Flowering Plants

Taxonomy and Phylogeny

Flowering Plants
Taxonomy and Phylogeny

B. Bhattacharyya
B.M. Johri

Springer-Verlag

Narosa Publishing House

Dr. Bharati Bhattacharyya
Department of Botany, Gargi College
University of Delhi South Campus
New Delhi 110 049, India

Prof. (Retired) B.M. Johri
Central Reference Library
University of Delhi
Delhi 110 007, India

ISBN 3-540-61303-X Berlin Heidelberg New York
ISBN 0-387-61303-X New York Berlin Heidelberg
ISBN 81-7319-079-8 Narosa Publishing House, New Delhi

Printed in India.

To

Botanists

Who have significantly advanced our knowledge of
Taxonomy and Phylogeny of Angiosperms

Preface

The present book is a fully illustrated (215 Figures), well-documented (ca. 1100 References) and critically analysed comprehensive comparative account of 62 Orders and 344 Families as revised by Melchior (1964) following A. Engler's (1909) *Syllabus der Pflanzenfamilien*. The relative placement of the Families in several other Systems of Classification is compared in Appendix 3.

Chapter 2 on Development of Classificatory Systems is a historical perspective to highlight the progress of the subject from ca. 370 B.C. to the present time. A comparative study of the different, frequently used classificatory systems is included. The relevant rules of the International Code of Botanical Nomenclature are discussed in Chapter 5.

The trends in evolution of different floral parts have been considered in Chapter 4.

Distinguishing features are mentioned for each order of the Dicotyledonae-Archichlamydeae (orders 37, families 227, figures 135, tables 26), Dicotyledonae-Sympetalae (orders 11, families 64, figures 30, tables 2) and Monocotyledonae (orders 14, families 53, figures 32, tables 5).

Today, the study of taxonomy requires a holistic approach with interdisciplinary input. For each family, therefore, attention has been drawn to geographical distribution, important vegetative and floral characters, anatomy, playnology, embryology, basic (or haploid or diploid) chromosome number, phytochemical data and economic importance.

Vegetative and floral characters are mainly based on the studies by Hutchinson (1973)—*The Families of Flowering Plants Arranged According to a New System Based on Their Probable Phylogeny*; Lawrence (1951)—*Taxonomy of Vascular Plants*; and Willis (1973)—*A Dictionary of the Flowering Plants and Ferns*. Personal observations and illustrations have been added for a number of families which we have studied during the past 25 years.

For anatomical details, the main source of information is the publication by Metcalfe and Chalk (1972)—*Anatomy of the Dicotyledons*, Vols. 1 and 2; and Tomlinson (1961, 1969)—*Anatomy of the Monocotyledons*, Vols. 1 and 3. The publications of Carlquist (1960, 1964, 1966, 1969, 1976 a, b, c, d, 1977a, b, 1985), Tomlinson (1962) and Tomlinson and Zimmerman (1969) have provided much useful information.

For the embryological features, data on pollen grains, ovules, embryo sac and endosperm have been added from the encyclopaedic work of Johri, Ambegaokar and Srivastava (1992)—*Comparative Embryology of Angiospersms*, Vols. 1 and 2. Data on the morphology of monocot pollen grains have been obtained from the work of Zavada (1983). Data on the pollen morphology of the Ranalean complex has been taken from the publication of Walker (1974). In addition, various other publications of different embryologists have been consulted to make the present book as informative as possible.

The data on the chromosome numbers is based on the work of Kumar and Subramanian (1987)—*Chromosome Atlas of Flowering Plants of the Indian Subcontinent*, Vols. 1 and 2, substantiated by the studies of Takhtajan (1987), *Systema Magnoliophytorum*, and other workers.

Chemosystematic studies of Harborne and Turner (1984)—*Plant Chemosystematics*; Harborne and Swain (1969)— *Perspective in Phystochemistry*; Heywood (1971)—*The Biology and Chemistry of the Umbelliflorae*; Jensen and Fairbrothers (1983)—*Protein and Nucleic Acids in Plant Systematics*; Swain (1966)—*Comparative Phytochemistry* and Young and Seigler (1981)—*Phytochemistry and Angiosperm Phylogeny*, have enriched the data on the chemical features of various families. Research articles of Dahlgren (1975b, 1980a, 1983a, b), Fairbrothers (1968, 1983), T.J. Mabry (1976), T.J. Mabry and Behnke (1976), T.J. Mabry and Turner (1964) and T.J. Mabry, Eifert, Chang, H. Mabry, Kidd and Behnke (1975) have also been very useful. Often, these data have been a decisive factor in determining the placement of disputed taxa. The histochemical and ultrastructural features have also been discussed following the studies of Behnke (1969, 1974, 1975, 1976a, b, 1981, 1982, 1985, 1986), Behnke and Turner (1971) and other contemporary workers.

The systematic assignment of the families has been discussed following the study by Johri et al, (1992). Attention has been drawn to the classificatory systems proposed by Cronquist (1968, 1981), Dahlgren (1980a, 1983a), Takhtajan (1980, 1987) and Thorne (1983).

Engler's (1909) systematic considerations also discuss the above data, but only marginally. The studies by Dutta (1988) and Singh and Jain (1989) have also attempted a similar approach, but without full justification.

In the present work, special features have been pointed out for selected taxa, including endemic and threatened species. The inter- and intrafamilial and inter- and intraordinal phylogenetic and evolutionary tendencies are discussed, especially for the taxa of disputed taxonomic assignments.

Data from various fields of study have been indicated to support or reject the existing/proposed relationship: wherever possible, taxa with indeterminate taxonomic assignment have been discussed. Families for which adequate information is either not available or has not been worked out at all are also listed.

It should be realized that a "final" classificatory system is simply not possible. Fashions of teaching and research change with the advance of knowledge and, in this race, taxonomy has become a victim. Advances in this area do not command the same attention or facilities as those in the fields of physiology, biochemistry, molecular biology and genetic engineering. However, taxonomy is a basic and fundamental discipline which should command the highest priority.

There is a great dearth of botanists, with broad-based knowledge, since most present-day research students are more enamoured in following only a few selected areas for specilisation. A physiologist, a biochemist, a molecular biologist or a genetic engineer hardly cares for taxonomic involvement in the advancement of plant sciences.

This challenging assignment was taken up keeping all these factors in view, and we sincerely hope that this book will be of help not only to taxonomists, but also to those interested in various other fields of study.

Procedures for field study, methods of collection and preparation of herbarium and fresh specimens in the laboratory have been added as Appendices.

A comprehensive List of References is cited in the text, and a Glossary of Technical Terms precedes the Plant Index.

The readers are welcome to convey their advice/suggestions to the authors, so that this book can be further improved.

B. Bhattacharyya
B.M. Johri

Acknowledgements

For the preparation of this book we have received much help from our colleagues in India and abroad, during the past 8 years.

We wish to express our gratitude to:

Dr. M.A. Rau (formerly Deputy Director, Botanical Survey of India) for critically examining most of the manuscript,

Dr (Mrs) Chhaya Biswas (formerly Principal, Gargi College, University of Delhi) for her keen interest in this work,

Dr. Virendra Kumar (Zakir Hussain College, University of Delhi) for providing us with the *Chromosome Atlas of Flowering Plants of the Indian Subcontinent*, Vols. 1 and 2,

Dr. M. Hakki (Botanical Garden and Museum, Berlin-Dahlem) for advising us to include new families such as Saccifoliaceae and others in this book, and for providing the publications of Professor Th. Eckardt on Dysphaniaceae and Centrospermae,

Professor P.K. Endress (Botanical Garden, Zürich), for providing his publications on Magnoliales and also for providing the photographs of some taxonomists,

Professor S. Carlquist (Claremont Graduate School, Claremont, California), for his publications on the Australian family, Stylidiaceae,

Professor H.Y. Mohan Ram (Department of Botany, University of Delhi) for his publications on Podostemaceae and Lentibulariaceae,

Dr. B. Mukherjee (B.C. Roy Post-Graduate Institute of Basic Medical Sciences, Calcutta) for his publications on the chemistry of the Annonaceae,

Mr. Arvind Mathur (Architect, New Delhi) for providing the semi-transparent non-glossy paper for the illustrations, and Mr. Arindam Bhattacharyya for sketching several illustrations,

Dr. (Mrs.) Atiya Habib (Jawaharlal Nehru University) and Mrs. Ruma Chanda for translating the French references into English,

Dr. A.M. Goswami (Head of the Division of Horticulture, Indian Agricultural Research Institute, New Delhi) and Dr. M.K. Roy (Principal Scientist, Nuclear Research Laboratory, Indian Agricultural Research Institute, New Delhi) for their help in bibliographic work,

Professor A.K. Sharma (Centre of Advanced Study, Department of Botany, University of Calcutta), Dr. P.K. Hazra (Director, Botanical Survey of India, Calcutta) and Dr. Prithipal Singh (Kirorimal College, University of Delhi) for their help in bibliographic work. Dr. Singh has also made available to us the information on the Tokyo edition of the International Code of Botanical Nomenclature.

Dr. (Mrs.) Krishna Kumar and Dr. (Mrs.) Kiran Prabha (both of Gargi College, University of Delhi), Mrs. Bijaya Sen Gupta (Delhi University Library) and Miss Somdutta Sinha Roy (Ph. D. Scholar, Department of Botany, University of Delhi) for helping us in many ways.

Special appreciation deserves to be expressed to Prof. A.K. Bhattacharyya, Mrs. Raj Johri and other members of the families of both the authors for their encouragement and moral support.

Professor Upendra Baxi (former Vice-Chancellar), Professor A.L. Nagar (former Pro-Vice-Chancellar), Professor A.P. Srivastava (former Librarian), and Mr. Sher Singh (Deputy Librarian), University of Delhi, for providing an office in the Central Reference Library where this work was completed.

Contents

1
Introduction

Taxonomy is one of the oldest fields of biology, with a long history of development. Man, from time immemorial, has known and been dependent on the plant world for many of his needs. This dependency made it necessary even for prehistoric man to identify the plants, and also to classify them into different groups, such as food plants, poisonous plants, medicinal plants, etc., which were important for him. This was the beginning of plant taxonomy which includes the *identification*, *nomenclature* and *classification* of plants. From the simple methods of recognition of economically useful plants by the earliest man, today it has become a highly complex and all-embracing biological science.

Plant taxonomy has long been treated as a primitive or orthodox science, in the sense that it is primarily based on morphology. It can even be called the most controversial, misunderstood and maligned subject. These adjectives can be applied because of the very nature of the subject. One of the aims of taxonomy is to provide help to non-taxonomists. Its principles and practices are thus more often scrutinised by non-specialists, as compared to other sciences. Much of the distrust and criticism of this subject is due to the lack of understanding of these users.

In fact, the concept of taxonomy has no obvious, finite, single aim or purpose. One important aspect of taxonomy is its fundamental property of being at the same time basic and all-embracing. Basic, because taxonomy is the basis of all other branches of biological science, providing the basic information—the identification, name and systematic position of the materials on which research can be carried out. At the same time, it is of supreme importance as the only truly synthetic science, based on data from different fields. It is the only unifying discipline.

The thirst for plant taxonomy may, on a small scale, be quenched simply by knowledge of the local flora and the ability to identify it; but this is not always enough. Data for taxonomic studies are accumulated by various methods—research performed in different herbaria of the world, in laboratories, in field and garden experiments, and also from libraries.

The two terms systematic botany and taxonomy are commonly used interchangeably by most authors of this science, as there is no agreed distinction, and they are so treated in this book also. Lawrence (1951) points out that the term taxonomy may be used for all objects of biological origin and may be defined as the study and description of the variation of organisms, the consequences of these variations, and the utilisation of this to produce a system of classification.

Duties of a taxonomist: The knowledge of plant taxonomy is helpful to the common man in various ways. The name of a wild plant may be of little general interest, but to a plant breeder it may prove to be a useful plant in having important genes for disease resistance or better yield. It is a great responsibility of plant taxonomists to provide the correct names and classification of such plants. Similarly, a chemist may find a new source of an important chemical, but the name of this source plant has to be provided by the taxonomist. Extensive use of this plant science has been made in many fields. In forestry, for example, trees must be named and classified because of their innumerable economic importance. Forest lands are often leased out for grazing. A taxonomist would be able to guide as to which plants are palatable to cattle, what methods of propagation must be adopted for the plants, or what should be the maximum grazing limit so that plants do not totally disappear due to overgrazing. In agriculture also the knowledge of taxonomy is useful in various ways. A taxonomist is ready to help a plant breeder with his broad knowledge of which plants can be used as suitable stock for hybridisation experiments. Even the introduction of seeds and plants from one country to another benefits from the expertise of a taxonomist. In range management also, plant taxonomists play an important role, by suggesting which plant species have better soil-binding capacity, or which plant can act as a suitable windbreak so that erosion does not take place.

Ecologists with sound knowledge of plant taxonomy can easily identify soil type or climatic condition of a region from the plants that are growing there. These are only some of the duties performed by plant taxonomists.

Aims and Objectives. Plant taxonomy is primarily concerned with the identification, nomenclature, classification and evolutionary relationship between diverse plant groups.

Identification is the determination of a specimen as being identical to or different from a previously known plant. A previously known specimen will not always be available, and the unknown specimen might prove to be a new one. The process of naming is usually not involved at this stage. For example, of the three specimens A, B and C, A is *Solanum nigrum* and B and C are unknown. Specimen B is similar to A and therefore it is *S. nigrum*; but specimen C is dissimilar and therefore further comparison with additional known specimens is advisable. However, there are also other methods of identification, such as consulting (a) a flora, monograph or revision, (b) a herbarium specimen, or (c) a garden specimen, and the comparison may provide the information required. If, however, a specimen shows no similarity to any known specimen, then it may be counted as a completely new taxon.

Nomenclature. Simple identification is not enough. These identified specimens must be given correct, valid names—a name by which these plants can be made known to the rest of the world. Nomenclature is an orderly application of names for which certain rules have to be followed. As it is of international importance, the rules for nomenclature are listed in the International Code of Botanical Nomenclature (ICBN). The basic rule is that binomial nomenclature should be followed for naming all specimens that are treated as plants. This means that each plant must have two names—the first is the generic name and the second the specific epithet. Apart from this basic rule, many others are proposed by the International Association for Plant Taxonomy (IAPT). These rules are finalized every 5 years during an International Botanical Congress.

Classification. Classification is an arrangement of plants or plant groups according to a particular plan. The habit of humanity is to classify whatever it comes across. As civilized man came to understand the utility of the plants, he classified them as food, fruit, medicine and many such groups. Originally, morphology was the basis of classification, but today classificatory systems utilise the information collected from various fields such as anatomy, embryology, cytology, biochemistry, ecology and phytogeography.

The various major units of classification are divisions, subdivisions, classes, subclasses, orders, suborders, tribes, genera and species. There are also infraspecific units, apart from these nine major ones, and all these together constitute the plant kingdom.

At different times, different bases have been used for classification. Many classificatory systems of earlier times were based on economic uses of plants or their habits; some of the systems suggested by herbalists were based on medicinal uses of plants. These systems were, however, incomplete in the sense that those plants which did not fit into these systems were ignored. The only system of classification that was complete, at least for the plants known until that time, was the one based on the natural relationship amongst the members of different plant groups. Even this "natural relationship" meant only similarity of plant parts based on the study of comparative morphology.

In fact, a natural classification should depict the natural relationship existing among plants as they evolved from the ancestral to the most advanced forms. All the later classificatory systems, such as those of Cronquist (1968, 1981), Dahlgren (1975a, 1980a, 1983a), Thorne (1968, 1983) and Takhtajan (1969, 1980, 1987), are both natural and phylogenetic.

Development of Classificatory Systems

Whenever we are confronted with a mass of material, evidence or data in our everyday life, we try to classify them into convenient groups for better appreciation of their nature, an analysis of their characteristics and for future reference. It can thus be said that classification of information is a basic human activity. Whatever we come across or observe, we try to classify. Man started classifying plants even when he was still only a food-gatherer. He classified which plants were edible and which were not, which could be used for medicinal purposes, and which were poisonous and so on.

Plant classification is the placement of plants or groups of plants in separate compartments at different levels according to phenetic similarities, phylogenetic relationships or mere artificial criteria. Thus, individuals resembling each other very closely may be grouped under a "species"; a few species sharing many common characteristics under a "genus"; similar "genera" under a "family" and so on.

These different groups show a box-within-box arrangement in which the sizes of the boxes may vary. Parallel with these different levels of groups is a series of categories forming a hierarchy. The hierarchy of categories is like a set of empty shelves arranged at different heights. Once formed, the groups are assigned to a particular category, and the category in which we place a group determines the name of the group.

Taxonomic categories are artificial, highly subjective and can be defined only by their position relative to other categories. The taxonomic groups are natural, can be defined and are objective. A taxonomic group is made up of lower taxonomic groups, but a category is not made up of lower categories. Groups like Angiospermae, Monocotyledoneae, Leguminales or Fabaceae are composed of larger and diverse types of plants. These groups are known as subdivision, class, order and family, and are together called major categories. The categories—genus (*Brassica*), species (*B. campestris*) and variety (*B. campestris* var. *oleracea*) are collectively known as minor categories and are of smaller magnitude. These major and minor categories may be divided into subordinate categories intermediate to the next lower category. ICBN recognises 21 different taxonomic categories. These are

Divisio (division)	Subfamilia (subfamily)	Series
Subdivisio (subdivision)	Tribus (tribe)	Subseries
Classis (class)	Subtribus (subtribe)	Species
Subclassis (subclass)	Genus	Subspecies
Ordo (order)	Subgenus	Varietas (variety)
Subordo (suborder)	Sectio (section)	Subvarietas (subvariety)
Familia (family)	Subsectio (subsection)	Forma (form)

Amongst these, the most frequently used categories are subclass, order, family, genus, section, species, subspecies and variety. At least one taxon is required for each category and subcategory, and at least one character is necessary to distinguish between two taxa belonging to the same category. Usually, as a rule, the lower the rank or category of the taxon, fewer are its members and higher is the number of common features shared by its members; the higher the rank, more numerous are its members and share fewer common characters.

In actual practice, classification deals more with the placing of a plant group in its proper position within the system than the placing of an individual plant in one of the several minor categories.

The history of plant classification is a fascinating subject. One learns how the different systems have evolved during the various stages of development, and also about the people responsible for them.

The **Pre-Linnaean Systems**. The systems of classification proposed by different systematists before Linnaeus can be broadly divided into two groups; (a) artificial classifications, and (b) mechanical classifications.

(a) The **artificial systems** of classification were based on the habit of the plants classified.

Theophrastus (370–285 B.C.) is considered the Father of Botany and was a student of the great Greek philosopher, Aristotle. He classified all plants on the basis of form and texture. He recognised trees, shrubs, undershrubs and herbs, and distinguished between annual, biennial and perennial plants. He also recognised determinate and indeterminate inflorescences; hypogynous, perigynous and epigynous flowers; apetalous, polypetalous and gamopetalous condition; the distinction between monocots and dicots; and the distinction between roots and rhizomes. He considered trees to be the most highly developed plants; his groups were strictly artificial and used no categories as we do today. His writings include *Enquiry into Plants* and *The Causes of Plants*. He had a botanic garden of his own, attached to his Lyceum near Athens, and all the classification work was the result of the studies conducted on the plants of this garden. In his *Historia Plantarum* he roughly classified ca. 480 types of different plants.

Dioscorides (60 A.D.) was a physician. He dealt with medicinal plants and identified ca. 600 different species. His book *De Materia Medica* included information on roots, stems, leaves and sometimes flowers of different plants of medicinal value.

Albertus Magnus (1193–1280 A.D.) was Bishop of Ratisbon. He recognised differences in stem structure of dicots and monocots anatomically, even with the help of the crude lenses available at that time. He also distinguished between leafy and non-leafy plants; leafy plants were further divided into monocots and dicots, and dicots into herbaceous and woody plants. In other major respects, he agreed with the classification of Theophrastus.

Otto Brunfels (1464–1534), a German, was one of the renowned herbalists of his time. He wrote an illustrated herbal based on material from the works of Theophrastus, Dioscorides and Pliny. However, real credit goes to him for recognising the Perfecti and Imperfecti groups of plants, i.e. plants with and without flowers, observable by keeping the plant on branch at arm's length. The three volumes of *Herbarium Vivae Eicones* by Otto Brunfels are well known for the illustrations of living plants. Contemporary with Brunfels came **Jerome Bock** (1498–1554), **Leonard Fuchs** (1501–1566), a Bavarian physician; **P. Mathiola** (1501–1577), an Italian herbalist; **Mathias de l'Obel** (1538–1616), a Dutch herbalist; **John Gerard** (1545–1612), an English surgeon and botanist, and a Flemish botanist, **Charles de l'Ecluse** (1526–1609). All these herbalists contributed to the descriptive phase of systematics and to the botany of medicinal plants.

(b) The **mechanical systems** are based on one or a few selected characters of mainly morphological nature. During this period, the hierarchy of categories was improved and more natural groups were recognised. These systems were published between 1580 and about 1760.

Andrea Caesalpino (1519–1603), an Italian botanist, has often been referred to as the first plant taxonomist. He classified plants first on the basis of habit,—woody or herbaceous—and further divided them on the basis of fruits, seeds and embryo characters. He recognised inferior and superior ovary, bulbs present or absent, sap milky or colourless, and the number of locules in an ovary. His famous work *De Plantis* (1583) included descriptions of some 1520 plant species arranged into woody and herbaceous groups. His most important conclusion was that the flowers and fruits were more reliable characters than the habit. However, he never ordered his ideas into outline or synopsis and they were therefore not adopted by his immediate successors; but many later workers, like Tournefort, John Ray and Linnaeus, were influenced by his work.

Jean Bauhin (1541–1631) and **Gaspard Bauhin** (1560–1624). Jean Bauhin is well-known for his comprehensive work carried out on about 5000 plants. The descriptions included were good diagnoses. *Historia Plantarum Universalis* was an excellently illustrated work, published in three volumes in 1650, after his death. His brother, Gaspard Bauhin, made the distinction between genera and species and described ca. 6000 species in his work *Pinax Theatri Botanici* (1623). He followed l'Obel and Gerard, the herbalists, for classifying the plants. To many of these plants, he gave a generic and specific epithet. Thus, binary nomenclature, with which Linnaeus' name is associated, was actually founded by Gaspard Bauhin about a century before its use in the *Species Plantarum* of Linnaeus.

John Ray (1628–1705) was an English philosopher, theologian and naturalist. His classificatory system was based on form relationship and about 18000 species were included. These were broadly divided into two groups—Herbae, i.e. herbaceous, and Arborae, i.e. woody. The Herbae were divided into Imperfectae, i.e. basically cryptogams, and Perfectae. The group Perfectae was further divided into dicots and monocots. The Arborae were also divided into dicots and monocots, which were further classified on the basis of fruit types as cone-bearing, nut-bearing, bacciferous, pomiferous, pruniferous and siliquous. He closely followed Caesalpino and pointed out that characters from all parts of the plant should be used for classification. He distinguished between inherited and imputed-differences, and noted the significance of the former for taxonomy. He also made use of anatomical characteristics that were collected by others like Grew and Malpighi. He recognised plant groups like the composites, umbels, mints, crucifers, legumes and a few others. In many respects, his system was superior to the Artificial System of Linnaeus.

Joseph Pitton de Tournefort (1656–1708) while working in Paris as a professor of botany, travelled widely in Europe and made collections of plant materials. He followed some of his predecessors in suggesting a system of classification based on form relationship. He broadly classified all the flowering plants into two categories—trees and herbs. Each of these categories was subdivided on the basis of apetalous/petalous/gamopetalous corolla, and then regular/irregular flowers. This system was used widely in different parts of Europe until 1760, when the system of Linnaeus appeared. In France it was used until 1780, when it was replaced by the system of de Jussieu. Tournefort is also given credit for being the father of the moden genus concept. He recognised 698 genera and 10146 species, and provided generic descriptions of these. Many of the generic names used by him are still in use, for example, *Salix, Populus, Fagus, Betula, Castanea, Quercus, Ulmus* and many others. However, this system was too artificial, as it was based more on vegetative characters, and was considered inferior to that of John Ray. No distinction was made between phanerogams and cryptogams or between monocots and dicots.

The Pre-Linnaean era ends here with the system of Tournefort.

Linnaeus' System of Classification and Thereafter

Carolus Linnaeus is regarded as the father of taxonomic botany, and also of zoology. He was Swedish, born on May 23rd, 1707. While working in the University of Uppsala as a student, he published *Hortus Uplandicus* in Latin, which was an enumeration of the plants in the Uppsala Botanical Garden. The plants were arranged according to Tournefort's system in the first edition. In the second edition, Linnaeus gave his own system of classification. He spent about 3 years, from 1737 to 1739, in Holland, and this was the most important period of his life. He pubished 14 treatises during this period, including *Genera Plantarum, Flora Lapponica* and *Hortus Cliffortianus*. He was also very popular as a teacher. He died on January 10, 1778 at the age of 70.

The Sexual System of Linnaeus, as it is usually called, appeared at a time when it had become absolutely necessary. For two centuries, plant materials of various forms had been collected, but they could not be identified or classified, as there was no logical system of classification available. Thus, when the scientific world was eagerly waiting, Linnaeus published his System by which these plants could be easily classified.

In short, the System has 24 classes for all the plants known at that time, based on number, union and length of stamens. Each class was then subdivided into orders on the basis of number of styles in the flowers. The system was first published in the second edition of *Hortus Uplandicus* (1732). Later, it was revised and formed the basis of *Genera Plantarum* (1737). Descriptions of 935 genera are included in this publication.

Species Plantarum (1753) is the most important work of Linnaeus in the systematics of vascular plants. In this book he included the identification and description of nearly 6000 species belonging to 1000 genera. Linnaeus is to be remembered as having given a precise system of plant nomenclature. *Species Plantarum*, published in May 1753, in two volumes, was the first book in which binomial nomenclature was used for describing all the plants. Each plant was given a generic name, followed by a trivial name and then a specific phrase-name. These were followed by references to previous publications or his herbarium specimens and then the native habitat of the plant concerned. The trivial name was in the true sense the specific name and, together with the generic name, formed the binomial for most plants. The third part of the name, that is the specific phrase-name, was considered by Linnaeus to be the specific name and was in the form of a polynomial descriptive phrase. The binomial nomenclature, formed of a generic name and a specific epithet, was formulated by Gaspard Bauhin in 1596. Bauhin, however, did not use this system in his work very consistently, whereas Linnaeus followed it carefully throughout his *Species Plantarum*. For this very reason, the taxonomists of today have chosen the 1st of May, 1753, as the starting point of current botanical nomenclature. Any name published before this date is not valid.

The post-Linnean period can also be divided into two parts.

(a) Period of Natural Systems (from 1760 to 1880).   Now that the easy method of identification was at hand (the Linnaean system of classification), many naturalists and explorers went on expeditions to collect plants from different parts of the world.

John Clayton (1685–1773) was probably the first taxonomist to go to America in 1705. A lawyer by profession, his interest in botany was unusual. He collected plants, wrote descriptions, and tried to identify them by corresponding with **J. Gronovius**, who was a professor at Leiden in the Netherlands. Later, in 1739, on the basis of these writings, *Flora Virginica* was published by **Gronovius**. Another Englishman, **Mark Catesby** (1680–1749), also explored parts of North America and published a voluminous *Natural History of Carolina, Florida, and the Bahama Islands* (1731–1743).

Almost at the same time another botanist, **George Eberhard Rumpf** (1628–1702), was studying the tropical plants of Indonesia and the island of Amboina. Unfortunately, both his collections and manuscript were destroyed, and it was only much later that the *Herbarium Amboineuse* could be published by **J. Burmann** (1741–1755) on the basis of a new manuscript compiled by him. About 1750 species of plants were described, of which 1060 were also illustrated in this book.

A large group of enthusiastic students of Linnaeus also explored different parts of the world. They were **Peter Kalm,** who went to North America, **Christopher Ternstrom**, who went to the East Indies, **Frederik Hasselquist**, who visited Palestine, **Carl Peter Thunberg**, who went to the Cape of Good Hope, and **Peter Forskal**, who went to Egypt and Arabia. Many other botanists explored North and South America, Africa and Australia. Quite a few of the European botanists also came to India and therefore most of the earlier literature on floristic explorations of different zones of this country was written by English botanists such as **Sir J.D. Hooker, D. Prain, J.F. Duthie, H.H. Haines** and **J.S. Gamble.**

Michel Adanson (1727–1806) was a French botanist who explored places such as tropical Africa. He rejected all the artificial classifications that had been published before his time, as he believed in giving equal significance to all the characters. This is known as the Adansonian Principle and became important in the classificatoy systems of the 20th Century. He is often referred to as the grandfather of numerical taxonomy.

By means of all these explorations throughout the globe, botanists came to know about many more plants than were included in Linnaeus' System. As a consequence, the search started for a natural system of classification. Adanson tried to form a new system, but was not successful.

Jean B.A.P.M. de Lamarck (1744–1829), also a French botanist, came up with a brilliant idea. This is known as Lamarckism, and means that changes in the environment bring about changes in the structure of the organisms.

All three brothers, **Antoine** (1686–1758), **Bernard** (1699–1776) and **Joseph** (1704–1779) **de Jussieu** were botanists. **Bernard de Jussieu** arranged the plants in the botanical garden at Le Trianon, Versailles, according to a new system of his own, which was neither based on the Aristotelian concept of habit nor as artificial as that given by Linnaeus. The work was published by another botanist, his nephew, **Antoine Laurent de Jussieu** in *Genera Plantarum* (1789). In this system, he divided all the plants into 15 classes, and these classes were subdivided into 100 orders. There was a solitary class Acotyledons that included all the known cryptogams. Class 15, i.e. Diclines Irregularis, was also not exclusively angiospermic, as Coniferae were included; but all the other 13 classes were of flowering plants alone.

A.P. de Candolle (1778–1841) coined the term taxonomy for the first time, meaning the arrangement of plants. His important works include *Théorie Elémentaire de la Botanique* (1813), in which his approach to plant classification was explained. According to him, anatomy should have been the basis of taxonomic classification rather than physiology. Later, while staying at Geneva, he wrote one of the classics in botany: *Prodromus Systematis Naturalis Regni Vegetabilis* (1806–1893). This was a compilation of the descriptions of the then known families, genera and species. It ran in 17 volumes, of which 7 volumes were written and produced by himself. The other 10 volumes were written by specialists and published under the supervision of his son, **Alphonse de Candolle**.

The first half of the 19th Century was important in the history of taxonomy, as a number of systems of classification were put forward during this time, and also numerous botanical explorations were undertaken.

Robert Brown (1773–1858): The main contribution of this Scottish scientist was the recognition of gymnosperms as a group distinct from angiosperms, and that they were characterised by the presence of naked ovules. He also carried out floral morphological studies of many families like Asclepiadaceae, Euphorbiaceae, Polygalaceae and others.

Bentham and Hooker: George Bentham (1800–1884), an Englishman, was a very critical and well-trained taxonomist. **Sir Joseph Dalton Hooker** (1817–1911), son of a botanist father, **Sir William J. Hooker**, was the then Director of Royal Botanical Garden at Kew, London. He was also a very enthusiastic plant explorer and phytogeographer. Bentham and Hooker's System of Classification first appeared in a three-volume work in Latin named *Genera Plantarum*. The system is based on groups of plant characters which are correlated with each other. For example, if a specimen consistently shows syngenesious stamens, inferior, bicarpellary, unilocular ovary with basal placentation, then it can confidently be assigned to the family Compositae. His work includes names, description and the classification of all the seed plants then known. Basically, it is on the same basis as that of de Candolle but with some modifications such as greater emphasis on the free and fused condition of petals.

The Dicotyledonae were divided into three groups—Polypetalae, Gamopetalae and Monochlamydeae. Polypetalae includes three series—Thalamiflorae, Disciflorae and Calyciflorae, of which the first and third are mentioned in de Candolle's work. Similarly, Gamopetalae and Monochlamydeae were the other two subclasses, Corolliflorae and Monochlamydeae in de Candolle's system. Gamopetalae is subdivided into three series—Inferae, Bicarpellatae and Heteromerae. Each of these series is subdivided into a number of cohorts, which are equivalent to the families in the present-day classificatory systems. Each of the cohorts comprises a few related families. The Monochlamydeae are arranged in eight series of unequal value. For example, the Curvembryeae was a natural group, but the Unisexuales included families which later showed quite different affinities.

The monocots are divided into seven series, starting with epigynous orders, Orchidaceae and Scitaminae, then the petaloid hypogynous Liliaceae, then orders where the perianth is not petaloid, and then orders like Padanaceae and Aroideae with aborted perianth. Next is the series Apocarpae, where the carpels are free. The last order is the Glumaceae, where the floral structure is highly reduced.

Although this system was accepted throughout the British Empire, in the United States of America and also in some Continental countries, its shortcomings were realised. Placing the Gymnospermae between the Dicotyledoneae and Monocotyledoneae had also been criticised, the critics trying to keep all the gymnosperms together instead of mixing them with the dicot families (as was done by de Candolle). Another point of criticism is the inclusion of certain families like the Chenopodiaceae in the Monochlamydeae, when they have clear affinities with Caryophyllaceae of Polypetalae.

Whether the simple flower structure is a primitive feature or a reduced variation of a highly evolved form has always been a matter of controversy. Monochlamydeae, a group showing reduced flower structure, should therefore be a homogeneous group of all the families showing reduced floral structure.

Among the monocotyledonous families, the relative position of ovary and perianth has been over-emphasised. For this reason, related families like Liliaceae and Amaryllidaceae are placed in two different orders and unrelated families like Juncaceae and Palmae are placed together.

However, in spite of all its disadvantages, the Bentham and Hooker system is still very popular in many countries, because identification of plants in the field is comparatively easy. It is followed in the Kew Herbarium for the arrangement of plant specimens, as also in many other herbaria, particularly those of the Commonwealth countries.

(b) Systems Based on Phylogeny. Towards the end of the 19th Century, Darwin's work on the theories of evolution and origin of species was also published coinciding, incidentally, with the publication of *Genera Plantarum* by **Bentham and Hooker** (1862–1883). Darwin's theories were so well-documented that the concept of natural classification was dropped. Even Hooker wanted to revise their system of classification, but Bentham deterred him from doing so, as he could not agree with Darwin's ideas.

Nevertheless, the classificatory systems that were written after the publication and general acceptance of Darwin's work were all based on theories of descent and evolution. Everyone waited to account for the true phylogenetic relationship of plants. Two schools of thought, therefore, developed and remained side by side for a considerable time, regarding the *nature of the primitive flower*. One of these was the Englerian school, proposed by a German taxonomist, **Alexander Braun**, in 1859 in his book *Flora der Provinz Bradenburg*. He classified monocotyledons and showed a progression from more simple to complex forms. He treated the naked, unisexual flowers of Lemnaceae as the most primitive, and gradually increasing in complexity, reached the most complex family, Orchidaceae. Amongst the dicots, he treated the Apetalae as the most primitive and, through gradually more complex forms like Sympetalae and Eleutheropetalae, ended with the Leguminosae.

His successor, **A.W. Eichler** (1875), published a modified form of this system in his *Blütendiagramme*. In 1883, he elaborated it further, and this system gradually replaced that of de Candolle.

Eichler divided the plant kingdom into two subgroups: Cryptogamae, containing flowerless plants, and Phanerogamae, including seed plants. The former was divided into three main parts: Thallophytes, including

Algae and Fungi; Bryophytes, including Hepaticae and Musci; Pteridophytes, including three classes—Equisetineae, Lycopodineae and Filicineae. The Phanerogamae were divided into two groups: Gymnospermae and Angiospermae. The latter were subdivided into two classes: Monocotyledoneae and Dicotyledoneae. In these two groups, plants were arranged according to increasing complexity of floral structure.

Adlof Engler (1844–1930) published a classificatory system (based on Eichler's system) in 1892 in the form of a guide to the plants of Breslau Botanic Garden. The phanerogams were named Embryophyta Siphonogama, and were divided into Gymnospermae and Angiospermae. The Angiospermae were further divided into the classes Monocotyledoae and Dicotyledonae. In both these classes, flowers with simple structures were treated as primitive forms. The class Monocotyledonae and the two subclasses of the Dicotyledonae were next divided into orders, each order comprising a few related families. The Monocotyledonae were divided into 11 orders and 45 families. Subclass Archichlamydeae of the Dicotyledonae comprised 37 orders and 227 families; subclass Metachlamydeae includes 11 orders and 64 families (as revised by **Melchior** 1964).

Engler himself did not consider his system of classification to be a phylogenetic one; but he maintained and expanded the concept of the primitive flower as proposed by **Alexander Braun** (1859). In addition to the fact that the unisexual and naked flowers borne on catkin-like inflorescences were the primitive flowers, Engler also pointed out that these flowers were wind-pollinated, as in the gymnosperms. It was assumed that in all probability the primitive angiospermous flowers originated from a gymnospermous ancestor bearing a unisexual strobilus. This was recognised as the Englerian concept of primitive flower. The Englerian school further proposed a polyphyletic origin for angiosperms.

Engler and Prantl applied this system to plants all over the world, and published a 23-volume work, *Die Natürlichen Pflanzenfamilien*, during the years 1887 and 1915. In this work, identification of all known plants from algae to angiosperms was taken up, and all the taxa were fully described. The original work was subsequently revised by many botanists. The latest revision was by Melchior (1964), who retained the basic structure of the system and accepted the concept of primitive flower. The descriptions of individual taxa now include the compilation of data from various fields other than morphology and embryology (considered in earlier revisions), so that, the position of some of the orders and families have been altered. Details of Engler and Prantl's system of classification as revised by Melchior form the basis of our book.

In many American and Continental European herbaria, Engler and Prantl's system is followed for the arrangement of plant specimens.

A.B. Rendle (1865–1938) is known for his two-volume work, *Classification of Flowering Plants* (1904, 1925). It is again based on Engler and Prantl's system of classification, with certain minor modifications. He, too, did not regard his system as a phylogenetic one, but rather a system of convenience. The monocots were treated as more primitive than the dicots. In this group the taxon Palmae had a controversial position. Engler placed it in the order Principes, but Rendle treated it as a family under Spathiflorae. The dicots, on the basis of flower structure, were divided into three groups: Monochlamydeae, Dialypetalae and Sympetalae. Salicales was the most primitive order, and Umbelliferae the most advanced amongst the dicots. Rendle is also known for his taxonomic studies of the Gramineae, Orchidaceae and Naiadaceae. His descriptions are very clear and complete, with beautiful diagrams, a record of exceptional features and discussion of the relationship of families with each other.

Richard von Wettstein (1862–1931), an Austrian botanist, proposed a classificatory system which, although based on Englerian concepts, was more a phylogenetic system. Like Engler, he also presumed that unisexual perianth-less flowers were more primitive than the bisexual flowers with elaborate perianth. He believed in the monophyletic origin of angiosperms from *Gnetum* or a *Gnetum*-like ancestral form of gymnosperms. He also treated the dicots as more primitive than the monocots and stated that the monocots were probably derived from Ranalian stock of the dicots. Further, according to Wettstein, herbaceous plants were more advanced than woody plants, and numerous flowers in an inflorescence was more primitive than a solitary or few-flowered inflorescence. Naturally, therefore, his arrangement of individual taxa was quite different from that of Engler or Rendle, inclining more towards phylogenetic classification. Moreover, his descriptions included data from a much larger range of fields, even from serology. In general, he considered (a) the presence of abortive reproductive parts in a flower as evidence of reduction; (b) adaptive modifications as indicating advancement, and (c) the spiral arrangement of flower parts as primitive as compared to cyclic arrangement.

Although this system was not accepted on a large scale, it is known that many of Wettstein's ideas were adopted in other phylogenetic systems of later years. A revision of his work was published posthumously in 1935: *Handbuch der Systematischen Botanik* from Leipzig.

August A. Pulle (1878–1950), from the Utrecht Botanical Museum, The Netherlands, published in 1938, a phylogenetic system of classification based essentially on Englerian concepts. In this system, however, the division Spermatophyta comprised four subdivisions (instead of two). These subdivisions were:

(a) Pteridospermae, a subdivision of two extinct families,
(b) Gymnospermae, comprising the classes (i) Cycadinae, (ii) Bennettitinae (extinct), (iii) Cordaitinae (extinct), (iv) Ginkgoinae and (v) Coniferae;
(c) Chlamydospermae (Gnetales and Welwitschiales), and
(d) Angiospermae (Monocotyledoneae and Dicotyledoneae).

The Monocotyledoneae was subdivided into ten orders, each comprising a number of families. The Dicotyledoneae was divided into eight series; each series comprised a few orders, and each order had a few families.

Pulle's classification of Gymnospermae was more acceptable, as it was in closer accord with phylogenetic trends. Although the arrangement of orders and families of the Angiospermae was more or less on the same lines as that of Wettstein, it showed some originality also. He regarded Sympetalae as a polyphyletic group, and rearranged its components. Also, Wettstein did not consider Sympetalae and Monochlamydeae to be natural taxa.

Carl Skottsberg (1880–1963), a Swedish professor, gave a modified Englerian classification, utilising some of the concepts of Wettstein as well (1940). He considered the monocots as having originated from some unknown dicot. Amongst the dicots he considered the Casuarinaceae to be the most primitive, but not more primitive than other dicots. In his view, apocarpy and polycarpy were primitive; syncarpy and monocarpy were advanced; apetalous forms polyphyletic. He adopted some of the concepts of the Benthamian-Bessey school, and placed the taxon Amentiferae after the Rosales. He differed from Pulle, as he retained the Primulales in the Sympetalae.

Charles E. Bessey (1845–1915), a student of **Asa Gray**, was the first American to bring out a system of classification. He is well-known also as the originator of the Ranalian concept of primitive flower, which was later modified by Hutchinson (1926).

According to this concept, the primitive angiosperm flower bore a resemblance to a certain gymnosperm with a bisexual strobilus having microsporophylls below and megasporophylls above. In the course of evolution, the lower sporophylls developed into sepals and petals by progressive sterilization, and the upper sporophylls into stamens and carpels. The central axis became shortened into the thalamus. Eventually, the primitive flower has numerous free perianth lobes and free stamens, and carpels spirally arranged on the receptacle (or the thalamus). Amongst the living angiosperms, the genus *Magnolia* of the family Magnoliaceae is the nearest to this structure.

Bessey's idea was to formulate a classificatory system for the flowering plants that would reflect their evolutionary relationship. To achieve this, he proposed a set of dicta based on empirical evidence, to distinguish between primitive and advanced features. Primitive features were those which were expected in ancient plants, and advanced features were expected in the more recent taxa. However, it must be made clear that primitive and advanced are not the same as simple and complex. An advanced taxon may be structurally simple due to reduction, or it may have an elaborate structure and therefore be highly complex.

Bessey's classificatory system resembled that of Bentham and Hooker. It may be said to be a reorganised system of Bentham and Hooker, on the basis of phylogeny and evolution. He considered the seed plants to have had a polyphyletic origin and to comprise three distinct phyla, of which only the Anthophyta (Angiosperms) was dealt with. Anthophyta was divided into two classes: Alternifoliae (or monocots) and Oppositifoliae (or dicots). Both these classes were divided into two subclasses: Strobiloideae, including orders with superior ovary, and Cotyloideae, including orders with inferior ovary. In the dicots, each subclass was further divided into superorders, then orders, and thereafter families.

Bessey's system received increasing support from all over the world. He may be criticied for assuming perigynous and epigynous forms to constitute a single evolutionary line.

Hans Hallier (1868–1932), a German botanist from Hamburg, independently developed ideas similar to those of Bessey. His system of classification was based on data synthesised from various fields like palaeobotany, anatomy, serology and ontogeny, herbarium material and previous literature. He did not agree with the Englerian concept of a primitive flower. Instead, he also proposed the idea of a strobiloid ancestory for a primitive angiospermous flower, as did Bessey. The Ranalian concept is therefore also known as the Besseyan-Hallierian concept. He considered the angiosperms as having originated monophyletically from a Bennettitalean type of ancestor. Dicots were regarded as more primitive than monocots. Polycarpy and spiral arrangement were considered primitive, and syncarpy and cyclic arrangement advanced. His realignment of certain families displayed much thoughtfulness, e.g. Salicaceae and Flacourtiaceae are related to each other; similarly, Cactaceae and Aizoaceae are aligned, and so also Cucurbitaceae and Passifloraceae. He also recognised the family Amaryllidaceae as an unnatural assemblage of plants, and divided it into the Agavaceae, Alstroemeriaceae and Amaryllidaceae.

In his 1912 publication, 213 families were recognised. However, in spite of such an advanced outlook, his system did not receive much recognition.

John Hutchinson (1884–1972) was the Keeper of the Royal Botanic Gardens, Kew, England. He was concerned with the phylogeny of angiosperms alone, and a classificatory system for this plant group was published in his two-volume work: *The Families of Flowering Plants* (1926, 1934), and later revised in his *British Flowering Plants* (1948), *Evolution and Phylogeny of Flowering Plants* (1969) and *The Families of Flowering Plants arranged according to a New System Based on Their Probable Phylogeny* (1973).

This classificatory system has close affinities with those of Bentham and Hooker, and Bessey. According to Hutchinson, the gymnosperms formed a lineal monophyletic series starting from the primitive Cycadaceae and developing into an ascending order of Ginkgoaceae, Taxaceae, Pinaceae and finally Cupressaceae.

Hutchinson considered the angiosperms to have originated monophyletically from the hypothetical Pro-angiosperms. The angiosperms were divided into two major groups: Dicotyledons and Monocotyledons. A cleavage of the group dicotyledons into predominantly woody Lignosae and predominantly herbaceous Herbaceae was the next step. These two groups were considered to be the two different evolutionary lines, Lignosae stemming from Magnoliales, and Herbaceae stemming from Ranales, and they developed parallel to each other.

There are 83 orders and 349 families included in the dicotyledons, and 29 orders and 69 families in the monocotyledons in the latest revision of his system. No one order was directly derived from another;

derivation was shown to be from the ancestral stock. The monocots separately formed another evolutionary line and were derived from Ranales at a very early stage, starting with the Butomales and ending in the Graminales. On the basis of the nature of the perianth, the monocots were divided into three distinct groups: Calyciferae, Corolliferae and Glumiflorae.

The main objection to this system is the early distinction of the dicotyledons into arborescent and herbaceous groups. The splitting up of many families is also inacceptable to some botanists.

Hutchinson's system was not followed on a large scale, but it encouraged a number of phylogenists to revise their systems of classification.

Oswald Tippo, a American from Illinois University, published an outline of a system in 1942 which was claimed to be phylogenetic. It was based on two previously published works—one by G.M. Smith (1938) for non-vascular plants, and the other by A.J. Eames (1936) for vascular plants. In dealing with the angiosperms, Tippo considered Magnoliales to be a very primitive order, and Amentiferae a highly advanced taxon.

Karl Christian Mez (1866–1944), a professor of botany at the University of Koenigsberg, Germany, proposed a classificatory system based on serological studies of various families of angiosperms. According to him, the relationship between the different groups of angiosperms could be ascertained by analysis of their protein reactions. This experimental method, also called serum diagnosis or serodiagnosis, consisted of mixing an extracted plant protein with blood serum from some experimental animal. This would result in the formation of antibodies in the inoculated serum. To this was added the protein extract of the second plant whose relationship with the first was to be established. Depending upon the quantity of precipitate formed, the degree of relationship between different plants could be judged.

This method is very useful to determine the relationship of plants of unknown or doubtful affinities.

Lyman Benson of California published his work in 1957 in his book *Plant Classification*. It was basically a Besseyan-type system and the relationship between the different groups was indicated in a unique manner. Instead of producing a phylogenetic tree, he represented the whole system in the form of a two-dimensional horizontal chart. The Dicotyledoneae were divided into five groups which were informal polyphyletic grades representing stages in floral evolution. The woody members of the Ranales were regarded as the most primitive taxa, and it was presumed that the ancestors to the angiosperms also retained the features of this group, in addition to other primitive features. The position of any order in this chart shows its approximate degree of specialisation and also the degree of departure from the features characteristic of the order Ranales amongst the dicots, and of the order Alismatales amongst the monocots. This system was not accepted widely because of certain demerits such as: (a) the primitive families under Ranales do not appear to be a natural group; (b) division of Sympetalae into Corolliflorae and Ovariflorae is artificial, and (c) inclusion of Proteales, Cactales and some other taxa in the Calyciflorae is similarly artificial.

Armen Takhtajan of the Soviet Academy of Sciences, Leningrad, USSR, proposed his system of classification in 1954 (translated into English in 1958). Some modifications were made in his later revisions in the years 1969, 1973 and 1980. Hallier's synthetic evolutionary classification of flowering plants had made a great impression on him. According to Takhtajan (1980), Hallier had a deep insight into the morphological evolution and phylogeny of flowering plants, and his system was a synthetic one.

In Takhtajan's system of classification, Magnoliales was suggested to be the most primitive taxon that has given rise to all the other groups of angiosperms—both dicotyledons and monocotyledons. It is based on all data available from various fields of study and therefore is a better phylogenetic system. Instead of the conventional terms, he used Magnoliophyta for the angiosperms, Magnoliopsida for the dicots and Liliopsida for the monocots. Magnoliopsida is divided into 7 subclasses, 20 superorders, 71 orders and

333 families. Liliopsida is divided into 3 subclasses, 8 superorders, 21 orders and 77 families (1980). He introduced a supplementary rank of superorder between the subclass and order. The superorder names have—anae endings. He derived Magnoliophyta monophyletically from Bennettitalean ancestors.

Arthur Cronquist, of the New York Botanical Garden, published his classificatory system in his book, *Evolution and Classification of Flowering Plants* (1968). It was later revised in 1981 in the book *An Integrated System of Classification of Flowering Plants*. In the later system, Cronquist replaced the usual terminology—Dicotyledoneae and Monocotyledoneae—with Magnoliatae and Liliatae, respectively.

The class Magnoliatae is further divided into six subclasses, of which the subclass Magnoliidae is assumed to be the basal complex from which all other subclasses have been derived. This is the most primitive subclass. The members of subclass Hamaelididae are mostly wind-pollinated families with reduced, chiefly apetalous flowers, often borne on catkins. The Caryophyllidae members are characterized by free-central or basal placentation and the occurrence of betacyanins in many of them. The Rosidae and Dilleniidae are supposed to be parallel groups with very little morphological distinction. The difference between the two groups lies in the centripetal development of stamens, uniovulate locules and the presence of a nectariferous disc in the Rosidae, and centrifugal development of stamens, multiovulate locule and absence of nectariferous disc in the Dilleniidae. The Asteridae are the higher sympetalous families with stamens, usually as many as corolla lobes. These are mostly derived from Rosidae.

The class Liliatae consists of four subclasses, and is derived from primitive Magnoliatae of herbaceous habit with apocarpous flowers, ordinary perianth, and uniaperturate pollen.

Takhtajan's (1969) four subclasses of Liliopsida were adopted in Cronquist's system with some changes in the Liliidae and Commelinidae. The most primitive Alismatidae is an aquatic group. The Arecidae are characterized by the large petiolate leaves, arborescent habit, and flowers in spadix. Commelinidae members have reduced floral structure accompanied by wind pollination. The most advanced Liliidae have modified stem and insect pollination.

Soo (1975) presented a system slightly modified from those of Takhtajan and Cronquist, chiefly based on phytochemical studies. In this system the subclass Dilleniidae was replaced by the Malvidae, and a few families were excluded from the former. Also, he changed the terminology—Dicotyledonopsida for dicots, Monocotyledonopsida for monocots, and Eucomiidae for Hamamelididae.

Robert Thorne presented a synopsis of his classificatory system in 1968, and the detailed system was published in 1976. This system was claimed to be considerably different from those of Takhtajan and Cronquist, and included 21 superorders, 50 orders, 74 suborders, 321 families and 432 subfamilies. His aim was to stress relationships rather than differences. However, there have been criticisms also from various quarters.

R.M.T. Dahlgren (1975a, 1977a) published a system of classification which agreed with the major features of Takhtajan's and Cronquist's systems. He, however, tried to represent the system in the form of a three-dimensional phylogenetic shrub in transection instead of the traditional two-dimensional representation. In this system (as revised in 1980a), the angiosperms or Magnoliopsida have two subclasses— Magnoliidae (dicots) and Liliidae (monocots). The former is divided into 24 superorders, 80 orders and 346 families, and the latter into 7 superorders, 26 orders and 92 families. An attempt was made to utilise as much data as possible to bring out the distinctions between different groups. He used the ending -florae instead of -anae for the nomenclature of superorders.

In the recent past, many workers have published classificatory systems based on phytochemical data (**Dahlgren** 1983a, b, **Gornell et al.** 1979, **Soo** 1975, **Young** 1981). Cronquist (1983) warned against constructing phylogenetic classificatory systems based on "concealed evidence" such as serological data, amino acid sequence data, biosynthetical pathways of secondary metabolites, and minute details in embryology and other microstructural evidence. In doing so, we may deviate to such an extent from a phenetically useful classification that the practical taxonomist would be happy to completely ignore our classification in favour of an artificial one that will prove useful in the field and in a modestly equipped laboratory. There may some day be one type of classification for pheneticists and another for phylogenists. In fact, this is already happening—in floras and practical handbooks, mostly Englerian classification is followed, while in academic institutions the approach is more phylogenetic.

All classificatory systems, however phylogenetic they may be at the moment, are only temporary or transient in character. They are not permanent because new findings will always bring about changes. However, it is always useful to study the historical background to aquaint oneself with the basic principles of the modern classificatory systems, which may otherwise appear to be arbitrary.

Classificatory Systems: A comparative study

3

It is an impossible task for any one (person or group) to study all the plants that occur on this earth. Even if we consider the angiosperms alone, the situation is no better, because this is the most dominant group on the global surface. We therefore have to classify them into groups on the basis of their similarities and dissimilarities, and then arrange these groups in different levels or categories. This arrangement—or classification, as it is termed—is exceedingly important.

From time to time, taxonomists have proposed different classificatory systems: Artificial Systems, Natural Systems, and Phylogenetic Systems.

An **Artificial System** classifies organisms usually by one or a few characters, irrespective of any relationship amongst them. It is only for the sake of convenience and as an aid to identification.

A **Natural** or **Formal System**, as it is sometimes called, is based mainly on morphological features which can be examined with the naked eye. The use of correlated characters is also important in these systems. Another important feature is the use of as many characters as possible so that the closely related taxa can be placed together or close to each other. This so-called Adansonian Principle (after Michel Adanson 1727–1806) is the basis of the Natural System of Classification. A natural classification aims to arrange all known plants into groups, which are graded according to the degree of resemblance so that each species, genus, tribe, family and order stands next to those it resembles in most respects. This system is based on data available during a particular period and the relationship amongst the taxa is a function of the overall similarities and dissimilarities. This is also known as Phenetic Relationship and is understood easily without referring to ancestral groups. The advantages of the Natural System are: (*a*) plants alike in hereditary constitution are grouped together, (*b*) a great deal of information is obtained, and (*c*) additional information can be easily incorporated.

Phylogenetic System. Darwin (1859) proposed the theory of evolution, in which he brought forward the fact that the present-day plants originated from some ancestral ones after undergoing periodical modifications due to environmental changes; therefore, all living plants today are related to each other in one way or the other. Thus, later classificatory systems are mostly phylogenetic, showing the presumed evolution of different plants or plant groups, and usually constructed on the basis of natural classification.

Natural Classification is often known as horizontal classification, i.e. based on data available during a particular time period. Phylogenetic classification, on the other hand, is known as vertical classification, as it mainly depends upon evolutionary relationship, or presumed ancestry. Sometimes this is designated Evolutionary Classification.

The major goal of taxonomic studies is to have a truly phylogenetic classificatory system. One of the best pre-evolutionary natural systems of classification was that of George Bentham (1800–1884) and Sir Joseph Dalton Hooker (1817–1911). The generic descriptions written from actual herbarium specimens were models of completeness and precision. Even now, this system is the most convenient for identification of plants in the field.

Two German botanists, Adolf Engler (1844–1930) and Karl Prantl (1842–1893), published their classical work *Die Natürlichen Pflanzenfamilien* during the post-evolution era. This system was widely accepted over a long period of time. It is a well-illustrated work with phylogenetic arrangement and modern keys and provides data for identification of all the known genera of plants from the primitive algae to the advanced seed plants. However, they did not recongnise the significance of reduction and, therefore, simplicity of structure was equated with primitiveness. The catkin-bearing families of "Amentiferae" were treated as primitive and placed before the petaliferous families like Ranunculaceae, Magnoliaceae and others. Another criticism is the placement of the monocotyledons before the dicotyledons.

John Hutchinson (1884–1972) considered the angiosperms as having originated monophyletically from the Hypothetical Pro-angiosperms. Though similar to natural systems, an attempt was made to arrange the plant groups according to their routes of descent (Jones and Luchsinger 1987); but Hutchinson made the error of splitting the dicotyledons into two linear evolutionary lines, the Herbaceae and the Lignosae. This unnatural division resulted in separating closely allied families. For example, Saxifragaceae was separated from the related family Rosaceae, and the Araliaceae from Umbelliferae. Also, families like the Euphorbiaceae and Papilionaceae include a large number of herbs and shrubs as well as trees. Hutchinson was, however, well-versed in the families of angiosperms all over the world, and therefore his treatments of these families are most informative.

Armen Takhtajan's System was more natural than any other system published so far and the inclusion of evidence from all branches of botany (anatomy, embryology, palynology, chromosome number, vegetative and floral morphology, chemical features and geographical distribution) made it a phylogenetic system. With further progress in our knowledge, this system is liable to change. Derivation of the monocots from the Nymphaeales is another point of criticism. Similarities between the two taxa are probably due to convergent evolution, and the ancestors of both are completely obscured in geological history (Stebbins 1974).

Dahlgren's System is a modification of Takhtajan's system and has similar approach. This system of classification was revised in 1975, 1977, and 1980, and yet, according to Dahlgren, it is still provisional. Monocots (Liliidae), according to him, are possibly a monophyletic group due to (a) single cotyledon, and (b) characteristic triangular protein bodies in the sieve tube plastids. Use of the termination -florae (e.g. Rosiflorae) for superorders is a point of criticism. Like Takhtajan's System, with further increase in our knowledge, this system, too, is liable to change.

A comparison of the placement of families in Bentham and Hooker's, Hutchinson's, Cronquist's Dahlgren's and Takhtajan's Systems in relation to Melchior's (1964) system is provided in Appendix 5.

A comparative account (Table 3.1) of some important classificatory systems follows:

Table 3.1 Comparative account of five classificatory systems—Bentham and Hooker, Engler and Prantl, Hutchinson, Takhtajan and Dahlgren.

Bentham and Hooker's System
1. First published between 1862 and 1883 as *Genera Plantarum* in Latin, a three-volume work.
2. This system essentially deals with seed plants, and around 97000 species are described.
3. It was modelled directly on the system developed by de Candolle but is refined.
4. A system based on form relationship and corelated characters.
5. The seed plants are divided into three major classes—Class I-Dicotyledons, Class II–Gymnosperms, and Class III–Monocotyledons. Monocotyledons are treated as the most advanced.
6. Each class is further divided into subclasses, each subclass into series, and each series into cohorts; the cohorts include the families.
7. One of the unusual features of this system is the position of the gymnosperms between the dicotyledons and the Monocotyledons.
8. Although the system was more natural than that of the earlier workers like de Candolle, a number of taxa could still not be classified satisfactorily. Such orders as could not be accommodated anywhere were placed under "Ordines anomali".

(Contd)

Engler and Prantl's System
1. First published as *Die Natürlichen Pflanzenfamilien,* a 23-volume work, between 1887 and 1915.
2. The entire plant kingdom, i.e. from algae to angiosperms has been described.
3. The classification of Eichler was adopted and modified.
4. This system emphasised that the incomplete or unisexual flowers were primitive. The subsequent addition of other floral whorls was an advancement.
5. In this system the angiosperms were treated as a division—Angiospermae—divided into classes—Monocotyledons and Dicotyledons. Monocots treated as primitive.
6. The Class Monocotyledonae includes 11 orders, some of which are further divided into suborders and families. The class Dicotyledonae is divided into two subclasses—Archichlamydeae (33 orders) and Sympetalae (11 orders) which in turn are divided into orders and families.
7. The Gymnospermae were placed before the monocots and were presumed to be the progenitor of the catkin-bearing Amentiferae. Angiosperms were considered to be polyphyletic; derived from seed ferns as well as the gymnosperms through Amentiferae.
8. Although intended otherwise, the system proved to be more natural and less phylogenetic. Interpretation of simple unisexual flowers (Amentiferae) as primitive is one of the main demerits.

Hutchinson's System
1. First published as *The Families of Flowering Plants,* in two volumes—Vol. I Dicotyledons in 1926 and Vol. II Monocotyledons in 1934.
2. This system mainly concerns the angiosperms.
3. Based on the principles of Charles Bessey's dicta on the relative primitivity and advancement of plant characters.
4. Based on the principles that evolution is both progressive and retrogressive and that all parts of a plant may not be involved in evolution at the same time.
5. The angiosperms were considered to originate from hypothetical Pro-angiosperms, and were classified into two major groups—Dicotyledons and Monocotyledons. A cleavage of the Dicotyledons into a herbaceous (Herbaceae) and a woody (Lignosae) group; two evolutionary lines were suggested. The Monocots have been derived (at an early stage) from the Ranales of Herbaceae.
6. Amongst Herbaceae, the most primitive order is the Ranales, and the most advanced order is the Lamiales; amongst Lignosae the most primitive order is the Magnoliales and the most advanced is Verbanales. In the Monocots, three groups—Calyciflorae, Corolliferae, and Glumiflorae—are based on the nature of the perianth; Butomales is the most primitive and Graminales the most advanced order.
7. Gymnosperms form a linear monophyletic series beginning with the primitive Cycadaceae and ending in the most advanced Cupressaceae.
8. This system is phylogenetic in nature, but has not been followed widely. The controversial taxon Amentiferae is regarded as an advanced group.

Takhtajan's System
1. First published as *Die Evolution der Angiospermen,* in one volume, in 1959.
2. Mainly concerned with angiosperms.
3. Influenced by Hallier's work, who attempted to create a synthetic evolutionary classification of flowering plants.
4. The important features of this system are—(*a*) Magnoliales s.l. is the most primitive group that gave rise to all the branches of angiosperms, and (*b*) Monocotyledons and Nymphaeales are derived from a hypothetical common dicotyledonous ancestor with vesselless wood and monocolpate pollen.
5. The angiosperms are considered monophyletic in origin, from primitive fossil orders like Bennettitales. Angiosperms are termed Magnoliophyta, which is classified into classes Magnoliopsida and Liliopsida; Liliopsida or the monocots are derived from the Nymphaeales.
6. Magnoliopsida includes 7 subclasses, 20 superorders, 71 orders and 342 families, and Liliopsida includes 3 subclasses, 8 superorders, 21 orders and 77 families (1980). Amentiferae is considered to be an advanced group; the naked, unisexual inflorescences are derived from multiwhorled, bisexual flowers and inflorescences.

(Contd)

arranged that it represents the sequence of repeated branching or "cladogenesis" and the degree and character of evolutionary modifications of branches (Takhtajan 1980).

Angiosperms are the dominating vegetation of most terrestrial ecosystems and consist of ca. 250,000–300,000 extant species. But this dominant plant group's origin has remained shrouded in mystery because of the apparently uninformed fossil record, uncertain relationships among the living members of this group and probably the inseparable morphological 'gaps' between the angiosperms and other seed plants, i.e. the gymnosperms.

The angiosperms could have originated variously—from some extinct gymnosperms, from the Bennettitales, from the Pteridosperms (seed-ferns) and more or less all the groups of extinct and extant gymnosperms could be the potential ancestors. Recent studies based on morphological and molecular evidences support monophyletic origin of the angiosperms and also the earlier ideas that the extinct Bennettitales and extant Gnetales are the two taxa most closely related to the angiosperms. This group (Bennettitales, Gnetales and Angiosperms) has been termed 'anthophyta', i.e., they possess flower-like reproductive structures. But even then, in many respects the morphological gap between the various members of anthophyta is rather wide (Crane et al. 1995).

The earliest fossil angiosperms include relatives of the modern *Liriodendron* and *Magnolia*, as well as some members of the Amentiferae. This has led to controversy amongst the phylogenists: according to some, the Amentiferae should be considered as among the most primitive angiosperms, and according to others the Ranales should be treated as the basal group in the evolution of the angiosperms. According to Bessey (1915), within any one phylad (an evolutionary line), the evolution of different organs may proceed at different rates. At any one time, any particular group may present both relatively advanced as well as relatively primitive characters. As a result, no one family will have all primitive features; instead, they will have some comparatively advanced features in addition. On the other hand, it is also true that a family with highly primitive features will never show highly advanced, but only comparatively advanced, characters. For example, in the primitive subclass Magnoliidae (sensu Cronquist 1968, 1981), some members are primitive in certain features but advanced in certain others. The family Winteraceae is vessel-less (primitive character) but has sieve tubes and companion cells (advanced character). Austrobaileyaceae have vessels but very primitive, gymnosperm-like phloem, without any companion cells (Cronquist 1968).

The hypothetical primitive angiosperm was an evergreen tree or large shrub with alternate, simple, entire, stipulate, pinnately net-veined leaves. The flowers were borne singly in the axils of leafy bracts, were bisexual, large and showy with numerous, spirally arranged perianth lobes which were not distinguishable as calyx and corolla, or only sepals were present and petals absent. Numerous stamens were spirally arranged and were laminar, i.e. without any differentiation into filament and anther. Carpels were numerous and free, usually stipitate and with unsealed margins which were covered with glandular hairs, which formed the elongated stigma. Placentation in many taxa was laminar, i.e. unilocular ovary bearing ovules over the whole abaxial surface. Ovules were anatropous, bitegmic and crassinucellate, and seeds endospermous.

Recent studies emphasize great diversity of floral forms in the Magnoliidae—variation in number and arrangement of floral parts is extreme and both large, multiparted, bisexual flowers as well as small, simple, often unisexual flowers quite common amongst the members. Current morphological and molecular evidences favour phylogenetic models with small, trimerous flowers (or even simpler) to represent the basal angiospermous form. The available fossil record (from Portugal and North America) also reveals that these flowers are generally few-parted and often with undifferentiated perianth; stamens with small pollen sacs, valvate dehiscence and apically extended connective; and carpels with poorly differentiated stigmatic surface (Crane et al. 1995).

The trends in evolution of the flower have taken place by (1) reduction in number, (2) fusion, (3) specialisation of parts, and (4) changes in symmetry.

Flower. The primitive angiosperm flower might be considered as having numerous, spirally arranged, prominent tepals, numerous, spirally arranged stamens and numerous unsealed carpels. All these characters occur in one or the other member of the present-day Magnoliales. Flowers of *Degeneria* and Winteraceae are of medium size, and with a moderately elongated receptacle. According to Stebbins (1974), the early angiosperms had flowers of moderate size, which supports the hypothesis that they were small woody plants, inhabiting pioneer habitats exposed to seasonal drought. "Under these ecological conditions, rapid development of flowers and seeds would have had an adaptive advantage and would be most easily acquired by reduction in size of the reproductive shoots" (Stebbins 1974). Hallier (1912) and Parkin (1914), however, had suggested large-sized flowers as the most primitive.

According to Takhtajan (1980), large flowers, as seen in many Magnoliaceae and Nymphaeaceae, as well as large flowers like *Rafflesia arnoldii* are of secondary origin, and evolved in response to selection pressure for different methods of pollination. Small flowers like those of Monimiaceae and Amborellaceae are also derived and may be correlated with specialisation of the inflorescence or reduction of the whole plant.

As evolution progressed, shortening of the receptacle brought the floral parts closer together, in a series of whorls. The number of floral parts in each whorl is reduced, and they become connate or adnate.

Other tendencies are towards elaboration and differentiation of parts. These two apparently opposing tendencies may sometimes be expressed in the same structure. Sympetalous corolla has evolved from polypetalous corolla through union and at the same time, many sympetalous flowers are also irregular, e.g. many members of the Scrophulariaceae (*Antirrhinum majus*).

However, these evolutionary tendencies are independent of each other and, as a result, a flower selected at random will show some relatively primitive and some relatively advanced features, e.g. sympetalous flowers may be polysepalous as in the Sapotaceae, or vice versa, as in some Caryophyllaceae; in Papilionaceae the calyx is synsepalous and regular but the corolla is polypetalous and irregular.

Pentamerous vs. Trimerous Flowers. Trimerous flowers are one of the unifying traits of the monocotyledons. In a few cases where they are absent (Pandanaceae and Sparganiaceae) in this group, some hypotheses relate the simpler flower structure to the trimerous ground plan. The same trimerous flowers are also seen in the Magnoliales, Laurales, Ranunculales and Piperales, and to some extent in the Polygonales.

The evolutionary origin and systematic significance of trimerous flowers is debatable. Based on the distribution of trimerous flowers and their correlation with other characters, Dahlgren (1983a) concludes that this trait is ancient and must have appeared in the angiosperms before differentiation of the monocotyledons, and that the trimerous flowers presumably occurred side by side with flowers with helically arranged parts. Referring to the frequent occurrence of trimerous flowers in the Ranunculales, Dahlgren (1983a) goes on to say that the pentamerous condition has most likely evolved out of a trimerous state in this order, and a similar situation is also observed in the Polygonales. In Nymphaeaceae, he presumes that the large polymerous flowers are derived from small or medium-sized trimerous flowers as observed in the present-day *Cabomba*. Dahlgren and Clifford (1982) also indicate the possible derivation of large, polymerous flowers with helical phyllotaxy (Nymphaeaceae, Magnoliaceae and Illiciaceae) from fairly small and oligomerous flowers in which there was "a tendency for the whorls to be trimerous."

This would mean a return from trimerous condition to pentamerous condition and spiral anthotaxy.[1]

---

[1]Anthotaxy—a term used as against "phyllotaxy" for the floral parts (Kubitzki 1987).

Burger (1978) proposes the derivation of trimerous flowers from the phylogenetic fusion of three simple flowers by contraction of internodes. This concept—and claiming the phylogenetic primacy of the monocotyledons (Burger 1981)—would turn the situation upside down, because ultimately trimerous flowers would have given rise to pentamerous or even polymerous flowers.

A transition from spiral to trimerous anthotaxy has been demonstrated in the Magnoliaceae by Erber and Leins (1982, 1983). In *Magnolia stellata* the flowers exhibit fully spiral anthotaxy, whereas in *M. denudata* and *Liriodendron tulipifera* the floral envelope consists of three whorls, one considered as calyx and two as corolla. This condition is derived from two situations: (1) after the inception of three floral primordia, there is no space left for any further primordia in the same whorl, and (2) a pause after inception of every third element of the floral envelope leads to its trimerous arrangement. Tucker (1960) and Hiepko (1965) also presented similar findings in *Michelia* and *Magnolia,* respectively. It is a fact that the pentamerous condition is the most common amongst the dicotyledons. In the Ranalean complex, however, co-occurrence of the petamerous, trimerous, as well as dimerous condition has been observed, and sometimes even in the same individual. For example, almost all genera of the Berberidaceae have a trimerous flower, but in *Berberis vulgaris* the terminal flower of the inflorescence is often pentamerous. Another genus, *Epimedium*, has a dimerous flower. This feature reflects a close association between trimery, pentamery and dimery. Similar examples are also reported amongst the members of Lauraceae (Mez 1889, Kasapligil 1951). Dimery frequently occurs with trimery in members of Annonaceae, Lauraceae, Berberidaceae and Papaveraceae (Kubitzki 1987). In Monimiaceae, many primitive genera like *Hortonia* (Endress 1980a), *Laurelia* (Sampson 1969), *Trimenia* (Endress and Sampson 1983) have spiral anthotaxy, while in the more advanced genera like *Wilkiea, Kibara, Tetrasynandra, Steganthera* and *Tambourissa* (Endress 1980b), floral envelopes are dimerous. The trimerous condition is absent altogether. This makes it clear that dimerous and trimerous conditions have both originated from spiral anthotaxy, and the pentamerous condition might be the evolutionary equivalent of two trimerous or dimerous whorls (Kubitzki 1987). The transition from spiral to whorled anthotaxy is documented in several extant families of the Ranalean complex and must have taken place in various parallel evolutionary lines. Except for its predominance amongst the monocots and frequent occurrence in the Ranalian complex, trimery is relatively rare in other dicots. Kubitzki (1987) proposes that trimery is a morphological constraint which offers only very limited possibilities for meristic variation, with no possible return to pentamery or spiral anthotaxy.

Unisexual vs. Bisexual Condition.   The primitive flower was bisexual with both stamens and pistils, and unisexual flowers are derived from the bisexual ones. Sometimes, there is an intermediate stage with both types of sex organs well developed, but only one is functional, e.g. in Sapindaceae, some members have an apparently bisexual but functionally unisexual flower. Some members of Compositae exhibit a unique case of unisexuality where the disc florets are functionally staminate. In these florets, the ovary is phyletically lost, but pollen from the anther tube is still pushed out by the elongating style.

Monoecious and dioecious groups are also derived from bisexual ancestors through many intermediate stages.

Hypogynous vs. Epigynous Flower.   The primitive flower was hypogynous and with free sepals, petals and stamens arising from below the ovary. The basal parts of the different floral whorls are frequently fused to form a structure—the hypanthium, which in some plants apparently resembles a calyx. In such plants, the petals and stamens appear to be inserted on the calyx tube, e.g. members of the Papilionaceae. In some others, however, such as Cucurbitaceae, the hypanthium is distinct, so that the calyx lobes, corolla and stamens appear to be arising from the apex of the hypanthium. Flowers with hypanthium and a superior ovary are perigynous, which is intermediate between hypogynous and epigynous conditions. The epigynous condition is derived from the perigynous when the hypanthium becomes adnate to the ovary, or the bases of the outer floral whorls may be fused with the ovary wall, as in the Rubiaceae, or the ovary itself may become submerged in the receptacle, as in the Santalaceae.

Perianth. Although the primitive angiospermous flowers had no corolla, and the perianth consisted of the sepals only, in modern angiosperms, the presence of a corolla is a primitive feature and its absence is derived. Petals (or perianth) might have originated either from bracts (as in the Magnoliales, Illiciales and Paeoniales) or from stamens as in the Nymphaeales, Ranunculales, Papaverales, Caryophyllales and Alismatales (Takhtajan 1980). The extra petals in "double" flowers of many species are staminodial in origin.

The usual tendency of the perianth to consist of an outer protective part, the calyx, and an inner attractive part, the corolla, is essentially functional. There is, however, much variation and deviation from this usual pattern, e.g. in *Delphinium* and *Aconitum* of Ranunculaceae, the sepals are large and showy like petals, and the petals function as nectaries. In another example, *Mirabilis jalapa* of Nyctaginaceae, the calyx is corolloid and is subtended by a calyx-like involucre of five basally united bracts. In the most advanced families, usually there is only one whorl each of calyx and corolla. Two sets of calyx and two sets of corolla are much less frequent (Cronquist 1968).

Androecium. Comparative studies of the stamens of living angiosperms show that the most primitive type of stamen is a broad, laminar, three-veined organ which is not differentiated into filament or anther. It develops two pairs of elongated microsporangia embedded in the abaxial or adaxial surface, between the midvein and lateral veins (Fig. 4.1). Canright (1962) regards the stamen of *Degeneria* as "the closest of all known types to primitive angiosperm stamen". In *Degeneria, Galbulimima, Lactoris, Beliolum* (Winteraceae), Annonanceae and *Liriodendron,* the microsporangia are situated on the abaxial surface and therefore the stamens are extrorse. On the other hand, in the Magnoliaceae (except *Liriodendron*), Austrobaileyaceae and Nymphaeaceae, the position of the microsporangia is on the adaxial surface, and the stamens are introrse (Takhtajan 1980). It is not certain, however, which of these two conditions is more primitive. According to Takhtajan (1980), both the abaxial and adaxial position of microsporangia have been derived from a common ancestral type, which could only have been marginal.

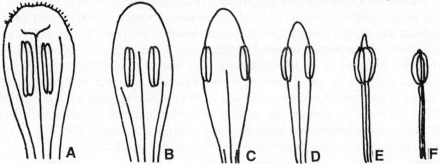

Fig. 4.1 Evolutionary Trends in Androecium.

Many taxonomists, including Moseley (1958), Eames (1961), Canright (1962), and Cronquist (1968) consider that deeply sunken microsporangia in the staminal tissue, as in *Degeneria* and *Galbulimima,* is a primitive feature. Other evolutionary features observed in androecia include the reduction in the number of stamens from many to a few. This feature sometimes coexists within the same genus. In *Hibbertia,* of the family Dilleniaceae, all conditions from numerous free stamens, to five fascicles of stamens to five separate stamens and lastly to a single stamen are reported (Wilson 1964). In Guttiferae the genus *Hypericum* has a similar series from numerous separate stamens to five fascicles of three stamens each.

Once the stamens have been reduced to a single whorl, the number of stamens usually does not increase. On the other hand, there is a possibility of further reduction in their number, as in Bignoniaceae, Labiatae and Scrophulariaceae. In these families, the number of corolla lobes is five, but the number of stamens is two or four, the missing stamens often represented by staminodia. However, according to some workers (see Cronquist 1968), in certain families of the Guttiferales and Malvales, progression has been from a few stamens in a single whorl to fascicled stamens and to numerous separate stamens. Eames (1961) interpreted the anatomical evidence in these families as indicating a reduction series. Cronquist (1968) also supports this view.

Pollen Grains. Most primitive angiosperm pollen is considered to have a single distal germinal furrow or colpus or sulcus in the sporoderm (Cronquist 1968, Sporne 1972, Stebbins 1974, Takhtajan 1980). Angiosperm pollen grains are mainly of two types: uniaperturate and triaperturate. The former has a single germinal pore or furrow usually on the surface opposite the contact point of the pollen grains in a tetrad. The latter has three germinal furrows which do not closely approach each other.

Uniaperturate grains are characteristic of the monocotyledons and some members of the Magnoliales and Ranales, and triaperturate grains are known in most of the dicotyledons. Apart from these two main types, multiaperturate and nonaperturate grains are also known. These are the derived ones and, hence, more advanced.

Pollination. Discoveries of the early- and mid-Cretaceous fossils of angiosperm flowers indicate that the early members of this group were insect-pollinated. Stamens in these fossil flowers have small anthers with valvate dehiscence, lesser amount of pollen grains; generally poorly developed and unelaborated stigmatic surface and smaller pollen grains (than those for wind dispersal). These early flowers were most probably pollinated by pollen-collecting or pollen-eating insects. Fossils of flowers pollinated by nectar-collecting insects appeared much later (Crane et al. 1995).

Gynoecium. The most primitive type of carpel is essentially a megasporophyll and this concept of megasporophyll-like carpel fits very well with the Ranalian theory of angiosperm evolution.

It was assumed for a long time that the primitive pre-angiospermous carpel had a row of ovules along each margin of the laminar structure, and with the closure of such a carpel, the two rows of ovules were brought together and formed a single row. Bailey and Swamy (1951) and Eames (1961) are of the opinion that the most primitive carpels are unsealed, conduplicate, and more or less stipitate structure with a large number of ovules scattered on the adaxial surface. This type of conduplicate carpel is seen in *Degeneria* and *Tasmannia* and also in some primitive monocotyledons. Another important characteristic of the primitive carpel is the absence of a true style, and stigmas occurring along the margin of the conduplicate carpels. These stigmatic margins are not fused at the time of pollination and, in the course of evolution, this primitive stigma was transformed into subapical and then apical stigmas (Fig. 4.2). All these transitional stages can be seen amongst the members of the Magnoliales and Ranales.

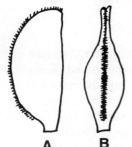    

A   B   C   D   E   F

Fig. 4.2 Evolutionary Trends in Carpels.

Primitive angiosperm gynoecium consisted of a large number of carpels, arranged spirally on the surface of a more or less elongated receptacle. The carpels, each containing a large number of ovules, spread open when mature and let out the seeds. Amongst both primitive dicots and monocots, this type of carpel is common. Sometimes, these spirally arranged free carpels become fused to form a "pseudo-syncarpous" condition, as in some species of *Magnolia*, but this line of evolution did not progress any further.

In the normal course of evolution, the number of carpels became less and they became restricted to a single whorl. A syncarpous gynoecium emerged from an apocarpous gynoecium by lateral fusion of closely connivent carpels arranged in a whorl. The primitive form amongst these still shows free upper portions of the fertile region of the carpels but, eventually, a compound style with one compound stigma is achieved.

There may be further reduction in the number of carpels in some families, which is in accord with the usual trend of evolution. The genus *Linnaea* of the Caprifoliaceae has a trilocular ovary but only one of these locules contains a normal ovule which develops into the seed. The other two locules contain several abortive ovules. A similar feature is seen in *Valerianella* of the Valerianaceae, in which two locules of the trilocular fruit are sterile. In another member of this family *Valeriana*, the ovary is tricarpellary with a trilobed stigma but two of the carpels are vestigial. In addition, there are several other families (Leitneriaceae, Krameriaceae) with "pseudomonomerous" gynoecium, i.e. an apparently monocarpellary ovary which is phyletically derived from a many-carpelled ovary.

Placentation. According to Takhtajan (1980), "the main directions of evolution of the gynoecium determine the main trends of evolution of placentation". The most primitive type of placentation is the laminar-superficial type, i.e. ovules scattered over the adaxial surface. Ovule position restricted only to the margin of these laminar structures was an early evolutionary step, and is termed laminar-marginal. From this stage with complete closure, marginal plancentation was derived. When several carpels in a whorl, each with marginal placentation, fuse at their margins, axile placentation is derived. If, however, there is incomplete closer, fusion of such carpels in a whorl, this will give rise to parietal placentation (Fig. 4.3). From axile placentation, through the process of reduction, both apical and basal placentation are derived (Fig. 4.3).

In many Apocynaceae (*Catharanthus roseus*) and Asclepiadaceae (*Calotropis procera*), there are two separate ovaries but style and/or stigma are fused. This condition is secondary in origin and not a primitive feature. There is no evidence to show that carpellary fusion has proceeded from the stigmatic end. Whatever the reason of the secondary splitting of the ovaries may be, the style and stigma have remained fused so that both the carpels may be pollinated by the same agent from a common pollinating surface and this is, no doubt, advantageous.

Most of the advanced families show an axile, parietal, free-central, basal and apical type of placentation.

Ovules. Anatropous, orthotropous, amphitropous and campylotropous ovules are known amongst the angiosperms. Orthotropous is the most primitive type and is typical of the living gymnosperms. Most angiosperms have anatropous ovules, which means that this type of ovule must have formed very early in the evolution of the angiosperms. Campylotropous and amphitropous conditions are derived from the anatropous type. Although orthotropous ovules are known amongst the gymnosperms, in angiosperms they are believed to have been derived from anatropous ovules (Eames 1961, Cronquist 1968, Corner 1976, Takhtajan 1980).

The importance of the number of integuments in an ovule has long been recognized. The bitegmic ovules are considered to be more primitive as compared to the unitegmic ones. In particular, the multilayered integuments in the orders Magnoliales, Laurales, Aristolochiales, Piperales and Illiciales are generally

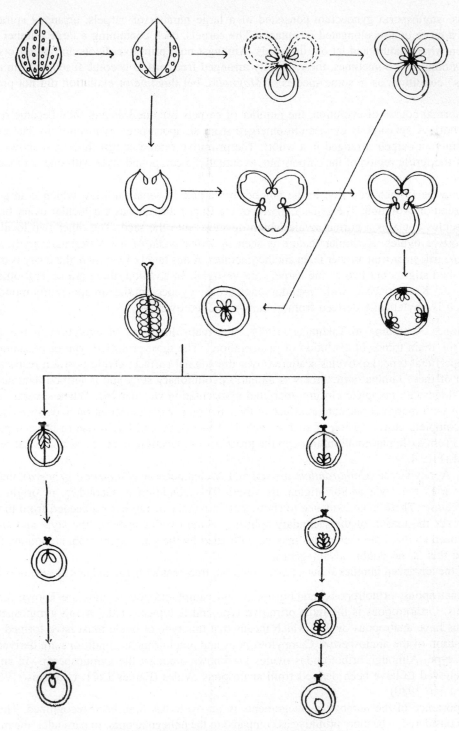

Fig. 4.3 Evolutionary Trends in Placentation.

regarded as the most primitive (Dahlgren 1975b). Bitegmic ovules are predominant amongst the monocotyledons and in most of the Polypetalae of the dicotyledons. Ovules are exclusively unitegmic in most of the gamopetalous orders of the dicotyledons. Unitegmic ovules sometimes occur in a number of isolated families of the orders with otherwise predominantly bitegmic ovules, proving that there are independent lines of evolution.

A unitegmic condition is a derived one and is formed either by fusion of the two integuments or by reduction of either one of them.

An interesting point of coincidence is that iridoids are almost 100% restricted to groups with unitegmic (and generally tenuinucellate) ovules (Jensen et al. 1975, Dahlgren 1975a). Usually, the primitive families amongst the angiosperms have crassinucellate ovules, and more advanced ones have tenuinucellate ones. In many groups of angiosperms, the development of the nucellus is directly related to the number of integuments and the type of endosperm formation (Dahlgren 1975b). Thus, the most primitive ovules are bitegmic and crassinucellate, and the most advanced ovules are unitegmic and tenuinucellate. The studies of Philipson (1974, 1975) have, however, shown that there are a large number of intermediate types, e.g. the bitegmic-tenuinucellate condition in Theaceae and Primulaceae, and the unitegmic-crassinucellate condition in Cornaceae and Araliaceae.

Seed. Seeds of primitive angiosperms are medium-sized, with abundant endosperm and a minute embryo which is often undifferentiated (Takhtajan 1980). In advanced seeds, however, the embryo is large but the endosperm is highly reduced and often absent. Both small and large seeds are derived.

Chemistry. Ellagitannins, and leucoanthocyanins, also called the proanthocyanins, are usually present in the primitive angiosperms. On the other hand, iridoids and glucosides are more common amongst the more advanced members, e.g. iridoids are distributed in most of the gamopetalous families, such as Gentianaceae, Apocynaceae and others.

To summarise, the trends of floral evolution are reduction in number, adnation or fusion of the floral parts, specialisation of these parts, and changes in symmetry. It has also been observed that the different parts of a plant do not evolve at the same rate, and as a result certain taxa show both primitive and advanced features. This phenomenon of unequal rate of evolution of different features within one lineage is known under various names: "chevauchement des spécialisations" (Dollo 1893) and "mosaic evolution" De Beer 1958). Different evolutionary stages or grades of different characters of the given taxon are the result of such mosaic evolution. Takhtajan (1959, 1966, 1980) named this difference in grades "heterobathomy" (Greek bathmos—step, grade).

Because of heterobathomy, an organism may present a mosaic combination of characters of quite different evolutionary levels. For example, the genera *Trochodendron* and *Tetracentron* have primitive vesselless wood and, at the same time, rather specialised flowers. On the other hand, the genus *Magnolia*, with primitive flowers, has rather advanced wood anatomy. Heterobathomy can also be expressed within the flower, e.g. in *Delphinium* the flower is zygomorphic and therefore advanced, but the stamens are arranged spirally, which is a primitive feature.

The more strongly heterobathomy is expressed, the more contradictory is the taxonomic information provided by different sets of characters. The "more heterobathomic a taxon, more complete and alround must be its study" (Takhtajan 1980).

Nomenclature

The purpose of giving a name to a plant is to provide an easy means of reference. Giving names to the newly acquired plants, or determining the correct name of already known plants follows a set of rules of nomenclature.

The elemental rules of nomenclature were first suggested by Linnaeus (1737, 1751; see Lawrence 1951). Augustine de Candolle's *Théorie Élémentaire de la Botanique* includes a detailed account of the rules for plant nomenclature (1813; see Lawrence 1951), and was the first significant work since the publications of Linnaeus. Later, these rules (also called de Candolle's rules) were adopted by the *International Code of Botanical Nomenclature (ICBN)*.

The first International Botanical Congress was held in Paris in August 1867, when many botanists from various countries met and adopted a set of rules for the naming of plants. Subsequent Congresses made significant contributions in modifying and amending some of these rules. It was only at the Cambridge Congress in 1930, that "for the first time in botanical history, a code of nomenclature came into being that was international in function as well as in name" (Lawrence 1951). The rules adopted at the Cambridge Congress (1930) were modified and amended, and the *International Code of Botanical Nomenclature (ICBN)*, presently in use, appeared in 1978. It was adopted at the 12th International Botanical Congress held in Leningrad (Russia) in August 1975.

The aim of the ICBN is to provide a suitable procedure of naming various taxonomic groups and also to avoid or reject such names as are contrary to rules and not valid.

Binomial Nomenclature

Man, from time immemorial, has been a "nomenclaturist". For his convenience, he has given names to everything that has come his way—animals, plants, birds or anything else. To begin with, the names given to plants were long descriptive sentences, e.g. *Grevillea robusta grandiflora australiana*. Although, these names were meaningful, it became impossible to remember such long plant names when the number of plants increased. Hence, this polynomial system did not continue for long. Gaspard (or Casper) Bauhin (1560–1624) came up with the novel idea of having only two names for every plant. He made a distinction between the generic name and specific epithet of the plants. This binary or binomial nomenclature, with which Carl Linnaeus's name is always associated, was, in fact, suggested by Casper Bauhin at least 100 years earlier.

However, the Swedish Botanist Linnaeus (1707–1778) was responsible for naming all living things from buffalo to buttercup, methodically applying two names to each, i.e. the binomial system. The vegetable kingdom named in this fashion was introduced in the book entitled *Species Plantarum* in 1753. The scientific names are in Latin, not the classical Latin, but a more or less popular Latin spoken by common people during the Middle Ages. One may raise an eyebrow and ask: "Why only Latin?" Well, Latin because: (1) It is specific, i.e. gives the precise meaning. (2) It is precise and concise and, therefore, (3) it is pertinent to the needs of descriptive phases of natural sciences. (4) Latin is written in the Roman alphabet and the confusion that will be created by the use of any other language of different scripts such as Chinese, Greek or Sanskrit, can be avoided. (5) Being a "dead" language now, it cannot arouse political controversy. Objections, however, have been voiced from many quarters against the use of Latin for plant names. Kelsey and Dayton (1942) in their *Standardised Plant Names* tried to introduce an English nomenclature using mostly common English names for plants, or anglicised Latin names. Using any spoken language and that, moreover, the national language of any country, is quite difficult. It is still more difficult, as no equivalent world flora has been published where English names or any other "vernacular" names for plants have been used.

Why not use the common names? There is no dearth of common names in any language. Benson (1962) pointed out the reasons why vernacular or common names cannot replace the Latin or Latinised botanical names:

1. Names in a common language are ordinarily applicable in only a single language; they are not universal.
2. In most parts of the world, relatively few species have common or vernacular names in any language.
3. Common names are applied indiscriminately to genera, species or varieties.
4. Often two or more unrelated plants are known by the same name, and frequently even in one language a single species may have two to several common names applied either in the same or different localities.

The same plant, *Piper nigrum*, is variously known as black pepper, white pepper, kali mirch, gole mirch, etc. On the other hand, the common name lily is used for many genera of the Liliaceae (*Lilium, Erythronium, Hemerocallis*), Iridaceae (*Belamcanda, Nemastylis*), Amaryllidaceae (*Zephyranthes, Crinum*) and Zingiberaceae (*Hedychium*). Moss is a group of plants belonging to the Bryophytes, but reindeer moss is a lichen, Spanish moss is an angiosperm, and bogmoss is *Sphagnum*. *Spathodea campanulata* is known by four different names in the English language alone: squirt tree, scarlet bell, fountain tree or African tulip.

Every binomial consists of two parts. The first part is the generic name, the second the specific epithet. The two parts should be in italics when in print, and underlined separately when typed or hand-written. The generic name always starts with a capital letter, whereas a specific epithet usually starts with a small letter, except in a few cases where the use of capital letter is permissible, e.g. *Pinus Roxburghii*. It is always a noun, also singular, and denotes the nominative case. There can be various sources, such as:

1. Names from many vernacular languages, e.g. *Salmalia* from shalmali (Sanskrit), *Madhuca* from madhukaha (Sanskrit), *Populus* from poplar (English), and others. Some names are based on the names in local languages of the areas where they occur, e.g. *Ginkgo* from the Chinese, *Tsuga* from the Japanese, *Nelumbo* from the Ceylonese and *Ravenala* from the Madagascarian.
2. Some names reflect the botanical character, e.g. *Trifolium* (with three leaves), *Cephalanthus* (flowers in heads), *Callicarpa* (with beautiful fruits), and *Liriodendron* (tree with lily-like flowers).
3. Many genera are named in honour of some famous botanist, e.g. *Bauhinia* (for Bauhin), *Hookera* (for Hooker), *Linnaea* (for Linnaeus); well-known scientists, e.g. *Einsteinia* (for Einstein); famous heads of state, e.g. *Victoria* (for Queen Victoria) and *Washingtonia* (for George Washington).
4. Some generic names are mythological in origin, e.g. *Narcissus* is after the famous Greek god Narcissus, *Circaea* or enchanter's nightshade refers to Circe, the famous enchantress, and *Nymphaea* refers to the water nymphs.
5. Some others are named after planets, e.g. *Mercurialis* after Mercury and *Neptunia* after Neptune.
6. Some are named after the name of country: *Salvadora* after El Salvador.

Specific epithets may likewise be derived from any source; it may be in honour of a scientist: *Pinus Roxburghii*; depicting some character of the plant: *Casuarina equisetifolia* (*Equisetum*-like leaves), *Jacaranda mimosifolia* (*Mimosa*-like leaves); or geographical distribution: *Ocimum americanum, Camellia sinensis*;

or simply after a vernacular name: *Psidium guajava* after guava. Often, the specific epithet is made up of two hyphenated words: *Hibiscus rosa-sinensis, Alisma plantago-aquatica.*

The two parts of the plant name usually belong to the same gender and often have similar endings. When the specific epithet is an adjective, it must agree in gender with the generic name. Usually the generic ending *-us* is masculine, the ending *-a* is feminine and the ending *-um* is neuter. By convention, all trees are considered feminine for nomenclatural purposes but exceptions are permissible: *Quercus rubra, Pinus nigra.* Here the generic names are both masculine but the specific epithets are feminine. In Latin there are four main sets of adjectives used as name-endings (see Table 5.1).

Table 5.1 Four Different Sets of Name-Endings used in Latin

	Masculine	Feminine	Neuter
1	-us *sativus*	-a *sativa*	-um *sativum*
2	-er *niger*	-ra *nigra*	-rum *nigrum*
3	-er *sylvester* *campester*	-ris *sylvestris* *campestris*	-re *sylvestre* *campestre*
4	-is *humilis* *occidentalis*	-is *humilis* *occidentalis*	-e *humile* *occidentale*

Given below are some common specific epithets and their meanings:

aphylla	-	leafless	nigra	-	black
alba	-	white	ochroleucus	-	yellowish-white
aquatica	-	in water	purpureus	-	purple
aureus	-	golden	palustris	-	of marshes or swamps
borealis	-	northern	repens	-	creeping
cerifera	-	wax-bearing	rara	-	rare
coccineus	-	scarlet	roseus	-	rose-coloured
communis	-	gregarious	serrata	-	with serrate margin
decumbens	-	reclining	scaposus	-	having a scape
dulcis	-	sweet	sulcatus	-	furrowed
edulis	-	edible	tridentata	-	with three spines
foetida	-	ill-scented	tenellus	-	slender, tender, soft
fluitans	-	floating	terrestris	-	growing on dry ground
grandiflora	-	large-flowered	tuberosus	-	tuberous
humilis	-	dwarf	uncinatus	-	hooked
linearis	-	narrow, linear	virens	-	green
magnus	-	large	viridis	-	green
minutus	-	very small	velutinous	-	velvety
mexicana	-	of Mexico	zeylanicus	-	of Ceylon

Another interesting feature of the specific epithets in honour of persons is that the proper ending is *i* or *ii* if the person honoured is a man: *agharkarii, baileyi*; it is *ae* if the person honoured is a woman: *margaratae, piersonae.*

Citation of Author's Name

To have a complete botanical/scientific name for a particular plant, it must be followed by the name of the person who identified and described the plant and suggested the name on the basis of this description, e.g. *Sesamum indicum* was identified, described and named by Linnaeus and, hence, should be written as *Sesamum indicum* L. In all systematic/taxonomic work, it is essential to cite the authority of the scientific names. Citation of the author's name is helpful if ever confusion results from two persons giving the same name. Author's names may be cited as full names or in abbreviated from, e.g. Roxb. for Roxburgh, Ait. for Aiton, Buch. -Ham. for Buchanan Hamilton, Cav. for Cavanilles, All. for Allioni, Wall. for Wallich, Bl. for Blume, Willd. for Willdenow. For two or more than two persons of the same family, different methods are adopted, e.g. William Hooker's name is abbreviated as Hook. and his son Joseph Dalton Hooker's as Hook. f., where f. stands for *filius* meaning son. In the de Candolle family, the father Augustin de Candolle is cited as DC., the son Alphonse as A. DC. and the grandson Casimir as C. DC.

If two persons have named a plant together, their names are joined by et or &, e.g. *Antigonon leptopus* Hook. & Arn.

When a name is proposed by one author and not validly published, and a second author has it published validly at a later date and ascribes it to the former author, the name of the former author, followed by the word ex, should be inserted before the name of the second author. For example, in *Cassia montana* Heyne ex Roth, Heyne proposed the name but did not publish it validly and the valid publication was done by Roth. The meaning of ex is "validly published by".

When a name proposed, described and diagnosed by one author is published in the work of another, the two names of the two author's are linked together by the word in. For example, *Hygrophila salicifolia* (Vahl) Nees in Wall.; *Euonymus indicus* Heyne ex Wall. in Roxb.

International Code of Botanical Nomenclature

Modern botanists all over the world use the *International Code of Botanical Nomenclature* (ICBN) which in a simple and precise manner deals with (1) terms which denote the ranks of taxonomic groups or units, and also (2) the scientific names which are applied to the individual taxonomic groups of plants (Greuter 1988). This code aims to provide a stable method of naming taxonomic groups, avoiding and rejecting the names that are ambiguous or create confusion.

Six principles form the basis of botanical nomenclature. The detailed provisions are divided into *Rules and Recommendations*. The objective of the rules is to put the nomenclature of the past into order and to reject the names that do not comply with the rules. The objective of the recommendations is to try to bring about greater uniformity and clarity, particularly in future nomenclature. Names that are recommended should await the formation of rules for their application. The *Rules and Recommendations* apply to all living organisms treated as plants (including fungi, but excluding bacteria), and also to fossils. The nomenclature of bacteria is governed by the *International Code of Nomenclature of Bacteria*.

The *principles* are the guidelines for the legitimate naming of any taxon:

I. "*Botanical nomenclature is independent of zoological nomenclature. The Code (Greuter 1988) applies equally to names of taxonomic groups treated as plants whether or not these groups were originally so treated.*" For the purpose of this code, "plants" do not include bacteria. As the code provides only for the nomenclature of plants, the same name may sometimes be assigned both to a plant and to an animal, e.g. *Cecropia* is the name of a moth, according to zoological nomenclature, and at the same time it refers to a tree belonging to the Moraceae.

II. "*The application of names of taxonomic groups is determined by means of nomenclatural types.*"

According to this principle, the name of each species is permanently associated with a particular specimen, the nomenclatural type. The type for a genus is a species, for a family it is a genus, and for an order it is a family. The following types are recognised:

The *holotype* is the one specimen or other element used or designated by the author in the original publication as the main nomenclatural type. Any type selected after the original publication is not to be regarded as a holotype. At present, it is essential that a holotype designated for a newly described species be deposited in a national herbarium.

An *isotype* is a duplicate specimen of a holotype. These are plants forming part of the same gathering as the holotype or growing with it and gathered at the same time. A *syntype* is one or two or more specimens studied and cited by the author, when the holotype is not designated by him. A *paratype* is a specimen cited with the original description in addition to the holotype. When the author fails to designate a holotype or the holotype is missing, a *lectotype* or a *neotype* is selected to serve as a nomenclatural type. A *lectotype* is a specimen selected from those cited by the author with the original description. A *neotype* is a specimen selected from the material that was not cited by the author with the original description. A *neotype* is selected only when all the original specimens collected and cited by the author are missing.

Impatiens thomsonii Hook. is a member of the Balsaminaceae and its description is given in the *Flora of British India*. The author has cited three specimens on which the description was based:

1. Collected by Thomson from Piti and Kunawur (Inner ranges of temperate Himalayas).
2. Collected by Strach and Wint from the Kumaon Hills.
3. Collected by Hooker from Sikkim.

For each, specimen-number, place and date of collection, and the name of the collector are given. Hooker stated that specimen no. 3 is the nomenclatural type and therefore holotype. Specimens 1 and 2 are the paratypes. If Hooker had not designated the 3rd specimen as holotype, then all three would have been syntypes. One of these syntypes can serve as a lectotype, if the holotype is missing. If all three specimens are destroyed for some reason, then a fourth specimen (collected by Wallich from Sikkim), which does not find mention in Hooker's description, will be treated as a neotype. Duplicate specimens collected by Hooker, along with the holotype, are treated as isotypes.

III. *The nomenclature of a taxonomic group is based upon "Priority of Publication."* According to this principle, each taxon should bear only one correct name and that should be the earliest published name. The rule of priority states: "For any taxon from family to genus inclusive, the correct name is the earliest legitimate one with the same rank, except in cases of limitation of priority by conservation." It also states: "The principle of priority does not apply to names of taxa above the rank of family."

To avoid confusion caused by the strict application of this rule of priority, certain specific, generic and family names are conserved in preference to the earlier published names, by resolution of the International Botanical Congresses. Conserved names are known as *nomina conservanda*. For example, Sterculiaceae Lindl. 1830 is a conserved name and not published earlier. The earlier published name for the same family is Byttneriaceae R. Br. 1814 (see Lawrence 1951). Conservation of specific names is restricted to names of species of major economic importance (Greuter 1981) and was adopted at the International Botanical Congress held at Sydney, Australia, in 1981.

Any association of a specific epithet with a generic name is known as *combination*. All those names which are presented in accordance with the rules of nomenclature are termed legitimate and the names contrary to the rules of nomenclature are termed illegitimate.

Priority of nomenclature for vascular plants is applicable from May 1st, 1753 which is the date of publication of Linnaeus' *Species Plantarum*.

When taxonomic study indicates that a species described in one genus is to be transferred to another genus, the specific epithet, if legitimate, should be retained. In the following example:

Sida cordata (Burm.f.) Borssum 1966
Melochia cordata Burm.f. 1768
Sida veronicifolia Lamk. 1787
Sida multicaulis Cav. 1785,

According to the rule of priority, *Melochia cordata* Burm.f. is the valid name. A later study by Borssum revealed that the genus was *Sida* and not *Melochia*. As the specific epithet was legitimate, *cordata* was retained and a new combination *Sida cordata* was made by Borssum. *M. cordata* Burm.f. is now known as a *basionym*. To write the present name correctly, the author's name of the basionym is placed in parentheses, followed by the name of the author who made the new combination.

When two or more taxa of the same rank are united, the earliest legitimate name or epithet is selected. For example, if the genera *Sloanea* L. 1753, *Echinocarpus* Blume 1825, and *Phoenicosperma* Miq. 1865 are combined, *Sloanea* L. should be the correct name to be used, as it is the earliest name and the other two would be the *taxonomic synonyms*.

If the taxonomic revision of a genus reveals that it should be divided into two or more genera, the original generic name should be retained for the genus that includes the species designated as the type. For example, the genus *Aesculus* is divided into four sections; *Aesculus* sect. *Aesculus*, sect. *Pavia*, sect. *Macrothyrsus* and sect. *Calothyrsus*. If the last three are regarded as distinct genera, i.e. the four sections are treated as separate genera, then the name *Aesculus* should be retained for the first one—*Aesculus* sect. *Aesculus*, which includes the type species *A. hippocasatanum* L. Similarly, when a species is broken into two or more than two parts, the specific epithet should be retained for the one which includes the type specimen/figure/description. For example, *Acer saccharum* Marshall was first described by Marshall. Later, Michaux considered it to comprise two species and, according to the rule, he retained the name *Acer saccharum* for the specimen that was described by Marshall and named the other specimen *Acer nigrum* Michx.f.

IV. "*Each taxonomic group with a particular circumscription, position, and rank can bear only one correct name, that is a validly and effectively published name.*"

All those names that are published in printed form in scientific journals and are available in botanical institutions with libraries, which are accessible to botanists in general, are the effectively published names. On the contrary, names that are published in nursery catalogues, newsprint or seed-exchange lists are not effectively published names. A plant name is not effectively published if printed on a label attached to herbarium specimens even if the specimens are widely distributed (Jones and Luchsinger 1987). For valid publication, a name must be effectively published; it must be accompanied with a description or a reference to a previously published description of that taxon. From January 1st, 1935, names of new taxa of recent plants (with the exception of algae and fossil) must be accompanied by a Latin diagnosis for valid publication. The description itself need not be in Latin, although it is recommended. The description and diagnosis (the distinguishing features of the taxon as mentioned by the author) of new taxa, published before January 1st, 1935, are treated as valid, even if they were in any modern language including Japanese, Russian or any other where Roman alphabets are not used.

The name of a taxon is not validly published if it is cited merely as a synonym.

Phalaris arundinacea Linn. Sp. Pl. 55, 1753. This name, given by Linnaeus, published in his *Species Plantarum* on p. 55 in 1753, has a Latin diagnosis and therefore a valid name.

Digitaria sanguinalis (L.) Scop. Fl. Carn. ed. 2, 1:52, 1772
Panicum sanguinale L. Sp. Pl. 57, 1753.

Scopoli discovered that the type specimen had the characters of *Digitaria* and, hence, the new combination was made by him and published in *Flora Carniolica*. He did not give a Latin diagnosis, as it had already been given by Linnaeus in *Species Plantarum* in 1753. The second name now becomes the *nomenclatural synonym*.

V. *Scientific names of taxonomic groups are treated as Latin regardless of their derivation.* According to this rule, generic names, specific epithets, as well as other names should be Latin or Latinised with the addition of prefixes and suffixes, whatever source they might have been taken from (cf. p. 29, 30).

VI. *The rules of nomenclature are retroactive unless expressly limited.* In connection with this principle, the principle of *Later Homonym* emerged. A name is a *later homonym* if it is spelled like a name previously and validly published for a taxon of the same rank, based on a different type specimen. Different genera (of the same family or different families) and different species of the same genus cannot have the same name. In such instances, the later-formed name or the later homonym is illegitimate and has to be rejected. For example, *Tapienanthus Boiss.* ex Benth. 1848 of Labiatae is a later homonym of *Tapienanthus* Herb. 1837 of Amaryllidaceae, and must be rejected; *Viburnum fragrans* Bunge 1831 and *Viburnum fragrans* Lois. 1824 both belong to the same family Caprifoliaceae, but the type specimens for them are different. Therefore, the later homonym, *V. fragrans* Bunge 1831, should be rejected. To indicate that a plant has a later homonym, the word non is used before the author's name of the later homonym and placed after the early homonym, e.g. *Viburnum fragrans* Lois 1824 non Bunge 1831 will indicate that this plant has a later homonym. If the plant name itself is a later homonym, the word nec is used before the author's name of the early homonym and placed after the later homonym, e.g.

Viburnum farreri Stearn 1966
V. fragrans Bunge 1831 nec Lois. 1824

The fifteenth International Botanical Congress was held at Yokohama, Japan, in 1993. The International Code of Botanical Nomenclature (called the Tokyo Code) adopted at this Congress is significantly different from the earlier Code (called the Berlin Code) which was adopted at the fourteenth Botanical Congress. Some of the important changes are as follows:

   (i) the rules on Typification and Effective publication have been clarified by creating a logical arrangement of the Articles 7–10, and 29–31, respectively;
  (ii) the proposals for (a) Conservation of species names, and (b) rejection of any name which would cause a disadvantageous nomenclatural change, were accepted by overwhelming majority;
 (iii) an entirely new concept has been incorporated in the Tokyo code. This concerns the recognition of "Interpretative Type" to serve the requirement of typification when an established name cannot be reliably identified for the purpose of precise application of a name.
  (iv) this Code permits the use of the term "phylum" as an alternate to "divisio".
   (v) an extensive revision of Article 46 has clarified the use of the prepositions '*ex*' and '*in*' in author citations;
  (vi) for valid publication of a new taxon of fossil plants, on or after January 1,1996, there must be an accompanying description or diagnosis in Latin or English, (or a reference to such earlier publication) and not in any language as before;
 (vii) The 15th International Botanical Congress proposed that after January 1, 2000 and after approval by the 16th International Botanical Congress (in 1999), new names must be registered.

Nomenclature of Hybrids

A hybrid is an offspring of two different genera, or species of plants. Hybrids between two species of the same genus or interspecific hybrids are usually shown by a formula: *Verbascum lychnite* × *V. nigrum* or *Verbascum lychnite* × *nigrum*, or may be given a formal name: *Verbascum* × *nigralychnites*. Intergeneric hybrids also can be named in the same fashion: either by a formula, e.g. *Cochloidea* × *Odontoglossum*; *Cooperia* × *Zephyranthes*, or by a formal name × *Cooperanthes*; × *Triticale*.

Some Basic Definitions Related to Nomenclature

Synonyms are different names for the same plants. It is a rejected name due to wrong application or difference in taxonomic judgement. When two or more than two names are given to the same taxon, based on the same type specimen, they are *nomenclatural synonyms*. For example, *Chilocarpus malabaricus* Bedd., and *Hunteria atrovirens* DC. are the nomenclatural synonyms of *Chilocarpus atrovirens* (G. Don) Blume, as the type specimen is the same. In another example, *Vernonia leiocarpa* DC. and *Vernonia melanocarpa* (Gleason) DC. are taxonomic synonyms, as the two are based on different type specimens.

Basionym. When a species is described in one genus but transferred to another later, the specific epithet, if legitimate, should be retained. *Myrobalanus bellirica* Gaertner 1791 is now known as *Terminalia bellirica* (Gaertner) Roxb. 1805. *Myrobalanus bellirica* is the *basionym* of *Terminalia bellirica*.

Homonym. When two or more identical names are given based on different type specimens, they are homonyms and the earliest published one amongst these is legitimate and should be retained. *Centranthera indica* (L.) Gamble 1924, if shifted to *Limnophila*, cannot be named *Limnophila indica* because there is already a species with the same name, i.e. *Limnophila indica* (L.) Druce 1914. In the event of doing so, these will be homonyms.

Tautonym is an illegitimate binomial where the generic name and specific epithet are exactly the same, e.g. *Sassafras Sassafras* (L.) Karst 1882, is a tautonym and therefore a rejected name.

Autonym is a legitimate, automatically created tautonym for infrageneric or infraspecific taxa. For example, *Hypericum* subgenus *Hypericum* section *Hypericum*; *Sesbania sesban* var. *sesban*.

Nomen nudum is a name without description and so should be rejected. In some cases, the names are nomenclaturally *superfluous*, i.e. published as a substitute for an already published legitimate name. For example, *Sassafras triloba* Raf. 1840 was a superfluous name for *Laurus Sassafras* L. 1753, and therefore rejected.

Dicotyledons

Subclass Archichlamydae

6

Order Casuarinales

The order Casuarinales comprises a solitary family, the Casuarinaceae. Evergreen trees or shrubs with whorled scale-like leaves, monoecious or dioecious. Male flowers in catkins and females in spherical heads.

Casuarinaceae

A unigeneric family with the genus *Casuarina* (65 species) occurs mostly in Australia, Malaya and New Caledonia. At present many species are grown worldwide in arid and semi-arid regions, particularly in the developing countries.

Vegetative Features. Evergreen, woody trees or shrubs with jointed, verticillate and striate branches (Fig. 6.1A). Leaves in whorls of 4 to 16, scale-like, usually linear to lanceolate, leaf tips appear only as teeth; basally connate forming a sheath around the twig (Fig. 6.1 B), internodes with striate grooves, as many as the leaves.

Floral Features. Plants dio- or monoecious, flowers unisexual. Staminate flowers (Fig. 6.1 C, D) comprise a single stamen subtended by 4 bracteoles (Fig. 6.1D), bracteoles obovate with serrate margin and arise from within the leaf sheaths (Fig. 6.1C); anther basifixed, bicelled, longitudinally dehiscent (Fig. 6.1E). Pistillate flowers in subglobose to ovoid catkins (Fig. 6.1F), each flower represented by a single bicarpellary gynoecium, subtended by a small bract and two bracteoles (Fig. 6.1G). Ovary superior, bilocular, but later becomes unilocular due to the suppression of 1 locule. Ovules 2 to 4, on parietal placenta; sytle 1, stigmas 2, filiform (Fig. 6.1G). Fruit a one-seeded winged samara. The pistillate catkin (at maturity) is woody, cone-like (Fig. 6.1H) and indehiscent. Seeds (Fig. 6.1I) non-endospermous, embryo straight.

Anatomy. In a transverse section the stem shows alternate ridges and furrows. Rubiaceous type of stomata and simple or branched stomatal hairs occur in the furrows. The stomata are oriented at right angles to the longitudinal axis of the branch. Two whorls of vascular bundles—the outer leaf-trace bundles and the inner cauline vascular bundles—occur in the stem. Vessels solitary, with simple perforation plates, occasionally scalariform. Parenchyma apotracheal, diffuse and in narrow bands. Rays small, uniseriate, homogeneous.

The root nodules—which occur in many species of *Casuarina*—are modified lateral roots which branch profusely to form "coralloid" masses (Metcalfe and Chalk 1972). They contain bacteria similar to those of the root nodules of the Leguminaceae, which fix atmospheric nitrogen.

Embryology. Pollen grains acolpate, 3-porate, with a smooth exine, shed at 2-celled stage. Ovules orthotropous, with multiple embryo sacs (P. Maheshwari 1950), bitegmic, crassinucellate; Polygonum type of embryo sac, 8-nucleate at maturity. Polyembryony and parthenocarpy is reported (Johri et al. 1992). This is the only family with nucellar vascularisation (Bouman 1984). Endosperm formation of the Nuclear type.

Chromosome Number. The basic chromosome numbers are x = 8,9, 11–14. In *C. distyla* group (of 13 closely related species), although sexual tetraploids and apomicts are formed, apparently they do not characterise any particular morphological grouping. The tetraploids occur only within the species and are

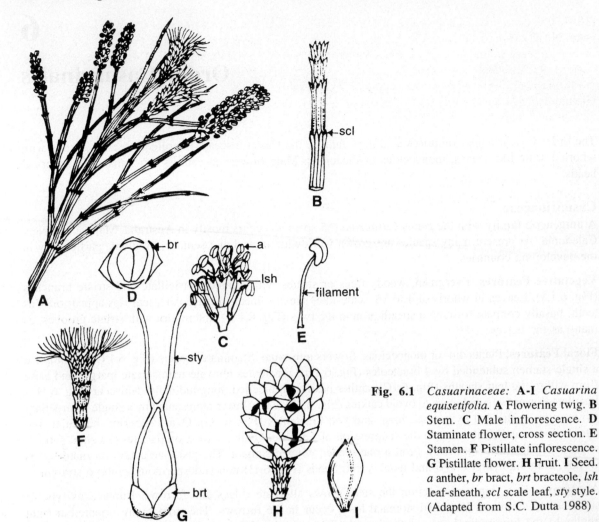

Fig. 6.1 *Casuarinaceae:* **A-I** *Casuarina equisetifolia.* **A** Flowering twig. **B** Stem. **C** Male inflorescence. **D** Staminate flower, cross section. **E** Stamen. **F** Pistillate inflorescence. **G** Pistillate flower. **H** Fruit. **I** Seed. *a* anther, *br* bract, *brt* bracteole, *lsh* leaf-sheath, *scl* scale leaf, *sty* style. (Adapted from S.C. Dutta 1988)

not morphologically separable from their respective diploids. Species evolution has been entirely at the diploid level and polyploidy has hardly any evolutionary significance (Barlow 1959).

Chemical Features. A group of chemicals related to the flavonoids, called biflavonyls, have been reported from *C. stricta*. The biflavonyls were originally discovered amongst various gymnosperms (Baker and Ollis 1961). Embryo rich in oil, bark rich in tannins; polyphenolics including ellagic acid, catechin and leucoanthocyanins also present in various species of *Casuarina*.

Important Genera and Economic Importance. *Casuarina* is the only genus with ca. 65 species distributed in the southern hemisphere. It is grown as a successful windbreak, particularly along seashores. It is an important timber tree.

Taxonomic considerations

Casuarinaneae is a highly isolated family. Because of the simple anemophilous flowers, catkin-like inflorescences and large rays in the xylem, it was treated as the most primitive dicot family by Engler

(1887d). Rendle (1925) and Wettstein (1935) included it in the Amentiferae and treated it as the most primitive angiosperm derived from the Ephedraceae. These ideas were abandoned as soon as the concepts of evolutionary trends in angiosperms changed. Bessey (1915), Hallier (1912), Hutchinson (1926) and Tippo (1938) treated Casuarinaceae as one of the highly reduced types and highly advanced. Moseley (1948) derived this family from the Hamamelidales, on the basis of its anatomy and floral morphology. Hjelmquist (1948), on the basis of floral morphology, treated this family as a primitive dicot but not as a member of the Amentiferae.

Palynologically, Casuarinaceae is allied to Betulaceae, Juglandaceae, Corylaceae and Myricaceae (Chanda 1969, Kuprainova 1965). Chanda (1969), however, does not consider this family as a primitive one, and does not derive it from the Hamamelidales.

More recent workers like Cronquist (1968, 1981), Dahlgren (1975a, 1980a, 1983a) and Takhtajan (1980, 1987) presume that the Casuarinaceae has its origin from the Hamamelidaceae.

7

Order Juglandales

Juglandales comprises two families: Myricaceae and Juglandaceae, mostly distributed in temperate regions of the northern hemisphere.

Deciduous or evergreen trees, rarely shrubs, leaves estipulate, entire in Myricaceae, and digitate or imparipinnate in Juglandaceae. Inflorescence unisexual catkins, flowers highly reduced, anemophilous. Pollen grains 3-porate, shed at 2-celled stage, chalazogamy common. Gynoecium syncarpous, bicarpellary, ovary superior, unilocular or basally 2-4-loculed as in Juglandaceae. Ovules atropous or orthotropous, unitegmic, crassinucellate; embryo sac of Polygonum type. Endosperm formation of the Nuclear type.

Myricaceae

A small family of two genera—*Myrica* and *Comptonia*—according to Hjelmquist and Small (see Lawrence 1951). Engler, Asa Gray and Hutchinson (see Lawrence 1951) treat Myricaceae as an unigeneric family, i.e. only *Myrica,* with ca. 40 species. Takhtajan (1987) includes three genera—*Myrica, Canacomyrica* and *Comptonia*—with ca. 50 species. Commonly called the sweet gale or wax myrtle family, it is distributed in the cool areas of the north temperate zone and in South Africa. *Myrica* occurs in xeric or swampy areas.

Vegetative Features.   Deciduous or evergreen, aromatic trees, rarely shrubs, monoecious or dioecious. Leaves alternate, simple, entire (Fig 7.1 A), rarely pinnatifid, usually short-petioled, coriaceous, with yellow resin glands (Fig 7.1 B), estipulate, stipulate in *Comptonia*.

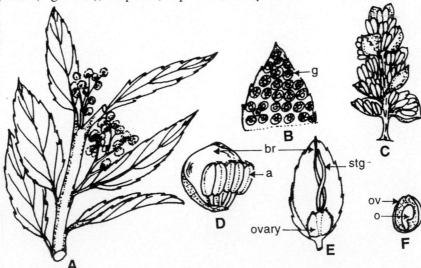

Fig 7.1   Myricaceae: **A, B** *Myrica arborea,* **C–F** *M. pensylvanica.* **A** Flowering twig. **B** Glands on leaf surface. **C** Staminate inflorescence. **D** Staminate flower. **E** Pistillate flower. **F** Ovary, vertical section. *a* anther, *br* bract, *g* gland, *o* ovule, *ov* ovary, *stg* stigma. (**A, B** adapted from Hutchinson 1969, **C–F** from Lawrence 1951)

Floral Features. Inflorescence axillary, densely flowered, unisexual spikes. When bisexual, the male flowers are at the basal and female at the apical end. Staminate flowers bracteate, bract 1, perianth absent, stamens usually 4 to 8 (Fig. 7.1 C, D) (may be 2 to 20); filaments free or connate at the base, anthers bicelled, basifixed, dehisce by vertical slits (Fig. 7.1 D). Pistillate flowers bracteate (Fig. 7.1 E), bract 1, bracteolate or ebracteolate; gynoecium syncarpous, bicarpellary; ovary superior, unilocular with a single basal ovule (Fig. 7.1 F); style reduced, stigmas 2 (Fig 7.1 E). Bisexual flowers have a central pistil surrounded by 3 or 4 stamens, and are subtended by a bract. Fruit a drupe covered by waxy coating.

Anatomy. Roots bear root tubercles which contain N_2-fixing bacteria. According to Bottomley (1911, 1912), the nodules arise as lateral branches and clusters of nodules are formed. Youngken (1919) identified the organism in the nodules as an Actinomyces. This aspect requires further study.

Wood ring-porous; vessels small, solitary; perforation plates exclusively scalariform, or scalariform and simple; intervascular pitting intermediate to alternate. Parenchyma apotracheal, diffuse, rays uniseriate or up to 4 to 8 cells wide, heterogeneous. Leaves nearly always dorsiventral with characteristic peltate glands which secrete an aromatic waxy material. The leaves have multiple layers of palisade tissue. Stomata of Ranunculaceous type, confined to the lower surface.

Embryology. The embryology has not been fully investigated. Pollen grains pororate, similar to those of Betulaceae and Fagaceae. Pollen tube often enters through the chalaza. In *Juglans regia* the entry of pollen tube into the ovule is chalazogamous or porogamous, depending on the extent of development of the single orthotropous ovule (Luza and Polito 1991). Ovules basal, orthotropous, unitegmic, crassinucellate in *Myrica;* pendulous and anatropous in *Canacomyrica* (Johri et al. 1992). Embryo sac of Polygonum type, 8-nucleate at maturity. Endosperm formation of the Nuclear type.

Chromosome Number. Basic chromosome number is $x = 8$.

Chemical Features. The conspicuous peltate, resinous glands on leaf surfaces give a characteristic smell. Fruit a small drupe, with wax-secreting glands on its surface. The endosperm is rich in oil and protein.

Important Genera and Economic Importance. Of the three genera—*Myrica, Comptonia* and *Canacomyrica*—*Myrica* is the most important. The waxy material obtained from the fruit of several species is used in the manufacture of bayberry candles and soap. Bayberry bark from the roots of *M. cerifera* is used medicinally, due to its astringent nature. Tannic acid is obtained from *M. gale.* Fruits of *M. nagii* are edible.

Taxonomic Considerations. Due to the catkin-bearing character, Myricaceae was considered as one of the primitive families by Engler (1887). Rendle (1925) suggested Myricaceae to be a member of the order Juglandales, on the basis of some common features like the single unilocular ovary with a large seed, orthotropous ovule, and aromatic compounds in the leaves. Possibly, Myricaceae and Juglandaceae have a common ancestor among the Hamamelidaceae of the Rosales. Hjelmquist (see Lawrence 1951) treated Myricaceae as the most primitive of his Juglandales, and retained it within the Amentiferae. According to Hutchinson (1969, 1973), Myricaceae is more advanced than Fagales and is derived from the Hamamelidales. Cronquist (1968, 1981) states that anatomically the two families—Juglandaceae and Myricaceae—are similar, but the inclusion of some other families like Rhoipteleaceae and Picrodendraceae in the Juglandales prevents it from being merged with the Myricales. Accordingly, it is more appropriate to retain Myricaceae in an order of its own. Dahlgren (1975a, b, 1980a, 1983a) places this family next to the order Juglandales, in the subclass Rutanae. The chemistry of Myricaceae is allied to that of Juglandales, in which it could be included. According to Takhtajan (1980), the family Myricaceae has much in common on the one hand with the Casuarinales and Betulaceae of the Fagales, and on the other with the Juglandales. Takhtajan (1987) includes this family in a separate order Myricales, emphasising a common origin from the Hamamelidales.

From the above discussion it is concluded that the family Myricaceae should be placed in a distinct order, Myricales, derived from Hamamelidales.

Juglandaceae

A small family of only 6 genera and 60 species, abundantly distributed in eastern Asia and Atlantic North America. There is one distributional zone extending through Central America along the Andes to Argentina, and another from temperate Asia extending down to Java and New Guinea.

Vegetative Features.   Trees or shrubs, monoecious or rarely dioecious, e.g. *Engelhardtia* spp. Leaves alternate or opposite as in *Alfaroa* and *Engelhardtia;* compound, digitate or imparipinnate or paripinnate as in *Juglans nigra,* usually estipulate. Leaflets mostly resinous-dotted beneath and aromatic.

Floral Features.   Inflorescence variable, staminate flowers usually in catkins (Fig. 7.2A, B), pistillate solitary or only a few in a cluster (Fig. 7.2A). Staminate flowers subtended by 1 primary bract and 2 secondary bracts (or bractlets). Perianth of 4 or fewer tepals (or sepals), or perianth totally absent as in *Carya;* both secondary bracts and perianth absent in *Platycarpa.* Stamens 3 to 300 in one or more series, filaments short, free; anthers erect, dithecous, basifixed, dehisce longitudinally (Fig. 7.2 C). Pistillate flowers subtended by 1 primary bract, entire or occasionally trilobed, and 2 secondary bracts; epigynous, incomplete. Perianth of 1 whorl of tepals or sepals, may be absent or modified in some members. Gynoecium syncarpous, bicarpellary, ovary inferior, unilocular above but 2- to 4-loculed below; ovule one, erect, in the centre, at the apex of the incomplete partition, appears basal in young flowers (Fig. 7.2 D); styles 2 or 1 with 2 lobes, stigmas 2. Fruit a large nut (Fig. 7.2 F) with a dehiscent or indehiscent

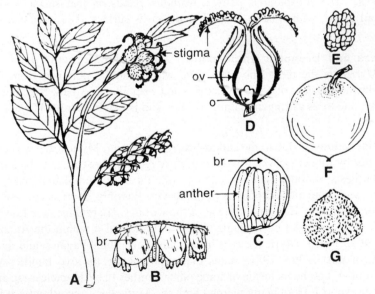

Fig. 7.2   Juglandaceae: **A–D, F, G** *Juglans nigra,* **E** *J. regia.* **A** Branch with pistillate (upper) and staminate (lower) inflorescence. **B** Staminate inflorescence. **C** Staminate flower. **D** Pistillate flower, vertical section. **E** Seed of *Juglans regia.* **F,G** Fruit (**F**) and seed (**G**) of *J. nigra. br* bract, *o* ovule, *ov* ovary, (Adapted from Lawrence 1951)

leathery or fibrous outer cover derived from the bracts, bracteoles; perianth surrounds the ovary; fruit sometimes a winged nutlet, incompletely 2- to 4-celled. Seed 2- to 4-lobed (Fig. 7.2 E, G), non-endospermous and with a large embryo.

Anatomy.   Wood often ring-porous; vessels typically few, medium-sized to large; intervascular pitting alternate. Parenchyma either predominantly apotracheal or as broken bands intermediate between apo- and paratracheal. Rays 2 to 7 cells wide, heterogeneous. Spirally thickened vascular tracheids occur in *Platycarya*. Short-stalked peltate glands present on leaf surfaces which secrete ethereal oil. Stomata of the Ranunculaceous type, confined to the lower surface.

Embryology.   Pollen grains usually 3- to 7-porate, shed at 2-celled stage. Ovules atropous or orthotropous, unitegmic, crassinucellate and chalazogamous; Polygonum type of embryo sac, 8-nucleate at maturity. Endosperm formation of the Nuclear type.

Chromosome Number.   Basic chromosome number is x = 16.

Chemical Features.   Members rich in polyphenols (including various tannins, myricetin and ellagic acids), naphthaquinones and citrullin.

Important Genera and Economic Importance.   Various species of *Juglans, Carya* and *Engelhardtia* are useful sources of timber. *J. regia* and *C. illinoensis* yield valuable food nuts. Many species of *Juglans, Carya, Pterocarya* and *Platycarya* are grown as ornamental trees. The wood of *C. ovata* is strong and elastic and is used for axe handles in North America. Hickory nuts are the seeds of this plant. Pecan nuts are obtained from *C. pecan*. Walnut wood from *Juglans regia* is highly prized for cabinet-making, for furniture, carving and inside work. In European countries, at the beginning of the 19th Century, it was prized for gun-stocks.

Taxonomic Considerations.   Engler and Diels (1936) considered Juglandaceae to be primitive, and included it in the order Juglandales of the Amentiferae. Bessey (1915) and Hutchinson (1959) presumed that it evolved from one of the taxa in Sapindales. Benson (1970) supported this view and further suggested that the Anacardiaceae of the order Sapindales could be the probable ancestor. Hallier (1908) included Juglandaceae members in Terebinthaceae, as derived from the Rutaceae. Heimsch (1942), on the basis of anatomy, concluded that the Juglandaceae were unrelated to the Anacardiaceae. Cronquist (1968) included this family in his subclass Hamamelidae and pointed out its direct evolution from the Hamamelidales. However, he treated the order as comprising three families: Rhoipteleaceae, Picrodendraceae and Juglandaceae. All three families have pinnately compound leaves, which sets them apart from the rest of the members of the Hamamelidae. He also stated that although Juglandaceae has often been considered to be related to the Anacardiaceae, their wood is distinctly more primitive than that of Anacardiaceae and the pollen is similar to that of several families of Hamamelidae. They also share the presence of peltate glands on their leaves, with Myricaceae and Fagaceae.

Hutchinson (1969, 1973) discarded his earlier conclusion and regarded Juglandaceae as a pinnate-leaved derivative of the Corylaceae, a member of his Fagales. Dahlgren (1975 a, b), however, re-established the affinity between the Sapindales and Juglandaceae, placing both in the same subclass, Rutanae. Takhtajan (1980, 1987) observes that this family has much in common with the Myricaceae, particularly serological similarities (Chupov 1978, Peterson and Fairbrothers 1979). He considers them to have evolved from a common Hamamelidalean ancestor.

Dahlgren (1980a, 1983a) recognizes the two separate orders, Juglandales and Myricales, for Juglandaceae and Myricaceae. According to him, serological data are decisive in placing them next to the Fagales.

Taxonomic Considerations of the Order Juglandales

The order Juglandales comprises two families, Myricaceae and Juglandaceae, according to Melchior (1964). Most other taxonomists treat these two families as belonging to two distinct orders—Myricales for Myricaceae, and Juglandales for Juglandaceae alone (Lawrence 1951), or along with Rhoipteleaceae and Picrodendraceae (Hutchinson 1969, Cronquist 1968, 1981). Dahlgren (1975a, 1980a, 1983a) accepts two independent orders under the subclass Rutanae. Takhtajan (1980, 1987) erects the subclass Juglandanae to accommodate the two orders, Myricales and Juglandales.

There are a number of resemblances between the two families as well as some differences. Embryologically, the two families show close affinities, and the order is homogeneous (Johri et al. 1992). At the same time, both families also resemble members of the order Fagales.

Dahlgren (1983a) includes the families Rhoipteleaceae and Juglandaceae in Juglandales, and places Myricaceae in a separate unifamilial order, Myricales. The flavonoid chemistry favours the position of Juglandaceae in the superorder Hamamelidae (Gornall et al. 1979). Peterson and Fairbrothers (1979) support the placement of Myricales and Juglandales along with Fagales, on the basis of serological evidence. Ehrendorfer (1983a) also indicates the assignment of this order close to the Hamamelididae as "they tend towards anemophily and reduction of polymerous to oligomerous monochlamydeous flowers".

The two families, Myricaceae and Juglandaceae, should be treated as belonging to the same order, Juglandales, and their placement next to the Fagales.

8

Order Balanopales

A monotypic order with only one family, Balanopaceae; dioecious shrubs or trees with simple leaves. Flowers borne in catkins.

Balanopaceae

A unigeneric family with the genus *Balanops* (nine species) distributed in New Caledonia and tropical Australia.

Vegetative Features.   Shrubs or trees, leaves alternate, simple, entire, estipulate.

Floral Features.   Plants dioecious; staminate flowers in catkins, each flower consists of 2 to 12 stamens, perianth absent. Pistillate flowers solitary, subtended by a multibracteate involucre. Gynoecium syncarpous, bicarpellary, ovary superior, partly divided into two locules, often incomplete; ovules 4, erect, basal or nearly so. Fruit a drupe surrounded by a persistent involucre, sometimes divisible into two parts. Seeds with straight embryo and fleshy endosperm.

Anatomy.   Epidermal cells of stem cortex with thick outer walls. Xylem vessels small, isolated, perforation plates scalariform with numerous bars. Leaves dorsiventral, stomata Ranunculaceous type, confined to the lower surface; clustered crystals abundant in the mesophyll.

Embryology.   Embryology has not been fully investigated. Pollen grains 3–4-colporidate, minutely spinulose. Ovules erect, unitegmic, crassinucellate; endosperm formation of the Nuclear type.

Chemical Features.   Bark is rich in tannins and triterpenes.

Important Genus and Economic Importance.   *Balanops* does not have any economic importance.

Taxonomic Considerations

According to Benson (1970), Balanopaceae is a member of a separate order of its own near the families Leitneriaceae, Myricaceae and Batidaceae, and therefore shows some affinities with them. Cronquist (1968, 1981) places this family in the order Fagales along with the families Fagaceae and Betulaceae, sharing common characters like more than one ovule, bitegmic or unitegmic, and anatropous. Plants not aromatic. Cronquist also changed the terminology, i.e: Balanopaceae from Balanopsidaceae, on the basis of the type genus *Balanops*.

Hutchinson (1969, 1973) places Balanopaceae in a separate order of its own, Balanopales, derives it directly from the Hamamelidales, and places it near the other two orders, Leitneriales and Myricales. Dahlgren (1975a) also places Balanopaceae in Balanopales; although he includes it in subclass Hamamelidanae near the Fagales, he is not sure of its position. In 1980a Dahlgren removes this order in the Rosiflorae, without disturbing its relative position. Thorne (1977) places Balanopaceae near the family Daphniphyllaceae, in his order Pittosporales.

Takhtajan (1980) reports that Balanopales is an isolated order consisting of unigeneric family, the Balanopaceae, which some authors include in the Fagales. However the genus *Balanops* differs from the members of Fagales in its basal, anatropous ovule, an obturator-like enlargement of the funicle, a thin layer of endosperm around the large embryo, and drupaceous fruit. He is of the opinion that it is probably derived directly from the Hamamelidales; but in 1987 Takhtajan derived this family from the Daphniphyllaceae, as the two taxa have a number of common anatomical, embryological and palynological features. There appears to be an apparent relationship between the Daphniphyllaceae of Buxales and the order Balanopales.

In our opinion, further studies should be undertaken to determine the appropriate assignment and relationship of the Balanopaceae.

9

Order Leitneriales

The order Leitneriales comprises two families; Leitneriaceae with only one genus and one species—*Leitneria floridana*, and Didymelaceae with the only genus *Didymeles*, with two species. Members of both the families are highly localised in distribution. Mostly dioecious trees bearing catkins of highly reduced flowers; pollination anemophilous. Gynoecium monocarpellary with an unilocular superior ovary, and only one style.

Leitneriaceae

A monotypic family; the only species, *L. floridana*, is distributed mainly in the swampy areas of the southeastern United States from southern Missouri to northern Georgia, and southward to Texas and Florida.

Vegetative Features.   Deciduous, dioecious shrubs or small trees. Leaves alternate, simple, entire (Fig. 9.1 A), estipulate or with rudimentary stipules, more or less coriaceous.

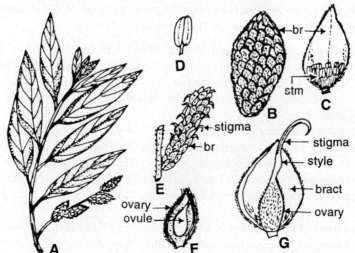

Fig. 9.1   Leitneriaceae: A–F *Leitneria floridana*. **A** Branch with staminate catkins. **B** Staminate catkin. **C** Staminate flower. **D** Anther. **E** Pistillate catkin. **F** Pistillate flower. **G** Ovary, vertical section. *br* bract, *o* ovule, *ov* ovary, *stm* stamen. (**A–D** adapted from Lawrence 1951)

Floral Features.   Flowers unisexual, staminate catkins drooping, pistillate catkins erect. Staminate inflorescences composed of 40 to 50 large, imbricate, glandular-pubescent bracts (Fig. 9.1 B), each subtends a staminate cymule. Perianth absent, stamens 3 to 12 (Fig. 9.1 C), filaments distinct, curved inwards, anthers erect, basifixed, introrse, bicelled, dehisce longitudinally (Fig. 9.1D). Pistillate inflorescences (Fig. 9.1 E) with a primary axis bearing large, primary bracts which enclose a pair of much smaller secondary bracts; the two series together subtend the pistillate cymule. Each female cymule is subtended by three bracts and a perianth of 4 tepals. Gynoecium monocarpellary, ovary superior, unilocular with a single subapical ovule, i.e. parietal but attached near the apex (Fig. 9.1 G); style one, stout, linear, constricted at base and with a groove along the side (Fig. 9.1 F), stigma one. Fruit a leathery compressed drupe, 1 to 2 cm long, 5 to 7 mm in diameter; usually many fruits remain aggregated in a cluster, each subtended by a persistent bract. Seed with a large, straight embryo and a thin fleshy endosperm.

Anatomy.   Secretory canals with yellow resinous contents, very conspicuous at the margin of pith. Sometimes these canals extend up to the finest veins of the leaves. The young stem is covered with uniseriate hairs. Vessels small, in diagonal or zigzag rows, semi-ring-porous, with spiral thickening, perforation plates simple, intervascular pitting scalariform to opposite. Parenchyma scanty, paratracheal to vasicentric and terminal. Rays exclusively uniseriate. Leaves dorsiventral, stomata superficial (not in pits).

Embryology.   Pollen grains 3–6-colporate with reticulate surface, shed at 2-celled stage. In *Leitneria* there is an excessive development of the upper portion of the integuments so that the micropylar canal lies in folds over the nucellus. Ovules bitegmic, amphitropous, crassinucellate; Polygonum type of embryo sac; 8-celled at maturity. Division of zygote is either transverse or vertical. Endosperm formation of the Nuclear type.

Chromosome Number.   Haploid chromosome number of *Leitneria* is n = 16.

Chemical Features.   The bark is rich in tannins.

Important Genera and Economic Importance.   The only member, *Leitneria floridana,* commonly known as Florida corkwood, is a small tree, native to muddy saline swamps of the southeastern United States. The tree yields light wood used as floats for fishing nets.

Taxonomic Considerations.   Bentham and Hooker (1965c) placed the family Leitneriaceae in the Unisexualis. Engler and many other workers grouped the family Leitneriaceae with others which have catkin inflorescence. Melchior (1964) also presumed its close relationship with the Myricales and Juglandales, because of similar chromosome number, i.e. n =16. Hutchinson (1926) derived it from the Rosales through Hamamelidaceae.

Bessey (1915) considered Leitneriaceae to belong to the family Ranunculaceae. Hjelmquist (see Lawrence 1951) included this family in the Amentiferae, although the anatomical characters are contradictory. From the anatomical and floral morphological studies, Abbe and Earle (1940) concluded that the monocarpellary condition and the perianth-like structure of the female flowers are similar to the sepals. On the basis of these observations, they point out that this family should be placed in the Geraniales or Rosales.

Cronquist (1968) observed that the family Leitneriaceae is a highly isolated group and its relationship is uncertain. Its gynoecium is pseudomonomerous rather than strictly unicarpellate, as shown by the occasional occurrence of bicarpellate pistils with two styles. The well-developed resin canals in the stem and leaves are comparable with those of the Juglandaceae.

The multicellular clavate glands of the leaves might, or might not, be homologous with the peltate glands of the Juglandales, Myricales and Fagales. However, several authors feel that it has originated by reduction from Hamamelidaceae (Cronquist 1981). According to Hutchinson (1969, 1973), this family might have been derived directly from an apocarpous family, possibly Rosaceae, and not through the intermediate stock of the Hamamelidales, in which syncarpy with reduction is the general rule.

Hjelmquist (see Lawrence 1951) includes Leitneriaceae in the Amentiferae, in the line of Myricaceae. Benson (1970) states that the study of vascular bundles of the pistillate flowers suggests earlier presence of several carpels and several ovules to indicate that the bract-like structures in the female flowers are the sepals. Therefore, the relationship could be with the Myricales or, more likely, with the Hamamelidaceae (order Rosales).

Dahlgren (1975a) places Leitneriaceae next to the Myricales, in the subclass Rutanae. Dahlgren (1980a, 1983a), however, changed its position to the Sapindales (Rutiflorae) near Anacardiaceae. Takhtajan (1980) included this family as the most advanced order of his superorder Hamamelidanae of subclass Hamamelididae. According to him, this unigeneric family is evidently one of the anemophilous derivatives of the Hamamelidales. On the basis of serotaxonomic studies, Peterson and Fairbrothers (1983) sought its alliance with the Simaroubaceae, which is again a member of Rutales. Takhtajan (1987) also placed Leitneriaceae in a distinct order Leitneriales next to Rutales. Morphological, anatomical and embryological studies suggest a relationship of Leitneriaceae with Anacardiaceae or Simaroubaceae of the Rutales (or Sapindales).

The position of Leitneriaceae, and the order Leitneriales, should be close to the Rutales. This view is also supported by chemical data (Peterson and Fairbrothers 1983).

Didymelaceae

A unigeneric family, the genus *Didymeles* has 2 species, endemic to Madagascar.

Vegetative Features.   Dioecious trees with simple, entire, alternate, estipulate, coriaceous leaves; yellowish-green when dry.

Floral Features.   Inflorescence catkins of highly reduced flowers, dioecious, axillary or supra-axillary. Male Flowers shortly paniculate subtended by 0–2 scales (or barcts or sepals?). Stamens 2, united by the filaments, anthers sessile, cuneate, tetralocular (bilocular according to Hutchinson 1973). Female flowers spicate, with thickened rachis, subtended by 0–4 scales; gynoecium monocarpellary, ovary unilocular, cylindrical, with large, oblique decurrent stigma with median groove, often recurved at the apex. Ovule one, seed with apical embryo in copious endosperm.

Anatomy.   Xylem parenchyma absent, rays uniseriate; tissues with abundant sclereids.

Embryology.   Embryology has not been studied in detail. Ovules hemianatropous, bitegmic, and crassinucellate. Endosperm formation of the Nuclear type.

Taxonomic Considerations.   Didymelaceae has not been recognised as a distinct family by many workers like Bentham and Hooker (1965c), Engler and Diels (1936), Bessey (1915), Hallier (1908), Hutchinson (1959) and Benson (1970). Leandri (1937) erected the family Didymelaceae for the genus *Didymeles*. Melchior (1964) includes this family in Leitneriales. The family Leitneriaceae, however, is anatomically more advanced, with simple perforation plates, although in floral morphology and embryology Leitneriaceae and Didymelaceae show resemblances. Cronquist (1968) included Didymelaceae in the Hamamelidales. He mentions that Didymelaceae has the most reduced and specialised inflorescences and flowers amongst the members of this order. Takhtajan's (1966) observation that it has a very primitive carpel and wood led to its treatment as a separate order derived from Hamamelidales (Cronquist 1981, Takhtajan 1969, 1980, 1987). Dahlgren (1980a), although placing it in the Euphorbiales, was uncertain about its position. Therefore he suggested an alternate position in Buxales; again a derivative from the Hamamelidales. It is true [as Cronquist (1968) pointed out] that Didymelaceae "is so distinctive that the problem is to find any relative at all, rather than to choose among different possibilities".

Taxonomic Considerations of the Order Leitneriales

Melchior (1964) treats Leitneriales as one of the primitive orders, which is unlikely. Most of the other workers have considered Leitneriales to be more highly evolved by reduction either from the Hamamelidales (Cronquist 1968, 1981, Benson 1970) or Rutales/Sapindales (Dahlgren 1980a, 1983a, Peterson and Fairbrothers 1983, Takhtajan 1987). Evolution by reduction from Rutales or Sapindales seems to be more likely because of similarities (cf. Leitneriaceae), as also corroborated by serological data. Leitneriaceae show affinities with the Geraniales in tricolporate pollen grains, bitegmic, crassinucellar ovule, porogamous entry of pollen tube, Nuclear type of endosperm, and albuminous seed with perisperm (Johri et al. 1992). However, its placement close to the order Rutales would be appropriate (cf. Leitneriaceae).

10

Order Salicales

Salicales is a monotypic order with the family Salicaceae, mostly of woody trees or shrubs with stipulate leaves and catkin inflorescences, capsular fruits, and comose seeds.

Salicaceae

A small family of two genera—*Salix* and *Populus*—with 300 species of almost cosmopolitan distribution. They are abundant in four centres of the temperate zone: Pacific North America, the area around the Behring's Sea, Central Europe and the Himalayas. They are absent in Australia and the Malayan Archipelago (Dutta 1988).

Vegetative Features.   Trees or shrubs, certain species of the arctic and alpine regions are dwarf, carpet-like plants. Leaves simple, alternate, petiolate (Fig. 10.1A), stipulate, stipules sometimes persistent and foliaceous.

Floral Features.   Most members are dioecious, with unisexual flowers in pendulous or erect catkins (Fig. 10.1 B, D). Flowers sessile, bracteate, bracts fringed or hairy, perianth absent or vestigial. Male flowers consist of a small group (sometimes only 2) of stamens set in a cup-shaped, often glandular disc (Fig. 10.1 C). Female flower consists of a single pistil in a concave disc (Fig. 10.1 E). Gynoecium syncarpous, bicarpellary or tetracarpellary; ovary superior, unilocular, with 2 to 4 parietal placentae, ovules numerous; style short, stigma 2 or 4 (Fig. 10.1 E). Fruit, a capsule, seeds comose, embryo straight, endosperm scanty or none.

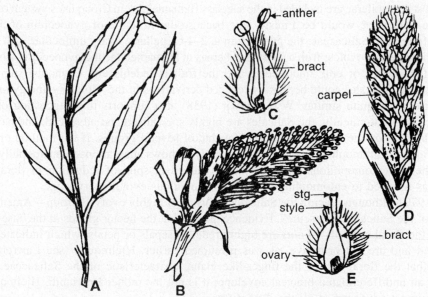

Fig. 10.1   Salicaceae: **A-E** *Salix tetracarpa*. **A** Twig. **B** Staminate inflorescence. **C** Staminate flower. **D** Pistillate inflorescence. **E** Pistillate flower. *stg* stigma. (Adapted from Gangulee and Kar 1987)

Anatomy.   Vessels moderately small, perforations simple, intervascular pitting alternate and large; parenchyma only terminal. Rays exclusively uniseriate, hetero- or homogeneous. Crystalliferous fibres in phloem is a characteristic feature. Leaves dorsiventral, or sometimes isobilateral as in *Populus nigra, Salix alba, S. babylonica* and *S. purpurea.* Stomata Rubiaceous type, frequently on both surfaces of the leaf.

Embryology.   Pollen grains 3-colpate with reticulate exine in *Salix,* inaperturate, and granulose in *Populus* (Johri et al. 1992); shed at 2-celled stage. Ovules unitegmic, crassinucellate, anatropous or campylotropous; Polygonum type of embryo sac, 8-celled at maturity. Endosperm formation of the Nuclear type.

Chromosome Number.   Basic chomosome numbers are x = 11, 12, 19.

Chemical Features.   Common flavonols, flavones and O-methyl derivatives are present. C-glycoflavones and myricetin are also prominent (Gornall et al. 1979). Tannins and phenolic glycosides like salicin and populin are present (Dahlgren 1975a). Cyanidin 3-glucoside and delphinidin 3-glucoside occur in the bark of *Salix purpurea* and some other species (Bridle et al. 1973).

Important Genera and Economic Importance.   Members of Salicaceae thrive in sunshine but not in a hot climate. With adequate water supply they can grow on almost any type of soil. No other tree but *Salix* and *Populus* can grow in the semiarid prairie regions and in the cold marshlands. Wood of *Salix* and *Populus* is a source of cellulose and, therefore, used for paper pulp, rayon, cellophane and lacquer manufacture. Cricket bats are manufactured from the wood of *Salix* spp. The foliage is not used as cattle feed as such, but the shrubby willows (*Salix*) save many deer and other wild animals from starvation when grass is covered with heavy snow (Swingle 1962).

Taxonomic Considerations.   The flowers of the ancestral forms of Salicaceae in all probability possessed a perianth, which is now represented by the cup-like gland in *Populus*. This feature and the simplicity of the flowers in this family are largely due to extreme reduction (Fisher 1928).

   Cronquist (1968) stated that the family Salicaceae is taxonomically rather isolated. Previous workers referred it to the Amentiferae mainly because of the unisexual flowers (without perianth) that are borne in catkins. Most Amentiferae are included in the subclass Hamamelidae in Cronquist's system of classification. However, the Salicaceae would be a misfit here because the structure of gynoecium of the two groups is entirely different. In Salicaceae, the gynoecium is 2–4-carpellary, ovary unilocular with 2 to 4 parietal placentae and numerous ovules; fruit a capsule, whereas in Hamamelidae the gynoecium, when compound, tends to be few-seeded or only one-seeded, and the fruit is indehiscent. Cronquist suggested another possibility, i.e. the Salicales could be florally reduced derivatives of the Violales—the gynoecial structure of the two groups is quite similar. W.H. Brown (1938) also supports this suggestion on the basis of nectary structure. Anatomically, the Salicales are highly specialised and advanced, and have no parallel. Pollen morphology, on the other hand, resembles that of several groups. If the method of pollination is considered, *Salix* is entomophilous although apparently it shows adaptations to anemophily. It is difficult to decide whether entomophilous habit has been retained in spite of reduction in floral structure, or whether it has reverted to entomophily after an anemophilous evolutionary stage.

   Benson (1970), although placing the Salicales in his most highly evolved group—Amentiferae, stated: "The origin of the Salicales is obscure". Evidence proves that the nectar glands at the base of the flowers of *Salix* and the floral cups of *Populus* are highly reduced sepals or petals, which indicates that they are much reduced and are not primitive, as was presumed earlier. Hjelmquist (see Lawrence 1951) also emphasised that the floral cup or the finger-like gland, characteristic of the Salicaceae, is formed by reduction of an undifferentiated bracteal envelope; it is not the reduced perianth. Hjelmquist treats this family as the most advanced of all the Amentiferae.

Hutchinson (1969, 1973) retained Salicaceae with other amentiferous families of presumed Hamamelidaceous affinities, and treated it as the most primitive of them. He stated that the supposed affinity with Tamaricaceae is entirely superficial.

Dahlgren (1975b, 1980a, 1983a) placed the Salicaceae (Salicales) in the subclass Violanae, mainly because of the similarity in the basic structure of gynoecium. Gornall et al. (1979), on the other hand, observed that on the basis of chemical data, the Salicales is isolated from the Violiflorae. Although its removal from the Violiflorae on this basis presents some morphological problems, it is important to note that many of the highly methylated compounds occur in bud secretions rather than in leaf or floral tissue, and are therefore of less taxonomic significance. Cronquist (1981) treats Salicaceae as the only member of Salicales and derives it from the Violales. According to Takhtajan (1980), Salicaceae can be derived from the family Flacourtiaceae of the order Violales. Corner (1976) emphasises: "The small plumose seeds (of Salicaceae) have the same structure as those of the Tamaricales, except that the hairs arise from the placenta instead of the testa itself. The affinity with the exotegmic Flacourtiaceae and Violaceae is ruled out." Takhtajan (1987) derives Salicales and the Salicaceae from the Tamaricales.

Taxonomic Considerations of the Order Salicales

From the above discussion, it is concluded that the order Salicales comprises the only family Salicaceae which is related more closely to and probably derived from the Tamaricales, and not from the members of the Violales.

11

Order Fagales

The order Fagales includes two families—Betulaceae with 6 genera and 150 species and Fagaceae with 8 genera and 900 species. Monoecious trees or shrubs distributed in the temperate zones of the New and the Old World. Leaves alternate, simple, stipulate and deeply-lobed. Staminate flowers in catkins, but pistillate flowers may be solitary, in small clusters, cymules or rarely in catkins. Staminate flowers are subtended by pairs of bracts followed by a free or connate perianth. Number of stamens varies from 2 to 12, sometimes up to 40. Pistillate flower is also bracteate, with or without perianth, 2- to 6-carpellary; 2 ovules per locule. Chalazogamy is common. Fruit a nut or a nutlet subtended by persistent bracts. Seeds non-endospermous with a straight embryo.

Both flavones and flavonoids are present in Betulaceae members and common flavonols in Fagaceae.

Betulaceae

A small family comprising 6 genera and 150 species, distributed in the temperate and arctic alpine regions of Eurasia, North and South America.

Vegetative Features.   Deciduous, monoecious trees and shrubs, often with bark peeling off, as in *Betula utilis*. Leaves alternate, simple, stipulate, stipules often deciduous; margin serrate or sometimes deeply lobed.

Floral Features.   Male inflorescences are pendulous catkins (Fig. 11.1A), female inflorescences are

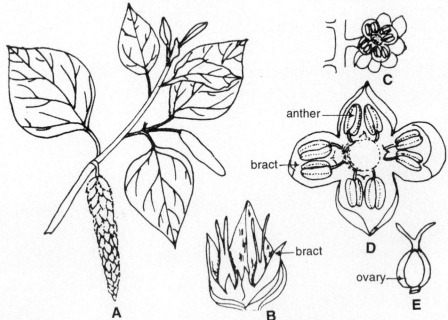

Fig. 11.1 Betulaceae: **A-E** *Alnus rugosa*. **A** Twig with staminate inflorescence. **B** Pistillate cymule. **C** Staminate inflorescence. **D** Staminate flower. **E** Gynoecium. (Adapted from Lawrence 1951)

cymules of 2 or 3 flowers (Fig. 11.1B). In each staminate catkin there is primary axis with spirally arranged cymules. Each cymule comprises a primary bract, in its axil is the secondary axis. The secondary axis gives rise to a pair of secondary bracts and ends in a flower. From the axil of the two secondary bracts, two tertiary axes arise, each terminating in a flower, and a pair of tertiary bracts are borne on each. Staminate cymules of all the genera consist of these three flowers although the bracts in many genera have been lost. A perianth of 4 minute, free tepals subtends 1 to 4 stamens of each flower in *Alnus* and *Betula* (Fig. 11.1C, D). Tepals absent in *Carpinus, Corylus, Ostrya* and *Ostryopsis*. Number of stamens per cymule varies from 2 to 20; filaments short, free or basally connate, anthers bithecous, longitudinally dehiscent. Because of the absence of both bracts and tepals in certain genera, the stamens of all three flowers of a cymule appear to be of a single flower.

Pistillate flowers usually 2 per cymule (3 in *Betula*) (Fig. 11.1B). In some genera, the bracts are absent and perianth may or may not be present. Gynoecium syncarpous, bicarpellary, ovary inferior (when perianth present) or "nude" (when perianth absent) (Fig. 11.1E), bilocular below, unilocular above where the placental septum ends. Each locule with one ovule or only one ovule in an ovary; styles 2 each ending in a single stigma. Fruit a small nut or a winged samara with a single seed. Seeds with a straight, large embryo, endosperm absent; seeds of *Betula davurica* contain a large embryo with scanty endosperm.

Anatomy.   Vessels small, with scalariform perforation plates, intervascular pitting moderately large and opposite to minute and alternate. Parenchyma diffuse and terminal; rays up to 3 or 4 cells wide, exclusively uniseriate, homogeneous. Occasional tracheids reported. Leaves usually dorsiventral, hairs simple, unicellular or sometimes uniseriate as in some species of *Alnus*. Epidermis frequently mucilaginous, stomata usually confined to the lower surface, Ranunculaceous type.

Embryology.   Pollen grains with 3–, 4– or 5–porate, shed at 2–celled stage. Ovules anatropous, uni– or bitegmic, crassinucellate. Polygonum type of embryo sac, 8- nucleate at maturity. Endosperm formation of the Nuclear type.

Chromosome Number.   Basic chromosome number is x = 7 in *Alnus, Betula* and *Corylus* and x = 8 in *Carpinus, Ostrya* and *Ostryopsis*.

Chemical Features.   Both flavones and flavonols present; B-Ring deoxyflavonoids are distinctive (Gornall et al. 1979). Embryo rich in oils and fats as well as in starch.

Important Genera and Economic Importance.   Some of the important genera of Betulaceae are *Alnus, Betula* (birch), *Corylus* (hazel), *Carpinus* (hornbean) and *Ostrya* (hop hornbean) and *Ostryopsis*. Wood of various species of *Betula* is used in making paper pulp. *B. pendula* and *Alnus glutinosa* are used for making plywoods in northern Europe. Wood of *B. lenta* and *B. lutea* is used for making furniture. *B. alnoides* is a tree distributed in the temperate and subtropical Himalayas, Khasi Hills and Manipur in eastern India. Its good quality wood is used for plywood, interiors of furniture, cabinet work and turnery (Singh et al. 1983). In ancient India the bark of *B. papyrifera* was used as a writing material in place of paper. It is known as *Bhurjapatra* (Sans.). Bark of *Alnus* is rich in tannin and hence used in the tanning industry. Wood of *A. nepalensis*, from the temperate Himalayas, is used for making tea chests and superior pencils. *A. nitida* (Himalayan black cedar) is a large tree from northwestern Himalayas and the Punjab. Its wood is used for making matchsticks and for tanning, and twigs are woven into baskets. Hazel nuts and filberts are obtained from two species of *Corylus* and are consumed fresh or after roasting. These nuts are rich in oils and fats, some amount of starch and also vitamins and minerals. Birch-beer is made from the sugary sap of *Betula* spp.

Taxonomic Considerations.   Bessey (1915) placed Betulaceae in the Sapindales along with others like

Juglandaceae, Fagaceae ad Myricaceae. Hutchinson (1926) and Tippo (1938) pointed out that although highly specialised, this family is a derivative of the Hamamelidaceae. Hjelmquist (see Lawrence 1951) treats Betulaceae as a separate order, Betulales, arising from the Fagales. Cronquist (1968, 1981), Benson (1970), Dahlgren (1975a, 1980a), Takhtajan (1969, 1980) and Thorne (1983) include this family in the Fagales. According to Cronquist (1968), the two families Fagaceae and Betulaceae are fairly closely related and are derived from Hamamelidaceae-like ancestors by reduction in floral structures. The Betulaceae is apparently, more advanced than the Fagaceae but their pistillate flowers are so different that one cannot be said to be ancestral to the other. Abbe (1935) reported that the carpellary situation of this family could have originated from an ancestral form with 3-carpellary condition, a situation also true for Hamamelidaceae.

Takhtajan (1987), while retaining both Betulaceae and Fagaceae in the superorder Hamamelidanae, segregated them in two orders—Betulaceae in Betulales and Fagaceae in Fagales.

It appears that these two families originated from Hamamelidales-like ancestors along different lines, but they are also closely related. Therefore, they can well be retained in the order Fagales.

Fagaceae

A small family of 8 genera and 900 species, distributed in the temperate and tropical regions of both the New and the Old World.

Vegetative Features.   Evergreen or deciduous monoecious trees and shrubs. Leaves alternate, petiolate, simple, entire (Fig. 11.2 A) or deeply lobed as in *Castania*, stipulate, stipules deciduous.

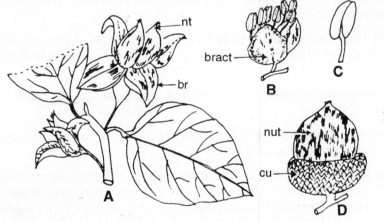

Fig. 11.2 Fagaceae: **A, B** *Fagus sylvatica*, **C, D** *Quercus borealis*. **A** Fruiting branch. **B** Staminate flower. **C** Stamen. **D** Acorn. *br* bract, *cu* cupule, *nt* nut. (After Lawrence 1951)

Floral Features.   Male inflorescences are simple spikes or thyrses except in *Fagus* and *Nothofagus*. In *Fagus* the much-condensed and several-flowered pendulous head is actually a dichasium without a primary terminal flower. *Nothofagus* has single flowers, or 2- or 3-flowered dichasia in the axils of foliage leaves (Fey and Endress 1983). The female inflorescences are also spikes or thyrses. *Fagus* has 2–flowered dichasia (or 2-flowered spikes). Flowers unisexual; staminate flowers (Fig. 11.2 B) with a perianth of 4-(6)-7 tepals arranged imbricately, free and enclose 4–40 stamens; filaments filiform, anthers basifixed (Fig. 11.2 B, C), bicelled, the two lobes merging together or confluent, dehiscence longitudinal. Pistillate flowers usually enclosed by an involucre of many adnate or imbricate bracteoles. Perianth lobes 4 to 6, free. Sometimes, however, 2 or 3 female flowers occur together in a cymule, usually surrounded by an involucre (Fig. 11.2 A). Gynoecium syncarpous, 3- to 6-carpellary, ovary inferior, 3- to 6-loculed, placentation axile with 2 ovules per locule; styles as many as the number of locules. Fruit, a one-seeded nut surrounded by the persistent hardened, cup-like involucre or cupule which is spiny, bristly or tuberculate at maturity (Fig. 11.2 D). Seeds non-endospermous with a large straight embryo.

The cupule, best interpreted by Fey and Endress (1983): "is composed of a more or less highly branched system of modified (compact, sterile) ultimate parts of the inflorescential cyme. The cupular valves do not correspond to just one branch, but to a branching system of several to many axial orders".

Anatomy.   Vessels moderately small, medium-sized or large, usually solitary; few and with a marked radial or oblique pattern; perforation plates simple to predominantly scalariform. Intervascular pitting intermediate, opposite or alternate. Parenchyma apotracheal; rays either wholly uniseriates or of two distinct sizes, the large 20–60 cells wide and very high; homo- or heterogeneous. Leaves dorsiventral to isobilateral, with simple, unicellular hairs or clustered or glandular peltate hairs. Stomata Ranunculaceous type confined to the lower surface.

Embryology.   Pollen grains 3- to 6- or 7–aperturate (colpate), and 2–celled at dispersal stage. Ovules anatropous, bitegmic, unitegmic in *Nothofagus,* crassinucellate. Embryo sac of Polygonum type, 8-nucleate at maturity. Endosperm formation of the Nuclear type.

Chromosome Number.   Basic chromosome number is x = 12.

Chemical Features.   Common flavonols present; C-glycoflavones distinguish this family (Gornall et al. 1979). Quercitol (of carbohydrate group) is common in the members of the Fagaceae except in *Castanea* and *Fagus.*

Important Genera and Economic Importance.   Some of the important genera of this family are *Quercus* (oak) with ca. 450 species, *Fagus* (beech) with 10 species and *Castanea* (chestnut) with 12 species. Economically, Fagaceae is important; *Quercus, Fagus and Castanea* yield high-quality timber. *Quercus suber* is the source of commercial cork. Tannic acid, used in tanning industry, is obtained from various insect galls occurring on some species. Fruits of *Castanea sativa* (European chestnut), *Castanopsis hystrix, Fagus sylvatica* (European beech) and the acorns of *Quercus ilex* (holly or holm oak) are edible. *Q. infectoria* (dyer's oak) is a shrub or small tree, indigenous to Greece, Syria and Iran. The galls—variously known as Aleppo gall, Mecca gall, Turkey gall—are used for tanning, dyeing, mordanting and in the preparation of ink. The wood of *Q. lamellosa*, an evergreen tree from the eastern Himalayas, Assam, Manipur and Darjeeling, is used for house building, bridges and poles (Singh et al. 1983).

Taxonomic Considerations.   It has already been stated earlier that the Fagaceae and the Betulaceae are closely related (cf. Betulaceae). Bessey (1915) treated Fagaceae as a family belonging to the order Sapindales. Rendle (1925), Hutchinson (1926) and Wettstein (1935) retained both the families (Betulaceae and Fagaceae) in the Fagales. Hjelmquist (see Lawrence 1951) raised both families to the rank of orders and assigned them to Amentiferae. Hjelmquist considered it to be highly advanced. Berridge (1914) and Tippo (1938) derived Fagaceae from the same ancestral stock from which Hamamelidaceae have been derived; Cronquist (1968, 1981) agrees with this view (cf. Betulaceae). According to Dahlgren (1975a, 1980a), the Fagaceae belong to the order Fagales (along with two other families) and are closely related to the Hamamelidales; both the orders are assigned to superorder Rosiflorae (Dahlgren 1980a). He supports the views expressed by other taxonomists. Takhtajan (1969, 1980, 1987) also includes Fagaceae in the Fagales, which is directly derived from the order Hamamelidales.

Taxonomic Considerations of the Order Fagales
The Betulaceae and Fagaceae are closely related and derived from ancestors similar to the Hamamelidaceae. In some respects, the Betulaceae are advanced over the Fagaceae; the pistillate flowers in the two families

are so different that one cannot be directly ancestral to the other (Cronquist 1968). Bessey (1915) grouped these two families with the Juglandaceae and treated the three as members of the order Sapindales. Hjelmquist (1948, see Lawrence 1951) segregated the two families in two separate orders—Betulales and Fagales. This view is not supported by others.

Endress (1977a) stresses that the Fagales and Hamamelidales are undoubtedly related. Cronquist (1968) and Takhtajan (1959) included a third family, Balanopaceae, in this order. Hutchinson (1969) and Dahlgren (1975a) included a much-reduced and therefore more advanced Corylaceae in this order. Both Cronquist (1981) and Takhtajan (1980) removed Balanopaceae from Fagales and retained only two families. However, Takhtajan (1987) made a further change and erected two monotypic orders—Fagales (with Fagaceae) and Betulales (with Betulaceae).

It may be concluded that the Fagales is an advanced order because of: (a) highly reduced flowers borne on catkin-like inflorescence, (b) syncarpous ovary, and (c) non-endospermous seeds. Also, this order must have originated from some Hamamelidaceous ancestor by further reduction in floral structure.

12

Order Urticales

The order Urticales comprises five families—a monotypic family Rhoipteleaceae, Ulmaceae, unigeneric Eucommiaceae, Moraceae and another large family, Urticaceae. Predominantly woody plants, a few herbs and climbers. Flowers simple, with undifferentiated perianth or perianth absent, ovary bicarpellary, produces only one seed. Bisexual flowers occur in some Ulmaceae (*Ulmus*). Wind pollination is the general rule.

Rhoipteleaceae

The family comprises a single genus with one species—*Rhoiptelea chilantha*—restricted to Indo-China.

Vegetative Features.   Woody, aromatic trees, ca. 20 m high. Imparipinnately compound, stipulate, deciduous leaves, thickly covered with numerous shield-shaped glands.

Floral Features.   Inflorescence large, pendulous, terminal panicles. Flowers bisexual, in triplets. The three flowers are surrounded by a large unlobed bract and 2 bracteoles. The middle flower is perfect and fertile with 4 uninerved, scarious, persistent sepals, 6 episepalous stamens, erect and persistent; bicelled, longitudinally dehiscent anthers. Gynoecium syncarpous, bicarpellary, ovary superior, transversely bilocular; one locule aborts, the other with a single ovule attached to the partition wall. The two stigmatic lobes are lamelliform and subrhombic, often bi- or tricuspidate. The lateral 2 flowers are female and sterile, each subtended by 2 small bracts. Fruit a samara, 5 to 8 mm in diameter. Seed non-endospermous, embryo straight with thick cotyledons.

Anatomy.   Wood ring-porous, vessels solitary, rarely in groups of 2 or 3, angular in cross section, perforation plates scalariform, end walls oblique with numerous alternate intervascular pittings. Rays heterogeneous, wood parenchyma abundant.

Embryology.   Pollen grains triangular, ca. 22 μm in diameter, tricolpate, with three pores arranged equatorially at the angles of the grains. Ovules semianatropous, bitegmic. Seed exalbuminous with a straight embryo which has a slender radicle and thick cotyledons (Johri et al. 1992).

Taxonomic Considerations.   The family Rhoipteleaceae was described by Handel-Mazzetti in 1932. He suggested that it may belong to the Urticales or to the Juglandales. Wettstein (1935) and Engler and Diels (1936) include it in the Urticales. Tang (1932) does not support this assignment, but suggests no other placement of the family.

Withner (1941), on the basis of anatomical resemblances, treats Rhoipteleaceae as a primitive member of the Juglandales. Lawrence (1951) includes it in the Urticales. Cronquist (1968) considers it as the most primitive family of the Juglandales and therefore a good connecting link with the Urticales. According to Hutchinson (1969), *R. chilantha* is an interesting monotype, probably representing a primitive stock leading up the more advanced Juglandaceae. Dahlgren (1975a, 1983a) and Takhtajan (1980) also include it in the Juglandales. However, Takhtajan (1987) erects a separate order, Rhoipteleales, to accommodate this family and places it before Juglandales. In our opinion also, Rhoipteleaceae should be raised to the rank of order, primitive to the Juglandales.

Ulmaceae

A family of 18 genera and ca. 200 species, mainly confined to temperate and subtropical zones of northern hemisphere.

Vegetative Features.   Mostly trees or shrubs with watery exudates. Leaves simple, alternate, petiolate, stipulate, stipules caducous; serrate margined (Fig. 12.2A, D), with unequal or oblique base,

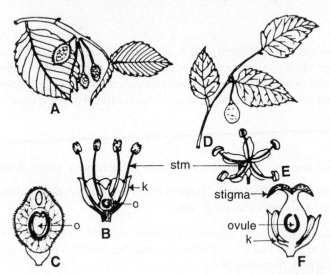

Fig. 12.2   *Ulmaceae*: **A–C** *Ulmus americana*, **D–F** *Celtis occidentalis*. **A** Fruiting twig. **B** Flower, vertical section. **C** Fruit, vertical section. **D** Fruiting twig. **E** Staminate flower. **F** Bisexual flower, vertical section. *k* calyx, *o* ovule, *stm* stamen. (Adapted from Lawrence 1951)

Floral Features.   Inflorescence cymose or in axillary groups or bunches or fascicles or solitary. Flowers bi- or unisexual (Fig. 12.2B, E, F), actinomorphic, tetramerous, apetalous. Sepals 4 to 8, united to form a more or less campanulate calyx (Fig. 12.2B, E). Stamens as many as the sepals, erect in bud, inserted at the base of calyx and opposite to the calyx lobes (Fig. 12.2F). Filaments distinct, anthers bicelled (Fig. 12.2B), longitudinally dehiscent. Gynoecium syncarpous, bicarpellary, ovary superior, unilocular and 1-ovuled (Fig. 12.2B, F), ovule pendulous, styles 2, linear, stigmatic surface papillose. Fruit a flat membranous samara (Fig. 12.2A), or a nut or drupe (Fig. 12.2D). Seeds (Fig. 12.2C) with a straight embryo, non-endospermous.

Anatomy.   Wood ring-porous or diffuse-porous. Vessels predominantly solitary in radical multiples, sometimes spirally thickened and with simple perforations. Intervascular pittings alternate, small to moderately large. Parenchyma scanty, paratracheal; rays hetero- or homogeneous. Leaves usually dorsiventral, subcentric in *Celtis*. Hairs glandular as well as non-glandular, stomata usually confined to the lower surface, subsidiary cells apparently absent or not well defined.

Embryology.   Pollen grains tricolpate, shed at 2-celled stage; the generative cell in many genera (*Holoptelea*) divides on the surface of the stigma. Often 2-pollen tubes enter the same embryo sac. Entry of pollen tube may be micropylar or chalazal. Ovules anatropous, bitegmic, crassinucellate. Development of embryo sac in *Ulmus* is interesting; it is transitional between Adoxa type and a modified Drusa type (Walker 1950). Endosperm formation of the Nuclear type. Polyembryony is known in some species of *Ulmus*.

Chromosome Numbers.   Basic chromosome numbers are x = 10, 11, 14.

Chemical Features.   Rich in flavonols and glycoflavones (Giannasi 1978). L-quebrachitol, a type of cyclitol, is reported in *Celtis* but not in *Ulmus* (Harborne and Turner 1984). Fossil leaves of *Zelkova* reveal the chemical preservation of kaempferol and dihydrokaempferol, the two "fossil flavonoids" (Niklas and Giannasi 1977a). Niklas and Giannasi (1977b) conclude from further experiments that the flavonoids concerned did not experience post-depositional temperatures above 80 °C or extreme pH changes. Similar extracts from fossil leaves of *Celtis* and *Ulmus* contained quercetin 3-0-glycosides, apigenin and luteolin carbon glycosides (Giannasi and Niklas 1977). These chemicals are also reported in the extant species of these genera.

Important Genera and Economic Importance.   *Ulmus* and *Celtis* are the two important genera. *Trema, Holoptelea, Aphananthe, Zelkova* and *Planera* are also well-known. Timber from various species of *Ulmus* is very strong and durable under water and used for shipbuilding, panelling and furniture making. Some species of *Aphananthe, Celtis* and *Holoptelea* also yield timber used locally in European countries. Because of its fragrance, the wood of *Planera abelica,* known as false sandalwood, is used in cabinet making. Many species of *Ulmus* and *Celtis* are grown as ornamentals. The gelatinous inner bark of *U. fulva* is used medicinally. The seeds of some species of *Celtis* are edible.

Taxonomic Considerations.   Hallier (1912) included Ulmaceae in the Urticales and pointed out its derivation from the family Terebinthaceae. Bessey (1915) treated this family as a member of the Malvales. Hutchinson (1948) treated it as a member of the Urticales, derived from the Hamamelidales. According to Cronquist (1968, 1981) also, Ulmaceae belongs to the Urticales and is derived from the Hamamelidales. Dahlgren (1983a) and Takhtajan (1987) support this assignment.

Eucommiaceae

This unigeneric family is reported from China only and has been preserved due to cultivation. *Eucommia ulmoides* does not occur in the wild state.

Vegetative Features.   A deciduous, dioecious, elm-like tree characterized by rubber latex and lamellate pith. Leaves are alternate, simple, petiolate, estipulate, serrate-margined (Fig. 12.3 A).

Floral Features.   Flowers unisexual (Fig. 12.3B, C), actinomorphic, solitary in axils of bracts at the base of twigs and without perianth. Staminate flowers pedicellate, stamens 4 to 10; filaments short and distinct, anthers linear, bicelled, mucronate, dehiscence longitudinal. Pistillate flowers with shorter pedicels, bracteate (Fig. 12.3C); gynoecium syncarpous, bicarpellary, ovary apparently superior, bilocular, one carpel aborts, placentation axile, ovule one, pendulous (Fig. 12.3D). Styles 2, short and reflexed, inner

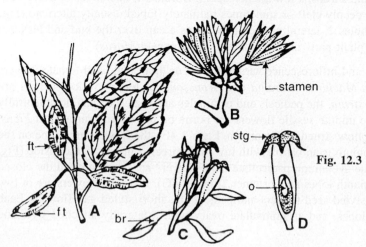

Fig. 12.3 Eucommiaceae: **A–D** *Eucommia ulmoides.* **A** Fruiting twig. **B** Staminate inflorescence. **C** Pistillate inflorescence. **D** Pistillate flower, vertical section. *br* bract, *ft* fruit, *o* ovule, *stg* stigma, *stm* stamen. (Adapted from Lawrence 1951)

surface stigmatic (Fig. 12.3C, D). Fruit a 1-seeded samara (Fig. 12.3A). Seed with a large straight embryo, endosperm abundant.

Anatomy.   Wood semi-ring-porous, vessels very small and nearly all solitary with spiral thickening, perforations simple, intervascular pitting opposite to alternate. Parenchyma diffuse and terminal. Rays up to 3 or 4 cells wide, a few uniseriate, almost homogeneous. Leaves dorsiventral, hairs simple, unicellular; stomata of Ranunculaceous type, confined to the lower surface. Laticiferous cells and tubes are of common occurrence in stem as well as in leaf.

Embryology.   Pollen grains tricolpate, shed at 2-celled stage. Ovules anatropous, unitegmic, weakly crassinucellate; Polygonum type of embryo sac, 8-nucleate at maturity. Endosperm formation of the Cellular type. Suspensor polyembryony is reported (Johri et al. 1992).

Chromosome Number.   Haploid chromosome number is n = 17.

Important Genus and Economic Importance.   The only genus, with only one species, is *Eucommia ulmoides,* the bark furnishes a drug, much valued by the Chinese. The inferior quality of rubber is scanty and extraction has not been tried.

Taxonomic Considerations.   Phyletic position of Eucommiaceae has been dealt with under taxonomic considerations of the Urticales. Eucommiaceae was originally placed in the Magnoliaceae. According to Hutchinson (1969), it is a "near relative" of the Ulmaceae. According to Lawrence (1951), it belongs to the Rosales, suborder Saxifragineae. Cronquist (1968), Dahlgren (1975a) and Takhtajan (1980) place it in a distinct order, Eucommiales, and treat it as primitive to the Urticales. Thorne (1983) includes this family in a much higher order, the Hamamelidales. On the basis of morphological and anatomical features, Eucommiaceae appears to be more closely related to the Urticales than to the Hamamelidales, or of stocks ancestral to them. The reduced flowers, presence of secretory ducts, the Cellular type of endosperm, and samaroid fruit of *Eucommia* are some of the features allied to Hamamelidaceae.

Moraceae

Moraceae is a large family of 53 genera and 1400 species distributed in the tropical and subtropical zones of the world.

Vegetative Features.   Mostly trees and shrubs, a few are herbs, e.g. *Dorstenia,* deciduous or evergreen, mono- or dioecious. Leaves entire or deeply cleft or sometimes palmately lobed, usually alternate (Fig. 12.4A), rarely opposite, stipulate, stipules 2, lateral or each pair forms a cap over the bud and leaves a cylindrical scar on the stem. All the plant parts usually contain latex (except *Morus*).

Floral Features.   Flowers unisexual and inflorescence varied. In *Morus,* both male and female flowers in pendulous catkins (Fig. 12.4B), in *Maclura, Cudrania* and *Broussonetia,* female inflorescences are condensed into globose heads. In *Dorstenia,* the pedicels and peduncles are coalesced and dorsiventrally flattened into a laminate structure, the minute sessile flowers are borne on the ventral surface. In *Ficus,* the receptacle has developed into a hollow structure (syconium, Fig. 12.4F), the flowers are borne on the inner surface. At the apex of the syconium is an ostiole with inwardly directed bracts on the inside (Fig. 12.4F). On the inner fleshy wall of the syconium, pedicellate male flowers are borne around the closed ostiole; each flower has 3 hooded perianth lobes and a stamen (Fig. 12.4G). Pistillate flowers are of two types: sessile or almost sessile, long-styled seed flowers and pedicellate, short-styled gall flowers. Each pedicellate flower has five perianth lobes, and an uniovulate ovary with a laterally attached style and papillate stigma (Fig. 12.4H, I).

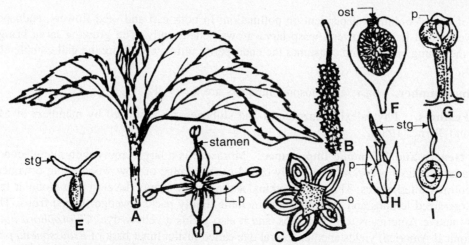

Fig. 12.4 Moraceae: **A–E** *Morus alba*, **F–I** *Ficus glomerata*. **A** Twig. **B** Pistillate inflorescence. **C** Cross section. **D** Staminate flower. **E** Gynoecium. **F** Inflorescence, vertical section. **G** Staminate flower. **H** Pistillate flower. **I** Vertical section. *o* ovule, *ost* ostiole, *p* perianth, *stg* stigma. (**A–E** by Arindam Bhattacharyya, **F–I** adapted from Benson 1970)

In other members, flowers are minute, perianth of 4 tepals in 2 whorls, free or slightly connate at the base (Fig. 12.4D). Number of stamens in male flowers (Fig. 12.4D), as many as the tepals or reduced, opposite the perianth lobes, filaments short, thick and prominent, incurved in bud; anthers versatile and longitudinally dehiscent. The female flowers with or without a perianth, when present, tepals 4; gynoecium syncarpouns, bicarpellary, one carpel often abortive; ovary superior or inferior, unilocular and 1-ovuled (Fig. 12.4C). Styles 2, filiform, stigmas 2 (Fig. 12.4E). Fruit usually compound or aggregate. Numerous small achenes or drupes are adnate to the perianth and the receptacles of the other adjacent flowers, forming an aggregate fruit. Seeds with curved embryo and fleshy endosperm.

Anatomy. Wood occasionally ring-porous, vessels usually medium-sized to large and predominantly solitary; spirally thickened, perforation simple, intervascular pitting minute to large. Parenchyma paratracheal, rays 1 to 15 cells wide, slightly heterogeneous to homogeneous, with a few uniseriate rays. Latex tubes present in rays of many genera. Leaves usually dorsiventral, hairs both glandular and non-glandular, stomata distribution variable, raised above or depressed below the leaf surface. Cruciferous type, as in *Conocephalus*, or Ranunculaceous type, or surrounded by a rosette of subsidiary cells, as in some species of *Ficus*.

Embryology. Pollen grains elliptic and biporate as in *Ficus* or spheroidal as in *Morus*, exine smooth, shed at 2-celled stage. Ovules hemianatropous (*Ficus*) or anatropous (*Morus*, *Dorstenia*), bitegmic, crassinucellate. Polygonum type of embryo sac, 8-nucleate at maturity. Endosperm formation of the Nuclear type. Seed coat is formed by both the integuments.

In different species of *Ficus*, the most interesting feature is the study of pollination by different fig pollinators. These insects have specific organs for collecting pollen, located on various parts of their body. *Ceratosolen arabicus*—pollinator of *F. sycomorus*, and *Blastophaga quadraticeps*—pollinator of *F. religiosa*—have thoracic pollen pockets closed from outside by a movable lid. Some other types of wasps have coxal corbiculae in the form of longitudinal depressions along the coxae, i.e. proximal joints of forelegs. In *B. psenes* pollen is stored in the intersegmental and pleural invaginations on the body of the wasp. Wasps that have no pollen pockets usually pollinate such species of *Ficus* as produce abundant pollen grains.

Growth of the syconium is dependent on pollination. In both gall and seed flowers, endosperm and embryo develop and in gall flowers, wasp larva grows concurrently. The growing larva brings about abortion of the young embryo and consumes the endosperm, ultimately filling the gall completely (Johri et al. 1992).

Chromosome Number.   Basic chromosome numbers are x = 7, 12, 13.

Chemical Features.   Phytoalexin benzofurans and stilbenes are produced by members of Moraceae (Harborne and Turner 1984).

Important Genera and Economic Importance.   Moraceae is a large family with many economically important taxa. *Maclura pomifera,* commonly called osage orange or bow wood, is an ornamental tree from the central United States. The round catkins of the female plants develop into globular fruits that look like large-sized lemons and consist of innumerable tightly packed wedge-shaped fruits. The wood was used by native Americans to make clubs, and it also yields a yellow dye. *Chlorophora tinctoria* (a tree from tropical America) yields another natural dye called fustic. Inner bark of *Broussonetia papyrifera* stem is the source of a good fibre used in the manufacture of paper. It is widely grown in Central Asia for making tapa cloth. Most species of the genus *Dorstenia* (from tropical South America and Africa) are highly poisonous. Various species of *Morus* (*M. alba, M. nigra*) are source of the mulberry fruit. Leaves of these plants are also used as feed for rearing silk moths. Other important fruit trees are *Artocarpus communis* (bread fruit), *A. heterophyllus* (jackfruit) and *Ficus carica* (fig). African bread fruit is the fruit of *Treculia africana. Castilla elastica* from Mexico and Central America gives Panama rubber. *Ficus elastica* from India is another source of rubber. The genus *Ficus* has more than 1000 species. Some of these, like *F. benghalensis, F. religiosa* and *F. krishnae,* are shade trees and also have religious importance. *F. aurea* is an epiphyte, *F. pumila* and *F. radicans* are both climbers with mottled leaves from Japan. These three species are grown as indoor plants. *F. carica* and *F. glomerata* are grown for their edible fruits. *Ficus benghalensis* is a tree with large canopy and numerous aerial roots produced from the branches grow downwards. These roots reach the ground and help in supporting the tree. *F. pumila,* known as Indian ivy, is a root climber and the climbing roots secrete a gummy substance by which the roots become attached to the support.

Taxonomic Considerations.   Bentham and Hooker (1965c) treated this family as a highly advanced one and placed it in series Unisexuales of the Monochlamydeae along with other families like Euphorbiaceae, Urticaceae, etc. Hallier (1912), however, kept all its members in the family Urticaceae. Bessey (1915) treated it as a member of the Malvales. Cronquist (1968, 1981) also treats this family as a member of the Urticales and derives it from the Hamamelidales. Benson (1970) places it in the Urticales of his Thalamiflorae. Dahlgren (1983a) treats the Urticales (including the family Moraceae) as a member of the superoder Dillenianae. Takhtajan (1980, 1987) includes Moraceae in the suborder Urticinae of order Urticales. The two herbaceous genera *Cannabis* and *Humulus* (containing only watery sap and no latex) are kept separately in the family Cannabaceae by Hutchinson (1948).

Cannabinaceae (=Cannabaceae)

This family is represented by only two genera, *Cannabis* and *Humulus,* and 4 species, distributed in temperate Eurasia and North America. This family is often treated as a constituent part of Moraceae, although both genera are herbaceous and without any latex.

Vegetative Features.   Annual or perennial herbs and climbers. Leaves simple and palmately lobed or compound, palmately veined, alternate, stipulate, stipules persistent.

Floral Features.   Flowers unisexual, borne on dioecious plants. Male flowers in loose racemes or panicles with 5 polysepalous sepals, corolla absent. Stamens 5, alternate with sepals, filaments distinct, anthers bicelled, basifixed, longitudinally dehiscent. Female flowers in dense clusters, sepals 5, completely fused to form a cup-like structure enclosing the ovary. Gynoecium syncarpous, bicarpellary, one carpel abortive so that the ovary becomes one-loculed with one ovule. Fruit achene enclosed in persistent calyx.

Anatomy.   Vessels medium-sized, solitary or in small irregular clusters, pits simple, perforation plates simple. Leaves dorsiventral, crystals and cystoliths present, stomata Ranunculaceous type.

Embryology. Pollen grains shed at 2-celled stage. Ovules anatropous, bitegmic, crassinucellate; nucellar cap present; Polygonum type of embryo sac, 8-nucleate at maturity. Endosperm formation of the Nuclear type with chalazal, free-nuclear haustorium, polyembryony reported. Inner integument collapses before maturity of seed. Mature seed with a rudimentary seed coat.

Chromosome Number.   Basic chromosome number is x = 8 in *Humulus* and x = 10 in *Cannabis*.

Chemical Features.   The resinous exudate contains Δ-tetrahydrocannabinol, an active hallucinogenic compound. Other chemical constituents are cannabidiolic acid, tetrahydrocannabinol-carboxylic acid, cannabigerol and cannabichromene (Kochhar 1981).

Important Genera and Economic Importance.   There are only two genera, *Cannabis* and *Humulus*, and both are economically important. *Cannabis* is a multipurpose plant. It gives hemp fibre, hemp seed oil and narcotics. It is also used in indigenous medicines. The narcotic constituents are mainly concentrated in the resin produced by the glandular hairs, more abundant on the bracts surrounding the female flowers. Three forms of intoxicants available from this plant are bhang, ganja, and charas. Of these, charas or hashish is the strongest form. *Humulus lupulus* is used as a flavouring material in beer. Here also, the glands occurring on the female inflorescences are important.

Taxonomic Considerations.   Earlier workers like Bentham and Hooker (1965c), Hallier (1912) and Bessey (1915) included Cannabinaceae in Moraceae. Lawrence (1951) placed the genera *Cannabis* and *Humulus* in the subfamily Cannaboideae of the Moraceae. Rendle (1925) and Hutchinson (1948) raised this subfamily to family rank and designated it as Cannabinaceae. This view is supported by most of the recent workers like Dahlgren (1983a), Cronquist (1981) and Takhtajan (1987).

   The family name has, however, been changed from Cannabinaceae to Cannabaceae, on the basis of the type genus *Cannabis*. It is distinct from other members of the family Moraceae as it is herbaceous, contains watery sap instead of latex, differs also in chemical constituents and embryological features.

Urticaceae

A large family of 45 genera and 850 species, mostly distributed in the tropical and subtropical regions of the world. Nearly 40% of the members are in the New World.

Vegetative Features.   Mono- or dioecious herbs and undershrubs, sometimes small trees, rarely climbers. The vegetative parts of many genera are covered with stiff 'stinging hairs' (Fig. 12.6B) (*Urtica, Fleurya*). The lumen of the hairs is filled with an acidic sap. The hairs have a sharp glassy-pointed tip, which is easily broken off if touched slightly, and injects the 'poison' that causes a burning sensation. Leaves alternate or opposite (Fig. 12.6A), simple, stipulate or estipulate (*Parietaria*).

Floral Features.   Inflorescence basically bracteate, axillary cymes, sometimes pendulous and catkin-like (Fig. 12.6A), sometimes highly condensed to form a head. Flowers unisexual (Fig. 12.6C, D) or imperfectly bisexual, actinomorphic, minute, inconspicuous, monochlamydeous or naked. Perianth, if present, biseriate of 4 or 5 distinct tepals (Fig. 12.6D, E), green, free or connate; staminate flowers mostly with 4 stamens, borne opposite the tepals (Fig. 12.6C). In bud condition stamens bent inwards and downwards but, at the time of anthesis, they spring back elastically, releasing a cloud of pollen grains; anthers bicelled, longitudinally dehiscent; rudimentary pistil often present in staminate flowers. Pistillate flowers with a perianth similar to staminate flowers, gynoecium monocarpellary, ovary superior or inferior, unilocular with a solitary basal ovule (Fig. 12.6F); style one, stigma one, often a brush-like tuft or scale-like staminodes present at the base of the pistil. Fruit an achene (Fig. 12.6G) or drupe, often enclosed in the persistent perianth. Seed with a straight embryo, surrounded by an oily endosperm.

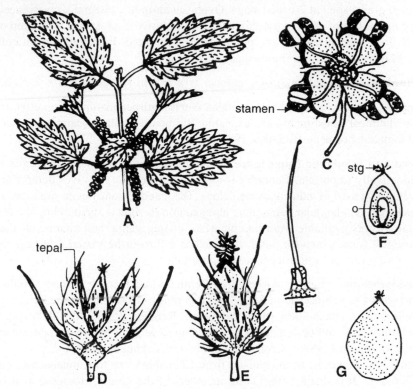

Fig. 12.6 Urticaceae: **A–G** *Urtica urens*. **A** Twig with infloresecence. **B** Stinging hair. **C** Staminate flower. **D** Pistillate flower. **E** Inner sepal. **F** Gynoecium, vertical section. **G** Achene. *o* ovule, *stg* stigma. (Adapted from Benson 1970)

Anatomy. Vessels medium-sized to large, usually solitary or 2 to 4 together; perforations simple. Parenchyma of 2 types: (a) normally lignified cells associated with the vessels, and (b) unlignified cells occur as bands or islands which, on disintegration, appear as phloem islands. Rays 5 to 10 cells wide, heterogeneous. Leaves dorsiventral, stinging hairs abundant in some genera, stomata much varied in distribution, Ranunculaceous type as in *Boehmeria* and *Pilea* or Cruciferous as in *Elatostema*, *Laportea*, *Pellonia* and others.

Embryology. The pollen grains are 2-celled at the dispersal stage. Ovules basal, orthotropous, bitegmic and crassinucellate with the inner integument forming the micropyle. Polygonum type of embryo sac, 8-nucleate at maturity. Development of endosperm of the Nuclear type.

Chromosome Number. Basic chromosome numbers are x = 6, 7, 11, 12, 13.

Important Genera and Economic Importance. The genus *Urtica* has 35 species. This and nine other genera have stinging hairs, e.g. *Girardinia, Pouzolzia, Fleurya* and *Laportea*. Genera without stinging hairs are: *Pilea, Parietaria, Boehmeria, Helxine, Debregeasia,* etc. The well-known ramie fibre is obtained from *Boehmeria nivea. Pilea cadiera* from Indo-China and *P. microphylla* from America make excellent pot plants. The fibres of stinging nettle, *Urtica dioica,* are remarkable for their high tensile strength, but difficulties in cultivation and extraction, as well as the small amount of fibre in the stems, has prevented any commercial use. *Pilea umbrosa* and *P. scripta* are popularly known as artillary plants or gun powder plants because of their explosive stamens. Also, in this genus, the perianth segments are connate and thrown off in the form of a cup. *Laportea crenulata* is commonly called fever or devil nettle. Its stinging hairs are so poisonous that it may cause fever to the person who is stung by it.

Taxonomic Considerations. The family Urticaceae is very distinct from other members of the order Urticales, because of the stinging hairs in several taxa, monocarpellary ovary with a single style, solitary basal orthotropous ovule, and cymose inflorescence on short axillary shoots. According to Cronquist (1968), the differences between Moraceae and Urticaceae are not very distinct. A subfamily of the Moraceae, Conocephaloideae, is more or less transitional between the two. The monotypic herb *Fatoua* is always retained in the Moraceae, but externally it looks very much like any Urticaceae member except for the presence of an inconspicuous vestigial second style. It is also debatable whether Urticaceae is derived from Moraceae, or whether both have diverged from a common ancestor.

Affinities with the Euphorbiaceae were suggested by Guérin (1923) because of the presence of latex ducts in *Laportea* and *Urera*. Other taxonomists like Dahlgren (1975a, 1980a, 1983a) and Takhtajan (1980, 1987), treat this family as the most advanced amongst all the members of the Urticales.

Taxonomic Considerations of the Order Urticales
According to Engler and Diels (1936), the Urticales included the families Ulmaceae, Rhoipteleaceae, Moraceae, (Cannabinaceae) and Urticaceae. The Urticales can be readily distinguished from its allies by (a) the presence of bicarpellary, unilocular, superior/inferior ovary with a single basal or pendulous ovule; and (b) stamens usually equal in number and opposite the perianth lobes.

Bessey (1915) placed Ulmaceae, Moraceae and Urticaceae in Malvales. Whereas, Hallier (1912) included them in his Terebinthales. Withner (1941) presumed that Rhoipteleaceae should be removed to Juglandales. Cronquist (1968, 1981) raises Eucommiaceae to ordinal rank, Eucommiales, and does not include it in the Urticales. The differences between these two orders are (a) Eucommiales has 2 ovules instead of one, (b) ovules unitegmic, and (c) stipules absent. Of these 5 (or 6) families, Ulmaceae is the least evolved. Whether the Urticaceae is the most advanced, and is derived from the Moraceae, or the two have diverged from a common ancestor is controversial. Cronquist (1981) places the Rhoipteleaceae in the Juglandales, mainly because of the presence of pinnately compound leaves. Hutchinson (1973) states that, as the stem of many herbaceous taxa of the Urticaceae are fibrous, they might have been derived from a ligneous stock similar to the Malvaceae, though on the basis of floral morphology, he considers it to have been derived from a Hamamelidaceous stock.

Benson (1970) includes 5 families in the Urticales: Ulmaceae, Eucommiaceae, Moraceae, Cannabinaceae and Urticaceae. He includes Rhoipteleaceae in the Juglandales. Dahlgren (1975a, 1980a, 1983a) does not include Eucommiaceae or Rhoipteleaceae in the Urticales, and places both of them next to the Malvales in subclass Dillenianae. Eucommiales is a distinct order in subclass Cornanae, next to the Sarraceniales. Rhoipteleaceae is regarded as the least advanced family in the Juglandales, subclass Rutanae.

Takhtajan (1980, 1987) also does not include these two controversial families in the Urticales.

Eucommiaceae is in a distinct order of its own, Eucommiales. It is related to the Urticales and has a common origin from the Hamamelidaceae. The family Rhoipteleaceae—though included in the Juglandales— has a hypothetical intermediate position between Hamamelidaceae and Juglandaceae.

Table 12.7.1   Status of the families Eucommiaceae and Rhoipteleaceae

Order/family	Placement	Reference
Eucommiales	An independent order	Cronquist (1968, 1981), Dahlgren (1980a, 1983a), Takhtajan (1980, 1987)
Eucommiaceae	A family in the Urticales	Hutchinson (1969, 1973), Benson (1970)
Eucommiaceae	A family in the Rosales	Engler and Diels (1936), Lawrence (1951)
Rhoipteleaceae	A family in the Juglandales	Withner (1941), Cronquist (1968, 1981), Hutchinson (1969, 1973), Benson (1970), Dahlgren (1975a, 1980a, 1983a), Takhtajan (1987)
Rhoipteleaceae	A family in the Urticales	Engler and Diels (1936), Tippo (1940), Lawence (1951), Melchior (1964)

From the above analysis (Table 12.7.1), it may be concluded that the family Eucommiaceae in the order Urticales should indeed be treated as a distinct order—Eucommiales. The distinctions between these two orders are enumerated in Table 12.7.2. Also, the family Rhoipteleaceae might well be placed in the Juglandales on the basis of pinnately compound leaves. On the other hand, however, this family also shows many features that are common to Urticales. As Cronquist (1968) suggests, Rhoipteleaceae may be a link between the Urticales and the Juglandales.

Table 12.7.2   Differences between the Urticales and Eucommiales

Urticales	Eucommiales
1 Stipules present	Stipules absent
2 Anthers short and thick without mucron	Anthers linear, mucronate
3 Number of ovules per locule 1	Number of ovules per locule 2
4 Ovules bitegmic	Ovules unitegmic

An over-all analysis confirms that the family Urticaceae is the most advanced amongst all the Urticales, because: (a) members mostly herbs or undershrubs, (b) monocarpellary ovary with 1 style, and (c) basal ovule.

Therefore, the Urticales may be regarded as a natural one, and only four families—Ulmaceae, Moraceae, Cannabaceae and Urticaceae—are included in it. Rhoipteleaceae and Eucommiaceae are treated separately.

13

Order Proteales

Proteales is a monotypic order with only one family, Proteaceae.

Proteaceae

A fairly large family of 62 genera and 1050 species distributed in tropical Asia, Malaysia, Australia, New Zealand, tropical South America, Chile, mountains of tropical Africa, South Africa and Madagascar. Many species, such as *Grevillea robusta*, are cultivated in the Indian subcontinent.

Vegetative Features.   Trees or shrubs, rarely herbs; sometimes dioecious. Leaves alternate, rarely opposite or whorled as in *Macadamia*, simple, entire or bipinnately dissected, estipulate, petiolate.

Floral Features.   Inflorescence usually showy, bracteate, head, spike or raceme; flowers in an inflorescence borne in pairs as in Grevilleoideae, or are solitary as in Persoonioideae. Flowers bisexual or unisexual due to abortion as in *Leucadendron*, actionomorphic or zygomorphic as in *Lomatia,* perianth uniseriate (only calyx according to some workers); tepals 4, petaloid, often very colourful, usually fused to form a tube with valvate lobes. Stamens 4, opposite the tepal lobes, filaments adnate or fused to the tube, free in *Symphyonema,* anthers bicelled, in some species of *Conospermum* and *Synaphea* 2 stamens with bicelled anthers and 2 only 1-celled; dehiscence longitudinal. Gynoecium monocarpellary, ovary superior, raised on a short gynophore, unilocular with 1 to many ovules, placentation parietal or the single ovule pendulous, style 1, filiform, stigma 1, often bulbous. Fruit a follicle, capsule, nut or drupe, seeds often winged, without endosperm.

Anatomy.   Vessels of moderate size with simple perforations, intervascular pitting alternate, minute to moderately large. Parenchyma in narrow to wide bands, sometimes enclose the vessels. Rays of 2 types; the larger up to 10 to 30 cells wide and high, usually homogeneous; sizes transitional between the largest and the smallest are common in *Leucadendron, Macadamia and Persoonia.* Leaves leathery, structure correlated with the arid nature of their habitat; dorsiventral or isobilateral to centric, hairs when present unicellular and often thick-walled. Stomata on all parts of the acicular and narrow leaves, confined to the lower surface in dorsiventral leaves; often in small cavities, Rubiaceous type.

Embryology.   Pollen grains triangular, 2- or 3-porate and angulaperturate; in *Aulax* and *Franklandia* spherical, triporate in *Banksia.* Ovules ortho-, hemiana- or anatropous, bitegmic, crassinucellate; micropyle formed by inner integument only. Polygonum type of embryo sac, 8-nucleate at maturity. Endosperm formation of the Nuclear type, differentiated into an upper cellular and a lower nuclear haustorial tube in Grevilleoideae.

Chromosome Number.   Basic chromosome number is x = 5, 7, 10–13.

Chemical Features.   The cotyledons of mature embryo are rich in protein and fat. Flavonol derivatives, leucoanthocyanins, arbutin, tannins and cyanogenic compounds are characteristic of the members. L-quebrachitol is reported in *Grevillea robusta* and *Hakea laurina,* and monomethyl ethers of myo-inositol in *Banksia integrifolia, Macadamia ternifolia* and *Stenocarpus sinuatus* (Plouvier 1963). Tropane alkaloids are also reported (Harborne and Turner 1984).

Important Genera and Economic Importance.   Many taxa are grown for their ornamental value, e.g. *Banksia, Leucadendron, Embothrium, Hakea* and *Protea*. Fruits of *Macadamia ternifolia* and *Gevuina avellana* are edible. Timber from *Faurea macnaughtonii* and *Knightia excelsa* is used for decorative work, cabinet-making, and wood panelling. *Grevillea robusta* grows on wastelands and is grown for reforestation. It is also grown as a shade tree in coffee plantations. Bark of *Leucospermum carpodendron* is used for tanning leather. *Hakea epiglottis* is a Tasmanian endemic with variable forms. Dioecious and bisexual populations are reported in this species (Lee 1987).

Taxonomic Considerations.   Engler and Diels (1936), Hallier (1912), and Melchior (1964) treat Proteaceae as a primitive family. The reduction of petals into nectariferous scales, occasional zygomorphy, ornithophilous flowers and the occurrence of four types of ovules prove beyond doubt that this family is an advanced one (Haber 1959, 1961). Hutchinson (1973) considered Proteaceae to have originated from the Thymelaeales. Benson (1970) derives it from the Rosales. Cronquist (1981) observes that the most likely origin of the Proteaceae is near the Thymelaeaceae in Myrtales. The Thymelaeaceae show a clear progression from syncarpous to pseudomonomerous condition of gynoecium. The flowers of the Thymelaeaceae show pseudomonomerous condition like that of Proteaceae: (a) a prominent hypanthium or calyx tube, (b) tetramerous calyx, and (c) petals reduced or totally absent. As suggested by Cronquist (1968), the monocarpellary ovary of Proteaceae could also be pseudomonomerous. As the Myrtales are probably derived from the Rosales, possibly the ancestry of Proteaceae lies therein (Cronquist 1968, 1981). Johnson and Briggs (1975) also accept the affinity with Rosales s. l. According to Takhtajan (1980, 1987), in all probability, Proteaceae is derived from some Saxifragalean ancestor which was closely related to the modern Cunoniaceae (in Rosidae).

   From the above discussion, it may be concluded that the Proteaceae, of the Proteales is an advanced taxon and its ancestry lies in one of the Rosales s. l. members (or Rosidae of Cronquist and Takhtajan).
   The family Proteaceae comprises 2 subfamilies:

1. Grevilleoideae—flowers in pairs borne on apparent racemes or spikes. Ovules many or sometimes 2. Fruit a many-seeded follicle, e.g. *Banksia, Hakea, Lomatia*.
2. Persoonioideae—flowers single in axil of bracts. Ovules 1, 2 or more. Fruit usually a one-seeded drupe, e.g. *Leucadendron, Persoonia, Protea*.

Taxonomic Considerations of the Order Proteales
The order Proteales includes only one family, Porteaceae. This order is an advanced one and most likely has a Rosalean ancestry.

14

Order Santalales

The order Santalales includes 7 families: Olacaceae, unigeneric Dipentodontaceae, Opiliaceae, unigeneric Grubbiaceae, Santalaceae, unigeneric Myzodendraceae (= Mîsodendraceae) and Loranthaceae (Melchior 1964). All the members, except Dipentodontaceae and Grubbiaceae, are parasitic or semiparasitic and mostly distributed in the tropics and subtropics, with a few extending to the temperate regions.

Mostly trees, shrubs or vines, often epiphytes; leaves alternate (except in Loranthaceae), simple, entire, often coriaceous. Flowers inflorescent, uni- or bisexual (Olacaceae), perianth smaller and opposite the adnate stamens; distinct calyx and corolla in Olacaceae. Ovules often indistinct; when distinct, anatropous as in Olacaceae, Opiliaceae and Grubbiaceae. Seeds endospermous, without testa in Santalaceae and Loranthaceae.

Vessels with simple perforations in all the families, schizogenous resiniferous ducts occasionally present (Grubbiaceae); cells with mucilage and tannin usual (Dahlgren 1975a). Polyacetylenes occur in Olacaceae, Opiliaceae, Santalaceae, Loranthaceae and Viscaceae (Bohlmann et al. 1973, Dahlgren et al. 1981). Tannins common in some families; sometimes leucoanthocyanins, myricetin and ellagic acid are also present as in Santalaceae. Triterpenes common in Loranthaceae.

Olacaceae

A family of 25 genera and 250 species distributed in tropical and subtropical South Africa, Madagascar, Indomalaysia and Australia, Indian subcontinent and tropical America. The Olacaceae of Malaysia are connected with those of south and southwest Asia on the one hand and the Pacific on the other. Some of the Malaysian genera are represented in Africa and Madagascar, and even in Central and South America (Sleumer 1980).

Vegetative Features. Trees, shrubs or vines, leaves simple, alternate, mostly entire, estipulate (Fig. 14.1A).

Floral Features. Inflorescence cymose or thyrse. Flowers usually bisexual (Fig. 14.1B), but some unisexual and plants polygamodioecious. Perianth biseriate; sepals 4–6, imbricate or open in bud, calyx highly reduced, oftern resembling the calyclus of Loranthaceae (Fig. 14.1B), often accrescent. Petals 4–6, polypetalous or variously connate at base, valvate. Stamens 4–12, opposite the petals when the number is same as the number of petals, free or rarely monoadelphous, staminodes occasional; anthers bicelled, dehiscence longitudinal (Fig. 14.1B,C) or poricidal, disc present, mostly annular. Gynoecium syncarpous, 3- to 4-carpellary, ovary superior, often appearing as inferior due to adnation to the surrounding disc, 3- to 4-locular or unilocular due to incomplete septation; placentation axile or free-central from locule apex when unilocular, ovule solitary in each locule (Fig. 14.1 D, E), pendulous; style 1 with a 2- to 5-lobed stigma (Fig. 14.1D). Fruit a drupe or a berry, often surrounded by the persistent enlarged calyx. Seeds endospermous, embryo very small within the apex of fleshy endosperm.

Anatomy. Vessels small, occasionally medium–sized, solitary or with numerous radial multiples, perforation plates scalariform or simple; intervascular pitting scalariform, opposite or alternate. Parenchyma apotracheal, diffuse to uniseriate; rays 1–4 cells wide and usually 1.5–4 mm high. Leaves usually dorsiventral, centric in *Olax stricta* and *Ximenia coriacea*. Stomata mostly confined to the lower surface, but on both surfaces in *Ximenia* and *Olax*; Rubiaceous type. Stellate hairs, resin canals and/or latex occur in some taxa.

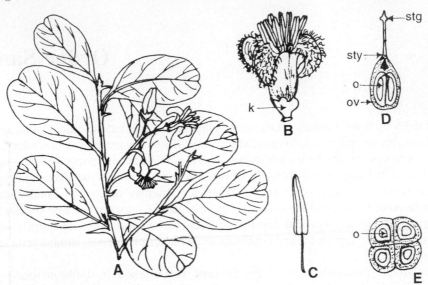

Fig. 14.1 Olacaceae: **A-E** *Ximenia americana*. **A** Flowering twig. **B** Flower. **C** Stamen. **D** Gynoecium, vertical section. **E** Ovary, cross section. *k* calyx, *o* ovule, *ov* ovary, *stg* stigma, *sty* style. (Adapted from Lawrence 1951)

Embryology. Pollen grains triangular, 3-colporate, 2- or 3-celled at the time of shedding; occasionally square-shaped with 4 germ pores (Agarwal 1963). Ovules anatropous, undifferentiated as in *Olax*, unitegmic as in *O. wightiana* or bitegmic as in *Coula, Ximenia*; nucellus reduced. Embryo sac of Allium type, 8-nucleate at maturity. Endosperm formation of the Cellular or Helobial type; a uni- or multinucleate haustorium present.

Chromosome Number. Basic chromosome number is x = 19 or 20 (Takhtajan 1987).

Chemical Features. Fatty acids like triglycerides occur in seed oils (Gershenzon and Mabry 1983). The unusual fatty acid, ximenynic acid, and the glycoside, sambungrin, are reported in *Ximenia americana* (Sørensen 1963, Paris 1963).

Important Genera and Economic Importance. Some of the important genera are *Olax* with 55 species, *Schoepfia* with 35 species and *Ximenia* with 10 to 15 species, all distributed in the tropics and subtropics of both hemispheres. The genus *Olax* is distributed in diversity in Africa including Madagascar, where all the 4 sections recognised by Engler (1909) occur. One of these, section Triadrae, extends from Africa to the far East, ending with an endemic species in New Caledonia (Sleumer 1980). The genus *Ximenia* is scattered from India and Ceylon to Burma, Indochina, Thailand, Hainan, Andaman Islands, Malaysia, North and North-eastern Australia, tropical America and Africa (Sleumer 1980).

Economically, the family is not so important. Kernels of *Coula edulis* or Gaboon nut and *Ximenia americana* or wild olive are edible. The wood does not have much commercial importance. The timber of *Minquartia* is used in Central America for its durability for railway sleepers, posts and poles. The wood of *X. americana* is sometimes used as a substitute for sandalwood (Metcalfe 1935) and is also suitable for carving.

Taxonomic Considerations. The Olacaceae, in combination with the Icacinaceae, constitute the Olacineae in the classificatory system of Bentham and Hooker (1965a). The Olacaceae, together with Opiliaceae and other families, are included in a separate order, the Olacales in Hutchinson's system (1973).

Embryological data (Shamanna 1954, 1961, Agarwal 1961, 1963) and floral morphology (F.H. Smith and E.C. Smith 1942) suggested a close alliance between Olacaceae and Santalaceae. Most of the later taxonomists include this family in the Santalales (Dahlgren 1975a, 1980a, 1983a, Cronquist 1968, 1981, Takhtajan 1980, 1987). This assignment is satisfactory.

Dipentodontaceae

A monotypic family with the genus *Dipentodon* distributed in northeastern India and China.

Vegetative Features. Small deciduous trees, leaves alternate, ovate, denticulate with deciduous stipules.

Floral Features. Inflorescence globose, long-peduncled axillary umbels; flowers with 4 or 5 deciduous bracts at base, small, bisexual, pedicels slender, articulate mid-way, hypogynous. Calyx of 5 to 7 sepals, polysepalous, imbricate, linear, shortly united at the base, pubescent. Petals 5 to 7, polypetalous, imbricate, linear, slightly narrower than sepals or indistinguishable; disc-glands 5 to 7 opposite the petals. Stamens 5 to 7, opposite calyx lobes, erect with small anthers. Gynoecium syncarpous, tricarpellary, ovary superior, unilocular above, with two axile ovules per locule on free basal placenta, style simple, stigma capitate. Fruit a small, oblong, unilocular, 1-seeded, tomentose capsule with persistent style, tardily dehiscent, surrounded by persistent calyx, corolla and stamens.

Anatomy. Vessels usually small, solitary and in twos and threes, occasionally with spiral thickening, perforations simple and oblique; parenchyma absent or very sparse. Rays up to 3–5 cells wide, markedly heterogenous; fibres septate. Leaves dorsiventral, stomata confined to the lower surface, Rubiaceous type. Cluster crystals abundant in the cortical region of petiole, but infrequent in the cortex of *Dipentodon sinicus*.

Embryology. The embryology has not been investigated (see Johri et al. 1992).

Important Genus. The only genus *Dipentodon sinicus* is of no economic importance.

Taxonomic Considerations. Dipentodontaceae is a highly controversial family: included in the Olacales by Hutchinson (1969, 1973). Melchior (1964) and Cronquist (1968) consider it in the order Santalales. Thorne (1983) places it in the Cistales, whereas Dahlgren (1975a, 1983a) and Takhtajan (1987) consider it in the Violales. Lobreau (1969) and Lobreau-Callen (1982) report Dipentodontaceae to be allied to the tribe Homalieae of the Flacourtiaceae, on the basis of pollen grains. T.A. Sprague (cited in Metcalfe and Chalk 1972) confirms this assignment on the basis of presence of stipules and the nature of perianth.

Cronquist (1968), however, points out that *Dipentodon,* and also *Medusandra,* should be included in the Santalales although this would introduce two new characters into the order—stipules and dehiscent fruits. "The fruits are, however, one-seeded like those of the other families of Santalales, and if they merely failed to dehisce, they would be perfectly normal for the order".

The above discussion reflects that the Dipentodontaceae may well be retained in the Santalales, although it shows certain features common with the genus *Homalium* of the Flacourtiaceae of the Violales.

Opiliaceae

A small family of 8 genera and 60 species, distributed in the Asiatic tropics, Indo-China, Java, New Guinea and tropics of Africa.

Vegetative Features. Shrubs, trees or woody vines (*Opilia*). Leaves simple, alternate, entire, estipulate, turning a characteristic yellow-green on drying.

Floral Features. Inflorescence axillary racemes (Fig. 14.3A) or compact cymes as in *Melientha*. Flowers bisexual, epigynous or perigynous, actinomorphic (Fig. 14.3.B). Calyx of 4 or 5 minute (gamosepalous) sepals, cupular or obsolete. Corolla of 4 or 5 petals, gamo- or polypetalous, valvate, united into a short tube in *Lepionurus* and *Cansjera*. Stamens 4 or 5, oppositipetalous, often epipetalous, anthers bicelled, introrse, longitudinally dehiscent (Fig. 14.3C); disc of 4 or 5 scales at the base of the ovary. Gynoecium syncarpous, monocarpellary, ovary superior or half-inferior, unilocular with one basal (Fig. 14.3D) or apical ovule; style and stigma one. Fruit a drupe with a crustaceous stone (Fig. 14.3E), seed without testa, embryo short or long in relation to the amount of the fleshy endosperm, endosperm oily.

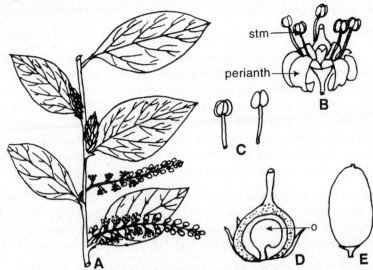

Fig. 14.3   Opiliaceae: **A-E** *Opilia amentacea*. **A** Twig bearing inflorescences. **B** Flower. **C** Stamens. **D** Gynoecium, vertical section. **E** Fruit. *o* ovule, *stm* stamen. (Adapted from Hutchinson 1969)

Anatomy.   Vessels exclusively solitary or in multiples, perforations simple, intervascular pitting rare, alternate and small. Parenchyma sparse, diffuse, strands usually of 2 cells. Rays up to 2–5 cells wide, with very few uniseriates, homogenous. Cystoliths in both leaf and stem, and branched, lignified cells in leaf mesophyll are characteristic features. Leaves usually dorsiventral but centric in *Opilia amentacea*. Stomata usually on both surfaces, Rubiaceous type.

Embryology.   Pollen grains triangular, tricolpate, shed at 2-celled stage. Ovules anatropous, unitegmic, nucellus reduced; Polygonum type of embryo sac, 8-nucleate at maturity. Endosperm formation of Cellular type with an uninucleate haustorium.

Chromosome Number.   Diploid chromosome number of *Lepionurus sylvestris* is 2n = 20 (Kumar and Subramanian 1987).

Chemical Features.   Polyacetylenes present in *Cansjera* and *Opilia* (Bohlmann et al. 1973, Dahlgren et al. 1981), fatty acid triglycerides occur in seed oils (Gershenzon and Mabry 1983).

Important Genera and Economic Importance.   Some of the important genera are *Cansjera, Lepionurus, Melientha* and *Opilia* but they have no economic importance.

Taxonomic Considerations.   Bentham and Hooker (1965a) did not recognise Opiliaceae as a distinct family, and included it in the family Olacaceae. It is distinct from the Olacaceae due to superior or half-inferior and unilocular ovary, suppressed calyx and the nature of the disc. Hutchinson (1969, 1973) included it in the Olacales along with Olacaceae and other families. However, most other systematists include Opiliaceae in the Santalales (Cronquist 1981, Dahlgren 1975a, 1980a, Melchior 1964, Takhtajan 1987). This assignment is also supported by chemotaxonomic (Bohlmann et al. 1973) and embryological studies (Shamanna 1955, Swamy 1960, Johri and Bhatnagar 1960). Johri and Bhatnagar (1960) suggested merging Opiliaceae with the Santalaceae as a tribe, Opilieae.

Grubbiaceae

A unigeneric family with the genus *Grubbia* with 3 species which occur only in the Mediterranean type flora of the Cape Province, South Africa.

Vegetative Features.   Small, heath-like, much-branched, woody shrubs. Leaves ericoid or flat, simple, linear or lanceolate, opposite decussate, estipulate, coriaceous; autotrophic plants of ericoid habit.

Floral Features.   Inflorescence small, axillary, 3-flowered dichasia or many-flowered strobiloid compound dichasia, subtended by 2 or more bracts. Flowers sessile, very small, bisexual, epigynous, actinomorphic. Perianth in a single whorl, tetramerous, calyx-like, and hairy on the outer surface, smooth and suffused pink to red on the inner surface, valvate (Carlquist 1977a). Stamens 8 in 2 whorls of 4 each, free, oppositisepalous, liguliform, laterally compressed filaments and 2-loculed basifixed anthers dehiscing laterally with valvular reflexion of the thecae wall; disc annular, hairy or papillate, epigynous. Gynoecium syncarpous, bicarpellary, ovary inferior, bilocular, with axile placentation at early stage but unilocular later. Ovules one in each locule, pendulous; style filiform, stigma bifid. Fruit indehiscent, with a fleshy exocarp and a sclerenchymatous endocarp. Each fruit united with adjacent fruits in an inflorescence forming a syncarp. Seed with abundant oily endosperm, embryo straight, hypocotyl and radicle longer than the two cotyledons.

Anatomy.   Vessels very small in diameter, solitary, perforation plates scalariform with 40 or more bars, parenchyma sparse. Rays both multi- and uniseriate; rhomboidal crystals and dark-staining deposits of unidentified gummy materials occur sometimes in ray cells of stem and root (Carlquist 1977b). Leaves dorsiventral, surface covered with simple, unicellular hairs. Stomata Ranunculaceous type, confined to the lower surface.

Embryology.   Pollen grains tricolporate, prolate. Ovules anatropous, unitegmic, tenuinucellate, with a long micropyle; Polygonum type of embryo sac, 8-nucleate at maturity. Endosperm formation of the Cellular type and with micropylar and chalazal haustoria.

Important Genus.   The only genus, *Grubbia*, is not known to have any economic importance.

Taxonomic Considerations.   The family Grubbiaceae was not recognised by Bentham and Hooker (1965c) and was considered as the tribe Grubbieae of the Santalaceae. On the basis of vegetative anatomy, Metcalfe and Chalk (1972) suggested a separate family for *Grubbia*, as it was very distinct from the Santalaceae (Table 14.4.1).

Table 14.4.1   Comparative Data for Grubbiaceae and Santalaceae

Grubbiaceae	Santalaceae
1. Vessels solitary with scalariform perforation	Vessels solitary with simple perforation or vessels numerous in small multiples
2. Parenchyma sparse	Parenchyma diffuse
3. Rays with rhombiodal crystals and dark staining deposits of gummy substance	Rays with prominent solitary crystals only; dark-staining deposits absent
4. Leaf surface covered with unicellular hairs	Leaf surface with striated cuticle
5. Stomata Ranunculaceous type, only on lower surface	Stomata Rubiaceous type, on both or only lower surface

Fagerlind (1947), on the basis of embryological features, suggested a close relationship between Grubbiaceae and Ericaceae, and its transfer to the Ericales. Erdtman (1952), however, observed that although the pollen grains of Grubbiaceae resemble those of Ericaceae, in most Ericaceae they occur in tetrads. The embryological features referred to by Fagerlind (1947) are also known in other members of Santalales. Most systematists include Grubbiaceae in Santalales (Cronquist 1968, Hutchinson 1973, Melchior 1964). According to Carlquist (1977b), the Grubbiaceae has a generalised rosoid position along with Bruniaceae and Geissolomataceae, mainly on the basis of similarity in wood anatomy.

Dahlgren (1980a) also placed Grubbiaceae in the Cunoniales near Bruniaceae. Cronquist (1981) and Takhtajan (1987), however, include the Grubbiaceae in the Ericales. The position of this family is still disputed and further comparative studies are necessary to assign it to its correct position.

Santalaceae
A large family of 35 genera and 400 species, distributed widely in the temperate as well as tropical zones of the world and concentrated in the relatively drier regions.

Vegetative Features.   Trees, shrubs or herbs, non-parasitic or parasitic on roots of higher plants. Leaves alternate or opposite, sometimes in pseudowhorls, simple, petiolate, estipulate, glabrous (Fig. 14.5A).

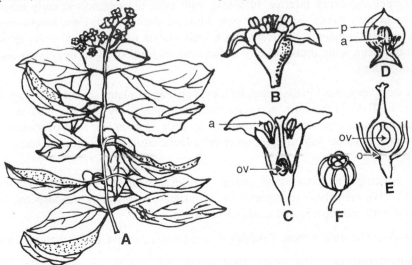

Fig. 14.5 Santalaceae: **A-F** *Santalum album*. **A** Flowering twig. **B** Flower. **C** Vertical section. **D** Perianth lobe with stamen attached at base. **E** Carpel, vertical section. **F** Fruit. *a* anther, *o* ovule, *ov* ovary, *p* perianth. (Adapted from Saldanha and Nicholson 1976)

Floral Features.   Inflorescence various: a raceme, spike, head, dichasium or solitary axillary flowers. Flowers minute (Fig. 14.5A), bi- or unisexual, actinomorphic, mostly epigynous, sometimes perigynous. Perianth only one-whorled, sepaloid or petaloid, 4- to 5-lobed, free, valvate or basally connate to form a tube (Fig. 14.5 B, C). Stamens 4 or 5, opposite the perianth lobes, attached at the base (Fig. 14.5 C, D) or the rim of the tube or cup-like structure. Filaments short, anthers bicelled, basifixed, dehiscence longitudinal (Fig. 14.5D). Gynoecium syncarpous, 3- to 5- and rarely bicarpellary, ovary unilocular, mostly inferior and embedded in receptacular disc or superior and borne on or surrounded by a nectar-secreting receptacular disc, which may have lobes projecting between the stamens, as in *Buckleya distichophylla*. Ovules 1-5 but usually 3 (only 1 matures), borne pendulously on a basal placenta; style 1, terminal stigma 1 and capitate or 3- to 5-lobed (Fig. 14.5E). Fruit an indehiscent nut (Fig. 14.5F) or drupe with a single seed. Seed without testa, with copious fleshy endosperm and a straight embryo.

Anatomy. Wood anatomy of two types:

(a) Vessels exclusively solitary with simple perforations. Parenchyma diffuse with numerous chambered crystals, strands of 2–4 cells. Rays 2–4 cells wide, with numerous uniseriates, slightly heterogeneous to homogeneous.

(b) Vessels numerous in small multiples, some tending to be ring-porous; spiral thickening may or may not be seen; perforations simple. Parenchyma paratracheal, strands mostly of two cells; rays 3–6 cells wide with very few uniseriates, hetero- or mostly homogeneous, usually with prominent solitary crystals.

Leaves dorsiventral, hairs mostly unicellular; cuticle frequently striated. Stomata on both surfaces or confined to only lower surface, mostly Rubiaceous type.

Embryology. Pollen grains tricolpate, globose, triangular or oblong, exine mostly smooth but may be granulate in some taxa (*Santalum album*) (Bhatnagar 1965); shed at 2–3-celled stage. In situ germination of pollen grains is common in *Santalum album* and *Thesium wightianum*. Ovules much reduced, hemianatropous or anatropous; no distinction between nucellus and integument except in *Comandra*, *Iodina* and *Thesium*, where ovules are unitegmic. Usually Polygonum type of embryo sac; sometimes Allium or Endymion type (Johri et al. 1992). Endosperm formation of the Cellular or Helobial type with chalazal haustorium, uni- or multicellular.

Chromosome Number. Basic chromosome numbers are x = 6, 7, 12, 13+.

Chemical Features. Endosperm rich in oils; anthocyanins, myricetin, and ellagic acid present (Dahlgren 1975a). Triglycerides of acetylinic fatty acids are also reported (Bohlmann et al. 1973, Dahlgren et al. 1981, Gershenzon and Mabry 1983).

Important Genera and Economic Importance. The sandal-wood tree (*Santalum album*) is the most renowned plant of this family. These trees were formerly abundant in the Fiji Islands, but the busy trade of this important commodity (between 1804 and 1816) made it nearly extinct. Its use dates back to the 5th Century B.C. and it is still used in India and China, often in performing certain religious ceremonies and funeral rites. In India, the wood is used in the manufacture of decorative articles like small boxes, hand fans, and others. The oil obtained on distillation is used in perfumery, and also in medicine. The powdered wood is used in cosmetics. The sawdust from the heartwood is made into scented cakes and sticks to be burnt as incense.

Calpoon compressum, Eucarya spicata, Santalum freycinetianum and *S. yasi* yield valuable wood. The fruits of *Acanthosyris falcata* and *Exocarpus cupressiformis*, and tubers of *Arjona tuberosa*, are edible. Species of *Buckleya* and *Pyrularia* are often cultivated for their showy ornamental flowers.

Taxonomic Considerations. In Bentham and Hooker's (1965c) classification Santalaceae is included in series Achlamydosporae of the subclass Monochlamydeae. Engler and Diels (1936) included Santalaceae in suborder Santalineae of the Santalales under Archichlamydeae. Other taxonomists also place Santalaceae in the order Santalales, although the circumscription of the order varies. The Santalaceae is closely related to the Loranthaceae in some characters of perianth, ovary and seeds; there are also some significant differences. According to some taxonomists, Santalaceae is also related to the Myzodendraceae. Both Santalaceae and Loranthaceae show: (a) Uni- or bisexual flowers, (b) a poorly or well-developed calyculus, (c) inferior ovary, (d) Polygonum type of embryo sac, and (e) seeds without testa.

The differences between these two families are tabulated in Table 14.5.1.

Table 14.5.1[1]   Comparative Data for Santalaceae and Loranthaceae

Santalaceae	Loranthaceae
1. Pollen grains spherical or oblong	Pollen grains triradiate
2. Distinct ovules present on central placenta	Ovules (in the usual sense) absent
3. Cellular type of endosperm; develops from a single embryo sac	Cellular type, composite endosperm; develops from all the endosperms of the different embryo sacs in the same ovary
4. Division of zygote transverse	Division of zygote vertical
5. Pericarp consists of the parenchymatous epicarp, stony mesocarp, and parenchymatous endocarp; endocarp is consumed by the endosperm	Pericarp consists of the outer leathery coat followed by the viscid layer and parenchymatous zone traversed by vascular strands

[1]Adapted from Johri and Bhatnagar (1960).

The two controversial genera of the Santalaceae are *Calyptosepalum* and *Exocarpus*. Anatomical features like trilacunar nodes, enlarged vessel members, predominantly uniseriate rays, and apotracheal parenchyma support the placement of *Calyptosepalum* in the Santalaceae (Swamy 1949). The other genus, *Exocarpus*, was placed in a separate family, Sarcopodaceae, near Taxaceae of the gymnosperms (Gagnepain and Boreau 1946). Both anatomical and embryological features confirm its assignment in Santalaceae (Ram 1959, P. Maheshwari and Kapil 1966).

Myzodendraceae

A unigeneric family of the genus *Myzodendron* with 12 species endemic to subantarctic forests of Chile and Argentina (Feuer 1981).

Vegetative Features.   Short semi-parasitic green shrubs attached to their hosts by suckers or haustoria; mostly parasitic on stems and branches of *Nothofagus*. Leaves green or scale-like, alternate, entire or crenate margined, estipulate.

Floral Features.   Flowers very small, unisexual, dioecious, naked in compound spikes. Male flowers with 2 or 3 stamens and a central disc; anthers terminal, monothecous, dehiscing by terminal tangential slit. Female flowers with 3 staminodes situated in longitudinal furrows of the ovary, lengthening into long plumose bristles in fruits. Gynoecium syncarpous, tricarpellary, ovary superior, trilocular, with 1 pendulous ovule per locule and 3 subsessile stigmas surrounded by an apical annular disc. Fruit a 3-angled or winged achene with three persistent plumose staminodes; seed endospermous, without testa.

Anatomy.   The two sections—Eumyzodendron and Gymnophyton—show different anatomical features (Metcalfe and Chalk 1972).

In the former the vascular system appears during the first year as a loose ring of vascular bundles separated by broad, unlignified, medullary rays; the vascular ring is supported externally by mechanical tissue. A second vascular ring may or may not be formed during the second year. Xylem vessels with spiral and scalarifom thickening and simple perforations are embedded in ground tissue of unlignified cells.

In the other section, Gymnophyton, xylem forms a continuous cylinder, traversed by narrow primary or sometimes secondary rays. Vessels with scalariform pitting and simple perforations.

Crystals of calcium oxalate are frequent in both the groups.

Embryology.   Pollen grains apolar, spheroidal, polyporate with (3–6–) 7–19 pores scattered randomly over the surface. Sculpturing uniformly echinate, extremely thickened near apurtures (Feuer 1981). Ovules ategmic, anatropous; Polygonum type of embryo sac, 8-nucleate at maturity. Endosperm formation of the Cellular type; chalazal haustorium present.

Important Genera and Economic Importance.   *Myzodendron* is the only genus and has no economic importance.

Taxonomic Considerations.   The family Myzodendraceae was considered by Rendle (1925) and Skottsberg (1935) as a reduced ally of the Santalaceae. Engler and Diels (1936) presumed that Myzodendraceae has closer links with the Loranthaceae. Hallier (1912) suggested direct derivation of Myzodendraceae from the Olacaceae. Kuijt (1968, 1969) suggested a possible derivation from *Chauochiton*-like ancestor fron the family Olacaceae. Ram (1970) reported—on the basis of embryological data—that Myzodendraceae differ from the Loranthaceae but bear a close resemblance to the Santalaceae and its allied families (Table 14.6.1).

Table 14.6.1   Comparative Data for Myzodendraceae and Loranthaceae

Myzodendraceae	Loranthaceae
1. Semi-inferior ovary, trilocular at the base, unilocular above	Inferior ovary, mostly unilocular
2. Free-central placenta	Placenta rudimentary or lacking
3. Polygonum type of embryo sac	Polygonum or Allium type of embryo sac
4. Cellular endosperm with a chalazal haustorium	Cellular endosperm 'composite'

Myzodendraceae resemble Santalaceae in: (a) semi-inferior ovary, trilocular at the base, unilocular above, (b) free-central placenta with 3 anatropous ovules, (c) Polygonum type of embryo sac, (d) inverted J-shaped form of mature embryo sac, (e) Cellular endosperm with a chalazal haustorium and (f) seeds without testa.

Takhtajan (1980, 1987) retains it close to Santalaceae, especially to the South American genera *Arjona* and *Quinchamalium,* and probably derived from them. However, they differ significantly from both Santalaceae and Loranthaceae. The data available on their floral, vegetative, anatomical and embryological features suggest erection of a separate family, the Myzodendraceae.

Loranthaceae

A large family of 65 genera and 850 species distributed all over the tropics, extending into the temperate zone. Commonly known as mistletoes, these plants occur in Southeast Asia, including the islands of Sunda- Shelf and Papua-New Guinea, in Central and Southern America, and in Africa. The subfamily Loranthoideae (or family Loranthaceae s.s.) is essentially southern in distribution, which probably indicates Gondwanan origin. The subfamily Viscoideae (or family Viscaceae) is more extensive in the northern hemisphere and, through the genus *Korthalsella*, is well represented in the Pacific basin, where this genus has a secondary center of diversity in the Hawaiian Islands. The genus *Arceuthobium* (dwarf mistletoes) is distributed exclusively in the northern hemisphere of both the New and Old World. These plants cause serious damage to the forest trees in North America, which affect the soft-wood timber industry (Calder 1983).

Vegetative Features.   Mostly stem parasites, rarely root parasites (*Atkinsonia ligustrina, Nuytsia floribunda* and *Gaiadendron punctatum*) which resemble non-parasitic trees and shrubs (Hocking and Fineran 1983). Evergreen shrubs, grow upon trees and attached to them by haustoria; rarely terrestrial trees and shrubs

(*Nuytsia, Gaiadendron*); branching dichotomous with swollen and jointed nodes. Leaves oppcsite or whorled, estipulate, simple, entire, often coriaceous, or sometimes reduced to mere scales; leaves alternate in *Nuytsia.*

Floral Features. Inflorescence axillary or terminal racemes (Fig. 14.7A) or spikes. Flowers actino- or slightly zygomorphic, uni- or bisexual (e.g. *Nuytsia*; Fig. 14.7D), often brightly coloured as in *Dendrophthoë falcata*; axis cup-like or disc-like. Sepals 2 or 3, free or united or poorly developed; often adnate to the ovary. Petals 2 or 3, free (*Helixanthera*), valvate or united into a tube and then the tube splits. Stamens 2 or 3, opposite and inserted on the petals; anthers bicelled or l-celled by confluence or multiloculate, dehiscing longitudinally (Fig. 14.7 B, D) or by transverse slits or by terminal pores; in staminate flowers, a rudimentary pistil may be present. Gynoecium syncarpous, 3–4-carpellary, ovary inferior, appears to be embedded in the receptacle, unilocular; ovules, as ordinarily expected, are absent. Several genera show a conical projection arising from the central portion of the ovary. This is termed mamelon. Some of the genera (*Amyema, Elytranthe, Lepeostegeres, Macrosolen*, etc.) have placenta and some others (*Dendrophthoë, Helixanthera, Scurrula*, etc.) do not. Placenta may be lobed or unlobed, and the ovary is accordingly uni-, tri- or tetra-locular. In *Amyema gravis* and *Helicanthes*, the placenta is unlobed, ovary unilocular. In *Elytranthe* and *Macrosolen* placenta 3-lobed, ovary unilocular. In *Nuytsia* placenta stalked and 3-lobed, ovary trilocular only at the base. In *Amylotheca* and *Lepeostegeres* placenta 3- or 4-lobed, ovary tri- or tetralocular. In *Lysiana* and *Peraxilla* placenta 4-lobed, ovary tetralocular. In *Barathranthus, Dendrophthoë falcata, Helixanthera, Tapinanthus* and others placenta absent, ovary unilocular (Johri and Bhatnagar 1972, Bhatnagar and Johri 1983).

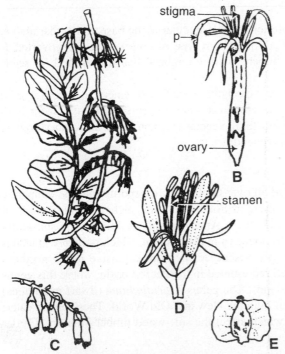

Fig. 14.7 Loranthaceae: **A-C** *Dendrophthoë falcata,* **D,E** *Nuytsia floribunda.* **A** Flowering twig. **B** Flower. **C** Fruit. **D** *Nuytsia,* inflorescence. **E** Fruit. *p* perianth. (**A-C** adapted from J.K. Maheshwari 1963, **D, E** after Johri and Bhatnagar 1972)

Style simple, terminal or absent, stigma 1, often sessile; ovule 1, indistinct. Fruit baccate or drupaceous, indehiscent (Fig. 14.7C, E); seed solitary, endospermous, without testa, embryo often 2 or 3 within each seed, and when paired, are united by cotyledons.

Anatomy. Vessels very small, greatly variable in arrangement, often with spiral thickenings; perforations simple, intervascular pitting alternate. Parenchyma mostly abundant, scattered or in short uniseriate lines and around the pores, typically storied. Rays 1–4 cells wide and without uniseriates, mostly heterogeneous. Leaves dorsiventral, isobilateral or with lamina mainly composed of isodiametric cells, as in *Loranthus europaeus* and *Tupeia antarctica*. In *Viscum album* the mesophyll cells of the biennial leaves consist of isodiametric cells in the first year and a layer of palisade tissue develops towards both the surfaces during the second year. Hairs rarely present; stomata Rubiaceous type, present on both surfaces of isobilateral and confined to the lower surface of dorsiventral leaves. In species with persistent leaves, the mesophyll sometimes contains strongly thickened stone cells which are often branched; solitary crystals freqently present in these elements.

Embryology. In Loranthoideae the pollen grains are triangular with 3 or sometimes 4, slightly curved, concavo-convex arms with the nucleus more or less centrally situated; each arm grooved. Exine smooth and thicker in the centre (Johri and Bhatnagar 1972); but uniformly thick in *Amylotheca dictyophleba* (Raj 1970), *Dendrophthoë falcata* (Singh 1952) and *Macrosolen cochinchinensis* (P. Maheshwari and Singh 1952). Germ pores present in the middle or slightly towards the tip of each arm; shed at 2-celled stage. Ovules undifferentiated; 1 to 4 hypodermal archesporial cells differentiate in each lobe of placenta or, as in *Helicanthes*, the entire hypodermal tissue of 30 to 40 cells become sporogenous (Johri and Bhatnagar 1972). Embryo sac of Polygonum type, 8-nucleate at maturity. The embryo sacs elongate at the 4-nucleate stage and their tips invade the style and stigma up to various levels. Endosperm formation of the Cellular type; a composite structure formed by the fusion of different endosperms developing in the same ovary.

In Viscoideae, the pollen grains are spherical, shed at 2-celled stage. Two or more archesporial cells differentiate in the central ovarian papilla and form a dyad. The upper dyad is larger and produces the embryo sac; the lower one degenerates. Embryo sac of Allium type, 8-celled at maturity. Endosperm foration of the Cellular type; develops individually in each embryo sac and does not fuse (Johri and Bhatnagar 1960, 1972). The presence of chlorophyll in the endosperm has been reported by Kuijt (1960).

Chromosome Number. Basic chromosome number is $x = 12$ and there has been progressive aneuploid reduction to $x = 8$. Stocks with $x = 12$ and 11 became established in temperate South America and Australia (Barlow and Wiens 1971).

Chemical Features. Monoethyl ethers of D I. inositol and L-quebrachitol occur in *Viscum album*. Triterpenoid saponins in Loranthaceae are the viscotoxin—a toxic substance.

Important Genera and Economic Importance. Amongst the Loranthoideae, the important genus is *Loranthus*, a tropical genus with about 500 species. Nomenclature of *Loranthus* has been cotroversial. The International Botanical Congress in 1930, rejected *Loranthus americanus* Linn. 1753 which was later included in the American genus *Psittacanthus*. In 1762 Linnaeus included four more species in this genus including *L. scurrula*. *L. scurrula* was earlier designated *Scurrula parasitica* Linn. 1753. In the same Botanical Congress, *Loranthus* 1762 was conserved with the type-species *L. scurrula*. This means that *Scurrula*, which was adopted by Danser (1929), should be called *Loranthus*. However, there has been an attempt to conserve the generic name *Scurrula* also. Balle et al. (1960) made proposals to prevent the alteration of *L. europaeus* Jacq. 1762 to *Hyphear europaeum* Danser 1929 and also for preservation of *Scurrula* as a separate genus.

Another important genus in *Dendrophthoë*, which alone parasitises 343 different hosts belonging to 57 different angiospermous families (Narasimha Rao and Ravindranath 1964, Singh 1962). Moore and Inamdar (1976) reported *D. falcata* as a parasite even on the leaf of *Mangifera indica*. *Desmaria mutabilis* is unique with dimorphic shoots, the short shoots produce a terminal inflorescence. It is a monotypic genus restricted to southern Chile (Kuijt 1986).

The genus *Psittacanthus* with 40–50 species includes most of the large-flowered neotropical Loranthaceae extending from northwest Mexico to the Buenos Aires region of Argentina (Feuer and Kuijt 1979). Hardly any tree or shrub is immune to the attack of the loranthaceous plants, commonly known as mistletoes. Many economically important plants such as rubber and kapok are attacked by *Loranthus* spp. and other related genera in Indonesia (Johri and Bhatnagar 1972). *Citrus, Castanea, Camphora, Diospyros, Liquidamber, Psidium* and *Pyrus* in China and Phillippines are also attacked by these parasites (Gill and Hawksworth 1961). Bamber (1916) reported *Dendrophthoë falcata* to be destructive to *Acacia, Mangifera, Melia, Morus, Pyrus* and *Quercus.*

Gaiadendron punctatum, a root parasite, is a medium-sized tree, bears racemose inflorescences of 30 or more yellow flowers. It is confined to high elevations in Bolivia, Colombia, Costa Rica and Peru. Another root parasite *Atkinsonia ligustrina,* a shrub with orange-red flowers, is endemic to the higher regions of the Blue Mountains, 100 km west of Sydney, New South Wales, in Australia. *Atkinsonia* is now listed as a rare and possibly endangered species (Hocking and Fineran 1983).

Amongst the Viscoideae, *Phoradendron,* an American genus with about 135 species is important. *P. californicum* are hemiparasitic, dioecious shrubs which occur in the southwestern United States of America and the Mexican deserts, and two of its hosts are *Prosopis glandulosa* var. *correyana* and *Acacia greggii* (Glazner et al. 1988). *Arceuthobium*, a scaly-leaved member, is parasitic on conifers. *A. minutissimum* parasitises the twigs of *Pinus wallichiana;* is a very minute plant with explosive fruits. Twigs of *Viscum album* and *Phoradendron flavescens* are used as Christmas decorations.

Taxonomic Considerations.   The family Loranthaceae is characterised by its semi-parasitic habit, plants attached to their hosts by haustoria, estipulate, opposite leaves, presence of calyculus (a characteristic rim below the perianth), considered by some as an aberrant calyx and by others as an outgrowth of axis (Nayar 1985), typical trilobate pollen grains, an inferior ovary and numerous ovules attached to the central placental zone.

Bentham and Hooker (1965c) included Loranthaceae in the Achlamydosporeae, whereas Cronquist (1968, 1981), Dahlgren (1975a, 1980a, 1983a), Hutchinson (1969, 1973), Melchior (1964), Takhtajan (1987) and Thorne (1968, 1983) place this family in the Santalales. Features like estipulate, opposite, thick, coriaceous leaves, inferior ovary and occurrence of fatty acids such as triglycerides in the seed oils support the placement of Loranthaceae in the Santalales.

Danser (1929) traced the complex nomenclatural history of the Loranthaceae s.l. and provided the basic classificatory structure of the entire group. He presumed the concept of a single family, the Loranthaceae, including both viscoid and loranthoid mistletoes (Calder 1983).

The family Loranthaceae has been usually divided into two subfamilies—Loranthoideae and Viscoideae. P. Maheshwari (1954, 1958) commented that the subfamily Loranthoideae is embryologically quite distinct from the Viscoideae and the two should perhaps be placed in separate families. P. Maheshwari et al. (1957), Johri and Bhatnagar (1960, 1972), Barlow (1964), Kuijt (1968, 1969) and Bhandari and Vohra (1983) have also suggested that these two subfamilies should be given the rank of separate families as they are distinct from each other in many features (see Table 14.7.1).

Some members of the Loranthaceae are now removed to another small family—Eremolepidaceae (Takhtajan 1987). Cytologically Eremolepidaceae members (n = 10, 13) are distinct from other Loranthaceae members (x = 8–12). Palynologically also they are distinct, with pollen grains typically echinate, tricolporate and oblate-spheroidal (Feuer and Kuijt 1978). Eremolepidaceae is probably a very early offshoot of Loranthaceae, with sessile, axillary flowers and simple spikes and racemes (Kuijt 1981). According to Bhandari and Vohra (1983), Eremolepidaceae members deserve the rank of a tribe within the Viscaceae.

Table 14.7.1   Comparative Data for Loranthoideae and Viscoideae

Loranthoideae	Viscoideae
1. Usually stem parasites except *Atkinsonia, Gaiadendron,* and *Nuytsia*—terrestrial root parasites.	All stem parasites
2. Flowers large, uni- or bisexual	Flowers minute, unisexual
3. Calyculus always present	Calyculus generally absent
4. Ovary may contain: a lobed placenta, free or fused with the inner wall of ovary in between the lobes, or may be unlobed and conical, or may be absent (Johri and Bhatnagar 1960)	A central placenta develops at the base of ovary in most species, placenta reduced in *Viscum*; ovules do not differentiate at all
5. Anther dehiscence longitudinal	Anthers with poricidal dehiscence
6. Pollen grains triradiate; round in *Atkinsonia, Ixocactus, Oryctanthus* and *Tupeia* (Johri and Bhatnagar 1972); 2-celled	Pollen grains spherical; 2-celled
7. Embryo sac of Polygonum type	Embryo sac of Allium type
8. Endosperm of Cellular type, develops from all the endosperms of the different embryo sacs growing in the same ovary, and fuse to form a composit structure	Endosperm of Cellular type, develops from a single embryo sac.
9. Pericarp distinguishable into 3 zones; outermost leathery coat is followed by the viscid and parenchymatous zones; viscid layer situated outside the vascular supply to corolla	Pericarp distinguishable into 3 zones; outer fleshy coat is followed by the viscid and parenchymatous zones; viscid layer situated internal to the vascular supply to perianth
10. Basic chromosome number x = 8–12	Basic chromosome number x = 10–15

They also share a few common features with the Loranthaceae, apart from several resemblances with the Viscaceae, and therefore are a possible link between these two families.

The taxonomic distinction of Loranthaceae and Viscaceae as two families of independent origin is supported by their geographic histories (Barlow 1983). The Loranthaceae probably originated in the Cretaceous era in the Gondowanaland region and the Viscaceae originated in the Tertiary period in Laurasian land mass. According to Barlow (1983), therefore, the similarities amongst the two families, Loranthaceae and Viscaceae, are due to parallelisms or convergences.

In view of the above dissimilarities, there is every justification in raising Loranthoideae and Viscoideae to the rank of independent families, Loranthaceae and Viscaceae; the third family, Eremolepidaceae, should be included in the suborder Loranthineae.

Taxonomic Considerations of the Order Santalales

The order Santalales includes two suborders—Santalineae and Loranthineae. In Santalineae (with 5 families) the plants are autotrophs (Olacaceae) or root parasites.

Leaves alternate, sometimes reduced to scales as in *Exocarpus*. Flowers uni- or bisexual, mono- or dichlamydeous (Olacaceae); "calyculus" sometimes present (*Choretrum* and *Mida* of Santalaceae); ovary superior, semi-inferior or inferior, 2- to 5-loculed at base and unilocular above.

In Loranthineae the plants are mostly stem parasites (except for *Atkinsonia, Gaiadendron* and *Nuytsia,* which are root parasites). Leaves opposite, often coriaceous, persistent. Flowers uni– or bisexual, calyculus present in Loranthaceae and not in Viscaceae. Ovary inferior, uni- or 3- or 4-chambered.

Embryologically, pollen grains spherical or oblong, 2- or 3- celled in Santalineae and triradiate or spherical, 2-celled in Loranthineae. The ovules absent only in the reduced forms of Santalineae such as *Exocarpus* (Santalaceae) and *Agonandra* (Opiliaceae), but in all members of Loranthineae. Embryo sac of Polygonum type except in *Buckleya* and *Iodina* (Santalaceae) where it is of the Allium type. In Loranthaceae of the Loranthineae, the embryo sacs conform to Polygonum type and in Viscaceae these are of the Allium type. Endosperm Cellular of Helobial in Santalineae, only Cellular in Lorathineae and a composite structure in Loranthaceae.

Morphological as well as embryological data support the formation of the two suborders of the Santalales, which, according to Johri and Bhatnagar (1972), should not be raised to the rank of separate orders as suggested by Van Tieghem (1896). The occurrence of acetylenic fatty acids as triglycerides in the seed oils of 5 families of the Santalales—Loranthaceae, Olacaceae, Opiliaceae, Santalceae and Viscaceae (Bohlmann et al. 1973, Dahlgren et al. 1981)—also helps unify this order. Recognition of the two subfamilies of Loranthaceae as distinct families—Loranthaceae and Viscaceae—is fully justified (see Table 14.7.1). In addition, Loranthieae may include a third family, Eremolepidaceae (Feuer and Kuijt 1978, Takhtajan 1987).

Although the Santalales is more or less a natural taxon, phylogenists are divided as to its phyletic position. Bessey (1915) included all these families in the Celastrales, deriving them from the Rosaceous ancestors. Hutchinson (1969) also considered them to have been derived from the Celastrales. Rendle (1925) presumed them to have originated from the Proteales. According to Cronquist (1968), Santalales, Celastrales, Rhamnales and Proteales have all been derived from a common Rosaceous ancestor with several cycles of stamens and a fully plurilocular ovary with more or less numerous ovules in each locule. Dahlgen (1975b, 1980a, 1983a) retains Santalales in the Celastranae close to the Celastrales. According to Takhtajan (1980), Santalales are not direct descendants of the Celastrales but share a common origin with them. The most primitive member of the Santalales, that is Olacaceae, is very close to the primitive families of the Celastrales, especially Icacinaceae.

From the above discussion it is apparent that the Santalales is closely allied to the Celastrales, and probably have a common origin with them.

15

Order Balanophorales

A monotypic order with the family Balanophoraceae: all taxa obligate parasities with reduced plant body and club-shaped inflorescences.

Balanophoraceae

A family of 18 genera and 120 species, distributed mainly in the tropical and subtropical regions of the world.

Vegetative Features. Annual or perennial, terrestrial, fleshy herbs, parasitic on the roots of woody plants; without chlorophyll. Stems succulent, rhizomes tuberous, leaves mostly absent or reduced to fleshy scales. Roots are usually absent except in *Corynaea*. Numerous light brown, short, easily breakable, unbranched roots more or less evenly distributed all over the surface of tuber (Kuijt and Bruns 1987).

Floral Features. Flowering stalks erect, short with scale leaves, at least when young. Inflorescence of *Corynaea crassa* is a fleshy structure consisting of an ovoid portion bearing minute flowers and dense indumentum of white hairs initially covered by massive, orange-brown, peltate scales and a stout, light brown stalk which lacks any leafy organ (Kuijt and Bruns 1987). Flowers uni- or rarely bisexual. Staminate flowers with 3–4–8-lobed perianth, stamens as many as perianth lobes. Pistillate flowers without perianth, hypogynous. Gynoecium 1–3-carpellary, syncarous, ovary superior, unilocular, ovules 1 to 3. Fruit a nut, seeds with abundant endosperm and an undifferentiated embryo.

Anatomy. Rhizomes and peduncles are covered with unicellular, thin-walled trichomes in *Longsdorffia* and *Thonningia*; filiform, multicellular hairs as well as glandular hairs also reported. Epidermis of small cells, devoid of stomata. In many genera, vascular bundles always collateral, much-branched, and pursue an irregular course in the tuberous rhizome and peduncles. Vascular bundles often arranged in 1, 2 or more rings. Vessels with reticulate or striate thickenings, performation plates simple. In some genera pith contains numerous sclerenchymatous fibres.

Embryology. The family is highly multipalynous. Pollen grains may be inaperturate or colpate or porate. According to Davis (1966), the pollen grains are shed at 3-celled stage but in genera studied by Brewbaker (1967) there are only 2 cells at the time of shedding. Ovules much reduced, ategmic, normally not well differentiated from the inner tissue of the ovary, the climax is the extremely small, archegonium-like female flowers in *Balanophora* (Hansen 1972); tenuinucellate. Embryo sac is of Polygonum type. Endosperm formation of the Cellular type. Embryo 4–12-celled, undifferentiated (Natesh and Rau 1984). The embryogeny conforms to the Piperad type. The first, as well as second, division is vertical in *Balanophora* (Zweifel 1939). In *B. dioica* (Ekambaram and Panje 1935) the third division is periclinal.

Chromosome Number. Diploid chromosome number of *Balanophora abbreviata* is 2n = 46 (Kumar and Subramanian 1987).

Chemical Features. A waxy or resinous substance called balanophorin is present in the tissues of *Balanophora, Langsdorffia* and *Thonningia*. A similar substance is present in the tuber of *Helosis* (Metcalfe and Chalk 1972). Common flavonols, dihydroflavonols and flavanones make up the flavonoid profile (Gornall et al. 1979).

Important Genera and Economic Importance. The Balanophoraceae are attached to their host plants in a very remarkable way. The subterranean tuberous rhizome of the plant is directly connected with the root of the host. At this connecting point the cortical tissues of the host root disorganize and the xylem strands penetrate the tissue of the parasite. In *Balanophora,* the tissues of the host and the parasite become intermingled in such a way that they cannot be distinguished easily (Metcalfe and Chalk 1972). In Java waxes are extracted from the plants of this family and are used for lighting. Sometimes the plants are reported to have aphrodisiac properties (Dutta 1988).

Taxonomic Considerations. Bentham and Hooker (1965c) considered Balanophoraceae to be a member of Achlamydosporae while Harms (1935) and Melchior (1964) treated it as Balanophorales. On the basis of highly reduced ovular structure and undifferentiated embryo, and copious endosperm, Cronquist (1981) placed Balanophoraceae in the Santalales, Dahlgren (1975a, 1980a) and Takhtajan (1980, 1987) erected a separate order Balanophorales. According to most phylogenists the nearest taxon allied to Balanophoraceae appears to be Cynomoriales. However, Takhtajan (1987) considers its nearest relatives in the Hydnoraceae and Rafflesiaceae, which are also total parasites.

Harms (1935) classified the following subfamilies under Balanophoraceae: Balanophoroideae (rhizome containing balanophorin), Mystropetaloideae, Dactylanthoideae, Sarcophytoideae, Helosioideae and Lophophytoideae (rhizome of all these subfamilies contains starch). Takhtajan (1987) has treated all these subfamilies as distinct families of the same order.

Taxonomic Cosiderations of the Order Balanophorales

The semi-parasitic taxa are included in the Santalales and the total parasites in the Balanophorales. In the latter, the ovule has undergone much reduction. Ovules of the other allied taxon, Cynomoriaceae, are unitegmic but in Balanophoraceae the ovules are ategmic and normally not well differentiated from the inner tissue of the ovary, as is the condition in extremely small archegonium-like pistillate flowers in *Balanophora.* The affinities of Balanophorales are controversial. Accordig to Dahlgren (1983a), Balanophorales could be allied to Gunneraceae, or possibly the Ericales, or even the Asterales. However, in our opinion, the Balanophoraceae is closer to Cynomoriaceae, at least on the basis of embryological features and their wholly parasitic nature (see Johri et al. 1992).

16

Order Medusandrales

A monotypic order with the fmaily Medusandraceae, comprising a single African genus of trees with secretory canals throughout the plant body.

Medusandraceae

A family of a single genus *Medusandra* with only one species in tropical West Africa.

Vegetative Features. Trees, 15–18 m high, numerous shoots arise from the base; branches smooth, with numerous shallow vertical cracks pealing off in thin irregular woody scales; dark green with grey-green patches (Brenan 1952). Leaves alternate, simple, crenate, filiform-stipulate, long-petioled; petiole pulvinate-geniculate at apex (Fig. 16.1 A) (Willis 1973.)

Floral Features. Inflorescence dense, pendulous, axillary racemes (Fig. 16.1 B). Sepals 5, open in bud (Fig. 16.1 C), persistent in fruit. Flowers small, bisexual, actinomorphic, hypogynous. Petals 5, imbricate, green. Stamens 5, opposite the petals, short with more or less large, 4-celled, latrorse anthers; staminodes 5, oppositisepalous, closely folded in bud, elongated, serpentine, long-exerted, densely papillose-pubescent, with abortive, terminal anthers (Fig. 16.1 D, E). Gynoecium syncarpous, tricarpellary (Fig. 16.1 F), ovary superior, unilocular, with slender central column and 6 pendulous ovules attached to the roof of the ovary, close to the column (Fig. 16.1 D, J, K); styles 3, short, conical, divergent (Fig. 16.1 F). Fruit, a 3-valved coriaceous capsule (Fig. 16.1 G), silky-fibrous within, subtended by reflexed accrescent calyx (Fig. 16.1 G, H); seed one, large, pulviniform, often with 6 radiating ribs above (Fig. 16.1 I). Endosperm copious, slightly ruminate, embryo small, straight.

Embryology. Pollen grains radio-symmetrical, occasionally bilateral, usually spheroidal, 2–3-colporoidate.

Anatomy. Vessels small to medium-sized, mostly solitary, tangential or oblique pairs also occur. Pitting on lateral walls mostly scalariform; perforation plates scalariform with usually 20, sometimes 50 or more very fine bars. Rays uniseriate, homogeneous. Parenchyma scanty, paratracheal and a little diffuse. Leaves dorsiventral, hairs and stomata confined to lower surface; hairs uni- or rarely bicellular, stomata infrequent, anomocytic. A secretory canal containing a lemon yellow amorphous substance accompanies every vascular bundle in leaf and petiole, and also at the other periphery of pith in stem (Metcalfe 1952a).

Important Genus. The only genus, *Medusandra* does not have any economic importance.

Taxonomic Considerations

Bentham and Hooker (1965a), Engler and Diels (1936), and Benson (1970) did not recognise the family Medusandraceae. According to Willis (1973), *Medusandra* is an interesting genus with possible affinities with many groups. Unusual placentation, anatomy and pollen structure suggest its connection with Olacaceae and Icacinaceae. Brenan (1952) observed that Medusandraceae may be remotely related to the Olacaceae. Affinity may be with Flacourtiaceae—particulaly the genus *Soyauxia* which was earlier included in this family (Brenan 1953). The habit, foliage and inflorescence are those of the genera *Baccaurea*, *Thecacoris* and *Mesobotrya*, which are referred to the Euphorbiaceae, particularly Phyllanthoideae. Anatomical evidence such as: (a) vessels often in tangential or oblique pairs, (b) perforation plates scalariform, sometimes with

Fig. 16.1 Medusandraceae: **A-H, K** *Medusandra richardsiana,* **I, J** *Soyauxia.* **A** Flowering twig. **B** Inflorescence. **C** Flower bud. **D** Vertical section. **E** Flower. **F** Gynoecium. **G** Fruit. **H** Dehisced fruit. **I** Seed. **J** *Soyauxia,* cross section gynoecium. **K** Vertical section. *br* bract, *c* corolla, *k* calyx, *o* ovule, *ov* ovary, *stm* stamen, *stn* staminode. (After Brenan 1952, 1953)

50 or more very fine bars, (c) Ranunculaceous stomata, and (d) secretory canal containing a lemon-yellow amorphous substance which accompanies every vascular bundle in leaf and petiole, support a somewhat isolated position. *Medusandra* has certain distinctive anatomical characters in common with the families Dipterocarpaceae, Lacistemaceae and Leitneriaceae, but there is complete lack of agreement as far as other features are concerned (Metcalfe 1952a). Hutchinson (1969) placed Medusandraceae in the Santalales, and later, in 1973, in Olacales close to the Santalales. Cronquist (1968, 1981) and Takhtajan (1987) also treat Medusandraceae as a member of the Santalales. Dahlgren (1980a), however, recognises it as a member of the Cornales and places it near the Icacinaceae. Metcalfe (1952a) also observed that there is definite affinity between *Medusandra* and the Icacinaceae although morphologically the two are very distinct.

Cronquist (1968) suggested the formation of a separate order, Medusandrales, for Medusandraceae and Dipentodontaceae (another controversial family of the Santalales). This order should stand alongside the Santalales.

Melchior (1964) had already recognised a distinct order Medusandrales in the revision of Engler's Syllabus der Pflanzenfamilien. This genus is so different from the members of the Santalales, that it should be treated as a separate order—Medusandrales—near the Santalales.

17

Order Polygonales

The Polygonales comprises only one family, Polygonaceae. The plants are herbs, shrubs or climbers, distributed mostly in north temperate regions of both New and Old Worlds. Leaves stipulate, stipules ochreate, nodes swollen. Pollen grains 3-celled. Ovary superior, unilocular, uniovular; fruit a nutlet or achenes. Ovules bitegmic, crassinucellate and usually orthotropous. Plants rich in anthocyanin pigments.

Polygonaceae

A family of 30 to 35 genera and about 1000 species, distributed mostly in temperate regions, especially in the northern hemisphere, a few taxa are subtropical; *Oxyria digyna* occurs in North Arctic and subarctic regions. *Koenigia islandica* occurs in Arctic, subarctic and Himalayan regions. Most of the genera are localized endemics and have a limited distribution. The genus *Polygonum* is distributed worldwide.

Vegetative Features. Annual (*Polygonum plebeium*) or perennial herbs, rarely shrubs or small trees (*Calligonum polygonoides*), sometimes climbing, e.g. *Antigonon leptopus. P. hydropiper* (Fig. 17.1 A) and *P. amphibium* are amphibian plants; *Muehlenbeckia platyclados* is xerophytic. Stems often with swollen nodes, occasionally geniculate (bent knee-like); stems and branches flattened and form ribbon-like cladodes or phylloclades jointed at nodes in *Homalocladium platycladium* and *M. platyclados*. Leaves petiolate, stipulate, stipules often ochreate (formation of a nodal sheath by fusion of 2 stipules) (Fig. 17.1 A) except in *Eriogonum* and *Chorizanthe;* alternate, opposite or verticillate in *Eriogonum, Lastarriaea, Chorizanthe, Pterostegia* and *Koenigia;* entire or sometimes lobed or serrulate, e.g. *Rheum* sp. and *Rumex acetocella;* generally smooth but sometimes, particularly in species growing on hills, woolly or covered with thick hairy growth.

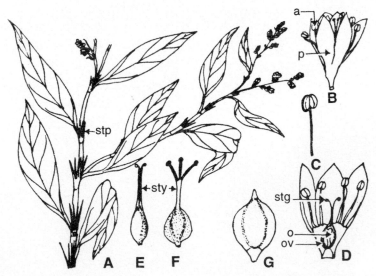

Fig. 17.1 Polygonaceae: **A-G** *Polygonum hydropiper*. **A** Twig; leaves with ochreate stipules. **B** Flower. **C** Stamen. **D** Flower, vertical section. **E, F** Gynoecium. **G** Fruit. *a* anther, *o* ovule, *ov* ovary, *p* perianth, *stg* stigma, *stp* stipule, *sty* style. (Adapted from Saldanha and Nicolson 1976)

Floral Features. Inflorescence raceme, spike, panicle, or head (the smaller units are cymules or cymes). Flowers solitary or 2 to 3 per axil, sessile or shortly pedicellate as in *Polygonum plebeium*. In *Fagopyrum esculentum* the flowers are in terminal subcapitate cymes. Flowers usually bisexual (when unisexual, plants dioecious or monoecious), actinomorphic, hypogynous, often bracteolate. Perianth biseriate, with 3 to 6, free, undifferentiated tepals; 2 whorls of 3 each is the normal condition; the 5-tepal condition represents a fusion of 2 tepals, one from each whorl; often persistent, enlarged and membranous in fruit.

The flowers are trimerous, rarely dimerous and may be cyclic with regular, alternate, isomerous whorls (Fig. 17.1 B, D) or acyclic. In cyclic flowers the usual floral formula is $P_{3+3} A_{3+3} G_{(3)}$. Acyclic flowers have P_5, rarely P_4, arranged quincuncially, $A_{5-8} G_{(3)}$. Stamens 6 to 9, in 2 series; the outer 6 often introrse and the three inner extrorse. *Pterostegia* has 2 whorls of 3 stamens each, in *Rumex* and *Koenigia* there is only one whorl of stamens. Stamens numerous in *Calligonum* and *Symmeria*. Filaments free or basally adnate, anthers bicelled, basifixed, dehiscence longitudinal (Fig. 17.1 C). Gynoecium syncarpous, (2–) 3 (–4–) carpellary (Fig. 17.1 D-F), ovary unilocular, superior, with one ovule, placentation basal, 3-angled, sessile, situated on an annular or lobed nectariferous disc; style 1 and stigmas (2–) 3(–4), corresponding to the number of carpels (Fig. 17.1 D-F). Fruit a 3-sided or biconvex, dry 1-seeded nut (Fig. 17.1 G) or a flat, angled or winged achene, e.g. *Triplaris*. Seed with copious mealy endosperm; embryo curved.

Anatomy. Vessels mostly medium-sized, often with moderately numerous multiples of 4 or more cells, sometimes tend to be ring-porous, with spiral thickenings; perforations simple, intervascular pitting alternate. Parenchyma paratracheal, scanty to narrowly vasicentric. Rays typically uniseriate or up to 2–3 cells wide, homo- or slightly heterogeneous. Fibres usually crystalliferous. Anomalous secondary growth such as formation of medullary bundles, or intraxylary phloem or interfascicular phloem strands have been observed. Leaves usually dorsiventral with hairs of various types; extrafloral nectaries occur sometimes. Stomata nearly always Ranunculaceous type, Rubiaceous in *Oxytheca* and *Triplaris*, present on both the surfaces or confined to lower surface.

Embryology. The family is highly multipalynous, colpate, colporate or porate. In some genera, the exine is elaborately reticulate, in others it is foveolate and in still others it may be granulose or spinuloid (P.K.K. Nair 1970); shed at tricelled stage. Ovules bitegmic, crassinucellate, orthotropous. Polygonum type of embryo sac, 8-nucleate at maturity. Endosperm formation of the Nuclear type; ruminate endosperm in *Coccoloba*.

Chromosome Number. Basic chromosome number is variable: may be $x = 4$–13, 17.

Chemical Features. Polygonaceous plants contain a high quantity of polyphenons such as tannin and anthraquinones; are also rich in flavonoids, commonly anthocyanins, and also paeonidin and malvidine glycosides (Kawasaki et al. 1986), anthraquinones are particularly common in *Rheum, Emex* and *Rumex* (Fairbairn and El-Muhtadi 1972). Indican (glucoside of indoxyl), which yields indigo, is reported in *Polygonum tinctorium* (Paris 1963).

Important Genera and Economic Importance. *Polygonum* is the largest genus with ca. 200 species and shows much variation in habit and habitat. *P. plebeium* of tropical regions is an annual prostrate herb with minute pink flowers. *P. hydropiper* is a perennial erect amphibian herb. *P. aubertii*, commonly called silver-lace vine, is a tall vigorous hardy perennial climber from Western China and Tibet. *P. cuspidatum*, Japanese knotweed, from Japan and *P. sachaliense* from the Sakhalin Islands are stout perennials, 4 to 12 ft high. Another important genus, *Rheum*, with ca. 25 species is a stout perennial herb from Asia bearing clumps of large radical leaves. *Rumex* with 125 species is mostly perennial weed of temperate regions. *Coccoloba uvifera* (sea grape) is a shrub or small tree with orbicular leaves and fruits resembling bunches of grapes; grown as an ornamental. In *Homalocladium platycladium* the branches are flat and jointed and form ribbon-like phylloclades or cladodes, native of Solomon Islands and grown as an ornamental. *Muehlenbeckia* is yet another ornamental taxon grown for its widely different looks and habit. The achenes of this plant are ovoid, 3-angled and enclosed in fleshy perianth forming a berry-like structure. Other important genera are *Fagopyrum, Chorizanthe, Eriogonum* and *Antigonon* (a climber).

The fruits of *Coccoloba uvifera,* starchy seeds of *Fagopyrum esculetum* (buck wheat) and succulent acid petioles of *Rheum rhaponticum*, a large herb native of south Siberia and China, are edible. Rootstocks of *R. officinale* and *R. palmatum* are the source of drug rhubarb. Roots and rhizomes of *Rheum emodi* (Indian rhubarb), a herb reported from subalpine Himalayas, are the source of a drug used as a laxative, tonic and purgative. Roots with madder and potash are used for dyeing fabrics red. *Polygonum tinctorium* is the source of a blue dye. *Rumex hymenosepalus* yields canaigre, used in tanning. Many taxa are grown as ornamentals.

Taxonomic Considerations of Polygonaceae and Polygonales

Because of monochlamydeous flowers and coiled or curved embryo, Bentham and Hooker (1965c) included this family in Curvembryeae under Monochlamydeae. According to Hallier (1912), it is included in Caryophyllales and origiated directly from the Ranales, on a line parallel to the Papaveraceae. Bessey (1915) considered Polygonaceae to be an advanced member of the Caryophyllales. Engler and Diels (1936), Lawrence (1951) and Melchior (1964) treat Polygonaceae family as the sole member of the order Polygonales and placed it before the Centrospermae (= Caryophyllales). According to Hutchinson (1973), the Polygonales is a degraded and reduced type of the Caryophyllales, descending from the Ranales. Cronquist (1968, 1981), Dahlgren (1975a, 1980a, 1983a) and Takhtajan (1980, 1987) also recognise the affinity of the Polygonales with the members of the Caryophyllales. In floral structure Polygonaceae resembles Amaranthaceae, Chenopodiaceae and Nyctaginaceae, and differs from them in having ochreate stipules, swollen nodes, triangular ovary, solitary, erect ovule, development of endosperm instead of perisperm, and occurrence of anthocyanin pigments instead of betalain pigments. Cronquist (1968, 1981) includes Polygonaceae in the Caryophyllales and derives it from the Caryophyllaceae, whereas Takhtajan (1987) derives the order Polygonales from the Caryophyllales.

From the distinctive features of Polygonaceae, it is clear that this family should preferably be treated as a member of a separate order, Polygonales, derived from the Caryophyllales.

18

Order Centrospermae (= Caryophyllales)

The order Centrospermae comprises 4 suborders and 13 families.

Suborder Phytolaccineae: Phytolaccaceae, Gyrostemonaceae, Achatocarpaceae, Nyctaginaceae, Molluginaceae and Aizoaceae;
Suborder Portulacineae: Portulacaceae and Basellaceae;
Suborder Caryophyllineae: Caryophyllaceae;
Suborder Chenopodiineae: Dysphaniaceae, Chenopodiaceae, Amaranthaceae and Didiereaceae.

Takhtajan (1987) recognized unigeneric Barbeuiaceae and Stegnospermataceae (from Phytolaccaceae), unigeneric Tetragoniaceae (from Aizoaceae), Hectorellaceae with 2 genera, and Halophytaceae (from Portulacaceae) and added Cactaceae to this order.

The plants are herbs, shrubs, trees or vines. The leaves alternate (Chenopodiaceae) or opposite-decussate (Caryophyllaceae), usually estipulate, frequently show anomalous secondary growth in both stem and root (except in Portulacaceae, Basellaceae and Didiereaceae). P-type plastids have been reported. Perianth typically biseriate; pollen grains 3-celled; ovules campylo- or amphitropous, bitegmic, crassinucellate with nucellar cap. Embryo generally coiled or curved. Most families (excluding Gyrostemonaceae, Caryophyllaceae and Molluginaceae) contain betalain pigments. The two families Caryophyllaceae and Molluginaceae contain anthocyanin pigments, and Gyrostemonaceae lacks both anthocyanin and betalain.

Phytolaccaceae

A small family of 16 genera and 75 species, distributed largely in the tropics and subtropics of America. Three genera—*Phytolacca americana*, *Rivina humilis* and monotypic *Petiveria*—are indigenous to North American tropics.

Vegetative Features. Herbs, shrubs, trees or vines. Leaves alternate, simple, entire, usually petiolate, estipulate, often with crystals bulging out on leaf blades or appearing as translucent dots (Rogers 1985).

Floral Features. Inflorescence racemose, paniculate or spicate, frequently bear simple or compound lateral dicasia. Flowers bisexual (Fig. 18.1A, D), actino- or zygomorphic (*Anisomeria*), hypogynous. Perianth mostly uniseriate, biseriate in *Stegnosperma* (because of one whorl of petaloid staminodes). Tepals 4 or 5, inconspicuous or showy, greenish or white, usually more or less connate at the base (Fig. 18.1A). Stamens 3 or 4 to numerous, sometimes vary within the same species, in one or two whorls, often borne on a hypogynous disc, filaments free or basally connate; anthers elongated, dorsifixed, bicelled, longitudinally dehiscent (Fig. 18.1A, D). Gynoecium syncarpous or apocarpous, 1- to 16-carpellary, ovary unilocular with one basal ovule as in *Rivina* (if monocarpellary) (Fig. 18.1C, E), or when multicarpellary, each carpel 1-loculed and 1-ovuled (*Phytolacca americana,* Fig. 18.1A); superior (inferior in *Agdestis*). When multicarpellary and syncarpous, placentation axile and each ovary is multilocular with one ovule in each locule (Fig. 18.1B). Style short or none, stigmas as many as carpels, usually linear to filiform, peltate in *Rivina* (Fig. 18.1D, E). Fruit variable depending upon the gynoecial condition, may be a berry or a drupe, dry and indehiscent utricle, achene or schizocarp. Seeds erect, compressed, circular to reniform, usually without aril (minutely arillate in *Rivina* and strongly so in *Stegnosperma*). Embryo annular or sharply bent, endosperm present.

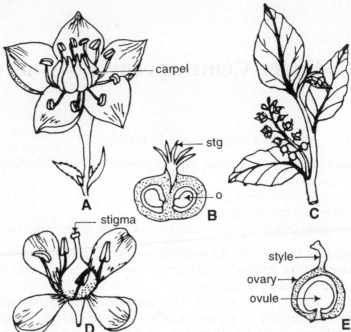

Fig. 18.1 Phytolaccaceae: **A, B** *Phytolacca americana*, **C-E** *Rivina humilis*. **A** Flower. **B** Vertical section of flower. **C** Fruiting twig of *R. humilis*. **D** Flower. **E** Carpel, vertical section. *o* ovule, *stg* stigma (Adapted from Lawrence 1951)

Anatomy. Vessels small to large, usually without any marked pattern, perforations simple, intervascular pitting alternate. Parenchyma predominantly paratracheal, vary from a few cells to complete sheaths around the vessels. Rays up to 2 to 10 cells wide, heterogeneous. Included phloem develops in many genera. Leaves usually dorsiventral, sometimes centric or isobilateral. Hairs infrequent, stomata Ranunculaceous or Rubiaceous type, mostly confined to the lower surface.

Embryology. Pollen grains tri- or hexacolpate or polycolporate; pantoporate or polyporate; exine spinulose, anulopunctate or punctate (Skvarla and Nowicke 1976, Behnke 1976); mature grains 2- or 3-celled at shedding stage. Ovules campylo- or amphitropous, bitegmic, crassinucellate, nucellar cap present. Polygonum type of embryo sac, 8-nucleate at maturity. Endosperm formation of the Nuclear type; totally consumed during embryo development (Eckardt 1976); mature embryo coiled or curved.

Chromosome Number. The basic chromosome number is x = 9.

Chemical Features. Plants with betalain pigments, also rich in saponins.

Important Genera and Economic Importance. The largest genus is *Phytolacca* with ca. 35 species. The Asiatic species *P. acinosa* produces dark red berries which yield a tasteless red juice used for colouring sweetmeats. The young shoots are used as greens. *Rivina humilis,* indigenous to South America, produces beautiful red berries, and is cultivated as an ornamental. Species of *Agdestis, Ervilia* and *Petiveria* are also grown as garden plants.

Taxonomic Considerations. Bentham and Hooker (1965c) assigned Phytolaccaceae to the order Curvembryae (because of the coiled or curved embryo). Engler and Diels (1936) placed it in the order Centrospermae (on the basis of embryological features). Cronquist (1981), Dahlgren (1983a) and Takhtajan

(1987) include Phytolaccaceae in the order Caryophyllales. Chemical data, i.e. presence of betalains, and anatomical features, i.e. presence of P-type plastids in the sieve elements, also support this viewpoint. It is included in the order Chenopodiales by Hutchinson (1973) and Thorne (1977). They recognised two orders—Chenopodiales (betalain present) and Caryophyllales (betalain absent). According to Ehrendorfer (1976a), Phytolaccaceae is the core member of the Caryophyllales from which the rest of the families have evolved.

The position of this family in the order Caryophyllales is quite appropriate.

Gyrostemonaceae

An endemic Australian family, including 5 genera and 16 species.

Vegetative Features.   Trees, shrubs or undershrubs, stem with normal secondary growth. Leaves semi-succulent, simple, entire, alternate, stipules very small or absent.

Floral Features.   Inflorescence racemes or spikes of staminate flowers, pistillate flowers solitary axillary. Flowers unisexual, actinomorphic, hypogynous, apetalous. Calyx of 4 or 5 sepals, polysepalous, shallowly lobed or entire, usually discoid or cupular; petals absent. Stamens 9 to 12 or numerous in one or many whorls in male flowers; anthers oblong to subcuneate, tetragonal, sessile or subsessile, arising from the edge of a convex or flat receptacle. In female flowers sepals 3 to 6, gynoecium syncarpous, 3- to 5-(9)-carpellary, stigma lobes as many as carpels, usually sessile and form a corona, rarely subulate, 3 to 4 mm long; ovary 3- to 5-(9)-loculed, each locule with one axile ovule. Fruit dry, each carpel usually dehisces dorsally or ventrally, or both, and separate from central column. Seeds brown, faintly rugose, arillate, non-endospermous, and with a coiled embryo.

Anatomy.   Vessels small in numerous clusters, perforations simple, intervascular pitting alternate, rather large. Parenchyma paratracheal. Rays up to 10 cells wide, almost homogeneous and with some uniseriates. Anomalous secondary growth absent. Leaves centric, mucilage-containing cells reported in epidermis and subepidermis (Metcalfe and Chalk 1972). Stomata of Ranunculaceous type. S-type of sieve element plastids, epicuticular wax rare (Engel and Barthlott 1988).

Embryology.   Pollen grains 3-colpate, 3-celled when shed. Ovules campylotropous, bitegmic, crassinucellate. Endosperm formation of the Nuclear type, embryo curved.

Chromosome Number.   Basic chromosome number is x = 14 (Keighery 1975).

Chemical Features.   Pigments do not usually occur in this family; glucosinolates occur in *Tersonia brevipes* (Kjaer and Malver 1979).

Important Genera and Economic Importance.   This family is important taxonomically, because of its disputed position. The five genera included are: *Codonocarpus, Cypselocarpus, Gyrostemon, Tersonia* and *Walternanthus*. The genus *Didymotheca* was segregated from *Gyrostemon* on the basis of number of carpels. George (1982) combined it with *Gyrostemon* once again, noting that there are several species with overlapping number of carpels. The genus *Walternanthus* has been added to the previous list of 4 genera by Keighery (1985).

Taxonomic Considerations.   The family Gyrostemonaceae has been treated differently by various workers. Most of the earlier authors included it in Phytolaccaceae. Eckardt (1964) and Hora (1979) referred Gyrostemonaceae to the Centrospermae. Ehrendorfer (1976a) suggested its removal from this order on the basis of chemical (absence of betalains), palynological (psilate-scabrate exine) and cytological (basic chromosome number x = 14) features. Anatomically too it is different from Centrosperms in having S-type sieve element plastids.

Ehrendorfer (1976b) stated that "apparently Gyrostemonaceae belong to Capparales". Takhtajan (1987) includes it in Batales along with Bataceae, and places this order near the Capparales. Cronquist (1981), Dahlgren (1980a, 1983b) and Jørganson (1981) also support this assignment.

From the above discussion it is evident that the correct position for Gyrostemonaceae is near the Capparales and not in Centrospermae.

Achatocarpaceae

Achatocarpaceae is a small family of only 2 genera and 10 species distributed from Texas and northwest Mexico to Paraguay and Argentina in South America.

Vegetative Features.   Dioecious, thorny shrubs or small trees with normal growth in thickness. Leaves simple, entire, alternate, estipulate.

Floral Features.   Flowers unisexual, dioecious, in axillary branched, bracteolate cymes. In *Achatocarpus* flowers bracteolate and with 5 sepals; in *Phaulothamnous* ebracteolate and only the terminal flower with 5 sepals and others with 4 sepals only; imbricate, persistent. Petals absent. Stamens 10 to 20, with filiform filament and basifixed elongate-oblong anthers. Gynoecium syncarpous, bicarpellary, unilocular with one basal ovule and 2 conspicuous, simple, free, subulate, divergent, uncinate styles. Fruit a small 1-seeded berry; seed exarillate with copious mealy perisperm.

Anatomy.   Vessels small, in multiples of 2 or 3 cells, perforations simple, intervascular pitting alternate, minute. Parenchyma paratracheal, a few cells around the vessels. Rays not more than 3 or 4 cells wide, heterogeneous. Phloem includes strongly thickened fibres and stone cells. Anomalous secondary growth not reported. Tannin abundant in cortex, medullary rays and xylem. Leaves usually dorsiventral, stomata of Ranunculaceous type. P-type sieve element plastids are reported. The family is characterized by parallel-oriented platelets of wax with irregularly undulate edges (Engel and Barthlott 1988). Anatomically, Achatocarpaceae is distinct from Phytolaccaceae for the occurrence of: (a) fibres and stone cells in phloem, (b) druses, sphaero- and solitary crystals in the same genus, and (c) absence of anomalous thickening.

Embryology.   Pollen grains 6-porate, exine verrucate; shed at 2- or 3-celled stage. Ovules campylotropous, bitegmic, crassinucellate. Seed exarillate, solitary, erect, lenticular, with strongly bent embryo surrounding the mealy endosperm (Johri et al. 1992).

Chemical Features.   C-glycosylflavones (vitexin and isovitexin) are reported (Richardson 1981). Betalain pigments are also present.

Important Genera and Economic Importance.   The two genera included are *Achatocarpus* and *Phaulothamnus*; no economic importance.

Taxonomic Considerations.   Achatocarpaceae is a small family included in Phytolaccaceae. Cronquist (1981) and Hutchinson (1973) also did not consider it as a distinct family. On the basis of anatomy, Heimerl (1934b) considered Achatocarpaceae as a distinct family. Takhtajan (1987), however, treats it as a separate family derived from Phytolaccaceae. Achatocarpaceae differs from Phytolaccaceae in unisexual flowers, normal secondary growth, and 6-porate pollen grains with verrucate exine. The presence of betalain pigments supports its placement in Centrospermae.

Nyctaginaceae

A family of 30 genera and 300 species, distributed in the temperate and tropical regions of the New World. Three of the genera are indigenous to the Old World—*Boerhavia, Oxybaphus* and *Pisonia.*

Vegetative Features.   Herbs (*Boerhavia*), shrubs or trees (*Pisonia*), sometimes scandent (*Bougainvillea*). Leaves usually opposite, simple, entire, estipulate (Fig. 18.4A). *Tricycla* is a spiny shrub and lateral shoots are spine-tipped in *Phaeoptilum.*

Floral Features.   Inflorescence cymose; flowers bracteate, ebracteolate, bisexual or sometimes unisexual (*Phaeoptilum*), actinomorphic, hypogynous, incomplete. Bracts 2 to 5, foliaceous, resembling a calyx or coloured and subtend a flower as in *Bougainvillea* (Fig. 18.4B). Perianth uniseriate, calyx of 5 sepals, connate to form a tube which resembles a sympetalous corolla (Fig 18.4A, C), plicate or contorted in bud; true corolla absent. Stamens vary from 1 to 30, usual number 5, filaments connate basally in a tube, unequal; anthers bicelled, longitudinally dehiscent (Fig. 18.4D). Gynoecium monocarpellary, ovary unilocular with 1 basal erect ovule (Fig. 18.4C), superior, style 1, slender (Fig. 18.4F), stigma 1. Fruit an achene (Fig. 18.4E), often enveloped by the persistent calyx which may be variously modified to facilitate dissemination. Seeds endospermous, embryo straight or curved.

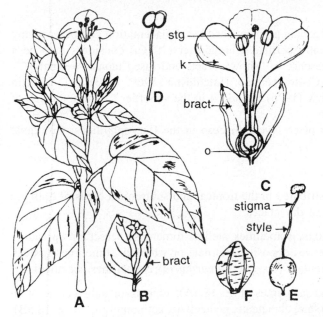

Fig. 18.4   Nyctaginaceae: A, C-F *Mirabilis jalapa.* B *Bougainvillea spectabilis.* **A** Flowering twig. **B** Flower of *B. spectabilis*, subtended by a leafy bract. **C** Flower, vertical section. **D** Stamen. **E** Carpel. **F** Seed. *k* calyx, *o* ovule, *stg* stigma. (Original)

Anatomy.   Vessels small, typically in radial groups and clusters behind the phloem strands, perforations simple. Parenchyma scanty and limited to a few cells around the vessels, sometimes storied; rays small and 1 or 2 cells wide and hetero- or homogeneous or absent. Included phloem always present; anomalous growth in thickness of the axis, by the development of successive rings of collateral vascular bundles. Leaves dorsiventral or isobilateral, hairs of various types occur, stomata confined to the lower surface in most species, Ranunculaceous or Rubiaceous type.

Embryology.   Pollen grains 3- to 4-colpate with coarse, reticulate or spinulose exine; polyporate (Nowicke 1970, Behnke 1976b). Ovules anatropous (*Pisonia*) or campylotropous, bitegmic, crassinucellate. Embryo sac of Polygonum type, 8-nucleate at maturity. Endosperm formation of the Nuclear type. Perisperm present.

Chromosome Number.   Basic chromosome number is x = (13, 17) 20–29 (Behnke 1976b) or x = 10, 13, 17, 29 and 33 (Takhtajan 1987).

Chemical Features.   The pigment betalain is present in members of Nyctaginaceae.

Important Genera and Economic Importance.   *Mirabilis multiflora* is an interesting genus in which the apparent calyx is only an involucre of bracts and the actual calyx is modified into an attractive and coloured structure that resembles a sympetalous corolla (Fig. 18.4A, B). Species of the genus *Neea* often show cauliflory. Bracts petaloid and calyx less conspicuous in *Bougainvillea* and *Coligonia. Okenia* species produce geocarpic fruits. In *O. hypogaea* the peduncles elongate up to as much as 11 inches and dig deep into the sand, where the fruits mature. Although the plant appears to be perennial, it is actually annual. Every season the fruits buried deep down into the sand give rise to new plants. *Okenia* also shows anisophylly. *Pisonia grandis,* a large tree, is widely distributed by various sea birds as the fruits are covered by sticky glands and adhere to the feathers of the birds. Similar fruits are also seen in *Boerhavia diffusa*, a small prostrate herb with pink flowers. In *Ramisia brasiliensis*, the persistent calyx spreads out like wings in mature fruits to help in dissemination.

A few taxa are ornamentals like *Bougainvillea, Mirabilis jalapa* (the 4 o'clock plant) and *Abronia* (the sand verbenas).

Taxonomic considerations.   The presence of betalain pigments and basal placentation do not leave any doubt about the inclusion of this family in Centrospermae. Ehrendorfer (1976a) considers it to be a derivative of Phytolaccaceae, because of the presence of "showy bracts and fused tepals (= sepals)".

Most taxonomists include this family in the Centrospermae; Hutchinson (1969, 1973) includes it in the Thymelaeales. Mabry et al. (1963) and Mabry (1976) include it in the suborder Chenopodiineae of the order Caryophyllales, as it contains betalain.

Cronquist (1981) and Takhtajan (1987) also place Nyctaginaceae in the Caryophyllales; Cronquist derives this family from the Phytolaccaceae.

Molluginaceae

A small family of 13 genera and 100 species distributed in the tropical and subtropical areas of both the New and the Old World; mostly woody plants of dry places.

Vegetative Features.   Mostly herbaceous plants, sometimes stellate-tomentose, erect or spreading (Fig. 18.5A), often dichotomously branched. Leaves opposite, alternate or whorled, sometimes radical, linear to obovate or spathulate, stipulate or estipulate (*Orygia, Macarthuria*); sometimes succulent.

Floral Features.   Inflorescence axillary dischasial cymes (Fig. 18.5A), umbels or subraceme, often axillary and fascicled. Flowers bisexual, ebracteolate, ebracteate, pedicellate, actinomorphic (Fig. 18.5B), hypogynous. Sepals 5, persistent, green. Petals 5 or 0 (*Mollugo verticillata*) or numerous (*Orygia*). Stamens 3 to 5, rarely more (*Orygia*), anthers bicelled, linear-oblong (Fig. 18.5C), longitudinally dehiscent. Gynoecium syncarpous, 2- to 5-carpellary; ovary 2- to 5-locular, with axile placentation (Fig. 18.5D,E), superior; unilocular with only one ovule in *Adenogramma*. Fruit a loculicidal capsule. Seeds (Fig. 18.5F) non-endospermous, embryo curved.

Anatomy.   Xylem vessels with simple perforation plates and varying proportion of parenchyma. Leaves bilateral or dorsiventral, stomata generally Ranunculaceous type.

Embryology.   Pollen grains tricolpate with spinulose, tubuliferous or punctate exine; 3-celled at shedding stage. Ovules campylotropous, bitegmic, crassinucellate. Polygonum type of embryo sac, 8-nucleate at maturity. Endosperm formation of the Nuclear type; mature embryos coiled or curved, perisperm present.

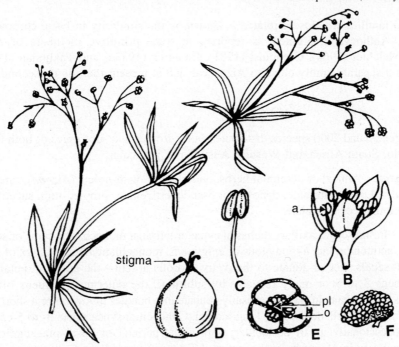

Fig. 18.5 Molluginaceae: **A-E** *Mollugo pentaphylla*. **A** Flowering branch. **B** Flower. **C** Stamen. **D** Carpel. **E** Ovary, cross section. **F** Seed. *a* anther, *o*, ovule, *pl* placenta. (Adapted from Saldanha and Nicolson 1976)

Chromosome Number. Basic chromosome number is x = 9. Hsu (1967) reported the diploid chromosome number of *Mollugo pentaphylla* to be 2n = 18, 36.

Chemical Features. Members of Molluginaceae contain C-glycosylflavones—a character which distinguishes them from herbaceous members of Aizoaceae (Richardson 1981). Almost all the members contain anthocyanins instead of betalains; the genus *Limeum* is without any pigment.

Important Genera and Economic Importance. Most of the members of this family are common weeds.

Taxonomic Considerations. The Molluginaceae is one of the core Centrospermous families. It bears close affinities to Aizoaceae and Phytolaccaceae. Bentham and Hooker (1965a) and Thorne (1983) recognize this family only as a part of the Aizoaceae. Hutchinson (1973) recognizes it as a distinct family—Molluginaceae. The differences between Molluginaceae and Aizoaceae are tabulated below (Table 18.5.1)

Table 18.5.1 Comparative Data of Molluginaceae and Aizoaceae

Molluginaceae	Aizoaceae
1. Mainly herbaceous	Both herbs and shrubs
2. Contains only anthocyanin or pigment-free	Contains betalains
3. Contains C-glycosyl-flavones	C-glycosyl flavones absent
4. Ovules campylotropous	Ovules anatropous
5. Ovary 2–5-carpellary, placentation axile	Ovary 3–5-carpellary, placentation axile, parietal or free-central

That these two families are closely related is shown by the similarity in basic chromosome number and pollen type. Anthocyanin pigment is reported in some primitive members of Aizoaceae and Phytolaccaceae. Melchior (1964), Cronquist (1981), Dahlgren (1983a) and Takhtajan (1987) recognize Molluginaceae as a separate family close to Aizoaceae and at present there is no second opinion about it.

Aizoaceae

A family of 130 genera and 2000 species distributed in the tropics and subtropics of both the New World and the Old World, South Africa and Western Australia in particular.

Vegetative Features.   Annual or perennial herbs, undershrubs or shrublets (*Aizoon*), rarely woody, e.g. *Gliscrothamnus ulei*. Leaves alternate, opposite or pseudoverticellate, simple, often succulent or reduced to scale leaves, stipulate or estipulate.

Floral Features.   Inflorescence axillary dichasial cyme or terminal monochasial cyme or solitary flowers. Flowers bisexual, actinomorphic, hypogynous or epigynous, monochlamydeous. Calyx of 5 to 8 connate, herbaceous, green sepals, free or adnate to the ovary. Corolla absent—the apparent petals are modified staminodes. Stamens 3 to 5 or many (usually by splitting), the outermost stamens often sterile and petaloid; filaments free, in pairs in *Galenia,* basally connate into bundles in *Aizoon* or a short monadelphous sheath; anthers small, bicelled, dehiscence longitudinal. Gynoecium syncarpous, 3- to 5-carpellary, ovary 1- to 5-locular, occasionally up to 20-locular, with axile, parietal or basal placentation, superior or inferior, ovules many as in *Mesembryanthemum* and *Tetragonia*, rarely one; style 1 or absent, stigmas 2 to 20 and usually radiating. Fruit a loculicidal capsule as in *Galenia, Plinthus* and *Acrossanthes* (all South African) or septicidal capsule as in Australian *Gunnia* spp. or circumscissile capsule as in *Sesuvium*. Seeds 2 or more per locule, small, with long funicles, compressed, subreniform; embryo large, cylindrical and curved enclosing the mealy endosperm.

Anatomy.   The epidermis of both leaf and stem generally includes large bladder-like water storage cells arranged between much smaller normal epidermal cells, e.g. in *Aizoön, Cryophytum, Mesembryanthemum*. Xylem includes vessels with bordered pits, simple perforation plates and varying proportion of parenchyma. Leaves nearly always centric but relatively flat leaves isobilateral as in *Sesuvium portulacastrum;* dorsiventral in *Gisekia*. Stomata generally Raunculaceous type, in some Rubiaceous type. Abnormal secondary growth in the form of successive rings of vascular bundles in the inner parenchymatous portion of the pericycle. Included phloem of the concentric type present in many genera (Metcalfe and Chalk 1972).

Embryology.   Pollen grains 3-colpate, exine spinulose and tubuliferous-punctate; 3-celled at shedding stage. Ovules anatropous or campylotropous, bitegmic, crassinucellate. Nucellar cap present. Embryo sac of Polygonum type, 8-nucleate at maturity. Endosperm formation of the Nuclear type, embryo coiled.

Chromosome Number.   Basic chromosome number is x = 8, 9.

Chemical Features.   The members of Aizoaceae are rich in betalains. Occurrence of betalains and leucoanthocyanidins were detected in *Carpobrotus edulis* (Kimler et al. 1970)

Important Genera and Economic Importance.   Some of the common plants of this family are *Glinus lotoides* a prostrate herb, *Trianthema portulacastrum* a creeping herbaceous weed in dry areas, and *Mesembryanthemum* sp., a garden ornamental. Fruits of *M. acinaciforme* and *M. edule* of South Africa (Hottentot figs) are edible. Tender shoots of *Tetragonia tetragonioides* are used as green vegetable.

Taxonomic Considerations.   Bentham and Hooker (1965a) included Aizoaceae in the order Ficoidales.

Engler and Diels (1936) placed it in the Centrospermae. Most taxonomists include Aizoaceae in the order Caryophyllales (Mabry et al. 1963, Behnke and Turner 1971, Cronquist 1968, 1981, Mabry 1976, Dahlgren 1980a, 1983a, Takhtajan 1980, 1987).

The family Aizoaceae is closely related to Molluginaceae morphologically as well as in many other features. However, it can be differentiated from the Molluginaceae on the basis of betalain pigments. It is considered that the Aizoaceae is derived from ancestral stocks of Phytolaccaceae.

Portulacaceae

A family of 30 genera and 600 species, distributed along two centres—Pacific coast states of North America and South America.

Vegetative Features. Herbs (*Portulaca grandiflora*), shrubs (*Portulacaria* sp.) or even small trees (*Calyptrotheca*); suffruticose, more or less decumbent as in *Calandrinia*, perennial herbs with prostrate stems or stolons, e.g. *Neopaxia, Claytonia* (Carolin 1987). Roots bear tubers in *Grahamia* and *Portulaca* spp.; solitary tuber in *Erocallis* (Carolin 1987). Leaves alternate, opposite (Fig. 18.7A) or spirally arranged or in rosettes, fleshy, simple, stipulate, stipules scarious or setaceous (absent in *Claytonia*).

Floral Features. Inflorescence cincinnus, or dichasial cyme; monochasia resemble raceme or spike in *Monocosmia* and *Calandrinia;* or flowers solitary (Fig. 18.7A) surrounded by numerous bract-like structures in *Grahamia.* Flowers bracteate, ebracteolate, bisexual, actinomorphic, hypogynous (Fig.18.7B), biseriate. Calyx of two green sepals (Fig. 18.7C), more than two in *Leurisia* and *Grahamia*, free or basally connate, often caducous, imbricate. Corolla of 4 to 6 petals (Fig. 18.7D) (only 2 in *Calyptridium* and 3 in *Montia*), free or basally connate. Stamen number varies widely, usually as many as and opposite the petals (Fig. 18.7B, E), or 2- to 4-times as many (by splitting), rarely fewer as in *Calyptridium*; free or adnate to the base of the corolla or connate with each other forming a tube or a dome-shaped structure that arches over the summit of the ovary and obscures the nectary below, e.g. *Portulaca* (Carolin 1987); anthers bicelled, introrse, dehiscence longitudinal. Gynoecium syncarpous, 2- to 3-carpellary (Fig. 18.7F), ovary unilocular,

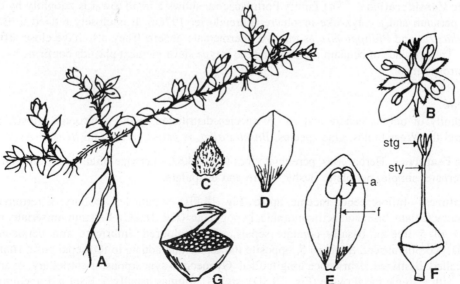

Fig. 18.7 Portulacaceae: **A-G** *Portulaca oleracea*. **A** Flowering plant. **B** Flower. **C** Sepal. **D** Petal. **E** Petal with anther attached at the base. **F** Carpel. **G** Capsule with seeds. *a* anther, *f* filament, *stg* stigma, *sty* style. (Original)

with basal placentation and numerous ovules, superior (half-inferior in *Portulaca*); styles and stigmas 2 to 5. Ovary base may be stalked or just narrowed. Fruit mostly dehiscent capsules, indehiscent in *Silvaea*, *Portulacaria* and *Ceraria*. Capsule of *Portulaca* is circumscissile (Fig. 18.7G) and opens through terminal pore in *Calandrinia*.

Anatomy.   Vascular bundles collateral, distinct, arranged in a ring in (trasections) stem. Vessels small with simple perforations, sometimes in radial rows with simple pits in *Ceraria gariepina*, and with spiral thickenings in *Montia perfoliata*, *Portulaca grandiflora* and *P. oleracea*. Dark brown mucilage cells commonly present. Leaves with thick or thin cuticle, Stomata mostly on both the surfaces, only on lower surface in *Montia;* usually Rubiaceous type. Anomalous secondary growth not known.

Embryology.   Pollen of three different types: (a) pancolpate with a few broad colpi and numerous irregular papillae scattered on the surface of aperture, (b) pancolpate with many narrow colpi often with a single line of papillae on the surface of aperture, and (c) trizonocoplate. Ovule orthotropous, crassinucellate, bitegmic; hemi-campylotropous or campylotropous. Embryo sac of Polygonum type, 8-nucleate at maturity. Endosperm may be well developed, or both perisperm and endosperm reduced, as in *Grahamia,* or perisperm well developed and endosperm reduced, as in *Calandrinia*, or perisperm as well as endosperm reduced with massive embryo, as in *Calyptrotheca* (Carolin 1987). Aril or strophiole may or may not develop.

Chromosome Number.   Basic chromosome number varies, x = (4–) 8–11 (–12).

Chemical Features.   Portulacaceae members contain betalain pigments.

Important Genera and Economic Importance.   *Portulacaria* of South Africa and *Philippiamra* of Chile are two important genera as they are the link between this family and its closest allies, Basellaceae. Some other interesting genera are *Grahamia* and *Portulaca*, which bear root tubers. Economically the family is important for some ornamentals, such as *Portulaca grandiflora* (rose moss) and many species of *Talinum*, *Leurisia* and *Calandrinia*.

Taxonomic Considerations.   The family Portulacaceae shows a trend towards zoophily by developing a petaloid perianth and a calyx-like involucre (Ehrendorfer 1976a). It is closely related to Basellaceae, with *Portulacaria* and *Philippiamra* as the two intermediate genera. They also have close affinity to the Aizoaceae. The presence of betalain pigment and the P-type sieve-element plastids confirms its assignment in the Centrospermae.

Basellaceae

A small family of only 3 genera and ca. 20 species distributed in the neotropics, mostly in tropical America and the West Indies. One species, *Basella alba*, is native to the Old World (Asia).

Vegetative Features.   Herbaceous, perennial vines (Fig. 18.8A); herbage often fleshy and mucilaginous. Leaves alternate, simple, petiolate, fleshy, entire and estipulate.

Floral Features.   Inflorescence raceme, spike (Fig. 18.8B) or panicles, axillary or terminal. Flowers bracteate, ebracteolate, bisexual, actinomorphic, hypogynous (Fig. 18.8C). Perianth uniseriate, represented by a calyx of 5 free or basally connate sepals, often coloured, imbricate and persistent in fruit (Fig. 18.8D); corolla absent. Stamens 5, opposite the sepals and adnate to the sepal base; filaments free, anthers bicelled, basifixed, dehiscence longitudinal. Gynoecium syncarpous, tricarpellary, ovary superior, unilocular with a single basal ovule (Fig. 18.8D), style 1, stigmas usually 3. Fruit a drupe, enveloped by persistent fleshy calyx; seeds with a large, annular or spirally twisted embryo.

Anatomy.   The stem shows (in t.s.) a ring of isolated bundles of unequal size. Xylem includes vessels

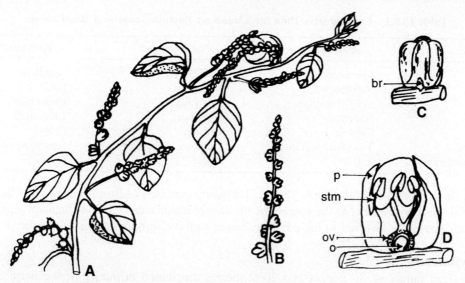

Fig. 18.8 Basellaceae: **A-D** *Basella rubra*, **A** Flowering branch. **B** Inflorescence. **C** Flower. **D** Vertical section of flower. *br* bract, *o* ovule, *ov* ovary, *p* perianth, *stm* stamen. (After Lawrence 1951)

that are 200 µm or more in diameter and with simple perforations. Intraxylary phloem develops in larger bundles of stem which are therefore bicollateral. Mucilage cells are common in both stem and leaf. Leaves dorsiventral, stomata Rubiaceous type, present on both surfaces.

Embryology.   Pollen grains 6- to polycolpate, exine spinulose and tubuliferous-punctate, and cuboidal in shape. Ovules campylotropous, bitegmic, crassinucellate. Embryo sac of Polygonum type, 8-nucleate at maturity. Endosperm formation of the Nuclear type. Seeds exarillate and contain perisperm.

Chromosome Number.   Basic chromosome number is x = (11-) 12.

Chemical Features.   Basellaceae members contain betalain pigments.

Important Genera and Economic Importance.   The members of this family often perennate by underground tubers, or bear tuberous branches though apparently they are annual (Ulbrich 1934b). In these plants mucilage enables them to survive very adverse conditions. Whilst in a plant press, one species of *Ullucus* gives rise to axillary tubercles. The tuberous starchy roots of *U. tuberosus* are consumed as potato substitute in the Andes in Central America. The fleshy mucilaginous leaves and twigs of *Basella* spp. are eaten as spinach. The Madiera vine or *Boussingaultia* is an ornamental.

Taxonomic Cosiderations.   The systematic position of Basellaceae is disputed. Engler and Diels (1936), Lawrence (1951), and Melchior (1964) include Basellaceae in the Centrospermales close to the Portulacaceae. Bessey (1915), Gunderson (1950), Cronquist (1981) and Takhtajan (1987) use an alternate name, Caryophyllales, for the order. Hutchinson (1973) included Basellaceae in a separate order, Chenopodiales, along with other advanced Caryophyllaceous families. Mabry (1976) placed it in suborder, Chenopdiineae, of order Caryophyllales. Hooker (1965) assigned it to Chenopodiaceae as the subfamily Baselleae. Franz (1908) treated it as a tribe of subfamily Montioideae under Portulacaceae. According to Ehrendorfer (1976b), it is linked with Aizoaceae. Basellaceae and Chenopodiaceae differ from each other in embryological as well as anatomical features, and cannot be placed together. A comparative study between Aizoaceae, Portulacaceae and Basellaceae shows that there are differences in morphological and palynological features,

Table 18.8.1 Comparative Data for Aizoaceae, Portulacaceae and Basellaceae

Feature	Aizoaceae	Portulacaceae	Basellaceae
Flowers	Apetalous	Petaloid perianth	Apetalous
Stamens	Numerous	Numerous	Five
Ovary	1–5–20-locular	Unilocular	Unilocular
Ovules	Numerous	Numerous	Uniovulate
Fruit	Capsule	Capsule	Drupe
Pollen grains	3-colpate, Spherical	3-colpate, Spherical	6-polycolpate, Cuboidal
Chromosome number	8, 9	(4–) 8–11 (–12)	(11)–12

and also in chromosome numbers (Table 18.8.1). Therefore, members of Basellaceae cannot be included in either of these two families. At the same time the family Basellaceae contains betalain pigments and P-type plastids and it is justified to include Basellaceae in the Caryophyllales as a distinct family.

Caryophyllaceae

A medium-sized family of 80 genera and 2000 species distributed primarily in the north temperate regions, a few genera in the south temperate regions and higher altitude areas of the tropics. The Mediterranean area is the centre of distribution. Two genera—*Colobanthus* and *Lyallia*—occur in Antarctica, and many genera such as *Silene, Lychnis, Spergula. Spergularia, Sagina, Arenaria, Paronychia* and *Scleranthus* grow in the cold Arctica.

Vegetative Features.   Annual or perennial herbs, sometimes suffrutescent shrubs. Stem herbaceous with swollen nodes, leaves opposite, rarely alternate, simple, mostly linear to lanceolate, often sessile and basally connected by a transverse line or by a shortly connate-perfoliate base as in *Dianthus* (Fig. 18.9H), stipulate with scarious stipules or estipulate.

Floral Features.   Inflorescence simple or complex dischasial cyme (Fig. 18.9A) or solitary terminal flowers, globose spiny heads in *Sphaerocoma*. Flowers ebracteate, often bracteolate (Fig. 18.9H) as in *Dianthus,* bisexual, actinomorphic, hypogynous, dimorphic in some *Stellaria* species; sterile flowers petalliferous and fertile ones apetalous. Calyx of 5 sepals, free as in *Cerastium, Stellaria, Spergularia* (Fig. 18.9B) and *Polycarpaea* or united as in *Dianthus* (Fig. 18.9H), *Silene* and *Lychnis*. Corolla with 5 petals, polypetalous, with a distinct limb and claw (Fig. 18.9I), often deeply bifid, e.g. in *Silene* and *Stellaria* (Fig. 18.9B, C) or limb multifid as in *Dianthus* (Fig. 18.9H, I). Corolla absent in *Cerdia, Colobanthus* and *Microphyes*. Stamens in 1 or 2 whorls, same as or double the number of petals, filaments free, anthers bicelled, longitudinally dehiscent (Fig. 18.9B, H), petaloid staminodes sometimes present. Stamens alternate with sepals in apetalous flowers of *Colobanthus*. Gynoecium syncarpous, 2- to 5-carpellary (Fig. 18.9D). Ovary unilocular with free-central placentation (Fig. 18.9E, F, J), or basally 3–5-loculed with axile placentation in the lower part and unilocular with free-central condition above as in some Silenoideae; or unilocular ovary with solitary, basal ovule as in *Acanthophyllum* and *Drypis;* ovules 1 to numerous. Ovary usually on a gynophore which is a stipe-like torus. Styles and stigmas 2 to 5 or as many as the carpels; fruit a capsule dehiscing apically by valves or teeth or circumscissilely or indehiscent utricle or achene. Seed usually with a hard endosperm (soft in *Agrostemma*) and a curved embryo (straight in *Dianthus*), often winged as in *Spergula arvensis* (Fig. 18.9G).

Anatomy.   In transverse section of stem, xylem and phloem appear to be in the form of a continuous cylinder or as distinct bundles separated by broad rays. Vessels are with simple perforations. Anomalous secondary growth occurs more frequently in roots of certain genera. Leaves dorsiventral or centric;

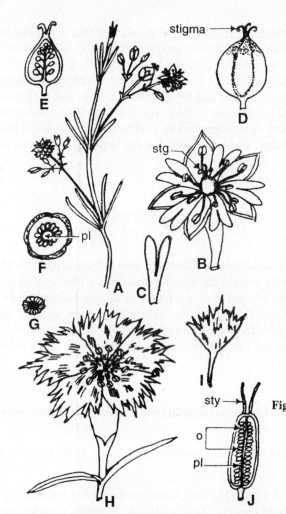

Fig. 18.9 Caryophyllaceae: **A-G** *Spergula arvensis,* **H-J** *Dianthus caryophyllus.* **A** Flowering branch. **B** Flower. **C** Petal. **D** Carpel. **E** Vertical section of carpel. **F** Ovary, cross-section. **G** Seed. **H** Flower of *D. caryophyllus.* **I** Petal with limb and claw. **J** Vertical section of carpel. *o* ovule, *pl* placenta, *stg* stigma, *sty* style. (Original)

central tissue serves as water storage tissue in *Sphaerocarpos.* Various types of hairs—both uniseriate and multiseriate and glandular—occur on leaf surface. Wax is thickly deposited on the leaf surface of certain species of *Cometes, Dicheranthus* and *Pteranthus;* stomata generally of the Caryophyllaceous type.

Embryology. Pollen grains 3-colpate, pantoporate, exine spinulose and tubuliferous-punctate or finely reticulate, shed at 3-celled stage. Ovules campylotropous, bitegmic, crassinucellate. Embryo sac of Polygonum type, 8-nucleate at maturity. Endosperm formation of the Nuclear type. Seed exarillate and perispermous.

Chromosome Number. Basic chromosome number is x = (5-) 9-15 (-19).

Chemical Features. Rich in anthocyanin pigments, some taxa also contain saponin.

Important Genera and Economic Importance. A large number of genera have rather unusual distribution from South America to Africa to Australia and New Zealand. Many genera grow at an altitude of 2000 to 4000 ft in the Himalayas.

This family is economically important for the ornamentals, e.g. *Dianthus, Gypsophila, Saponaria, Silene, Lychnis, Arenaria* and *Cerastium.* The roots of *Vaccaria pyramidata* yield saponin, which forms lather with water.

Taxonomic Considerations.   The origin of Caryophllaceae is disputed. One view is that it originated from the Phytolaccaceae, where the outer whorl of stamens becomes converted to petals, and the outer whorl of carpels to stamens. This view is supported by Eichler (1875; see Lawrence 1951), Pax (1927), Rendle (1925) and Wettstein (1935). The second view is that it originated from the Ranalean ancestors and is the source of origin for Amaranthaceae, Chenopodiaceae and the Primulales. This was proposed by Wernham (1911) and is supported by Bessey (1915), Lawrence (1951), Hutchinson (1973), Cronquist (1981), Dahlgren (1983a) and Takhtajan (1987). A third view, suggested by Dickson (1936), is that it has originated from the Geraniales. The present knowledge about Caryophyllaceae supports the view of Wernham.

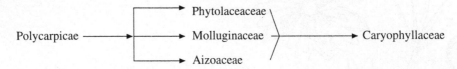

According to Ehrendorfer (1976a), Caryophyllaceae has originated from the woody Polycarpicae (with showy perianth, primary polyandry and anthocyanin pigments) through Phytoaccaceae-Molluginaceae-Aizoaceae, although it has diverged from this group in course of evolution. They have retained the anthocyanin pigments from the ancestral "Polycarpicae" although anthocyanins have been replaced by betalain pigments in the immediate ancestors or the core families. Mabry (1976), on the basis of DNA-RNA hybridisation data, is also of the opinion that the Caryophyllaceae is derived from a common "Centrospermous" ancestor.

Dysphaniaceae

A unigeneric family of the genus *Dysphania* with 3 or 4 species (Eckardt 1969), restricted to sandy soils in Australia.

Vegetative Features.   Small, branched, more or less prostrate, perennial herbs (Fig. 18.10A); leaves simple, entire or crenate, alternate estipulate.

Floral Features.   Inflorescence dense axillary fascicles (Fig. 18.10B), sometimes crowded into leafless, false spikes, indeterminate; bracteoles absent in flower clusters. Tepals usually 3, rarely 1, imbricate, membranous, persistent, accrescent (Fig. 18.10C). Stamens normally 3, sometimes 1, opposite the tepals, exerted, with basally thickened straight filaments and ovoid, introrse, bicelled anthers (Fig. 18.10D); disc absent. Gynoecium syncarpous, bi- or tricarpellary, ovary unilocular with 1 ovule (Fig. 18.10E); 1 or 2 filiform styles. Fruit a small achene or a nutlet with a very thin pericarp. Seed with a slightly curved embryo and a furrow above the embryo (Eckardt 1968).

Anatomy.   Uniseriate hairs with a glandular cell at the tip, centric mesophyll and crystal-sand in leaves, and P-type sieve element plastids are the important features.

Embryology.   Pollen grains polyporate with spinulose or tubuliferous-punctate exine; shed at 3-celled stage. Ovules campylotropous, bitegmic, crassinucellate (Eckardt 1968).

Chromosome Number.   Basic chromosome number is x = 9.

Chemical Features.   Plants rich in betalains, mainly flowers.

Important Genera and Economic Importance.   The only genus is *Dysphania;* economic importance not known.

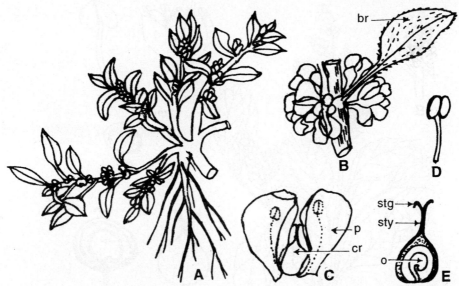

Fig. 18.10   Dysphaniaceae: **A-E** *Dysphania myriocephala*. **A** Flowering plant. **B** Inflorescence. **C** Flower. **D** Stamen. **E** Vertical section of carpel. *br* bract, *cr* carpel, *o* ovule, *p* perianth, *stg* stigma, *sty* style. (After Eckardt 1969)

Taxonomic Considerations.   Dysphaniaceae is allied with Caryophyllaceae-Illecebraceae (Bentham 1870), or as tribe Dysphanieae (Pax 1889; see Eckardt 1976) or with Chenopodiaceae (Hutchinson 1973, Cronquist 1981, Takhtajan 1987). It was raised to a unigeneric family Dysphaniaceae by Pax and Hoffmann (1934), Takhtajan (1959) and Eckardt (1964). However, Eckardt (1967) concluded that *Dysphania* is very similar to *Chenopodium* in flower development and morphology and did not deserve a family status. Mabry and Behnke (1976) are of the opinion that the presence of betalains and P-type sieve-element plastids in this genus indicate that it belongs to the group of betalain-containing Centrospermous families. Takhtajan (1987) agrees with Eckardt (1976) and includes *Dysphania* in tribe Dysphanieae of subfamily Chenopodioideae, family Chenopodiaceae.

   From the above discussion it is apparent that *Dysphania* should be treated as a genus in the family Chenopodiaceae. Apart from floral morphology and anatomy, this genus also shows alliance to Chenopodiaceae in palynology and basic chromosome number.

Chenopodiaceae

A family with 105 genera and 1600 species, more or less cosmopolitan in distribution, especially in the xeric environment and halophytic areas. The family is well represented on the prairies and plains of North America, the pampas of South America, the shores of the Red, Caspian and Mediterranean seas, the Central Asiatic region, the South African karroo and the salt plains of Australia (Lawrence 1951). Many genera are indicators of saline habitats.

Vegetative Features.   Annual or perennial herbs (Fig. 18.11A) or shrubs, rarely trees, e.g. *Haloxylon* of Central Asiatic Steppes. Plants mostly halophytes, *Salsola kali* grows near seashore and *Salicornia herbacea* in salt marshes, or in steppes and deserts which at one time were covered with sea water but are dry and supersaturated with salt at present. These plants exhibit typical xerophytic characters to reduce the rate of transpiration. Stems sometimes fleshy, jointed and nearly leafless. Leaves alternate

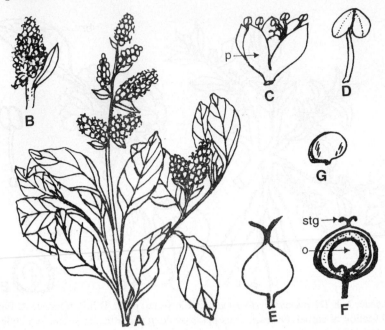

Fig. 18.11 Chenopodiacese: **A-G** *Chenopodium album*. **A** Flowering twig. **B** Inflorescence. **C** Flower. **D** Stamen. **E** Carpel. **F** Vertical section of carpel. **G** Seed. *o* ovule, *p* perianth, *stg* stigma. (Original)

(Fig. 18.11A), rarely opposite as in *Nitrophila* and *Salicornia*, simple, estipulate, fleshy and terete in some and reduced to scales in others.

Floral Features.   Inflorescence dichasial or unilateral cymes (Fig. 18.11B). Flowers bracteate, ebracteolate, bi- or unisexual and the plants are dioecious (as in *Grayia*) or monoecious (as in *Sarcobatus*), actinomorphic, mostly hypogynous, epigynous in *Beta*, minute and greenish. Perianth uniseriate, pentamerous, sepaloid, basally connate (Fig. 18.11C), persistent (sometimes absent in male flowers, only 2 in *Atriplex,* and 3 or 4 in *Salicornia*), imbricate. Stamens as many as perianth lobes and opposite the tepals, inserted on a staminal disc or a hypogynous disc. Filaments distinct, incurved in bud, anthers bicelled, introrse, dehisce longitudinally (Fig. 18.11D). Gynoecium syncarpous, 2- or 3-carpellary, ovary unilocular with a solitary basal ovule (Fig. 18.11E, F), superior or inferior, styles and stigmas 1 to 3. Fruit an indehiscent nut, nutlet or achene, enclosed in a persistent perianth; sometimes a large number aggregate together by connation of fleshy perianth. Seeds (Fig. 18.11G) small with endosperm surrounded by a peripheral or coiled embryo. Endosperm scanty or none in *Salsola, Sarcobatus* and *Suaeda* (Lawrence 1951).

Anatomy.   Vessels small, typically in clusters on the inner side of the phloem strands, sometimes tending to be ring-porous; with spiral thickening, perforations simple; intervascular pitting alternate. Parenchyma conjunctive, linking the strands of phloem in broad irregular bands and scattered round and among the vessel groups; rays absent. Typical flattened leaves develop in only a few genera such as *Atriplex, Beta, Chenopodium, Hablitzia, Obione* and *Rhagodia*. Leaf highly reduced in size in other genera. Various types of trichomes occur (Carolin 1982). Epicuticular wax of platelet type is reported in Chenopodiaceae members (Engel and Barthlott 1988). Stomata occur on all parts of the surface of both cylindrical and flattened leaves; generally Ranunculaceous type, sometimes Rubiaceous, as in some species of *Camphorosma, Salicornia, Salsola* and *Suaeda*. Anomalous secondary growth recorded in stem and root. Mature stems contain numerous vascular bundles, laid down together with the conjunctive tissue around them, by a succession of rings or arcs of cambium. Although, usually situated in the pericycle, these may also originate in phloem (Metcalfe and Chalk 1972).

Embryology.   Pollen grains polyporate with spinulose or tubuliferous punctate exine and shed at the 3-celled stage. Ovules campylotropous, bitegmic, crassinucellate. Polygonum type of embryo sac, 8-nucleate at maturity. Endosperm formation of the Nuclear type. Embryo annular or conduplicate in Cyclobeae and spirally coiled in Spirolobeae.

Chromosome Number.   Basic chromosome umber is x = (6–) 9.

Chemical Features.   Chenopodiaceae is a betalain-containing family. The flavonoid chemistry (Young 1981) of this family is allied to Dilleniiflorae and Malviflorae than to Magnolliiflorae. According to Hartley and Harris (1981), ferulic acid is present in the cell walls of Chenopodiaceae.

Important Genera and Economic Importance.   Some of the important genera are *Beta, Chenopodium* and *Spinacia. Beta* is the only genus (in this family) with inferior ovary. *B. vulgaris,* commonly called garden beet or sugar beet, is largely cultivated for its roots. The white variety is the source of sucrose or beet sugar and the red variety is used as a vegetable. *Chenopodium anthelminticum* yields an essential oil, "oil of wormwood", used as a vermifuge. *C. album* and *C. murale* are often used as pot herb. The seeds of *C. quinoa* are boiled and eaten like rice. *Spinacia oleraceae* is used as a green vegetable. *Kochia indica* is an ornamental plant. *Suaeda, Salsola* and *Salicornia* are plant indicators for saline soil. *Salsola* is a good fodder for camels.

Taxonomic Considerations.   The Chenopodiaceae is a member of the Centrospermae, according to Engler and Diels (1936), Wettstein (1935) and Melchior (1964). Bentham and Hooker (1965c) treated it as a member of the Monochlamydeae (because of uniseriate perianth), Hutchinson (1973) recognises a separate order Chenopodiales—advanced over the Caryophyllales; Cronquist (1981) includes it in his Caryophyllales (= Centrospermae). Takhtajan (1987) also includes all the betalain-containing families in the order Caryophyllales. However, he includes the two allied families—Amaranthaceae and Chenopodiaceae—in the same suborder Chenopodiineae of the order Caryophyllales.

Amaranthaceae

A family of 65 genera and 900 species distributed widely, but abundant in tropical America and tropical Africa. Nearly one-third of the genera are monotypic.

Vegetative Features.   Annual or perennial herbs (Fig. 18.12A), rarely shrubs or trees; *Alternanthera aquatica* is an aquatic herb. Stem often angular or ridged, green or sometimes reddish, as in some species of *Amaranthus*. Leaves alternate or opposite, simple, entire, estipulate, often covered with adpressed hairs as in *Aerva tomentosa* and *Achyranthes aspera*.

Floral Features.   Inflorescence a simple or branched spike or raceme, the ultimate branches often dichasial cymes. Flowers minute and often densely crowded in the inflorescence to give an attractive appearance, e.g. *Amaranthus* and *Celosia;* bracteate, bracteolate, bisexual, less commonly unisexual, actinomorphic, hypogynous. Perianth uniseriate, usually of 4 or 5 perianth lobes (Fig. 18.12B), free or basally connate, dry, membranous, white or coloured, often hairy. Stamens same as the number and opposite the tepals, filaments usually partially connate along their entire length into a membranous tube (Fig. 18.12C); lobed or fringed petaloid outgrowths may alternate with the anthers (Fig. 18.12C); anthers 4-celled at anthesis in Amaranthoideae and bicelled in Gomphrenoideae, dehiscence longitudinal (Fig. 18.12D). Gynoecium syncarpous, 2- or 3-carpellary (Fig. 18.12E); ovary unilocular with a solitary basal ovule, superior; ovules several on a seemingly single basal funicle in Celosieae. Styles and stigmas 1

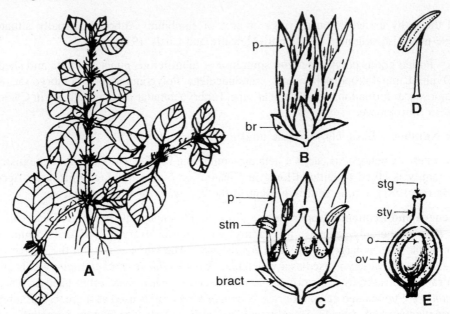

Fig. 18.12 Amaranthaceae: **A-E** *Alternathera pungens*. **A** Flowering branch. **B** Flower. **C** Vertical section of flower. **D** Stamen. **E** Carpel, vertical section. *br* bract, *o* ovule, *ov* ovary, *p* perianth, *stg* stigma, *stm* stamen, *sty* style. (Original)

to 3. Fruit a circumscissile capsule as in *Celosia* or a utricle or nutlet, rarely a drupe or berry. Seed with embryo enveloping the mealy endosperm, usually disc-shaped and with a shiny testa.

Anatomy. Vessels small to medium-sized, perforations simple, intervascular pitting alternate and moderately large. Parenchyma paratracheal, scanty to vasicentric, sometimes storied. Rays absent and replaced by radial sheets of conjunctive parenchyma. Included phloem common. Stem frequently angular with collenchyma well-developed in the ribs. Leaves dorsiventral (isobilateral in *Celosia argentea*). Woolly or silky covering of uniseriate hairs is common in addition to various special types. Stomata present on both surfaces but usually more numerous on the lower than the upper surface in *Achyranthes, Aerva, Allmania, Amaranthus, Celosia, Gomphrena* and *Pupalia*. Vascular bundles of both large and small veins are surrounded by sheaths of usually green, almost cubical, parenchymatous cells. Cluster crystals of very large size recorded in *Iresine* (Schinz 1934). Crystal-sand reported in the axis as well as leaf of *Acnida, Allmania, Amaranthus, Celosia, Cyathula, Deeringia* and *Pupalia* (Metcalfe and Chalk 1972).

Embryology. Pollen grains polyporate with spinulose or tubuliferous-punctate exine, shed at the 3-celled stage. Ovules campylotropous, bitegmic, crassinucellate. Endosperm formation of the Nuclear type.

Chromosome Number. Basic chromosome number is x = 7–9 (–13).

Chemical Features. Amaranthaceae members contain betalain pigments.

Important Genera and Economic Importance. The family is not very important except for a few species of *Amaranthus*, whose leaves and seeds are edible. A few species of *Amaranthus, Celosia* and *Gomphrena* are grown as ornamentals. A number of genera like *Achyranthes, Alternanthera, Aerva, Digera* and *Pupalia* grow as weeds.

Taxonomic Considerations. Anatomically, Amaranthaceae is affiliated to Nyctaginaceae but it does not contain any raphides or styloids. It is also close to Chenopodiaceae because of the similar type of anomalous secondary growth, and trichomes (Carolin 1982).

According to Lawrence (1951), Amaranthaceae was earlier presumed to be a primitive family but recent studies of the bracts and bractlets provide evidence that the basic inflorescence is a dichasium of 3 flowers, of which 2 have been lost, and only the bractlets remain. Each flower represents an ancestral dichasium and this is certainly an advanced feature. Hutchinson (1969) relates Amaranthaceae to the tribe Polycarpeae of the Caryophyllaceae in which the calyx is similarly dry and scarious. Cronquist (1981) and Takhtajan (1987) include Amaranthaceae in Caryophyllales and refer to its alliance with Chenopodiaceae.

Didiereaceae

A small family of 4 genera and 11 species endemic to southwestern Madagascar (Melchior 1964).

Vegetative Features. Perennial trees or shrubs with habit like that of cacti or calciform euphorbias, often armed with solitary, paired or fascicled spines. Leaves simple, alternate, estipulate.

Floral Features. Inflorescence cymes or fascicles. Flowers unisexual, both male and female flowers in the same inflorescence or rarely bisexual and female flowers in an inflorescence as in *Decaryia.* Involucre 2-leaved, more or less produced or decurrent at base. Perianth lobes 4 in 2 series of 2 each, imbricate. Stamens 8 to 10, staminodes in female flowers, shortly united at base. Gynoecium syncarpous, 2- to 4-carpellary, ovary with 1 fertile locule and 1 basal ovule. Style 1, usually bearing an expanded 3- to 4-lobed stigma, scarcely stigma elongated as in *Decaryia.* Pistillode present in male flowers. Fruit dry, indehiscent, usually loosely enclosed in persistent involucre. Seed endospermous or non-endospermous, with a curved embryo and a small aril.

Anatomy. Vessels solitary and in irregular clusters and groups of multiples; intervascular pitting transitional, opposite and alternate; perforations simple. Parenchyma paratracheal, scanty and terminal. Rays up to 4 cells wide, more than 1 mm high, heterogeneous. Large cluster crystals and mucilage cavities present in the cortex. Leaves isobilateral, with a homogeneous mesophyll, stomata infrequent, slightly depressed.

Embryology. Pollen grains 3-7-colpate with spinulose, anulopunctate exine (Behnke 1976). Ovules hemicampylotropous, bitegmic, crassinucellate. Embryo sac of Polygonum type, 8-nucleate at maturity (Johri et al. 1992). Endosperm formation of the Nuclear type.

Chromosome Number. Basic chromosome number is x = 24; represented exclusively by polyploids (Ehrendorfer 1976a, Schill et al. 1974).

Chemical Features. Didiereaceae contain betacyanin pigments.

Important Genera and Economic Importance. The four genera of this family are *Alluaudia, Alluaudiopsis, Decaryia* and *Didierea.* No economic importance is known.

Taxonomic Considerations. Bentham and Hooker (1965a) included Didiereaceae in the Sapindaceae. Hallier (1912) placed it in Caryophyllales and derived it from the Portulacaceae. According to Ehrendorfer (1976b): "Both Didiereaceae and Cactaceae may be regarded as parallel evolutionary lines, originating not far from Portulacaceae and Aizoaceae". Bessey (1915) did not recognise Didiereaceae as a distinct family. Engler and Diels (1936) and Lawrence (1951) included it in Sapindales. Melchior (1964), however, placed it in Caryophyllales (= Centrospermae). Willis (1973) treated it as a Centrospermous family on the basis of pollen morphology, floral features, anatomy, embryology and the presence of betalain pigments. This view is supported by later workers like Cronquist (1981), Dahlgren (1980a, 1983a) and Takhtajan (1987). On the basis of available data, the family Didiereaceae is correctly placed in the Centrospermae (= Caryophyllales).

In addition to the 13 families (Melchior 1964), Takhtajan (1987) added 6 more families—Berbeuiaceae, Tetragoniaceae, Stegnospermaceae, Hectorellaceae, Halophytaceae and Cactaceae. Of these, Cactaceae has been dealt with under Cactales, Berbeuiaceae and Stegnospermaceae are often included in Phytolaccaceae, Tetragoniaceae in Aizoaceae, Hectorellaceae in Portulacaceae, and Halophytaceae in Chenopodiaceae.

Taxonomic Considerations of the Order Centrospermae

The Centrospermae or the Caryophyllales comprise 13 families distributed in 4 suborders. These 13 families are quite homogeneous and have many common embryological features. They have campylotropous or amphitropous (rarely anatropous), bitegmic crassinucellate ovules, 3-celled pollen grains at shedding stage, seed with a curved peripheral embryo with scanty or no endosperm and the food storage tissue is perisperm.

Amongst all the constituent families, Molluginaceae, Aizoaceae and Phytolaccaceae show closer affinities. These plants are more or less woody with normal secondary growth, non-succulent, alternate leaves and sieve element plastids with round crystalloides; thyrsoid inflorescence, pentamerous flowers with a single whorl of perianth, 3-colpate pollen grains and 5 partly-fused carpels, numerous ovules on axile placentation. All these primitive features suggest that these three families are close to the ancestors of this order (Ehrendorfer 1976b). Nyctaginaceae with fused tepals and showy green or coloured bracts and Achatocarpaceae with dioecious, reduced flowers, are derivatives of Phytolaccacae. The position of Gyrostemonaceae in Centrospermae is doubtful. This family is more allied to the Bataceae and the two together are placed in Capparales by Takhtajan (1987). Portulacaceae and Basellaceae are derivatives of Aizoaceae. Caryophyllaceae with mostly herbaceous members is an advanced family and it contains anthocyanin instead of betalain. Molluginaceae is another such family. Most advanced families amongst the Caryophyllales are those belonging to the suborder Chenopodiineae—Dysphaniaceae (very often included in Chenopodiaceae), Chenopodiaceae, Amaranthaceae and Didiereaceae. These families are more or less herbaceous, show typical Centrospermous embryological features, abnormal secondary growth, uniseriate, sepaloid or scarious perianth, unilocular ovary with a single basal ovule and have betalain pigments. Three of the families, Aizoaceae, Portulacaceae and Didiereaceae, and to some extent Basellaceae, are succulent and many Chenopodiaceae are halophytes. Both these characters are common to the plants of xeric environment.

According to Ehrendorfer (1976b), Centrospermae are monophyletic and data on morphology and sieve-element ultrastructure point to woody "Polycarpicae" as the probable ancestral form. The ancestral forms of Centrospermous families or the "Proto-Centrospermae" were plants of open, warm, dry and windy habitats and were anemophilous. They either still had the original anthocyanin pigments (from the Polycarpicae) or lost them, possibly in the course of their trend towards anemophily. *Limeum*, a relic Molluginaceae, contains neither of the pigments. More advanced groups have radiated either into (a) more humid and forested areas with a large number of pollinating insects, or (b) in further open and increasingly dry areas where the insects have expanded in number later resulting in xeromorphic plants. For this reason many advanced Centrospermae such as Cactaceae have developed showy petals from stamens or have colourful bracts (e.g. Nyctaginaceae). These pigmentations are due to betalains and not anthocyanins. It is assumed, therefore, that this reversed selection pressure in favour of zoophily is responsible for the origin of the new betalain pigments after anthocyanins were lost. Other advanced families like Chenopodiaceae and Amaranthaceae (with compact inflorescences) also show zoophily, and members such as Caryophyllaceae and Molluginaceae retained the anthocyanin pigments from the ancestral forms although they are also pollinated by insects.

Mabry (1976) concluded (on the basis of DNA-RNA hybridization data) that the two groups—one containing anthocyanins and the other betalains—are derived from a common "Centrospermous" ancestor.

The two groups are best treated as members of one order, Caryophyllales, consisting of one betalain suborder, Chenopodiineae, and one anthocyanin suborder, Caryophyllineae. Mabry et al. (1963) and Behnke and Turner (1971) also classified the Centrospermous familes in the same way. Cronquist (1968) does not include Gyrostemonaceae in his Caryophyllales. Cronquist (1968) also does not recognise Achatocarpaceae and Dysphaniaceae as distinct families. He includes Cactaceae in this order, and the total number of families is then 11. Takhtajan (1969) included Bataceae and Gyrostemonaceae in this order and also recognised a few smaller families like Dysphaniaceae, Tetragoniaceae, Halophytaceae and Hectorellaceae. Hutchinson (1973) includes Elatinaceae, Molluginaceae, Caryophyllaceae, Aizoaceae and Portulacaceae in his Caryophyllales, and Barbeuiaceae, Phytolaccaceae, Gyrostemonaceae, Agdestidaceae, Petiveriaceae, Chenopodiaceae, Amaranthaceae, Theligonaceae, Batidaceae and Basellaceae in his Chenopodiales. Obviously he does not attach much importance to the pigments and broke up some families into many smaller units. Hershkovitz (1989) points out that two characters—the P_3-type sieve tube plastid, and the presence of bound ferulic acid in unlignified cell walls—are common in all Centrospermous families but not in any other dicot.

19

Order Cactales

The order Cactales comprises only one family, Cactaceae, herbs or shrubs, distributed mostly in xeric habitats of the world. Most cacti are native to the United States of America except Maine, New Hampshire and Vermont. These are fleshy herbs or subshrubs, rarely woody. Stem green and succulent, variously shaped and therefore of much value as ornamentals. Flowers usually bisexual, highly ornamental, and show primitive features.

Cactaceae

The fleshy parts of the plants and their spiny nature render them difficult to make an adequate number of herbarium specimens and, therefore, one has to depend mainly upon a limited number of vegetative parts for identification. Britton and Rose (1923) recognised 100 genera, Parish (1936) reduced these to only 26 (see Lawrence 1951). Takhtajan (1987) reports 105 and Dutta (1988) 87 genera. There are about 2000 species. The main centre of distribution is the dry, xeric habitats of America, especially Mexico and the adjoining parts of the United States of America.

Vegetative Features.   Succulent herbs, shrubs or sometimes woody plants, perennials. Stems simple or cespitose, fleshy, globular, cylindrical or flattened, angular or ribbed, frequently constricted and jointed; sap watery or rarely milky, as in *Coryphantha*. Leaves alternate, simple, flat leaflike in appearance and fleshy in *Pereskia* and *Pereskiopsis;* cylindrical, scale-like or absent in others; often reduced to tufts of bristles or spines or glochidia arranged in aeroles (Fig. 19.1A).

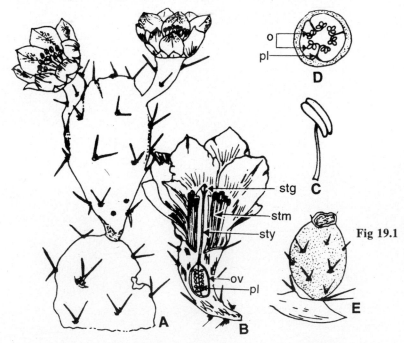

Fig 19.1   *Cactaceae*: **A–E** *Opuntia dillenii.* **A** Phyllodes with flowers. **B** Flower, vertical section. **C** Stamen. **D** Cross section of ovary. **E** Fruit. *o* ovule, *ov* ovary, *pl* placenta, *stg* stigma, *stm* stamen, *sty* style. (Adapted from Arachi 1968)

Floral Features.   Flowers usually solitary, sometimes in clusters as in *Pereskia;* bisexual, rarely unisexual, actinomorphic, epigynous, the hypanthium adnate to the ovary (Fig. 19.1B). Perianth usually weakly differentiated into sepals and petals, gradual transition from sepals to petals and often merged into bracts, free or united to form a tube. Stamens numerous arising spirally or in clusters from the inner face of hypanthium (Fig. 19.1A, B); anthers bicelled, dehiscence longitudinal, filaments long, slender (Fig. 19.1C). Gynoecium syncarpous, 4- to many-carpellary, ovary unilocular with parietal placentation, inferior, ovules numerous (Fig. 19.1D); style usually 1 (as many as carpels in *Pereskia*), stigmas as many as carpels, radiating. Fruit a berry, often glochidiate, spiny or bristly (Fig. 19.1E); seeds with (*Rhipsalis*) or without endosperm, and a straight or curved embryo.

Anatomy.   Stems, generally spiny and succulent in most genera, form the main part of the plant body and serve as principal photosynthetic organ. Vessels small, except in *Pereskia,* solitary or in clusters, perforations simple; intervascular pitting scalariform to opposite. Parenchyma paratracheal, a few cells to a complete sheath around the vessels. Rays mostly 6 to 10 cells wide and without uniseriates. Mucilage cells in the ground tissue of most taxa, and laticiferous canals in some genera occur. Spines on stem surface vary in length and thickness; in some species barbed and called glochidia; usually grouped in small circular areas called areoles on stem surface and covered with dense hairs at the base. In some taxa the hairs are as long as or longer than the spines and form a definite flossy (silky fibrous) coating on the stem. Stomata of different types, abundant on stem surface. Rudimentary leaves of *Opuntia* provided with abortive stomata; numerous normal stomata present in *Pereskia* (Metcalfe and Chalk 1972). Leaves reduced to scales except in *Pereskia.*

Embryology.   Pollen grains 3-4-colpate, 3-colporate, 6- to polyrugate or multiforate and 3-celled at shedding stage. Exine often spinulate and reticulate. Ovules anatropous, hemicircinotropous or circinotropous, bitegmic, crassinucellate with a nucellar cap. Endosperm formation of the Nuclear type; nucellar polyembryony occurs in *Mammillaria, Opuntia* and *Pereskia* (Tyagi 1970). Seed is perispermous.

Chromosome Number.   Basic chromosome number is $x = 9 - 11$.

Chemical Features.   Two types of betalain pigments (betanin and phyllocactin) are present in flowers and fruits of Cactaceae. Flavonol-glycosides and sugar-free flavonols have been isolated from the tepals of several species of cactaceous plants (Iwashina et al. 1986). Many of the hallucinogenic principles from Cactaceae members are isoquinoline alkaloids; anhalonin has been isolated in the dried crowns (mescal buttons) of the peyote cactus, *Lophophora williamsi.*

Important Genera and Economic Importance.   Many genera of this family are grown as ornamentals throughout the world, e.g. *Cereus, Echinocactus, Epiphyllum, Mammillaria* and many others. The fruits of *Cereus variabilis, Opuntia elatior, O. ficus-indica* and *O. megacantha* are edible. Cochineal insects thrive on *Nopalea cochinillifera* and are the source of cochineal dye. Hallucinogenic drug is obtained from the dry heads or mescal buttons from *Ariocarpus, Astrophytum, Aztekium, Obregonia, Pelecyphora* and *Solisia*—popularly known as peyotes. A variety of alkaloids are obtained from these taxa. Many species of *Opuntia* are locally grown as hedge plants.

   The symbol of American desert is the Saguaro cactus, *Cereus giganteus,* largest of all the cacti, but one of the slowest-growing plants. Saguaro cactus is the state emblem of Arizona and it can attain a height of about 10 m with a life span of about 200 years. Other well-known cacti are Queen of the Night (*Selenicereus grandiflorus*) and Princess of the Night (*S. nycticalus*) (Nayar 1984).

Taxonomic Considerations.   The family Cactaceae shows heterobathmy; it has primitive, unspecialised flowers with highly advanced vegetative parts. Ecologically, it is interesting as all the members survive under adverse climatic conditions.

Bentham and Hooker (1965a) placed Cactaceae in Ficoidales, next to the Passiflorales, because of parietal placentation. Engler and Diels (1936) agreed to its alliance with Aizoaceae of the Centrospermae, but placed it next to the Parietales as the sole member of Opuntiales. Besşey (1915) and Hutchinson (1973) also included it in the Cactales and placed it between the Cucurbitales and Theales. Gunderson (1950) and Mitra (1956) derived Cactaceae from the Ranales, on the basis of primitive floral features such as spirocyclic flowers, transition from sepals to petals, and numerous fasciculated stamens. Many toxonomists suggest Cactaceae to have a position in or near Centrospermae (Wettstein 1935, Buxbaum 1944, P. Maheshwari 1945, and Martin 1946), on the basis of anatomy, floral morphology, and embryology. Melchior (1964) placed it in a separate order, Cactales, next to the Centrospermae. The occurrence of betalain pigments further supports their inclusion in the Centrospermae (Smith 1976). Cronquist (1981), Dahlgren (1983a) and Takhtajan (1987) treat it as a member of Centrospermae. Behnke (1976), Ehrendorfer (1976a, b) and Mabry (1976) also include it in the Centrospermae. According to Ehrendorfer (1976b), it is one of the advanced families of the Centrospermae which has developed from the core families and has radiated in increasingly dry habitats. According to Johri et al. (1992), Cactaceae forms a bridge between Aizoaceae and Portulacaceae.

This family is correctly placed in an independent order Cactales, and is derived from the Centrospermae.

20

Order Magnoliales

A large order of 22 primitive families, mostly distributed in the tropics and subtropics of the southern hemisphere. Some of the small, isolated relic families of this order are common in the tropical and subtropical rain forests in the Western Pacific region: Austrobaileyaceae, Eupomatiaceae, Himantandraceae, Idiospermaceae-Calycanthaceae are confined to Eastern Australia and Eastern Malaysia (Endress 1983).

Plants evergreen or deciduous, large to medium-sized trees or shrubs. Leaves simple, mostly alternate. Flowers usually solitary, large and showy, sometimes in inflorescence as in Trochodendraceae. Numerous whorls of sepals and petals and indefinite stamens are arranged spirally; anthers introrse. Gynoecium monocarpellary or many-carpelled and apocarpous. Wood anatomy shows primitive vessels with scalariform end-walls. Pollen grains are shed at 2-celled stage; shedding in permanent tetrads in Winteraceae, Monimiaceae and rarely in Annonaceae and Magnoliaceae. Ovules anatropous, bitegmic, crassinucellate. Many of the families are rich in alkaloids.

Magnoliaceae

A small family of about 14 genera and 240 species, distributed mainly in the tropics and subtropics of both the New and the Old World. Some members occur in North temperate zones also. The geological record shows that the family was at one time much more widely distributed in the Northern hemisphere.

Vegetative Features.   Large to medium-sized trees or shrubs; evergreen or deciduous. Leaves simple, alternate, entire, stipulate, sometimes thick, coriaceous and shiny as in *Magnolia*; stipules often large and protective to the young buds, deciduous leaving a circular scar around the node as the leaves expand.

Floral Features.   Flowers usually solitary, terminal or axillary; hermaphrodite (unisexual in *Kmeria*), actinomorphic, bracteate, ebracteolate, mostly very large and ornamental, often with fragrance as in *Michelia champaca* (Fig. 20.1A, B). Calyx distinct or indistinct; when distinct, sepals 3, cyclic, green. Corolla of 6 petals or more, spirally arranged, often around the base of an elongated receptacle as in *Magnolia*. Stamens numerous, hypogynous, distinct, arranged spirally; anthers bicelled, introrse, dehiscence longitudinal (Fig. 20.1C). Gynoecium sessile or borne on an elongated axis or gynophore; carpels numerous, free, arranged spirally on the axis (Fig. 20.1E). Ovary monocarpellary, unilocular, placentation parietal, ovules 1 to numerous (Fig. 20.1D), style 1, stigma 1. Fruit a follicle, berry or samara; an etaerio of follicles in *Magnolia*.

Anatomy.   Vessels usually medium-sized but small (less than 100 μm) in *Alcimandra, Liriodendron, Magnolia, Michelia* and a few other taxa; solitary or in small groups, sometimes with spiral thickening; perforation plates typically scalariform with a few, widely spaced bars; but perforations simple in *Magnolia acuminata*. Intervascular pitting scalariform to opposite; parenchyma terminal; rays usually up to 3 or 4 cells wide, hetero- or homogeneous. Leaves dorsiventral, stomata mostly confined to the lower surface; usually Rubiaceous sometimes Ranunculaceous type. Calcium oxalate crystals not very common.

Embryology.   Pollen grains uni- or triaperturate, monocolpate; mature grains oval-shaped with thick walls; shed at 2-celled stage in *Magnolia*, 3-celled in *Liriodendron tulipifera*. Ovules anatropous, bitegmic, crassinucellate; Polygonum type of embryo sac, 8-nucleate at maturity Endosperm formation of the Cellular type.

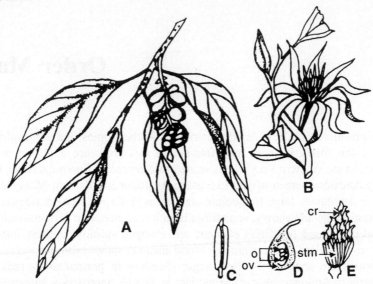

Fig. 20.1   Magnoliaceae: **A–C, E** *Michelia champaca,* **D** *Magnolia grandiflora.* **A** Fruiting twig. **B** Flower. **C** Stamen. **D** Carpel, vertical section. **E** Stamens and carpels arranged spirally. *cr* carpel, *o* ovule, *ov* ovary, *stm* stamen. (**A, B** after DeWit 1963)

Chromosome Number.   Basic chromosome number is x = 19.

Chemical Features.   D-pinitol, a methyl ether of inositol, has been isolated from *Magnolia* spp. A dimethyl ether of myo-inositol, liriodendritol, occurs in only two species of *Liriodendron—L. chinense* and *L. tulipifera* (Plouvier 1963). An alkaloid, liriodendrine, occurs in the heartwood of *L. tulipifera* (Hegnauer 1963). Essential oils of different types have been reported in various species of *Magnolia* and *Michelia* (Thien et al. 1975). Sesquiterpene lactones have been recorded in *Liriodendron* and *Michelia.*

Important Genera and Economic Importance.   A large number of trees are ornamental. *Michelia champaca* is a large tree and bears solitary axillary flowers with an elongated cone-like torus or gynophore; perianth petaloid in many whorls of 3 each, followed by spirally arranged free stamens and carpels. It is cultivated for its fragrant flowers. The timber of *M. nilagirica* is very handsome and used for furniture, railway sleepers; bark yields essential oil. Various species of *Magnolia* are large or medium-sized trees which bear beautiful, large, fragrant flowers and timber that is used for cabinet work. *Liriodendron tulipifera* is a cultivated deciduous tree of the eastern United States of America. It yields a commercial timber called Canary white wood. Timber of *Magnolia acuminata* is similar to that of *Liriodendron* and is soft and easy to work with. *Aromadendron* is another interesting genus in which the carpels are concrescent (fused), fleshy and indehiscent, and the ovules in each ovary are reduced to 2. In *Pachylarnax* also the carpels are fused but they open completely along their abaxial suture and partially along their line of junction. Therefore, the fruit in more or less like a woody loculicidal capsule (Hutchinson 1969). *Magnolia acuminata, Manglietia hookeri, Michelia baillonii, M. deltsopa* and *Pachylarnax pleiocarpa* produce valuable timber.

Taxonomic Considerations.   The family Magnoliaceae is considered to be the most primitive family amongst the dicots (Bentham and Hooker 1965a, Hallier 1905, Hutchinson 1973, Stebbins 1974). Hallier (1905) even compared the elongated floral axis (torus) bearing the spirally arranged floral whorls with the sporophyll-bearing axis of the Bennettitales. Some members of this family show very primitive features, e.g. *Magnolia stellata* and *Aromadendron* have tepals, stamens and carpels, all spirally arranged, with large, more or less petaloid tepals that are not distinguishable into sepals and petals. On the other hand, *Liriodendron* has three series of three tepals each, the outermost series is comparable to sepals. Anatomically, too, Magnoliaceae members show primitive structures such as xylem vessels with scalariform perforation plates and spiral thickening. The tree-like habit with large, showy, solitary flowers is also a primitive feature.

Magnoliaceae is indeed a primitive family, but certainly not the most primitive. Melchior (1964), Benson (1970), Cronquist (1981), Dahlgren (1980a, 1983a) and Takhtajan (1987) agree to this suggestion.

Degeneriaceae

Degeneriaceae is a monotypic family with the genus *Degeneria* with only one species endemic to the Fiji Islands.

Vegetative Features. A mature tree of *Degeneria vitiensis* is extremely slender and with a freely branching crown and dense foliage; about 18–30 m high. Leaves simple, petiolate and estipulate (Fig. 20.2A).

Floral Features. The solitary axillary or supra-axillary flowers are bisexual, actinomorphic, hypogynous, pedicellate, and complete. Calyx of usually 3, rarely 4 sepals, polysepalous (Fig. 20.2B). Petals occasionally up to 18 and arranged in 3 or 4 series, polypetalous; differ from sepals in size and form but resemble in texture and cellular composition. Stamens 20 to 30, in 3 or 4 series; laminar, without any differentiation into filaments, connective and anther (Fig. 20.2C); the 4 microsporangia are embedded on the abaxial surface. A single indehiscent carpel, rarely 2. The carpel resembles an adaxially-folded 3-veined sporophyll (Bailey and Smith 1942). Ovules vary from 20–32, carpel devoid of any style and stigma and, instead, has a decurrent stigmatic crest (Fig. 20.2D, E). Mature fruits are rich-pink to purple, falcate-oblong-ellipsoid (Fig. 20.2F) and dehisce along ventral suture.

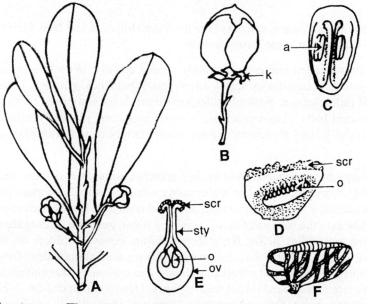

Fig. 20.2 Degeneriaceae: **A–F** *Degeneria vitiensis*. **A** Flowering twig. **B** Flower. **C** Laminar stamen. **D, E** Longisection of carpel. **F** Fruit. *o* ovule, *ov* ovary, *scr* stigmatic crest, *sty* style. (Adapted from Melchior 1964)

Anatomy. The primary vascular cylinder is a much dissected dictyostele of numerous bundles. The vessels in secondary xylem are thin-walled and angular, occur singly or in small radially oriented clusters. Perforation plates scalariform and intervascular pitting also scalariform; nodes pentalocular.

Embryology. Pollen grains broadly ellipsoid, monocolpate with a single narrow furrow which broadens at the poles. Exine smooth; 2-celled at shedding stage. Ovules anatropous, bitegmic, crassinucellate; micropyle formed by the inner integument; Polygonum type of embryo sac, 8-nucleate at maturity. Endosperm of the Cellular type; mature endosperm ruminate. Mature embryo well differentiated with tri- or tetracotyledonous condition.

Chromosome Number. Basic chromosome number is x = 6 (Takhtajan 1987).

Important Genera and Economic Importance. *Degeneria vitiensis* has no economic importance. It is important as one of the primitive or relict families of angiosperms.

Taxonomic Considerations. The family Degeneriaceae is a member of the magnolian complex and exhibits close similarity to the Magnoliaceae in:

a) Internal structure of vegetative organs,
b) type of pollen grains,
c) vasculature of stamen, and
d) Polygonum type of embryo sac.

The family Degeneriaceae was established by Bailey and Smith (1942) and has been recognised as a distinct taxon. However, Hutchinson (1969, 1973) treats it as a member of his Winteraceae, because of its resemblances with *Exospermum* and *Zygogynum*. According to Bailey et al. (1943) and Smith (1949), Degeneriaceae is closely related to Magnoliaceae and Himantandraceae. Melchior (1964), Bhandari (1971), Cronquist (1981), Dahlgren (1980a, 1983a) and Takhtajan (1987) support this assignment.

This monotypic family is correctly placed in Magnoliales close to Magnoliaceae and Himantandraceae. Although the three families are closely related, they are distinct from each other.

Himantandraceae

A monotypic family with the only genus *Himantandra*, which occurs from the Moluccas and New Guinea to northeastern Australia. There are 3 species of these aromatic trees.

Vegetative Features. Trees; branchlets slender, subterete or faintly angled distally, densely lepidote (covered with small scurfy scales); peltate, membranous scales cover young branches, petioles, lower surface of leaves and external parts of inflorescence. Stalk of scales very short, body composed of 30 to 56 radiating, flattened, laterally coalescent hairs. Leaves alternate, simple, estipulate, pinnately veined; petiole slender, rugulose (irregularly ridged), leaf blades coriaceous, entire, faintly recurved or plain at margin.

Floral Features. Flowers usually solitary and terminal on short axillary branches, sometimes accompanied by one or two lateral flowers (Fig. 20.3A), bracteate, bracts 2 or 3, alternating with each other, subcoriaceous, oblong, 1–3 mm long, occasionally foliaceous. Calyx of 2 subcoriaceous ovoid-conical or obtuse, calyptrate sepals that rupture along an irregular line near the base and fall off, leaving a small, undulate or irregularly lobed calycine remnant attached to the torus (Fig. 20.3A, B); glabrous within, densely lepidote on the outer surface. Corolla similar to calyx in texture, lanceolate, followed by the petaloid staminodes. Torus carnose (fleshy), flaring to the attachment of calyx and corolla, columnar and copiously staminiferous, concave on the distal surface and give rise to a conical carpel-bearing apex. Outer staminodes ca. 7–23, in 1 or 2 series, fleshy, sharply reflexed at anthesis. Stamens numerous, in many series, closely appressed, similar to outer staminodes in texture and shape, rapidly elongating and reflexed after anthesis; pollen sacs 4, extrorse immersed in the sporophyll tissue, linear, obtuse at base and apex (Fig. 20.3C), dehiscence

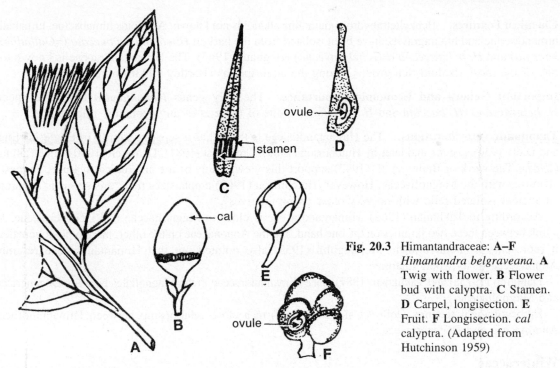

Fig. 20.3 Himantandraceae: **A–F** *Himantandra belgraveana.* **A** Twig with flower. **B** Flower bud with calyptra. **C** Stamen. **D** Carpel, longisection. **E** Fruit. **F** Longisection. *cal* calyptra. (Adapted from Hutchinson 1959)

longitudinal. Inner staminodia ca. 13 to 20 in 1 to 3 series, similar to outer staminodes in texture, linear-subulate, narrowed to an acute apex, usually erect and coherent at base. Carpels 7 to 10 spirally arranged on the conical apical portion of the torus, free but laterally appressed. Ovary ovoid or oblong-ellipsoid, densely lepidote on the thick outer surface, glabrous on the thinner lateral surfaces, gradually narrowed distally into a subulate style; styles plumose, soft, glandular. Ovary unilocular, usually uniovulate (rarely biovulate, but the second ovule develops rarely), attached to the ventral margin (Fig. 20.3D) at various levels in different carpels, often pendulous. Fruit a gall-like, ellipsoid or subglobose syncarp (Fig. 20.3E, F), up to 25 mm in diameter at maturity. Pericarp coriaceous, 0.5–1.5 mm thick, red, rugulose when dried, lepidote outside. Endocarp cartilaginous. Seeds have a small embryo with oily endosperm (Hutchinson 1973).

Anatomy. Vessels of mature wood usually with simple perforation plates; young stems with simple and scalariform perforations. Alternate intervascular pitting is reported. Wood parenchyma occurs in concentric circles and rays heterogeneous. Lower surface of leaves covered with a dense indumentum of peltate scales , those of *Himantandra baccata* smaller and less crowded than those of *H. belgraveana* . Upper surface of mature leaves glabrous but Bailey et al. (1943) report scales or stellate hairs on the upper surface of immature leaves in some species. Stomata arranged in approximately circular clusters around each of the peltate scales on the lower surface. Pairs or small cluster of crystals present in the cells of lower epidermis, and in the strands accompanying the sheaths of sclerenchyma around the vascular bundles of the veins and veinlets; secretory cells noted in leaf as well as stem tissue.

Embryology. Mature pollen grains are single, monosulcate, scabrate, atectate, spheroidal and shed at the 2-celled stage. Ovules anatropous, bitegmic, crassinucellate; Polygonum type of embryo sac, 8-nucleate at maturity. Endosperm formation of the Cellular type.

Chromosome Number. Basic chromosome number is x = 6.

Chemical Features.   Benzyltetrahydroisoquinoline alkaloids not known. Alkaloids himabacine, himabeline, himandravine and himangravine have been isolated from the bark of *Himantandra baccata* (=*Galbulimima baccata*) and *H. belgraveana* (=*G. belgraveana*) (Hegnauer 1963). The Himantandraceae is known to be one of the most alkaloid-rich groups among the angiosperms (Hartley 1973, Endress 1983).

Important Genera and Economic Importance.   The only genus *Himantandra* has three species: *H. belgraveana, H. baccata* and *H. parviflora.* None of these is of any economic importance.

Taxonomic Considerations.   The Himantandraceae is treated as a separate family in Engler's (Engler and Diels 1936) system, and later by Hutchinson (1969), Cronquist (1981), Dahlgren (1980a) and Takhtajan (1987). The work of Bailey et al. (1943) support this view. Many of the anatomical features suggest its affinities with the Magnoliaceae. However, Hutchinson (1969) emphasizes that "this genus is apparently an ancient isolated relic with no very close living relatives".

According to Takhtajan (1969), Himantandraceae is close to Magnoliaceae and Degeneriaceae. It is a link between these two families on the one hand, and the Annonaceae on the other, though phylogenetically it represents a lateral branch. McLaughlin (1933) also pointed out that Himantandraceae resembles Annonaceae in its wood anatomy.

Dahlgren (1975a) and Takhtajan (1987) place Himantandraceae in the Magnoliales between Degeneriaceae and Magnoliaceae.

Embryological features (Johri et al. 1992) confirm a close relationship between Himantandraceae, Annonaceae and Degeneriaceae.

Winteraceae

Winteraceae is a primitive family of 6 genera and ca. 90 species distributed in Australia, Tasmania, New Guinea, the Philippines, Borneo, Celebes and Amboina in the Old World. In the New World it has been reported from South America. *Drimys* is the only genus which occurs in both the New and the Old World.

Vegetative Features.   Trees or large shrubs; dwarfing of the entire plant or only its leaves in reported. When growing in very exposed condition the plants tend to be gnarled and often have leaves closely crowded towards the apex of the branches. Branchlets sometimes glaucous when young, subterete and usually longitudinally striate. Leaves alternate or irregularly crowded, petiolate, petiole strong; simple, entire, glabrous, estipulate, pinnately veined, variously shaped—oblong to elliptic to obovate. Leaves microphyllous in some epiphytic species of *Drimys.*

Floral Features.   Inflorescence variable, usually clustered at apices of branchlets; umbellate (Fig. 20.4A), peduncle 7 to 40 mm long with 3 to 6 flowers in each inflorescence; solitary axillary in *Drimys,* solitary terminal in *Zygogynum.* Variable number of sepals, gamosepalous, membranous as in *Drimys* or subcoriaceous in *Bubbia, Bellonia* and *Exospermum;* pellucid-glandular, suborbicular to ovate, 6–8 mm long. The sepals show three conditions: (1) Reduction in number of partly concrescent sepals to 2 as in *Drimys* and *Pseudowintera*, (2) elimination of lobes to form entire or rotate calyces, or lastly (3) complete concrescence of sepals to form a calyptrate calyx. Corolla of 7 to 8, rarely 12 petals, polypetalous (Fig. 20.4B), membranous, sparsely and obscurely pellucid glandular, oblong. Stamens numerous (30 to 40), free, broadly truncated sporophyll-like bearing transversely oriented apical sporangia in *Bubbia, Exospermum, Pseudowintera* and *Zygogynum;* apically constricted sporophylls bear markedly protuberant, laterally attached subapical sporangia in *Drimys* (Fig. 20.4B); filaments terete, comparatively slender in *Drimys;* flattened in *Bubbia* and *Bellonia.* Gynoecium 3- to 6-carpellary, apocarpous; carpels obovoid (Fig. 20.4B, C), about 3 mm at the time of anthesis, sporophyll-like, stipitate, with conduplicate, palmately 3-veined lamina, 2 closely approximated stigmatic crests; ovary unilocular with numerous ovules on parietal

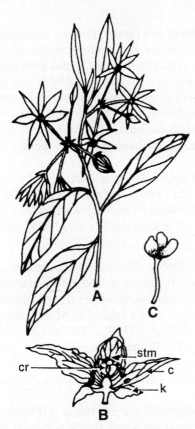

Fig. 20.4 Winteraceae: **A–C** *Drimys winteri*. **A** Flowering twig. **B** Flower, longisection. **C** Fruit. *c* corolla, *cr* carpel, *k* calyx, *stm* stamen. (Adapted from Hutchinson 1969)

placenta. Carpels free even in buds in *Bubbia,* firmly concrescent (syncarpous) in *Zygogynum,* free only after anthesis in *Exospermum.*

Anatomy. The wood of Winteraceae is unique. Xylem is vesselless with diffuse parenchyma, primitive heterogeneous type of rays, up to 10 cells wide and occasionally with "oil" cells. Tracheids with a row of circular pits. Leaves dorsiventral, glabrous; lower epidermis papillose. Stomata Rubiaceous type confined to lower surface and restricted to depressions filled with a white granular material different from wax reported in Magnoliaceae and Schisandraceae.

Embryology. Pollen permanently in tetrahedral tetrads, each pollen grain is monocolpate with a distally oriented germ pore, intine thick and protruded in the region of germ pore, exine coarsely reticulate; shed at 2-celled stage. Ovules anatropous, bitegmic, crassinucellate. Polygonum type of embryo sac, 8-nucleate at maturity. Endosperm formation of the Cellular type. The embryo is minute and undifferentiated in *Drimys,* and dicotyledonous in *Pseudowintera* (Johri et al. 1992).

Chromosome Number. Basic chromosome number of all Winteraceae is x = 43 except *Drimys* sect. *Tasmannia* where the base number is x = 13. *Drimys* sect. *Tasmannia* is therefore at a lower ploidy level (Ehrendorfer et al. 1968).

Chemical Features. Stem bark of *Drimys winteri* contains sesquiterpenoids of the drimane type, in addition to cryptomeridiol, cirsimaritin, quercetin, astilbin and quercitrin. Leaf surface contains a chemical which is different from wax and has a higher melting point, and is insoluble in boiling alcohol, hot ether and other non-polar substances (Bailey and Nast 1945a). Presence of phenolic compounds reported in anther, ovule and fruit wall (Bhandari 1971).

Important Genera and Economic Importance.   The type genus of this family, *Drimys,* is an important taxon. It has ca. 20 species distributed in both Australasia and Central and South America. These are evergreen trees and shrubs with pellucid-punctate leaves. It is a typically montane genus growing up to a height of 3800 m but at sea level also. Microphyllous *D. microphylla* is an epiphyte. Old World *Drimys* was named *Tasmannia.* Later, it was treated as section *Tasmannia* of genus *Drimys.* However, Ehrendorfer et al. (1968) consider that it should be treated as a distinct genus because: (1) dioecious flowers, narrow filaments, carpels not utriculate, (2) different pollen morphology and anatomy, (3) occurrence of flavones in addition to flavonoids, and (4) chromosome number $2x = 26$. *Tasmannia*, at a lower ploidy-level, is more advanced both morphologically as well as chemically.

In *Zygogynum vieillardii* the flower is solitary, terminal and sessile. In other species of *Zygogynum,* and in *Exospermum* there are three terminal flowers; in *Bubbia* and *Drimys* there is a terminal cluster of cymes. Carpels syncarpous in tribe Exospermeae; reduced to only one in *Drimys dipetala* and *Zygogynum pomiferum* (Hutchinson 1969). The genus *Takhtajania* is an almost or completely extinct Madagascarian relic (Vink 1978, Tucker and Sampson 1979).

Taxonomic Considerations.   In Bentham and Hooker's system (1965a), Winteraceae was not recognised as a distinct family. Hallier (1912) and Bessey (1915) also did not recognise it. The cytological findings of Whitaker (1933) supported the segregation of family Winteraceae and some others from Magnoliaceae (s.l.). Hutchinson (1973) also treats Winteraceae as a very primitive family related to Magnoliaceae and retains it in the Magnoliales. Cronquist (1981), Stebbins (1974), Dahlgren (1980a, 1983a) and Takhtajan (1980) agree with this assignment. According to Takhtajan (1980), Winteraceae is the most primitive family of the Magnoliales and belongs to the monotypic suborder Winterineae. Némejc (1956) and Smith (1971) place it in a distinct order Winterales (cf. Takhtajan 1980); Takhtajan (1987) erected a distinct order Winterales to accommodate this family.

In most characters, Winteraceae resembles the members of Magnoliaceae; the major difference is the presence of vesselless xylem in some members of this family. This is not a convincing argument to erect a separate order Winterales, and Winteraceae should still be treated as a member of the Magnoliales.

Annonaceae

The Annonaceae are a common pantropical woody family comprising about 130 genera and 2300 species (Okada and Ueda 1984). A large family of aromatic trees, shrubs or climbers, which occur in tropical and subtropical regions. In the tropics of the Old World, they are usually of climbing or straggling habit, and occur in lowland dense evergreen forest. In tropical America, they are nearly all shrubby or arboreal, and grow mostly in the open grassy plains (Hutchinson 1964, Leboeuf et al. 1982). The only genus extending into the temperate zone is *Asimina,* which occurs in North America as far north as the Great Lakes. According to Takhtajan (1969), 51 genera and ca. 950 species are confined to Asia and Australasia, 40 genera and ca. 450 species in Africa and Madagascar and 38 genera and 740 species on the American continent. Thus Asia, together with Australasia, is the basic centre of distribution of the Annonaceae. However, on the basis of phytogeographical and palynological data, Le Thomas (1981) hypothesizes South America or Africa to be the centre of origin of this family.

Vegetative Features.   The members of Annonaceae are trees (*Polyalthia longifolia*), shrubs (*Annona squamosa, A. reticulata*) or vines such as *Artabotrys odoratissima* (Fig. 20.5A), *Uvaria* spp. with aromatic wood and foliage. Leaves simple, alternate, entire, estipulate; deciduous or persistent, gland-dotted.

Floral Features.   Flowers usually solitary (Fig. 20.5A), rarely inflorescent, bisexual (unisexual and plants dioecious in *Ephedranthus, Stelechocarpus,* and *Thonnera*), actinomorphic, hypogynous, spirocyclic,

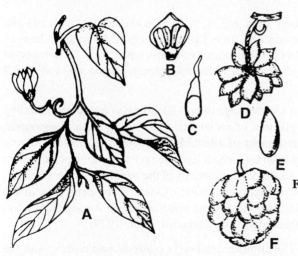

Fig. 20.5 Annonaceae: **A–E** *Artabotrys odoratissima*, **F** *Annona squamosa*. **A** Twig with flower. **B** Stamen. **C** Carpel. **D** Fruit. **E** Seed. **F** Aggregate fruit of *A. squamosa*. (**A–E** adapted from DeWit (1963)

trimerous, usually scented. Perianth triseriate, the outer whorl, a calyx of 3 basally connate or free, valvate, persistent sepals, and two inner whorls, a corolla of usually 6 distinct, similar or dissimilar petals, imbricate or valvate. The axis usually extends and enlarges beyond the point of perianth attachment. Corolla sympetalous in *Asteranthe, Disepalum, Enneastemon;* outer petals with long dorsal appendages in *Rollinia;* petals 6 in *Haplostichanthus, Monanthotaxis* and *Monocyclanthus* and only 3 in *Xylopia.* Stamens numerous, free, spirally arranged, only 9 to 12 arranged in whorls in *Mezzettia;* filaments short and thick, anthers 4-celled at anthesis (Fig. 20.5B), extrorse, but introrse in *Mezzettia parviflora;* anther lobes linear with a prolonged connective; anthers transversely locellate in *Cardiopetalum, Hornschuchia* and *Porcelia,* and connective is not elongated in *Alphonsea.* Gynoecium of a few to numerous carpels (basally connate in *Monodora),* arranged spirally; carpel solitary in *Kingstonia, Mezzettia, Monocarpia* and *Tridimeris.* Ovary superior, unilocular (Fig. 20.5C), but multilocular in *Pachypodanthium;* ovules 1 to many, typically parietal but often appear to be basal; rarely all the carpels are united to form a unilocular ovary with parietal placentation as in *Monodora.* Style 1, very short or absent, stigma 1. Fruit a berry or etaerio of berries (Fig. 20.5D) or the mature pistils are connate and adnate to the floral axis to form a fleshy, aggregate fruit as in *Annona squamosa* (Fig. 20.5F). Seed large (Fig. 20.5E) with a small embryo and copious ruminate endosperm, often arillate.

Anatomy. Vessels very small (less than 50 μm mean tangential diameter), as in some species of *Malmea, Orophea, Oxandra* and *Popowia,* or large (more than 200 μm diameter), as in some species of *Cananga, Cleistopholis, Guatteria, Rollinia* and *Unona;* usually a few, with simple perforations. Parenchyma apotracheal, in numerous fine lines, often storied. Rays typically wide and high, 3-16-celled wide, slightly heterogeneous to homogeneous. Leaves generally dorsiventral; stomata Rubiaceous type, confined to the lower surface. Simple, stellate and peltate hairs on both surfaces. Both solitary and clustered crystals occur in epidermal cells.

Embryology. Pollen grains either inaperturate or monocolpate and provided with a proximal germ pore, Generally mature pollen grains show a tendency to remain in groups in tetrads in *Annona;* in pairs in *Cananga,* and are shed as tetragonal or rhomboidal tetrads (Parulekar 1970). Exine sculpturing varies from psilate, verrucate, reticulate to echinate (Canright 1962); shed at 2-celled stage. Ovules anatropous, bi- or tritegmic, crassinucellate. Embryo sac of Polygonum type, 8-nucleate at maturity. Endosperm formation of the Cellular type and becomes horny and ruminate in the mature seed.

Chromosome Number.   Basic chromosome numbers are x = 7, 8, 9, 14 (Takhtajan 1987). Okada and Ueda (1984) report that among Asian genera of Annonaceae, 2n = 14 occurs only in *Mezzettia;* 2n = 16 is widespread and is also reported in *Anaxagorea* (with some primitive characters). 2n = 18 is reported for 11 genera, and tetraploidy (2n = 36) has been observed in *Polyalthia* species. Basic no. x = 8 or 9 has been suggested for the Asian genera.

Chemical Features.   Annonaceae members produce a wide range of non-alkaloidal compounds including carbohydrates, lipids, amino acids and proteins, polyphenols, essential oils, terpenes, aromatic compounds, and a host of other substances in addition to a large number of alkaloids (Leboeuf et al. 1984). Among the alkaloidal components, it is interesting that isoquinolines, which have benzylisoquinolines as in vivo precursors (Shamma and Moniot 1978), are the main alkaloidal constituents of the members of Annonaceae. Isolation of isoquinolines is significant from the chemotaxonomic point of view. The occurrence of the alkaloids—anonaine, xylopine, liriodenine and lanuginosine—in *Xylopia brasiliensis* and *Annona squamosa* suggests a chemotaxonomic relationship between these two taxa (Bhaumik et al. 1979).

Important Genera and Economic Importance.   The climber *Artabotrys odoratissima* climbs with the help of strong hooks borne on its flower stalks. *Polyalthia longifolia* is an ornamental tree with long, lanceolate, shiny waxy leaves and yellowish-green flowers in dense umbels. *Monodora myristica* of tropical Africa is the only genus of this family with polycarpellary, syncarpous ovary. Its flowers are also zygomorphic. The habit of the genus *Geanthemum* is remarkable; the flowers are borne on subterranean sucker-like shoots.

Economically, the fleshy fruits of various species of *Annona* are edible. The white custard-like pulp of custard apple (*Annona squamosa*) is sweet. The fruits of *A. reticulata* (bullock's heart) are so named because of their shape like a heart; these are sour to taste. The fruit of *A. muricata* is also sour to taste. *Artabotrys odoratissima, Unona discolor* and *Cananga odorata* are cultivated for their fragrant flowers. The essential oil, "ylang-ylang", is obtained from the flowers of *Cananga odorata,* and has widespread use in perfumery. Oils and fatty acids have also been isolated from fruits and/or seeds of various species of *Annona, Asimina triloba, Dennettia tripetala, Xylopia aethiopica, X. brasiliensis, X. longifolia* and from the leaves of *Annona muricata* and *A. senegalensis* (Hegnauer 1964, Leboeuf et al. 1984). These could possibly be used as edible oils after refining. Some products are used as spices, e.g. seeds of *Monodora myristica* as nutmegs, and *Xylopia aethiopica* as the source of Ethiopian pepper. The beautiful leaves of *Polyalthia longifolia* are often used for festive decorations.

Taxonomic Considerations.   The family Annonaceae is closely related to the Magnoliaceae; it differs from Magnoliaceae in having estipulate leaves, valvate corolla, and ruminate endosperm. There is general agreement that the Annonaceae is a derivative from Magnoliaceous stock. Hutchinson (1973) erected a separate order Annonales and included in it Annonaceae and Eupomatiaceae; he also pointed out that Annonales was related to, but more advanced than, the Magnoliales.

According to Cronquist (1968), within the Mangoliales, 5 families—Magnoliaceae, Winteraceae, Degeneriaceae, Himantandraceae and Annonaceae—form a cluster and share some common primitive characters. Later, in 1981, Cronquist changed the position of Annonaceae and placed it with the Myristicaceae and Canellaceae. Dahlgren (1980a) treats Annonales including Annonaceae, Myristicaceae, Eupomatiaceae and Canellaceae as the most primitive dicot order. Takhtajan (1969, 1980) at first placed Annonaceae in the Magnoliales, but in 1987 he also erected a separate order Annonales, comprising Annonaceae, Canellaceae and Myristicaceae. Takhtajan treats Annonales as an advanced group over the Magnoliales and Eupomatiales.

The families Magnoliaceae and Annonaceae have many common features: (a) tree habit (a few exceptions in Annonaceae), (b) multilacunar nodes, (c) Rubiaceous type of stomata (Ranunculaceous type in some Magnoliaceae), (d) bisexual, trimerous flowers, (e) embedded microsporangia, (f) monocolpate pollen grains (acolpate in some Annonaceae), (g) free carpels arranged spirally (exceptions are *Monodora* and *Isolona* of Annonaceae where gynoecium is syncarpous), (h) anatropous, bitegmic, crassinucellate ovules, (i) Polygonum type of embryo sac, (j) Cellular type of endosperm, and (k) follicular fruit. All these features strongly support the close relationship of Magnoliaceae and Annonaceae. At the same time, the differences between them are no less significant (see Table 20.5.1).

Table 20.5.1 Comparative Data for Magnoliaceae and Annonaceae

	Magnoliaceae	Annonaceae
1	Stamens with 3–7 vascular trace	Stamens with 1 vascular trace
2	Pollen grains released as monads	Pollen grains released in permanent tetrads
3	Endosperm not ruminate	Endosperm ruminate
4	Seeds not arillate	Seeds arillate
5	Basic chromosome number x = 19	Basic chromosome number x = 7, 8, 9, 14
6	Leaves stipulate	Leaves estipulate
7	Fruits follicular	Fruit a fleshy aggregate in which the follicles fuse with the floral axis
8	Vessels with scalariform perforation plates	Vessels with simple perforation plates

There is no doubt that the Annonaceae belong to the Magnolian stock, but as the Annonaceae have several features advanced over the Magnoliaceae, the erection of a separate order Annonales is justified.

Eupomatiaceae

Eupomatiaceae is a unigeneric family. The genus *Eupomatia*, with two species is distributed in New Guinea and eastern Australia. *Eupomatia laurina* has very wide distribution. It ranges from temperate Victoria in southern Australia to tropical Queensland and New Guinea. In New Guinea it occurs from sea level up to 1300 m altitude and is one of the common shrubs in Australian rain forests. *E. bennettii* is much less common and is distributed from New South Wales to southern Queensland. It does not occur in New Guinea.

Vegetative Features.   Shrubs or small trees, sometimes with tuberous roots. Leaves alternate, simple, entire, estipulate.

Floral Features.   Flowers fairly large, solitary, terminal, perigynous, bisexual. Sepals and petals indistinct and form a deciduous calyptra on the rim of the expanded concave torus (Hutchinson 1959). According to Willis (1973), the calyptra is a modified bract and perianth is absent. Stamens numerous, perigynous, the inner ones sterile and petaloid, and the few outer ones fertile with two linear extrorse anther lobes and acuminate connective. Gynoecium many-carpellary, syncarpous, enclosed in a hollow receptacle or immersed in the turbinate receptacle (Hutchinson 1959), with flat top. From the areolate upper part the stigmas project shortly. Ovules several, ventrally attached. Fruit an urceolate-turbinate berry, apex truncate and margined by annular base of deciduous calyptra with 1 or 2 seeds per locule. Seeds angular, pale-brownish with copious ruminate endosperm, and a very small embryo.

Anatomy.   Vessels small, solitary and in short to long radial multiples; perforation plates scalariform, intervascular pitting scalariform to opposite. Parenchyma sparse, diffuse and paratracheal; rays very broad and high; fibres septate. Leaves weakly dorsiventral, almost glabrous. Stomata of Rubiaceous type confined to the lower surface. Secretory cells with amorphous contents scattered in the ground tissue of the petiole and cortex.

Embryology.   The two-celled pollen grains may germinate in situ (Johri et al. 1992). Ovules anatropous, bitegmic and crassinucellate with the micropyle formed by the inner integument in *Eupomatia laurina* (Mohana Rao 1983); Polygonum type of embryo sac, 8-nucleate at maturity. Endosperm formation of the Cellular type, embryo small, embedded in abundant oily endosperm.

Chromosome Number.   Haploid chromosome number is n = 10.

Important Genera and Economic Importance.   *E. laurina* is a shrub up to 5 m high and yields valuable timber. Fruits are edible (A.B. Cribb and J.W. Cribb 1975) and taste aromatic, like *Psidium guajava* (guava). Webb and Tracey (1981) regard *Eupomatia* as adapted to bird dispersal. However, as the fruits show a syndrome of mammalichory—strong smell, colour not brilliant, dropping when ripe or remaining but still accessible from the ground by larger mammals—they are probably dispersed by both mammals and birds (Endress 1983).

Taxonomic Considerations.   Bentham and Hooker (1965a) included Eupomatiaceae in the Annonaceae. Although *Eupomatia* has some features in common with the Annonaceae (ruminate endosperm), it is generally accepted as the only genus of a distinct family Eupomatiaceae. The Eupomatiaceae differ from Annonaceae in wood anatomy: absence of diaphragms composed of stone cells in the pith, the bast not stratified into hard and soft zones, and presence of bordered pits in wood parenchyma (Garatt 1933). Lemesle (1936, 1937), on the basis of floral morphology and wood anatomy, supported this alliance. He further pointed out that the Eupomatiaceae are among the most primitive of the angiosperms, more primitive than even Magnoliaceae. The immersion of otherwise free carpels in the expanded receptacle is probably a parallel to the similar condition in the Nymphaeaceae (Hutchinson 1959).

According to Willis (1973), Eupomatiaceae is an isolated and ancient group without close relatives, except perhaps distantly to the Calycanthaceae. Endress (1983) considers Eupomatiaceae to be one of the "small archaic relic families of the angiosperms". With less than 4 years, *Eupomatia* has the shortest life-span amongst all these members. Apart from rain forests, two species of *Eupomatia* also occur in the transition region between rain forests and sclerophyll forests, and they seem to be adapted to temporal droughts by a xylopodium with accessory buds (*E. laurina*) or by root tubers (*E. bennettii*).

The large indehiscent fruits are considered as typical of primitive families of the Magnoliales, and also good adaptations in rain forest habitats. For this reason, Endress (1983) considers Eupomatiaceae to be close to other archaic families like Austrobaileyaceae, Himantandraceae, Calycanthaceae, etc. Cronquist (1981) and Takhtajan (1980, 1987) consider Eupomatiaceae to be allied to the Himantandraceae. Cronquist (1968), however, pointed out that Eupomatiaceae is anomalous in both Magnoliaceous and Lauraceous groups, and included in the latter group with some reservation. Takhtajan (1987) erected a separate order, Eupomatiales, to accommodate this family between Magnoliales and Annonales.

This family is more allied to the Himantandraceae and should be placed close to it. The alliance with Annonaceae is only superficial.

Myristicaceae

For a long time the family Myristicaceae was considered to be unigeneric; Warburg (1897) recognised 15 genera and ca. 260 species. These genera are widely distributed in the tropics of both the New and the Old World.

Vegetative Features.   Dioecious or monoecious, evergreen trees or shrubs with usually aromatic wood and foliage; inner bark often exudes a reddish liquid. Leaves alternate, simple, entire, coriaceous, often with glandular dots, estipulate in *Myristica fragrans*. Leaves white or glaucous beneath.

Floral Features. Inflorescence racemose, corymbose, fasciculate or capitate, axillary, supra-axillary or sometime from the axils of old leaves. Flowers unisexual, actinomorphic, hypogynous, pedicellate, bracteate, bracteolate; bracteoles embrace the base of calyx. Perianth represented by a saucer- to funnel-shaped and usually 3-lobed calyx, gamosepalous. Staminate flowers with 2–30 stamens, filaments united into a solid column; anthers bicelled, extrorse, distinct or connate, dehisce longitudinally. Pistillate flowers with only one pistil; ovary superior, monocarpellary and unilocular with one basal ovule; style 1 or almost absent, stigmas connate into a sulcate mass. Fruit a dehiscent drupe; seed partially or completely enveloped by a brightly coloured laciniate fleshy aril. Seed large, embryo small, endosperm copious and ruminate.

Anatomy. Vessels mostly medium-sized, solitary and in small multiples, perforation plates both simple and scalariform; intervascular pitting varies from scalarifrom to alternate. Parenchyma paratracheal, scanty to almost vasicentric, occasionally with oil cells. Rays up to 2–3 cells wide, usually markedly heterogeneous, oil or mucilage cells sometimes present. Leaves dorsiventral, hairs are sympodially branched uniseriate trichomes of various forms. Stomata Rubiaceous type, confined to the lower surface.

Embryology. Pollen grains monosulcate, 2-celled at shedding stage. Ovules anatropous, bitegmic, crassinucellate; micropyle constituted by the inner integument alone; Polygonum type of embryo sac, 8-nucleate at maturity. Endosperm formation of the Nuclear type, but later it becomes cellular and ruminate. The aril develops from the outer integument.

Chromosome Number. Basic chromosome numbers are x = 18, 21, 25. For *Myristica fragrans,* haploid chromosome number is n = 21 and for *Pycnanthus angolensis* it is n = 19 (Bhandari 1971).

Chemical Features. *Myristica fragrans* is rich in an aromatic volatile substance called myristicin. This is a biologically interesting molecule, since it reputedly has hallucinogenic activity in man (Harborne and Turner 1984).

Important Genera and Economic Importance. *Myristica fragrans,* a handsome aromatic evergreen tree, grows to a height of 9 to 12 m (under cultivation). The flowers are small, pale yellow and fleshy. The fruits are like large apricots and orange-yellow when ripe. It is cultivated in both East and West Indian Islands for its fruits called nutmegs. The large shiny black seeds constitute the commercial nutmeg and the orange much dissected aril, the commercial mace. Both are used as a spice.

The nutmeg from the East Indian Islands are better in essential oil content and flavour, but poorer in fatty oil content. The essential oil of nutmeg and mace contains a highly toxic substance myristicin and therefore should be used carefully. The seeds and aril also contain a fatty acid, myristic acid $C_{14}H_{28}O_2$, which is used in the manufacture of soaps, perfumes and ointments. Nutmeg is reported to have hallucinogenic properties. As few as two nutmeg seeds may be fatal (Kochhar 1981).

Taxonomic Considerations. Earlier workers like Hallier (1912) and Bessey (1915) placed Myristicaceae in the Ranales. Wettstein (1935) indicated its close relationship with the Annonaceae. This alliance has been accepted by Joshi (1946), Lawrence (1951), Melchior (1964), Dahlgren (1975a, 1980a), Cronquist (1968, 1981) and Takhtajan (1980, 1987). However, Hutchinson (1969, 1973) includes the Myristicaceae in the Laurales, segregating it from Annonaceae, which he includes in the Annonales. Wood anatomical features also suggest its closer relationship with the Laurales (Garrat 1933, Metcalfe and Chalk 1972). The two families—Annonaceae and Myristicaceae—are closely related and resemble each other in a number of features: (a) uniaperturate pollen grain, (b) stamens not laminar, (c) leaves estipulate, (d) perianth mostly trimerous, (e) aromatic wood, (f) tri- or multilacunar nodes, (g) endosperm ruminate, and (h) embryo small. These families may have originated from the same ancestral stock.

Canellaceae

The Canellaceae is a small family of 5 genera and ca. 11 species distributed in the tropical regions of both the hemispheres—America, East Africa and Madagascar.

Vegetative Features. Medium-sized glabrous, aromatic trees. Leaves alternate, simple, entire, coriaceous, glandular-punctate, estipulate.

Floral Features. Inflorescence solitary axillary or in small groups. In *Canella* it is a terminal raceme (Fig. 20.8A) or cyme; in *Cinnamodendron* the racemes are axillary. Flowers bisexual, hypogynous, actinomorphic, often subtended by three persistent imbricate bracts. Calyx of 3 (rarely 4 or 5) distinct fleshy imbricate sepals. Corolla of variable number of petals. *Canella* has 5 petals, fused at the base and arranged quincuncially (Fig. 20.8B); *Pleodendron* has 12 petals in 2 series of 6 each; outer petals larger

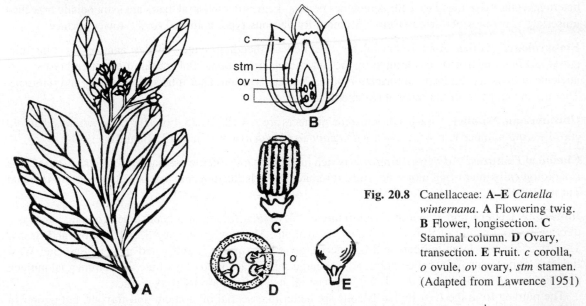

Fig. 20.8 Canellaceae: A–E *Canella winternana.* **A** Flowering twig. **B** Flower, longisection. **C** Staminal column. **D** Ovary, transection. **E** Fruit. *c* corolla, *o* ovule, *ov* ovary, *stm* stamen. (Adapted from Lawrence 1951)

and thicker than inner ones. Within each series, there is alternation of large and small petals. *Capsicodendron* has 6 to 8 petals in 2 series of 3 or 4 each. Petals 6 or 10 in *Cinnamodendron,* in 2 series; outer larger ones alternate with inner smaller ones. *Warburgia* has 10 petals in 2 series of 5 each—outer quincuncial and inner rotate. In *Cinnamosma* 5 or 6 petals are united into a tube at least half their length (Wilson 1966). Petals thick, fleshy and dark red or purple. Stamens variable (12–20), united into a tube by their filments (Fig. 20.8B, C) which almost surround the pistil; anthers bilobed, each lobe attached by its entire length to the filament and has two microsporangia; dehiscence by longitudinal slits, extrorse. Gynoecium syncarpous, 2- to 5-carpellary, ovary superior, unilocular with parietal placentation (Fig. 20.8B, D); style short and thick, stigma lobes as many as carpels (except in *Cinnamodendron*). Long glandular stigmatic hairs and stylar canals correspond to the number of carpels. Stylar canals are open and unobstructed from the tip of style up to the ovary. Just before flowering the canals become filled with a dense hairy growth. These hairs prevent the pollen grains from penetrating the ovary directly (Wilson 1964). Ovules 2 to many on each placenta. Fruit a berry (Fig. 20.8E); seeds with oily endosperm and a straight or slightly curved embryo.

Anatomy. Vessels moderately small, solitary and few; perforations scalariform. parenchyma diffuse and contains numerous oil or mucilage cells. Rays uniseriate or up to 3 or 4 cells wide, usually homogeneous or of square and upright cells only. Fibres with conspicuous bordered pits, moderately to very long. Leaves mostly dorsiventral, stomata confined to the lower surface; Rubiaceous type in *Canella* and *Cinnamodendron*, and Ranunculaceous in *Cinnamosma fragrans*. Clustered crystals of calcium oxalate usually present in both stem and leaf.

Embryology. Pollen grains monocolpate, 2-celled at shedding stage. A large number of pollen grains germinate in situ. Ovules hemianatropous, bitegmic, crassinucellate; Polygonum type of embryo sac, 8-nucleate at maturity. Endosperm formation of the Cellular type. Mature endosperm copious and ruminate; ruminations caused by the seed coat, non-vascularized.

Chromosome Number. Haploid chromosome numbers are n = 11, 13, 14 (Takhtajan 1987).

Important Genera and Economic Importance. *Canella* is an important genus of Canellaceae. The bark of *C. alba* is used medicinally. The wood of *Cinnamosma fragrans* is scented and often used in religious ceremonies (Metcalfe and Chalk 1972).

Taxonomic Considerations. The flowers of Canellaceae are distinct and have: (a) thick fleshy sepals and petals, (2) oil cells, and (3) monadelphous stamens, extrorse longitudinally dehiscent anthers, and (4) syncarpous ovary with parietal placentation.

Most treatments ally the Canellaceae either with the Parietales or the Ranales. They resemble the Parietales in parietal placentation and trilacunar nodes. Common Ranalian features are: (1) relatively low level specialization of wood, (2) unspecialised phloem, (3) spiral insertion of some floral parts, and (4) thick, fleshy sepals and petals (Wilson 1966). On the basis of wood anatomy, Vestal (1937) suggested the placement of Canellaceae close to the Myristicaceae and other arboreal Ranales. From the study of wood of Myristicaceae, Garratt (1933) concluded that the Canellaceae is far removed from Myristicaceae. The presence of oil cells or mucilage cells, thick petals and sepals, and parenchyma suggest affinity with Magoliaceae and Lauraceae (Metcalfe and Chalk 1972).

Although the family Canellaceae has been placed close to Violaceae in the Parietales (Lawrence 1951), it is more allied to Magnoliales (in spite of syncarpous ovary). "The trilacunar nodes, hypogynous flowers, mostly inaperaturate pollen and endospermous seeds of Canellaceae ally them to the Magnoliaceous rather than Lauraceous cluster of families in the Magnoliales" (Cronquist 1968). Takhtajan (1969, 1980, 1987) also includes Canellaceae in the Magnoliales. Some doubt has been expressed about this alliance because of the syncarpous ovary in Canellaceae. A similar condition is met with in two genera of Annonaceae also—*Isolona* and *Monodora* of Annonaceae—have syncarpous ovary.

Schisandraceae

Schisandraceae is a very small family of only 2 genera, *Schisandra* and *Kadsura*, and 47 species, well distributed and indigenous to eastern and southeastern Asia and Malaysia. Only one species, *S. glabra*, is American, found on steep slopes of ravines in a few places like South Carolina, Georgia, northwestern Florida, Alabama, Tennessee and Louisiana.

Vegetative Features. Mostly clambering or twining (Fig. 20.9A) woody monoecious or dioecious vines. Leaves alternate, simple, deciduous, estipulate.

Floral Features. Flowers solitary, axillary, pedicellate (Fig. 20.9A), unisexual, actinomorphic, ebracteate, ebracteolate. Perianth of few to many, undifferentiated, free tepals, in 2 to many series (Fig. 20.9C). Staminate flowers with 4 to 80 stamens, filaments basally connate into a modified column (Fig. 20.9D) or wholly united into a fleshy globose mass. Anthers bicelled, basifixed, dehiscence longitudinal. Pistillate

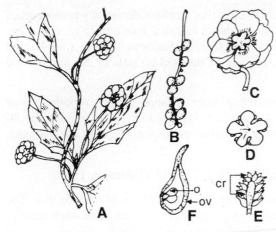

Fig. 20.9 Schisandraceae: **A–F** *Schisandra coccinea* **A** Flowering twig. **B** Fruiting peduncle. **C** Staminate flower. **D** Androecium, top view. **E** Gynoecium, vertical section. **F** Carpel, longisection. *cr* carpel, *o* ovule *ov* ovary, (Adapted from Lawrence 1951)

flowers with 12 to 300 free, acyclic pistils arranged in a head (Fig. 20.9E). Later, in fruit, the pistils spread along an elongated axis (Fig. 20.9B). In *Kadsura*, however, the carpels in fruit remain in a globose or ellipsoid head. Each ovary superior, unilocular, monocarpellary, conduplicate; ovules mostly 2 to 5, ventrally attached and pendulous (parietal) (Fig. 20.9F). Ovary is not fully closed along the ventral margin and the 2 parallel edges bear a ciliate stigmatic crest, its distal end usually projects into an unvascularised pseudostyle. Fruit aggregate, baccate, composed of an elongated axis and sessile drupe-like fruits attached to it (Fig. 20.9B). Seed usually 1 to 5, with a small embryo and copious endosperm.

Anatomy.   Vessels large and solitary, 1 to 15 per mm^2 in *Kadsura* and ca. 60 per mm^2 in *Schisandra*; with spiral thickenings. Perforation plates scalariform with up to 15 bars in *Schisandra* and up to 7 bars in *Kadsura*. Parenchyma in terminal bands, 1 to 3 cells wide. Rays also 1 to 3 cells wide, up to 1 mm high and heterogeneous, with uniseriate margins of 1 to 7 cells. Enlarged oil cells occur in the marginal rows. Stomata present on both surfaces in *Kadsura* and only on lower surface in *Schisandra*.

Embryology.   Pollen grains tri- or hexacolpate, with 3 small and 3 large meridianly arranged furrows. The large colpi fuse at one end to form a triradiate mark; shedding at 2-celled stage. Ovules anatropous, bitegmic, crassinucellate; Polygonum or Allium type of embryo sac, 8-nucleate at maturity. Both the integuments form the micropyle. Endosperm formation of the Cellular type.

Chromosome Number.   The haploid chromosome number is n = 14.

Important Genera and Economic Importance.   The two genera, *Kadsura* and *Schisandra* have no economic importance.

Taxonomic Considerations.   Bentham and Hooker (1965a), Engler and Prantl (1887–1915), and Rendle (1925) included *Schisandra* and *Kadsura* in a tribe Schisandreae, or a subfamily Schisandroideae of the Magnoliaceae. On the basis of morphology, wood anatomy and chromosome numbers, McLaughlin (1933), Whitaker (1933), Lemesle (1945), Ozenda (1946), Smith (1947) and Bailey et al. (1948) concluded that Schisandreae should be raised to family rank as Schisandraceae. This suggestion has been accepted by Lawrence (1951), Hutchinson (1969, 1973), Melchior (1964), Cronquist (1968, 1981), Dahlgren (1975a, 1983a) and Takhtajan (1980, 1987).

Presently, the families Illiciaceae and Schisandraceae are included in the order Illiciales, even though they differ in a number of characters (see Table 20.9.1).

Kapil and Jalan (1964) pointed out that as these two families differ from each other in many features, there cannot be a close relationship between them, as suggested by Whitaker (1933), Smith (1947),

Table 20.9.1 Comparative Data for Illiciaceae and Schisandraceae

Illiciaceae	Schisandraceae
1 Shrubs or small trees	Straggling or twining vines
2 Mostly dioecious	Mostly monoecious
3 Pseudosiphonostelic stem with poorly developed pericycle	Eustelic stem with well-developed pericycle
4 1-traced, unilacunar nodes	3-traced, unilacunar nodes
5 Pseudoverticillate leaves	Alternate leaves
6 Pitted sclereids without crystals	Non-pitted sclereids with crystals
7 Stamens free; hexacolpate pollen	Stamens monadelphous; tricolpate pollen
8 Whorled carpels with styles	Spirally arranged carpels, without styles
9 Fruit a follicle	Fruit a berry

Bailey and Nast (1948). According to Eames (1951), the three families—Magnoliaceae, Illiciaceae and Schisandraceae—have probably been derived from a common ancestral form and the last two families are more specialized. The view of Smith (1947) that "*Illicium* has no close allies other than *Schisandra* and *Kadsura*" is corroborated by the chromosome number of n = 14; however, the karyotypic analysis of the two families does not support this statement (Stone and Freeman 1968, Stone 1968).

The above discussion indicates that Scisandraceae and Illiciaceae should be removed from the Magnoliales because: (1) stipules are absent, and (2) the basic chromosome number is x = 14 (and not x = 19) and assigned to a separate order Illiciales (see also Illiciaceae).

Illiciaceae

Illiciaceae is a monotypic family with the genus *Illicium* (ca. 42 species) distributed in both the hemispheres. It occurs in the northern hemisphere from Assam and Sikkim in Northeastern India to China, Japan and Korea, and in the southern hemisphere to Borneo in the Old World and Florida, Mexico and the West Indies in the New World.

Vegetative Features. Shrubs or small trees. Leaves alternate, simple, entire, pinnately nerved, often aromatic, estipulate; sometimes clustered at distal nodes.

Floral Features. Flowers not inflorescent, mostly solitary (Fig. 20.10A) or in 2's or 3's, axillary or supra-axillary, rarely on the main stem. Flowers bisexual, actinomorphic (Fig. 20.10B), hypogynous, ebracteate, ebracteolate, pedicellate. Perianth undifferentiated into calyx and corolla. Tepals 7 to 33, in many series, imbricate, inner gradually larger and transitional towards formation of stamens. Stamens rarely as few as 4, usually numerous, 1- to many-seriate. Anthers bicelled, basifixed, dehiscence longitudinal (Fig. 20.10C); connective sometimes enlarged and extends beyond the thecae. Gynoecium apocarpous, polycarpellary, number of carpels vary from 7-15-21, and arranged in one whorl. Ovary superior, unilocular, ovule 1 (Fig. 20.10D) near the base, on parietal placenta; style tapering, conduplicate and its ventral surface distally stigmatic. Fruit a follicetum (a group of lightly cohering follicles) of one to numerous seeded dehiscent follicles borne on short, thick peduncles; seed glossy, light-brown, with copious endosperm and a minute embryo. Roberts and Haynes (1983) reported ballistic seed dispersal in *Illicium*. The dehisced fruits often persist on the peduncle after seed ejection.

Anatomy. Vessels solitary, perforation plates scalariform, with numerous fine bars; intervascular pitting scalariform to opposite. Parenchyma sparse, paratracheal and sometimes scattered along the ring-boundary.

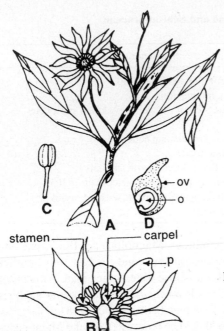

stamen — — carpel

Fig. 20.10 Illiciaceae: **A–D** *Illicium floridanum*. **A** Flowering twig. **B** Flower, longisection. **C** Stamen. **D** Carpel, longisection. *o* ovule, *ov* ovary, *p* perianth. (Adapted from Lawrence 1951)

Rays up to 3 cells wide, with numerous uniseriates, heterogeneous. Leaves glabrous with thick cuticle on both surfaces. Stomata confined to the lower surface and without clearly diffentiated surrounding cells.

Embryology.   Pollen grains tricolporate, shed at 2-celled stage. Ovules anatropous, bitegmic, crassinucellate; Polygonum type of embryo sac, 8-nucleate at maturity. Endosperm of the Cellular type.

Chromosome Number.   Haploid chromosome number n = 13 for *Illicium floridanum* and n = 14 for *I. parviflorum* (Stone and Freeman 1968). Origin of n = 13 in *I. floridanum* is by aneupolidy from n = 14 of *I. parviflorum*.

Important Genera and Economic Importance.   *Illicium verum* (Chinese star) bears fruits and seeds with a flavour similar to that of anise or *Pimpinella anisum* of Umbelliferae (Hutchinson 1969). The extracted oil is used as a flavouring agent and carminative.

Taxonomic Considerations.   The Illiciaceae is treated as a tribe of the Magnoliaceae (Bentham and Hooker 1965a, Engler and Diels 1936, Rendle 1925). Hutchinson (1926) included *Illicium* in the Winteraceae. Smith (1947) and Bailey et al. (1948) do not agree with this assignment.

According to Cronquist (1968), Illiciaceae and Schisandraceae form a closely knit group which differs from the rest of the Magnoliales in unilacunar nodes—a primitive feature—triaperturate or more advanced pollen, and stamens with well-differentiated filament and anther. At the same time, neither of the characters is ancestral to the other. Hutchinson (1969, 1973) recognised Illiciaceae as a family distinct from the Winteraceae; he included both Illiciaceae and Schisandraceae in the Magnoliales.

Dahlgren (1980a, 1983a) and Takhtajan (1980, 1987), however, prefer to include these two families in a separate order, Illiciales. On the basis of differences between these two families and other members of the Magnoliales, these two families should be placed in a distinct order, Illiciales.

Austrobaileyaceae

Austrobailleyaceae is a monotypic family with the genus *Austrobaileya* which has a very restricted distribution in tropical rain forests of North Queensland in Eastern Australia, at altitudes between 380 and 1100 m. It is centred around Bartle Frere (Boonjee), Mt. Koolmoon and Bellenden Ker (around 17.30ft), farther north on Mt. Levis and Mt. Spurgeon (including Mt. Misery, around 16.30ft). Although 2 species—*A. scandens* and *A. maculata*—were recognised earlier, according to Endress (1983), *A. maculata* described from Mt. Spurgeon (White 1948) is synonymous with *A. scandens*.

Vegetative Features. A woody vine, leaves opposite-decussate to start with but later (during subsequent growth) their position may change; entire, glabrous, pinnately veined, coriaceous at maturity, petiolate, estipulate.

Floral Features. Single flowers borne in the axils of foliage leaves, more rarely of bracts; not inflorescent but sometimes aggregation of 2 or 3 flowers occurs. Open flowers are pendant. Perianth lobes numerous, free, undifferentiated into calyx and corolla. Tepals overlap each other and the larger ones embrace the flower bud right over the axis. There is a gradual transformation—in size, shape, coloration pattern, vascular pattern and histology—from the outermost to the innermost tepal. Outer tepals glossy green, the inner ones yellowish-green with red dots. Androecium consists of stamens and inner staminodes which vary in shape and size. They change from relatively flat to strongly boat-shaped, and from almost acute or truncate to rounded at the apex. Innermost stamens resemble the inner staminodes. Sometimes the last stamen may have only one theca. The surface of stamens and staminodes are papillose; on both surfaces of stamens scattered stomata differentiate. Each theca opens by a longitudinal slit. The floral centre appears dark due to numerous black-purplish spots on the yellowish stamens and more so on the inner staminodes. Carpels bright yellow but not visible from the outside, as they are covered by the staminodes. Gynoecium apocarpous, 6-8-14-carpellary. Each carpel has roughly the form of an oblique wine bottle (Endress 1980a). Styles representing the bottle neck are excentric; styles and stigmas of all the carpels, therefore, converge. Each carpel has a massive solid base taking up about half the length of the ovary. Ovules arranged alternately in two longitudinal rows at the adaxial side of the carpel. Before anthesis the stylar canal and the bilobed carpellary tip (or stigma) become papillose and secretory. A mass of mucilage (or a massive cap of mucilage) is produced in which all styles and stigmas become embedded. This helps in transmitting the pollen tubes. The number of ovules per carpel varies from 4 to 10; and only a few develop into fruit. Each forms a stalked ellipsoid or globose fleshy berry, up to 8 cm long and 4 cm broad, indehiscent; skin of the berry orange and the flesh yellow. Ripe fruits have a thin, tough and leathery skin and a thin layer of mealy-pulpy flesh around the seeds (Endress 1983). Seeds lenticular with a small dicot embryo and a massive starchy and ruminate endosperm.

Anatomy. Vessels solitary or in tangential or oblique pairs, with spiral thickenings, angular; perforation plates scalariform or scalariform and simple, intervascular pitting scalariform to opposite. Parenchyma only terminal, rays up to 3 cells wide, heterogeneous. Leaves dorsiventral, stomata occur more on the lower surface, have large oval-shaped guard cells but subsidiary cells are absent; nodes unilacunar.

Embryology. Pollen grains monosulcate, mature grains spherical with a single germinal furrow running nearly from pole to pole; exine thick, finely pitted, shed at 2-celled stage. Ovules anatropous, bitegmic, crassinucellate. At the time of anthesis the embryo sac is mature, presumably of the Polygonum type (Johri et al. 1992) and is 8-nucleate (Endress 1980a); micropyle is formed either by the inner integument alone or by both.

Chromosome Number. Somatic chromosome number of $2n = 44$ for *Austrobaileya* is reported by Rüdenberg (1967). Chromosomes are highly asymmetrical and of unequal size.

Important Genera and Economic Importance.   The single monotypic genus *Austrobaileya scandens* is a vine, a dominant plant locally in the rain forests of Eastern Australia. The fruits smell like pumpkin and taste unpleasant; the ruminate seeds resemble those of *Asimina* of the Annonaceae (Endress 1983).

Taxonomic Considerations.   White (1933) included *Austrobaileya* in the Magnoliaceae; Croizat (1940)— misled by some superficial resemblances—with *Hibberetia scandens*, removed it to the Dilleniaceae, but later erected a new family Austrobaileyaceae (without any further comments).

Following the work of Bailey and Swamy (1949), the family has almost unanimously been placed at the beginning of the Laurales near the Monimiaceae s.l. (Hutchinson 1969, Thorne 1976, 1983, Dahlgren 1975a, 1980a, 1983a, Takhtajan 1969, 1980). Cronquist (1968) emphasised its isolated position and did not include it in either Magnoliales or Laurales. Takhtajan (1987) erected a separate order Austrobaileyales for this family in between Magnoliales and Laurales.

According to Endress (1980a), the flowers and fruits of this genus certainly support its isolated position among the woody Ranales, but at the same time they have closer relationship with the Magnolialian and not the Lauralian families, mainly with Annonaceae. However, it should definitely not be included in the Annonaceae. A comparison of *Austrobaileya* with Magnoliales and Laurales would make it clear (see Table 20.11.1)

Table 20.11.1   Comparative Data for *Austrobaileya*, Magnoliales and Laurales

	Austrobaileya	Magnoliales	Laurales
1	Flower large, solitary	Same	Same, except *Idiospermum*
2	Flower sapromyophilous	Only some in Annonaceae	Flowers not sapromyophilous
3	Floral cup absent	Same	Deeply-urceolate hypanthium present
4	Pollen anasulcate	Same	Pollen not anasulcate
5	Carpels extremely ascidiate	Only in some Magnoliaceae	Only in Trimeniaceae
6	Carpellary apex transversely bilobate	In *Goniothalamus* (Annonaceae) only	Carpellary apex not bilobate
7	Stigma papillate and heavily secretory	In Annonaceae only	In Monimiaceae; stigma plumose, hairs tanniniferous in Trimeniaceae
8	Pollen tube transmitting tissue hairy papillate	In Annonaceae only	Not so
9	Several ovules in each carpel	Same	One erect or pendulous ovule in each carpel
10	Seeds ruminate	Same	Seeds not ruminate
11	Fruitlets berries with a long stalk	In some Annonaceae only	Fruitlets in some Monimiaceae are drupes not berries

The characters that separate *Austrobaileya* from Annonaceae are:

a) Perianth spiral (alternate whorls in Annonaceae)
b) Tepals imbricate (at least outer tepals valvate)
c) Carpels ascidiate (epeltate in Annonaceae)
d) Stamens extremely laminar (± terete or weakly laminar in Annonaceae)
e) Pollen anasulcate, exine reticulate-rugulate (different from this) in Annonaceae

Ehrendorfer et al. (1968), on the basis of chromosome pattern, and Walker (1976), on the basis of palynological features, include Austrobaileyaceae in the Laurales as a primitive family.

Austrobaileyaceae has an extremely isolated position within the woody Ranales, where they form an independent line between Magnoliales and Laurales, but approach the former more than the latter. Its monotypic relic nature, with a very limited distribution, is rather interesting. *Austrobaileya* shows very primitive pollen morphology. The pollen resembles the earliest known angiosperm pollen fossils, *Clavatipollenites* from the lower Cretaceous (Endress and Honegger 1980).

Austrobileyaceae are one of the most prominent "living fossils" among the angiosperms (Endress 1983).

Monimiaceae

Monimiaceae is a family of 32 genera and 330 species (Takhtajan 1987), distributed mainly in the tropics of the southern hemisphere. They are absent from India proper and are very sparsely represented in Africa (Hutchinson 1969); mostly Australia, Polynesia, Madagascar and Oceania, South America and Mexico are the main areas of distribution.

Vegetative Features.   Evergreen trees or shrubs. Leaves entire, mostly opposite-decussate, rarely alternate as in *Tambourissa* spp; simple, pinnately veined, commonly aromatic, coriaceous; estipulate.

Floral Features.   Inflorescence cymose or flowers solitary; short axillary cymes in *Monimia ovalifolia* (Fig. 20.12A); terminal cyme in *Boldea boldus* (Fig. 20.12B). Flowers bisexual in most genera but unisexual in some e.g. *Boldea* (Fig. 20.12C, D); actinomorphic, perigynous, ebracteate, ebracteolate, pedicellate. Perianth, when present, inconspicuous (mostly absent), of only one whorl and calyciform or biseriate and inner whorl petaloid, borne on the rim of convex to deeply urceolate hypanthium (Fig. 20.12C, D); number of tepals 4–8–numerous. Stamens numerous (7–22) in 1, 2 or many series, or scattered over the inner hapanthium surface in staminate flowers. At anthesis the stamens have a linear oblong anther and an extensive, elongated filament; lateral glandular appendages occur on filaments. Anthers bicelled, basifixed (Fig. 20.12F), slightly latrorse or introrse, dehiscence longitudinal or transverse or by 2 flap-like upturned valves, as in *Doryphora sassafras* (Fig. 20.12E). Gynoecium apocarpous, many carpellate; ovaries superior or half-inferior, unicarpellary, unilocular with one erect or pendulous ovule (Fig. 20.12G), placentation parietal; style 1, usually short and thick; stigma 1, terminal. Staminodes often present in pistillate flowers. Fruit variable, an achene, or drupe enclosed by the hypanthium which may become fleshy, or a drupe enclosed in persistent calyx, or pistils distinct and separate in fruits. Seed with minute embryo and copious oily endosperm.

Anatomy.   Vessels usually moderately small, mostly solitary, with slight to very oblique end-walls, rarely simple but usually scalariform perforation plates with 10 or more bars; intervascular pitting scalariform to alternate. Rays with moderately numerous uniseriates, usually up to 3 to 4 cells wide, weakly to distinctly heterogeneous; parenchyma mostly apotracheal. Leaves dorsiventral, hairs exclusively non-glandular; stomata usually Ranunculaceous type, confined to the lower surface.

Embryology.   Pollen grains vary in form from spherical to ellipsoidal to ovoid. Size varies from 10 to 15 μm to between 40 to 50 μm. There is variation in size and form of pollen grains even within the same anther. Exine structure variable. Grains mono-, di- or acolpate. Pollen grains of *Peumas* are echinate, those of *Hortonia* have spiral bands on surface. Ovules anatropous, unitegmic (*Siparuna*) or bitegmic (*Peumas*), crassinucellate; nucellar cap and hypostase present (Bhandari 1971). Embryo sac of Polygonum type, 8-nucleate at maturity. Endosperm formation of the Cellular type in *Peumas boldas*.

Chromosome Number.   Basic chromosome number is x = 11.

Chemical Features.   Secondary metabolities like allyphenols are reported (Gottlieb et al. 1989).

Important Genera and Economic Importance.   The members of this family exhibit an unusually wide range of variability. In *Glossocalyx,* opposite to each normal foliage leaf is a reduced filiform leaf which

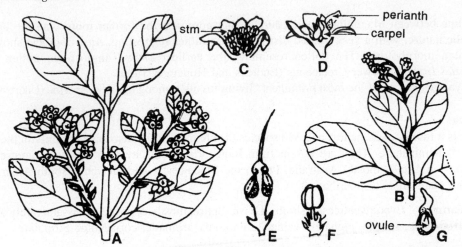

Fig. 20.12 Monimiaceae: **A** *Monimia ovalifolia*, **B-D, F, G** *Boldea boldus*, **E** *Doryphora sassafras*. **A** Fruiting branch. **B** Flowering branch. **C** Staminate flower, longisection. **D** Pistillate flower, vertical section. **E, F** Stamens. **G** Carpel, vertical section. *stm* stamen. (**A** after Hutchinson 1969, **B-D, E-G** after Lawrence 1951)

is caducous. Therefore, the overall appearance is as if it has alternate leaves. In some genera the flowers are highly specialised (Endress 1979, 1980c, Johri and Ambegaonkar 1984). In *Tambourissa*, the unisexual flowers are enclosed in an urceolate cup, which appears to be fig-like. The carpels are situated around the floral cup embedded in the fleshy tissue. The secretory transmitting tissue covering the carpels is continued into the channel at the mouth of the floral cup. This constitutes the "hyperstigma" (Johri et al. 1992). The hyperstigma gives protection against predation damage caused by pollinating insects (Favre-Duchartre 1984). Similar feature is also found in *Hennecartia, Wilkiea* and *Kibara* (Endress 1980c).

Laurelia has one species in New Zealand—*L. novae-zelandae* (Sampson 1969), and the other is endemic to Chile. There are very few economic products. Boldo leaves from the Chilean shrub *Peumas boldus* are fragrant and contain some essential oil. The fruit is sweet and edible. The wood of *Doryphora sassafras* and *Atherosperma moscata* is attractive and used for furniture making (Hutchinson 1969).

Taxonomic Considerations. Money et al. (1950) divided Monimiaceae s.l. into 4 subfamilies. All four subfamilies have been elevated to family status (at one time or another) by different phylogenists. Walker (1976) treated Atherospermatoideae as a distinct family, while retaining the other three as subfamilies of the Monimiaceae. *Hortonia* of Sri Lanka is the only member of the subfamily Hortonioideae which has been raised to family level, by Smith (1971). The subfamily Monimioideae is treated as Monimiaceae s.s. and is the largest including 24 genera. Another subfamily, Siparunoideae, was raised to family rank by Schodde (1970).

Atherospermataceae is very close to the Monimiaceae, especially to *Hortonia*, from which it differs in: (a) disulcate pollen grains (Walker 1976), (b) anthers that open by valves, and (c) basal, erect ovules with micropyle directed upwards. In Monimiaceae the ovules are apical and pendulous, with micropyle directed downwards (Takhtajan 1987).

Siparunaceae is another small family of 3 genera, *Siparuna, Bracteanthus* and *Glossocalyx* with 170 species distributed in tropical America and West Indies (the first two genera) and tropical west Africa (the last genus). Schodde (1970) proposed separating these three genera as family, Siparunaceae, from the Monimiaceae on the basis of anatomical, morphological, embryological features, chromosome number, and geographical and ecological distribution (see Table 20.12.1). Subsequent taxonomists like Walker (1976), Cronquist (1981) and Thorne (1983) do not recognise it as a separate family. Takhtajan (1987) treats Siparunaceae as a distinct family of his Laurales although he observes that this family is very close to Atherospermataceae; it is more closely related to Monimiaceae than to any other Lauralean family (Schodde 1970).

Table. 20.12.1 Comparative data for Monimiaceae and Siparunaceae

Monimiaceae	Siparunaceae
1 Small trees, shrubs or woody climbers, occur in tropical to tropical-montane and subtropical rain forests from Madagascar to the Malaysian region, Oceania, Australasia, central and southern America	Monoecious or dioecious, often aromatic, evergreen trees, occur in lowland and montane tropical rain forests of central America and western Africa
2 Vessels with simple perforation plates	Vessels with scalariform perforation plates
3 Rays multiseriate; wood fibres nonsepatate; parenchyma in apotracheal bands	Rays uniseriate; wood fibres septate; parenchyma apotracheal
4 Node trace single, arc-shaped	Node traces of 3 or more strands
5 Flowers unisexual, perigynous, syntepalous; perianth 4–8-lobed in 1–2 series, valvate	Flowers mostly bisexual, perigynous, choritepalous; perianth 4–8-many lobed, multiseriate, imbricate
6 Anthers bisporangiate with introrse valvular dehiscence	Anthers tetrasporangiate, dehisce by vertical fissures
7 Carpels 4-many, free, uniovulate, embeded in hypanthium wall and disc, styles elongated, stigmas clustered; placentation basal	Carpels numerous, free, uniovulate, borne exposed on an enlarged fruiting hypanthium; placentation parietal
8 Ovule unitegmic	Ovule bitegmic
9 Base chromosome number x = 22 (Ehrendorfer et al. 1968)	Base chromosome number x = 11

With the exclusion of Atherospermataceae, Siparunaceae and Trimeniaceae the Monimiaceae would be a more natural family.

Although referred to Magnoliales by Engler and Diels (1936), Lawrence (1951), Melchior (1964) and Cronquist (1968), the family Monimiaceae is of considerable phylogenetic importance as it shows both Magnolian and Lauralian affinities. Several phylogenists therefore, include Monimiaceae in a separate order, Laurales (Cronquist, 1981, Dahlgren 1980a, 1983a, Endress and Sampson 1983, Endress 1980b, c Schodde 1970, Takhtajan 1980, 1987, Walker 1976).

The family Monimiaceae is characterised by decussate toothed leaves, trends to unisexuality of flowers, varied development of floral hypanthium, basically tetramerous perianth, apocarpous gynoecia with numerous carpels, copious endosperm and relatively unspecialized secondary xylem. At the same time, there is marked diversity in the structure of secondary xylem, flower and fruit, as well as in pollen (Money et al. 1950) and chromosome numbers (Ehrendorfer et al. 1968). Pichon (1948) separated the *Atherosperma* group at familial level on account of its valved anthers and basal ovule. In addition, he erected a new family Amborellaceae which included the vesselless *Amborella*. Money et al. (1950) recognised Amborellaceae, and also treated *Trimenia* group as a distinct family Trimeniaceae. Schodde (1970) proposed a family rank for the *Siparuna* group, and the family Siparunaceae has been recognised by Takhtajan (1987).

Schodde (1970), on the basis of a detailed study has shown that Monimiaceae s.l. are not a unified family as had been presumed by previous workers. Splitting of this family into Atherospermataceae, Amborellaceae, Siparunaceae, Trimeniaceae and Monimiaceae s.s. (including 32 genera and 330 species) should make Monimiaceae s.s. a natural taxon.

Amborellaceae

Amborellaceae is a monotypic family with the genus *Amborella* (*trichopoda*) indigenous to New Caledonia.

Vegetative Features.   Shrubs of slightly scandent habit. Leaves alternate, distichous (Willis 1973), pinnately veined, estipulate, very variable in shape.

Floral Features.   Inflorescence complex dichasial cyme. Flowers hypogynous, unisexual, floral axis differentiated into pedicel and receptacle, bear spirally arranged bracteoles and tepals of progressively increasing size; staminate flowers with numerous or 10–12 stamens. Outer stamens larger than inner ones; with narrow, short, cylindrical filaments, basal parts bear unicellular hairs, anthers basifixed, longitudinally dehiscent. Staminate flowers show incipient perigynous tendencies due to (a) concavity of torus, and (b) adnation of parts. Pistillate flowers more or less bracteolate, bracteole 1 to 4, tepals 6 to 8, slightly connate at the base (Fig. 20.13A), in 2 series; inner ones broader with membranous margin. Staminodes 1 or 2, resemble stamens but sterile (Fig. 20.13B), basally adnate to subtending tepals. Gynoecium pentacarpellary (Fig. 20.13A), the basal parts of the carpels sealed, whereas the extensively developed paried stigmatic crests of the upper part of conduplicate carpels unfolded, each carpel with one ovule, attached to the ventral side of locule (Fig. 20.13C). Fruit drupaceous, with slightly compressed apex and sides, stipitate, conspicuously pitted-reticulate when dry with vestiges of stigmatic crests subterminal. Seeds endospermous, with a small embryo.

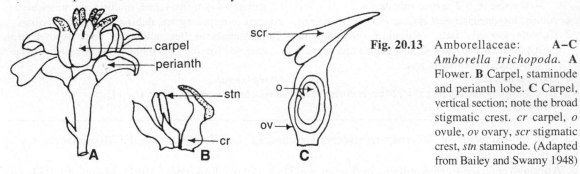

Fig. 20.13   Amborellaceae: **A–C** *Amborella trichopoda*. **A** Flower. **B** Carpel, staminode and perianth lobe. **C** Carpel, vertical section; note the broad stigmatic crest. *cr* carpel, *o* ovule, *ov* ovary, *scr* stigmatic crest, *stn* staminode. (Adapted from Bailey and Swamy 1948)

Anatomy.   Tracheids of vesselless xylem occur in relatively uniform radial seriation; numerous conspicuous bordered pits tend to be more or less closely crowded in the radial walls of the tracheids; smaller and of sporadic occurrence in tangential walls. Rays uniseriate or biseriate or even 3 to 5 cells wide multiseriates. Parenchyma strands few and diffuse. Nodes unilacunar; ethereal oil cells absent. Leaves dorsiventral, stomata confined to lower surface, Rubiaceous type, or pore surrounded by numerous ordinary parenchyma cells.

Embryology.   Pollen grains inaperturate, globose-spherical, exine finely papillose. Ovules anatropous, bitegmic, crassinucellate. Embryo sac type not known. Endosperm formation of the Cellular type.

Chromosome Number.   Haploid chromosome number is n = 13.

Important Genera and Economic Importance.   The only genus and species is *Amborella trichopoda*.

Taxonomic Considerations.   The family Amborellaceae was included in the Monimiaceae in earlier systems of classification. Pichon (1948) erected the new family including the only vesselless genus *Amborella*. Money et al. (1950) recognised this family, and most of the later phylogenists.

The family Amborellaceae is referred to the Magnoliaies although ethereal oil cells and mucilage cells are absent. According to Cronquist (1968): "no other order could be stretched to accommodate it; and it is not sufficiently distinctive, amongst a group of primitive relic families, to warrant separation as a monotypic order". However, most of the later authors include Amborellaceae in the Laurales along with other Lauralian families. It is closely related to the Monimiaceae in: (1) unisexual flowers, (2) similar pollen morphology, (3) uniovulate carpel, (4) unilacunar node, and (5) absence of druses and other large crystals of calcium oxalate. On the other hand, absence of vessels and septate fibres in xylem, absence of fibrous parenchyma in pith and cortex, and absence of ethereal oil cells distinguish it from the Monimiaceae.

From the above discussion, it is clear that the Amborellaceae is best treated as a member of the Laurales or as a Lauralian family in the Magnoliales, and is placed close to Monimiaceae.

Trimeniaceae

Trimeniaceae is a small western Pacific family of 2 genera and 5–9 species, distributed in eastern Australia, Celebes, the Molluccas, New Guinea, New Britain, Bougainville, New Caledonia, Fiji, Samoa and the Marquesas Islands.

Vegetative Features. *Trimenia papuana* is a tree of tropical mountain rain forests.

Floral Features. Inflorescence paniculate in leaf axils and at the end of leafy shoots. Flowers small, inconspicuous, scentless, mostly bisexual but some male flowers with gynoecium reduced or lacking. Perianth of 13 to 28 tepals, gradually changing in size and shape from the smaller, more or less round or depressed-obovate outermost ones to the larger, obovate to spathulate inner ones. The spathulate tepals form hoods over the androecium (because of their apically broadened shape). Tepals 2 to 11 in *Piptocalyx.* Stamens 11 to 15 in *Trimenia neocaledonica*, 7–16 in *Piptocalyx moorei,* 4 mm long and slender. Anther and filament equal in size. Mature anthers latrorsely dehiscent by 2 lateral slits and capped by a short ligniform connective apex. Gynoecium mono- or rarely bicarpellary, and apocarpous. At anthesis ovary barrel-shaped, slightly assymmetric, topped by a capitate tuftlike stigma, style not distinct. Ovule 1, in ventral median position, pendulous, almost completely filling the ovary, the small gaps between the ovary wall and ovule contain a mucilage-like secretion. The plumose stigma consists of long, multicelled, multiseriate, tanniniferous hairs. Fruit a 1-seeded berry (Endress and Sampson 1983).

Anatomy. Vessels usually moderately small, exclusively scalariform perforation plates with 20 fine bars. Rays not more than 3–4 cells wide, parenchyma usually sparse. Leaves dorsiventral, stomata confined to the lower surface, Rubiaceous type in *Trimenia.*

Embryology. Pollen grains inaperturate and spherical in *Trimenia papuana,* biaperturate in *T. neocaledonica* and *Piptocalyx moorei;* polyporate also reported (Endress and Sampson 1983); shed at 2-celled stage. Ovule anatropous, bitegmic, crassinucellate; embryo sac type not known, micropyle directed upward. Endosperm formation of the Cellular type.

Chromosome Number. Haploid chromosome number is n = 8.

Chemical Features. Secondary metabolites neolignans and piptoside are reported (Gottlieb et al. 1989).

Important Genera and Economic Importance. The two genera *Trimenia* and *Piptocalyx* do not have any economic importance.

Taxonomic Considerations. The Trimeniaceae is an archaic family characterised by acyclic flower and absence of floral calyx. Endress and Sampson (1983) comment that although this family occupies a comparatively isolated position within the Ranales,. it shows closest relationship with the Lauraceae, Monimiaceae, Chloranthaceae and Hernandiaceae. However, in its pollen morphology, the Trimeniaceae have no close relationship to these families, and remains isolated (Sampson and Endress 1984). According to Takhtajan (1987), they are closer to the Monimiaceae, with which they were associated earlier.

Although there are a few differences to indicate that the Trimeniaceae should be treated as a separate family from the Monimiaceae (mono- or rarely bicarpellary gynoecium, Rubiaceous type stomata, inaperturate or biaperturate pollen grains, and haploid chromosome number n = 8), yet the two families are closely related, as reflected by their distribution pattern, anatomical and embryological features.

Calycanthaceae

Calycanthaceae is a small family of 3 genera and 7 species, distributed in California and the southeastern United States of America in the New World, and China and Japan in the Old World.

Vegetative Features.   Deciduous or evergreen shrubs of mostly tropical forests. Leaves opposite, simple, entire, petiolate, petiole short, pinnately veined, estipulate.

Floral Features.   Flowers not inflorescent, solitary axillary on leafy lateral branches. Flowers bisexual, actinomorphic, perigynous, fragrant. Perianth lobes undifferentiated into sepals and petals, free, numerous, arranged spirally on the outer rim of a thickened cup-like or tubular receptacle. Stamens 5–30, inserted on the receptacle rim, filaments short and distinct; anthers bicelled, laterally extrorse (or latrorse); inner ones often sterile; connective broad and above the anthers. Gynoecium apocarpous, carpels numerous, borne on the inside of the urceolate receptacle. Each ovary superior, unilocular with 1 or 2 ovules; placentation parietal, style linear to filiform, stigma decurrent. Fruits 1-seeded dry, indehiscent achenes enclosed within the enlarged and fleshy receptacle. Seed without endosperm and a large embryo (Fig. 20.15B) with foliaceous cotyledons. Fruits show reticulate surface, caused by the enlarged attachment fields of the fallen tepals (Fig. 20.15A, C) and the "crown" of persistent staminodes at the apex of the receptacle (Fig. 20.15A, C; Endress 1983).

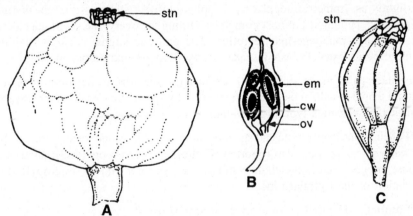

Fig. 20.15   Calycanthaceae: **A** *Idiospermum australiense*, fruit. **B** *Chimonanthus praecox*, fruit, longisection. **C** *Calycanthus floridus*, several-carpellate fruit. *cw* carpellary wall, *em* embryo, *ov* ovary, *stn* staminodes. (After Endress 1983)

Anatomy.   Vessels very small, with a marked oblique or flame-like pattern; ring-porous, with spiral thickening; perforations simple. Parenchyma limited to a few cells round the vessels. Rays up to 4 cells wide and composed almost entirely of square and upright cells. Leaves dorsiventral with unicellular hairs, surrounded by a rosette of cells with silicified walls. These appear as dark or translucent dots in dried specimens. Stomata Rubiaceous type confined to the lower surface.

Embryology.   Pollen grains monocolpate, biaperturate or rarely triaperturate; shed at 2-celled stage, rarely in tetrads as in *Chimonanthus fragrans;* exine reticulate. Ovules anatropous, bitegmic, crassinucellate; micropyle formed by the inner integument alone; Polygonum type of embryo sac, 8-nucleate at maturity. Endosperm of the Cellular type.

Chromosome Number. The haploid chromosome numbers are n = 11, 12. Sax (1933) reported 2n = 22 in *Calycanthus fertilis* and *C. floridus,* and *C. floridus* var. *ovatus* is a triploid.

Important Genera and Economic Importance. Only two genera are included in this family—*Calycanthus* and *Chimonanthus,* the former distributed in southeastern North America and the latter in China (Takhtajan 1969). Takhtajan (1987) reports a third genus *Sinocalycanthus.* Endress (1983) suggests the inclusion of *Idiospermum* also in this family, although many taxonomists treat it as a distinct family—Idiospermaceae (Cronquist 1981, Dahlgren 1983a, Takhtajan 1987).

The family is not of much economic importance except that the plants are grown for their fragrant flowers.

Taxonomic Considerations. The Calycanthaceae in Bentham and Hooker's (1965a) system were placed in the Ranales between the Dilleniaceae and Magnoliaceae; similarities to the Rosaceae, Combretaceae and Monimiaceae have been pointed out. In Engler and Diels (1936) system also, this family is placed near the Magnoliaceae. Hutchinson (1969, 1973) includes Calycanthaceae in the Rosales, on the basis of their affinities with the Rosaceae and Dichapetalaceae. In Magnoliaceae and other allied families, there is always abundant endosperm and a small embryo in the seed, but in Calycanthaceae the endosperm is absent and embryo is large with foliaceous cotyledons, the androecial and gynoecial elements are spirally arranged and there is a cup-like receptacle. All these features support their position in or near the Rosales.

However, the anatomical and embryological features of the Calycanthaceae resemble those of the Monimiaceae and their allies. This justifies the present position of this family in the Laurales which is supported by most of the taxonomists (Cronquist 1981, Dahlgren 1980a, 1983a, Takhtajan 1987).

Gomortegaceae

Gomortegaceae is a monotypic family of the genus *Gomortega* with only one species (*G. nitida*) which occurs in South America.

Vegetative Features. Large trees containing aromatic oil; wood heavy, durable and beautifully figured. Leaves simple, entire, opposite, petiolate, shiny and aromatic; evergreen, narrowly elliptic, pinnately-veined, estipulate.

Floral Features. Inflorescence axillary and terminal racemes or panicles. Flowers bisexual, actinomorphic, subtended by two opposite bracts; sepals 6–10, spirally arranged, epigynous; petals absent. Stamens 2–11, epigynous, filaments free, anthers bilocular, introrse, dehisce by valves; inner stamens with two shortly-stalked glands at the base of each filament. Gynoecium 2–3-carpellary, syncarpous; ovary inferior, 2- to 3-locular, ovule one per locule, pendulous; style 2–3-lobed. Fruit a drupe with a bony endocarp and fleshy exocarp. Seed with a large embryo in abundant oily endosperm.

Anatomy. Vessels small, perforation plates scalariform with numerous bars; scanty wood parenchyma and narrow heterogeneous rays. Spherical secretory cells with yellow resinous contents present both in stem and leaf. Leaves dorsiventral, stomata confined to the lower surface; Rubiaceous type, the sudsidiary cells sometimes secondarily divided by a wall at right angles to the pore.

Embryology. Pollen grains inaperturate, globose-spherical, echinate, shed at 2-celled stage. The spines of the pollen of *Gomortega* are reminiscent of those in the Lauraceae (Walker 1976). Embryology has not been fully studied (Johri et al. 1992).

Chromosome Number. The haploid chromosome number is n = 12

Chemical Features. Secondary metabolite dimethoxylated coumarin occurs in Gomortegaceae members (Gottlieb et al. 1989).

Important Genera and Economic Importance.   *Gomortega nitida* is the only member of this family and has no economic importance.

Taxonomic Considerations.   *Gomortega* was included amongst the Lauraceae by Bentham and Hooker (1965a), the woody Ranales by Engler and Diels (1936), as a distinct family and as a monotypic family within the Laurales by Hutchinson (1959, 1969, 1973). The occurrence of secretory cells in all parts of the plant, and of scalariform perforation plates confirms its alliance to the Lauralian families. The fibres with distinct bordered pits in the ground tissue of the wood distinguish it from the rest of the Laurales. According to Stern (1955): "most likely Gomortegaceae is closely allied to the Monimiaceae through a *Hortonia*-like forbearer with valvular (=valvate) anthers". Based on floral morphology, pollen grains and xylem anatomy, Cronquist (1968) included Gomortegaceae in the cluster of 6 families—Amborellaceae, Trimeniaceae, Monimiaceae, Calycanthaceae, Lauraceae and Gomortegaceae—under the Magnoliales which conforms to the Laurales of Cronquist (1981), Dahlgren (1983a) and Takhtajan (1987). All these families are characterised by : (1) unilacunar nodes, (2) neither uni- or triaperturate pollen grains, (3) perigynous to epigynous flowers, (4) only 1, rarely 2 ovules per ovary, and (5) absence of endosperm.

Lauraceae
A medium-sized family of 32 genera and 2500 species, with main centres of distribution in Southeast Asia and Brazil.

Vegetative Features.   Trees (*Cinnamomum, Persea*) or shrubs (*Laurus*), rarely twining, leafless parasitic herb e.g. *Cassytha* (Fig. 20.17A). All parts have aromatic oil glands. Leaves alternate, simple, evergreen, coriaceous, entire, or sometimes lobate (e.g. *Sassafras albidum*), rarely palmately compound, estipulate.

Floral Features.   Inflorescence cymose or racemose (Fig. 20.17A), bisexual, polygamous or unisexual. Flowers small, greenish or yellowish, acyclic, spirocyclic or cyclic (Fig. 20.17B), mostly with a hypanthium. Calyx of usually 6 sepals, inferior, tube sometimes enlarges in fruit, limbs imbricate; petals absent.

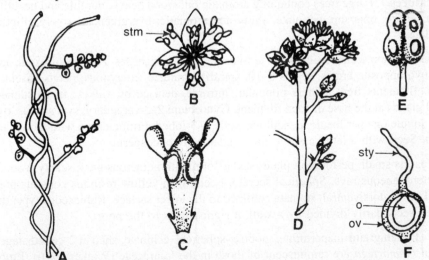

Fig. 20.17   Lauraceae: **A–C** *Cassytha filiformis*, **D–F** *Persea americana*. **A** Twinning branches bearing flowers and fruits. **B** Staminate flower, face view. **C** Stamen. **D** Flowering branch of *P. americana*. **E** Stamen. **F** Carpel, longisection. *o* ovule, *ov* ovary, *stm* stamen, *sty* style. (**A–C** adapted from Hutchinson 1969, **D–F** from Lawrence 1951)

Stamens (3-)-5-numerous, spiral or cyclic in 4 whorls, often the 4th row is suppressed or reduced to staminodes; filaments sometimes glandular at the base, rarely the glands fuse into a disc; anthers more or less ribbon-shaped, continuous with the filaments, dehiscence by valves; valves superimposed or more or less collateral, opening from base upwards by flaps (Fig. 20.17C, E), introrse, or sometimes the third whorl extrorse. Gynoecium monocarpellary, ovary superior or inferior, unilocular with a solitary, pendulous ovule (Fig. 20.17F); style terminal, simple, stigma small. Fruit a berry or drupaceous, surrounded by persistent perianth-base seated upon an enlarged receptacle. Seeds non-endospermous, and embryo straight.

Anatomy. Vessels mostly medium-sized, perforation plates exclusively simple; sporadic occurrence of scalariform plates; intervascular pitting alternate. Parenchyma paratracheal, scanty to vasicentric; rays 2-3 cells wide, uniseriates occur, weakly heterogeneous to homogeneous, often with oil cells. Leaves generally dorsiventral, hairs mostly unicellular, frequently thick-walled. Stomata confined to the lower surface, frequently sunken, Rubiaceous type. Secretory cells contain oil or mucilage which is a characteristic feature.

Embryology. Pollen grains monocolpate (*Cinnamomum, Laurus, Litsea*) or acolpate (*Cassytha*), exine minutely spinescent; shed at 2-celled stage (Johri et al. 1992). Ovules anatropous, bitegmic, crassinucellate; micropyle formed by both the integuments. Embryo sac of Polygonum type, 8-nucleate at maturity. Endosperm formation of Nuclear type; Cellular type in *Cassytha*.

Chromosome Number. Haploid chromosome number is n = 12, 18, 21, 24.

Chemical Features. Seeds of *Laurus nobilis* and *Cinnamomum camphora* are rich in lauric acid; fruit coat fats are palmitic, oleic and linoleic acids (Shorland 1963). Aliphatic polyol—D-perseitol reported only in *Laurus persea,* and dulcitol in *Cassytha filiformis* (Plouvier 1963). Alkaloids anibine and 4-methoxyparacotoin occur in *Aniba rosaeodora* and *A. duckei;* cryptopleurine and pleurospermine, respectively, in bark and leaves of *Cryptocarya pleurosperma* (Hegnauer 1963). Both saturated and unsaturated fatty acids occur in fruits of *Persea gratissima*. Alkanes are reported in *Persea* and *Beilschmiedia* (Scora et al. 1975). Sesquiterpene lactones have been recorded in *Neolitsea, Lindera* and *Laurus* (Harborne and Turner 1984). Subfamily Cassythoideae is particularly rich in benzylisoquinolines (Gottlieb et al. 1989).

Important Genera and Economic Importance. Some of the important genera are *Cinnamomum* (ca. 250 spp.), *Persea* (150 spp.), *Sassafras* (3 spp.), *Lindera* (100 spp.), *Nectandra* (100 spp.), *Cassytha* (20 spp.) and *Laurus* (2 spp.).

According to a Greek legend, the large evergreen shrub *Laurus nobilis,* commonly called laurel, was sacred to the Greek God Apollo. Its branches were woven into crowns for the winners in the ancient Olympic Games. Later, laurel became the symbol of triumph in Rome. Another interesting plant is *Cassytha filiformis,* a parasitic twiner. In a family of mostly woody representatives, this plant appears to be anachronistic, but the two resemble each other in floral morphology and seed structure. *Cryptocarya floribunda, Dehaasia kurzii, Machilus villosa,* and *Phoebe lanceolata* are some of the trees.

Plants of Lauraceae are important sources of aromatic oils, fruits and timbers. A perfume "bois de rose" is obtained from *Aniba panurensis* of Guyana and *A. rosaeodora* var. *amazonica* of Brazil. *Cinnamomum burmanni* of Indonesia, *C. cassia* of Burma, *C. massoia* of New Guinea, *C. oliveri* of Australia and *C. tamala* of India are the source of "Cassia bark"—a flavouring agent. Distillation of wood of *Cinnamomum camphora* yields camphor. Many of the genera yield good timber, e.g. Nan-Mu wood is obtained from *Persea nanmu* of China. *Beilschmiedia roxburghiana, Nectandra rodiaei, Ocotea bullata* and *Umbellularia california* are the other timber-yielding plants. The leaves of *Laurus nobilis* constitute the commercial sweet bay much used in cooking. Bark of *Cinnamomum zeylanicum* gives the commercial cinnamon and is cultivated in India and Sri Lanka. *Sassafras albidum* yields oil of sassafras, and the bark from the roots is boiled for tea. Avocado or alligator pear is the edible fruit from *Persea gratissima* of tropical America.

Taxonomic Considerations. Lauraceae has been treated as one of the woody Ranales by Engler and Diels (1936), Bessey (1915), and Lawrence (1951); Melchior (1964) and Cronquist (1968) retain it in the Magnoliales. Hallier (1912) included it in his Annonales along with other woody members of the Ranalian complex. Hutchinson (1973) placed this family (together with Monimiaceae, Hernandiaceae, Myristicaceae and other allied families) in his Laurales, an order he considered as reduced perhaps from Winteraceous ancestors of the Magnoliales.

Most taxonomists agree to the erection of a separate order, Laurales, including Lauraceae and its allies (Cronquist 1981, Dahlgren 1983a, Takhtajan 1987, Walker 1976). Thorne (1983), however, treats Lauraceae as a member of the suborder Laurineae of his order Annonales.

The family Lauraceae (and also Amborellaceae, Trimeniaceae, Monimiaceae, Calycanthaceae, Gomortegaceae and Hernandiaceae) is characterised by unilacunar nodes, usually with two traces, specialised pollen grains that are neither uni- nor triaperturate, usually peri- or epigynous, highly reduced flowers, presence of floral cups, only 1 (rarely 2) ovule in each carpel and non-endospermous seeds. This group of families, therefore, is so different from other Magnolian families that it would be better to treat them as belonging to a distinct order, Laurales.

Although Hutchinson (1948) referred to Winteraceous ancestry for this group, Cronquist (1968) observed that they cannot be considered either ancestral to or derived from either of the other two groups amongst the Magnoliales (Magnoliaceae-Winteraceae-Degeneriaceae-Himantandraceae-Annonaceae complex, and Schisandraceae-Illiciaceae complex), since it combines both primitive and advanced characters. According to Takhtajan (1969), however, Laurales is derived "from some vesselless member of the Magnoliales", probably because of the presence of the vesselless family Amborellaceae in the Laurales.

Hernandiaceae

Hernandiaceae is a family of only four genera and ca. 23 species (only two genera and 42 species according to Takhtajan 1987), distributed in the tropics and subtropics of Asia and Africa. Amongst these 4 genera, *Gyrocarpus* has been raised to the rank of a family, Gyrocarpaceae (see Takhtajan 1987).

Vegetative Features. Trees (Fig. 20.18A) or shrubs, sometimes scandent; leaves simple or digitately compound as in *Illigera*, alternate, ovate, often peltate, estipulate, petiolate, with oil cells.

Floral Features. Inflorescence axillary corymbs or paniculate cymes, bracteate or ebracteate. Flowers bi- or unisexual as in *Hernandia* (Fig. 20.18B, E, F), sometimes polygamous by abortion, actinomorphic, perigynous, bracteate, ebracteolate. In *Hernandia* the partial inflorescence of 3 flowers is surrounded by an involucre of 4 or 5 foliaceous bracts; the central flower pistillate, flanked by two staminate flowers on both sides. Perianth usually biseriate with 4–8 subequal segments in each series, sepaloid. Stamens 3–5, in one series, alternating with the outer sepals; anthers bicelled, dehisce introrsely or laterally by two valves (Fig. 20.18D); connective broad (Fig. 20.18C, D), often reaching beyond the length of anthers. Gland-like staminodes often present in 1 or 2 whorls outside the stamens (Fig. 20.18B). Gynoecium in pistillate flowers, monocarpellary; ovary inferior (Fig. 20.18E), unilocular, with only one pendulous ovule (Fig. 20.18F), placentation parietal; style 1, short and thick, stigma 1. Fruit is an achene, incompletely enclosed by the expanded or inflated receptacle as in *Hernandia,* or with 2–4 wings as in *Illigera.* or with 2 terminal wings formed by enlarged perianth segments as in *Gyrocarpus.* Seeds without endosperm and a straight embryo with thick, lobed subruminate cotyledons.

Anatomy. Vessels large, mostly solitary, perforations simple, intervascular pitting alternate. Parenchyma

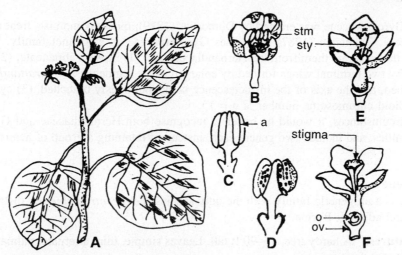

Fig. 20.18 Hernandiaceae: **A–F** *Hernandia ovigera*. **A** Flowering branch. **B** Staminate flower. **C** Stamen. **D** Stamen after dehiscence. **E** Pistillate flower. **F** Same, vertical section. *a* anther, *o* ovule, *ov* ovary, *stm* stamen, *sty* style. (Adapted from Lawrence 1951)

paratracheal, rays up to 3–5 cells wide, with very few to moderately numerous uniseriates, homogeneous. Leaves mostly dorsiventral, hairs nearly always unicellular. Stomata confined to the lower surface, Ranunculaceous type in *Gyrocarpus* and *Sparattanthelium,* and Rubiaceous type in *Hernandia* and *Illigera;* crystals and cystoliths present.

Embryology.   Pollen grains nonaperturate or rarely uniaperturate, monocolpate; shed at 2-celled stage. Ovules anatropous, bitegmic, crassinucellate. Embryo sac of *Gyrocarpus* is 8-nucleate at maturity. Endosperm formation is of the Cellular type.

Chromosome Number.   Basic chromosome number is x = 20.

Chemical Features.   Some common lactones are reported from various members (Gottlieb et al. 1989).

Important Genera and Economic Importance.   *Hernandia* is the only genus that is cultivated as an ornamental tree in some places. Other genera are *Gyrocarpus, Illigera* and *Sparattanthelium;* of these, *Gyrocarpus* has already been raised to family level and *Sparattanthelium* is also included in this family, Gyrocarpaceae (Takhtajan 1987).

Taxonomic Considerations.   The family Hernandiaceae was accepted by Wettstein (1935) and Engler and Diels (1936). The genus *Hernandia* was formerly included in the Lauraceae by Bentham and Hooker (1965a). According to Lawrence (1951), this family lacks natural affinities, homogeneity and "it is likely that its genera will be further segregated into additional families".

Hutchinson (1973) placed Hernandiaceae in the Laurales because of absence of petals, more or less perigynous flowers, stamens in whorls and dehiscence of anthers by valves. Cronquist (1968) placed Hernandiaceae in Magnoliales, although he recognised its affinities with Lauralian families. Later, in 1981, he separated Laurales from the Magnoliales and Hernandiaceae as the most advanced family of the Laurales. Dahlgren (1983a) also treats it as a member of Laurales.

Takhtajan (1980, 1987) also accepted Hernandiaceae as a member of the Laurales, but he retained only two genera—*Hernandia* and *Illigera*—in this family and recognised a separate family Gyrocarpaceae, which included the genera *Gyrocarpus* and *Sparattanthelium*.

The family Gyrocarpaceae was erected by Dumortier (1829); most taxonomists treat it as included in the Hernandiaceae. Takhtajan (1987) recognises Gyrocarpaceae as a distinct family. The features of Gyrocarpaceae that separate them from the Hernandiaceae are: (1) flowers ebracteate, (2) in *Gyrocarpus* fruits crowned by two terminal wings formed by enlarged calyx segments; in *Sparattanthelium* the fruits are dry and ribbed, and the axis of the infructescence is dichotomously branched, (3) cystoliths present, and (4) the haploid chromosome number is n = 15.

Under the circumstances, it would be better to recognise both Hernandiaceae and Gyrocarpaceae as two distinct families, and include two genera each, instead of retaining a group of heterogeneous genera in one family.

Tetracentraceae

Tetracentraceae is a unigeneric family with the only species, *Tetracentron sinense,* distributed in south-central China and adjacent Burma.

Vegetative Features.   A hardy tree, 15–90 ft tall. Leaves simple, thin, alternate, palmately nerved, and with stipular flanges.

Floral Features.   Inflorescence a pendulous spike (Fig. 20.19A), each flower in the axil of a subtending bract (= pherophyll according to Briggs and Johnson 1979), bisexual, hypogynous, actinomorphic, sessile (Fig. 20.19B). Perianth of 4 tepals which cover the inner floral organs in bud, arranged in two alternate pairs, polytepalous, imbricate, persistent; the torus flattened and inconspicuous. Stamens 4, opposite the tepals (Fig. 20.19B), filaments short and thick; anthers basifixed, dehiscence by valves (Endress 1986).

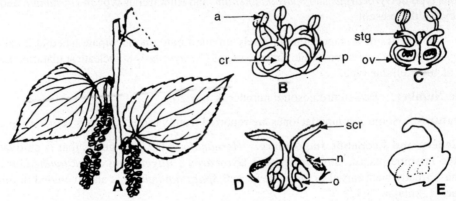

Fig. 20.19   Tetracentraceae: **A-E** *Tetracentron sinense*. **A** Twig with inflorscences. **B** Flower. **C** Vertical section. **D** Carpels, longisection, note the nectary (*n*). **E** A single pistil. *a* anther, *n* nectary, *o* ovule, *ov* ovary, *stm* stamen. (**A-C, E** adapted from Lawrence 1951, **D** after Endress 1986)

Gynoecium tetracarpellary, concrescent below but free above, i.e. the margin of the carpels is free except at the base of the syncarpous gynoecium; each carpel unilocular, with 5 or 6 ovules borne on the placentae only in the middle (Fig. 20.19C, D). The carpels have a conspicuous dorsal bulge that serves as a nectary (*n* in Fig. 20.19D). The epidermis of the nectary is covered with numerous slightly sunken stomata surrounded by a ring of epidermal cells with prominent cuticular folds (Endress 1986). Style conduplicate (folded), stigmatic along most of the ventral suture (Fig. 20.19E). Fruit, a folliceta with strong ventral development. Seeds with a spongy and wing-like outer integument.

Anatomy.   Vessels absent; much of the xylem tissue formed of extremely long tracheids with circular to scalariform pits on the radial walls. Parenchyma diffuse, rays heterogeneous, up to 4 cells wide.

Leaves dorsiventral, stomata confined to the lower surface; two subsidiary cells surround the pore, or 4/5 well-defined subsidiary cells surround the pore. Large, branched, secretory idioblasts with resinous contents reported from leaf mesophyll and outer part of cortex in stem.

Embryology.   Pollen grains very small (10–15 μm diameter), simple, spheroidal, tricolpate; exine rugulate-reticulate but striate near the apertures; shed at 2-celled stage. Ovules anatropous, bitegmic, crassinucellate (Johri et al. 1992). Embryo sac type not known. Endosperm formation of the Cellular type, embryo minute.

Chromosome Number.   The haploid chromosome number is n = 19.

Important Genera and Economic Importance.   *Tetracentron sinense* is often grown as an ornamental.

Taxonomic Considerations.   This family is treated as a member of the order Trochodendrales (Walker 1976, Dahlgren 1980a, Cronquist 1981, Endress 1986, Takhtajan 1987), or of Trochodendrineae of the Hamamelidales (Thorne 1963). The genus *Trochodendron* of the Trochodendraceae has many features common with *Tetracentron*, therefore the taxonomic features of Trochodendraceae may also be referred to Tetracentraceae.

Trochodendraceae

Trochodendraceae is a unigeneric family and *Trochodendron aralioides* is a native of South Korea, Japan, Taiwan and Formosa.

Vegetative Features.   A small evergreen tree; branchlets terete, marked at the nodes by leaf scars. Leaves estipulate, pseudoverticillate in clusters of 4, 6 or 12 at the tips of branches; petiole canaliculate; simple, coriaceous, serrulate, pinnately veined (Fig. 20.20A).

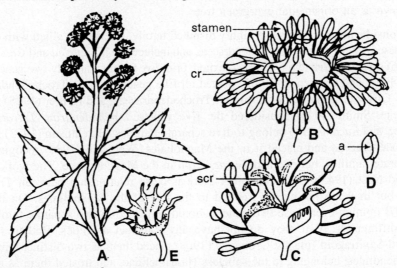

Fig. 20.20   Trochodendraceae: **A–E** *Trochodendron aralioides*. **A** Twig with flowers. **B** Flower. **C** Longisection. **D** Stamen. **E** Fruit. *cr* carpel, *o* ovule, *ov* ovary. (**A, C–E** adapted from Lawrence 1951, **B** after Endress 1986)

Floral Features.   Inflorescence dichasial cyme, terminal at inception, later appears axillary. Flowers bracteate, bracts caducous, bracteolate, incomplete, bisexual, actinomorphic, pedicellate. Perianth 0, sometimes bracteoles considered as remnants of perianth (Lawrence 1951). Stamens numerous, in 3 or 4 superimposed series on an expanded torus, the lower ones caducous; filaments filiform, anthers basifixed, oblong, bicelled, mucronate (Fig. 20.20B–D), each theca opens by valves (Endress 1986). Gynoecium of (4-)-6-11 sessile carpels fused to the torus at the base (Fig. 20.20B, C); in one whorl, and laterally concrescent. Ovary unilocular, with 2 rows of ovules borne near the ventral suture; styles free, bent outwardly with decurrent stigmatic surfaces (Fig. 20.20C); above the ovary there is a short zone where the neighbouring contiguous carpels are postgenitally fused. As in *Tetracentron*, there is a dorsal bulge on the carpels which serves as a nectary; the ovary is partially inferior (Fig. 20.20B, C). Fruit a follicetum of laterally coalescent follicles (Fig. 20.20E); ventral loculicidal dehiscence.

Anatomy. The wood is vesselless; much of the xylem tissue formed of extremely long tracheids with circular to scalariform pits on the radial walls; parenchyma diffuse. Rays of 2 distinct sizes, the larger up to 12 cells wide; heterogeneous. Leaves dorsiventral, upper epidermis composed of cells with very thick outer walls. Stomata confined to the lower surface; subsidiary cells horse-shoe-shaped and contain deposits of a gum-like substance. Large, branched, sclerenchymatous idioblasts occur in leaf mesophyll and primary cortex of stem.

Embryology. Pollen grains simple monads, spherical with three pores, i.e. tricolpate; exine rugulate, shed at 2-celled stage. Ovules anatropous, bitegmic, crassinucellate; micropyle formed by inner integument. Embryo sac of Polygonum type, 8-nucleate at maturity. Endosperm formation of the Cellular type.

Chromosome Number. Basic chromosome number is x = 19.

Chemical Features. Polyphenolics like leucodelphinidin and quercitin usually present (Dahlgren 1975a).

Important Genera and Economic Importance. The only member, *Trochodendron aralioides*, is sometimes grown as an ornamental evergreen tree.

Taxonomic Considerations. This is a highly disputed family. It has been allied with the Magnoliaceae. Prantl (1888) described the family Trochodendraceae and included in it *Euptelea* and *Cercidiphyllum*. Engler and Diels (1936) separated Cercidiphyllaceae—a more primitive family with only two genera, *Cercidiphyllum* and *Euptelea*. Finet and Gagnepain (1905) placed all five genera in tribe Trochodendreae of the family Magnoliaceae, Oliver (1895) in a single family, Trochodendraceae, and Hallier (1905) in a single family, Hamamelidaceae. Smith (1945) considered the five genera *Trochodendron, Tetracentron, Euptelea, Cercidiphyllum,* and *Eucommia* to belong to five separate families. Hutchinson (1973) included *Euptelea* also in Trochodendraceae and placed it in the Magnoliales. Wettstein (1935) recognised four of these genera as separate families, but retained *Tetracentron* in the Magnoliaceae. The works of Smith (1945) and Bailey and Nast (1945b, c) indicate that there is close relationship between *Trochodendron* and *Tetracentron,* but they are not closely related to the rest of the Magnoliales. On the other hand, in Croizat's (1947) opinion even these two genera should be placed in two separate families as they show a number of differences. Also, they do not have any Ranalian affinities. Rather, they are closer to "Hamamelidoid-Saxifragoid" plexus. Cronquist (1968) placed them as two distinct families in a separate order, Trochodendrales, belonging to the subclass Hamamelidae, and treated them as the most primitive taxa in this group. Members of this order have conduplicate, scarcely sealed carpels, and vesselless xylem; are derived from the Magnoliales. On the other hand, the poorly developed perianth clearly reflects a reduction in structure; structure of stamens and pollen grain are also not primitive.

Dahlgren (1975a) treats this order at the base of superorder Hamamelidanae; he includes the families Eupteleaceae and Cercidiphyllaceae in the Trochodendrales. Later in 1980a and 1983a, Dahlgren includes the families Tetracentraceae, Trochodendraceae, Eupteleaceae and Cercidiphyllaceae in the Rosiflorae.

Eupteleaceae and Cercidiphyllaceae are included in the order Trochodendrales in 1980a and in Cercidiphyllales in 1983a.

According to Takhtajan (1980), in many respects the order Trochodendrales occupies an intermediate position between Magnoliales and Hamamelidales, but in totality, they stand nearer to the latter. Both Dahlgren and Takhtajan recognise the families Trochodendraceae and Tetracentraceae.

According to Endress (1986), *Trochodendron* and *Tetracentron* are very closely related (see Table 20.20.1) and more so than the two other genera, *Euptelea* and *Cercidiphyllum,* of this group, and therefore the two should be retained in the same family, Trochodendraceae.

Table 20.20.1 **Comparative Data for** *Trochodendron, Tetracentron, Euptelea* **and** *Cercidiphyllum*

	Tetracentron	*Trochodendron*	*Euptelea*	*Cercidiphyllum*
Leaves	Alternate, deciduous	Alternate, evergreen	Alternate, deciduous	Opposite, deciduous
Flowers	Bisexual, sessile	Bisexual, pedicelled	Bisexual, pedicelled	Unisexual, sessile
Stamens	Four	Numerous	Numerous	Numerous
Pollen grains	10–15 μm diameter	10–15 μm diameter	20–30 μm diameter	20–30 μm diameter
Exine	Striate-rugose	Striate-rugose	Finely reticulate	Finely reticulate
Anther dehiscence	Valvate	Valvate	Longitudinal	Longitudinal
Pollination	Entomophilous	Entomophilous	Anemophilous	Anemophilous
Carpels	Four, sessile	Numerous, sessile	Numerous, stipitate	1, stipitate
Gynoecium	Syncarpous	Syncarpous	Apocarpous	Unicarpellate
Ovary	Semi-inferior	Semi-inferior	Superior	Superior
Fruits	Dehiscent	Dehiscent	Indehiscent	Dehiscent
Seeds	With appendages	With appendages	Without appendages	With appendages
"Oil" cells	Present in floral organs	Present in floral organs	Absent	Absent
Wood	Without vessels	Without vessels	With vessels	With vessels
Nectaries	Present on dorsal surface of carpel	Present on dorsal surface of carpel	Absent	Absent

According to Endress (1986: p. 321), the Trochodendrales are closely related to the Magnoliales, and at the same time to the Hamamelidales. The Trochodendrales are intermediate between the core Magnoliidae and Rosidae/Hamamelidae. Some genera have already acquired tricolpate pollen grains and valvate anthers of Hamamelidian type. Under the circumstances, it would be best to treat this whole group as a separate order between Magnoliales and Hamamelidales, or as the most primitive group amongst the Hamamelidales (Takhtajan 1987). In Takhtajan's classificatory system (1987), superorder Trochodendranae of the Hamamelididae, includes three orders: Trochodendrales with *Trochodendron* and *Tetracentron*, Eupteleales with *Euptelea,* and Cercidiphyllales with *Cercidiphyllum.*

Eupteleaceae

Eupteleaceae is also a unigeneric family with the genus *Euptelea*; of the two species, one is endemic to Japan (*Euptelea polyandra*) and the other to China and northeastern India (*E. pleiosperma*).

Vegetative Features.   Deciduous trees or shrubs; branchlets alternate, with lenticels; marked at the base of each year's growth by the numerous concentric scars of bud scales. Buds always axillary; terminal buds abort. Leaves alternate, estipulate; 3–10 per season on the longer branchlets, often crowded and pseudo-whorled on short lateral branches; first-formed leaves in a season are small and underdeveloped. Petiole of mature leaves slender, canaliculate, often dilated to form a sheathing, bud-subtending base. Leaf-blade smooth with acuminate apex, serrate margin, pinnately veined (Fig. 20.21A).

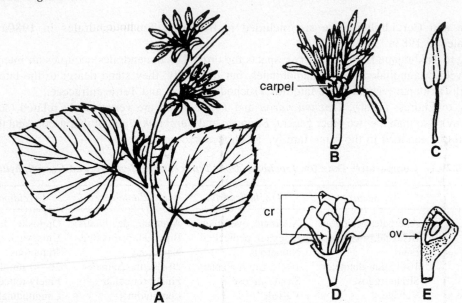

Fig. 20.21 Eupteleaceae: **A–E** *Euptelea polyandra*. **A** Flowering branch. **B** Flower. **C** Stamen. **D** Carpels. **E** Longisection of carpel. *cr* carpel, *o* ovule, *ov* ovary. (**A, C–E** adapted from Lawrence 1951, **B** after Endress 1986)

Floral Features. Inflorescence of about 6–12 flowers borne in the axils of bracts around the growing point and subsequently become lateral due to the vegetative development and proliferation of the axis (Troll 1964); auxotelic raceme (Briggs and Johnson 1979). Bracts caducous. Flowers bisexual (Fig. 20.21B), protandrous, anemophilous; pedicels long, subterete or slightly flattened, straight, slender; perianth absent; stamens borne in a single whorl on the margin of the flattened torus; filaments filiform or slightly flattened, often twisted, anthers basifixed, linear-oblong, dehiscence longitudinal, connective produced into an apical acute appendage (Fig. 20.21B, C). Carpels numerous, free, borne in a single whorl, just on the inner side of the stamens (Fig. 20.21B, D). Ovary flattened, oblong or elliptic with a ventral stigmatic margin; the stigmatic zone is covered with minute tangled sticky processes. Locule 1, ovules 1–3 (rarely 4), suborbicular, attached to the ventral edge of the locule (Fig. 20.21E). Fruit is a cluster of samaras, each conspicuously stipitate; stipes filiform at base, gradually swollen and flattened distally and expanded into the wing of the fruits. Seeds 1–3 or 4, ellipsoid or obovoid. Endosperm oily, granular, copious, embryo small, near the basal end of seed.

Anatomy. The thin-walled, more or less angular vessels of *Euptelea* are numerous, solitary and diffusely distributed. These are relatively long with overlapping end and have scalariform perforation plates with numerous bars. Wood parenchyma in terminal bands 1–2 cells wide; rays up to 10 cells wide, heterogeneous. Leaves dorsiventral, lower epidermis frequently papillose in *E. pleiosperma*. Hairs on young leaves simple, uniseriate with single or several basal cells; stomata restricted to lower surface, Ranunculaceous type.

Embryology. Pollen grains tricolpate in *E. pleiosperma* or pluriaperturate (Endress 1986), exine finely reticulate; 2-celled at dispersal stage. Ovules bitegmic, crassinucellate, immature at the time of anthesis. Polygonum type of embryo sac, 8-nucleate at maturity. Endosperm formation of the Cellular type.

Chromosome Number. Basic chromosome number is x = 14.

Important Genera and Economic Importance. The two species of *Euptelea* are cultivated (to some extent) as ornamentals.

Taxonomic Considerations. Most taxonomists including Hutchinson (1969, 1973), considered *Euptelea* to be a member of Trochodendraceae in the Magnoliales. Nast and Bailey (1946) and Smith (1946) considered this genus to be an isolated taxon amongst the woody Ranales because of the presence of tricolpate pollen grains and absence of ethereal oil cells. Cronquist (1968, 1981) placed *Euptelea* in a distinct family, Eupteleaceae, in the Hamamelidales; he also mentioned that its position is debatable. It has a wood anatomy similar to that of *Illicium* (Illiciales) as well as Hamamelidaceae (Hamamelidales); its flower is, with numerous stamens, primitive to that of other Hamamelidales but, at the same time, it also shows the general pattern of reduction associated with anemophily. Dahlgren (1975a, 1980a), Endress (1986) and others placed Eupteleaceae in the Trochodendrales. Takhtajan (1969, 1980, 1987), however, always retained this family in a distinct order—Eupteleales—placing it near the Hamamelidales. Because of some of its distinct features, he did not include Eupteleaceae in the Hamamelidales. Although, the family Eupteleaceae has a number of features common to both the Magnoliales and the Hamamelidales, yet, because of certain specific features of its own, like pluriaperturate pollen grains, stipitate carpels without style, indehiscent fruits and seeds without appendage, it is justifiable to treat it as a distinct order, Eupteleales. This order may be placed between Trochodendrales and Hamamelidales.

Cercidiphyllaceae

Cercidiphyllaceae is another unigeneric family with the genus *Cercidiphyllum* and only one species *C. japonicum*, distributed only in China and Japan. This genus was widely distributed in the northern hemisphere in the Tertiary (Brown 1962, Hummel 1971, Becker 1973, Chandrasekharan 1974, Iljinskaja 1972, Jähnicen et al. 1980, Scott and Wheeler 1982, Basinger and Dilcher 1983, Hickey et al. 1983, Stockey and Crane 1983), and perhaps back to the Upper Cretaceous (Hickey et al. 1983). In the Paleocene, some other genera, such as *Joffrea* (Crane and Stockey 1985), and with less certainity *Jenkinsella* (Chandler 1964, Crane 1978), were affiliated with *Cercidiphyllum*. Similarities between early *Cercidiphyllum* and the mid-Cretaceous genus *Prisca* were pointed out by Retallack and Dilcher (1981).

Vegetative Features. A dioecious, deciduous tree; branches bear short shoots. Leaves dimorphic— broadly cordate or reniform, palmately veined leaves with crenate margin borne on short shoots. Leaves on long shoots fluctuate from elliptic to deltoid or broadly ovate in shape, with entire or finely rounded-serrate margin; leaves opposite, stipulate, stipules deciduous.

Floral Features. The reproductive axis is highly condensed and bears either staminate or pistillate flowers (Fig. 20.22A). The male reproductive axis terminates in a cluster of stamens subtended by four membranous bracts (Fig. 20.22B). Stamens 8–13, filaments filiform, anthers bicelled, basifixed, dehiscence longitudinal (Fig. 20.22C). The pistillate inflorescence (Fig. 20.22D) are pedunculate, of 2–6 sessile flowers, each subtended by a bract; gynoecium of a single pistil, ovary monocarpellary, unilocular, superior, ovules numerous, in 2 rows along the ventral edge of the locule (Fig. 20.22E); style linear, bears 2 stigmatic ridges. Fruit a follicle, seeds winged.

Anatomy. Vessels very small and numerous, perforation plates scalariform with numerous bars. Parenchyma sparse, terminal, sometimes diffuse. Rays 1 or 2 cells wide, heterogeneous and often fused vertically. Leaves dorsiventral, lower epidermis sub-papillose; margin with glandular emergences which exude a gummy substance; stomata Ranunculaceous type, confined to the lower surface; large cluster of crystals occur in the mesophyll.

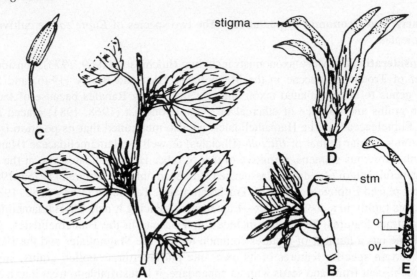

Fig. 20.22 Cercidiphyllaceae: **A–E** *Cercidiphyllum japonicum.* **A** Flowering twig. **B** Staminate flower. **C** Stamen. **D** Pistillate flower. **E** Carpel, longisection. *o* ovule, *ov* ovary, *stm* stamen. (**A, C–E** adapted from Lawrence 1951, **B** after Endress 1986)

Embryology.   Pollen grains spherical, tricolpate, with very broad furrows and finely pitted exine; shed at 2-celled stage. Ovules bitegmic, crassinucellate, hemianatropous; micropyle formed by both integuments; Polygonum type of embryo sac, 8-nucleate at maturity. Endosperm formation of the Cellular type. Wing of the seed formed from the outer integument.

Chromosome Number.   Haploid chromosome number is n = 19.

Chemical Features.   Some polyphenolics like leucodelphinidin and quercitin and also ellagic acid are reported to occur (Dahlgren 1975a).

Important Genera and Economic Importance.   *Cercidiphyllum japonicum* is a timber tree of China and Japan. It grows on the slopes of low hills in a moist deep rich soil. It attains a height of about 100 ft and often develops clusters of stems 8–10 ft wide. The timber is valuable, soft, straight-grained and light-yellow. The foliage is bright red in spring and yellow in autumn (Hutchinson 1969).

Taxonomic Considerations.   Most taxonomists include Cercidiphyllaceae within 'the Magnoliaceae. The studies of Swamy and Bailey (1949) brought to light that it has a pistillate inflorescence of 2 to 6 apetalous flowers with monocarpellary, multiovulate ovaries. They are of the opinion that Cercidiphyllaceae does not have a close affinity with any other family, and that it does not belong to any known angiosperm order. Cronquist (1968) placed it in the Hamamelidales because of the unusual structure of the pistillate flowers. However, he also agrees that the Cercidiphyllaceae is a taxonomically isolated family. According to Hutchinson (1969): "it shows some distant relationship with *Liriodendron,* but also approaches Hamamelidaceae", although he himself placed it in the Magnoliales.

Dahlgren (1975a, 1983a) treats Cercidiphyllaceae as a member of order Trochodendrales of superorder Hamamelidanae. He, too, agrees to its relationship with the Hamamelidaceous group.

Takhtajan (1980) commented that the Cercidiphyllales, an isolated unigeneric order with the only family Cercidiphyllaceae, probably shares a common origin with Trochodendrales. Both these orders belong to superorder Hamamelidanae. Takhtajan (1983, 1987) and Cronquist (1981) include this family in Hamamelidae, although Takhtajan places it in an order of its own, i.e. Cercidiphyllales; Cronquist places it in the Hamamelidales. Endress (1986) includes Cercidiphyllaceae in the Trochodendrales along with *Euptelea,* Tetracentraceae and Trochodendraceae; he places Trochodendrales in between Magnoliidae and Hamamelidae.

In view of its reduced floral structure, tricolpate pollen grains and monocarpellary ovary, it is justified to place it near Hamamelidales or even within Hamamelidales as a distinct family but not in the Magnoliales.

Taxonomic Considerations of the Order Magnoliales

The morphology, anatomy, palynology, cytology and embryology of the order Magnoliales has been extensively studied. The interrelationships between various members have been ascertained by many taxonomists (see Bhandari 1971). Embryologically the order is quite homogeneous.

Undoubtedly, the family Magnoliaceae is most primitive in this order, and Lauraceae with unisexual flowers is more advanced. Takhtajan (1987) treats Lauraceae and related families (Amborellaceae, Trimeniaceae, Monimiaceae, Atherospermataceae, Gomortegaceae, Hernandiaceae, Calycanthaceae, Idiospermaceae and Gyrocarpaceae) in a distinct order, Laurales. Scandent habit of many genera, unisexual highly reduced flower, typical floral structure are some of the differences from Magnolian families.

Cercidiphyllaceae with stipulate, palmatelyveined leaves, tricolpate pollen grains, wood anatomy similar to that of *Corylopsis* (Hamamelidaceae), and several embryological features do not suggest a relation to Magnoliaceae. Cronquist (1968), Takhtajan (1969) and Thorne (1968, 1983) include this family, as well as Eupteleaceae, Trochodendraceae and Tetracentraceae, in the Hamamelidales or as distinct orders, Cercidiphyllales, Eupteleales, Trochodendrales and Tetracentrales in subclass Hamamelididae (Takhtajan 1987). Swamy and Bailey (1949) comment that "transference of such genera as *Tetracentron, Trochodendron, Euptelea* and *Cercidiphyllum* into close relationship with the Hamamelidaceae or Saxifragaceae would merely serve to expand another order into a less homogeneous assemblage". To this, Bhandari (1971) states that evidence available from various fields "suggests that retention of this family (Cercidiphyllaceae) within the Magnoliales would not make the order more heterogeneous than would its exclusion". The other three genera are also best treated as three separate families in the Magnoliales.

The present trend is to include all four genera in subclass Hamamelididae (Cronquist 1981, Takhtajan 1987, Smith 1972). Dahlgren (1980a) places all the four genera in Trochodendrales. Every genus in a separate family under three orders— Trochodendrales, Cercidiphyllales and Eupteleales—is the conclusion of Takhtajan (1987).

According to Endress (1986), the Trochodendrales are closely related to the Magnoliales and, at the same time, to the Hamamelidales. It is intermediate between the core Magnoliidae and Rosidae/Hamamelididae (Table 20.23.1).

Some of the genera have retained vesselless wood and "oil" cells, but have already acquired tricolpate pollen grains and valvate dehiscence of anthers of the Hamamelidian type. In the Magnoliales, the perianth is not yet differentiated into typical sepals and petals (Hiepko 1965). In Trochodendrales, the perianth is reduced, whereas in Rosales/Hamamelidales, it is often differentiated into sepals and petals, sometimes also reduced (at least partly from a double perianth; Endress 1977a, Ehrendorfer 1977).

It is assumed, therefore, that the Trochodendrales evolved form an ancestral group with perianth not differentiated into sepals and petals as in the Magnoliales. Also, they are not ancestral to the Hamamelidales. This is, therefore, a conservative isolated group with common ancestry with the Hamamelidales and have retained more Magnolian features than have the Hamamelidales.

Table 20.23.1   Comparative Data of Trochodendrales*, Magnoliidae and Hamamelididae

	Magnoliidae	Trochodendrales	Hamamelididae
Perianth	Not differentiated into calyx and corolla	Not differentiated into calyx and corolla	Differentiated into calyx and corolla
Wood	Vesselless in some members	Vesselless in Tetracentraceae, and Trochodendraceae	Vessels present
'Oil' cells	Present	Present (Te, Tr)	Absent
Anther dehiscence	Longitudinal	Valvate (Te, Tr)	Valvate in some members
Filaments	Short	Long	Long
Carpel number	Broadly variable	Broadly variable	Not variable
Carpel	Stipitate, style absent	Stipitate, style absent (E)	Not stipitate
Stigma	With unicellular papillae	With unicellular papillae	Without papilla
Ovary	Superior	Semi-inferior (Te, Tr)	Semi-inferior
Ovules	Mature at anthesis	Immature at anthesis (E, C)	Immature at anthesis
Fruits	Indehiscent	Dehiscent	Dehiscent
Seeds	Not winged	Winged (Te, Tr, C)	Winged
Myricetin	Absent	Present	Present

C—Cercidiphyllaceae, E—Eupteleaceae, Te—Tetracentraceae, Tr—Trochodendraceae

*Trochodendrales has certain features common with Magnoliidae and certain others common with Hamamelididae.

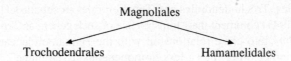

The above discussion makes it clear that the Magnoliales sensu lato is a heterogeneous assemblage of a large number of families, and it would probably be better to recognise three separate orders—Magnoliales s.s., Laurales and Trochodendrales—instead of one. Many taxonomists also support this view.

21

Order Ranunculales

The order Ranunculales comprises 2 suborders and 7 families.

The suborder Ranunculineae includes Ranunculaceae, which is the largest and most primitive. Berberidaceae, Sargentodoxaceae, Lardizabalaceae and Menispermaceae are the other families. All the families have herbaceous members or soft woody climbers or climbing shrubs.

The two families of the suborder Nymphaeineae—Nymphaeaceae and Ceratophyllaceae—are both aquatic herbs with submerged rhizomes and long-petioled, floating or submerged leaves.

Flowers inflorescent in Ranunculineae but solitary axillary in Nymphaeineae. Flowers of both suborders have spiral floral organs. Sepals, petals and stamens are all spiral. Anthers adnate, bicelled, longitudinally dehiscent. Carpels few to many, free or fused; when free, spirally arranged. Ovaries uni- or multiovulate. Nectaries at the base of the petals occur in Ranunculaceae, Berberidaceae and Lardizabalaceae. The members of this order are characterised by triaperturate pollen, anatropous, uni- or bitegmic ovules

The isoquinoline alkaloid, berberine, is reported in a few families of this order.

Ranunculaceae

A moderately large family with ca. 35 genera and 2000 species, chiefly distributed in the cooler temperate zones of the northern hemisphere. In the subtropics and tropics, they are fewer and occur at higher altitudes.

Vegetative Features.   Annual or perennial herbs, a few are woody climbers such as *Clematis*; some aquatic, e.g. *Ranunculus aquatilis*; some are amphibious, e.g. *R. sceleratus* (Fig. 21.1A). *Naravalia zeylanica* is

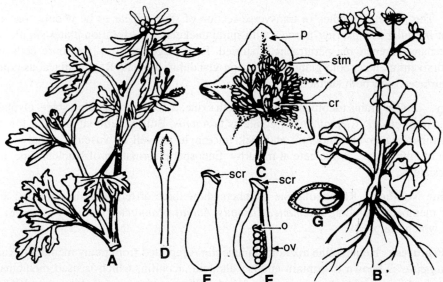

Fig. 21.1   Ranunculaceae: **A** *Ranunculus sceleratus*, **B-G** *Caltha palustris*. **A** Flowering twig. **B** Habit of *C. palustris*. **C** Flower. **D** Stamen. **E** Carpel. **F, G** Carpel, longisection and transection. *cr* carpel, *o* ovule, *ov* ovary, *p* perianth, *scr* stigmatic crest, *stm* stamen. (**A** original, **B-G** adapted from Radford 1987)

a common climbing shrub in the tropical forests at the foothills of the Himalayas. Stem usually herbaceous, rarely woody at the base. The perennial herbs persist by their rhizomes.

Leaves usually alternate, petiolate, estipulate, opposite in *Clematis*; compound [entire in *Caltha* (Fig. 21.1B), *Coptis*], pinnately compound in *Xanthorrhiza* and *Actaea*, palmately compound in *Nigella* and *Delphinium*. In *Clematis aphylla* the entire leaf is modified into tendril. In *Naravelia* leaves trifoliate with the terminal leaf modified to tendril. Leaves radical in *Anemone* and *Callianthemum*. Leaf base broad and sheathing. Heterophylly is seen in *Ranunculus aquatilis*, the submerged leaves much dissected and the floating ones lobed.

Floral Features.   Inflorescence varied: dichasial cyme in *Ranunculus* (Fig. 21.1A), raceme in *Delphinium*, solitary axillary in *Clematis cadmia*, solitary terminal in *Trollius*, *Nigella* and a panicle in *Clematis nutans*. Flowers bracteate, bracteolate, bisexual (unisexual in some species of *Thalictrum*), actinomorphic or zygomorphic as in *Delphinium*. Calyx of 5 sepals, polysepalous, green or petaloid, imbricate or quincuncial; induplicate-valvate in *Clematis* and *Naravelia*, more or less half-imbricate and half-valvate in *Clematopsis*; spurred or saccate in *Myosurus*. Corolla of 5 or more petals, usually polypetalous, imbricate (Fig. 21.1C), pocket-like nectaries at the base of each petal; *R. pinguis* of New Zealand has 2 or 3 nectaries on the petals, petals represented by petaloid staminodes in some species of *Clematis*; totally absent in *Anemone* and most species of *Clematis*. Stamens indefinite, polyandrous, spirally arranged, hypogynous, distinct (Fig. 21.1C); anthers bicelled, basifixed or adnate, dehiscence longitudinal (Fig. 21.1D). Gynoecium apocarous or syncarpous, 3- to many-carpellary, very rarely monocarpellary, e.g. *Delphinium*, when polycarpellary and apocarpous, arranged spirally on the receptacle (Fig. 21.1C), ovaries superior, unilocular with one basal ovule in *Ranunculus*; in monocarpellary gynoecium of *Delphinium*, and each carpel of polycarpellary gynoecium in *Caltha* (Fig. 21.1E, F), ovules 1 to many, placentation parietal along the ventral suture (Fig. 21.1F, G); in polycarpellary *Nigella* ovary as many loculed as the number of carpels, ovules many, placentation axile. Style and stigma 1 (many in *Nigella*). Fruit typically a follicle. Sometimes achene as in *Ranunculus*, or berry as in *Actaea*, or a capsule as in *Nigella*. Seed with minute embryo and copious endosperm.

Anatomy.   The vascular bundles in transverse section of stem appear to be widely spaced. Vessels in tangential or irregular groups, ring-porous, with spiral thickening, perforation plates simple, intervascular pittings alternate; parenchyma paratracheal, storied. Rays large, up to 12 or more cells wide. Leaves generally dorsiventral, hairs both glandular and non-glandular. Stomata Ranunculaceous type confined to the lower surface or on both surfaces.

Embryology.   Pollen grains tricolporate with smooth exine; 2-celled at shedding stage. Ovules anatropous, unitegmic as in *Anemone*, *Clematis* or bitegmic as in *Adonis*. Both functional and non-functional ovules occur. Non-functional ovules ategmic and lack micropyle as well as vascular supply; Polygonum or Allium type of embryo sac, 8-nucleate at maturity. Endosperm formation of Nuclear type, later becomes cellular, persistent.

Chromosome Number.   Ranunculaceae members show three different basic numbers: x = 7, 8 or 9. Most of the members, including *Anemone*, *Clematis*, *Adonis*, *Ranunculus*, etc. show x = 8; *Coptis* is the only genus which has x = 9.

Chemical Features.   Isoquinoline alkaloid, ranunculin is reported from many members (Ruijgrok 1968). *Aconitum napellus* is known to contain another alkaloid, aconitin, which is used medicinally. The two alkaloids, magnoflorine and berberine, have been isolated from some species (Hegnauer 1963). Ecdysones, an insect-moulting hormone, has been detected in *Helleborus* sp., along with saponins and bufadienolides (Hardman and Benjamin 1976).

Important Genera and Economic Importance. *Ranunculus* is the largest genus, with about 250 species, amongst which hydrophytes, amphibians, and mesophytes are known. Often it is a troublesome weed in the cropfields. *Clematis* is another large genus of woody climbers with about 200 species. Many species are cultivated as ornamentals because of the clusters of fragrant flowers. *Delphinium* with 250 species and *Anemone* with 100 species are other large genera. *Aquilegia, Thalictrum, Myosurus, Cimicifuga* and *Coptis* are sometimes grown as ornamentals. Various members are used medicinally. Dried tuberous roots of *Aconitum napellus*, *A. heterophyllum*, *A. chasmanthum* and *A. deinorrhizum* contain some very toxic alkaloids like aconitine, aconine and benzoylaconine. These alkaloids are used externally for neuralgia and rheumatism, and internally to relieve pain and fever (Kochhar 1981). *Cimicifuga* or black snakeroot, also used in medicine, consists of dried rhizomes and roots of *C. racemosa* (Metcalfe and Chalk 1972). Seeds of *Nigella damascena* are used as spice and also medicinally.

Taxonomic Considerations. As early as 1783, De Jussieu (see Lawrence 1951) concluded that the Ranunculaceae was the most primitive amongst the dicot families. This fact has been accepted by most later workers like Bentham and Hooker (1965a), Hallier (1912), Bessey (1915) and Hutchinson (1948). Engler and Diels (1936) and later Lawrence (1951), Melchior (1964) and Stebbins (1974) did not agree with this assignment. On the basis of wood anatomy of Ranunculaceae members, and those of some monocots like Alismataceae, Metcalfe and Chalk (1972) conclude it to be the most primitive dicot family.

According to Cronquist (1968, 1981), the order Ranunculales is the herbaceous equivalent of the Magnoliales. Ranunculaceae is ancestral to all other families in this order but is itself a derivative of the Magnoliales. Dahlgren (1983a) and Takhtajan (1987) also accept this view.

It is evident that, although anatomically Ranunculaceae is primitive, in floral features, it is advanced. There is no vesselless genus in this family and members with zygomorphic flowers, a derivative feature, are not known in more than one genus. Ranunculaceae should be treated as advanced over Magnoliales.

Glaucidium, a monotypic genus, is usually included in Ranunculaceae (Melchior 1964, Cronquist 1981). Hutchinson (1973) includes this genus in Helleboraceae, a family separated from Ranunculaceae. Langlet (1928) and Miyaji (1930) suggested its inclusion in Berberidaceae. Tamura (1963, 1972), Takhtajan (1966), Dahlgren (1975a, 1983a), Tobe (1981) and Thorne (1983) treat it as a member of an independent family, Glaucidiaceae. Tamura (1972) includes it in the Hypericales. Tobe (1981) places it together with Paeoniaceae. Dahlgren (1983a) and Thorne (1983) are also of the same opinion. Takhtajan (1987), however, erected a separate order, Glaucidiales, and placed it between Ranunculales and Paeoniales. Embryologically, *Glaucidium* is so distinct that it does deserve a family rank (Tobe 1981). It is so distinct from other nearby families morphologically, anatomically, chemically and in chromosome numbers, that it deserves an ordinal rank too.

Berberidaceae

A small family of 10–12 genera (only 2 according to Hutchinson 1969), and about 650 species. It is mainly distributed in northern temperate regions. In tropics and subtropics they grow at higher altitudes.

Vegetative Features. Perennial herbs or shrubs; rootstocks sometimes of creeping rhizomes (e.g. *Vancouveria hexandra*) or tubers. Leaves simple (*Epimedium*), unifoliolate (*Berberis*, Fig. 21.2A) or pinnately compound (*Mahonia*), basal or cauline and alternate; deciduous as in *Berberis*, or evergreen, estipulate (stipulate in *Epimedium*). There are two types of shoots in *Berberis*—short shoots (Fig. 21.2A) and long shoots; the long shoots bear leaves metamorphosed into thorns; in their axils appear short shoots. Petiole terete, swollen and hairy at junction and at insertion of leaflets in *Epimedium*.

Floral Features. Inflorescence axillary cymes (Fig. 21.2A), a raceme terminating the long or short shoots, or a thyrse. Flowers bisexual, actinomorphic, hypogynous, ebracteate and ebracteolate. Sepals 4

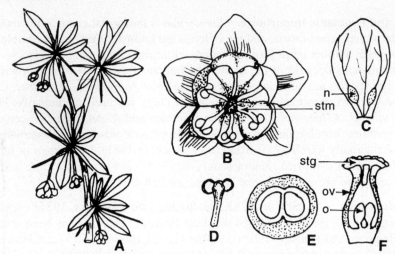

Fig. 21.2 Berberidaceae: **A-F** *Berberis thunbergii*. **A** Flowering twig. **B** Flower. **C** Petal with nectaries. **D** Stamen. **E** Carpel, cross section. **F** Carpel, longisection. *n* nectary, *o* ovule, *ov* ovary, *stg* stigma, *stm* stamen. (Adapted from Lawrence 1951)

to 6, free, often petaloid. Petals 4 to 6, free and distinct (Fig. 21.2B); sepals and petals often caducous. Sometimes one or two whorls of petaloid nectaries (Fig. 21.2C) occur between corolla and androecium. Stamens 4 to 18, distinct, generally in two whorls, those of the outer whorl opposite the petals; anthers bicelled, basifixed, mostly dehisce by flap-like valves (Fig. 21.2B, D), recurve from the base of the thecae or longitudinal dehiscence as in *Podophyllum*; filaments sometimes expanded. Gynoecium monocarpellary (derived from 2- to 3-carpellary condition), unilocular (Fig. 21.2 E, F) (2-loculed in some species of *Epimedium*); superior, ovules many on parietal placentation, or a few and basal. Style short and thick or absent, stigma 1. Fruit usually a berry as in *Mahonia* and *Berberis*, or a follicle as in *Jeffersonia*. In *Caulophyllum* the pericarp ruptures and withers just after fertilisation and the two naked seeds mature independently. Seed with copious fleshy endosperm and a small embryo. Seeds oblong, straight or slightly curved, sometimes arillate.

Anatomy. Vessels very small, commonly in irregular clusters, ring-porous in *Berberis aristata* and *B. vulgaris*; semi-ring-porous in some other species of *Berberis* and *Nandina*; with spiral thickenings. Perforations simple, slightly oblique; scalariform plates with many bars; parenchyma absent; rays up to 6 to 12 cells wide, homogeneous. Leaves usually dorsiventral; glandular hairs only in *Epimedium*. Stomata confined to lower surface, Ranunculaceous type. Solitary or cluster crystals of calcium oxalate often occur.

Embryology. Pollen triaperturate, exine spiny in *Diphylleia*. 2-celled at shedding stage. Ovules anatropous, bitegmic, crassinucellate, nucellar cap present in *Mahonia* and *Berberis*; parietal cell absent in *Jeffersonia*, *Podophyllum* and *Epimedium*. Funicular obturator and hypostase present in *B. aristata*. Polygonum type of embryo sac, 8-nucleate at maturity. Endosperm formation of the Nuclear type.

Chrmosome Number. Basic chromosome numbers are x = 6, 7, 8, 14. In *Podophyllum, Jeffersonia, Epimedium, Bongardia* and *Achlys*, x = 6, in *Ranzania* x = 7, in *Caulophyllum, Gymnospermium* and *Leontice* x = 8, and in *Berberis* and *Mahonia* x = 14 (Nowicke and Skvarla 1981). *Berberis turcomania* has 2n = 56, a tetraploid form. Hybridization between *Berberis* and *Mahonia* results in a sterile hybrid plant x *Mahoberberis neubertii* (Derman 1931).

Chemical Features.   Berberine, an aromatic alkaloid of benzylisoquinoline nature, has been obtained from the roots of *Berberis*. Besides berberine, bases of the lupin type have been obtained from the genera *Caulophyllum* and *Leontice*. Sparteine, lupanine, methylcytisine and leontine have all been isolated from these. Aglycone amurensin has been isolated from *Epimedium*. Another alkaloid protopine has been obtained from the seeds of *Nandina domestica*. About 40 alkaloids have been identified in ca. 8 genera of this family.

Important Genera and Economic Importance.   *Podophyllum hexandrum*, commonly called Indian podophyllum, is a succulent erect herb with a creeping rootstock, two leaves, 15 to 25 cm diameter and a single white or pinkish flower; occurs at high altitudes in the inner ranges of the Himalayas, usually at about 3000 to 4000 m. The rhizome of this plant is used medicinally as a very strong purgative. Its action is slow but severe. The rhizome contains a resin called podophyllin. *Podophyllum* is also reported to be useful in many skin diseases and tumerous growth (Jain 1968). *P. peltatum*, an American species, is sometimes cultivated for its rhizomes in the Himalayan ranges in Kashmir, Himachal Pradesh, Punjab, Uttar Pradesh and Sikkim in India.

Berberis is another important genus; *B. vulgaris* acts as the obligate host of the aecidial stage of wheat rust (not in India), *Puccinia graminis*. Dried roots of *B. aristata, B. asiatica* and *B. lycium* have medicinal importance. The root bark, roots and stem on boiling with water give a semi-solid substance which is used externally against eye diseases. Root and stem of *B. aristata* yield a yellow dye used for tanning and colouring leather (Jain 1968). Another important genus is *Mahonia*, which was formerly often included in *Berberis*. Their chromosome number is also same and they form intergeneric hybrids. One such hybrid is x *Mahoberberis neubertii*, it is not known to bear flowers. In *Mahonia* there are only the long shoots; leaves on these shoots are normal and either trifoliate or imparipinnate and evergreen. There are no thorns.

Taxonomic Considerations.   The family Berberidaceae (= Berberideae) is a member of the most primitive order, Ranales (Bentham and Hooker 1965a, Cronquist 1968, 1981, Dahlgren, 1983a, Takhtajan 1987). Hutchinson (1969, 1973) segregated Berberidaceae in a separate order Berberidales. According to him, the members of Berberidales are advanced over those of the Ranales. The flowers in this group are advanced in structure, as they are of reduced types of the order Ranales. However, Berberidaceae can be retained in the Ranales and is derived from the Ranunculaceae.

Dahlgren (1983a) and Takhtajan (1987), although include Berberidaceae in the Ranunculales, segregate the families Podophyllaceae, Nandinaceae and Circaeastraceae from Berberidaceae. According to them, Berberidaceae is the most highly evolved family of this order. Meacham (1980) also suggest the separation of genus *Nandina* to a monotyic family Nandinaceae.

Takhtajan (1980) suggests the alternate name Berberidales for the Ranunculales; but as Cronquist (1968) observed, more than half the species of this order belong to the Ranunculaceae and therefore it is better to use the terminology Ranunculales. Takhtajan (1987) includes Berberidaceae in suborder Berberidineae under order Ranunculales and recognises three families—Hydrastidaceae, Berberidaceae and Nandinaceae.

The genus *Hydrastis* has usually been included in the Ranunculaceae. Hallier (1905) and Wettstein (1935), however, retained it in the Berberidaceae. It is more closely related to the latter family because: (a) ovule has a longer outer integument, (b) exine morphology of pollen grains (Nowicke and Skvarla 1981), and (c) presence of the alkaloid berberin (Takhtajan 1987). However, the differences are also not less significant. In *Hydrastis*, the basic chromosome number is x = 13, and the gynoecium consists of 8 to 15 free conduplicate carpels. Therefore, it has been treated as a separate family, Hydrastidaceae, connecting link between the Ranunculaceae and Berberidaceae.

Takhtajan (1987) also recognised Nandinaceae as a distinct family, although the genus *Nandina* is very close to the Berberidaceae. It differs in basic chromosome number, x = 10, numerous sepals, morphology of pollen grains, flowers without nectary and longitudinal dehiscence of anthers.

Berberidaceae is divided into 4 subfamilies: (a) Podophylloideae includes: *Podophyllum*, *Dysosma*, *Diphylleia*, (b) Caulophylloideae *Caulophyllum*, *Gymnospermium*, *Leontice*, (c) Epimedioideae *Epimedium*, *Vancouvaria*, *Jeffersonia*, *Achlys* and *Bongardia*, and (d) Berberidoideae *Ranzania*, *Mahonia* and *Berberis*. A fifth subfamily Nandinoideae should include *Nandina* as well rather than treating it as a distinct family.

Sargentodoxaceae

Sargentodoxaceae is a family with the only taxon *Sagrentodoxa cuneata*, reported from Central China.

Vegetative Features.   Similar to members of Lardizabalaceae. Stem climbers with soft woody stem.

Floral Features.   Inflorescence similar to that of the Lardizabalaceae. Male flowers are also similar in structure to those of Lardizabalaceae. Female flowers are very much like those of Schizandraceae, the only difference is that the carpels are uniovulate.

Anatomy.   The vascular system consists of 4 large bundles alternating with smaller bundles. Vessels large; the ground tissue of the xylem consists of radial rows of fibres with circular bordered pits, obliquely crossed apertures on both the radial and tangential walls. Secretory cells (in association with vascular bundles) contain tannin.

Chemical Features.   Plants are rich in tannins.

Taxonomic Considerations.   The family Sargentodoxaceae shows characters partly of the Lardizabalaceae and partly of Schizandraceae. As long as the floral features were not known, it was included in Lardizabalaceae under the name *Holboellia cuneata*. Hutchinson (1926) gave it the status of a distinct family in the order Berberidales. According to Lemesle (1943), it is a very primitive family, on the basis of its floral features and the nature of fibres which constitute the ground tissue of xylem.

Dahlgren (1983a), Cronquist (1981), Thorne (1983) and Takhtajan (1987) also treat Sargentodoxaceae as a distinct family in the Ranunculales. It is closely related to Lardizabalaceae.

Lardizabalaceae

Lardizabalaceae is a small family of 7 (or 8) genera and ca. 50 species distributed in the northeastern Himalayas, China, Japan and in Chile in South America.

Vegetative Features.   Mostly scandent woody climbers or lianas, except *Decaisnea* which is an erect shrub. Stem soft woody. Leaves alternate, estipulate, petiolate, palmately or pinnately compound as in *Decaisnea* (Takhtajan 1969). Foliaceous stipules occur in *Lardizabala biternata*.

Floral Features.   Inflorescence racemose or solitary flowers. Flowers actinomorphic, cyclic, trimerous, hypogynous, and functionally unisexual; bisexual or polygamous only in *Decaisnea*. Calyx of 3 to 6 sepals, polysepalous, often petaloid. Corolla may or may not be present, when present, 6 small distinct petals. Staminate flowers with 6 distinct, free or basally connate stamens; anthers bicelled, basifixed, extrorse, dehisce longitudinally, nectaries usually present between perianth and filaments. Pistillate flowers with 3 to 15 free, distinct and divergent pistils. Gynoecium apocarpous, 3- to 15-carpellary, ovary superior, unilocular with usually numerous ovules on parietal placenta; stigma 1, oblique, subsessile. Fruit a berry, splitting longitudinally when mature, seed with a small embryo and a copious, firm-fleshy endosperm.

Anatomy.   Vessels of two distinct sizes in *Akebia quinata*, very large and solitary and very small and in tangential multiples or clusters. Spiral thickenings in *Lardizabala biternata*. Perforations simple but scalariform plates in *Decaisnea fargesii*; parenchyma very sparse or absent. Crystals abundant in pericyclic fibres and pith. Leaves dorsiventral, stomata confined to the lower surface and of Ranunculaceous type.

Embrylogy.   Pollen grains 3-colpate, shed at 2-celled stage. Ovules anatropous, bitegmic, crassinucellate; micropyle formed by inner integument; Polygonum type of embryo sac, 8-nucleate at maturity. Endosperm formation of the Cellular type, with large micropylar and a small chalazal chamber (Johri et al. 1992).

Chromosome Number.   The haploid chromosome numbers are x = 14, 15, 16.

Important Genera and Economic Importance.   Some of the important genera of this family are *Lardizabala, Akebia, Decaisnea, Stanutonia* and *Sinofranchetia.* The fruits of *Akebia lobata* are eaten in Japan. The most primitive genus *Decaisnea* occurs in eastern Himalayas and in western China (Takhtajan 1969).

Taxonomic Considerations.   The small family Lardizabalaceae is closely related to the Berberidaceae in which it was included by Bentham and Hooker (1965a). Bessey (1915) and Hallier (1912) treated it as a separate family, Lardizabalaceae. Most taxonomists recognise it as a separate family with close relationship with Berberidaceae and other members of the Ranales. Hutchinson (1973) places it in his Berberidales. On the basis of comparative embryological data, Lardizabalaceae deserves the status of an independent family (Johri et al. 1992), and need not be included as a tribe under the Berberidaceae of Bentham and Hooker. A comparative account on the basis of other features also support this viewpoint (see Table 21.4.1).

Table 21.4.1   Comparative Data for Berberidaceae and Lardizabalaceae

	Berberidaceae	Lardizabalaceae
Habit	Perennial herbs or shrubs	Scandant woody climbers or lianas
Flowers	Bisexual	Unisexual (except in *Decaisnea*)
Calyx	Sepals 4 to 6	Sepals 3 to 6
Corolla	Petals 4 to 6	Present or absent
Anthers	Dehiscence by flap-like valves recurving from the base of thecae (except in *Podophyllum*)	Dehiscence longitudinal
Gynoecium	Monocarpellary	Apocarpous, 3- to 15-carpellary
Anatomy	Vessels very small	Vessels very small and very large
Chromosome number	Basic chromosome no. x = 6, 7, 8, 14	Haploid chromosome no. n = 14, 15, 16
Endosperm	Nuclear type	Cellular type
Berberine	Present	Absent

It is, therefore, fully justified that Lardizabalaceae is treated as an independent family from the Berberidaceae.

Menispermaceae

A large family of 80 genera and ca. 370 species distributed in the tropics and subtropics of both the hemispheres.

Vegetative Features.   Perennial scandent shrubs or herbaceous vines (Fig. 21.5A,C), woody at least at the base; exceptions—stem tree-like in *Cocculus laurifolius*; herbaceous, low and erect from a woody rhizome in some American species of *Cissampelos*. Sometimes with a colourless bitter exudate; mostly dioecious. Leaves alternate, petiolate, simple, rarely trifoliate as in *Burasaia*; usually estipulate, entire or palmately lobed, mostly palmately veined.

Floral Features.   Inflorescence commonly supra-axillary, bracteate, many-flowered racemes, panicles (Fig. 21.5A) or umbels; rarely solitary flowers. Flowers unisexual, minute, greenish, yellowish or whitish, actinomorphic (Fig. 21.5B, D); perianth cyclic, mostly many-seriate, often deciduous. Sepals 4 to 9 in 3-merous cycles; petals 6 to 8, mostly biseriate, free and distinct, sepals united into a campanulate or globose calyx in *Cyclea*; the inner ones valvate in *Triclisia* and *Limacia*; the inner 3 sepals connate and valvate into a corolla-like structure and the 6 petals very small in *Synclisia*. The staminate flowers with usually 6 stamens but sometimes only 3, and sometimes numerous, filaments free or monadelphous; anthers 4-locular become 1- or 2-locular at anthesis; dehiscence longitudinal. The pistillate flowers with a gynoecium of 3–6 or 32 free carpels in one series, often on a gynophore. Ovary superior, unilocular with 2 ovules, one aborts, placentation parietal (Fig. 21.5F); style very short or none; stigma terminal, capitate or discoid, entire or lobed (Fig. 21.5E). Fruit, a drupe or an achene, when drupe, fleshy and indehiscent. Seed usually curved, with or without endosperm.

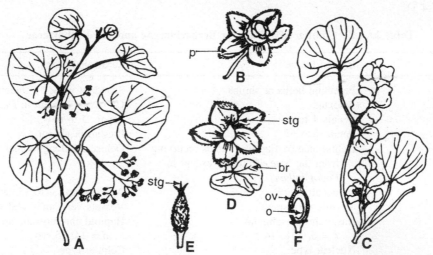

Fig. 21.5   Menispermaceae: **A-F** *Cissampelos pareira*. **A** Male plant. **B** Staminate flower. **C** Female plant. **D** Pistillate flower. **E** Carpel, **F** Carpel, longisection. *br* bract, *o* ovule, *ov* ovary, *p* perianth, *stg* stigma. (Adapted from J.K. Maheshwari 1963)

Anatomy.   Vascular bundles in young stems are individually distinct (even to naked eye), as they are separated by broad primary medullary rays which may or may not be lignified. Vessels small to medium-sized, very small in *Cocculus* (25–50 μm). Mostly solitary, but occasionally in groups of 2 to 4, seldom in contact with the rays. Perforation plates simple, slightly oblique. Intervascular pitting rather scarce, alternate and moderate-sized. Parenchyma of two types: (a) Conjunctive, between successive layers of xylem and phloem bundles including layers of isodiametric or radially elongated stone cells. (b) Apotracheal, diffuse, and in short tangential lines. Rays up to 10-23 cells wide. Anomalous secondary growth occurs in both stem and root.

Leaves generally dorsiventral, centric in *Cocculus leaeba*. Stomata mostly on lower surface, on both surfaces in *Antizoma* and *C. cebatha*; exhibits no characteristic feature—may be of Rubiaceous type (e.g. *C. carolinus* and *C. laurifolius*). Rosettes of subsidiary cells seen in certain species of *Albertisia, Anamirta, Cocculus, Fawcettia, Limacia, Macrococculus* and others (see Metcalfe and Chalk 1972). Various types of hairs occur on leaf surfaces. Crystals in the form of needles, prisms, solitary or clustered raphides and crystal sand have been recorded.

Embryology.   Pollen grains triporate, 2-celled at shedding stage; dimorphic in *Tripodandra grandiflora* and filled with starch grains in *Tiliacora*. Ovules hemianatropous to amphitropous, bitegmic, as in *Cocculus villosus* and *Tiliacora*, or unitegmic, as in *Cissampelos, Stephania* and *Tinospora*, crassinucellate; micropyle formed by inner integument in *Tiliacora*, zigzag in *Tinospora cordifolia*; nucellar cap conspicuous. Polygonum type of embryo sac, 8-nucleate at maturity. Endosperm formation of the Nuclear type; ruminate.

Chromosome Number.   Basic chromosome numbers are x = 11–13, 19, 25. Diploid chromosome number has been reported to be 2n = 18, 20, 24, 26, 38, 48, 50, 52, 54 and 78 from about 10 genera. In *Cocculus carolinus* it is 2n = 78, in *Cissampelos pareira*, 2n = 24, in *Menispermum dahuricum*, 2n = 52–54 and in *M. canadense*, 2n = 52.

Chemical Features.   Members of Menispermaceae are rich in various types of alkaloids. Benzylisoquinoline alkaloids (Thornber 1970) and sesquiterpene lactones, known as picrotoxins (Seigler 1981b), have been reported. Commercial curare (intocostrin) for medicinal use is prepared from a species of *Chondrodendron*. Potions of certain taxa are used as poison by some aborigins. Tubocurarine, the principle alkaloid of bamboo-tube curare (so named from the Brazilean packing method) is a benzylisoquinoline from *Chondrodendron tomentosum*. The members are also rich in polyalcohols like D-quercitol which occurs in bark and roots of many species (Harborne and Turner 1984). Seed fat of *Menispermum canadense* is rich in stearic acid.

Important Genera and Economic Importance.   *Tinospora cordifolia* is a large climber with succulent stems and gives out aerial roots like the trees of *Ficus benghalensis* (Moraceae). Stems and branches speckled with white glands. The unisexual flowers are minute—male flowers in cluster in the axil of bracts and female flowers solitary. Dried stem of this plant (with the bark intact) constitutes a drug useful as a tonic. The starch from its roots and stems is useful against diarrhoea and dysentry (Jain 1968). The stem of *Coscinium fenestratum* contains the alkaloid berberine. It is a robust woody, climbing shrub, with yellowish bark, and densely hairy when young. Drug obtained from dried stems is said to be effective against fever, general debility and dyspepsia. Stems yield a yellow dye which is used singly or in combination with turmeric. The sliced root of *Jateorhiza palmata*, commercially known as Calumba or Colombo root, is used medicinally for treating indigestion. The roots of *Chondrodendron tomentosum* have toxic and diurectic properties (Metcalfe and Chalk 1972). The fruits of *Anamirta paniculata* are also used medicinally in preparation of certain ointments. *Cissampelos pareira* is a high-climbing shrub with petiole as long as leaf blade; leaf blades reniform, unisexual flowers borne in axils of bracts.

Taxonomic Considerations.   The Menispermaceae have been considered as a constituent of the family Berberidaceae, Magnoliaceae or Annonaceae. De Candolle recognised it as a distinct family, a view that was upheld by Eichler and his successors (see Lawrence 1951). Most authors treat this family with a close affinity with the Berberidaceae and Lardizabalaceae (Hutchinson 1959, 1969, 1973, Cronquist 1968, 1981, Benson 1970, Stebbins 1974, Dahlgren 1975a, 1980a, 1983a, Takhtajan 1980, 1987). According to Metcalfe and Chalk (1972), the anatomical features of this family clearly suggest its affinity with the woody members of Berberidaceae. This is also confirmed by the presence of the alkaloid berberine in many members of Menispermaceae.

The placement of Menispermaceae in the Ranunculales near Berberidaceae is undisputed.

Nymphaeaceae

A small family of only 8 genera and ca. 90 species in aquatic habitats all over the world. A new genus, *Odinea*, has been added to this family (Den Hartog 1970).

Vegetative Features.   Annual (*Euryale*) or perennial aquatic herbs. These genera are common in freshwater pools and ponds. The stem is mostly rhizomatous, erect (*Victoria*) or creeping (*Nymphaea*); rhizome absent in *Cabomba*. Leaves alternate, petiolate, simple, usually floating as in *Nymphaea* or emerged as in *Nelumbo*, or heterophyllous leaves as in *Cabomba*; submerged leaves repeatedly dichotomously divided into 60–150 segments; aerial ones entire, peltate, oval or slightly cordate. Usually smooth, but prickly lower surface in *Victoria* and *Euryale*; milky latex often present. Venation actinodromous, suprabasal, reticulate type in *Cabomba* (Hickey 1973).

Floral Features.   Flowers solitary, with a long pedicel; hermaphrodite, actinomorphic, hypogynous, ebracteate, ebracteolate, large and showy, often fragrant. Calyx of 3 sepals in Cabomboideae, 4 or 5 sepals in Nymphaeoideae and indefinite in Nelumboideae, polysepalous, usually green and as large as petals; larger than petals and yellow in *Nuphar*. Petals usually showy, numerous, spirally arranged but only 3 and small, scale-like in *Nuphar*, often bear nectaries. In Nymphaeoideae the innermost petals are petaloid staminodes. Stamens 3 to 6 and cyclic in *Nuphar*, more commonly numerous and spirally arranged; introrse, filaments often extend as a sterile appendage beyond the anther lobes; anthers bicelled, dehiscence longitudinal. Gynoecium is very variable: 2- to 3-carpellary, apocarpous, ovary superior, unilocular, ovules 1 or 2 per carpel in *Cabomba* and *Brasenia*. Indefinite, free (apocarpous) carpels, each buried in a pit in the spongy torus and contains one ovule in each carpel, as in *Nelumbo*. Indefinite carpels, syncarpous gynoecium with a multilocular ovary occurs in *Nymphaea;* each locule contains several ovules scattered over the carpellary surface; the back of each carpel prolonged into a stylar process. Fruit is a pod containing 3 seeds in *Cabomba*, a spongy berry with numerous seeds in *Nymphaea*, and an aggregate of indehiscent nutlets in *Nelumbo*. Seeds with straight embryo and a starchy endosperm, exalbuminous in *Nelumbo* (Johri et al. 1992).

Epignous flowers with inferior ovaries occur in *Euryale* and *Victoria*.

Anatomy.   Vascular bundles closed and scattered resembling those of the monocots. Being aquatic plants, there are numerous intercellular spaces in the parenchymatous tissues. True vessels usually absent; present only in roots, rhizomes and young aerial stems of *Cabomba*. These are narrow, long with spiral side-wall thickening, and with simple, obliquely placed perforation plates at opposite ends (Inamdar and Aleykutty 1979). Branched sclerenchymatous idioblasts common. Laticiferous tubes or sacs present in parenchymatous tissues of all organs. Rhizomes and stolons polystelic. Leaves dorsiventral, uniseriate hairs present only on younger parts and restricted to lower surface. Stomata Ranunculaceous type, confined to the upper surface only. Hydathodes present; idioblasts of various shapes are also common.

Embryology.   Pollen grains monosulcate in *Nuphar*, *Brasenia* and *Cabomba*; zonasulcate in *Nymphaea*, *Euryale*, *Victoria* and *Odinea*; and tricolpate in *Nelumbo*; permanent tetrads in *Victoria* (Ito 1987); 2-celled at shedding stage. Ovules anatropous, bitegmic, crassinucellate; Polygonum type of embryo sac, 8-nucleate at maturity. Endosperm formation of Cellular, Helobial or occasionally of Nuclear type (Dahlgren 1975b). Seeds with endosperm and perisperm and a small embryo.

Chromosome Number. The haploid chromosome number of *Cabomba* is n = 12, that of *Brasenia* n = 40, 52; and of *Hydrostemma* (= *Barclaya*) n = 17, 18. Basic chromsome number for *Nymphaea*, *Odinea*, *Nuphar*, *Euryale* and *Victoria* is x = 10, 12, 14, 17, 29. Diploid chromosome number of *Nelumbo nucifera* is 2n = 16. Subrahmanyam and Khoshoo (1984) report that the genus *Nymphaea* supports a wide range of ploidy level from 2n = 28 to 16n = 224, both at inter- and intraspecific levels based on x = 14. The diploids occur in India and Africa; while polyploids are common in areas of extreme climatic stresses (Nayar 1984a).

Chemical Features. The occurrence of pseudoalkaloids nupharidine and deoxynupharidine in this family is reported. The recent identification of the aporphine derivatives, roemerine, nuciferine, nornuciferine and armepavine, in *Nelumbo nucifera* is taxonomically more important since it shows relationship between the Nymphaeaceae and Polycarpieae of Wettstein (Hegnauer 1963). Sesquiterpene pseudoalkaloid (Desoxynupharidine) has been reported from *Nuphar japonicum* (Hegnauer 1963).

Important Genera and Economic Importance. Most of the genera of Nymphaeaceae are grown as ornamentals. A striking member is *Victoria amazonica* with its enormous floating leaves, upturned at the margins and prickly below. It is a native of the backwaters of the Amazon River. Plants of *Euryale* and *Victoria* are armed with prickly projections. *Nuphar luteum* is the yellow water lily and is common in most parts of the northern hemisphere. Its leaves are dimorphic and flowers yellow arising from the axils of small bracts. *Nelumbo nucifera* is the sacred lotus of the Hindus. Its flowers are white or pink and often with fragrance. Its rhizome, stem as well as fruits, are edible.

Euryale ferox, a stemless, prickly aquatic herb whose fruits are also edible, occurs in Eastern China, Japan and the Vladiovostock region of Russia (Voroshilov and Nekrasov 1954). *Odinea* is a genus newly added to this family. It may be regarded as an apetalous form of *Nymphaea* (Den Hartog 1970).

Taxonomic Considerations. The family Nymphaeaceae is divided into three subfamilies:

a) Cabomboideae includes two genera *Cabomba* and *Brasenia*, flowers trimerous with 3 free sepals, 3 free petals, 3–6 stamens and 2–3 free carpels with a few ovules on marginal/parietal placenta. Fruit a follicle or pod containing usually 3 seeds.

b) Nelumboideae includes a single genus *Nelumbo*. Flowers with a perianth of indefinite, free, petaloid parts arranged spirally; stamens indefinite, spirally arranged on the basal part of a large flattened receptacle. In the flat upper portion of this spongy receptacle, indefinite free carpels are buried, each in a distinct round pit. In each carpel there is one pendulous ovule. The carpels mature into indehiscent nuts with very hard pericarp and remain buried within the spongy torus, until it decays and the nuts are set free.

c) Nymphaeoideae is the largest subfamily comprising six genera—*Nymphaea*, *Nuphar*, *Euryale*, *Odinea*, *Victoria* and *Barclaya*. The typical genus, *Nymphaea*, has flowers with a 4- to 5-segmented free sepals, followed by two whorls of petals (number of segments same as that of sepals) within which are present eight series of spirally arranged segments which depict a gradual transition from petals to stamens. These are further followed by numerous spirally arranged stamens and in the centre is the gynoecium. Gynoecium polycarpellary, syncarpous, superior, multilocular with numerous ovules scattered over the carpellary surface. Ovary inferior in *Euryale* and *Victoria*. Fruit is a spongy berry that matures below the surface of water and dehisces.

The above division was supported by Bentham and Hooker (1965a). Bessey (1915) raised these subfamilies to the rank of families, retained Cabombaceae and Nelumbonaceae in the Ranales, and placed Nymphaeaceae s.s. in Rhoeadales because of the occurrence of syncarpous ovary. Hutchinson (1973) treats Cabombaceae as a distinct family but retains the other two subfamilies under Nymphaeaceae. Both were placed in the Ranales. Cronquist (1968, 1981) recognised a separate order, Nymphaeales, including two families, Nymphaeaceae and Nelumbonaceae. Cabombaceae is included in Nymphaeaceae. According to him, the Nymphaeales might (with some justification) be treated as a suborder of the Ranunculales on purely phenetic grounds, but phylogenetically they are far apart from each other. The Ranunculales, with triaperturate pollen, appear to be related to the families of the Magnoliales. The origin of Nymphaeales must be sought amongst the families with uniaperturate pollen grains.

The absence of vessels (except *Cabomba*) in Nymphaeaceae could be interpreted as either a primitive feature or a reduction due to aquatic habitat. If it is considered to be a primitive feature, Nymphaeaceae should be treated as a separate order—Nymphaeales. Hallier (1905) suggested that the Nymphaeaceae were: "the ancestors of Helobiae and of the whole division of Monocotyledons". According to Arber (1925), the family Nymphaeaceae descended from an ancient stock which gave rise to the monocotyledons. In Takhtajan's system (1969), Nymphaeales is indicated as a branch of the monocotyledonous stock. Haines and Lye (1975) proposed that the Nymphaeales should be placed among the moncotyledons, closely allied to the Helobiae. Stebbins (1974) recognises Nymphaeales with three families: Ceratophyllaceae, Nymphaeaceae (including Cabombaceae) and Nelumbonaceae.

Dahlgren (1980a, 1983a) recognised a superorder Nymphaeiflorae, including the orders Piperales and Nymphaeales. The families in the order Nymphaeales were Cabombaceae, Ceratophyllaceae and Nymphaeaceae (including Barclayaceae). The order Nelumbonales with Nelumbonaceae were included in superorder Magnoliiflorae on the basis of: (1) vessels with scalariform perforation plates in rhizome, (2) pollen grains tricolpate, (3) receptacle obconical with monocarpellate cavities on upper side, (4) each carpel with one subapical ovule, (5) endosperm Cellular type, (6) seeds lack endosperm and perisperm, and (7) benzylisoquinoline alkaloids are present. Dahlgren (1983a) aligns the Piperales and Nymphaeales as they resemble the monocotyledons. Huber (1977) mentioned a continuous gradation of characters from advanced monocotyledons through dicotyledon-like monocots (Dioscoriales), monocotyleon-like dicots (Aristolochiales) to dicotyledons. The absence of secondary thickening due to the lack of cambial tissue in vascular strands is reported in the Nymphaeales, many Piperales and also several members of Magnoliiflorae.

Takhtajan (1980) also recognises a distinct superorder Nymphaeanae, but includes in it two orders— Nymphaeales and Nelumbonales. In his oinion the Nymphaeales are probably derived from some ancient vesselless stock of the order Magnoliales. Takhtajan (1987), however, raises the order Nelumbonales to the rank of a superorder—Nelumbonanae, and thus includes two orders Nymphaeales and Ceratophyllales in superorder Nymphaeanae. Both the superorders are treated as the most highly evolved amongst the Magnoliidae.

Nymphaeaceae s.l., including the 3 subfamilies Cabomboideae, Nymphaeoideae and Nelumboideae (Melchior 1964), is a well-accepted circumscription of this family. Also, the entire family has a distinct ecological niche. All the members grow in quiet fresh water with floating and/or submerged leaves. Cronquist (1968) observes that "the similarities in habitat, habit, aspect of flowers and placentation might conceivably reflect mere parallelism, but there is no reason to believe that they must".

Nelumbonaceae are indeed taxonomically distinct from other members of Nymphaeaceae in anatomical, morphological, embryological as well as chemical features. The distinctness is also supported by serological data (Simon 1970). A separate order, Nelumbonales, has been recognised by many authors (Bukowiecki et al. 1972, Dahlgren 1980a, Khanna 1965, Li 1955).

The Nymphaeaceae should indeed be raised to the rank of an order, distinct from the Ranales, and should include the families Cabombaceae, Nymphaeaceae and Ceratophyllaceae. The family Nelumbonaceae should be included in the Ranales.

Ceratophyllaceae

Ceratophyllaceae is a small unigeneric family of the genus *Ceratophyllum* with 3 species, of cosmopolitan distribution. This plant occurs throughout the world in lakes, ponds and slow streams.

Vegetative Features.   Submerged, perennial, aquatic plants; roots absent; stem thin, soft and chlorophyllous. Leaves whorled, dichotomously divided, sessile, the filiform or linear segments with serrulate margins, estipulate (Fig. 21.7A, B).

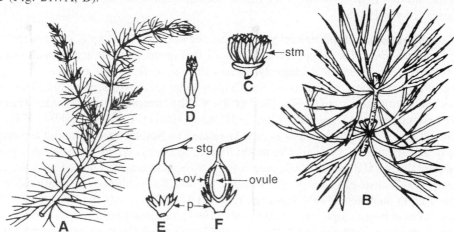

Fig. 21.7   Ceratophyllaceae: **A-F** *Ceratophyllum demersum*. **A** Vegetative plant. **B** Portion of twig. **C** Staminate flower. **D** Stamen. **E** Pistillate flower. **F** Longisection. *ov* ovary, *p* perianth, *stg* stigma, *stm* stamen. (Adapted from Saldanha and Nicolson 1976)

Floral Features.   Flowers solitary axillary, unisexual, actinomorphic, minute, inconspicuous and apetalous. Staminate flowers (Fig. 21.7C) with 10–15 sepals, basally connate to form a gamosepalous calyx; stamens 10-20, spirally arranged on a flat receptacle; each stamen with very short filament and erect linear-oblong bicelled anthers; dehiscence longitudinal (Fig. 21.7D). A thickened, often coloured connective projects beyond the anther lobes. Pistillate flowers (Fig. 21.7E, F) also with a similar perianth (or calyx) which surrounds a single pistil. Gynoecium monocarpellary, ovary unilocular, superior with one pendulous ovule (Fig. 21.2F), placentation parietal; style slender and acute, stigma 1, often not distinguishable from style. Fruit a nut, terminated by the persistent style. Seed without endosperm, embryo large and straight.

Anatomy.   Epidermal cells of stem contains chlorophyll; stomata absent from both stem and leaf. In stem the peripheral part of the primary cortex collenchymatous, but inner part including a ring of air spaces, separated from one another by radial plates, each consisting of a single layer of cells. Endodermis well defined, containing starch and provided with casparian thickenings. There is a solitary vascular strand. Xylem is represented by vertically elongated amyliferous cells which surround the centrally placed air canal in the vascular strand. Vessels absent. Secretory cells tanniniferous, occur in both leaf and stem (Metcalfe and Chalk 1972).

Embryology.   Pollen grains uniaperturate, 2-celled at shedding stage. Occasionally the pollen may germinate before being released from the anther (Sehgal and Mohan Ram 1981). Ovules orthotropous, unitegmic, crassinucellate. Polygonum type of embryo sac, 8-nucleate at maturity. Endosperm formation of the Cellular type.

Chromosome Number.   Haploid chromosome numbers are n = 12, 20, 36. In *Ceratophyllum demersum* the diploid number is 2n = 24 (Subramanyam 1962).

Chemical Features.　The glandular hairs at the apices of leaf segments contain a rose-coloured oily liquid called myriophyllin (Metcalfe and Chalk 1972). The older glands and the secretory cells in mesophyll and cortical tissue of stem contain tannin. Benzylisoquinoline alkaloids absent.

Important Genera and Economic Importance.　*Ceratophyllum* is the only genus and it has no economic importance. However, the adaptation for cross-pollination under water is interesting. The anthers break off, and, with the aid of a float at the tip of the theca, float through water till they establish on a stigma of another flower (Nayar 1984b).

Taxonomic Considerations.　The Ceratophyllaceae are regarded as monotypic, comprising the single extant genus, *Ceratophyllum*. The origin and affinities of this genus have attracted much interest, as it shows many features reminiscent of a relatively ancient lineage of flowering plants.

Historically, the phylogenetic affinities of the Ceratophyllaceae have been vague. The genus was variously associated with such unlikely families as the Callitrichaceae, Cloranthaceae, Haloragaceae, Lythraceae, Podostemaceae and Urticaceae (Gray 1848, Aboy 1936 cf. Les 1988, Gibbs 1974). Asa Gray (1848) pointed out that Ceratophyllaceae was closely related to Nelumbonaceae and Cabombaceae.

In Bentham and Hooker's system, Ceratophyllaceae was placed in Ordines Anomali under Monochlamydeae (1965c). Hallier (1912) derived it from Nymphaeaceae and placed both the taxa in the Ranales. Bessey (1915) also placed it in the Ranales but did not recognise its relationship with the Nymphaeaceae. Engler and Diels (1936), however, retained it in the Ranales and next to Nymphaeaceae. The family was treated in the same way by Melchior (1964) and Hutchinson (1973). Stebbins (1974), Cronquist (1981), Dahlgren (1983a) and Takhtajan (1987) recognise Ceratophyllaceae as a member of the Nymphaeales, particularly the genus *Cabomba*. They also agree with its close relationship with the Ranales.

Les (1988) confirmed the relationship of *Cabomba* with *Brasenia*, *Euryale* with *Victoria*, and *Nymphaea* with *Nuphar*, *Odinea* and *Barclaya*, based on an evaluation of critical character distributions by cluster analyses. Les did not support a close affinity of *Ceratophyllum* to any Nymphaealean genus. It is also argued that if, on the basis of phenetic analysis, *Nelumbo* (which is also relatively distant from other Nymphaealean members) can be treated as belonging to a separate order, Nelumbonales, a similar disposition is also warranted for *Ceratophyllum*. The isolation of these two genera from other Nymphaeales is further supported by studies on their seed anatomy (Kak and Durani 1986).

Takhtajan (1980) presumed that *Cabomba* and *Ceratophyllum* had a common origin. His views were supported by Cronquist (1981), who regarded similarities in the submerged foliage of *Cabomba* as indicating a "link" between *Ceratophyllum* and the rest of the Nymphaeales. A close relationship of *Ceratophyllum* has also been proposed by Batygina et al. (1980) on the basis of embryological characters, Okada and Tamura (1981), on the basis of chromosomal, and Sundari et al. (1982) on the basis of chemotaxonomic evidences. Young (1983), on the other hand, observed that the Nymphaeales was not entirely monophyletic and that a common ancestry of *Ceratophyllum* to other Nymphaeales is unlikely (in Les 1988).

Ito (1987), on the basis of cladistic studies, concluded that the Nymphaeales were a monophyletic group and that *Ceratophyllum* and *Cabomba* are closely related phylogenetically. Les (1988), in his analysis, has shown that none of the embryological, chromosomal, chemosystematic and cladistic evidence is strong enough to establish a link between *Ceratophyllum* and *Cabomba*. He regards the Ceratophyllaceae as a truly "living fossil" representing a group that probably diverged from some of the earliest angiosperm progenitors.

As the Ceratophyllaceae shows no close affinity to any extant angiosperm group, the family is viewed as a vestige of ancient angiosperms that became separated quite early from the line, giving rise to most of the other modern taxa. The occurrence of typically moncotyledonous features in *Ceratophyllum* and

Nymphaeales (Dahlgren et al. 1985, Haines and Lye 1975) may indicate a common gene pool somewhere in the remote ancestry of all these groups (Les 1988).

To conclude, it may be pointed out that this family is highly isolated and its affiliations to other primitive families are yet to be ascertained. In such a situation, "to better reflect the isolated position of Ceratophyllaceae in phylogenetic schemes" (Les 1988), a proposal for a new order Ceratophyllales may not be incorrect.

Taxonomic Considerations of the Order Ranunculales

The circumscription of this order has been very variable. It was originally termed Ranales (Bentham and Hooker 1965a) and included only 8 families, of which Ranunculaceae was the most primitive and Magnoliaceae comparatively advanced. Hallier (1912) did not include Magnoliaceae in his Ranales. Bessey (1915) included 24 families in the order Ranales, amongst which Magnoliaceae was treated as most primitive. Engler and Diels (1936), however, did not consider the Ranales as the most primitive dicot. Cronquist (1981) prefers to term it Ranunculales and includes 8 families. According to him, the family Ranunculaceae is the most primitive. Hutchinson (1973) considers the Ranales to be the most primitive group of the herbaceous flowering plants. Stebbins (1974) treats Ranunculales as the 7th order, comprising only 5 families. He recognised a separate order, Nymphaeales, distinct from the Ranunculales, and family Ranunculaceae as the most primitive.

More recent workers like Dahlgren (1983a, b) consider superorder Magnolianae to be the most primitive dicot and not the superorder Ranunculanae. It comprises three orders—Nelumbonales, Ranunculales and Papaverales. In Takhtajan's (1980, 1987) system also the order Ranunculales is preceded by 8 other orders. Ranunculales is included in subclass Ranunculidae and superorder Ranunculaneae. Ranunculales is related to the order Illiciales of Magnolianae and has a common origin with them. Many taxonomists, however, do not agree with this view. Metcalfe and Chalk (1972), on the basis of resemblance in wood anatomy between Ranunculaceae members and monocotyledonous family Alismataceae, regard it as the most primitive dicot.

According to Cronquist (1981), the order Ranunculales is the herbaceous equivalent of the Magnoliales. Dahlgren (1975a, 1980a, 1983a) and Takhtajan (1987) also have similar views. .

Although anatomically Ranunculales is primitive, in floral features it is more advanced. There is no vesselless genus in Ranunculales (except some Nymphaeaceae) and many genera have zygomorphic flowers—an avanced feature. Therefore, Ranunculales should be treated as an advanced taxon over Magnoliales.

Recent studies on Ceratophyllaceae (Les 1988) reveal that it is not closely affiliated to any Nymphaealean genera, such as *Cabomba* in particular. Under the circumstances, the proposal of raising this family to ordinal level and creating a new order Ceratophyllales is worth consideration.

22

Order Piperales

The order Piperales includes the families Saururaceae, Piperaceae, Chloranthaceae, and the monotypic family Lactoridaceae.

The members are predominantly herbaceous, but shrubs and trees are also known. Leaves simple, entire, stipulate (stipules absent in Piperaceae). Ethereal oil cells occur in the vegetative parts.

Flowers mostly in spikes or racemes (axillary in Lactoridaceae), bisexual, without perianth but with bracts. The seeds of Piperaceae and Saururaceae have perisperm; endospermic seeds are formed in the other two families. Often the embryo remains merely a mass of undifferentiated cells at the time of seed-shedding.

Saururaceae

This is a small family of 5 genera and 7 species distributed in Southeastern Asia, the United States of America and Mexico. *Saururus cernuus* occurs in eastern North America. A single species *Houttuynia cordata* occurs from the Himalayas to Japan.

Vegetative Features. Saururaceae has mostly perennial herbs growing in moist shady places. Leaves simple, alternate, stipulate, stipules adnate to petiole.

Floral Features. Inflorescence dense racemes or spikes (Fig. 22.1A, B). Flowers hermaphrodite, hypo- or perigynous, bracteate (lower bracts often petaloid and coloured), perianth absent. Stamens 2 to 8 (Fig. 22.1C), filaments distinct, anthers basifixed, dehiscence longitudinal. Pistil 1– to 3-4–carpellary

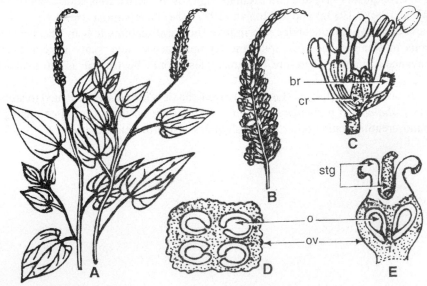

Fig. 22.1 Saururaceae: **A-E** *Saururus cernuus*. **A** Flowering branches. **B** Inflorescence. **C** Flower. **D** Ovary, cross-section. **E** Gynoecium, vertical section. *br* bract, *cr* carpel, *o* ovule, *ov* ovary, *stg* stigma. (After Lawrence 1951)

(Fig. 22.1D, E), ovary superior or half-inferior, mostly syncarpous, placentation parietal (axile in *Saururas*); ovules 1 to 10 per carpel. Fruit a fleshy follicle or capsule dehiscing apically.

Anatomy. Vascular bundles widely spaced but arranged in a single row. Crystals and secretory cells abundant in the ground tissue. Stomata numerous on the lower surface, occur sporadically on the upper surface. Stomata are surrounded by a rosette of epidermal cells.

Embryology. Pollen grains uniaperturate, sulcate. Ovules uni- or bitegmic, orthotropous, crassinucellate; Polygonum type of embryo sac, 8-nucleate at maturity. Seeds with copious perisperm, scanty endosperm, and a small embryo. Endosperm of the Cellular type, embryo minute and lacks differentiation when seed is shed.

Chromosome Number. Haploid chromosome number is n = 11, 26.

Important Genera and Economic Importance. The five genera of this family are: *Saururus, Gymnotheca, Anemopsis, Houttuynia* and *Circaeocarpus. Anemopsis* is a monotypic genus, the only species, *A. californica*, occurs in northern Mexico and parts of United States. *Houttuynia* is interesting as there is only a single highly polypoid species *H. cordata,* a single apomictic clone. Economically, the family is not of much importance.

Taxonomic Considerations. Saururaceae bears catkins and reduced flowers. It resembles the family Piperaceae, in similar type of inflorescence, and seeds with copious perisperm and a small embryo. According to Engler and Diels (1936), it is a primitive group. Hallier (1912) placed this family in his Piperales along with Piperaceae; they resemble each other in their inflorescence type. Bessey (1915) treated it as one of the families of Ranales. A study of the origin and gradual evolution of the involucre of bracts in the members of Saururaceae indicates that in the highest evolved *Anemopsis californica*, the leaves are nearly all radical stem-leaves reduced to stipule-like organs and the inflorescence looks like a Ranunculaceous flower. Hutchinson (1959) also included it in Piperales and derived it from the family Piperaceae. Cronquist (1968), Dahlgren (1980a), as well as Takhtajan (1987), treat it as a member of Piperales, superorder Magnolianae.

Piperaceae
A small family of 10 genera and ca. 1000 species (9 genera and 3100 species according to Takhtajan 1987), distributed in the moist tropical belt of both the hemispheres.

Vegetative Features. Erect (Fig. 22.2A) or scandent herbs, shrubs or rarely trees. Mostly evergreen herbaceous plants, sometimes succulent. Stem in many species twining (Fig. 22.2D). Leaves petiolate, simple, entire, sometimes succulent, alternate, rarely opposite or whorled, stipules absent.

Floral Features. Inflorescence dense, fleshy, catkin-like spikes (Fig. 22.2A, D). Flowers hermaphrodite, minute, hypogynous, bracteate, bracts often peltate (Fig. 22.2B, C), perianth absent. Stamens 1–10, filaments distinct, anthers bithecous, longitudinal dehiscence (Fig. 22.2C). Pistil 1, ovary 2- to 5-carpellary, syncarpous, unilocular, superior. Stigma 1–5; fruit a small drupe.

Anatomy. Vascular bundles in stem are scattered as in the monocots. Vessels with simple perforations, or scalariform plates with a few bars. Secretory cells occur in both stem and leaf. Stomata of Ranunculaceous type, or surrounded by rosettes of epidermal cells, and are more or less confined to the lower surface of the leaves.

Embryology. Pollen grains monocolpate; shed at 2-celled stage. Ovules bitegmic (unitegmic in *Peperomia*), crassinucellate. Frittilaria type of embryo sac; in *Peperomia* it is of the Peperomia type, 16 (synergid,

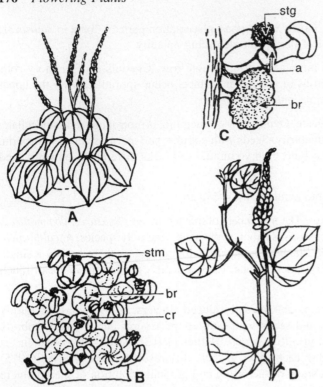

Fig. 22.2 Piperaceae: **A-C** *Peperomia sandersii* var. *argyreia;* **D** *Piper nigrum.* **A** Flowering plant. **B** Part of inflorescence. **C** Flower. **D** Fruiting twig of *P. nigrum. a* anther, *br* bract, *stg* stigma, *stm* stamen. (Adapted from Lawrence 1951)

egg, 6 antipodals, 8 polars)- or 8 (synergid, egg, 6 polars)-nucleate at maturity. Cellular type of endosperm dominates, but Nuclear type in *Piper*. Endosperm scanty in mature seed. Zygote divides vertically (Swamy 1953a). Perisperm copious and embryo minute.

Chromosome Number.   The haploid chromosome number for the family is n = 8, 11, 14, 16. High degree of polyploidy is known in *Peperomia* spp.

Chemical Features.   *Piper nigrum* contains a volatile oil in pericarp cells which gives the aroma. The pungent taste is due to a nonvolatile oleoresin and various alkaloids. Piperine, $C_{17}H_{19}NO_3$ is the chief alkaloid. Other alkaloids like chavicine, piperidine and piperettine are present in small quantities (Kochhar 1981).

Important Genera and Economic Importance.   *Piper* and *Peperomia* are the important genera. *Piper nigrum* is indigenous to the Malabar coast of southwest India. At present it is grown in the tropics of both the eastern and western hemispheres. However, the main production is from India, Indonesia and Malaysia. Sri Lanka, Cambodia, Thailand, Ivory Coast, Jamaica, Brazil and Haiti are the minor producers. *Piper betle* or betel leaf is cultivated in India, the Philippines, Indonesia and Malaysia. Leaves of this plant are chewed along with other spices, lime and tannin from *Acacia catechu* wood. *Piper methysticum* or Kava, used as a hypnotic drug, is indigenous to Fiji and other Pacific islands. *Peperomia* is predominant from Mexico to South America and also in peninsular Florida. The genus *Pothomorphe* is from South America; *Nematanthera* from Guianas, and *Verhuellia* from Cuba.

The dry fruits of *Piper nigrum* are the source of commercial black pepper. White pepper is obtained when the seeds are ground after removing the pericarp. *Piper cubeba* is the source of cubeb, another spice. *Piper longum* is yet another member. Its seeds are used as spice and also as medicine against cold and cough.

Taxonomic Considerations. On the basis of anatomical evidence, the three families, Piperaceae, Saururaceae and Cloranthaceae are related to each other (Metcalfe and Chalk 1972). Rousseau (1928) pointed out that these three families should be amalgamated into a single family, the Piperaceae. According to him, Piperaceae s.s. is the main line from which by minor modifications *Saururus* and *Chloranthus* have evolved, and by major modifications *Peperomia* has been derived. Engler and Diels (1936) treat this family as one of the most primitive dicots. Hallier (1912), Bessey (1915) and Hutchinson (1948) regard Piperaceae to be an independent and terminal offshoot of Ranalian ancestry. Rendle (1925) and later Benson (1970) considered it to be related to the Polygonales, as both the taxa have solitary orthotropous ovules. On the basis of embryological evidence, P. Maheshwari and Kapil (1966) conclude that Piperaceae is derived from Saururaceae with *Peperomia* as the connecting link. Hutchinson (1969) regards it as being derived from Ranunculaceae (Ranales) after reduction in floral whorls. According to Cronquist (1968), the entire order Piperales is derived from the Magnoliales. Dahlgren (1975a, 1980a) and Takhtajan (1987) agree with this view.

Chloranthaceae

A tropical and south-temperate family of 5 genera and 75 species (4 genera and 70–75 species according to Takhtajan 1987).

Vegetative Features. The members are aromatic herbs, shrubs and trees. Stem and branches jointed at the nodes (Fig. 22.3A). Leaves opposite, simple, stipulate; bases of opposite leaves connate (Fig. 22.3B).

Floral Features. Inflorescence terminal, slender spikes (Fig. 22.3A), cymes (Fig. 22.3G), panicles or head. Flowers hermaphrodite or unisexual, minute, bracteate, actinomorphic, epigynous or hypogynous. Flowers without sepals or petals (Fig. 22.3C-E), or with only a calyx of 3 sepals in pistillate flowers. Bisexual flowers with 1 or 3 stamens, coalescent and adnate to one side of the base of the ovary. Anther lobes 2 or sometimes 1, dehiscence longitudinal. Pistil 1, ovary monocarpellary, inferior, or superior, unilocular, with one pendulous ovule (Fig. 22.3E); style absent or highly reduced, stigma flattened, subsessile, truncate. Fruit a small ovoid or globose drupe-like berry (Endress 1987).

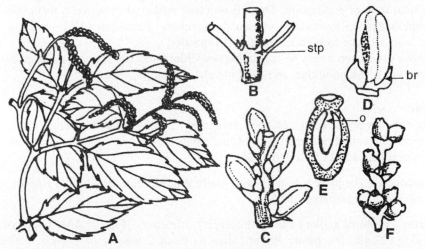

Fig. 22.3 Chloranthaceae: **A-F** *Ascarina lucida*. **A** Flowering branch. **B** Node with stipule. **C** Staminate inflorescence. **D** Staminate flower. **E** Pistillate inflorescence. **F** Pistillate flower, vertical section. *br* bract, *o* ovule, *stp* stipule. (**A-F** after Lawrence 1951).

Anatomy.   Vessels small, almost exclusively solitary; perforation plates highly elongated, scalariform and with numerous fine bars. Vessels absent in *Sarcandra*. Plants aromatic due to the occurrence of secretory cells in the leaf and stem. Stomata confined to the lower surface of the leaves and surrounded by variously arranged epidermal cells.

Embryology.   Of the three genera of this family, *Ascarina* and *Hedyosmum* have anacolpate pollen grains, and *Chloranthus* has more or less zonocolpate pollen grains. The solitary ovule is pendulous and orthotropous, bitegmic, crassinucellate. Polygonum type of embryo sac, 8-nucleate at maturity. Endosperm of the Cellular type; seed with copious endosperm and without perisperm. In most genera, the embryo is merely a mass of cells, undifferentiated at seed-shedding stage.

Chromosome Number.   Haploid chromosome number is n = 8 (*Hedyosmum*), 14 (*Ascarina*) and 15 (*Sarcandra, Chloranthus*).

Important Genera and Economic Importance.   The two most pimitive genera are *Sarcandra* and *Chloranthus; Sarcandra* has vesselless xylem. Two other genera are *Ascarina* and *Hedyosmum*. *C. glaber*, a low ornamental shrub, is sometimes grown for its foliage and bright red fruits. *Sarcandra* and *Chloranthus* are apparently entomophilous and *Ascarina* and *Hedyosmum* are wind-pollinated (Endress 1987).

Taxonomic Considerations.   Various views have been expressed about the interrelationship of the Chloranthaceae. Bentham and Hooker (1965c) regarded it as a member of Microembryeae (Monochlamydeae) because of apetalous flowers and minute embryo. Engler and Diels (1936) and Hutchinson (1959, 1969) recognised Chloranthaceae as a member of Piperales, a primitive order. Bessey (1915) considered it related to the Ranales.Takhtajan (1960, 1980) and Thorne (1968) include this family in Laurales and Annonales, respectively, whereas Cronquist (1968) retains it in the Piperales along with Saururaceae and Piperaceae. In the light of anatomical evidence, Chloranthaceae and Saururaceae have more characters common with each other than either family has with Piperaceae. According to Bhandari (1971), the presence of monocolpate pollen grains, unilacunar node and ethereal oil cells indicate Ranalian affinities. Chloranthaceae also differs from Saururaceae and Piperaceae in unilacunar nodes, pollen with reticulate exine, and Cellular type of endosperm. There is no close relationship between them. Dahlgren (1980a) includes it in the Magnoliales amongst the woody members. Takhtajan (1987) erects a distinct order, Chloranthales, with alliance to both Laurales and Piperales.

   From the above discussion it may be concluded that Chloranthaceae should be placed between Laurales, the woody members of Magnoliidae, and the Piperales of Ranunculidae.

Lactoridaceae

A small family of a monotypic genus, *Lactoris*, restricted to Juan Fernandez Islands off the western coast of South America. Currently fewer than 6 individual plants are known from Masatierra, the largest island of this Archipelogo (Zavada and Taylon 1986).

Vegetative Features.   The plant is a shrub with small, simple, alternate, stipulate leaves. Secretory cells on the leaves appear as transparent dots.

Floral Features.   Flowers axillary (not inflorescent), trimerous (Fig. 22.4A), hypogynous. Sepals 3, polysepalous (Fig. 22.4B, D); petals absent. Stamens 6, in 2 whorls of 3 each. Anthers with short filaments, extrorse; fixation adnate (Fig. 22.4C), dehiscence longitudinal. Pistil 1, 3-carpellary, nearly apocarpous, alternate with the sepals. (Fig. 22.4D,E). Ovules 6 in each carpel arranged in 2 vertical rows (Fig. 22.4F). Fruit a capsule. Seeds endospermous (Fig. 22.4G), perisperm absent.

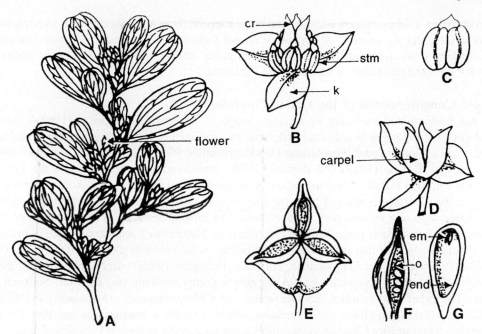

Fig. 22.4 Lactoridaceae: **A-G** *Lactoris fernandeziana*. **A** Flowering twig. **B** Flower. **C** Stamen. **D** Sepals and carpels. **E** Fruit. **F** Carpel (opened), note the ovules. **G** Seed, vertical section. *cr* carpel, *em* embryo, *end* endosperm, *k* calyx, *o* ovule, *stm* stamen. (**A-C, E, F** after Takhtajan 1969, **D** and **G** after Hutchinson 1959)

Anatomy.   Wood contains small vessels with bordered pits and simple perforations; wood parenchyma also with bordered pits. Epidermal cells of leaves papillose. Secretory cells present in the spongy mesophyll.

Embryology.   Pollen grains uniaperturate, 2-celled at shedding stage, shed in permanent tetrads (Zavada and Taylon 1986). Ovules anatropous, bitegmic, crassinucellate. Inner integument forms the micropyle. Endosperm formation of the Nuclear type (Cronquist 1981). The mature seed has copious oily endosperm, and a minute undifferentiated embryo.

Chromosome Number.   The haploid chromosome number of *Lactoris* is n = 20.

Important Genus.   The only genus and species of this family, *Lactoris fernandeziana*, is of very restricted distribution.

Taxonomic Considerations.   Bentham and Hooker (1965c) included the family Lactoridaceae in the family Piperaceae. McLaughlin (1933) supported this placement on the basis of wood anatomy, and assigned it to the order Piperales. Lawrence (1951) treated it as a member of suborder Magnoliineae (Ranales). Cronquist (1968) and Hutchinson (1969) placed it in the order Magnoliales. Hutchinson pointed out that the trimerous nature of the flowers indicated an affinity with the monocotyledons. In 1973, Hutchinson regarded it to be closely related to the Winteraceae, of which it is probably a reduced form. Stebbins (1974), Dahlgren (1975a) and Takhtajan (1980) prefer to retain Lactoridaceae in the order Laurales, although Dahlgren (1975a) is not very certain about its position. Later, Dahlgren (1980a) placed it in the Magnoliales near Chloranthaceae.

Takhtajan (1980) erects a distinct suborder Lactoridineae and comments that Lactoridaceae occupies a rather isolated position within the order Laurales, and apparently stands nearest to Chloranthaceae. On further study, Takhtajan (1987) regards it as a member of a separate order, Lactoridales, between Laurales and Chloranthales.

The woody habit, axillary flowers with well-developed sepals, more or less apocarpous ovaries, numerous anatropous ovules and the absence of perisperm do not support the placement of Lactoridaceae in the Piperales. At best it should be treated as an isolated group amongst the Laurales, or, better still, as a distinct order Lactoridales between Laurales and Piperales.

Taxonomic Considerations of the Order Piperales

Piperales has been treated variously by different workers. Bentham and Hooker (1965a) and Bessey (1915) did not treat Piperales as a distinct order, but incorporated it in the order Ranales. Hallier (1912) recognised it as a distinct order. Hutchinson (1948) presumed Piperales to be the terminal offshoot of Ranalian ancestry. Rendle (1925) and Benson (1970) considered it to be related to the Polygonales because of the common feature of solitary, orthotropous ovule in both these orders. According to Cronquist (1968), the order Piperales is derived from the Magnoliales. All the features in which they differ from the Magnoliales represent evolutionary advancement. The differences are: (a) most species of Piperales are herbaceous, (b) absence of perisperm in the members of Magnoliales, and (c) presence of orthotropous ovules in Piperales. Within this order, the Lactoridaceae is the probable ancestor. Dahlgren (1975a) at first included Piperales in the Magnolianae, but later (Dahlgren 1980a) placed this order in the super-order Nymphaeflorae near Nymphaeaceae. Also, now it comprises only two families: Saururaceae and Piperaceae. The other two families, Lactoridaceae and Chloranthaceae, are included in Magnoliales (Magnoliiflorae). Takhtajan (1980, 1987) includes only two families, Saururaceae and Piperaceae in the order Piperales. The families Chloranthaceae and Lactoridaceae are placed in Laurales (Takhtajan 1980) and later in two distinct orders. These two orders, Lactoridales and Chloranthales, are positioned between Laurales and Piperales (Takhtajan 1987).

The most primitive taxon in the group is the family Lactoridaceae and is probably the ancestral form. The Lactoridaceae and the Chloranthaceae are distinct from the Saururaceae and the Piperaceae and should therefore be treated as members of separate orders. The woody members, and seeds without perisperm, of these two families are closer to the Magnoliales s.l. (Magnolianae/Magnoliiflorae). They are also different from Magnoliales s.l. since the flowers are minute and apetalous. These two families should be grouped together in a single order, and placed between Magnoliales s.l. and Piperales.

23

Order Aristolochiales

The order Aristolochiales (as circumscribed by Melchior 1964), comprises 3 families—Aristolochiaceae, Rafflesiaceae, and Hydnoraceae. Of these, Aristolochiaceae members are all autotrophs, herbaceous or sometimes woody, scandent shrubs, but the members of both Rafflesiaceae and Hydnoraceae are parasitic in nature, mostly total root parasites. Flowers large to very large, often brightly coloured, e.g. *Cytinus* sp. and with one whorl of petaloid calyx lobes that are united below. However, other floral structures and fruits of Aristolochiaceae are quite different from those of Rafflesiaceae and Hydnoraceae.

Aristolochiaceae

A small family of only 6 genera and about 400 species distributed primarily in the tropics (12 genera and ca. 625 species according to Takhtajan 1987).

Vegetative Features. Herbs or woody scandent shrubs (Fig. 23.1A). Leaves alternate, simple, entire, petiolate, estipulate, palmately veined.

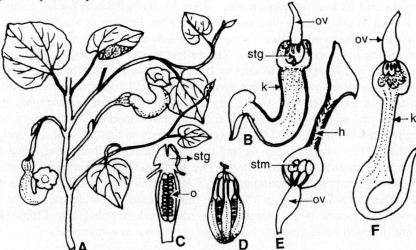

Fig. 23.1 Aristolochiaceae: **A-C** *Aristolochia durior*, **D** *A. indica*, **E,F** *A. clematitis*. **A** Flowering twig. **B** Flower, longisection. **C** Ovary, vertical section. **D** Fruit of *A. indica*. **E, F** Flower of *A. clematitis*, longisection before (**E**) and after pollination (**F**). *h* hair, *k* calyx, *o* ovule, *ov* ovary, *stg* stigma, *stm* stamen. (**A-C** adapted from Lawrence 1951, **D-F** from Gangulee and Kar 1987)

Floral Features. Flowers solitary, in axillary clusters or racemes, e.g. *Pararistolochia triactina*. Flowers bisexual, mostly zygomorphic (actinomorphic in *Pararistolochia, Asarum*). Perianth usually represented by a petaloid gamosepalous calyx (Fig. 23.1B), variously trilobed, often coloured and with a disagreeable odour. An inner whorl of 3 minute teeth-like structures (comparable to a vestigial corolla) is sometimes present, as in *Asarum;* conspicuous petals in *Saruma*. Stamens 6 to 36, free or adnate to style to produce a gynostegium-like structure. Filaments short and thick, anthers free and adnate to style, bicelled, dehiscence longitudinal. Gynoecium syncarpous, 4- to 6-carpellary, ovary inferior or half-inferior, 4- to 6-loculed, placentation axile (Fig. 23.1C), each locule with numerous ovules; style 1, short and stout, stigmas equal to the number of carpels. Fruit a septicidal capsule, often dehisce basally, i.e. parachute-like (Fig. 23.1D); an elongated, ribbed, hard, indehiscent fruit in *Pararistolochia*. Embryo minute and endosperm copious.

Anatomy.   Vessels large, and tend to be solitary in the twining species; perforations simple, intervascular pitting alternate, and small to large. Parenchyma variable, as uniseriate apotracheal lines or scanty paratracheal. Rays in climbing species wide and very high, heterogeneous and contain oil or mucilage cells. Leaves generally dorsiventral, rarely isobilateral in some *Aristolochia* species. Hairs mostly simple, uniseriate; stomata usually confined to the lower surface, Ranunculaceous type. Secretory cells occur in all plant tissues; crystals fairly frequent.

Embryology.   Pollen grains non-aperturate, exine smooth or sculptured, shed at 2-celled stage. Ovules with massive funiculus (Johri and Bhatnagar 1955); anatropous, circinotropous in *Bragantia*, bitegmic, crassinucellate; micropyle formed by inner integument; Polygonum type of embryo sac, 8-nucleate at maturity. Endosperm formation of the Cellular type.

Chomosome Number.   Basic chromosome number much variable, x = 4–7, 12, 13; in *Saruma henryi*, 2n = 52 (Takashi 1987).

Chemical Features.   Aristolochic acid is reported from different members of this family. Magnoflorine and related bases of the benzylisoquinoline type have also been isolated from some species.

Important Genera and Economic Importance.   *Aristolochia* is the best known genus of this family. It is variously known as Dutchman's pipe and pelican flower, because of the shape of calyx and fruit. *Aristolochia clematitis,* commonly called birthwort, was formerly used in medicine by midwives. This plant has an interesting pollination mechanism. The yellow flowers are upright when young. Above the inferior ovary, a spherical chamber is formed by the calyx which elongates in a funnel-like structure (Fig. 23.1E). Attracted by aroma and colour, small flies enter the tube and the chamber through the funnel-like opening. A day later, the stigmas dry up and the 6 anthers surrounding the head of the columnar style open and shower their pollen over the captured insect/s. As the hairs bend inwards or downwards, the fly cannot escape. Next, the flower stalk bends so that the formerly upright flowers tilt forward (Fig. 23.1F). The hairs also dry up and the fly escapes, but only after being showered with pollen, and is captured in another younger flower (De Wit 1963). *Pararistolochia* (a genus recognised by Hutchinson 1969) has a perfectly actinomorphic, 3-lobed calyx and a highly elongated, ribbed, hard, indehiscent fruit. The genus *Saruma* is an interesting link with the Berberidaceae as it has distinct petals and separate pistils. *S. henryi* is an evergreen, perennial herb distributed in China (Takashi 1987). Economically not of much importance; some species are grown as ornamentals.

Taxonomic Considerations. Bentham and Hooker (1965c) treated this family as a member of Monochlamydeae. Engler and Diels (1936) placed it in the Aristolochiales, next to the Santalales. Wettstein (1935) considered it as one of the 25 families of his Polycarpicae and derived it from Myrtaceous and Annonaceous ancestors. Melchior (1964) also placed Aristolochiaceae in the Aristolochiales along with Rafflesiaceae and Hydnoraceae.

   The systematic position of the family Aristolochiaceae in the Magnoliales is supported by the presence of nitrocompounds, aristolochic acid, which is closely related to benzylisoquinoline alkaloids (Gershenzon and Mabry 1983). Hegnauer (1963) also commented that "the occurrence of these chemicals justifies its (Aristolochiaceae) placement in Wettstein's classification".

   Others, like Cronquist (1968, 1981), pointed out: the "ancestry of this family must be sought within Magnoliidae rather than in any more advanced group". Amongst Magnoliidae, the Magnoliales are considered by him to be the direct ancestor because of: (a) uniaperturate pollen grains, and (b) ethereal oil cells in both the groups. Ranunculales and Piperales have also been suggested to be the probable ancestors. However, the two features mentioned above rule out the possibility of the Ranunculales being the ancestor. A certain similarity of *Asarum* to species of *Piper* and *Peperomia* of the Piperales, and also the similarity of pollen grains of *Saruma* and some Chloranthaceae members, have led some phylogenists to consider Piperales to be a probable ancestor. However, the members of the Piperales are highly advanced as far as floral reduction is concerned, and they cannot be ancestral to Aristolochiaceae.

According to Hutchinson (1973), the Aristolochiaceae is very closely related to some petaliferous families like Menispermaceae. As the stem anatomy of this family also is similar to that of the Menispermaceae, he presumes that the ancestral group for the Aristolochiaceae is a herbaceous group like the Ranales. Another similarity with the Ranales is the stomata of the leaves without any special subsidiary cells.

Dahlgren (1983a) places Aristolochiales (with the single family Aristolochiaceae) next to Annonales as one of the most primitive families amongst the Magnoliidae. Takhtajan (1987) treats this family as the sole member of the order Aristolochiales and places it next to the Piperales, probably because of the genus *Saruma*. Amongst the few dicots which have monocolpate pollen grains are the Piperales and the genus *Saruma* of the Aristolochiaceae. Also, its gynoecium is semi-apocarpous and semi-inferior and its fruit a semi-syncarpous multifolliculus. These affinities suggest multiple ancestry—polyphyletic origin of Aristolochiaceae.

Neither the Ranunculales nor the Piperales could be the ancestral form for the Aristolochiaceae. Similarity, in uni-aperturate pollen grains and presence of ethereal oil cells supports Magnoliales as the ancestral form. The primitive genus *Saruma* of Aristolochiaceae with its perigynous flowers, well-developed petals and sepals, almost free carpels and follicular fruits, could have evolved from a Magnoliales member.

Rafflesiaceae

A small family of 7 genera and 27 species distributed in the tropical and subtropical regions of the world (only 3 genera and 15 species according to Takhtajan 1987).

Vegetative Features.   Usually non-green, rootless, dioecious or monoecious, fleshy herbs, parasitic on roots and branches of various hosts (Fig. 23.2A). Vegetative body thalloid or reduced to mycelium-like tissues that invade the host; leaves usually scale-like and alternate.

Floral Features.   Flowers minute to very large; *Rafflesia arnoldii* of Malaya produces flowers that are about 1 meter in diameter and weigh up to 20 pounds. These are the largest flowers of the vegetable kingdom. Flowers solitary, unisexual, sessile, ebracteate, ebracteolate, actinomorphic, epi- or perigynous. Calyx of 4–10, free or basally connate sepals, arranged in one whorl mostly, petaloid (Fig. 23.2B); corolla absent. Staminate flowers (Fig. 23.2E) with an indefinite number of stamens, anthers sessile, bicelled, dehisce by longitudinal slits or apical pores. Pistillate flowers (Fig. 23.2C, D) with a 4-6-8-carpellary, syncarpous gynoecium; ovary unilocular, inferior or half-inferior, placentation parietal or apical, ovules numerous (Fig. 23.2D, F); style 1 or absent altogether, stigma discoid, capitate or of many lobes or surfaces. Fruit a berry with numerous seeds. Seeds minute with undifferentiated embryo and oily endosperm.

Anatomy.   The plants are devoid of chlorophyll. A vascular system exists only in the members with a massive thalloid body. The floral axis usually arises from endogenously-formed large masses of parenchyma cells called floral cushions, from within the host plant. The thallus usually consists of simple or branched rows of cells resembling fungal hyphae which penetrates the host tissue. Vascular system is highly

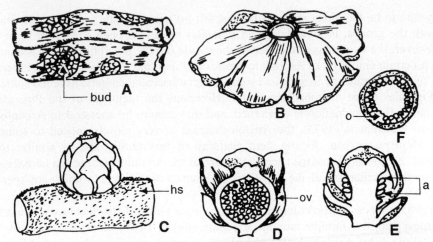

Fig. 23.2 Rafflesiaceae: **A, B** *Rafflesia arnoldii*, **C-F** *Pilostyles thurberi*. **A** Flower buds on roots of host plant. **B** Flower. **C** Pistillate flower of *P. thurberi*. **D** Longisection. **E** Staminate flower, longisection. **F** Ovary, cross section. *a* anther, *hs* host, *o* ovule, *ov* ovary. (**C-F** adapted from Lawrence 1951)

reduced, mostly reported to occur in the floral cushion and the floral axis (Metcalfe and Chalk 1972). The conducting elements are usually absent in filamentous thallus, but definite vascular strands are present in more massive thallus. Vascular bundles are present in floral cushions and also the flower-bearing axis. The conducting bundle elements are mostly broad, dense, spirally thickened tracheids without contents. The older elements are usually devoid of transverse walls, and therefore appear like vessels, as in *Rafflesia patma*.

Embryology.   Pollen grains 3- or 4-colpate or 2- or 3-porate. Ovules are not fully developed in open flowers and development continues after pollination. Ovules ana- or orthotropous (*Cytinus*), bi- or unitegmic (*Rafflesia*) and tenuinucellate. Polygonum type of embryo sac, 8-nucleate at maturity. Endosperm formation of the Nuclear type. Embryo with 4 to 10 cells in 3 to 5 tiers (Natesh and Rau 1984). Seed coat endotegmic (Boesewinkel and Bouman 1984) and sclerechymatous (Johri et al. 1992).

Chromosome Number.   Haploid chromosome number is n = 12.

Chemical Features.   Benzylisoquinoline alkaloids present.

Important Genera and Economic Importance.   Amongst the genera included in this family, *Rafflesia* is noteworthy. It is an East Indian species of parasitic plants named after its discoverer, Sir Thomas Stanford Raffles, a British officer and naturalist. About 6 species of this genus have been reported from Malaysian forested mountains. *R. arnoldii* of the Island of Sumatra, in particular, has the largest flower in the vegetable kingdom although the parasitic plant body is only thalloid in nature. The flower emits a foetid smell of tainted meat and attracts swarms of carrion flies, which are believed to bring about pollination. *Cytinus hypocistus*, a parasite on roots of *Cistus* spp., produce brilliant scarlet flowers. None of the members is of any economic importance.

Taxonomic Considerations.   Melchior (1964) and Hutchinson (1969, 1973) treated Rafflesiaceae as a member of the Aristolochiales, presumably because of similarity in the perianth. According to Cronquist (1968, 1981), however, the two taxa are very different. Rafflesiaceae members are all parasitic in nature, whereas none of the Aristolochiaceae are. Aristolochiaceae pollen grains are of uniaperturate series, whereas Rafflesiaceae has pollen grains of triaperturate series. Ethereal oil cells of the Aristolochiaceae find no parallel in the Rafflesiaceae. Placentation is axile in syncarpous members of Aristolochiaceae but parietal or apical in Rafflesiaceae. Therefore, the Rafflesiaceae is misplaced in the Aristolochiales. It should be placed in an order of its own with some other allied family like the Hydnoraceae which is also parasitic in nature.

According to Dahlgren (1980a, 1983a), Rafflesiaceae (including Cytinaceae and Mitrastemonaceae) and Hydnoraceae belong to the same order, Rafflesiales, next to the Aristolochiales.

Takhtajan (1987), however, splits this family and recognises three other families, Apodanthaceae, Cytinaceae and Mitrastemonaceae, presumably because of different haploid chromosome numbers. Apodanthaceae is distributed from southeast part of the USA to the Strait of Magellan, South Africa, Western Asia and West Australia. The members included are *Apodanthes, Pilostyles* and *Berlinianche*.

Cytinaceae is distributed in tropical South Africa and Madagascar, the Mediterranean region, Asia Minor, West Caucasus (*Cytinus*), Mexico and El Salvador (*Bdallophytum*). The third member of this family is *Botryocytinus*. Flowers unisexual, in racemes. Stamens in one ring. Ovary inferior with 8 to 14 placentae. Haploid chromosome number is n = 16.

Mitrastemonaceae is distributed in east and southeast Asia, Mexico and Central America. The only genus included is *Mitrastemon*, with bisexual, solitary flower with a superior ovary and haploid chromosome number n = 10. The family Mitrastemonaceae is not recognised by Hallier (1912) and Bessey (1915). Hutchinson (1969, 1973) and Dahlgren (1975a, 1980a) include *Mitrastemon* in Rafflesiaceae. However, Cronquist (1968, 1981) recognises Mitrastemonaceae as a distinct family as it has bisexual flowers with superior ovary and opposite scale leaves. It has been placed between Rafflesiaceae and Hydnoraceae.

Hydnoraceae

A small family of only 2 genera, *Hydnora* and *Prosopanche*, with 20 species, distributed in Africa, Malagasy Islands and South America, particularly the Pampas of Argentina.

Vegetative Features. Thick, creeping, branched, underground "rhizomatoids", terete, or angled or ribbed and devoid of roots. Unlike real rootstock, they are not divided into nodes and internodes and are provided with hood-shaped formation which protects the growing apex. These modified "roots" serve the purpose of searching new host plants.

Floral Features. Flowers thick, fleshy, sessile or rhizomatoids, bisexual, actinomorphic, apetalous, epigynous. Calyx thick, fleshy, 3-(4-) 5-lobed, sometimes tubular, lobes valvate. Stamens 3-(4-)5, opposite the calyx lobes, usually sessile, united to form a thick sinuose-annular or ovoid synandrous structure, with numerous parallel thecae; anthers dehisce longitudinally. Gynoecium 3–5-carpellary, syncarpous, ovary inferior, wholly or partially embedded in soil, unilocular with numerous placentae; may be laminar or parietal, ovules numerous; stigma sessile, truncate-pulvinate. Fruit a large thick-walled berry with fleshy pulp; seeds numerous, minute with small undifferentiated embryo, surrounded by copious endosperm and a thin layer of perisperm.

Anatomy. Absorption of nutrients from the host plants is through the 'rhizoids' which may be angular or circular (in transection) in different species. Outer surface of these rhizoids is covered by a layer of cork cells which protect the parenchymatous vascularized cylinder. The vascular bundles vary in number and are arranged in one or two rings; sieve tubes in the phloem of *Prosopanche* (Metcalfe and Chalk 1972) and simple perforation in the vessels (Takhtajan 1987) have been reported. Strands of elongated pith cells, fibrous in *Prosopanche* and prismatic in *Hydnora*, are present in the central region of the 'rhizoids'. The cells of the parenchymatous ground tissue contain brown tanniniferous mucilaginous contents, starch and sometimes also crystals.

Embryology. Pollen grains mono- or 2-colpate; 2-celled at shedding stage. Ovules orthotropous, unitegmic, tenuinucellate. Allium type of embryo sac, 8-nucleate at maturity. Endosperm formation of the Cellular type.

Imortant Genera and Economic Importance. The two genera *Hydnora* and *Prosopanche* are both total root parasities. *Hydnora* survives on the roots of *Acacia* and *Euphorbia* in Africa and Madagascar; *Prosopanche* is parasitic on the roots of the Argentinian algarobba, a member of the genus *Prosopis* (De Wit 1963).

Taxonomic Considerations. The family Hydnoraceae is closely related to the Rafflesiaceae as well as to the Aristolochiaceae, especially to the genus *Asarum*. Although most phylogenists retain these three families in the order Aristolochiales (Engler and Diels 1936, Melchior 1964, Cronquist 1968, 1981, Dahlgren 1980a), Bentham and Hooker (1965c) do not recognise Hydnoraceae as a distinct family. Takhtajan (1980) and Thorne (1983) include it in the Rafflesiales along with Rafflesiaceae. The two families are close to each other in their parasitic habit, but Hydnoraceae differ from Rafflesiaceae in having root-like structures, ebracteolate bisexual flowers, and stamens borne on a column-like structure. Therefore, the erection of a distinct order Hydnorales (Takhtajan 1987) for the family Hydnoraceae alone is justified.

Taxonomic Considerations of the Order Aristolochiales

Engler and Diels (1936) and Melchior (1964) place the three families Aristolochiaceae, Rafflesiaceae and Hydnoraceae in this order. In our opinion the two parasitic families cannot be included in Aristolochiales (see Table 23.4.1).

Table 23.4.1   Comparative Data for Aristolochiaceae, Rafflesiaceae and Hydnoraceae

Aristolochiaceae	Rafflesiaceae and Hydnoraceae
1. Autotrophic plants	Total root parasites
2. Herbaceous or scandent shrubs	Thalloid plant body
3. Ethereal oil cells present	Ethereal oil cells absent
4. Pollen of uniaperturate series	Pollen of triaperturate series
5. Ovary, when present, syncarpous, axile placentation	Ovary syncarpous, parietal or apical placentation

Therefore, the Aristolochiaceae alone should be included in the order Aristolochiales. It has been pointed out earlier (p. 185) that although Rafflesiaceae and Hydnoraceae are both total root parasites, they differ from each other in many features. Consequently, these two families should also be placed in distinct orders—Rafflesiales and Hydnorales (Takhtajan 1987).

24

Order Guttiferales

The order Guttiferales comprises 4 suborders and 16 families.

The suborder Dilleniineae includes Dilleniaceae, unigeneric Paeoniaceae, Crossosomataceae, unigeneric Eucryphiaceae, monotypic Medusagynaceae, and Actinidiaceae.

The suborder Ochnineae includes Ochnaceae, Dioncophyllaceae, unigeneric Strasburgeriaceae and Dipterocarpaceae.

The suborder Theineae includes Theaceae, Caryocaraceae, Marcgraviaceae, Quiinaceae and Guttiferae.

The suborder Ancistrocladineae includes unigeneric Ancistrocladaceae.

The member are trees, shrubs or woody climbers, rarely herbs, with simple or occasionally compound, stipulate or estipulate leaves; mostly alternate, opposite or whorled in Quiinaceae, Medusagynaceae and Guttiferae. Flowers showy, mostly hypogynous with imbricate calyx and mostly polypetalous corolla; stamens numerous, centrifugal, or less frequently few and cyclic; pollen is usually shed at 2-celled stage. Gynoecium 2- to many-carpellary, syncarpous, rarely apocarpous as in Dilleniaceae, and mostly axile, seldom parietal placentation. Ovules bitegmic except in Actinidiaceae. Endosperm present or absent. Many of the families are rich in alkaloids, tannins, terpenes, myricetin, quercetin and other chemicals.

Dilleniaceae

A family of 12 genera and 350 species (Takhtajan 1987), distributed in the tropics of both the New and the Old World, mostly in Australia and South America. Occurrence of the genus *Hibbertia* in Madagascar and Australasia is an example of discontinuous distribution.

Vegetative Features.   Trees (*Dillenia*), shrubs (*Hibbertia*) or woody climbers; rarely herbaceous (*Acrotrema thwaitesii*, Fig. 24.1A). Leaves simple, petiolate, spirally arranged or alternate, stipulate, stipules caducous, winged and adnate to petiole as in *Davila vaginalis*, *D. wormifolia* (Fig. 24.1B) or absent; venation seemingly parallel.

Floral Features.   Flowers sessile, or borne on short pedicels, singly or rarely in inflorescence. Flowers small to large, ebracteate, ebracteolate, actinomorphic, bi- or unisexual and then the plants are dioecious or less commonly monoecious, hypogynous, without intrastaminal disc. Sepals 5, polysepalous, imbricate, persistent, often thick and fleshy as in *Dillenia indica*. Petals free, 5, imbricate, often crumpled in bud; only 3 in *Trisema* (Fig. 24.1E). Stamens generally numerous, free or variously basally fasciculate, mostly persistent; anthers bicelled, linear (Fig. 24.1F), often of two types—the inner erect and introrse and the outer recurved and extrorse; dehiscence longitudinal or by apical pores as in *Hibbertia miniata*, *H. montana* and *H. serrata*. Gynoecium of numerous free carpels adherent to central axis (Fig. 24.1C, D, H), rarely only one as in *Trisema* (Fig. 24.1G), *Delima* and *Doliocarpus* spp. Ovary superior, unilocular with parietal placentation, ovules one or more in each ovary; styles as many as carpels, free. Fruit a follicle or berry-like, often enclosed in persistent sepals as in *Dillenia indica* (Fig. 24.1I). Seed with minute embryo, copious fleshy endosperm and with funicular aril.

Anatomy.   Vessels predominantly solitary, medium-sized, perforation plates scalariform or simple and scalariform, intervascular pitting opposite to scalariform. Parenchyma predominantly apotracheal, diffuse or in uniseriate lines with a few cells around the vessels; rays 8 to 10 cells wide, heterogeneous.

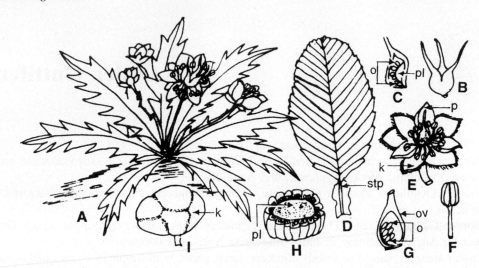

Fig. 24.1 Dilleniaceae: **A-C** *Acrotrema thwaitesii,* **D** *Davilla wormifolia,* **E-G** *Trisema coriacea,* **H,I** *Dillenia indica.* **A** Flowering plant. **B** Gynoecium. **C** Carpel, vertical section. **D** Leaf of *D. wormifolia.* **E** Flower of *T. coriacea.* **F** Stamen. **G** Carpel, longisection. **H** Carpel, cross section. **I** Fruit. *k* calyx, *o* ovule, *ov* ovary, *pe* petal, *pl* placenta, *stp* stipule. (**A-H** adapted from Hutchinson 1969, **I** after Lawrence 1951)

Parenchyma and ray cells may contain raphides. Leaves mostly dorsiventral, stomata usually Ranunculaceous but Rubiaceous type in *Tetracera oblonga.* Sunken stomata in xeromorphic Hibbertieae, especially in the phylloclades of *Pachynema.*

Embryology.   Pollen grains are monads, spheroidal, sometimes subprolate or suboblate; tricolpate, tricolporate, tetracolpate or sometimes inaperturate; exine finely punctate or coarsely reticulate. Ovules anatropous, campylo-amphitropous, bitegmic and crassinucellate. Polygonum type of embryo sac, 8-nucleate at maturity. Endosperm formation of Nuclear type.

Chromosome Number.   Basic chromosome numbers variable: x = 4, 5, 8–10,12,13.

Chemical Features.   *Davilla, Curatella, Doliocarpus* and *Hibbertia* are rich in myricetin and/or quercetin and kaempferol, in varying combinations. Myricetin is rare in *Tetracera* and *Dillenia;* methylated flavonols occur frequently. Both taxa have isorhamnetin and rhamnetin; in addition, azaleatin and rhamnocitrin present in *Tetracera;* ellagic acid and leucoanthocyanins are also common (Kubitzki 1968).

Important Genera and Economic Importance.   Some of the important genera of this family are *Acrotrema, Dillenia, Tetracera, Curatella* and *Hibbertia.* The fleshy calices of *Dillenia indica* and *D. pentagyna* are edible. Leaves of *Curatella americana* from Central America, and some species of *Tetracera* are used as a substitute for sand paper in polishing work. Species of *Dillenia, Hibbertia* and *Tetracera* are important ornamentals.

Taxonomic Considerations.   The family Dilleniaceae is included in the order Ranales by Bentham and Hooker (1965a), in the Parietales by Engler and Diels (1936), in the Theales by Thorne (1968, 1983) and in the Dilleniales by Cronquist (1968, 1981), Dahlgren (1975a, 1980a, 1983a), Hutchinson (1959, 1969, 1973) and Takhtajan (1980, 1987). According to Stebbins (1974), Dilleniaceae is one of the primitive families which exhibits a wide range of ecological preferences. It provides "an admirable example of ancient adaptive radiation from intermediate habitats both towards more xeric and more mesic adaptations."

On the basis of pollen morphology, Dickison (1976b) concluded that more or less tricolpate, tricolporate, triporate, tetracolpate and reticulated pollen of Dilleniaceae can have resemblances with those of Theales (or Guttiferales) members but not any Ranalean member.

Dilleniaceae resemble the members of Paeoniaceae and Crossosomataceae and all three taxa can be placed in a distinct order, Dilleniales.

Paeoniaceae

A unigeneric family; the genus *Paeonia* has 33 species distributed in temperate countries of both the Old and the New World. There are 3 sections: (1) Moutan—4 species of shrubs, native to western China, (2) Paeonia—27 species of perennial herbs, occur from Spain and north Africa through temperate Asia into Japan, and (3) Onaepia—2 species of perennial herbs native to western North America (Keefe and Mosely 1978).

Vegetative Features. Perennial rhizomatous herbs, occasionally shrubby, roots sometimes tuberous. Leaves alternate, biternate, estipulate.

Floral Features. Flowers solitary, terminal or axillary; bisexual, ebracteate, ebracteolate, pedicellate, complete, large, white. Calyx of 3 to 5 sepals, free, rounded or subfoliaceous, imbricate, persistent. Corolla of 5(–10) petals, polypetalous, large, imbricate. Stamens numerous, centrifugal with oblong extrorse anthers; nectariferous disc fleshy, or separate glands or forming a large subglobose envelope around the gynoecium. Gynoecium apocarpous, 2- to 5-carpellary, ovary more or less fleshy, bearing numerous biseriate ovules, style inconspicuous, stigma thick, falcate. Fruit of 2 to 5 large, leathery, ventrally dehiscent follicles. Seeds large, at first red, later black and shiny, arillate, with copious endosperm.

Anatomy. Vessels extremely small to moderately small (16–88 μm) in diameter, with scalariform thickenings and scalariform perforation plates with a few to intermediate number of bars; distinctly ring-porous. Rays only up to 3 cells wide, uni- or multiseriate, mostly heterogeneous, rarely homogeneous; parenchyma apotracheal and sparse, and occurs in a scattered diffuse pattern. Calcium oxalate crystals present. Leaves mostly dorsiventral, stomata usually confined to the lower surface, Ranunculaceous type.

Embryology. Pollen grains 3-colporate, prolate, spheroidal with reticulate exine; 2-celled at shedding stage. Ovules anatropous, bitegmic, crassinucellate; both the integuments form the micropyle. Polygonum type of embryo sac, 8-nucleate at maturity. The aril at the base of each ovule is a prominent feature. The young ovule has a prominent vasculated hypostase which may be a relatively primitive character. Endosperm develops at the chalazal end (Camp and Hubbard 1963). Endosperm formation of the Nuclear type.

The embryogeny of *Paeonia* is a controversial issue. According to Yakovlev and Yoffe (1957, 1961), the nucleus of the zygote undergoes repeated free-nuclear divisions to form a coenocyte and the nuclei become distributed around a large central vacuole. Centripetal wall formation occurs, producing a proembryonal mass. Several peripheral cells develop meristematic centres, and produce embryonal primordia. Only one of these primordia matures into a dicotyledonous embryo.

According to Murgai (1962), the division of the zygote nucleus is followed by wall formation. The basal cell becomes coenocytic with a vacuole in the centre. The embryo develops from one of the peripheral groups of actively dividing cells in the coenocyte. According to Johri et al. (1992), these observations require confirmation.

Chromosome Number. Basic chromosome number is x = 5. *Paeonia californica* and *P. brownii* show a high degree of meiotic irregularity in natural populations, such as chromosome fragmentation (Walters 1956).

Chemical Features.   Paeoniflorin, benzoylpaeoniflorin, oxypaeoniflorin, paeonol, paeonoside, paeonolide and apiopaeonoside are the seven compounds that have been isolated from the root of *Paeonia suffruticosa* and *P. lactifera* (Yu Jin and Xiao Pei-Gen 1987). Amongst these, the paeonol compounds are restricted to the woody section Moutan, and absent in the herbaceous section Paeonia.

Important Genera and Economic Importance.   The only genus, *Paeonia*, bears large, ornamental flowers. Although a number of alkaloids have been isolated from various species of this genus, these have not been put to any use so far.

Taxonomic Considerations.   *Paeonia* is a highly controversial genus and has been assigned to various taxonomic disciplines, as shown in Table 24.2.1. It has been variously placed in Berberidaceae, Ranunculaceae, Paeoniaceae or the Guttiferales. Worsdel (1908a) placed this unique genus in a separate family, Paeoniaceae, and pointed out its anatomical resemblance to the Magnoliaceae. Worsdel (1908b) recognised its resemblance to the Calycanthaceae also, as the "floral bracts" pass imperceptibly into the sepals in both taxa.

Table. 24.2.1   Taxonomic Position of Paeoniaceae

Order	Family	Reference
Ranales	Berberidaceae	Hallier (1905), Lotsy (1911), Langlet (1928)
Ranales	Ranunculaceae	Engler and Diels (1936), Lawrence (1951), Benson (1970), Melville (1983)
Guttiferales	Paeoniaceae (near Dilleniaceae)	Melchior (1964)
Dilleniales	Paeoniaceae	Dickison (1967a, b), Cronquist (1968, 1981), Keefe and Moseley (1978), Corner (1946), Takhtajan (1969), Gornall et al. (1979), Dahlgren (1975a)
Theales	Paeoniaceae	Thorne (1974)
Paeoniales	Paeoniaceae	Dahlgren (1980a), Takhtajan (1980, 1987)

On the basis of centrifugal stamens, persistent sepals, aril, hard testa, woody tendency and estipulate leaves, Corner (1946) pointed out the relationship of *Paeonia* to the Dilleniaceae. Melchior (1964) placed it in a distinct family, Paeoniaceae, near Dilleniaceae in the order Guttiferales. Keefe and Moseley (1978) also pointed out close relationship between these two taxa: "*Paeonia* is most probably best placed as a valid and distinct unigeneric family with the Dilleniales". Most phylogenists treat Paeoniaceae as a unigeneric family (Cronquist 1981, Dahlgren 1983a, Dickison 1967a,b, Hutchinson 1973, Takhtajan 1987). Takhtajan (1969) separated it from Dilleniaceae because of: (1) thick, fleshy carpels, (2) broad stigmas, (3) a lobed and unusually prominent, staminal nectariferous disc, (4) a massive outer integument, (5) seed coat character, and (6) its peculiar embryogeny (Yakovlev 1957, Johri et al. 1992).

Takhtajan (1987) even recognises a distinct order, Paeoniales, and places it between Dilleniales and Theales. The Paeoniaceae probably arose from Dilleniaceous ancestors in the Indo-Malayan region, and migrated northwards into western China.

Crossosomataceae

A small family of 3 genera and 10 or 11 species, distributed in the American tropics. *Crossosoma californica* is reported from California and New Mexico.

Vegetative Features.   Shrubs or small trees with rough bark. Plants are often xerophytic. Leaves alternate, or clustered on short shoots, entire, simple and leathery.

Floral Features.   Flowers solitary, terminal on short shoots, showy, bisexual, actinomorphic, perigynous. Sepals 5, polysepalous but basally connate, imbricate; petals 5, poypetalous, imbricate, white or purplish, deciduous, orbicular to oblong in shape. Stamens numerous in 3 or 4 whorls, free, arranged on a disc lining the hypanthium. Anthers oval-shaped to oblong, bicelled, basifixed; dehiscence longitudinal. Gynoecium of 3- to 5-carpellary, apocarpous, perigynous ovaries; each ovary unilocular, style highly shortened or obsolete, stigma bilobed (?); placentation parietal, ovules many in rows. Fruit a follicle, seeds globose to reniform, arillate; embryo curved, endosperm scanty.

Anatomy.   Vessels very small and angular, ring-porous, with simple perforations; intervascular pittings alternate; parenchyma sparse, paratracheal. Rays up to 6 cells wide, heterogeneous; fibres with large bordered pits. Leaves isobilateral; stomata on both surfaces but more numerous on the lower surface; Ranunculaceous type. Minute, yellow, acicular crystals closely packed in clusters of cells in the phloem have been observed. These are different from raphides and crystal sand reported in the Dilleniaceae members (Metcalfe and Chalk 1972).

Embryology.   Pollen grains tricolporate or bicolporate, prolate with reticulate exine; 2-celled at the shedding stage. Ovules amphitropons, bitegmic and crassinucellate. Polygonum type of embryo sac, 8-nucleate at maturity. Endosperm formation of the Nuclear type.

Chromosome Number.   Basic chromosome number is x = 6.

Important Genera and Economic Importance.   Most taxonomists report it as a family with the only genus *Crossosoma. Apacheria* is the other genus, and *Forsellesia* was added to this family by Thorne and Scogin in 1978. The family has no economic importance.

Taxonomic Considerations.   Crossosomataceae has usually been treated as a distinct family allied to the Rosaceae (Engler and Diels 1936). It was placed near Dilleniaceae by Hutchinson (1969, 1973), and Cronquist (1968, 1981). The resemblances between the two taxa are: (1) centrifugal development of stamens, (2) similar floral anatomy, and (3) arillate seeds. However, there are also differences between the Dilleniaceae and Crossosomataceae: (1) perigynous flowers, (2) apocarpous ovaries, (3) presence of hypanthium, (4) more advanced xylem, and (5) absence of raphides and crystal sand.

Features like perigynous flowers and apocarpous gynoecium help us to place it in the Rosales, as has been suggested by Dahlgren (1980a). Takhtajan (1987), however, places it in a separate order, Crossosomatales, in Rosidae, thereby supporting Dahlgren's view. Additional data are necessary for a correct assignment of this family.

Eucryphiaceae

A unigeneric (*Eucryphia*) family with 4 species (6 species, according to Takhtajan 1987). This family is of interest because of its disjunctive distribution, endemic to Southeast Australia, Tasmania and southern Chile.

Vegetative Features.   Evergreen trees or shrubs, often resinous. Leaves opposite, simple or pinnately compound, petiolate, stipulate, stipules minute.

Floral Features.   Solitary axillary flowers, bisexual, actinomorphic, showy, white, hypogynous, ebracteate, ebracteolate. Sepals 4, imbricate, joined together at the apex forming a calyptra; it gets detached from the base at anthesis. Petals 4, polypetalous, imbricate, white. Stamens numerous, free, centripetal, arising in many whorls from a thin disc at the base of the ovary; staminodes may or may not be present. Anthers bicelled, small, basifixed, dehiscing longitudinally. Gynoecium syncarpous, 5- to 12- (or 18-) carpellary, ovary as many-loculed, superior, placentation axile, ovules numerous in each locule, pendulous; styles as many as carpels. Fruit a woody or leathery capsule; mature ovary separates by dehiscence along ventral sutures into distinct follicle-like units. Seeds winged, flattened, embryo small, endosperm copious.

Anatomy. Vessels small, mostly solitary but with some multiples; spiral thickenings present in some species, absent in *Eucryphia cordifolia.* Perforation plates scalariform or scalariform and simple, intervascular pitting transitional between scalariform and opposite. Parenchyma apotracheal, diffuse to slightly banded and sometimes in terminal bands. Rays uniseriate, or 2–3 cells wide, heterogeneous. Leaflets dorsiventral, stomata Rubiaceous type, confined to the lower surface. Solitary as well as cluster crystals are sometimes present.

Chromosome Number. Haploid chromosome numbers are n = 15,16.

Chemical Features. Flavonol 5–methyl ethers are present (Harborne 1969). The flavonoids caryatin, azaleatin and quercetin are reported in the two species from South America. Species from Australia have a simpler flavonoid pattern (Bate Smith et al. 1967). They lack azaleatin and caryatin.

Important Genera and Economic Importance. *Eucryphia* is a highly ornametal tree with its showy flowers. Strong, hard and close-grained wood of *E. cordifolia* is used for flooring, furniture, telegraph poles, etc. It is also used for making canoes. Flowers are an excellent source of honey (Hutchinson 1969).

Taxonomic Considerations. There are diverse opinions regarding the taxonomic position of Eucryphiaceae. Bentham and Hooker (1965a) considered it to be a member of the Rosales. Engler and Diels (1936) and Lawrence (1951) treated it as a member of the suborder Theineae of the order Parietales and allied it to Dilleniaceae. Hallier (1912) included it within the Theaceae and considered it to have been derived from the Malvales. Wettstein (1935) accepted it as a distinct family allied to Dilleniaceae and Ochnaceae. Cronquist (1968, 1981) commented that pollen morphology and wood anatomy of this family suggest its alliance to the Cunoniaceae. Also, the centripetal stamens are more common in the Rosidae and not in the Dilleniideae. Hutchinson (1973) considered Eucryphiaceae to be an advanced opposite-leaved group derived from the Theaceae rather than a member of Rosales, as suggested by some earlier workers like Bentham and Hooker (1965a) and Cronquist (1968). Dahlgren (1980a) and Takhtajan (1987) also prefer to place it in the Cunoniales of Rosiflorae.

The centripetal stamens, pollen morphology and wood anatomy of the Eucryphiaceae suggest its alliance more towards Cunoniaceae than Dilleniaceae.

Palynological data (Hideux and Ferguson 1976) and comparative anatomical results (Dickison 1978) suggest its close relationship within the Rosales (or Cunoniales). Harborne (1969) supported the earlier view of Melchior (1964), i.e. its position is within the Dilleniales, close to the Dilleniaceae or Theaceae. The P-type sieve element plastids of the form present in Eucryphiaceae have not been recorded in Dilleniales/Theales nor in Rosales/Cunoniales. These unusually small plastids correspond to the equally small S-type plastids of Cunoniaceae (Behnke 1985). On this basis, its position was suggested to be between P-type family Connaraceae and S-type family Cunoniaceae, as proposed by Cronquist (1981).

Similarity between Eucryphiaceae and Ochnaceae has also been observed, primarily in the structure of fruit. At maturity, the carpels separate and remain attached to the axis, but in wood anatomy Eucryphiaceae is more primitive. It is probably an offshoot of Dilleniaceae, and the similarity in fruit structure with Ochnaceae is due to parallel evolution (Decker 1966).

Medusagynaceae

This family is represented by a single monotypic genus, *Medusagyne*, from the Seychelles Islands. The species included is *M. oppositifolia.*

Vegetative Features.   Shrubs or small trees, leaves opposite, entire, simple with a crenate margin, estipulate.

Floral Features.   Flowers in terminal panicles; bisexual, actinomorphic, hypogynous. Calyx of 5 sepals united at the base, imbricate, persistent. Corolla of 5 petals, polypetalous, twisted. Stamens numerous, free; filaments often shorter than petals, anthers basifixed, longitudinally dehiscent; often anthers are arranged at different heights. Gynoecium 17- to 25-carpellary, carpels nearly free from the central axis; ovary superior, 17- to 25-locular, ovules 2 per locule, 1 erect and the other pendulous; styles as many as loculi, short, free and arranged in a ring on the shoulders of the carpels; stigma capitate. Fruits capsules, dehiscence septicidal from base and diverging like the ribs of an umbrella. Seeds winged, without endosperm.

Anatomy.   Young stem shows a fairly broad, parenchymatous cortex followed by a pericycle layer containing a closed ring of sclerenchyma. Phloem stratified into alternate concentric rings of sieve tissue and strongly thickened cells. Xylem ring interrupted by primary rays that are mostly 5–8 cells wide. A characteristic feature is the presence of cortical bundles in members of Medusagynaceae. Vessels mostly solitary, only 15 to 20 μm in diameter and with simple perforations. Leaves dorsiventral, stomata Ranunculaceous type, confined to the lower surface; mucilage cells present in mesophyll. Clustered crystals occur in the petiole and around vascular strand of the midrib.

Embryology.   Pollen grains (2)-3-(4)-porate. The anatropous ovule has a long funicle in *Medusagyne oppositifolia* (Johri et al. 1992). Detailed embryology has not been studied.

Taxonomic Considerations.   The taxonomic position of Medusagynaceae is somewhat uncertain. Earlier workers like Bentham and Hooker (1965a), Bessey (1915) and Hallier (1912) did not recognise Medusagynaceae as a family. Engler and Melchior (1925) placed it in the Parietales. Melchior (1964), while revising Engler's work, included it in the Guttiferales. It has been variously ascribed to the Guttiferae and Theaceae. Anatomically, the genus *Medusagyne* differs from Guttiferae and Marcgraviaceae by the absence of resin canals, and from the Theaceae in the absence of sclerenchymatous idioblasts. It resembles the Ochnaceae in cortical bundles. Hutchinson (1969, 1973) placed this family in his Theales. According to him, it has a close connection with the order Magnoliales. Cronquist (1968, 1981) also includes it in the Theales. Dahlgren (1980a), although accepting this assignment, was not very sure of its position. Takhtajan (1987) observes that: "Medusagynaceae is an almost extinct family and occupies such an isolated position that it was necessary to place it in an independent order". Detailed data on the taxon *Medusagyne* are not available, except for exomorphology and some anatomy. He, therefore, concluded that: "it probably belongs to subclass Dilleniidae and possesses certain features common with the Paracryphiales, Theales and Ochnales". It is therefore, advisable to follow the observations of Takhtajan, until more data about this family, particularly in serology become available.

Actinidiaceae

A small family of 4 genera: *Actinidia, Saurauria, Clematoclethra* and *Sladenia*, and about 285 species distributed mainly in the Asiatic tropics (3 genera and 300 species according to Dickison et al. 1982, Takhtajan 1987). *Sladenia* was raised to the rank of a family by Airy Shaw in 1965. *Actinidia* spp. occur in the subtropics and temperate regions of eastern and southern Asia; *Saurauria* is a much larger genus and occurs in both the New and the Old World tropics.

Vegetative Features.   *Actinidia* species are mostly woody climbers, whereas *Saurauria* sp. are trees and shrubs. Leaves alternate, simple, entire or with serrate margin, petiolate, stipulate or with minute caducous stipules.

Floral Features.   Flowers solitary axillary or fascicled, cymose or paniculate; bisexual or unisexual and then plants dioecious, actinomorphic, hypogynous, pedicellate, ebracteate and ebracteolate. Calyx of 5 sepals, free, imbricate. Petals 5, polypetalous, imbricate. Stamens 10 to numerous, free or adnate to the base of the petals; anthers bicelled, versatile, dehiscence poricidal or longitudinal. Gynoecium syncarpous, 3- to 5- or more-carpellary; ovary with as many locules, superior, placentation axile; ovules numerous in each locule; styles as many as carpels, free or connate, generally persistent. Fruit a berry or leathery capsule, seeds without aril, with a large embryo and copious endosperm.

Anatomy.   Vessels small or small and large; with spiral thickenings, perforation plates simple or scalariform. Parenchyma apotracheal, diffuse or in irregular bands; rays up to 6 cells wide, heterogeneous. Leaves dorsiventral, hairs simple, uniseriate or multicellular glandular. Stomata Ranunculaceous type, confined to the lower surface of leaves. Raphides reported in leaves as well as stems.

Embryology.   Pollen grains usually monads, oblate-spheroidal to prolate; 3-colporate, colpi usually long. In *Saurauria* the mature pollen is shed in tetrads; 2-celled at shedding stage. Ovules anatropous, unitegmic, tenuinucellate; Polygonum type of embryo sac, 8-nucleate at maturity, hypostase present, conspicuous during post-fertilization stages. Endosperm formation of the Cellular type.

Chromosome Number.   Basic chromosome number is x = 15.

Important Genera and Economic Importance.   The 36 species of *Actinidia* are all woody climbers or lianas distributed in the eastern and southern Asiatic tropics. Fruits of *A. chinensis, A. polygama* and *A. callosa* are edible and highly esteemed in southeast Asia, used as baked products. The genus *Saurauria* is interesting as the corolla in the same species is polypetalous, in some gamopetalous, and in others apetalous. In *S. cauliflora* the flowers are borne on the main stem. The terminal flower of *S. roxburghii* is often pistillate and in *S. conferta* the flowers are borne in dense bracteate heads.

Taxonomic Considerations.   *Actinidia*, treated by most taxonomists as a member of the Dilleniaceae, was raised to family rank mainly on the basis of presence of unitegmic ovules against bitegmic ovules of other Dilleniaceae members (Vijayaraghavan 1970a). Engler and Diels (1936) placed this family along with Dilleniaceae under Theineae in the order Parietales. Melchior (1964) treated Actinidiaceae along with the Dilleniaceae and Theaceae in the Guttiferales. Lawrence (1951) pointed out that it was not a phylogenetic taxon and there could be some realignment in future. Airy Shaw (1965) raised *Sladenia* to a family rank—Sladeniaceae. Similarly, Hutchinson (1959) segregated Saurauriaceae. He treated Actinidiaceae as distinct from Dilleniaceae and included *Actinidia, Clematoclethra* and *Sladenia*.

As reported by Vijayaraghavan (1965), *Actinidia* is very different from Dilleniaceae in exomorphic and embryological characters.

Bentham and Hooker (1965a) and Hutchinson (1959) included Actinidiaceae in the Theales. Serological studies also support its relationship with the Theaceae (Vijayaraghavan 1970a). Embryologically, also, the Actinidiaceae resemble Theaceae (Kapil and Sethi 1963, Vijayaraghavan 1965). Many taxonomists, therefore, place it in or near Theales (Cronquist 1968, 1981). Hutchinson (1973) also places it in the Theales. Takhtajan (1987) places this family in a distinct order, Actinidiales, belonging to superorder Theanae.

In view of the above facts, the Actinidiaceae should be placed in or near Theales. It has more resemblances with the members of Theaceae than the members of Dilleniaceae.

Ochnaceae

A family of 21 genera and 375 species distributed in the tropics and subtropics of the Old World and northeastern South America and Mexico. *Indovethia* and *Neckia* are native to Borneo, and *Neckia* spreads to Sumatra and Malaya Archipelago.

Vegetative Features.   Ochnaceae members are mostly trees or shrubs, rarely herbs such as *Sauvagesia erecta* (Fig. 24.7A) and *Lavradia glandulosa*. Accordingly, the stem may be woody, or green and herbaceous. Leaves alternate, simple, pinnately compound in some species of *Godoya*, leathery, stipulate, stipules paired, axillary; leaves with close parallel venation.

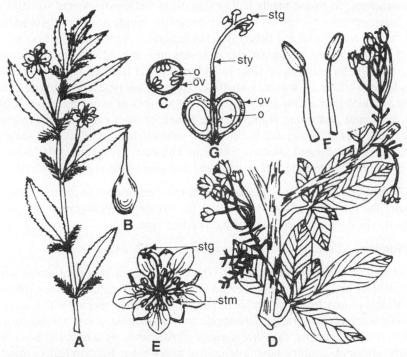

Fig. 24.7   Ochnaceae: **A-C** *Sauvagesia erecta*, **D** *Ochna obtusata*, **E-G** *O. multiflora*. **A** Flowering plant. **B** Carpel. **C** Cross section. **D** Flowering twig of *O. obtusata*. **E** Flower. **F** Stamens. **G** Gynoecium, longisection. *o* ovule, *ov* ovary, *stg* stigma, *stm* stamen, *sty* style. (**A-C** adapted from Hutchinson 1969, **D** after Saldanha and Nicolson 1976, **E-G** from Lawrence 1951)

Floral Features.   Inflorescence panicles, racemes (Fig. 24.7D) or cymes, mostly from scaly buds arising from below the 1-year-old leaves. Flowers bisexual, actinomorphic, hypogynous, pedicellate, pedicel jointed in *Ochna* (Fig. 24.7D); ebracteate, ebracteolate. Sepals 4 or 5 or 10 (*Blastemanthus*), polysepalous or basally connate, usually imbricate. In the tribe Blastemantheae, the sepals are spirally arranged and cone-like in bud. In *Lophira* the outer 2 or 3 sepals are large. Petals 4 or 5, rarely 10, polypetalous, contorted or imbricate in bud. Stamens 5, 10, or more, free (Fig. 24.7E), sometimes with 1 to 3 series of staminodes, inner staminodes petaloid in *Sauvagesia*; often borne on an elongated androphore (or anthers are connate into a unilateral mass) as in *Luxembergia*. Stamens numerous in many whorls in *Brackenridgea* sect. Notochnella and *Ochna* (Kanis 1968). *Testulea* has only one fertile stamen, anthers bicelled, basifixed, each cell dehisces longitudinally (Fig. 24.7F) or with an apical pore. Gynoecium 2- to 5-carpellary, syncarpous. Pseudoapocarpous in *Ochna atropurpurea*, i.e. to start with there are separate primordia but later they fuse (Pauže and Sattler 1979). Ovary superior, entire (*Elvasia*) or deeply-lobed (*Ochna*), 2- to 5- or rarely 10- to 15-loculed; unilocular in *Sauvagesia* (Fig. 24.7C); placentation axile, ovule 1 to many per locule, erect or rarely pendulous. Style 1, gynobasic, central; stigmas 1 to 5 (Fig. 24.7B, E, G). Fruit baccate and borne usually on an enlarged torus, the carpels separate into fleshy cocci. Seed with a straight embryo; endosperm present or absent. A winged capsule occurs in *Lophira*; and an indehiscent 1-seeded nut in Elvasieae. Seeds winged in *Tyleria* (Decker 1966).

Anatomy.   Vessels extremely small to rather large, almost exclusively solitary, occasionally with some radial pattern, perforations typically simple, intervascular pitting alternate; members of medium length. Parenchyma paratracheal, in broad bands in *Lophira*. Rays uniseriate or multiseriate, up to 2–8 cells wide, markedly heterogeneous except in *Lophira*. Fibres with simple to distinctly bordered pits. Cortical vascular bundles occur throughout the family. Leaves generally dorsiventral but occasionally centric in certain species of *Ouratea*. Hairs infrequent, when present, uni- or multicellular and uniseriate. Glandular shaggy hairs recorded on the stipules of *Godoya* and on the leaf teeth of *Lavradia glandulosa*. Stomata confined to the lower surface of leaves; sometimes Rubiaceous type and without subsidiary cells in *Lophira*. Stomata arranged in crowded groups between the network of veins in *Godoya*. Cells with strongly thickened inner tangential and radial walls, and the lumen of each cell filled completely with a large cluster or more rarely a solitary crystal, occur in the parenchymatous tissues of some genera such as *Ochna, Brackenridgea, Elvasia* and *Ouratea*. These are known as cristarque cells and are in the outer or inner part of the cortex of the petiole and young stem, and also in the region of lateral veins of the leaf.

Embryology.   Pollen grains 3-colpate and 2-celled at the shedding stage. Ovules anatropous, bitegmic and tenuinucellate; the integuments are free only at the micropyle. Polygonum type of embryo sac, 8-nucleate at maturity. Endosperm formation of the Nuclear type.

Chromosome Number.   Basic chromosome number is x = 12, 14.

Important Genera and Economic Importance.   *Ochna* is the most important genus, with ca. 80 species distributed in the tropics and subtropics of the Old World. A samll genus, *Godoya* (of tropical America) is interesting as its leaves may be either simple or pinnate in different species which seem to be otherwise closely related. The advanced genera are *Sauvagesia* and *Vausagesia*, a few species of the former are so far advanced as to become annuals. *Lophira*, a genus of African trees inhabiting both the moist tropical forest and the dry savannah country have a wing-like calyx lobe. Economically, this family is not of much importance. The kernels of *Lophira alata* are the source of meni oil and niam fat (Metcalfe and Chalk 1972). Its wood, called African oak, has strong, hard, interlocked grain and is used for railway sleepers and some construction work. The wood of *Ochna arborea* is used in South Africa for the handles of tools.

Taxonomic Considerations.   The family Ochnaceae belongs to the order Geraniales in Bentham and Hooker's system (1965a). Bessey (1915) removed it to the Guttiferales near Theaceae, Cistaceae, Eucryphiaceae and other families. Engler and Diels (1936) placed it in the order Parietales along with families like Dilleniaceae, Eucryphiaceae and Theaceae. Cronquist (1968, 1981) included it in the Theales, and Hutchinson (1969, 1973) raised it to a separate order Ochnales, including a number of new families like Strasburgeriaceae, Diegodendraceae and others. Dahlgren (1980a), however, placed it back in Theales. Takhtajan (1987) prefers to retain it in the Ochnales of superorder Theanae, but he segregated the genus *Lophira* to Lophiraceae, and *Sauvagesia* and 25 other genera in Sauvagesiaceae.

   Although there is much difference of opinion about the position of Ochnaceae, this family is allied to the Theaceae. It seems better, however, to retain Ochnaceae and other relaed families in Ochnales so that Theales too remains a natural taxon.

Ochnaceae occupies a position amongst the most primitive dicots (Decker 1966) and probably originated from Dilleniaceae.

Dioncophyllaceae
A small family of 3 genera and 3 species distributed in tropical Africa.

Vegetative Features. Plants woody climbers with stout, hooked branch tips or leaf tips. In *Triphyophyllum peltatum* the leaves are of 3 types: (a) on long shoots typical assimilatory leaves bear 2 hooks, (b) on axillary short shoots normal leaves without hooks, and (c) a third type occurs on some nonflowering shoots and have highly reduced lamina and on midrib bear glandular hairs.

Floral Features. Inflorescence a lax, more or less supra-axillary cyme. Flowers bracteate, bisexual, actinomorphic, hypogynous, pedicellate. Sepals 5, free, valvate, persistent. Petals 5, free, contorted, caducous, thick or delicate, white; disc absent. Stamens 10 (–30), equal or unequal, short or elongated; anthers elongated, linear or shortly ellipsoid, opening lengthwise. Gynoecium syncarpous, 2- or 5-carpellary; ovary superior, unilocular, opens loculicidally at a very early stage and exposes the numerous anatropous ovules on parietal placentae. Styles 2–5, free, filiform, stigma capitate or plumose. Fruit an ovoid to globose capsule, opens early to expose young developing seeds. Seeds few, large, flat, disc-like and winged, peltately attached to the elongated funicle; embryo large, discoid-obconic, mostly surrounded by copious endosperm.

Anatomy. Vessels very large (*Triphyophyllum*) or small as in *Habropetalum,* mostly solitary or arranged in radial multiples of up to 5 cells; perforations simple, intervascular pitting alternate, large; parenchyma paratracheal as narrow vasicentric sheaths around the vessels. Rays uniseriates and rather few. Interxylary phloem common. Hooks are leafy structure. Stomata numerous, confined to the lower surface, actinocytic, i.e. surrounded by a circle of subsidiary cells (Metcalfe 1952b).

Embryology. Pollen grains 3- to 4-colporate. Ovules anatropous, bitegmic, crassinucellate with inner integument longer than the outer. Polygonum type of embryo sac, 8–nucleate at maturity. Embryology has not been studied in sufficient detail.

Chromosome Number. Basic chromosome number is x = 9.

Important Genera and Economic Importance. The three genera of this family are *Dioncophyllum, Triphyophyllum* and *Habropetalum,* with one species each. No economic importance.

Taxonomic Considerations. The family Dioncophyllaceae was proposed by Airy Shaw in 1952. Melchior (1964) included it in the Guttiferales. Hutchinson (1969, 1973), however, included it in the Flacourtiaceae. It has been recognised as a separate family but placed in different positions. Cronquist (1968, 1981) preferred to include it in the Violales, on the basis of unilocular ovary. Schmid (1964) indicated that the Dioncophyllaceae may have closer affinity with the Violales, and it is also more convincing to refer them to the Violales, because of their unilocular ovary. However, it has also been referred to the related order, Theales (Dahlgren 1980a). Takhtajan (1987) includes it in a separate order of its own, i.e. Dioncophyllales, which belongs to the superorder Violanae. It appears that with unilocular ovary, parietal placentation, and seeds with copious endosperm, the family Dioncophyllaceae is more closely affiliated to the members of Violanae, and it may be treated as a member of the order Violales or of a distinct order Dioncophyllales.

Strasburgeriaceae

This is a unigeneric family with only one species, *Strasburgeria*. It is an interesting isolated survival from the primitive reservoir of forms endemic to New Caledonia (Willis 1973).

Vegetative Features. Trees with large obovate, spathulate, alternate, remotely dentate, stipulate leaves, stipules paired, connate and intrapetiolar.

Floral Features. Flowers solitary axillary, bisexual, actinomorphic, short-pedicelled. Sepals 8–10, imbricate, spirally arranged, outermost smallest and persistent. Corolla lobes or petals 5, polypetalous, imbricate, thick. Stamens 10, in one series, with stout, subulate filaments and large, oblong, dorsifixed, versatile, introrse anthers; interstaminal disc thick, annular, sinous, 10-lobed. Gynoecium syncarpous, 5-carpellary, ovary superior, 10-ribbed, narrowed into a subulate style with small capitate stigma. Placentation axile, with 2 superposed descending ovules per locule. Fruit a large, globular, multilocular, indehiscent corky-woody structure with 1 or 2 seeds per locule, or fruit is 1-seeded. Seeds trigonous, endosperm fleshy.

Anatomy. Vessels long with scalariform perforation plates with 20–30 bars; tracheids highly elongated with numerous bordered pits; rays heterocellular with elongated ends. Parenchyma diffuse. Nodes 3–lacuner. Stomata Ranunculaceous type.

Embryology. The embryology has not been studied fully. Ovules anatropous, bitegmic, crassinucellate.

Taxonomic Considerations. Most taxonomists include the Strasburgeriaceae in Ochnaceae (Cronquist 1981, Dickison 1981, Hutchinson 1973, Wettstein 1935). Thorne (1983) and Takhtajan (1987) recognise this family as a member of the Theales and Ochnales, respectively. On the basis of scanty data available, this family is closely linked with the Ochnaceae and both the families should be placed in the same order Ochnales.

Dipterocarpaceae

A family of 14 genera and 570 species; palaeotropically distributed, chiefly in Indo-Malaya. In Sri Lanka 44 of the 45 species are endemic, and they belong to *Cotylelobium, Hopea, Dipterocarpus, Shorea, Stemonoporus, Vateria* and *Vatica* (Bandaranayake et al. 1977). According to Meher-Homiji (1979), Borneo is considered to be the centre of origin of this family. In India, the drier climate due to the uplift of the Ghats and the Himalayas caused much extinction in the Late Tertiary. Kostermans (1985) observes that a palatable assumption is that the Dipterocarpaceae evolved in Gondowanaland and/or Laurasia and spread from there to other places.

Vegetative Features. Trees, usually tall with large spreading, sympodially branched emergent crowns; rarely shrubs. Growth is gregarious in areas with marked dry season or on poor soils; often dominate the evergreen, mixed lowland tropical rain forests (Willis 1973); also known as dipterocarp forests of humid tropical Indo-Malaya.

Leaves simple, entire, leathery, alternate, stipulate, stipules simple, sometimes surround the internode; prominently veined, frequently with domatia (depressions, pockets, sacs or tufts of hairs) (Metcalfe and Chalk 1979) in axils of veins.

Floral Features. Inflorescence axillary or terminal racemes or panicles, rarely cymose (*Upuna* and *Vatica* spp.). Flowers bisexual, actinomorphic, pentamerous, fragrant, bracteate, bracts caducous; floral axis broad and saucer-shaped (Fig. 24.10A). Calyx of 5 sepals, polysepalous, imbricate or valvate with short or long tube, free or adnate to ovary, persistent in fruit. Corolla of 5 petals, polypetalous, convolute, often connate at base. Stamens numerous (Fig. 24.10A) or 5 as in *Monoporandra*, or from 10 to 5 as in

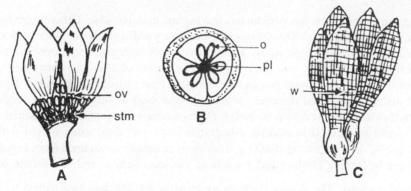

Fig. 24.10 Dipterocarpaceae: **A-C** *Shorea robusta.* **A** Flower, longisection. **B** Carpel, cross section. **C** Fruit. *o* ovule, *ov* ovary, *pl* placenta, *stm* stamen, *w* wing. (Adapted from Hutchinson 1969)

Stemonoporus and *Balanocarpus,* or 10, centrifugal, sometimes with androgynophore; in one or more whorls, usually connate, frequently adnate to the base of the petals; connective prolonged into a terminal process; anthers bicelled, basifixed as in *Dipterocarpus* or versatile, longitudinally dehiscent. Gynoecium syncarpous, tricarpellary, ovary superior, trilocular (Fig. 24.10B), placentation axile, ovules 2 per locule; style entire or trifid, frequently on a stylopodium; stigma usually obscure or 3- to 6-lobed. Fruit a l-seeded nut usually enclosed in persistent calyx of which 2 (*Dipterocarpus*), 3 or all 5 sepals (*Shorea*) form wings (Fig. 24.10C) aiding in dispersal. Seeds without endosperm and cotyledons often twisted, lobed or laciniate.

Anatomy. The most outstanding anatomical characteristic is the occurrence of a branched system of resin canals. Vessels medium-sized, solitary or with radial multiples of 2 or 3 cells; perforations simple; intervascular pitting alternate. Parenchyma abundant, both paratracheal and apotracheal. Rays up to 4–8 cells wide; uniseriate in *Marquesia* and *Monotes*, often more than 1 mm high; heterogeneous to homogeneous. Leaves dorsiventral, hairs of both uni– and multicellular types. Disc-shaped extrafloral nectaries recorded on the surface of foliage leaves and stipules of *Shorea*. Stomata confined to the lower surface; with distinct subsidiary cells; Rubiaceous type in *Balanocarpus, Shorea* and *Vateria* species. Mucilage cells and resin canals common. Cortical bundles and triangular phloem groups are also characteristic of this family.

Embryology. Pollen grains tricolpate, rarely tetracolpate, spheroidal; exine 2-3-layered; shed at 2-celled stage. Ovules anatropous, bitegmic, crassinucellate. Polygonum type of embryo sac, 8-nucleate at maturity. Endosperm formation of the Nuclear type.

Chromosome Number. Basic chromosome numbers are x = 7 and 11. The haploid chromosome number of *Shorea robusta* is n = 7.

Chemical Features. Bark, timber and resin of the Dipterocarpaceae members contain various types of triterpenoids and sesquiterpenoids (Bisset et al. 1967) and their presence or absence is helpful in classification at the generic level. They also contain abundant tannin, leucodelphinidin, leucocyanidin, myricetin, quercetin and kaempferol glycosides. Seeds of *Shorea robusta* are rich in stearic acid. The polyphenol "hopeaphenol" has been isolated from *S. talura* and *S. robusta,* which was originally reported in *Hopea odorata* and *Balanocarpus heimii* (Madhav et al. 1967).

Important Genera and Economic Importance. *Shorea*, with a number of timber-yielding species, is an important genus. *Dipterocarpus* is a genus of tall resinous trees with large stipules at the end of the shoots, enclosing the bud, which are deciduous, leaving an annular scar. Other important genera are *Hopea odorata*, a garden ornamental, *Dryobalanops aromatica* yields yellow camphor or Borneo camphor, and *Vateria indica* which yields edible seeds. The family is economically very important as its members are the source of timber, resin, fatty oils and camphor. The timber of *Shorea* is very strong and durable and is used for house and ship building, gun carriage manufacture, railway sleepers, etc. *Shorea koodarsii* yields turpentine and a resin called dammer resin or dhoona used as incencse. Oily seeds of *S. aptera* yield edible fat used as a substitute for cocoa butter. *Dipterocarpus laevis* and *D. aromatica* are the source of Gurjan oil, a liquid resin used in making lithographic ink, and also, when mixed with dammer, to prevent white ant attack. Another strong-smelling, thick resin or balsam is obtained from *Isauxis lancaefolia*, a very common tree in Bangla Desh. Starchy seeds of *Vateria, Vatica* and *Doona* are edible.

Taxonomic Considerations. The family Dipterocarpaceae is divided into two tribes:

(1) Dipterocarpoideae—resin and balsam ducts present in pith; anthers basifixed and filaments short. Distributed in the Indo-Malayan region, *Dipterocarpus, Shorea* and others.

(2) Monotoideae—no resin or balsam ducts. Anthers versatile with long filaments. Distributed only in Africa, *Marquesia* and *Monotes*.

Bentham and Hooker (1965a), Engler and Diels (1936) and Bessey (1915) placed Dipterocarpaceae along with the families like Guttiferae and Theaceae. Hallier (1912) placed it in his Columnales, i.e. Malvales near Tiliaceae. Vestal (1937), on the basis of anatomical features, commented that resemblance of Dipterocarpaceae with the Guttiferae is due to parallel evolution. Hutchinson (1969, 1973) treated this family as a member of the Ochnales, and derived all the families of this order from Theaceae. The alliance with the Theaceae is agreed to by Cronquist (1968, 1981). However, Willis (1973), Dahlgren (1975a, 1980a, 1983a) and Takhtajan (1980, 1987) agree to Hallier's view and place Dipterocarpaceae in the Malvales near Tiliaceae, on the basis of certain morphological and anatomical features.

Kostermans (1978, 1985), who examined Monotoideae in detail concluded that Monotoideae (including *Pakaraimaea*) alone is more closely allied to the Tiliaceae and not the Old World dipterocarps. His reasoning is based on some morphological, anatomical and chemical characteristics of this subfamily. Maury (1981; cf. Takhtajan 1987) separated *Monotes, Marquesia* and *Pakaraimaea* to form a distinct family, Monotaceae, with alliance to the Dipterocarpaceae. Gottwald and Parameswaran (1966) also observed that the presence of uniseriate rays and absence of resin canals are sufficient reason to raise both the subfamilies to family level.

It is evident, therefore, that the Dipterocarpaceae s.s. are more closely related to Guttiferae and Theaceae. Subfamily Monotoideae is better treated as a distinct family Monotaceae. These two families are quite closely related, although geographically they are distant.

Theaceae

A family of ca. 30 genera and 500 species distributed mainly in the tropics and subtropics of the world. A few genera and species extend up to the temperate regions.

Vegetative Features.   Mostly erect trees or shrubs; leaves alternate, simple, coriaceous as in *Camellia* or membranous, evergreen, entire or with serrate margin, estipulate (Fig. 24.11A).

Floral Features.   Solitary, axillary flowers or fasciculate. Flowers normally bisexual, unisexual and dioecious plants in *Eurya*; actinomorphic, hypogynous, often subtended by a pair of bracts. Sepals 5 or 6, gradually transformed from bracteoles to petals; free, usually persistent, imbricate; petals 5 (rarely 4 or more than 5), free or connate at the base, contorted or imbricate. Stamens numerous, rarely 15 or less (Fig. 24.11B), free or basally monadelphous or fascicled in 5 bundles that are opposite and often basally

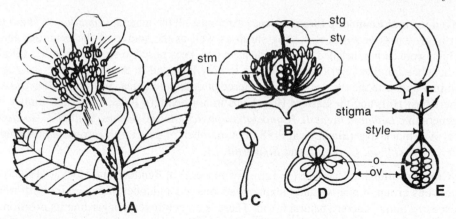

Fig. 24.11 Theaceae: **A-F** *Camellia* spp. **A** *C. sasanqua*, flower. **B** Flower, longisection. **C** Stamen. **D** Carpel, cross section. **E** Carpel, longisection. **F** Fruit. *o* ovule, *ov* ovary, *stg* stigma, *stm* stamen, *sty* style. (**A, B** adapted from Lawrence 1951)

adnate to the petals. Anthers bicelled, dehiscing longitudinally, basifixed (Fig. 24.11B, C) or versatile. Gynoecium syncarpous, 3- to 5-carpellary, ovary superior, 3- (Fig. 24.11D) to 5-loculed, sessile, with axile placentation (Fig. 24.11D, E), ovules usually 2 or more in each locule; styles and stigmas as many as carpels, free or connate. Fruit is a woody or subligneous, loculicidal capsule (Fig. 24.11F). In members of the tribe Ternstroemieae the capsule is fleshy and indehiscent. Seeds wingless, non-arillate and with a straight or curved embryo, and scanty or no endosperm.

Anatomy.   Vessels typically small, predominantly solitary, semi-ring-porous in some species, with spiral thickenings; oblique perforation plates often with many bars; intervascular pitting scalariform or opposite. Parenchyma apotracheal, diffuse or in short lines. Rays usually only 2 or 3 cells wide but may be 1–8 cells wide also, heterogeneous. Leaves generally dorsiventral, covered with a thick cuticle. Hairs infrequent, but when present usually unicellular, acuminate and thick-walled, as in *Camellia*, *Eurya*, *Gordonia*, etc. Stomata confined to lower surface, surrounded by 2–5 narrow subsidiary cells in certain species of *Adinandra*, *Camellia*, *Ternstroemia* and others. Stomata Rubiaceous type in *Archytaea*, Ranunculaceous in *Sladenia celastrifolia*. Sclerenchymatous idioblasts in the parenchymatous tissue of leaf and cortex and pith of stem is characteristic of the family.

Embryology.   Pollen grains tricolporate. triangular and 2-celled at shedding stage. Ovules anatropous, bitegmic, tenuinucellate; micropyle formed by inner integument, hypostase present. The mature embryo has two large cotyledons that grow backwards on either side of the radicle (Sethi 1965). Polyembryony occurs in some species of *Camellia*. The endosperm fomation is of the Nuclear type.

Chromosome Number.   Basic chromosome numbers are x = 15, 18, 21, 25. The haploid chromosome number of *Camellia sinensis* is n = 15.

Chemical Features.   Theaceae is unique in making 8-oxygenated flavonols (which may indicate a link with Primulaceae) and 3-deoxyanthocyanidins; C-glycoflavones and flavonoids also occur in Theaceae (Parks and Kondo 1974). The three principal constituents of *Camellia sinensis* are the essential oils, the alkaloidal fraction, and polyphenols or the tannins. The flavour of tea is due to the oil theol and its stimulating and refreshing qualities is due to theine. Sulfated flavonoids also occur in Theaceae members (Gurni and Kubitzki 1981, Gurni et al. 1981).

Important Genera and Economic Importance.   Amongst all the genera, *Camellia* (= *Thea*) is the most important. It is grown as beautiful flowering shrub as well as the source of the beverage tea. Some of the genera which were earlier included in Theaceae have now been raised to separate families (Sladeniaceae), or included in other families, e.g. *Clematoclethra* is now included in Actinidiaceae.

Dried terminal bud with the two leaves is the source of commercial tea. Wild tea, *Camellia kissi* grows in Assam, the Khasi Hills and the Eastern Himalayas in northeastern India. Important timber is obtained from some genera like *Laplacea brenesii, Haploclathra paniculata* and *Schinea wallichii. Gordonia excelsa* is the source of a black dye (Singh et al. 1983). A number of genera are grown as ornamentals, e.g. *Cleyera, Eurya, Franklinia, Gordonia* and *Stewartia.*

Taxonomic Considerations.   Theaceae (= Ternstroemiaceae) of Bentham and Hooker (1965a) comprised a nonhomogeneous group of plants. Later, taxonomists merged a number of genera into different families to make Theaceae a homogeneous, natural taxon. There is no controversy regarding its position. Theaceae has been treated as a member of the Theales by most workers.

Caryocaraceae

A very small family of only two genera, *Caryocar* and *Anthodiscus,* with 23 species distributed in Latin American countries.

Vegetative Features.   Trees or shrubs, stem surface of *Caryocar* covered with abundant short hairs when very young but shed later on (Metcalfe and Chalk 1972). Leaves opposite as in *Caryocar* or alternate as in *Anthodiscus,* petiolate, digitately compound, 3- to 5-foliolate, stipulate; extrafloral nectaries occur on the margin of the young leaf and on the stipules, stipules deciduous.

Floral Features.   Inflorescence a raceme. Flowers bisexual, actinomorphic, hypogynous. Calyx lobes 5 or 6, gamosepalous, imbricate or open; corolla of 5 or 6 petals. polypetalous as in *Caryocar,* imbricate or petals connate into a calyptra as in *Anthodiscus.* Stamens numerous, united into a ring or in 5 bundles. Gynoecium syncarpous, 4–8–20-carpellary, ovary with as many locules, superior, placentation axile, 1 pendulous ovule per locule; styles as many as the number of carpels. Fruit usually a drupe with oily mesocarp, and woody endocarp which splits into 4 mericarps; sometimes a leathery schizocarp; seeds with scanty or no endosperm.

Anatomy.   Vessels moderately small in *Anthodiscus,* large in *Caryocar;* perforations simple; intervascular pitting alternate, medium-sized; parenchyma predominantly apotracheal but with some paratracheal in *Caryocar;* only paratracheal in *Anthodiscus.* Rays biseriate. Roots of *Caryocar* show anomalous structure. Leaves dorsiventral or sometimes centric, e.g. *C. glabrum.* Stomata mostly confined to the lower surface; a few in depressions beside the veins on the upper surface in *C. nuciferum;* Ranunculaceous type in *Caryocar.* Branched sclerenchymatous idioblasts present in the mesophyll and the ground tissue of petiole. Crystals numerous.

Embryology.   The embryology has not been fully studied. The ovule is anatropous and bitegmic. The endosperm, in mature seed, is scanty (Johri et al. 1992).

Important Genera and Economic Importance.   Only two genera, *Caryocar* and *Anthodiscus,* comprise this family. *C. amygdaliferum* is an interesting species with large pouch-like glands at the base of the leaflets which probably harbour ants. In *C. nuciferum* dark spots on the lamina represent tanniniferous cells. Both the genera have a spirally coiled, enormous radicle with small, inflexed cotyledons. Economically an important genus is *Caryocar.* Butter nuts or souari nuts are the fruits of *C. nuciferum,* and are rich in edible fat. The timber of some species of *Caryocar* is used for ship building, flooring and other purposes.

Taxonomic Considerations. Bentham and Hooker (1965a) treated Caryocaraceae as the first tribe of family Ternstroemiaceae (= Theaceae). Engler and Diels (1936) recognised it as a distinct family in the Parietales. However, it differs from the family Theaceae in morphological, anatomical and palynological features, and also the typically twisted structure of hypocotyl with small inflexed cotyledons and drupaceous fruit.

Later workers treat it as a distinct family, Caryocaraceae (Melchior 1964, Hutchinson 1969, 1973, Cronquist 1968, 1981, Dahlgren 1975a, 1980a, Takhtajan 1980, 1987), of the order Theales.

The placement of this family near the Theaceae is justified. The two families resemble each other in being woody plants, and have hypogynous flowers, bitegmic, crassinucellate ovules and a large number of stamens.

Marcgraviaceae

A small family of 5 genera and 100 species distributed in tropical America.

Vegetative Features. Climbing or scandent shrubs usually grow upon other trees and shrubs as epiphytes; sometimes trees. Leaves alternate, stipulate, often heterophyllous.

Floral Features. Inflorescence pendulous racemes. Flowers bisexual, actinomorphic, bracteate, bracts often modified into pitcher-like vessels which secrete nectar; ebracteolate, hypogynous. Sepals 4 or 5, free, imbricate; petals 4 or 5 gamo- or polypetalous, drop off as a cap when the flower opens. Stamens 3-numerous, free or connate at base, epipetalous; anthers longitudinally dehiscent. Gynoecium syncarpous, 2- to 5-carpellary, ovary superior, originally unilocular with parietal placentation but later 2-numerous locules formed by ingrowth of placentae. Ovules numerous per locule; style 1, short, with inconspicuous or shortly 5-lobed stigma. Fruit a fleshy, leathery, tardily dehiscent, many-seeded capsule. Seeds small, numerous, non-endospermous.

Anatomy. Anatomically, the two genera, *Marcgravia* and *Norantea*, show certain differences. In *Marcgravia* the vessels are medium-sized to large, with simple perforation plates; intervascular pitting alternate; parenchyma paratracheal, vasicentric. In *Norantea* the vessels are moderately small with scalariform perforation plates and many fine bars; intervascular pitting opposite; parenchyma apotracheal, diffuse. Rays 10 or more cells wide in both the genera. Raphides and sclerenchymatous idioblasts are two of the most characteristic anatomical features. Leaves dorsiventral, hairs absent; stomata confined to lower surface. Bundles of raphides in special cells in the mesophyll.

Embryology. Pollen grains 3-colpate, and germinate in situ. Ovules anatropous, bitegmic, tenuinucellate; the inner integument forms the micropyle. Polygonum type of embryo sac, 8-nucleate at maturity. Polyembryony unknown; seed coat formed by only the epidermis of outer integument, its cells become filled with tannin. Endosperm formation of the Cellular type.

Important Genera and Economic Importance. Amongst the 5 genera of this family, *Marcgravia* and *Norantea* are important. Economically the family is not of much importance.

Taxonomic Considerations. The family Marcgraviaceae was not recognised by Bentham and Hooker (1965a). Engler and Diels (1936) placed Marcgraviaceae in the Parietales along with families like Theaceae, Guttiferae and others. It shows much resemblance with Theaceae. Melchior (1964) removed these families to the order Guttiferales. The name of the order was later changed to Theales (Cronquist 1981, Dahlgren 1975a, 1980a, Hutchinson 1973, Takhtajan 1987).

Families Ochnaceae, Marcgraviaceae, Theaceae and Guttiferae are embryologically similar, yet the signs of fusion of the integuments, Cellular type of endosperm and arillate albuminous seed distinguish Marcgraviaceae from the rest of the families.

The family Marcgraviaceae differs from Theaceae and other related families also by its epiphytic and scandent habit, considerably advanced flower structure, more specialised wood, and the presence of raphides in leaves.

Marcgraviaceae is an advanced family and is correctly included in the order Theales.

Quiinaceae

A small family of only 4 genera and 50 species indigenous to tropical South America.

Vegetative Features.   Woody trees or shrubs, sometimes scandent. Leaves opposite or veticellate, simple or pinnatifid, entire or with crenate margin, stipulate, stipules interpetiolar, 1–4 pairs; elegant plumose-reticulate venation pattern (Foster 1950); heterophylly known (Foster 1951). Younger leaves of a seedling are pinnatifid, but the mature leaves are simple and entire.

Floral Feature.   Inflorescence racemes or panicles. Flowers small, regular, bisexual or rarely unisexual. Sepals 4 or 5, free, imbricate; corolla 4-5-8-lobed, imbricate or contorted. Stamens 15–30-numerous, may be 160–170, free or adnate to corolla at the base. Gynoecium 3- or (2–3) or (7–1)–carpellary, with as many filiform finally reflexed styles with obliquely peltate stigmas. Ovules 2 per locule, erect or ascending, on axile placentation. Fruit baccate, dehisce by valves, usually unilocular, 1-(4)–seeded. Seed often with densely velutinous covering; without endosperm.

Anatomy.   Vessels mostly solitary, sometimes with radial or oblique pattern, perforations simple; intervascular pitting alternate and very small, pits of the ray cells similar. Parenchyma apotracheal, diffuse. Rays up to 2–4 or 5–6 cells wide, heterogeneous. Leaves dorsiventral, stomata Rubiaceous type, confined to the lower surface. Members of Quiinaceae lack schizogenous canals with resinous contents and, instead, have lysigenous intercellular spaces filled with mucilage particularly in the vascular strand of midrib and petiole. A sheath of very thick-walled fibres surrounds the vascular bundles of the leaves which give feathery appearance of the veins. Crystals and druses common.

Important Genera and Economic Importance.   The genera included in this family are *Quiina, Touroulia, Lacunaria* and *Froësia*, with no known economic importance.

Taxonomic Considerations.   Taxonomic position assigned to Quiinaceae is much varied but it belongs to the Thealean complex. Most phylogenists include it in the Theales (Engler 1925, Cronquist 1981, Dahlgren 1983a, Thorne 1968, 1983). According to Hutchinson (1969, 1973) and Cronquist, Quiinaceae is close to Guttiferae, and the former author placed it in his Guttiferales. Melchior (1964) also placed Quiinaceae near the Guttiferae. However, the two families differ anatomically (Metcalfe and Chalk 1972). According to Gottwald and Parameswaran (1967), Quiinaceae shows a close affinity to Ochnaceae in various anatomical features and in having SiO_2 particles in *Lacunaria*. Thorne (1983) and Takhtajan (1987) place the Quiinaceae close to the Ochnaceae, and Takhtajan (1987) includes it in the Ochnales.

Serological studies of these families would be helpful in understanding their correct position.

Guttiferae (= Clusiaceae)

A moderate-sized family with 47 genera and 950 species disributed in the tropical belt of southeastern Asia.

Vegetative Features.   Trees or shrubs, often epiphytic, rarely climbers, resinous. Leaves opposite (Fig. 24.15A) or whorled, simple, coriaceous, with a prominent midvein, estipulate.

Floral Features.   Inflorescence a panicle (Fig. 24.15A) or the flowers are solitary. Flowers usually bisexual, sometimes unisexual or polygamodioecious; actinomorphic, hypogynous. Sepals 2–10 or more,

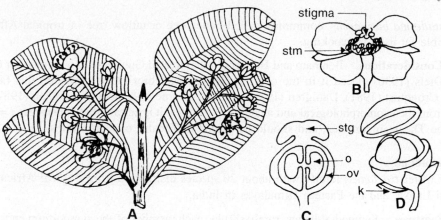

Fig. 24.15 Guttiferae: **A** *Calophyllum inophyllum*, **B-D** *Garcinia mangostana.* **A** Flowering branches. **B** Flower. **C** Longisection. **D** Fruit. *k* calyx, *o* ovule, *ov* ovary, *stg* stigma, *stm* stamen. (**A** adapted from Hutchinson 1969, **B-D** after Lawrence 1951)

decussate or imbricate; petals (Fig. 24.15B) usually imbricate, subvalvate or contorted. Stamens few or numerous, hypogynous, free or variously united into bundles; anthers bicelled, dehiscence longitudinal; pistillode often present in staminate flowers. Gynoecium 3–5 or more-carpellary, syncarpous, ovary with as many locules, or only unilocular, superior, placentation axile (Fig. 24.15C), basal or infrequently parietal; ovule 1 to many in each locule. Styles and stigmas as many as the number of carpels, often peltate or radiating (Fig. 24.15B,C). Fruit a septicidally dehiscent capsule (Fig. 24.15D) or sometimes a berry or drupe. Seeds non-endospermous, with a large, straight embryo (undifferentiated embryo in *Pentadesma*), arillate in *Garcinia*.

Anatomy. The wood structure of two extreme types has been noted in the two groups, Clusieae and Calophylloideae. Clusieae is more primitive with vessels in small multiples, simple or simple and scalariform perforation plates, paratracheal parenchyma, and 4 to 9 cells wide markedly heterogeneous rays. Calophylloideae members have exclusively solitary vessels with simple perforations, apotracheal parenchyma in broad bands or diffuse heterogeneous rays only 1 to 3 cells wide. Most of the other genera are intermediate. Leaves dorsiventral in most species, centric or subcentric in some others. Hairs rarely occur, when present, are simple, unicellular or uniseriate trichomes. Stomata confined to the lower surface; Rubiaceous type. Calcium oxalate crystals common in parenchymatous tissue.

Embryology. Pollen grains tricolpate and shed at 2-celled stage. Ovules anatropous or amphitropous, bitegmic, and tenuinucellate. Polygonum type of embryo sac, 8-nucleate at maturity. Endosperm formation of the Nuclear type (Johri et al. 1992).

Chromosome Number. Basic chromosome numbers are x = 7, 8, 9, 10.

Chemical Features. Flavones and biflavones have been reported (Gornall et al. 1979).

Important Genera and Economic Importance. An important genus is *Calophyllum* with a large number of species. Wood of various species characteristically strong and durable and used for general construction, ship building and furniture-making. Another important timber tree of this family is *Mesua ferrea. Garcinia mangostana,* native of the Moluccas is widely cultivated in the tropics for edible fruits (Fig. 24.15D). *Mammea americana* of the West Indies and northern South America also produces edible fruits. The seeds of *Pentadesma butyracea,* commonly called butter tree or tallow tree of tropical Africa, are the source of edible fat used for cooking.

Taxonomic Considerations.   Bentham and Hooker (1965a) treated Guttiferae as a member of Guttiferales. Engler and Diels (1936) placed it in the order Parietales. Melchior (1964) retained this family in the Guttiferales. Cronquist (1981), Dahlgren (1980a, 1983a), Thorne (1983) and Takhtajan (1987) include it in Theales. From the exomorphological and anatomical features, it is certain that Clusiaceae (= Guttiferae) belongs to the Thealean complex, and should be placed in Theales or Guttiferales.

Ancistrocladaceae

A unigeneric (*Ancistrocladus*) family with about 20 species distributed from tropical Africa to Western Malaysia, Sri Lanka, and the Eastern Himalayas in India.

Vegetative Features.   Sympodial lianas, rarely shrubs, each member of the sympodium ends in a watch-spring hook. Leaves alternate, simple, obovate or oblanceolate, entire with scattered minute glandular pits; stipules when present, minute and caducous.

Floral Features.   Inflorescence lax or condensed dichotomous cymes or apparent racemes. Calyx lobes 5, polysepalous, imbricate, adnate to ovary, persistent, outer surface often with glandular pits. Corolla of 5 petals, polypetalous, more or less fleshy, shortly connate basally, imbricate or contorted. Stamens 10 (rarely 5), uniseriate, slightly unequal, filaments short, fleshy, connate below, anthers basifixed, introrse or latrorse and opens longitudinally. Gynoecium tricarpellary, syncarpous, ovary inferior, unilocular, with one basal, erect ovule; 3 free or connate, articulate styles, thickened upwards with hippocrepiform or punctified stigma. Fruit dry, woody, indehiscent, surrounded by persistent sepals. Seed large with a small embryo and strongly ruminate endosperm.

Anatomy.   Leaf surface covered with peltate, multicellular glands, sunk in the epidermis. Stomata confined to lower surface, actinocytic, i.e. surrounded by a circle of more or less clearly-defined subsidiary cells. Xylem occupies a large part of the stem. Vessels exclusively solitary, small to medium-sized; perforations simple, oblique. Parenchyma apotracheal; pith wide; rays uni- or biseriate. The hook-like structure is essentially a stem structure (Metcalfe 1952).

Embryology.   The embryology has not been studied in detail. Ovules hemianatropous, bitegmic, crassinucellate. Endosperm formation of the Cellular type.

Important Genera and Economic Importance.   The only genus, *Ancistrocladus*, is an interesting plant, climbing with hooks that are modified branch tips.

Taxonomic Considerations.   Bentham and Hooker (1965a) included Ancistrocladaceae in Dipterocarpaceae. Engler and Diels (1936) placed it in the Parietales. Cronquist (1981), Dahlgren (1980a), Hutchinson (1973), Melchior (1964) and Takhtajan (1987) recognise it as a distinct family. Opinions differ regarding its position. Although most phylogenists include Ancistrocladaceae in the Theales/Guttiferales, Cronquist (1981) treats it as a member of the Violales, because of unilocular ovary and basal placentation (not axile as in Theales). He also observes that this family has no close relative (except Dioncophyllaceae) in either Theales or Violales, although both families belong to this alliance in all probability. Takhtajan (1987) erected a separate order, Ancistrocladales, to accommodate this family, and included it in the superorder Theanae.

   In the absence of any close relative in the orders mentioned above, it seems appropriate to place it in a distinct order of its own.

Taxonomic Considerations of the Order Guttiferales

It is a heterogeneous order comprising a large number of families. Most phylogenists prefer these members under more than one order. For example, Cronquist (1968, 1981) treats Dilleniaceae, Paeoniaceae and Crossosomataceae under Dilleniales; another 13 families, including Guttiferae, Theaceae, Ochnaceae, Actinidiaceae and others in the Guttiferales (or Theales), and two other families in the Violales. Takhtajan (1987) introduces still further cleavage and recognises a number of monotypic orders.

Whatever the circumscription, members of all these families are woody, lofty trees or shrubs and are characterised by the presence of numerous centrifugal stamens, pollen grains 2-celled at shedding stage, mostly syncarpous multicarpellary ovary with axile or parietal placentation and numerous ovules.

In this group the families Dilleniaceae, Paeoniaceae and Crossosomataceae differ from the rest in apocarpous gynoecium and presence of chemicals like myricetin, quercetin, kaempferol, various alkaloids, particularly in Paeoniaceae. It is therefore justified to place them in a separate order, Dilleniales.

According to Dahlgren (1980a), Dilleniales can be placed with the Malvales; Takhtajan (1980) and Thorne (1981) prefer to place it with the Theales; and Cronquist (1981) with both. With their stellate hairs, foliar sclereids and scalariform perforation plates, combined with hypogynous, coripetalous flowers without a disc, the multistaminal androecia occassionally with centrifugal succession, the flattened stamens, the mutually free carpels and the crassinucellate ovules, they combine the characters of Malvales as well as Theales.

Like Malvales and Theales, the Dilleniales also have flavonoid and polyphenol spectra which include myricetin and O-methylated flavonols.

The families Dioncophyllaceae and Ancistrocladaceae are distinct in being scandent shrubs climbing by hooks that are modified branch tips or leaf tips, and in having unilocular ovary with parietal or basal placentation (and not axile as in the rest of the members of this order). Takhtajan (1987) includes Dioncophyllaceae in superorder Violanae and retains Ancistrocladaceae in Theanae, although in an order of its own, Ancistrocladales, near Ochnales. Cronquist (1968, 1981), however, includes both these families in the Violales on the basis of unilocular ovary. This view appears to be more appropriate.

With the removal of these two groups of families, the order Guttiferales/Theales becomes more homogeneous and natural. If, however, these five families are retained in the Guttifeales (Melchior 1964), Dilleniaceae is naturally the most primitive family of this order and is the link with the ancestral order Magnoliales or the Magnolian complex. The Dilleniaceae itself is directly ancestral to the Theaceae and all other related families, which differ only in having syncarpous ovary.

25

Order Sarraceniales

Sarraceniales is an order comprising three families: Sarraceniaceae, unigeneric Nepenthaceae and Droseraceae. All the members are herbaceous and insectivorous plants with different insect-capturing devices. Flowers inconspicuous, actinomorphic and hermaphrodite; fruits usually loculicidal capsules.

Sarraceniaceae

A small family of 3 genera and 15 species distributed mainly in the United States of America and northern South America. These plants usually grow in water-logged soil that contains little or no soluble nitrates.

Vegetative Features.   Plants herbaceous, rarely shrubs, perennials. Leaves simple, alternate or in the form of rosette, often winged or tubular with usually a small terminal lamina (Fig. 25.1A); the tubular part retrorsely hairy within.

Floral Features.   Inflorescence a scape, bears a solitary flower (Fig. 25.1A) or a raceme as in *Heliamphora.* Flowers actinomorphic, hypogynous, hermaphrodite, pedicellate (Fig. 25.1B), bracteate in *Heliamphora*, ebracteate in the other two genera—*Darlingtonia* and *Sarracenia*. Sepals 4 or 5, polysepalous, imbricate, hypogynous, persistent, often coloured and showy. Petals 5 in *Darlingtonia* and *Sarracenia*, absent in *Heliamphora*, polypetalous. Stamens numerous (15 in *Darlingtonia*), free; anthers bicelled, introrse, dehiscence longitudinal (Fig. 25.1C), shortly adnate to the base of petals. Gynoecium syncarpous, 3- to 5-carpellary, ovary 3- to 5-loculed, and as many lobed (Fig. 25.1D, E), superior; placentation axile, ovules numerous in each locule (Fig. 25.1D). Style 1, with normal radiating stigmas in *Darlingtonia*; enlarged into an umbrella-like expansion with as many lobes as the carpels and each lobe tip stigmatic beneath in *Sarracenia* (Fig. 25.1A-C). Fruit a loculicidal capsule (Fig. 25.1E), seeds minute, broadly winged in *Heliamphora*; with small linear embryo and copious fleshy endosperm.

Anatomy.   Vascular bundles collateral, of different size and shape, arranged in a discontinuous ring and separated by rays of unequal width. Vessels few with spiral thickenings, perforation plates scalariform. The radical tubular leaves have specialised glands and hairs with which insects and other small organisms are entrapped. These are ultimately digested by the plant. From apex downwards, various zones can be recognised on the inner surface of the leaf: (1) Glandular zone occupies the inner surface of the lid of the leaf of *Heliamphora* and *Sarracenia*, and of the hood of *Darlingtonia.* (2) Slippery zone consists of epidermal cells with downwardly directed projections. (3) 'Eel-trap' zone is provided with downwardly directed needle-shaped hairs and numerous glands. Stomata Ranunculaceous type, occur only in the glandular zone.

Embryology.   Pollen grains in monads, shed at 2-celled stage. Ovules anatropous, uni- or bitegmic, tenuinucellate; Polygonum type of embryo sac, 8-nucleate at maturity. Inner integument forms the micropyle. Endosperm formation of the Cellular type.

Chromosome Number.   The basic chromosome number is different for all the three genera—$x = 13$ in *Sarracenia*, $x = 15$ in *Darlingtonia* and $x = 17$ in *Heliamphora*.

Chemical Features.   A derivative of seco-iridoids is reported in *Sarracenia purpurea* and *Darlingtonia*

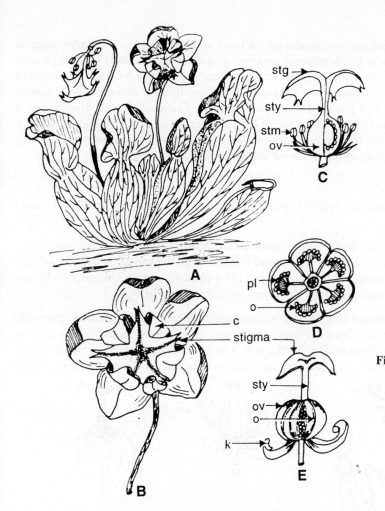

Fig 25.1 Sarraceniaceae: **A-E** *Sarracenia purpurea.* **A** Flowering plant. **B** Flower. **C** Vertical section of flower, perianth removed. **D** Ovary, cross section. **E** Dehisced capsule. *c* corolla, *k* calyx, *o* ovule, *ov* ovary, *pl* placenta, *stg* stigma, *stm* stamen, *sty* style. (**A** adapted from Takhtajan 1969, **C**, **D** from Lawrence 1951)

californica (Jensen et al. 1975). Anthocyanins and tannins of the condensed type and kaempferol, cyanidin and caffeic acid occur in the extracts of leaves.

Important Genera and Economic Importance.   The three important genera are *Darlingtonia, Heliamphora* and *Sarracenia.* None of these is economically important.

Taxonomic Considerations.   Bentham and Hooker (1965a) included Sarraceniaceae in the order Parietales and placed it before the Papaverales. Bessey (1915) included Sarraceniaceae and Nepenthaceae in the order Sarraceniales. Engler and Diels (1936) included the family Droseraceae also in the same order and placed it next to the Rhoeodales. Melchior (1964) supported this view. Cronquist (1968) also retains this family in the order Sarraceniales along with the other two families—Droseraceae and Nepenthaceae—mainly because of similar exomorphic features and similar mode of insect-catching. In 1981, Cronquist changed the order name to Nepenthales.

Hutchinson (1973) places Sarraceniaceae in the order Sarraceniales along with Droseraceae. According to him, Engler's Sarraceniales is a polygeneous group (Reihe) which seems to depend almost exclusively on the presence of insectivorous leaves. The three families are otherwise not related. De Buhr (1975) presents floral and embryological data to support the relationship between Sarraceniaceae and Theales.

Dahlgren (1975b) places Sarraceniaceae next to Ericales on the basis of morphological and embryological features. The presence of seco-iridoids in Sarraceniaceae supports a position close to the Cornales (Jensen et al. 1975). Dahlgren (1980a) treats this family as a monotypic member of the order Sarraceniales and places it in Corniflorae near Ericales and Cornales. Takhtajan (1987) also treats it as a monotypic family belonging to the order Sarraceniales and places it near the superorder Ericanae. Sarraceniaceae has a number of characters common with the family Pyrolaceae of Ericales (Jensen et al. 1975), and chemically also it is close to the Cornales. Keeping these points in view, we think that a position between Ericales and Cornales is justified for Sarraceniaceae, as has been shown by Dahlgren (1980a).

Nepenthaceae

A unigeneric family of the genus *Nepenthes* with 72 species distributed in the tropical rain forest areas of the Old World.

Vegetative Features.   The members are insectivorous, herbaceous, suffrutescent or shruby plants, often climbers. Leaves cauline, alternate, composed of 4 different zones: (1) an expanded or winged petiole, followed by (2) a constricted, often coiled or tendriller zone, (3) a pendant, highly coloured, cylinder-shaped pitcher with a recurved rim, terminated by (4) a variously margined lid.

Floral Features.   Inflorescence racemose (Fig. 25.2A) to paniculate. Flowers inconspicuous, greenish, unisexual, actinomorphic, ebracteate, ebracteolate, hypogynous and monochlamydeous. Sepals 3 or 4, polysepalous, nectariferous, Staminate flowers (Fig. 25.2C) with 4 to 24 stamens, filaments monadelphous to form a column (Fig. 25.2C,D); anthers free, bicelled, dehiscence longitudinal. Pistillate flowers with

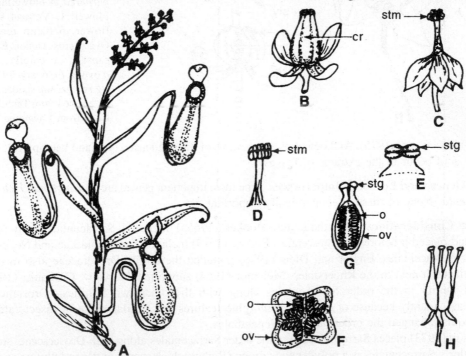

Fig. 25.2   Nepenthaceae: **A-H** *Nepenthes* sp. **A** Flowering twig. **B** Pistillate flower. **C** Staminate flower. **D** Stamens. **E** Stigmatic disc. **F** Ovary, cross section. **G** Carpel, longisection. **H** Fruit. *o* ovule, *ov* ovary, *stg* stigma, *stm* stamen. (Adapted from Lawrence 1951)

a single pistil; gynoecium syncarpous, 3- to 4-carpellary (Fig. 25.2B); ovary 3- to 4-loculed, superior, placentation axile (Fig. 25.2F, G), ovules numerous; style 1, thick and stout or almost inconspicuous terminated by a discoid stigma (Fig. 25.2B, E). Fruit an elongated, leathery loculicidal capsule (Fig. 25.2H); seeds numerous. Embryo straight, surrounded by fleshy endosperm.

Anatomy.   The vessel elements have simple perforation plates and transverse end walls. The imperforate elements have bordered pits and some diffuse parenchyma; ray cells upright, square and procumbent. Lianas and vines have more specialised vessels (De Buhr 1977).

Embryology.   Embryology has not been fully investigated. Pollen grains in permanent tetrads. Ovules anatropous, bitegmic, crassinucellate.

Chemical Features.   Iridoids are reported to be absent (Jensen et al. 1975). The waxy interior (containing alkanes) of the pitcher effectively helps to trap insects (Eglinton and Hamilton 1963).

Important Genera and Economic Importance.   The only genus *Nepenthes*, is commonly known as the pitcher plant. The leaf blade portion is modified into a flask-like structure which captures the victims. Once inside, they cannot come out because of the downward-facing hairs on the inner wall of the pitcher. It does not have any economic importance.

Taxonomic Considerations.   Most authors have placed the Nepenthaceae with the Sarraceniaceae and the Droseraceae in the same order (Engler and Diels 1936, Cronquist 1981). However, Hutchinson (1969) is of the opinion that the Nepenthaceae is more closely related to the Aristolochiaceae. Thorne (1968) presumed that the Nepenthaceae has common ancestry with the Theales but not directly with the Sarraceniaceae. Dahlgren (1980a) includes it in Theales. Takhtajan (1980) includes Nepenthales comprising Nepenthaceae alone in the superorder Rosanae. Later, in 1987, he changed his opinion and placed this order in the superorder Nepenthanae near Rafflesianae.

Chemically, the two families—Sarraceniaceae and Nepenthaceae—are much apart from each other. Morphologically, also, they are distinct. Therefore, it is better to keep *Nepenthes* in a distinct family and order of its own—Nepenthaceae and Nepenthales.

Droseraceae

A medium-sized family of 4 genera and 100 species of insectivorous plants.

Vegetative Features.   The members are herbaceous, annual or perennial, mostly grow in marshy areas. *Aldrovanda* is a submerged aquatic plant. Leaves alternate, usually in basal rosettes (Fig. 25.3A); both surfaces covered with stalked glandular hairs (Fig. 25.3A, B) that help in entrapping insects (except in *Dionaea*). In *Dionaea* the two halves of the leaf on two sides of the midvein snap together when an insect lands on its surface.

Floral Features.   Inflorescence a raceme or a panicle. Flowers hermaphrodite, actinomorphic, hypogynous (Fig. 25.3C), bracteate. Sepals 4 or 5, connate only at the base, persistent, imbricate. Petals 5, polypetalous, convolute. Stamens 5–20 in 1 or more pentamerous whorls, free or rarely filaments connate at the base; anthers bicelled, dehiscence longitudinal (Fig. 25.3D); anthers often with a broad connective, extrorse. Gynoecium syncarpous, 3- to 5-carpellary, ovary unilocular, superior or half-inferior, with 3–5 parietal placentae (Fig. 25.3E, F), ovules few to many on each placenta. In *Dionaea* and *Drosophyllum* several ovules are situated at the base of the unilocular ovary; styles 3 to 5, free in *Drosophyllum*, connate in *Dionaea*; stigmas as many as stylar tips. Fruit a loculicidal capsule; seeds numerous, endospermous, embryo straight.

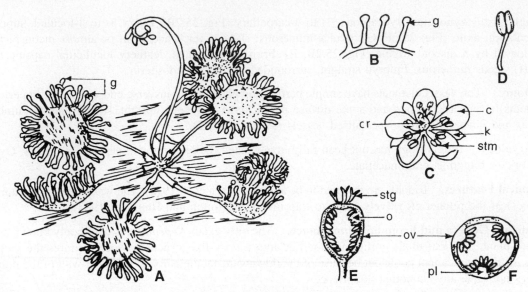

Fig. 25.3 Droseraceae: **A-F** *Drosera rotundifolia.* **A** Habit. **B** Glandular leaf-margin hairs. **C** Flower, front view. **D** Stamen. **E** Ovary, longisection. **F** Cross section. *c* corolla, *cr* carpel, *g* gland, *k* sepal, *o* ovule, *ov* ovary, *stg* stigma, *stm* stamen. (Adapted from Lawrence 1951)

Anatomy. Anatomically, the four genera are quite different from each other. Detailed anatomical descriptions are given in Metcalfe and Chalk (1972). In *Drosera*, the leaf surface is covered with sessile glandular hairs often filled with a red or purple fluid. In addition, there are tentacles on the margin and upper surface of the leaves which consist of a stalk with a large number of tracheids that terminate in a head of glandular cells.

General morphology of *Drosophyllum* is similar to that of *Drosera* but they are more shrub-like in habit. Tentacles and glands similar to those of *Drosera*. Vascular system of the scape consists of 2 rings of vascular bundles.

In *Dionaea*, the distal ends of the leaves are differentiated as traps. The marginal cilia of the leaves interlock when the leaf is stimulated to close. Petiole portion is winged.

Aldrovanda is a monotypic genus, the only species is *A. vesiculosa*. It is a rootless floating aquatic. Leaves somewhat like those of *Dionaea*. Outer surface of the trap bears short 2-armed trichomes and also almost sessile glands. Petiole shows air cavities below the epidermis and a centrally situated single vascular bundle.

Embryology. Pollen grains in tetrads; 2- or 3-celled at shedding stage, e.g. *Drosera* (Brewbaker 1967). Sometimes, they germinate in situ (Johri et al. 1992). Ovules anatropous, bitegmic, tenuinucellate or crassinucellate (Dahlgren 1975b). Polygonum type of embryo sac, 8-nucleate at maturity. Development of endosperm is of the Nuclear type.

Chromosome Number. There is much variation in chromosome number. The haploid number may be n = 6, 10, 11, 14–17, 20, 30, 36, and 40 (Takhtajan 1987).

Chemical Features. Iridoids have been reported to be absent from *Drosera rotundifolia* and *Drosophyllum lusitanicum* (Jensen et al. 1975). Naphthoquinone 'plumbagin' occurs in a free state in *Drosera* (Paris 1963).

Important Genera and Economic Importance. All the 4 genera are important as they are specimens of academic interest. *Aldrovanda vesiculosa* is one of the rare and threatened plants. Although south European in origin, it now occurs only in a few saline ponds near Calcutta, India. With the conversion of these water bodies into land, these plants are slowly disappearing. Economically, none of these plants is important.

Taxonomic Considerations. Many authors place the family Droseraceae in the Parietales near Violaceae and Ochnaceae, on the basis of parietal placentation. Bentham and Hooker (1965a) place them next to Saxifragaceae because of determinate inflorescence, often half-inferior ovary, and similarity between the two groups in androecium, perianth and ovule morphology. Because of aquatic habit of *Aldrovanda* and presence of glandular hairs, it was also presumed to be similar to Lentibulariaceae (Lawrence 1951).

Melchior (1964) and Cronquist (1968) place Droseraceae in Sarraceniales along with 2 other families.

None of the three families of the order can be considered ancestral to anyone of the others. In 1981, Cronquist changed the ordinal name from Sarraceniales to Nepenthales.

Hutchinson (1973) treats Droseraceae as a member of Sarraceniales along with the Sarraceniaceae and are closely related to the members of Saxifragaceae. Dahlgren (1980a) places Droseraceae in a distinct order Droserales and includes 2 other families—Lepuropetalaceae and Parnassiaceae—and derives it from the Theales. Takhtajan (1980) treats this family as a member of Saxifragales of subclass Rosidae. However, in 1987, he erects a separate order, Droserales, to include this family alone, and derives it from the Saxifragales. From the above discussion, it is clear that the family Droseraceae is more akin to the Saxifragales and should be placed near it.

Taxonomic Considerations of the Order Sarraceniales

Cronquist (1981), Engler and Diels (1936), Lawrence (1951), and Melchior (1964) consider Sarraceniales as an order comprising 3 families of insectivorous plants. Dahlgren (1980a), Hutchinson (1973), and Takhtajan (1987) do not agree to this view. Hutchinson (1973) includes Droseraceae and Sarraceniaceae in his Sarraceniales with their nearest relatives in the Saxifragaceae, and segregates Nepenthaceae in the Aristolochiales. Cronquist (1968) is of the opinion that the similarities between *Ancistrocladus* and *Dioncophyllum* of the Violales and members of the Sarraceniales reflect a common ancestry in the Theales. Dahlgren (1980a) derives only Droseraceae from the Theales. Sarraceniales sensu Dahlgren includes only Sarraceniaceae, placed between the Ericales and Cornales, as Sarraceniaceae resembles both these orders (see under Sarraceniaceae). Takhtajan (1987) places Sarraceniales s.s. near Ericanae, Nepenthales next to Balanophorales, and Droserales next to Saxifragales. This means that the insectivorous adaptations in the three families are due to parallel evolution and the families are derived from different ancestors. Therefore, these families have different taxonomic assignment.

26

Order Papaverales

The order Papaverales comprises 6 (or 7) families under 4 suborders:
   The suborder (a) Papaverineae includes Papaveraceae,
   (b) Capparineae includes Capparaceae, Cruciferae (= Brassicaceae) and Tovariaceae,
   (c) Resedineae includes Resedaceae, and
   (d) Moringineae includes monotypic Moringaceae (Takhtajan 1987).

The members are mostly herbaceous, except *Moringa* of Moringaceae. Flowers bisexual, actino- or zygomorphic, hypogynous; number of stamens variable—numerous in Papaveraceae, to 6 (in most Brassicaceae) or even 2, as in *Coronopus didymus* of Brassicaceae. Gynoecium syncarpous with parietal placentation (axile in Tovariaceae). Pollen grains shed at 2-celled stage, ovules anatropous, bitegmic, crassinucellate, endosperm mostly absent (except in Tovariaceae). Most members have benzylisoquinoline (Papaverineae) or methylglucosinolates.

Papaveraceae

A medium-sized family with 24 genera and 250 species (26 genera and 200 spp. according to Kumar and Subramanian 1987), distributed chiefly in north-temperate zone and at higher altitudes in the tropics (between 2640 and 5610 m in the Himalayas, Nepal and Bhutan). Some cultivated, a few naturalized.

Vegetative Features.   Herbaceous annuals or perennials, rarely shrubs (*Dendromecon*) or trees (*Bocconia*). Stem soft and herbaceous, usually with coloured (*Argemone*) or milky sap (*Papaver*), watery sap in *Eschscholzia* and *Hunnemannia*. Leaves alternate, radical or cauline, simple, entire or pinnately (*Papaver*) or palmately (*Eschscholzia*) cleft, petiole often winged, estipulate. Leaf margin spinescent in *Argemone*.

Floral Features.   Flowers mostly solitary, axillary (Fig. 26.1A) or terminal as in *Argemone*; inflorescence compound raceme in *Bocconia* and *Macleaya*. Flowers ebracteate, ebracteolate, bisexual, actinomorphic, hypogynous, showy. Calyx of 2 or 3 sepals (fused and calyptrate in *Eschscholzia*), polysepalous, imbricate, caducous. Corolla of 4 to 6 or 8 to 12 petals (absent in *Macleaya*) in 1 or 2 whorls, polypetalous (Fig. 26.1B), imbricate, often crumpled in bud, deciduous. Stamens numerous in several whorls (4 stamens in *Pteridophyllum* and *Hypecoum*), filaments often petaloid, anthers basifixed, bicelled, dehiscence longitudinal (Fig. 26.1C). Gynoecium syncarpous, 2- to many-carpellary (pistils several and coherent in *Platystemon*), ovary superior, unilocular, 2 to many ovules on parietal placentae (Fig. 26.1D); style usually 1 or indistinct, stigma capitate. Fruit a capsule (follicular in *Platystemon*) with poricidal or valvular dehiscence; seeds with minute embryo and copious endosperm.

Anatomy.   The stem (in transverse section) exhibits a single ring of widely-spaced vascular bundles. Vessels small, tend to be ulmiform or medium-sized and without any specific pattern; sometimes ring-porous and with spiral thickenings, intervascular pitting alternate. Parenchyma vasicentric, rays multiseriate, up to 5 to 12 cells wide, heterogeneous. Leaves dorsiventral, hairs scanty—uniseriate, biseriate or multiseriate; stomata Ranunculaceous type, generally confined to the lower surface; on both the surfaces in *Papaver pilosum, P. spicatum* and *Roemeria dodecandra*.

Embryology.   Pollen grains 3-colpate, colporate, pantoporate and pantocolpate, shed at 2-celled stage. Ovules anatropous, bitegmic, crassinucellate. Endosperm formation of the Nuclear type.

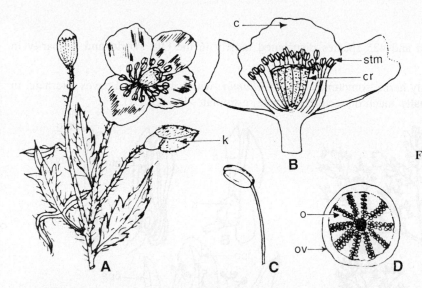

Fig. 26.1 Papaveraceae: **A-D** *Papaver rhoeas.* **A** Flowering twig. **B** Flower, longisection. **C** Stamen. **D** Carpel, cross section. *c* corolla, *cr* carpel, *k* calyx, *o* ovule, *ov* ovary. (Original)

Chromosome Number. Basic chromosome number is x = 6, 7. Induced polyploidy is reported in some *Papaver* spp. (Mary and Malik 1973).

Chemical Features. Rich in alkaloids like morphine and codeine; berberin occurs in *Argemone mexicana.*. Glucosinolates absent and benzylisoquinolines present (Hegnauer 1963). Meconic acid can be a taxonomic marker in this family (Fairbairn and Williamson 1978).

Important Genera and Economic Importance. *Argemone* is an important genus with 10 species distributed mainly in Mexico, West Indies, western and eastern United States of America and Asiatic tropics. The oil from the seeds of *A. mexicana* is used as an adulterant for mustard oil. *Eschscholzia californica* (Californian poppy) is an ornamental. *Meconopsis*, with its 42 species from the Himalayas to western China and one species in western Europe, is another important genus. It grows in the temperate and subalpine Himalayas between 2640 and 5610 m altitude. *Papaver* (100 species) has a wide distribution. The plants of opium poppy or *Papaver somniferum* are erect annuals, often grown as an ornamental for their large and showy, white, scarlet, or purple flowers. The capsules are large, ovoid and globular; incisions are made on the surface of these capsules in late evening and the exudate collected before sunrise. Opium in its crude commercial form is the air-dried coagulated latex from unripe capsules. Opium is a habit-forming drug, though in lower doses it can be a pain reliever. In recent years, opium addiction has diminished due to strict governmental control of its cultivation. *P. somniferum* is the source of the alkaloids morphine, codeine and thebaine. The first two have both uses and abuses and the third can be used safely in medicine. The seeds are devoid of latex. *P. rhoeas* is a cultivated ornamental—the garden poppy.

Taxonomic Considerations. Bentham and Hooker (1965a) placed Papaveraceae in the Parietales and Engler and Diels (1936) in Rhoeadales. Cronquist (1968, 1981), Dahlgren (1980a, 1983a), Melchior (1964), and Takhtajan (1987) include it in the Papaverales. Dahlgren (1983a) also points out that it is undisputedly related to the Ranunculales.

Fumariaceae

A small family of 17 genera and 425 species distributed mostly in the Old World and primarily in temperate Eurasia.

Vegetative Features.   Mostly herbs, sometimes lianas (*Adlumia*) with watery sap. Leaves alternate, in basal rosettes or cauline, usually much dissected, petiolate, estipulate (Fig. 26.2A).

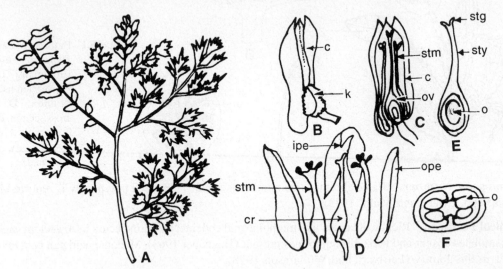

Fig. 26.2   Fumariaceae: **A-E** *Fumaria indica*, **F** *Dicentra spectabilis*. **A** Flowering twig. **B** Flower. **C** Flower, longisection. **D** Dissected, whole-mount. **E** Carpel, vertical section. **F** Ovary, cross section, of *D. spectabilis*. *c* corolla, *cr* carpel, *ipe* inner petal, *k* calyx, *o* ovule, *ope* outer petal, *ov* ovary, *stg* stigma, *stm* stamen, *sty* style. (Adapted from G.L. Chopra 1973)

Floral Features.   Inflorescence raceme (Fig. 26.2A). Flowers bisexual, zygomorphic, hypogynous, ebracteate and ebracteolate. Calyx of 2 minute caducous sepals (*k* in Fig. 26.2B); corolla of 4 petals, more or less coherent and sometimes basally connate, in 2 whorls, one or both outer petals saccate or spurred at the base (Fig. 26.2C, D), inner ones narrower, crested and united over the anthers. Stamens 6, in 2 bundles, three on each side of the pistil, opposite the inner petals; each bundle has a single broad filament which divides into three parts at the apex (Fig. 26.2D), the central part bears a bicelled anther and the lateral parts bear one-loculed half-anther. One or two nectaries usually present at the base of the stamens. Gynoecium syncarpous, bicarpellary, ovary superior, unilocular (Fig. 26.2E), ovules 2 to many on parietal placenta (Fig. 26.2F); style 1 and slender, stigma one or bilobed, stigmatic surfaces 2, 4 or 8. Fruit, a transversely septate capsule dehiscing by valves or a 1-seeded (only one ovule matures) indehiscent nut in *Fumaria*. Seeds with minute embryo and copious endosperm. Cotyledons usually 2, 1 in some species of *Corydalis* and *Dicentra*.

Anatomy.   Similar to that of the Papaveraceae herbs.

Embryology.   The pollen grains in *Dicentra* are prolate-spheroidal to spheroidal, 3-colpate, number of apertures variable (Stern 1962). In *Fumaria*, pollen grains are pantoporate, 2-celled at the time of dehiscence (Johri et al. 1992). Ovules campylotropous, bitegmic, and crassinucellate. Embryo sac of Polygonum type, 8-nucleate at maturity. Cytokinesis occurs in *Fumaria*, after several hundred endosperm nuclei are formed (Johri et al. 1992).

Chromosome Number.   Basic chromosome number is x = 6–8, 16.

Chemical Features.   The presence of spirobenzylisoquinoline alkaloids is characterstic of only this family (Preisner and Shama 1980). Fahselt (1971) reported the occurrence of flavones with 5-hydroxyl and 4-hydroxyl groups and a glycoside of cyanidin.

Important Genera and Economic Importance.   Some of the genera of this family are *Dicentra, Corydalis, Fumaria.* Economically, the Fumariaceae is of little importance, except for some ornamentals like *Dicentra* (bleeding heart). *Corydalis ramosa* is useful in the treatment of eye diseases. The pea-like tubers of *Dicentra canadensis* form the commercial corydalis. *Fumaria indica* is a weed in cultivated fields. Alkaloids in some species of *Dicentra* cause livestock poisoning (Fahselt an Ownbey 1968).

Taxonomic Considerations.   Bentham and Hooker (1965a), Fedde (1936), and Thorne (1977) do not recognise Fumariaceae as a distinct family. Hutchinson (1973) includes it in the order Rhoeadales, and Cronquist (1981), Dahlgren (1980a, 1983a) and Takhtajan (1987) in the Papaverales. Fedde (1936) placed the Fumariaceae within Papaveraceae.

The family Fumariaceae differs from the Papaveraceae in: (1) 4 or 6 stamens, (2) zygomorphic flowers, (3) saccate or spurred petals, (4) absence or occassional presence of latex system by scattered secretory cells. According to Cronquist (1968): "most of the characters which mark Fumariaceae as a separate group within the order Papaverales, represent phylogenetic advances over the characters of the Papaveraceae".

It is concluded that, on the basis of differences in floral morphology as well as chemistry, Fumariaceae should be treated as a separate family in the order Papaverales.

Capparaceae

Capparaceae, a medium-sized family of 42 to 45 genera and 850 species (Takhtajan 1987) is distributed palaeotropically in both hemispheres.

Vegetative Features.   Annual or perennial herbs (*Cleome*), shrubs (*Capparis*) or trees (*Crataeva*), sometimes lianas (*Maerua arenaria*), often xerophytic such as *Capparis decidua*. Stem herbaceous or woody, without latex. Leaves alternate, simple (*Maerua*) or palmately compound, pentafoliate, trifoliate or unifoliate (*Steriphoma peruviana*); stipulate or estipulate, stipules minute, glandular or spinose (*Capparis decidua, C. sepiaria*).

Floral Features.   Inflorescence terminal (Fig. 26.3A, E) or axillary racemes or cymes; flowers solitary in *Niebuhria*. Flowers bisexual, complete, sometimes unisexual—the plants are then monoecious, e.g. *Podandrogyne;* actinomorphic (Fig. 26.3B) or zygomorphic (*Capparis decidua,* Fig. 26.3G), bracteate, bracteolate. Flowers borne on old wood in *Bachmannia*. Calyx of 4 sepals, polysepalous, valvate, outer surface often covered with glandular hairs as in *Cleome viscosa*; sepals unequal in *Capparis decidua*, the posterior one forms a hood-like structure; corolla of 4 petals, polypetalous, sometimes connate, the two posterior ones form a large hood-like structure in *Emblingia*, imbricate, sometimes absent; all equal or 2 posterior ones larger, clawed or sessile. Stamens 4 to many, never tetradynamous, but are derived from tetrandrous condition by splitting of the 4 primordia and then many filaments often lack anthers; anthers bi- or tetracelled, dehisce longitudinally (Fig. 26.3C). In some genera the androecium and gynoecium are borne on androgynophore as in *Cleome gynandra* (Fig. 26.3B). Gynoecium syncarpous, bicarpellary, ovary usually borne on a long or short gynophore (Fig. 26.3C), superior, unilocular with parietal placentation (Fig. 26.3D) and numerous ovules; style 1, short or filiform and elongate, stigma bilobed or capitate (Fig. 26.3B, C). Fruit a capsule dehiscing by valves as in *Cleome* or elongated or torulose berry as in *Capparis* (Fig. 26.3H) and with transverse constrictions, as in *Maerua* (Fig. 26.3F), sometimes indehiscent nuts as in *Emblingia*. Seeds reniform with a curved, folded embryo and fleshy endosperm.

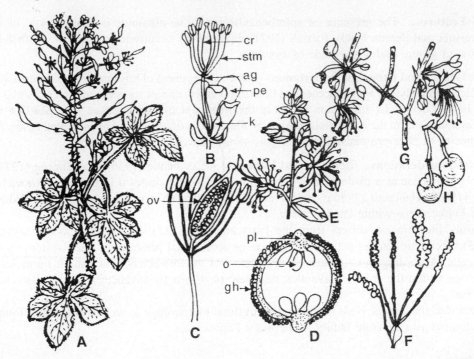

Fig. 26.3 Capparaceae: **A-D** *Cleome gynandra,* **E, F** *Maerua arenaria,* **G, H** *Capparis decidua.* **A** Flowering twig. **B** Flower. **C** Longisection, sepals and petals removed. **D** Ovary, cross section. **E** Flowering twig of *M. arenaria.* **F** Fruits. **G** Flowering twig of *C. decidua.* **H** Fruits. *ag* androgynophore, *cr* carpel, *gh* glandular hairs, *k* calyx, *o* ovule, *ov* ovary, *pe* petal, *pl* placenta, *stm* stamen. (Original)

Anatomy. Vessels very small to medium-sized, often in clusters or long, radial multiples, perforations simple, intervascular pitting alternate, small. Parenchyma paratracheal, sparsely vasicentric, sometimes storied, rays up to 2 to 5 cells wide, homogeneous. Included phloem occurs in some genera like *Boscia, Cadaba, Forchhammeria, Maerua* and *Stixis.* Leaves dorsiventral, isobilateral or centric, hairs of various types; stomata Ranunculaceous type, on both the surfaces in isobilateral and centric leaves and confined to lower surface in dorsiventral leaves.

Embryology. Pollen grains spherical, tricolpate, with small spine-like structures on the thick exine; shed at 2- or 3-celled stage. Ovules campylotropous (anatropous in *Crataeva*), bitegmic, crassinucellate. Polygonum type of embryo sac, 8-nucleate at maturity. Endosperm formation of the Nuclear type.

Chromosome Number. Basic chromosome numbers are x = 8 to 17. As both *Podandrogyne* and *Cleome* have a very high basic chromosome number x = 29, Cochrane (1978) suggested phyletic relationship amongst the species of these two genera.

Chemical Features. Methylglucosinolates prominent (Ettlinger and Kjaer 1968). Ellagitannins usually absent. Common flavonols and their O-methyl derivatives also occur (Gornall et al. 1979).

Important Genera and Economic Importance. Amongst the important genera are *Cleome* (150 spp.), *Capparis* (250 spp.), *Crataeva* (9 spp.), *Cadaba* (30 spp.), *Maerua* (100 spp.) and *Roydsia*. Economically, the Capparaceae is not so important. *Capparis spinosa* is grown in the Mediterranean area for the unopened flower buds called capers which are useful in seasoning foods. Fruits and flower buds of *C. decidua* are edible either raw or pickled. The bark of *Crataeva nurvala* finds application as a remedy for bladder stones; the fruit is edible and the rind is used as a mordant in dyeing. *Cleome gynandra, C. brachycarpa* and *C. viscosa* are noxious weeds. *Capparis sepiaria* is a hedge plant and climbs by hooked spines. *Oceanopapaver* is a new monotypic genus from New Caledonia (Schmid et al. 1984).

Taxonomic Considerations.   The family Capparaceae has been placed in Rhoeadales by the earlier authors (Lawrence 1951), in Papaverales by Melchior (1964) and in Capparales by Cronquist (1968, 1981), Dahlgren (1980a, 1983a) and Takhtajan (1980, 1987). It is intermediate between the families Papaveraceae and Cruciferae, but more closely allied to the Cruciferae. A comparison between the two families is given in Table 26.3.1.

Table 26.3.1   Comparative Data for Capparaceae and Cruciferae

Capparaceae	Cruciferae
Mostly shrubs and trees; very few herbs	Mostly herbs
Leaves simple or palmately compound, not auriculate	Leaves simple, lobed or pinnatifid; auriculate
Stamens 6–8 to numerous, never tetradynamous	Stamens 6, tetradynamous
Androphore and/or gynophore present	Androphore and gynophore absent
Fruit capsule or berry, rarely siliqua	Fruit siliqua/silicula, never a capsule

Capparaceae is a natural taxon and well placed as the most primitive family amongst the Capparales. That it is a primitive taxon is further proved by the fact that Capparaceae is distinct, mainly due to the presence of alanine-derived methyl glucosinolates. The other families of the Capparales have more complex glucosinolates which are derived from it (Harborne and Turner 1984).

Its placement is more desirable in the Capparales than in the Papaverales.

Aleykutty and Inamdar (1978) considered it unnecessary to separate the tribe Cleomoideae (from the Capparaceae) as a distinct family Cleomaceae.

Cruciferae (= Brassicaceae)

A large family of 350 genera and 2500 species distributed primarily in northern hemisphere and in colder alpine regions of the tropics. Some of the genera, such as *Cardamine, Lepidium, Sisymbrium* are cosmopolitan.

Vegetative Features.   Annual, biennial or perennial herbs, rarely undershrubs. Stem herbaceous, soft and green, sometimes modified for storage of food, e.g. *Brassica caulorapa*. Roots of some genera are also modified for food storage, e.g. *Armoracia rusticana, Brassica rapa* and *Raphanus sativus*. Leaves alternate, radical or cauline, simple, auriculate, often lyrate as in *R. sativus*, and *Sisymbrium irio* (Fig. 26.4A); estipulate.

Floral Features.   Inflorescence racemes, spikes or corymbs. Flowers bisexual, actinomorphic, zygomorphic in *Iberis* and *Teesdalia*, owing to the enlargement of two outer petals, hypogynous, ebracteate, ebracteolate. Calyx of 4 sepals, polysepalous (Fig. 26.4B, H), in 2 whorls of 2 each; corolla of 4 petals, polypetalous (Fig. 26.4J), with a distinct limb and claw, cruciform; often petals very small as in *Coronopus didymus* (Fig. 26.4J) or absent as in *Lepidium*. Stamens 6 in 2 whorls (Fig. 26.4C), often reduced to 4 (*Cardamine hirsuta*) or 2 (*Coronopus didymus*, Fig. 26.4G, I) or sometimes numerous as in *Megacarpaea* (16 stamens); usually tetradynamous, filaments of each inner pair sometimes connate, those of outer pair may be winged or toothed. Nectariferous glands present in between the filaments, anthers bicelled, dorsifixed, dehiscence longitudinal. Gynoecium syncarpous, bicarpellary, ovary unilocular but bilocular later due to

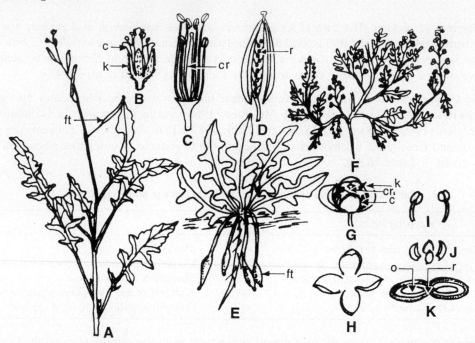

Fig. 26.4 Brassicaceae: **A-D** *Sisymbrium irio*, **E** *Geococcus pusillus*, **F-K** *Coronopus didymus*. **A** Flowering twig. **B** Flower. **C** Tetradynamous stamens and carpel. **D** Replum with seeds on it. **E** Fruiting plant of *G. pusillus*. **F** Flowering plant of *C. didymus*. **G** Flower. **H** Dissected calyx lobes. **I** Stamens. **J** Corolla. **K** Ovary, cross section. *c* corolla, *cr* carpel, *ft* fruit, *k* calyx, *o* ovule, *r* replum. (**E** adapted from Hutchinson, 1969, **A-D** and **F-K** Original)

the formation of a false septum called 'replum' (*r* in Fig. 26.4D, K), superior, placentation parietal, ovules numerous or only a few; style one or obsolete, stigmas 2, capitate. Fruit a siliqua or a silicula. Generally, the ovary is sessile on the receptacle but there is a stipe between the ovary and receptacle in *Stanleya*. Fruit a 1-seeded indehiscent nut in *Bunias* and *Isatis*. Seeds with scanty or no endosperm and a large embryo.

Anatomy.   Vessels small, perforations simple, intervascular pitting alternate with horizontal apertures. Parenchyma paratracheal, extremely sparse, rays up to 2 to 4 cells wide, heterogeneous. Leaves dorsiventral or isobilateral, stomata Cruciferous type, hairs variable. Myrosin cells distributed throughout leaf parenchyma.

Embryology.   Pollen grains mostly 3-colpate but inaperturate in *Matthiola* (P.K.K. Nair 1970); shed at 3-celled stage. Ovules bitegmic, tenuinucellate. Polygonum type of embryo sac, 8-nucleate at maturity. Endosperm formation of Nuclear type (Dahlgren 1980a).

Chromosome Number.   Basic chromosome numbers are x = 3–13.

Chemical Features.   The seeds of Brassicaceae are rich in fatty acids. Erucic acid, a rare fatty acid, occurs in higher quantities in the seed oil from *Brassica napus*. They are also rich in isothiocyanates and glucosinolate sinigrins (Ettlinger and Kjaer 1968).

Important Genera and Economic Importance.   Important genera include *Brassica* with 50 species and a number of varieties, common in Europe, Asia and Mediterranean region, 8 species are cultivated in warmer and temperate regions. *Raphanus* is another important genus with modified roots for storage and are edible. *R. sativus* has numerous strains that are cultivated. Economically also this family is very important. A number of winter vegetables such as cabbage, *Brassica oleracea* var. *capitata*, cauliflower, *B. oleracea* var. *botrytis*, Brussels sprouts, *B. oleracea* var. *gemmifera*, Kohl rabi, *B. caulorapa*, radish, *R. sativus*, cress, *Lepidium sativum* are obtained from this family. Fatty oils used for cooking purposes are obtained from *Brassica juncea* var. *sarson*. Fatty oil from *Eruca sativa* is used for adulteration of mustard oil, has unpleasant taste and aroma. Table mustard is made from powdered seeds of *B. nigra* and *B. alba*. In *Geococcus pusillus* (Fig 26.4E) the fruits mature underground.

Isatis tinctoria is the source of a blue dye 'woad'. Dame's violet—*Hesperis matronalis*, sweet alyssum—*Lobularia maritima* (= *Alyssum maritimum*), wallflowers—*Matthiola incana* and *M. tristis*, candytuft—*Iberis amara*, and stock—*Cheiranthus* are some of the cultivated ornamentals. A few, like *Capsella bursa-pastoris*, *Coronopus didymus* (Fig. 26.4F), *Nasturtium* sp. and *Sisymbrium irio*, are weeds in waste places.

Taxonomic Considerations. Cruciferae (= Brassicaceae) was placed in the order Parietales (Bentham and Hooker 1965a), Rhoeadales (Rendle 1925, Lawrence 1951, Melchior 1964), Cruciales (Hutchinson 1969), Brassicales (Hutchinson 1973) and Capparales (Cronquist 1981, Dahlgren 1983a, Takhtajan 1987). It is a natural taxon. The only controversy about its origin is whether it is derived from Papaveraceae or Capparaceae. On the basis of androecial and gynoecial morphology and anatomy, it is derived from Capparaceous ancestors (Lawrence 1951). Chemical features also support this view; isothiocyanates are reported in both Capparaceae and Cruciferae (Harborne and Turner 1984). From the Papaveraceae it differs chemically and in endospermous seeds, although there are a few resemblances in androecial and gynoecial features and the tetramerous perianth.

On the basis of studies of different taxonomists, it is apparent that the Cruciferae has originated from Capparaceous ancestors and are better placed in the order Capparales.

Tovariaceae

A unigeneric family, *Tovaria* with two species distributed in tropical America, Mexico, Jamaica and the tropical Andes.

Vegetative Features. Shrubs or annual herbs with alternate, trifoliate, stipulate leaves, leaflets entire. Plants emit the smell of *Apium* or *Cestrum* when fresh, and of the chemical coumarin when dried (Willis 1973).

Floral Features. Inflorescence loose, elongated, many-flowered, terminal racemes. Flowers bisexual, actinomorphic, hypogynous. Calyx of 8 sepals, polysepalous, imbricate. Corolla of 8 petals, each with a short claw, imbricate. Stamens 8, opposite the sepals with lanceolate, long-papillate filament, and basifixed, introrse anthers. Gynoecium syncarpous, 6- to 8-carpellary, ovary superior, 6- to 8-loculed with axile placentation and numerous ovules, short style and peltate, lobulate stigma. Fruit a globose, green, slender-pedicelled, many-seeded berry, mucilaginous when young; pericarp membranous; seeds with curved embryo and well-developed endosperm.

Anatomy. Similar to Capparaceae members, but myrosin cells absent. Leaves dorsiventral, stomata Ranunculaceous type, confined to the lower surface.

Embryology. Pollen grains 2-celled at dispersal stage. Ovules campylotropous, bitegmic, crassinucellate. Polygonum type of embryo sac, 8-nucleate at maturity. The endosperm formation of the Nuclear type.

Chromosome Number. Haploid chromosome number is n = 14.

Chemical Features. Glucosinolates present (Gershenzon and Mabry 1983).

Important Genus. *Tovaria* is the only genus; economic importance not known.

Taxonomic Considerations. The family Tovariaceae was included in Capparaceae by Bentham and Hooker (1965a) and Metcalfe and Chalk (1972). It is treated as a distinct family by Benson (1970), Cronquist (1969, 1981), Dahlgren (1980a, 1983a) and Takhtajan (1980, 1987).

Originally placed in Capparaceae, Tovariaceae is the smallest and the most distinctive family of this group. With axile placentation, absence of myrosin cells, well-developed endosperm and floral parts in eights, Tovariaceae stands apart from the remaining Capparales members. In spite of this, most taxonomists retain Tovariaceae in the Capparales, mainly because of its similar palynological and embryological features. The elongate true racemes are very suggestive of those of Cruciferae and many of the Capparaceae (Cronquist 1968). The endospermous seeds and axile placentation are primitive features, but the number of floral parts may reflect a secondary increase which is an advanced feature. According to Willis (1973), it is related to Cleomoideae of the Capparaceae. Therefore, Tovariaceae is a distinct family of the Capparales and is allied to the subfamily Cleomoideae of the Capparaceae.

Resedaceae

A small family of 6 genera and 70 to 75 species, mainly distributed in the Mediterranean region. The genus *Oligomeris* is primarily African, but *O. linifolia* is widely distributed in arid and saline habitats of the warmer northern hemisphere. The natural origin of most Resedaceae is in South Arabia and Somalia. Some of the 13 spp. recognized from Jordan are endemic to Jordan and Palestine (Al-Eisawi 1988), e.g. *Reseda alopecuros*, which has very limited geographical distribution.

Vegetative Features.   Annual or perennial herbs, e.g. *Reseda*, *Oligomeris*, suffrutescent or rarely shrubs such as *Ochradenus* and *Randonia*. Vegetative parts soft, herbaceous, with a watery sap. Leaves alternate, simple or pinnately dissected, stipulate, stipules minute, glandular.

Floral Features.   Inflorescence racemes or spikes (Fig. 26.6A). Flowers bisexual, rarely unisexual and then the plants monoecious, zygomorphic, hypogynous. Sepals and petals usually 4 or 8, valvate, petals

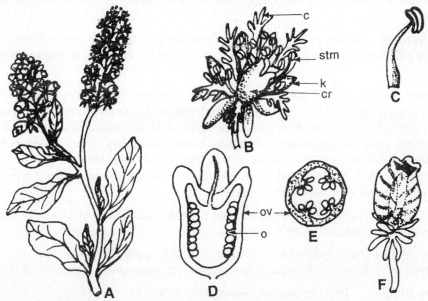

Fig. 26.6   Resedaceae: **A-F** *Reseda odorata*. **A** Flowering twig. **B** Flower. **C** Stamen. **D** Ovary, longisection. **E** Cross section. **F** Fruit. *c* corolla, *cr* carpel, *k* sepal, *o* ovule, *ov* ovary, *stm* stamen. (Adapted from Lawrence 1951)

sometimes absent or only 2; the posterior ones much larger and often laciniate (Fig. 26.6B). Stamens 3 to 40, usually borne on a disc, anthers bicelled, dehiscence longitudinal, introrse (Fig. 26.6C). Gynoecium syncarpous, 2- to 6-carpellary, ovary superior, unilocular (Fig. 26.6D, E), usually open at the apex; styles absent, each ovary with its own stigma, ovules numerous, placentation parietal (Fig. 26.6E). Fruit a capsule (Fig. 26.6F) or berry, seeds reniform, with a curved embryo and fleshy endosperm. In *R. odorata*, fruit 4-carpelled, ovary partially closed by 4 incurved triangular flaps formed from the fused margins of two adjacent carpels (Arber 1942).

Anatomy.   Xylem and phloem form closed cylinders traversed by narrow primary medullary rays. Vessels with small lumina and simple perforations. Rays 1 or 2 cells wide. In the leaves, the mesophyll is not clearly differentiated into spongy and palisade regions. Hairs simple, unicellular; stomata Ranunculaceous type, often present on both the surfaces. Myrosin cells present.

Embryology.   Pollen grains 2–3-colporate, shed at 2-celled stage. Ovules ana- or campylotropous, bitegmic, crassi- or tenuinucellate. Polygonum type of embryo sac, 8-nucleate at maturity. Endosperm formation of the Nuclear type.

Chromosome Number.   Basic chromosome number is x = 6–15.

Chemical Features.   Flavones at low frequency recorded (Gornall et al. 1979).

Important Genera and Economic Importance.   Genera belonging to Resedaceae are *Oligomeris*, *Ochradenus*, *Reseda*, *Randonia*, *Caylusea* and *Astrocarpus*. Some species of *Reseda* are grown as a garden novelty. *Oligomeris linifolia* is a small weedy annual.

Taxonomic Considerations.   Resedaceae has been placed in Papaverales (Benson 1970), Rhoeadales (Lawrence 1951), Capparales (Cronquist 1981, Dahlgren 1983a, Takhtajan 1987). Hutchinson (1973) erected a separate order, Resedales, for this family. Some morphological features, such as the inflorescence type and parietal placentation, embryology, palynology and the presence of myrosin cells, suggest its relationship with the families of the Capparales. According to Cronquist (1968), if the ancestor of Resedaceae was known, it would have been placed in Capparaceae.

From the studies taken up so far, Resedaceae is better placed in Capparales than in the Rhoeadales or Papaverales. Its members do not show any affinity with the Papaveraceae and Fumariaceae.

Moringaceae

A unigeneric (*Moringa*) family with 14 species native to the Old World tropics.

Vegetative Features.   Large deciduous trees; leaves alternate, bi- or tripinnately compound or decompound, stipules and stipels modified to extrafloral nectaries, as in *Moringa oleifera*.

Floral Features.   Axillary terminal cymose panicles (Fig. 26.7A). Flowers (Fig. 26.7B) bisexual, zygomorphic, complete, hypogynous, bracteate, ebracteolate. Calyx of 5 sepals, polysepalous, borne on a short hypanthium (Fig. 26.7C) and reflexed, unequal, imbricate. Petals 5, polypetalous, borne on the hypanthium, the 2 posterior ones smaller and reflexed with laterals ascending and the anterior one larger (Fig. 26.7B, C). Stamens 5, of unequal length and alternating with an outer whorl of 3 to 5 filiform staminodes, borne on the rim of the hypanthium; filaments broad, anthers 1-celled, dehiscence longitudinal, lying side by side to form a structure similar to staminal column through which the style protrudes at the time of anthesis (Fig. 26.7B). Gynoecium syncarpous, 3-carpellary, ovary superior, shortly stipitate (Fig. 26.7C), hairy, curved, unilocular, ovules numerous, placentation parietal (Fig. 26.7D); style 1, slender, stigma 1, truncate. Fruit an elongated siliqua-like, triquetrous capsule (Fig. 26.7E), 3-valved,

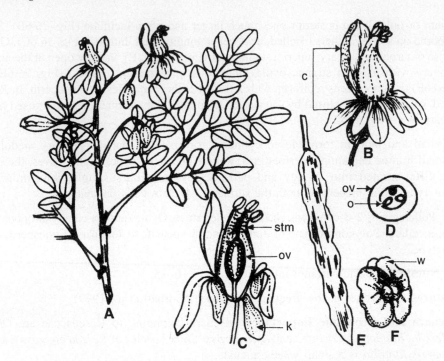

Fig. 26.7 Moringaceae: A-F *Moringa oleifera*. **A** Flowering twig. **B** Flower. **C** Longisection. **D** Ovary, cross section. **E** Fruit. **F** Seed. *c* corolla, *k* calyx, *o* ovule, *ov* ovary, *stm* stamen, *w* wing. (Original)

beaked. Seeds numerous, large, ovate, winged or wingless (3-winged in *M. oleifera*; Fig. 26.7F). Seeds non-endospermous, embryo straight.

Anatomy.   Vessels medium-sized with numerous multiples of 3 or 4 cells, perforations simple; intervascular pitting alternate, large. Parenchyma paratracheal, vasicentric or slightly aliform, sometimes storied. Rays 2 to 3 cells wide, homogeneous. Leaves compound, leaflets dorsiventral with unicellular hairs; stomata Ranunculaceous type. Starch, myrosin and cluster crystals of calcium oxalate abundant.

Embryology.   Pollen grains tricolpate, 2-celled at shedding stage. Ovules anatropous, bitegmic, crassinucellate. Polygonum type of embryo sac, 8-nucleate at maturity. Endosperm formation of Nuclear type; cells filled with abundant starch (H.S. Narayana 1962).

Chromosome Number.   Haploid chromosome numbers are n = 11, 14.

Chemical Features.   Glucosinolates reported (Gershenzon and Mabry 1983).

Important Genera and Economic Importance.   The only genus is *Moringa*; fruits of *M. oleifera* are used as a vegetable.

Taxonomic Considerations.   Moringaceae is a controversial family. It has been placed in the orders Leguminales (Jussieu 1789, see H.S. Narayana 1970, Hallier 1905, 1912), Violales (Dutta and Mitra 1947), Rhoeadales (Pax 1936, Lawrence 1951), Capparales (Hutchinson 1973, Dahlgren 1983a, Cronquist 1981) and Moringales (Takhtajan 1987). It differs from the Leguminales by the absence of monocarpellary ovary, marginal placentation, and chalazal endosperm haustorium (H.S. Narayana 1970). Moringaceae is also different from Violales as the calyx, corolla and stamens grow in a ring on the hypanthium and staminodes alternate with stamens, 1-celled anthers, short gynophore and non-endospermous seeds. Similarly, Moringaceae has no morphological and chemical resemblance with the Rhoeadales, particularly with Papaveraceae.

Presence of short gynophore, anatropous ovule and hypostase indicate the early divergence of this group from the Capparaceae. Kolbe's (1973, 1978) serological studies do not support affinities between Moringaceae and the other families of Capparales. Under the circumstances, Takhtajan's (1987) suggestion may be followed to treat this family as the only member of the order Moringales. He places Moringales next to the Capparales which is appropriate.

Taxonomic Considerations of the Order Papaverales

Many earlier taxonomists (Lawrence 1951, Melchior 1964) included 6 families belonging to 4 suborders in this order. Some taxonomists (Cronquist 1981, Dahlgren 1980a, 1983a, Takhtajan 1987) do not agree with this view and, seggregate it into 2 or 3 orders, such as Papaverales (including Papaveraceae and Fumariaceae), Capparales (including Capparaceae, Cruciferae, Resedaceae, and Tovariaceae) and Moringales (Moringaceae).Hutchinson (1973) places Resedaceae also in a distinct order, Resedales.

Various morphological, anatomical and chemical studies reveal that the order Papaverales should be divided into two separate orders—Papaverales s.s., including Papaveraceae and Fumariaceae, and Capparales, including the rest of the families. Table 26.8.1 gives a comparative account of the two orders.

Table 26.8.1   Comparative Data for Papaverales s.s. and Capparales

Papaverales s.s.	Capparales
Perfect, hypogynous flowers	Same
Bicarpellary, syncarpous ovary	Same
Parietal placentation	Same
Usually with a latex system	Commonly have myrosin cells
Stamens centripetal	Stamens centrifugal
Benzylisoquinoline present	Methylglucosinolates present

Melchior (1964) does not recognise Fumariaceae as a distinct family and includes it in the Papaveraceae. From the present study it is apparent that Fumariaceae should be treated as a separate family. The genera *Hypecoum* and *Pteridophyllum* with regular flowers and spurless petals—but otherwise resembling the Fumariaceae—form the connecting link between the two families, Papaveraceae and Fumariaceae.

27

Order Batales

Batales is a monotypic order with the family Bataceae; it includes the genus *Batis*. Dioecious shrubs with fleshy leaves, axillary ament-like or catkin-like inflorescences and naked flowers.

Bataceae

A family of one genus, *Batis* with two species, *B. argillicola* and *B. maritima* (Goldblatt 1976). These are widespread in littoral zones of the tropics and subtropics, *B. maritima* in the New World and naturalized in the Hawaiian Islands, and *B. argillicola* in New Guinea.

Vegetative Features.   Succulent, halophytic, low, straggling subshrubs. Stem spreading or prostrate, strongly scented. Leaves opposite, simple, sessile, entire, fleshy, semiterete, stipulate; glaucous and fleshy, cylindrical at the distal end, subcylindrical in the middle, adaxially grooved and stem-clasping at the base (Johnson 1935).

Floral Features.   Inflorescence axillary, sessile, cone-like or ament-like erect spikes of unisexual flowers, borne on separate plants in *B. maritima;* or solitary staminate and pistillate flowers borne on the same plant as in *B. argillicola.* Flowers unisexual, reduced, tetramerous, actinomorphic, hypogynous.

Staminate flowers many in a spike, subtended by persistent imbricate bracts, calyx shallowly 2-lipped, campanulate and membranous; stamens 4 or 5, alternate with as many clawed and rhomboidal staminodes (or petals); filaments free, anthers dorsifixed, bicelled, introrse, longitudinally dehiscent. Pistillate flowers 4 to 12 in an inflorescence, each subtended by small, deciduous, non-imbricate bracts; calyx and corolla absent. Gynoecium syncarpous, tetracarpellary, ovary tetraloculed, each locule with a solitary basal ovule; style highly reduced, stigma 1, cushion-like. Fruit a berry; all the fruits of an inflorescence connate into a multiple structure. Seeds without endosperm and perisperm and with a large spathulate erect embryo.

Anatomy.   Stem angular; vascular system in an internodal zone comprises 3 to 5 bundles corresponding to each angle of the stem, separated from each other by multiseriate rays; a smaller bundle is present opposite each lateral surface of stem. Vessels very small (up to 30 μm in diameter), circular and thick-walled with simple perforations. Parenchyma paratracheal, rays broad. Leaves subterete, stomata Rubiaceous type, with semilunar subsidiary cells. Phloem with S-type sieve element plastids.

Embryology.   Pollen grains 3- or 4-colporate, with scabrate exine; shed at 2-celled stage. Ovules anatropous, bitegmic, crassinucellate. Endosperm formation is of the Nuclear type (Dahlgren 1975b).

Chromosome Number.   Basic chromosome number is x = 11; the diploid number for *B. maritima* is 2n = 22.

Chemical Features.   It contains neither anthocyanin nor betalain; contains glucosinolates (common in Cruciferae and Capparaceae).

Important Genera and Economic Importance.   The only genus, *Batis,* has no economic importance.

Taxonomic Considerations.   The phyletic position of Bataceae is much disputed. Bessey (1915) placed it in his Caryophyllales, but only doubtfully. Buxbaum (1961), Hutchinson (1973) and Takhtajan (1969) also include Bataceae in the Caryophyllales (= Centrospermae). It has been allied with Buxaceae, Empetraceae, Polygonaceae and Juglandaceae (McLaughlin 1959), with Salicaceae (Wettstein 1935) and with Capparales (Dahlgren 1983a, Goldblatt 1976). Melchior (1964), Cronquist (1981), Jørgensen (1981) and Takhtajan (1987) place it in a separate order, Batales, near Capparales. Cronquist (1968) remarked: "The habitual resemblance of *Batis* to *Sarcobatus* of the Chenopodiaceae can hardly be overlooked and the maritime habitat of *Batis* is in accord with the alkaline, inland habitat of *Sarcobatus* and other Chenopodiaceae".

The absence of both anthocyanin and betalain pigments supports its exclusion from the Centrospermae. Mabry et al. (1963), Behnke and Turner (1971), and Mabry (1976) exclude Bataceae from Caryophyllales (= Centrospermae). Ehrendorfer (1976a) also suggests that the Bataceae apparently belongs to the Capparales. The basic chromosome number x = 11 is further evidence to exclude Bataceae from the Centrospermae and Salicaceae; x = 11 is consistant with the basic chromosome number of Capparales and Violanae in general. The presence of the chemical glucosinolates also suggests its placement in or near the Capparales.

Taxonomic Considerations of the Order Batales

Earlier taxonomists did not have a clear idea about the correct position of this order. According to Lawrence (1951), Batidales have little or no relationship with either Juglandales or Fagales as interpreted earlier. Cronquist (1968) remarks on its alliance with the Caryophyllales (= Centrospermae) particularly the Chenopodiaceae, on the basis of similar morphology of the plant body and similar habitat. The absence of betalains and anthocyanins in the members of the order Batales contradicts this view. Both the presence of the glucosinolates and basic chromosome number x = 11 reflect its relationship with the Capparales. Melchior (1964), Ehrendorfer (1976a) and Takhtajan (1987) place the order Batales next to these glucosinolate-containing groups and this is the correct position.

28

Order Rosales

A large order comprising 19 families distributed in four suborders:

1. Suborder Hamamelidineae with unigeneric Platanaceae, Hamamelidaceae, and unigeneric Myrothamnaceae.
2. Suborder Saxifragineae includes Crassulaceae, unigeneric Cephalotaceae, Saxifragaceae, monogeneric Brunelliaceae, Cunoniaceae, unigeneric Davidsoniaceae, Pittosporaceae, unigeneric Byblidaceae and Roridulaceae, and Bruniaceae.
3. Suborder Rosineae includes Rosaceae, Neuradaceae, and Chrysobalanaceae.
4. Suborder Leguminosineae includes Connaraceae, Leguminosae, and Krameriaceae.

The members are trees, shrubs and herbs, of cosmopolitan distribution. Flowers generally cyclic, typically pentamerous, hypo-, peri- or epigynous, stamens mostly in many whorls; gynoecium apo- to syncarpous, style and stigma distinct; ovules bitegmic, seeds endospermous. The taxa of this order are characterised by the presence of common flavonols and, to a lesser extent, by flavones together with their O-methyl derivatives.

Platanaceae

A unigeneric family of the genus *Platanus* with 10 species distributed in many parts of the northern hemisphere, exclusive of Africa.

Vegetative Features.   Trees with deciduous or exfoliating bark; vegetative parts with stellate hairs. Leaves alternate, simple, palmately 3- to 9-lobed, petiolate; petiole sheathing, encloses the axillary bud (Fig. 28.1A); stipulate, stipules membranous, deciduous.

Floral Features.   Several dense globose heads of unisexual flowers on separate pendulous peduncles. Male flowers ebracteate, actinomorphic, with a cupular calyx, 3–7 tridentate, minute petals alternate with

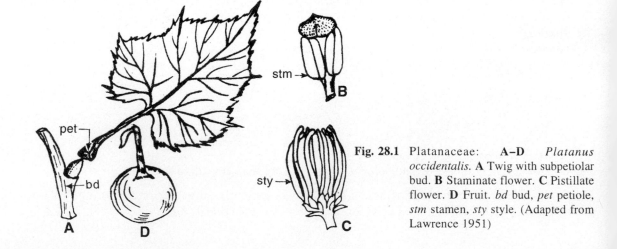

Fig. 28.1 Platanaceae: **A–D**   *Platanus occidentalis*. **A** Twig with subpetiolar bud. **B** Staminate flower. **C** Pistillate flower. **D** Fruit. *bd* bud, *pet* petiole, *stm* stamen, *sty* style. (Adapted from Lawrence 1951)

a whorl of 3 to 7 stamens; stamens free, filaments short or absent, anthers bicelled, elongated with a large, peltate, apically placed connective (Fig. 28.1B), dehiscence longitudinal; rudimentary pistils as well as staminodes often present. Female flowers ebracteate, calyx 3- to 5-lobed and cup-shaped or 3 to 5 free sepals, petals usually absent (present in *Platanus racemosa*); gynoecium apocarpous, 5- to 9-carpellary (Fig. 28.1C), ovary superior, unilocular, placentation parietal, ovules 1 or 2, pendulous; style 1, linear, usually recurved (Fig. 28.1C), stigma extends most of the inner face of the style length. Fruit (Fig. 28.1D) a 1-seeded, linear, quadrangular achene, rarely follicular; seeds with a straight embryo and a thin but fleshy layer of endosperm.

Anatomy. Vessels small, mostly solitary, perforation plates simple and scalariform, intervascular pitting usually opposite. Parenchyma apotracheal, diffuse or in uniseriate bands, rather sparse. Rays 10 to 15 cells wide, with a few or no uniseriates, homogeneous. Leaves isobilateral, young leaves covered with a dense woolly tomentum of deciduous "candelabra" hairs; poorly branched hairs and capitate glands have also been observed. Stomata are somewhat raised and surrounded by rather numerous cells without any special arrangement.

Embryology. Pollen grains 3-colpate or polyporate, and 2-celled at dispersal stage. Ovules orthotropous, bitegmic and crassinucellate. Micropyle is formed by both integuments. Polygonum type of embryo sac, 8-nucleate at maturity. Endosperm formation of the Nuclear type, or occasionally of the Cellular type.

Chromosome Number. Basic chromosome number is $x = 7$.

Chemical Features. Endosperm rich in oil and protein. Tannins, leucoanthocyanins and myricetin are typical.

Important Genera and Economic Importance. *Platanus* is the only genus with about 10 species. These are ornamental trees and often grown as avenue trees. They are pollution indicators of various atmospheric gases.

Taxonomic Considerations. Bentham and Hooker (1965c) considered Platanaceae to be a member of the Unisexuales, near Urticaceae, because of the similarities in inflorescences, simplicity of pistil and variability in the number of floral parts. Bessey (1915), Boothroid (1930), Wettstein (1935), Tippo (1938) and Lawrence (1951), placed Platanaceae in the Rosales, because of its closer affinities with Rosaceae and Hamamelidaceae. Hutchinson (1969, 1973) includes Platanaceae in Hamamelidales and derives this order from the Rosales. Cronquist (1968, 1981) also places it in the Hamamelidales, but treats this order as a primitive one deriving it from the Magnoliales. Takhtajan (1987) accepts this assignment. Although Dahlgren (1980a, 1983a) places it in the same order, he treats the order Hamamelidales as primitive to the Rosales.

Platanaceae, just as many other members of the Hamamelidaceae, has both primitive as well as advanced features, and is correctly placed as a primitive taxon in the advanced order Rosales.

Hamamelidaceae

A small family of 26 genera and ca. 100 to 130 species with a discontinuous subtropical and temperate distribution. The family appears loosely knit with many apparently relict genera.

Vegetative Features. Deciduous or evergreen trees or shrubs, often with stellate hairs on vegetative parts. Leaves usually alternate, simple or palmately lobed, petiolate, stipulate.

Floral Features. Inflorescence racemose, often a spike or head, sometimes with an involucre of coloured bracts, e.g. *Rhodoleia*. Flowers bi- or unisexual (the plants then monoecious, e.g. *Liquidamber,* or dioecious), actinomorphic, rarely zygomorphic, hypo- or epigynous. Calyx of 4 or 5 sepals, basally connate, more or less adnate to the ovary; corolla of 4 to 5 petals, sometimes absent, as in the central flowers of *Rhodoleia,* polypetalous, imbricate or valvate. Stamens 2 to 8 in 1 whorl, free, anthers bicelled, dehiscing longitudinally or by upturning valves, connective often exerted. Gynoecium syncarpous, bicarpellary, carpels diverging and often separating apically; ovary half-inferior to inferior, bilocular, placentation axile; ovules 1, 2 or more in each locule, pendulous; styles and stigmas 2. Fruit a loculicidal or infrequently septicidal capsule, often with a woody to leathery exocarp. Seed sometimes winged, endospermous, embryo large and straight.

Anatomy.   Vessels small, often entirely solitary, sometimes with scanty spiral thickening; perforation plates scalariform with a few to many bars; intervascular pitting scalariform to opposite. Parenchyma apotracheal, mostly diffuse, sometimes in bands. Rays 2 to 3 cells wide, exclusively uniseriate in a few genera, heterogeneous. Leaves always dorsiventral, hairs mostly tufted or stellate. Glandular leaf teeth, each containing a bundle termination, cells filled with tanniniferous mucilage and stomata on the upper surface occur in *Liquidambar styraciflua* (Metcalfe and Chalk 1972). Cuticle thick in *Bucklandia, Corylopsis, Dicoryphe, Rhodoleia* and *Trichocladus.* Stomata Rubiaceous type, with 1 or more subsidiary cells on either side of pore, confined to the lower surface. Secretory cells as well as intercellular secretory canals are present.

Embryology.   Pollen grains usually 3-colpate, rarely polyporate, shed at 2-celled stage. Ovules anatropous bitegmic, crassinucellate; chalazogamy common. Polygonum type of embryo sac, 8-nucleate at maturity. Endosperm formation of the Nuclear type; Cellular type of endosperm formation in *Parrotiopsis jacquemontiana* (Kaul 1969).

Chromosome Number.   Basic chromosome number is x = 8, 12. Goldblatt and Endress (1977) reported x = 16 for *Liquidambar* and x = 32 for *Exbucklandia.*

Chemical Features.   Seeds rich in oil and protein. Iridoids usually absent but present in *Liquidambar.* Tannins, leucoanthocyanins and myricetin are present. Ellagic acid, shikimic acid and quinic acid also occur. Serological similarity between Hamamelidaceae and Saxifragaceae is very high. *Hamamelis* is very similar to *Saxifraga.* The subclasses Hamamelididae and Rosidae contain polyphenols and tannins, and are phylogenetically related (Grund and Jensen 1981).

Important Genera and Economic Importance.   The sweet gum or storax— *Liquidambar*— is a large tree with beautiful star-like palmately lobed leaves, which turn red and yellow in autumn. Witch hazel extract prepared from *Hamamelis* bark is used as a liniment. Storax is the balsam produced when the bark of *L. orientalis* is wounded, and is used against certain skin ailments. Various members are cultivated as ornamentals, such as *Hamamelis* spp., *Parrotia persica, Corylopsis, Fothergilla* and *Loropetalum.* Useful timbers are produced by *Altingia* and *Bucklandia* in the Indo-Malayan region, and *Liquidambar* in North America.

Taxonomic Considerations.   The family Hamamelidaceae is allied to the Platanaceae and Myrothamnaceae. Endress (1977a) linked Hamamelidales with Fagales and Betulales through families Myrothamnaceae and Platanaceae. Meeuse (1975a) and Ehrendorfer (1977) have advocated reasons for considering wind-pollinated flowers as relatively ancestral. Takhtajan (1969) and Cronquist (1968) combine Hamamelidales with the ancient amentiferous orders, while Dahlgren (1983a) and Thorne (1983) indicate that the present-day Hamamelididae are remnants of an ancestral order of dicotyledonous families, and they are segregated into different subclasses. According to Takhtajan (1969), Hamamelidales includes Hamamelidaceae,

Platanaceae, and Myrothamnaceae; in 1987 he separates Rhodoleiaceae and Altingiaceae from Hamamelidaceae s.s., and does not include Myrothamnaceae in this order.

Myrothamnaceae

A unigeneric family of the genus *Myrothamnus* with only 2 species, distributed in South Africa and Madagascar.

Vegetative Features.   Small xeromorphic, resinous shrubs, twigs quadrangular, rigid, more or less spiny. Leaves simple, cuneate to fan-shaped, plicately folded along the nerves, opposite, stipulate.

Floral Features.   Inflorescence erect, terminal catkin-like spikes. Flowers bracteate, unisexual (plants dioecious), actionmorphic, hypogynous. Calyx and corolla absent, stamens commonly 5 but sometimes 4 to 8, filaments coherent, free above; anthers bithecous, longitudinally dehiscent. Gynoecium syncarpous, tricarpellary, ovary superior, trilocular with axile placentation, ovules numerous in each locule. Fruit a small leathery capsule, apices of carpels free and spreading. Seeds endospermous.

Anatomy.   Vessels small, mostly solitary, perforation plates scalariform. Parenchyma absent, rays uniseriate, heterogeneous. Fibres with bordered pits. Leaves isobilateral, resin cells occur in leaf epidermis, hairs absent, cuticle fairly thick; stomata Ranunculaceous type.

Embryology.   The binucleate pollen grains remain in permanent tetrads in *Myrothamnus flabellifolia* and *M. moschata* (Jäger-Zürn 1966). Ovules anatropous, bitegmic, crassinucellate. Allium type of embryo sac, 8-nucleate at maturity. Endosperm formation of the Nuclear type.

Chromosome Number.   Haploid chromosome number is n = 10.

Chemical Features.   Myricetin and common flavonols present (Gornall et al. 1979). Iridoids absent.

Important Genera and Economic Importance.   *Myrothamnus* is the only genus with two species of restricted distribution: economic importance is not reported.

Taxonomic Considerations.   Myrothamnaceae is closely linked with Hamamelidaceae and Platanaceae, although in pollen structure it resembles Monimiaceae of the Laurales to some extent. Most authors place the three families in the same order, Hamamelidales (Wettstein 1935, Tippo 1938, Hutchinson 1969, 1973, Takhtajan 1969, Dahlgren 1975a, 1977a). Takhtajan (1987) removes Myrothamnaceae to a distinct order, Myrothamnales. It appears that these three families should be retained in the same order; Melchior (1964) retains them under the suborder Hamamelidineae in Rosales.

Crassulaceae

A family of 35 genera and 1500 species of wide geographical distribution, absent from Australia and Oceania, and only a few representatives in South America. It is widely distributed in South Africa.

Vegetative Features.   Annual or perennial, succulent herbs, shrubs or rarely scandent. Leaves opposite, alternate as in *Sempervivum* and *Sedum*, or whorled (Fig 28.4A), mostly persistent, simple and entire, fleshy, estipulate, often form compact rosettes. Leaves peltate in *Umbilicus rupestris*. Leaves of *Bryophyllum* have adventitious buds along their margin which grow into new plants (Fig. 28.4E).

Floral Features.   Inflorescence cymose, start as a dichasial and later a monochasial. Flowers bracteate, bracteolate, or ebracteolate, hermaphrodite, actinomorphic, hypogynous (Fig. 28.4B). Calyx of usually 4 or 5 sepals, poly- or gamosepalous forming a tube, rarely with 3 sepals as in *Sedum* or with 30 sepals as in *Sempervivum*. Corolla of same number of petals as the sepals, polypetalous and imbricate or united

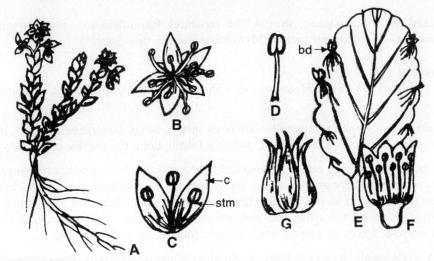

Fig. 28.4 Crassulaceae: **A–D** *Sedum acre*, **E–G** *Bryophyllum* sp. **A** Flowering plant. **B** Flower. **C** Vertical section. **D** Stamen. **E** Leaf of *Bryophyllum* with adventitious buds along the margin. **F** Corolla, longisection. **G** Carpels. (**A–D** adapted from Lawrence 1951, **E–G** Original) *bd* bud, *c* corolla, *stm* stamen.

only at base; exceptions are *Cotyledon* and *Kalanchoë* with completely united corolla lobes forming a tube. Stamens as many or twice as many as petals in 2 whorls (Fig. 28.4C), free except in *Cotyledon,* generally hypogynous, unless petals basally connate and then usually borne on the corolla tube, the ante-petalous stamens are slightly above the antesepalous ones (Fig. 28.4F); anthers bicelled, introrse, dehiscence longitudinal (Fig. 28.4D). Gynoecium apocarpous or only basally united, 3-, 4- or 5-carpellary (Fig. 28.4G) (usually as many as petals), each bears a scale-like or petaloid nectary at the base (absent in *Greenovia*); ovary superior, unilocular, with marginal placentation, ovules numerous, rarely few arranged in 2 rows; style 1, short or elongated, stigma 1. Fruit a membranous or leathery follicle. Seeds minute, with a straight embryo, endospermous or without endosperm.

Anatomy. Xylem nearly always in the form of a continuous cylinder, consists either wholly of prosenchyma without vessels, or small groups of vessels accompanied by elongated, unlignified, parenchyma cells. Vessels with simple pits or reticulate thickenings, perforations simple. Leaves usually centric, glandular hairs common; stomata surrounded by a girdle of 3 subsidiary cells and occur all over the leaf. Anthocyanin pigment common in many members.

Embryology. Pollen grains 3- or rarely 4-colporate, colpi tenuimarginate, exine smooth, starch grains occassionally present; shed at 2-celled stage. Ovules anatropous, bitegmic, crassinucellate. Polygonum or Allium type of embryo sac, 8-nucleate at maturity. Endosperm formation of the Cellular type. Endosperm with uninucleate chalazal haustorium is common (Johri et al. 1992).

Chromosome Number. Basic chromosome number is x = 4–22, 31.

Chemical Features. Plants rich in calcium oxalate and many organic acids—crassulacic acid, malic acid, etc, (Harborne and Turner 1984). Waxy coating on the leaves of subfamily Sempervivoideae is rich in alkanes.

Important Genera and Economic Importance. Some of the important genera are *Sedum* (350 spp.), *Sempervivum* (30 spp.), *Crassula* (250 spp.), *Kalanchoë* (125 spp.) and *Echevaria* (80 spp.). Most of these plants are grown as ornamentals because of their succulent foliage. *Sedum spectabilis*—commonly called live-forever—is prized for its fine clusters of pink flowers. *Sempervivum tectorum*—the house leek— is often grown as a border plant in gardens for its radical, closely crowded, rosette-forming leaves. *Kalanchoë* species are of medicinal value.

Taxonomic Considerations. The family Crassulaceae is included in the order Rosales by Bentham and Hooker (1965a), Engler and Diels (1936), Melchior (1964), Thorne (1977) and Cronquist (1981), whereas Hutchinson (1969, 1973), Dahlgren (1980a, 1983a) and Takhtajan (1980, 1987) consider it to be included in the order Saxifragales. The Crassulaceae is closely related to the Saxifragaceae, although its members are succulent, their carpels apocarpous and the same number as the petals. A haustorium develops from megaspores, synergids and antipodals.

While considering the family Crassulaceae in the order Saxifragales, Dahlgren (1983a) mentions that this order is a heterogeneous assemblage. According to Takhtajan (1980, 1987), Crassulaceae is very close to Saxifragaceae, especially to the monogeneric Penthoroideae (= unigeneric family Penthoraceae), and had a common origin. In some respects, including embryological features (Subramanyam 1962), the Crassulaceae are more primitive than the Saxifragaceae. Melchior (1964) also supports this and treats Crassulaceae as more primitive to the Saxifragaceae. Krach (1977) indicates that Crassulaceae have probably evolved as part of or parallel with the Saxifragaceae in North America and Eastern Asia. Many taxa of these two families are herbaceous, grow in extreme habitats, are succulent, accumulate sedoheptulose and have CAM metabolism (Grund and Jensen 1981). Therefore, Crassulaceae have been placed with Saxifragaceae, in the order Saxifragales, by Dahlgren (1975a).

However, its placement in the order Rosales s.l. is also appropriate, as it is a very large and heterogenous order.

Cephalotaceae

A unigeneric family of the genus *Cephalotus* with only one species (*C. follicularis*) endemic to marshy lands in King George's Sound in southwestern Australia.

Vegetative Features. An interesting plant with "pitchers", as in *Nepenthes* and *Sarracenia* though not related to either of them. Perennial herbs with underground woody rhizomes. Leaves in basal rosettes; the lower leaves of the rosette form pitchers, the upper are flat and green, the woody rhizome annually produces both types of leaves. The pitcher is more or less similar to that of *Nepenthes* and catches insects in the same way.

Floral Features. Inflorescence a small, adpressed-hairy thyrse borne on a long scape. Flowers bisexual, apetalous, actinomorphic, hypogynous. Sepals 6, valvate, stamens 12, in 2 whorls of 6 each, connective large, globose; disc broad, thick, green, papillose. Gynoecium 6-carpellary, syncarpous or polycarpous, each ovary unilocular, with 1 or rarely 2, basal, erect ovules with dorsal raphe and short, recurved, subulate style with simple stigma. Fruit a follicle with one seed, embryo small in a fleshy endosperm.

Anatomy. The epidermis of the young rhizome has long unicellular hairs, the broad cortex amyliferous or tanniniferous. Vasculature of a ring of phloem and xylem interrupted by foliar traces. Of the 3 types of leaves—(a) the scale leaves are covered with unicellular hairs and glands; (b) the foliage flat leaves with unicellular hairs, glands and stomata on both surfaces and also on petiole; (c) in the third type of leaf that forms the pitcher, the epidermis on the outer surface of pitcher is of isodiametric cells, perforated by stomata, bears depressed glands and unicellular hairs. Various types of glandular hairs are borne inside the pitcher.

Chromosome Number. Haploid chromosome number is n = 10.

Important Genera and Economic Importance.   *Cephalotus* is the only genus which is not of any economic importance.

Taxonomic Considerations.   The family Cephalotaceae has often been presumed to be closely related to the families Sarraceniaceae and Nepenthaceae, because of similar pitcher-like adaptation of the leaves. However, the floral morphological studies reveal that its placement in Rosales is appropriate, as has been pointed out by Bentham and Hooker (1965a) and Engler and Diels (1936). Hutchinson (1969, 1973) includes this family in his Saxifragales of the Herbaceae. Cronquist (1968, 1981) and Takhtajan (1980, 1987) also include this family in the Rosales. Keighery (1979) observes that the Cephalotaceae are related to Saxifragaceae and the Crassulaceae, but differ markedly from these families in being insectivorous. The formation of pitcher-like leaves as in two other families—Nepenthaceae and Sarraceniaceae—could be due to adaptation towards similar ecological conditions in which they grow.

Saxifragaceae

A fairly large family with ca. 30 genera and 600 species, more or less cosmopolitan, but more abundant in the north-temperate regions.

Vegetative Features.   Herb, shrubs or small trees; herbs tend to be succulent in temperate and cold regions. Leaves alternate, sometimes opposite, as in *Bauera*, simple (Fig. 28.6A, B) or compound, usually deciduous, mostly estipulate.

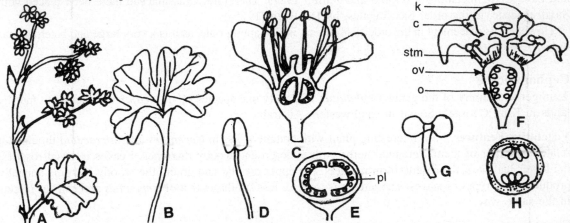

Fig. 28.6   Saxifragaceae: **A, B** *Saxifraga granulata*, **C–E** *S. macnabiana*, **F–H** *Ribes sativum*. **A** Flowering twig. **B** Leaf. **C** Flower, longisection. **D** Stamen. **E** Ovary, cross section. **F** Flower, vertical section. **G** Stamen of *R. sativum*. **H** Ovary, cross section. *c* corolla, *k* calyx, *o* ovule, *ov* ovary, *pl* placenta, *stm* stamen. (Adapted from Lawrence 1951)

Floral Features.   Inflorescence basically cymose (Fig. 28.6A), sometime racemose to paniculate. Flowers bracteate, ebracteolate, bisexual, actinomorphic (zygomorphic in *Saxifraga*), hypo-, peri- or epigynous. Calyx of 4 or 5 sepals, often gamosepalous, sometimes highly coloured and petaloid. Petals as many as sepals, borne on a receptacle or hypanthium (Fig. 28.6C), sometimes smaller than calyx (Fig. 28.6F) or absent. Stamens usually the same number as and alternate with the petals or twice as many with the outer whorl opposite the petals (only 3 stamens in *Tolmiea*), distinct; anthers usually bicelled (Fig. 28.6D, G), 1-celled in *Leptarrhena,* dehisce longitudinally, staminodes or nectariferous glands often present. Gynoecium apocarpous or syncarpous, 2- to 5-carpellary; when syncarpous, may be unilocular with parietal placentation (Fig. 28.6F, H) or 2- to 5-loculed with axile placentation (Fig. 28.6E); when apocarpous, the 2 to 5 separate pistils are often basally connate, styles and stigmas as many as carpels. Ovary superior, half-inferior or inferior, sometimes within the hypanthium; ovules numerous on swollen placenta. Fruit a capsule or berry, seeds endospermous with a small embryo, rarely winged as in *Sullivantia*.

Anatomy.   In a transection of stem, phloem and xylem usually appear as a ring of individual collateral, bundles. Vessels small with simple perforations. Leaves dorsiventral or centric, both glandular and non-glandular hairs are common. Stomata Ranunculaceous type and their distribution is of diagnostic value at specific level. Chalk glands or hydathodes secrete water-containing calcium carbonate, which is deposited on leaf surface as solid carbonate.

Embryology.   Highly multipalynous family. Pollen grains basically 3-colpate but there are variations in the number and distribution of apertures; shed at 2- or 3-celled stage. Ovules anatropous, bitegmic, crassinucellate (tenuinucellate in *Bistella*). Micropyle is formed by both the integuments. Polygonum type of embryo sac, 8-nucleate at maturity. Endosperm formation of the Cellular, Helobial or Nuclear type (Johri et al. 1992).

Chromosome Number.   Basic chromosome numbers are x = 6–15.

Chemical Features.   Iridoids absent. Hydrangeic acid is reported in *Hydrangea* (H. Erdtman 1963).

Important Genera and Economic Importance.   Most of the genera are grown as ornamentals: *Saxifraga* (saxifrages), *Philadelphus* (mock orange), *Heuchera* (coral bells), *Hydrangea* (hydrangeas) and many others. Fruits of *Ribes* (commonly called currants and gooseberries) are edible.

Taxonomic Considerations.   The Saxifragaceae in Bentham and Hooker (1965a) system included plants belonging to Cunoniaceae, Grossulariaceae, Escalloniaceae and Hydrangeaceae. Engler (1930b) recognised it as a very large family of 15 subfamilies. Hutchinson (1973) treats the Saxifragaceae s.s. as evolved along a different line from the Cunoniales, in which he includes the smaller families: Escalloniaceae, Grossulariaceae, Greyiaceae and Hydrangeaceae. Cronquist (1968, 1981) treats Saxifragaceae as a member of the Rosales, and includes most of the smaller families.

According to Dahlgren (1983a), Saxifragaceae s.s. belongs to Saxifragales along with the smaller families: Iteaceae, Frankoaceae, Vahliaceae and Greyiaceae. He includes Hydrangeaceae in Cornales, and Roridulaceae in Ericales. Takhtajan (1987) recognises a large number of segregates, like Baueraceae, Greyiaceae, Frankoaceae, Lepuropetalaceae, Parnassiaceae, Eremosynaceae, Rousseaceae, Iteaceae, Penthoraceae and Grossulariaceae. All these are unigeneric families (except Frankoaceae), closely related to Saxifragaceae, but distinct from it. He places Baueraceae and Davidsoniaceae in Cunoniales (in Rosanae), Hydrangeaceae and Roridulaceae in Hydrangeales, Byblidaceae in Byblidales, and Pittosporaceae in Pittosporales, all (except Cunoniales) included in a separate superorder, Cornanae.

According to Cronquist (1968), Saxifragaceae is connected to the Crassulaceae by an intermediate genus *Penthorum*. It has 5 separate follicular carpels as in the Crassulaceae, but without succulence. *Penthorum* is treated as a member of Crassulaceae or Saxifragaceae. Takhtajan (1987) treats it as a distinct family, Penthoraceae.

There is diverse opinion about the circumscription of the family Saxifragaceae. According to Bentham and Hooker (1965a) and Engler (1930b), it is a large family embracing herbs, shrubs as well as trees but Dahlgren (1980a), Takhtajan (1987) and others include only herbaceous members in Saxifragaceae s.s, and the rest in separate families.

Brunelliaceae

A family of a single American genus, *Brunellia*, with 51 species, distributed in the Andes, central America and the Caribbean (Ehrendorfer et al. 1984).

Vegetative Features. Tall trees, usually tomentose all over the plant body. Leaves opposite or verticillate, simple or pinnately compound, often dentate, with small caducous stipules.

Floral Features. Inflorescence axillary or terminal spikes. Flowers actinomorphic, unisexual, dioecious, hypogynous. Calyx of 4–5(-7) sepals, valvate, shortly connate below and free above; corolla absent, disc cupular, 8- to 10-lobed, adnate to calyx. Stamens 8–10(-14), with filiform, pubescent filament and dorsifixed, versatile, introrse anthers, staminodes occur in female flowers. Gynoecium apocarpous, 5-, 4- or 2-carpellary, ovaries superior, unilocular, gradually produced into long subulate style with punctiform stigmas and with 2 ventrally collateral ovules per locule; small pistillode occurs in male flowers. Fruit of 5 to 2 ventrally dehiscent, 1- to 2-seeded follicles; seeds with mealy endosperm.

Anatomy. Vessels small, with numerous multiples, perforation plates simple and scalariform, intervascular pitting scalariform. Parenchyma absent; rays up to 6 cells wide or exclusively uniseriate, heterogeneous. Leaves dorsiventral, stomata not easily observed, but apparently confined to portions of the lower epidermis above the spongy tissue.

Embryology. Pollen grains 3-colporate, shed at 2-celled stage. Ovules bitegmic, crassinucellate. Endosperm formation of the Nuclear type.

Chromosome Number. Polyploids, numbers 2n = 28 in *Brunellia comocladiifolia* and *B. mexicana* (Ehrendorfer et al. 1984).

Taxonomic Considerations. The genus *Brunellia* was included in Simaroubaceae (Bentham and Hooker 1965a), but the scalariform perforation plates differentiate it from this family. The family Brunelliaceae is included in Rosales due to its alliance with Cunoniaceae (Willis 1973, Takhtajan 1980, 1987). Cronquist (1968) includes this genus in Cunoniaceae. Later, Cronquist (1981) treats it as a separate family Brunelliaceae, the most primitive amongst the Rosales. Dahlgren (1980a, 1983a) places it in Cunoniales under superorder Rosiflorae.

In our opinion the Brunelliaceae should be treated as a distinct family of the Rosales, close to Cunoniaceae.

Cunoniaceae

A small family of 21 genera and 350 species, restricted to southern hemisphere; distributed in Australia, New Guinea and New Caledonia; *Lamanonia* (=*Belangera*) and *Caldcluvia* are from South America, and *Platylophus* from South Africa.

Vegetative Features. Perennial woody shrubs or trees; leaves pinnately or trifoliately compound, rarely simple; opposite or whorled, margin often glandular-serrate, stipulate, stipules sometimes large and connate.

Floral Features. Inflorescence capitate or paniculate. Flowers bi- or unisexual (the plants are then dioecious), actinomorphic, small, pentamerous, hypogynous. Calyx of (3-) 4-5 (-6) sepals, free or basally connate. Corolla lobes as many as sepals, smaller than sepals, basally connate, sometimes absent. Stamens mostly numerous, frequently obdiplostemonous, sometimes a few and then opposite the sepals, annular nectariferous disc present; distinct anthers and filaments, each anther with a connective and a 4-celled anther lobe, dehisce longitudinally. Gynoecium 4- or 5-carpellate, free or more commonly opposite concrescent carpels fused in the region of the ovary, i.e. synovarious (Dickison 1975); ovary superior, often in hypanthium, glabrous to densely pubescent with numerous ovules on axile placentation; when bi- or uniovulate, pendulous; styles and stigmas as many as carpels, terminal, capitate. Fruit a capsule or nut. Seeds endospermous, with a small embryo.

Anatomy.   Vessels moderately small, solitary or with some multiples, perforation plates scalariform only or simple with a few scalariform plates; intervascular pitting scalariform to opposite. Parenchyma apotracheal, diffuse or in bands, 1 to 4 cells wide. Rays up to 2 to 4 cells wide, markedly heterogeneous. Leaves dorsiventral, stomata small, almost circular in *Cunonia* and *Platylophus.* Secretory cells with amorphous, presumably tanniniferous contents present in the unlignified tissues of the petiole in *Weinmannia trichosperma.*

Embryology.   Unipalynous family with 3-colporate pollen grains (P.K.K. Nair 1970); 2-celled at shedding stage. Ovules anatropous, bitegmic, crassinucellate. Polygonum type of embryo sac, 8-nucleate at maturity. Endosperm formation of the Nuclear type.

Chromosome Number.   Basic chromosome number is x = 12, 15.

Important Genera and Economic Importance.   *Weinmannia* is the largest genus with 126 species. *W. pubescens* of northern Andes yields tannin. *Ceratopetalum apetalum* of New South Wales (Australia) is the source of timber. Many species of *Ackama, Callicoma, Ceratopetalum* and *Weinmannia* are cultivated as ornamentals.

Taxonomic Considerations.   Cunoniaceae members are allied to the Saxifragaceae, but differ in woody tree habit, leaves always opposite or whorled, 4-celled anthers and unipalynous condition.

Bentham and Hooker (1965a), and Schulze-Menz (1964) place Cunoniaceae in Rosales, Cronquist (1968, 1981) between Eucryphiaceae and Davidsoniaceae in the large and heterogeneous order, Rosales. Thorne (1968) favours a close association of Cunoniaceae, Brunelliaceae, Eucryphiaceae and Davidsoniaceae. Takhtajan (1969) supports this arrangement, but includes them in Saxifragales. Hutchinson (1969, 1973) regards the Cunoniales as composed of many taxa which Engler (1930b) included in the Saxifragaceae. In his opinion, Cunoniales is allied to primitive members of the Dilleniales and Rosales. Dahlgren (1975a, 1980a, 1983a) and Takhtajan (1980, 1987) also recognise a separate order, Cunoniales.

Rosales is a very large and heterogeneous order; it should be split into smaller groups to include a few larger families. The order Cunoniales with family Cunoniaceae could be one such order.

Davidsoniaceae

A monotypic family of the genus *Davidsonia* with one species—*D. pruriens* distributed in northeastern Australia.

Vegetative Features.   Small slender trees, bear irritant hairs. Leaves alternate, pinnate, elongated, stipulate, stipules large, reniform; rachis winged, dentate, leaflets dentate-serrate.

Floral Features.   Inflorescence axillary or supra-axillary, large, lax, pedunculate panicle or a dense pedunculate spike. Flowers bracteate, bracts large, amplexicaule, ebracteolate; bisexual, actinomorphic, hypogynous. Sepals 4, valvate, thick, united approximately 3/4th their length. Petals absent. Stamens 10 or 20, in 2 whorls of 10 each, opposite and alternate with sepals, nectariferous scales inserted on disc; filaments more or less tumid (swollen, inflated) below, anthers oblong, versatile. Gynoecium syncarpous, bicarpellary, ovary superior, bilocular, placentation axile, ovules ca. 7 per locule; styles basally fused but filiform and free above. Fruit a large, 2-pyrened drupe, red velvety when young, glaucous and pruinose (with a bloom on the surface) when ripe. Seeds 2, large, pendulous, non-endospermous.

Anatomy. Anatomy is similar to that of Cunoniaceae members.

Important Genus and Economic Importance. *Davidsonia* is the only genus with a single species; no economic importance.

Taxonomic Considerations. The family Davidsoniaceae is allied to the Cunoniaceae but differs because of indehiscent fruits and non-endospermous seeds. Both the families are distributed in the southern hemisphere. Bentham and Hooker (1965a) and Hutchinson (1969) included the genus *Davidsonia* in the family Cunoniaceae. However, Melchior (1964), Cronquist (1968, 1981) and Takhtajan (1980) place it in a distinct family, and Takhtajan (1987) recognises a separate order Cunoniales. Therefore, a distinct order Cunoniales with the families like Cunoniaceae, Davidsoniaceae and other allied taxa is advisable, rather than including these families in Rosales s.l.

Pittosporaceae

A small family of 9 genera and ca. 350 species, distributed in the tropics and subtropics of the Old World. The genus *Pittosporum* occurs in Asia, Australia and Africa. Eight genera are endemic to Australia (Nayar 1986a).

Vegetative Features. Trees, shrubs and lianas, sometimes spiny such as *Billardiera, Sollya*. Leaves simple, alternate or whorled, often evergreen, sometimes subverticillate at the tip of the branches; often coriaceous, estipulate.

Floral Features. Inflorescence cymose or paniculate or flowers solitary. Flowers bisexual (Fig. 28.10A), actinomorphic or zygomorphic (*Cheiranthera*), campanulate or rotate. Sepals 5, polysepalous or basally connate and free above. Petals 5, sometimes basally cannate or coherent, imbricate. Stamens 5, alternate with the petals (Fig. 28.10A), free or monadelphous as in *Marianthus*, anthers lanceolate to linear, bicelled (Fig. 28.10B), introrse, dehiscence poricidal as in *Sollya,* or by longitudinal slits. Gynoecium syncarpous, 2- to 5-carpellary, ovary 2- to 5-locular with axile placentation (Fig. 28.10C) or unilocular with parietal and sometimes basal placentation; ovules numerous in two ranks; style 1 and short, stigma as many as carpels. Fruit a loculicidal capsule, as in *Pittosporum* (Fig. 28.10D), or a berry usually containing a viscous pulp, as in *Billardiera,* seeds with abundant endosperm and a minute linear embryo.

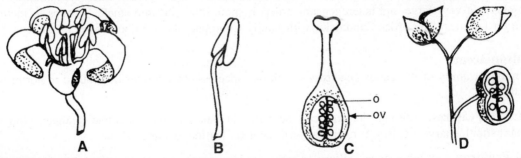

Fig. 28.10 Pittosporaceae: **A–D** *Pittosporum undulatum.* **A** Flower. **B** Stamen. **C** Carpel, longisection. **D** Fruits. *o* ovule, *ov* ovary. (Adapted from Lawrence 1951).

Anatomy. Vessels small, in numerous small multiples and clusters, often with a diagonal pattern; perforation plates simple (Meylan and Butterfield 1975); intervascular pitting alternate and small. Parenchyma sparse, vasicentric; rays up to 3 to 7 cells wide, with a few uniseriates and almost homogeneous. Septate fibres are an interesting feature. Leaves mostly dorsiventral, centric in a few species. The hairs may be uniseriate, two-armed trichomes as well as club-shaped glandular hairs. Stomata with 2 to many subsidiary cells parallel to the pore; confined to the lower surface.

Embryology.   Pollen grains 3-colporate and 2-celled at dispersal stage in 7 species of *Pittosporum* (Gardner 1975). Brewbaker (1967) recorded 3-celled condition in one species. Ovules anacampylotropous, unitegmic, tenuinucellate, Polygonum type of embryo sac, 8-nucleate at maturity. Endosperm formation of the nuclear type (Johri et al. 1992).

Chromosome Number.   Basic chromosome number is x = 12.

Chemical Features.   Various essential oils, ursene-type saponins, polyacetylenes and caffeic acid esters have been isolated from the members of this family (Hegnauer 1971). Etheric oils and resins occur in excretory canals (Grund and Jensen 1981).

Important Genera and Economic Importance.   *Pittosporum* is the largest genus with about 160 species that grow all over the Old World tropics. *Pittosporum rhombifolium, P. viridiflora, P. tobira, Hymenosporum flavum, Bursaria spinosa* and *Sollya* are grown as ornamentals. *Billardiera longiflora* of Tasmania is grown for its blue edible berries.

Taxonomic Considerations.   The family Pittosporaceae has been treated under Pittosporales by most workers (Hutchinson 1969, 1973, Dahlgren 1975a, 1980a 1983a). Bentham and Hooker (1965a) included this family in Polygalinae along with Tremandraceae and Polygalaceae. Lawrence (1951), Melchior (1964) and Cronquist (1968, 1981), however, place it in Rosales s.l. According to Lawrence (1951), the phyletic position of Pittosporaceae is close to the Saxifragaceae, particularly to the genus *Escallonia* and its relatives.

Some of the anatomical features such as the septate fibres and paratracheal parenchyma in Pittosporaceae suggest a relationship with Araliaceae of Araliales (Metcalfe and Chalk 1972). Dahlgren (1980a) considers the Pittosporales to be allied to the Araliales. The Apiaceae, Araliaceae and Pittosporaceae are linked by the production of similar polyacetylenes (Nayar 1986a). Hegnauer (1971) also points out that there are close chemical ties between Araliales and Pittosporales. The occurrence of flavonoids, caffeic acid and sinapinic acid suggests similarity with Saxifragaceae, but other characters indicate relationship with Araliaceae and Apiaceae (Dahlgren 1975a).

Takhtajan (1966) retained Pittosporaceae in Saxifragales, but later (1987), he allied it with the Araliales and placed it close to this order as a distinct order Pittosporales.

Although Pittosporaceae is chemically related to the Araliales, it cannot be placed in this order because of differences in other features. Pittsporaceae should rather be treated as a member of a separate order Pittosporales close to the Araliales as suggested by Takhtajan (1987).

Byblidaceae

A unigeneric family of the genus *Byblis,* distributed in tropical northern Australia. Of the two species, *B. linifolia* and *B. gigantea,* the latter occurs on white sandy areas that are swampy in winter but dry in summer. They show preference for establishment on recently burned or otherwise disturbed areas (Carlquist 1976a).

Vegetative Features.   *B. linifolia* is a small insectivorous herb; *B. gigantea* attains a height of up to 0.5 m and is a moderately woody herb. Leaves alternate, elongate-linear, estipulate, crowded, circinate vernation, and bears both sessile and stalked capitate glands.

Floral Features.   Flowers solitary, axillary, ebracteolate, bisexual, actinomorphic, long-pedicelled. Calyx of 5 sepals, connate at base, free above, imbricate, persistent. Corolla of 5 petals, contorted, broad-cuneate, apically fimbriate. Stamens 5, alternipetalous, sometimes unequal or declinate, anthers basifixed, introrse, dehisce by apical pores or short slits. Gynoecium syncarpous, bicarpellary, ovary superior, bilocular with numerous erect ovules on axile placentation. Fruit a 2-loculed, 2- to 4-valved, many-seeded capsule; seeds endospermous with coarsely verrucose testa.

Anatomy. Vessels small in diameter, mostly solitary, intervascular pitting alternate. Perforation plates mostly simple, but scalariform or multiperforate plates with 1 to 3 bars. Parenchyma sparse, diffuse. Rays mostly bi- or triseriate. Leaf triangular in transverse section with rounded angles, but nearly cylindrical towards the apex, terminating in a hydathode. Sessile glands on leaf surface confined to longitudinal furrows with glandless ridges between them. Stomata Rubiaceous type situated in the glandless regions inbetween the furrows of sessile glands.

Embryology. Embryology is not sufficiently known. Ovules anatropous, bitegmic, tenuinucellate. Endosperm formation of the Cellular type (Johri et al. 1992).

Chemical Features. Iridoid compounds present (Dahlgren 1975b).

Important Genus and Economic Importance. *Byblis* is the only important genus; no economic importance.

Taxonomic Considerations. Bentham and Hooker (1965a) included *Byblis* in Droseraceae, because of its insectivorous habit. Cronquist (1981), Diels (1930a), Hutchinson (1973), Melchior (1964) and Takhtajan (1987) recognise a separate family Byblidaceae. Cronquist (1981) and Hutchinson (1973) include *Roridula* also in this family. Vani-Hardev (1972) and Gornall et al. (1979) do not agree with this assignment. The wood anatomy of *Byblis* is consistent with relationship to Roridulaceae as well as placement in the orders Rosales and Pittosporales (Carlquist 1976a). Takhtajan (1987), however, raises this family to an order rank, Byblidales and places it close to the Pittosporales.

Table. 28.11.1   Comparative Data for Byblidaceae and Rosales

Byblidaceae	Rosales
1. Distributed in tropical northern Australia, in recently burned or otherwise disturbed area	Distribution cosmopolitan
2. Moderately woody herbs	Trees, shrubs, and herbs
3. Leaves in circinate vernation and bear sessile and stalked capitate glands on the surface	Leaves normal, without capitate hairs on the surface
4. Stomata Rubiaceous type	Stomata of various types: Rubiaceous, Ranunculaceous, and often with more than two subsidiary cells.
5. Iridoid compounds present, flavonols and flavones absent	Iridoid compounds absent, common flavonols and flavones together with their O-methyl derivatives occur

   Although Diels (1930a) and Melchior (1964) placed Byblidaceae in the Rosales, a number of morphological, anatomical and chemical features of this family differ from those of the other Rosalean members (Table 28.11.1). Its resemblance with the Droseraceae is also superficial. Therefore, its treatment as a separate order by Takhtajan (1987) may be considered valid until further study.

Roridulaceae

A unigeneric family of the genus *Roridula* with 2 species distributed in the rocky regions of the south and southwestern parts of South Africa.

Vegetative Features. Insectivorous shrublets (Willis 1973), with stalked, capitate, viscous glands of various length on stems, leaves and calyx. Leaves simple, alternate, elongated, linear-lanceolate, entire (in *Roridula gorgonius*) or pinnatifid, circinate vernation, estipulate; leaf margin studded with numerous tentacular glands.

Floral Features. Inflorescence long, slender racemes. Flowers ebracteate, bibracteolate, bisexual, actinomorphic, hypogynous. Calyx of 5 sepals, basally connate and free above, imbricate, persistent, thickly coated with glands. Corolla of 5 obovate or broad-elliptic petals, imbricate, glabrous. Stamens 5, alternipetalous, free with downwardly pointing anthers (Vani-Hardev 1972). Connectives swollen, forming massive nectaries at the base of the anthers, anthers basifixed, inverted in bud, but during dehiscence undergo a curvature of 180° and become erect, dehiscence by apical pore or small slits. Gynoecium syncarpous, tricarpellary, ovary superior, trilocular, with several pendulous ovules on axile placentation in each locule; ovary wall devoid of glands. Style simple and stigma funnel-shaped and papillate. Fruit a 3-valved loculicidal capsule; seeds rather large, 4-sided and sticky with spongy seed coat turning mucilaginous when placed in water; endospermous.

Anatomy. In the stem, a ring composed of fibre-like cells adjoins the vascular bundles. Vessels isolated, spirally thickened, perforation plates scalariform with many bars. Rays 1 to many cells wide. Leaves provided with tentacular glands of varying length, each consists of multiseriate stalks bearing ellipsoidal heads. Stomata confined to the lower surface.

Embryology. Pollen grains single, 3-colpate, shed at 2-celled stage. Ovules pendulous, anatropous, unitegmic, tenuinucellar; the nucellar epidermis degenerates with the growth of the gametophyte (Vani-Hardev 1972, Gornall et al. 1979). Polygonum type of embryo sac, 8-nucleate at maturity. Endosperm formation of the Cellular type and without any haustorium.

Chemical Features. Plants are rich in tannins, calcium oxalate crystals and iridoids.

Important Genera and Economic Importance. *Roridula* is the only genus. Commonly known as Vliegebos or fly bush, the plant is hung up in country houses to catch flies, the sticky leaves acting as fly-catchers. *Synoema marlothi,* a spider, which is immune to the balsam-like secretion, moves on the plant freely and thrives on the insects caught in the viscid secretion (Vani-Hardev 1972). *R. gorgonius* is a small shrub.

Taxonomic Considerations. The genus *Roridula* has been regarded as a member of the family Droseraceae in many classificatory systems (Bentham and Hooker 1965a, Diels 1930b, Netolitzky 1926 in Takhtajan 1987, Wettstein 1935). It was placed in Clethraceae by Hallier (1912). Hutchinson (1969, 1973) and Cronquist (1968) place *Roridula* in Byblidaceae along with *Byblis* and point out that the family is related to Pittosporaceae. In a study of the embryological characters of *Roridula gorgonius,* Vani-Hardev (1972) compared the genus with *Byblis* and pointed out a number of differences that warranted placing the two genera in separate families. It is also well separated from Droseraceae anatomically, embryologically and chemically. Plumbagin and 7-methyljuglone, the two naphthaquinones, occur in Droseraceae, but are absent from both *Byblis* and *Roridula* (Harborne and Turner 1984).

The presence of iridoids in combination with morphological, anatomical and embryological features supports placing the family Roridulaceae in or near the Ericales (Jensen et al. 1975, Dahlgren 1975a, b, 1980a, 1983a). Serological studies with seed proteins (Grund and Jensen 1981) reveal that Roridulaceae is closely related to Hydrangeaceae. Takhtajan (1987), therefore, includes this family in Hydrangeales near the family Hydrangeaceae.

Under the circumstances, it is best to treat Roridulaceae as a distinct family; there is controversy about its phyletic position. Further research on chemical data would be helpful in the assignment of an appropriate position to the family Roridulaceae.

Bruniaceae

A family of 12 genera and 75 species form South Africa.

Vegetative Features. Heath-like shrubs, with simple, entire, alternate leaves, estipulate or with rudimentary stipules.

Floral Features. Inflorescence spicate or capitate. Flowers small, bisexual, actinomorphic, pentamerous, generally perigynous. Calyx of 4 or 5 sepals, gamo- or polysepalous, imbricate, persistent. Corolla of 4 or 5 petals, rarely connate below and free above, imbricate, often persistent. Stamens 4 or 5, free, alternipetalous, rarely connate with corolla into a tube, often persistent; anthers dorsifixed, versatile or introrse, longitudinally dehiscent, connective sometimes extended, disc rare (present in *Audouinia, Thamnea* and *Tittmannia*). Gynoecium syncarpous, 2- or 3-carpellary (*Audouinia*), ovary mostly inferior or half-inferior, rarely superior, with as many locules as the number of carpels or unilocular (then pseudomonomerous). Ovules mostly 1 or 2 per locule, 4 in *Thamnea* and 10 in *Lonchostoma;* 1 ovule in unilocular ovary in *Mniothamnea* and *Berzelia,* more or less united stylodium with small apical stigma. Fruit a capsule with 2 seeds or nut with only 1 seed, usually with a persistent calyx, ventrally dehiscent into 2 (in *Lonchostoma* 2 to 4) mericarps or indehiscent. Seeds often arillate (*Staavia, Liconia*) with endosperm and a very small, straight embryo.

Anatomy. Vessels with scalariform perforations and numerous bars. Stomata of different types and are often surrounded by 4 to 7 or more epidermal cells.

Embryology. Pollen grains 3-colporate or 6- to 11-colporate. Ovules anatropous, unitegmic, crassinucellate. Polygonum type of embryo sac, 8-nucleate at maturity. Endosperm formation of the Nuclear type.

Chromosome Number. Haploid chromosome number of *Staavia* is $n = 8$ (Takhtajan 1987).

Chemical Features. Iridoids absent.

Important Genera and Economic Importance. The 12 genera are grouped under 4 subfamilies. *Linconia, Raspalia, Staavia, Mniothamnea, Lonchostoma, Brunia* are some of the genera. No economic importance is reported.

Taxonomic Considerations. Hallier (1912) had placed Bruniaceae together with Cunoniaceae and Saxifragaceae (although with some doubt) in the Rosales. Hutchinson (1959, 1969) and Novák (1961, see Takhtajan 1987) included Bruniaceae in Hamamelidaceae. Dahlgren (1975a, 1980a, 1983a) includes it in the Cunoniales. Cronquist (1968, 1981) groups this family in the Rosales. According to Takhtajan (1987), the family Bruniaceae is undoubtedly related to the Cunoniaceae and other allied families but as it differs from them in the structure of flower, pollen morphology, unitegmic ovules, and seed structure, it would be preferable to place Bruniaceae in a distinct order, Bruniales (see also Nakai 1943 in Takhtajan 1987).

Rosaceae

A large family of 100 genera and 3000–3350 species, cosmopolitan in distribution, abundant in eastern Asia.

Vegetative Features. Trees (*Malus, Pyrus*), shrubs (*Prunus*) or herbs (*Fragaria, Potentilla*) often thorny

(*Rosa*), sometimes climbing (*Spiraea, Rosa*); low growing herbs, e.g. *Fragaria vesca,* and scrambling bushes, e.g. *Rosa* sp. Both root and stem concerned in vegetative propagation; suckers develop in *Rubus idaeus* and runners in *Fragaria vesca.* Leaf buds are commonly produced on the roots of *Prunus cerasus.* Leaves alternate, rarely opposite, simple or pinnately compound (*Rosa*), usually stipulate, stipules sometimes caducous or adnate to petiole.

Floral Features. Inflorescence determinate or indeterminate, Flowers usually bisexual, rarely unisexual and then the plants are dioecious as in *Aruncus;* actinomorphic (zygomorphic in *Hirtella*). The receptacle is generally hollow and there is a concavity. Accordingly ovary fused with the hollow receptacle is hypo-, peri-, or epigynous. Calyx of 5 sepals, almost invariably greenish and sepaloid, foliaceous sepals in *Rosa* (Fig. 28.14D). *Fragaria* has an epicalyx formed from the stipules of the sepals. Sepals basally connate but free above. Corolla of 5 to numerous petals, or absent as in *Alchemilla, Sanguisorba, Acaena, Poterium* and others; generally large and of different colours, imbricate, convolute in *Gillenia*; the hypanthium often bears a nectariferous glandular disc. Stamens considerably variable in number—in *Alchemilla* sp. stamens only 4, alternate with the 4 sepals; *A. arvensis* has only one stamen. Generally, the stamens are 2-, 3- or 4-times the number of corolla lobes, incurved in bud. Filaments elongated, filiform, anthers small, bicelled, dehisce longitudinally, rarely transversely or by pores. Gynoecium syncarpous or apocarpous, polycarpellary; ovaries cyclically or spirally disposed, situated usually within a hypanthium or the hypanthium adnate to a compound ovary, superior, semi-inferior or inferior; 2- to 5-loculed, when compound, with axile placentation and a few to several ovules per locule; terminated by as many styles and stigmas as the carpels. When apocarpous, each ovary unilocular with a pendulous ovule.

The shape of the torus and the relative position and number of stamens and carpels, as well as the structure of fruit, vary widely. On this basis, there are a number of subfamilies:

1. Spiraeoideae—Plants generally unarmed shrubs with simple or pinnately-compound, often estipulate leaves. The torus of the flower form a shallow cup (Fig. 28.14A). Carpels generally 2 to 5, from a central whorl at the base of this shallow cup; the gynoecium is apocarpous. Fruit a dehiscent capsule with 2 or more ovules in each ovary, e.g. *Spiraea decumbens* and *S. hypericifolia.*

2. Pomoideae—Shrubs or trees with simple or pinnately compound, stipulate leaves. The torus forms a deep cup (Fig. 28.14B) in which there are 2 to 5 carpels. Carpels more or less completely united with each other, i.e. syncarpous, and also totally fused with the receptacle. The ovary, therefore, is inferior. Thalamus becomes fleshy in nature in mature fruits, carpels are covered with a cartilaginous sheath (Fig. 28.14B, C). Fruit a pome, e.g. in *Pyrus communis* and *Malus sylvestris.*

3. Rosoideae—Plants generally shrubs or prostrate herbs with suckers or runners. Under this subfamily the torus shows two conditions:

   a) The torus is very much hollowed out. Number of carpels indefinite, they never fuse with the torus and are apocarpous. The carpels are completely enclosed inside the torus, each with a separate style and stigma. Fruits are a collection of achenes, e.g. *Rosa canina* (Fig. 28.14D).

   b) The torus is convex and the indefinite apocarpous carpels are borne on the surface of the convex torus (Fig. 28.14E). The torus may be fleshy, juicy and edible, e.g. *Fragaria vesca* (Fig. 28.14G), or may remain dry as in *Rubus* sp. where the fruit is an etaerio of achenes (Fig. 28.14F).

4. Prunoideae—Deciduous or evergreen trees or shrubs with simple stipulate leaves, stipules small and caducous. Flowers showy, generally in racemes. Torus hollow, gynoecium monocarpellary, ovary superior with a terminal style, free from the torus (Fig. 28.14H). Fruit a drupe with stony endocarp, e.g. *Prunus domestica, P. amygdalus, P. armeniaca, P. persica* and P. *cerasus.*

Two other subfamilies, Chrysobalanoideae and Neuradoideae, have been raised to the rank of families: Chrysobalanaceae and Neuradaceae.

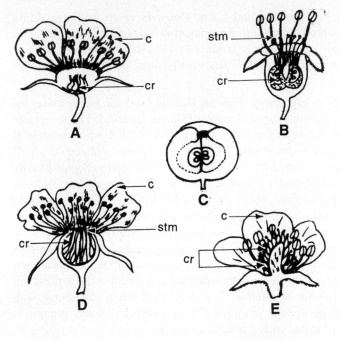

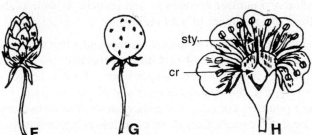

Fig. 28.14 Rosaceae: **A** *Spiraea,* flower longisection. **B** *Malus,* flower longisection; note carpels covered with cartilaginous sheath. **C** Same, fruit longisection. **D** *Rosa canina,* flower longisection. **E** *Fragaria vesca,* longisection of flower and fruit. **F** *Rubus* sp, fruit. **G** *Ribes* sp, fruit. **H** *Prunus domestica,* longisection of flower. *c* corolla, *cr* carpel, *stm* stamen, *sty* style. (Adapted from Rendle 1925)

Anatomy.   Vessels in temperate species mostly small and numerous, with a tendency to form ring-porous condition, with oblique or radial arrangement in the Prunoideae—*Laurocerasus, Nuttallia, Padus* and *Prunus;* perforation plates mostly simple, multiperforate in some species of *Amelanchier, Crataegus, Padus, Pyrus, Sorbaria, Sorbus* and *Spiraea;* intervascular pitting small, alternate. Parenchyma apotracheal, diffuse or in short bands, rarely scanty paratracheal as in *Kerria, Laurocerasus, Padus, Prunus, Rubus* and *Spiraea.* Rays mostly 2 to 5 cells wide, hetero- or homogeneous. Calcium oxalate crystals are either solitary or in clusters. Leaves usually dorsiventral, centric in *Crataegus azarolus.* Hairs commonly unicellular but variations occur. Nectaries present on the petiole of some species of *Prunus,* and also on leaf teeth. Stomata Ranunculaceous type, mostly confined to lower surface.

Embryology.   Pollen grains tricolporate, occasionally tricolpate or stephanocolporate; 2-celled at dispersal stage. Ovules ana- to hemianatropous, uni- or bitegmic and crassinucellate. Mostly Polygonum type of embryo sac, rarely Allium type, 8-nucleate at maturity. Endosperm formation of the Nuclear type.

Chromosome Number.   Basic chromosome numbers are x = 9 in Spiraeoideae, x = 7 in Rosoideae, x = 17 in Pomoideae and x = 8 in Prunoideae.

Chemical Features.   Many species are rich in cyanogenic compounds (*Prunus*), phenolic compounds

(*Malus*), flavonoids (*Pyrus*). *Crataegus* of Rosoideae contains flavone-C-glycosides which has been reported in quite a few genera of Spiraeoideae, Pomoideae and Rosoideae (Challice 1974). Gum containing various types of sugars such as arabinose, rhamnose, galactose and glucoronic acid reported in *Prunus* spp. (Harborne and Turner 1984). Carotenoid pigment rubixanthin occurs in the fruits (hips) of *Rosa* spp. The sugar alcohol sorbitol, isolated first from *Sorbus aucuparia*, has been observed in all Pomoideae and Prunoideae, and most Spiraeoideae. The subfamily Rosoideae is heterogeneous. Only three genera— *Rhodotypos, Kerria* and *Neviusia* of tribe Kerrieae have sorbital (Wallaart 1980). Sorbitol is a potential taxonomic marker for the family Rosaceae (Harborne and Turner 1984). Wallaart (1980) has also shown a correlation between sorbitol accumulation and chromosome number within the family.

Important Genera and Economic Importance.   There are a large number of important genera in this family, such as *Fragaria* or strawberry, *Potentilla* an annual herb of temperate climate, and *Pygeum acuminatum* a tree with two-seeded fruits. Brambles or *Rubus hexagynus* is a shrubby plant. Roses— various species of the genus *Rosa*—have been under cultivation from time immemorial for their beautiful scented flowers. Some genera like *Pyrus, Prunus, Malus* and many others are important fruit trees.

The Rosaceae is economically important also for edible seeds such as almond from *Prunus amygdalus* var. *dulcis,* and edible fruits such as loquat or *Eriobotrya japonica,* strawberry or *Fragaria chiloensis, F. vesca* and *F.virginiana,* apple or *Malus pumila,* medlar or *Mespilus germanica,* apricot or *Prunus armeniaca,* cherry or *Prunus cerasus,* plum or *Prunus domestica,* peach or *Prunus persica,* pear or *Pyrus communis,* blackberry or *Rubus fruticosus* and raspberry or *R. idaeus.* The dried inner bark of *Quillaja saponaria* forms the commercial soapbark (Dutta 1988). *Sorbus aucuparia* is the source of sorbitol, a sweet-tasting substance that can be used as a sugar substitute in the diet for diabetics. The detection of sorbitol in a fruit product derived from fruits not containing sorbitol, indicates its adulteration with cheap apples and pears (Wallaart 1980). It also yields valuable timber. A number of genera are grown as ornamentals: *Cotoneaster, Crataegus, Geum, Kerria, Potentilla, Rosa* and *Spiraea.* An essential oil, 'Otto of roses' is obtained by distillation from the petals of *Rosa damascena.* Rose petals soaked in concentrated sugar syrup are eaten with betel leaf.

Taxonomic Considerations.   The Rosaceae is a large and diverse family. Sometimes, it is split into a number of subfamilies (Dahlgren 1980a). Two of these subfamilies are recognised as separate families: Chrysobalanaceae and Neuradaceae (Cronquist 1981, Kalkman 1988, Melchior 1964, Schulze Menz 1964, Takhtajan 1980, 1987, Willis 1973). Thorne (1983) excludes Chrysobalanaceae but includes Neuradaceae, while Hutchinson (1973) retains both subfamilies in the Rosaceae. Sometimes the inclusion of these two families in the order Rosales is doubted. Willis (1973) suggests placing Neuradaceae close to the Malvaceae and alliance of Chrysobalanaceae with the Geraniales and Sapindales. Dahlgren (1983a) treats them as belonging to a separate order, Chrysobalanales, in the Myrtiflorae. However, Prance (1972)—in his monograph on Chrysobalanaceae—retains them in the Rosales. Hutchinson (1969) presumed the Rosaceae to be "derived from the same stock as the Dilleniaceae" and tribe Quillajeae as the link between them. Hutchinson (1973) and Takhtajan (1959, 1980) consider Rosaceae to be the most primitive family in this order and that the order Rosales is linked with the primitive Saxifragales through subfamily Spiraeoideae (Takhtajan 1980, 1987). Dahlgren (1980a) allied Rosales with the Cunoniales.

According to Kalkman (1988), an ancestor to Rosaceae should have had the following characters:

1. Woody plants with compound, alternate, stipulate leaves,
2. relatively unspecialised flowers,
3. hypanthium not well-developed,
4. perianth biseriate,

5. stamens numerous, free, whorled,
6. pistils numerous, free, with numerous ovules per carpel, and
7. fruits dehiscent, dry, many-seeded.

On the basis of these features, the possible ancestral groups could be either Connaraceae or Cunoniaceae. Cunoniaceae s.s., i.e. without Davidsoniaceae, could be the closest relative of the Rosaceae. This is also supported by chemical features such as the occurrence of tannins, proanthocyanins, ellagic acid, and calcium oxalate crystals, and absence of alkaloids and iridoids (Cronquist 1981).

According to ICBN, Pomoideae should be renamed Maloideae, but the original name is still in use (Challice 1974). Pomoideae are postulated to have arisen by allopolyploidy between different forms of Rosaceae, one of x = 8 (a prunoid) and the other of x = 9 (a spiraeoid). It is presumed that the primitive forms with x = 8 and x = 9 were alike as compared to the present-day taxa (Challice 1974).

Neuradaceae

A small family of 3 genera and 10 species distributed from the Mediterranean to India and South Africa.

Vegetative Features.   Annual prostrate tomentose herbs, more or less woody below. Leaves alternate, variously lobed or pinnatifid, stipulate, stipules minute or absent.

Floral Features.   Flowers solitary, axillary; bisexual, actinomorphic, sometimes with an epicalyx of 5 bracteoles, often showy. Calyx of 5 sepals, gamosepalous, calyx tube broad and flat, lobes valvate. Corolla lobes 5, polypetalous, convolute, inserted in the throat of calyx tube. Stamens 10, in 2 whorls of 5 each, filaments elongated, subulate, persistent or caducous; anthers small, ovoid, longitudinally dehiscent. Gynoecium syncarpous, 3- to 10-carpellary. The torus is deeply cup-shaped; the carpels are united with each other as well as with the base of the cup-shaped torus, which does not become fleshy but enlarges and forms a dry covering around the developing fruits. Ovary horizontal, verticillate, 3- to 10-loculed, with 1 pendulous ovule per locule and 3 to 10 short, persistent styles with a small capitate stigma. Fruits orbicular, depressed-conical, laterally membranous, winged or spinulose-muricate. Carpels dehisce ventrally, styles sometimes spinescent; seeds horizontal, without endosperm.

Anatomy.   Anatomy is comparable to that of the Rosaceae.

Chromosome Number. Basic chromosome number is x = 6.

Important Genera and Economic Importance.   The three genera are *Grielum, Neuradopsis* and *Neurada*. An interesting feature is that the yellow corolla of the first two genera change colour to bluish black on drying.

Taxonomic Considerations. The systematic position of the family Neuradaceae is doubtful. According to Nayar (1984c), Neuradaceae is "a segregate of the family Rosaceae and its alliance with Malvaceae requires study". This family is included in the family Rosaceae as a subfamily Neuradoideae by Bentham and Hooker (1965a), Rendle (1925), Lawrence (1951), Hutchinson (1969, 1973) and Thorne (1968, 1983). Melchior (1964), Cronquist (1969, 1981) and Takhtajan (1980, 1987) include it in the order Rosales.

Morphologically and cytologically it is different from the Rosaceae s.s. Therefore, it is justified to treat it as a distinct family.

Chrysobalanaceae

The family Chrysobalanaceae comprises 17 genera and 450 species (Tobe and Raven 1984) of worldwide distribution in tropical and subtropical regions, especially well-represented in the New World tropics.

Vegetative Features.   Small trees, shrubs or subshrubs. Leaves simple, entire, alternate, stipulate, pinnately veined.

Floral Features.   Inflorescence terminal and subterminal cymose, panicles or terminal and axillary cymules. Flowers bisexual, rarely unisexual and polygamous; perigynous, actino- or zygomorphic. Calyx of 5 sepals, gamosepalous, tube turbinate (top-shaped) or campanulate, more or less unequal or calcarate (spurred) at base; segments free or more or less connate, imbricate. Corolla of 5 petals, rarely apetalous, inserted on the mouth of calyx tube, shortly unguiculate (clawed), imbricate. Disc forms a lining to floral tube (Prance 1970). Stamens 2 to numerous, inserted around the margin of the disc, often larger and fertile, opposite the larger calyx segments and shorter and fertile or more or less sterile on opposite side (Willis 1973); filaments filiform, exerted, anthers extrorse. Gynoecium monocarpellary. The torus shows two conditions (Rendle1925):

a) the torus is shallow, gynoecium represented by a single carpel with a lateral style on the shallow torus (Fig. 28.16A). It does not fuse with the torus; fruit a dry drupe as in *Chrysobalanus icaco*.
b) The torus is tubular with the monocarpellary gynoecium, has a lateral style (Fig. 28.16B).

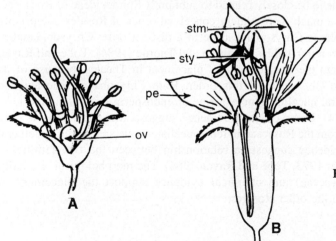

Fig. 28.16 Chrysobalanaceae: **A** *Chrysobalanus icaco,* longisection of flower. **B** *Hirtella*, longisection of flower. *ov* ovary, *pe* petal, *stm* stamen, *sty* style. (Adapted from Rendle 1925)

Ovules usually 2, rarely 1, erect, basal, collateral; style simple, filiform, lateral with simple stigma. Fruit a sessile or stipitate drupe, rarely a crustaceous berry; embryo with thick fleshy cotyledons, endosperm none.

Anatomy.   Vessels moderately to very large, almost exclusively solitary, few to moderately numerous, typically in oblique lines, spiral thickenings absent, perforation plates simple. Parenchyma apotracheal, predominantly in uni- to triseriate bands. Rays uniseriate, heterogeneous. In leaves the entire mesophyll consists of palisade tissue, except in *Leucostemon*. Normally unicellular hairs are common, but stellate hairs occur in *Chrysobalanus,* and peltate in *Leucostemon*. Stomata Rubiaceous type, confined to lower surface.

Embryology.   Pollen grains medium-sized with three furrows, oblate-spheroidal to subprolate, triangular in polar view; exine scabrous to verrucose; 2-celled at shedding stage (Prance 1970). Ovules anatropous, tenuinucellate, bitegmic; modified Polygonum type of embryo sac, 4-nucleate at maturity. Endosperm formation of the Nuclear type.

Chromosome Number.   Basic chromosome number is x = 10 or 11.

Chemical Features.   Cyanidins are absent in Chrysobalanaceae, ∝-elaeostearic acid occurs in the family, as well as in *Prunus* (of Rosaceae) (Hegnauer 1973).

Important Genera and Economic Importance.   The genus *Chrysobalanus* with 4 species is distributed in tropical America, the West Indies, and tropical Africa. In this genus the syle is basal. Fruits of *C. icaco* are edible. Genus *Hirtella* has 95 species in Central and tropical South America and West Indies and 3 species in tropical East Africa and Madagascar. The flowers in this genus are zygomorphic, axis deeply hollowed on one side, stamens and carpels are not enclosed in the hollow, but develop on the other side of the surface of the axis. The third genus, *Licania* has one species in Malaysia and 135 in southeastern United States of America to tropical South America and the West Indies. Sixteen species of *Magnistipula* occur in tropical Africa.

Taxonomic Considerations.   Close affinities of Chrysobalanaceae have been proposed with Dichapetalaceae and Trigoniaceae (Hallier 1923), Tropaeolaceae and Geraniaceae (Hauman 1951), and Connaraceae (Gutzwiller 1961). Prance (1972) placed it between the Rosaceae and the Fabaceae. Takhtajan (1980, 1987) has placed this family in his order Rosales, comprising only Rosaceae, Chrysobalanaceae and Neuradaceae and considers Chrysobalanaceae to be closely related to subfamily Spiraeoideae of Rosaceae. Cronquist (1981) has placed this family in a much more widely conceived order of Rosales comprising 24 families. Dahlgren (1983a), however, places Chrysobalanaceae in a distinct order Chrysobalanales, within the superorder Myrtiflorae, a position reiterated by Dahlgren and Thorne (1984). Tobe and Raven (1984) proposed removal of this family from the Rosales and its placement in Theales, because of its similarities with Theaceae, Ochnaceae and Clusiaceae of this order. Both Chrysobalanaceae and the Theales have tenuinucellate ovule, ephemeral nucellus, Nuclear type of endosperm formation, and non-endospermous seeds (Tobe and Raven 1984). Embryological evidence suggests that this family is only distantly related to the Rosaceae. It differs from the Rosaceae in tenuinucellate ovule and scanty ephemeral nucellus. On the other hand, chemical evidence suggests a relationship between the two families—Rosaceae and Chrysobalanaceae—(Hegnauer 1973, Tobe and Raven 1984). The morphological similarity with the subfamily Spiraeoideae (of Rosaceae) and chemical evidence support the placement of Chrysobalanaceae close to the Rosaceae in the order Rosales.

Connaraceae

A family of 25 genera and 200 species distributed in the tropics of America, Africa, Asia and also in Malaysia, southeastern Asia and Madagascar.

Vegetative Features.   Trees (*Ellipanthus, Cnestis, Connarus*) or twining shrubs; trees of *Jollydora* are palm-like in habit. Leaves alternate, compound, sometimes unifoliate, estipulate.

Floral Features.   Inflorescence axillary or terminal panicles; flowers small, actinomorphic or slightly zygomorphic, bisexual, or rarely unisexual, hypogynous. Calyx of 5 sepals, poly- or rarely gamosepalous, imbricate' or valvate, mostly persistent; corolla of 5 petals, polypetalous or sometimes adherent to the base, imbricate, rarely valvate. Stamens 10 or rarely 8 in 2 whorls, sometimes joined below, inner whorl may be staminodal, filaments free or fused at base to form a short tube; anthers longitudinally dehiscent. Gynoecium apocarpous or united only at the base, unicarpellary, tetra- or pentacarpellary, each carpel with 2 erect, orthotropous ovules; stigma bulbous. Fruit usually a follicle with 1 seed. Seeds endospermous or non-endospermous, often arillate.

Anatomy. Vessels medium-sized to large, almost exclusively solitary or with numerous multiples, occasionally with spiral thickening; perforations simple, intervascular pitting alternate. Parenchyma absent or only a few cells round the vessels. Rays mostly uniseriate, sometimes biseriate, numerous, hetero- or homogeneous. Latex tubes, gum cysts and intercellular canals present in some members. Included phloem present in *Rourea*. Leaves dorsiventral; hairs mostly unicelluar, occasionally biseriate with considerable variation in form; stellate hairs common in *Agelaea*. In *Paxia calophylla,* the stellate hairs form a golden-yellow clay-like coating on young leaves and stem. Stomata mostly confined to the lower surface, Rubiaceous type in *Bernardinia, Paxia, Rourea* and others, and Caryophyllaceous type in *Jollydora*.

Embryology. Pollen grains mostly 3-colporate, sometimes 3-colpate, rarely 4-colpate; 2-celled at shedding stage. Ovules anatropous to hemi-anatropous, collateral, bitegmic, crassinucellate. Endosperm formation of Nuclear type.

Chromosome Number. Basic chromosome number is x = 6.

Taxonomic Considerations. The family Connaraceae has been included in Rosales by Schellenberg (1938), Melchior (1964) and Thorne (1983), in the order Dilleniales by Hutchinson (1969, 1973), in the order Sapindales by Cronquist (1981) and Dahlgren (1983a) and in the order Connarales (Takhtajan 1980, 1987).

According to Nayar (1984d), the family Connaraceae is allied to Leguminosae and Averrhoaceae. On the basis of anatomical and embryological features, Connaraceae resembles several members of Sapindales (Cronquist 1968, 1981, Corner 1976, Metcalfe and Chalk 1972). Embryologically, Connaraceae is similar to Cunoniaceae, but they are almost unique in Sapindales because of apocarpous ovaries. On the other hand, their advanced wood anatomy and arillate seed are comparable to that of Sapindales. According to Corner (1976): "With estipulate, pinnate leaves, apocarpous flowers and arillate follicles, Connaraceae fits well as a side-branch of Meliaceous-Sapindaceous ancestry".

A separate order Connarales, between Sapindales and Cunoniales is appropriate.

Leguminosae

A large taxon, mostly treated as distinct order comprising 3 families: Papilionaceae (Fabaceae), Caesalpiniaceae and Mimosaceae. Of all these families, Papilionaceae is predominantly herbaceous with a few shrubs and trees, but both the Caesalpiniaceae and Mimosaceae are chiefly arborescent. The flowers in racemose inflorescence, bisexual, actino- or zygomorphic, usually highly ornamental. Stamens few to numerous, basifixed, mostly dehisce longitudinally. Ovary monocarpellary, superior, unilocular, placentaion marginal. Fruit a dehiscent or indehiscent legume.

Mimosaceae

A family of ca. 56 genera and 2800 species, more or less confined to the tropics and subtropics of both the hemispheres.

Vegetative Features. Mostly trees and shrubs, often with spiny outgrowths on stem, xerophytes common, hydrophytes also reported (*Neptunia*). Leaves bipinnate, unipinnate in *Affonsea* and *Inga,* sometimes reduced to phyllodia (Fig. 28.18.1A); rachis pulvinate, generally gland-bearing (Fig. 28.18.1A), stipulate, stipules spiny; leaves show sleeping movement. Leaves of *Mimosa pudica* are sensitive to touch.

Floral Features. Flowers in spike or head (condensed racemes) inflorescences, involucre common. In *Dichrostachys* upper part of the spike is bisexual, the lower neutral with long staminodes. Flowers bisexual, actinomorphic, hypogynous. Sepals 4 or 5, inconspicuous, cup-shaped (Fig. 28.18.1B, C), valvate, odd sepal anterior. Petals 4 or 5, polypetalous, sympetalous in *Acacia* (Fig. 28.18.1D) and *Albizia,* valvate. Stamens 10 or equal to the number of petals, free or monadelphous (Fig. 28.18.1E) as in *Inga,* all fertile. Anthers gland-tipped in some genera—*Acacia* (Fig. 28.18.1D, E), *Adenanthera, Parkia* and *Prosopis;* bithecous and longitudinally dehiscent; pollen granular or agglutinated into tetrads or polyads (Fig. 28.18.1F). Gynoecium monocarpellary, ovary unilocular, superior, placentation marginal (Fig. 28.18.1I); ovules numerous; style long, filiform, coiled in bud (Fig. 28.18.1G), stigma truncate (Fig 28.18.1H). Fruit a legume or lomentum as in *Entada, Pseudoentada* and *Plathymania.* Seeds dorsiventrally flattened, funicle long and coiled, pleurogram present (absent in *Pithecellobium*); embryo straight, endosperm present.

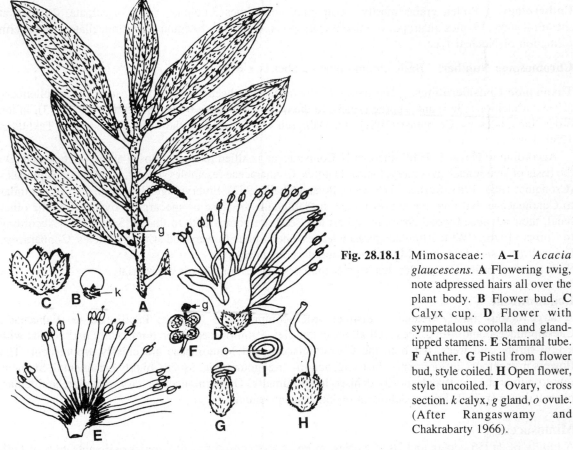

Fig. 28.18.1 Mimosaceae: **A–I** *Acacia glaucescens.* **A** Flowering twig, note adpressed hairs all over the plant body. **B** Flower bud. **C** Calyx cup. **D** Flower with sympetalous corolla and gland-tipped stamens. **E** Staminal tube. **F** Anther. **G** Pistil from flower bud, style coiled. **H** Open flower, style uncoiled. **I** Ovary, cross section. *k* calyx, *g* gland, *o* ovule. (After Rangaswamy and Chakrabarty 1966).

Anatomy.   Wood diffuse-porous, vessels medium-sized to large, typically solitary, spiral thickenings absent; parenchyma abundant, paratracheal, rays 2 to 5 cells wide, homogeneous, mostly of small cells. Fibres with few, small, simple pits. Anomalous structure rarely seen. Leaves dorsiventral, isobilateral or centric. Hairs of both glandular and non-glandular types occur. Stomata Rubiaceous type, confined to the lower surface in Adenanthereae, Ingeae and Parkieae and uniformly distributed on both sides of leaf in Acaciae, Eumimoseae and in *Dichrostachys* and *Neptunia.*

Embryology.   Pollen simple, granular in *Neptunia, Leucaena, Prosopis* and *Desmanthus;* shed as tetrads or polyads in other genera. Ovules anatropous, campylotropous or amphitropous, bitegmic, crassinucellate Polygonum type of embryo sac, 8-nucleate at maturity. Endosperm formation of the Nuclear type; chalazal haustorium common.

Chromosome Number. Diploid chromosome numbers for various genera are 2n = 16, 22, 24, 26, 28, 36, 44, 52, 56 and 104 (Kumar and Subramanian 1987).

Chemical Features. Rich in tannins; a glycoside, dihydroacacipetalin reported in *Acacia* sp. Non-protein amino acid albizzine occurs in seeds of *Albizia julibrissin, Acacia* (except series Gummiferae; Seneviratne and Fowden 1968), and *Mimosa.* Carotenoids present in yellow-flowered *Acacia decurrrens* var. *mollis, A. discolor* and *A. linifolia.* Cyanogenic glucosides reported in some *Acacia* spp. (Secor et al.1976, Seigler et al. 1978). Gum containing sugars arabinose, rhamnose, galactose and glucoronic acid present in *Acacia.*

Important Genera and Economic Importance. The genera *Acacia, Albizia, Adenanthera, Inga, Entada, Enterolobium, Mimosa* and *Pithecellobium* are important. Leaves of *M. pudica,* commonly called touch-me-not or the sensitive plant are sensitive to touch. Leaves of two aquatic plants *Neptunia oleracea* and *N. plena* also show similar characteristics. *Entada scandens* is an immensely woody climber with almost 1-meter-long fruits, and round seeds that are 2" in diameter.

Many genera are economically important. *Acacia nilotica* var. *gangeticus* yields fuel and gum. The wood is also used for making agricultural implements, tent pegs, etc. *A. senegal* is the source of gum arabic. The heartwood of *A. catechu* (on boiling) gives a tannin known as katha and is commonly used with betel leaf. A yellowish dye obtained from this heartwood is also used in dyeing khaki cloth. Pods and bark of *A. farnesiana* are used medicinally and for tanning; the flowers yield a perfume. *Xylia dolabriformis* or iron-wood tree from Burma yields valuable timber.

Acacia auriculiformis, the phyllode-bearing Australian *Acacia* tree, *Albizia lebbeck, A. procera, Enterolobium,* etc. are grown as avenue trees. *Calliandra haematocephala* is an ornamental shrub.

Taxonomic Considerations. The family Mimosaceae is allied to the Rosaceae and shares common characters: actinomorphic flowers and numerous stamens. However, it can readily be distinguished because of the hypogynous stamens, anteriorly-placed odd sepal, superior ovary and the fruit a legume or lomentum. Occurrence of trees and shrubs and rarely herbs, actinomorphic flowers and numerous stamens are primitive characters of this family. On the other hand, predominantly bipinnate leaves and the existence of xerophytes and hydrophytes are the advanced features.

It is considered to be the most primitive family/subfamily of the Leguminosae.

Caesalpiniaceae

A large family of ca. 152 genera and over 2800 species (Kumar and Subramanian 1987), distributed in the tropics and subtropics of both hemispheres, abundant in America.

Vegetative Features. Predominantly arborescent, xerophytes less common (*Parkinsonia aculeata*; Fig. 28.18.2A), hydrophytes not known. Stem mostly glabrous. Leaves unipinnate (*Cassia*), or bipimnate (*Delonix*) or rarely simple (*Bauhinia*); rachis pulvinate, rarely gland-bearing (*Cassia*), stipulate, stipules sometimes foliaceous, e.g. *C. auriculata.* Leaflets exhibit sleeping movement; sometimes with mucronate tip.

Floral Features. Flowers in corymb or simple raceme (Fig. 28.18.2A), sometimes pendulous as in *Cassia fistula;* bracteate, ebracteolate, pedicellate, bisexual, zygomorphic. Sepals 5 (4 in *Amherstia*) polysepalous, rarely gamosepalous as in *Bauhinia;* descendingly imbricate, odd sepal anterior and outermost. Petals 5 (Fig. 28.18.2B, D), rarely fewer (3 in *Tamarindus* and *Amherstia*; 1 in *Afzelia*), and absent in *Saraca*; petals dissimilar, often clawed. Stamens 10 or less (3 in *Tamarindus*, 3 to 8 in *Saraca*), free as in *Cassia* and *Caesalpinia* or monadelphous as in *Tamarindus* or diadelphous as in *Amherstia*. Stamens in *Cassia* unequal in size and staminodes present. In *Caesalpinia, Delonix,* and *Parkinsonia* (Fig. 28.18.2D), all stamens fertile. Anthers basifixed, bithecous, longitudinally dehiscent or poricidal; filaments free,

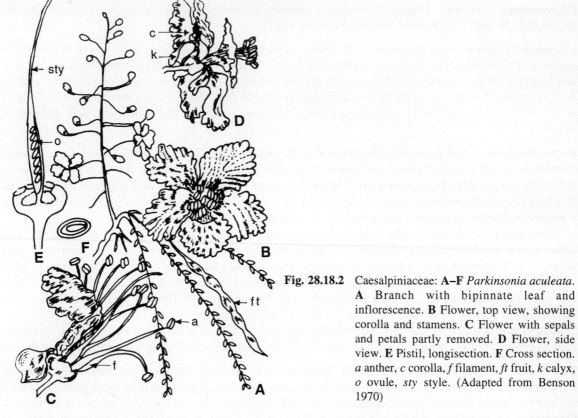

Fig. 28.18.2 Caesalpiniaceae: **A–F** *Parkinsonia aculeata.* **A** Branch with bipinnate leaf and inflorescence. **B** Flower, top view, showing corolla and stamens. **C** Flower with sepals and petals partly removed. **D** Flower, side view. **E** Pistil, longisection. **F** Cross section. *a* anther, *c* corolla, *f* filament, *ft* fruit, *k* calyx, *o* ovule, *sty* style. (Adapted from Benson 1970)

massive, dilated at the base (Fig. 28.18.2C). Gynoecium monocarpellary, ovary superior, unilocular, placentation marginal (Fig. 28.18.2E, F), ovules numerous; style massive, often slightly recurved as in *Cassia*, stigma capitate. Fruit a legume or large woody pod, dehiscent or indehiscent, cylindrical (*Cassia fistula*), or flattened (*Delonix regia*). Seeds dorsiventrally flattened, funicle longer than seed; endosperm present.

Anatomy. Wood diffuse-porous, vessels typically medium-sized, solitary, sometimes with spiral thickenings; parenchyma moderately abundant, paratracheal; rays 1 to 3 cells wide, heterogeneous. Leaves generally dorsiventral except in certain species of *Hoffmanseggia* and *Hymenaea*. Hairs glandular or nonglandular; stomata mostly Rubiaceous or Ranunculaceous type, variations common. Abnormal anatomy reported in many species of the genus *Bauhinia*.

Embryology. Pollen grains basically 3-colpate in this highly multipalynous family; shed singly or as tetrads as in *Afzelia*, at 2-celled stage. Ovules anatropous, campylotropous, or amphitropous, bitegmic, crassinucellate. Polygonum type of embryo sac, 8-nucleate at maturity. Endosperm formation of the Nuclear type; chalazal haustorium present.

Chromosome Number. Diploid chromosome numbers are 2n = 16, 18, 22, 24, 26, 28, 42, 48, 52 and 56.

Chemical Features. Plants rich in tannins; carotenoids in *Delonix regia* flowers and ánthraquinones in *Cassia* are reported. Non-protein amino acids occur in the seeds of some species of *Caesalpinia* (Evans and Bell 1978).

Important Genera and Economic Importance.   Important genera of this family are *Bauhinia, Cassia, Caesalpinia, Delonix, Amherstia, Haematoxylon, Hardwickia, Humboldtia, Parkinsonia* (Fig. 28.18.2A), *Saraca,* and *Tamarindus.* Economically, the family Caesalpiniaceae is quite important. Fruits of *Cassia fistula* are used medicinally, and those of *Tamarindus indica* have carminative and laxative properties. The seeds of *Cassia occidentalis* are powdered and mixed with coffee powder as an adulterant. The heartwood of *Haematoxylon campechianum* is the source of the dye heamatoxylin, used as nuclear stain in biological sciences. *Hardwickia binata* is the source of valuable timber. The wood of *Parkinsonia aculeata* is good for making charcoal. Many genera are grown as ornamentals or as avenue trees.

Bauhinia vahlii is a large woody climber with stem tendrils. The simple apically notched leaves are almost 30 cm in diameter and are used as substitute for plates; ropes made from the bark of this plant are very tough and used for making suspension bridges over small rivers and rivulets in the Himalayan region. *B. anguinia* is another such climber with flat, ribbon-like and twisted stem giving the appearence of a snake, and the common name is 'nagpat'.

The flower buds of *Bauhinia variegata* are used as a vegetable. *Caesalpinia bonducella* or fever nut tree is also used medicinally. The wood of *C. sappan* yields a red dye used for dyeing wool and silk. The red colour mixed with starch powder is used during the Holi festival in India.

Taxonomic Considerations.   According to Hutchinson (1969,. 1973), the Caesalpiniaceae is the most primitive amongst the members of the Leguminosae, and is therefore closest to the Rosales, from which it has been derived. The vertical sections of the flowers of *Parinari* of Rosaceae and *Bauhinia* of Caesalpiniaceae resemble each other. The position of this family should be between the Mimosaceae and Papilionaceae. The family Caesalpiniaceae has been retained as a subfamily Caesalpinioideae in the family Leguminosae by Stebbins (1974) and Takhtajan (1980, 1987), as a family Caesalpiniaceae in the order Fabales by Dahlgren (1977a, 1980a, 1983a); in the order Leguminales by Jones (1955), Hutchinson (1973) and Rangaswamy and Chakrabarty (1966).

Papilionaceae (Fabaceae)

A very large family of about 482 genera and 12000 species, cosmopolitan, abundant in tropics and subtropics and some in temperate zones.

Vegetative Features.   Chiefly herbs and climbers, some are shrubs (*Sesbania sesban*), trees (*Pongamia pinnata, Sophora* sp.) and woody climbers (*Abrus precatorius*); xerophytes (*Alhagi pseudalhagi*) and hydrophytes (*Aeschynomene aspera*) rarc. Bacterial root nodules are commonly present. The stem surface mostly hairy except in woody species. Leaves simple (*Indigofera cordifolia*) or unipinnate, often trifoliate as in *Cajanus cajan, Rhynchosia minima;* rachis pulvinate, grooved, lamina gland-dotted in *Rhynchosia minima;* stipulate, stipules spiny in *Robinia;* sleeping movement common in many genera. Leaflets often modified to tendrils, as in *Pisum sativum* (Fig. 28.18.3.A), *Vicia hirsuta* and others.

Floral Features.   Flowers in a raceme or spike, sometimes highly condensed to form heads, e.g. *Medicago lupulina, Trifolium pratense;* bracteate, ebracteolate, pedicellate, bisexual, hypogynous, zygomorphic (Fig. 28. 18.3B, C). Sepals 5, connate, campanulate (Fig. 28.18.3C), odd sepal anterior and inferior, often coloured, as in *Pongamia.* Corolla papilionaceous, petals 5, the posterior odd petal outermost and is called standard, two lateral ones the wings, and the two anterior ones fused to form a keel or carina (a boat-shaped structure (Fig. 28.18.3D, E). Stamens 10, diadelphous (Fig. 28.18.3G), rarely 9 and monadelphous, as in *Abrus;* in *Erythrina* 10 monadelphous stamens, all fertile, filaments fused to form staminal column but free near the apex. Anthers basifixed, bithecous, longitudinally dehiscent, pollen granular. Gynoecium monocarpellary, superior, unilocular, ovary with marginal placentation and one row of ovules (Fig.28.18.3I, J); style thick and curved or reflexed at base, stigma brushy (Fig. 28.18.3F) or

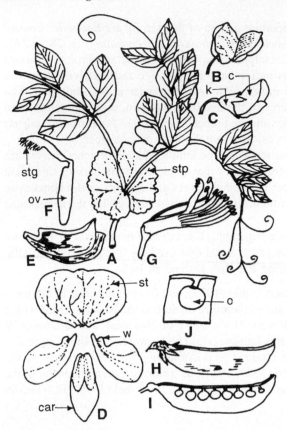

Fig. 28.18.3 Papilionaceae: **A–J** *Pisum sativum*. **A** Flowering twig. **B** Flower. **C** Flower bud. **D** Papilionaceous aestivation of petals. **E** Keel. **F** Pistil. **G** Stamens and pistil. **H, I** Pod (**H**) l.s (**I**). **J** Seed and placenta. *c* corolla, *car* carina, *k* calyx, *o* ovule, *ov* ovary, *st* standard, *stg* stigma, *stp* stipule *w* wing. (Original)

capitate. Fruit a dehiscent (Fig. 28.18.3H) or indehiscent legume. Seeds reniform or rounded (Fig. 28.18.3I J); funicle shorter than seed. Embryo pleurorhizal, endosperm present.

Anatomy. Wood ring-porous, vessels medium-sized to very small, spiral thickenings occasional; parenchyma moderate to abundant, paratracheal. Rays as in Caesalpiniaceae. Anomalous growth occurs in many genera, climbers in particular. Leaves usually dorsiventral, less frequently isobilateral. Hairs both glandular and non-glandular; glandular leaf-teeth in *Myroxylon pubescens* and extrafloral nectaries are present on the stipules of *Canavalia, Dolichos, Erythrina* and *Vicia*. Stomata variable in structure and distribution. Stomata are present on both surfaces of the leaf in many species of *Alysicarpus, Arachis, Argyrolobium, Canavalia, Crotalaria, Rhynchosia, Smithia* and many others; confined to upper surface in *Coelidium, Dillwynia, Ormocarpum, Eutaxia, Geoffraea*, and *Pultenaea;* confined to lower surface in *Aeschynomene, Chadsia, Clitoria, Derris, Desmodium, Dioclea, Dumasia, Millettia, Mucuna, Strongylodon* and some others. The stomata may be Rubiaceous type as in *Alysicarpus, Arachis, Bowdichia, Cicer* and many others, Rubiaceous type but with two pairs of subsidiary cells parallel to the pore, e.g. species of *Aotus, Brachysema, Dillwynia* and *Oxylobium;* surrounded by 3 or more subsidiary cells as in most Galegeae, Hedysareae, Podalyrieae and Sophoreae; approximating to the Cruciferous type in species of *Borbonia, Crotalaria, Lebeckia, Lotononis, Priotropis, Rafnia* and *Viborgia;* surrounded by a rosette of cells in *Anarthrophyllum, Genista, Lebeckia* and *Templetonia;* Ranunculaceous type in most Loteae and Vicieae.

Embryology. Highly multipalynous family with the fundamental form as 3-colpate; pollen grains shed singly at 2-celled stage; usually smooth, spinuliferous in *Dolichos* (Dnyansagar 1970). Ovules anatropous, campylotropous or amphitropous, bitegmic, crassinucellate, Polygonum type of embryo sac most common, Allium type and Oenothera type are also known, 8- or 4-nucleate at maturity. Endosperm formation of the Nuclear type; chalazal part forms a haustorium.

Chromosome Number. Haploid chromosome numbers are n = 7, 8, 10, 11, 12 and 13 of which 7 and 8 are more common. Natural hybridization occurs amongst various species of *Baptisia* (Alston and Turner 1963).

Chemical Features. Many non-protein amino acids are present, e.g. canavanine in *Canavalia ensiformis* (Jackbean), dopa or tyrosine in *Mucuna prurita* and *Vicia faba* seeds, lathyrine in seeds of *Lathyrus tingitanus,* and pipecolic acid in the seeds of *Phaseolus vulgaris.* Isoflavones occur profusely only in the Papilionaceae (amongst all the plant groups): in the flowers, leaves, seeds, roots and heartwood of the genera *Cytisus, Ulex, Trifolium* and *Lathyrus* (Harborne and Turner 1984). *Baptisia*, a genus with 18 species of perennial herbs from North America, contains 9 flavones, 16 flavonols and 18 isoflavone glycosides (Markham et al. 1970). Another related genus *Thermopsis* is also rich in flavonoids (Dement and Mabry 1972). Glycosides are present in the flower pigments of the tribe Vicieae of Papilionaceae (Harborne and Turner 1984). Quinolizidine in lupins (*Lupinus*) deters the feeding of herbivores and inhibits the growth and development of bacteria and fungi and also inhibits the germination of grass seeds. Alkaloid-free lupins have higher incidence of herbivory and disease (Wink 1985).

Important Genera and Economic Importance. Papilionaceae includes numerous important genera. Many are ornamental trees such as *Butea monosperma, Erythrina* spp., *Sophora* sp., *Pongamia pinnata, Sesbania grandiflora* and others; some are climbers, viz. *Derris elliptica, Lathyrus odoratus* and *Wisteria chinensis.* Various members of Papilionaceae provide many essential commodities. Food from *Pisum sativum, Dolichos lablab, Cajanus cajan, Lens esculenta, Phaseolus* spp., *Vigna* spp., *Canavalia* sp., *Cyamopsis tetragonoloba,* etc., fatty oils from *Arachis hypogaea;* fodder from *Trifolium* and *Trigonella* spp., dye from *Indigofera tinctoria* and *Butea monosperma;* timber from *Dalbergia sissoo, D. latifolia* and *Pterocarpus santalinus;* medicines from *Glycyrrhiza glabra* used for sore throat and cough; and an insecticide from *Derris elliptica.* Dried flowers of *Butea monosperma* yield yellow colour, used for dyeing, and also during the festival of colour—"Holi". Cowage or *Mucuna prurita* is a climber whose pods are covered with stinging hairs. It is useful as green manure and cover crop. Seeds have medicinal value. Some are ornamental herbs like *Lupinus* and climbers like *Lathyrus odoratus* and *Clitoria ternatea.* There are many herbs of wild growth: *Alysicarpus, Indigofera, Heylandia, Medicago, Melilotus, Tephrosia, Vicia* and *Zornia.* The seeds of the stout liana *Abrus precatorius* are the source of the protein, abrin. The seeds of this plant are bright red with a black spot and oval-shaped. Each seed has such accurate weight that they are used for weighing gold and silver. *Adenanthera pavonia* (Mimosaceae) is a tree with similar bright red disc-shaped seeds that are used as curios after scooping out the cotyledons and filling the empty space with miniature animals made of ivory. Another interesting plant is *Aschynomene aspera*, the light spongy wood is of ivory colour and used for making decorative articles. The decorations for the foreheads of brides and bridegrooms in Bengal are made of this wood.

Taxonomic Considerations. This family is the most advanced of the three families—Mimosaceae, Caesalpiniaceae and Papilionaceae—with predominantly herbaceous members, some xerophytes and hydrophytes, zygomorphic flowers and fewer and diadelphous stamens. The status of Papilionaceae has changed repeatedly according to the changes in the status of the order Leguminales (= Leguminosae). The presence of phenylated flavones, flavonoids with a methylenedioxy group and 5- and 7-deoxy-flavonoids in Fabaceae (= Papilionaceae) and Rutaceae indicate their close association (Wollenweber 1982). On the other hand, cyanogenic glycosides are reported to occur (Seigler 1977) in both Fabaceae and Rosaceae, supporting their alliance.

Taxonomic Considerations of the Order Leguminales

The order Leguminales/Fabales comprises three families: Mimosaceae, Caesalpiniaceae and Fabaceae or Papilionaceae. Whether the taxon Leguminales (= Leguminosae) represents a family of 3 subfamilies or an order embracing 3 families is still disputed (see Table 28.18.3). Bentham and Hooker (1965a), Rendle (1925), Wilber (1963), Cronquist (1968) and Takhtajan (1980, 1987) consider the Leguminosae as a family of 3. subfamilies.

That the 3 subfamilies enjoy the rank of individual families and, therefore, the Leguminosae constitute an order is not, however, the latest view. As early as 1814, Brown reported that: "this extensive tribe, i.e. Leguminosae, may be considered as a class (that is, an order in present-day terminology) divisible into at least 3 orders (that is families) namely Mimosae, Lomentaceae or Caesalpiniaceae, and Papilionaceae". Hutchinson (1926, 1969, 1973), Stebbins (1974) and Dahlgren (1975a, 1977a, 1980a, 1983a) also treat the Leguminosae as an order. Following the International Code of Botanical Nomenclature, Jones (1955) proposed the ordinal name Leguminales. Stebbins (1974) and Dahlgren (1980a, 1983a) have further changed the name to Fabales based on the type Family Fabaceae (= Papilionaceae).

Table 28.18.3 Taxonomic Status of Leguminosae

Leguminosae as a family	Leguminosae as an order	Leguminales	Fabales
Bentham and Hooker (1965a)	Brown (1814)	Jones (1955)	Stebbins (1974)
Rendle (1925)	Hutchinson (1926)	Hutchinson	Dahlgren (1980a)
Wilber (1963)	Hallier (cf.	(1959, 1969)	1983a)
Cronquist (1968)	Lawrence 1951)		
Benson (1970)	Dahlgren (1975a, 1977a)		
Stebbins (1974)			
Takhtajan (1980, 1987)			

On the basis of the above discussion, the three taxa should be treated as distinct families: Mimosaceae, Caesalpiniaceae and Papilionaceae, and included in a distinct order, Leguminales, independent of the Rosales, but next to and derived from it. As Fabaceae is the alternate name for Papilionaceae, according to ICBN, the term Fabales may also be used alternatively for the Leguminales.

Krameriaceae

An unigeneric family of the genus *Krameria* with 25 species distributed from Mexico to Chile.

Vegetative Features.   Shrubs or perennial herbs, mostly pubescent or sericeous (silky); leaves simple or unifoliate or rarely 3-foliolate as in *K. cytisoides*, entire, alternate, estipulate.

Floral Features.   Inflorescence axillary or terminal racemes. Flowers bisexual, bracteate, usually with two opposite foliaceous bracts, zygomorphic, hypogynous. Calyx of 4 or 5 sepals, polysepalous, unequal, imbricate, the three larger sepals nearly enclose the entire flower along with the two smaller ones; brightly coloured with a reddish pigment giving them a petaloid appearance. Corolla of 5 petals, polypetalous, very unequal; a pair of short, thick, fleshy petals subtend the pistil on the adaxial side of the flower and the other three petals are long-clawed and sometimes partly united at their bases, into a single stalk-like structure. Stamens usually 4, united at their base and sometimes adnate to claws of upper petals; anthers tetra-sporangiate, conical, dehiscence by a single, terminal pore. Gynoecium syncarpous, bicarpellary—

1 fertile and 1 sterile carpel; ovary superior, unilocular with 2 collateral, pendulous ovules per locule; style simple, stigma discoid. Fruit globose, indehiscent, 1-seeded, covered with bristles or spines (often barbed); seed non-endospermous, embryo straight with thick cotyledons.

Anatomy.   Xylem and phloem in the form of a broad or narrow continuous cylinder traversed by inconspicuous rays. In different species, vessels vary in size, frequency and arrangement usually, with spiral thickening and simple perforation. The hard ground tissue of the wood is composed of fibers with bordered pits. Cluster crystals and secretory cells with presumably tanniniferous contents are common. Leaves with relatively larger lamina are isobilateral and with smaller lamina are centric. Stomata present on both the surfaces, mostly Rubiaceous type.

Embryology.   Pollen grains monads, isopolar, spheroidal, 3-colporate to 3-porate; exine thick, surface striate; no starch and therefore rarely collected by bees (Simpson and Skvarla 1981). Ovules anatropous, bitegmic; endosperm formation of the Nuclear type (Johri et al. 1992).

Chromosome Number.   Basic chromosome number $x = 6$.

Chemical Features.   3-acetoxy fatty acids occur in free state, in floral glands or nectaries of *Krameria* spp. (Simpson et al. 1977, 1978, Seigler et al. 1978, Harborne and Turner 1984).

Important Genera and Economic Importance.   The only genus is *Krameria,* without any economic importance.

Taxonomic Considerations.   The genus *Krameria* has been treated variously (see Table 28.19.1). Rendle (1925) treated *Krameria* as a solitary genus of the tribe Kramerieae of subfamily Caesalpinioideae. In the posteriorly placed odd petal and the monocarpellary gynoecium, *Krameria* conforms to the floral pattern of the Leguminosae (Rangaswamy and Chakrabarty 1966). Bentham and Hooker (1965a) placed this genus in the Polygaleae of the order Polygalinae. Benson (1970) considers it to be the sole member of the fourth subfamily, Kramerioideae, of family Leguminosae. Small (1903; see Lawrence 1951) and Jones (1955) raised it to a family, Krameriaceae, in the order Leguminosae/Leguminales. Melchior (1964) also treats it as a distinct family, but places it in Rosales along with Leguminosae.

On the basis of serotaxonomical studies *Krameria* is closer to Polygalaceae members. Floral, morphological and floral-anatomical studies of Milby (1971) and Verkerke (1985) also support this view. The syncarpous, bicarpellary gynoecium of *Krameria* is pseudomonomerous (Leinfellner 1971) and although one carpel is suppressed, this fact provides the strongest evidence for placing Krameriaceae in the Polygalales (Simpson and Skvarla 1981).

However, all species of *Krameria* investigated so far are semi-parasitic root parasites (Musselman 1977, Musselman and Mann 1978). No member of Rosales or Polygalales is parasitic. Detailed palynological and cytological study (Turner 1958) also fails to provide any significant data as to the phylogenetic affinities.

Most of the contemporary phylogenists treat this genus as a member of the Polygalales (Hutchinson 1973, Thorne 1976, Cronquist 1968, 1981, Stebbins 1974, Dahlgren 1980a, 1983a, Takhtajan 1987).

The Krameriaceae is indeed closer to the Polygalaceae and should, therefore, be included in the order Polygalales.

Taxonomic Considerations of the Order Rosales

Rosales is a heterogeneous group of families, held together by a complex pattern of overlapping similarities, although the order is morphologically diffuse and difficult to define. According to Cronquist (1981), the Rosales are derived from within or near the Magnoliales.

Table 28.19.1 Status of the Genus *Krameria*

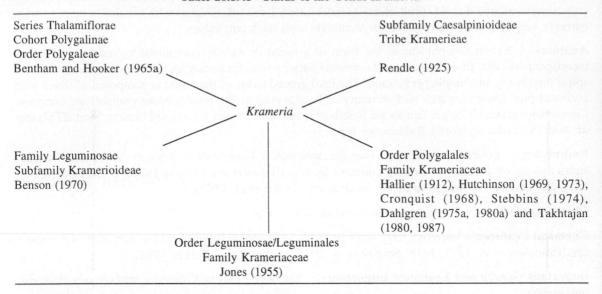

Series Thalamiflorae
Cohort Polygalinae
Order Polygaleae
Bentham and Hooker (1965a)

Subfamily Caesalpinioideae
Tribe Kramerieae
Rendle (1925)

Krameria

Family Leguminosae
Subfamily Kramerioideae
Benson (1970)

Order Polygalales
Family Krameriaceae
Hallier (1912), Hutchinson (1969, 1973),
Cronquist (1968), Stebbins (1974),
Dahlgren (1975a, 1980a) and Takhtajan
(1980, 1987)

Order Leguminosae/Leguminales
Family Krameriaceae
Jones (1955)

There is much disagreement about the limits of families within the Rosales. Takhtajan (1969) recognises Hamamelidales, a distinct order including Platanaceae, Hamamelidaceae and Myrothamnaceae. Takhtajan (1987) separates Myrothamnaceae to Myrothamnales. Saxifragaceae, according to earlier taxonomists is a large family comprising herbs and shrubs as well as trees. Recent phylogenists include only the herbaceous members in Saxifragaceae s.s, and the rest in separate families.

Cunoniaceae is regarded by Hutchinson (1969) to comprise many taxa which were earlier included in the Saxifragaceae by Engler (1930a). Cunoniaceae, along with the unigeneric Davidsoniaceae, is often treated as member of a separate order, Cunoniales. Pittosporaceae is yet another controversial family of this order. Close chemical ties between Araliales and Pittosporaceae (Hegnauer 1971, Nayar 1986a) form the basis of placing this family near Araliales as a distinct order Pittosporales (Takhtajan 1987).

The two monotypic families—Byblidaceae and Roridulaceae—are often merged together (Cronquist 1968). Although included in the Rosales in this treatment, both Byblidaceae and Roridulaceae differ from the rest of the Rosalean members in a number of features. Takhtajan (1987) raises Byblidaceae to an order Byblidales and places it near Pittosporales. Serological studies on Roridulaceae reveal its close relationship with Hydrangeaceae and it is placed close to this family in the order Hydrangeales.

The Leguminosae is also a disputed taxon. It is recognised as a separate order, Leguminales, distinct from Rosales, although closely related and probably derived from it. The studies on Krameriaceae suggest its inclusion in the Polygalales, and not in the Rosales. Still another family of this order, Connaraceae, appears to be a connecting link between the Rosales and the Sapindales. Embryologically, the Connaraceae is similar to the Cunoniaceae; it is strictly apocarpous but its wood structure and the arillate seeds are comparable to those of the Sapindales.

Morphological diversity of the Rosales and its large families has resulted in lack of ecological unity. Crassulaceae, a family of succulents, is the only family that is ecologically distinct.

The circumscription of the order Rosales will be very different, given the possibility of removing some of the families, such as Cunoniaceae, Davidsoniaceae, Pittosporaceae, Byblidaceae, Roridulaceae, Krameriaceae, Connaraceae and even Leguminosae, that are included today.

Order Hydrostachyales

Hydrostachyales is a monotypic order with the family Hydrostachyaceae from South Africa and Madagascar.

Hydrostachyaceae

A unigeneric family with the genus *Hydrostachys* and 25 species.

Vegetative Features.   Submerged aquatic herbs, stems tuberous, leaves in rosettes, simple to bi- or tripinnately divided, partly covered with scale-like excrescences (Dahlgren 1975a).

Floral Features.   Inflorescence spicate, borne on unbranched, leafless peduncle; flowers usually without sepals and petals, unisexual, sessile, each in the axil of a bract, usually with a tuft of hairs on each side. Male flowers with one extrorse stamen, dehiscence longitudinal. In female flowers the gynoecium is syncarpous, bicarpellary; ovary unilocular with parietal placentation; ovules numerous on each placenta. Fruit a capsule with numerous seeds; seeds non-endospermous.

Anatomy.   *Hydrostachys natalensis* has leaves with pointed emergences. The structures of all the organs of this species are remarkably homogeneous. A ring of isolated vascular bundles, along with medullary and cortical strands, occurs in the inflorescence axis of *H. imbricata.* Clustered crystals of calcium oxalate are reported.

Embryology.   Pollen grains are shed in permanent tetrads. The ovules are anatropous, unitegmic, and tenuinucellate. Polygonum type of embryo sac, 8-nucleate at maturity. Endosperm formation is of the Cellular type and has micropylar haustorium.

Chromosome Number.   Haploid chromosome numbers are n = 10–12.

Chemical Features.   Druses of calcium oxalate occur in vegetative parts.

Important Genus.   *Hydrostachys* is the only genus.

Taxonomic Considerations of Hydrostachyaceae and the Order Hydrostachyales

Taxonomists have treated this family in different ways. Hydrostachyaceae is member of a distinct monotypic order, Hydrostachyales, of Amentiferae (Lawrence 1951). Melchior (1964) placed it just before the Geraniales along with the Podostemales. According to Cronquist (1968), this family shows more resemblances with the Scrophulariales and is an aquatic, apetalous offshoot of this order. Hutchinson (1969, 1973) and Benson (1970), however, included Hydrostachyaceae in the Podostemales. Cronquist (1981) assigns this family to the Callitrichales along with Hippuridaceae and Callitrichaceae, another aquatic family with reduced flowers. Dahlgren (1975a) treats Hydrostachyales as a still more advanced family by placing it in Lamianae. Dahlgren (1980a, 1983a) has a different opinion and points out that this family approaches Myrothamnaceae and allied families of the Rosiflorae; he retains it in the Lamiales mainly on the basis of embryological features.

Takhtajan (1969, 1980) includes Hydrostachyaceae in Scrophulariales with its probable origin from Scrophulariaceae. Takhtajan (1987) treats it as a member of the monotypic order, Hydrostachyales, and derives it from Plantaginaceae of the Scrophulariales.

This family and order are both advanced, although there is considerable reduction in floral structures. The taxonomic position requires further studies.

Order Podostemales

Podostemales is a monotypic order and comprises the family Podostemaceae.

Podostemaceae

It is a medium-sized family of 49 genera and 240 species, distributed mostly in the tropics of both the hemispheres. Some members occur in temperate regions. Of the 20 species that occur in India, 18 are endemic and grow mostly in the Western Ghats (Sehgal et al. 1993).

Vegetative Features.   Aquatic herbs, perennial, usually submerged and resemble algae, mosses or liverworts (Fig. 30.1A). They grow on stones and rocks, usually attached to them by adhesive, polymorphic, photosynthetic, creeping, dorsiventral flat roots. The rocks are often submerged and sometimes even under waterfalls. Primary axis is reduced in size, often dorsiventrally flat, branching and forming a thallus-like structure with filiform to laminate branches or segments. The leaves are borne on special secondary leaf-bearing branches. Vegetative parts have milky latex in some members. Leaves simple, alternate, linear or broad, often with sheathing leaf-base.

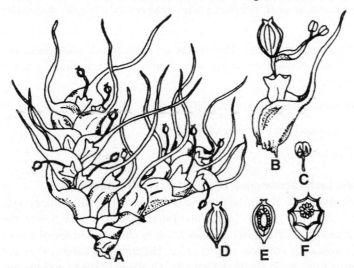

Fig. 30.1 Podostemaceae: **A-F** *Podostemon subulatus*. **A** Portion of plant. **B** Flower. **C** Stamen. **D** Gynoecium. **E** Vertical section. **F** Cross section of gynoecium. (After Subramanyam 1962).

Floral Features.   Flowers are produced only when the plants are exposed because of low level of water. Flowers bisexual, minute, zygomorphic, fragrant, entomophilous, solitary and terminal (Fig. 30.1A, B) on an elongated pedicel or in cymes, flower buds enclosed within capsule-like structure called "spathella". Tepals absent or 2 to 5, hypogynous, basally connate. Stamens (Fig. 30.1B, C) 1, 2, 4 or more, sometimes in two whorls; if only two, may be fused by their filaments (Fig. 30.1B); anthers 4-celled at the time of anthesis, introrse, with 1 staminode on each side of the stamens. Gynoecium (Fig. 30.1D) syncarpous, bi- or tricarpellary, ovary superior, bi- or trilocular, with axile placentation (Fig. 30.1E, F), rarely unilocular with free-central placentation; ovules numerous in each locule; styles distinct, 2 or 3, filiform or shortened (Fig. 30.1B, D), often strongly papillate; stigmas as many as styles. Fruit a septicidal capsule, seeds numerous, endosperm absent, embryo thick and straight.

Anatomy. Internal structure of mature thallus is divided into 3 regions—upper and lower epidermis and ground tissue. Intercellular air spaces occur very rarely although the members are aquatic. Mechanical tissue mostly in the form of collenchyma, in some members sclerenchyma has been reported. Silica bodies occur in most members and prevent the plants from collapsing in time of drought. Nagendran et al. (1980) report facultative sunken stomata on upper epidermis of thallus in *Griffithella* and the not-so-well differentiated epidermis contains chloroplasts. Vessels poorly developed and large air canals develop where they collapse. The vessels with spiral and annular thickenings; sieve tubes absent. In *Griffithella* several cells of lower epidermis develop into rhizoids.

Embryology. Anthers 4-celled at the time of anthesis; pollen grains free or in dyads, shed at 2-celled stage. Ovules anatropous, bitegmic, tenuinucellate. Embryo sac of various types, the most common one is Podostemum type. The interpretation of nuclei of the embryo sac is controversial (see Johri et al. 1992). In the embryo sac only the upper polar nucleus differentiates, the single fertilization gives rise to true embryo (with a suspensor haustorium) which is able to develop without the endosperm. This is so because the nucellar cells underlying the embryo sac stretch longitudinally, their walls break down, the protoplast unite and a pseudo-embryo sac is formed in which the embryo develops (Favre-Duchartre 1984). It is also known as "nucellar plasmodium" (Johri et al. 1992).

Chromosome Number. Basic chromosome number is x = 10.

Chemical Features. Silicate bodies, laticiferous or resin ducts are common. Salts accumulate in the plant body. Cubical crystals of calcium oxalate have been reported in *Mourera fluvitalis*. The gummy substance produced by the plant *Griffithella hookeriana* shows 5 monosaccharide sugars, namely arabinose, galactose, rhamnose, glucose and mannose. Galacturonic acid occurs only in traces (Vidyashankari, unpubl. data).

Important Genera and Economic Importance. Of the 20 species of this family growing in India, the majority occur in the Western Ghats, South Canara, Madhya Pradesh and Kerala. *Dicraeia dichotoma* is an algiform herb reported from the Nilgiris and Malabar Hills up to 25000 m. *Griffithella hookeriana* is a rootless, polymorphic, thalloid plant with a funnel- or goblet-shaped thallus, growing in the rivers Netravati (Karnataka, India) and Bhima (Maharashtra, India) (Vidyashankari and Mohan Ram 1987). *Podostemon ceratophyllum* is a water plant, an indicator of clean streams. It also indicates the impact of dam building, strip mining, and all kinds of water pollution in the rivers of eastern North America (Meijer 1976). It is difficult to collect these plants once abundant in Ottawa river, Canada, because pollution is believed to have exterminated them.

Taxonomic Considerations of Podostemaceae and the Order Podostemales

The family Podostemaceae has been variously treated by different taxonomists. Engler (1930a) considered it to be close to Saxifragaceae. Hutchinson (1926) placed Podostemaceae and Hydrostachyaceae in the order Podostemales and pointed out that these are highly reduced and apetalous forms of Saxifragaceae. P. Maheshwari (1945), on the basis of embryological features, said that it is "almost certain that the Podostemaceae are much reduced apetalous derivatives of the Crassulaceae".

Cronquist (1968) retained only one family in his order Podostemales, and considered Podostemaceae to be related to the Crassulaceae. However, Crassulaceae may seem to be an unusual starting point for a group of aquatics; one member of this family, *Tillaea aquatica*, is semi-aquatic. This plant may not be the direct ancestor of the Podostemales, but it shows that the Crassulaceae has the potentiality to adapt to aquatic habit.

Dahlgren (1975a) recognised 2 families—Tristicaceae and Podostemaceae—and derived them from the Saxifragales, Takhtajan (1980, 1987) derived Podostemaceae from Crassulaceae-like ancestors. He divided the family into two subfamilies—Tristichoideae and Podostemoideae.

Subramanyam and Sreemadhavan (1971) and Nayar (1986b) consider the family Tristicaceae to be distinct from Podostemaceae (Table 30.2.1).

Table 30.2.1 Comparative Data for Tristicaceae and Podostemaceae

Tristicaceae	Podostemaceae
1. Simple branches	Branches in the axil of lower stipule
2. Simple estipulate leaves	Complex stipulate leaves
3. Simple or slightly thalloid secondary shoots	Secondary shoots complex with leaves on their margin
4. Spathe absent	Spathe present
5. Perianth present	Perianth absent
6. Staminodes absent	Staminodes present

Dahlgren (1980a, 1983a) regarded the Podostemales to be so specialized that they could not be associated with any other superorder and therefore erected a distinct superorder Podostemiflorae. The specialized habit and lack of endosperm and presence of silica bodies are some of the interesting features of this family. Close relationships have been proposed with Hydrostachyaceae (Hutchinson 1973), Saxifragaceae (Engler 1930a) and Crassulaceae (Mauritzon 1933). Embryologically, it is close to Crassulaceae (Herr 1984).

The above account clearly indicates that the Podostemaceae are closely linked with the Saxifragales, and Crassulaceae in particular. Datta (1988) recognises three subfamilies—Podostemoideae, Tristicoideae and Weddelinoideae—which differ from each other in the presence or absence of spathella, and number and union of perianth segments.

In our opinion, formation of three subfamilies is unwarranted. Various features examined justify the erection of an independent order, Podostemales, with a single family, Podostemaceae.

31

Order Geraniales

The order Geraniales comprises three suborders and nine families: Limnanthaceae, Oxalidaceae, Geraniaceae, Tropaeolaceae, Zygophyllaceae, Linaceae, Erythroxylaceae, Euphorbiaceae and Daphniphyllaceae.

Most families are predominantly herbaceous except some members of Erythroxylaceae, Euphorbiaceae and Daphniphyllaceae. The members of Geraniales are characterised by obdiplostemonous stamens, pendulous ovules with a ventral raphe, and the micropyle pointing upwards or erect ovules with a dorsal raphe and the micropyle pointing downwards. Ovary syncarpous, the number of carpels vary, styles often persistent and seeds normally without endosperm.

Limnanthaceae

A small family of only 2 genera, *Limnanthes* and *Floerkea*, and 11 species, all indigenous to North America. *Floerkea* is a monotypic genus.

Vegetative Features. Mostly annual herbs of aquatic habitat. Stem weak and flaccid, remains green, and rarely branched in *Floerkea*; stouter and more branched near the base in *Limnanthes*. Leaves alternate, petiolate, pinnately dissected, estipulate, pinnules also deeply cut (Fig. 31.1A).

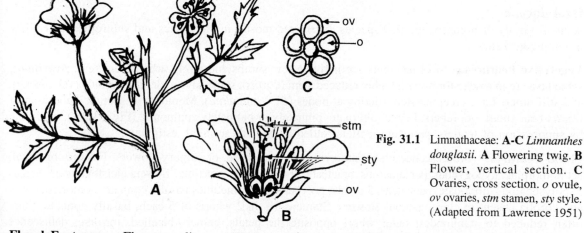

Fig. 31.1 Limnathaceae: **A-C** *Limnanthes douglasii*. **A** Flowering twig. **B** Flower, vertical section. **C** Ovaries, cross section. *o* ovule, *ov* ovaries, *stm* stamen, *sty* style. (Adapted from Lawrence 1951)

Floral Features. Flowers solitary axillary on long pedicels (Fig. 31.1A); bisexual, actinomorphic, hypogynous, ebracteate, ebracteolate. Sepals 3- to 5-merous, valvate, free or connate at the base, persistent; petals 3 or 5, contorted, polypetalous, usually with distinct limb and claw, sometimes alternate with nectaries; stamens twice as many as the petals, in two whorls, outer one alternate with the petals; anthers bicelled, dehisce longitudinally. Gynoecium 3- to 5-carpellary, ovaries apocarpous, styles gynobasic, connate into a common one from as many branches as the number of ovaries (Fig. 31.1B), stigma capitate; ovaries superior, unilocular, placentation basal-parietal and only one ovule (Fig. 31.1C). Fruit a l-seeded achene, seed non-endospermous, with straight embryo.

Anatomy. Stem (in a transection) shows 6- or 7-layered cortex of large round cells with thin walls.

Vascular bundles 8–10 in a ring. Phloem poorly developed; interfascicular cambium absent. Leaves dorsiventral, stomata on both the surfaces but more on the lower surface.

Embryology.   Pollen grains tetracolporate and shed at 2-celled stage. Ovules anatropous, unitegmic, tenuinucellate (Johri 1970a). An unusual type of tetrasporic embryo sac, the mature embryo sac contains only the egg apparatus and the polar nuclei. Endosperm formation of the Nuclear type; in *Floerkea* a caenocytic pouch is formed adjacent to the funicular side. Embryo straight, with large upwardly extended cotyledons; cotyledons bifurcate in *Floerkea* (Johri et al. 1992).

Chromosome Number.   Basic chromosome number is x = 5.

Chemical Features.   Unsaturated fatty acids present in seed oil of *Limnanthes douglasii* (Shorland 1963). The seeds also contain a thioglucoside, glucolimnanthin (Kjaer 1963).

Important Genera and Economic Importance.   There are only two genera, *Limnanthes* and *Floerkea*, and these have no economic importance.

Taxonomic Considerations.   The genus *Limnanthes* was included in the Geraniaceae by Bentham and Hooker (1965a), but later was raised to the family rank—Limnanthaceae. It was included in the Geraniales (Hutchinson 1969, 1973, Melchior 1964), or in the Sapindales (Lawrence 1951). The assignment in the Geraniales is accepted by most taxonomists. However, on the basis of embryological features P. Maheshwari and Johri (1956) and Johri (1970a) suggested the erection of the order Limnanthales. Takhtajan (1987) also treats Limnanthaceae as the only member of the order Limnanthales. This is quite appropriate because (1) Limnanthaceae are succulent marshy herbs, (2) flowers always actinomorphic, bisexual, (3) with gynobasic style and free ovaries, and (4) ovules unitegmic, tenuinucellate.

Oxalidaceae

A small family of 6 genera and 900 species distributed mostly in the tropics and subtropics, and a few in temperate zones.

Vegetative Features.   Most members are herbs, a few shrubs and rarely arborescent (e.g. *Averrhoa*), some taxa are in rosette forms with highly reduced stem (*Oxalis*), some have underground bulbs (*O. bowiei*), and still others have creeping stem rooting at nodes (*O. corniculata*). Members of section Tuberosae of *Oxalis* bear small root tubers. Leaves alternate, pinnately or palmately compound (Fig. 31.2A), or simple by suppression of leaflets, petiolate; leaflets folded in bud and at night, estipulate.

Floral Features.   Inflorescence umbellate cymes, or racemose, or solitary flowers. Flowers bisexual, actinomorphic, hypogynous, pentamerous, bracteate, ebracteolate; sometimes flowers cleistogamous. Sepals 5, polysepalous, imbricate, persistent; 5 petals, polypetalous, sometimes basally connate, contorted; calyx and corolla absent in cleistogamous flowers. Stamens 10, in 2 whorls of 5 each, basally connate, 5 are often reduced to staminodes; outer whorl opposite the petals, anthers bicelled, introrse, dehiscence longitudinal. Gynoecium syncarpous, pentacarpellary, ovary superior, pentalocular with axile placentation, ovules 1 or more in each locule; styles 5, free, persistent, stigmas terminal and mostly capitate. Fruit a loculicidal capsule or a berry (*Averrhoa*); seeds sometimes arillate, the aril separates from testa elastically and expels the seed explosively from the capsule, e.g. *O. acetocella* (Fig. 31.2B, C); embryo straight, endosperm present.

Anatomy.   A transection of stem shows a ring of collateral bundles. Vessels in multiples of 4 or more cells, perforations simple, intervascular pitting alternate and large. Parenchyma typically vasicentric and scanty. Rays uniseriate, heterogeneous or almost homogeneous. Leaves usually dorsiventral, stomata Rubiaceous type, calcium oxalate crystal and crystal sand occur.

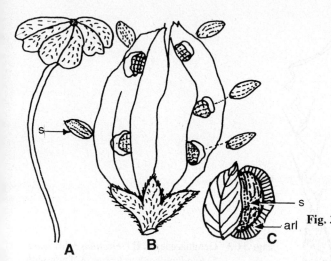

Fig. 31.2 Oxalidaceae: **A-C** *Oxalis acetocella*. **A** Leaf. **B** Fruit. **C** Seed with aril. *arl* aril, *s* seed. (Adapted from Hutchinson 1969)

Embryology. Pollen grains 3-colporate and 2-celled at the shedding stage (L.L. Narayana 1970a). Ovules anatropous, bitegmic, tenuinucellate; crassinucellate in *Averrhoa bilimbi* and *A. carambola* (Johri et al. 1992). Embryo sac of Polygonum or Allium type, 8-nucleate at maturity. Endosperm formation of the Nuclear type. Both the integuments take part in the formation of seed coat.

Chromosome Number. Basic chromosome number is x = 5 – 12.

Chemical Features. Rich in oxalic acid.

Important Genera and Economic Importance. *Oxalis* is one of the important genera with 800 species; some are weeds in the garden and some others are grown as ornamentals. *Averrhoa* spp. are large trees cultivated for their sour edible fruits.

Taxonomic Considerations. The Oxalidaceae was treated as a subfamily of the Geraniaceae by Bentham and Hooker (1965a). Wettstein (1935), Hutchinson (1973) and Takhtajan (1987) consider this family to be of independent status. However, the Oxalidaceae resembles the Geraniaceae in anatomical and many embryological features, and should be retained in the order Geraniales.

Geraniaceae

A family of 11 genera and 750 species, distributed mainly in the temperate zone.

Vegetative Features. Mainly herbaceous, sometimes suffrutescent or shrubby (e.g. *Sarcocaulon*). Stems often fleshy, hairy (thorny in *Sarcocaulon*). Leaves alternate or opposite, compound or simple and lobed, venation mostly palmate, margin dentate; petiolate, stipulate, surface hairy (Fig. 31.3A).

Floral Features. Inflorescence umbellate cyme of a few flowers, often reduced to a 1-flowered peduncle. Flowers actinomorphic or zygomorphic (*Pelargonium* and *Erodium*), bisexual, hypogynous, bracteate, ebracteolate. Sepals 5, polysepalous, imbricate (sepals 4 in *Viviania* and 8 in *Dirachma*); corolla usually of 5 petals, polypetalous, normally alternate with nectaries (number of petals may vary from 8, 4, 2 or none), imbricate. Stamens typically 5–15, in 1 to 3 whorls of 5 each, with stamens of 1 or 2 whorls sometimes reduced to staminodes, often basally connate. Gynoecium syncarpous, 3–5-carpellary (8 in *Dirachma*); ovary 3–5-locular with axile placentation, ovules usually 1 or 2 in each locule (many in *Balbisia*), pendulous; styles 3 to 5, slender, beak-like, stigmas as many as the styles and ligulate, rarely capitate. Fruit septicidal or loculicidal (*Viviania*) capsule, separate into 1-, 2- or many-seeded, usually

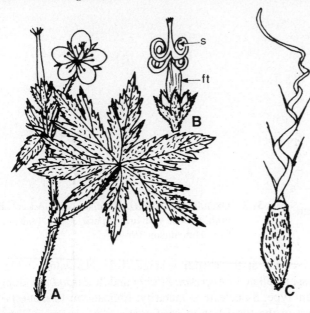

Fig. 31.3 Geraniaceae: **A, B** *Geranium sylvaticum*, **C** *Erodium pachyrrhizum*. **A** Twig with flower and fruit. **B** Dehisced fruit. **C** Seed. *ft* fruit, *s* seed. (Adapted from Hutchinson 1969)

dehiscent mericarps, the styles remain attached to the ovarian beak and the basal portion recurves elastically from the base upwards (Fig. 31.3B) and sometimes spiral as in *Erodium* (Fig. 31.3C).

Yeo (1985) reported three main types of seed discharges:

a) *Geranium* subgenus *Geranium*—Ballistic expulsion of seed from mericarp.
b) *Geranium* subgenus *Robertium*—Forcible discharge of the mericarp with the seed in it, separately from the awn—'carpel projection'.
c) *Geranium* subgenus *Erodioideae*—Seed-containing mericarp is thrown off with the attached awn which becomes helically coiled—*Erodium* type.

Seeds with curved embryo, the radicle incumbent on the folded or convolute cotyledons; endosperm usually absent in mature seed (present in *Biebersteinia*).

Anatomy.   The vascular bundles in stem are widely separated and arranged in 1 or 2 rings in different species. The xylem and phloem form close cylinders in mature stems of shrubby species. Vessels solitary and in multiples and clusters; sometimes ring-porous, simple or multiperforate plates and alternate to opposite pitting. Parenchyma scanty, paratracheal. Rays, when present, heterogeneous; fibres commonly septate. Both simple and glandular hairs occur on leaf and stem surface. Glandular leaf teeth seen in *Geranium robertianum*. Leaves isobilateral or dorsiventral. Stomata Ranunculaceous type, present on both surfaces or confined to lower surface only.

Embryology.   Pollen grains tricolporate, 2- or 3-celled at the shedding stage (L.L. Narayana 1970b). Ovules ana- to campylotropous, bitegmic, crassinucellate. In *Geranium grevellianum*, *G. nepalense* and *G. ocellatum* the ovule is anatropous when young and campylotropous at maturity (L.L. Narayana 1970b). Only outer integument takes part in the formation of micropyle. Polygonum type of embryo sac, 8-nucleate at maturity. Endosperm formation of the Nuclear type.

Chromosome Number.   Basic chromosome number is x = 7–14.

Chemical Features.   Leaves and stems of some members are rich in an essential oil—geranium oil. Tannin is also reported.

Important Genera and Economic Importance. The genus *Sarcocaulon*, a shrubby plant from the deserts of southwest Africa, bears thorns formed from the petioles after the lamina falls off at the end of the rainy season. The stem is covered with a yellow, elastic, inflammable layer of cork, its cells are filled with resinous contents, and this layer serves as protective layer during the dry season. *Hypseocharis* is another interesting plant. The loculicidal capsule of this taxon does not show the specialized method of dispersal of the beaked fruits of other Geraniaceae members. This taxon may be a relic of the ancestors of modern Geraniaceae that lived in the Gondowana continent before the separation of South America and Africa (Boesewinkel 1988). Eight species are reported in the subalpine zone of the Andes from Peru to Bolivia into Northern Argentina at 2000-4000 m altitute. Other important genera include *Geranium* and *Pelargonium*. Florist's geranium, *P. zonale*, and other species and varieties of this genus, are grown for their aromatic foliage and flowers. *Geranium* spp. or crane's bill and *Erodium* spp. or stork's bill, so called because of the shape of the fruits, are grown as garden ornamentals. The leaves of *P. graveolens* and *P. odoratissimum* yield geranium oil, on distillation. The roots of *G. nepalense* and *G. wallichianum* have astringent properties and are used as tanning material.

Taxonomic Considerations. Geraniaceae has been included in the order Geraniales by all phylogenists. It is a homogeneous group and shows affinities with the families Balsaminaceae and Oxalidaceae. Takhtajan (1987) removed the genera:

1. *Biebersteinia* to Biebersteiniaceae on the basis of presence of endosperm and basic chromosome number x = 5.
2. *Dirachma* to Dirachmaceae as it has 8 calyx lobes and a 8-carpellary ovary.
3. *Balbisia* (= *Ledocarpon*) and *Wendtia* to Ledocarpaceae on the basis of numerous ovules and basic chromosome number x = 9.
4. *Viviania* (of Geraniaceae s.l.) and *Araeoandra*, *Caesarea* and *Cissarobryon* to Vivianiaceae as they have 4 sepals and petals, a loculicidal capsule and the basic chromosome number x = 7.

Therefore, Geraniaceae s.s. comprises only 6 genera according to Takhtajan (1987). According to Bortenschlager (1967), palynologically only *Pelargonium*, *Monsonia* and *Sarcocaulon* belong to Geraniaceae proper. He, too, recognised the families Biebersteiniaceae, Ledocarpaceae and Vivianiaceae.

To recognize Geraniaceae as a natural and homogeneous taxon, therefore, the above-mentioned genera should be removed from this family.

Tropaeolaceae

A small family of 3 genera and 92 species distributed in the Andes and other parts of Central and South America.

Vegetative Features. Succulent, often scandent herbs with watery acrid sap with an unpleasent smell. Stem twining or clambering. Leaves alternate, petiolate, simple, peltate, sometimes lobed or dissected, estipulate (Fig. 31.4A).

Floral Features. Flowers solitary axillary on long pedicels, bisexual, zygomorphic, hypogynous, ebracteate, ebracteolate and spurred, Calyx of 5 sepals, bilabiate, the dorsal one forms the spur, and the other 4 free, imbricate. Corolla of 5 petals, polypetalous, with distinct limb and claw, imbricate, upper 2 petals differ from the lower 3 in size and pattern. Stamens 8 in 2 whorls of 4 each, free, unequal; anthers bicelled, longitudinal dehiscence. Gynoecium syncarpous, 3-carpellary, ovary trilobed (Fig. 31.4B), 3-loculed with axile placentation (Fig.31.4C) and 1 pendulous ovule per locule; style 1, apical, stigmas 3, linear. Fruit a 3-seeded schizocarp (Fig. 31.4D), each mericarp separates from the central axis, and is indehiscent, usually rugose. Seed with straight embryo and without endosperm.

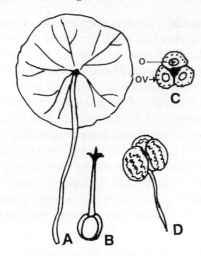

Fig. 31.4 Tropaeolaceae: **A-D** *Tropaeolum majus*. **A** Peltate leaf. **B** Gynoecium. **C** Ovary, cross section. **D** Fruit. *o* ovule, *ov* ovary. (Original)

Anatomy.   Vascular bundles in young stems separate and arranged in a single ring. Vessels up to 80 μm in diameter, those formed earlier with spiral thickenings, but later ones show small bordered pits; perforations simple, sometimes reticulate. Pith occupies a large area. Leaves dorsiventral, stomata always on the lower surface, rarely on upper surface; Ranunculaceous type.

Embryology.   Pollen grains tricolpate, 3-celled at the shedding stage (L.L. Narayana 1970c). Ovules anatropous, bitegmic, tenuinucellate. Polygonum type of embryo sac, 8-nucleate at maturity. The endosperm formation is of the Nuclear type. The two uppermost cells of the proembryo undergo repeated divisions to produce a basal mass of cells which form an extensive suspensor haustorial system (Johri et al. 1992).

Chromosome Number.   Basic chromosome number is x = 12–14.

Chemical Features.   Myrosin cells are present in roots and axis and to some extent in the leaves.

Important Genera and Economic Importance.   The most important genus is *Tropaeolum*, which is grown as an ornamental.

Taxonomic Considerations.   Bentham and Hooker (1965a) included the genus *Tropaeolum* in the family Geraniaceae, but later workers treat it as a distinct family. Anatomically, the true Geraniaceae are different from the Tropaeolaceae. The well-developed ring of mechanical tissue in the pericycle—a characteristic feature of Geraniaceae—is absent in the Tropaeolaceae. Also, the myrosin cells reported from this family are absent in the Geraniaceae. Thus, anatomically and chemically the two families are distinct. Takhtajan (1987) prefers to place the Tropaeolaceae in a separate order of its own, Tropaeolales, which is not necessary.

Zygophyllaceae

A family of 22 genera and 220 species, distributed mainly in the tropics and subtropics, arid zones in particular, extending into temperate regions.

Vegetative Features.   Mostly perennial herbs and shrubs, rarely trees, e.g. *Guaiacum* spp.; often prostrate, e.g. *Tribulus terrestris* (Fig. 31.5A). Branches fleshy, often jointed at nodes. Leaves opposite, or alternate, pinnately compound, sometimes 2-foliolate as in *Zygophyllum fabago* (Fig. 31.5D), or simple as in *Peganum*, *Tetradiclis*, *Malacocarpus*, *Nitraria* and *Sericodes*; often fleshy to coriaceous, stipulate, stipules coriaceous, hairy, fleshy or spinescent, persistent.

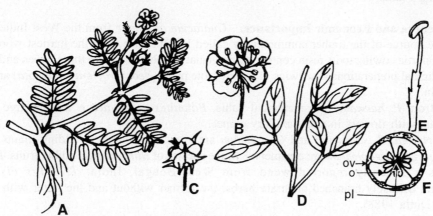

Fig. 31.5 Zygophyllaceae: **A-C** *Tribulus terrestris*, **D-F** *Zygophyllm fabago*. **A** Flowering twig. **B** Flower. **C** Fruit. **D** Flowering twig of *Z. fabago*. **E** Stamen. **F** Ovary, cross section. *o* ovule, *ov* ovary, *pl* placenta. (**A-C** Original, **D-F** adapted from Hutchinson 1969)

Floral Features. Inflorescence cymose, or flowers solitary or in pairs. Flowers bisexual (unisexual and then plants dioecious as in *Neoluederitzia*), actinomorphic or rarely zygomorphic, hypogynous, ebracteate, ebracteolate (Fig. 31.5B). Sepals 5–(4), polysepalous, sometimes basally connate, imbricate. Petals as many as sepals, rarely absent, as in *Miltianthus* and *Seetzenia*, polypetalous, a convex or depressed disc present. Stamens numerous, in 1, 2 or 3 whorls of 5 stamens each, often unequal, outer whorl opposite the petals; filaments usually with basal scales (Fig. 31.5E), anthers bicelled, introrse, dehiscence longitudinal. Gynoecium syncarpous, 4- to 5-carpellary, ovary as many loculed, superior, with axile placentation (Fig. 31.5F), ovules 2 to many (rarely 1) on each placenta, pendulous; ovary furrowed, angled or winged; style 1, angular or furrowed, stigma 1 and simple. Fruit a 4- or 5-angled or winged, indehiscent drupaceous berry (Fig. 31.5C) or breaking septicidally into 5 cocci, or loculicidally 5-valved. Seed 1 or more per locule, pendulous, with a straight or curved embryo and scanty endosperm; endospermous seeds observed in *Larrea*, *Malacocarpus*, *Metharme*, *Viscainoa*, *Bulnesia*, *Fagonia*, *Peganum* and a few others. Seeds dark brown, spongy and albuminous in *P. harmala* (Kapil and Ahluwalia 1963).

Anatomy. Vessels very small to moderately large, solitary, as in *Guaiacum* and *Larrea*, or in large clusters of solitary vessels, as in *Balanites*, or with a few multiples, as in *Bulnesia* and *Sericodes*, perforations simple. Parenchyma apotracheal, diffuse or in uniseriate bands, rays narrow, short and storied or all large and not storied; homogeneous.

Embryology. Pollen grains tricolpate and shed at 2- or 3-celled stage. In *Peganum* and *Guaiacum* some of the pollen grains germinate in situ. Ovules ana-, hemiana-, or campylotropous, bitegmic, crassinucellate, with a prominent integumentary tapetum. Polygonum type of embryo sac, 8-nucleate at maturity. Endosperm formation of the Nuclear type and wall formation proceeds from micropylar end. Ruminate endosperm occurs in *Guaiacum* (Masand 1970). Both the integuments form the seed coat.

Chromosome Number. Basic chromosome number is x = 6, 8–13.

Chemical Features. Benzylisoquinoline alkaloids are in general absent in Zygophyllaceae members, except in *Peganum harmala* which contains quinazoline, hermaline, yageine and harmine, which are used as drugs and are psycomimetics, i.e. act as hallucinogens. Piritol occurs in *Zygophyllum fabago* (Plouvier 1963). Zygophyllaceae seeds are rich in fats with palmitic, oleic and linoleic acids (Shorland 1963). Presence of *P. harmala* indicates accumulation of potassium nitrate in the soil (Kapil and Ahluwalia 1963).

Important Genera and Economic Importance. *Guaiacum officinale* from the West Indies and South America is the source of the timber commercially called lignum vitae. It is the hardest wood with self-lubricating properties owing to its resin content. Gum guaiacum obtained from this species and *G. sanctum* is used in medicinal preparations. The twigs of the creosote plant (*Neoschroetera tridentata*) are the source of yellow resin.

Alkaloids from *P. harmala* have medicinal value. *Balanites aegyptiaca* is a spiny tree of arid and semi-arid areas, with flowers in small axillary cymes.

Bulnesia arborea, *B. sarmientii* and *G. sanctum* are some other timber-yielding plants. Species of *Larrea, Tribulus* and *Zygophyllum* are ornamental. Some species of *Nitraria* yield edible fruits. *Kallstroemia pubescens* is a recently reported weed from West Bengal, India. *Tribulus cistoides* and *T. terrestris* are diffusely branched prostrate herbs, the former without and the latter with spiny fruits (Fig. 31.5C) (Dutta 1988).

Taxonomic Considerations. Most of the earlier taxonomists have retained Zygophyllaceae in the Geraniales (Lawrence 1951). Hutchinson (1973) places it in the Malpighiales, and Takhtajan (1987) in the Rutales. Cronquist (1968) observed that "although the Zygophyllaceae apparently have no very close allies, their affinities clearly lie with the Sapindales-Geraniales-Linales-Polygalales complex". He prefers to retain this family in the primitive and already somewhat heterogeneous order Sapindales, the other three orders being more homogeneous. However, in Sapindales this family occupies an isolated position.

Balanites and *Peganum* are two controversial genera. Hutchinson (1973), Willis (1973), Corner (1976), Dahlgren (1983a) and Takhtajan (1987) separate *Balanites* in a distinct family, Balanitaceae. Some authors treat it as a member of Simaroubaceae (Cronquist 1981, Datta 1988). According to Record (1921), wood structure of *Balanites* has no resemblance to *Bulnesia, Guaiacum* and *Porlieria* of Zygophyllaceae but it has many less obvious features in common with these genera. Heimsch (1942) considered that, apart from the rays, the wood anatomy of *Balanites* suggests affinity with the Zygophyllaceae rather than with the Simaroubaceae. Masand (1970) observes that while there are several embryological resemblances between this genus and the Zygophyllaceae, *Balanites* has certain distinct features of its own.

The family Peganaceae includes the only genus *Peganum*. This assignment is accepted by Souéges 1953), Rau (1962), Dahlgren (1983a) and Takhtajan (1987). According to Masand (1970), this genus also has some distinct embryological features. On the basis of floral anatomy also, Nair and Nathawat (1958) favour its placement in the Zygophyllaceae.

It appears, however, that neither the anatomical nor the embryological features of these genera are significant enough to treat them as separate families. Therefore, it would be better to recognise them as two separate subfamilies—Balanitoideae and Peganoideae—as envisaged by Melchior (1964).

Linaceae

A family of 6 genera and 250 species, distributed mostly in the temperate regions of both northern and southern hemispheres.

Vegetative Features. Annual herbs, rarely shrubs. Stem herbaceous, soft, smooth. Leaves alternate or opposite, sometimes whorled (*Linum grandiflorum*), simple, entire, sessile or shortly petioled, stipules present or absent.

Floral Features. Inflorescence dichasial cyme or cincinnus, sometimes appear racemose. Flowers bisexual, actinomorphic, hypogynous, bracteate, ebracteolate. Calyx of 5 free or basally connate sepals, imbricate; corolla of 5 petals, rarely 4, free, contorted in bud, often with distinct limb and claw, claw naked or crested, early deciduous. Stamens 5, alternate with 5 or 10 toothlike staminodes; anthers bicelled, introrse, dehiscence longitudinal. Gynoecium syncarpous, pentacarpellary, ovary 5-locular, rarely 10-locular by intrusion of carpel midrib; superior, placentation axile, ovules typically 2 in each locule. Styles as many as ovary locules, free, filiform, each terminated by a capitate stigma. Fruit is a septicidal capsule or drupe (as in *Hugonia*) surrounded by persistent calyx. Seed with a straight embryo, endosperm absent or scanty (present in *Linum*).

Anatomy. Vessels exclusively solitary except in *Hugonia*, perforation plates scalariform or simple. Parenchyma much varied in different genera, rays up to 2–5 cells wide, markedly heterogeneous. Leaves generally dorsiventral, hairs of different types observed. Stomata Rubiaceous type, usually confined to the lower surface but present on both surfaces in some taxa.

Embryology. Pollen grains 3-celled at shedding stage. Exine variously ornamented in different genera. Ovules anatropous, bitegmic, crassinucellate. Embryo sac of Polygonum type, 8-nucleate at maturity. Endosperm formation generally of the Nuclear Type; Helobial type has been reported (Dorasami and Gopinath 1945) in *Linum mysorense*. Chalazal endosperm haustoria reported in *L. grandiflorum, L. rubrum, L. perenne, Reinwardtia trigyna* and *Anisadenia saxatilis*. Both integuments take part in the formation of seed coat (only outer integument in *Hugonia*).

Chromosome Number. Basic chromosome number is x = 6–11.

Chemical Features. *L. usitatissium* is rich in cyanogenic glycosides, linamarin and lotaustralin (Harborne and Turner 1984). The seeds of *Linum* spp., linseed, are rich in linolenic acid, and seed mucilage contains acidic polysaccharides (Harborne and Turner 1984).

Important Genera and Economic Importance. *Linum usitatissimum* yields the flax fibre, and seeds of some varieties yield linseed oil. The oil cake is used as cattle feed. *Ctenolophon parviflorum* of Malaya yields a hard and durable timber. *Hugonia obtusifolia* and *H. platysephala*, both from Africa, yield edible fruits. Some species of *Linum* and *Reinwardtia* are cultivated as ornamentals.

Taxonomic Considerations. The phyletic position of Linaceae is controversial. Bessey (1915), Rendle (1925), Wettstein (1935), and Lawrence (1951) placed it in the Geraniales. Hallier (1912) included it in the Guttales, Hutchinson (1973) in the Malpighiales and Cronquist (1968, 1981) in the Linales. Melchior (1964) places it in the Geraniales.
Takhtajan (1987) also places it in the Linales but separates the following genera in distinct families:

1. *Indorouchera, Roucheria, Philbornea, Hebepetalum* and *Hugonia* in Hugoniaceae. There are anatomical differences between the two families and Hugoniaceae also lacks a floral disc and has drupaceous fruits; basic chromosome number is x = 13.
2. *Ctenolophon* in Ctenolophonaceae. It is anatomically more primitive and bears a closer relationship to the Humiriaceae. This genus has opposite leaves, nut-like fruit and an extra-staminal disc. L.L. Narayana and Rao (1971), however, consider this genus to be allied to the Linaceae, due to exomorphic, floral and wood anatomical resemblances.
3. *Cyrillopsis, Ixonanthes, Ochthocosmus* and *Phyllocosmus* in Ixonanthaceae. According to L.L. Narayana (1970d), the two genera, *Ochthocosmus* and *Ixonanthes*, are different from other members of the Linaceae only in having a massive nucellus, and need not therefore be placed in a separate family. It has an intra-staminal disc and drupaceous fruits.

With these segregates, Linaceae is a heterogeneous family. Linaceae including only the 6 genera— *Anisadenia, Hesperolinon, Linum, Radiola, Reinwardtia* and *Tirpitzia*—is a natural group (see Takhtajan 1987). Its nearest related family is Erythroxylaceae.

Erythroxylaceae

A small family of 4 genera and 260 species distributed largely in the American tropics, mostly in South America and a few in Africa.

Vegetative Features.   Shrubs or small trees, no herbs. Stem woody, sometimes with small tubercles on surface. Leaves alternate (opposite in *Aneulophus*), involutely folded in bud, simple, entire, short petioled and stipulate, stipules interpetiolar.

Floral Features.   Inflorescence thyrse or axillary fascicles, or solitary axillary flowers. Flowers bisexual, actinomorphic, hypogynous, bracteate. Calyx of 5, rarely 6, gamosepalous, campanulate, imbricate, persistent sepals. Corolla of 5 petals, rarely 6, rotate, polypetalous, convolute or imbricate, clawed with bifid ligulate appendages on inner side. Stamens 10, in two whorls of 5 each, unequal in size, monadelphous to form a basal tube; anthers bicelled, dehiscence longitudinal. Gynoecium syncarpous, tricarpellary, ovary superior, trilocular, but usually only 1 locule develops into fruit; placentation axile, ovules 1 or 2 per locule; styles 3, free or basally connate, stigma capitate, clavate, or stigmatic surface obliquely depressed; heterostyly reported in *Erythroxylon coca* (Ganders 1979). Fruit drupaceous or berry. Seed with a straight embryo, endosperm present and fleshy, rarely absent.

Anatomy.   Vessels small to medium-sized, occasionally tending to form long radial multiples, perforations simple, intervascular pitting alternate and very small. Parenchyma paratracheal; rays 2 to 5 cells wide, heterogeneous. Leaves generally dorsiventral; epidermal cells mucilaginous in some species of *Aneulophus* and *Erythroxylon*; lower epidermis papillose in some species. Stomata Rubiaceous type, confined to the lower surface; crystals abundant.

Embryology.   Pollen grains tricolpate and 3-celled at shedding stage; exine pattern uniform in the family (stenopalynous). Ovules anatropous, bitegmic, crassinucellate. Embryo sac of Polygonum type, 8-nucleate at maturity. Endosperm formation of the Nuclear type, ultimately becoming cellular. Both integuments take part in the formation of seed coat. Ganders (1979) reports *E. coca* to be a distylous species with strong self-compatibility linked with floral dimorphism.

Chromosome Number.   Basic chromosome number is x = 12.

Chemical Features.   *Erythroxylon coca* of this family is the source of the drug cocaine, which is a mixture of the alkaloids cocaine ($C_{17}H_{21}O_4N$), tropococaine, cinnamylcocaine, truxillines, and benzoylecgonine. It has anaesthetic properties.

Important Genera and Economic Importance.   Two species of *Erythroxylon—E. coca* or Bolivian coca and *E. truxillense* or Peruvian coca are important sources of cocaine. Chewing of the leaves of these plants is habit-forming and it enables the person concerned to work hard for long periods, without taking any food, drink or rest. It is an evergreen shrub or small tree, growing best at 900–2750 m altitude. Leaves dark-green, leathery and ovate. The drug cocaine is used medicinally as an anaesthetic, particularly in dentistry. Overconsumption of this substance is harmful.

Taxonomic Considerations.   Bentham and Hooker (1965a) and Hallier (1912) included Erythroxylaceae in the Linaceae and Hutchinson (1973) in the Malpighiales. Most other taxonomists treat it as a distinct family and place it in the Geraniales (Engler and Diels 1936, Lawrence 1951, Melchior 1964, Takhtajan 1966). Cronquist (1981) and Takhtajan (1987) regard it as a member of the order Linales. According to Cronquist (1968), the Linales is a simple-leaved offshoot of the Sapindales, parallel to the Geraniales and Polygalales.

Heimsch (1942) remarks that on anatomical basis—general lack of scalariform perforation plates—the Erythroxylaceae should not be united with the Linaceae, even though the two families are closely related on the basis of other characters.

Therefore, a separate order Linales need not be erected. Instead, both Linaceae and Erythroxylaceae should be retained in the Geraniales.

Euphorbiaceae

A large family of 300 genera and 7500 species of cosmopolitan distribution except in the arctic and antarctic regions.

Vegetative Features.   Annual or perennial herbs, shrubs or trees, sometimes xerophytic. The genus *Euphorbia* has prostrate herbs (*E. thymifolia, E. hirta*), shrubs (*E. tirucalli, E. neriifolia*) and trees (*E. nivulia*). Other tree members are *Bischofia javanica, Mallotus philippensis, Putranjiva roxburghii, Emblica officinalis* and *Cicca acida.* Shrubs include *Jatropha gossypifolia, Poinsettia pulcherrima* and *Antidesma ghesaembilla*, and herbs are *Phyllanthus niruri, P. simplex, Croton bonplandianum* (Fig. 31.8A), *Acalypha indica* and others.

Many genera are latex-bearing (except *Bridelia, Phyllanthus, Baccaurea* and *Poranthera*); stem usually soft, herbaceous and green, sometimes modified to phyllodes as in *Xylophylla*; woody in tree members. Leaves mostly alternate, sometimes opposite, as in *E. hirta*, or whorled as in *Acalypha indica*, entire or lobed, as in *Ricinus communis* and *Jatropha* spp., stipulate, stipules modified to glandular hairs (*J. gossypifolia*) or spines (*E. milli*; Fig. 31.8B).

Floral Features.   Inflorescence shows variations. Usually the first branching of racemose type is followed by cymose types. Catkins or pendulous racemes are seen in *Acalypha indica*, 1 or 2 axillary flowers have been observed in *Emblica officinalis* and *Phyllanthus niruri* (Fig. 31.8F). An erect raceme is known in *Ricinus communis*; terminal dichasial cymes in *Jatropha* (Fig. 31.8N); simple or compound racemes in *Manihot* and in *Trewia*, the male flowers in drooping catkins and a large female flower is solitary on a long peduncle. In *Euphorbia* spp. the inflorescence gives the appearance of a single flower, cyathium. In a cyathium, a centrally situated, highly reduced female flower is surrounded by a large number of male flowers, each represented by a single stalked stamen subtended by a bract. All these flowers are enclosed within an involucre formed of 5 bracts alternating with 5 nectaries and there are 1 or 2 large bracts forming the outermost layer (Fig. 31.8J-L). The reduced male flowers are in 2–5 groups and arranged in scorpioid cymes, the oldest is nearest to the female flower.

Flowers unisexual (Fig. 31.8G, H, K, L), plants may be mono- or dioecious, male and female flowers may be borne on the same inflorescence or on separate ones, complete or incomplete, zygo- or actinomorphic, hypogynous, bracteate, ebracteolate. Caulifloral inflorescences seen in *Baccaurea* sp. Both calyx and corolla are present in *Jatropha* (Fig. 31.8N). Sepals 5, polysepalous, imbricate; petals 5, polypetalous, valvate or contorted. Stamens usually 10, in two whorls of 5 each, filaments basally connate. In female flowers the gynoecium is syncarpous, 3-carpellary, ovary superior, trilocular with axile placentation and one ovule per locule (Fig. 31.8M). In *Croton*, the female flowers may be with or without a conspicuous corolla (Fig. 31.8D, E). In *Phyllanthus* and *Ricinus*, both male and female flowers are apetalous and only a sepaloid perianth is present (Fig. 31.8G, H). In *Manihot*, the calyx is petaloid and in *Euphorbia* the flowers are without any calyx and corolla (Fig. 31.8J-L). Stamens may be 1 to numerous in male flowers (Fig. 31.8C), free or variously branched. In *Euphorbia* spp., each male flower is represented by one stalked stamen (Fig. 31.8K) which is bracteate. In *Ricinus*, there are 5 stamens that are profusely branched and each branch terminates in an anther. In *Jatropha, Crozophora* and *Phyllanthus*, the stamens are basally connate or monadelphous (Fig. 31.8H). Pistillodes are sometimes present. Anthers bicelled, dehiscence longitudinal, transverse or poricidal.

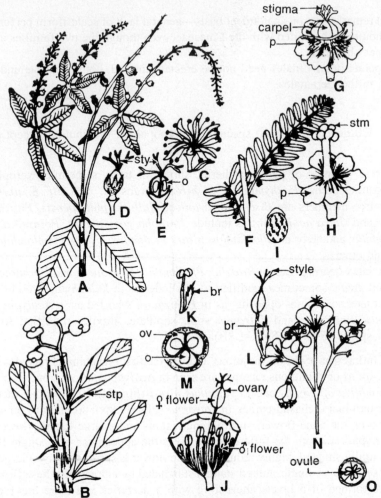

Fig. 31.8 Euphorbiaceae: **A, C-E** *Croton bonplandianum*, **F-H** *Phyllanthus niruri*, **B, J-M** *Euphorbia milli*, **I** *Ricinus communis*, **N, O** *Jatropha* sp. **A** Flowering twig of *C. bonplandianum*. **B** Of *Euphorbia milli*. **C, D** male (**C**) and female (**D**) flowers of *C. bonplandianum*. **E** Female flower, longisection. **F** Flowering twig of *P. niruri*. **G** Pistillate flower. **H** Staminate flower. **I** Seed of *R. communis*. **J** Cyathium cut open of *E. milli*. **K, L** Staminate and pistillate flower. **M** Ovary, cross section. **N** Inflorescence of *Jatropha* sp. **O** Ovary, cross section. *br* bract, *o* ovule, *ov* ovary, *p* perianth, *stm* stamen, *stp* stipule, *sty* style (Sketched by Arindam Bhattacharyya)

Gynoecium syncarpous, tricarpellary, ovary superior, trilocular with 1 or 2 ovules per locule (Fig. 3.8M, O), placentation axile; styles three, free or basally connate (Fig. 31.8D, E, G, L), each often bilobed, stigmas 3 or 6, linear or broadened, often papillate or dissected into filiform segments. Fruits usually 3-valved schizocarpic capsule, splitting into three l-seeded cocci that dehisce ventrally; seeds with straight or curved embryo and fleshy endosperm. Seeds with a caruncle in some members, e.g. *Jatropha*, *Ricinus* (Fig. 31.8I).

Anatomy. The anatomical structure exhibits a wide range of variation with the diversity of habit, and there is no important character throughout the numerous tribes into which this family is divided (Metcalfe and Chalk 1972). Vessels variable in size, sometimes even within the same genus, with simple (Crotonoideae and Glochidion type of Phyllanthoideae) or scalariform perforations (Aporusa type of Phyllanthoideae). Parenchyma abundant, apotracheal in Crotonoideae, diffuse in Phyllanthoideae (Aporusa type); absent or only a few cells about the vessels in Phyllanthoideae (Glochidion type). Rays mostly of two distinct sizes; typically 2 to 3 cells wide or exclusively uniseriate as in Crotonoideae. Leaves may be of ordinary laminate types with a distinct dorsiventral mesophyll, rolled and furrowed forms which are often centric or sometimes much reduced as in succulent species. Hairs of glandular, non-glandular and stinging types known; extrafloral nectaries also common. Stomata usually Rubiaceous type, generally Cruciferous type in *Andrachne, Aporusa, Baccaurea* and *Richria*, predominantly of Ranunculaceous type in European species of *Euphorbia*; usually confined to the lower surface, rarely on both the surfaces. Latex tubes are sometimes present in the rays.

Embryology. Pollen grains 2- or 3-celled at the time of shedding, have smooth or reticulate exine; generally triporate, nonaperturate in *Baliospermum montanum* and *Croton bonplandianum*; quadriporate in *Acalypha indica, A. alnifolia* and *Micrococca mercurialis*; 10- to 12-porate in *Melanthesa rhamnoides*. Ovules mostly ana-, hemiana- (*Chrozophora*) or orthotropous (*Breynia patens*), bitegmic, crassinucellate and have an obturator, prominent nucellar beak, hypostase and vascular supply in the integuments of some species. Embryo sac of Polygonum type (8-nucleate at maturity) is seen in majority of species. Monosporic 5-nucleate condition occurs in *Codiaeum variegatum*; three nuclei at the micropylar end form the egg apparatus, and the 2 polar nuclei form secondary nucleus. Bisporic, Allium type occurs in *E. amygadaloides* and *E. mauritanica*; Penaea type in *E. procera* and *E. palustris*; "Acalypha indica" type in *A. indica*. Peperomia type is reported in *A. lanceolata* and *E. dulcis* shows Fritillaria type; Drusa type in *Mallotus japonicus*; Adoxa type in *E. pulcherrima*; Chrysanthemum parthenium type (16-nucleate) and C. cineraraefolium (12-nucleate) type are reported in *E. epithymoides* (Johri et al. 1992). Endosperm formation of the Nuclear type and eventually becomes cellular. Seeds usually with a caruncle. Polyembryony occurs in *Alchornea* and *Euphorbia dulcis*.

Chromosome Number. Basic chromosome number is $x = 6–12$.

Chemical Features. The seeds of many Euphorbiaceae are rich in linolenic, linoleic and oleic acids such as *Antidesma diandrum, Bischofia javanica, Euphorbia heterophylla, E. marginata, Mercurialis annua* and others (Shorland 1963). Natural polyols have been reported in *E. pilulifera* and in the latex of *Hevea* (Plouvier 1963). Coumarin glycoside with aglycone aesculetin occurs in *E. lathyris* (Paris 1963); triterpenes in resins and bark of trees, and in the latex of *Euphorbia* and *Hevea*. The acrid, milky or colourless juice of most members contain triterpenoids, flavonoids and alkaloids, coumarins, cyanogenic compounds and tannins (Rizk 1987).

Important Genera and Economic Importance. *Euphorbia* is the most important genus of this family, showing much variation in habit and habitat. Many of the xerophytic euphorbias bear so much resemblance to cacti that they can be separated only by the presence of latex and pairs of stipulary spines. The genus has a remarkable floral structure not shared by many genera. There are other important genera like *Ricinus communis, Hevea braziliensis, Emblica officinalis, Croton, Jatropha, Phyllanthus* and others.

Hevea braziliensis is an important tree from the Amazon River valley, and its latex is the source of rubber. Its mature fruit is a hard, woody, trilobed capsule that dehisces violently into 3 pieces (when dry) throwing the seeds to a distance away from the mother tree. The seeds are recalcitrant due to their high moisture content and loose viability if stored under open-air condition (Thomas et al. 1996). *Croton tiglium* and *Ricinus communis* are two other important plants which yield croton oil and castor oil, respectively. Both the oils have many commercial uses. *Hura crepitans*, commonly called the sandbox tree, is a large or medium-sized tree. Its trunk is usually covered with short, sharp spines. The male flowers are in a dense spike, and the female flower is a solitary one borne on the side of the stalk of the male inflorescence. The capsule looks like a small pumpkin and consists of about fifteen 1-seeded woody chambers; when ripe, it explodes with a loud report. The milky sap of the plant is poisonous and is often mixed with meal to stupefy fish. Another poisonous plant is *Hippomane mancinella* from Panama, Venezuela, the West Indies and South Florida.

Aleurites fordii of China is the source of tung oil used in varnishes; *A. moluccana* yields candlenut oil used as a preservative for the hulls of vessels. The bark of *Bischofia javanica* and *Bridelia retusa* are useful in tanning. Candelilla wax is extracted from the stems of *Euphorbia antisyphylitica* and *Pedilanthus pavonis*, both from Mexico and Texas. The oil from the seeds of *Givotia rottleriformis* is used as a lubricant. Ink is prepared from the ripe fruits of *Kirganelia reticulata*. Chinese tallow tree or *Sapium sebiferum* is a native of subtropical China and has been cultivated for at least 14 centuries as a seed oil crop (Seibert et al. 1986).

Some plants yield edible fruits and roots. Fruits of *Aleurites moluccana*, *Baccaurea sapida*, *Bridelia squamosa*, *Hemicyclia andamanica*, *H. sepiaria* and *Trewia nudiflora* are used for culinary purposes. The fruits of *Cicca acida* and *Emblica officinalis* are rich source of vitamin C. Starchy roots of *Manihot esculenta* are the commercial tapioca.

Some taxa are cultivated as ornamentals: *Acalypha*, *Codiaeum*, *Dalechampia*, *Jatropha*, *Euphorbia*, *Poinsettia* and others. *Baccaurea* sp. with its colourful caulifloral inflorescences is a spectacular tree from Kerala and some parts of Western Ghats in South India.

Taxonomic Considerations.   There is much controversy regarding the systematic position of Euphorbiaceae. Hallier (1912) regarded it as a member of his Passionales; Lawrence (1951), Melchior (1964) in Geraniales; Rendle (1925) and Wettstein (1935) in Tricoccae; and Bentham and Hooker (1965c) placed it in the Unisexuales under Monochlamydeae, on the basis of its floral structures. Hutchinson (1969, 1973) placed Euphorbiaceae in a separate order of its own Euphorbiales, next to the Malpighiales. According to him, it is a highly evolved family, almost comparable to the Asteraceae because of so much reduction in the floral structure. Also, the family Euphorbiaceae comprises a group of genera which have been derived from different stocks—like Tiliaceae, Sterculiaceae, Malvaceae and Celastraceae. It also bears a relationship with the Geraniales and Sapindales, on account of the nature of ovules. There is no doubt that the Ephorbiaceae has a polyphyletic origin.

Cronquist (1981) includes Euphorbiaceae in the Euphorbiales along with four other families—Buxaceae, Daphniphyllaceae, Aextoxicaceae and Pandaceae. All these families have unisexual, mostly monochlamydeous flowers, 1 or 2, bitegmic ovules per locule and copious endosperm. Takhtajan (1987) includes Dichapetalaceae in Euphorbiales, in addition to Pandaceae and Aextoxicaceae, and removes Buxaceae and Daphniphyllaceae to separate orders. The families Buxaceae, Dichapetalaceae and Daphniphyllaceae have earlier been treated as different tribes of Euphorbiaceae. They are all embryologically distinct from Euphorbiaceae, but they show common features such as: monochlamydeous, unisexual flowers, bitegmic ovules and copious endosperm. Three genera of the 4 members of the Pandaceae were earlier included in the Euphorbiaceae and the Pandaceae is a monotypic family. Forman (1966) moved these three genera back to Pandaceae. The two families are closely related. Webster (1967) agreed to this assignment and expressed similar views about the families Aextoxicaceae, Buxaceae and Daphniphyllaceae.

In our opinion, there should be a separate order, Euphorbiales to include all these 6 families—Buxaceae, Euphorbiaceae, Dichapetalaceae, Daphniphyllaceae, Pandaceae, and Aextoxicaceae.

Daphniphyllaceae

A unigeneric family of the genus *Daphniphyllum* with about 9 or 10 species, distributed in eastern Asia and Malaysia.

Vegetative Features. Trees or shrubs with alternate or sometimes closely arranged leaves at the apex of a branch, giving the appearance of a verticillate type; long-petioled, entire, oblong-lanceolate, simple, pinnately veined and estipulate leaves, more or less glaucous below.

Floral Features. Inflorescence axillary racemes with deciduous bracts. Flowers unisexual, actinomorphic, apetalous, and sometimes also without calyx, hypogynous, bracteate, pedicellate. Calyx of 3 to 6 sepals, gamosepalous, minute, imbricate, sometimes missing; corolla absent; stamens (5) 6 – 12 (14), free, with short filaments; anthers large, oblong, basifixed, latrorse, bicelled, longitudinally dehiscent. In *Daphniphyllum himalayense* (Bhatnagar and Garg 1977), the anthers are fused and the connective projects slightly beyond the anther lobes. Short staminodes present in female flowers. Gynoecium syncarpous, bicarpellary, ovary superior, imperfectly 2-loculed and with 2 to 4 parietal, pendulous ovules per locule (Bhatnagar and Kapil 1982); stylodium present, stigmas 2, thick, subsessile and recurved; stigmatic surface on the inner side. Fruit a 1-seeded (rarely 2-seeded), glaucous drupe with persistent stigmas; seeds with minute, apical, straight embryo and copious endosperm and perisperm.

Anatomy. Vessels small, solitary and numerous, perforation plates scalariform with 20 to 30 or more cross-bars; intervascular pitting scalariform to opposite but rare; parenchyma diffuse, heterogeneous, rays 2 cell wide (Bhatnagar and Garg 1977). Leaf dorsiventral, stomata confined to the lower surface; Rubiaceous type; crystals abundant.

Embryology. Pollen grains spherical, tricolpate, 2-celled at shedding stage, with a reticulate exine and granular area of membrane at the colpa (Bhatnagar and Garg 1977). Ovules anatropous, or epitropous with ventral suture, bitegmic, crassinucellate. Embryo sac of Polygonum type, 8-nucleate at maturity. Endosperm formation of the Cellular type (Bhatnagar and Kapil 1982).

Chromosome Number. Basic chromosome number is x = 8.

Chemical Features. A unique alkaloid, daphniphillin, is reported in this plant. Crystalline protein reserves are common in perisperm.

Important Genera and Economic Importance. *Daphniphyllum himalayense* is a large tree, timber is coloured and used by the tribals for marking their foreheads.

Taxonomic Considerations. Daphniphyllaceae has only one genus, *Daphniphyllum*, which was formerly included in the family Euphorbiaceae under the tribe Phyllantheae. Lawrence (1951) and Melchior (1964) include it in the Geraniales and next to the Euphorbiaceae. Hutchinson (1973) regards it as a member of the Hamamelidales and close to the Buxaceae. It is placed in the Euphorbiales by Cronquist (1981). However, Daphniphyllaceae differs from the Euphorbiaceae: (a) morphologically—absence of stipules, and drupaceous fruits; (b) embryologically—abundant endosperm and a minute embryo, endotegmic seed coat, and perisperm with abundant albuminous crystals; (c) chemically—occurrence of the alkaloid daphniphillin; and (d) anatomically—small, solitary, numerous vessels, abundant crystals, and only Rubiaceous type of stomata.

Takhtajan (1987) treats Daphniphyllaceae as the only member of the Daphniphyllales, and is of the opinion that this family is more closely related to the Hamamelidaceae than to the Euphorbiaceae. As early as 1929, Janssonius observed that the difference between Euphorbiaceae and *Daphniphyllum* in wood anatomy alone is sufficient to doubt its position in the Euphorbiaceae. He, too, considered it to be more closely related to "a large group of families that includes, for example, the Theaceae (s.l.) and Hamamelidaceae". Bhatnagar and Garg (1977) report its resemblance to Hamamelidaceae in floral, embryological and palynological features.

Therefore, Daphniphyllaceae should be treated as a separate family, in a separate order, as also suggested by Takhtajan (1987). Its position near Euphorbiaceae is also doubtful.

Taxonomic Considerations of the Order Geraniales

The order Geraniales (sensu Melchior 1964) comprises 9 families, some of the families are referred to separate orders—Linales, Polygalales, Sapindales, Euphorbiales and Daphniphyllales—by Hutchinson (1973), Cronquist (1981), Dahlgren (1983a) and Takhtajan (1987).

The family Limnanthaceae is so distinct from other members of the Geraniales that it would be appropriate to place it in a separate order, the Limnanthales. The mutual relationship of Oxalidaceae, Geraniaceae and Tropaeolaceae is generally accepted. The first two families have many overlapping features, although the typical forms are distinct enough. The Tropaeolaceae is distinct from other members, both anatomically and chemically, and is justifiably placed in a separate order, Tropaeolales. Linaceae and Erythroxylaceae are closely related and should be retained in one order. The family Daphniphyllaceae has often been included in the Euphorbiales but they are distinct enough to be treated as a separate order (Bhatnagar and Garg 1977, Bhatnagar and Kapil 1982). The Euphorbiaceae, with tricarpellary and few-ovuled ovary, simple leaves, unisexual flowers which are often highly reduced, is probably out of place in the Geraniales.

According to Cronquist (1968), who includes only herbaceous members in his Geraniales, this order is a herbaceous offshoot of the more primitive Sapindales. The Geraniales, as recognised by Melchior (1964), is highly unnatural and it appears that it should be broken up into smaller units:

a) Limnanthales including Limnanthaceae.
b) Geraniales s.s. including Oxalidaceae and Geraniaceae.
c) Tropaeolales with Tropaeolaceae.
d) Linales with Linaceae and Erythroxylaceae.
e) Monotypic Euphorbiales.
f) Monotypic Daphniphyllales.

We agree with the above arrangement.

32

Order Rutales

The order Rutales comprises 3 suborders:

(a) Suborder Rutineae includes 6 families, (b) suborder Malpighiineae 3 families, and (c) suborder Polygalineae 2 families.

Members of the Rutales are mostly trees and shrubs, a few climbers and a few herbs. Leaves simple or pinnately or palmately compound, mostly estipulate, often glandular-punctate (as in Rutaceae). Inflorescence much variable, flowers usually actinomorphic, scented and with a nectariferous interstaminal disc (absent in Polygalineae). Embryologically and anatomically, these families are allied. Pollen grains are monads and usually shed at 2-celled stage. Ovules mostly bitegmic, crassinucellate, and anatropous, hemianatropous or epitropous.

Rutaceae

A large family comprising 150 genera and 1500–1600 species, widely distributed in both temperate and tropical zones of the New as well as the Old World.

Vegetative Features. Mostly shrubs (*Citrus*) and trees (*Aegle marmelos, Feronia elephantum*), sometimes climbers, e.g. *Paraminya scandens*, a woody climber with strong axillary, recurved spines; herbs rare, e.g. *Ruta graveolens,* (Fig. 32.1A), *Boenninghausenia* sp. Stem herbaceous in *Boenninghausenia, Monnieria* and *Dictamnus*; woody below and herbaceous above in trees and shrubs. Leaves alternate or opposite, simple (*Boronia, Pitavia, Phelline*), or palmately (*Citrus*) or pinnately (*Ruta*) compound (Fig. 32.1A), sometimes reduced to spine, estipulate. In *Citrus* spp. the petiole is winged and separated from the lamina

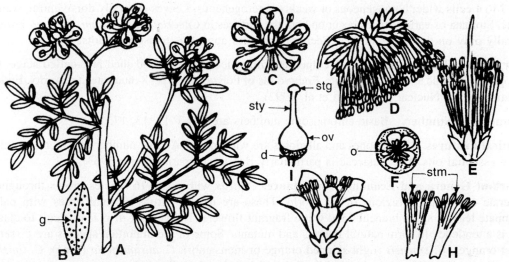

Fig. 32.1 Rutaceae: **A-C** *Ruta graveolens*, **D-F** *Diplolaena dampieri*, **G-I** *Citrus limon*. **A** Flowering twig. **B** Glandular-punctate leaflet. **C** Flower, face view. **D** Inflorescence of *D. dampieri*. **E** Flower. **F** Ovary, cross section. **G** Flower of *Citrus limon*. **H** Polyadelphous stamens. **I** Gynoecium. *d* disc, *g* gland, *o* ovule, *ov* ovary, *stg* stigma, *stm* stamen, *sty* style. (**A-F** adapted from Hutchinson 1969, **G-I** original)

by a distinct joint—often considered as a unifoliate, palmately compound leaf. Leaves glandular-punctate (Fig. 32.1B).

Floral Features. Inflorescence usually cymose, e.g. *Toddalia*, sometimes racemose—a raceme or corymb (*Murraya paniculata*), or sometimes solitary axillary as in *Triphasia aurantiola*. In the Autsralian genus *Diplolaena*, the flowers are densely grouped into a head with a 3- or 4-seriate involucre of bracts, the inner is petaloid (Fig. 32.1D,E). Epiphyllous flowers are borne in *Erythrochiton hypophyllanthus* from South America.

Flowers mostly bisexual (Fig. 32.1C,E,G), rarely unisexual as in *Evodia* and *Zanthoxylum*, actinomorphic (zygomorphic in *Dictamnus*), usually pentamerous, may be tetra- or trimerous also, e.g. in *Ruta*, the terminal flower is pentamerous and the lateral ones tetramerous. Calyx mostly of 3 to 5 sepals. poly- or gamosepalous (*Citrus*), imbricate or quincuncial; corolla of 3 to 5 petals, polypetalous (gamopetalous and campanulate in *Correa speciosa*), imbricate. Stamens 3 to 10 or more, obdiplostemonous or in 2 whorls—outer ones opposite the petals; a disc present between stamens and ovary. In *Citrus* stamens numerous and polyadelphous (Fig. 32.1G,H); in *Aegle marmelos* the number of stamens is 50 or more, in *Murraya* 10, in *Skimmia* 5, and in *Zanthoxylum* 3 to 5. All stamens attached at base or rim of the nectariferous disc (Fig. 32.I C,I), some occasionally reduced to staminodia, free or basally connate, rarely adnate to petals, usually straight and unequal. Anthers bicelled, introrse, dehiscence longitudinal (Fig. 32.1C,E,G,H), connective often with glandular apex. Gynoecium syncarpous, 5- or 4-carpellary, or carpels weakly connate or only basally or apically connate (Lawrence 1951); ovary superior, usually deeply lobed, typically 4- to 5-loculed with axile placentation (Fig. 32.1F) (unilocular with parietal placentation in *Feronia limonia*); ovules 1, 2 or more in each locule; styles as many as carpels and free or only 1, stigma 1, capitate. Fruit various—a valvate capsule or a hesperidium, or separating into mericarps or a winged berry or drupe or samara. Seed with a large, straight or curved embryo and fleshy endosperm or endosperm absent.

Anatomy. Wood ring-porous or semi-ring-porous; vessels small to medium-sized, typically in multiples and sometimes with a distinct radial or oblique pattern. Perforation mostly simple, intervascular pitting alternate. Parenchyma terminal, paratracheal, usually vasicentric, often include crystal cells. Rays uniseriate, up to 2 to 4 cells wide; homogeneous or weakly heterogeneous. Leaves generally dorsiventral, sometimes centric. Stomata of various types occur on both the surfaces in *Cneoridium dumosum* and *Ruta graveolens*, generally only on the lower surface. Secretory cavities appear as transparent dots.

Embryology. Pollen grains 2- to 8-colporate with reticulate exine and shed at 2-celled stage. Ovules anatropous, bitegmic and crassinucellate. Embryo sac of Polygonum type, 8- nucleate at maturity. Endosperm formation of the Nuclear type (Johri et al. 1992)

Chromosome Number. Basic chromosome numbers are x = 7–11, 13, 17, 19.

Chemical Features. Flavanones and alkaloids are widespread. A large number of genera are relatively rich in essential oils. The Rutaceae is particularly rich in coumarins (Price 1963).

Important Genera and Economic Importance. *Citrus* with about 16 species grows throughout the temperate and tropical regions of the world. These are mostly shrubs or small trees with palmately unipinnate leaves, spiny branches and white fragrant flowers. The genus *Citrus* is difficult to classify as there is a tendency to form natural hybrids and mutants. Some of the important species are *C. reticulata* (sweet orange), *C. sinensis* (tight-skinned orange or mousambi), *C. aurantiifolia* (lime), *C. limellioides* (sweet lime), *C. maxima* (shaddock), *C. medica* (citron) and *C. limon* (lemon). Apart from being used as fruits, they produce a large number of commercial products such as essential and fixed oils, citric acid and pectin. Fruits also find use in preparation of juices, squashes, marmalades and jellies. The juice of

C. limon is rich in vitamin C. Seeds contain varying amount of fixed oil, protein and limonin; the oil is used in soap industry. The waste pulp finds use in the production of food yeast, industrial alcohol and ascorbic acid.

Aegle marmelos is important for its medicinal use. It is a medium-sized spiny tree with tripinnate leaves and fragrant flowers. Pulp of the ripe fruit is a good laxative; unripe fruits after boiling or roasting is often used against diarrhoea and dysentry. The mucilage around the seeds is used as an adhesive. The leaves are beleived to be sacred in Hindu mythology and are offered in prayers to Lord Shiva. *Ruta graveolens* is a strongly smelling ornamental herb with yellow flowers; 'oil of rue' is distilled from its leaves. Trees of *Chloroxylon swietenia* and *Zanthoxylum flavum* yield useful timbers, commonly called satinwoods. Timber from *Flindersia brayleana* is an important hard wood from Australia, used for cabinet work, veneers, aeroplane construction and rifle stocks. Various plants have medicinal value—*Cusparia febrifuga* gives cusparia bark used as a substitute for quinine; the dried leaves of *Barosma betulina*, *B. crenulata* and *B. serratifolia* form the drug buchu.

The violet-scented 'oil of boronia' used in perfumery is derived from *Boronia megastigma*. 'Mexican elemi', an oleoresin, is obtained from *Amyris balsamifera* and *A. elemifera*. The twigs of *Glycosmis pentaphylla* and *Zanthoxylum alatum* are used as chewsticks or toothbrushes. 'Chinese box', *Murraya paniculata* is a garden ornamental, and *M. koenigii* is grown extensively for its leaves, that are used as a condiment. The roots of *Toddalia asiatica*, a spiny shrub, are the source of a yellow dye.

Taxonomic Considerations.   Bentham and Hooker (1965a) included Rutaceae in Geraniales and so did Engler and Diels (1936). But Melchior (1964) regarded it as a member of a separate order Rutales and placed it next to the Geraniales. Rendle (1925), and Hutchinson (1973) also place Rutaceae in Rutales, but the circumscription of the order varied.

Cronquist (1981) treats it as a member of the Sapindales although Takhtajan (1980, 1987) agrees with the views of earlier workers. The family Rutaceae is more allied to Meliaceae, Sapindaceae and Anacardiaceae in their exomorphic and anatomical features, and it appears that this family is better placed in an order distinct from the Geraniales. Rutales could be an appropriate order for this family.

Tetradiclis.   This genus has been included in various families like Crassulaceae, Elatinaceae, Zygophyllaceae and Rutaceae. Fenzl (1841), Hallier (1908, 1912), and Takhtajan (1966) placed it in the Rutaceae, but many taxonomists place it in Zygophyllaceae. According to Takhtajan (1987), *Tetradiclis* differs sufficiently from both Rutaceae and Zygophyllaceae to deserve the rank of an independent family, Tetradiclidaceae. This family is somewhat closer to the Rutaceae.

Cneoraceae

A small family of two genera, *Cneorum* and *Neochamaelea*, and three species distributed in Cuba, the Canary Islands and the Mediterranean region.

Vegetative Features.   Shrubs with alternate, simple, entire, leathery, estipulate leaves; glandular-punctate but inconspicuous. Bark and wood without prominent resin ducts.

Floral Features.   Flowers solitary or in corymbs; bisexual, actinomorphic with elongated torus or bolster-like (swollen or fluffy) disc, tri- or tetramerous. Stamens 3 or 4, free, latrorse; intrastaminal disc present. Gynoecium syncarpous, 3- to 4-carpellary; ovary 3- to 4-loculed, superior, ovules 2 in each locule; style 1. Fruit schizocarpic; seeds endospermous.

Anatomy.   Vessels small, mostly in clusters, with radial chains or irregular clusters; perforations simple, intervascular pitting alternate and minute. Parenchyma predominantly paratracheal, scanty in some, often storied. Rays up to 2 or 3 cells wide, with a few uniseriates, homogeneous in mature material. Leaves dorsiventral, multicellular external glands often present. Stomata Ranunculaceous type, occur sporadically on the upper and abundantly on the lower surface. Secretory cells with oily or resinous content present in stem cortex and mesophyll, rather inconspicuous.

Embryology. Pollen grains 2-celled at the time of shedding. Ovules anatropous, bitegmic, crassinucellate (Boesewinkel 1984a). Embryo sac of Polygonum type, 8-nucleate at maturity. Endosperm formation is initially Nuclear but subsequently cell formation takes place. Mature, strongly campylotropous seed is enclosed in a thick and liquified fruit wall. The cellular endosperm and embryo are rich in lipids and poor in starch (Boesewinkel 1984a)

Chromosome Number. Basic chromosome number is x = 9.

Important Genera and Economic Importance. *Cneorum* and *Neochamaelea* are the two genera of this family; economic importance not known.

Taxonomic Considerations. Bentham and Hooker (1965a) included *Cneorum* in the Simaroubaceae. Bessey (1915) placed it between Zygophyllaceae and Rutaceae in his Geraniales. Hallier (1912) derived this family from Rutaceae. Engler and Diels (1936) and Lawrence (1951) also retained Cneoraceae in Geraniales. Melchior (1964) and Cronquist (1981) place it in the Sapindales, and Takhtajan (1987) includes Cneoraceae in the Rutales. On the basis of wood anatomy, Cneoraceae are closer to the Rutaceae than to the Zygophyllaceae. Pollen-morphological studies and embryological data (Lobreau-Callan et al. 1978, Boesewinkel 1984a) reveal that this family is closely linked with the Rutaceae. Campylotrophy of seeds is rather rare but in Rutaceae it occurs in *Ruta* and *Thamnosma*. The subfamily Rutoideae, especially the tribe Ruteae, shows maximal resemblance with Cneoraceae in ovule and seed characters (Boesewinkel 1984a).

As Cneoraceae is more allied to Rutaceae and Simaroubaceae than to Geraniaceae, it is justifiable to regard it as a member of the Rutales.

Simaroubaceae

A family comprising 22 genera and 165 species, disributed mostly in the pantropical regions and a few in temperate regions.

Vegetative Features. Shrubs or trees, bark contains some bitter principles. Leaves alternate, pinnately compound (simple in *Suriana*), without pellucid dots, usually estipulate, often very large (as much as 4 ft long); plants leafless in *Holacantha*.

Floral Features. Inflorescence terminal or axillary panicles. Flowers unisexual by abortion and then the plants dioecious or less commonly bisexual, actinomorphic, hypogynous. Calyx of 3 to 8 sepals, gamosepalous, imbricate or valvate; corolla absent, e.g. *Alvaradoa* or of 5 petals, polypetalous, imbricate. Stamens as many as or twice the number of petals, borne on or at the base of the disc; filaments with a short scale at the base, anthers dorsifixed, bicelled, dehiscence longitudinal. Gynoecium on a short, broad gynophore; ovary comprises 2–5 unilocular simple pistils (only one pistil in *Guilfoylia*) with free styles in *Suriana, Cadellia* and *Rigiostachys*. In most other genera the pistils are basally connate or connate by styles into a lobed 2- to 8-carpellary and 2- to 8-loculed ovary with axile placentation and 1 or 2 ovules (rarely more) in each locule; styles 2 to 8 (1 in *Guilfoylia*), free or connate. Fruit a capsule, schizocarp or samara, or rarely drupe as in *Simarouba*; seed with straight or curved embryo and scanty or no endosperm.

Anatomy. Vessels mostly medium-sized and evenly distributed, a few species slightly ring-porous; perforations simple. Parenchyma extremely sparse or absent to abundant, in broad bands, usually with at least some parenchyma associated with the vessels. Rays multiseriate (uniseriate in *Aeschrion, Guilfoylia,* and *Quassia*), 2 to 4 cells wide, mostly homogeneous, Leaves usually dorsiventral, sometimes isobilateral or centric. Hairs mostly simple, unicellular or uniseriate. Glandular hairs, sunken glands and extrafloral nectaries also occur. Stomata mostly confined to the lower surface, usually Ranunculaceous type, (Rubiaceous in *Castela, Irvingia, Klainedoxa* and *Picrodendron*).

Embryology. Pollen grains mostly 3-colporate and 2- or 3-celled (sometimes with supernumerary nuclei as in *Balanites roxburghii*) at shedding stage; exine smooth. High percentage of pollen sterility known in *Samadera indica.* Ovules anatropous, hemianatropous or campylotropous, bitegmic and crassinucellate. Ovules unitegmic in *Suriana.* Endosperm development of the Nuclear type. Seed coat formed by both the integuments.

Chromosome Number. Basic chromosome numbers for the family are x = 8, 13–15, 18, 25 and 31.

Chemical Features. Flavonols, leuco-anthocyanins, cinnamic acids and akaloids occur in dried leaves of some Simaroubaceae members (Nooteboom 1966).

Important Genera and Economic Importance. Many genera and species are important timber trees. The bark of *Ailanthus excelsa, Simarouba amara* are often used as medicines. Some taxa such as *Balanites, Irvingia* and *Picrodendron* are taxonomically disputed genera.

Taxonomic Considerations. The Simaroubaceae is a highly heterogeneous family. It was placed in the Geraniales by Bentham and Hooker (1965a), Bessey (1915) and Engler and Diels (1936). Hallier (1912) placed it in his Terebinthinae and Wettstein (1935) in Terebinthales. Hutchinson (1973), Melchior (1964) and Takhtajan (1987) include this family in Rutales, and Cronquist (1981) in Sapindales.

Embryologically, Simaroubaceae resemble other members of the Rutales such as Burseraceae, Rutaceae and Meliaceae. It differs from Rutaceae in unisexual flowers and leaves without glandular dots, from Burseraceae in eurypalynous pollen grains and absence of lysigenous resin ducts in the bark (Webber 1936, 1941); from Meliaceae in the absence of monadelphous stamens and in bisexual flowers (rarely). These differences are important enough to treat it as a distinct family. However, the circumscription of the family Simaroubaceae is often different according to different taxonomists. Takhtajan (1987) recognises as distinct families: Surianaceae (including *Suriana, Cadellia, Rigiostachys* and *Guilfoylia*), Balanitaceae (including *Balanites*), Irvingiaceae (including *Irvingia, Ixonanthes, Allantospermum,* etc.) and Picrodendraceae (including *Picrodendron*).

Balanites aegyptiaca from Egypt is a thorny shrub or tree of very slow growth, with alternate, compound (rarely simple), coriaceous leaves, each with two leaflets. Flowers in axillary corymbs, small, green, pentamerous. Fruit an edible drupe. Six flavonoid glycosides have been identified from the leaves and branches of this plant (Maksoud and EL Hadidi 1988). In floral anatomical features and chemical features *Balanites* resembles Zygophyllaceae. On the other hand, some anatomical and palynological features also support its position near Simaroubaceae. Although some authors retain *Balanites* in the family Simaroubaceae (Datta 1988), in view of the differences recorded, it should be treated as a subfamily of Zygophyllaceae (see Zygophyllaceae).

Suriana. *Suriana* and three other genera (formerly of Simaroubaceae) have simple leaves, free styles and stigmas, unitegmic ovules and some distinct anatomical features such as stalked glands on the leaves, numerous stomata on both the surfaces of leaves, 3 to 5 subsidiary cells or typically Cruciferous type of stomata and absence of secretory canals.

It appears, therefore, that the genera *Suriana, Cadellia, Rigiostachys and Guilfoylia* can be segregated in a separate family Surianaceae.

Allantospermum. *Allantospermum* is yet another disputed taxon. The two families Irvingiaceae and Ixonanthaceae include 4 genera each. The former is allied to Simaroubaceae and the latter to Linaceae (Takhtajan 1987). A comparison of the two families is listed in Table 32.3.1

Table 32.3.1 Comparative Data for Irvingiaceae and Ixonanthaceae

Irvingiaceae	Ixonanthaceae
1 Fruit a drupe or samara	Fruit a capsule
2 Seed without wing or aril	Seed winged or arillate
3 Secretory canals present	Secretory canals absent
4 Petals deciduous	Petals persist on the fruit
5 Ovules 1 per locule	Ovules 2 per locule
6 Filaments free from the disc	Filaments attached to the disc
7 Stipules intra-petiolar	Stipules lateral

With the addition of *Allantospermum* to Ixonanthaceae, the gross morphological difference between the two families, Irvingiaceae and Ixonanthaceae, come down to (a) drupaceous or samaroid fruit (in Irvingiaceae) and capsular fruit (in Ixonanthaceae), and (b) presence or absence of secretory canals. According to Forman (1965), these characters are not sufficient to justify the two small groups being treated as distinct families. He recognised only Ixonanthaceae with two subfamilies—Ixonanthoideae and Irvingioideae. Nooteboom (1967), however, placed *Allantospermum* in Irvingioideae under Simaroubaceae because of true aril, interpetiolar stipules, wood anatomy and phenolic substances in the dry leaves. Although Hallier (1923) and Kool (1980) advocated Linaceous affinity for *Irvingia* and *Ixonanthes,* respectively, both the taxa have many features in common with those of Simaroubaceae members. With correct placement of the genus *Allantospermum,* the two small families—Ixonanthaceae and Irvingiaceae—can be combined together into a single family, Ixonanthaceae, and placed near Simaroubaceae.

Picrodendron is the only member of the family Picrodendraceae with 3 species, confined to the West Indies. These are trees with trifoliate, minutely stipulate leaves. The male inflorescences are in slender, interrupted simple or slightly branched catkins borne in the axils of leaves of the previous season's growth; female flowers are solitary axillary on slender pedicels, on the new shoots. Flowers unisexual, highly reduced, actinomorphic, hypogynous, incomplete. Female flowers with 1 to 4 sepals or calyx teeth; in male flowers stamens numerous, anthers basifixed, bicelled, longitudinally dehiscent. Gynoecium syncarpous, bicarpellary; ovary superior, bilocular with 2 ovules in each locule, apical, pendulous. Fruit drupaceous.

Bentham and Hooker (1965a), Engler and Diels (1936), Lawrence (1951), and Takhtajan (1987) do not recognize Picrodendraceae as a separate family. Melchior (1964) retained it in the Rutales (probably) because of its trifoliate leaves and drupaceous fruits. Unisexual flowers are also known in Simaroubaceae and Burseraceae of the Rutales. Cronquist (1968) included it in Juglandales because of some features common to Rhoipteleaceae, and Juglandaceae. Hutchinson (1973) placed it near Euphorbiaceae because of extremely different male and female inflorescence and flowers. According to Hayden et al. (1984), *Picrodendron* conforms more closely to the Euphorbiaceae, as supported by gross morphology, pollen structure, chromosome number and anatomical features. Takhtajan (1987) includes *Picrodendron* in Euphorbiaceae. The position of *Picrodendron* in Rutales, near Simaroubaceae, may be erroneous and its correct placement must await further investigations.

Burseraceae

A small family of 20 genera and 500 to 550 species distributed both in the Old World and the New World tropics.

Vegetative Features. Large trees or shrubs, the inner bark with resin ducts. Leaves alternate, usually unipinnate, deciduous or persistent, estipulate or stipules represented by a basal pair of small leaflets inserted at the very base of the petiole or subulate or flat and auricle-shaped (Leenhouts 1959).

Floral Features. Inflorescence axillary or terminal, cymose panicles. Flowers small, usually unisexual by abortion, sometimes bisexual, actinomorphic, hypogynous, 3-merous or 5-merous. Sepals 3 to 5, more or less basally connate, imbricate or valvate; petals 3 to 5, alternate with sepals, polypetalous, imbricate or valvate; annular or cup-shaped, intrastaminal, nectariferous disc present. Stamens 6 to 10, in 1 or 2 whorls, outer whorl opposite the petals, free, inserted below the disc; anthers versatile, introrse, bicelled, longitudinally dehiscent, sterile in female flower. Gynoecium syncarpous, 3-(2–5)-carpellary; ovary as many loculed, superior, with axile placentation and 2 ovules per locule; style 1, short or highly reduced, stigma lobes as many as carpels, female flowers with staminodes. Fruits usually drupaceous (Brizicky 1962) or 1- to 5-seeded berry, with ± dry or fleshy exocarp and mesocarp, sometimes tardily dehiscent capsule; seeds 1 per locule with a straight or curved embryo and no endosperm.

Anatomy. Vessels moderately small to medium-sized, perforations simple, intervascular pitting alternate, with large, hexagonal, bordered pits. Parenchyma paratracheal varying from scanty to vasicenric; rays up to 2 to 4 cells wide with a few uniseriates, heterogeneous. Leaves generally dorsiventral, occasionally isobilateral. Hairs simple or stellate clothing trichomes or glandular; stomata Ranunculaceous type, frequently confined to the lower surface.

Embryology. Pollen grains isopolar, 3-zonocolporate with tectate exine, prolate to prolate-spheroidal or simply spheroidal (Mitra et al. 1977) and shed at 2-celled stage, 3-celled in *Garuga pinnata* and *Boswellia serrata* (L.L. Narayana 1960a). Ovules hemianatropous, bitegmic, crassinucellate [unitegmic in *Canarium oleosum, C. asperum, Santiria rubiginosa* and *S. commiphora* (L.L. Narayana 1960b)]. Embryo sac of Polygonum type, 8-nucleate at maturity. The endosperm development follows the Nuclear type.

Chromosome Number. Basic chromosome number is x = 11, 13, 23.

Chemical Features. The members of this family are rich in gums and resins.

Important Genera and Economic Importance. The divergent genera are *Tetragastris* and *Trattinnickia* with their petals united into a tube and valvate. *Triomma* is an interesting genus from Malaya having fruits with thick, broad wings. In *Garuga* adnation of calyx, corolla and androecium results in a cup-like structure, from the rim of which these organs separate (L.L. Narayana 1960a). This feature is known in *Scutinanthe* and *Commiphora* also (Hutchinson 1969).

Many members are rich in resins and gums of commercial value. The resinous substance 'myrrh' from *Commiphora molmol* and *C. abysinica* and 'franckincense' from *Boswellia carteri* are used medicinally and also as incense. A resin, 'copal', used as a cement and varnish in Mexico, is obtained from *Bursera glabrifolia* and *B. penicillata. Aucoumea klaineana* is a very important timber tree giving 'Gaboon mahogony' or 'okoume'.

Taxonomic Considerations. The family Burseraceae is closely allied to the Simaroubaceae, Rutaceae, Meliaceae and Anacardiaceae, and is variously placed in the Geraniales (Bentham and Hooker 1965a, Bessey 1915, Engler and Diels 1936), in the Terebinthales (Hallier 1935), Rutales (Hutchinson 1973, Melchior 1964, Takhtajan 1987). It is a homogeneous family and there is no controversy about its phyletic position in the Rutales. It is distinguished from Rutaceae and Simaroubaceae by the presence of lysigenous or schizogenous resin ducts in the bark (Webber 1941).

Meliaceae

A medium-sized family comprising 53 genera and 1350 species distributed in the tropics and warm temperate region of both Old and New World.

Vegetative Features. Shrubs or trees, stem more or less herbaceous in *Munronia* and *Naregamia* (Fig. 32.5A). Leaves usually alternate, pinnately compound (simple in *Turraea, Nymania, Vavaea* and *Nurmonia*) or decompound, lack pellucid dots, estipulate.

Floral Features. Inflorescence axillary panicles (Fig. 32.5F). Flowers bisexual (Fig. 32.5B,G), rarely unisexual and then the plants polygamodioecious, actinomorphic, hypogynous. Calyx of 4 or 5 sepals, usually basally connate, imbricate, petals 4 or 5, polypetalous, contorted or imbricate, sometimes adnate to the staminal column (*Munronia*) and then valvate. Stamens 8 to 10, rarely 5 or numerous as in *Vavaea*, mostly monadelphous (Fig. 32.5B,C,G), filaments free in *Cedrela* and *Walsura;* a disc usually present between stamens and ovary (*d* in Fig. 32.5 E); anthers bicelled, dehisce longitudinally, introrse, often with apical appendage (Fig. 32.5C,D). Gynoecium syncarpous, 2- to 5-carpellary, ovary as many loculed, superior, with axile placentation (Fig. 32.5H), ovules 2 per locule, rarely more (12 in *Swietenia*), style 1 or absent, stigma capitate or discoid. Fruit a berry, capsule or rarely a drupe (Fig. 32.5I). Seeds winged, endosperm fleshy or none.

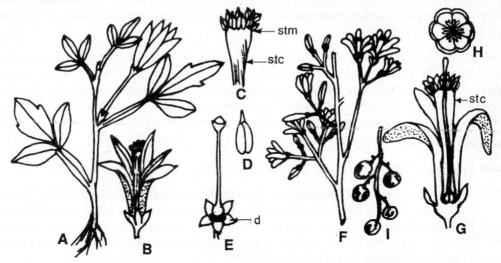

Fig. 32.5 Meliaceae: **A-E** *Naregamia alata*, **F-I** *Melia azadarach*. **A** Flowering plant. **B** Flower. **C** Stamens and staminal column. **D** Anther. **E** Gynoecium. **F** Part of inflorescence of *M. azadarach*. **G** Flower, longisection. **H** Ovary, cross section. **I** Fruits. *d* disc, *stc* staminal column, *stm* stamen. (**A-E** adapted from Hutchinson 1969, **F-I** original)

Anatomy. Vessels moderately small to medium-sized, radial multiples of 2 or 3 cells common; perforations simple, intervascular pitting typically minute. Parenchyma paratracheal, diffuse parenchyma sometimes present and often consists of crystalliferous cells. Rays usually exclusively uniseriate or 2 to 4 cells wide, mostly heterogeneous. Leaves dorsiventral, hairs of various types (Metcalfe and Chalk 1972). Stomata Ranunculaceous type, confined to the lower surface. Peg-like pneumatophores occur in *Amoora cuculata* and *Carapa moluccensis*.

Embryology.   Pollen grains 3- to 5-colporate (3- to 9-colporate in *Azadirachta indica*), with thick, smooth exine and abundant reserve food (N.C. Nair 1970a) and 2- or 3-celled when shed. Ovules anatropous, bitegmic (unitegmic in *Dysoxylum alliaceum*), crassinucellate. Embryo sac of Polygonum type, 8-nucleate at maturity. Endosperm formation of the Nuclear type. Both integuments take part in the formation of seed coat. Adventive embryony known in 'duku' and 'langsat' varieties of the southeast Asian fruit tree *Lansium domesticum*; ca. 60% seeds degenerate during early development but the edible fleshy aril continues to grow independently (Prakash et al. 1977).

Chromosome Number.   Basic chromosome number is x = 10 – 14 (Takhtajan 1987).

Chemical Features.   Seed fats are rich in stearic acid. Alkaloids absent .

Important Genera and Economic Importance.   Members of the Meliaceae are important as a source of hardwood timber. Mahogony from *Swietenia mahogani*, African mahogany from *Khaya senegalensis*, sapele mahogony from *Entandrophragma candollei*, African walnut from *Lovoa* species and Spanish cedar from *Cedrela* species are all commercially known timbers. Other timber-yielding plants are *Chloroxylon*, *Dysoxylum malabaricum*, *Swietenia macrophylla*, *Toona ciliata* and *Walsura robusta*. True mahogony wood is medium- to fine-textured, very hard and resistant to indentation. Its beauty is due to light reflecting qualities of the rays (Kochhar 1981). It is used for furniture, aeroplane propellers, panelling, veneering, cabinet-making, printer's block, musical instrument and ship-building.

The leaves of *Azadirachta indica* are used as a pot-herb and as an insect-repellant and the twigs are used for toothbrushes; seeds yield margosa oil, used in soap industry. Carapa oil from seeds of *Carapa guianensis* and *C. moluccensis* is used as an illuminant. A red dye is obtained from the flowers of *Chickrasia tabularis* and its astringent bark is of medicinal value. The fruits of *Lansium domesticum* and *Sandoricum koetjape* are edible (Dutta 1988). Some plants such as *Amoora*, *Melia*, *Swietenia* and *Turraea* are grown as ornamentls. *Azadirachta indica* and *Melia azadarach* are grown as avenue trees.

Taxonomic Considerations.   The family Meliaceae was included in the order Terebinthales by Hallier (1912), derived from the Rutaceae. Bessey (1915), Bentham and Hooker (1965a) and Engler and Diels (1936) retained it in the Geraniales, while Hutchinson (1973) segregates it as the only family of his Meliales. Melchior (1964) and Takhtajan (1987) include it in Rutales and near Burseraceae, Simaroubaceae and Rutaceae. Cronquist (1981) includes this family in the Sapindales. Meliaceae resembles Rutaceae anatomically as well as embryologically, and differs from them in having leaves without gland dots and a staminal column. Various features of Meliaceae support its placement in the Rutales along with other closely allied families.

Akaniaceae

A monotypic family including the genus *Akania* with only one species *A. lucens*, from eastern Australia.

Vegetative Features.   Trees with alternate, imparipinnate, estipulate leaves.

Floral Features.   Inflorescence paniculate; flowers bisexual, actinomorphic, hypogynous. Calyx of 5 sepals, polysepalous, imbricate; petals 5, polypetalous, contorted , disc absent. Stamens typically 8 in two whorls, 5 outer ones opposite the sepals. Gynoecium syncarpous, 3-carpellary, ovary superior, trilocular, ovules epitropous, 3 in each locule, pendulous. Fruit a loculicidal capsule. Seeds with straight embryo and fleshy endosperm.

Anatomy.   Vessels moderately small, solitary and in radial, tangential or oblique multiples and clusters. Perforation plates simple and oblique or rarely scalariform. Parenchyma paratracheal, scanty. Rays up to

16 cells wide, heterogeneous; no uniseriates. Leaves dorsiventral, hairs infrequent; stomata confined to special depressions on the lower surface of the leaf.

Important Genera and Economic Importance.   *Akania* is the only genus; no economic importance.

Taxomomic Considerations.   The family Akaniaceae has been included in the Sapindales (Cronquist 1981, Metcalfe and Chalk 1972, Takhtajan 1987), in the Rutales (Melchior 1964) or in the Geraniales (Lawrence 1951). In Bentham and Hooker's (1965a) system, the genus *Akania* was placed in the family Sapindaceae. Willis (1973) also supported this view. The broad rays, absence of disc and presence of endosperm in mature seeds are uncommon in the Rutales. Pentamerous flowers and 8 stamens and the pollen morphology of *Akania* are similar to that in the Sapindaceae. Wide medullary rays, absence of nectariferous disc, presence of epitropous ovules and ample fleshy endosperm are the differences from Sapindaceae (Takhtajan 1987). However, *Akania* does not have any special feature for which it can be retained in the Rutales, and it would be better to shift Akaniaceae to Sapindales near Sapindaceae.

Malpighiaceae

A large family of 60 genera and 1200 species distributed primarily in the tropics and subtropics of the New World.

Vegetative Features.   Trees or shrubs, usually woody, climbing plants, plant body covered with hairs with 1,2 or more horizontal branches (Fig. 32.7F,G). Leaves opposite (Fig. 32.7A), rarely alternate or ternate; often with petiolar glands and jointed petioles, stipulate, stipules variable, margin entire or spiny dentate.

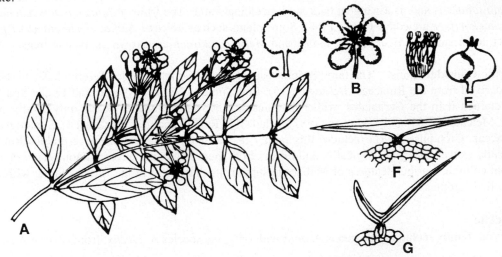

Fig. 32.7   Malpighiaceae: **A-E** *Malpighia glabra,* **F** *M. urens,* **G** *Mascagnia cordifolia.* **A** Flowering twig. **B** Flower, face view. **C** Petal. **D** Stamens with basally connate filaments. **E** Gynoecium. **F,G** Hairs. (**A-E** adapted from Hutchinson 1969, **F,G** from Metcalfe and Chalk 1972)

Floral Features.   Inflorescence axillary and terminal, fasciculate or umbellate corymbs (Fig. 32.7A), rarely solitary flowers. Flowers bracteate, bisexual, actinomorphic (Fig. 32.7B) or obliquely zygomorphic, hypogynous, often cleistogamous. Calyx of 5 sepals, polysepalous, imbricate, rarely valvate, some or all with large, sessile or stalked glands. Petals 5, polypetalous, clawed, fringed or toothed (Fig. 32.7B,C), convolute. Stamens 10, in 2 whorls of 5 each, all perfect or some often reduced to staminodes, filaments usually basally connate (Fig. 32.7D); anthers bicelled, introrse, often with enlarged connective, dehiscence longitudinal. Gynoecium syncarpous (Fig. 32.7E), (2-) 3- (4-5)-carpellary, ovary as many loculed, superior, axile placentation. Ovule 1 in each locule, pendulous; styles as many as carpels, stigmas entire or minutely lobed. Fruit a samara, schizocarp, capsule, berry or rarely a drupe; seeds without endosperm and with a large embryo.

Anatomy.   Vessels mostly medium-sized in radial multiples of 2 to 3 and sometimes with a definite radial pattern; perforations simple, intervascular pitting alternate. Parenchyma usually paratracheal, scanty to confluent, occasionally abundant and banded, sometimes diffuse. Rays 2 to 3 cells wide, markedly heterogeneous. Anomalous secondary growth occurs. Leaves dorsiventral, sometimes isobilateral or centric; stomata Rubiaceous type, confined to the lower surface.

Embryology.   Pollen grains are 2-celled at dispersal stage. Ovules semiana- or anatropous, bitegmic and crassinucellate. Embryo sac mostly of Penaea type, sometimes of Allium type; 8- or 16-nucleate at maturity. Endosperm formation of the Nuclear type. Failure of fertilization is a common feature. Nucellar polyembryony is frequent (Johri et al. 1992).

Chromosome Number.   Basic chromosome number is $x = 6,9-12$ (Takhtajan 1987).

Chemical Features.   Gums and tannins present.

Important Genera and Economic Importance.   The most important genus is *Malpighia* with 35 species. *Hiptage benghalensis* and *Malpighia* spp. are grown as ornamentals. *Banisteria laevifolia*, a climbing shrub with yellow flowers, and *Heteropteris leona* from west tropical Africa with yellow flowers in panicles and reddish samaras are other ornamentals. A hallucinogenic drug 'caapi' or 'ayahuasca' is obtained from *Banisteriopsis caapi*; the active constituents are harmaline, harmine and tetrahydroharmine (Kochhar 1981).

Taxonomic Considerations.   Most phylogenists have allied Malpighiaceae with the Geraniales. Hutchinson (1973) places it in his Malpighiales along with 11 other families and derives it from Tiliaceae stock. Lawrence (1951) includes Malpighiaceae in Geraniales along with Trigoniaceae and Vochysiaceae in suborder Malpighiineae. Melchior (1964) places it in the Rutales. Cronquist (1981) treats this family as the most primitive in the order Polygalales, which includes 8 other families. Takhtajan (1987) also includes it in the Polygalales but along with only five other families.

   Malpighiaceae is closely related to the families Tremandraceae and Polygalaceae. Anatomically, it has a more highly specialized wood structure than Linaceae, Humiriaceae and Erythroxylaceae, with which it is often associated (Heimsch 1942). On the basis of pollen ultrastructure (Simpson and Skvarla 1981), the families Polygalaceae, Malpighiaceae and Krameriaceae may have evolved convergently. However, there is not much difference in the phyletic position of the Malpighiaceae in the Rutales.

Trigoniaceae

A small family of 4 genera and 35 species, distributed in tropical America, Madagascar and Malaysia.

Vegetative Features.   Shrubs, often climbing, rarely trees. Leaves simple, entire, alternate or opposite, stipulate or estipulate; stipules of opposite leaves sometimes connate.

Floral Features.   Inflorescence terminal or axillary racemes or thyrses, rarely 3-flowered cymes. Calyx lobes 5, gamosepalous, imbricate; corolla of 5 or 3 petals, unequal, imbricate or contorted, rarely valvate or subimbricate. Stamen 5 to 12, usually include 3 to 6 staminodes; filaments connate into longer or shorter staminal tube, split posticously; anthers introrse, disc glands 1 to 3, usually adjoining the split.

Gynoecium syncarpous, tricarpellary, ovary superior, 3- or 1-locular; ovules 1 to numerous in each locule, biseriate, placentation axile; style simple, stigma capitate. Fruit usually a septicidal capsule, sometimes winged, rarely a samara. Seeds usually long-pilose, rarely glabrous, with or rarely without endosperm.

Anatomy.   Vessels solitary, perforations simple; parenchyma apotracheal or paratracheal. Rays up to 4 cells wide, uniseriates moderately numerous, heterogeneous. Leaves dorsiventral, stomata commonly of the Rubiaceous type, confined to the lower surface.

Embryology.   Embryology has not been fully studied. Ovules anatropous, bitegmic and tenuinucellate. The 8-nucleate embryo sac contains starch and probably conforms to the Polygonum type. The scanty nucellus collapses so that the inner integument surrounds the embryo sac (Johri et al. 1992). Endosperm formation of the Nuclear type.

Chromosome Number.   The haploid chromosome number is n = 10.

Important Genera and Economic Importance.   The four genera of this family are *Trigoniastrum, Trigonia, Humbertiodendron* and *Euphronia* (= *Lightia*). None of these has any economic importance. *Trigoniastrum* has often been included in the Polygalaceae but systematically it is closest to *Trigonia* (Takhtajan 1987).

Taxonomic Considerations.   The family Trigoniaceae in Bentham and Hooker's (1965a) system was included in the Vochysiaceae. Engler and Diels (1936) also placed it near Vochysiaceae in the order Geraniales. Most taxonomists agree to the alliance of the three families: Vochysiaceae, Trigoniaceae and Malpighiaceae. Hutchinson (1973), however, included Trigoniaceae and Vochysiaceae in Polygalales and the Malpighiaceae in Malpighiales. Takhtajan (1987) places Trigoniaceae in the Polygalales along with Malpighiaceae, Vochysiaceae, Polygalaceae, Krameriaceae and Tremandraceae. The ovule and seed anatomy of *Trigonia* (Johri et al. 1992) is similar to that of *Linum* and related taxa. They share several other embryological features with Linaceae and Erythroxylaceae as well as with Vochysiaceae, Malpighiaceae and Tremandraceae. According to Corner (1976), the family Trigoniaceae cannot be classified near the Polygalaceae because of the strongly lignified exotegmen. Pollen morphology suggests some affinity with the Malpighiaceae, but Trigoniaceae differs in the absence of glands at the base of calyx lobes, stylodium united with style, and a 16-nucleate embryo sac (Takhtajan 1987).

After comparing the different views about this family, it appears that the Trigoniaceae should be placed together with the Vochysiaceae and Malpighiaceae. On the basis of seed structure, these families correspond fairly well with the Linaceae complex (see Johri et al. 1992) and therefore may be included in the Geraniales.

Vochysiaceae

A small family of 6 genera and 200 species, distributed in tropical America and western Africa.

Vegetative Features.   Trees or shrubs, rarely herbs; leaves opposite or vertical or rarely alternate, entire, simple, stipulate or estipulate (Fig. 32.9 A).

Floral Features.   Inflorescence compound raceme of cincinni (Fig. 32.9B). Flowers bisexual, obliquely zygomorphic, hypogynous. Calyx of 5 sepals, polysepalous, imbricate, connate at the base, posterior lobe often spurred (Fig. 32.9C,E). Corolla usually of 3 to 1, unequal petals, rarely 5, contorted. Stamen 1, fertile, introrse (Fig. 32.9E), staminodes 2 to 4. Gynoecium syncarpous, tricarpellary, ovary superior, trilocular, with 1 to numerous ovules per locule, placentation axile; rarely a monocarpellary gynoecium with unilocular ovary and 2 lateral ovules; style simple, stigma small. Fruit a loculicidal capsule or samaroid and indehiscent (Fig. 32.9D). Seeds often winged, sometimes pilose, non-endospermous.

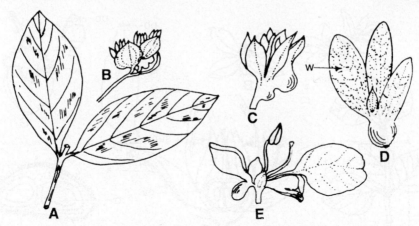

Fig. 32.9 Vochysiaceae: **A-D** *Erismadelphus exsul*, **E** *Qualea multiflora*. **A** Twig with leaves. **B** Inflorescence. **C** Flower. **D** Fruit. **E** Flower of *Q. multiflora*. *w* wing. (**A-E** adapted from Hutchinson 1969)

Anatomy.   Vessels medium-sized to large and few, perforation plates simple, intervascular pitting alternate. Parenchyma paratracheal; rays up to 2 to 8 (usually 3 to 5) cells wide, hetero- to homogeneous. Included phloem sometimes present. Leaves dorsiventral (isobilateral in *Qualea glauca*); glandular hairs absent. Extrafloral nectaries occur above the stipules in *Erisma laurifolium* and *Vochysia* sp. Stomata of Ranunculaceous type, confined to the lower surface, Rubiaceous type in *Callisthene* and *Qualea*. Characteristic features are the intercellular canals and banded parenchyma.

Embryology.   The embryology has not been investigated fully. Ovules hemianatropous, bitegmic and crassinucellate (Johri et al. 1942). Development of embryo sac has not been studied . Endosperm formation of the Nuclear type.

Chromosome Number.   The haploid chromosome number is n = 11.

Chemical Features.   Usually accumulates aluminium (Takhtajan 1987).

Taxonomic Considerations.   The family Vochysiaceae is closest to the Trigoniaceae, from which it differs chemically, anatomically, and in some exomorphological features like obliquely zygomorphic flowers, and structure of androecium. There is no second opinion about its alliance with Trigoniaceae and Malpighiaceae, and most taxonomists place the three families together.

Tremandraceae
A small family of 3 genera, *Tremandra, Tetratheca* and *Platytheca,* and 43 species, distributed mainly in western Australia and Tasmania.

Vegetative Features.   Suffrutescent herbs or small shrubs, often stellate-tomentose or with glandular hairs; stems and branches broadly winged in *Tetratheca affinis* var. *platycauda*. Leaves opposite, whorled (Fig. 32.10A) or alternate, simple, estipulate, Leaves of many species have a tendency to roll their margins.

Floral Features.   Flowers solitary, axillary; bisexual, actinomorphic, hypogynous (Fig. 32.10A-C). Calyx of 4 or 5 sepals, polysepalous, valvate; corolla of 4 or 5 petals, polypetalous, induplicate-valvate; a lobed glandular disc sometimes present between stamens and corolla. Stamens 8 to 10 (rarely 6), mostly in two whorls; anthers 2- to 4-celled, dehiscence by a transverse terminal valve or prolonged into a beak with

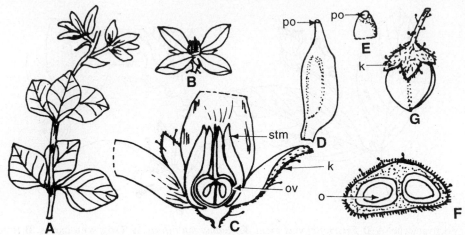

Fig. 32.10 Tremandraceae : **A-G** *Tetratheca ciliata.* **A** Flowering twig. **B** Open flower. **C** Vertical section. **D** Anther. **E** Tip with apical pore. **F** Ovary, cross section. **G** Fruit. *k* calyx, *o* ovule, *ov* ovary, *po* pore, *stm* stamen. (Adatped from Lawrence 1951)

terminal pores (Fig. 32.10D,E). Gynoecium syncarpous, bicarpellary; ovary superior, bilocular, placentation axile, ovules 1 or 2 in each locule (Fig. 32.10F); style and stigma solitary, simple. Fruit a compressed loculicidal or septicidal capsule (Fig. 32.10G); seeds hairy with an aril-like appendage from the chalaza; endosperm copious, embryo small and straight.

Anatomy. Vessels small, sometimes with spiral thickenings; perforations simple. Parenchyma scanty paratracheal or absent. Rays 1 or 2 cells wide, composed of upright cells. Leaves flat or linear with margins frequently curved inwards. Simple, stellate and glandular hairs recorded. Stomata confined to the lower surface, Ranunculaceous type.

Embryology. Embryology has not been investigated in sufficient details. Ovules anatropous, bitegmic, crassinucellate (Johri et al. 1992).

Important Genera and Economic Importance. The three genera are *Tetratheca, Tremandra* and *Platytheca.* The family is not of much economic importance; a species of *Tetratheca* is sometimes grown as an ornamental.

Taxonomic Considerations. Bentham and Hooker (1965a) included the Tremandraceae in the Polygalales. Bessey (1915) retained it in the Geraniales. Wettsein (1935) placed this family in his Terebinthales along with Polygalaceae. Melchior (1964) included Tremandraceae in the Rutales along with Polygalaceae in the suborder Polygalineae. Hutchinson (1973) placed it in his Pittosporales.

Cronquist (1981) and Takhtajan (1987) include Tremandraceae in Polygalales, Even though different workers assign separate position to this family, everyone recognises its affinity with the Polygalaceae. This family is different from the rest of the Polygalales in actinomorphic flowers, but they are alike in wood anatomy and poricidal dehiscence of anthers. Pollen morphology is to some extent similar to the Malpighiaceae (G. Erdtman 1952).

The position of the family Tremandraceae close to Polygalaceae and Malpighiaceae is justified. However, these families are included in different orders—Rutales, Malpighiales or Polygalales.

In our opinion, the family Tremandraceae is closer to Polygalaceae, Trigoniaceae, Malpighiaceae, Zygophyllaceae and Vochysiaceae than to the other members of the Rutales. These 5 or 6 (also Krameriaceae) families should preferably be kept under a separate order, Polygalales (or Malpighiales).

Polygalaceae

A family of 15 genera and 900 species, distributed widely except in arctic regions and New Zealand.

Vegetative Features. Herbs, shrubs or trees of medium height, sometimes climbing or twining. *Epirrhizanthes* is a non-chlorophyllous saprophyte from Malayan region. Leaves usually alternate (Fig. 32.11A), sometimes opposite or whorled, simple, sometimes scale-like, mostly estipulate or with small stipular glands.

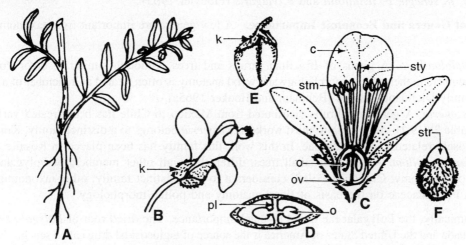

Fig. 32.11 Polygalaceae: **A** *Polygala chinensis,* **B,D** *P. paucifolia,* **E, F** *P. cabrae.* **A** Plant bearing fruits. **B, C** Flower of *P. paucifolia* (**B**) and of *P. cabrae,* longisection (**C**). **D** Ovary, cross section. **E, F** Fruit (**E**) and seed (**F**) of *P. cabrae. c* corolla, *k* calyx, *o* ovule, *ov* ovary, *pl* placenta, *stm* stamen, *str* strophiole, *sty* style. (**A** original, **B-F** adapted from Hutchinson 1969)

Floral Features. Inflorescence a spike, raceme or crowded into "heads", sometimes solitary, rarely paniculate. Flowers bracteate, bracteolate, bisexual, zygomorphic, hypogynous (Fig. 32.11B). Calyx of 5 sepals, sometimes varies from 4 to 7; the 2 lower united or polysepalous with the 2 inner ones petaloid and wing-like. Corolla of 5 petals, polypetalous but usually only 3 present—the 2 posterior and 1 anterior, more or less basally adnate to the androecium; anterior petal often concave with or without a fringed crest or keel. Stamens basically 10 in 2 pentamerous whorls, usually only 8, monadelphous (Fig. 32.11C), sheath split up to 3/4th the length; anthers basifixed, usually confluently (merging or blending together), one-celled, dehisce by an apical or subterminal pore, sometimes by a transverse slit. Gynoecium syncarpous, bicarpellary; ovary bilocular with axile placentation, superior (Fig. 32.11C,D); ovules solitary on each placenta, pendulous; style 1, stigmas as many as carpels. Fruit usually a loculicidal capsule (Fig. 32.11E), rarely a nut, samara or drupe; seeds often pilose, with a conspicuous micropylar aril or strophiole (Fig. 32.11F); endosperm soft, fleshy and embryo straight.

Anatomy. Vessels small to large, usually solitary, perforations simple; intervascular pitting alternate. Parenchyma usually paratracheal but predominantly apotracheal, diffuse or banded in tribes Montabeae and Xanthophylleae. Rays uniseriate or up to 2 or 3 cells wide, heterogeneous. Included phloem develops in climbers. Leaves dorsiventral, stomata generally Ranunculaceous type, rarely Rubiaceous (some species of *Securidaca* and *Xanthophyllum*), confined to both surfaces or only the lower surface. Mostly solitary but sometimes clustered crystals usually present in leaf and axis.

Embryology. Pollen grains 3-colpate and 2- or 3- celled at dispersal stage. Ovules anatropous, bitegmic and crassinucellate. Embryo sac of Polygonum type, 8-nucleate at maturity. Endosperm formation of the Nuclear type.

Chromosome Number.   Basic chromosome numbers are x = 5–11 (Takhtajan 1987).

Chemical Features.   A phytosterol, called spinasterol, occurs in the roots of *Polygala senega* (Harborne and Turner 1984). Aliphatic polyols or polyalcohols have been isolated from different species of *Polygala*— *P. amara, P. senega, P. tenuifolia* and *P. vulgaris* (Plouvier 1963).

Important Genera and Economic Importance.   A few genera are important from taxonomic point of view:

a) *Diclidanthera*. A genus of Brazilian shrubs and trees, although similar to other Polygalaceae members in their pollen morphology and wood anatomy is often treated as a member of a monotypic family Diclidantheraceae (Bentham and Hooker 1965a).

b) *Krameria*. A genus of shrubs distributed from Mexico to Chile has been treated variously (see Table 28.19.1). According to most workers, *Krameria* belongs to a distinct family, Krameriaceae, closely related to Polygalaceae. In this work this family has been placed in Rosales.

c) *Xanthophyllum*. A genus of small trees, differs from all other members of Polygalaceae in its wood anatomy; Cronquist (1968) considers it to be a distinct family; other taxonomists retain it in Polygalaceae on the basis of floral anatomy and pollen morphology.

Economically, the Polygalaceae is not of much importance. The dried root of *Polygala senega* from South Canada and the United States of America is the source of a glucosidal drug called senega. *P. butyracea* of tropical Africa and *Securidaca longipedunculata* of east Africa produce fibres. A few species of *Polygala* yield dyes. Some species of *Bredemeyera, Polygala* and *Securidaca* are grown as ornamentals.

Taxonomic Considerations.   There are diverse opinions regarding the phylogenetic position of the Polygalaceae. In Bentham and Hooker's (1965a) system this family is placed between the Parietales and Caryophyllinae. Bessey (1915) and Engler and Diels (1936) placed it in the Geraniales, Wettstein (1935) in his Terebinthales. Hallier (1912), Hutchinson (1973), Cronquist (1981) and Takhtajan (1987) treat this family as a member of the Polygalales.

Melchior (1964) retained it in the Rutales along with Tremandraceae, Vochysiaceae and Malpighiaceae. On the basis of wood anatomy, Polygalaceae is a member of an interrelated group, the other members of this group are Trigoniaceae, Tremandraceae, Zygophyllaceae, Malpighiaceae and Vochysiaceae. Although more specialised in their wood structure, these families are related to the Linaceae, Humiriaceae and Erythroxylaceae (Metcalfe and Chalk 1972). Melchior (1964) includes the Polygalaceae in Rutales along with all the above families (and a few others), and treats it as the most advanced taxon of this order. Zygomorphic flowers, specialised wood anatomy and numerous herbaceous members (in this family) support this view.

In our opinion, the family Polygalaceae (along with Trigoniaceae, Tremandraceae, Zygophyllaceae, Malpighiaceae and Vochysiaceae) should be treated as a distinct order Polygalales. Another family which is close to the Polygalaceae is—Krameriaceae (see p. 256) and should be included in the same order, Polygalales.

Taxonomic Considerations of the Order Rutales

Rutales comprises 12 families distributed in three suborders, according to Melchior (1964). Engler and Diels (1936) and Lawrence (1951) place all these families (except Picrodendraceae) in the Geraniales in three suborders.

Cronquist (1981) includes Rutaceae, Cneoraceae, Akaniaceae, Burseraceae, Simaroubaceae and Meliaceae in the order Sapindales, Picrodendraceae in Juglandales and the rest of the families in the Polygalales. The family Akaniaceae is allied to Sapindaceae in the order Sapindales. The family Picrodendraceae is more allied to the Euphorbiaceae in pollen structure, chromosome number, and some morphological and anatomical features. Takhtajan (1987) includes the genus *Picrodendron* in Euphorbiaceae. Cronquist (1981), on the other hand, places Picrodendraceae in the Juglandales.

The order Rutales is large and highly heterogeneous. In would be better to break up this large order into three smaller ones: Rutales s.s., Picrodendraceae in Euphorbiales, and the Polygalales including Trigoniaceae, Vochysiaceae, Malpighiaceae, Tremandraceae and also Krameriaceae. Rutales s.s. should include the families Rutaceae, Cneoraceae, Burseraceae, Simaroubaceae and Meliaceae. These families and the Akaniaceae are included in the Sapindales by some workers (Cronquist 1981). The family Akaniaceae may, however, be included in the Sapindales near Sapindaceae.

33

Order Sapindales

This order comprises 10 families[1] belonging to 4 suborders, distributed both in the New and the Old World. Mostly shrubs and trees, rarely herbs; unisexual, bisexual or polygamous flowers. Ovules pendulous with dorsal raphe and micropyle upwards, or erect with ventral raphe and micropyle downwards.

Coriariaceae

Coriariaceae is a unigeneric family, the genus *Coriaria* has 20 species, discontinuous distribution from western South America to islands near Australia including New Zealand and China, Japan, Fiji, Samoa, New Hebrides and other islands nearby. A few members are indigenous to South America.

Vegetative Features.   Suffrutescent, perennial herbs or shrubs with angular twigs. Leaves opposite or whorled, simple, entire, estipulate, with unicostate, reticulate venation (Fig. 33.1A).

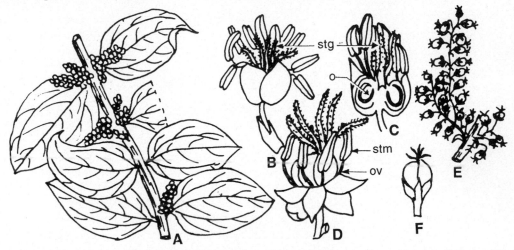

Fig. 33.1   Coriariaceae: **A-F** *Coriaria nepalensis*. **A** Flowering twig. **B** Flower. **C** Longisection. **D** Perianth spreadout and a few stamens removed. **E, F** Fruits. *o* ovule, *ov* ovary, *stg* stigma, *stm* stamen   (After V.K. Sharma 1968).

Floral Features.   Inflorescence long, lax, axillary racemes; 3 or 4 pairs of sterile minute bracts below the flower-bearing region. Flowers bracteate, ebracteolate, bisexual (*C. ruscifolia*) or unisexual (*C. japonica*), actinomorphic, hypogynous, pentamerous (Fig. 33.1B–D). Calyx of 5 sepals, polysepalous, imbricate; petals 5, polypetalous, imbricate, smaller than the sepals, with a keel on the inner face, fleshy, persistent, enlarged and adhering to the fruiting carpels. Stamens 10, free, in 2 alternating whorls of 5 each, those opposite the petals are adnate to the petal keel; filaments long, slender, anthers exerted, often coloured (coral red in *C. nepalensis*), bicelled, introrse and attenuated, longitudinally dehiscent. Gynoecium syncarpous, 5- to 10-carpellary, basally fused but free above; ovary unilocular, one pendulous ovule (Fig. 33.1C) from

---

[1]The family Aextoxicaceae has not been included in the text.

parietal placenta, superior; styles 5, free, terminal, linear, thick, stigmatic along ventral side, usually divergent (Fig. 33.1B–D). Fruit composed of 5, 1-seeded carpels; the petals enlarge, become purple and adhere to the fruiting carpel, and produce a pseudo-drupe (Fig. 33.1E, F). Fruiting carpels separated by the projecting keel of petals. Seeds with scanty endosperm and straight embryo.

Anatomy. Vessels small, commonly grouped tangentially, sometimes ring-porous; perforations simple, intervascular pitting alternate, small. Parenchyma paratracheal, scanty, vasicentric or confluent, storied. Rays up to 15 cells wide and without uniseriates. Leaves dorsiventral; stomata present on both the surfaces, more on the lower surface; Rubiaceous type.

Embryology. Pollen grains globose, 3-colporate, and 3-celled at shedding stage. Ovules anatropous, bitegmic, crassinucellate, and with a nucellar cap. Embryo sac of Polygonum type, 8-nucleate at maturity. The outer integument forms a flap-like covering on either side of the funicule. Endosperm formation of the Nuclear type. Seed coat formed from the outer integument (Sharma 1968).

Chromosome Number. The basic chromosome number is x = 10.

Chemical Features. Plants rich in tannin. Flavonoids occur in leaf tissue (Bohm and Ornduff 1981).

Important Genera and Economic Importance. The genus *Coriaria* is interesting because of its discontinuous distribution (see p. 296) and the taxonomic dispute about its position. Leaves of *C. myrtifolia* are often used as a source of tannin. Young shoots and seeds are said to be poisonous to cattle (Hutchinson 1969). The mature fruits are fermented into a pleasant drink by the Maoris of New Zealand.

Taxonomic Considerations. The taxonomic position of the family Coriariaceae is highly disputed. Both Bentham and Hooker (1965a) and Engler and Diels (1936) placed this family next to the Anacardiaceae in the order Sapindales. Mauritzon (1936) and Takhtajan (1959) placed it near the Salvadoraceae in the Celastrales, Wettstein (1935) placed it in his Terebinthales, along with Cyrillaceae. According to Lawrence (1951), the Empetraceae is most closely related to this family and both belong to the Sapindales. However, this is not tenable, as Empetraceae has already been removed to Ericales (Melchior 1964). Takhtajan (1966) included it in the Rutales, which also retained the Anacardiaceae. Hutchinson (1969, 1973) treats it as the only member of the order Coriariales and positioned it next to the Dilleniales, "for want of a better position". He was doubtful about its affinity with the Sapindales because of its discontinuous distribution, a primitive feature. Cronquist (1968, 1981) treats Coriariaceae as a primitive family and includes it in the Ranunculales. According to him: "their wood anatomy suggests that they may be only secondarily woody" and "the glycosides of the Coriariaceae are very much like some of those in the Menispermaceae". According to Garg (1981), palynologically, *Coriaria* seems to have closest affinities with the Sapindaceae. Takhtajan (1987) includes Coriariaceae in a separate order, Coriariales, in the superorder Rutanae, along with the orders Sapindales, Rutales and others. According to Carlquist (1985). "One could imagine *Coriaria* as a derivative of a herbaceous or semi-herbaceous line allied to the Simaroubaceae and other Sapindalean—Ranalean families". It perhaps represents an isolated ancient stock with a remote connection with some of the arborescent families of the Sapindales (Sharma 1968, Rau and Sharma 1970). Coriariaceae is best treated as the sole member of a distinct order, Coriariales, and placed near the Sapindales.

Anacardiaceae

Anacardiaceae is a family of 80–85 genera and 600 species, distributed in both northern and southern hemispheres, in the tropics as well as in temperate zones.

Vegetative Features. Trees or shrubs, usually with resinous bark. *Tapirira* is a climber and *Schinus* is often spinescent. Leaves usually alternate (opposite in *Bouea* and *Dobinea*), simple, trifoliate or pinnate (*Comocladia dentatus, Rhus* spp. (Fig. 33.2B), estipulate or stipules obscure.

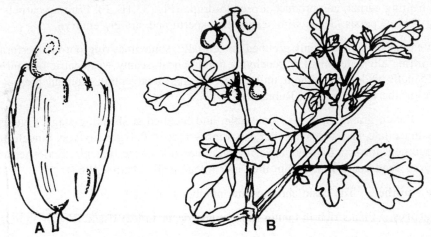

Fig. 33.2   Anacardiaceae : **A** *Anacardium occidentale,* fruit. **B** *Rhus mysurensis,* fruiting twig

Floral Features.   Inflorescence paniculate in the tribe Spondieae and thyrsoids in tribes Anacardieae and Rhoeae (Barfod 1988). *Toxicodendron* of Rhoeae is an exception and has axillary panicles. Flowers bisexual or unisexual by abortion, actinomorphic, small, hypogynous. Calyx of 3 to 5 sepals, basally connate and sometimes adnate to the gynoecium along with other whorls; petals 3 to 5, polypetalous or rarely basally connate; sepals and petals both absent in *Pistacia chinensis* (Copeland 1955). Stamens 10 in 2 whorls of 5 each (one stamen in *Anacardium* along with 6 to 9 staminodes). Anthers bicelled, dehiscence longitudinal. Gynoecium syncarpous, 3-carpellary but functionally monocarpellary; ovary unilocular with a single functional ovule (ovary 5-carpellary and 5-locular in *Buchanania*), placentation axile, ovules solitary in each locule, style usually 1, stigmas as many as carpels. Fruit usually a drupe with a resinous mesocarp. Seeds with a curved embryo and scanty endosperm.

Anatomy.   Wood both diffuse-porous and ring-porous, as in *Rhus* (Heimsch 1940). Vessels small to moderately large, sometimes with an oblique pattern; perforation plates mostly simple, intervascular pitting alternate, large. Parenchyma paratracheal, scanty, vasicentric, often rather sparse; with terminal or irregularly distributed broad bands in some genera. Rays mostly 2 or 3 cells wide but sometimes 8 to 10 cells wide; uniseriates occur, heterogeneous; intercellular canals in the rays of many genera. Leaves dorsiventral, stomata mostly Ranunculaceous type; hairs of various types common.

Embryology.   Pollen grains spherical, binucleate and with finely pitted exine in *Pistacia chinensis;* very large (22 μm to 40 μm), spherical to elliptical and tricolpate in *Rhus* (Heimsch 1940). Ovule anatropous, unitegmic, crassinucellate. Embryo sac of Polygonum type, 8-nucleate at maturity. Endosperm formation of the Nuclear type.

Chromosome Number.   Basic chromosome number much variable: x = 7, 10, 12, 14–16.

Chemical Features.   The seed coat of Anacardiaceae members contains fat, e.g. *Rhus,* commonly called sumac tallow. The seeds of *Mangifera indica* contain palmitic, oleic, and linoleic acids. Rhamnose, a methyl derivative of monosaccharide pentose occurs free in *Rhus* spp. Stellacyanin, a glycoprotein, is obtained from the Japanese lacquer tree, *Rhus vernicifera* (Harborne and Turner 1984).

Important Genera and Economic Importance. *Mangifera* is an important genus in the tropical regions of the world. *M. indica* has numerous varieties of edible mangoes. A native of southeastern Asia, mango has been cultivated in India for more than 4000 years, and finds an important place in many religious observances and festivals. From its place of origin, it was introduced into East Africa by the Persians in the 10th Century A.D. Later it was introduced to many other countries. It has two types of seeds, the monoembryonic form with one true zygotic embryo, and the polyembryonic form with several adventitious embryos from the nucellus, in addition to the zygotic embryo.

Anacardium occidentale, another important taxon, is native to tropical America from Mexico to Peru and Brazil, and to the West Indies. It reached India in the 16th century, and today 80% of the total world production is from India. It is an evergreen tree with gnarled or twisted stem. The fruit, commonly called cashew apple is the pear-shaped swollen peduncle; at its top is borne the kidney-shaped, single-seeded nut (Fig. 33.2A). It can be grown either from seeds or by various vegetative means.

Rhus toxicodendron, commonly called poison sumac, is yet another important taxon.

The tribal divisions in the Anacardiaceae are supported by the inflorescence morphology. Tribe Spondieae has paniculate inflorescence, tribe Anacardieae and Rhoeae have thyrsoids. *Toxicodendron* in tribe Rhoeae is exceptional in having axillary panicles and usually a specialised liana-like growth habit (Barfod 1988).

Economically, the family Anacardiaceae is very important. The seeds of *Anacardium occidentale*, *Pistacia vera*, *Buchanania lanzan* are edible and rich in proteins and fats. The juice of cashew apple is fermented into a pleasant beverage called kaju wine. *Mangifera indica*, *Spondias dulcis*, *S. pinnata*, *Harpephyllum caffrum* and *Dracontomelon mangiferum* are some of the fruit trees. Timber from *Cotinus coggygria* is used for making picture frames and cabinet work. The wood of *Drimicarpus racemosus*, *Schinopsis* and *Astronium* spp. is also important. A number of genera yield resins such as the sap of *Gluta* sp. (Malaya), trunk and fruit rind of *Holigarna arnottiana* (used as a varnish), *Melanorrhoea usitata* (Burmese lacquer) and *Rhus vernicifera* (Japanese lacquer). *Pistacia cabulica* and *P. lentisens* yield a resin called mastic. *P. integerrima*, *Rhus coriaria* and *Schinopsia lorentzii* are cultivated for tannin. Leaves of *P. lenticus* are also rich in tannin. The gum exuding from the trunk of *Lannea coromandelica* is used for calico-printing. *Semecarpus anacardium* for marking nut tree has an almost round nut seated on the top of a pyriform fleshy peduncle; the tannin in the nut is used for putting marks on clothes by laundries. *Rhus toxicodendron* and *R. vernix*—also known as poison ivy—secrete a resinous substance which is highly poisonous and may cause severe dermatitis. Other species of *Rhus*, such as *R. typhina* are ornamentals and grown in temperate countries for their glorious autumn colour.

Taxonomic Considerations   There is a general agreement among the taxonomists to include the Anacardiaceae in the Sapindales (Lawrence 1951). Although Anacardiaceae members resemble Sapindaceae members in habit and floral structure, they can be distinguished easily by the presence of intrastaminal disc, usually unilocular ovary, drupaceous fruit and presence of resin canals. Takhtajan (1987) places it in the Rutales, near Burseraceae. Although there are resemblances between the two families (Anacardiaceae and Burseraceae) in having resin canals and wood anatomical structures, there are differences in general habit and floral morphology. These resemblances could be due to parallel evolution.

In our opinion, the Anacardiaceae is correctly placed in the Sapindales.

Aceraceae

Aceraceae is a small family of only 2 genera—*Acer* and *Dipteronia*, and about 120 species. *Acer* is indigenous to mountainous regions of the northern hemisphere and *Dipteronia* is indigenous to Central China.

Vegetative Features   Trees or shrubs; leaves simple, opposite, pedicellate, usually with palmate venation or pinnately compound, sap often milky, estipulate.

Floral Features.   Inflorescence corymb, racemes or panicles. Flowers unisexual or less commonly bisexual; when unisexual, plants may be monoecious, dioecious or polygamodioecious, i.e. functionally dioecious but has a few flowers of the opposite sex or a few bisexual flowers on all plants at flowering time. Flowers actinomorphic, hypogynous. Calyx of 5 or 4 sepals, polysepalous or basally connate, imbricate; petals 5 to 4 or none, polypetalous, imbricate; extra- or intrastaminal disc usually present, mostly flat, sometimes lobed, divided or reduced to teeth. Stamens 4 to 10 (mostly 8), usually arising from the margin of the disc; anthers bicelled, filaments short and thick; rudimentary ovary often present in staminate flowers. Gynoecium syncarpous, bicarpellary, ovary bilocular, superior, usually compressed at right angles to the septum; placentation axile with 2 ovules per locule. Styles 2, divergent, free or basally connate, stigmas 2, terminal. Fruit a samaroid schizocarp separating into 2 single-winged mericarps (Fig. 33.3A) or a double samara (Brizicky 1963). Seeds non-endospermous with a straight embryo.

Fig. 33.3   Aceraceae: **A** *Acer maple,* winged fruits.

Anatomy.   Vessels moderately small, with spiral thickening, perforations simple, intervascular pitting alternate and moderately large. Parenchyma in terminal bands, or absent or with a few cells round the vessels. Rays 5 to 7 cells wide, homogeneous. Leaves usually dorsiventral; hairs of various types occur. Stomata confined to the lower surface; Ranunculaceous type.

Embryology.   Pollen grains tricolporate and shed at 2-celled stage. Ovules anatropous, bitegmic, crassinucellate. Embryo sac of Polygonum type, 8-nucleate at maturity. Endosperm formation of the Nuclear type. Suspensor polyembryony is reported in *Acer platanoides* (Johri et al. 1992).

Chromosome Number.   The basic chromosome number is x = 13.

Chemical Features.   Quebrachitol occurs in 20 species of *Acer;* but not recorded from *Acer carpinifolium* and *Dipteronia sinensis.*

Important Genera and Economic Importance.   *Acer,* with about 120 species, is an important genus, occurs mostly in temperate or high altitude regions. The other genus, *Dipteronia,* is confined to Central China.

The bark of most species of *Acer* contains sugar and occurs in sufficient quantities to be extracted, as in *Acer saccharinum.* The sap containing the sugar is obtained by tapping the trunks. Several species of *Acer* such as *A. rubrum, A saccharum* and *A pseudoplatanus* are sources of valuable timber.

Taxonomic Considerations.   In Bentham and Hooker's (1965a) system, the two genera, *Acer* and *Dipteronia,* are included in the Sapindaceae. They are distinct from other Sapindaceous members due to opposite and mostly palmately veined leaves, actinomorphic flowers, schizocarpic fruits and septate fibres. Most other taxonomists consider it as a distinct family Aceraceae, belonging to the order Sapindales.

The family Aceraceae is most closely allied to the Anacardiaceae. The two families resemble each other in a number of morphological, floral, anatomical and embryological features such as: (a) trees or shrubs with simple, estipulate or stipulate leaves, (b) inflorescence of various types: racemes, corymbs, panicles or thyrsoids, (c) flowers often unisexual, small, actinomorphic and hypogynous, (d) Onagrad type of development of embryo, (e) exalbuminous seeds, (f) vessels with simple perforation plates, (g) alternate intervascular pitting, (h) leaves dorsiventral, and (i) Ranunculaceous type of stomata.

There is no doubt that Anacardiaceae and Aceraceae are closely related and belong to the same order, Sapindales.

Bretschneideraceae

Bretschneideraceae is a monotypic family, the genus *Bretschneidera* has one species earlier known to occur only in southwestern China to northern Vietnam (Fu and Fu 1984 in Tobe and Peng 1990); recently discovered in Taiwan (Hsu and Lu 1984 in Tobe and Peng 1990, Lu et al. 1986).

Vegetative Features.   Trees; leaves alternate, compound, imparipinnate, estipulate.

Floral Features.   Inflorescence a terminal raceme. Flowers large, slightly zygomorphic, hermaphrodite and hypogynous. Calyx of 5 sepals, gamosepalous, aestivation open; corolla of 5 petals, polypetalous, imbricate, inserted on the calyx tube. Stamens 8, filaments hairy, anthers versatile, bicelled, longitudinally dehiscent, disc absent. Gynoecium syncarpous, tricarpellary, ovary trilocular, superior with axile placentation; ovules pendulous, 2 per locule; style long, simple and curved, stigma small, capitate. Fruit an obovate, thick walled, 3-valved, dehiscent capsule. Seeds red, without endosperm.

Anatomy.   Wood anatomy is similar to that of Sapindaceae.

Embryology.   Embryology has not been fully investigated. Ovule campylotropous, bitegmic, crassinucellate; outer integument thick. Embryo sac formation of Allium type, 8-nucleate at maturity. Seed exalbuminous (Tobe and Peng 1990).

Chromosome Number.   Basic chromosome number is x = 9 (Lu et al. 1986, Yang and Hu 1985).

Chemical Features.   Myrosin cells and glucosinolates occur in *Bretschneidera* (Tobe and Peng 1990).

Important Genera and Economic Importance.   The only genus is *Bretschneidera,* and economic importance is not known.

Taxonomic Considerations.   Earlier authors did not recognise Bretschneideraceae as a distinct family. Radlkofer (1908) treated the genus *Bretschneidera* as related to Capparales or Moringales. Zygomorphic flower, tricarpellary and trilocular ovary suggest the relationship with the Moringaceae. But according to Heimsch (1942), the wood anatomy of the Bretschneideraceae is closely related to the Sapindaceae. Cronquist (1968) and Hutchinson (1973) include *Bretschneidera* in the family Sapindaceae. Dahlgren (1980a) is uncertain about the position of this family although he retains it in the Sapindales. Metcalfe and Chalk (1972) include the genus *Bretschneidera* in the family Hippocastanaceae, and Takhtajan (1987) recognises it as a distinct family close to Hippocastanaceae in the Sapindales. On the basis of embryological data, Bretschneideraceae resembles both Hippocastanaceae and Sapindaceae more closely than Moringaceae and Capparaceae (Tobe and Peng 1990), although it is known to be one of the glucosinolate-producing families like the Capparaceae.

It appears that the family Bretschneideraceae needs to be studied in more details to come to any conclusion.

Sapindaceae

A large family of 150 genera and 2000 species distributed in the tropics and subtropics of Asia and America.

Vegetative Features. Trees (*Koelreuteria*), shrubs (*Dodonaea*), or sometimes tendril-bearing climbers such as *Cardiospermum halicacabum* (Fig. 33.5A), perennial or annual. Thorns represent reduced axillary branches in *Stocksia*. Some of the erect members of this family are branchless with a terminal collection of leaves, and resemble a palm tree, e.g. *Pseudima, Talisia, Tripterodendron,* etc. Leaves alternate (opposite in *Valenzuelia*), simple (*Dodonea*) or more commonly compound (*Koelreuteria, Sapindus*); bipinnate in *Macphersonia, Cardiospermum* and *Paullinia,* trifoliolate in *Hypelate;* usually estipulate (stipulate in climbing species). Leaves covered with resiniferous glands in *Dodonaea*.

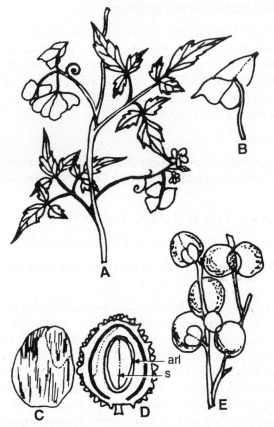

Fig. 33.5 Sapindaceae: **A, B** *Cardiospermum halicacabum,* **C, D** *Litchi chinensis,* **E,** *Sapindus laurifolius.* **A** Twig with flowers and fruits. **B** Fruit. **C** Aril of *L. chinensis.* **D** Fruit, longisection. **E.** A bunch of fruits of *Sapindus laurifolius. arl* aril, *s* seed. (**A, B** Original, **C-E** after G.L. Chopra 1973)

Floral Features. Inflorescence terminal or axillary racemes or panicles of cymes. Flowers bisexual or unisexual, plants monoecious, dioecious or polygamodioecious; actinomorphic (*Litchi, Nephelium* and *Sapindus*) or zygomorphic (*Allophylus, Cardiospermum* and *Erioglossum*), hypogynous. Calyx of 5 (in zygomorphic flowers) or 4 (in actinomorphic flowers), sepals mostly polysepalous, imbricate (valvate in *Koelreuteria*); petals 5 or 4 or absent as in *Dodonaea*, polypetalous, imbricate, equal or unequal, often with scaly or hair-tufted nectaries on lower inner side; extrastaminal glandular disc present; horn-like disc glands occur in *Xanthoceras sorbifolia*. Stamens 10, in 2 whorls of 5 each (12 or more in *Deinbollia*), free, inserted within or on the disc; anthers vesatile, introrse or extrorse as in *Melicoccus,* longitudinally dehiscent. Gynoecium syncarpous, usually 3-, rarely 2- or 4(–6)-carpellary (Brizicky 1963); ovary usually

3-loculed, rarely 2-, 4- or 6-loculed, superior, placentation axile, rarely parietal, ovules 1 or 2 per locule, ascending; style 1, short or elongated, stigma simple or lobed. Fruit variable: drupe, berry, capsule or samara-like; lobes of fruit inflated in *Cardiospermum* (Fig. 33.5B). Seed mostly 1 per locule, non-endospermous, arillate (Fig. 33.5D) and with curved or coiled embryo. Fruit, a dry, wing-like, papyraceous capsule in *Koelreuteria paniculata*; seeds with a bony seed coat (Meyer 1976).

Anatomy.   Wood occasionally ring-porous; vessels typically small and numerous, with many multiples of 2 or 3 cells but without a definite pattern, perforations simple, intervascular pitting alternate and small to minute. Parenchyma paratracheal, sparse, sometimes vasicentric. Rays exclusively uniseriate, numerous, homogeneous. Various types of anomalous secondary growth occurs in different lianas. Leaves dorsiventral, hairs of various types reported. Stomata Rubiaceous or Ranunculaceous type, confined to the lower surface.

Embryology.   Pollen grains single, spheroidal, polar view triangular, 3-colporate, exine finely striate. Ovules campylotropous or anatropous, bitegmic and often with a third integument developing into aril. Embryo sac of Polygonum type, 8-nucleate at maturity. Endosperm formation of the Nuclear type.

Chromosome Number.   Basic chromosome number is variable: $x = 10$ to 16. In *Koelreuteria* $x = 11$ and the diploid number is $2n = 22$ (Meyer 1976).

Chemical Features.   The Sapindaceae is the only family in which cyanolipids occur in seed oils (Seigler and Kawahara 1976). Seed fats of various members of this family are rich in arachidic, behenic or lignoceric acids (Shorland 1963); the aliphatic polyol, quebrachitol, has been isolated from *Heterodendron oleaefolium* (Plouvier 1963). Triterpenoid saponins have also been isolated from various Sapindaceous members, e.g. *Sapindus* spp.

Important Genera and Economic Importance.   *Cardiospermum halicacabum* is an interesting plant in which the pair of tendrils for climbing is formed by the modified pedicels at the lowest part of each raceme. In *Dodonaea viscosa* the leaves are simple and encrusted with a viscid, resinous secretion, and the fruits are 3-winged. The edible part of the fruits of *Litchi chinensis* (Fig 33.5 C, D) is the aril, which is the modified third integument. Many members of this family contain saponin in their stem and/or fruit, e.g. *Sapindus laurifolius* (Fig. 33.5E) and *S. mukorossi*. Pericarp of the fruits of these plants makes lather with water, and is largely used as a substitute for soap for washing silk and woollen fabrics.

The trees of *Koelreuteria paniculata,* commonly called varnish tree, and shrubs of *Xanthoceras sorbifolia* are cultivated as hardy ornamental plants. Other ornamental plants include *Euphoria* (Longan), *Melicoccus* (Spanish lime) and *Ungnadia* (Mexican buckeye). *Dodonaea viscosa* is often grown as a hedge plant as its leaves are not eaten by cattle. *Litchi chinensis* is cultivated in the tropical and subtropical regions, for its soft, fleshy, sweet and juicy edible aril. *Blighia sapida* of tropical Africa, *Melicoccus bijugatus* and *Euphoria longan* of tropical America, *Nephelium lappaceum* of Malaya and *Pappea capensis* of South Africa are other important fruit trees. Lac tree or *Schleichera oleosa* is a large deciduous tree in India. It is considered to be one of the best hosts for the culture of lac insects. The seeds of this plant also yield commercial 'kusum' oil, and the flowers are the source of a dye.

Taxonomic Considerations.   The Sapindaceae is a member of the order Sapindales. The placement of the order differs in various systems of classification. In habit and general floral structures, Sapindaceae is related to the Anacardiaceae, although it differs from the latter in having irregular flower and in the absence of intercellular canals in the stem tissue. Occurrence of septate fibres links the Sapindaceae with the Meliaceae. The cyanolipids is a taxonomic marker for this family; it has not been recorded in any other family of the order Sapindales.

Hippocastanaceae

The Hippocastanaceae is a small family of only 2 genera and 15 species distributed mostly in North and South America. *Aesculus indica* is abundant in the northwestern Himalayas from India to Nepal between 4000 and 10000 ft. altitude. *A. hippocastanum* or horse chestnut is occasionally cultivated in northern India as an ornamental tree and also for its edible fruits.

Vegetative Features.   Trees or shrubs; leaves opposite, palmately compound, digitately 3- to 9-foliolate; leaflets with serrate or entire margin and pinnately nerved; estipulate.

Floral Features.   Inflorescence terminal panicles or thyrses. Flowers hermaphrodite, sometimes the upper ones unisexual and staminate, zygomorphic, hypogynous. Sepals 4 or 5, basally connate, polysepalous in *Billia*, imbricate; petals 4 or 5, polypetalous, unequal, clawed, imbricate; extrastaminal disc present. Stamens 5 to 9, free, filaments elongated; anthers bicelled, basifixed, dehiscence longitudinal. Gynoecium syncarpous, tricarpellary; ovary trilocular, superior, placentation axile with 2 ovules in each locule; style and stigma 1. Fruit usually a single-seeded and single-loculed leathery loculicidal capsule. Seeds large, non-endspermous and non-arillate, with a curved embryo.

Anatomy.   Vessels small and numerous, often angular, as in *Aesculus,* with numerous multiples and clusters; perforations simple, intervascular pitting alternate; solid deposits and tyloses often present. Parenchyma terminal, scanty paratracheal, sometimes very sparse or absent, often storied. Rays exclusively uniseriate, homogeneous or slightly heterogeneous.

Embryology.   The embryology has not been investigated fully. The pollen grains are 2-celled at dispersal stage. Ovules amphitropous, bitegmic, crassinucellate. Polygonum type of embryo sac, 8-nucleate at maturity. Endosperm formation of the Nuclear type (Johri et al. 1992).

Chromosome Number.   Basic chromosome number is x = 10. Interspecific hybrids produced readily within the genus *Aesculus.*

Chemical Features.   Flavonoids occur in *Aesculus.* Coumarin glycoside aesculin is reported from *A. hippocastanum* and fraxin from *A. turbinata* (Paris 1963).

Important Genera and Economic Importance.   Of the two genera, *Billia* and *Aesculus, Aesculus* spp. are known for their timber. *A. hippocastanum* is a native of northern Greece and is grown all over the temperate regions as an ornamental tree. The timber is fine-grained, light but fairly tough.

Taxonomic Considerations.   The Hippocastanaceae are included in the Sapindaceae (Hallier 1912, Hutchinson 1926). It is treated as a distinct family near Sapindaceae by Bentham and Hooker (1965a), Engler and Diels (1936), Hutchinson (1969, 1973), Dahlgren (1980a, 1983a) and Takhtajan (1980, 1987). It is distinct from the Sapindaceae in palmately compound leaves, thyrses of large, ornamental flowers, and a leathery capsule with a very large, solitary seed and in the absence of the chemical cyanolipids.

Sabiaceae

A small family of 4 genera and 120 species distributed mostly in the tropical eastern Asia. The genus *Meliosma* is occasional throughout the tropics.

Vegetative Features.   Trees, shrubs or vines (*Meliosma*); leaves alternate, simple or pinnate, estipulate; surface hairy in *Meliosma.*

Floral Features.   Inflorescence terminal panicles; flowers bisexual, zygomorphic, hypogynous, ebracteate, ebracteolate, pedicellate. Sepals 3 to 5, free or basally connate, imbricate; petals 4 or 5, sometimes basally connate, the outer ones usually broad and imbricate, with the inner 2 much reduced; annular disc small. Stamens 3 to 5, free, opposite the petals, free or adnate to petals, all or only 2 fertile and the rest reduced to staminodia; anthers bicelled, the loculi separated by a thick connective and dehisce by a transverse slit or deciduous cap. Gynoecium syncarpous, bicarpellary, ovary bilocular (bicarpellary derived from a tricarpellary condition still seen in some members), superior, placentation axile, ovules usually 2 in each locule, pendulous, horizontal or ascending, styles 2 but often connate, stigmas 2. Fruits berries, sometimes dry and leathery; seed usually non-endospermous with a large embryo.

Anatomy. Vessels mostly medium-sized, exclusively solitary or with some groups, perforation plates simple or simple and scalariform, sometimes reticulate, intervascular pitting alternate. Parenchyma paratracheal, rather scanty as in *Meliosma* or absent as in *Sabia,* occasionally vasicentric as in *Ophiocaryon.* Rays usually up to 3–9 cells wide, sometimes 15–20 cells high, heterogeneous. Fibres sometimes septate. Leaves dorsiventral, stomata confined to the lower surface, Ranunculaceous in some species and Rubiaceous in others.

Embryology. Pollen grains tricolporate and 2-celled at dispersal stage. Ovules hemianatropous, unitegmic, crassinucellate. Embryo sac of Polygonum type, 8-nucleate at maturity. Endosperm formation of the Helobial type (Johri et al. 1992).

Chromosome Number. The haploid chromosome number is n = 12 in *Sabia* and n = 16 in *Meliosma.*

Important Genera and Economic Importance. Most of the species belong to *Meliosma* and *Sabia; Phoxanthus* and *Ophiocaryon* are monotypic. Some species are cultivated as ornamentals.

Taxonomic Considerations. Sabiaceae is recognised as a distinct family and is allied to the Sapindaceae by all phylogenists (Lawrence 1951, Melchior 1964, Cronquist 1981, Takhtajan 1987).

Melianthaceae

The Melianthaceae is a small family of 2 genera (*Melianthus* and *Bersama*) and 15 species confined to tropical west Africa. One species, *Melianthus major,* is almost naturalised in the Kumaon Hills, Darjeeling and the Nilgiri Hills in India and in Bhutan (Kumar and Subramanian 1987). A third genus, *Greyia,* with 3 species, was segregated as a distinct family, Greyiaceae by Hutchinson (1926) and placed in Cunoniales.

Vegetative Features. Trees, shrubs, or infrequently suffrutescent herbs. Leaves alternate, pinnately compound, imparipinnate, stipulate; stipules interpetiolar; leaflet margin serrate or dentate.

Floral Features. Inflorescence terminal or axillary racemes. Flowers hermaphrodite, sometimes interspersed with staminate and pistillate flowers in the same inflorescence; zygomorphic, hypogynous; pedicel of flowers twisted by 180° by the time of anthesis. Sepals 5 or 4 by connation, free or basally connate, unequal, imbricate; petals 4 or 5, free, clawed, unequal, imbricate. Disc cresent-shaped, extrastaminal. Stamens 4 or 5, or 10, free, or shortly basally connate, alternipetalous; anthers bicelled, dehiscence longitudinal. Gynoecium syncarpous, 4- or 5-carpellary; ovary 4- or 5-locular, superior, usually deeply lobed, placentation axile, ovules 1 to many on each placenta, erect or pendulous, style 1, stigma 4- to 5-lobed (sometimes truncate, capitate or toothed). Fruit a loculicidal or apically dehiscent capsule; seeds often arillate, endospermous, embryo straight.

Anatomy. Vessels rather small, perforations simple, intervascular pitting alternate, and very small. Parenchyma paratracheal, scanty with scattered crystalliferous cells as in *Bersama.* Rays 3–9 cells wide, homogeneous, no uniseriates. Fibres extremely short and with simple pits. Leaves dorsiventral, stomata confined to the lower surface; Ranunculaceous type. Styloids (elongated prismatic crystals) often present.

Embryology.   The elliptical pollen grains are monads and tricolporate and shed at 2- or 3-celled stage. Ovules anatropous, bitegmic and crassinucellate. Embryo sac of Polygonum type, 8-nucleate at maturity. Endosperm formation of the Nuclear type (Johri et al. 1992).

Chromosome Number.   The basic chromosome number is x = 19.

Chemical Features.   Seigler and Kawahara (1976) report the absence of cyanolipids from seed oils of Melianthaceae, although it is closely related to the Sapindaceae.

Important Genera and Economic Importance.   *Melianthus* is sometimes cultivated as an ornamental tree.

Taxonomic Considerations.   The genera comprising the Melianthaceae are included in Sapindaceae by Bentham and Hooker (1965a). Melianthaceae has been recognised as a distinct family by Lawrence (1951), Hutchinson (1973), Cronquist (1981), and Takhtajan (1987). Dahlgren (1983a), however, includes it in the Sapindales with some reservation. According to Takhtajan (1987), Melianthaceae is allied to Sapindaceae and Hippocastanaceae.

Greyia.   The genus *Greyia* has sometimes been included in the Melianthaceae (Lawrence 1951). Hutchinson (1973) recognised a distinct family Greyiaceae and placed it in the order Cunoniales, between Grossulariaceae and Escalloniaceae. Greyiaceae differs from Melianthaceae in a few morphological, floral, embryological and anatomical features as shown in Table 33.8.1.

Table 33.8.1   Comparative Data for *Greyia* and Melianthaceae

Greyia	Melianthaceae
1. Leaves simple, stipules absent	Leaves pinnate, stipules interpetiolar
2. Flowers actinomorphic	Flowers zygomorphic
3. Sepals free, petals equal, disc cupular within the petals	Sepals partly united, petals unequal, disc unilateral lining the calyx
4. Stamens 10, obdiplostemonous	Stamens 5, haplostemonous
5. Placentation parietal, ovules numerous	Placentation axile, ovules 2–4
6. Capsule septicidal	Capsule loculicidal
7. Intervascular pitting scalariform.	Intervascular pitting alternate
8. Rays heterogeneous; raphides present	Rays homogeneous; styloides present

Heimsch (1942) considers the anatomical differences sufficient to justify the segregation of the genus *Greyia* as a separate family next to Melianthaceae.
The two taxa *Greyia* and Melianthaceae resemble each other in: (a) usually perfect flowers, (b) ovules more than two per locule or carpel, (c) similar pollen structure, and (d) extrastaminal disc.

In our opinion, the genus *Greyia* can be associated with the family Melianthaceae, in view of the above mentioned resemblances between the two taxa. However, because of the differences between them, the two groups should be treated as two closely related but distinct families, in the Sapindales.

Balsaminaceae

A small family of 4 genera (only two genera according to Lawrence 1951) and 600 species distributed widely, although more common in the tropics of Asia and Africa.

Vegetative Features.   Plants herbaceous, sometimes aquatic (*Hydrocera triflora*) or suffrutescent. Stem succulent in *Impatiens,* often with adventitious supporting roots which grow from the basal part of the stem. Leaves alternate, opposite or in whorls of three, simple, petiolate, with serrate margin; extrafloral nectaries on petiole and a few lower serrations or teeth of leaf margin, usually estipulate.

Floral Features. Flowers solitary or several on axillary peduncles; bisexual, zygomorphic, hypogynous, often spurred, nodding, pentamerous. Calyx of 3–5 sepals, often petaloid, imbricate, the posterior sepal very large, sac-like and gradually prolonged backward into a tubular spur which contains nectariferous glands. Petals 5, alternate with sepals, free or connate and appear as 3 petals, lower ones larger than the upper ones. Stamens 5, more or less syngenesious, filaments flattened and closely cover the ovary and often the style, like a hood, anthers connate at the top of the ovary in *Hydrocera* (Venkateswarlu and Lakshminarayana 1957); anthers bicelled. Gynoecium syncarpous, pentacarpellary, ovary pentalocular, superior, placentation axile; ovules 3 to many on each placenta, pendulous; style 1, short, stigmas 1 to 5. Fruit a fleshy, 5-valved capsule with explosive dehiscence; the valves coil up elastically and the tension distributes the seeds forcibly in *Impatiens;* a drupe in *Hydrocera.* Seeds endospermic with a straight embryo (non-endospermic in *Hydrocera*).

Anatomy. About 12 vascular bundles distinct and arranged in a circle (as seen in a transection of stem). Vessels with simple perforations and spiral thickenings. Large cells filled with mucilage are scattered in the ground tissue. Raphides present in both leaf and stem. Leaves dorsiventral, extrafloral nectaries recorded on the petiole and stem of some species. Stomata present on both surfaces or only on the lower surface as in *Impatiens sultani;* mostly Ranunculaceous type, some tend to be Cruciferous type.

Embryology. Pollen grains 3- to 5-colpate and shed at the 2-celled stage; exine reticulate. Ovules anatropous, bitegmic, tenuinucellate. Integuments fused along their length, are free for a short distance only at the top. Embryo sac of Allium type, 8-nucleate at maturity. Endosperm formation is of the Cellular type, with both micropylar and chalazal haustoria.

Chromosome Number. The basic chromosome number is x = 13.

Chemical Features. *Impatiens* spp. contain anthocyanin pigments 3, 5-diglucoside, pelargonidin and paeonidin. The seeds of *I. balsamina* and *I. glandulifera* contain fatty acids of the nature of linoleic, oleic, and linolenic acids.

Important Genera and Economic Importance. *Hydrocera triflora* is an aquatic member bearing red globose succulent fruits (Venkateswarlu and Lakshminarayana 1957). Many species of *Impatiens* are cultivated for their beautiful flowers.

Taxonomic Considerations. The family Balsaminaceae has been treated as a subfamily of the Geraniaceae by Bentham and Hooker (1965a) and as an independent family by Lawrence (1951), Hutchinson (1973), Melchior (1964), Takhtajan (1966) and Cronquist (1968). Lawrence (1951) and Melchior (1964) place it in the Sapindales, on the basis of similar ovule attachment. Dahlgren (1980a, 1983a) and Takhtajan (1987) prefer to place it in a separate order, the Balsaminales, close to Geraniales, Oxalidales and Tropaeolales. As reported by L.L. Narayana (1970e), Balsaminaceae resembles Oxalidaceae, Geraniaceae and Tropaeolaceae in some exomorphological and embryological features. According to L.L. Narayana, anatomical, embryological and cytological features indicate its alliance with the Geraniales and others, but not with Sapindales.

Under the circumstances, the family Balsaminaceae is best treated as a member of a distinct order Balsaminales, as suggested by Takhtajan (1987).

Taxonomic Considerations of the Order Sapindales

The circumscription of the order Sapindales differs according to various phylogenists. Engler and Diels (1936) and Lawrence (1951) included 23 families in this order, whereas Melchior (1964) includes only 10 families, Cronquist (1968) 17 families, and Takhtajan (1987) 12 families. There is, however, a general agreement that the Sapindales should include the families Sapindaceae, Hippocastanaceae, Sabiaceae, Aceraceae, Melianthaceae and Anacardiaceae.

The family Coriariaceae is interesting because of its discontinuous distribution from South America to some smaller islands near Australia, New Zealand, China, and Japan. The monotypic Bretschnideraceae is also distributed in China. The position of this family is uncertain (Dahlgren 1980a, 1983a) and some taxonomists presume its alliance with the Capparales or Moringales. The rest of the families show close resemblance to each other. Balsaminaceae is the only herbaceous member of this order and reportedly has some alliance with the Geraniales (L.L. Narayana 1970e). Takhtajan (1987) places it in a separate order Balsaminales.

The order Sapindales has some resemblance to and is probably derived from the Rutales.

34

Order Julianiales

A monotypic order comprising the family Julianiaceae.

Julianiaceae

Julianiaceae is a small family of only 2 genera; *Amphipterygium* from Mexico and *Orthopterygium* from Peru.

Vegetative Features.   Resinous shrubs and small trees. Leaves alternate, pinnate (Fig. 34.1A), bear resin glands, estipulate. Plants dioecious.

Floral Features.   Flowers reduced, unisexual. Staminate flowers in catkins, consist of 3 to 9 stamens surrounded by 3 to 9 sepals (Fig. 34.1B). Pistillate flowers in groups of 3 or 4 enclosed by an involucre (Fig. 34.1C), perianth and disc absent; gynoecium syncarpous, tricarpellary, ovary unilocular with only one ovule, stigma broad and elongated (Fig. 34.1C). Fruit a samara, nutlike, 3 or 4 in a cluster, enclosed in and attached to the persistent involucre (Fig. 34.1A). Seeds non-endospermous, embryo large, cotyledons plano-convex.

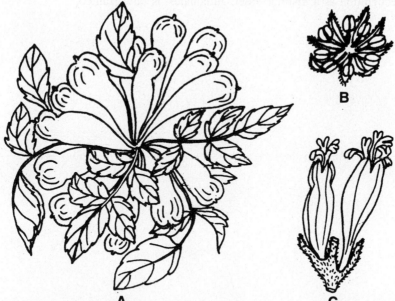

Fig. 34.1   Julianiaceae: **A-C** *Amphipterygium adstringens.* **A** Fruiting twig. **B** Staminate flower. **C** Pistillate flowers
(Adapted from Hutchinson 1969).

Anatomy.   Vessels small, perforations simple, intervascular pitting opposite to alternate. Parenchyma paratracheal, scanty; rays up to 6 cells wide, very few uniseriates, almost homogeneous. Leaves dorsiventral, stomata confined to lower surface, Ranunculaceous type. Secretory canals contain resin, common in phloem tissue of veins and petioles of leaves, and the stem.

Embryology.   Pollen grains small and spheroidal. Ovules semi-anatropous, unitegmic and partly covered by a large fleshy structure which develops from the funiculus.

Chemical Features.   Embryo rich in oils, protein and starch; tannins occasionally present in older stems; triterpene saponins and quebrachitol and polygalitol are also common (Dahlgren 1975a).

Important Genera and Economic Importance.   The two genera of this family are *Amphypterygium* (= *Juliania*) and *Orthopterygium*. Bark of *A. adstringens* is used medicinally as an astringent.

Taxonomic Considerations of the Order Julianiales

Bentham and Hooker (1965a) considered *Juliania* as a member of Anacardiaceae. Hemsley (1907; see Hutchinson 1969) created the family Julianiaceae and, while admitting that they resembled the Anacardiaceae, he considered their affinities with Juglandales and Cupuliflorae (= Fagales). Anatomical studies (particularly that of secondary xylem) support its alliance with Anacardiaceae (Fritsch 1908, Kramer 1939, Heimsch 1942). Hutchinson (1926) and Wettstein (1935) considered Julianiaceae as a member of Juglandales, but later (1969, 1973) Hutchinson supported its resemblance with the Anacardiaceae and placed Julianiaceae in the Sapindales. Dahlgren (1975a) also supports this alliance. However, Dahlgren (1980a, 1983a) includes the genus *Juliania* in Anacardiaceae. Takhtajan (1980) showed its affinities with this family and actually treats it as a subfamily Julianioideae of Anacardiaceae (1987).

Name of the type genus *Juliania* Schlechtd is illegitimate and has been changed to *Amphipterygium* Standley (Hutchinson 1969, Takhtajan 1987).

It is clear that Julianiaceae is so closely related to the Anacardiaceae that it can even be included in this family; its recognition as a distinct order, Julianiales, is unwarranted.

35

Order Celastrales

The order Celastrales comprises 13 families: Cyrillaceae, unigeneric Pentaphylacaceae, Aquifoliaceae and Corynocarpaceae, Pandaceae, Celastraceae, Staphyleaceae, Hippocrateaceae, Stackhousiaceae, Salvadoraceae, Buxaceae, Icacinaceae, and unigeneric Cardiopteridaceae.

The members are generally woody plants, except some Stackhousiaceae; leaves opposite or alternate, mostly estipulate. Flowers usually with a disc except in Salvadoraceae, tetra- or pentamerous, haplostemonous. Ovules in most members anatropous, unitegmic, tenui- or crassinucellate. Seeds with or without endosperm.

Cyrillaceae

A small family of only 3 genera—*Cyrilla*, *Cliftonia* and *Purdiaea*—and 14 species, distributed in North and South America. The first two taxa are monotypic and indigenous to the southeastern United States of America.

Vegetative Features. Deciduous or evergreen shrubs or small trees. Leaves alternate, simple, entire, coriaceous, estipulate.

Floral Features. Inflorescence a raceme. Flowers ebracteate, ebracteolate, hermaphrodite, actinomorphic, hypogynous (Fig. 35.1A). Sepals 5, polysepalous, basally connate, imbricate, rarely valvate, persistent, often enlarged in fruit. Petals 5, polypetalous or slightly basally connate, imbricate or contorted. Stamens 10 in 2 whorls of 5 each (*Cliftonia*, *Purdiaea*) or the inner whorl absent or reduced to staminodes as in *Cyrilla* (Fig. 35.1A); filaments shorter than petals, free and dilated, anthers bicelled, dehiscence longitudinal. Gynoecium syncarpous, 2- to 4-carpellary, ovary superior, 2- to 4-loculed with axile placentation and usually 1 or 2 or rarely 4 pendulous, collaterally disposed ovules per locule (Fig. 35.1B). Style 1, short or nearly obsolete, stigmas 2 and linear-ovate. Fruit broadly 3- to 4-winged, as in *Cliftonia*, or wingless, as in *Purdiaea*; a dehiscent capsule or a leathery or fleshy drupaceous berry as in *Cyrilla* (Fig. 35.1C). Seed with fleshy endosperm and a small straight embryo.

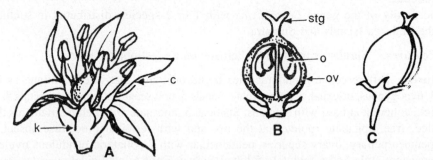

Fig. 35.1 Cyrillaceae : **A-C** *Cyrilla racemiflora*. **A** Flower. **B** Ovary, longisection. **C** Fruit. *k* calyx, *c* corolla, *o* ovule, *ov* ovary, *stg* stigma. (Adapted from Lawrence 1951)

Anatomy. Vessels very small and numerous especially in *Cliftonia* (Metcalfe and Chalk 1972); occasionally with a tendency to be semi-ring-porous. Perforation plates exclusively scalariform with numerous, fine, closely spaced bars; intervascular pitting scalariform to opposite. Parenchyma diffuse or diffuse to scanty paratracheal. Rays mostly 4 or 5 cells wide with a few uniseriates that are 1 or 2 cells high, heterogeneous. Leaves sometimes with irregularly arranged ridges of cuticle on the lower side, as in *Cliftonia nitida*. Stomata confined to the lower surface; surrounded by numerous ordinary epidermal cells. Mesophyll consists wholly of palisade tissue. Both solitary and clustered crystals are reported.

Embryology.   Pollen grains monads, shed at 2-celled stage. Ovule anatropous, unitegmic, tenui- or crassinucellate. Polygonum type of embryo sac, 8-nucleate at maturity. In *Cliftonia monophylla* both normal and aposporous embryo sacs develop in the same ovule (Vijayaraghavan 1969). Endosperm formation of the Cellular type. Both micropylar and chalazal endosperm haustoria are formed. The mature embryo has a small suspensor.

Chromosome Number.   Basic chromosome number is x = 10.

Important Genera and Economic Importance.   *Cyrilla* and *Cliftonia* are the two important genera. *Cyrilla recemosa* or leatherwood and *Cliftonia monophylla* or buckwheat are often cultivated as ornamentals for their fragrant, white flowers and showy autumn coloration of the foliage.

Taxonomic Considerations.   The phyletic position of this family is disputed. Bentham and Hooker (1965a) negate any relationship with the Ericales. Some taxonomists suggest its association with Aquifoliaceae and Celastraceae (Lawrence 1951, Melchior 1964, Hutchinson 1969, 1973). Hallier (1912) included it in his Bicornes (= Ericales). Copeland (1953), also on the basis of embryological evidence, relates it to the Bicornes (= Ericales). Vijayaraghavan (1969, 1970) compared the embryological features of Aquifoliaceae, Celastraceae, Cyrillaceae and Ericaceae.

Cyrillaceae show close affinities with Ericaceae in unitegmic, tenuinucellate ovules, Polygonum type of embryo sac with a broad micropylar and narrow chalazal end, uninucleate persistent antipodals, and Cellular endosperm with chalazal and micropylar haustoria. Cyrillaceae also resembles Clethraceae (another member of the Ericales) in : (a) presence of fibrous endothecium, (b) druses in the cells of connective, and (c) pollen grains shed as monads (Vijayaraghavan and Dhar 1978). Cyrillaceae differ from Celastraceae because of: (a) unitegmic, tenuinucellate ovules, (b) Cellular type of endosperm, and (c) absence of nucellar polyembryony.

Cronquist (1981), Dahlgren (1983a, b) and Takhtajan (1987) include Cyrillaceae in the Ericales near Clethraceae. The position of Cyrillaceae in the Celastrales is doubtful.

Pentaphylacaceae

An unigeneric family of the genus *Pentaphylax* with 1 or 2 species, distributed in southeastern Asia, particularly the Malacca Islands and Sumatra.

Vegetative Features.   Shrubs or trees with alternate, entire, estipulate leaves.

Flower Features.   Inflorescence pseudoracemes borne on twigs that often continue as leafy shoots. Flowers small, hypogynous, bisexual, actinomorphic. Sepals 5, polysepalous, imbricate; petals 5, polypetalous, imbricate, thick, coherent at base with stamens. Stamens 5, alternate with petals, filaments thick and flat, anthers bicelled, free, apiculate (pointed at the tip) and with terminal pore; disc absent. Gynoecium syncarpous, pentacarpellary, ovary superior, pentalocular with 2 collateral pendulous ovules per locule; one simple persistent style and a minutely 5-lobed stigma. Fruit a loculicidal capsule; seeds with very little endosperm, and an embryo.

Anatomy.   Vessels solitary, perforation plates and intervascular pitting scalariform; vessels extremely long. Parenchyma diffuse; rays up to 5 cells wide, heterogeneous. Leaves dorsiventral, epidermis mucilaginous. Stomata confined to lower surface, mostly Rubiaceous type.

Embryology. Embryology has not been investigated fully. The ovules are ana- or campylotropous, bitegmic and crassinucellate. The seeds are more or less winged, with scanty endosperm and horse-shoe-shaped embryo (Hutchinson 1973).

Important Genera and Economic Importance. *Pentaphylax* is the only genus; it has no economic importance.

Taxonomic Considerations. *Pentaphylax* was treated as a member of Ternstroemiaceae (= Theaceae) by Bentham and Hooker (1965a). It was retained in the order Theales by Hutchinson (1969).

Willis (1973) also mentions its relationship with Theaceae. However, in the Englerian system it is treated as a distinct family with affinities with Cyrillaceae and Celastraceae. Lawrence (1951) and Melchior (1964) relate it with Cyrillaceae and Celastraceae. Cronquist (1968, 1981) includes it in Theaceae. Takhtajan (1987) places it in the Theales. The Pentaphylacaceae is related to Theaceae and its placement in the Celastrales is untenable; the correct assessment is possible only when this taxon has been fully investigated.

Aquifoliaccae

A small family of only 2 genera—*Ilex* (incl. *Byronia*) and *Nemopanthus*—with about 400 species distributed in tropical and temperate zones of eastern United States, Central America, Asia and Australia.

Vegetative Features. Evergreen or deciduous trees or shrubs with alternate, simple, thick, coriaceous, stipulate leaves, margin entire, dentate or spinous, stipules minute.

Floral Features. Inflorescence a fascicle or axillary cyme of a few flowers, or solitary axillary flowers. Flowers bisexual or unisexual (then the plants dioecious or polygamodioecious), actinomorphic, hypogynous, small and greenish. Sepals 3 to 6, more or less basally connate, imbricate; corolla of 4 to 9 petals, free or slightly connate basally, rotate, imbricate. Stamens 4 to 9, free, alternipetalous, sometimes adherent basally to petals; filaments thick and short; anthers basifixed, bicelled, longitudinally dehiscent, disc absent; sterile anthers bear staminodes; usually present in pistillate flowers and a rudimentary pistil in staminate flowers. Gynoecium syncarpous, 3- to many-carpellary, ovary superior, 3- to many-loculed with axile placentation, ovules 1 or 2 on each placenta, pendulous; style 1, terminal or absent, stigma lobed or capitate. Fruit a berry or drupe, usually with 4 pyrenes. Seeds with copious fleshy endosperm and a minute straight embryo.

Anatomy. Vessels small, often in marked radial groups or lines, spiral thickenings usually present in *Ilex* and *Byronia*, perforation plates exclusively scalariform and with many bars; oblique end walls with (11)-14-(20) bars in *I. collina* (Baas 1984); intervascular pitting usually opposite. Parenchyma more or less abundant, diffuse; rays of 2 size, the larger 5–15 cells wide and 2–5 mm high, heterogeneous. Leaves dorsiventral, stomata mostly confined to the lower surface, Ranunculaceous type. Leaf margins strengthened mechanically by cuticular thickenings, or sclerenchymatous tissue that contains chloroplast (*I. insignis*) or by a sheath of sclerenchyma (*I. aquifolius*).

Embryology. Pollen grains tricolporate with poorly defined pores, variable exine ornamentations, and prolate or oblate as in *Ilex* (Martin 1977); 2-celled at shedding stage. Ovules apotropous, unitegmic, crassinucellate. Polygonum type of embryo sac, 8-nucleate at maturity. Antipodals persist after fertilization. Endosperm formation is of the Cellular type; no endosperm haustoria. In *Ilex* the embryo undergoes continuous development until germination which may take 2 to 8 years (Martin 1977).

Chromosome Number. Basic chromosome number is x = 9, 10.

Important Genera and Economic Importance.   *Ilex* is an important genus with about 295 species that are widely distributed in the temperate regions. It is common to the eastern United States and Asia, with its chief centre of world distribution in Central and South America. Commonly known as holly, its leaves with prickly margin and fleshy drupaceous fruits are used in Christmas decorations. Various species of *Ilex* are also important for the white hard wood. The other genus, *Nemopanthus*, is not known to be of any economic use. Leaves of 60 different species of *Ilex* are used for making beverages in South and North America, South Africa and the Chinese-Tibetan border area. *I. paraguariensis* is cultivated extensively for tea in Brazil and Paraguay. It is poorly represented in Australian flora but fossil pollen of *Ilex* proves its earlier wide distribution (Martin 1977).

Taxonomic Considerations.   Most phylogenists include Aquifoliaceae in the suborder Celastrineae of the order Sapindales (Lawrence 1951). Melchior (1964) erected a separate order Celastrales for this family; accepted by Hutchinson (1969, 1973), Cronquist (1968, 1981), Dahlgren (1980a, 1983a) and Takhtajan (1987).

Corynocarpaccac

A unigeneric family of the genus *Corynocarpus* with 5 species distributed in New Zealand and some of the Pacific Islands. *C. laevigatus* occurs in coastal and lowland forests (Patel 1975).

Vegetative Features.   Trees or shrubs; leaves alternate., simple, entire, estipulate; young leaves with hairy surface.

Floral Features.   Inflorescence terminal panicles. Flowers bisexual, complete, actinomorphic, hypogynous. Calyx of 5 sepals, polysepalous, imbricate; corolla of 5 petals, polypetalous, imbricate. Stamens 5, antipetalous and epipetalous, alternate with 5 petaloid staminodes; an intrastaminal disc of 5 large cushion-like glands are present opposite to the staminodes. Gynoecium syncarpous, 1- or 2-carpellary, ovary superior, with only 1 fertile locule with 1 pendulous ovule (Philipson 1987), the other locule sterile; styles and stigmas 1 or 2, free. Fruit a globular fleshy drupe, seed exalbuminous.

Anatomy.   Vessels moderately small, mostly in small multiples and irregular clusters; perforations simple, intervascular pitting alternate and small. Parenchyma abundant in discontinuous paratracheal bands and vasicentric, storied; rarely scanty paratracheal (Patel 1975). Rays multiseriate, up to 16 cells wide and very high; uniseriates absent, heterogeneous. Leaves dorsiventral with globular multicellular hairs on both surfaces of young leaves which fall off in older leaves. Stomata confined to the lower surface; Ranunculaceous type.

Embryology.   Pollen grains shed at binucleate stage (Gardner 1975). Ovules anatropous, bitegmic, and crassinucellate. Development of embryo sac not investigated. Endosperm formation of the Nuclear type.

Chromosome Number.   The haploid chromosome number is n = 22.

Chemical Features.   The seeds of *Corynocarpus laevigatus* contain a poisonous substance resembling digitalin. According to Hegnauer (1964), polyphenols of *Corynocarpus* are similar to those in some Myrtales (e.g. Melastomataceae), and karakin is probably the same as hiptagin (from *Hiptage*, Malpighiaceae).

Important Genera and Economic Importance.   *Corynocarpus* is the only genus, with 4 or 5 species. The fruits of *C. laevigatus* are edible after culinary treatment, but not very palatable. The Maoris of New Zealand are quite fond of these fruits, which they consume after prolonged soaking or steaming to remove the toxic substance (Patel 1975).

Taxonomic Considerations. The position of the genus *Corynocarpus* is uncertain and it has been included variously in the Berberidaceae, Myrsinaceae and Anacardiaceae (Metcalfe and Chalk 1972). Bentham and Hooker (1965a) included it in the Anacardiaceae. Krause (1942a) raised it to family rank and placed Corynocarpaceae in the Sapindales. Cronquist (1968) included this family in the Ranunculales. According to him, the wood anatomy of *Corynocarpus* suggests that they are secondarily woody. Metcalfe and Chalk (1972) also pointed out that there are many common wood-anatomical features between Corynocarpaceae and Berberidaceae (also of the Ranunculales). However, Cronquist (1981) regrouped it in the Celastrales. Melchior (1964), Dahlgren (1980a) and Takhtajan (1987) also include it in the Celastrales. Philipson (1987) reported that the absence of resin cavities, occurrence of leaves with sheathing stipules, and presence of bitegmic crassinucellate ovules are some of the features that suggest the affinity of Corynocarpaceae with the Celastrales.

Pandaceae

A small family of 4 genera and 28 species, distributed in west tropical Africa and Asia.

Vegetative Features. Dioecious trees with alternate or distichous simple leaves; often serrate, stipulate.

Floral Features. Inflorescence axillary fascicles as in *Microdesmis*, or cymes as in *Centroplacus*, or in terminal (*Galearia*) or cauliflorus (*Galearia* and *Panda*) racemiform thyrses. Flowers unisexual, borne on separate plants, ebracteate, ebracteolate, actinomorphic, hypogynous. Calyx of 5 sepals, gamosepalous, imbricate or open, cupular; petals 5, polypetalous, imbricate or valvate. Stamens 5 to 10 or 15, sometimes unequal; filaments long, anthers basifixed, introrse, small staminodes sometimes present as in *Centroplacus*; staminodes and disc absent from female flowers. In male flowers a linear-subulate rudimentary ovary is present. Gynoecium syncarpous, 2- to 5-carpellary, ovary superior, 2- to 5-loculed; ovules usually 1 or rarely 2 per locule as in *Centroplacus*, apical, pendulous. Style short with 2-5-(10)-lobed stigma. Fruit drupaceous, flattened, rarely a capsule (e.g. *Centroplacus*), usually variously tuberculate or muricate or pitted or ridged, sometimes dehisce by valves. Seeds endospermous, usually flattened-concave, rarely ovoid as in *Centroplacus*.

Anatomy. Similar to that of Euphorbiaceae.

Embryology. Embryology has not been fully investigated. Pollen grains similar to some Euphorbiaceae members. Ovules ortho- or anatropous, bitegmic. Seeds usually flattened, rarely ovoid, with copious fatty endosperm and a straight embryo (Johri et al. 1992).

Chromosome Number. Diploid chromosome number is $2n = 30$ in *Microdesmis* (Takhtajan 1987).

Important Genera and Economic Importance. Four genera—*Panda, Galearia, Microdesmis* and *Centroplacus*—are included in this family; none of these is of any economic importance.

Taxonomic Considerations. The family Pandaceae is included in the Euphorbiaceae by most authors. Melchior (1964) and Willis (1973) include it in the Celastrales. However, the presence of dioecious flowers justifies its placement in the Euphorbiales.

Cronquist (1968, 1981), Hutchinson (1969, 1973) and Takhtajan (1987) include this family in the Euphorbiaceae. The family Pandaceae was regarded as a unigeneric family till 1966, when Forman showed the relationship of *Panda* to *Galearia* and *Microdesmis* (traditionally regarded as members of the Euphorbiaceae). At present, Pandaceae includes a fourth genus *Centroplacus* (Takhtajan 1987).

Celastraceae

A family of 58 genera and 860 species, distributed widely except in the Arctic regions.

Vegetative Features. Trees or shrubs, sometimes climbers or twining (e.g. *Maytenus*); branches spiny in *Acanthothamnus* and *Gymnosporia*. Leaves alternate or opposite (as in *Euonymus maackii*), simple, deciduous or evergreen, stipulate or estipulate, stipules small and caducous. In *Sarawakodendron* young branches, petiole, midrib and floral parts contain sulphur-yellow kautchuk particles (Hou 1967).

Floral Features. Inflorescence cymose, often borne on the midrib, e.g. *Polycardia baroniana* (Fig. 35.6A). flowers bisexual (Fig. 35.6B), sometimes functionally unisexual and then plants are polygamodioecious; actinomorphic, hypogynous. Calyx of 4 or 5 sepals, polysepalous but basally connate, imbricate (Fig. 35.6 B); petals 4 or 5, rarely absent, e.g. *Microtropis*; polypetalous, imbricate or valvate as in *Caryospermum* and *Perrottetia*. Stamens usually 4 or 5, sometimes 10, as in *Glossopetalous*, arise from the rim of an intrastaminal disc that surrounds the ovary; anthers bicelled, dehiscence longitudinal. Anther-connective very thick with the loculi divergent at the top in *Glyptopetalum*; connective protrudes beyond anther lobes in *Kokoona*. Gynoecium syncarpous, 2- to 5-carpellary, ovary superior, 2- to 5-loculed, with axile placentation; ovules usually 2 on each placenta, erect (Fig. 35.6C); style 1, short, stigma capitate or indistinctly 2- to 5-lobed. Fruit a loculicidal capsule, berry, samara or drupe, a thick, coriaceous, hard capsule in *Sarawakodendron*; seed usually covered with a bright-coloured pulpy aril; endospermous, embryo enveloped by the endosperm.

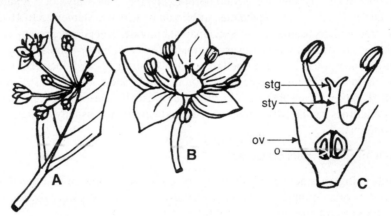

Fig. 35.6   Celastraceae : **A** *Polycardia baroniana,* **B,C** *Celastrus orbiculatus.* **A** Inflorescence. **B** Flower of *C. orbiculatus.* **C** Longisection, sepals and petals removed. *o* ovule, *ov* ovary, *stg* stigma, *sty* style. (**A** after Hutchinson 1969, **B, C** after Lawrence 1951)

Anatomy. Vessels typically small, mostly solitary, sometimes ring-porous or semi-ring-porous, occasionally with spiral thickening; perforations simple (scalariform in *Kurrimia* and *Parrottetia*). Parenchyma very sparse or absent, diffuse or in broad multiseriate bands in some genera. Rays mostly uniseriate, homogeneous and with prominent intercellular spaces. Leaves generally dorsiventral but isobilateral in *Gymnosporia*, *Maytenus* and *Mortonia*; hairs infrequent, but when present, unicellular or uniseriate. Stomata usually confined to the lower surface, mostly Cruciferous and Ranunculaceous type (Rubiaceous in *Kurrimia*); gaurd cells surrounded by a rosette of smaller epidermal cells in *Mortonia*. Sometimes stomatal type of even a single leaf may vary (Hartog and Baas 1978).

Embryology. Pollen grains 3-colporate, shed at 2-celled stage; exine smooth, granulate or reticulate;

in *Lophopetalum* pollen grains are shed in tetrads or polyads, triporate or tricolpate (Hou 1969). The ovules anatropous, bitegmic, crassinucellate, Polygonum type of embryo sac, 8-nucleate at maturity. Endosperm formation of the Nuclear type, endosperm haustoria absent. Polyembryony common.

Chromosome Number.   Basic chromosome number is x = 12 or 16+.

Chemical Features.   An active principle, maytansine, with significant tumour-inhibiting activity in vivo has been obtained from *Maytenus* sp. and *Putterlickia verrucosa* (Sebsebe 1985).

Important Genera and Economic Importance.   Some of the Celastraceae are cultivated as ornamentals for their handsome foliage and decorative fruits, for example, *Euonymus alatus* and *E. japonicum*. The spindle tree, yielding wooden spindles used in textile machinery is *E. europaeus*. Leaves of khat plant, *Catha edulis*, are used in making Arabian tea. Trees of *E. atropurpureus* commonly called burning bush are so called because the leaves turn purplish-red during autumn. The genus *Maytenus* is important in traditional medicine. They contain a potential anticancer agent (Sebsebe 1985).

Taxonomic Considerations.   The family Celastraceae is included in the Celastrales by Bentham and Hooker (1965a), Loesener (1942), Hutchinson (1973), Cronquist (1981), Dahlgren (1983a) and Takhtajan (1987). Thorne (1968), however, treats this family under Santalales, and Lawrence (1951) under Sapindales. The family is allied to the Aquifoliaceae from which it differs in having the intrastaminal glandular disc and brightly-coloured arillate seeds.

According to Hartog and Baas (1978), the family comprises ca. 90 genera and 1000 species if Hippocrateaceae is included on the basis of macromorphology. anatomy and pollen morphology.

Staphyleaceae

A small family of 5 genera and 50 to 60 species (3 genera and 46 species according to Takhtajan 1987), distributed mostly in the north-temperate regions. Earlier phylogenists included *Huertea* and *Tapiscia* also in this family. These are now included in a separate family, Tapisciaceae.

Vegetative Features.   Shrubs or trees; leaves opposite, pinnately compound, pinnately trifoliate in *Staphylea trifolia*, stipulate, stipules linear, leaflets stipellate.

Floral Features.   Inflorescence drooping racemes or panicles. Flowers bisexual or polygamous, rarely imperfect, actinomorphic, hypogynous, bracteate, bracteolate. Calyx of 5 sepals, polysepalous, imbricate, deciduous; petals 5, polypetalous, imbricate, both sepals and petals arise from or below the hypogynous disc, often connate at the base. Stamens 5, free, alternipetalous, inserted on or below the large cupular disc; anthers bicelled, dehiscence longitudinal. Gynoecium syncarpous, 2- or 3-carpellary, ovary superior, 2- or 3-loculed, with axile placentation, ovules numerous, in 2 rows on each placenta; styles as many as carpels and distinct. Fruit, a membranous, often lobed, inflated, apically dehiscent capsule. Seeds few, subglobose, sometimes arillate, endospermous and with a straight embryo. Endosperm stores fatty food reserve. In *Staphylea* the number of capsules per inflorescence is low, compared to the number of flowers per inflorescence (Sponberg 1971).

Anatomy.   Vessels small, solitary, numerous, sometimes with spiral thickenings; perforation plates oblique, scalariform. Wood diffuse-porous with scanty paratracheal parenchyma; rays 4–7 cells wide and usually more than 1 mm high, heterogeneous. Leaves dorsiventral, with glandular leaf teeth; glandular hairs absent. Stomata Cruciferous type, confined to lower surface.

Embryology.   Pollen grains 2- or 3-colporate or polyporate, 2-nucleate at shedding stage. Ovules apotropous, bitegmic, crassinucellate. Embryo sac of Polygonum type, 8-nucleate at maturity. Ovule abortion and low seed formation are common. Low percentage of seed may be due to self-incompatibility (Sponberg 1971). Endosperm formation of the Nuclear type.

Chromosome Number.   Basic chromosome number is x = 13.

Important Genera and Economic Importance.   The type genus *Staphylea* is a very important plant. Species of *Staphylea* are deciduous shrubs, rarely trees with terete branches and grey to black bark that is often mottled (Sponberg 1971). Leaves trifoliate as in *S. trifolia* or pinnately compound. Commonly called as bladder nuts because of the inflated bladdery fruits.

Taxonomic Considerations.   Bessey (1915), Rendle (1925), Wettstein (1935) and Melchior (1964) placed Staphyleaceae in the Celastrales. On the basis of wood anatomy, Heimsh (1942) supported this assignment. Hallier (1908, 1912) and Van Steenis (1959, 1960) included it in the Cunoniales, on the basis of strong morphological and anatomical similarities. Davis (1966) also supported this view on the basis of embryological data. Van Steenis (1960) pointed out that "it represents a marked northern counterpart of the Cunoniaceae which is largely a southern hemisphere family". Most other phylogenists, however, include this family in the Sapindales (Bentham and Hooker 1965a, Krause 1942b, Hutchinson 1973, Cronquist 1981, Dahlgren 1983a and Takhtajan 1987). According to Cronquist (1981), the presence of compound leaves supports the inclusion of this family in the Sapindales. If retained in the Celastrales, Staphyleaceae is to be treated as the most primitive family.

Johri et al. (1992) support its position in the Celastrales, on the basis of embryological features. In our opinion also, Staphyleaceae should be retained in the Celastrales, keeping in view the morphological, floral, embryological and anatomical similarities: (a) trees or shrubs with mostly stipulate leaves, (b) bisexual, actinomophic, hypogynous flowers with imbricate aestivation of floral parts, (c) syncarpous, superior ovary with axile placentation, (d) arillate, endospermous seeds, (e) numerous small vessels with spiral thickening, (f) generally dorsiventral leaves with Cruciferous type of stomata most common, (g) pollen grains shed at 2-celled stage, (h) bitegmic, crassinucellate ovules, and (i) Nuclear type of endosperm.

Staphyleaceae is distinct from Celastraceae because of: pinnately compound leaves, 2- to 3-carpellary ovary with numerous ovules in 2 rows on each placenta, heterogeneous rays, apotropous ovules and low percentage of seeds due to self-incompatibility.

Hippocrateaceae

A small family of 18 genera and 225 species distributed in the tropical zones of both the New and the Old World.

Vegetative Features.   Trees, shrubs or lianas; vegetative parts with well-developed latex system. Leaves usually opposite, simple, entire, minutely stipulate or estipulate.

Floral Features.   Inflorescence thyrsoid, cymose, racemose, paniculate or fasciculate (Fig. 35.8A). Flowers

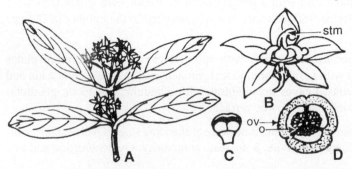

Fig. 35.8 Hippocrateaceae: **A-D** *Hippocratea africana*. **A** Flowering shoot. **B** Flower. **C** Stamen, note transverse dehiscence. **D** Ovary, cross section. *o* ovule, *ov* ovary, *stm* stamen. (Adapted from Hutchinson 1969)

ebracteate, bracteolate, bisexual, actinomorphic, hypogynous. Calyx of 5 sepals, polysepalous or fused up to the middle, imbricate, persistent. Petals usually 5, polypetalous (Fig. 35.8B), imbricate or valvate. Stamens usually 3 (rarely 2, 4 or 5), filaments dilated and often connate basally, inserted within the disc; the disc extrastaminal, usually annular and continuous, sometimes discontinuous and form staminiferous pockets; anthers bicelled, extrorse (introrse in *Tristemonanthus*), dehiscence transverse (Fig. 35.8C). Gynoecium syncarpous, usually 3-carpellary, ovary superior, trilocular (Fig. 35.8D), the disc sometimes adnate to its base and conceal it; placentation axile, ovules 2 to 14 in each locule, usually 2-ranked. The style trifid and usually short, stigmas may or may not be prominent, mostly 3, entire or bifid. Fruit a schizocarp, a 3-valved capsule, e.g. *Hippocratea*, or a berry. Seeds compressed, angular or winged, non-endospermous.

Anatomy. Vessels medium-sized to large, usually solitary, with simple perforations. Parenchyma paratracheal, sparse; rays exclusively uniseriate or of 2 distinct sizes, the larger up to 30 cells wide and very high. Included phloem present. Leaves generally dorsiventral, rarely centric in some species of *Hippocratea* and *Salacia*. Stomata usually confined to the lower surface and mostly Ranunculaceous type. Laticiferous canals present in both leaf and axis.

Embryology. Pollen grains tricolpate, rarely tetracolpate; with reticulate exine, 2-celled at dispersal stage. Ovules anatropous, bitegmic, tenuinucellate. Embryo sac of Polygonum type, 8-nucleate at maturity. Endosperm formation of the Nuclear type (Johri et al. 1992).

Chemical Features. Sugar dulcitel is reported in *Toutelea brachypoda*, roots of *Pristimera indica* and *Salacia prinoides* (Plouvier 1963).

Important Genera and Economic Importance. Hippocrateaceae has no economic importance.

Taxonomic Considerations. Most taxonomists treat Hippocrateaceae as a distinct family in the Celastrales. Hallier (1912) and Takhtajan (1987), however, regard it as a subfamily of Celastraceae. Smith (1940) acknowledged the affinity of this family with the Celastraceae but he also pointed out the differences between the two families: (a) the disc in Celastraceae is intrastaminal and merely surrounds the ovary and in Hippocrateaceae the disc is extrastaminal with the stamens seated on the disc, (b) the anther is transversely dehiscent in Hippocrateaceae and longitudinally dehiscent in Celastraceae.

Because of these differences, it would be better to treat Hippocrateaceae as a distinct family in the Celastrales and near Celastraceae.

Stackhousiaceae

A small family of 3 genera: *Macgregoria*, *Stackhousia* and *Tripterococcus* with 27 species, distributed in Malaysia, Australia and New Zealand.

Vegetative Features. Annual or perennial, more or less xerophytic, rhizomatous herbs. Leaves alternate, simple, entire, estipulate.

Floral Features. Inflorescence racemose or cymose, rarely umbellate. Flowers bisexual, actinomorphic, perigynous, bracteolate or ebracteolate. Calyx 5-lobed or 5-cleft, imbricate; corolla of 5 perigynous petals; petals long-clawed, claws elongated, either free at base and fused above to form a tube or totally free as in *Macgregoria*. Stamens 5, inserted on the margin of the interstaminal disc, equal and with short filaments as in *Macgregoria* or 3 long and 2 short stamens with elongated filaments, as in *Stackhousia* and *Tripterococcus*. Anthers with apical appendage in *Macgregoria*, obtuse or shortly mucronate in *Stackhousia* and *Tripterococcus*. Gynoecium syncarpous, 2- to 5-carpellary, ovary half-inferior, 2- to 5-

locular and 2- to 5-lobed with one erect ovule per locule; style 2 to 5, free or connate, with a discoid collar beneath stigmatic lobes in *Macgregoria* and without a collar in the other two genera. Fruit a schizocarp, covered with hooked hairs in *Macgregoria*; cocci winged in *Tripterococcus*.

Anatomy. In many species of *Stackhousia* the outer cortex of young stems contains abundant pitted fibres corresponding to the ribs on the surface of the stem. Tanniniferous cells common in cortex. Xylem vessels with simple perforations, a small amount of parenchyma and fibres with bordered pits; rays absent. Phloem cells of the rhizome contain secretory cells filled with dark-brown contents. Leaves dorsiventral or centric, stomata in *S. spathulata* are on both the surfaces; Ranunculaceous type.

Embryology. Pollen grains 3-celled, tricolpate with smooth exine in *Macgregoria* and lamellate-arelate exine in *Stackhousia* and *Tripterococcus*; occasionally double-sized pollen grains occur. Ovules anatropous, bitegmic, tenuinucellate. Polygonum type of embryo sac, 8-nucleate at maturity. Endosperm formation is of the Nuclear type; embryo with small undivided suspensor cell.

Chromosome Number. Basic chromosome number is much variable—x = 9, 10 or 15.

Important Genera and Economic Importance. Some members are grown as ornamentals.

Taxonomic Consideration. The family Stackhousiaceae is included in the Celastrales by most phylogenists: Bentham and Hooker (1965a), Bessey (1915), Hutchinson (1973), Cronquist (1981), Melchior (1964), and others. However, Hallier (1912) placed it in the Rosales.

The studies of Narang (1953) reveal that, embryologically, Stackhousiaceae is closest to Celastraceae and Hippocrateaceae, rather than to the Scrophulariaceae, Lobeliaceae or Selaginaceae. Morphological and anatomical features also support its inclusion in the order Celastrales.

Salvadoraceae

A small family of 3 genera—*Azima*, *Dobera* and *Salvadora*—with 11 or 12 species distributed in arid and semi-arid, often saline regions of Africa, Madagascar and tropical and subtropical Asia.

Vegetative Features. Trees, shrubs or scramblers, often armed with axillary spines as in *Azima*. Leaves opposite (Fig. 35.10A), simple, entire, petiolate, stipules minute, rudimentary, caducous or entirely absent. The entire plant usually olive-green in colour.

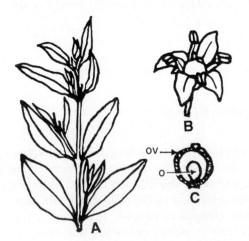

Fig. 35.10 Salvadoraceae : **A-C** *Salvadora oleoides*. **A** Twig. **B** Flower. **C** Ovary, longisection. *o* ovule, *ov* ovary. (Adapted from J.K. Maheshwari 1965)

Floral Features. Inflorescence a terminal raceme or spike, or dense axillary panicles. Flowers small, bisexual or polygamodioecious, actinomorphic, hypogynous. Calyx of 2 to 5 sepals, gamosepalous, imbricate or more or less valvate, campanulate; corolla of 4 or 5 petals, polypetalous in *Azima* and *Dobera*, or partly connate as in *Salvadora* (Fig. 35.10 B), imbricate or contorted, usually with teeth or glands on the inner surface. Stamens 4 or 5, epipetalous, inserted at or near the base of the petals, alternipetalous; sometimes free, as in *Azima*, filaments sometimes connate into a tube, e.g. *Dobera*; disc absent but stamens sometimes alternate with nectariferous glands. Anthers bicelled, dehiscence longitudinal. Gynoecium syncarpous, bicarpellary, ovary superior, unilocular with 1 ovule in *Salvadora persica* (Fig. 35.10 C) and bilocular with 2 ovules (1 in each locule) on basal placenta in *Azima tetracantha*; style 1 short with bifid stigma. Fruit a berry or drupe; seeds non-endospermous.

Anatomy. Vessels small, in radial multiples of 2 to 4 and in irregular clusters. Growth rings absent, wood diffuse-porous; perforation plates simple with horizontal end walls; intervascular pitting alternate. Parenchyma abundant, vasicentric and scantily diffuse; rays 10 cells wide, uniseriates few, sometimes storied. Included phloem present in *Dobera* and *Salvadora* (Singh 1944). Leaves dorsiventral or isobilateral, hairs infrequent. *Dobera* and *Salvadora* have thick and centric leaves, stomata are abundant all over the surface; in *Azima* only on the lower surface of the dorsiventral leaf. Stomata Rubiaceous type in *Salvadora*.

Embryology. Pollen grains triporate, 2-celled at shedding stage. Ovules anatropous, bitegmic, crassinucellate. Micropyle formed by inner integument alone in *Salvadora persica* and by both in *Azima tetracantha* (Maheswari Devi 1972). Endosperm development of the Nuclear type.

Chromosome Number. Basic chromosome number is x =12.

Chemical Features. Plants rich in glucosinolate or isothiocyanate (Dahlgren 1980a). Glucotropaeolin, a thioglucoside, is reported in *Salvadora persica*.

Important Genera and Economic Importance. *Salvadora*—commonly known as 'salt bush' as it grows on saline soils, or 'mustard tree' as the leaves taste like mustard, or 'toothbrush tree' as its twigs are used for brushing teeth—is the most important genus of this family. Young shoots and leaves are eaten as salad, and they also provide fodder for camels. It is also the source of vegetable salt called Kegr, derived from the ash of the plant. Seed fat finds application in the manufacture of candles. Root of *Azima tetracantha* is used medicinally against dropsy and rheumatism.

Taxonomic Considerations. Salvadoraceae is a small family but with an unusual assemblage of characters. Bentham and Hooker (1965a) and Bessey (1915) placed it in the order Gentianales, because of sympetaly in some of its members. Rendle (1925) preferred to treat it along with Oleaceae in the Oleales. These authors have ignored the importance of polypetalous corolla and hypogynous stamens in *Azima* and *Dobera*. Engler and Diels (1936) included this family in the Polypetalae, in the suborder Celastrineae of order Sapindales, owing to the free petals and hypogynous stamens in 2 of the 3 genera. Cronquist (1968, 1981), Gunderson (1950), Hutchinson (1973) and Melchior (1964) also place it in the Celastrales. Dahlgren (1980a), however, suggested a separate order, Salvadorales, in the superorder Violiflorae to accommodate this family, because of bicarpellary ovary, predominantly polypetalous condition, and presence of glucosinolates.

Embryological data (Maheswari Devi 1972) indicate that the Salvadoraceae stand apart from Gentianales and Oleales, and support its inclusion in the Celastrales. This is also supported by floral anatomical studies (Kshetrapal 1970). Biochemical data (Gibbs 1958) and pollen grain studies (Lobreau 1969) also suggest its relationship with Aquifoliaceae of the Celastrales. Den Outer and Van Veenandal (1981) accept this position of the family Salvadoraceae.

Gamopetaly in *Salvadora* should be regarded as an exceptional feature as the polypetalous condition in *Fraxinus* is exceptional in otherwise gamopetalous Oleaceae. Melchior (1964) and Takhtajan (1980, 1987) also regard Salvadoraceae as a member of the Celastrales.

Buxaceae

A family of 6 genera and 80–90 species indigenous to the tropical and subtropical regions of the Old World mainly, and also in mountainous regions of West Virginia, south to western Florida and Louisiana (Lawrence 1951). Distribution of the Buxaceae is discontinuous. *Pachysandra* occurs in the southeast United States of America (1 species) and in China and Japan (3 species). *Buxus*, the largest genus, is widespread but is absent from Australasia. *Buxus sempervirens* is widely scattered from southern England, Azores and Morocco to Asia Minor, with some very closely related species in China and Japan.

Vegetative Features.   Monoecious or dioecious, evergreen herbs, undershrubs or trees. Vegetative parts without latex. Leaves alternate (*Styloceros*) or opposite (*B. sempervirens*), simple, entire or toothed margin (*Pachysandra*), coriaceous, estipulate.

Floral Features.   Inflorescence axillary or terminal spike, or raceme with the terminal female flower and the rest male flowers. Flowers bracteate. ebracteolate, unisexual, actinomorphic, hypogynous, with or without perianth. Perianth of only one whorl of calyx; sepals usually 4 (sometimes 4–12 in pistillate flowers), basally connate. Staminate flowers with 4 stamens opposite to the sepals; filaments thick, anthers dorsifixed, introrse, longitudinally dehiscent; rudimentary ovary sometimes present. Pistillate flowers fewer than staminate ones; gynoecium syncarpous, tricarpellary; ovary superior, 3-loculed, deeply lobed, with axile placentation and 2 collateral and pendulous ovules per locule. Styles as many as carpels, simple, basally connate or widely separated and divergent. Fruit an ovoid 3-horned loculicidally dehiscent capsule. Seeds black, shiny, usually with a caruncle; endospermous and with a straight embryo.

Anatomy.   Vessels extremely small, exclusively solitary, usually with scalariform perforation plates. In *Simmondsia* the vessels have spiral thickening and simple perforation plates. Intervascular pitting usually very small, opposite to transitional (alternate in *Simmondsia*). Anomalous secondary growth observed in *Simmondsia*. Parenchyma apotracheal, diffuse; rays 2–4 cells wide, with numerous uniseriates, markedly heterogeneous. In *Simmondsia* parenchyma rare except for the bands of conjunctive tissue in the anomalous secondary growth. Leaves dorsiventral in *Buxus*, *Pachysandra*, *Sarcococca* and *Styloceros*. Hairs simple, thick-walled 1- to many-celled. Stomata usually confined to the lower surface, surrounded by a well-defined rosette of subsidiary cells; on both sides and of Ranunculaceous type in *Simmondsia*.

Embryology.   Pollen grains 2- or 3-celled at the dispersal stage. Ovules anatropous, bitegmic, crassinucellate. Embryo sac of Polygonum type, 8-nucleate at maturity. Endosperm formation is of the Nuclear type (Johri et al. 1992).

Chromosome Number.   Basic chromosome number is x = 10, 14.

Chemical Features.   β-diketones have been reported in waxes of certain species of *Buxus* (Behnke 1982). Seed wax of *Simmondsia californica* contain eicosenoic acid as a major constituent. Oleic acid is almost absent. Benzylisoquinoline alkaloids are also present (Dahlgren et al. 1981).

Important Genera and Economic Importance.   Amongst the genera of this family, *Buxus* is widely distributed. Its wood, commonly called boxwood, is used for engraving blocks, rulers, etc. *Simmondsia californica* seeds are rich in waxes.

Taxonomic Considerations.   The family Buxaceae has been treated variously. Some taxonomists consider it to be a member of the Celastrales (Melchior 1964), or of Sapindales (Lawrence 1951) or of Euphorbiales (Cronquist 1981), or even of Buxales (Takhtajan 1987).

The two genera, *Simmondsia* and *Styloceros*, are often treated as members of Simmondsiaceae and Stylocerotaceae, respectively (Willis 1973, Takhtajan 1987). Anatomically, these two genera are so different from other members of Buxaceae that it is justified to treat them as distinct families.

Buxaceae have many features in common with the Euphorbiaceae: unisexual, monochlamydeous flowers, tricarpellary, trilocular, superior ovary, axile placentation, seeds with caruncle and occurrence of benzylisoquinoline alkaloids. This family may be placed near the Euphorbiaceae, either in the Geraniales (Melchior 1964) or in the Euphorbiales (Cronquist 1968, 1981). Takhtajan (1987) considers Buxales to be a fairly primitive order near Balanopales and Casuarinales which is doubtful. As suggested by Cronquist (1968, 1981), the origin of Buxaceae (along with other members of his Euphorbiales) is in or near the Celastrales is probable.

Icacinaceae

A medium-sized family of 56 genera and 400 species (Takhtajan 1987), distributed in tropical regions of southern hemisphere—South Africa, southwestern India, northeastern Himalayas, Malaysia and northeast Australia.

Vegetative Features.   Trees, shrubs or tall climbers (*Icacina senegalensis*). Stems woody, sometimes climbing; tuberculate in *Trematosperma cordatum* and *Pyrenacantha* spp. Leaves usually alternate, opposite in *Cassinopsis*, *Iodes*, *Mappianthus*, *Polyporandra* and *Tridianisia*; simple, entire, spinulose-dentate in *Cassinopsis*; toothed in *Villaresia*, *Natsiatum*, *Natsiatopsis*, and *Hosiea*; lobed in *Natsiatum* and *Phytocrene*; often coriaceous, estipulate, palmately veined in *Natsiatum*, *Natsiatopsis*, *Phytocrene*, *Pyrenacantha* and *Polycephalium*.

Floral Features.   Inflorescence paniculate, usually terminal, leaf-opposed in *Gomphandra*, interpetiolar in *Cassinopsis ilicifolia*, extra-axillary in *Leptaulus*, and cauliflorous in *Pseudobotrys* and *Lavigeria*. Flowers hermaphrodite (rarely unisexual by abortion), actinomorphic, hypogynous, bracteate, ebracteolate. Calyx of 4 or 5 sepals, smaller than petals, polysepalous; petals 4 or 5, polypetalous, sometimes basally connate, valvate. Stamens as many as petals, alternipetalous (Fig. 35.12A), free; anthers bicelled, usually introrse, dehiscing longitudinally; locellate and opening by numerous pores in *Polyporandra*. Gynoecium syncarpous, 3- or 5-carpellary, all except 1 carpel are usually lost by suppression or compression; ovary unilocular, superior, placentation apical and usually 2 pendulous ovules from locule apex (Fig. 35.12B). Style 1, stigma usually 3, rarely 2 or 5. Fruit usually a drupe, rarely dry and winged; seeds endospermous, with a straight or curved embryo.

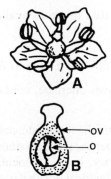

Fig. 35.12   Icacinaceae: **A,B** *Villaresia mucronata*. **A** Flower. **B** Ovary, longisection. *o* ovule, *ov* ovary. (After Lawrence 1951)

Anatomy.   Vessels show an unusual range from unspecialised to specialized forms. Unspecialised vessels are typically numerous, small, solitary, with exclusively scalariform perforation plates, scalariform or alternate pitting; vessels highly elongated. Specialised vessels are small to medium-sized, often in radial multiples and irregular groups, with simple perforation plates, and alternate pitting. In genera with scalariform perforation plates, parenchyma is apotracheal, mostly diffuse, sometimes in definite bands; in more advanced woods parenchyma varies from banded apotracheal to predominantly paratracheal. Rays 3 to 10 cells or more wide, heterogeneous and often with sheath cells. Included phloem present in climbers. Leaves dorsiventral, stomata Ranunculaceous and   Cruciferous type (Rubiaceous in *Apodytes dimidiata*). Hairs of various types reported on leaf surfaces; laticiferous tubes and crystals have also been observed.

Embryology.   Pollen grains shed at 2-celled stage (Johri et al. 1992). Ovules unitegmic, tenuinucellate; Polygonum type of embryo sac, 8-nucleate at maturity. Endosperm formation of the Nuclear type and endosperm haustoria present.

Chromosome Number.   Basic chromosome number is x = 10, 11.

Chemical Features.   Iridoids present (Dahlgren 1975b).

Important Genera and Economic Importance.   Some of the important genera are *Irvingbaileya*, *Icacina*, *Apodytes* and *Citronella*. A substitute of maté tea (*Ilex paraguariensis*) is obtained from *Citronella gongonha*. Starch and oil are obtained from fruit and seed of *Poraqueiba* in the Para region of Brazil. Tubers of *Humirianthera* are edible, but they have to be washed properly (before use) to remove toxicity. A blue dye is obtained from bark, leaves and fruits of *Calatola*. The wood of this family is not of much use except that of *Apodytes dimidiata* (South Africa), *Dendrobangia* (tropical America) and *Ottoschulzia* (West Indies) used locally. A few species of *Pennantia* and *Villaresia* are cultivated as ornamentals.

Taxonomic Considerations.   Many genera of this family were included in the tribes Icacineae and Phytocreneae of the Olacineae (= Olacaceae) in Bentham and Hooker's (1965a) system. Sleumer (1942b) treated Icacinaceae as a separate family under the Sapindales. Hutchinson (1973) and Wettstein (1935) included it as a distinct family under the Celastrales. Cronquist (1981) and Takhtajan (1987) also treat it as a member of Celastrales. However, Dahlgren (1975a, b, 1980a) includes it in the Cornales, because of similar embryological, anatomical and chemical features. He observed that the presence of iridoids is correlated with embryological characters such as unitegmic, tenuinucellate ovules, and Cellular type of endosperm with haustoria. On the other hand, according to Baas (1975), Icacinaceae is allied to Aquifoliaceae of the Celastrales, on the basis of embryological and exomorphological features, though iridoids are not reported from the Aquifoliaceae.

The position of Icacinaceae can be ascertained only after additional data becomes available.

Cardiopteridaceae

A monogeneric family, *Peripterygium*, with 3 species distributed in southeastern Asia to Australia.

Vegetative Features.   Climbing herbs with abundant milky latex. Leaves alternate, cordate, entire or lobed, membranous, estipulate.

Floral Features.   Inflorescence axillary, scorpioid cymes. Flowers very small, ebracteate, actinomorphic, hypogynous. Calyx of 5 sepals, gamosepalous, imbricate; corolla of 5 petals, gamopetalous, imbricate. Stamens 5, epipetalous, disc absent. Gynoecium syncarpous, bicarpellary, ovary unilocular, superior, with 2 apical pendulous ovules; styles 2, dissimilar, one longer, thicker, cylindrical or subclavate, persistent

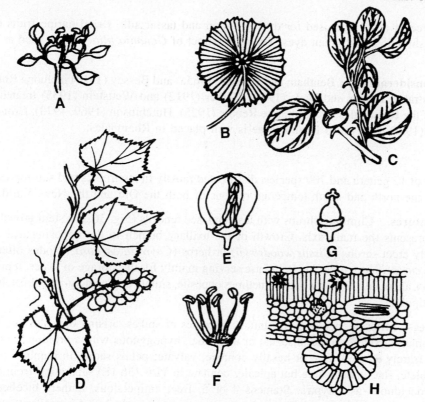

Fig. 36.1 **A–H. A–C** Rhamnaceae, **D–G** Vitaceae, **H** Leeaceae. **A** *Ceanothus americanus*, flower. **B** *Paliurus orientalis*, fruit. **C** *Ziziphus jujuba*, twig bearing fruit. **D** *Vitis vinifera*, fruiting twig. **E** Flower. **F** Staminate flower. **G** Pistillate flower. **H** *Leea aequata*, leaf section with pearl gland. (**A, E–G** adapted from Lawrence 1951, **B** after Hutchinson 1969, **H** after Metcalfe and Chalk 1972)

Chromosome Number. Basic chromosome numbers are x = 9–13 and 23.

Chemical Features. All the 23 species of *Rhamnus* and *Berchemia racemosa* contain L-bornesitol, one of the natural cyclitols. A glycoside, franguloside, is obtained from *R. frangula*; xanthorhamnin, a flavonoid, from *R. infectoria*. The anthaquinone glycoside, frangulin, occurs in *R. frangula* and *R. purshiana*. Seed fat of *Rhamnus* spp. is rich in linolenic acid.

Important Genera and Economic Importace. One important genus is *Ziziphus*, with 100 species distributed in tropical America, Africa, the Mediterranean region, Indo-Malaysia and Australia, and also the tropical parts of India, Nepal, Pakistan, Bhutan, Bangla Desh, and Sri Lanka. The taxa are undershrubs, shrubs as well as small trees. The genus *Rhamnus*, with 150 species of cosmopolitan distribution, is also important. *Ventilago*, with 37 species, is common in tropical Africa, Madagascar, India and China to New Guinea. *V. denticulata* is a strong woody climber with winged fruits. *Gouania tiliaefolia* is a climber with watch-spring-like tendrils on the inflorescence and inferior three-winged fruits.

Many Rhamnaceae members produce valuable timber, such as *Frangula alnus*, *Maesopis eminii* and *Ziziphus chloroxylon*. *Krugiodendron ferreum* of the West Indies yields black ironwood, one of the heaviest woods with a specific gravity of 1.3. Fruits of various species of *Ziziphus* and the fruit axis of *Hovenia dulcis* are edible. Berries of *Rhamnus cathartica*, *R. frangula* and *R. purshiana* are cultivated for their bark of medicinal value. The bark of *Ventilago denticulata* is used for making cordage and its root is the source of a red dye used for dyeing cotton and tassar silk. Dried unripe fruits of *Rhamnus infectoria* furnish yellow and green dyes. The root extract of *Gouania tiliaefolia* is used as a substitute for soap.

Taxonomic Considerations. Bentham and Hooker (1965a) and Bessey (1915) included Rhamnaceae in the order Celastrales. Engler and Diels (1936), Hallier (1912) and Wettstein (1935) treated this family as a member of the Rhamnales. According to Rendle (1925), Hutchinson (1969, 1973), Cronquist (1981) and Takhtajan (1980, 1987) also Rhamnaceae is well-placed in Rhamnales.

Vitaceae

A small family of 12 genera and 700 species, distributed mostly in the tropics and subtropics, sometimes extending into the north and south temperate regions of both the Old and the New World.

Vegetative Features. Climbing shrubs with leaf-opposed tendrils (Fig. 36.21); stem growth sympodial, and tendril represents the main axis. Growth of the axillary branch in the opposing leaf axil is more vigorous. Rarely erect shrubs (*Cissus woodrowii*) or herbs (*Cayratia carnosa*). Nodes often swollen or jointed. *Cissus currori* has a tuberous stem base serving mainly for the storage of water, it produces erect branches. Leaves alternate, the lower ones sometimes opposite, simple or pinnately- or palmately-compound, stipulate or estipulate.

Floral Features. Inflorescence cymes, panicles, racemes or spikes, arising opposite to a leaf at node. Flowers very small, actinomorphic, bisexual or unisexual, hypogynous with a prominent disc, greenish. Sepals 4 or 5 (rarely 3 to 7), free or basally connate, valvate; petals same (in number) as the sepals, minute or obsolete, flat, usually free but apically connate in *Vitis* (36.1E) separating from each other at the base and deciduous as calyptra. Stamens 4 or 5, free, antipetalous, anthers bicelled, dorsifixed (Fig. 36.1F), longitudinally dehiscent. Gynoecium syncarpous, 2- (rarely 3- to 6-carpellary); ovary superior or more or less united with a nectariferous gland (Fig. 36.1G), number of locules same as the number of carpels, ovules 1 or 2 per locule, on axile placentation; style 1 and short, stigma discoid or capitate. Fruit a berry; seeds small with cartilaginous endosperm and a small, straight embryo.

Anatomy. Vessels large, or both large and small, commonly in radial multiples; perforation plates simple; intervascular pitting usually scalariform, sometimes opposite or alternate. Parenchyma paratracheal, usually sparse, sometimes moderately abundant. Rays broad and very high, up to 20 cells wide in some species, uniseriate rays less common. Leaves dorsiventral, hairs of various types and deciduous, pearl glands (in *Vitis*) present; stomata Ranunculaceous type. Raphides and raphide sacs common in parenchyma tissue. P-type plastids known (Behnke 1974, 1975).

Embryology. Pollen grains tricolporate, with smooth exine; 2- or 3-celled at the time of shedding. A high degree of degeneration of pollen grains has been reported (N.C. Nair 1970c). Ovules anatropous, epitropous, bitegmic and crassinucellate; micropyle formed by inner integument alone. Embryo sac of Polygonum type, 8-nucleate at maturity. The endosperm formation is of the Nuclear type and the endosperm is ruminate.

Chromosome Number. Basic chromosome number varies—it is x = 11–13, 15, 19. 2n = 48 and 50 in some species of *Cissus;* 2n = 60 in *Cayratia japonica*.

Chemical Features. Phytoalexin viniform occurs in *Vitis*. Other chemicals like 3-glucoside, 3, 5-diglucoside, cyanidin, paeonidin, delphinidin, petunidin, malvidin and p-coumaric acid also occur in the fruits of *Vitis*.

Important Genera and Economic Importance.   The most important genus is *Vitis* with 60 to 70 species and numerous varieties and cultivers, and it is cultivated extensively for its edible fruit. Dried fruits are made into raisins; they are also fermented into wine. *Parthenocissus tricuspidata* (Boston ivy) and *P. quinquefolia* (Virginia creeper) are cultivated as ornamentals. *Cissus quadrangularis* is a climber with a jointed quadrangular herbaceous, sympodial stem, grown as an ornamental. Stems of *V. papillosa* and *V. sicyoides* are used as cordage. Grape seeds yield an oil used in the manufacture of soaps and paints, the waste is used as cattle feed. Some plants like *Cissus setosa* and *Cayratia* are used in indigenous medicines.

Cissus with 350 species is grown mostly in the tropics.

Taxonomic Considerations.   Most authors link the family Vitaceae with the Rhamnaceae and place the two in the order Rhamnales (Engler and Diels 1936, Wettstein 1935, Lawrence 1951, Melchior 1964, Cronquist 1981). Hutchinson (1973) includes two other families—Elaeagnaceae and Heteropyxidaceae— also in this order. Takhtajan (1987), however, includes Vitaceae along with Leeaceae in a separate order Vitales.

Rhamnaceae and Vitaceae have many morphological, floral-anatomical and embryological features in common. But there are some striking differences in the anatomical and embryological characters: (1) P-type plastids in Vitaceae and S-type in Rhamnaceae, (2) raphide sacs in parenchyma of Vitaceae which are absent in Rhamnaceae, and (3) ruminate endosperm of Vitaceae in not known in Rhamnaceae (N.C. Nair 1970b, c, Takhtajan 1987). Because of these differences, the two families—Rhamnaceae and Vitaceae— have often been treated as members of two distinct orders. This is not necessary, and the two families can be retained is the order Rhamnales.

Leeaceae

A unigeneric (genus *Leea*) family with 70 species distributed in the tropics of Asia, Africa and Australia.

Vegetative Features.   Trees, shrubs or herbs, occasionally prickly. Leaves pinnate to tripinnate, rarely ternate or simple, alternate, usually dentate, estipulate, petiole usually with 2 auricles or sheathing expansions near base.

Floral Features.   Inflorescence usually corymbose, many-flowered, erect and terminal (rarely axillary and pendulous), often ferrugino-tomentose. Flowers ebracteate, ebracteolate, hermaphrodite, actinomorphic, hypogynous, small, greenish. Calyx lobes 5 or 4, gamosepalous, cupular, shortly dentate or lobate. Corolla lobes same as number of sepals, valvate, reflexed, polypetalous, united with the staminal tube at the base. Stamens 5 or 4, oppositipetalous and epipetalous; staminal tube short or long, conical or subglobose, variously 5-lobed, lobes entire or bifid, alternate with stamens, with an internal free, membranous tube, pendent midway from the staminal tube. Anthers extrorse, longitudinally dehiscent. Gynoecium 3- to 8-carpellary, syncarpous; ovary superior, somewhat immersed in the receptacle, with simple style and axile placentation; ovules one per locule. Fruit 3- to 8-loculed dry berry; seeds with ruminate endosperm and small erect embryo.

Anatomy.   Vessels small (less than 100 μm mean tangential diameter), perforation plates simple; parenchyma paratracheal, sparse and scattered among the fibres in *Leea sambucina;* rays uniseriate or broad. Leaves usually dorsiventral, often with deciduous pearl glands (Fig. 36.1H) on the surface. These glands are spherical with a short few-celled stalk and a group of polygonal cells surrounded by an epidermis perforated by stomata; druses common in *Leea*. P-type plastids are characteristic.

Embryology.   Pollen grains triangular, 3-colporate with smooth (N.C. Nair 1970d) or scalariform (netted) exine; 2- or 3-celled at shedding stage. Ovules anatropous, apotropous, bitegmic, crassinucellate. Polygonum type of embryo sac, 8-nucleate at maturity. Endosperm formation of the Nuclear type; mature endosperm ruminate.

Chromosome Number.   Basic chromosome number is x = 11, 12.

Important Genus.   *Leea* is the only genus; it has no economic importance.

Taxonomic Considerations.   Bentham and Hooker (1965a) and Hutchinson (1973) included the genus *Leea* in family Vitaceae. The erection of the family Leeaceae was suggested by Dumortier 1829 (cf Süssenguth 1953), and accepted by Lawrence (1951), Süssenguth (1953), Melchior (1964) and Takhtajan (1987). Dahlgren (1983a), however, includes *Leea* in the family Vitaceae. Periasamy (1962) based on the development of ruminations in the seed concluded that *Leea* should remain a member of Vitaceae. The family Leeaceae differs from the Vitaceae in many respects (Table 36.3.1):

Table 36.3.1   Comparative Data for Leeaceae and Vitaceae

Feature	Leaceae	Vitaceae
Habit	Erect	Scandent
Inflorescence	Terminal	Axillary, leaf-opposed
Stipules	Absent	Present
Stamens	Epipetalous	Free
Staminal tube	Present	Absent
Staminodes	,,	,,
Obdiplostemony	,,	,,
Ovary	4- to 8-loculed	3- to 8-loculed
Ovules	1 ovule/locule	2 ovules/locule
Rays (Vascular)	Uniseriate rays present	Uniseriate rays absent

In view of the above differences, *Leea* should be included in a separate family, Leeaceae. Whatever resemblances they have indicate that the two families are closely related (N.C. Nair 1970b, c).

Taxonomic Consideration of the Order Rhamnales

The order Rhamnales comprises three families—Rhamnaceae, Vitaceae and Leeaceae (Lawrence 1951, Melchior 1964). The three families have many morphological, floral-anatomical and embryological features in common which support their inclusion in the same order.

However, certain other features help to distinguish them also (as mentioned under Vitaceae). Dahlgren (1977a) points out: "Rhamnales should be divided, Vitaceae and Leeaceae being placed in a separate order". This is justified on the basis of embryology also (Johri et al. 1992).

It is apparent, therefore, that the three families should not be treated under the same order. Because of the striking differences between Rhamnaceae and Vitaceae, Takhtajan (1987) treats them as belonging to two distinct orders—Rhamnales, including only Rhamnaceae, and Vitales including two families Vitaceae and Leeaceae. A comparative study of the two families, Leeaceae and Vitaceae, shows both resemblances as well as differences between the two. This emphasises further that the two families—Leeaceae and Vitaceae—are distinct from each other but, at the same time, are closely related, to be placed in the same order.

37

Order Malvales

The Malvales is a large order comprising 4 suborders and 7 families; Elaeocarpaceae, Sarcolaenaceae, Tiliaceae, Malvaceae, Bombacaeae, Sterculiaceae and Scytopetalaceae.

The taxa are predominantly woody plants of the tropics and subtropics of both the northern and southern hemisphere. Vegetative parts often stellate-pubescent and mucilage-producing. Flowers bisexual, actinomorphic, hypogynous and mostly pentamerous. Calyx valvate, corolla valvate or contorted. Stamens numerous, in one or more than one whorl, often monadelphous. Ovary multicarpellary, multilocular and usually with axile placentation. Nectary glands are characteristic multicelluar hairs packed close together to form cushion-like growths.

Elaeocarpaceae

A small family of 12 genera and 350 species, distributed in the tropics and subtropics of both the New and Old World. The genus *Muntingia* with 3 species occurs in tropical South America and the West Indies; the 200 species of *Elaeocarpus* are distributed in East Asia, Indo-Malaysia, Australia and the Pacific Islands.

Vegetative Features.   Trees or shrubs; leaves simple, entire, alternate or opposite, stipulate.

Floral Features.   Inflorescence raceme, panicle, or dichasial cyme. Flowers bisexual, actinomorphic, hypogynous, without any involucre. Sepals 4 or 5, polysepalous or slightly connate at the base, valvate; petals 4 or 5 or 0, polypetalous, rarely basally connate, often incised or hairy, usually valvate. Stamens numerous, free, arising from a disc; anthers bicelled, dehiscence by two apical pores; interstaminal disc present and sometimes develops into an androphore. Gynoecium syncarpous, 2- to many-carpellary, ovary as many loculed as the number of carpels, superior, with axile placentation, ovules 2 to many in each locule; style 1, mostly simple or shortly lobed. Fruit a capsule or drupe; seed with a straight embryo and copious endosperm.

Anatomy.   Wood semi-ring-porous in *Muntingia*, vessels solitary, usually small, with pronounced radial multiples in some genera. Perforations simple. Parenchyma paratracheal, very sparse, sometimes terminal, rays uni- or multiseriate, heterogeneous (homogeneous in *Muntingia*). Leaves usually dorsiventral; simple, unicellular and glandular hairs recorded in *Sloanea*. Stomata confined to the lower surface. Mucilage cavities and canals absent.

Embryology.   Pollen grains smooth-walled and triporate (Venkata Rao 1953a); shed at 2-celled stage. Ovules anatropous, bitegmic, crassinucellate. Polygonum type of embryo sac, 8-nucleate at maturity. The outer integument, at an early stage, overgrows the inner one and at the time of fertilisation the two integuments form a zigzag micropyle (Wunderlich 1967). Endosperm formation of the Nuclear type.

Chromosome Number.   Basic chromosome number is x = 12, 14 or 15.

Chemical Features.   Tropane alkaloids occur in various members of this family (Harborne and Turner 1984). Common flavonols are major compounds; myricetin prominent (Gornall et al 1979).

Important Genera and Economic Importance.   The genus *Elaeocarpus* with 200 species is distributed in Eastern Asia, Indo-Malaysia, Australia and the Pacific Islands. *E. sphaericus* is cultivated for the ornamental seeds. that are used as beads. Three species of *Muntingia* occur in tropical South America and the West Indies. A few species of this genus and of *Sloana* also occur in Mexico. The genus *Sericola* (15 species) is endemic to New Guinea (Balgooy 1982) and includes shrubs, scramblers, trees or treelets with spreading, distally drooping branches. Berries of *Aristotelia macqui* are edible. Species of *Aristotelia, Crinodendron, Elaeocarpus* and *Muntingia* are cultivated as ornamentals.

Taxonomic Considerations. Bentham and Hooker (1965a) and Hutchinson (1969) included Elaeocarpaceae in the family Tiliaceae; Lawrence (1951), Cronquist (1981), Takhtajan (1980, 1987) and Thorne (1983) include this family in the Malvales. It is, related to the Tiliaceae and Malvaceae. Wood anatomy, floral morphology and absence of nectary suggest that it is a primitive family of the Malvales. Although some characters are common with the Tiliaceae, the Elaeocarpaceae can be distinguished by the absence of mucilage cavities and canals.

Sarcolaenaceae

A small family of 8 genera and 33 species, endemic to Madagascar.

Vegetative Features. Trees or shrubs, with simple, entire, alternate, evergreen, stipulate leaves, stipules often large, similar to that of *Ficus* of Moraceae, extra- or intra-petiolar, caducous.

Floral Features. Flowers single or 2 together in an involucre of various forms, in cymose inflorescence. Flowers bisexual, actinomorphic, bracteolate, bracteoles in most genera are united to form an involucel or epicalyx which often persists and encloses the fruit. Calyx of 3 to 5 sepals, polysepalous, imbricate, equal or unequal. Corolla of 5 to 6 petals, polypetalous, contorted, large, disc present. Stamens numerous, rarely 5–10, sometimes fasciculate, inserted within the disc; anthers basi- or dorxifixed, introrse or extrorse. Gynoecium syncarpous, 1- to 5-carpellary, ovary 1- to 5-loculed, superior, placentation basal, apical or axile, ovules few to several per locule, ascending or descending. Style usually thick, more or less elongated and with a lobed stigma. Fruit a many-seeded loculicidal capsule or 1-seeded and indehiscent, enclosed in a woody sac or surrounded by lignified bract or a cupule. Seeds with endosperm and a straight embryo with flat and plaited cotyledons; endosperm fleshy or horny.

Anatomy. Vessels of medium size, exclusively solitary, perforations simple, intervascular pitting alternate. Parenchyma apotracheal, diffuse, rays exclusively uniseriate. Leaves dorsiventral with various types of epidermal hairs, stomata surrounded by a large number of ordinary epidermal cells, often in depressions. Mucilage cells occur in both stem and leaf.

Embryology. Pollen grains in tetrads or 4- to 16-celled pollinia, mature grains 2-celled at dehiscence (Carlquist 1964). Ovules bitegmic, anatropous, crassinucellate.

Important Genera. Some of the important genera are *Sarcolaena, Leptolaena, Perrierodendron* and *Eremolaena.*

Taxonomic Considerations. Sarcolaenaceae is a disputed family. Bentham and Hooker (1965a), Lawrence (1951), Melchior (1964) and Takhtajan (1987) consider it as a member of the Malvales. Cronquist (1981) retains it in the Theales near Dipterocarpaceae, because of the presence of mucilage cells in the stem cortex. It differs from Dipterocarpaceae by the absence of resin cells. On the basis of vegetative, floral and anatomical features, Sarcolaenaceae is more closely related to the members of Malvales.

Tiliaceae

A medium-sized family of 46 genera and 450 species (Takhtajan 1987), mostly restricted to tropical regions with a few members distributed in temperate zones. The genus *Carpodiptera* has 3 of its 5 species in the West Indies, one in the coastal zone of east tropical Africa, and the fifth in the Comoro Islands.

Vegetative Features.   Trees or shrubs, rarely herbs, e.g. *Corchorus* (Fig. 37.3A), *Triumfetta;* with mucilage. Stem mostly woody, leaves usually alternate, rarely opposite as in *Plagiopteron,* simple, entire, stipulate, stipules deciduous; margin dentate, serrate or lobed, often oblique at the base.

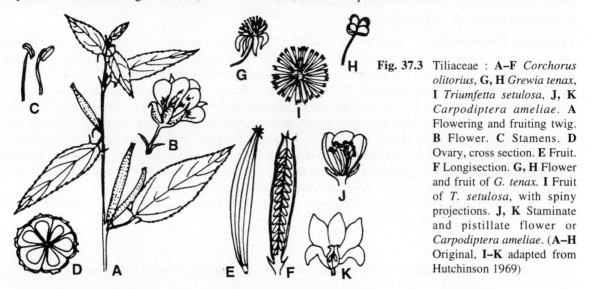

Fig. 37.3 Tiliaceae : **A–F** *Corchorus olitorius,* **G, H** *Grewia tenax,* **I** *Triumfetta setulosa,* **J, K** *Carpodiptera ameliae.* **A** Flowering and fruiting twig. **B** Flower. **C** Stamens. **D** Ovary, cross section. **E** Fruit. **F** Longisection. **G, H** Flower and fruit of *G. tenax.* **I** Fruit of *T. setulosa,* with spiny projections. **J, K** Staminate and pistillate flower or *Carpodiptera ameliae.* (**A–H** Original, **I–K** adapted from Hutchinson 1969)

Floral Features.   Axillary or terminal cymes, often highly reduced so as to appear as a solitary flower, e.g. *Corchorus;* often peduncles winged up to the middle with the adnate, foliaceous or coloured bract. Flowers ebracteate, ebracteolate, actinomorphic, bisexual or rarely unisexual as in *Vasivaea* and *Carpodiptera* (Fig. 37.3G, K), hypogynous. Calyx of 4 or 5 sepals, free or slightly united at base, usually valvate. Corolla of 4 or 5 petals (Fig. 37.3B) or absent as in *Prockia,* sometimes sepaloid, polypetalous, valvate or imbricate. Stamens 10 or more, free or basally connate or polyadelphous as in *Grewia* (Fig. 37.3G); anthers introrse, bicelled, dehisce longitudinally (Fig. 37.3C) or by apical pores. Gynoecium syncarpous, 2- to 10-carpellary, ovary superior, 2- to 10-loculed, placentation axile (Fig. 37.3D), ovules 1 to numerous in each locule; style 1, simple, stigmas usually as many as locules. Fruit fleshy, berry-like or drupaceous (*Grewia,* Fig. 37.3 H) or dry, capsule (*Corchorus,* Fig. 37.3 E, F), dehiscent or indehiscent. Seeds small, usually endospermic, with straight embryo; sometimes covered with stellate hairs as in *Triumfetta* (Fig. 37.3I).

Anatomy.   Vessels small to medium-sized with radial multiples of 4 or more cells in some genera, semi-ring-porous, perforations simple; parenchyma predominantly apotracheal but paratracheal and intermediate types also reported. Rays uni- or multiseriate, 2–3 cells or 4–15 cells wide. Leaves generally dorsiventral but consist wholly of palisade tissue in some species of *Apeiba, Berrya, Corchorus, Diplodiscus* and *Grewia.* Hairs unicellular, uniseriate, stellate, tufted; stomata generally confined to the lower surface, Ranunculaceous type. Mucilage cells and cavities observed in cortex and pith.

Embryology.   Pollen grains shed at 2-celled stage (Sharma 1969). Ovules anatropous, bitegmic, crassinucellate. Polygonum type of embryo sac, 8-nucleate at maturity. Endosperm formation of the Nuclear type.

Chromosome Number.   Basic chromosome number is x = 7–10, 41.

Chemical Features.   Phytoalexin sesquiterpene occurs in *Tilia* and C-glycoflavones in members of the Tiliaceae (Gornall et al. 1979).

Important Genera and Economic Importance.   *Berrya cordata, B. cordifolia, Erinocarpus nimmonii, Tilia americana* and *T. vulgaris* are important timber-yielding trees. Linden or basswood is obtained from *Tilia americana. Grewia tiliafolia* is another tree, its berries are edible. *Corchorus* is yet another important genus, two species—*C. capsularis* and *C. olitorius*—yield the bast fibre, jute. *C. aestuans* and *Triumfetta rhomboidea* are weeds of wastelands.

Taxonomic Considerations.   Most taxonomists treat Tiliaceae as a member of the Malvales. Hutchinson (1973) includes Tiliaceae in an independent order, Tiliales. *Pakaraimaea* is a disputed genus. Although Maguire et al. (1977) placed it in Dipterocarpaceae, according to Kostermans (1978, 1985) it belongs to the Tiliaceae. The anatomical structure of *Pakaraimaea* indicates that it should be retained in the order Theales of the Dilleniidae; Sarcolaenaceae and Dipterocarpaceae are its closest allies. Along with two African genera, *Marquesia* and *Monotes, Pakaraimaea* easily forms a group separate from the true Dipterocarps. The numerous stamens on androgynophore, tricolpate pollen grain, gum sacs instead of resin canals, uniseriate wood ryas, and absence of glandular hairs bring this group closer to the Tiliaceae. Takhtajan (1987) treats this group as a separate family, Monotaceae, of the Malvales and places it next to the family Tiliaceae. Takhtajan's Malvales also include Dipterocarpaceae.

In our opinion also, *Pakaraimaea, Marquesia* and *Monotes* should be included in a distinct family, Monotaceae, and placed near the Tiliaceae.

Malvaceae

A medium-sized family with 75 to 85 genera and 1500–1600 species, distributed predominantly in the tropics.

Vegetative Features.   Trees (*Thespesia populnea, Kydia calycina*), shrubs (*Hibiscus mutabilis, H. rosa-sinensis*) or herbs (*Malvastrum, Sida, Urena, Malva*—(Fig. 37.4A), the entire plant contains mucilaginous sap. Leaves alternate, simple, petiolate, stipulate, stipules free-lateral; margin dentate, crenate, entire or deeply lobed as in *Abelmoschus esculentus,* usually palmately veined.

Floral Features.   Inflorescence basically cyme (Fig. 37.4A) but often solitary, axillary as in *H. rosa-sinensis* (Fig. 37.4D). Flowers bracteate, bracteolate, bracteoles usually form a whorl of epicalyx (exceptions *Sida, Abutilon*), pedicellate, actinomorphic, hypogynous, bisexual, pentamerous. Epicalyx of 7 to 10 lobes, free, green, hairy. Calyx of 5 sepals, gamosepalous, valvate, outer surface hairy; corolla of 5 petals, polypetalous (Fig. 37.4B), often basally adnate to the staminal column, convolute or twisted, contain mucilage. Stamens numerous, monadelphous, from a staminal column around the style, fused with the basal parts of the corolla. Upper part of the branched filament is free, each bears a monothecous, reniform, half-anther which opens by a transverse slit; extrorse, pollen formation profuse. All the stamens are derived by copious branching of 5 antipetalous stamens; the outer antisepalous whorl of stamens is lost, though in *Hibiscus* it is represented by 5 teeth of the staminodes on the summit of the staminal tube. Gynoecium syncarpous, 2- to many-carpellary, ovary superior, 2- to many-loculed (Fig. 37.4C), usually in a ring or infrequently superposed as in *Malope* (Fig. 37.4E); placentation axile, ovules 1 to many in each locule, style 1 and apically branched or as many as the carpels; stigmas as many or twice as many as the carpels (Fig. 37.4D, E), capitate or discoid. Fruit typically a loculicidal capsule (Fig. 37. 4F) or schizocarpic or carcerulus (Fig. 37.4H), or rarely a berry as in *Malvaviscus* or a samara. Seeds mostly reniform (Fig. 37.4I); often pubescent or comose as in *Gossypium* (Fig. 37.4G).

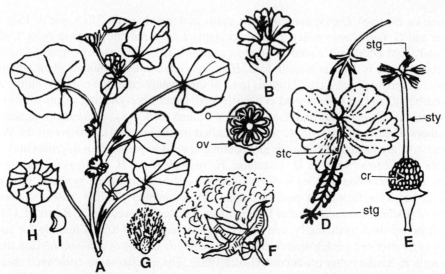

Fig. 37.4 Malvaceae : **A–C** *Malva parviflora*, **D** *Hibiscus rosa-sinensis*, **E** *Malope trifida,* **F,G** *Gossypium hirsutum*, **H, I** *Malvastrum coromandelianum.* **A** Twig bearing flowers and fruits. **B** Flower. **C** Ovary, cross section. **D** Flower of *H. rosa-sinensis*. **E** Gynoecium of *M. trifida*. **F,G** Boll and seed of *G. hirsutum* **H, I** Fruit and seed of *M. coromandelianum. cr* carpel, *o* ovule, *ov* ovary, *stc* staminal column, *stg* stigma, *sty* style. (**A, B, H, I** after J. K. Maheshwari 1965, **E** after Hutchinson 1969)

Anatomy. Vessels small to medium-sized, semi-ring-porous; perforations simple, intervascular pitting small and alternate. Parenchyma predominantly paratracheal in tribes Malveae and Ureneae, and predominantly apotracheal in Hibisceae. Rays may be of 2 types: (a) multiseriate rays usually up to 4–9 and rarely up to 23 cells wide, low to high, and (b) uniseriate; typically heterogeneous. Leaves dorsiventral, centric in *Malva parviflora*, covered with both glandular and non-glandular hairs. Extrafloral nectaries also occur in some members. Stomata of the Ranunculaceous type, always on the lower surface. Mucilage cells and secretory cavities common.

Embryology. Pollen grains spherical, pantoporate or multiporate (Venkata Rao 1954a), spinous and very large. Pollen sterility is reported in *Thespesia populnea*. Ovules campylotropous, bitegmic, crassinucellate; outer integument longer than the inner one. Polygonum type of embryo sac, 8-nucleate at maturity. Endosperm formation of the Nuclear type.

Chromosome Number. Basic chromosome number is variable—$x = 5–13, 15–17, 19–23, 29, 33, 39$.

Chemical Features. In the members of Malvaceae, the seed fats contain palmitic, oleic and linoleic acids as major components. Unusual fatty acids like malvalic and sterculic acid occur in many members. Aglycone anthocyanin pigments such as cyanidin and malvidin are present in many flowers. Cotton fibres are rich in cellulose.

Important Genera and Economic Importance. *Hibiscus* with 300 species and *Pavonia* with 200 species each are the largest genera. *Hibiscus* species, mostly with the large showy flowers, are distributed in the tropical and subtropical regions of both the New and the Old World. *Pavonia* and *Sida* also occur in the tropics and subtropics. Economically, the Malvaceae is very important. The commercial cotton is derived from the densely hairy seeds of *Gossypium*. Cotton fibres have been traced to the Indus Valley civilization (ca. 3000 BC). The Greeks found this crop when they invaded India with Alexander the Great and they called it 'lamb on the tree'. *Gossypium hirsutum,* cultivated mainly in the USA, yields long staple cotton, *G. arboreum* and *G. herbaceum* which yield short staple cotton, are cultivated in Asia. Cotton seeds are rich in fat and yield edible oil called the cotton seed oil. It is also used in manufacturing soaps and lubricants; the oil cake is used as a cattle feed. *Hibiscus cannabinus* is another plant which yields fibre, used widely for cordage, ropes, etc.; fatty oil used in the manufacture of linoleum, paints and varnishes and also as edible oil and oil cake is used as a cattle feed. Other fibre-yielding plants of minor importance are *Abutilon indicum, A. persicum* and *A. theophrastii* (China jute), *Sida cordifolia, S. acuta, Urena lobata* and many others. *Hibiscus sabdariffa* (commonly called roselle) is a shrub, native to the West Indies; its epicalyx and calyx are fleshy, rich in acids and pectin and used in preparation of jellies and confectionery. Many species of *Hibiscus* such as *H. mutabilis, H. rosa-sinensis, H. schizopetalous* are cultivated as ornamentals; *H. elatus* (blue mahoe) is the national flower of Jamaica, *H. syriacus* growing in eastern Mediterranean has large flowers of pinkish colour and is commonly called 'rose of Sharon'.

The fruits of *Abelmoscus esculentus* or lady's finger are used as a vegetable. Roots of *Althea officinalis* or marshmallow are used medicinally, considered to be a very useful herbal remedy for cough. *A. rosea* or hollyhock is a cultivated garden ornamental. The mucilaginous root of *Pavonia hirsuta* from Zimbabwe is added to milk to hasten butter production (Dutta 1988). The mucilaginous substance obtained from the stem of *Kydia calycina* is used for clarifying sugar; its wood is used for matches, packing cases, pencils, shoe heels, picture frames, veneers, plywood, paper and rayon-grade pulp (Singh et al. 1983). The Portia tree or *Thespesia populnea* is an avenue tree. *Anoda hastata, Pavonia odorata,* various species of *Malva, Malvastrum, Decaschistia, Sida* and *Urena* are weeds of waste-lands. *Sidalcea nelsoniana* is an endemic species in Oregon, USA, and is a threatened species. Its seed coat is very hard, and must be softened to such an extent that water and oxygen enter and the embryo is allowed to expand and break out of the seed coat (Halse and Mishaga 1988).

Taxonomic Considerations.   There is no controversy about the assignment of the family Malvaceae in the Malvales. The distinguishing features include presence of mucilage cells, monadelphous stamens and monothecous anthers, spinous and polyporate pollen grains, and pentacarpellary ovary with the style passing through the staminal tube.

Bombacaceae

A small family of 25–31 genera and ca. 200 species, distributed in the tropics, especially the American tropics.

Vegetative Features.   Tall trees with unusually thick trunks often covered with large woody thorns or spines, as in *Bombax ceiba, Chorisia speciosa.* Leaves alternate, simple or palmately compound (Fig. 37.5A), deciduous, stipulate, stipules caducous; often with slime cells and the vesture of stellate hairs or peltate scales.

Floral Features.   Flowers solitary or fasciculate in leaf axils or opposite a leaf, showy, hermaphrodite, actinomorphic, hypogynous, commonly bracteate, often appearing before the leaves. Calyx of 5 sepals, polysepalous or basally connate, valvate, often with epicalyx. Corolla of 5 petals, polypetalous, contorted or convolute in bud. Stamens 5 to many, free or monadelphous (Fig. 37.5B); anthers mono- or bithecous, longitudinally dehiscent, pollen grains smooth. Gynoecium 2- to 5- carpellary, syncarpous; ovary superior, 2- to 5-loculed, placentation axile, ovules 2 to many in each locule, style simple, stigma 1- to 5-lobed or capitate. Fruit a loculicidal pod (rarely indehiscent) or berry-like. Seeds smooth, often embedded in hairs springing from inner surface of fruit wall, endosperm scanty or none.

Anatomy.   Vessels medium-sized to large, perforations simple, intervascular pitting alternate. Parenchyma

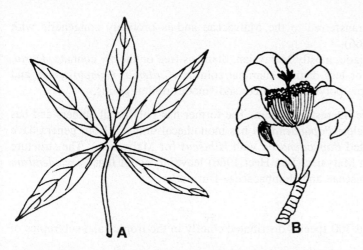

Fig. 37.5 Bombacaceae : **A, B** *Salmalia malabaricum.* **A** Leaf, **B** Flower. (Original)

abundant and sometimes forms the ground tissue of the wood, vasicentric. Rays 4–10 cells wide in tribes Adansonieae and Matisieae, heterogeneous. Leaves generally dorsiventral, extrafloral nectaries occur on the lower side of the midrib and on dorsal surface of petiole in some species of *Adansonia, Ceiba, Chorisia* and *Pachira.* Stomata confined to the lower surface.

Embryology. Pollen grains smooth-walled, triporate (Venkata Rao 1954b, Sharma 1970), shed at 2-celled stage. Ovules erect, anatropous, bitegmic, crassinucellate; outer integument longer than the inner one. Polygonum type of embryo sac, 8-nucleate at maturity. Endosperm formation of the Nuclear type.

Chromosome Number. Diploid chromosome number is very variable—2n = 56, 72, 80, 88, 96, 140, 150.

Chemical Features. Seed fats rich in palmitic, oleic and linoleic acid, as in *Ceiba acuminata* (Shorland 1963).

Important Genera and Economic Importance. Some of the larger genera are *Bombax, Ceiba* and *Chorisia. Adansonia digitata,* commonly called baobab tree, is an unusual tree with a short trunk of considerable size close to the ground and tapering gradually upwards. *Bombax ceiba* or the red silk cotton tree is remarkable, as its leaves are shed in December and it remains leaf-less until April, but flowers in January. White silk cotton tree is *Ceiba pentandra.* The hairy floss from the pericarp of the fruits of these two plants are used for stuffing pillows and cushions. Dug-out canoes are made from the soft wood of *Bombax ceiba.* The acid pulp of the fruits of *Adansonia digitata* makes good beverage. *Durio zibethinus,* a common plant of Indo-Malaya, yields Durian fruit eaten by the Burmese. The wood of *Ochroma pyramidale* or balsa is the lightest wood of the world, with a specific gravity of 0.12.

Taxonomic Considerations. The nearest relative of the Bombacaceae is the Malvaceae, from which it differs in exine ornamentation of the pollen grains and the number of carpels. It is a natural taxon and all phylogenists include it in the Malvales.

Borssum Waalkes (1966) divided the Bombacaceae into 3 tribes:

a) Bombaceae—Members both American and Asian. Stamens show all transitions between free stamens and distinct staminal column, and monothecal (*Adansonia, Bombax*) or polythecal anthers and smooth pollen.

b) Matisieae—All members are American. Stamens mostly free or shortly connate. *Montezuma,* formerly placed in Matisieae, has been transferred to the Malvaceae and is probably congeneric with *Thespesia* (Borssum Waalkes 1966).

c) Durioneae—All members are Asiatic, mostly Malaysian. Stamens free or shortly connate (*Durio, Neesia, Coelostegia*) or a more or less distinct staminal column (*Cullenia, Camptostemon* and *Papuodendron*). Most genera show polythecal anthers and smooth pollen grains.

Camptostemon and *Papuodendron* are interesting genera. The former has polythecal anthers and has therefore been excluded from the Malvales; *Papuodendron* has monothecal anthers. Both genera have echinate pollen. Kostermans (1960) united *Papuodendron* with *Hibiscus* (of Malvaceae). The structure and ontogeny of the staminal column in Malvales (Van Heel 1966) leave no doubt that *Papuodendron* is the only link between Malvaceae-Hibisceae and Bombacaceae-Durioneae.

Sterculiaceae

A medium-sized family of 60 genera and 700 species distributed chiefly in the tropics and subtropics of both the New and the Old World.

Vegetative Features. Trees, shrubs (*Dombeya*), a few herbs (*Pentapetes*) and rarely climbers, e.g. *Ayenia*. Leaves alternate, simple, entire or infrequently palmately lobed or compound, stipulate, stipules caducous.

Floral Features. Inflorescence usually cymes of various types, often cauliflorous (Fig. 37.6A). Flowers actinomorphic or rarely zygomorphic (*Helicteres*), hypogynous, bi- or unisexual and plants monoecious as in *Sterculia* and *Cola*; bracteate, ebracteolate. Calyx of 3 to 5 sepals, slightly united at base, valvate; petals 5 or absent as in *Sterculia* and *Heritiera*. In *Theobroma cacao* each petal bulges out at the base, followed by a considerably narrow and constricted zone that ends in an expanded tip (Fig. 37.6B, C); sometimes adnate to the base of androecium, contorted in bud. Stamens numerous, in two whorls, outer

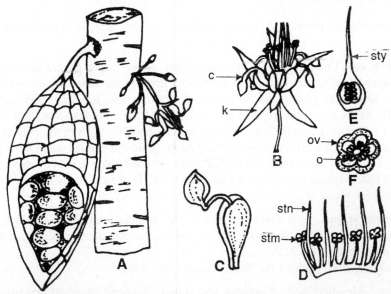

Fig. 37.6 Sterculiaceae : **A–F** *Theobroma cacao*. **A** Twig bearing fruit and inflorescence on the stem surface. **B** Flower. **C** Petal. **D** Stamens and staminodes. **E** Ovary, vertical section. **F** Cross section. *c* corolla, *k* calyx, *o* ovule, *ov* ovary, *stg* stigma, *stm* stamen, *stn* staminode, *sty* style. (Adapted from G.L. Chopra 1973)

antisepalous and reduced to staminodes, or scales or absent (*Eriolaena*), the inner whorl of stamens antipetalous and fertile (Fig. 37.6D); free or monadelphously connate in a single tube; anthers bicelled, longitudinally dehiscent, rarely cohering apically. Gynoecium syncarpous, 4- to 5-carpellary; ovary superior, 4- to 5- or rarely 10- to 12-loculed, placentation axile, ovules 2 to numerous in each locule (Fig. 37.6E, F). Ovary often raised on an androgynophore except in *Helicteres, Mansonia* and some others. Fruit leathery, or fleshy, rarely woody, dehiscent or indehiscent, sometimes the carpels split into cocci. Seeds arillate, often winged, endospermous, with a straight or recurved embryo.

Anatomy. Vessels medium-sized to large, few (1 to 5 per mm), perforations simple, intervascular pitting alternate, small. Parenchyma of 2 types: (a) diffuse, slightly vasicentric, and (b) broad, apotracheal to confluent bands; storied. Rays uniseriate or multiseriate, 3 to 20 cells wide, heterogeneous. In Buettnerioideae vessels in multiples of 4 or more cells, mostly 5 to 20 per mm, with a tendency to be ring-porous in many genera. Parenchyma either diffuse, with scanty vasicentric parenchyma or exclusively paratracheal. Rays of 2 distinct sizes; the larger usually 4 to 10 cells wide, typically heterogeneous; the smaller rays less than 5 cells wide, often storied. Leaves generally dorsiventral with stellate hairs or even unicellular, uniseriate, glandular and peltate hairs. Stomata mostly Ranunculaceous type, confined to the lower surface. Mucilage commonly present in special cells, cavities or canals. Tannin is abundant.

Embryology. Pollen grains spinescent and triporate (Venkata Rao 1949), shed at 2-celled stage. Ovules anatropous, bitegmic, crassinucellate, outer integument longer than the inner one. Polygonum type of embryo sac, 8-nucleate at maturity. Endosperm formation of the Nuclear type.

Chromosome Number. Basic chromosome numbers are x = 5–23.

Chemical Features. Anthocyanidin glycosides, 3-arabinoside and 3-galactoside are reported in *Theobroma cacao* pods. Aglycones, cyanidin and apigeninidine, occur in the calyx of *Chiranthodendron*. The alkaloid theobromine is also obtained from *Theobroma cacao*. Cola nuts from *Cola vera* and *C. acuminata* contain caffein and theine.

Important Genera and Economic Importance. *Byttneria herbacea* is a herb of this family. *Heritiera fomes,* commonly called 'sundri', is a tree from which the Sunderbans—the famous mangrove forests on the coast of the Bay of Bengal—takes its name. Follicles of the shrub *Helicteres isora,* on dehiscing, twist spirally and expel their seeds. *Pterospermum acerifolium* is an avenue tree with large white bisexual flowers and long fleshy sepals.

Firmiana colorata, Kleinhovia hospita, Pterygota alata and *Sterculia foetida* are some of the important avenue trees.

Seeds of *Theobroma cacao* are the source of cocoa powder, cocoa butter and chocolate. A number of other genera are grown for ornamental purposes, e.g. *Abroma, Bracychiton, Dombeya, Firmiana, Fremontodendron, Pterospermum* and *Reevesia. Cola acuminata* is the source of cola nuts, which produce caffeine and theine used in preparation of soft drinks. Good quality timber is obtained from *Heritiera fomes*. The bark of *Guazuma ulmifolia, Helicteres isora, Sterculia urens* and *S. villosa* yield fibres.

Taxonomic Considerations. The family Sterculiaceae is closely related to the Malvaceae and Bombacaceae. Most taxonomists include this family in the Malvales, although Hutchinson (1973) retains it in the Tiliales along with other woody members of this complex. Recent workers like Takhtajan (1980, 1987) place Sterculiaceae in the Malvales which also includes Tiliaceae.

Scytopetalaceae

A small family of 5 genera and 20 species distributed in west tropical Africa.

Vegetative Features.   Trees or shrubs, sometimes scandant. Leaves simple, alternate, often distichous, entire or dentate, estipulate, often asymmetrical at base.

Floral Features.   Inflorescence axillary or terminal panicles or racemes or in fascicles on old wood. Flowers bisexual, actinomorphic, often long-pedicelled, ebracteate and ebracteolate. Calyx of 3 or 4 or more sepals, gamosepalous, often petalliferous or cupuliferous with entire margin, persistent. Corolla of 3 to 16 petals, polypetalous, valvate, sometimes do not separate, and then falling off as an entire cap, mostly thick and reflexed at anthesis; inconspicuous, annular disc present. Stamens numerous, in 3 to 6 series, sometimes connate below. Anthers bicelled, sometimes poricidal. Gynoecium syncarpous, 3- to 8-carpellary; ovary superior, with as many locules as the number of carpels, often locules incomplete above; placentation axile, with 2-6-numerous, biseriate ovules per locule; style 1, simple, stigma reduced. Fruit usually a tardily dehiscent capsule, more rarely drupaceous, unilocular and 1- to 8-seeded. Seeds sometimes covered with agglutinated mucilaginous hairs, endospermous; endosperm sometimes ruminate.

Anatomy.   Vessels medium-sized, solitary or in multiples of 2 or 3 cells, perforations mostly simple, but some scalariform plates with up to 12 bars are present; intervascular pitting alternate. Parenchyma apotracheal, in numerous uniseriate bands; rays up to 6 cells wide, few uniseriates, heterogeneous. Leaves usually dorsiventral, stomata Cruciferous type, either on both surfaces or only on the lower surface.

Embryology.   Pollen grains smooth-walled, triporate; shed at the 2-celled stage. Ovules pendulous, anatropous, bitegmic, tenuinucellate; the inner integument is longer than the outer one. Polygonum type of embryo sac, 8-nucleate at maturity. Development of endosperm has not been investigated.

Chromosome Number.   Basic chromosome numbers are x = 9, 11.

Important Genera and Economic Importance.   The five genera, *Oubanguia*, *Scytopetalum*, *Rhaptopetalum*, *Brazzeia* and *Pierrina*, do not have any economic importance.

Taxonomic Considerations.   Scytopetalaceae is a controversial family. Engler and Diels (1936) included it in the Malvales, and Gunderson (1950), Melchior (1964) and Cronquist (1968) support this assignment. The genus *Rhaptopetalum* has been treated in the Bentham and Hooker's (1965a) system as an anomalous genus under Olacinae. Anatomical studies suggest its affinities with both the Tiliaceae and the Ochnaceae. It is distinct from other members of the Malvales because of tenuinucellate ovules and the inner integument longer than the outer one. Embryological studies (Vijayaraghavan and Dhar 1976) suggest its affinities with the Tiliaceae, Elaeocarpaceae and Bombacaceae, but not Malvaceae. Cronquist (1981) and Thorne (1983), however, suggest its relationship with the Quiinaceae and Ochnaceae; Takhtajan (1987) includes Scytopetalaceae in the Ochnales. Observations of Vijayaraghavan and Dhar (1976) support Hutchinson's concept of two separate orders, Malvales, including Malvaceae, and Tiliales, including Scytopetalaceae and other closely related families.

Taxonomic Considerations of the Order Malvales

The order Malvales is a natural taxon as indicated by the presence of stellate hairs and mucilage cells, sacs, canals or cavities (except in Elaeocarpaceae and Scytopetalaceae) in its members. Most phylogenists follow the concept of the order as adopted by Engler and Diels (1936). Hutchinson (1973) restricts the Malvales to include only the Malvaceae; he considers the Malvales to be more advanced than the rest of the families usually included in this order. Tiliaceae, Bombacaceae and Sterculiaceae have been retained in a separate order, Tiliales. Melchior (1964) includes seven families in the Malvales— Elaeocarpaceae, Tiliaccae, Sarcolaenaceae, Sterculiaceae, Bombacaceae, Malvaceae and Scytopetalaceae.

Cronquist (1968, 1981) prefers to include Sarcolaenaceae in his Theales, as this family is distinct from other members of this order in: (a) imbricate calyx (valvate in the Malvales), and (b) pollen grains in permanent tetrads, although they have mucilage cells and the typical phloem structure of the Malvales, i.e. phloem stratified into alternate bands of fibrous and non-fibrous tissue.

Malvales is a more or less homogeneous order because of hypogynous flowers with valvate calyx, mostly separate petals that are often convolute in bud condition, numerous centrifugal stamens that are connate by their filaments and basically syncarpous ovary with axile or rarely parietal placentation.

Inclusion of Sarcolaenaceae will make the Malvales a heterogeneous assemblage. Therefore, Cronquist (1981) includes this family in the already heterogeneous Theales. He removed Scytopetalaceae also from the Malvales to include it in his Theales, probably because of its wood anatomy and embryological features which suggest affinities with the Ochnaceae, also a member of the Theales. Dahlgren (1980a) includes 13 families in his Malvales, many of the families, belonging to the orders like Ochnales, Theales, Dilleniales or Guttiferales. Takhtajan (1987) considers a few other families like Plagiopteridaceae, Monotaceae, Dipterocarpaceae, Rhopalocarpaceae and Huaceae also to be the members of his Malvales, along with the usual six families. The monotypic African genus *Hua* is a tree with alternate entire leaves, flowers with valvate sepals, long-clawed petals with induplicate-valvate aestivation, 10 stamens and a unilocular ovary with 1 basal ovule and a fruit dehiscing into 5 valves. This genus has been variously associated with the Malvales, Ebenales and Linales (Cronquist 1968). It is now treated as a distinct family belonging to the Violales (Cronquist 1981, 1983), or Malvales (Baas 1972, Thorne 1983, Takhtajan 1987).

The family Huaceae includes *Hua* with 1 species and *Afrostyrax* with 2 species distributed in tropical Africa. Baas (1972) observed that the 2 genera have many anatomical and other features in common with the Malvales; he placed Huaceae in the Malvales. According to Cronquist (1983), because of the presence of unilocular ovary with basal placentation and non-stratified phloem and absence of mucilage cells and cyclopropenoid fatty acids, Huaceae cannot be assigned to the Malvales.

The Malvaceae is a rather homogeneous family in contrast to other families of the order Malvales. The common features of all the members are: (a) calyx valvate, (b) stamens united into a staminal column, (c) monothecal anthers, (d) echinate pollen grains and (e) septate ovaries.

The Bombacaceae can be separated from the Malvaceae on the basis of : (a) stamens free or shortly connate, (b) anthers of 2 or more thecae, and (c) smooth pollen grains (Borssum Waalkes 1966).

Although the circumscription of the Malvales varies in different systems of classification, Malvaceae, Sterculiaceae, Bombacaceae and Tiliaceae have been retained in Malvales by most taxonomists, as they are very closely allied to each other. Elaeocarpaceae, Scytopetalaceae and Sarcolaenaceae are somewhat different from the rest of the families of this order. Elaeocarpaceae has sometimes been allied with the Tiliaceae (cf. Cronquist 1968). It differs from the Tiliaceae in the absence of stellate hairs and mucilage canals, and presence of tenuinucellate ovules instead of crassinucellate ones.

The Malvales, with a large number of stamens and/or carpels, is probably derived from the less modified members of the Theales, from which they differ in valvate calyx (not imbricate) and in the occurrence of stellate hairs and mucilage canals in most of the members of this order. The order Violales in also closely related to the Malvales as, in all probability, both these orders have evolved from the Theales.

38

Order Thymelaeales

Thymelaeales is an order comprising 5 families: monotypic Geissolomataceae, Penaeaceae, Dichapetalaceae, Thymelaeaceae, and Elaeagnaceae distributed mostly in warm temperate regions of the world.

Most members are shrubs of xerophytic nature, some are trees and a very few herbs. Leaves simple, coriaceous, entire and estipulate. Inflorescence mostly racemose. Flowers bisexual or unisexual and then the plants are monoecious or polygamodioecious; mostly tetramerous, actinomorphic, hypogynous. Perianth mostly uniseriate (biseriate in Dichapetalaceae), stamens as many or twice as many as sepals. Gynoecium 4-, 2- or monocarpellary. Ovules anatropous, bitegmic, crassi- or tenuinucellate. Fruit a drupe, nut, achene or berry, often enclosed in a persistent receptacle as in the Thymelaeaceae. Seeds with scanty or no endosperm.

Geissolomataceae

Geissolomataceae is a monotypic family with the taxon *Geissoloma marginatum*, endemic to South Africa. This taxon occurs on the mid-elevation slopes of the Langeberge, a range of sandstone mountains in southwestern Cape Province in South Africa (Carlquist 1976). It grows in small colonies on moist south-facing sites often beside sandstone outcrops that provide shade, a greater degree of water retention and some protection from fire.

Vegetative Features. Small, xerophytic shrubs which grow in colonies. Underground stems tend to develop as lignotubers and are capable of sprouting after a shrub has been burnt. Leaves opposite, entire, stipulate, evergreen.

Floral Features. Flowers solitary axillary, bracteate, subtended by 6 persistent bracts, hermaphrodite, actinomorphic, hypogynous. Calyx of 4 sepals, polysepalous, petaloid, imbricate, persistent; corolla absent. Stamens 8, in 2 whorls of 4 each, with slender filaments and ellipsoid, dorsifixed anthers; anthers bicelled, longitudinally dehiscent. Gynoecium syncarpous, tetracarpellary, ovary superior, 4-locular with 2 pendulous ovules per locule; styles 4, subulate, free below but coherent above. Fruit a 4-loculed capsule with 4 seeds. Seeds shiny, endospermous.

Anatomy. Vessels mostly solitary, small, angular (in transections); perforation plates scalariform. Intervascular pitting opposite or scalariform. Axial parenchyma diffuse, in strands of 4 or 5 very long cells. Rays multiseriate as well as uniseriate. Gummy deposits which stain darkly with safranin are common in ray cells (Carlquist 1976c).

Embryology. Pollen grains 3-colporate, 3-celled at shedding stage (Carlquist 1976c). Ovules anatropous, bitegmic and crassinucellate. Embryo sac of Polygonum type, 8-nucleate at maturity. Endosperm formation of the Cellular type (Johri et al. 1992).

Chemical Features. C-glycofavones are present, myricetin absent.

Important Genus. The only genus and species, *G. marginatum* is not known to be economically important.

Taxonomic Considerations. The family Geissolomataceae has been variously treated by different workers. Genus *Penaea* was placed in the Myrtales (Linnaeus 1771; see Dahlgren and Rao 1969). Similarity of

Geissoloma to Penaeaceae is based on (a) tetramerous flowers of both, and (b) decussate nature of small sclerophyllous leaves. Hutchinson (1973) places this family in his Thymelaeales along with the Penaeaceae. Cronquist (1981) includes Geissolomataceae in the Celastrales. According to Carlquist (1976c): "although primitive wood is found in the Celastrales and the capsules (of *Geissoloma* and Celastrales members) also resemble to some extent, the totality does not suggest a close link". According to Dahlgren and Rao (1969), it is closer to the Oleaceae and Salvadoraceae. However, on the basis of mainly wood anatomy, Carlquist (1976c) suggested its position near Bruniaceae and Grubbiaceae. Takhtajan (1987) erects a separate order, Geissolomatales, and places it next to Bruniaceae of Bruniales.

With the available data, this family is best treated when placed in a separate order, as suggested by Takhtajan (1987). For a better treatment, detailed chemical, cytological and embryological data are essential.

Penaeaceae

Penaeaceae is a small family of 6 genera and 25 species distributed in South Africa. Two species, *Brachysiphon rupestris* and *Sonderothamnus petraeus*, belong to a habitat form peculiarly characteristic of the Table Mountain sandstone: diminutive subshrubs restricted to crevices in sandstone cliffs and outcroppings. Most species grow in highly restricted localities on the Table Mountain sandstone and are endemic within Cape Province (Dahlgren 1967a, 1971). The genus *Penaea* extends eastwards from Cape Peninsula as far as the Port Elizabeth Division in Cape Province and most species occur within a 200 km radius of Cape Town in the southwesternmost portion of this region (Carlquist and De Buhr 1977).

Vegetative Features. Shrubby xerophytes of ericoid habit, with several branches from woody base; *Stylapterus fruticulosus* has notably succulent bark on massive underground stems. Leaves opposite, entire, evergreen, slightly keeled with cuneate base in *Glischrocolla formosa;* sessile, coriaceous and rather closely set in *Endonema*; stipules minute or absent.

Floral Features. Inflorescence a compact raceme (*Glischrocolla*) or spike (*Endonema*). Flowers bracteate, bracteolate, bracts and bracteoles often coloured, purple or carmine at least on apical half as in *Glischrocolla;* bisexual, actinomorphic, hypogynous. Calyx of 4 sepals, gamosepalous, valvate, persistent, corolla absent (according to Dahlgren 1967a, 1971; this is a cylindrical perianth tube with 4 narrow ovate, valvate lobes). Stamens 4, alternisepalous, filaments short, inserted in the throat of the calyx (or perianth), incurved or not so in bud, disc absent; anthers introrse, bicelled, medifixed as in *Endonema*, longitudinally dehiscent. Gynoecium syncarpous, tetracarpellary; ovary superior, 4-loculed with 2 to 4 ovules per locule—2 upper ovules ascending and the two lower ones pendulous; style simple with 4-lobed stigma. Fruit a loculicidal capsule, often 4-seeded; seeds non-endospermous.

Anatomy. Vessels small (22-46 μm in diameter), perforation plates simple, alternate pitting on lateral walls. Axial parenchyma scanty and diffuse in distribution, rays predominantly uniseriate; crystals less common. A dark-staining amorphous deposit is present mostly in ray cells and axial parenchyma cells. Leaves usually dorsiventral, occasionally isobilateral or centric. Stomata present on both surfaces or confined only to the lower surface, mostly Ranunculaceous type. Sclerenchyma fibres, with or without spiral markings present in mesophyll.

Embryology. Pollen grains 3- to 5-colporate, ellipsoidal, almost hexalobate in polar view as in *Endonema*, almost square in equatorial view as in *Glischrocolla* (Dahlgren 1967a,b); 2-celled at shedding stage. Ovules ascending, descending or both, anatropous, bitegmic and crassinucellate. Embryo sac of Penaea type, 16-nucleate at maturity. Endosperm formation has not been investigated.

Chromosome Number. Basic chromosome number is x = 10.

Important Genera and Economic Importance. Of the 6 genera, *Penaea* is the largest. All the genera occupy a wide range of ecological habitats, which experience various degrees of drought with approaching summer heat. *Brachysiphon rupestris* and *Sonderothamnus petraeus* are the most xerophytic representatives and grow in rock crevices on cliffs or outcrops. *Stylapterus fruticulosus* grow on sand flats, east and north of Cape Peninsula, are also xerophytic. Open slopes on Table Mountain sandstone, where *Endonema lateriflora* and *E. retzioides* grow, are more mesic. The river valleys or the wet forest where *Penaea cneorum* grows is most mesic.

Taxonomic Considerations. Bentham and Hooker (1965c) and Hallier (1912) regarded Penaeaceae as a member of Daphnales of Monochlamydeae. Engler and Diels (1936) treated it as a member of Myrtales, with which it has anatomical resemblances like presence of intraxylary phloem, crystalliferous strands in axial xylem, both uni- and multiseriate rays, and amorphous deposits in ray cells.

Melchior (1964) removed Penaeaceae to the Thymelaeales along with Geissolomataceae and others. Hutchinson (1973) also included it in the Thymelaeales which differs from Thymelaeaceae because of a more advanced (i.e. valvate) aestivation. Cronquist (1968) regarded it to be a member of the Myrtales mainly because of similarity in anatomical features. Takhtajan (1980, 1987) is also of the same opinion. Dahlgren and Thorne (1984) commented: "The ancestors of Penaeaceae could have had common origin with Rhyncocalycaceae", also a member of the Myrtales. Embryologically, too, Penaeaceae are closer to the Myrtales (Carlquist and De Buhr 1977, Dahlgren and Thorne 1984, Johri et al. 1992). The family Penaeaceae has a controversial position, though on the basis of wood anatomy and embryology they are closer to the Myrtales.

Dichapetalaceae

A small family of 3 genera and 155 species distributed in most parts of the tropics extending into southeast Africa.

Vegetative Features. Mostly shrubs, but some trees and lianas (Fig. 38.3A,B) also reported; sometimes stems with grey pubescence. Leaves alternate, simple, entire, stipulate, with a few flat glands near the base.

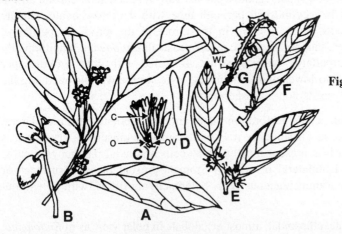

Fig. 38.3 Dichapetalaceae : **A-D** *Dichapetalum toxicarium*, **E-G** *Tapura amazonica*. **A,B** Flowering and fruiting twig. **C** Flower, vertical section. **D** Petal. **E,F** Epiphyllous flowers, and fruit of *Tapura amazonica*. **G** Unicellular hair with wartlike papillae. *c* corolla, *o* ovule, *ov*, ovary, *wr* wart. (**A-F** adapted from Hutchinson 1969, **G** after Metcalfe and Chalk 1972).

Floral Features. Inflorescence axillary dichotomous cymes or fascicles, with frequent concrescence of the peduncle and petiole (Fig. 38.3E,F) as in *Tapura amazonica*, the inflorescence then appears to arise from the base of the leaf blade. Flowers ebracteate, ebracteolate, pedicellate (*Dichapetalum*) or sessile (*Tapura*), bi- or unisexual, actinomorphic, or zygomorphic as in *Tapura*. Calyx of 5 sepals, poly- or gamosepalous, imbricate, outer surface hairy; corolla mostly of 5 petals, usually polypetalous but gamopetalous in *Stephanopodium* and *Tapura*, petals mostly bifid (Fig. 38.3C,D) or bilobed, often black on drying. Stamens 5, oppositisepalous, epipetalous as in *Tapura* or free as in *Dichapetalum*, anthers introrse, bithecous, longitudinally dehiscent; disc cupular or lobed. Gynoecium syncarpous, bi- or tricarpellary, ovary bi- or trilocular, with 2 apical pendulous ovules per locule; styles 2 or 3, basally connate, stigma capitate (?). Fruit a drupe, usually pubescent, 1- to 3-lobed, and loculed and locules 1-seeded; seeds non-endospermous, sometimes with a caruncle.

Anatomy.   Vessels small, perforations simple or simple and scalariform, intervascular pitting usually alternate and minute. Parenchyma predominantly paratracheal, often vasicentric along with some diffuse parenchyma. Rays up to 5 or 6, occasionally 8 to 10 cells wide, markedly heterogeneous. Leaves usually dorsiventral. Simple, unicellular hairs with conical or wart-like papillae in *Dichapetalum* and *Tapura* (Fig. 38.3G). Stomata Rubiaceous type, confined to the lower surface.

Embryology.   Pollen grain structure not studied. Ovules anatropous, bitegmic and tenuinucellate. Embryo sac of Polygonum type, 8-nucleate at maturity. Endosperm formation of the Nuclear type (Johri et al. 1992).

Chromosome Number.   The diploid chromosome numbers are reported to be 2n = 20, 24.

Chemical Featurres.   Organic fluoride compounds and flavonoids and leucoanthocyanins common (Dahlgren 1975a).

Important Genera and Economic Importance.   The three genera included in the family are *Dichapetalum* (= *Chailletia*), *Stephanopodium* and *Tapura*. *Dichapetalum* is a large genus common to most parts of the tropics, *D. cymosum* and other species of this genus are poisonous. The genus *Tapura* is rather interesting. The peduncle of the inflorescence is often adnate to the petiole of the subtending leaf; the entire structure appears as if the inflorescence is arising from the base of the leaf blade. There is marked variation in the floral structures of various members, e.g. in this small family, polypetalae, gamopetaleae, actinomorphy and zygomorphy are reported.

Taxonomic Considerations.   A highly controversial family, has been regarded variously by different workers. Bentham and Hooker (1965a) used the name Chailletiaceae and included it in the Geraniales. Engler and Diels (1936) changed the name to Dichapetalaceae but retained it in the Geraniales. Melchior (1964) includes it in the Thymelaeales and Hutchinson (1969,1973) in the Rosales. Cronquist (1968,1981) states that "this family is reminiscent in some respects of the Euphorbiaceae and sometimes has been associated with that family; they have perfect flowers which are not nearly so much reduced as those of Euphorbiaceae". According to him, Dichapetalaceae with their perfect dichlamydeous flowers and non-endospermous seeds appear to be much better placed in the Celastrales. Their inclusion in the Euphorbiales "would vitiate the conceptual unity of the order".

According to Takhtajan (1987), however, this family is closely linked with the Euphorbiaceae, especially with the subfamily Phyllanthoideae. Dahlgren (1980a,1983a) includes it in Thymelaeales but according to him also this family has a closer link with the Euphorbiaceae. Gornall et al. (1979) observed that the absence of myricetin in this family might support its link with myricetin-poor Euphorbiaceae.

A more detailed study is essential for proper placement of the family Dichapetalaceae.

Thymelaeaceae

A medium-sized family with 50 genera and 500 species distributed in both the temperate as well as tropical areas, particularly in Africa. Many members occur in the Himalayas up to 3000 m height. Its members are rather cosmopolitan except in the polar regions, with concentration in Australia, South Africa. Mediterranean region and the steppes of Central and Western Africa.

Vegetative Features.   Trees and shrubs, rarely herbs; leaves entire, alternate or opposite, estipulate, simple, persistent or deciduous (Fig. 38.4A).

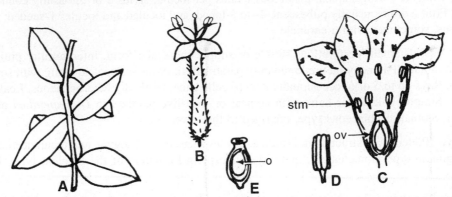

Fig. 38.4   Thymelaeaceae : **A** *Pimelea ferruginea*, **C-E** *Daphne mezereum.* **A** Vegetative shoot. **B** Flower of *Daphne mezereum.* **C** Opened out. **D** Stamen. **E** Ovary, longisection. *o* ovule, *ov* ovary, *stm* stamen. (Adapted from Lawrence 1951)

Floral Features.   Inflorescence raceme or spike or umbel; sometimes flowers are solitary. Flowers hermaphrodite, or rarely unisexual (and then plants dioecious), actinomorphic, hypogynous (or perigynous?). Calyx 4 or 5, imbricate, petaloid, connate to form a cuplike or tubular structure with spreading lobes (Fig. 38.4B). This so called calyx tube may be a true hypanthium of appendicular origin as is evidensed by the reduced petals and the stamens borne on it (Fig. 38.4C). Petals 4 to 12, often reduced to scale-like appendages, or absent, arising usually from near the mouth of the tube. Stamens variable: as many as and alternating with the sepals or twice as many or reduced to only 2 as in *Pimelea ferruginea* (Fig. 38.4B); anthers bicelled, introrse, dehiscence longitudinal (Fig. 38.4D); a hypogynous nectariferous disc, annular, cupular, or of scales often present. Gynoecium syncarpous, bicarpellary, one carpel markedly reduced; ovary superior, unilocular due to reduction of one carpel, placentation axile or parietal, ovule solitary and pendulous (Fig. 38.4E), when bilocular the ovary has 1 ovule per locule; style 1 or highly reduced, stigma typically discoid. Fruit a drupe, nut, achene or berry, often enclosed in the persistent receptacle. Seeds carunculate or arillate; endosperm scanty or none, embryo straight.

Anatomy.   Vessels small, sometimes in radial or oblique groups and such genera often with spiral thickening and tending to be ring-porous; perforations simple. Parenchyma paratracheal, rays either predominantly uniseriate or up to 2–4 cells wide with a few uniseriates, mostly homogeneous. Included phloem develops in many genera. Leaves usually dorsiventral, stomata on both surfaces or confined to only the lower surface; Ranunculaceous type.

Embryology.   Pollen grains 3-colporate to polyporate, 3-celled at dispersal stage. Ovules pendulous, anatropous, bitegmic, crassinucellate; Polygonum type of embryo sac, 8-nucleate at maturity. Obturator usually present; endosperm formation of the Nuclear type.

Chromosome Number.   The basic chomosome number is variable: x = 9, 13, 14, 15 and 45.

Chemical Features.   Toxic substances and coumarine derivatives like daphnin occur commonly in this family; leucoanthocyanins also reported, but lacks ellagic acid.

Important Genera and Economic Importance.   *Daphne* and *Wikstroemia* are the two large genera with about 70 species each. The former is common in Europe, North Africa, temperate and subtropical Asia, Australia and the Pacific Islands. In India it occurs in the Himalayas up to 3000 m. *Wikstroemia* occurs in South China, Indo-China, Australia and the Pacific Islands. In India it abounds in the temperate Himalayas up to 2700 m and northeastern India. The genus *Edgeworthia* with its 3 species is distributed from the Himalayas to Japan.

The family is economically important. Paper is made from *E. tomentosa* and *W. canescens*. The fibrous inner bark of *D. cannabina* is utilized for preparing Nepal paper, and the lace-like inner bark of *Lagetta lintearia* is suitable for making dresses. The wood of *Aquilaria malaccensis* is rich in resin and is the source of an incense called agaru, often burnt during Hindu religious festivals in India.

Taxonomic Considerations.   The position of the family Thymelaeaceae is disputed. According to earlier workers like Bentham and Hooker (1965c) and Hallier (1912), it belongs to the Daphnales of the Monochlamydeae. Engler and Diels (1936) and Rendle (1925) treated if as a member of the Myrtiflorae. Lawrence (1951) and Cronquist (1981) include Thymelaeaceae in the Myrtales. Melchior (1964) erected a separate order Thymelaeales. Dahlgren (1980a, 1983a) treats this family as a member of the order Thymelaeales derived from the Euphorbiales. Takhtajan (1987) also considers it to belong to a distinct order, Thymalaeales, and places this order next to the Euphorbiales. Heinig (1951) and Hutchinson (1969,1973) suggested its derivation from the family Flacourtiaceae, and Euphorbiaceae (partly) is also derived from the same line.

Various floral, anatomical and embryological features show a close relationship between Thymelaeaceae and the next family Elaeagnaceae. These two families alongwith a few others can form a separate order, Thymelaeales.

Elaeagnaceae

A small family of 3 genera and 50 species, distributed mostly as steppe and rock plants with the centres of abundance in southern Asia, Europe and North America.

Vegetative Features.   Mostly erect, much branched shrubs, rarely trees; vegetative parts densely covered with silvery brownish or golden-coloured lepidote or stellate hairs, branches sometimes spiny. Leaves alternate, opposite or whorled, coriaceous, simple, entire, estipulate (Fig. 38.5A).

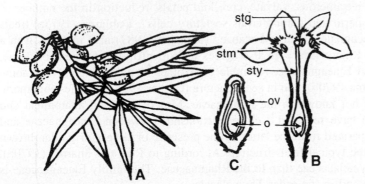

Fig. 38.5 Elaeagnaceae : **A-C** *Elaeagnus umbellata*. **A** Fruiting twig. **B** Flower, longisection. **C** Ovary, longisection. *o* ovule, *ov* ovary, *stg* stigma, *stm* stamen, *sty* style. (Adapted from Lawrence 1951)

Floral Features.   Inflorescence racemose. Flowers bisexual or unisexual (the plants then dioecious as in *Shepherdia* and *Hippophäe*, or polygamodioecious as in *Elaeagnus*), actinomorphic, hypogynous. Perianth uniseriate, petaloid and developed into a hypanthium or a receptacle. The receptacle in male flowers is flat, and in female flowers and the bisexual ones it is tubular and may be fused with the ovary. Stamens as many or twice as many as the perianth lobes; anthers introrse, bithecous, longitudinally dehiscent (Fig. 38.5B). Gynoecium monocarpellary, ovary superior, unilocular, with a solitary basal ovule (Fig. 38.5B,C); style and stigma 1. Fruit a dry, indehiscent achene surrounded by the fleshy persistent perianth and called a pseudo-drupe. Seeds with scanty or no endosperm, and a straight embryo.

Anatomy.   Vessels small, solitary, ring-porous, with simple perforations. Parenchyma diffuse, sometimes sparse. Rays up to 3 to 20 cells wide, hetero- or homogeneous, with a few uniseriates. Crystals of various kinds and crystal sand and tanniniferous substances abundant. Leaves dorsiventral; stomata of the Ranunculaceous type and confined to the lower surface.

Embryology.   Pollen grains usually 3-colporate and shed at 3-celled stage. Ovules anatropous, bitegmic, crassinucellate; glandular obturator in micropylar region. Polygonum type of embryo sac, 8-nucleate at maturity. Endosperm of the Nuclear type.

Chromosome Number.   Basic chromosome number variable: $x = 6, 10, 11, 13, 14$.

Chemical Features.   The family Elaeagnaceae is characterised by the presence of L-quebrachitol, not reported in neighbouring families. Ellagic acid, quercetin and other polyphenolics also present. Accumulation of simple indole bases, sinapic acid and saponins common.

Important Genera and Economic Importance.   The genus *Elaeagnus* is the largest taxon with 45 species which occur in different parts of Europe, Asia and North America. In India they grow up to 2300 m altitude in the Himalayas. Species of *Hippophäe* occur between 1600 m and 4000 m in the Himalayas. Economically, the family is also important. Different species of *Elaeagnus* are cultivated as deciduous or evergreen ornamental shrubs. The fruits of *E. angustifolia* and *Shepherdia argentea* are edible; those of *Hippophäe rhamnoides* are made into jelly or sauce. Its hardwood is used in turnery.

Taxonomic Considerations.   Most systematists treat family Elaeagnaceae as an ally of Thymelaeaceae (Bentham and Hooker 1965c, Lawrence 1951, Melchior 1964), although Melchior (1964) removed these two families along with some others to a separate order Thymelaeales. Hutchinson (1973) places Elaeagnaceae in a much advanced order Rhamnales. In his opinion, it "is a more" advanced group than and related to the Rhamnaceae.

Cronquist (1981) has placed this family in the Proteales along with Proteaceae, both with "strongly perigynous flowers, estipulate leaves, tetramerous, valvate, corolloid petals, reduction in the number of carpels to one, with or without endosperm and tanniniferous secretory cells". Dahlgren (1975a) treats Elaeagnaceae as a separate order, Elaeagnales, under Myrtanae. Takhtajan (1987) also includes it in a distinct order, Elaeagnales, but under Rhamnaceae. He also agrees with its affinities to the Proteaceae.

The two families, Rhamnaceae and Elaeagnaceae, resemble each other in palynological and some embryological features (Rau and Sharma 1970b), and in seed stucture (Corner 1976); but the monomeric ovary of Elaeagnaceae is a character not known in the Rhamnaeae. Moreover, floral anatomical and embryological studies (Sharma 1966) have revealed a close relationship betweeen Elaeagnaceae and Thymelaeaceae. The former can be separated from the latter by the presence of silvery or golden-brown lepidote on the vegetative parts, and the typical fruit structure. According to Rau and Sharma (1970b), the Elaeagnaceae are closer to the Thymelaeaceae than to the Rhamnaceae. The family Elaeagnaceae is better placed near the Thymelaeaceae and in the order Thymelaeales.

Taxonomic Considerations of the Order Thymelaeales

The order Thymelaeales comprises 5 families which have been assigned different positions by different authors as shown in Table 38.6.1.

Table 38.6.1   Comparative Data for the Five Families

Family	Lawrence 1951	Hutchinson 1973	Dahlgren 1983a	Cronquist 1981	Takhtajan 1987
Geissolomataceae	Myrtales	Thymelaeales	Celastrales	Celastrales	Geissolomatales
Penaeaceae	,,	,,		Myrtales	Myrtales
Dichapetalaceae	Geraniales	Rosales	Thymelaeales	Celastrales	Euphorbiales
Thymelaeaceae	Myrtales	Thymelaeales	Thymelaeales	Myrtales	Thymelaeales
Elaeagnaceae	Myrtales	Rhamnales	Elaeagnales	Proteales	Elaeagnales

The above analysis shows the assignment of each of the 5 families. Geissolomataceae is probably related to the Celastraceae but its pollen morphology is similar to that of the Euphorbiaceae members. Dahlgren and Rao (1969) and Carlquist (1976c) have different views about the position of this family. Penaeaceae closely resembles Oliniaceae of the Myrtales and Dichapetalaceae is closer to Euphorbiaceae. Elaeagnaceae shares many common features with the Rhamnales, including seed anatomy (Corner 1976) and uredinological data (Holm 1979, Savile 1979), but unicarpellate gynoecium does not indicate a direct derivation from the Rhamnales. Geissolomataceae, Penaeaceae and Thymelaeaceae (Order Thymelaeales of Hutchinson) are quite close to each other. The family Dichapetalaceae also has features common to the Euphorbiaceae. The family Elaeagnaceae resembles the Thymelaeaceae, particularly in floral, anatomical and embryological features (Rau and Sharma 1970b). As suggested by Hutchinson (1969,1973), the Flacourtiaceae is the probable ancestor of this order. Its resemblances with the Euphorbiaceae may be due to parallel development/evolution.

39

Order Violales

The order Violales comprises 6 suborders and 20 families:

1. Suborder Flacourtiineae includes Flacourtiaceae, Peridiscaceae, Violaceae, unigeneric Stachyuraceae, and Scyphostegiaceae, Turneraceae, unigeneric Malesherbiaceae, Passifloraceae, and Achariaceae.
2. Suborder Cistineae includes Cistaceae, unigeneric Bixaceae, Sphaerosepalaceae and Cochlospermaceae.
3. Suborder Tamaricineae comprises Tamaricaceae, Frankeniaceae and Elatinaceae.
4. Suborder Caricineae includes only Caricaceae.
5. Suborder Loasineae is monotypic and includes Loasaceae.
6. Suborder Begoniineae comprises Datiscaceae and Begoniaceae.

Herbs, shrubs or trees, sometimes climbers, e.g. Passifloraceae and Datiscaceae; annual or perennial, some xerophytic, e.g. Tamaricaceae, and some hydrophytic, e.g. Elatinaceae. The members of Violales have a basic floral construction with 5+5 perianth members, a variable number of stamens and 3 syncarpous carpels, with parietal placentation; ovules numerous. Embryological features are anatropous (orthotropous in Flacourtiaceae and Cistaceae), bitegmic (unitegmic in Loasaceae and Begoniaceae), crassinucellate ovules, Nuclear type of endosperm (Cellular in Loasaceae) and frequently copious endosperm in seed. The taxa of some families are rich in cyanogenic compounds and those of the Caricaceae contain glucosinolates.

Flacourtiaceae

A medium-sized family of 70–75 genera and 1200–1250 species from pantropical and subtropical regions of the world—South Africa, New South Wales and Chile; *Idesia* grows in Japan.

Vegetative Features. Small trees or shrubs, seldom climbers; leaves alternate, rarely opposite, simple, coriaceous, persistent, stipulate, stipules caducous.

Floral Features. Inflorescence lateral or terminal cymose. Flowers hermaphrodite, sometimes unisexual and then the plants are monoecious or dioecious as in *Pangium*; actinomorphic, hypogynous, ebracteate and ebracteolate (?). Sepals 2–15, polysepalous, free, imbricate, sometimes undifferentiated from corolla. Petals, when present, are usually equal in number to the sepals or more numerous, with or without a basal scale, imbricate. Stamens numerous, hypogynous, free or in groups alternating with sepals; anthers bicelled, dehiscence longitudinal, often alternate or appendaged. Gynoecium 2- to 10-carpellary, syncarpous, ovary unilocular with parietal placentation; superior, rarely half-inferior, ovules numerous on each placenta; style 1 or as many as the number of carpels. Fruit a loculicidal capsule or berry, seeds arillate with a straight embryo and abundant endosperm.

Anatomy. Vessels usually moderately small, solitary and in 2s and 3s. Perforation plates simple, rarely scalariform (*Hydnocarpus*), intervascular pitting scalariform, opposite or alternate; parenchyma sparse, paratracheal. Rays usually 3–5 cells wide, heterogeneous. Septate fibres present. Leaves dorsivenral, hairs of various types; stomata Rubiaceous or Cruciferous type, usually confined to lower surface.

Embryology.   Pollen grains 2-celled at the dispersal stage. Ovules orthotropous (*Casearia tomentosa*), hemiana- (*Idesia polycarpa*) or anatropous (*Kiggelaria*), bitegmic, crassinucellate. Polygonum type of embryo sac, 8-nucleate at maturity. Endosperm formation of the Nuclear type (Johri et al. 1992).

Chromosome Number.   Basic chromosome number is x = 0–12.

Chemical Features.   The cyclopentoid cyanogenic glycosides are reported (Saupe 1981). Gynocardin, a cyanogen with cyclopentene ring, is obtained from *Gynocardia odorata*.

Important Genera and Economic Importance.   Many of the genera are grown as ornamenals such as *Azara, Berberidopsis, Xylosma* and *Idesia. Hydnocarpus kurzii* is an important tree with a tall trunk, drooping branches, small yellow flowers and round, brown fruits. The seeds are the source of 'chalmoogric' and 'hydnocarpic' oil, used against leprosy. Medicinal oil is obtained from species of *Taraktogenos*. Fruits of *Flacourtia indica* and *Doryalis* spp. are edible.

Taxonomic Considerations.   The Flacourtiaceae is included in the order Violales by Cronquist (1968, 1981), Dahlgren (1983a), Melchior (1964) and Takhtajan (1980,1987); in the order Bixales by Hutchinson (1973) and in the Cistales by Thorne (1968,1983). According to Willis (1973), the Flacourtiaceae is related to the Euphorbiaceae, Tiliaceae and Passifloraceae. Tendencies towards perigyny and epigyny, unisexuality, reduction of stamens, fusion of filaments, development of corona, reduction in the number of carpels, fusion of styles and loss of endosperm from the seeds can all be observed in the members of this family (Cronquist 1968). The Flacourtiaceae is the most primitive family of the Violales. Kolbe and John (1979) reported that, serologically, this family is more closely related to the Violalean families with hypogynous flowers such as Passifloraceae, Violaceae and Turneraceae, and less closely to and highly distinct from those with epigynous flowers such as Datiscaceae, Begoniaceae and Cucurbitaceae (presently placed in a separate order Cucurbitales).

Peridiscaceae

A small family of 2 genera and 2 species only. *Peridiscus* is disributed in Amazonian region of Brazil and Venezuella, and *Whittonia* in Guiana.

Vegetative Features.   Trees, leaves alternate, petiolate, subobliquely elliptic, acuminate, entire, shining above, strongly 3-nerved from the base with a large pit in the axil of each basal nerve below. Stipules intrapetiolar, early deciduous and have a narrow oblique scar.

Floral Features.   Inflorescence in axillary cluster of very short racemes covered with short branched hairs or flowers fasciculate (as in *Whittonia guianensis*). Bracts fairly large, persistent; pedicels short. Sepals 5–7, imbricate, very hairy. Petals absent. Stamens numerous, inserted outside a large, cupular, several-lobulate or annular fleshy disc; filaments partially connate at base, anthers very small, monothecous, opening from the middle by lateral valves, introrse. Ovary sessile, and half-immersed in or subtended by the disc, depressed, unilocular,with numerous ovules hanging from the top (free-central apical placentation). Styles 4 or 3, short, free, spreading. Fruit stipitate (contracted at the base), ovoid, unilocular and 1-seeded, indehiscent. Seeds with curved embryo and without endosperm.

Anatomy.   Similar to Flacouriaceae, but different in having non-septate fibres. In *Whittonia* the vessels are in long radial multiples, the tangential walls of individual vessels in a multiple assemblage, flattened when in contact with one another. Lateral pitting scalariform; parenchyma abundant, scattered. Stomata absent in *Whittonia*, on abaxial surface in *Peridiscus*; Ranunculaceous type (Metcalfe 1962).

Important Genera and Economic Importance.   There are only two genera—*Peridiscus* and *Whittonia*; no economic importance.

Taxonomic Considerations. The family Peridiscaceae was not recognised by Bentham and Hooker (1965a) who included the four taxa of Peridiscaceae in Flacourtiaceae. Melchior (1964) places it in the Violales near Flacourtiaceae. Hutchinson (1969,1973) and Willis (1973) also support this view. According to Takhtajan (1987), the family Peridiscaceae is close to the Flacourtiaceae, but differs by the ovules pendant from apex of the ovary and anatomically by the unseptate fibrous elements.

Violaceae

A family of 15 genera and 850 species of wide distribution.

Vegetative Features. Shrubs or perennial herbs, rarely climbers, e.g. *Anchietea, Corynostylis, Noisettia* and *Schweiggeria. Hybanthus havanensis* of Cuba and *Hymenanthera dentata* of Australia and New Zealand are spiny shrubs. Stem woody and climbing (in climbers) but soft and herbaceous in annual herbs. Leaves alternate (opposite in *Hybanthus* and *Rinorea*), simple, rarely lobed or divided, stipulate, stipules minute, or foliaceous, persistent; often pellucidpunctate as in *Leonia*.

Floral Features. Mostly solitary axillary flowers (Fig. 39.3A); sometimes a raceme or a spike, subspicate in *Paypayrola* and borne on main trunk in *Leonia glycycarpa*. Flowers bisexual, rarely unisexual as in *Melicytus* and *Hymenanthera*, hypogynous, irregular (regular in *Rinorea*). Sepals 5, more or less free, imbricate, usually persistent; petals 5, the lowermost often spurred or saccate and larger than others, imbricate or contorted. Stamens 5, closely arranged around the pistil, anthers bicelled, introrse, often spurred, dehiscence longitudinal, connective protrudes above anther lobes. In *Viola* the stamens have a finger-like curved, nectar-secreting horn (Fig. 39.3B) that projects back from the connective of each of the two lower anthers into the spur of the lower petal. Cleistogamous flowers known in *Viola* and *Hybanthus*. Gynoecium 3- to 5-carpellary, syncarpous, ovary unilocular with parietal placentation (Fig. 39.3B,C); ovules 1, 2 or numerous on each placenta, style and stigma one. Fruit a loculicidal capsule (Fig. 39.3D) or a berry. Seeds winged in some lianas, e.g. *Anchietea*; embryo straight, endosperm copious, fleshy.

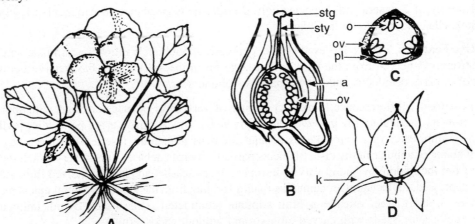

Fig. 39.3 Violaceae: **A-D** *Viola tricolor*. **A** Flowering plant. **B** Flower, vertical section, outer whorls partly removed. **C** Ovary, cross section. **D** Capsule with persistent calyx. *a* anther, *k* calyx, *o* ovule, *ov* ovary, *pl* placenta, *stg* stigma, *sty* style. (Original)

Anatomy. Vessels small and numerous, occasionally with a radial pattern, sometimes semi-ring-porous and with spiral thickening; perforation plates simple, scalariform or both; intervascular pitting usually opposite or scalariform. Parenchyma absent or as rare cells about the vessels. Rays absent from *Viola*; in other genera of 2 distinct size. Leaves generally dorsiventral, hairs of simple unicellular or uniseriate trichomes. Stomata Cruciferous or Rubiaceous type, on both surfaces or confined to the lower surface.

Embryology. Pollen grains 3- to 5-colporate, prolate or spheroidal, shed at 2-celled stage. Ovules anatropous, bitegmic, crassinucellate. Embryo sac of Polygonum type, 8-nucleate at maturity. Endosperm formation of the Nuclear type (Johri et al. 1992).

Chromosome Number. Basic chromosome number is x = 6–13,17,21,23. Polyploidy is widely distributed. A polyploid series is known for section Chamaemelanium of *Viola* with a base number x = 6 (2n = 24, 36,48,60,72).

Important Genera and Economic Importance. *Viola* is the largest genus with about 400 spp, mostly herbs, often cultivated as ornamentals; roots tuberous in *V. arborescens, Hybanthus havanensis* and *Hymenanthera dentata* are erect shrubs. *Corynostylis, Anchietea* and *Agatea* are climbers. In *Anchietea* of South America and *Decorsella* of Africa, capsules dehisce almost immediately after pollination and the seeds mature uncovered (Brizicky 1961a).

Taxonomic Considerations. Phylogenetic position of the family Violaceae is uncertain. Bentham and Hooker (1965a) placed the Violaceae in Parietales along with Papaveraceae, Resedaceae and others, on the basis of parietal placentation.

Hallier (1912) considered the Violaceae to be close to other families with spurred corolla like Balasaminaceae, Flacourtiaceae and Turneraceae, and included it as a primitive member in his Polygalinae. It has also been placed in the Guttales (Bessey 1915) and Parietales (Rendle 1925, Wettstein 1935, Lawrence 1951). Brizicky (1961a) considered Violaceae to be closely related to Flacourtiaceae and through this with Turneraceae, Malesherbiaceae and Passifloraceae.

Violaceae seems to be a climax family; the primitive members are woody with actinomorphic flowers, and the more advanced ones are herbaceous with zygomorphic flowers. They exhibit a range of structural variability in secondary xylem; some genera with strikingly primitive wood and others with more advanced but not highly evolved features (Taylor 1972). As treated by most phylogenists, this family is now regarded as a member of the Violales (Hutchinson 1973, Cronquist 1981, Dahlgren 1980a, Takhtajan 1987, Thorne 1983).

Stachyuraceae

A unigeneric family of the genus *Stachyurus* and 16 species, indigenous to eastern and central Asia.

Vegetative Features. Shrubs or small trees, leaves alternate, simple, deciduous, minutely stipulate or stipules absent.

Floral Features. Inflorescence axillary racemes, appearing before the leaves (Fig. 39.4A). Flower hermaphrodite, hypogynous, actinomorphic, ebracteate, ebracteolate and sessile (Fig. 39.4B). Sepals 4, polysepalous, imbricate; petals 4, polypetalous, imbricate; stamens 8, free, anthers bicelled (Fig. 39.4B, C), dehiscence longitudinal. Gynoecium 4-carpellary, syncarpous, ovary superior, unilocular, with 4 intruding parietal placentae that meet in the centre, so that the ovary appears to be tetralocular with axile placentation (Fig. 39.4D); ovules numerous. Style 1, thick and short with usually 4-lobed, capitate-peltate stigma (Fig. 39.4C). Fruit a berry, seeds arillate, with large embryo and copious oily endosperm.

Anatomy. Vessels small, solitary, commonly with spiral thickening, perforation plates scalariform with

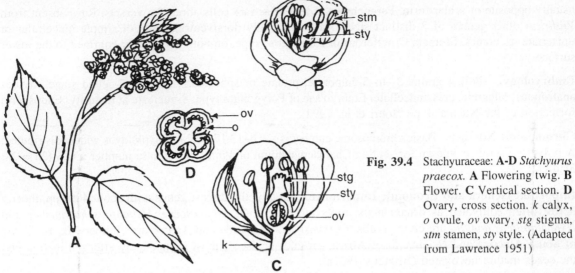

Fig. 39.4 Stachyuraceae: **A-D** *Stachyurus praecox*. **A** Flowering twig. **B** Flower. **C** Vertical section. **D** Ovary, cross section. *k* calyx, *o* ovule, *ov* ovary, *stg* stigma, *stm* stamen, *sty* style. (Adapted from Lawrence 1951)

many bars; parenchyma apotracheal, scattered among the fibres, rays up to 2–4 cells wide, markedly heterogeneous. Fibres with distinctly bordered pits, sometimes with spiral thickening. Leaves dorsiventral, stomata confined to the lower surface; Ranunculaceous type. Cluster crystals present in the spongy mesophyll and the parenchyma cells of the petiole.

Embryology.   Development of anther and pollen grains has not been investigated. Ovules anatropous, bitegmic, crassinucellate. Embryo sac of Polygonum type, 8-nucleate at maturity. Endosperm formation of the Nuclear type (Johri et al. 1992).

Important Genus.   The only genus, *Stachyurus* has no economic importance.

Taxonomic Considerations.   Stachyuraceae is a small family of uncertain taxonomic position. It has been variously placed in the Violales (Melchior 1964), Guttiferales (Metcalfe and Chalk 1972), Hamamelidales (Hutchinson 1973) and Theales (Cronquist 1968, 1981, Dahlgren 1983a, Takhtajan 1980, 1987, Thorne 1976,1983). Stachyuraceae closely resembles Hamamelidaceae members in anatomical features. It also resembles the Flacourtiaceae. Plants characterised by tanniniferous tissues, somewhat spongy cortex in the young stem, xylem with small, somewhat angular vessels with scalariform perforations and large cluster crystals in the parenchymatous tissues occur in both the Flacourtiaceae and the Hamamelidaceae. It was once united with Theaceae, from which it differs in: fewer stamens, unbranched style and parietal placentation. The absence of fibrous exotegmen in the seeds forbids the alliance with Flacourtiaceae as well (Corner 1976, Takhtajan 1987).

A more detailed study is necessary for assignment of correct position of the family Stachyuraceae.

Scyphostegiaceae

A unigeneric family with the only member *Scyphostegia borneensis*, endemic to Mount Kinabalu, Borneo, at about 2000 ft altitude.

Vegetative Features.   Small dioecious trees with soft wood and slender lenticular branchlets that are 3- to 4-angled when young. Leaves simple, alternate, distichous, shortly petioled with very small caducous stipules, close transverse venation and serrulate margin.

Floral Features. Inflorescence axillary and terminal racemes or panicles, usually leafy below (Fig. 39.5A,B); paniculate branches bare below, bears 3 to 4 (in male inflorescence) or 2 (in female inflorescence) tiers of infundibuliform or tubular bracts; the upper bracts each shortly exerted from the lower, with truncate margins (Fig. 39.5A,C), each bract subtends a single flower (Fig. 39.5D).

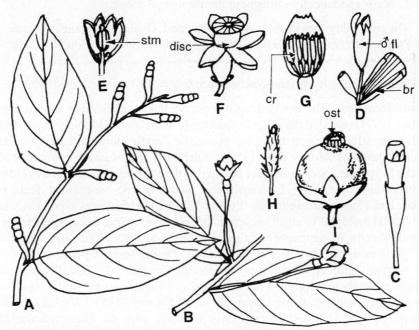

Fig. 39.5 Scyphostegiaceae : **A-I** *Scyphostegia borneensis.* **A, B** Male and female shoot bearing inflorescences. **C** Staminal inflorescence. **D** Bract and staminate flower. **E** Longisection. **F** Pistillate flower. **G** Vertical section of the disc with free carpels. **H** Carpel. **I** Fruit. *br* bract, *cr* carpel, ♂ *fl* staminate flower, *ost* ostiole, *stm* stamen. (After Hutchinson 1969)

Staminate Flowers: pedicellate, pedicels 2-nerved, flattened, perianth lobes 6, in 2 whorls of 3 each, imbricate, stamens 3, opposite the inner tepals, filaments fused into a column with clavate common apical connective (Fig. 39.5E), anthers 4-locular, extrorse, oblong (Fig. 39.5E; *stm*), column surrounded at the base by 3 short fleshy glands opposite the 3 stamens (Fig. 39.5E); disc absent.

Pistillate Flowers: perianth of 6 tepals in one whorl enclose a large fleshy globular disc with several inflexed stigma-like rays (Fig. 39.5F) (but no apparent stigmas), and an aperture (opening) in its centre. Gynoecium of numerous free, cylindrical, fusiform, puberulous, stipitate carpels (Fig. 39.5G,H) which stand erect on a broad, shallow receptacle within the disc (Fig. 39.5G), each subtended by 3 hyaline, partly adnate lobules from the receptacle at the base and a subterminal, oblique stigma; ovule 1, basal, erect and anatropous.

Fruit a bunch of narrow achenes enclosed in the enlarged fleshy disc (Fig. 39.5B, I) which tends ultimately to separate into 8–12 segments; the entire structure is subtended by a persistent perianth. Seeds with a large oily embryo and scanty or no endosperm.

Anatomy. Vessels sometimes solitary, but mostly in radial multiples of 2–4 or occasionally more; 40–100 μm in diameter, perforation plates simple, very oblique; lateral pitting alternate. Fibres constitute almost complete ground tissue of the wood. Parenchyma absent, rarely a few paratracheal cells observed.

Rays mostly uniseriate, a few partly bi- or triseriate; uniseriate rays of tall narrow cells; partly bi- or triseriate rays are 9–28 cells high, markedly heterogeneous.

Leaves dorsiventral, hairs absent; stomata Rubiaceous type, abundant, confined to the lower surface. Some of the mesophyll cells are filled with amorphous, probably tanniniferous contents. Large solitary clustered crystals occur sporadically throughout the mesophyll tissue.

Embryology.   The embryology has not been fully investigated. Ovules anatropous, bitegmic, crassinucellate. Embryo sac of Polygonum type, 8-nucleate at maturity. Endosperm formation of the Nuclear type (Johri et al. 1992).

Chromosome Number.   Basic chromosome number is x = 9.

Taxonomic Considerations.   The family Scyphostegiaceae has been placed in various orders by different taxonomists. Stapf (1894) included the genus *Scyphostegia* in the family Monimiaceae; Hutchinson (1926) in the Urticales near Moraceae, on the basis of similar female flowers. Later, in 1959 Hutchinson removed it to the Celastrales. Money et al (1950), on the basis of anatomy, morphology and palynology, concluded that the Scyphostegiaceae could not be allied to the Monimiaceae. Metcalfe (1956) and Melchior (1964) place this family near the Flacourtiaceae. Some of the anatomical features common to Scyphostegiaceae and Flacourtiaceae (of the Violales) are : (a) Rubiaceous type of stomata and adaxial hypodermis, (b) radial multiples of small vessels in the wood, (c) sparse wood parenchyma, (d) uniseriate rays of high, upright cells, (e) septate wood fibres, and (f) especially large cluster crystals, Heel (1967) ruled out its position near the Ranales or the Celastrales. On the basis of placentation, Heel decided the position of the Scyphostegiaceae to be nearer the Flacourtiaceae and the Tamaricaceae of the Parietales. Most phylogenists place these families (Flacourtiaceae and Tamaricaceae) in the Violales. On the basis of anatomical evidence and the floral interpretation by Swamy (1953), Violales is the order where Scyphostegiaceae fits best. It has some taxonomic affinities with the Flacourtiaceae of the Violales, parlicularly in anatomical features (as mentioned above); embryological features: bitegmic, crassinucellate ovules, Polygonum type of embryo sac, and Nuclear type of endosperm; and exomorphological features: tendency towards perigyny, unisexuality and reduction in the number of stamens.

In our opinion, inclusion of Scyphostegiaceae in the Violales is fully justified.

Turneraceae

A small family of 6 genera and 110 species, distributed mostly in tropical America, tropical Africa and Madagascar. *Turnera ulmifolia* is a native of the West Indies and tropical America (Johnson 1958). It also grows wild along coastal regions of south India (Mudaliar and Rao 1951).

Vegetative Features.   Herbs, shrubs or rarely trees, usually pubescent. Leaves alternate, simple, entire or toothed or lobed, often with two extrafloral nectaries or glands at the base. *Turnera subnuda* of Brazil and *T. guianensis* of Guyana and Trinidad are sometimes leafless.

Floral Features.   Flowers solitary or fasciculate, hermaphrodite, actinomorphic, hypogynous, bracteate and typically bibracteolate. Sepals 5, borne on a hypanthium, imbricate, deciduous; petals 5, inserted at the throat of the calyx, clawed, polypetalous, contorted. Stamens 5, opposite the sepals, borne on the hypanthium, anthers bicelled, dehiscence longitudinal. Gynoecium syncarpous, 3-carpellary, ovary superior or half-inferior (e.g. *Turnera*), unilocular, with parietal placentation and numerous ovules; styles 3, free, linear or flattened, stigmas flabellately lobed or fringed. Fruit a loculicidal capsule, ovoid or oblong, valves with usually numerous seeds in the middle. Seeds curved, sulcate, pitted or rough, arillate, with straight embryo and a fleshy endosperm.

Anatomy. Vessels very small to medium-sized with numerous multiples, perforations simple or simple and scalariform, intervascular pitting alternate and very small. Parenchyma apotracheal, diffuse or reticulate; rays up to 5 cells wide, heterogeneous or composed entirely of square and upright cells. Leaves generally dorsiventral, hairs of various types. Stomata Rubiaceous, sometimes Cruciferous and Ranunculaceous type.

Embryology. Pollen grains spheroidal or triangular, tricolpate, with a thick exine; shed at 2-celled stage. Ovules anatropous, bitegmic, crassinucellate. Polygonum type of embryo sac, 8-nucleate at maturity. Endosperm formation of the Nuclear type (Vijayaraghavan and Kaur 1966); aril present.

Chromosome Number. The basic chromosome number is x = 7,10.

Chemical Features. The genera *Erblichia standleyi, Piriqueta odorata* and *Turnera* spp. contain hydrocyanic acid (HCN). The cyclopentenoid cyanogenic glycosides are also reported from some genera (Saupe 1981).

Important Genera and Economic Importance. The genera *Erblichia* and *Piriqueta* have a corona, as in some Passifloraceae members; in *Mathurina* the peduncle bears a pair of large leafy bracts and an arillate seed clothed with long thread-like hairs, With its yellow, clawed petals and siliqua-like fruits, *Wormskioldia* bears a striking resemblance to some Brassicaceae members. None of the members is economically important.

Taxonomic Considerations. The family Turneraceae is closely related to the Passifloraceae and Malesherbiaceae (Brizicky 1961b), belonging to the same order Violales. Taylor (1938), on the basis of anatomical studies, allied it with the Flacourtiaceae. Hutchinson (1973) places Turneraceae in the Loasales along with Loasaceae. Embryological evidence (Vijayaraghavan and Kaur 1966) supports its close affinities with the Passifloraceae and to some extent with Flacourtiaceae, and not with the Loasaceae.

Malesherbiaceae

A unigeneric family of the genus *Malesherbia* with about 35 species distributed in southwestern America.

Vegetative Features. Herbs and undershrubs with alternate, often deeply lobed, estipulate leaves, sometimes very hairy.

Floral Features. Inflorescence racemes or cymes; flowers bisexual, actinomorphic, hypogynous, pentamerous except the gynoecium. Calyx of 5 sepals, long, tubular, polysepalous, lobes valvate, or occasionally imbricate. Stamens 5, borne on androphore. Gynoecium syncarpous, tricarpellary, ovary superior, unilocular with parietal placentation and numerous ovules. Styles 3 or 4, free from each other and widely separated at the base, appear below the apex of the ovary. Fruit capsule, seeds without aril, and with pitted testa.

Anatomy. Xylem includes numerous, radially arranged vessels, mostly with simple, circular or elliptical perforations, but scalariform plates with a few bars have also been recorded. Wood fibres consist of fairly short elements with very small, slit-shaped or elliptical pits. Rays mostly 1–2 or rarely 3 cells wide. Leaves isobilaeral; hairs of two types—moderately stiff, unicellular trichomes, and long, filiform, multiseriate, frequently glandular hairs, secreting a substance with an unpleasant smell.

Embryology. The embryology has not been fully investigated. Ovules anatropous, bitegmic, crassinucellate. Seeds albuminous with aleurone grains and fat globules (Johri et al. 1992).

Chemical Features. Cyanogenic glycosides with a cyclopentenoid ring structure present (Saupe 1981, Spencer and Seigler 1984).

Important Genera and Economic Importance. *Malesherbia* is the only genus with 35 species. These are erect or procumbent herbs or undershrubs, occur in dry habitats in the Andes from Peru to Bolivia and in the Argentine.

Taxonomic Considertions. Engler and Diels (1936) and Lawrence (1951) included Malesherbiaceae in the order Parietales, Melchior (1964) in the order Violales and Hutchinson (1969,1973) treats it as a member of the Passiflorales and near Passifloraceae. *Malesherbia* differs from Passifloraceae in: the styles more deeply inserted and widely separated at the base. Most taxonomists include Malesherbiaceae in the Violales which also includes the Passifloraceae. Malesherbiaceae is undoubtedly a member of the Violales on the basis of morphological, anatomical, embryological and chemical features. It is derived from the most primitive family of this order, i.e. Flacourtiaceae, as observed by Hutchinson (1969).

Passifloraceae

A family of 16 genera and 650–700 species (12 genera and 500 species, according to Brizicky 1961b), distributed predominantly in the tropical America but also found in New Zealand, Africa, Madagascar and Asia.

Vegetative Features. Herbs or shrubs, often climbing with the help of axillary tendrils (Fig. 39.8A); perennial. Leaves alternate, simple, entire, lobed or partite, rarely compound (pinnately compound in *Deidamia;* trifoliate or pentafoliate in *Efulensia*); stipulate, petiolate, petiole often glandular (Fig. 39.8A, B); tendrils opposite the leaves correspond to the terminal flower of a dichasium or to the first flower of a monochasium, unbranched.

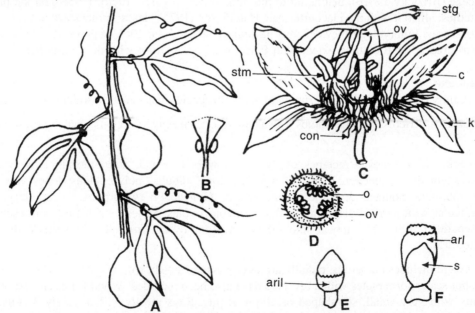

Fig. 39.8 Passifloraceae : **A-F** *Passiflora incarnata.* **A** Portion of stem with fruits. **B** Glands on leaf base. **C** Flower with calyx and corolla partly removed. **D** Ovary, cross section. **E** Young seed with aril. **F** Older seed with aril. *arl* aril, *c* corolla, *con* corona, *k* sepal, *o* ovule, *ov* ovary, *stg* stigma, *stm* stamen. (Original)

Floral Features. Flowers usualy solitary or axillary pairs; usually bisexual (when unisexual, plants monoecious or rarely dioecious as in *Adenia*), actinomorphic, hypogynous, bracteate, bracteolate. Sepals

5 (or 4), free or basally connate, often petaloid or fleshy, imbricate, persistent; lobes often horned on the back below the apex. Petals 5 (or 4) or absent, free or basally connate; often smaller than sepals, imbricate. Corona (Fig. 39.8C; *con*) single or double, fleshy, concave or cup-like structure lying between corolla and androecium. Stamens 5 or more, usually oppositipetalous, numerous in *Barteria* and *Smeathmannia*; filaments usually adnate to the elongated gynophore forming an androgynophore; anthers versatile as in *Passiflora* (Fig. 39.8C; *stm*) or basifixed, bicelled, dehisce longitudinally, staminodes present in some genera. Gynoecium syncarpous, 3- to 5-carpellary, raised on a gynophore or an androgynophore; ovary superior, unilocular, with parietal placentation, ovules numerous on each placenta (Fig. 39.8D). Styles as many as carpels, free or all fused together, stigmas 3 to 5, capitate or discoid. Fruit a berry or a loculicidal capsule with arillate seeds (Fig. 39.8 E,F); seeds with straight embryo and fleshy endosperm.

Anatomy.   Vessels extremely small to large, few to numerous, sometimes in radial multiples of 4 or more cells, or with a radial or oblique pattern; perforations usually simple, rarely with a few scalariform plates, intervascular pitting alternate. Parenchyma typically apotracheal, diffuse to numerous uniseriate bands; rays up to 1 to 8 cells wide, heterogeneous. Leaves usually dorsiventral, rarely isobilateral. Hairs unicellular or uniseriate, stomata Ranunculaceous type.

Embryology.   Pollen grains 3- to 12-colporate in *Passiflora;* with reticulate exine and are shed at the 2-celled stage. Ovules anatropous, rarely orthotropous, bitegmic, crassinucellate. Polygonum type of embryo sac, 8-nucleate at maturity. Endosperm formation of the Nuclear type.

Chromosome Number.   Basic chromosome number is x = 6, 9–11.

Chemical Features.   Anthocyanins of the nature of 3-glucosyl glucoside occur in fruit rind of *Passiflora* (Harborne 1963). Some members of Passifloraceae yield HCN (Gibbs 1963). The fruits of *Passiflora edulis* contain an essential oil that is rich in 250 volatile components (Murray et al. 1972). Saupe (1981) reported cyanogenic glycosides in various members of this family.

Important Genera and Economic Importance.   The primitive genera *Smeathmannia* and *Barteria* are erect shrubs without tendrils and bear numerous stamens and poorly developed corona. *Adenia venetata*, distributed from Arabia to Tanganyika is remarkable for its enormous fleshy rootstock and spiny leafless branches. Many *Passiflora* species are grown as ornamentals, and fruits of some species are edible.

Taxonomic Considerations.   The family Passifloraceae has been treated variously by different taxonomists. It has been placed in the Parietales by Engler and Diels (1936) and Lawrence (1951); in the Passiflorales by Bentham and Hooker (1965a), Hutchinson (1969, 1973) and Takhtajan (1966); in the Cistales by Thorne (1968, 1983); in the Violales by Cronquist (1968,1981), Dahlgren (1983a), Melchior (1964) and Takhtajan (1980,1987). The Passifloraceae resemble the members of the Violales in unilocular ovary with parietal placentation. Chemically, it is closer to Turneraceae as they both contain HCN. It is quite an advanced family, as most of the members are herbaceous or climbers, with few stamens and corolline corona, Its inclusion in the Violales is fully justified.

Achariaceae

A small family of only 3 genera and 3 species, confined to South Africa.

Vegetative Features.   Habit various: woody shrublets, e.g. *Acharia tragioides,* stemless or acaulescent herbs, e.g. *Guthriea capensis* or a slender, scandent herb, e.g. *Ceratiosicyos ecklonii.* Leaves alternate, simple, estipulate with crenulate, serrulate or lobed leaves (radial cordate leaves in *Guthriea*).

Floral Features.   Inflorescence axillary fascicle of a few flowers or raceme or often solitary flowers.

Flowers unisexual, male or female, plants monoecious, flowers actinomorphic, hypogynous. Sepals 3 or 5, open; petals 3 to 5, gamopetalous, campanulate, lobes valvate. Stamens 3 to 5, epipetalous, connective expanded, anthers introrse; staminodes 3 to 5, short, fleshy and alternate with stamens. Gynoecium syncarpous, 3- to 5-carpellary, ovary unilocular with parietal placentation; ovules 2-numerous on parietal placentae. Fruit, a globose to linear 3- to 5-valved capsule. Seeds arillate and with copious endosperm, and a small straight embryo.

Anatomy.   Vessels usually with simple perforations, narrow, radially arranged, with bordered pits, embedded in a ground mass of thick-walled, septate fibres with simple pits in *Acharia;* wide and scattered, embedded in ground tissue consisting of septate, prosenchymatous, simple-pitted elements intermixed with parenchyma. Rays broad, parenchyma mostly around the vessels. Leaves dorsiventral, hairs usually simple, multicellular; stomata confined to lower surface, Ranunculaceous type.

Embryology.   Embryology has not been studied; ovule anatropous (Johri et al. 1992).

Important Genera and Economic Importance.   Three genera are included in this family—*Acharia, Ceratiosicyos* and *Guthriea. G. capensis* is a stem-less herb with fleshy roots arising from a rhizome. No economic importance is known.

Taxonomic Considerations.   Bentham and Hooker (1965a) included Achariaceae in the Passifloraceae. Melchior (1964) places Achariaceae in the Violales, Hutchinson (1973) in the Passiflorales, Cronquist (1981), Thorne (1983), Dahlgren (1983a) and Takhtajan (1980,1987) in the Violales. The Achariaceae is very close to the Passifloraceae, and differs from it mainly in the absence of stipules and presence of sympetalous, campanulate corolla. It also shows certain similarities with the Cucurbitaceae in the structure of leaves, flowers, pollen and seed structure (Takhtajan 1987).

   However, the presence of syncarpous, unilocular ovary with parietal placentation in the Achariaceae makes it undoubtedly a member of the Violales.

Cistaceae
A family of 8 genera and 200 species distributed in the warmer parts of the northern hemisphere, and abundant in the Mediterranean region.

Vegetative Features.   Perennial or annual, herbs or shrubs, often covered with stellate hairs. Leaves opposite, simple, petiolate, stipulate or estipulate.

Floral Features.   Inflorescence cymose or cymose racemes, often solitary axillary. Flowers bisexual, actinomorphic, hypogynous, often cleistogamous. Sepals 3–5, polysepalous, convolute, unequal in some genera; petals 5, rarely 3 or absent, convolute, caducous. Stamens numerous, borne on an elongated and projecting receptacle, centrifugal, free; anthers introrse, bicelled, dehiscence longitudinal. Gynoecium syncarpous, 3- or 5- to 10-carpellary, unilocular or falsely 5- to 10-loculed due to intrusion of placentae; placentation parietal, ovules 2 to many on each placenta. Style 1, stigma 1 or 3 to 5. Fruit a leathery or woody capsule, loculicidally dehiscent, seeds small, angular; endosperm present, embryo curved.

Anatomy.   Vessels small, numerous, in radial or tangential rows or semi-ring-porous; perforations simple, intervascular pitting alternate and small. Parenchyma usually absent; rays mostly unseriates, sometimes up to 3 cells wide, heterogeneous. Leaves dorsiventral or centric with different types of hairs; stomata Ranunculaceous type, mostly present on both sides.

Embryology.   Pollen grains with three germ pores, 2-celled at dispersal stage. Ovules orthotropous, rarely anatropous, bitegmic, crassinucellate. Polygonum type of embryo sac, 8-nucleate at maturity. Endosperm formation of the Nuclear type.

Chromosome Number. Basic chromosome number is x = 5–9, 11.

Chemical Features. Members of the the Cistaceae lack cyanogenic compounds as well as glucosinolates, but accumulate sulphated flavonoids (Gershenzon and Mabry 1983).

Important Genera and Economic Importance. A few species and hybrids of the genus *Cistus* and some species of *Helianthemum* and *Hudsonia* are cultivated as ornamentals.

Taxonomic Considerations. The position of Cistaceae is controversial. It was included in the order Cistineae along with Bixaceae by Engler and Diels (1936). Bessey (1915) treated Cistineae (= Cistaceae) and Bixaceae as members of the Guttiferales. Rendle (1925), Wettstein (1935), and Lawrence (1951) retained Cistaceae in the Parietales. Hutchinson (1973), Cronquist (1968,1981) and Takhtajan (1980,1987) place this family in the Bixales. Melchior (1964), however, includes Cistaceae in the Violales. Dahlgren (1983a) and Thorne (1983) point out that the Cistaceae along with Bixaceae and Cochlospermaceae should be retained in the Malvales. Chemical data (Gornall et al. 1979, Young 1981), unilocular ovary, parietal placentation, and the exotegmic seed coat relate Cistaceae to Bixaceae.

Both Cistaceae and Bixaceae lack cyanogenic compounds as well as glucosinolates but they do accumulate sulphated flavonoids, a character shared with Tamaricaceae and Frankeniaceae (Harborne 1975b), which are also treated in or very close to the Violales. Dahlgren (1980a,1983a) and Thorne (1981,1983), however, include these two families in the Malvales, because of the pesence of myricetin and 8-OH-flavonoids. Also, they lack cyclopropenoid fatty acids—another characteristic of most of the Malvales members (Gibbs 1974, Shorland 1963).

The appropriate placement of these two families is controversial, and further investigations are necessary to ascertain their taxonomic affinity.

Bixaceae

A unigeneric family of the genus *Bixa* with only 1 species (*B. orellana*), widely naturalized in the tropical belt.

Vegetative Features. Shrubs or small trees with reddish sap, stem surface hairy. Leaves alternate, simple, entire, petiolate, stipulate.

Floral Features. Inflorescence a panicle. Flowers bisexual, actinomorphic, hypogynous, pedicellate, pedicels with 5 glands below the calyx. Calyx of 5 sepals, polysepalous, imbricate, caducous. Corolla of 5 petals, polypetalous, large, showy, imbricate and twisted in bud. Stamens numerous, free, in many whorls, anthers horseshoe-shaped, bicelled, introrse, longitudinal dehiscence. Gynoecium syncarpous, bicarpellary, ovary superior, unilocular, with parietal placentation, ovules numerous. Style 1, slender, stigmas 2, flattened. Fruit a loculicidal capsule, outer surface spinous, 2-valved; seeds with a bright red, fleshy testa, endosperm present, embryo large with broad cotyledons.

Anatomy. Vessels moderately small, with numerous small multiples, perforations simple, intervascular pitting alternate and minute. Parenchyma diffuse, storied; rays mostly 1–3 cells wide, heterogeneous. Leaves dorsiventral, hairs tufted or peltate, stomata Ranunculaceous type, confined to the lower surface.

Embryology. Pollen grains 2- or 3-celled, tricolpate with smooth exine. Ovules anatropous, bitegmic, crassinucellate. Polygonum type of embryo sac, 8-nucleate at maturity. Endosperm formation of the Nuclear type. Seed coat formed by both integuments.

Chromosome Number. The basic chromosome number is x = 6,7,8.

Chemical Features. Rich in sulphated flavonoids (Gershenzon and Mabry 1983, Harborne 1975a), myricetin and 8-OH-flavonoids (Gornall et al. 1979).

Important Genera and Economic Importance. *Bixa orellana* is the only member of this family. Commonly called annatto, it is a shrub or small tree of Central America and often cultivated in South India. Colouring matter, bixin, is the main constituent of pigment mass surrounding each individual seed. Annatto is an important vegetable colouring material used in dairy and food industries. Also used in ice cream, bakery products and edible oils. Apart from food colouring, annatto is also used in floor wax, furniture and shoe polishes, nail gloss, brass lacquer, wood stains and hair oils (Singh et al. 1983).

Taxonomic Considerations. The family Bixaceae is closely related to the Cistaceae and the Cochlospermaceae. Like Cistaceae, the position of this family is also disputed. The composition of the orders in which Bixaceae have been placed by various authors varies, but the affinity with the families Cistaceae and Cochlospermaceae is recognised.

Sphaerosepalaceae

A small family of 2 genera, *Dialyceras* and *Rhopalocarpus* and 14 species, reported mostly from the Madagascar region.

Vegetative Features. Trees or shrubs with alternate, simple, entire, sometimes trinerved, stipulate leaves; stipules interpetiolar, caducous, leave prominent scar on the stem.

Floral Features. Inflorescence terminal and axillary panicles of umbelliferous cymes. Flowers hermaphrodite, actinomorphic, hypogynous, bracteate, ebracteolate, usually tetramerous. Calyx of 4 sepals (rarely 3 + 3), polysepalous, unequal, imbricate; corolla of 4 petals (rarely 3), unequal, imbricate, slightly clawed, densely streaked with short resinous lines. Stamens numerous, filaments filiform, more or less connate at base, resin-dotted, anthers small, bicelled, locules widely separated by a broad, glandulous connective; disc (interior to stamens) large, cupular, wrinkled, denticulate. Gynoecium syncarpous, bi- or rarely tricarpellary, ovary superior, bilocular and deeply-bilobed, with a simple, geniculate style (arising between the lobes) and entire stigma; ovules 1 to 3 or more per locule, erect, basal. Fruit globose or didymous, densely muricate, 1- (or rarely 2)-seeded; seeds large, reniform with a horny, more or less ruminate endosperm, and minute embryo with foliaceous cotyledons.

Anatomy. Somewhat similar to that of Cochlospermaceae.

Embryology. Embryology has not been studied in sufficient detail. Ovules anatropous with a massive tanniniferous chalaza (Johri et al. 1992).

Important Genera and Economic Impotance. It is a small family of Madagascar trees—of the genera *Dialyceras* and *Rhopalocarpus*. Economic importance not reported.

Taxonomic Considerations. A controversial family, has been termed as Rhopalocarpaceae. The family is taxonomically isolated (Cronquist 1968) and has been variously referred to the Theales, Malvales and Violales. Cronquist (1968, 1981), Thorne (1983) and Takhtajan (1987) are of the opinion that this family is best accommodated near the Sarcolaenaceae, in the Theales or Ochnales (Hutchinson 1969,1973). Syncarpous, bilocular ovary with basal placentation and partly connate stamens are an unusual combination of characters.

The name of the genus, *Sphaerosepalum*, has been changed to *Rhopalocarpus*, and hence the family name Rhopalocarpaceae (Cronquist 1968,1981, Thorne 1983, Takhtajan 1987).

Cochlospermaceae

A small family of only 2 genera, *Amoreuxia* and *Cochlospermum*, and 20–38 species (Takhtajan 1987) distributed in tropical countries. *C. religiosum* is native to India.

Vegetative Features. Essentially woody trees, shrubs or rarely herbs, sometimes rhizomatous (*C. vitifolium, C. regium*) with reddish or orange-coloured sap. Leaves alternate, stipulate, stipules inconspicuous, subulate, pubescent, 1–2 cm long, in *Amoreuxia malvifolia* the leaves are more conspicuous and chestnut-coloured (Poppendieck 1980). Leaves simple but palmately lobed and veined.

Floral Features. Inflorescence axillary or terminal racemes or panicles. Flowers bisexual, actinomorphic (zygomorphic in *Amoreuxia*), large, showy, hypogynous, bracteate, ebracteolate. Sepals and petals 5 (or 4), free, deciduous, imbricate or quincuncial. Stamens numerous, free, in several whorls, equal or unequal, anthers bicelled, linear, dehisce by apical slits; anthers of *Amoreuxia* are dimorphic. Gynoecium syncarpous, 3- to 5-carpellary, ovary superior, unilocular with parietal placentation (*Cochlospermum*) or falsely 3- to 5-loculed (*Amoreuxia*) due to intrusion of the placentae; style 1, stigmas as many as carpels, dentate, fruit a capsule with 3–5 valves. Seeds with curved embryo and an oily endosperm; densely covered with woolly hairs.

Anatomy. Vessels medium-sized to large and rather few, perforations simple, intervascular pitting alternate. Parenchyma in broad apotracheal bands. Rays up to 5 or 6 cells wide, uni- or multiseriate, heterogeneous. Leaves with simple, unicellular, elongated hairs and stomata confined to the lower surface. A conspicuous feature is the fine red stripes on the leaves, sepals and petals brought about by idioblasts containing a red fluid. There are secretory canals in the pith and cortex of *Cochlospermum*, and secretory cells in *Amoreuxia*.

Embryology. Embryology has not been fully investigated. Ovules ana-campylotropous, bitegmic and crassinucellate. Endosperm formation of the Nuclear type (Johri et al. 1992).

Chromosome Number. Basic chromosome number is x = 6 (Takhtajan 1987); 2n = 12 in *C. tinctorium* and *C. planchonii*.

Chemical Features. Harborne (1975b) reports ellagic acid, kaempferol, quercetin and myricetin in 5 species of *Cochlospemum*.

Important Genera and Economic Importance. *Cochlospermum religiosum* is a medium-sized tree which occurs in dry forests of India. Commonly called white silk cotton, the fibrous covering of its seeds is used as stuffing material. An edible gum obtained from this plant is used as a substitute for gum tragacanth. *C. vitifolium*, with showy yellow flowers is often grown as an ornamental. A yellow dye is obtained from the roots of *C. tinctorium*. Some species, e.g. *C. vitifolium, C. angolense, C. planchonii* and *C. tinctorium*, are reported to be used as a cure for jaundice. Roots of some species are used as vegetables, e.g. *Amoreuxia palmatifida;* young fruits are also eaten raw (Poppendieck 1980).

Taxonomic Considerations. Melchior (1964) placed this family in his Violales which comprises the major part of Engler's Parietales. Takhtajan (1966) accepts this placement. Dahlgren (1977a, 1980a, 1983a) included Cochlospermaceae in the Malvales. Keating (1969), on the basis of wood anatomy, allied it with the Malvaceae, Tiliaceae and Sterculiaceae of the Malvales. Cochlospermaceae has been grouped along with Bixaceae and Cistaceae by all workers. The features common to these families are: (a) mostly woody perennials, (b) numerous stamens in many whorls, (c) unilocular ovary, and (d) parietal placentation. Chemically, however, Cochlospermaceae and Bixaceae are not very closely related (Harborne 1975b), and also because of palmately lobed leaves and oily endosperm. Dahlgren (1983a) and Thorne

(1968,1983) include this family along with the Bixaceae and Cistaceae, in the Malvales. Cronquist (1981) and Takhtajan (1987) recognise a separate order, Bixales, for these three families and place it in the Malvanae.

Although included in the Violales, the family Cochlospermaceae is more closely related to the Malvales because of: (a) presence of digitately veined and palmately lobed leaves, (b) numerous stamens, (c) hairy seeds, and (d) presence of myricetin. These families are better placed in the order Malvales, or in a separate order, Bixales, under the Malvanae.

Tamaricaceae

A small family of 4 genera and ca. 100 species, mostly of the Mediterranean region and semi-arid and saline regions of the Old World.

Vegetative Features. Small trees or shrubs of shorter height, halophytic or xerophytic. Leaves minute, scaly, sometimes subulate, appressed to the stem and branches (Fig. 39.14A).

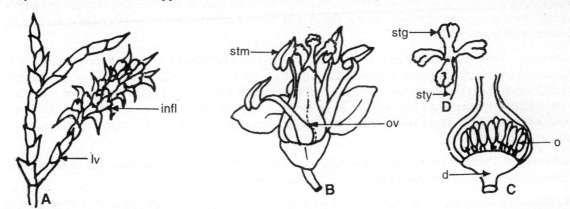

Fig. 39.14 Tamaricaceae: **A-D** *Tamarix dioica*. **A** Twig bearing inflorescence. **B** Flower. **C** Ovary, vertical section. **D** Style and stigma. *d* disc, *infl* inflorescence, *lv* leaves, *o* ovule, *ov* ovary. *stg* stigma, *stm* stamen, *sty* style. (**A, B** after J.K. Maheshwari 1965, **C, D** after Lawrence 1951)

Floral Features. Inflorescence dense, spike-like racemes (Fig. 39.14A) or solitary flowers as in *Reaumuria*. Flowers minute, bisexual, actinomorphic (Fig. 39.14B), ebracteate, hypogynous, tetra- or pentamerous. Sepals 4 or 5, polysepalous, imbricate; petals 4 or 5, polypetalous, imbricate, persistent in fruit. Stamens as many or twice as many as the petals; anthers bicelled, dehiscence longitudinal. Gynoecium syncarpous, 3- to 4-carpellary, ovary superior, unilocular, with parietal, or basal-parietal (Fig. 39.14C) placentation; ovules 2 to many on each placenta; styles as many as carpels (Fig. 39.14D), free or basally connate or absent with as many sessile stigmas. Fruit a capsule, sometimes falsely multilocular, seeds densely hairy all over or only at distal end, rarely winged; embryo straight, endosperm absent in *Tamarix* but present in *Reaumuria, Hololachna,* etc.

Anatomy. Vessels small to medium-sized , solitary, in short multiples and in clusters; semi-ring-porous, perforations simple, intervascular pitting alternate. Parenchyma scanty, paratracheal to vasicentric. Rays very wide, heterogeneous. Stomata present on both surfaces or only on upper surface; Ranunculaceous type.

Embryology. Pollen grains (2)-3(4)-colpate; 2-celled at the dispersal stage. Ovules anatropous, bitegmic, crassinucellate. The inner integument grows faster than the outer one and forms the micropyle. Endosperm formation of the Nuclear type (Johri et al. 1992). Five different types of embryo sacs have been reported: Fritillaria, Chrysanthemum cinerariaefolium, Drusa, Adoxa and Plumbagella, and, accordingly, the embryo sac may be 8-nucleate, 16-nucleate or 10- or 12-nucleate at maturity. Polyembryony reported in *Tamarix ericoides* (Johri et al. 1992).

Chromosome Number.   Basic chromosome number is x = 6.

Chemical Features.   Sulfated flavonoids reported (Gershenzon and Mabry 1983).

Important Genera and Economic Importance.   *Myricaria, Hololachna, Reaumuria* and *Tamarix* are the four genera of this family. *Myricaria* has seed character similar to *Tamarix,* but it has monadelphous stamens. *Tamarix* species are often grown as wind-break along the sea shore; many species of *Tamarix* and *Myricaria germanica* are also grown as ornamentals because of their pink bloom.

Taxonomic Considerations.   A family of controversial affinities, Tamaricaceae has been grouped in several orders. Bessey (1915) included it in the Caryophyllales. Engler and Diels (1936), Lawrence (1951), Rendle (1925) and Wettstein (1935) placed Tamaricaceae in the Parietales. Melchior (1964) included it in the Violales, mainly because of the presence of unilocular ovary with parietal placentation. Hutchinson (1969,1973) recognised a distinct order, Tamaricales, for the families Tamaricaceae, Frankeniaceae and Fouquieriaceae. Cronquist (1968,1981) and Thorne (1983), however, treat Tamaricaceae as a member of the Violales. On the other hand, Corner (1976), Dahlgren (1983a), Gunderson (1950), Novak (1961, see Takhtajan 1987) and Takhtajan (1987) consider Tamaricaceae and Frankeniaceae to be included in a separate order, Tamaricales, with affinities with the Violales.

Chemically, Tamaricaceae and Frankeniaceae are related and the only families (of the Violales) with this feature (Cistaceae, Bixaceae and Cochlospermaceae), are more closely related to the Malvales. Hence, in our opinion, these two families—Tamaricaceae and Frankeniaceae—should be assigned a separate order Tamaricales.

Frankeniaceae

A small family of 4 genera and 34 species, of cosmopolitan distribution but more common in the Mediterranean region.

Vegetative Features.   Perennial herbs or shrubs. Leaves opposite, decussate, simple, entire, the pairs joined at the base by a ciliated line on the stem surface, estipulate.

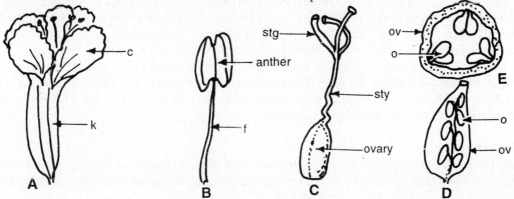

Fig. 39.15   Frankeniaceae : **A-E** *Frankenia grandifolia.* **A** Flower. **B** Stamen. **C** Gynoecium. **D** Ovary, vertical section. **E** Cross section. *an* anther, *c* corolla, *f* filament, *k* calyx, *o* ovule, *ov* ovary, *stg* stigma, *sty* style. (After Lawrence 1951)

Floral Features. Inflorescence terminal or axillary cymes or solitary flowers. Flowers bisexual, actinomorphic, hypogynous. Sepals 4–7, gamosepalous, form a tubular basal portion with short calyx teeth (Fig. 39.15A). Petals 4–7, polypetalous, imbricate, each petal long-clawed and with a ligule-scale at the base of each limb. Stamens usually 6 (4–7) in 2 whorls, free or basally connate, often didynamous; anthers bicelled, extrorse, versatile, longitudinally dehiscent (Fig. 39.15B). Gynoecium syncarpous, 2- to 4-carpellary, ovary superior, unilocular, with parietal placentation (Fig. 39.15D, E), ovules 2 or 3 or many; style 1, slender, stigmas three (Fig. 39.15C). Fruit a loculicidal capsule, enclosed in the persistent calyx. Seeds with straight embryo and a mealy endosperm.

Anatomy. Phloem and xylem usually in the form of continuous cylinders; vessels small, scattered, with simple perforations; ground tissue composed of prosenchyma with simple pits. Characteristic feature is the universal occurrence of epidermal salt glands of a special type. Leaves generally dorsiventral, sometimes subcentric; stomata of the Ranunculaceous type, usually sunken, in some species confined to grooves.

Embryology. Pollen grains spherical, tricolpate, 3-celled at shedding stage. Ovules anatropous, bitegmic, pseudo-crassinucellate. Polygonum type of embryo sac, 8-nucleate at maturity. Endosperm formation of the Nuclear type (Walia and Kapil 1965).

Chromosome Number. Basic chromosome numbe is x = 10, 15.

Chemical Features. In six different species of *Frankenia,* flavonol conjugates were isolated. Most of the flavonoids in the leaves appear to be bound as bisulphates (Harborne 1975b).

Important Genera and Economic Importance. *Frankenia,* with about 70 species, are low shrublets of ericoid habit. *F. laevis* has rose-coloured blooms. *Hypericopsis* is distributed in south Iran. Five species of *Anthobryum* occur at high altitudes in western South America, and *Niederleinia juniperoides* is endemic to Patagonia. *N. juniperoides* differs from other members of the family by its dioecious or polygamo-monoecious flowers and fruits with 1-seeded placenta (Hutchinson 1969). There is also a considerable concentration of endemic species in west and southwest Australia.

Taxonomic Considerations. The family Frankeniaceae has been associated with the Tamaricaceae and the Elatinaceae. Niedenzu (1925b) and Lawrence (1951) included Frankeniaceae in the Parietales, Melchior (1964) in the Violales. Bentham and Hooker (1965a) placed Frankeniaceae near Caryophyllaceae in the Caryophyllineae. Gundersen (1927) also accepted the placement, although the placentation differs in the two families. According to Kapil and Walia (1965), this family should be placed close to Elatinaceae in the Parietales. Cronquist (1981) included it in the Violales, and Dahlgren (1983a, b), Hutchinson (1969, 1973), Takhtajan (1980, 1987) and Thorne (1983) in the order Tamaricales.

The Frankeniaceae is allied to the Tamaricaceae, but differs in opposite leaves, free sepals, and one style; while in Tamaricaceae the leaves are alternate, sepals connate below and styles three or four. According to Meeuse (1975b), the order Tamaricales (of Tamaricaceae and Frankeniaceae) is allied to the Violales and Salicales. Therefore, these two families may be retained in the Violales or may be treated as members of a smaller order, Tamaricales.

Elatinaceae

A small family of only 2 genera, *Bergia* and *Elatine* with about 30 species of cosmopolitan distribution. *Elatine* grows in freshwater habitats of both temperate and tropical regions and *Bergia* in the tropics and subtropics of both the Old and the New World.

Vegetative Features. Annual or perennial aquatic or terrestrial herbs, or small shrubs with creeping or erect stems rooting at the nodes (Fig. 39.16A, B), sometimes succulent, e.g. *Bergia capensis*. Leaves opposite or whorled, simple, stipulate, gland-dotted, entire or serrate or crenate; stipules caducous.

Floral Features.   Inflorescence small axillary cymes or small solitary axillary flowers (Fig. 39.16B). Flowers bisexual (Fig. 39.16C, D), actino- or zygomorphic, hypogynous. Sepals 5 in *Bergia,* 2–4 in *Elatine* (Subramanyam 1970), free or basally connate, imbricate with membranous margin, persistent. Petals same as number of sepals, polypetalous, imbricate, persistent. Stamens as many as or twice the number of petals, in 1 or 2 whorls, free; anthers versatile, bicelled, dehiscence longitudinal (Fig. 39.16D). Gynoecium syncarpous, 3- to 5-carpellary, ovary superior, 3- to 5-locular with axile placentation (Fig. 39.16E); ovules numerous in 2 or more rows on each placenta, styles 3–5, free and short, stigmas capitate. Fruit a septicidal capsule (Fig. 39.16F), seeds straight or curved, often rugose or pitted, with cylindrical or curved embryo and scanty or no endosperm.

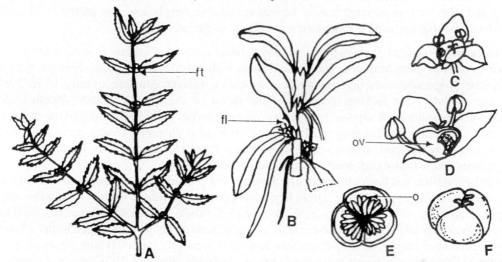

Fig. 39.16   Elatinaceae: **A** *Bergia ammanioides,* **B-F** *Elatine americana.* **A** Fruiting twig. **B** Flowering twig of *E. americana,* **C, D** Flower, vertical section (**D**). **E** Ovary, cross section. **F** Fruit. *fl* flower, *ft* fruit, *o* ovule, *ov* ovary. (After Lawrence 1951)

Anatomy.   Vessels mostly somewhat quadrangular and arranged in radial rows, perforations simple. Growth rings seen in *Bergia.* Stomata apparently Ranunculaceous type and on both surfaces.

Embryology.   Pollen grains triangular with three germ pores; 2-celled at shedding stage in *Bergia* and 3-celled in *Elatine.* They germinate in situ in the cleistogamous flowers of *E. triandra* (Johri et al. 1992). Ovules anatropous, bitegmic, crassinucellate. Polygonum type of embryo sac, 8-nucleate at maturity. Endosperm formation of the Nuclear type.

Chromosome Number.   Basic chromosome number is x = 6, 9.

Important Genera and Economic Importance.   The two genera of this family are *Bergia* and *Elatine*. No economic importance is known.

Taxonomic Considerations.   The taxonomic position of the Elatinaceae is uncertain. Niedenzu (1925b) and Lawrence (1951) included it in the Parietales, Melchior (1964) in the Violales, and Hutchinson (1973) in Caryophyllales. It has also been placed in the Theales by Cronquist (1968), Takhtajan (1980), and Thorne (1983). Takhtajan (1987), however, placed it in a separate order, Elatinales, in the Theanae.

The family Elatinaceae is characterised by the presence of stipular, opposite, decussate or whorled leaves, superior, 3- to 5-loculed ovary with axile placentation, septifragally dehiscent, capsular fruits and pitted or otherwise sculptured seeds.

According to Willis (1973), Elatinaceae is a peripheral Centrosperm group, but is possibly also connected to Hippuridaceae, Haloragidaceae and Lythraceae.

It is rather difficult to find any single alliance for this family, and it is probably better to follow Takhtajan (1987) and place it in a separate order of its own, i.e. Elatinales.

Caricaceae

A small family of 4 genera and 30–35 species, distributed in tropical and subtropical America and Africa.

Vegetative Features. Small, soft-wooded, dioecious or monoecious trees with milky sap, sparsely branching and with a crown of large leaves. Leaves simple, but deeply lobed or partite as in *Carica* or compound and 5-7-foliate; petiole elongated and hollow, estipulate.

Floral Features. Staminate flowers borne on long pendant racemes. Pistillate flowers usually axillary-solitary or in pairs or even in threes. Staminate flowers with calyx 5-lobed, gamosepalous, very small, valvate; petals 5, gamopetalous but free above, coriaceous, valvate. Stamens usually 10 in 2 whorls, sessile, inner whorls often lacking, arranged at the throat of the corolla; anthers bicelled, dehisce longitudinally. Gynoecium syncarpous, 5-carpellary, ovary superior, unilocular, with parietal placentation; placenta often intruding and forming a falsely 5- to 10-loculed chamber, ovules numerous. Styles 5, broadly cuneate to fan-shaped, stigma simple or fimbriate. Fruit a large, many-seeded berry; seeds with straight embryo and a fleshy endospem, rich in fatty food reserves.

Floral polymorphism has been observed. The flowers on a plant may be (a) *staminate* with 10 sessile stamens in 2 whorls and a rudimentary, rarely functional pistil, (b) *pistillate*, solitary or in a few-flowered corymbs in leaf axils, with a large globose ovary and 5 sessile, fimbriate stigma, (c) *long-fruited*, multiflowered corymb inflorescence, flowers with 10 sessile stamens at the base of petals and a functional pistil; fruits cucumber-shaped, or (d) *polygamous* bearing flowers of 2 types: (i) with 10 sessile stamens, all at the throat of the corolla, petals connate to form an elongated tube, and (ii) with only 5 stamens with long filaments and attached at the base of the corolla, near the ovary, corolla tube very short.

Anatomy. Vessels large, solitary or in radial multiples of up to 5 or more numbers. Perforations simple or horizontal. Articulated laticiferous canals present in all parts of the ground tissue. Leaves dorsiventral; stomata Ranunculaceous type, confined to the lower surface.

Embryology. Pollen grains 3-colporate with reticulate exine; 2-celled at the dispersal stage. Ovules anatropous, bitegmic, crassinucellate. Polygonum type of embryo sac, 8-nucleate at maturiy. Endosperm formation of the Nuclear type (Johri et al. 1992).

Chromosome Number. Basic chromosome number is x = 9.

Chemical Features. Benzyl isothiocyanate is the only acrid oil present in various members of Caricaceae—*Carica, Jarilla* and *Jacaratia* (Tang et al. 1972). Analysis of macerated seeds of *Carica* (6 spp.), *Jarilla* (1 sp.) and *Jacaratia* (3 spp.) indicate that the first two genera are relatively rich in benzyl isothiocyanate (Tang et al. 1972). Since *Carica* and *Jacaratia* are otherwise difficult genera to distinguish, this qualitative chemical character is useful for their identification (Harborne and Turner 1984). *Carica papaya* contains the cyclopentene ring containing cyanogenic glucoside, tetraphyllin B as well as the aromatic prunassin. This is the first report of cyanogenic glucoside from a species known to produce glucosinolates (Gershenzon and Mabry 1983). Seed fats of *C. papaya* are rich in palmitic, oleic and linoleic acid (Shorland 1963). The latex of Caricaceae members is rich in papain—a protein-breaking enzyme.

Important Genera and Economic Importance.   The genus *Cylicomorpha* of tropical Africa are trees with prickly erect stems, and *Jarilla* of American tropics and subtropics are herbs with smooth prostrate stems. In *Jarilla* and *Jacaratia* the leaves are digitately 5- to 12-foliolate, often covered with bloom below as in *Jacaratia* (Hutchinson 1969). *Carica papaya* is well-known for the large, edible fruits, consumed both unripe and ripe. The latex of unripe fruits is useful in making meat tender while cooking; it is also used medicinally against liver ailments. *C. candicans* of Peru, *C. chrysophila* of Colombia and *C. pentagyna* of Equador are also cultivated for their edible fruits and seeds (Dutta 1988). *C. cundinamarcensis* can be grown successfully at higher altitudes in the tropics where *C. papaya* cannot be raised. Fruits of *Jacaratia mexicana* and *Jarilla caudata* are eaten in the Andes.

Taxonomic Considerations.   The Caricaceae was not treated as a distinct family and was retained under Passifloraceae by Bentham and Hooker (1965a). Takhtajan (1966) also considered this family as belonging to the Passiflorales. The relationship of the two families (Caricaceae and Passifloraceae) can be traced to morphological, embryological and chemical features. Van Tieghem (1869) considered Caricaceae as related to the Cucurbitaceae. Hutchinson (1969, 1973) also supported this view, and placed this family in the Cucurbitales. Cronquist (1968, 1981), Melchior (1964), Dahlgren (1983a) and Takhtajan (1987) consider the Caricaceae to be a member of the Violales, whereas Thorne (1983) includes it in his Cistales.

The presence of chemicals like cyclopentoid cyanogenic glycosides, restricted only to the families Flacourtiaceae, Passifloraceae, Turneraceae, Caricaceae and Malesherbiaceae, supports their treatment in the same order, Violales (Saupe 1981, Gershenzon and Mabry 1983). Caricaceae members also have glucosinolates, and the occurrence of both the classes of compounds suggests that the family Caricaceae might be intermediate between the Capparales and the Violales (Spencer and Seigler 1984).

According to Baker (1976), the ancestor of this family bore bisexual flowers. The staminate and pistillate flowers of Caricaceae are derived by the abortion of gynoecium in the former and the replacement of stamens in the latter.

Loasaceae

A family of 16 genera and 300 species (Takhtajan 1987, 14 genera and 200 species according to Ernst and Thompson 1963), mostly distributed in western South America, and a monotypic genus *Kissenia* occurs from Arabia to southwest Africa.

Vegetative Features.   Annual or perennial herbs or shrubs, sometimes woody lianas, e.g. *Fuertesia,* and rarely trees, e.g. *Mentzelia arborescens.* Leaves opposite or alternate, simple or compound, pinnatisect or pinnatifid (Fig. 39.18A), estipulate, hairy, petiolate; tendrils absent.

Floral Features.   Inflorescence axillary racemes or spikes or capitate or subcapitate or solitary, axillary flowers. Flowers bisexual, perigynous, actinomorphic (Fig. 39.18A, C), bracteate, ebracteolate, tetramerous or 6- to 7-merous. Sepals 4 to 6 or 7 (*Loasa*), polysepalous, imbricate, persistent; accrescent and wing-like in the fruit of *Kissenia* and *Fuertesia.* Petals 4 to 6 or 7, polypetalous as in *Mentzelia* or connate at base as in *Eucnide* and *Sympetaleia;* 3-lobed and laciniate in *Fuertesia,* induplicate-valvate, often alternate with an inner series of petaloid staminodes or nectar-scales. Stamens numerous, free or in bundles, opposite the petals (Fig. 39.18B, C) or all filaments basally fused to form a low ring or short cylinder; anthers bicelled (1-celled in *Sympetaleia*), dehiscence longitudinal; only 2 fertile stamens in *Petalonyx crenatus,* outer 5 petaloid staminodes in *Mentzelia;* filaments very short and connective produced into an elongated appendage in *Cevallia.* Gynoecium syncarpous, 3- to 7-carpellary, ovary half-inferior to inferior, 1- to 3-locular, placentation parietal (Fig. 39.18B, D) or axile, ovules solitary to numerous on each placenta; style and stigma 1. Fruit a loculicidal capsule, the valves usually spirally twisted (Fig. 39.18E,F); seeds endospermous with straight embryo.

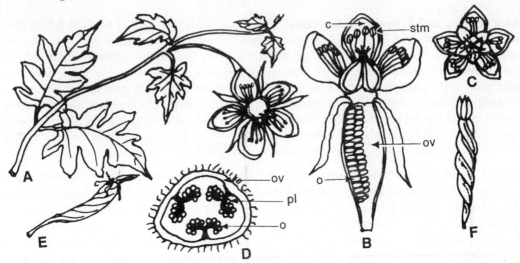

Fig. 39.18 Loasaceae : **A-E** *Caiophora lateritia*, **F** *Sclerothrix fasciculata*. **A** Flowering branch. **B** Flower, vertical section. **C** Face view. **D** Ovary, cross section. **E** Fruit. **F** Fruit of *S. fasciculata*. *c corolla*, *pl* placenta, *o* ovule, *ov* ovary, *stm* stamen. (**A-E** adapted from Lawrence 1951, **F** after Hutchinson 1969)

Anatomy. Stem of herbaceous species contains a ring of collateral vascular bundles and the woody species a cylinder of xylem as in *Mentzelia* and *Petalonyx*. Vessels various, with simple perforations. Leaves centric, hairs of various types; both uni- or multicellular, dorsiventral, e.g. *Caiophora* or isobilateral as in *Petalonyx;* stomata Ranunculaceous type.

Embryology. The development of anther and pollen has not been studied (Johri et al. 1992). Ovules anatropous, unitegmic, tenuinucellate. Polygonum type of embryo sac, 8-nucleate at maturity. Endosperm formation of the Cellular type; chalazal endosperm haustorium conspicuous (Garcia 1962).

Chromosome Number. Basic chromosome number is x = 7–15 (Takhtajan 1987).

Chemical Features. Iridoid glycosides present (Kooiman 1974, Dahlgren 1977b). Mentzeloside and decaloside have been isolated from *Mentzelia decapetala*. Soluble proteins are reported in the seeds of *Mentzelia* (Hill 1977).

Important Genera and Economic Importance. A predominantly American family with the only representative, *Kissenia*, in the Old World. The genus *Mentzelia* shows much phenotypic plasticity. The seed coat topography is characteristic for a species or species groups and therefore helpful in identification (Hill 1976, Glad 1976). Soluble seed protein analysis is helpful in deciding the evolutionary relationships amongst the species of this genus (Hill 1977). Economically, the family is not of much importance; a few species of *Blumenbachia, Caiophora, Eucnide, Loasa* and *Mentzelia* are grown as ornamentals.

Taxonomic Considerations. The family Loasaceae has controversial affinities. It has been included in the Parietales (Engler and Diels 1936, Lawrence 1951), in the Violales (Melchior 1964, Cronquist 1968, 1981) and in the Loasales (Dahlgren 1975a, 1983a, Cronquist 1983, Takhtajan 1980, 1987, Thorne 1983). Thorne (1983) places this new order in superorder Loasiflorae, just before Myrtiflorae, whereas Takhtajan (1987) places the Loasales in superorder Loasanae between the Gentiananae and Solananae.

Superficially, Loasaceae resembles the Cucurbitales in inferior or semi-inferior ovary with parietal placentation, but differs in embryological and chemical features, and in the absence of tendrils.

Kolbe and John (1979) reported that the Loasanae is serologically distinct from other members of the

Violales and shows affinities with certain sympetalous families like Polemoniaceae, Hydrophyllaceae, Campanulaceae, Ericaceae and Primulaceae.

Loasaceae has unitegmic, tenuinucellar ovules, with Cellular type of endosperm and extensive endosperm haustoria (Garcia 1962). These embryological features are typical for many Corniflorae and Lamiiflorae, and iridoid-containing Gentianiflorae. Iridoids are abundant in Loasaceae and include seco-iridoids, which are otherwise common in Gentianiflorae and Corniflorae (Cornales, Dipsacales).

Dahlgren (1980a, 1983a) considers the Loasaceae as the only member of the order Loasales in the superorder Loasanae. He proposes some connection with the Dipsacales, Cornales and woody Saxifrages. Alliance of Loasaceae with Cornaceae and Hydrangeaceae has been shown by Hufford (1992) on the basis of morphological and chemical data. Recent molecular evidence from restriction site comparisons of the chloroplast genome (Downie and Plamer 1992) suggest an alliance of this family within the Cornales s.1. Study on rbcL sequence data (Hempel et al. 1995) also support this view.

It is evident, therefore, that the Loasaceae are related to the Cornales s.1. of the Sympetalae.

Datiscaceae

A family of 3 genera—*Datisca , Tetrameles* and *Octomeles*—and 4 species. *D. glomerata* is a native of western Central America and *D. canabina* is Asiatic. The other 2 genera are monotypic. *Tetrameles* is a tall tree in India and Java, and *Octomeles* grows in the Philippines and other nearby countries.

Vegetative Features.   Perennial herbs, large shrubs or small trees, mostly dioecious. *Tetrameles* has characteristic large buttresses. Leaves alternate, simple or pinnately compound, estipulate, hairy, petiolate.

Floral Features.   Inflorescence raceme or spike or axillary fascicle of flowers. Flowers unisexual (Fig. 39.19A, C), actinomorphic, epigynous. Staminate flowers with 3- to 9-lobed calyx, polysepalous; 8-lobed corolla, polypetalous or petals absent. Stamens variable 4–25 (Fig. 39.19C), opposite the calyx lobes, free, filaments usually short, anthers bicelled, longitudinally dehiscent, sometimes with a rudimentary pistil. Pistillate flowers with a calyx tube adnate to the ovary, staminodes often present. Gynoecium syncarpous, tricarpellary, ovary inferior, unilocular, with parietal placentation and numerous ovules on

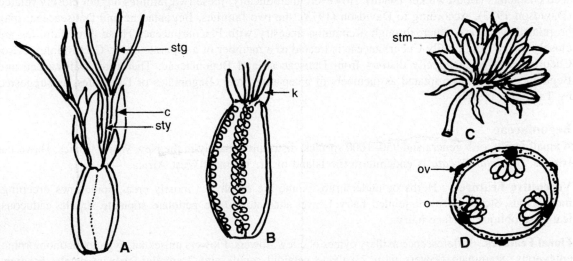

Fig. 39.19   Datiscaceae : **A-D** *Datisca glomerata.* **A** Pistillate flower. **B** Ovary, vertical section. **C** Staminate flower. **D** Ovary, cross section. *c* corolla, *k* calyx, *o* ovule, *ov* ovary, *stg* stigma, *stm* stamen, *sty* style. (Adapted from Lawrence 1951)

each placenta (Fig. 39.19B, D); styles 3, each bifid (Fig. 39.19A). Fruit a capsule, dehisces apically between the persistent styles; seeds small, probably dispersed by wind or water (Boesewinkel 1984b), with straight, oily embryo and scanty endosperm.

Anatomy.   Vessels large, perforations simple, intervascular pitting alternate. Parenchyma paratracheal, vasicentric; rays up to 4–7 cells wide, with a few uniseriates, heterogeneous. Leaves dorsiventral, hairs with multicellular stalk and glandular heads. Stomata Ranunculaceous type on both or only one surface.

Embryology.   The embryology has not been fully studied. Ovules anatropous, bitegmic, crassinucellate. Embryo sac of Allium type, 8-nucleate at maturity. Endosperm formation of the Nuclear type.

Chromosome Number.   Haploid chromosome number of *Datisca* is n = 11.

Important Genera and Economic Importance.   There are three genera in this family—*Datisca, Tetrameles* and *Octomeles*. *D. canabina* is a bushy herb distributed in the Western Himalayas from Kashmir to Nepal. The roots yield a dye which is used for colouring wool and cotton (Singh et al. 1983). *Tetrameles nudiflora* is a tree distributed in Sikkim Himalayas, Western Ghats and the Andamans. Decoction of the bark is used for rheumatism, dropsy and jaundice. *Datisca canabina* and *D. glomerata* are sometimes grown as ornamentals.

Taxonomic Considerations.   The family Datiscaceae has been placed in Parietales by Engler and Diels (1936) and Lawrence (1951), in Cucurbitales by Hutchinson (1969, 1973), in Violales by Melchior (1964), Cronquist (1968, 1981) and Thorne (1983), and in Begoniales by Takhtajan (1987). Dahlgren (1975a) provisionally included this family in the Violales, and indicated that its systematic position is uncertain. Dahlgren (1983a) included it in the order Cucurbitales along with Cucurbitaceae and Begoniaceae, based on the serological findings of Kolbe and John (1979).

The family shows some general affinities with the Cucurbitaceae: unisexual flowers, inferior ovary and presence of stipules. A close link with the Begoniaceae is suggested by embryological features and seed characters (Boesewinkel 1984b). However, anatomically, these two families are not closely related (Davidson 1973). According to Davidson (1973), the two families, Begoniaceae and Datiscaceae, may be related with each other, through a common ancestry with Flacourtiaceae. These three families are closely linked. The family Cucurbitaceae is treated as a member of a separate order, Cucurbitales. Also, Cucurbitaceae is chemically distinct from Datiscaceae and Begoniaceae. Therefore, Datiscaceae and Begoniaceae should be treated as members of a separate order, Begoniales or Datiscales, as suggested by Takhtajan (1987).

Begoniaceae

A small family of 5 genera and 950–1000 species, distributed mainly in the New World tropics, Hawaiian Islands. *Begonia baccata* is endemic in the Island of St. Thomas, West Africa.

Vegetative Features.   Herbs or undershrubs, some are epiphytic, mostly erect, sometimes creeping, monoecious. Stems succulent, jointed, hairy. Leaves alternate, simple, petiolate, stipulate, stipules caducous, leaf-base oblique, surface hairy.

Floral Features.   Inflorescence axillary cymes of a few flowers. Flowers unisexual, zygo- or actinomorphic, epigynous. Staminate flowers with 2 valvate petaloid sepals and 2 smaller, valvate petals. Stamens numerous in many whorls, free or basally connate, anthers basifixed, bicelled, dehiscence longitudinal, connective often exerted. Pistillate flowers with 2- to many petaloid tepals (calyx and corolla not distinct), free, imbricate. Gynoecium syncarpous, 3-(2- or 5)-carpellary, ovary inferior (half-inferior in *Hillebrandia*), trilocular with axile placentation or unilocular with 5 parietal placentae as in *Hillebrandia,* often 1–3-(or 6)-winged; ovules numerous. Styles 2–5, free or basally connate, stigmas papillose and often twisted. Fruit a loculicidal capsule or rarely a berry, e.g. *Begonia baccata;* seeds with straight, oily embryo and no endosperm.

Anatomy. Vascular bundles in one ring. Vessels in radial rows and progressively larger in diameter towards the exterior of the stem surrounded by parenchyma. Intervascular pitting scalariform, simple circular perforations or scalariform plates with many bars present. Leaves dorsiventral, with various types of hairs. Stomata solitary or in groups, always surrounded by 3–6 subsidiary cells.

Embryology. Pollen grains are 2-celled at the dispersal stage. Ovules anatropous, bitegmic, crassinucellate. Polygonum type of embryo sac, 8-nucleate at maturity. Endosperm formation of the Nuclear type (Johri et al. 1992).

Chromosome Number. Basic chromosome number is x = 10–21.

Important Genera and Economic Importance. Many species of *Begonia* are cultivated as ornamentals.

Taxonomic Considerations. Bentham and Hooker (1965a) placed Begoniaceae in the order Passiflorales, Engler and Diels (1936) and Lawrence (1951) in the Parietales, Melchior (1964) and Cronquist (1968,1981) in the Violales, Hutchinson (1969,1973) and Takhtajan (1987) in the Begoniales. There is a general agreement about its close relationship with the Datiscaceae. Both families have unisexual flowers, numerous stamens, inferior ovary with parietal placentation, oily embryo and scanty or no endosperm. The Begoniaceae are different from the Datiscaceae as they are: (a) monoecious, (b) have stipulate leaves and (c) winged or angled ovary. They also differ in anatomical features.

According to Davidson (1973), the two families may be related with each other through a common ancestry with Flacourtiaceae. As discussed earlier under Datiscaceae, these two families may be treated as the members of a separate order—Begoniales or Datiscales.

Taxonomic Considerations of the Order Violales

The Violales (as defined here) comprises 20 families. The designation Parietales, Bixales and Cistales have also been applied to this order, and the Cucurbitales, Datiscales, Passiflorales and Tamaricales have been used for certain families or groups of families.

The order Violales is a heterogeneous assemblage of a large number of families. Any attempt to try to make them more homogeneous will give rise to a large number of smaller orders, each comprising only one or two families.

The members of Violales have a basic floral structure of 5 + 5 perianth lobes, a variable number of stamens and three carpels, which is also common to the Theales, Euphorbiales and some Malvales. The stamens are rarely diplostemonous: *Casearia* of Flacourtiaceae, some Passifloraceae, and most Caricaceae are the only examples. They tend to be either numerous, as in most Flacourtiaceae and Begoniaceae, or haplostemonous. A dominant feature is the unilocular ovary with parietal placentation (in Begoniaceae, only in the genus *Hillebrandia*). Embryological features are the bitegmic, crassinucellate ovules, Nuclear types of endosperm, and endospermous seed.

Kolbe and John (1979) observed that the families with hypogynous flowers in this order, i.e. Flacourtiaceae, Violaceae, Passifloraceae and Turneraceae, are serologically separate from families with epigynous flowers such as Begoniaceae and Datiscaceae. Gynophores, androgynophores and corolline corona are particularly common in the former group of families.

Studies on pollen and floral morphology of Bixaceae, Cochlospermaceae and Flacourtiaceae by Keating (1972a,b, 1974) have, along with other evidences from stem anatomy (Keating 1969,1970) suggested that the families Bixaceae and Cochlospermaceae although closely related, are distinct. These two families—Bixaceae and Cochlospermaceae (along with Cistaceae)—show trends of specialisation quite different from the Flacourtiaceae. They are logically related with the Malvales (Dahlgren 1983b, Thorne 1983). Embryological studies (Dathan and Singh 1972) of *Bixa* and *Cochlospermum* as well as biochemical studies (Poppendieck 1980, Young 1981a) also draw the same conclusion. Dahlgren (1983b) also emphasises that the Bixaceae, Cistaceae and Cochlospermaceae should be retained in the Malvales. The flavonoid spectrum (Gornall et al. 1979, Young 1981a) and the exotegmic seed coat relate Cistaceae to Bixaceae, in addition to unilocular ovary and parietal placentation.

Scyphostegiaceae is yet another family with disputed position in this order. Though placed in various orders by different authors, on the basis of floral interpretation by Swamy (1953), and anatomical features, Scyphostegiaceae is closely allied to the Flacourtiaceae of the Violales.

Stachyuraceae and Elatinaceae are other disputed families. The characteristic anatomical features of Stachyuraceae are also met with in some members of both the families—Hamamelidaceae and Flacourtiaceae. Some taxonomists include this family in the Theales, from which it differs in fewer stamens, unbranched style, and parietal placentation. The position of Stachyuraceae is still uncertain. Elatinaceae is yet another family with possible alliance to too many families of different groups. It is best treated as a member of an order of its own, Elatinales.

The family Cucurbitaceae is treated as belonging to a separate order Cucurbitales, although some taxonomists include it in the Violales. Chemically it is distinct from the members of Violales.

The family Caricaceae, with glucosinolates as well as cyclopentenoid cyanogenic glycosides, may be intermediate between the Capparales and the Violales.

The family Loasaceae has unitegmic, tenuinucellar ovules with Cellular type of endosperm and extensive endosperm haustoria. These embryological features are typical of many sympetalous orders like Corniflorae and Lamiiflorae, and partly Gentianiflorae. Loasaceae members are rich in iridoids and seco-iridoids which are otherwise common in Gentianiflorae and some Corniflorae (Cornales, Dipsacales). These features, coupled with serological affinities with certain sympetalous families like Polemoniaceae, Hydrophyllaceae, Campanulaceae, Ericaceae and Primulaceae, make this family closely allied to Dipsacales, Cornales and related orders. Thorne (1981) and Cronquist (1981) do not accept this alliance but Thorne (1983) and Takhtajan (1987) recognise the monotypic superorders Loasiflorae as including Loasales and Loasaceae.

The more primitive members of the Violales are trees with alternate, stipulate leaves, bisexual, hypogynous, polypetalous flowers with numerous centrifugal stamens, a syncarpous ovary with free styles, parietal placentation, and seeds with copious endosperm. The family Flacourtiaceae, which is considered to be the most primitive family of the Violales, shows combination of these features in most of its members. However, the Flacourtiaceae is not directly ancestral to all the families of this order. Instead, there are groups of related taxa undergoing a series of parallel evolutionary changes.

40

Order Cucurbitales

The order Cucurbitales includes only one family—Cucurbitaceae. The plants are easily recognisable by their prostrate or climbing habit, often with the help of tendrils. Stem and branches herbaceous, covered with hairs; leaves simple, lobed or divided, hairy on both surfaces. Flowers unisexual, epigynous and ovary with parietal placentation. Many fruits are edible.

In transverse section of stem, the vascular bundles are bicollateral, and the pith is hollow. Seeds of many species are rich in amino acids, oils and storage proteins (Jeffrey 1980). Seed oils contain mostly linoleic, oleic, linolenic or conjugated polyethenoid acids as major components (Shorland 1963).

Cucurbitaceae

A large family with ca. 90 genera and 700 species distributed in the tropics and subtropics of both the Old and the New World.

Vegetative Features. Annual or sometimes perennial climbers or prostrate, monoecious or dioecious herbs, rarely trees, e.g. *Dendrosicyos,* a small tree on Socotra Islands. Stems cylindrical or more often pentangular, e.g. *Cucurbita, Luffa;* hairy, sometimes rooting at nodes, usually fistular. Leaves alternate, hairy, palmately 3- to 5-lobed, often with extrafloral nectaries embedded in leaf lamina, tendrils simple or multifid, usually in extra-axillary position, estipulate or stipulate.

Floral Features. Inflorescence axillary racemes or cymes of male flowers, female flowers solitary (Fig. 40.1A); unisexual (hermaphrodite in *Schizopepon*), actinomorphic, epigynous (Fig. 40.1B, C), incomplete, usually yellow or white. Calyx lobes 5, gamosepalous, outer surface hairy, imbricate, often with extrafloral nectaries on outer surface as in *Coccinia, Luffa* (Fig. 40.1A, G). Corolla of 5 petals, gamopetalous (Fig. 40.1B, C, F, H) (polypetalous in *Fevillea*), salver-form as in *Momordica* or campanulate as in *Cucurbita* (Fig. 40.1F, H), margin of upper parts of the petals often fringed as in *Trichosanthes;* valvate, imbricate or quincuncial. Staminate flowers show much variation in the form and structure of the 5 stamens, due to cohesion and fusion. Stamens 5, free in *Fevillea*; in *Thladiantha*, of the 5 stamens, 4 are in 2 pairs by the slight cohesion of the basal portion of the filaments, and the 5th is free. In *Bryonia* and *Momordica,* apparently there are 3 stamens—2 by the fusion of 2 each and the 3rd is the free 5th stamen. In *Cucurbita*, the situation is more complex; anthers of all the 5 stamens are spirally twisted into a central column (Fig.40.1F, G) and the filaments are connate except at the extreme base (Fig. 40.1F). In *Cyclanthera,* the stamens are monadelphous with the thecae of the anthers in 2 horizontal link-like rings around the edge of a peltate mass of connective and filament tissue, the thecal links dehisce (seemingly) transversely by a single suture. Rudimentary ovary may be present in a staminate flower.

In pistillate flowers staminodes often present, pistil single; 5-carpellary in *Fevillea,* mostly 3-carpellary, syncarpous, ovary unilocular with parietal placentation (Fig. 40.1D, E) (rarely trilocular with axile placentation), ovules usually numerous (unilocular with 1 large apically parietal ovule in *Sechium edule*); style 1, rarely 3 with as many branches as of stigma (Fig. 40.1D, H). Fruit a berry with soft or hard pericarp and numerous seeds (Fig 40.1J), also called a pepo. Seeds many, germinate within the fruit in *Sechium edule;* embryo straight with large cotyledons, endosperm absent.

Anatomy. Predominantly bicollateral vascular bundles separated from one another by broad strips of

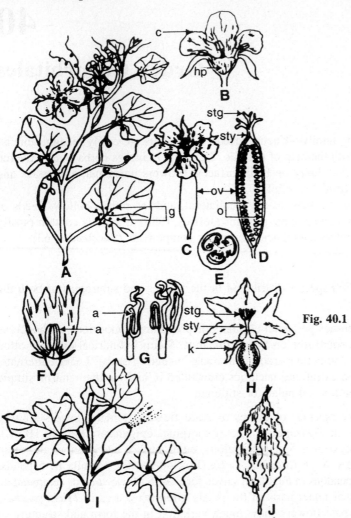

Fig. 40.1 Cucurbitaceae: **A–E** *Luffa cylindrica*, **F–H** *Cucurbita maxima*, **I** *Ecballium elaterium*, **J** *Momordica charantia*. **A** Plant with staminate inflorescence. **B** Staminate flower, vertical section. **C** Pistillate flower. **D** Carpel, vertical section. **E** Cross section. **F** Staminate flower of *C. maxima*, corolla cut open. **G** Anthers. **H** Pistillate flower, corolla cut open. **I** Plant of *E. elaterium* with squirting fruits. **J** Fruit of *Momordica charantia*. *a* anther, *c* corolla, *g* gland, *hp* hypanthium, *k* calyx, *o* ovule, *ov* ovary, *stg* stigma, *sty* style. (**I** adapted from Hutchinson 1969, **A–H**, **J** Original)

ground tissue and frequently arranged in two rings. Xylem vessels in older stems with wide lumina and simple perforations. Sieve tubes of the phloem are very large and conspicuous with transverse sieve plates. Leaves usually dorsiventral; various types of hairs recorded on leaf surfaces. Stomata on both surfaces of leaves or confined only to the lower surface; Ranunculaceous type.

Embryology. Pollen grains 2- or 3-celled at shedding stage, the family is eurypalynous, 3- to 10-colporate, 3- to 5-porate or 3-aperturate. Exine smooth or variously sculptured. Ovules anatropous, bitegmic, crassinucellate. Polygonum type of embryo sac, 8-nucleate at maturity. Seed coat is formed by the outer integument only. Endosperm formation of the Nuclear type; chalazal haustorium common. Endosperm haustoria of varied forms occur in a number of genera (Johri et al. 1992).

Chromosome Number. The basic chromosome numbers are x = 7–14.

Chemical Features. A group of bitter triterpenes, the cucurbitacins are almost characteristic components of the vegetative parts and fruits of Cucurbitaceae. *Lagenaria* fruits are the rich source of the enzyme catalase and the seeds of water melon (*Citrullus lanatus*) are rich in the enzyme urease. The seeds of Cucurbitaceous plants are rich in fatty acids of various kinds.

Important Genera and Economic Importance.   A large number of genera of this family are grown for their fruits that are consumed as vegetables, e.g. *Benincasa hispida* (white gourd), *Citrullus lanatus* (water melon), *Coccinia grandis* (scarlet gourd), *Cucumis melo* (melon), *C. sativus* (cucumber), *Cucurbita maxima* (pumpkin), *Lagenaria siceraria* (= *L. vulgaris*) (bottle gourd), *Luffa acutangula* (ribbed gourd), *L. cylindrica* (vegetable bath sponge), *Momordica charantia* (bitter gourd), *Sechium edule* (squash), *Trichosanthes anguina* (snake gourd) and *T. dioica* (parwal).

Some plants of this family are of medicinal value, e.g. bryony, used against common cold, is obtained from *Bryonia dioica*. A purgative called elaterium is produced from the fruits of *Ecballium elaterium* and a powerful laxative colocynth is obtained from dried fruit pulp of *Citrullus colocynthis*. The seed (testa) extract of *Momordica charantia* is usuful for diabetic patients. A few taxa are also grown as ornamentals, e.g. *Coccinia cordifilia, Ibervillea sonorae, Xerosicyos* sp., etc.

Interesting plants of this family are: *Acanthosicyos horridus,* an erect, shrubby and very spiny genus from Africa; *Dendrosicyos,* a tree genus from Socotra Islands; *Hodgsonia macrocarpa,* a gigantic climber attaining a length of ca. 25 m; *Ecballium elaterium,* the squirting cucumber (Fig. 40.1I) with explosively dehiscent fruits; *Alsomitra macrocarpa,* with winged seeds; and the Indomalayan *Hodgsonia* with ovules and seeds in pairs.

Taxonomic Considerations.   The systematic position of the Cucurbitaceae is controversial. Most phylogenists include this family amongst those with trilocular ovary and parietal placentation. The stamen and fruit morphology are unique to agree to this association. The family Cucurbitaceae is distinguished by the structure of its anthers; the anther cells in most cases are very long, winding up and down on the outer surface of the connective. The stamens may be monadelphous, or syngenesious and variously disposed upon the filaments, usually vertical but sometimes horizontal or inclined. The anther cells show various type of configurations: simple to curved, variously convoluted or with annular patterns. The structure of stamens, and the ovary, habit of the plants, and presence of tendrils together make this family unique amongst the dicot families.

On the basis of gamopetalous corolla, unusual stamen structure and inferior ovary, Wettstein (1935) and Eichler (1878; see Lawrence 1951) allied it to the Campanulales. Engler and Diels (1936) constructed a separate family, Cucurbitaceae, under Cucurbitales, and placed it next to the Campanulales.

According to Bentham and Hooker (1965a), the presence of unisexual flowers, inferior unilocular ovary with parietal placentation and stem tendrils, make it the nearest relative of the Passifloraceae. Hutchinson (1973) included this family in his Cucurbitales along with Begoniaceae, Datiscaceae and Caricaceae. Cronquist (1968, 1981) placed Cucurbitaceae in Violales along with all those families with unilocular ovary and parietal placentation. Mez (1926), by his serological studies, showed that the Cucurbitaceae are akin to the Campanulaceae and not to the Passifloraceae. Many embryological features also favour its inclusion in the Sympetalae but the constantly unisexual flowers, inferior ovary and presence of tendrils bring the family closer to the Passifloraceae (Dutta 1988).

Chakravarty (1966) observed that the Cucurbitaceae are highly evolved. There has been progression together with consequent metamorphosis of the structures associated with the reduction. amalgamation, connation, adnation and sterilization of the sex organs. It resembles the Passifloraceae in: placenta retained on the wall, single-chambered ovary, fleshy fruit, tendril structure and extrafloral nectaries. However, in spite of these resemblances, Cucurbitaceae deserves a separate rank by itself because of some of its typical features, especially of the male organs and the typical gourd-type fruit.

Takhtajan (1987) presumed its affinity with Begoniaceae and Datiscaceae, and therefore places the Cucurbitales near the Begoniales. Serologically, the Cucurbitaceae are related to Datiscaceae and Begoniaceae and do not show any affinity with the Loasaceae and Caricaceae of the Violales (Jeffrey 1980, Gershenzon and Mabry 1983).

The ovary structure, unisexual flowers, predominantly herbaceous nature and serological affinity indicate that the Cucurbitaceae is closely allied to some of the most advanced members of the Violales.

The family Cucurbitaceae is divided into 2 subfamilies and 9 tribes (Takhtajan 1987, Dutta 1988):

Subfamily 1. Cucurbitoideae—tendrils unbranched or branched (2–7) from lower part, spiralling above the point of branching. Pollen of various types; style 1, seeds not winged.

Tribe (i) Benincaseae—pollen sacs convolute and pollen grains with reticulate exine. Hypanthium in pistillate flowers short; ovules horizontal. Fruit usually smooth and indehiscent. e.g. *Benincasa, Citrullus* and others.

Tribe (ii) Cucurbiteae—pollen grains large, spiny, with many pores. Ovules erect or horizontal. Fruit fleshy, indehiscent and 1- to many-seeded, e.g. *Cucurbita.*

Tribe (iii) Cyclanthereae—stamen filaments united into a single column; pollen grains punctate. Ovules erect or ascending in a unilocular ovary. Fruit often spiny, usually dehiscent, often explosive, e.g. *Cyclanthera, Marah.*

Tribe (iv) Joliffieae—pollen grains reticulate. Hypanthium short. Petals fimbriate or with basal scales. Ovules usually horizontal, e.g. *Momordica, Telfairia.*

Tribe (v) Melothrieae—pollen-sacs straight or almost so and pollen grains reticulate. Hypanthium campanulate or cylindrical and alike in both sexes. Ovules horizontal, e.g. *Cucumis, Zehneria.*

Tribe (vi) Schizonpeponeae—stamens 3, free, pollen grains reticuloid. Ovules pendulous in a trilocular ovary. Fruit dehisce explosively into three valves, e.g. *Schizopepon.*

Tribe (vii) Sicyoeae—filaments united into a central column; pollen grains spiny. Ovules single, pendulous in a unilocular ovary. Fruit 1-seeded, indehiscent, usually hard or leathery or fleshy, e.g. *Sechium.*

Tribe (viii) Trichosanthieae—pollen grains striate, smooth or knobbled, never spiny. Hypanthium long and tubular in both sexes. Petals entire or with fimbriate margin. Ovules horizontal. Fruit fleshy or dry, dehiscing by three valves, e.g. *Trichosanthes.*

Subfamily II. Zanonioideae

Tribe (ix) Zanonieae—tendrils twice-branched from near the apex, spiralling above or below the point of branching. Pollen grains small, striate, uniform. Styles three. Seeds often winged, e.g. *Alsomitra, Zanonia.*

On the basis of vegetative and floral morphology, palynology and seed morphology, the erection of two subfamilies and nine tribes is fully justified.

41

Order Myrtiflorae

The order Myrtiflorae comprises 17 families in 3 suborders:

1. Suborder Myrtineae includes Lythraceae, unigeneric Trapaceae, Crypteroniaceae, Myrtaceae, unigeneric Dialypetalanthaceae, Sonneratiaceae, Punicaceae, Lecythidaceae, Melastomataceae, Rhizophoraceae, Combretaceae, Onagraceae, monotypic Oliniaceae and Haloragaceae.
2. Suborder Hippuridineae includes unigeneric Theligonaceae and Hippuridaceae.
3. Suborder Cynomoriineae comprises only Cynomoriaceae.

Mostly trees and shrubs, some herbs and a few aquatic herbs are included in this order. Members of Myrtales usually have opposite, simple, commonly entire and mostly estipulate (or with caducous stipules) leaves. Flowers cyclic, mostly tetramerous, usually with a hypanthium which may in some families be wholly adnate to the ovary; and shows a transition from perigyny to epigyny. Stamens often numerous and develop in a centripetal sequence. Gynoecium 2- to 5-carpellary, ovary with as many locules and axile or apical, rarely parietal placentation. Seeds with scanty or no endosperm.

Anatomically Myrtales is characterised by the presence of intraxylary phloem.

Lythraceae

A family of 28 genera and 600 species (Takhtajan 1987), abundant in the American tropics and in the tropics of the Old World, relatively few species in temperate regions.

Vegetative Features. Annual or perennial herbs (*Ammannia*), shrubs (*Lawsonia*) or trees (*Lagerstroemia, Lafoensia*), sometimes spinescent as in *Lawsonia*. Leaves opposite or whorled, simple, entire, estipulate or rarely with minute stipules.

Floral Features. Inflorescence mostly terminal racemes, spikes, or panicles, sometimes cymose. Flowers bisexual, actino- or zygomorphic, 4- or 6-merous, hypanthium present, hypogynous. Calyx a hollow tube, sepals 4 or 6, appear as lobes of the hypanthium. Corolla with an equal number of petals as sepals, polypetalous or absent as in *Peplis* and *Rotala,* apparently arising from the rim or upper inner surface of the hypanthium, crumpled in bud. Stamens usually twice as many as sepals or petals or more, inserted at different levels on the calyx tube. Anthers bicelled, introrse, dorsifixed, dehisce longitudinally. Gynoecium syncarpous, 2- to 6-carpellary, ovary superior, sessile or shortly stipitate, 2- to 6-locular, placentation axile, ovules numerous, ascending; style simple, heterostyly occasional, stigma usually capitate. Fruit a dry capsule, seeds non-endospermous and with a straight embryo.

Anatomy. Vessels small to medium-sized, solitary or in radial multiples of 2 to 8. Wood diffuse or semi-ring-porous to ring-porous (Baas and Zweypfenning 1979); perforations simple, intervascular pitting alternate. Parenchyma predominantly paratracheal, scanty or vasicentric to aliform and confluent, sometimes with numerous crystalliferous cells. Rays uniseriate or up to 2–3 cells wide, hetero- or homogeneous. Leaves usually dorsiventral, isobilateral in *Pemphis acidula*. Hairs may be simple, unicellular, bicellular, 2-armed, tufted or glandular; spherical glandular hairs with very short stalks, situated in pits and appearing as black dots on the lower surface of leaves in species of *Adenaria, Grislea, Lagerstroemia* and *Woodfordia*. Stomata Ranunculaceous type, usually confined to lower surface. Intraxylary phloem is a characterstic feature.

Embryology.   Pollen grains 3-colporate, shed at 2-celled stage. Ovules anatropous, bitegmic, crassinucellate; hypostase present. Polygonum type of embryo sac, 8-nucleate at maturity. Endosperm formation of the Nuclear type. Genus *Cuphea* is eurypalynous; pollen grains mostly tricolpate, oblate, triangular to oval-triangular in polar view (S.A. Graham and A. Graham 1971).

Chromosome Number.   Basic chromosome number is x = 5 to 11.

Chemical Features.   Capric acid is the main component of seed fat in *Cuphea llavea* (Shorland 1963). Anthocyanins of 3, 5-diglucoside type and pelargonidin and malvidin occur in *Lythrum* and *Cuphea* (Harborne 1963).

Important Genera and Economic Importance.   Some of the larger genera include *Cuphea* with 200 species, *Lagerstroemia*, commonly called crepe myrtle, and *Lythrum* with 30 species each, and *Rotala* with 50 species. *Ammannia baccifera* is a marshy herb. *Woodfordia fruticosa* is an ornamental shrub with racemes of red flowers. The leaves of *Lawsonia inermis* are the source of the dye henna, used largely as a hair dye and also for dyeing silk, wool and leather. *Cuphea*, *Lagerstroemia* and *Lythrum* are cultivated for their ornamental value. *Lawsonia inermis* is cultivated as a hedge plant; its flowers on distillation yield an aromatic oil useful in perfumery and embalming. Timber is obtained from *Lagerstroemia lanceolata*, *L. parviflora*, *L. speciosa*, *Physocalymma scaberrimum* and *Woodfordia fruticosa*.

Taxonomic Considerations. The family Lythraceae has been included in the Myrtales/Myrtiflorae by most phylogenists. Hutchinson (1969, 1973) included it in the Lythrales. However, its position in the Myrtales is justified on the basis of all the features studied.

Trapaceae

A unigeneric family of the genus *Trapa* with 3 species distributed mainly in the Old World tropics. One species, *T. natans* var. *bispinosa*, is naturalized in other tropical regions also.

Vegetative Features.   Floating annual herbs with reduced stem; leaves dimorphic: the floating ones in rosettes, stipulate, stipules finely dissected and caducous, with serrate margin and spongy, inflated petioles; the submerged leaves opposite, finely pinnatisect, resembling roots (Fig. 41.2A).

Floral Features.   Flowers solitary, axillary, bisexual, actinomorphic, perigynous, tetramerous; hypanthium present. Calyx of 4 sepals, valvate, persistent in fruit. Corolla of 4 petals, polypetalous, valvate. Stamens 4, in one whorl; anthers bicelled, dorsifixed, introrse (Fig. 41.2B, C). Gynoecium syncarpous, bicarpellary; ovary bilocular, placentation axile, with one pendulous ovule in each locule (Fig. 41.2C), one ovule usually aborts; style 1, thick, stigma capitate (Fig. 41.2C, D). Fruit is a 1-seeded, top-shaped drupe, with 2 to 4 spines representing persistent calyx (Fig. 41.2E). Seeds without endosperm and a large embryo with unequal cotyledons, one large and the other minute and scale-like. On germination, the radicle perforates the apex of the fruit (Fig. 41.2F).

Anatomy.   Anatomically, this family is quite similar to the family Onagraceae, where it was included earlier. The difference lies in the absence of crystal raphides in Trapaceae.

Embryology.   Pollen grains tricolpate, shed at 2-celled stage. Ovules anatropous, bitegmic, crassinucellate; differ from Onagraceae (4-nucleate embryo sac) in having 8-nucleate Polygonum type of embryo sac and in the absence of endosperm, presence of well-developed embryonal haustorial suspensor and extremely unequal cotyledons (Ram 1956).

Chromosome Number.   Diploid chromosome number is 2n = 48, 56 (Kumar and Subramaniam 1987).

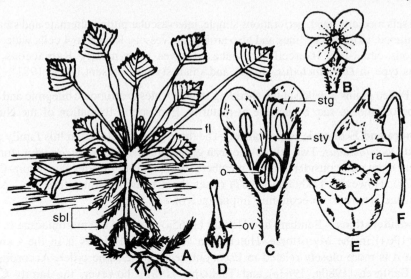

Fig. 41.2 Trapaceae: **A–F** *Trapa bispinosa*. **A** Flowering plant. **B** Flower. **C** Vertical section. **D** Gynoecium. **E** Fruit. **F** Germination. *fl* flower, *o* ovule, *ov* ovary, *ra* radicle, *sbl* submerged leaves, *stg* stigma, *sty* style. (Adapted from Dutta 1988)

Chemical Features.   Cotyledons rich in starch.

Important Genera and Economic Importance.   The only genus *Trapa,* an aquatic herb with rosettes of floating leaves, has about 3 species widespread in the tropical regions. The fruits are eaten raw or cooked, and are also a source of starch.

Taxonomic Considerations.   The family Trapaceae was included in the Onagraceae (Bentham and Hooker 1965a, Rendle 1925, Hutchinson 1959). It was placed in Hydrocaryaceae, but later this family name was discarded because: (a) there is no genus named *Hydrocarya,* and (b) it was a superfluous name. Most phylogenists, however, treat it as a family distinct from the Onagraceae (Hutchinson 1969, Cronquist 1968, Dahlgren 1980a, 1983a, Takhtajan 1980, 1987). Dahlgren (1975a, b) placed it in the order Trapales.

   It is now correctly concluded that Trapaceae is a distinct family derived from the Onagraceae. It is different from members of Onagraceae in anatomical, embryological and floral anatomical features, and it is justified to treat Trapaceae as a distinct family.

Crypteroniaceae

A small family of 3 genera and 10 species distributed in tropical Asia, Assam in India, southeastern Asia and Malaysia.

Vegetative Features.   Trees or shrubs; leaves opposite, entire, simple, coriaceous, estipulate.

Floral Features.   Inflorescence a true panicle of small subsessile flowers. Flowers bisexual or polygamodioecious, actinomorphic, hypo- or epigynous. Calyx of 4 or 5 sepals, gamosepalous, valvate. Corolla of 4 or 5 petals, polypetalous, imbricate, or absent. Stamens 10 or 4 to 5, alternisepalous, filaments short or elongated, anthers small, sometimes with a thickened connective and thecae diverged downwards; disc absent. Gynoecium 2- to 5-carpellary, syncarpous; ovary 2- to 5-locular, locules sometimes incomplete, ovules 2 on axile or basal placentation; style 1 and one capitate stigma. Fruit a loculicidal, 2- to 5-valved capsule, valves usually remain connected by style. Seeds many, sometimes winged, endospermous.

Anatomy. Vessels medium-sized, perforations simple, intervascular pitting alternate and small. Parenchyma apotracheal, scattered, in uniseriate lines and also around the vessels. Rays 2 to 4 cells wide, heterogeneous, with conspicuous canal-like intercellular spaces. Leaves with mostly Rubiaceous type stomata; Ranunculaceous type in *Dactylocladus*; druses and small styloids present (Baas 1981).

Embryology. Pollen grains 2-celled at the shedding stage. Ovules anatropous, bitegmic and crassinucellate. Polygonum type of embryo sac, 8-nucleate at maturity. Endosperm formation of the Nuclear type.

Important Genera and Economic Importance. The three genera included in this family are: *Axinandra, Crypteronia* and *Dactylocladus*. Two other American genera *Alzatea* and *Rhynchocalyx* (formerly included in the Lythraceae), are also considered to be members of this family by Van Beusekom-Osinga and Van Beusekom (1975). However, Takhtajan (1987) raises them to two distinct families—Alzateaceae and Rhynchocalycaceae. There is no economic importance of any taxon.

Taxonomic Considerations. Bentham and Hooker (1965a) included Crypteroniaceae in the Lythraceae and Melchior (1964) in the Myrtiflorae. Hutchinson (1969, 1973) places it in the Cunoniales on the assumption that it is more closely related to Escalloniaceae of the same order. According to Cronquist (1968, 1981), Dahlgren (1980a, 1983a) and Takhtajan (1987), however, the family Crypteroniaceae belongs to the Myrtiflorae.

The assignment of the genus *Rhynchocalyx* is rather controversial. It has been related to the family Crypteroniaceae which also includes the genera *Crypteronia, Dactylocladus, Axinandra* and *Alzatea*. Pollen morphology differs in all five members. *Rhynchocalyx* and *Alzatea* differ from the other three members of Crypteroniaceae in wood anatomy and share characteristics with Melastomataceae and Lythraceae (Van Vilet 1975) because of nodal anatomy. Although Willis (1973) includes it in Lythraceae, *Rhynchocalyx* is distinct in having petals and stamens on the hypanthium and sclereids in leaf petioles. Embryologically, it is closer to the Myrtales because of crassinucellate ovule, 2-layered inner integument, Nuclear type of endosperm and exalbuminous mature seed (Johri et al. 1992).

Alzatea is yet another controversial genus. This taxon has all the important features of Myrtales and in addition it shows bisporic Allium type of embryo sac. Hence, it is an exceptional member of the Myrtales (Johri et al. 1992).

Genus *Axinandra*, with its one species in Sri Lanka, and the other three in West Malaysia, also has a controversial position. It has been placed in the Lythraceae (Bentham and Hooker 1965a), Melastomataceae (Baillon 1876, Van Bakhuizen 1943) and Memecylaceae (Willis 1973). Meijer (1972) regarded *Axinandra* as a very old genus that stood near the roots of the families Lythraceae, Myrtaceae, Melastomataceae and Rhizophoraceae, as the genus shares one or the other significant characteristic with some members of each of these families.

Van Beusekom-Osinga and Van Beusekom (1975) and Van Beusekom-Osinga (1977) place *Axinandra* in Crypteroniaceae s.l. along with *Crypteronia, Dactylocladus, Rhynchocalyx* and *Alzatea*. Pollen morphology does not support any relationship amongst these genera (Mueller 1975). Wood anatomical studies (Van Vilet 1975) suggest that Crypteroniaceae sensu Van Beusekom-Osinga and Van Beusekom (1975) and Van Beusekom-Osinga (1977) is a heterogeneous family. A study of the vegetative anatomy (Van Vilet and Baas 1975) indicates that *Axinandra* and *Crypteronia* are similar and should be included in the same family.

Cronquist (1981) includes only *Crypteronia* in his Crypteroniaceae. According to Dahlgren and Thorne (1984), and Johnson and Briggs (1985), this family comprises the three Asiatic genera—*Crypteronia, Dactylocladus* and *Axinandra*—which form a monophyletic group. Van Vilet et al. (1981) regard the Crypteroniaceae (including the same three Asiatic genera) as a subfamily of the Melastomataceae.

Embryologically, although *Axinandra* is more or less similar to the other members of the Myrtales, it differs from them in: (a) wall formation in the zygote occurs when several thousand free endosperm nuclei have been formed, (b) an integumentary tapetum is present. *Axinandra* resembles Melastomataceae because of ephemeral (non-fibrous) endothecium but differs in having 2-nucleate tapetal cells (Tobe and Raven 1983a, b). The genus *Axinandra* has been considered by Dahlgren and Thorne (1984) to be closely related to the Melastomataceae. On the basis of available embryological data, Johri et al. (1992) also support this view.

It is apparent, therefore, that the position of *Axinandra* is still uncertain. Studies on chromosome numbers and chemical features may be helpful in assigning the correct position to this genus.

Myrtaceae

A large family of 145 genera and 3650–4000 species distributed in tropical and subtropical regions of the world. The two centres of distribution are Australia and America.

Vegetative Features.   Shrubs or trees, stems when sufficiently young have wings. Leaves usually opposite (Fig. 41.4A), whorled in *Callistemon lanceolatus* (Fig. 41.4E), simple, entire, coriaceous, gland-dotted, with intra-marginal venation.

Floral Features.   Inflorescence cymose (*Eucalyptus*), racemose (*Callistemon*) or paniculate (*Syzygium aromaticum*), rarely solitary axillary flowers as in *Psidium guajava* (Fig. 41.4A). Flowers bisexual, actinomorphic, mostly epigynous (Fig. 41.4B). Calyx usually 4 or 5, mostly inconspicuous or thrown off as a calyptra as in *Eucalyptus* (Fig. 41.4F), free or basally connate. Petals 4 or 5, free, but connate forming an operculum, as in *Eucalyptus,* imbricate. Stamens numerous, bent inwards in bud (Fig. 41.4F), free or polyadelphous; anthers dorsifixed or versatile (Fig. 41.4C), rarely basifixed as in *Calothamnus,* bicelled, introrse, dehiscing longitudinally or apically by 2 pores as in *Actinodium, Darwinia, Homoranthus* and many others; the connective often conspicuous and gland-tipped. Gynoecium syncarpous, 2- to 5-carpellary; ovary inferior or half-inferior, uni-, bi-, tri-, or tetralocular with parietal (when unilocular) or axile placentation (Fig.41.4B, D); ovules 2 to many on each placenta, obliquely pendulous; style and

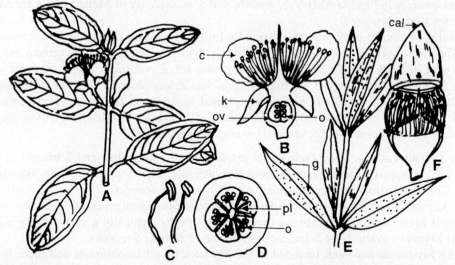

Fig. 41.4   Myrtaceae: **A–D** *Psidium guajava,* **E** *Callistemon lanceolatus,* **F** *Eucalyptus citriodora.* **A** Flowering twig. **B** Flower. **C** Stamens. **D** Ovary, cross section. **E** Twig of *Callistemon lanceolatus,* leaves gland-dotted. **F** Flower of *Eucalyptus citriodora. c* corolla, *cal* calyptra, *g* glands, *k* calyx, *o* ovule, *ov* ovary, *pl* placenta. (Original)

stigma 1. Fruit a berry (*Psidium*), or loculicidal capsule, rarely drupaceous or nutlike. Seeds usually a few, embryo variously shaped, endosperm scanty or absent.

Anatomy. Vessels small, numerous, solitary and without any definite arrangement in most genera (exception *Eucalyptus*), perforations simple, intervascular pitting alternate. Parenchyma diffuse or in uniseriate bands in the wood with solitary vessels, predominantly paratracheal in wood with numerous multiple and some intermediate forms with both diffuse and paratracheal types. Rays exclusively uniseriate or up to 2 to 6 cells wide. Intraxylary phloem is common to all the members; intercellular canals and cavities containing oil are also present. Leaves isobilateral or centric as in *Callistemon linearis*; hairs mostly unicellular, stomata usually Ranunculaceous type and occur on both surfaces of vertically placed leaves.

Embryology. Pollen grains shed at 2-celled stage. Ovules ana- to campylotropous, bitegmic (unitegmic in *Eugenia, Syzygium*), crassinucellate; micropyle formed by both integuments in *Wehlia*. Polygonum type of embryo sac, 8-nucleate at maturity. Endosperm formation of the Nuclear type.

Chromosome Number. Basic chromosome numbers are x = 6–11.

Chemical Features. Presence of ellagitannins is characteristic. Anthocyanins of 3-glucoside type and cyanidin and delphinidin are present in *Eucalyptus* and malvidin and delphinidin in *Metrosideros*. The glycoside prunasin occurs in *E. corynocalyx*; D-quercitol in fruits of *Syzygium cumini* and L-quercitol in *Eucalyptus populnea*. Triterpenoid saponins are also present in some Myrtaceae members.

Important Genera and Economic Importance. One important genus is *Eucalyptus*, some species are tall handsome trees bearing white or red flowers. *E. regnans* from Australia is considered as one of the tallest angiosperms, second only to redwoods (gymnosperm) of California (Nayar 1984f). Another interesting genus is *Callistemon*—commonly called bottle-brush tree because of the resemblance of its inflorescence, and it is cultivated on a large scale as an avenue tree. *Acmena acuminatissima* is a tree from the Andamans. *Decaspermum fruticosum* is a slender tree of Khasi Hills in Assam. *Psidium guajava* or guava, *Syzygium cumini*, black plum or Indian blackberry, *S. jambos,* and *S. malaccense* or Malaya apple are all cultivated as well-known fruit trees.

Economically, Myrtaceae is of great importance for fruits, oils, spices, timbers and ornamentals. Various species of *Eucalyptus* yield valuable timber, e.g. *E. diversifolia, E. leucoxylon, E. marginata* and *E. robusta*. The leaves of *E. globulus* on distillation yield eucalyptus oil, *E. maculata* and *E. rostrata* yield citron gum and red gum, respectively, and are used in medicine. *E. occidentalis* supplies mallet bark, rich in tannin. *Leptospermum laevigatum* is extensively planted in Australia for reclamation of moving sand. Cajeput oil is obtained from *Melaleuca leucadendron* of Malacca. Allspice and cloves are the spices derived from *Pimenta dioica* and *Syzygium aromaticum*, respectively.

Taxonomic Considerations. The Myrtaceae is divided into 2 subfamilies and 3 tribes:
  Subfamily I Leptospermoideae: leaves opposite or alternate; fruit a capsule or nut-like, dry.
  Tribe (i) Chamaelancieae: ovary unilocular, e.g. *Calytrix, Verticordia*.
  Tribe (ii) Leptospermae: ovary multilocular, e.g. *Callistemon, Eucalyptus, Melaleuca*.
  Subfamily II Myrtoideae: leaves always opposite, fruit fleshy, typically a berry, rarely a drupe.
  Tribe (iii) Myrteae: ovary 2- or 5-locular, e.g. *Myrtus, Psidium, Syzygium*.
  The family Myrtaceae has been included in the Myrtales by all taxonomists and there is no dispute about its position.

Dialypetalanthaceae

A unigeneric family of the genus *Dialypetalanthus* with only one species distributed in tropical South America.

Vegetative Features. Trees with white tomentose young branches. Leaves opposite, entire, simple, with large persistent, intrapetiolar stipules, members of opposite pairs connate, obscurely gland-dotted.

Floral Feature. Inflorescence, a terminal, many-flowered, bracteate thyrse. Flowers bisexual, conspicuous, scented, actinomorphic, epigynous. Calyx of 4 sepals, polysepalous, imbricate. Corolla of 4 petals, biseriate, polypetalous, white. Stamens 16 to 25, biseriate, free form corolla, equal, 2 linear thecae on flattened connective, introrse, dehiscence poricidal; disc annular, fimbriate. Gynoecium syncarpous, bicarpellary; ovary inferior, bilocular with axile placentation, ovules numerous; style simple, elongated with a shortly bifid stigma. Fruit a bilocular, loculicidal, many-seeded capsule, exerted from calyx at apex. Seeds numerous, fusiform, endospermous.

Anatomy. Anatomically, Dialypetalanthaceae resembles members of Rubiaceae, order Gentianales. Internal phloem is absent.

Embryology. Pollen grains are similar in structure to those of the Rhizophoraceae. Ovules anatropous, unitegmic, tenuinucellar. Embryology has not been fully investigated (Johri et al. 1992).

Chemical Features. Common flavones present.

Important Genera and Economic Importance. The only genus *Dialypetalanthus* is a tree, and has no economic importance.

Taxonomic Considerations. In spite of having some Myrtalean affinities, the family Dialypetalanthaceae has been retained in the Gentianales along with Rubiaceae by most phylogenists. Cronquist (1968) did include it in the Myrtales but, in 1981, he placed it back in the Gentianales. Dahlgren (1980a) included it in the Cornales. Dahlgren (1983a) and Dahlgren and Thorne (1984) are of the opinion that: "*Dialypetalanthus* is probably an aberrant early off-shoot of the Rubiaceae or a relict family, closely related to the Rubiaceae". Even Takhtajan (1987) includes this family in the Gentianales, which it resembles in: (a) interpetiolar stipules, (b) advanced wood anatomy, (c) absence of internal phloem, (d) inferior ovary, and (e) axile placentation. However, as stated by Cronquist (1968), the absence of internal phloem is not so important, as a few other Myrtalean genera also lack it. Even glandular-punctate leaves, polypetalous corolla and numerous stamens do not suggest any similarity with the Rubiaceae.

From the above discussion it is clear that further detailed study of this genus is essential to assign it to the correct position.

Sonneretiaceae

A small family of only one genus, *Sonneratia*, and 5 species distributed in coastal regions of tropical East Africa and Asia to northern Australia, New Caledonia and West Pacific Islands, New Hebrides and the Solomon Islands.

Vegetative Features. Trees or shrubs; leaves simple, entire, opposite, coriaceous, estipulate. In *Sonneratia* the aerial roots spring vertically out of the mud, arising as lateral negatively geotropic branches on the normal roots.

Floral Features. Inflorescence terminal cymes or corymbs of 1 to 3 flowers. Flowers bisexual, actinomorphic, hypogynous. Calyx of 4 to 8 sepals, gamosepalous, valvate, acute, coriaceous, often coloured inside. Corolla of 4 to 8 petals, polypetalous, sometimes broad and wrinkled, often absent. Stamens 12 to numerous, filaments filiform, inflexed in bud, anthers reniform, medifixed. Gynoecium syncarpous, 4- to 15-carpellary, seated on a broad base, ovary superior, multilocular with numerous ovules on axile placentae; long, robust, simple style and capitate stigma. Fruit a berry; seeds without appendages.

Anatomy.   Vessels medium-sized, solitary as well as in numerous multiples of 2 or 3 cells, perforation plates simple; intervascular pitting alternate. Parenchyma absent, rays exclusively uniseriate, rarely biseriate, homogeneous. Leaves isobilateral in *Sonneretia apetala*. Stomata deeply sunken and present on both surfaces in *S. acida;* Rubiaceous type and equally distributed on both surfaces in *S. apetala*. Intraxylary phloem present.

Embryology.   Pollen grains 3-porate, filled with starch grains in *S. apetala;* shed at 2-celled stage. Ovules anatropous, bitegmic, crassinucellate. Polygonum type of embryo sac. As the antipodals degenerate early and polar nuclei fuse before fertilisation, the mature embryo sac in 4-nucleate. Endosperm formation of the Nuclear type.

Chromosome Number.   The haploid chromosome number is n = 9.

Chemical Features.   Fatty acids, sterols and hydrocarbons are reported in the fresh leaves of *Sonneratia* (Hogg and Gillan 1984).

Important Genera and Economic Importance.   *Sonneratia* is the only genus, distributed mostly in the coastal regions of the tropics. The timber is sometimes used.

Taxonomic Considerations.   The genus *Sonneratia* was earlier included in the family Lythraceae along with another genus *Duabanga*. Dahlgren and Thorne (1984) included these two genera in the Lythraceae as two subfamilies : Sonneratioideae and Duabangoideae. Both genera differ from each other as well as from other members of Lythraceae, in floral morphology and wood anatomy. Therefore, three different families are now recognised : Sonneratiaceae, Duabangaceae and Lythraceae (Takhtajan 1987).

Punicaceae

A unigeneric family of the genus *Punica* with only 2 species, *P. granatum* and *P. protopunica,* a native of subtropical Asia but widely naturalised.

Vegetative Features.   Shrubs or small trees, with spiny branches. Leaves mostly opposite or fasciculate, simple, estipulate, glossy when young (Fig. 41.7A).

Floral Features.   Inflorescence terminal and cymose, or terminal or axillary solitary flowers (Fig. 41.7A). Flowers bisexual, actinomorphic, epigynous. Calyx of 5 to 8 sepals, valvate and fleshy; petals 5 to 7,

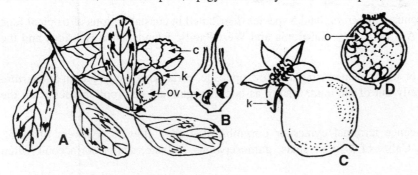

Fig. 41.7 Punicaceae: **A–D** *Punica granatum.* **A** Flowering twig. **B** Carpel, vertical section. **C** Fruit. **D** Vertical section. *c* corolla, *k* calyx, *o* ovule, *ov* ovary. (**A, C, D** Original, **B** after Lawrence 1951)

imbricate, emerging from the edge of the hypanthium, crumpled in bud. Stamens numerous, emerging in many whorls from within the upper half or more of the hypanthium; filaments free and nearly equal, anthers bicelled, dorsifixed, dehiscence longitudinal. Gynoecium syncarpous, 8- to 12-carpellary; ovary inferior, 8- to 12-locular, in early ontogeny of ovary the locules are in 2 concentric whorls, placentation axile to start with but as development progresses, the outer series of 5 to 9 locules become superposed over the originally inner series of usually 3 locules and the placentation of the upper series is transposed to a parallel position; ovules numerous, style and stigma 1 (Fig. 41.7B, D). Fruit a berry with persistent calyx lobes (Fig. 41.7C). Seeds with a fleshy testa that forms the pulp within the fruit, non-endospermous and with a straight embryo (Lawrence 1951).

Anatomy.   Vessels very small to moderately small, mostly solitary or in multiples of 2 or 3 cells, perforations simple; intervascular pitting alternate. Parenchyma absent; rays uniseriate, homogeneous. Leaves dorsiventral, stomata confined to the lower surface, mostly Ranunculaceous type. Intraxylary phloem present.

Embryology.   Pollen grains have not been studied. Ovules anatropous, bitegmic, crassinucellate. Polygonum type of embryo sac, 8-nucleate at maturity. Endosperm formation of the Nuclear type.

Chromosome Number.   Basic chromosome number is x = 8.

Chemical Features.   Rich in tannins. Anthocyanin of 3, 5—diglucoside type, pelargonidin and delphinidin are reported. Seed fats with linoleic, oleic, linolenic or conjugated polythenoid acid in *P. granatum* (Shorland 1963).

Important Genera and Economic Importance.   Only genus *Punica* is cultivated for its edible fruits and ornamental flowers.

Taxonomic Considerations.   A member of the Myrtales, it was earlier treated as a member of the Lythraceae. They differ in wood anatomy and in basic chromosome number. There is no dispute in accepting it as a distinct family.

Lecythidaceae

A family of 15 genera and 325 species endemic to tropical America.

Vegetative Features.   Trees; leaves in bunches at the end of twigs, alternate, simple, entire (Fig. 41.8A) or dentate, estipulate, usually dense at the end of branches.

Floral Features.   Inflorescence racemose or flowers solitary. Flowers very large, actino- or zygomorphic, peri- or epigynous, bisexual, always with complete fusion of receptacle and ovary; usually one intrastaminal disc and one under corolla and stamens. Calyx of usually 4 to 6 sepals, polysepalous, valvate, persistent in fruit. Corolla of 4 to 6 petals, rarely gamopetalous, imbricate. Stamens numerous in several whorls, free or basally connate; anthers usually versatile, bent inwards in bud. Sometimes stamens are of unusual appearance, because of one-sided development of the union and abortion of some anthers (Fig. 41.8A); numerous (up to 1200), rarely less (up to 10) in certain Napoleonaeoideae. Anthers longitudinally dehiscent, rarely by apical pore. Intrastaminal nectarine disc usually present. Gynoecium syncarpous, 2- to 6- or more-carpellary; ovary inferior, multilocular, 1 to numerous ovules in each locule on axile placentation (Fig. 41.8B); style simple or 3- or 4-lobed, bulbous or lobate stigma. Fruit a berry (Fig. 41.8C) or woody capsule, indehiscent or operculate; seeds non-endospermous.

Anatomy.   Vessels of various size—very small in *Grias,* very large in *Bertholletia, Couroupita* and others— but mostly medium-sized; sometimes with pronounced radial multiples or irregular clusters. Perforations

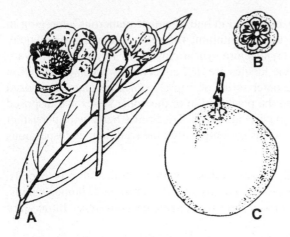

Fig. 41.8 Lecythidaceae: **A–C** *Couroupita guianensis*. **A** Leaf and inflorescence. **B** Ovary, cross section. **C** Fruit. (Adapted from Hutchinson 1969)

simple, intervascular pitting alternate. Parenchyma typically in apotracheal bands, predominantly aliform to confluent in a few genera, uniseriate and broken to wide and continuous in some others. Rays mostly 2 or 3 cells wide, but very broad in some genera, hetero- or homogeneous.

Leaves usually dorsiventral, centric in some species of *Asteranthos* and *Foetidia*. Hairs simple, unicellular or uniseriate, occasionally tufted. Stomata confined to lower surface; generally Cruciferous type.

Embryology. Pollen grains 3-colporate or 3-colpate; shed at 3-celled stage. Abnormal and multinucleate pollen grains known in *Napoleonaea imperialis*. Both single and compound pollen grains are reported in *Couroupita guianensis* (Jacques 1965). Ovules anatropous, bitegmic, tenuinucellate; micropyle formed by inner integument. Polygonum type of embryo sac, 8-nucleate at maturity. Endosperm formation of the Nuclear type.

Chromosome Number. The three tribes have different basic chromosome number. In Planchonoideae x = 13, in Napoleonaeoideae x = 16, and in Lecythidoideae x = 17 (Takhtajan 1987)

Chemical Features. Seed fat with palmitic, oleic and linoleic acid in *Bertholletia excelsa*. (Shorland 1963).

Important Genera and Economic Importance. Flowers of *Couroupita guianensis* are borne on old stems and the fruits are large spherical woody capsules, common name cannon-ball tree. It yields good timber. *Lecythis* flowers are more or less similar to those of *Couroupita,* but the stamens curving over the ovary are sterile. Capsules with woody lids are commonly known as monkey-pots, These fruits are used (with sugar) to catch monkeys, as they cannot withdraw the inserted hand. In *Bertholletia* the fruit is a large woody capsule, containing seeds with hard woody testa and oily endosperm, and are known as Brazil nuts. The fruits are indehiscent and the seeds are procured by opening it with an axe. It is closed by a plug formed of the hardened calyx, and during germination the seedlings emerge from this point.

Taxonomic Considerations. The systematic position of Lecythidaceae is not very clear. Traditionally, this family is included in the Myrtales. But it differs from the Myrtalean members and displays several common features with the Theales. Cronquist (1968, 1981)—for ample reasons—treats Lecythidaceae at an ordinal level and relates it to the Malvales through a commom ancestry in the Theales. Takhtajan (1987) also raises it to an ordinal rank and places it immediately after Theales. In any case, its position in the Myrtales is doubtful. Further studies are necessary for an appropriate placement of Lecythidaceae.

Melastomataceae

A large family of 200 genera and 4000–4500 species (Takhtajan 1987), distributed in the humid tropical regions of the world, the chief area of distribution is South America. It forms a characteristic feature of the flora of Brazil.

Vegetative Features. Herbs (*Sonerila tenera*), shrubs (*Tibouchina semidecandra*) or small trees (*Memecylon capitellatum*); erect, climbing or epiphytic (*Medinilla*). The only lianas are *Adelbotrys macrantha* and *Topobea alternifolia* (Ter Welle and Koek-Noorman 1981). Leaves mostly opposite and decussate, one of a pair is often smaller; rarely alternate by abortion of one of a pair; simple, entire; main veins 3-9, palmate and parallel, closely and transversely anastomosing, and without gland dots; hairy or glabrous (Fig. 41.9A).

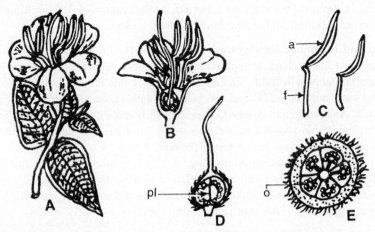

Fig. 41.9 Melastomataceae : **A–E** *Melastoma malabathricum.* **A** Flowering twig. **B** Flower, vertical section. **C** Stamens. **D** Carpel, vertical section. **E** Ovary, cross section. *a* anther, *f* filament, *o* ovule, *pl* placenta. (Adapted from S.C. Dutta 1988)

Floral Features. Solitary, terminal or axillary or cymose inflorescence. Flowers bisexual, actinomorphic, hypo-, peri- or epigynous. (Fig. 41.9A, B). Calyx of 4 or 5 sepals, polysepalous or connate into a calyptra-like hood, sometimes reduced to a rim on receptacle. Petals 5, polypetalous, convolute or imbricate. Stamens as many as, or twice as many as petals and then in 2 whorls; rarely more as in *Astrocalyx* (65) or only 3 as in *Sonerila* sp.; polyandrous, unequal and geniculate (bent like a knee) (Fig. 41.9C), inflexed in bud, anthers typically bicelled, sometimes seemingly unithecate or even 4-celled, as in *Conostegia* and some species of *Miconia*; basifixed, introrse, dehiscing with a subapical pore; connected often with sickle-form appendage. Gynoecium syncarpous, 4- to 14-carpellary; ovary inferior or half-inferior, number of locules same as the number of carpels; placentation basal, parietal or axile, ovules usually numerous on each placenta (Fig. 41.9D, E); style and stigma 1 and simple. Fruit a loculicidal capsule or berry, seeds non-endospermous (Willis 1973) with a small embryo.

Anatomy. Vessels usually small, occasionally medium-sized, small multiples and irregular clusters common; perforations simple, intervascular pitting alternate (Koek-Noorman et al. 1979). Parenchyma paratracheal and rather sparse or vasicentric in most genera, aliform to broad apotracheal bands in a few others, and diffuse in some others. Crystals of variable types such as druses, megastyloids and crystalline masses are present (Ter Welle and Koek-Noorman 1981). Rays mostly uniseriate, cells often with gummy content but no crystals. Leaves dorsiventral or centric; hairs of various complex types, glandular or not.

Stomata very variable in size, present only on the lower surface or both surfaces, mostly Ranunculaceous or Cruciferous type; Caryophyllaceous type in species of *Bertolonia, Marcetia, Medinilla, Sonerila* and some others; Rubiaceous type in *Memecylon, Miconia* and related genera. *Memecylon* contains globular silica bodies in leaf epidermis and styloids confined to the phloem of the midrib and major veins (Baas 1981).

Embryology.   Pollen grains 3-colpate, shed at 3-celled stage. Ovules anatropous or ana-campylotropous as in *Memecylon edule,* bitegmic, crassinucellate; hypostase present. Polygonum type of embryo sac, 8-nucleate at maturity. Endosperm formation of the Nuclear type. Polyembryony known in some species.

Chromosome Number.   Basic chromosome number is x = 7–17, 27.

Chemical Features.   Common flavonols are reported. Anthocyanins of 3, 5-diglucoside type, malvidin and p-coumaric acid are reported in *Tebouchina* (Harborne 1963).

Important Genera and Economic Importance.   In some parts of Brazil with tropical environment, the members of this family form a characteristic component of the vegetation. *Melastoma malabathricum* occurs in the Eastern Himalayas in India. The larger genera of this family include *Miconia* with ca. 900 species, *Tebouchina* with more than 200 species, *Leandra* with ca. another 200; *Clidemia* and *Medinilla* with 160 species each. Another genus, *Sonerila,* has ca. 175 species in tropical Asia. Twenty two species grow in the tropical evergreen moist lowland forests in southwestern parts of Sri Lanka and in the montane forests in the central parts, are endemic (Lundin 1983).

   Economically, this family is not so important. Species of *Bertolonia, Gravesia* and *Sonerila* are grown for their varigated leaves, and those of *Centradenia, Heterocentron, Medinilla* and *Rhexia* for their flowers. The wood of *Memecylon umbellatum* makes excellent firewood and charcoal. Its leaves and flowers contain dyes and are employed to dye mats and cotton fabrics.

Taxonomic Considerations.   The family Melastomataceae is divided into three subfamilies:
I. Astronioideae—placentation basal or parietal; fruit many-seeded, embryo minute, e.g. *Astronia, Kibessia.*
II. Melastomatoideae—placentation axile; fruit many-seeded, embryo minute, e.g. *Melastoma, Osbeckia.*
III. Memecyloideae—placentation free-central; fruit 1- to 5-seeded, embryo large with leafy cotyledons, e.g. *Memecylon.*
   Willis (1973) and Dahlgren (1983a) recognise the subfamily Memecyloideae as a distinct family Memecylaceae. It is, however, retained at subfamily rank by the rest of the phylogenists, including Takhtajan (1987).

   Melastomataceae is recognised as a member of the Myrtales or Myrtiflorae. According to Hutchinson (1969, 1973), this family is derived from the Myrtaceae but Van Vilet et al. (1981) presume its derivation from the Crypteroniaceae. The genus *Axinandra* may be the connecting link between the two families—Crypteroniaceae and Melastomataceae.

Rhizophoraceae

A family of 16 genera and 120 species distributed mainly in the coastal regions of the tropics of the Old World.

Vegetative Features.   Shrubs or small to medium-sized trees, branches with swollen nodes. Leaves opposite, simple, coriaceous, petiolate and stipulate (stipules caducous) or estipulate. The two leaves at each node are unequal in *Anisophyllea.*

Floral Features.   Inflorescence cymose or flowers solitary, axillary. Flowers actinomorphic, bisexual (rarely unisexuel by abortion and then plants monoecious). peri- to epigynous (Fig. 41.10A). Sepals 3 to

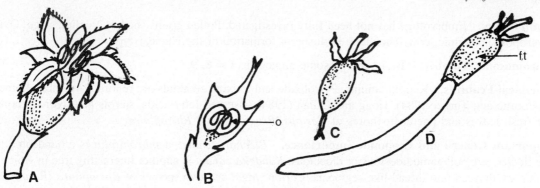

Fig. 41.10 Rhizophoraceae: **A–D** *Rhizophora mangle.* **A** Flower. **B** Vertical section, sepals and petals removed. **C** Fruit. **D** Germinated fruit. *ft* fruit, *o* ovule. (Adapted from Lawrence 1951)

16, usually 4–8, mostly basally connate, valvate, persistent. Petals as many as sepals or absent, often smaller, usually fleshy, free and often clawed, frequently lacerate or emarginate, convolute or inflexed in bud. Stamens 2- to 4-times as many as sepals, usually in 1 whorl, often in pairs opposite the petals, situated on edge of a lobed, peri- or epigynous disc. Filaments mostly very short, anthers introrse, 4-celled at anthesis (pollen sacs numerous in *Rhizophora*), dehiscence longitudinal. Gynoecium syncarpous, 2- to 4-carpellary; ovary superior, semi-inferior or inferior, placentation axile; ovules pendulous (Fig. 41.10B), usually 2- to 4-numerous on each placenta, style 1, stigma mostly as many lobed as carpels. Fruit usually a berry, terminated by the persistent calyx (Fig. 41.10C). Seeds occasionally viviparous (Fig. 41.10D), with a straight (often green) embryo, endosperm mostly present, sometimes absent (Willis 1973).

Anatomy. Wood anatomy is very variable in different genera.

In Group I Rhizophoreae (including *Bruguiera, Ceriops, Kandelia* and *Rhizophora*), vessels moderately small with fairly numerous multiples and clusters; perforation plates scalariform with a few bars; intervascular pitting scalariform. Parenchyma scanty, paratracheal. Rays mostly 3–6 cells wide, uniseriate rays rare, homo- or heterogeneous.

In Group II Gynotrocheae (including *Anisophyllea, Carallia, Combretocarpus, Crossostylis* and *Gynotroches*), vessels medium-sized to large and often very few; perforation plates simple, intervascular pitting alternate. Parenchyma typically banded or aliform, bands vary from narrow to broad, regular to irregular. Rays up to 10–15 cells wide, very high, uniseriates numerous, markedly heterogeneous.

In Group III Macarisieae (including *Anopyxis, Blepharistemma, Cassipourea, Macarisia* and *Sterigmapetalum*), vessels small to large, often exclusively solitary; perforation plates simple and scalariform with fine bars, intervascular pitting opposite to alternate. Parenchyma paratracheal and confined to the abaxial side of the vessels (apotracheal in *Blepharistemma*). Rays 3–4 cells wide, less than 1 mm high, uniseriates numerous, markedly heterogeneous.

Two genera—*Pellacalyx* and *Poga*—are different from all the three groups. In *Pellacalyx* the vessels are medium-sized, arranged in tangential groups, perforation plates simple. Parenchyma encloses the vessels and forms bands between large rays. Rays up to 25 cells wide and very high, uniseriates numerous, heterogeneous. In *Poga* vessels large, solitary, perforations simple and nearly horizontal; intervascular pitting alternate. Parenchyma surrounding the vessels forms narrow wings and apotracheal bands between the large rays. Rays of 2 sizes, the larger up to 20 cells wide, uniseriates numerous, heterogeneous.

Leaves usually dorsiventral, hairs mostly unicellular, cuticle often very thick. Stomata confined to the lower surface and often depressed. A mixture of Ranunculaceous, Cruciferous and Rubiaceous types of stomata recorded in *Anopyxis calaensis.* Subsidiary cells are usually not well differentiated.

Embryology.   Embryology has not been fully investigated. Pollen grains shed at 2-celled stage. Ovules anatropous, bitegmic, crassinucellate. Endosperm formation of the Nuclear type.

Chromosome Number.   Basic chromosome number is x = 8, 9.

Chemical Features.   Rich in tannins. Pyrrolizide and tropane alkaloids are reported in Rhizophoraceae (Harborne and Turner 1984). Hogg and Gillan (1984) observed fatty acids, sterols and hydrocarbons in the fresh leaves and pneumatophores of *Ceriops, Bruguiera* and *Rhizophora.*

Important Genera and Economic Importance.   *Blepharistemma membranifolia* is a handsome tree, the flowers are polygamodioecious or dioecious. *Kandelia candel* is another interesting tree in which the petals are divided into thread-like segments. *Ceriops tagal* and two species of *Rhizophora* (*R. apiculata* and *R. mucronata*) are the members of mangrove forests in Sunderbans area in West Bengal in eastern India. The wood of *R. mucronata* is used in making charcoal, and its bark for tanning and dyeing; the fruit is edible and the fruit juice is fermented for preparation of a drink. The wood of *Carallia integerrima* is also useful in various ways. Bark decoction of *Ceriops tagal* stops haemorrhage; tannin is obtained from roots and fruits and a dye from the bark.

Taxonomic Considerations.   The family Rhizophoraceae was placed in the Myrtales/Myrtiflorae by most of the earlier workers (Bentham and Hooker 1965a, Engler and Diels 1936, Wettstein 1935, Lawrence 1951 'and Melchior 1964).

Cronquist (1968) removed it from the Myrtales because of its anomalous position: lack internal phloem and has copious endosperm. He included this family in the Cornales because, in his opinion, it was "pretty clearly of Rosalean origin, parallel to the Myrtales" and this family could be included in the Cornales "as a near-basal side branch not far distant from the Myrtales". He also perceived a separate order for it.

Dahlgren (1980a) placed it in a distinct order Rhizophorales. According to him, "Rhizophoraceae s.s. deviate from the Myrtalean families to a variable extent" but their position next to the Myrtales is likely. Takhtajan (1987) also agrees with this view and treats it as a member of the order Rhizophorales.

In view of these treatments, the family Rhizophoraceae should be raised to the rank of an order, Rhizophorales.

Combretaceae

A family of 20 genera and 400 species (Takhtajan 1987), distributed mostly in the Old World tropics and subtropics.

Vegetative Features.   Trees (*Terminalia*) or shrubs (*Combretum fruticosum*), often scandent (*Calycopteris floribunda*), sometimes lianas (*Quisqualis indica*). Leaves alternate or opposite, simple, entire, stipulate (Fig. 41.11A).

Floral Features.   Inflorescence panicles, racemes or spikes. Flowers bracteate, ebracteolate (*Quisqualis*), or bracteolate, bracteoles often adnate to ovary as in *Lumnitzera* and *Laguncularia*, and even form wings as in *Macropteranthes* (Fig. 41. 11D); actinomorphic, bisexual or polygamodioecious; epigynous (perigynous in *Strephonema*). Calyx lobes 4 or 5 (to 8), persistent, valvate, adnate to the ovary to form hypanthium. Corolla lobes of the same number as that of sepals, or absent, imbricate or valvate, small (Fig. 41.11B). Stamens 2 to 5 or twice as many as calyx lobes and biseriate; filaments long, filiform as in *Combretum*, anthers versatile, bithecous, dehisce longitudinally (Fig. 41.11C), inserted on the tube or limbs of calyx (Fig. 41.11B). Gynoecium monocarpellary, ovary inferior, unilocular, ovules 2–6, pendulous on long funicle from locule apex; style slender and solitary with a capitate or not so obvious stigma. Fruit

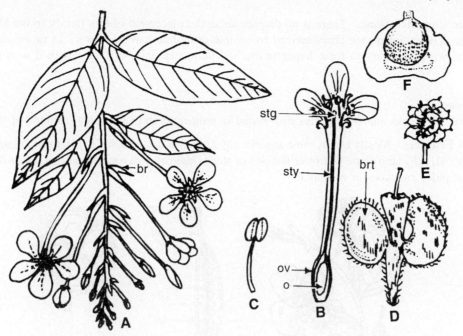

Fig. 41.11 Combretaceae: **A–C** *Quisqualis indica,* **D** *Macropteranthes kekwickii,* **E, F** *Anogeissus latifolia.* **A** Flowering twig. **B** Flower, vertical section. **C** Stamen. **D** Fruit of *M. kekwickii.* **E** Bunch of fruits of *Anogeissus latifolia.* **F** Single fruit. *br* bract, *brt* bracteole, *o* ovule, *ov* ovary, *stg* stigma, *sty* style. (**A–C** sketched by Arindam Bhattacharyya, **D** adapted from Hutchinson 1969, **E, F** after J.K. Maheshwari 1965)

a drupe, often winged as in *Terminalia arjuna;* crowded into a mass in *Ramatuela, Conocarpus, Anogeissus* (Fig. 41.11E, F) and *Finetia;* 1-seeded (by ovule abortion), seeds non-endospermous.

Anatomy. Vessels mostly medium-sized, exclusively solitary or with numerous multiples of 4 or more cells, perforations simple, intervascular pitting alternate. Parenchyma predominantly paratracheal, abundant, but apotracheal in *Terminalia bialata.* Rays exclusively uniseriate or occasionally biseriate as in *Anogeissus, Combretum, Conocarpus,* homo- or heterogeneous.

Leaves dorsiventral, rarely centric and with glandular or non-glandular hairs. Stomata Ranunculaceous type, usually confined to the lower surface.

Embryology. Pollen grains ridged, triporate, exine thick, smooth; shed at 2- or 3-celled stage. Ovules anatropous, bitegmic, crassinucellate. Polyembryony reported. Endosperm formation of the Nuclear type.

Chromosome Number. Basic chromosome numbers are x = 7,11,12,13.

Chemical Features. Bark and fruits of various members are rich in tannins.

Important Genera and Economic Importance. *Terminalia* spp., commonly called the myrobalans, are the most important plants of this family. The fruits of *T. chebula* and *T. bellirica* are used in Aurvedic medicines in India. Fruits and bark of different species of this genus are rich sources of tannin. Other important members are *Anogeissus, Combretum, Calycopteris* and *Quisqualis;* many of these are grown as ornamentals. *Lumnitzera racemosa* is a mangrove plant growing in muddy tidal areas along the coasts. These plants develop negatively geotropic roots.

Taxonomic Considerations. There is no dispute about the placement of this family in the Myrtales or Myrtiflorae. The members are characterized by simple estipulate leaves, flowers in racemose clusters, calyx fused with the ovary to form hypanthium, and unilocular inferior ovary with 2 to 6 pendulous ovules.

Onagraceae

A family of 17 genera and 680 species distributed in temperate and subtropical regions of the world.

Vegetative Features. Mostly herbs, some aquatic, e.g. *Ludwigia adscendens* (L.) Hara (= *Jussiaea repens* Linn.) (Fig. 41.12F), and some shrubs (*Fuchsia*) or trees (*Hauya*). Leaves alternate or opposite, simple, stipulate, stipules caducous or estipulate.

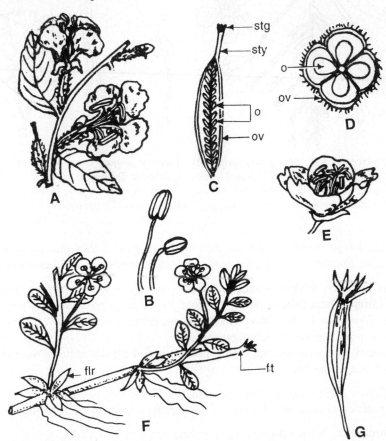

Fig. 41.12 Onagraceae : **A–D** *Clarkia elegans*, **E** *Oenothera pilosella*, **F, G** *Ludwigia adscendens*. **A** Flowering twig. **B** Stamen. **C** Ovary, vertical section. **D** Cross section. **E** Flower of *O. pilosella*. **F** Flowering branch of *L. adscendens*. **G** Fruit. *flr* floating roots, *ft* fruit, *o* ovule, *ov* ovary, *stg* stigma, *sty* style. (**A–E** after Lawrence 1951, **F, G** Original)

Floral Features. Inflorescence raceme, or spike or paniculate as in *Fuchsia,* or solitary axillary flowers. Flowers bisexual, actinomorphic, zygomorphic in *Lopezia,* typically tetramerous, epigynous (Fig. 41.12A, E). The biseriate perianth forms a hypanthium adnate to the ovary, sepals usually 4 but vary from 2–3–5 or

6, free or the hypanthium extends much beyond the ovary, valvate, persistent as in *Ludwigia* or deciduous. Petals mostly 4, sometimes 2 or more or absent as in *Ludwigia*, mostly with prominent limb and claw, convolute or imbricate. Stamens usually as many as corolla lobes or twice as many and then in 2 whorls, the outer whorl alternate with petals, free, arising from or near the rim of hypanthium. *Lopezia* is exceptional in having only 1 fertile stamen and one staminode. Anthers bicelled, longitudinally dehiscent (Fig. 41.12B), each cell transversely divided into 2, as in *Circaea*, or several chambers, as in *Clarkia;* versatile or innate. Gynoecium syncarpous, 4-carpellary, ovary inferior, tetralocular, ovules numerous in each locule, placentation axile (Fig. 41.12C, D); style simple, long filiform; stigma capitate, notched or 4-rayed. Fruit a berry (*Fuchsia*) or a capsule (*Clarkia*), often with persistent calyx (Fig. 41.12G). Seeds non-endospermous, embryo straight or nearly so. Fruit is bristly, 1- or 2-seeded nutlet in *Circaea* and a 1-seeded smooth nut in *Gaura*.

Anatomy. Vessels small often with numerous radial multiples and clusters, perforations simple, intervascular pitting alternate. Parenchyma paratracheal, scanty. Rays 2–9 cells wide, commonly more than 1 mm high, heterogeneous. Included phloem present.

Leaves dorsiventral or centric. Stomata usually Cruciferous type, occur on both surfaces or confined only to the lower surface. Raphides usually present.

Embryology. Pollen grains of different genera have distinctive morphology. In *Ludwigia* spp. pollen grains are triangular and connected together by viscous thread. Mature grains are 2-celled and triporate, shed singly or in tetrads. Ovules anatropous, bitegmic, crassinucellate and have a hypostase. Embryologically the family is of great interest as it is characterized by the universal occurrence of the monosporic 4-nucleate embryo sac designated 'Oenothera' type (Seshavataram 1970). Endosperm formation of the Nuclear type.

Chromosome Number. Basic chromosome numbers are $x = 7–17, 27$. A large number of allopolyploids are known in the genus *Clarkia* (Small et al. 1971).

Chemical Features. The chalcone, isosalipurposide, has been isolated from some Onagraceae members, but is absent in others (Dement and Raven 1973). Flavonoids luteolin and luteolin 7-glycoside and also apigenin-based c-glycosyl flavones have been obtained from stem and leaves of *Gaura triangulata* (Hilsenbeck et al. 1984).

Important Genera and Economic Importance. Some of the well-known garden plants of this family are: *Clarkia elegans* (Fig. 41.12A), *Oenothera biennis and O. grandiflora* (evening primrose), *Fuchsia splendens*, *F. corymbosa* and *F. magellanica* (dancing girl) from Central and South America, *Godetia grandiflora* (farewell to Spring) a native of California and *Circaea lutetiana* (enchanter's nightshade). *Ludwigia adscendens* is an aquatic herb bearing bunches of respiratory roots (Fig. 41.12F). From the perianth of *Fuchsia*, a red dye is extracted; this fuchsine is used as a staining reagent in botanical studies.

Taxonomic Considerations. This family was included in the Myrtiflorae by Engler and Diels (1936), Rendle (1925), Wettstein (1935). Hutchinson (1959) placed it in Lythrales along with Lythraceae, Trapaceae, Haloragidaceae and Callitrichaceae. Hutchinson (1969, 1973) included Lythraceae in the Myrtales and the rest of the families in a separate order, Onagrales, in the Herbaceae, deriving it from Caryophyllaceous stock through the subfamily Sileneoideae. According to him, it is an advanced family with perigynous to epigynous condition of ovary and many members with highly reduced structures.

Many morphological (tetramerous flowers), anatomical (presence of internal phloem) and embryological features (2-celled triporate pollen grains; anatropous, bitegmic ovules; monosporic embryo sac and Nuclear endosperm) support its inclusion in the Myrtales. Cronquist (1981), Dahlgren (1983a) and Takhtajan (1987) also treat the family Onagraceae as a member of the Myrtales which is appropriate.

Oliniaceae

A monotypic family of the genus *Olinia* with 10 species distributed in South and East Africa and St. Helena.

Vegetative Features.   Trees or shrubs. Leaves simple, entire, opposite, stipulate coriaceous.

Floral Features.   Inflorescence terminal or axillary cymes. Calyx of 5 sepals, gamosepalous, tubular, limb of 5 minute blunt teeth, or obsolete. Corolla of 5 petals, polypetalous, imbricate, spathulate, inserted on margin of calyx tube, alternate with 5 small, valvate, more or less cucullate, pubescent, coloured scales. Stamens 5, alternipetalous, inserted immediately below scales and at first hidden and enclosed by them; filaments very short, connective thickened, alternate with 5 small, thick subglobose, pubescent staminodes. Gynoecium syncarpous, 4- or 5-carpellary, ovary inferior, 4- to 5-locular with 2 (-3) superposed pendulous ovules per locule, placentation axile; style 1, simple, stigma clavate. Fruit a 3- to 5-locular false drupe, with 1 seed per locule. Seeds with spiral or convolute embryo, non-endospermous.

Anatomy.   Vessels small, in numerous small multiples, perforations simple, intervascular pitting alternate. Parenchyma very sparse, scanty, paratracheal. Rays 2 cells wide, markedly heterogeneous. Intraxylary phloem present.

Embryology.   Pollen grains 2-celled at dispersal stage. Ovules pendulous, campylotropous, bitegmic, crassinucellate. Polygonum type of embryo sac, the three antipodals are ephemeral and degenerate before fertilization. The mature embryo sac is 5-nucleate (Johri et al. 1992). Endosperm formation of the Nuclear type.

Chromosome Number.   Basic chromosome number is x = 12.

Important Genera and Economic Importance.   The only genus *Olinia;* one species *O. cymosa* produces a valuable timber called hard peer and is used for various purposes in South Africa.

Taxonomic Considerations.   In Bentham and Hooker's (1965a) system the genus *Olinia* was included in the Lythraceae. Hutchinson (1973) included this family in the order Cunoniales but with a pointer that it could have an affinity with the Lythraceae of Myrtales, due to the presence of intraxylary phloem. Other taxonomists have treated this family as a member of the Myrtales, which is appropriate.

Haloragaceae

A family of 10 genera and 100 species, widely distributed all over the world, but abundant in Australia. Haloragaceae members are distributed in temperate and subtropical regions of all continents and a few species of *Myriophyllum* occur in the Arctic. *Haloragis* with ca. 90 species, and *Gunnera* with 50 species occur mainly in the southern hemisphere. These two genera together constitute a large regional paleosubantarctic complex with a distribution concentrated around Australia, New Zealand and Tasmania (Praglowski 1970).

Most species of *Myriophyllum* occur in the northern hemisphere. The genus *Haloragis,* with the exception of two species from Juan Fernandez Islands and one species from New Caledonia, occurs only in the Australia-New Zealand-Tasmania area. These are terrestrial small herbs or subshrubs, usually growing in relatively humid areas. The genus *Gunnera* occurs mainly in the southern hemisphere. A few species are reported north of the equator in South America and in the Philippines. Subantarctic species of this genus are abundant in New Zealand, but they also occur in Chile, Argentina and Juan Fernandez Islands. It is vertically distributed from sea level to 4500 m altitude in the South American cordilliera (Praglowski 1970).

The genus *Myriophyllum* includes more or less submerged hydrophytic species and its centre of distribution is Australia; about 15 spp. grow here. Around 8 species grow in temperate North America and a few are arctic and subarctic.

Another interesting genus is *Laurembergia* with 20 species. These are small, prostrate, stoloniferous herbs which grow in humid but sunny areas. These species have an amphibic biology (Raynal 1965), i.e. each year the plants pass through a cycle of inundation and so change from hydrophytic to halophytic status. This genus has a large distributional area—Australia, New Zealand, tropical Asia, Madagascar, Africa and tropical America.

The distribution of the relatively infrequent findings of macro- and microfossils, originating from plants belonging to the Haloragaceae, and of sporomorphs from plants of "Haloragoideae-Oenotheraceae type" suggest that these taxa in the pre-Quarternary (= 2.5 million years ago) systems had a wider distribution than the family Haloragaceae has at present. Some of the recently isolated endemic species of this family from the main centers of their distribution, might be considered as representative of "conservative endemicity", i.e. regression from wider area of distribution towards a more limited occurrence.

Vegetative Features. Aquatic or terrestrial herbs, rarely suffrutescent. Leaves opposite or alternate (even within the same genus) or sometimes whorled; very variable in size, pectinately pinnatifid when submerged, estipulate. Adventitious roots well-developed (Fig. 41.14A, B).

Floral Features. Inflorescence simple raceme (*Haloragis*), corymbose or paniculate as in *Loudonia* or simple axillary clusters or flowers solitary axillary (Fig. 41.14B). Flowers unisexual (and then plants monoecious or polygamodioecious), minute, actinomorphic, mostly subtended by a pair of bracteoles, epigynous. Perianth 4 + 4, or 4 or 0, uniseriate or none at all, adnate to the ovary (Fig. 41.14C, E). Calyx of 2 to 4 or 0 sepals, petals usually 0, when present 2 to 4, early deciduous, free, usually larger than calyx lobes, imbricate or valvate. Stamens 4 or 8, the outer series (of stamens) when present, opposite the petals; anthers bithecous, longitudinally dehiscent (Fig. 41. 14C–E, G). Gynoecium syncarpous, bi- to tetracarpellary, ovary inferior, mono- to tetralocular, placentation axile, each locule with one pendulous ovule (Fig. 41.14D, I); when unilocular, the ovule is solitary, parietal and pendulous. Styles free, as many as the locules, stigmas often plumose (Fig. 41.14 H, I). Fruit a nut or drupaceous, sometimes winged; seeds endospermous and with a straight, cylindrical or obcordate embryo. Fruit a schizocarp in *Myriophyllum* (Fig. 41.14F).

Anatomy. The vascular structure of *Myriophyllum* is complex, and consists of an axile, fibro-vascular mass without a true pith. In *Haloragis alata* and *Loudonia aurea.* there is a ring of cambium and a continuous cylinder of secondary xylem traversed by narrow rays. Vessels with narrow lumina and simple perforations. Solitary and clustered crystals are common in the pith of *Haloragis alata.* Leaves dorsiventral or subcentric. A variety of hairs, warts and emergences are reported; stomata Ranunculaceous type and numerous in terrestrial genera.

Embryology. Pollen grains tetra- or pentaporate and are shed at 3-celled stage. Ovules anatropous, bitegmic, crassinucellate. Endosperm formation of the Nuclear or Cellular type.

Chromosome Number. Basic chromosome numbers are x = 7, 9, 21, 29.

Important Genera and Economic Importance. Some of the important genera are *Haloragis* with ca. 90 species and *Myriophyllum* also with a large number of species. Economically, it is not an important family. *Myriophyllum,* or water milfoil, is of some importance in limnological conservation practices and as an aquarium plant.

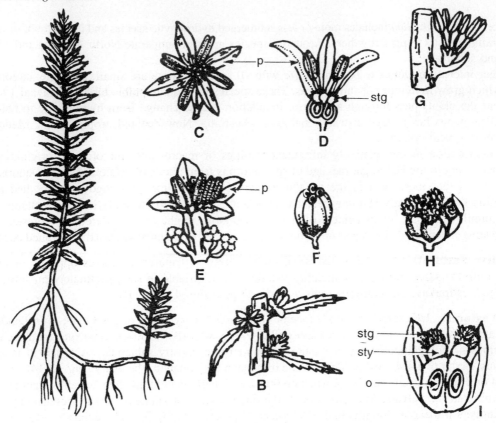

Fig. 41.14   Haloragaceae : **A–F** *Myriophyllum intermedium*, **G–I** *M. exalbescens*. **A** Plant. **B** Portion of plant. **C** Flower. **D** Vertical section. **E** Flower and fruits. **F** Fruit. **G** Staminate flower of *M. exalbescens*. **H** Pistillate flower. **I** Vertical section. *o* ovule, *p* perianth, *stg* stigma, *sty* style (**A–F** adapted from Subramanyam 1962, **G–I** from Lawrence 1951).

Taxonomic Considerations.   The famil· Haloragaceae has been included in the "order" Halorageae of the Polypetaleae among Bruniaceae, Rhizophoraceae, Combretaceae and Myrtaceae in Bentham and Haooker's (1965a) system. Wettstein (1935) placed it in the Myrtales, Onagraceae (= Oenotheraceae) and Lythraceae being the closest. Hutchinson (1973) reduced the large order Myrtales and established Lythrales in which Haloragaceae was placed along with Lythraceae, Onagraceae, Trapaceae and Callitrichaceae. Takhtajan (1959) recognised Myrtales and Haloragales in the Myrtiflorae, the latter included Haloragaceae, Gunneraceae and Hippuridaceae; Haloragaceae is the most primitive. This view is supported by floral morphology and embryology.

Chadefaud and Emberger (1960) placed Haloragaceae and Gunneraceae within the Myrtales, as two closely related but separate families. Among the 16 families of the Myrtales, Myrtaceae, Oenotheraceae (= Onagraceae), Hippuridaceae and Callitrichaceae are closely related to these two families.

Melchior (1964) recognized 3 suborders—Myrtineae, Hippuridineae and Cynomorineae. Haloragaceae is placed in the Myrtineae and is considered by Melchior (1964) as reduced but at the same time, as a more-developed fraction of the Oenotheraceae (= Onagraceae).

Cronquist (1981) places the family Haloragaceae in the Haloragales along with a separate family, Gunneraceae. Dahlgren (1983a) and Takhtajan (1987) consider this family to be the only member of the order Haloragales and treat Gunneraceae also as a member of a distinct order, Gunnerales. Thorne (1983) includes this family as well as Gunneraceae in the Cornales under suborder Haloragineae.

To conclude, Haloragaceae belongs to the Myrtales close to the Onagraceae, from which it differs due to the absence of intra-xylary phloem and presence of endospermous seeds. Wind-pollinated flowers and more or less aquatic habitat are some of the advanced features of this family. The family Haloragaceae, therefore, may be regarded as an advanced taxon of the Myrtiflorae/Myrtales.

Regarding the genus *Gunnera,* its separation into a distinct family Gunneraceae is justified (Praglowski 1970). The unigeneric family Gunneraceae was established by Meissner (1841, see Takhtajan 1987). Eichler (1880, see Praglowski 1970) included the genus *Gunnera* in the Haloragaceae. Gunneraceae as a family was re-established by Takhtajan (1959, 1969, 1980, 1987), followed by Dahlgren (1975a, 1980a, 1983a), Cronquist (1981) and Thorne (1976, 1981, 1983). *Gunnera* is distinct from other members of Haloragaceae in gross and floral morphology, embryology, and anatomy. Also, the genus is entirely stenopalynous, and its diploid chromosome number is 2n =24. Members of Gunneraceae are unique among the angiosperms in their peculiar symbiosis with blue-green algae (*Nostoc*) living in cortical cavities of the plant (Dahlgren 1983a). Also noteworthy are the presence of P-type sieve-tube plastids (Behnke 1986), unisexual flowers, a bicarpellary, unilocular ovary with a solitary ovule, tetrasporic embryo sac and Cellular type endosperm.

Dahlgren (1975a 1980a, 1983a) raised the three families—Haloragaceae, Gunneraceae and Hippuridaceae—to ordinal ranks and placed Haloragaceae in Myrtiflorae, Hippuridaceae in Lamiflorae and Gunneraceae in Saxifragalean alliance. Ehrendorfer (1983) and Dahlgren and Thorne (1984) also support this view. Studies on sieve-tube plastids (Behnke 1986) also support the alliance of Gunneraceae to Rosiflorae. Sieve-tube plastids are P-type in *Gunnera* and in many unrelated families of Rosidae, but they are S-type in Haloragaceae and Hippuridaceae.

In our opinion also, raising the three families—Haloragaceae, Gunneraceae and Hippuridaceae—to distinct orders is justified.

Theligonaceae

A unigeneric family of the genus *Theligonum* with 3 species distributed in the Mediterranean region, Western China and Japan.

Vegetative Features. Annual or perennial succulent herbs with fleshy, simple, entire leaves, alternate above and opposite below as in *Theligonum cynocrambe* (Kapil and Mohana Rao 1966a), by suppression of one leaf of each pair, stipulate, stipules interpetiolar.

Floral Features. Inflorescence axillary groups of small, sessile, uni- or bisexual cymes. Flowers unisexual; in male flowers the perianth or calyx is closed in bud, valvately 2- to 5-partite at anthesis, segments broad, revolute; stamens (2-) 7-12 (-30) in bundles, filaments filiform, short; anthers linear, erect in bud, pendulous at anthesis; pistillode absent. Female flowers have membranous perianth, more or less flask-shaped, extends above into a narrow tube, 2- to 4-dentate at mouth. Gynoecium monocarpellary, ovary unilocular, inferior, with one basal ovule and a gynobasic filiform style exerted from the mouth of the perianth. Fruit a subglobose, nut-like drupe with only one seed; endosperm fleshy.

Anatomy. Xylem forms a continuous cylinder, radial rows of numerous vessels. Pith contains more chlorophyll than the cortex. Idioblasts which contain raphides are present in the cortex. Leaves dorsiventral, upper epidermal cells papillose; stomata Rubiaceous type, more numerous on lower surface. Large, club-shaped glands occur in apical region of stem.

Embryology. Pollen grains 6-porate, 3-celled at shedding stage, packed with starch grains. Ovules campylotropous, unitegmic, tenuinucellate. Endosperm formation of the Nuclear type (Kapil and Mohana Rao 1966a).

Chromosome Number. Haploid chromosome numbers are n = 10,11.

Chemical Features. Iridoids asperuloside, desacetyl-asperuloside and paederoside are present in *Theligonum* (Jensen et al. 1975). Betalains absent.

Important Genera and Economic Importance. The only genus, *Theligonum,* is not economically important.

Taxonomic Considerations. The position of Theligonaceae is much disputed. Bentham and Hooker (1965c) include this family in the Urticaceae mainly because of the highly reduced unisexual flowers. Some taxonomists also include it in the Caryophyllales although it does not have the pigment betacyanin (Mabry et al. 1975). Although Melchior (1964) placed it in the Myrtales, there is no resemblance to the lofty Myrtalean trees. As Wunderlich (1971) pointed out, in Theligonaceae there are many features that are common to the members of the Rubiaceae. These include exomorphic features like simple, entire, opposite, stipulate leaves, and inferior ovary; anatomical features like Rubiaceous type of stomata and abundant raphides; embryological features like unitegmic, tenuinucellate ovules and Nuclear type of endosperm, and chemical features like the presence of iridoids.

The presence of unisexual flowers and monocarpellary ovary are somewhat unusual for its inclusion in the Rubiaceae. According to Wunderlich (1971), the monocarpellary ovary might have developed from originally 2 medianly placed carpels.

Cronquist (1968, 1981) has established its relationship with the members of his Haloragales and, according to him, their position in Caryophyllales would be anomalous in the absence of perisperm and betacyanin pigments, and presence of unitegmic ovules.

Dahlgren (1980a, 1983a) includes Theligonaceae in Rubiaceae. Kooiman (1971) and Takhtajan (1987) include it in the order Gentianales, and place it next to the Rubiaceae. This is the correct position of this family.

Hippuridaceae

A unigeneric family, the genus *Hippuris* has 3 species, almost cosmopolitan in distribution.

Vegetative Features. Aquatic herbs with creeping rhizome and erect sympodial shoots, upper parts usually project above the water surface. Leaves linear, entire, estipulate, verticellate, the submerged ones longer and more flaccid than the aerial ones (Fig. 41.16A, B).

Floral Features. Flowers solitary and sessile in the axil of each leaf (Fig. 41.16B), mostly bisexual or sometimes only female in upper axils, or on some stalks; protogynous. Perianth absent or a small, simple, 2- to 4-lobate; stamen 1, borne near the apex of the ovary, a large bilobed anther, longitudinally dehiscent (Fig. 41.16C, D). Gynoecium monocarpellary, ovary inferior, unilocular with 1 apical pendulous ovule (Fig. 41.16E); 1 long subulate sublateral style, stigmatic throughout; anemophilous. Fruit a smooth ovoid achene or a small drupe. Seeds non-endospermous, with a large embryo.

Anatomy. Internal structure of the inner branches are as follows: a single-layered epidermis of tangentially-thickened cells is followed by a broad cortex. The entire cortical tissue includes large intercellular spaces that are separated from one another by multicellular plates only 1-cell wide. Inner to the cortex is the layer of endodermis. The vascular system is reduced to a narrow ring of phloem surrounding a comparatively broader zone of xylem. Vessels small in diameter with spiral or reticulate thickenings. Leaves isobilateral

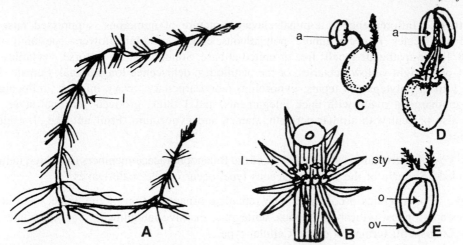

Fig. 41.16 Hippuridaceae : **A–E** *Hippuris vulgaris*. **A** Plant. **B** Node with flowers in axil of leaves. **C** Flower. **D** Style between anther cells. **E** Carpel, vertical section. *a* anther, *l* leaf, *o* ovule, *ov* ovary, *sty* style. (Adapted from Lawrence 1951)

to centric. Peltate hairs with an unicellular base and multicellular head are present. Stomata occur on both the surfaces with well-developed air cavities below these.

Embryology.   Pollen grains 4- to 6-colpate, trinucleate or 3-celled at the time of shedding. Ovules anatropous, unitegmic, tenuinucellate, or ategmic. Polygonum type of embryo sac, 8-nucleate at maturity. Endosperm formation of the Cellular type; no endosperm haustorium.

Chromosome Number.   Haploid chromosome number is n = 16.

Chemical Features.   Iridoids like aucubin and catalpol have been reported from *Hippuris* spp. (Jensen et al. 1975). Caffeic and ferrulic acids are also present (Dahlgren 1975a).

Important Genera and Economic Importance.   The only genus, *Hippuris,* is not known to have any economic importance.

Taxonomic Considerations.   Bentham and Hooker (1965a) and Hallier (1912) included *Hippuris* and also *Callitriche* (of Callitrichaceae) in the Haloragaceae but placed the family in the Ranales and considered it to have been derived from the Nymphaeaceae. Bessey (1915), Rendle (1925) and Wettstein (1935) accepted Hippuridaceae as a distinct family near Haloragaceae. However, later studies have shown that Hippuridaceae is very different from the Haloragaceae in morphology; embryology and chemistry, which support its placement in a distinct order, Hippuridales (Dahlgren 1975a, 1980a, 1983a, Jensen et al. 1975, Takhtajan 1987). All taxonomists are also of the opinion that this order should be placed near Scrophulariales in the Lamianae, particularly on the basic of chemical features.

Cynomoriaceae

A unigeneric family of the genus *Cynomorium* with 2 species distributed from Mediterranean region to Mongolia.

Vegetative Features.   Achlorophyllous, grassy, parasitic herbs, with thick brownish rhizome, which bears numerous short, haustorial appendages, and thick, simple, succulent flowering stems bearing numerous deltoid, alternate scale leaves. The plant is reddish brown.

Floral Features. Inflorescence a terminal clavate structure of numerous, suppressed false capitula interspersed with scales. Flowers minute, polygamous, epigynous. Male flowers—perianth of 1 to 5, rarely 6 to 8, linear-spathulate tepals, free or united at base. Stamen 1, with introrse, versatile, 4-locular anther and 1 or 2 bright-yellow nectaries, or the pistillodes; dehiscence longitudinal. Female flowers—perianth of 1 to 5, polytepalous tepals; gynoecium monocarpellary, ovary inferior, unilocular with 1 pendulous, submarginal ovule with thick integuments, and 1 thick, grooved, terminal style. Bisexual flowers are also present with similar perianth, stamen and gynoecium. Fruit nut-like, 1-seeded, seeds endospermous.

Anatomy. Vascular structure is similar to that of the Balanophoraceae members (in which family it was earlier included). Stomata of the Ranunculaceous type, occur on the scale leaves.

Embryology. Pollen grains 3-colporate with reticulate ornamentation of the exine, shed at 2-celled stage. Ovules apotropous, ortho-hemitropous, unitegmic, massive integument, crassinucellate. Nucellar cap present. Endosperm formation of the Cellular type.

Chromosome Number. Haploid chromosome number is n = 12.

Chemical Features. Endosperm contains oil.

Important Genera and Economic Importance. The only genus *Cynomorium* with its 2 species is a total root parasite.

Taxonomic Considerations. Taxonomic position of this family is disputed. More often it is related with the family Balanophoraceae, and has often been included in this family (Bentham and Hooker 1965c, Metcalfe and Chalk 1972). Dahlgren (1975a, 1980a, 1983a) included this family in an order of its own—the Balanophorales. It is more primitive than the Balanophoraceae in having a well-developed ovular integument, but more advanced in its inferior ovary (Cronquist 1968). In the absence of any apparent relative in any other order, and also because of its parasitic habit, and some other Santalean features, Cronquist (1968, 1981) includes Cynomoriaceae in Santalales and places it next to the Balanophoraceae. Takhtajan (1987) is of the opinion that although *Cynomorium* is similar to members of the Balanophorales (in habit), embryological features are so different that there is no connection between the two taxa. Similarly, any relationship between the Hippuridaceae and the Cynomoriaceae, as claimed by Melchior (1964), is also unacceptable.

On the basis of available data, it is probably better to erect a separate order Cynomoriales for this family, including it in the Rosidae, as has been suggested by Takhtajan (1987). However, its accurate position is yet to be settled. We must await further investigations.

Taxonomic Considerations of the Order Myrtiflorae
About the circumscription of this order, there are many views. The Mrytiflorae of Engler and Diels (1936) comprised 4 suborders and 23 families. Melchior (1964), in revising the work of Engler, included 17 families under 3 suborders in the Myrtiflorae. He treated the suborder Thymelaeineae as a separate order, Thymelaeales, and removed the families Heteropyxidaceae, Nyssaceae, Alangiaceae and Hydrocaryaceae to other orders. Instead, he added the families Dialypetalanthaceae, Trapaceae, Oliniaceae and Theligonaceae to the suborder Myrtineae. Earlier, there were doubts about the position of the Dialypetalanthaceae in this order. They are unusual in this order in lacking internal phloem, and the pollen is not Myrtalean. It was once presumed that it is related to the Rubiaceae, but numerous stamens and polypetalous corolla of *Dialypetalanthus* are not suggestive of this alliance. The family Lecythidaceae with alternate leaves, centrifugal stamens, no internal phloem, and a number of negative embryological features, should also be placed in the Dilleniiales (Cronquist 1968, 1981).

The family Theligonaceae, also, is more closely related to the Rubiaceae (Wunderlich 1971, Kooiman 1971). According to Takhtajan (1987), the family Hippuridaceae is closely linked to the Scrophulariaceae and this is supported by chemical data also (Gornal et al. 1979). He includes it in a separate order, Hippuridales, and also Cynomoriaceae in Cynomoriales. However, the position of the last family is still doubtful because of inadequate information.

Monotypic Hoplestigmataceae (sometimes included in Myrtales) also has a controversial position. It has been referred to the Violales (Cronquist 1981) and Boraginales (Takhtajan 1987). In *Hoplestigma pierreanum*, pollen is shed as monads, spheroidal to suboblate in equitorial view, 3-colporate, and the colpi bordered by meridional ridges. This type of pollen is not known in Begoniaceae (Presting et al. 1983), or Cistaceae (Nowicke 1989, unpubl. data) and all other families of the Violales (Nowicke and Millar 1989). Similar pollen morphology is seen in many Myrtales (Patel et al. 1985). Both taxa also have numerous stamens. However, there are also a number of differences, such as ovules bitegmic in Myrtales but unitegmic in *Hoplestigma*, and corolla polypetalous in Myrtales but united basally to form a short tube in *Hoplestigma*. In the absence of adequate data, it is not possible to establish its correct position. Hoplestigmataceae is treated in this work as a member of suborder Ebinineae of the order Ebenales (p. 448).

It appears that only the suborder Myrtineae (with the exception of Lecythidaceae and Theligonaceae) is the justified member of the order Myrtiflorae.

The ancestry of the order Myrtiflorae lies in the Rosales. The ancestral prototype could have been a woody plant with simple, opposite, minutely (or caducous) stipulate leaves, perigynous flowers with well-developed corolla, numerous centripetal stamens, a syncarpous gynoecium of many multiovulate carpels with a single style and capsule as a fruit. These characters are common to Sonneretiaceae. According to Cronquist (1968), such an ancestor could be traced only amongst the members of Rosaceae.

42

Order Umbellales

The order Umbellales comprises 7 families: Alangiaceae, Nyssaceae, monotypic Davidiaceae, Cornaceae, unigeneric Garryaceae, Araliaceae and Umbelliferae.

Members of this order are mostly trees and shrubs, herbs in Umbelliferae, and climbing vines in Araliaceae. Most of them are tropical and subtropical in distribution except the members of the Umbelliferae, which are cosmopolitan but prefer cold climate. Leaves are simple in Alangiaceae, Nyssaceae, Davidiaceae, Cornaceae and Garryaceae but pinnately compound or highly dissected in Araliaceae and Umbelliferae. The inflorescence mostly umbellate, capitate, spicate or racemose. Inflorescence of *Davidia* is different in the sense that the capitulum or head has numerous male flowers but only one perfect flower. Garryaceae and Nyssaceae have separate staminate and pistillate inflorescences. Flowers pentamerous, epigynous, often bracteate, complete or incomplete, uni- or bisexual. Calyx mostly adnate to the inferior ovary and represented by small teeth. Petals distinct, valvate or imbricate, sometimes absent. Stamens diplostemonous or obdiplostemonous or the number is same as the number of petals. Pollen grains 2- or 3-celled. Ovary inferior with varied number of carpels and locules, placentation axile; locules uniovulate, ovules pendulous, anatropous, usually unitegmic, rarely bitegmic.

Anatomically there are two groups:

1. Wood parenchyma diffuse, vessels with scalariform perforations, nodes trilacunar, secretory canals absent.

2. Wood parenchyma vasicentric, vessels with simple perforations, nodes multilacunar, schizogenous secretory canals present.

There are two distinct groups with respect to their chemical constituents. Nyssaceae-Cornaceae alliance contains iridoids and ellagitannins. Umbelliferae-Araliaceae group contains aromatic oils, unsaturated petroselinic acid in seed fats, and falcarinone alkaloids.

Alangiaceae

A small, unigeneric family of the genus *Alangium* with ca. 22 species in the tropics and subtropics of the eastern hemisphere. It extends from tropical Africa through Indo-Malaya, China and Japan to northeastern Australia.

Vegetative Features.   Trees or shrubs, sometimes spiny. Leaves alternate (Fig. 42.1A), entire or lobed, estipulate.

Floral Features.   Inflorescence axillary cymes. Flowers bisexual (Fig. 42.1B) with a jointed pedicel, bracteate, ebracteolate, pedicellate, epigynous. Calyx of 4 to 10 sepals, polysepalous, tooth-like or obsolete, adnate to the ovary. Corolla of 4 to 10 petals, polypetalous, valvate, linear or lorate, sometimes basally coherent; basal part of the corolla tubular but spreading above and densely hairy within (Fig. 42.1B). Stamens as many as or 2- to 4-times the number of petals. Anthers bicelled, dehiscence longitudinal, basifixed (Fig. 42.1C). Gynoecium syncarpous, bi- or tricarpellary, ovary inferior, uni- or rarely bi-locular; ovule solitary, pendulous (Fig. 42.1D); style 1, stigma 1- or 2- or 3-lobed. Fruit a drupe, usually with a crown formed by calyx lobes and the disc. Seed with a large embryo and copious endosperm.

Anatomy.   Vessels moderately small to medium-sized; perforation plates scalariform or simple; intervascular

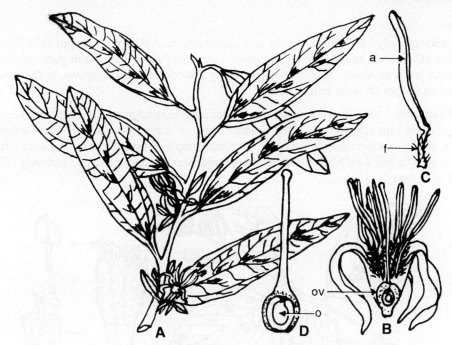

Fig. 42.1 Alangiaceae : **A-D** *Alangium salvifolium.* **A** Flowering twig. **B** Flower, vertical section. **C** Stamen. **D** Carpel, longisection. *a* anther, *f* filament, *o* ovule, *ov* ovary. (Adapted from Arachi 1968)

pitting alternate, parenchyma apotracheal, rays heterogeneous. Leaves dorsiventral, have hairs of different types. Stomata Ranunculaceous type, confined to the lower surface. Very large cluster crystals occur in the mesophyll.

Embryology. Pollen grains triporate, 3- or 4-colporate, shed at 2-celled stage; oblate spheroidal. Ovules anatropous, unitegmic, crassinucellate. Polygonum type of embryo sac, 8-nucleate at maturity. Endosperm formation of the Nuclear type (Johri et al. 1992).

Chromosome Number. Haploid chromosome number is n = 8, 11.

Chemical Features. Iridoids present (Jensen et al. 1975, Dahlgren 1983b).

Important Genera and Economic Importance. *Alangium*, the only genus, is not known to have any economic importance.

Taxonomic Considerations. The family Alangiaceae is closely related to the Nyssaceae-Cornaceae alliance. The distinctive features are: (a) presence of articulated pedicels of flowers, (b) valvate aestivation of petals, (c) corolla more or less hairy or villous within, and (d) ovary mostly unilocular with a single pendulous ovule.

Bentham and Hooker (1965a) do not recognise the family Alangiaceae. Bessey (1915) and Hallier (1912) included Alangiaceae in the order Santalales. Engler and Diels (1936), Wangerin (1910a), and Lawrence (1951) placed it in the Myrtiflorae. Melchior (1964) brought it back into the Umbellales. Hutchinson (1926) included it in his Umbelliflorae but later (1959, 1969) transferred it to the Araliales along with the rest of the woody families. Cronquist (1968, 1981) treats it as a member of the Cornales, along with members like Nyssaceae, Davidiaceae and others. Dahlgren (1975a) and Tákhtajan (1980, 1987) also place the Alangiaceae in the Cornales.

From the data available for Alangiaceae, Cornales is the appropriate order for it.

Nyssaceae

A small family of only two genera, *Nyssa* and *Camptotheca*. *Nyssa* with 6 spp. is indigenous to the United States of America and to Asia. *N. aquatica*, commonly known as cotton gum, grows in swamps and sometimes even in water. *N. sylvatica*, popularly called black gum, grows in the low woods, in swamps and on shores of lakes and marshes.

Vegetative Features.   Deciduous trees, shrubby under adverse conditions, branching markedly excurrent; bark grey, divided into segments by deep fissures. Base of trunk enlarged and roots sometimes form arches when grows in water. Leaves petiolate, alternate, simple, estipulate and deciduous; often crowded near the ends of branches. Leaf blade membranous or coriaceous, elliptic or ovate to narrowly (*Camptotheca*; Fig 42.2A) or broadly obovate.

Fig. 42.2 Nyssaceae : **A-F** *Camptotheca acuminata*. **A** Twig with inflorescences. **B** Flower (bisexual). **C** Stamen. **D** Carpel, longisection. **E, F** Bunch of fruits. *a* anther, *c* corolla, *o* ovule. (Adapted from Hutchinson 1969)

Floral Features.   Inflorescence umbellate, capitate (Fig. 42.2A), spicate or racemose. Flowers actinomorphic, staminate or bisexual, borne on different plants. In *Nyssa sylvatica* staminate flowers are in racemes and the bisexual ones in clusters of 2, 3 or 4 on slender penduncles. Male Flowers: Sepals 5, polysepalous, valvate; petals 5, fused below but free above, imbricate; stamens 5-10, arranged in two whorls of 5 each, anthers bicelled, basifixed, exerted (Fig. 42.2B, D), introrse, longitudinal dehiscence; central fleshy disc often present. Bisexual Flowers: have the same number of sepals and petals but stamens only 5 (10 in *Camptotheca*; Fig. 42.2 B). Gynoecium monocarpellary or bicarpellary and syncarpous; ovary inferior, unilocular, uniovulate, ovule pendulous (Fig. 42.2D). Style 1, often reflexed or coiled, sometimes cleft, stigma typically 1. Fruit a 1-seeded, ovoid or ellipsoid drupe; blue-black, purple or red, with corky epidermal spots, crowned by persistent disc (Fig. 42.2E, F). Seeds endospermous, with a straight embryo.

Anatomy.   Vessels small, usually solitary, perforation plates scalariform and usually with many fine bars. Parenchyma paratracheal; rays heterogeneous, fibres with bordered pits. Leaves dorsiventral, epidermal hairs of various types (sometimes mucilaginous). Stomata Rubiaceous type and confined to the lower surface.

Embryology.   Mature pollen grains are spheroidal, 3-colporate and shed at 2-celled stage. Ovules anatropous, unitegmic, pseudo-crassinucellate. Polygonum type of embryo sac, 8-nucleate at maturity. Endosperm formation of the Nuclear type (Johri et al. 1992).

Chromosome Number.   The haploid chromosome number is n = 22.

Chemical Features.   Linolenic-rich fatty acids are reported in both *Nyssa* and *Camptotheca*.

Important Genera and Economic Importance.   *Nyssa* and *Camptotheca* are the only two genera belonging to this family. Wood of *Nyssa* is a useful timber. Although it has a tendency to warp because of interlocking grain which makes the wood unsuitable for many purposes, it is used extensively for boxes and crates, as well as for furniture, flooring and paper pulp.

Taxonomic Considerations.   Bentham and Hooker (1865a), and Hallier (1912) placed *Nyssa* and its related genera in the family Cornaceae, on the basis of morphological features: inferior ovary, solitary seed in each locule and unitegmic ovule. Baillon (1873) and Wangerin (1910b) pointed out that *Nyssa* differs from other members of Cornaceae in diplostemony and in the mode of pollen development. Du Jussieu (1825; quoted in Takhtajan 1959) created a new family, Nyssaceae, on the basis of certain exomorphic features. Hutchinson (1973) supports this removal on the basis of the presence of imbricate petals in *Nyssa*. Eyde (1967) also treats Nyssaceae as a distinct family because of their larger number of petals and stamens, and their unique triangular germination valves.

Although there are some resemblances between *Nyssa* and the family Cornaceae, there are also several differences. Therefore, its removal from the Cornaceae and erection of a separate family Nyssaceae is justified.

Wangerin (1910b) considered *Nyssa* to have bitegmic ovules and therefore to be close to the family Combretaceae, other resemblances are: (a) tricolporate pollen grains, (b) albuminous seeds, and (c) Polygonum type of embryo sac. Bessey (1915) also held the same opinion, and placed Nyssaceae in the Myrtiflorae. Hutchinson (1969, 1973) retains it close to the Cornaceae in the order Araliales. Takhtajan (1959, 1969) and Melchior (1964) included it in the Umbelliflorae along with Cornaceae. Takhtajan (1987), however, includes Nyssaceae in the Cornales. On the basis of fatty acid analysis (Hohn and Meinschein 1974) and serological studies (Hillebrand and Fairbrothers 1970), the two families, Nyssaceae and Cornaceae, are closely related. Cronquist (1981) and Thorne (1983) also recognise Cornaceae and Nyssaceae to be closely related and group them together under the Cornales.

There is some controversy over the number of genera included in this family. Most taxonomists have included the genus *Davidia* in Nyssaceae. Li (1954) and Takhtajan (1959), on the basis of inflorescence, erected a separate family Davidiaceae. This view has later been supported by biochemical (Bate-Smith et al. 1975) and serological (Fairbrothers and Johnson 1964) studies. According to Eyde (1967), "*Davidia* is the nearest living thing to the common ancestor of *Cornus* and *Nyssa*".

The above discussion makes it sufficiently clear that the position of Nyssaceae is closer to the Cornaceae and its inclusion in the Umbellales s.s. is erroneous.

Davidiaceae

This is a newly formed unigeneric family comprising the single monotypic genus, *Davidia*. It occurs in northwestern Yunnan, Sikang, Szechuan, Kweichow and western Hupeh in China.

Vegetative Features.   Deciduous trees, grow in moist woods at medium to high altitudes; reach a height of ca. 60 ft and the girth of stem ca. 3 ft. Leaves alternate, simple, with dentate margin, estipulate.

Floral Features.   Inflorescence a subglobose head, densely beset with numerous, small staminate flowers and one bisexual flower situated laterally, above the middle, towards the tip; subtended by 2 large, white bracts. Flowers are andromonoecious and apetalous.

Staminate flowers numerous, scarcely distinct, consists of 5 or 6 stamens only, with long filaments and small anthers. The solitary bisexual flower is obliquely terminal. Perianth absent. Staminodes small, few to many, inserted half-way up the ovary. Gynoecium 6- to 9-carpellary, syncarpous; ovary as many loculed as the number of carpels, inferior, with a solitary, axile, pendulous ovule in each locule. Style columnar, with 6 to 9 stigmatic lobes that are bent towards periphery. Fruit a drupe, with granular mesocarp and bony 3- to 5-locular endocarp. Seeds with fleshy endosperm and straight embryo.

Anatomy. Anatomically, *Davidia* is not much different from *Nyssa*. However, the scalariform perforation plates of the vessels have a large number of bars in *Davidia*. The pericycle is a continuous ring of sclerenchyma. Petiole in transection shows a large crescent-shaped bundle and numerous large solitary crystals in the spongy tissue.

Embryology. Embryology has not been fully investigated. Pollen grains are tricolporidate. Ovules anatropous, unitegmic, crassinucellate. Polygomum type of embryo sac, 8-nucleate at maturity. Endosperm formation of the Cellular type (Johri et al. 1992).

Chromosome Number. Diploid chromosome number is $2n = \pm 40$ for *D. involucrata* (Dermen 1932).

Chemical Features. Rich in fatty acids; iridoids present (Jensen et al. 1975).

Important Genera and Economic Importance. *Davidia involucrata*, popularly called the 'handkerchief' tree, is the only member of this family. It is an interesting taxon, restricted to some areas in China. Fossil evidences of xylem and gynoecium characters have been studied and Eyde (1967) infers (from this) that *Davidia* is the nearest living representative to the common ancestors of *Cornus* and *Nyssa*

Taxonomic Considerations. Bessey (1915) and Hutchinson (1973) included *Davidia* in the family Nyssaceae. Titman (1949), on the basis of wood anatomy, preferred to retain *Davidia* in the Davidiaceae. G. Erdtman (1952) also agrees to this view because of differences in pollen morphology. Li (1954) and Takhtajan (1959) raised *Davidia* to a family rank, Davidiaceae, on the basis of inflorescence character and placed it next to the Nyssaceae. Eyde (1963), on the basis of fruit structure, pointed out that *Nyssa* is more allied to *Camptotheca* than to *Davidia*. He (Eyde 1967) observed that cytological evidence also supports this view.

On the basis of similarities in pollen, wood, and leaf character, Marcgraf (1963) placed it close to *Actinidia* (of Actinidiaceae); but the embryological data, like 3- to 5-carpellary ovary, tenuinucellar ovule, well-developed hypostase, and anatomical data like raphide-containing idioblasts in the leaves and floral appendages, and cytological data do not support this view.

Davidia involucrata has been variously placed in (a) *Nyssa*, (b) as a separate genus in Nyssaceae, (c) in its own family Davidiaceae, and (d) in Actinidiaceae. The fatty acid data reveal a close relationship between *Nyssa* and *Davidia*, in addition to some other similarities (Hohn and Meinschein 1976). In our opinion, *Davidia* should be placed in a family of its own, i.e. Davidiaceae, and its position should be next to the Nyssaceae.

Cornaceae

A family of 10 genera and 90 species distributed widely in the tropical and temperate regions of both the Old and the New World. Isolated in the Southern hemisphere are *Melanophylla* with 9 species and the monotypic *Kaliphora* in Madagascar. The monotypic *Curtisia* is confined to the coastal forests of South Africa and southeast tropical Africa. The genus *Griselinia* is common to New Zealand, Chile and Brazil. *Aucuba*, *Helwingia* and *Toricellia* are confined to temperate Asia.

Vegetative Features. Mostly trees and shrubs, or subshrubs; the genus *Chamaepericlymenum* is a herb with annual stems from a perennial rhizome and *Griselinia* is a woody vine. Leaves opposite, or less commonly alternate or fasciculate as in *Corokia*, *Mastixia*, *Toricellia*, *Kaliphora*, *Melanophylla* and

Griselinia, sometimes persistent, e.g. *Aucuba*. Usually petiolate, estipulate (stipulate in *Helwingia*); leaves palmately veined in *Toricellia*.

Floral Features. Inflorescence cymose heads or panicles, sometimes subtended by large, showy foliaceous bracts e.g. *Chamaepericlymenum, Cynoxylon, Dendrobenthamia, Cornus florida* (Fig. 42.4A) and *C. canadensis.* Flowers small, bisexual (Fig. 42.4 B) or unisexual; when unisexual, the plants are polygamodioecious[3] (or dioecious as in *Aucuba*), actinomorphic, epigynous. Sepals 4 or 5, polysepalous, adnate to the inferior ovary. Petals 4 or 5, polypetalous, rarely absent, distinct, valvate. Stamens same in number as the petals and alternate with them; filaments short, thick, with a pulvinate disc at the base (Fig. 42.4C); anthers bicelled, basi- or dorsifixed, dehiscence longitudinal. Gynoecium syncarpous, 2- to 4-carpellary, ovary inferior, 2- to 4-loculed, placentation axile (parietal in *Aucuba*); ovule solitary in each locule, pendulous (Fig. 42.4C, D). Style 1 or more arising from the epigynous, glandular disc, stigmas as many as styles and subcapitate. Fruit is typically a drupe with a bony stone, sometimes berry as in *Aucuba* and *Griselinia*, and rarely all the ovaries become united to form a fleshy syncarp as in *Dendrobenthamia*.

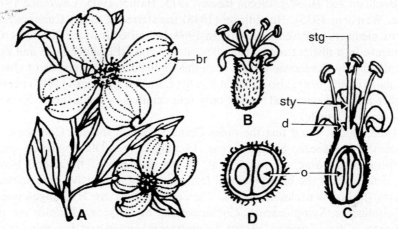

Fig. 42.4 Cornaceae : **A-D** *Cornus florida.* **A** Twig with inflorescences. **B** Flower. **C** Vertical section. **D** Ovary, cross section. *br* bract, *d* disc, *o* ovule, *stg* stigma, *sty* style. (**B,C** after Lawrence 1951)

Anatomy. Vessels small, exclusively solitary, numerous, sometimes with spiral thickening; perforation plates typically scalariform with many bars, sometimes simple; parenchyma sparse, typically diffuse. Rays heterogeneous. Leaves dorsiventral, simple unicellular hairs occur generally but 2-armed and glandular types are present in some members. The Ranunculaceous type of stomata are confined to the lower surface of leaves.

Embryology. Pollen grains spheroidal, 3-colporate and 2-celled at shedding stage. Ovules anatropous, unitegmic, crassinucellate. Polygonum type of embryo sac, 8-nucleate at maturity. Endosperm formation of the Cellular type. Nucellar apospory is reported.

Chromosome Number. Haploid chromosome numbers are n = 9, 10, 11, 12; diploid number 2n = 32, 36, 44, 72, 120 and 144 (Kapil and Mohana Rao 1966b).

---

[3]Polygamodioecious—functionally dioecious but with a few flowers of opposite sex or a few bisexual flowers on all plants at flowering time.

Chemical Features. Rich in iridoids; also contain ellagitannins and aucubins.

Important Genera and Economic Importance. The Cornaceae members are important ornamental shrubs or trees. *Aucuba* or gold-dust plant is grown in warm temperate regions. *Cornus florida*, *C. nuttallii*, *Benthamidia capitata* and *Mastixia rostrata* are also ornamentals. *Swida alternifolia* yields timber. Fruits of *C. mas* or cornelian cherry are edible.

Taxonomic Considerations. The family Cornaceae has been placed in the Umbellales, on the basis of its few-carpellate, inferior ovary, reduced calyx and tendency towards forming a many-flowered umbellate inflorescence. Anatomy of Cornaceae differs from Umbelliferae and Araliaceae, but the secretory ducts and extrorse micropyle of *Mastixia* form a link between Cornaceae and Araliaceae (Ferguson 1966b, c). Biochemical studies show that the chemical aucubin is present in Cornaceae and not in Umbelliferae and Araliaceae. Anatomically and morphologically, this family resembles Caprifoliaceae and in wood anatomy it resembles Hydrangeaceae.

Cornaceae has been treated variously by different workers. All earlier botanists placed it in Umbellales/ Umbelliflorae (Bentham and Hooker 1965a, Bessey 1915, Hallier 1905, Lawrence 1915, Rendle 1925, Wangerin 1910c, Wettstein 1935). Hutchinson (1948) transferred it to his Cunoniales, on the basis of woody habit of its members and stem anatomy. In 1969, he shifted it to the Araliales. Cronquist (1968) placed the Cornaceae in a distinct order, Cornales, along with Rhizophoraceae and other families. He pointed out that the family Cornaceae differs from the Umbellales in a number of characters. Dahlgren (1975a) and Takhtajan (1980, 1987) also included it in the Cornales. Cronquist (1981) treats Rhizophoraceae in a separate order Rhizophorales and retains only four families—Alangiaceae, Nyssaceae, Cornaceae and Garryaceae—in the Cornales.

As early as 1963, Huber stated that the order Cornales should include Cornaceae, Philadelphaceae (= Hydrangeaceae), Styracaceae, Symplocaceae, Escalloniaceae, Diapensiaceae and Aquifoliaceae. This circumscription was based on wood, leaf, flower and ovule characters. However, Dahlgren (1975b, 1980a) mentions that the embryological features are more decisive, and subsequently enlarged the order Cornales (in his own classification). The core families are Hydrangeaceae, Escalloniaceae, Icacinaceae, Aquifoliaceae, Symplocaceae, Cornaceae and Stylidiaceae. Iridoids are present in one or more species of each of these families, except Aquifoliaceae. Grund and Jensen's (1981) investigations show close mutual serological affinities between Hydrangeaceae, Escalloniaceae and Cornaceae.

As Cornaceae shows more differences than resemblances to the Umbellales, it is advisable to treat it separately as a member of the Cornales. The Cornales should include other iridoid-containing families like Nyssaceae, Alangiaceae, Davidiaceae and Garryaceae. The genus *Mastixia* may be considered as an exception, forming a link between the two orders, Cornales and Umbellales. Or, it can also be raised to family rank, as has been done with the genera *Aucuba*, *Toricellia* and *Helwingia* (Ferguson 1977, Takhtajan 1980, 1987). It may be noted that the circumscription of the Cornales is much enlarged and it should now include many more families which are chemically related (Dahlgren 1983b).

Garryaceae

A unigeneric family of the genus *Garrya*, distributed in Central America, extending from California through southwestern United States and Mexico to the West Indies. There are ca. 15 species.

Vegetative Features. Dioecious, evergreen trees or shrubs. Leaves opposite, entire, simple, coriaceous, petioles of opposite leaves basally connate.

Floral Features. Inflorescence staminate and pistillate (Fig. 42.5A, E), pendulous or drooping catkins that are silky-hairy; catkins upright when young and susequently pendulous (Kapil and Mohana Rao 1966b).

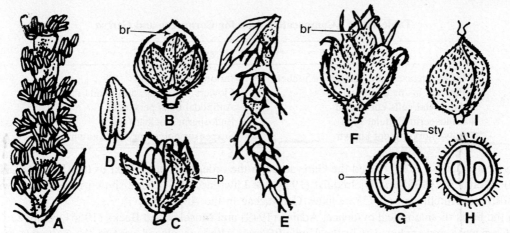

Fig. 42.5 Garryaceae : **A-I** *Garrya veatchii*. **A** Staminate inflorescence. **B** Bract embracing three male flowers. **C** Male flower. **D** Stamen. **E** Pistillate inflorescence. **F** Three pistillate flowers enclosed in a bract. **G** Carpel, vertical section. **H** Cross section. **I** Fruit. *br* bract, *o* ovule, *p* perianth, *sty* style. (Adapted from Kapil and Rao 1966b)

Staminate flowers in 2s or 3s or solitary in axils of bracts (Fig. 42.5B); the opposite bracts qualescent. Male flowers stalked, monochlamydeous; perianth of 4 sepals (Fig. 42.5C). Sepals often apically coherent, with stamens protruding between them. Stamens 4, distinct, alternate with the sepals; rudiment of pistil often present at the center of the flower. Anthers basifixed, longitudinally dehiscent (Fig. 42.5D). Pistillate flowers sessile in groups of 2 or 3, without perianth (Fig. 42.5F); gynoecium syncarpous, bi- or tri-carpellary, ovary unilocular, apparently superior, with parietal placentation and ovules pendant from near the apex of the locule (Fig. 42.5G, H). Styles 2, distinct, subulate and spreading, each with stigmatic papillae on the inside. Fruit a berry (Fig. 42.5I); seeds with copious endosperm and a minute embryo.

Anatomy.   Vessels very small, solitary, with spiral thickening and scalariform perforation plates; rays homogeneous. Leaf surface covered with a layer of unicellular hairs. Stomata confined to the hairy lower surface, protected by beak-like extentions of the guard or subsidiary cells.

Embryology.   Pollen grains spheroidal, 3-colporate, shed at 2-celled stage. Ovules anatropous, unitegmic, crassinucellate. Polygonum type of embryo sac, 8-nucleate at maturity. Endosperm formation of the Nuclear type. Suspensor polyembryony common (Mohana Rao 1963).

Chromosome Number.   Basic chromosome number is x = 11.

Chemical Features.   There is frequent occurrence of aucubin, an iridoid (Bate-Smith and Swain 1966).

Important Genera and Economic Importance.   Economically, this monotypic family is of little importance. A few species of *Garrya* are sometimes cultivated as ornamentals because of their glossy, evergreen foliage and reddish-purple drooping inflorescences.

Taxonomic Considerations.   Bentham and Hooker (1965a), Hallier (1912) and Bessey (1915) kept the genus *Garrya* in the family Cornaceae with which it shows the following resemblances: 1. spheroidal, 3-colporate pollen grains, 2. anatropous, unitegmic, crassinucellate ovules, 3. Polygonum type of embryo sac, 4. berry-like fruits, 5. occurrence of the chemical aucubin.
However, there are some differences also (see Table 42.5.1).

Table 42.5.1 Comparative Data for Cornaceae and *Garrya*

Cornaceae	*Garrya*
1 Inflorescence cyme, panicle or umbel	Inflorescence catkin
2 Flowers bi- or unisexual	Flowers unisexual, and plants dioecious
3 Antipodal cells ephemeral	Antipodal cells persistent
4 Endosperm Cellular type	Endosperm Nuclear type
5 Polyembryony not known	Suspensor polyembryony known

Engler and Diels (1936) placed the Garryaceae in the order Garryales, next to the Salicales. Hallock (1930), Wettstein (1935) and Hjelmquist (1948; see Lawrence 1951) have shown its affinities with the Umbelliferae. Hutchinson (1973) includes Garryaceae in the Araliales.

On the basis of anatomical evidence, Adams (1949) and Moseley and Beeks (1955), removed *Garrya* from Cornaceae and included it in the Umbelliflorae. Although placed next to the Salicales by many authors, on the basis of such features as: (a) pendulous, catkin-like inflorescences, (b) bicarpellary, 2-styled gynoecium, and (c) achlamydeous female flowers, there are hardly any resemblance between the two. The important differences between *Garrya* and Salicales are: (a) inferior ovary, (b) 2- or 3-nucleate tapetal cells, (c) 3-colporate pollen grains, (d) well-developed hypostase, (e) persistent antipodal cells, (f) suspensor polyembryony, and (g) scalariform perforation plates in xylem vessels.

According to Kapil and Mohana Rao (1966b), morphological and embryological data on *Garrya* confirms the family to be more closely allied to the Umbelliflorae (= Umbellales) than to the Amentiferae members. Hutchinson (1969, 1973) includes this family in the Araliales of his Lignosae. According to him, the family Garryaceae "is peculiar, and therefore not very easy to place in its true affinity". It, however, does not seem to be far removed from the Hamamelidaceae—the more advanced Amentiferous family. Cronquist (1968, 1981), Dahlgren (1975a, b 1983a), and Takhtajan (1980, 1987) place Garryaceae in the Cornales. According to Takhtajan (1980, 1987), it is related to *Aucuba* of the Aucubaceae (formerly Cornaceae) in similar pollen morphology (Eramian 1971). Both *Aucuba* and *Garrya* have decarboxylated iridoid aucubin and no tannins (Bate-Smith et al. 1975). Moreover, *Garrya* and *Aucuba* have been successfully grafted, showing their close relationship.

The affinities of Garryaceae with Cornaceae is supported by embryological, anatomical, palynological and phytochemical features. Garryaceae may be treated as an advanced family in the order Cornales.

Araliaceae

A large family with 65 genera and more than 800 species distributed primarily in the tropics. There are two centres of distribution, Indo-Malayan region and Tropical America. The genus *Tupidanthus*, the most primitive member of this family (Hutchinson 1969), grows only in the Khasi Hills of the Eastern Himalayan region. Other related genera are *Tetraplasandra* from the Pacific Islands and New Guinea, *Plerandra* from the Fiji Islands, and the monotypic *Indokingia* from the Seychelles. Three genera, *Aralia* (6 spp.), *Panax* (2 spp.), and the monotypic *Oplopanax* are indigenous to America. *Hedera helix* is of European origin but naturalised in many regions.

Vegetative Features.   Herbs, shrubs or trees, rarely lianas, e.g. *Hedera helix*. Stems solid, with a large pith, often prickly; herbaceous in *Panax* and *Stilbocarpa*, woody in others. Leaves alternate, rarely opposite as in *Arthrophyllum* or peltate as in *Harmsiopanax*; simple as in *Hedera*, or palmately compound as in *Tupidanthus*, or pinnately compound as in *Aralia*; surface covered with stellate hairs, stipulate, stipules modified as a membranous border of petiole base or liguliform.

Floral Features. Inflorescences are basically heads or umbels, sometimes borne on leaf blades as in *Helwingia chinensis* (Fig. 42.6.A, B). Some of the primitive genera show racemes or spikes with terminal umbels, for example *Reynoldsia sandwichensis* has a raceme with a terminal umbel; *Dipanax gymnocarpa* has a raceme of male flowers terminated by an umbel of bisexual flowers; *Cuphocarpus aculeatus* with abortive flowers in the raceme portion and fertile ones forming the umbel; *Cussonia spicata* with a spicate inflorescence. Much-branched panicles are seen in *Aralidium*. Flowers bi- or unisexual and then the plants are dioecious or polygamodioecious, actinomorphic, epigynous, minutely bracteate and often greenish. Calyx of 5 sepals, cupuliform, or inconspicuous, adnate to ovary and usually represented by 5 small teeth. Petals 5 to 10, rarely 3, e.g. *Helwingia* male flowers (Fig. 42.6C) or 4 as in female flowers of the same (Fig. 42.6D); broad at base, arising from the disc, valvate; caducous as calyptra-like cap in *Tupidanthus* or as distinct petals. Stamens usually as many as and alternate with the petals (Fig. 42.6C), rarely numerous as in *Plerandra*, *Tetraplasandra* and *Tupidanthus*, distinct, arising from disc; anthers bicelled, dorsifixed, longitudinally dehiscent; nectar-secreting disc, covering the apex of the ovary and usually confluent with style bases. Gynoecium syncarpous, 2- to 15-carpellary (nearly 160 carpels in *Tupidanthus*), ovary normally inferior, superior to half-inferior in *Dipanax*, 2- to 15-locular with axile placentation and 1 ovule in each locule (Fig. 42.6E), ovules pendulous. Styles as many as carpels, distinct and recurved or connate into a single cone or column, sometimes absent and then stigmas sessile. Fruit a berry (Fig. 42.6F) or rarely a drupe; broadly winged downwards in *Myodocarpus* and some species of *Astrotricha*. Seeds with small embryo and copious endosperm, sometimes ruminate as in *Hedera*.

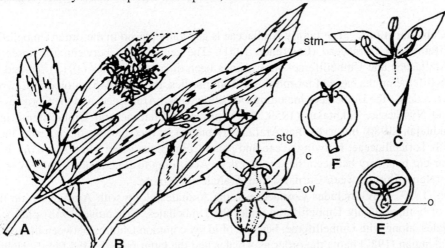

Fig. 42.6 Araliaceae : **A-F** *Helwingia chinensis*. **A, B** Twigs with staminate inflorescence and pistillate flowers borne on the leaf blades. **C** Staminate flower. **D** Pistillate flower. **E** Ovary, cross section. **F** Fruit. *c* corolla, *o* ovule, *ov* ovary, *stg* stigma, *stm* stamen. (Adapted from Hutchinson 1969)

Anatomy. Vessels very small to medium-sized, in short multiples and clusters, sometimes with a tangential or ulmiform arrangement; often ring-porous; sometimes with spiral thickenings, perforation plates usually simple; parenchyma paratracheal. Rays heterogeneous. Leaves usually dorsiventral with hairs of various types. Stomata Rubiaceous type, confined to the lower surface; secretory ducts and clustered crystals are reported.

Embryology. Spheroidal, 3-colporate pollen grains, 2-celled at shedding stage. Ovules anatropous, unitegmic, tenui- or crassinucellate. Polygonum type of embryo sac, 8-nucleate at maturity. Endosperm formation of the Nuclear type. Polyembryony not known.

Chromosome Number.   The haploid chromosome number is n = 11, 12.

Chemical Features.   Members of Araliaceae are rich in a polyacetelene called falcarinon (Sørensen 1963). Seed fats of *Aralia spinosa* are rich in petroselinic acid (Shorland 1963).

Important Genera and Economic Importance.   The important genera include *Aralia* with 35 species, *Hedera* with 15 species, *Fatsia* with only one species, *F. japonica, Panax* with 8 species, *Schefflera* with 150 species and monotypic *Tetrapanax. Hedera helix* commonly called English ivy grows wild in many parts of Europe. Many of its cultivars such as *H. canariensis* or Canary Island ivy, and various species of *Acanthopanax, Aralia, Dizygotheca, Fatsia, Polyscias* and *Schefflera*, and the intergeneric hybrid of *Fatsia* and *Hedera,* × *Fatshedera* (Jones and Luchsinger 1987) are grown as ornamentals. The wood of *Acanthopanax ricinifolium* of Japan is used for making hats. The light pith of *Tetrapanax papyriferum* is the source of Chinese rice paper. *Tieghemopanax murrayi* produces valuable timber.

Panax ginseng is considered a valuable drug in Korea, China and Japan. It has been shown that the root extracts inhibit growth of murine leukemia and sarcoma cells, and also inhibit DNA, RNA and protein synthesis in murine ascitic sarcoma (Hansen and Boll 1986). The active substance is panaquilon, a glycoside. It is also available from dried roots of *P. quinquefolia* and serves as a stomachic and stimulant. In Thailand *Schefflera* spp. extracts are used in treating asthma.

Several members of Araliaceae cause allergic contact dermititis and skin irritation, e.g. *Hedera helix* and *S. arboricola.* Falcarinol—the polyacetelene—is the major allergen from S*chefflera* (Hansen and Boll 1986).

Taxonomic Considerations.   The family Araliaceae is generally placed in the order Umbelliflorae along with Umbelliferae and Cornaceae (Lawrence 1951). The relationship between the three families—Cornaceae, Araliaceae and Umbelliferae—has always been disputed. Hallier (1912) merged Araliaceae with the Umbelliferae on the basis of morphological similarities. Ricket (1945; quoted in Lawrence 1951) recognised two orders—the Umbellales including Umbelliferae and Araliaceae, and the Cornales comprising Cornaceae and Nyssaceae. Takhtajan (1959, 1969) groups Araliaceae and Umbelliferae in the order Araliales. Takhtajan (1980) observes that Araliales (Apiales) is very closely linked with the Cornales, especially with Toricelliaceae, Helwingiaceae and some Cornaceous genera like *Mastixia*. It is difficult to draw a clear-cut boundary between these two orders. According to Bate-Smith et al. (1975), however, the chemical relationship between Cornales and Araliales is remote.

Hutchinson (1959, 1969) includes Cornaceae in the Araliales along with Araliaceae, on the basis of woody habit of both, and only Umbelliferae under the Umbellales. Cronquist (1968) places Araliaceae in the Umbellales, along with Umbelliferae, because of many common features between the two. Cronquist (1981) and Takhtajan (1987) name the order as Apiales and the latter includes the family Helwingiaceae. Thorne (1973, 1983) supports the merging of the Umbelliferae in the Araliaceae, on the basis of palynology, chemotaxonomy and fruit morphology. Roth (1977), after a detailed study of fruits, considers Umbelliferae as an over-developed tribe of Araliaceae. Dahlgren (1977a, b) treats Araliales as a distinct line of differentiation amongst the Sympetalae, along with Asterales, Campanulales, Pittosporales and Santalales. This group of families is primarily rich in polyacetelene and sesquiterpene lactone. According to Dahlgren (1983b), evidence from morphology, embryology and chemistry indicates that the Aralliales are very closely related to the Asteraceae and Campanulaceae.

Araliaceae and Umbelliferae have several common features: (a) sheathing leaf base, (b) pentamerous flowers, (c) inconspicuous calyx lobes, (d) solitary ovule in each locule, (e) epigynous disc, (f) subprolate, 3-colporate, 2-celled pollen grains, (g) abortive ovules, (h) Nuclear type of endosperm, (i) Rubiaceous type of stomata, (j) similar basic chromosome number and (k) similar chemical substance. The differences are: (a) fruit a berry or drupe in Araliaceae and schizocarpic in Umbelliferae, (b) ovary multicarpellary,

2- to 15-loculed in Araliaceae and bicarpellary, bilocular in Umbelliferae, (c) Onagrad or Chenopodiad type of embryogeny in Araliaceae, and Solanad type in Umbelliferae. Therefore members of the Araliaceae cannot be merged with the Umbelliferae; but the similarities do support the placement of the two families in the same order, Umbellales.

Umbelliferae

A very large family with 200 genera and 2900 species distributed widely all over the world, chiefly in the temperate zones but also reported from the tropics and subtropics. One species of *Azorella* is grows in the Antarctica.

Vegetative Features.   Mostly normal mesophytic plants except a few taxa that grow on wet soil, e.g. *Hydrocotyle* or water pennywort; *Actinotus bellidioides* grows in peaty soil at high altitudes in Tasmania. *Eryngium campestre* is a xerophyte.

Mostly biennial or perennial, aromatic herbs. Stem herbaceous, rarely woody, with prominent, swollen, solid nodes and hollow internodes, surface ridged, hairs when present, non-glandular, sometimes dendroid, e.g. *Xanthosia pilosa* and stellate, e.g. *Bowlesia tropaeolifolia*. Pith large, shrinks or dries at maturity.

Leaves alternate, rarely opposite as in *Apiastrum,* often basal, sometimes heteromorphic, e.g. *Coriandrum sativum* (Fig. 42.7A); usually pinnately or palmately compound or decompound, e.g. *Daucus carota*; sometimes simple and reniform as in *Centella asiatica* (Fig. 42.7H), petiole sheathing, estipulate.

Floral Features.   Inflorescence simple or compound umbel (Fig. 42.7A), in some genera, the pedicels are lost by reduction and the inflorescence is capitate, e.g. *Eryngium.* When compound, each primary ray or peduncle is terminated by an umbelet, whose pedicels are termed secondary rays. An umbel is often subtended by an involucre of distinct, simple or compound bracts. Similarly, an umbelet is subtended by an involucel of bractlets. The bracts may be caducous or persistent in fruit. Flowers generally bisexual but unisexual in *Echinophora;* in *Petagnia saniculifolia* (Fig. 42.7J, K), a central female flower surrounded by male umbels (Dutta 1988), plants dioecious in *Arctopus*, polygamous flowers in *Torilis anthriscus*. Polygamomonoecism exists when a few longer-pedicelled staminate flowers occur with a majority of bisexual ones as in *Astrantia*. Flowers actinomorphic/zygomorphic, pedicellate, bracteate, ebracteolate, epigynous (Fig. 42.7B, I).

Sepals 5, adnate to the ovary, persistent and mostly small, teeth-like, valvate (Fig. 42.7B, I); in *Pentapeltis peltigera* calyx lobes disc-like. Corolla of 5 distinct petals, polypetalous, inflexed in bud, often notched at the apex, more or less bifid in *Coriandrum sativum* (Fig. 42.7B, C) and *Foeniculm vulgare*, imbricate. Stamens 5, inflexed in bud but spreading later, alternipetalous, arising from an epigynous disc. Anthers bicelled, basi- or dorsifixed, longitudinal dehiscence (Fig. 42.7B, D). Gynoecium syncarpous, bicarpellary, ovary inferior, bilocular, placentation axile, ovules pendulous, 1 in each locule (Fig. 42.7E, F). Sytles 2, with a swollen base called stylopodium (Fig. 42.7F, I), stigmatic tip scarcely differentiated. Fruit a cremocarp composed of 2 mericarps (Fig. 42.7G), coherent and dehisce by their faces (commissure), flattened dorsally, i.e. parallel to the commissural face or laterally, i.e. at right angles to the commissural face. Each mericarp has 5 ribs distinguished as lateral, dorsal and intermediate; the ribs thin or corky and filiform or winged (*Thapsia villosa*) or spinescent as in *Daucus carota*. Oil ducts or vittae are present in the spaces between two ribs or under the ribs (Fig. 42.7F). Each mericarp is 1-seeded and usually suspended after dehiscence by a slender wiry stalk or carpophore. Seed with a minute embryo, endosperm abundant.

Anatomy. Vessels very small to medium-sized, often in clusters, perforations usually simple, rarely scalariform as in *Heracleum sphondylium.* Parenchyma paratracheal, rays heterogeneous to almost

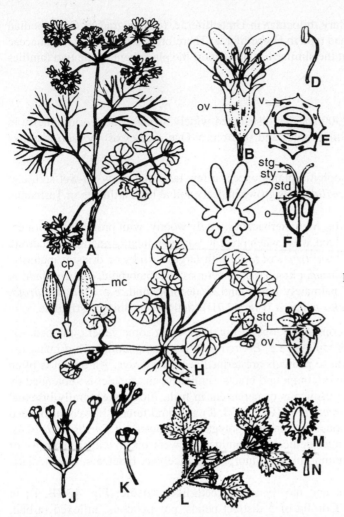

Fig. 42.7 Umbelliferae : **A-G** *Coriandrum sativum*, **H, I** *Centella asiatica*, **J, K** *Petagnia saniculifolia*, **L-N**, *Drusa oppositifolia*. **A** Plant with inflorescence. **B** Flower. **C** Petals, dissected. **D** Stamen. **E** Ovary cross section. **F** carpel, vertical section. **G** Cremocarp. **H** Plant of *C. asiatica*. **I** Flower. **J, K** Inflorescence and staminate flower of *P. saniculifolia*. **L** Flowering twig of *D. oppositifolia* **M** Fruit. **N** Hair, from the surface. *cp* carpophore, *mc* mericarp, *o* ovule, *ov* ovary, *std* stylopodium, *stg* stigma, *sty* style, *v* vittae. (**A-G** Original, **H-N** adapted from Hutchinson 1969)

homogeneous. Trichomes of various types common. Leaves dorsiventral, more stomata on lower surface, of Rubiaceous or Ranunculaceous type. Crystals absent, secretory ducts present.

Embryology.   Bilateral, 3-colporate and 3-celled pollen grains. Ovules pendulous, anatropous, unitegmic, tenuinucellate. Polygonum type of embryo sac most common; Allium, Drusa and Penaea types also recorded (Johri et al. 1992); 8- or 16-nucleate at maturity. Abortive ovules reported. Endosperm formation of the Nuclear type.

Chromosome Number.   The basic chromosome numbers are x = 3–12, most commonly 8 in Hydrocotyloideae and Saniculoideae, and 11 in Apioideae.

Chemical Features.   Umbelliferae seeds are rich in the unsaturated fatty acid, petroselinic acid (Hegnauer 1973). Monoterpenoids—carvone, limonene and fenchone— also occur. Polyacetelene falcarinose is common in most Umbelliferae members (Hegnaer 1971). The seeds contain an oligosaccharide, umbelliferose, in addition to various volatile oils. The bitter flavour of *Foeniculum vulgare* is due to fenchone and the sweet flavour is due to anethole. The pleasant delicate aroma of the fruits of *Coriandrum sativum* is due to coriandrol. The characteristic aroma and sharp taste of *Carum carvi* is due to the presence of caraway oil containing carvone, a ketonic substance (50–60%) and d-limonene. The main constituent of the volatile oil of cumin from *Cuminum cyminum* is cuminaldehyde (Kochhar 1981). Some of the aroma constituents of root, leaf or fruit are of phenylpropanoid type. Two specific constituents are parsleyapiole from *Petroselinium crispum* and dillapiole from *Anethum graveolens*. Another substance is myristicin, typical for Umbelliferae. This chemical is reported to have hallucinogenic effect in man (Shulgin 1966). It is reported in 14 genera of the subfamily Apioideae. Breeding and cultivation affects its concentration, and for taxonomic purposes the surveys should be limited to wild species only. Myristicin acts as a very interesting taxonomic marker in the tribe Caucalideae. It occurs in closely related genera *Daucus* and *Pseudorlaya*. In the latter, it occurs in both species, but in *Daucus* it occurs in only two polyploid members, *D. glochidiatus* and *D. montanus*. Another noteworthy aromatic volatile oil of this family is coumarin.

Important Genera and Economic Importance.   The family Umbelliferae is very important economically as they produce food, condiments, spices, and ornamentals. *Daucus, Pastinaca, Apium* and *Petroselinium* are cultivated for their food value. Fruits of *Coriandrum, Foeniculum, Carum, Anethum, Cuminum, Pimpinella* and *Anthriscus* are used as spices and condiments. A number of plants are grown as ornamentals such as *Ammi majus, Trachymene* spp., *Aegopodium* sp., *Angelica, Eryngium* and *Heracleum*.

Some taxa are poisoneus, e.g. *Conium maculatum*[4] (poison hemlock), *Cicuta* (water hemlock) and *Aethusa* (fool's parsley). The aromatic oil from *Foeniculum* and *Coriandrum* finds use in medicine as a carminative. In India the seeds of *Carum carvi* and *Cuminum cyminum* have long been used as stomachic and carminative. The residual mass after extraction of oil is a valuable cattle feed.

Chinese folk medicine have been making use of a crude extract of *Angelica acutiloba* var. *acutiloba-kitagawa*, due to its analgesic, sedative and antibacterial effects. Chemical examination proved the anaesthetic nature of the components.

Asafetida, *Ferula assafoetida*, is another member in which the flavouring constituents are present in the gummy, resinous exudate that collects on the surface of chopped roots or rhizomes. In western countries, liquors, particularly gin, are often flavoured with *Coriandrum* seeds. Apart from numerous economically important genera, this family also includes a few morphologically interesting plants. *Actinotus bellidioides* is a diminutive taxon which grows on wet peaty soil at high altitudes in Tasmania. *A. helianthi* of South Australia is so named because of its inflorescence resembling a sunflower (*Helianthus*). *Drusa oppositifolia*, an endemic of the Canary Islands, has opposite leaves, plant body covered with dense hairs and fruits with peculiar anchor-like processes on the margin (Fig. 42.7L-N). Another remarkable plant is *Petagnia saniculifolia* from Sicily; the flowers are unisexual; the female flower has attached to it 2 or 3 male flowers; their pedicels are partially adnate to a rib of the calyx (Fig. 42.7J, K). *Platysace deflexa* from Western Australia bears edible underground tubers. The leaves are terete and jointed in *Ottoa oenanthoides*, which grows from Central America to Ecuador.

Taxonomic Considerations.   There are no two opinions about the placement of the family Umbelliferae. The Araliaceae and Umbelliferae have always been retained in the same order i.e. Umbelliflorae of Bentham and Hooker (1965a); Umbellales of Engler and Diels (1936) and Melchior (1964); and Araliales of Hutchinson (1959), Dahlgren (1975a) and Takhtajan (1980). Hutchinson (1969) treats this family as the only member of his Umbellales. However, Thorne (1973, 1983) and Roth (1977) include Umbelliferae (= Apiaceae) in Araliaceae. Cronquist (1981) and Takhtajan (1987) use the alternate terminology, Apiales, for the order.

---

[4]The hemlock poison was served to Socrates who was condemned to death (for political reasons).

Because of the resemblances between the two families, Araliaceae and Umbelliferae, these should be retained in the same order, be it Araliales or Umbellales.

Taxonomic Considerations of the Order Umbellales

The order Umbellales (Umbelliflorae) was treated by Engler and his associates as composed of 3 families—Araliaceae, Umbelliferae and Cornaceae. This circumscription of the order was also retained by Rendle (1925). Wettstein (1935) added Garryaceae, whereas Bessey (1915) included the two genera *Nyssa* and *Garrya* in the Cornaceae.

Melchior (1964), in a revised version of Engler's work, increased the members of this order and included Alangiaceae, Nyssaceae, Garryaceae, Davidiaceae, Cornaceae, Araliaceae and Umbelliferae. The treatment of these six families by various workers has been very different. Bentham and Hooker (1965a) recognised only Araliaceae, Umbelliferae and Cornaceae, and placed them together in the Umbelliflorae. Engler and Prantl (as revised by Diels 1936) included Nyssaceae and Alangiaceae in the Myrtiflorae, Garryaceae in the Garryales, and Araliaceae, Umbelliferae and Cornaceae in the Umbelliflorae. *Davidia* was not raised to family rank. Hallier (1912) placed Nyssaceae and Cornaceae in the Cornales, Alangiaceae in Santalales and the other three families in the Umbelliflorae. Araliaceae was retained in the Umbelliferae. Bessey (1915) recognised only 3 families as belonging to the Umbellales; Garryaceae, Davidiaceae, Nyssaceae, and Alangiaceae were not mentioned in his system of classification. Hutchinson (1926) placed all the families (except Davidiaceae) in the order Araliales, but in 1948, restricted his Umbellales to include only Umbelliferae; Cornaceae was placed in the Cunoniales and Araliaceae in the Araliales. In his further revisions (1969, 1973), he retained only Umbelliferae in the Umbellales, and Cornaceae, Alangiaceae, Garryaceae, Nyssaceae and Araliaceae in the Araliales. According to Hutchinson, the segregation of *Davidia* in a separate family, was not desirable.

Cronquist (1968), although placed all these families in the subclass Rosidae, segregated them in two distinct orders. The Cornales included Davidiaceae, Nyssaceae, Alangiaceae, Cornaceae and Garryaceae, and the Umbellales included Araliaceae and Umbelliferae. According to him, the two orders show so many differences (Table 42.8.1) that it is unreasonable to retain them together.

Table 42.8.1 Comparative Data for Cornales and Umbellales

Umbellales	Cornales
1. Leaves always compound or lobed or dissected	Leaves always simple and entire or toothed
2. Sheathing leaf base or stiplate	Estipulate, base not sheathing
3. Nodes multilacunar	Nodes trilacunar
4. Pollen grains shed at 3-celled stage	Pollen grains shed at 2-celled stage
5. Schizogenous sercetory canals well-developed	Secretory canals usually absent
6. Wood parenchyma paratracheal, vasicentric	Wood parenchyma, when present, apotracheal
7. Vessels with simple terminal pore	Vessels mostly with scalariform perforations
8. Haploid chromosome numbers are n = 6–12	Haploid chromosome numbers are n = 8, 9, 10, 11
9. Contain aromatic volatile oils, polyacetelenes and unusual fatty acids like petroselinic acid	Contains iridoids, ellagitannins and aucubins

Dahlgren (1975a) includes Araliaceae and Umbelliferae in the Araliales of superorder Aralianae, and Nyssaceae, Davidiaceae, Garryaceae, Alangiaceae and Cornaceae in the order Cornales of superoder Cornanae. Takhtajan (1969) places all the families in the order Cornales but later, in his 1980 revision he removes Araliaceae and Umbelliferae to a separate order, Araliales, but retains them all in the same superorder, Cornanae. He also segregates a number of smaller families from the Cornaceae:

1. **Aucubaceae.** The genus *Aucuba* differs from all other Cornaceous genera in conspicuously oblique stigma, very small embryo at the apex of the endosperm, distinct intectate pollen grains and basic chromosome number x = 8. Therefore, a distinct family was suggested for it.

2. **Melanophyllaceae.** Includes *Melanophylla* and *Kaliphora* (both usually placed in the Cornaceae), which differ from other members of the Cornaceae in free and subulate styles, glandular hairs with spherical heads, and a very distinct pollen morphology.

3. **Griseliniaceae.** The genus *Griselinia* is very isolated within the order Cornales—both morphologically, cytologically and serologically. Rodriguez (1971) pointed out that *Griselinia* is a genus which shows the features of three different families—Cornaceae, Araliaceae and Escalloniaceae; and its basic chromosome number is x = 9.

4. **Toricelliaceae.** This family (only genus *Toricellia*) is also related to Cornaceae but differs in thick branches and thick *Sambucus*-like pith, palmatilobed leaves, multicellular hairs, simple perforation plates, lax pendulous thyrses, and the funicle is thick and forms an obturator. It is intermediate between Cornaceae and Araliaceae.

5. **Helwingiaceae.** The genus *Helwingia* has a controversial position. Bentham and Hooker (1965a) included it in the Araliaceae. Hutchinson (1959, 1969) also retained it in the Araliaceae. Eyde (1966, 1967) and Rodriguez (1971) suggest returning *Helwingia* to Araliaceae. *Helwingia* differs from other members of this family in the absence of secretory canals. It also differs from Cornaceae in the absence of iridoids, presence of stipulate leaves and pollen morphology.

Takhtajan (1980), therefore, treats it as a monotypic family Helwingiaceae. Its basic chromosome number is x = 19 (Raven 1975).

Takhtajan (1987) demarcates two other families, Curtisiaceae and Mastixiaceae. Subfamily Curtisioideae of the Cornaceae, including the monotypic genus *Curtisia*, was raised to the rank of a family. The Mastixiaceae includes only one genus, *Mastixia*, with 13 species. The family Toricelliaceae has been raised to the rank of a separate order, Toricelliales, and placed between Cornales and Apiales (Araliales).

From the above discussion, it is apparent that the order Umbellales comprises two distinct groups of families which are linked with each other by some intermediate families (or genera). One group includes Alangiaceae, Nyssaceae, Davidiaceae, Garryaceae and Cornaceae, which are rich in iridoids. The other group, comprising Araliaceae and Umbelliferae, contains volatile oils and resins in schizogenous cavities and a high quantity of petroselinic acid in seed fats. The two groups of families differ in some morphological, anatomical, embryological and cytological features also. Takhtajan (1987) treats these two groups as two separate orders—Cornales and Apiales—with the intermediate order, Toricelliales. Whether the seven families should be retained in one order or not is debatable. Serological studies of these families may be helpful in solving the problem. Until that time, these seven families may be retained in only one order, Umbellales, under the two suborders, Cornanae and Apianae, or they might be placed in two separate orders, Cornales and Apiales (or Umbellales), as suggested by Takhtajan (1987).

Dicotyledons

Subclass Sympetalae

43

Order Diapensiales

A monotypic order with the only family Diapensiaceae.

Diapensiaceae
A family of 6 genera and 18 species distributed in the temperate and arctic regions of the northern hemisphere.

Vegetative Features. Low-growing, evergreen, herbs or undershrubs, mostly grow in acidic soils; at high altitudes, cushion-like shrublets or suffruticose habit. Leaves alternate, spathulate or linear, simple, estipulate, sometimes reniform, e.g. *Galax aphylla.*

Floral Features. Inflorescence spicate, e.g. *Galax,* or flowers solitary, axillary, e.g. *Diapensia.* Flowers ebracteate, ebracteolate, hermaphrodite, actinomorphic, hypogynous, pedicellate, pentamerous. Calyx of 5 sepals, polysepalous, imbricate, persistent. Corolla of 5 petals, poly- or gamopetalous, campanulate or funnel-shaped, imbricate, interstaminal disc absent. Stamens 5, epipetalous, alternipetalous or hypogynous, free, or basally connate in a ring, often in 2 whorls, alternate with staminodes. Anthers bicelled, longitudinally or transversely dehiscent as in *Pyxidanthera;* staminodes absent in *Diapensia* and *Pyxidanthera. Galax* is an unusual genus with unilocular anthers transversely bivalved (Hutchinson 1969). Gynoecium syncarpous, tricarpellary, ovary trilocular with axile placentation, superior; ovules few to many per locule; style 1, short and thick, stigma 3-lobed. Fruit a loculicidal, 3-valved capsule; seeds with abundant endosperm and a straight or slightly curved embryo.

Anatomy. Vessels only a few, somewhat angular, up to about 20 μm in diameter, perforations simple in *Galax aphylla,* scalariform in *Diapensia lapponica;* rays ill-defined; ground tissue of xylem consists of fibres with numerous bordered pits. Leaves usually dorsiventral; stomata Ranunculaceous type, confined to the lower surface or on both the surfaces as in some species of *Diapensia, Galax, Pyxidanthera* and *Shortia;* hairs more or less absent. Crystals mostly clustered; tannin and oily substances present.

Embryology. Pollen grains single, 2-celled at shedding stage. Ovules anatropous or amphitropous, unitegmic, tenuinucellate. Embryo sac of Polygonum type, 8-nucleate at maturity. Endosperm formation of the Cellular type, haustoria absent.

Chromosome Number. The basic chromosome number in the family is x = 6. This low chromosome number is considered to be a conservative character (Scott and Day 1983).

Chemical Features. Diapensiaceae members have 8-oxygenated flavonols (Gornall et al. 1979).

Important Genera and Economic Importance. The members of this family are not of much economic importance but they are important in various other ways. These plants occur only in the colder belts of the northern hemisphere. *Diapensia lapponica* grows all around the Arctic circle and a few related species in the mountains of Japan, the White Mountains of New England, USA, and in the high mountains of Central Asia. According to Scott and Day (1983), there are two centres of distribution of this family— the Appalachian Mountain system of eastern North America, and eastern Asia. Poor seed dispersal is the reason for restricted distribution. Seeds often retained by ovary and sepals and they germinate in situ. In *D. lapponica* seed production is reduced due to fungal infection before and after capsule dehiscence.

The genus *Shortia* (Fig. 43.1.A) is represented by one species in the eastern part of the USA, another in Japan and a third one in southern Yunnan. Other important taxa are: *Galax aphylla, Shortia galacifolia, Pyxidanthera barbulata, Berneuxia* sp. and *Schizocodon.*

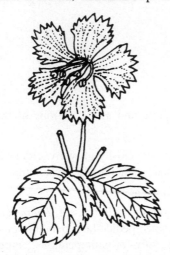

Fig. 43.1 Diapensiaceae: *Shortia uniflora.* Twig, flower shows polypetalous corolla and alternipetalous five stamens. (Adapted from Hutchinson 1969)

Taxonomic Considerations of Diapensiaceae and the Order Diapensiales.

The family Diapensiaceae was included within the Ericales until 1924 when it was separated as a distinct order by Engler and Gilg (1924). It was treated as an advanced family over the Ericaceae. Hallier (1912) included Diapensiaceae in his order Bicornes as its most advanced taxon. This view was accepted by Wettstein (1935). Diels (1914) considered this family to have been derived from Saxifragaceous ancestors in the Rosales.

Cronquist (1968, 1981) treats Diapensiaceae as the only member of the order Diapensiales. According to him, the presence of unitegmic, tenuinucellate ovules and glandular binucleate tapetum make this family distinct from the Ericales, in which it was placed earlier. These embryological features also suggest that the order Diapensiales is related to either the Asteridae or should be placed among the advanced orders of the Dilleniidae. Some of the floral features such as the presence of tricarpellary ovary with axile placentation, numerous ovules and frequent occurrence of staminodia strongly suggest their alliance with the advanced Dilleniidae. The Ericales (of Dilleniidae) could be the nearest allie. According to Takhtajan (1987), both Ericales and Diapensiales are advanced Dilleniidae. Two genera of Diapensiales have anthers with appendages that are similar to those in the Ericales. Wood anatomy and floral anatomy also suggest a common ancestor for these two orders. Dahlgren (1975a, 1980a, 1983a) and Takhtajan (1980), however, treat it as a member of the order Ericales. Hutchinson (1969, 1973) also considers Diapensiaceae as an advanced member of the Ericales, and presumed its polyphyletic origin from the families Ericaceae and Pyrolaceae.

The Diapensiales should be treated as a distinct order which orginated from a common ancestor along with the Ericales.

44

Order Ericales

The order Ericales includes 5 families: Clethraceae, Pyrolaceae, Ericaceae, Empetraceae and Epacridaceae.

The Ericales is represented by trees, shrubs, as well as herbs. Some of the herbs, like *Monotropa* and *Pyrola*, are dependent on mycorrhizal fungus (Cronquist 1944). The members of this order have simple, alternate, entire, estipulate, coriaceous leaves. Flowers mostly sympetalous, e.g. *Rhododendron*, and sometimes polypetalous, e.g. *Clethra*. Stamens typically twice as many as the petals and usually attached directly to the receptacle. Anthers mostly with prominent appendages and they open by terminal pores; pollen grains remain attached in tetrads. Gynoecium syncarpous, (2)-3–7(-10)-carpellary, ovary with locules as many as the number of carpels; mostly axile and rarely parietal placentation; sometimes unilocular in the upper portion. Ovary superior, half-inferior or inferior; ovules numerous, anatropous or campylotropous, unitegmic, tenuinucellar. Fruit a loculicidal or septicidal capsule, berry or drupe, calyx persistent in fruit. Seeds small, sometimes winged or tailed; embryo small, endosperm haustoria present.

Clethraceae
A small family of only 2 genera and ca. 31 species mostly American or paleotropic, extending from Maine (USA) to Brazil. The arborescent monotypic *Schizocardia* is native of British Honduras.

Vegetative Features. The members are deciduous, tall shrubs or low trees; *Schizocardia* is a large tree. Leaves alternate, simple, estipulate, serrate-margined.

Floral Features. Inflorescence terminal or axillary racemes or panicles. Flowers bracteate, ebracteolate, short-peduncled, hermaphrodite, actinomorphic, hypogynous, densely covered with simple stellate hairs. Calyx of 5 sepals, basally connate, persistent around the fruit. Corolla of 5 petals, polypetalous, disc absent. Stamens 10 or 12 in 2 whorls—outer whorl opposite corolla lobes, inner whorl opposite calyx lobes, free; filaments pubescent or glabrous; anthers bicelled, extrorse in bud, sagittate, inverted and introrse at anthesis, dehisce by slit-like apical pores. Pollen grains single. Gynoecium syncarpous, tricarpellary (pentacarpellary in *Schizocardia*), ovary trilocular with axile placentation, superior, ovules numerous. Style 1, thick, stigma 3-lobed.

Fruit a 3-valved, loculicidal capsule. Seeds small and ovoid with thin one-layered seed coat, irregular, extremely reticulate and without wings in *Clethra bodineri, C. barbinervis* and *C. cavaleriei* (Ganapathy 1970). The North American species of *Clethra* have highly compressed and winged seeds (Hu 1960).

Anatomy. Vessels moderately small, exclusively solitary, sometimes with fine spiral thickenings at the tips; perforation plates scalariform with 20–50 fine bars. Parenchyma apotracheal; rays markedly heterogeneous. Fibres with numerous bordered pits. Secretory cells and clustered crystals occur in unliginified tissue. Leaves dorsiventral with stomata confined to the lower surface: mostly Rubiaceous type, a few Cruciferous and Ranunculaceous type. Short, multicellular, stellate hairs often present.

Embryology. Pollen grains single, 3-colporate; 2-celled at shedding stage. Ovules anatropous, unitegmic, tenuinucellate. Embryo sac of Polygonum type, 8-nucleate at maturity. Endosperm formation of the Cellular type. Endosperm fleshy and oily, embryo cylindrical, straight; chalazal as well as micropylar endosperm haustoria present.

Chromosome Number. The haploid chromosome number is n = 8 in *Clethra arborea* and n = 16 in *C. alnifolia*. The basic chromosome number for the family is x = 8 (Thomas 1961).

Chemical Features. Iridoids are absent in *Clethra arborea* (Jensen et al. 1975).

Important Genera and Economic Importance. The genus *Clethra* has about 65 species; many of these are cultivated as ornamentals because of their fragrant flowers, e.g. *C. acuminatum*, *C. alnifolia*, *C. arborea*, *C. monostachya* and *C. tomentosa*.

Taxonomic Considerations. Bentham and Hooker (1965b) treated this family as a tribe of Ericaceae but later workers raised it to the rank of a distinct family. Hallier (1912) and Wettstein (1935) treated Clethraceae as a primitive taxon of the order Bicornes and derived it from the Ochnaceae. Bessey (1915) also considered it as the most primitive family of his Ericales. Lawrence (1951), Melchior (1964), and Hutchinson (1969, 1973) agreed with this assignment. Hutchinson (1973) derives Clethraceae from the Theales. The leaves and the polypetalous flowers of Clethraceae are strongly reminiscent of those of *Eurya* (Theaceae) but the flowers are borne in racemes or panicles and the anthers open by apical pores, as in the Ericaceae. The family Clethraceae, therefore, is a definite link between the Theales and the Ericaceae. Cronquist (1968,1981) also treats Clethraceae as one of the primitive families of the Ericales.

Benson (1970) recognises only one family, i.e. Ericaceae which is divided into 5 subfamilies: Clethroideae, Pyroloideae, Monotropoideae, Ericoideae and Vaccinioideae.

Dahlgren (1980a, 1983a) and Takhtajan (1980, 1987) include Clethraceae in the order Ericales.

In our opinion also, Clethraceae has correctly been treated as a primitive member of the Ericales. Its primitive features are: polypetalous corolla and actinomorphic flowers; embryologically, it shows close similarity with Ericaceae (Johri et al. 1992).

There is a controversy over the number of genera included in this family. According to Lawrence (1951), there are only 2 genera—*Clethra* and *Schizocardia*. Later workers recognise only one genus *Clethra* (in this family).

Pyrolaceae

A small family of 10 genera and 32 species confined mainly to the temperate zones of the northern hemisphere. In the troplical and subtropical regions they grow at high altitudes.

Vegetative Features. Perennial herbs with scaly rootstocks, without chlorophyll; fleshy saprophytes, sometimes suffrutescent. Stems more or less herbaceous and soft. Leaves radical or cauline, alternate, sometimes nearly opposite or in false whorls, simple, foliaceous or scale-like, persistent or deciduous, mostly coriaceous, toothed, stipulate.

Floral Features. Inflorescence variable—the genus *Moneses* has solitary flowers, *Chimaphila* has flowers in umbels or corymbs. Flowers bracteate (Fig. 44.2A), ebracteolate, hermaphrodite, actinomorphic, complete, pedicellate. Calyx of 5 sepals, polysepalous or basally connate, imbricate, persistent. Petals 5, polypetalous, hypogynous, disc present or absent, deciduous. Stamens 8–10, free, hypogynous; filaments often dilated basally, anthers dorsifixed, at first retroflexed but later erect, bicelled, the two "cells" slightly separated by the connective and each produced into a tube-like apex with a terminal pore (Fig. 44.2D, E), pollen remain attached in tetrads. In Monotropoideae the thecae open by longitudinal slits and pollen grains single. Gynoecium syncarpous, pentacarpellary, ovary pentalocular, 5-lobed, placentation axile with placentae intruded into each lobe (Fig. 44.2F) and bifurcate, superior; ovules numerous in each locule, style1, stigma 1, 5-lobed (Fig. 44.2 B, C). Fruit a loculicidal capsule, subglobose, 5-locular. Seeds very small with a loose testa and reticulate markings, abundant fleshy endosperm and small embryo.

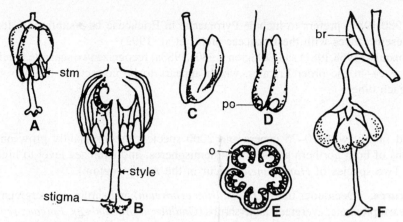

Fig. 44.2 Pyrolaceae: **A-F** *Ramischia secunda.* **A** Flower. **B** Vertical section. **C, D** Anthers. **E** Ovary, transection. **F** Fruit. *br* bract, *o* ovule, *po* pore, *stm* stamen. (Adapted from Lawrence 1951)

Anatomy. Similar to that of the Ericaceae.

Embryology. Pollen grains shed singly in *Monotropa*, otherwise in permanent tetrads in other members (Johri et al. 1992); 2-celled at dispersal stage. Ovules anatropous, unitegmic, tenuinucellate. Embryo sac of Polygonum type, 8-nucleate at maturiy. Endosperm formation of the Cellular type; haustoria short and 1-celled.

Chromosome Number. The diploid chromosome number is 2n = 32 in *Monotropa uniflora* and 2n = 46 in *Pyrola rotundifolia* (Kumar and Subramanian 1987).

Important Genera and Economic Importance. Some of the genera are parasitic or saprophytes without any chlorophyll, and leaves reduced to scales, e.g. *Allotropa, Monotropa, Pterospora* and *Sarcodes*. Some other genera have green leaves and are herbaceous, e.g. *Chimaphila, Moneses, Pyrola* and others. *Monotropa* is an interesting genus and shows infraspecific geographical difference in palynological characters. *M. hypopitys* is basically characterised by pollen grains without apertures in New World and with 3 apertures in the Old World, *M. uniflora* has 3-porate pollen in the New World and 3-colporate in the Old World (Takahashi 1987).

    M. uniflora and *Pyrola rotundifolia* grow in temperate Himalayas. None of the members is economically important.

Taxonomic Considerations. Bentham and Hooker (1965b) did not recognise the Pyrolaceae as a distinct family. Drude (1889) recognized it as a distinct family. Melchior (1964) and Hutchinson (1969) support this segregation. Many authors still hold the view that Pyrolaceae is not sufficiently distinct from Ericaceae to be treated as a separate family.

    According to Henderson (1919, see Metcalfe and Chalk 1972), the Pyrolaceae and Monotropaceae differ from the Ericaceae only in their gradually increasing saprophytism and the characters which go hand in hand are the loss of green colouring matter, the reduction from shrubs to herbs, the reduction of leaves to scales, the ovary from 5-celled with central placentae to almost completely 1-celled with parietal placentae, the increase in the number of seeds and the reduction in their size and in the number of cells in the endosperm and embryo.

    In addition there is a gradual decrease in the amount of wood formed and in the morphology of the leaves, from evergreen leathery leaves in Ericaceae to the chlorophyll-less scale leaves of *Monoropa hypopitys* and *M. uniflora*.

Takhtajan (1980,1987) prefers to include Pyrolaceae in Ericaceae as a subfamily. Its embryological features show resemblances with the Ericaceae (Johri et al. 1992).

However, Cronquist (1968,1981) and Dahlgren (1980a,1983a) recognize two separate families—Pyrolaceae and Monotropaceae—in the order Ericales, which appears to be correct, as the two genera are quite different from each other.

Ericaceae

A medium-sized family with 70–75 genera and 2000 species, predominantly grow on acidic soils of temperate regions of both northern and southern hemispheres, and from sea level to high altitude zones in the tropics. (Two species of *Harrimanella* occur in the Arctic region.)

Vegetative Features. Deciduous trees (*Oxydendrum arboreum*) and shrubs or evergreen shrubs (*Befaria racemosa*) or dwarf, prostrate, evergreen undershrubs (*Gaultheria procumbens, Epigaea repens*). *Cassandra calyculata* is a low evergreen shrub growing in *Sphagnum* bogs. Stem woody or herbaceous, erect or prostrate. Leaves of three different categories as reported by Hagerup (1953): (i) Rhododendron type— these are flat, generally broad leaves; stomata on the lower surface, as in *Andromeda, Arbutus, Arctostaphylos*. (ii) Needle-shaped leaves—narrow, pointed leaves with flat margins; stomata on both surfaces, as in *Erica, Cassiope* and *Harrimanella*. (iii) Ericoid leaves—these are small and linear with a hairy groove containing stomata on the lower surface, as in *Erica tetralix* and *Phyllodoce coerulea*. Leaves alternate, sometimes opposite or whorled, simple, often coriaceous and persistent, estipulate.

Floral Features. Inflorescence racemose (racemes, panicles or corymbs). Flowers bracteate, bracteolate hermaphrodite, rarely functionally unisexual, as in *Gaultheria* and *Epigaea;* actinomorphic, rarely zygomorphic, as in *Rhododendron* spp., commonly pentamerous, obdiplostemonous, pedicellate, showy and adapted for pollination by insects or birds (Wood 1961).

Calyx 4- to 7-lobed or polysepalous, usually persistent, valvate or imbricate. Corolla 4- to 7-lobed, usually gamopetalous, funnel-shaped, campanulate or urceolate, convolute or imbricate. Stamens as many or twice as many as petal lobes, usually inserted at the edge of a nectariferous, variously lobed disc, rarely epipetalous, free; filaments flattened, dilated or S-shaped, united in some tropical members of the Vaccinioideae. Anthers bicelled, becoming inverted during development; thecae often bulbous at the base, dehisce introrsely by longitudinal slits, clefts or pores, often with terminal awns or abaxial spurs; pollen grains in tetrads. Gynoecium syncarpous, (2)-3–7-(10)-carpellary, ovary superior or inferior, as in Vaccinioideae, number of locules same as the number of carpels, sometimes unilocular in the upper portion, placentation axile, ovules numerous in each locule. Style 1, stigma simple. Fruit a loculicidal or septicidal capsule, berry or drupe, sometimes enclosed within fleshy, persistent and adnate calyx, as in *Gaultheria*. Seeds small, sometimes winged or tailed, with fleshy endosperm and a straight embryo.

Anatomy. Vessels typically very small and numerous, often exclusively solitary; commonly semi-ring-porous, often with spiral thickenings; perforation plates usually scalariform or scalariform and simple. Parenchyma sparse or absent, moderately abundant and apotracheal in a few genera; rays heterogeneous. Leaves mostly dorsiventral, sometimes isobilateral or centric. Stomata generally occur only on the lower surface, confined to grooves (on the lower surface) in species with rolled leaves; generally Ranunculaceous type. Lower epidermis often papillose as in *Erica* spp., *Kalmia* spp. and *Rhododendron* spp.; upper epidermis sometimes papillose in *Erica*.

Embryology. Pollen grains usually in tetrads, 2-celled at dispersal stage. Ovules anatropous or campylotropous, unitegmic, tenuinucellar. Embryo sac of Polygonum type, 8-nucleate at maturity. Endosperm formaion of the Cellular type in *Daboecia, Kalmia, Phyllodoce,* and of the Nuclear type in

Rhododendron japonicum x *R. mucronatum*, and *Vaccinium* (Johri et al. 1992); both apical and basal haustoria present.

Chromosome Number. Basic chromosome number for the family is variable, x = 6, 8, 12, 13, 19, or 23.

Chemical Features. The members are rich in ellagitannins and iridoids, and seeds of *Arctostaphylos glauca* in linolenic fats.

Important Genera and Economic Importance. Several members of Ericaceae are important ornamentals: *Arctostaphylos, Calluna, Erica, Kalmia, Leucothoë, Rhododendron* and others. *R. hookeri* of the Eastern Himalayas are well-known for their clusters of bright red flowers.

Many species are the source of edible fruits, e.g. blueberries are obtained from several species of *Vaccinium*—*V. angustifolium, V. corymbosum, V. membranaceum, V. myrtilloides, V. ovatum,* and *V. vacillans,* cranberries from *V. oxycoccas,* cowberry from *V. vitis-idaea* and huckleberry from *Gaylussacia baccata.* The leaves of *Agapetes saligna* are often used as tea substitute; those of *Ledum palustre* furnishes Labrador tea, and *V. arctostaphylos* Broussa tea. Fruits of *Arbutus unedo* are used for making preserves and alchololic drinks, Briar pipes for smoking tobacco are made from the wood of *Erica scoparia.*

Leaves of *Gaultheria procumbens* and *G. shallon* are the source of the oil of wintergreen, used medicinally. Valuable timber is obtained from *Arbutus menziesii* (North America) and *A. unedo* (Mediterranean and South-west Ireland).

Rhododendron with 850 species, *Erica* with 600 species, and *Vaccinium* with 450 species are some of the important genera of this family. They often from the characteristic vegetation of areas with acid soils, particularly in the moors, swamps and mountain slopes of temperate countries of the world, and also on the mountain slopes of the tropical and subtropical regions.

Taxonomic Considerations. The family Ericaceae, according to some taxonomists, comprises two families—Ericaceae, with superior ovary and fruit typically a capsule, and Vacciniaceae, with inferior ovary and fruit typically a berry.

According to Engler and Diels (1936), the Ericaceae comprises 4 subfamilies—Rhododendroideae, Arbutoidae, Vaccinioideae and Ericoideae. They differ from each other in fruit types, position of ovary and morphological naure of anthers and seeds. Bessey (1915) and Lawrence (1951) treated Ericaceae as a member of the order Ericales. Melchior (1964), Hutchinson (1969, 1973), Dahlgren (1980a, 1983a) and Takhtajan (1980, 1987) support this view.

Hutchinson (1969, 1973) regards the genus *Vaccinium* with inferior ovary as belonging to a separate family Vacciniaceae. Takhtajan (1980, 1987), on the other hand, retains *Vaccinium* and also the members of Pyrolaceae in Ericaceae as subfamilies Vaccinioideae and Pyroloideae. The composition of the family is rather diversified, although it is universally regarded as a primitive family on the basis of: (i) superior ovary, (ii) hypogynous stamens that are free from corolla and (iii) the number of stamens double the number of petals.

The uniformity of embryological data of *Vaccinium* with other members of Ericaceae supports its retension as a subfamily, Vaccinioideae, of Ericaceae (Johri et al. 1992).

Empetraceae

A small family of 3 genera and 8 species distributed in the hilly regions of North and South America; also grow in the Arctic and Antarctic and circumpolar regions of the northern hemisphere. *Empetrum* and *Corema* are native of the Rocky Mountains and nearby mountain ranges of western North America. The monotypic *Ceratiola* (*C. ericoides*) occurs only in the Atlantic coastal plain from Florida to South Carolina and westward to Mississipi.

Vegetative Features. Dioecious, evergreen shrubs (Fig. 44.4A), resemble members of Ericaceae. Leaves simple, alternate, deeply-grooved beneath (Fig. 44.4B), linear, estipulate, sessile, small, rigid.

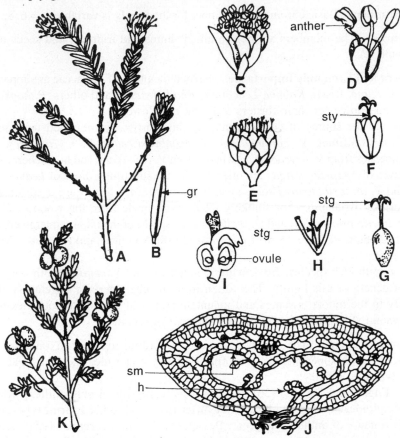

Fig. 44.4 Empetraceae: **A-J** *Corema conradii,* **K** *Empetrum nigrum.* **A** Twig with flowers. **B** Leaf. **C, D** Male inflorescence (**C**) and flower (**D**). **E, F** Female inflorescence (**E**) and flower (**F**). **G** Gynoecium. **H** Male flower (anthers removed) with vestigial gynoecium. **I** Ovary, longisection. **J** Leaf, transection. **K** Fruiting twig of *E. nigrum. gr* groove, *h* hair, *o* ovule, *sm* stomata, *stg* stigma, *sty* style. (**A-I** after Hutchinson 1969, **J** after Metcalfe and Chalk 1972, **K** after Lawrence 1951)

Floral Features. Inflorescence axillary or terminal heads. Flowers actinomorphic, bracteate and ebracteolate, usually hermaphrodite, sometimes unisexual (Fig. 44.4C-F), incomplete, hypogynous. Sepals 2–6, sometimes petaloid, imbricate, in 2 whorls. Petals usually absent, inner whorl of sepals often treated as petals by some authors. Stamens 2–4, hypogynous, distinct, anthers 2-celled, dehiscence longitudinal, disc absent. Gynoecium syncarpous, 2- to 9-carpellary, ovary 2- to 9-loculed with axile placentation, superior, one ovule per locule (Fig. 44.4I); style 1, short, variously lobed (Fig. 44.4G, H), fringed or divided; stigmatic branches as many as the number of carpels. Fruit a dry or fleshy berry containing 2 or more 1-seeded pyrenes. Seeds with fleshy endosperm and a long, straight embryo.

Anatomy. Vessels extremely small; almost exclusively solitary, tendency to become ring-porous in *Empetrum nigrum;* perforation plates scalariform in *Empetrum,* mostly simple and a few scalariform in *Ceratiola* and *Corema.* Parenchyma scanty; rays 1 or 2 cells wide, exclusively uniseriate in *E. nigrum.*

Fibres with numerous, small, bordered pits on both radial and tangential walls. Leaves deeply furrowed on the lower surface (Fig. 44.4B). The two margins eventually come closer and become almost closed by hairs, forming a central cavity (Fig. 44.4J). Hairs and stomata are confined to the epidermis lining this cavity; unicellular hairs are confined to the opposite leaf margins and are sufficiently interlocked to close the entrance to the cavity in *Corema conradii* and *E. nigrum* (Fig. 44.4J).

Embryology. Pollen grains in tetrads; shed at 2-celled stage. Ovules anatropous or nearly campylotropous, unitegmic, tenuinucellate. Polygonum type of embryo sac, 8-nucleate at maturity. Endosperm formation of the Cellular type; micropylar as well as chalazal endosperm haustoria present.

Chromosome Number. The basic chromosome number is x = 13.

Chemical Features. The members of Empetraceae have 8-oxyginated flavonols, but are devoid of any iridoid.

Important Genera and Economic Importance. The family is not of much economic importance. Species of *Corema* and *Empetrum* are sometimes cultivated as novelties or ornamentals (Fig. 44.4K).

Taxonomic Considerations. The family Empetraceae was considered to be closely allied to the Euphorbiaceae, because of: apetalous condition, imbricate aestivation of the sepals, number of sepals and stamens same, stamens opposite the sepals, bilocular anthers, superior ovary and lobed stigma. Bentham and Hooker (1965c) placed it in the Ordines Anomali of Monochlamydeae. Engler and Diels (1936) placed it in the Sapindales close to the Celastraceae and Buxaceae. Bessey (1915) and Lawrence (1951) also supported this view. Wettstein (1935) rejected the Sapindaceous resemblance (erect ovule and its ventral raphe) and adopted Ericaceous alliance due to embryological and endosperm affinities. This view is also supported by serological evidence.

According to Hutchinson (1969, 1973), its true affinities are rather doubtful and he places this family in the order Celastrales. Cronquist (1968, 1981) includes Empetraceae in the Ericales regarding it as a reduced apetalous form and polygamous or dioecious derivative of the latter. Samuelsson (1913), on the basis of embryological data emphasized that Empetraceae should be a member of the Ericales. Many other taxonomists also agree with this view (Dahlgren 1980a, 1983a, Takhtajan 1980, 1987).

The position of this family in the order Ericales is justified on the basis of its ericoid habit, rolled leaves, as in many other Ericales (Hagerup 1953), and embryological features.

Epacridaceae

A family of 23 genera and about 350 species, mainly distributed in the Australasian region of the world. Just as Ericaceae is a dominant family in South Africa, the Epacridaceae is equally so in Australia and Tasmania.

Vegetative Features. Mostly shrubs, shrublets or small trees; erect or prostrate; very small plants of only 5 cm height is *Leucopogon frazeri* whereas *Trochocarpa* sp. is 6–12 m tall. Stem woody. Leaves alternate, often crowded, jointed at the base, simple, small, usually stiff, xeromorphic, with very thick cuticle, estipulate; sometimes cordate with pointed tip. Leaves opposite in *Needhamiella*.

Floral Features. Flowers solitary axillary in upper leaf axils, sometimes in pairs, racemes or spikes; short axillary or terminal racemes in *Leucopogon* sp.; flowers solitary and terminal in *Sprengelia sprengelioides*. Flowers bracteate, bracts many, imbricate, cover the pedicel and overlap the calyx, and ebracteolate, normally hermaphrodite, actinomorphic, complete, hypogynous, pentamerous. Calyx of 4 or 5 sepals, polysepalous, imbricate, persistent. Corolla of 4 or 5 petals, gamopetalous, polypetalous in *Lysinema* and *Sprengelia*, imbricate or valvate. Stamens 5 or 4, epipetalous, filaments short, alternate with the corolla lobes, sometimes with alternating staminodes, represented by clusters of hairs or glands. Anthers introrse, 1-celled at anthesis, dehiscence longitudinal. Anthers appendaged in some species of *Leucopogon* and *Melichrus;* 2 perfect stamens, rest staminodes in *Oligarrhena;* anthers deeply slit in *Conostephium.* Gynoecium syncarpous, 4- or 5-carpellary, ovary superior, 1- to 10-loculed, placentation axile, ovules 1 to numerous per locule. Ovary often surrounded by a hypogynous glandular disc; style 1, stigma 1, capitate; style inserted in a tubular depression of the ovary in *Epacris* and *Sprengelia.* Ovary unilocular in *Monotoca* and *Leucopogon;* ovaries of *Leucopogon* and *Brachyloma* attenuate into the style.

Anatomy.   Vessels typically less than 50 μm in mean tangential diameter and sometimes even less than 25 μm as in *Drachophyllum* and *Richea;* often with spiral thickenings. Perforations partly or wholly simple or plates exclusively scalariform and sometimes irregularly reticulate. Parenchyma apotracheal, diffuse or in fine bands, sometimes very sparse; rays heterogeneous. Leaf lamina frequently dorsiventral, sometimes grooved, stomata confined to the grooves, of Ranunculaceous type, pores generally parallel to the long axis of the leaves.

Embryology.   The family Epacridaceae is unique and shows a variety of pollen grains (G. Erdtman 1952). These are mostly single, solitary and smooth in *Brachyloma*, and in *Epacris* sp. and *Leucopogon* the pollen is in tetrads. Ovules anatropous, unitegmic, tenuinucellate. Embryo sac of Polygonum type, 8-nucleate at maturity. Endosperm formation of the Cellular type.

Chromosome Number.   Basic chromosome number is x = 4, 6, 7, 9, 11, 12, 13.

Chemical Features.   Large amounts of tannin and calcium oxalate crystals occur freely in various members (Paterson 1960). Rimuene—a diterpene hydrocarbon—is reported in the epicuticular wax of *Richea continentis* (Salasoo 1985). The wax contains 41% rimuene and 25% hentriacontane, an alkane.

Important Genera and Economic Importance.   *Styphelia* with 175 species is the largest genus followed by *Epacris* with 34 species. Although most members have a normal 5-locular ovary, it is 1 to 10-locular in *Styphelia, Monotoca, Decatoca, Trochocarpa* and *Leucopogon;* 3-locular in *Wittsteinia* and 2-locular in *Needhamiella* and *Oligarrhena. Wittsteinia* is also interesting; it has pollen grains in tetrads and inferior ovary.

Species of *Cyathodes, Epacris* and *Leucopogon* are cultivated as ornamentals.

Taxonomic Considerations.   Bentham and Hooker (1965b) placed Epacridaceae (= Epacrideae) in the order Ericales, a member of series Heteromerae. Engler and Diels (1936) also treated it as a distinct family in the order Ericales. Hallier (1912) treated it as a member of his Bicornes along with Clethraceae, Pyrolaceae, Empetraceae, Ericaceae, Cyrillaceae and Diapensiaceae, and so did Bessey (1915) although he named the order Ericales. Lawrence (1951) mentions that "Epacridaceae is undoubtedly an advanced taxon of close alliance with the Ericaceae". Hutchinson (1969, 1973), Dahlgren (1980a,1983a) and Takhtajan (1980, 1987) are also of the same opinion. According to Takhtajan (1980), Epacridaceae stands very near the Ericaceae, especially to subfamily Ericoideae. the genus *Sprengelia* is one of the (most) primitive within the family and is closely related to the Ericaceae (Paterson 1961).

The Epacridaceae has much in common with the Ericaceae. Anatomically they are very similar, as reported by Metcalfe and Chalk (1972). Salasoo (1985) observes the occurrence of the same chemical rimuene in certain members of both the families—*Richea continentis* of the Epacridaceae and *Gaultheria* sp. of the Ericaceae. As Lawrence (1951) pointed out, it can also be regarded as the Australian counterpart of the family Ericaceae, although there are a few minor differences such as : (i) a single whorl of epipetalous stamens, (ii) one-celled anthers which dehisce by a single longitudinal slit, and (iii) anthers do not have appendages or tails that are so common in the Ericaceae.

Accordingly, the Epacridaceae should be treated as a distinct family and placed next to the Ericaceae.

Taxonomic Considerations of the Order Ericales

The Ericales—also Bicornes (Hallier 1912)—is characterised by generally pentamerous flowers with free or basally connate petals, obdiplostemonous stamens, pollen often in tetrads, placentation axile, ovules numerous, tenuinucellate and unitegmic.

Circumscription of this order has been much varied. Engler and Diels (1936) treated it as composed of four families: Clethraceae, Pyrolaceae, Ericaceae and Epacridaceae. Melchior (1964) includes Empetraceae also in this order. Hutchinson (1969, 1973) placed eight families is his Ericales separating Monotropaceae and Vacciniaceae from the Ericaceae. He further proposed the origin of the Ericales from the Theales, the common features are free petals in the primitive family Clethraceae and numerous stamens.

Dahlgren (1980a, 1983a) assigns Actinidiaceae, Cyrillaceae, Roridulaceae and Diapensiaceae in the order Ericales. However, the position of the Grubbiaceae may not be correct (Jensen et al. 1975). The family Roridulaceae has been included in this order on the basis of similar chemical constituents (iridoids) and also some of the morphological and embryological features. The families Actinidiaceae and Cyrillaceae have embryological features in common with other Ericales. Diapensiaceae was previously included in the Ericales as an advanced member as it has epipetalous stamens; but was separated as a distinct order by Engler and Gilg (1924). It has been placed back in this order by Dahlgren (1975a) mainly on the basis of some exomorphological features such as obdiplostemonous condition, epipetalous stamens, axile placentation and some anatomical features.

Takhtajan (1980, 1987) also includes Actinidiaceae, Diapensiaceae, Cyrillaceae and Grubbiaceae in his Ericales, in addition to Clethraceae, Empetraceae, Ericaceae and Epacridaceae. He considers the Pyrolaceae to be one of the subfamilies, Pyroloideae, of the family Ericaceae.

However, the position of the family Grubbiaceae and Actinidiaceae in the Ericales is doubtful. The embryological features of the Grubbiaceae are also common to the Santalaceae and allied families (Ram 1970). Lawrence (1951) and Hutchinson (1969), on the basis of morphological features, placed this family in the Santalales. The family Actinidiaceae has closer affinities with the family Theaceae and its inclusion in the order Theales is justified (Vijayaraghavan 1970).

The family Cyrillaceae shows close correspondence with the Ericaceae in most of the important embryological features. According to Vijayaraghavan (1970), this family should be retained in the Ericales, The family Diapensiaceae, though closely related to the Ericales, is quite distinct from them because of single pollen grains, tricarpellary ovary and bitegmic ovules (Lawrence 1951). In our opinion the order Ericales should comprise Clethraceae, Cyrillaceae, Empetraceae, Ericaceae, Epacridaceae and Pyrolaceae; the families Grubbiaceae, Actinidiaceae and Diapensiaceae may not be included in Ericales.

45

Order Primulales

The order Primulales comprises three families—Theophrastaceae, Myrsinaceae, and Primulaceae. According to Takhtajan (1987), there is a fourth family, Aegicerataceae, which Willis (1973) includes in the Myrsinaceae.

   Both Theophrastaceae and Myrsinaceae are predominantly arborescent. The Primulaceae are mostly perennial herbs, and most advanced amongst the three. All three families have sympetalous, pentamerous flowers with or without staminodes. Ovary unilocular, superior with free-central placentation; ovules mostly anatropous and bitegmic. None of the members appear to be adapted to any particular ecological niche.

Theophrastaceae

A family of 5 genera and 110 species distributed in tropical America and the West Indies.

Vegetative Features.   Monocaulous trees or shrubs, stem woody; branching extremely rare. Leaves alternate or opposite (as in *Jacquinia*), sometimes pseudoverticillate along the upper part of the stem; each pseudoverticel has about 20 leaves arranged in a condensed spiral. The apical meristem is protected by narrow bud scales persisting as spines on older parts of stem (Stähl 1987). Leaves entire or spinose-serrate, estipulate, often crowded at apex of stem, sometimes pungent.

Floral Features.   Inflorescence racemes, corymbs or panicles. Flowers ebracteate, ebracteolate, hermaphrodite or unisexual (the plants then dioecious), actinomorphic, hypogynous. Calyx lobes usually 5, polysepalous or connate at base, imbricate, persistent. Corolla 5-lobed, gamopetalous, imbricate, rotate, urceolate or funnel-shaped; sepals and petals sometimes gland-dotted or streaked. Stamens biseriate—an outer whorl of 5 alternipetalous petaloid staminodes, and an inner whorl of oppositipetalous epipetalous stamens; filaments usually free, rarely connate as in *Clavija*. Anthers bicelled, extrorse, dehiscence longitudinal. Gynoecium syncarpous, pentacarpellary, ovary unilocular, superior with numerous ovules on free-central placenta; style 1, short and thick, stigma discoid or conical, entire to irregularly lobed. Fruit usually a berry, rarely a drupe; seeds numerous to a few, rarely one, with a well-developed embryo and copious endosperm.

Anatomy.   Vessels moderately small, commonly in radial multiples of 3 or 4 cells and irregular clusters; perforation plates usually simple, intervascular pitting small and alternate. Parenchyma paratracheal and sparse, a few cells to narrow sheaths round the vessels. Rays multiseriate, 4 to 30 cells wide; usually heterogeneous. Leaves not dorsiventral; epidermal hairs often branched and sclerosed as in *Jacquinia armillaris*. Stomata confined to the lower surface mostly; of Cruciferous-Ranunculaceous type. Mesophyll in most members contains sclerenchymatous fibres beneath the upper and lower epidermis; resin ducts absent.

Embryology.   The embryology has not been fully investigated (Johri et al. 1992). Ovules numerous, ana- to campylotropous, bitegmic, tenuinucellate. Polygonum type of embryo sac, 8-nucleate at maturity. Development of the endosperm is of the Nuclear type. Endosperm copious in mature seed.

Chromosome Number.   The basic chromosome number is x = 9 (Takhtajan 1987).

Important Genera and Economic Importance.   All the members of Theophrastaceae have a very unusual appearance. *Clavija* sp. somewhat resembles a palm, since the plants have a long, slender, unbranched, leafless trunk bearing a terminal rosette-like head of leaves. *Theophrasta* is similar in habit but smaller in size whereas *Deherainia* is still smaller, bearing the rosette of leaves just above the soil level. *Clavija longifolia* is sometimes cultivated as an ornamental.

Taxonomic Considerations. The family Theophrastaceae has often been combined with the Myrsinaceae (Bentham and Hooker 1965b) as a subfamily, Theophrastieae. The family Theophrastaceae differs from the Myrsinaceae exomorphologically in having staminodes, absence of resin ducts, and the seeds large, yellow or orange.

Hutchinson (1969, 1973) included Theophrastaceae in Lignosae, along with Myrsinaceae, because of arborescent nature. Cronquist (1968, 1981) places it in the order Primulales and so does Takhtajan (1980, 1987).

The Theophrastaceae—along with Myrsinaceae and Primulaceae—are appropriately placed in the Primulales and the three families together form a natural group.

Myrsinaceae

The medium-sized family of about 35 genera (32–34 according to Takhtajan 1987), and 1000 species, mostly distributed in tropical and subtropical South Africa and New Zealand.

Vegetative Features. Trees or shrubs. Leaves mostly alternate, simple, entire or with serrate or crenate margin, usually coriaceous and persistent, estipulate, glandular-punctate or with linear resin ducts.

Floral Features. Inflorescence racemose—paniculate or corymbose or cymose. Flowers often fasciculate on scaly short shoots or on spurs in leaf axils (Lawrence 1951). In the genus *Myrsine*, flowers single or subumbellate, a few axillary or on previous season's wood (Hutchinson 1969). Flowers bracteate, ebracteolate, hermaphrodite or unisexual (the plants then dioecious or polygamodioecious), actinomorphic, hypogynous. Calyx of 4–6 sepals, polysepalous in *Embelia* and *Heberdenia* or more commonly basally connate, persistent, imbricate, glandular-punctate. Corolla 4- to 6-lobed, polypetalous (as in *Embelia*) or gamopetalous, rotate to salverform (Fig. 45.2A). Stamens as many as petals, epipetalous, opposite to petals, usually free but monadelphous in some taxa, syngenesious in *Amblyanthus*. Anthers introrse, bicelled, dehiscence longitudinal, or by apical slits (Fig. 45.2C, D) or pores, usually longer than filaments. Staminodes absent in male flowers, present in female flowers; often as large as stamens. Gynoecium syncarpous, 4- to 6-carpellary, ovary unilocular with free-central or basal placentation, or 4- to 6-locular with axile placentation (Fig. 45.2B), ovules numerous (rarely few) sunk in the placental tissue, usually superior (inferior to half-inferior in *Maesa*). Style and stigma simple (Fig 45.2D), stigma may sometimes be lobed. Fruit a berry with a fleshy exocarp and a stony endocarp, contains 1 to many seeds. Seeds with copious endosperm and a cylindrical embryo.

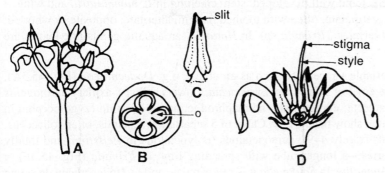

Fig. 45.2 Myrsinaceae: **A–D** *Ardisia crenata.* **A** Inflorescence. **B** Ovary, transection. **C** Anther, note apical slit. **D** Flower, longisection. *o* ovule. (After Lawrence 1951)

Anatomy. Anatomically, Myrsinaceae resembles Theophrastaceae. It has moderately small vessels with simple or scalariform perforation plates. Leaves dorsiventral and have secretory cells and cavities, with yellow or reddish-brown contents. Canals containing a similar secretion are also reported in the cortex or the pith of the axis.

Embryology. Pollen grains triporate and 2-celled at the dispersal stage (Johri et al. 1992). Ovules half-anatropous or half-campylotropous, bitegmic, tenuinucellate. Polygonum type of embryo sac, 8-nucleate at maturity. Endosperm formation is of the Nuclear type (Johri et al. 1992).

Chromosome Number. The basic chromosome number is x = 10 to 12 and 23.

Chemical Features. A polyalcohol called D-quercitol has been reported in the fruits of *Myrsine africana*, *M. semiserrata* and *Embelia ribes*.

Important Genera and Economic Importance. Among the larger genera of this family are *Ardisia* with 250 species, *Rapanea* with 140 species, *Maesa* with 100 species and *Embelia* with 60 species. Some species are cultivated as ornamentals. Many Myrsinaceae members show an interesting symbiotic association with bacteria (Johri et al. 1992).

Taxonomic Considerations. The family Myrsinaceae resembles the Theophrastaceae in more than one respect. Hallier (1912) treated the two as distinct families but related to each other. Wettstein (1935) considered them to be the most advanced of the Primulales because of frequency of dioecism, half-inferior ovary in Maesoideae and polyembryony in *Ardisia*. Hutchinson (1969, 1973) does not appreciate the view of Engler and Diels (1936), Melchior (1964) and Rendle (1925) and places Myrsinaceae in an order of its own, Myrsinales, in the group Lignosae. He also supports the separation of a small family Aegicerataceae which includes some mangrove swamp trees. Cronquist (1968, 1981) retains the three families in the Primulales, and Takhtajan (1987) includes the fourth family, Aegiceratceae, in this order.

   Except for the woody nature of Myrsinaceae and Theophrastaceae, the three families form a natural group and should be so maintained.

Primulaceae

A family of 23–27 genera and about 1000 species (28 genera and 800 species according to Lawrence 1951), widely distributed but more common in north temperate regions.

Vegetative Features. Mostly mesophytes, but a hydrophyte *Hottonia,* and a halophyte *Glaux maritima* are also reported. Mostly perennial herbs, perennate by rhizomes as in *Primula,* or by tuber as in *Cyclamen;* sometimes annual, e.g. *Anagallis.* The aerial stem internodes are often suppressed so that the leaves appear to be in the form of dense radical rosettes, e.g. *Primula, Dodecatheon* (Fig. 45.3A, E) and others; in *Lysimachia vulgaris* the stem is erect and well developed; stem creeping in *L. nummularia* and winged in *Anagallis.* Leaves usually simple, estipulate, often with toothed margin, alternate, opposite or whorled, obovate-spathulate; gland-dotted or farinose (*Primula* sp). In *Hottonia,* an aquatic genus, the leaves are submerged and finely dissected.

Floral Features. Inflorescence variable—racemose such as an umbel, e.g. *Dodecatheon* (Fig. 45.3A), *Cyclamen, Primula* (Fig. 45.3E) or racemes or spikes, e.g. *Lysimachia;* flowers solitary, axillary in *Anagallis* and *Trientalis borealis.* Flowers bracteate, ebracteolate, hermaphrodite, actinomorphic (zygomorphic in *Coris*), hypogynous, pentamerous, often show heterostyly. Calyx of 5 sepals, polysepalous, often foliaceous, persistent. Corolla mostly of 5 petals (rarely 4–9), gamopetalous (polypetalous in *Pelletiera* and totally absent in *Glaux,*), corolla shape varies—a longer tube with spreading limb in *Primula* (Fig. 45.3F), a shorter tube with spreading limb in *Anagallis.* In *Soldanella* it is campanulate and in *Dodecatheon* drooping

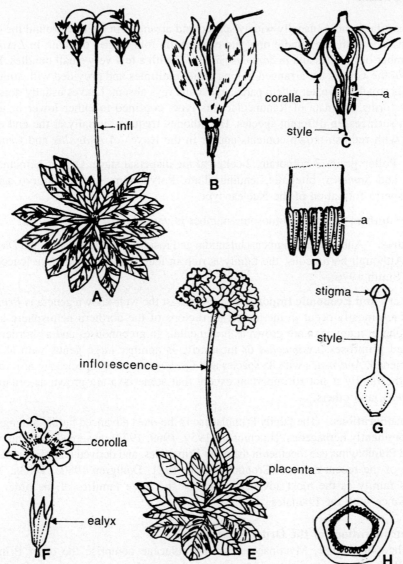

Fig. 45.3 Primulaceae: **A–D** *Dodecatheon meadia*, **E–H** *Primula denticulata*. **A** Plant with inflorescence. **B** Flower. **C** Same, vertical section. **D** Stamens (spread out). **E** Plant of *P. denticulata* with basal rosette of leaves and inflorescence. **F** Flower. **G** Gynoecium. **H** Ovary, transection. *a* anther, *infl* inflorescence. (Adapted from Lawrence 1951)

(Fig. 45.3A, B). *Pelletiera,* an exceptional genus, has 3-merous flowers. Stamens as many as corolla lobes, epipetalous and oppositipetalous in one whorl. The missing outer whorl sometimes represented by scale-like staminodes as in *Soldanella* and *Samolus*. Anthers bicelled, introrse, dehiscence longitudinal (Fig. 45.3C, D). Gynoecium syncarpous, 5-carpellary, ovary superior (half-inferior in *Samolus*), unilocular, ovules few to numerous, placentation free-central (Fig. 45.3H); style 1, heterostyly common, stigma capitate (Fig. 45.3G). Fruit usually a 5-valved or 5-toothed (sometimes 10-toothed) capsule or a pyxis (as in *Anagallis* and *Centunculus*); seeds small with transparent endosperm, and a small straight embryo.

Anatomy.   Vascular bundles usually widely spaced and arranged in a circle around the central pith, e.g. in *Anagallis arvensis* with one bundle opposite each of the four angles of stem; in *Lysimachia vulgaris* with a large number of bundles and in *Samolus valerandi* with a few very small bundles. In mature xylem of *L. hillebrandii* the vessels are arranged in long radial multiples and provided with simple perforations. Intervascular pits small, alternate; xylem parenchyma and rays absent. Leaves usually dorsiventral; centric in xeromorphic forms. Stomata of Ranunculaceous type, confined to either lower or upper surface or present on both surfaces, in different species. Hydathodes frequent, usually at the end of the midveins. Secretory cells with reddish-brown contents appear in the leaves of *Anagallis* and *Centunculus.*

Embryology.   Pollen grains 3-colporate, 2-celled at the dispersal stage. Ovules hemiana- or anatropous as in *Hottonia* and *Samolus,* bitegmic, tenuinucellate. Polygonum type of embryo sac, 8-nucleate at maturity. Endosperm formation of the Nuclear type.

Chromosome Number.   Basic chromosome number is variable x = 5, 8–15,19,22.

Chemical Features.   Anthocyanin pigments hirsutidin and rosidin occur in *Primula* and *Dionysia* (Harborne 1968, 1969b). Although herbaceous, the family is rich in leuco-anthocyanins like leucodelphinidin and leucocyanidin (Smith 1976).

Important Genera and Economic Importance.   Amongst the well-known genera is *Primula,* with about 400 species, which mostly occur in the temperate regions of the northern hemisphere and a few in the southern hemisphere; many taxa are grown as ornamentals in greenhouses and as borders and rockeries; commonly called primroses. *Lysimachia* or loosestrife is another large genus with about 100 species which are ornamental. *Androsace* with 85 species and *Cyclamen* with 20 species are also important genera. Economically, the family is not so important except that some taxa are grown as ornamentals, such as *Primula, Cyclamen* and others.

Taxonomic Considerations.   The family Primulaceae is the most advanced family amongst the Primulales which are predominantly herbaceous. Hutchinson (1959, 1969, 1973) disagreed with this view and placed Primulaceae and Plumbaginaceae together in the order Primulales, and derived them from the Caryophyllales. However, most of the recent authors (Cronquist 1968, 1981, Dahlgren 1980a, 1983a, Takhtajan 1980, 1987) treat this family as the most advanced amongst the three families of the order Primulales and presume its alliance with the Ebenales.

Taxonomic Considerations of the Order Primulales

The families Theophrastaceae, Myrsinaceae and Primulaceae comprise the order Primulales and are closely related. Both Theophrastaceae and Myrsinaceae are woody tropical groups (with a few exceptions), while the Primulaceae are herbaceous, apparently approaching the Myrsinaceae through a few woody species of *Lysimachia.* All three groups are held together: by the mostly pentamerous and sympetalous flowers; androecium of an outer antesepalous whorl of stamens; similar pollen; unilocular ovary and free-central placentation; bitegmic ovules.

   Although, in the Englerian sequence of families the Primulales are placed next to the Ericales and Diapensiales, the free-central placentation has led many taxonomists (Hutchinson 1969, 1973, Dickson 1936, Douglas 1936) to associate the Primulales with the Caryophyllales. The Plumbaginaceae is sometimes included in the Primulales (Bessey 1915, Hutchinson 1969), but there is evidence to indicate that the apparent similarities between them are due to parallelism. Melchior (1964) derived Primulales from the Ericales and so did Hallier (1912). Wettstein (1935) considered it to be allied to Bicornes (= Ericales).

   According to Cronquist (1968, 1981), the Primulales are probably related to the Ebenales because of the combination of sympetalous corolla and bitegmic ovules. They are not ancestral to each other but may have had a common ancestry in the Theales. Dahlgren (1980a) treats Ebenales and Primulales together in his Primuliflorae and suggests their derivation from Theiflorae on the basis of phytochemistry: "The chemical characters are more like those of the Theales". Iridoid and seco-iridoid compounds are absent in all the three orders: Primulales, Ebenales and Theales (Dahlgren 1977b). Takhtajan (1980, 1987) also retains the Primulales next to the Ebenales.

However, as the Primulales have none of the special features of the Caryophyllales other than free-central placentation, their derivation from the Caryophyllales is highly unlikely. The ancestral group for the Primulales should have been one that could give rise to Theophrastaceae and not the most advanced Primulaceae. The expected characters would be a tropical tree with hypogynous, sympetalous flowers with at least 2 whorls of epipetalous stamens, a syncarpous ovary and numerous bitegmic tenuinucellar ovules. The Theales are short of sympetalous corolla and could possibly be the ancestral group of this order.

46

Order Plumbaginales

A monotypic order comprising Plumbaginaceae with 22 genera and 600 species.

Plumbaginaceae

Distributed mainly in the semi-arid zones of the Old World, especially the Mediterranean and Central Asiatic regions and also on salt steppes and sea coasts.

Vegetative Features.   Perennial herbs or shrubs, sometimes scandent with alternate, simple, estipulate leaves, on the surface occur water glands or chalk glands.

Floral Features.   Inflorescence of various types, both racemose and cymose. In *Plumbago* raceme, raceme with dichasia as smaller units in *Ceratostigma*, a spike in *Limonium* and a capitulum of cincinna, surrounded by a whorl of bracts in *Armeria*. Flowers bracteate, ebracteolate, hermaphrodite, actinomorphic, hypogynous and pentamerous. Calyx 5-lobed, gamosepalous, odd sepal posterior, persistent, plicate, often 5- to 10-ribbed, -angled or -winged, sometimes membranous or scarious and showy or sometimes covered with glandular hairs on the outer surface. Corolla of 5 petals, gamopetalous, sometimes seemingly polypetalous, contorted or imbricate. Stamens 5, epipetalous and oppositipetalous, introrse; anthers bicelled, dehiscence longitudinal; pollen grains dimorphic in some members. Gynoecium syncarpous, 5-carpellary, ovary unilocular but 5-lobed or 5-ribbed, superior; ovule solitary, pendulous; styles 5, opposite the sepals, free or basally connate, often hairy or glandular; stigmas filiform; heterostyly in *Limonium vulgare*. Fruit a utricle or nut, often enclosed within the calyx; seed with a firm crystalline-granular or floury endosperm and straight embryo.

Anatomy.   Vessels very small, sometimes with a radial pattern, perforations simple; parenchyma paratracheal, scanty; rays up to 3 cells wide with numerous uniseriates. Included phloem reported in *Acantholimon, Aegialitis* and *Limoniastrum*. Epidermal glands secrete mucilage and/or calcium salts on stem and leaf surfaces. Leaves isobilateral, centric or dorsiventral; stomata generally on both the surfaces, Rubiaceous, Cruciferous or Ranunculaceous type, tannin abundant.

Embryology.   Pollen grains 3-, 5- or 6-colpate with spinulous exine, and 2- or 3-celled at the dispersal stage; pollen dimorphism reported in some species and genera. The family Plumbaginaceae has 2 tribes: in Plumbagineae heterostyly (when present) is connected with monomorphic pollen, and in Staticeae heterostyly is associated with dimorphic pollen (Weber-El Ghobary 1984). Ovules anatropous, bitegmic, crassinucellate. Plumbago type of embryo sac, 8-nucleate at maturity. Penaea type (in *Statice limonium*), Fritiliaria type (in *Armeria bupleuroides*) and Plumbagella type (in *Plumbagella micrantha*) of embryo sacs also occur (Johri et al. 1992). Endosperm formation of the Nuclear type.

Chromosome Number.   The basic chromosome number is x = 6–9.

Chemical Features.   Naphthoquinone plumbagin is obtained from *Plumbago*. Anthocyanin pigments of the forms cyanidin, delphinidin, malvidin and petunidin have been reported in various genera and species.

Important Genera and Economic Importance.   *Limonium* with more than 100 species is the largest genus, mostly occurs in steppes and salt marshes of the Old World. *Statice* with about 50 species occurs mainly in the alpine, arctic and maritime habitats in the north temperate zone and also in the Chilean Andes.

The Plumbaginaceae is not of much importance except for ornamentals like *Acantholimon, Armeria, Ceratostigma, Limonium* and *Plumbago,* various species are cultivated. *Plumbago zeylanica* is a widespread weed.

Taxonomic Considerations. Hallier (1912) included the Plumbaginaceae in the order Controspermae, and Hutchinson (1969, 1973) in the order Primulales. Most other taxonomists treat this family as a single member of the order Plumbaginales (Wettstein 1935, Lawrence 1951, Melchior 1964, Cronquist 1969, 1981, Dahlgren 1980a, 1983a, Takhtajan 1987).

The Plumbaginaceae resemble the Centrospermae members in unilocular ovary, basal ovule borne on a long funicle and a mealy endosperm.

The Plumbaginaceae resemble Primulales in similar floral structure such as stamens opposite petals, unilocular ovary, and bitegmic ovules. The differences are: a single ovule, 5 styles, mealy endosperm and absence of outer whorl of staminodes. Chemically, however, they are different from each other. It may be linked chemically with the Ebenaceae and Ericaceae. *Diospyros* roots have plumbagin and Ericaceae have azaleatin (Harborne 1967).

Mostly, however, this family has been derived from the Caryophyllales (= Centrospermae), with Primulaceae in an intermediate position.

Taxonomic Considerations of the Order Plumbaginales

This order comprises only one family, Plumbaginaceae. There is evidence that the Plumbaginales may have evolved from stocks ancestral to the Primulales, and most phylogenists also presume that they have close affinities with the Caryophyllaceous taxa. Takhtajan (1980) assigns his Caryophyllales and the Plumbaginales under the subclass Caryophyllidae. Melchior (1964) places Plumbaginales in Sympetalae but the Plumbaginaceae are strictly isolated within the Sympetalae, and correspond with the Centrospermae (= Caryophyllales) embryologically. However, Johri et al. (1992) point out that there are some important embryological differences between the two taxa.

Dahlgren (1980a, 1983a) is perhaps the only taxonomist to have placed this order in his Malviflorae, probably because of apparently free petals and glandular hairs on the vegetative parts.

The order Plumbaginales is an isolated taxon amongst the Gamopetalae as it shows alliance with a large number of taxa from both Poly- and Gamopetalae. Under the circumstances, further studies, particularly in the field of serology, will be helpful to assign a correct placement to this order.

47

Order Ebenales

The Ebenales include 7 families grouped in two suborders:

The suborder Sapotineae comprises Sapotaceae and unigeneric Sarcospermataceae.

The suborder Ebenineae comprises Ebenaceae, Styracaceae, Lissocarpaceae, Symplocaceae and Hoplestigmataceae.

The Ebenales are woody trees with simple alternate leaves, mostly of tropical distribution. They are characterized by estipulate leaves, sympetalous corolla and epipetalous stamens in usually 2 or 3 whorls, sometimes including staminodes. The ovary is basically septate with 1 to few ovules on each of the axile placentae. Ovules bitegmic.

There is some form or other of exudates from the members of this family, and they contain calcium oxalate crystals.

Sapotaceae

A family of (50) 55–60 (70) genera and 800 species (Takhtajan 1987), distributed mainly in the Old World and the American tropics.

Vegetative Features.   Woody trees or shrubs with milky sap in vegetative parts. Leaves simple, usually entire, coriaceous, alternate or rarely opposite, mostly estipulate.

Floral Features.   Inflorescence cymose or only solitary flowers in leaf axils or on old stems. Sepals 4–12, polysepalous, arranged in two whorls or spirals; basally connate, imbricate; petals as may as sepals, gamopetalous, imbricate, sometimes with lateral or dorsolateral appendages. Stamens 8–15, commonly in 1 to 3 series, the outer 1 or 2 series usually sterile or sometimes obsolete, often alternate with corolla lobes, sometimes petaloid; anthers bicelled, extrorse, dehisce longitudinally. Gynoecium syncarpous, 4- or 5-carpellary, ovary 4- or 5-locular, with axile placentation and 1 ovule per locule, superior; style 1, stigma bulbous or smoothly lobed. Fruit a berry; outer layer usually thin and horny or leathery. Seeds large with relatively broad raphe and thick shiny seed coat; more or less well-develpoed endosperm and a large embryo.

Anatomy.   Vessels medium-sized, in loose, radial or oblique lines, often in multiples of 4 or more cells; perforations simple, intervascular pitting alternate. Parenchyma apotracheal; rays 1–6 cells wide (typically 2–3 cells wide), heterogeneous. Leaves dorsiventral, sometimes centric; surface hairs thick- or thin-walled, unicellular, 2-armed and golden-yellow, rusty-brown or silvery-white, frequently secrete a resinous substance, especially on the lower surface of leaf. Stomata confined to the lower surface; on upper surface in a few species of *Chrysophyllum* and *Omphalocarpum*. Stomata of Ranunculaceous type, sometimes Rubiaceous type, deeply sunk in *Mastichodendron* and *Sideroxylon*, rarely elevated above the epidermis as in *Mimusops*. Laticifer ducts arranged in longitudinal rows, in the cortex of root, and cortex, phloem and pith of stem; and in leaf, accompany the veins.

Embryology.   Pollen grains 2- or 3-celled at dispersal stage, 3–4–(5–6)-colporate, and with reticulate exine. Ovules anatropous, to hemianatropous, apotropous, unitegmic, tenuinucellate. Embryo sac of Polygonum type, 8-nucleate at maturity. Endosperm formation of the Nuclear type.

Chromosome Number. Haploid chromosome number varies from n = 11 to 13.

Chemical Features. Sapotaceae members contain a group of latex products called 'balatas' which contain a high percentage of trans-polyisoprene and resin (Kochhar 1981). Lactose or 4-b-galactosylglucose, the sugar of cow's milk, has been detected in the chicle plant *Manilkara zapota* (Harborne and Turner 1984).

Important Genera and Economic Importance. Most of the genera are an important source of tropical fruits or timber. These include sapodilla or *Manilkara zapota*, star apple or *Chrysophyllum cainito*, and marmalade plum or *Lucuma mammosa*, chiefly from the West Indies and tropical America. The fruits of *Synsepalum dulcifolium*, commonly called miraculous berry, from West Africa, can give a sweet taste to even the sourest foodstuff. Shea butter is the fat from the seeds of *Butyrospermum parkii* of West Africa and Sudan. In Northern Africa and Morocco, a substitute for olive oil is extracted from the seeds of *Argania spinosa*. Many of the Sapotaceae members are the source of gutta-percha, which can be used as a substitute for natural rubber. The latex of *Palaquium gutta* is the main source of balata or gutta gum. A more or less similar substance is obtained from *Manilkara didentata*. Chicle gum, the base for the chewing gum industry, is obtained from the milky latex of *M. zapota*. Timber from various members of this family is very hard, heavy and durable but sometimes difficult to season and to work with tools. The most important genera are *Manilkara* and *Mimusops; Minusops elengi* with its sweet-scented flowers is planted as an avenue tree.

Taxonomic Considerations. There is not much controversy about the taxonomic position of the family Sapotaceae. Almost all the taxonomists place it in the order Ebenales (Bentham and Hooker 1965b, Engler and Diels 1963, Hutchinson 1969, 1973, Cronquist 1968, 1981, Dahlgren 1980a, 1983a, Takhtajan 1969, 1980).

However, Takhtajan (1987) raises the family to a distinct order, Sapotales.

There is much controversy over the nomenclature of various genera and species of this family which often leads to taxonomic problems.

Sarcospermataceae

A unigeneric family with 6 species distributed in southeastern Asia, from India and southern China to the Malayan Archipelago.

Vegetative Features. Trees or shrubs; sometimes laticiferous. Leaves simple, entire, opposite or subopposite, stipulate; stipules small, caducous.

Floral Features. Inflorescence axillary racemes, spikes, or panicles, or flowers are solitary. Flowers small, actinomorphic, hermaphrodite and hypogynous. Calyx of 5 sepals, polysepalous, imbricate. Corolla of 5 petals, gamopetalous with a short corolla tube; lobes imbricate, spreading, rounded. Stamens 10, in two whorls of 5 each, epipetalous; outer whorl of staminodia; filaments short, anthers basifixed, lateroextrorse. Gynoecium syncarpous, uni- or bicarpellary; ovary uni- or bilocular, superior, with a short and stout style and a more or less truncate stigma, and 1 basal-axile, apotropous, ascending ovule per locule. Fruit a drupe; 1- or 2-seeded; without endosperm but with a large embryo.

Anatomy. Anatomically, Sarcospermataceae is quite similar to Sapotaceae members, except that the wood fibres of *Sarcosperma* have conspicuous large bordered pits, and angular walls of fibre cells which occasionally contain crystals. On this basis Lam (1939) had proposed a distinct family.

Chromosome Number. The haploid chromosome number of *Sarcosperma arboreum* is n = 12 (2n = 24; Kumar and Subramanian 1987).

Taxonomic Considerations.   Most taxonomists treat *Sarcosperma* as a member of Sapotaceae (Bentham and Hooker 1965b, Hallier 1912, Bessey 1915). Melchior (1964) and Hutchinson (1969, 1973) consider it as a distinct family. Cronquist (1968, 1981), Dahlgren (1980a) and Takhtajan (1980, 1987), however, do not recognise it as a separate family.

In the absence of embryological and phytochemical data, we retain the family Sarcospermataceae in the Ebenales and near Sapotaceae, on the basis of exomorphological and anatomical features.

Ebenaceae

A family of only 3 genera and about 500 species distributed in the Indo-Malayan region. There are five genera according to Lawrence (1951), seven according to Hutchinson (1973); Willis (1973) and Kumar and Subramanian (1987) recognise three genera, and Takhtajan (1987) recognises only two genera.

Vegetative Features.   Dioecious trees or shrubs; heartwood often black, red or green; without any latex. Leaves alternate, opposite or whorled, simple, entire, coriaceous and estipulate.

Floral Features.   Inflorescence cymes or flowers solitary axillary (Fig. 47.3A). Flowers ebracteate, bracteolate, unisexual (Fig. 47.3B, C) (bisexual in *Diospyros*), actinomorphic, 3- to 7-merous, hypogynous. Calyx lobes variable, gamosepalous, persistent. Corolla 3- to 7-lobed, gamopetalous, campanulate (Fig. 47.3D) or tubular or urceolate; petals coriaceous, usually contorted and imbricate. Stamens as many as or 2 or 3 times the number of corolla lobes, separate or united in pairs; isomerous, or in 2 whorls, frequently numerous due to branching. Staminodes usually present in pistillate flowers; anthers bicelled, introrse, dehisce longitudinally. Gynoecium syncarpous, 2- to 16-carpellary, ovary with as many locules, superior, with axile placentation, ovules pendulous (Fig. 47.3D), typically 2 in each locule; styles and stigmas 2–8; styles free or connate at the base. Fruit usually a berry with fewer seeds than the number of ovules present. Embryo straight or slightly curved, in abundant cartilaginous endosperm.

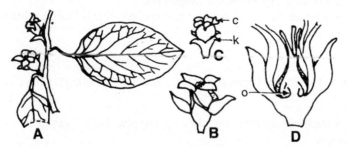

Fig. 47.3   Ebenaceae: **A–D** *Diospyros virginiana*. **A** Twig with flowers. **B, C** Flowers, male (**B**) and female (**C**). **D** Female flower, vertical section. *c* corolla, *k* calyx, *o* ovule. (Adapted from Lawrence 1951)

Anatomy.   Vessels small to medium-sized, with numerous multiples of 2 or 3 cells and sometimes with longer multiples of 4 or more cells in a radial or oblique pattern; perforation simple; intervascular pitting alternate and small. Parenchyma in numerous uniseriate, apotracheal bands, occasionally storied. Rays up to 2 cells wide or only uniseriate, heterogeneous. Leaves dorsiventral, tending to be isobilateral in some *Diospyros*, and *Euclea*. Epidermal hairs of various types, and sunken extrafloral nectaries reported. Stomata Ranunculaceous type and mostly on the lower surface.

Embryology.   Pollen grains 3-colporate with reticulate exine; 2-celled (*Diospyros virginiana*) at the dispersal stage. Ovules anatropous, bitegmic, tenuinucellate. Polygonum type of embryo sac, 8-nucleate at maturity. Endosperm formation of the Cellular type.

Chromosome Number.   The basic chromosome number is x = 15 (Takhtajan 1987). The somatic chromosome number for the genus *Diospyros* is reported as 2n = 30, 56 and 90 (Kumar and Subramanian 1987).

Chemical Features. The family Ebenaceae is particularly rich in pigments of naphthaquinone group (Harborne and Turner 1984); lycopene-based carotenoids are the fruit pigments in *Diospyros kaki* (Harborne and Turner 1984).

Important Genera and Economic Importance. *Diospyros* is the most important genus with ca. 240 species; many are of economic importance. *D. ebenum* of Sri Lanka and *D. melanoxylon* of India are the chief sources of the timber ebony. Completely black heartwood of these two plants is much sought for. Streaked or mottled heartwoods of *D. macassar* and *D. virginiana* are the sources of highly ornamental Macassar ebony and Andaman marblewood or zebrawood, respectively. The wood of *D. virginiana* is the standard wood for the heads of golf clubs. A gum is obtained (in India, Burma and Sri Lanka) from the pulp of the red fruits of *D. embryopteris*. The dried leaves of tendu or *D. melanoxylon* are used for wrapping bidi (commonly smoked in India), a tobacco product. The Japanese persimon is the edible fruit of *D. kaki*.

Taxonomic Considerations. The family Ebenaceae has been considered to be a member of the order Ebenales which is justified. The family has its ancestry in the Theales (Cronquist 1968).

Styracaceae

A small family of 12 genera and 180 species distributed in 3 centres of distribution: Eastern Asia to Western Malaysia, Southeastern United States and Mexico to tropical and subtropical South America (Morton and Dickison 1992); predominantly belongs to warmer regions. A single species (*Styrax officinale*) is Mediterranean.

Vegetative Features. Small trees or shrubs, with simple, estipulate, entire, soft, herbaceous to coriaceous, alternate leaves, often covered with stellate or peltate indumentum.

Floral Features. Inflorescence racemose to paniculate cymose, without any bracteole. In *Melliodendron* and *Halesia* the flowers are fasciculate or solitary at the nodes of the older shoots. Flowers bracteate, ebracteolate, hermaphrodite, rarely polygamodioecious as in *Bruinsmia,* actinomorphic, complete. Calyx 4- or 5-cleft or -toothed, tubular, gamosepalous, valvate or open, persistent. Corolla of 4 or 5 petals, gamopetalous, but often lobed up to the base, imbricate or valvate. Stamens 8–12 in one whorl, epipetalous, connate at base or into a tube. Anthers bicelled, oblong or linear, rarely rounded, dehiscence longitudinal. Gynoecium syncarpous, 3- to 5-carpellary, ovary mostly superior (half-inferior to inferior in *Halesia* and *Pterostyrax*); usually 3- to 5-locular below but unilocular above with 1 or a few pendulous ovules in each locule, placentation axile; style 1, simple, stigmas 1–5, capitate or lobed. Fruit drupaceous or capsular with fleshy or dry papery and dehiscent pericarp, and one or a few seeds. Embryo straight, endosperm present.

Anatomy. Vessels small to medium-sized; perforation plates scalariform, intervascular pitting scalariform or opposite; parenchyma apotracheal, diffuse and in regular uniseriate lines. Rays up to 2 to 4 cells wide, sometimes up to 6 cells, heterogeneous. Fibres with bordered pits. Leaves dorsiventral, bears scattered, brown tomentum of very small stellate hairs on the lower surface; stomata of Ranunculaceous type, confined to the lower surface. Both solitary and clustered crystals reported around the vascular bundles, in species of *Halesia* and *Styrax.*

Embryology. Pollen grains solitary, medium-sized, radially symmetrical; 2-celled at the time of shedding; 3-colporate and spinuliferous exine (Morton and Dickison 1992). Ovules anatropous, uni- or bitegmic (*Styrax*), tenuinucellate. Polygonum type of embryo sac, 8-nucleate at maturity. Endosperm formation of the Cellular type.

Chromosome Number.   The basic chromosome number is x = 8, 12.

Chemical Features.   Iridoids are absent in *Styrax japonicum* (Jensen et al. 1975). An aliphatic polyalcohol—L-styracitol—occurs only in the fruits of *Styrax obassia* (Plouvier 1963).

Important Genera and Economic Importance.   Some of the important genera are: *Styrax, Bruinsmia, Alniphyllum, Huodendron, Halesia, Pterostyrax, Parastyrax* and *Melliodendron. Pamphila* has only 5 stamens and is an isolated genus in Brazil; *Halesia* occurs in eastern North America and in eastern Asia. The sole representative in Africa is *Afrostyrax* confined to the Cameroons, West Africa.

The family members are the main source of the medicinally used benzoin. *Styrax benzoides* is the source of the resin Siam benzoin, and Sumatra benzoin is obtained from *S. tonkinensis;* storax, another resin, is obtained from *S. officinale. S. japonica* yields fine-grained timber used in making handles of umbrellas.

Taxonomic Considerations.   Most taxonomists treat the Styracaceae as a distinct family in the order Ebenales. It can be easily distinguished by the stellate pubescence, a single whorl of stamens, imperfectly septate ovary, the single style and the typical fruit. Hutchinson (1969, 1973) included Styracaceae in a separate order, Styracales, along with Lissocarpaceae and Symplocaceae, and placed the order next to the Cunoniales. Cronquist (1968, 1981) includes them in the Ebenales. However, Dahlgren (1983a) and Takhtajan (1980, 1987) include Styracaceae in the Theales.

With their sympetalous corolla, fewer stamens and axile placentation, the Styracaceae may be retained in Ebenales till further study.

The genus *Afrostyrax* is regarded as a member of Huaceae by some anthors such as Baas (1972) Cronquist (1981), Takhtajan (1980) and Willis (1973).

Lissocarpaceae

A unigeneric family with the only genus *Lissocarpa,* which is distributed in tropical South America. It is a segregate from Styracaceae.

Vegetative Features.   Small tree of the tropics of South America. Leaves alternate, estipulate.

Floral Features.   Calyx of 4 sepals, gamosepalous, imbricate. Corolla of 4 petals, sympetalous, lobes contorted. Stamens 8, inserted towards the base of the corolla tube; anthers apiculate. Gynoecium syncarpous, 4-carpellary, ovary 4-locular, inferior; ovules 2 in each locule, pendulous. Fruit more or less fleshy, indehiscent. Seeds with a straight embryo and copious endosperm.

Structurally the family Lissocarpaceae is similar to the family Styracaceae. It is different from Styracaceae in: (a) pubescence of simple hairs; (b) flowers with a corona, contorted corolla lobes and imbricate calyx; (c) stamens twice as many as the corolla lobes and (d) fruit fleshy.

Taxonomic Considerations.   Bentham and Hooker (1965b) treated *Lissocarpa* as a genus of the family Styracaceae, but other taxonomists recognize it as a family distinct from Styracaceae, on the basis of the differences listed above. Hutchinson (1969, 1973) placed it in his Styracales along with Styracaceae and Symplocaceae. But other taxonomists treat this family as a member of the Ebenales (Melchior 1964, Cronquist 1968, 1981, Dahlgren 1980a, 1983a, Takhtajan 1980, 1987).

Correct assignment of the position of family Lissocarpaceae must await detailed studies in various fields.

Symplocaceae

A unigeneric family of the genus *Symplocos* with about 300 species distributed in both the New World and the Old World tropics and subtropics, but absent from Africa.

Vegetative Features. Trees or shrubs, with alternate, simple, estipulate, often coriaceous leaves.

Floral Features. Inflorescence racemose or paniculate (Fig. 47.6A). Flowers ebracteate, bracteolate, hermaphrodite (Fig. 47.6B, C), actinomorphic, epi- or perigynous. Sepals 5, gamosepalous, imbricate, persistent. Corolla of 5 or 10 petals, gamopetalous but divided nearly to the base, in 1 or 2 series, imbricate. Stamens numerous (Fig. 47.6C) in 1–4 series; sometimes as few as 5 (Willis 1973) or 4 (Lawrence 1951, Hutchinson 1969), epipetalous or free of corolla or filaments variously connate; anthers ovate or rotund, bicelled, dehiscence longitudinal. Gynoecium syncarpous, 2- to 5-carpellary, ovary as many loculed, inferior or half-inferior, placentation axile, ovules 2 or 4 per locule (Fig. 47.6C); style 1, simple stigma capitate or 2- to 5-lobed. Fruit a berry (Fig. 47.6D) or drupe; seeds with copious endosperm and a straight or curved embryo.

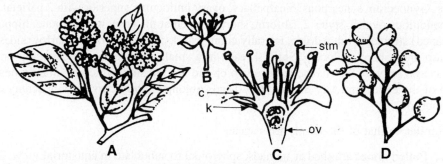

Fig. 47.6 Symplocaceae: **A–D** *Symplocos paniculata*. **A** Twig with flowers. **B** Flower. **C** Vertical section. **D** Fruits. *c* corolla, *k* calyx, *ov* ovary, *stm* stamen. (After Lawrence 1951)

Anatomy. Vessels small, solitary, perforation plates scalariform, with spiral thickenings; parenchyma apotracheal; rays up to 3–5 cells wide, markedly heterogeneous; fibres with bordered pits and often with spiral thickenings. Leaves dorsiventral, stomata Rubiaceous type, confined to the lower surface. Crystals solitary or clustered.

Embryology. Pollen grains 3- or 4-porate with spinuliferous exine and shed at 2-celled stage. Ovules anatropous, unitegmic, tenuinucellate. Polygonum type of embryo sac, 8-nucleate at maturity. Endosperm formation of the Cellular type.

Chromosome Number. The basic chromosome number is x = 11 to 14.

Chemical Features. Seed fats of *Symplocos paniculata* are rich in linoleic acids (Shorland 1963).

Important Genera and Economic Importance. *Symplocos* is the only genus, some of its species have medicinal importance, e.g. *S. racemosa,* commonly called lodh tree, its bark has astringent properties, and is the source of a dye that gives a red and yellow colour (Hutchinson 1969).

Taxonomic Considerations. Bentham and Hooker (1965b) treated *Symplocos* as a member of Styracaceae. Engler and Diels (1936) regarded it as a member of a distinct family, Symplocaceae. Cronquist (1968, 1981), Hutchinson (1969, 1973) and Takhtajan (1980, 1987) treat Symplocaceae as a family distinct from Styracaceae.

Although Hutchinson placed Styracales (including Symplocaceae) next to Cunoniales, Takhtajan (1987) and Thorne (1983) place it in the Theales. Symplocaceae usually has numerous stamens, often in five bundles and alternate with the petals; they do not seem to be particularly primitive in other respects (Cronquist 1968). In our opinion this family need not be included in the Theales. The position of Symplocaceae in the Ebenales is justified.

Hoplestigmataceae

A small unigeneric family of the genus *Hoplestigma* with only 2 species, endemic to tropical Africa.

Vegetative Features.   Trees with large simple, entire, alternate, estipulate leaves.

Floral Features.   Inflorescence terminal, ebracteate, brown-hirsute, subscorpioid cymes. Flowers hermaphrodite, actinomorphic, shortly pedicelled, hypogynous. Calyx entire in bud, globose, splitting into irregular lobes, persistent. Corolla of 11 to 14 petals, shortly connate, imbricate, irregular, 3- or 4-seriate. Stamens 20–35, free from corolla (Takhtajan 1987), 3-seriate, with filiform filament and elongate-oblong anthers. Gynoecium syncarpous, bicarpellary; ovary unilocular, superior with 2 parietal placentae, each bears 2 pendulous ovules; styles 2, filiform, shortly united at the base, with capitate, hippocrepiform (horseshoe-shaped) stigmas. Fruit a drupe, laterally compressed, channelled on the narrow sides, with soft leathery exocarp and a bony endocarp. Seeds 4, each in a small chamber in the endocarp on each side of the fruit, covered only by the channelled exocarp on the narrow side, which eventually becomes torn by the growth of the endocarp (Airy Shaw 1965). Seeds oblong slightly curved, with scanty endosperm and a large embryo.

Anatomy.   Similar to that of the family Ebenaceae.

Embryology.   Pollen grains are shed as monads, spheroidal to suboblate in equatorial view, 3-colporate and the colpi bordered by six meridional ridges (Nowicke and Millar 1989). Ovules anatropous, unitegmic, tenuinucellate. Development of embryo sac and endosperm have not been studied fully (Johri et al. 1992).

Taxonomic Considerations.   Hallier (1912) included *Hoplestigma* in Boraginaceae but it differs from this family: (a) many petals (11–14) in 2 or 3 series, (b) 20–35 stamens free from the corolla, and (c) fleshy fruit. Bentham and Hooker (1965b) and Bessey (1915) did not recognise it as a distinct family. Lawrence (1951) placed this family in the suborder Sapotineae of the Ebenales, along with the family Sapotaceae. Hoplestigmataceae is different from the Sapotaceae: (a) absence of laticiferous ducts, (b) presence of terminal inflorescences, (c) calyx entire in bud splitting into irregular lobes, (d) stamens free from corolla, and (e) unilocular ovary with parietal placentation. Melchior (1964) placed it in the suborder Ebenineae.

Hutchinson treated Hoplestigmataceae as a member of the Bixales (1973), probably because of parietal placentation. Wagenitz (1964) included it in the Ebenales, Cronquist (1981) in the Violales. According to Dahlgren (1980a, 1983a) and Takhtajan (1987), Hoplestigmataceae is a member of the Boraginales. It is (perhaps) closer to the Boraginaceae. The pollen grains are similar in *Bourreria* and *Ehretia* of the Boraginaceae (Nowicke and Miller 1989). Similar pollen morphology is also common in the Myrtales but not reported in the Ebenales or Violales. Bicarpellate ovary with few ovules and sympetalous corolla of Hoplestigmataceae is not typical of the Violales, although parietal placenta and numerous stamens suggest alliance to the Violales. Parietal placentation observed in the Hoplestigmataceae is not typical of the Ebenales but sympetalous corolla, stamens twice the number of the corolla lobes, and fewer ovules suggest alliance with the Ebenales. The inflorescence, sympetalous corolla, pollen morphology, bicarpellary ovary and unitegmic ovules suggest Boraginaceae but not the numerous stamens. *Hoplestigma* shares many features with the Myrtales, but differs in: 1) petals united at the base, and 2) unitegmic ovules. The placement of the family Hoplestigmataceae has never been resolved satisfactorily. In our opinion, this family may be retained in the suborder Ebenineae of the order Ebenales, till further investigations are carried out.

Taxonomic Considerations of the Order Ebenales

Most taxonomists agree that the Ebenales (including the families already mentioned, except Hoplestigmataceae) constitute a natural group, and that they have originated from the Theales. Ebenales also resemble the Ericales to some extent, but that is probably due to parallel evolution, both having a common ancestry in the Theales. The family Hoplestigmataceae has been included in Boraginales (Hallier 1912) or Bixales (Hutchinson 1973) or Violales (Cronquist 1968). "Hoplestigmataceae are an isolated group of uncertain affinities", according to Cronquist (1968).

In the present work, however, this family has been included in the Ebenales.

Order Oleales

This order comprises only one family—Oleaceae. The members are mostly trees or evergreen undershrubs with woody stems and opposite or alternate leaves, and flowers with gamopetalous, contorted corolla.

Oleaceae

A family or about 22 genera and about 500 species, distributed mostly in the temperate and paleotropical regions of the World.

Vegetative Features.   Mostly perennial climbing shrubs or trees (*Olea europaea*). Stem woody, herbaceous above, cylindrical, solid, with fine pubescence. Leaves opposite, alternate in *Jasminum humile, J. floridum* and *J. parkeri*; petiolate, estipulate, simple or pinnately compound.

Floral Features.   Inflorescence a raceme or a cyme—a compound raceme in *Syringa* spp. and *Fraxinus* spp.; a dichasial cyme in *Jasminum* spp. Flowers bracteate, ebracteolate, usually hermaphrodite (unisexual in *Fraxinus nigra* and *F. pennsylvanica*), actinomorphic, hypogynous, pedicellate. Calyx 4-lobed, rarely up to 15-lobed, valvate, absent in *Fraxinus* spp. and *Forestiera* spp., gamosepalous. Corolla basically with 4 or 5 petals, gamopetalous, valvate (*Syringa*) or imbricate (*Jasminum*); apetalous in *Fraxinus velutina, Forestiera* sp., and female flowers of *Olea dioica*. Stamens typically 2, rarely 4, as in *Hesperelaea* and *Tessarandra*.

Anthers distinct, bicelled, the cells usually back to back, dehiscence longitudinal; often apiculate due to extensive extension of the connective. Gynoecium syncarpous, bicarpellary; ovary bilocular, placentation axile, superior, ovules 2 per locule; style 1 or often absent; stigma 1 or 2. Fruit a berry in *Ligustrum*, drupe in *Olea*, loculicidal capsule in *Syringa* and *Forsythia*, circumsessile capsule in *Menodora* or a samara as in *Fraxinus*. Seed with fleshy endosperm and a straight embryo.

Anatomy.   Vessels typically small with variable arrangement; commonly ring-porous and with spiral thickenings, perforations simple, intervascular pitting typically very small to minute; parenchyma paratracheal, varying from a few cells round the vessels to aliform or locally confluent; absent in some genera. Rays 1–4 cells wide, heterogeneous or homogeneous. Leaves usually dorsiventral; stomata numerous, confined to the lower surface, generally of Ranunculaceous type. Groups of secretory type of hairs often form extrafloral nectaries in various species of *Forestiera, Fraxinus, Olea, Osmanthus, Phillyrea* and *Syringa*.

Embryology.   Pollen grains 3- or 4-colporate to colporoidate and 2-celled at shedding stage. Ovules anatropous, orthotropous in *Linociera intermedia*, unitegmic and tenuinucellate. Embryo sac of the Polygonum type, 8-nucleate at maturity. Endosperm formation of the Cellular type.

Chromosome Number.   The basic chromosome number is x = 10–14 and 23; in Oleoideae x = 23 and in Jasminoideae x = 11, 13 or 14. According to Harborne and Green (1980), all the species with x = 23 (except a few spp. of *Syringa* with x = 24) are of polyploid origin and are derived by allopolyploidy from 11- and 12-chromosome prototypes that are not in existence.

Chemical Features.   Fruit coat of *Olea europaea* is rich in fatty acids: palmitic, oleic and linoleic acids (Shorland 1963). Harborne and Green (1980) reported common occurrence of four flavonol glycosides and four flavone glycosides. The subfamily Oleoideae with higher basic chromosome number has complex flavonoid patterns; subfamily Jasminoideae with lower basic chromosome number has simple patterns based only on flavonols.

Important Genera and Economic Importance. The genus *Jasminum* is the largest with 300 species distributed in the Old World tropics and subtropics, chiefly in Eastern Himalayas and southwestern India and Nepal, Pakistan, Bhutan, Bangla Desh and Sri Lanka. Some species are cultivated for their scented flowers. *Ligustrum* with 40–50 species is another large genus distributed in Europe to North Iran, East Asia, Indo-Malaysia to New Guinea and Queensland, India, Bhutan, Bangla Desh and Sri Lanka. The genus *Olea* occurs mainly in the Mediterranean region, Africa, Mascarene, East Asia, Indo-Malaysia, East Australia, New Zealand and Polynesia. One species of *Olea* (*O. ferruginea*) is cutivated in many parts of India, Nepal, Pakistan, Bangla Desh and Sri Lanka for its edible fruits. Other important genera include *Fraxinus, Osmanthus, Schrebera* and *Syringa*.

Olive fruits from *Olea europaea* are edible and also the source of olive oil which has medicinal value. The timber from *Fraxinus excelsior* is very durable and used for cabinet work. Many genera are cultivated for their ornamental value such as *Fraxinus, Forsythia, Syringa*, and *Jasminum*.

Taxonomic Considerations. The family Oleaceae comprises two subfamilies:
(a) Jasminoideae—seeds generally erect, fruits vertically constricted if both the carpels develop equally— *Jasminum* and *Menodora;*
(b) Oleoideae—seeds pendulous, fruits not vertically constricted—*Olea, Fraxinus, Forestiera* and others.

Bentham and Hooker (1965b) placed the family Oleaceae in the order Gentianales, because of: a superior bicarpellary ovary, gamopetalous corolla with contorted or convolute aestivation, and stamens adnate to the base of the corolla. Other families included in the Gentianales are Apocynaceae, Asclepiadaceae, Loganiaceae, Gentianaceae and Salvadoraceae. Lingelsheim (1920) retained all these families—except Salvadoraceae—in the Gentianales and changed the name to Contortae. There is no justification to change the name because it causes confusion.

Wettstein (1935) and Rendle (1925) segregated Oleales as a distinct order including only the Oleaceae. Cronquist (1968, 1981) distinguishes Oleaceae from the Gentianales as the former do not have internal phloem, but do have integumentary tapetum and Cellular type of endosperm. He includes Oleaceae in his Scrophulariales along with the family Buddlejaceae. Hutchinson (1969, 1973) places Oleaceae and Loganiaceae in the order Loganiales. Dahlgren (1980a, 1983a) retains it in a distinct order Oleales of Superorder Gentiananae and gives weightage to both embryological as well as chemical features. According to Gornall et al. (1979): "Oleaceae is unusual in having both myricetin and flavones. As such, it perhaps could be intermediate in the myricetin-loss/flavone-gain trend. Thus, the family fits well between the myricetin-bearing Corniflorae and the myricetin-free Lamiiflorae". Takhtajan (1980, 1987) treats Oleaceae as the only family of the order Oleales, and places it next to the Gentianales.

On the basis of morphological, embryological and chemical data, the family Oleaceae appears to be best placed in a separate order, but because of some of the common features with the Gentianales members, this order should be placed near the Gentianales.

Taxonomic Considerations of the Order Oleales

The Oleales, as mentioned earlier, is segregated as an independent order by Rendle (1925) and Wettstein (1935). This order includes only one family, Oleaceae, The Oleales differ from the Gentianales in the absence of intraxylary phloem, and the presence of two stamens, integumentary tapetum, and the Cellular type of endosperm. The Oleales also contain both myricetin and flavones, the myricetin is absent from the Gentianales.

Melchior (1964), Gornall et al. (1979), Dahlgren (1980a, 1983a) and Takhtajan(1980, 1987) treat Oleales as an order separate from the Gentianales, which is justified.

49

Order Gentianales

The order Gentianales (sensu Melchior 1964) includes 7 families—Loganiaceae, unigeneric Desfontainiaceae, Gentianaceae, Menyanthaceae, Apocynaceae, Asclepiadaceae and Rubiaceae. In addition, Takhtajan (1987) considers Saccifoliaceae, Carlemanniaceae, Dialypetalanthaceae, Theligonaceae, Spigeliaceae and Plocospermataceae also as members of the order Gentianales.

The plants have opposite, simple or pinnately compound, estipulate leaves; internal phloem; hypogynous, actinomorphic flowers with contorted aestivation; 1 whorl of alternipetalous stamens; unitegmic, tenuinucellate ovules, mostly Nuclear type of endosperm, and many genera contain alkaloids or glucosides. The family Menyanthaceae, however, differs from the other families as they are aquatic plants, have alternate leaves, integumentary tapetum and Cellular type of endosperm. They also lack internal phloem.

Many of the Gentianalian genera are medicinally important as they contain alkaloids, glycosides and other chemicals.

Loganiaceae
A family of about 32 genera and 800 species distributed predominantly in the warmer and tropical regions of the world. A large number of genera are confined to the Old World tropics.

Vegetative Features.   Herbs, shrubs, trees or lianas. Stem woody or herbaceous; leaves simple, usually opposite and stipulate.

Floral Features.   Inflorescence cymes or thyrses. Flowers bracteate, bracteolate, hermaphrodite, actinomorphic, hypogynous (perigynous in *Mitreola*). Calyx of 4 or 5 sepals, polysepalous or gamosepalous; in *Usteria* one calyx lobe is larger and petaloid, Corolla of 4 or 5 petals, each petal often bilobed, gamopetalous, aestivation various; mouth of the corolla tube often hairy. Stamens 4 or 5, alternate with petals, epipetalous (rarely 1 stamen as in *Usteria);* anthers bicelled, longitudinally dehiscent, anther loculi confluent and peltate in open flowers of *Peltanthera*. Gynoecium syncarpous, bicarpellary, ovary bilocular (unilocular in *Strychnos* and incompletely bilocular in *Fagraea*), axile placentation, superior, (half-inferior in *Mitreola*); ovules many; style 1 or 2 with a single common stigma as in *Cynoctonum*; stigmas 1- or 2-, rarely 4-cleft. Fruit a capsule, rarely a berry or a drupe. Seeds sometimes winged, as in *Gelsemium* and *Coinochlamys*; endosperm fleshy or bony, embryo small and straight.

Anatomy.   Vessels variable in size and number, often with numerous multiples of 4 or more cells and with spiral thickenings; perforation plates simple; ring-porous. Parenchyma absent or only a few cells to a distinct sheath round the vessels or in narrow or broad bands; rays 3–12 cells wide, hetero- or homogeneous. Intraxylary as well as interxylary phloem reported in *Logania* and *Strychnos*. Leaves isobilateral in *Gomphostigma, Logania, Nuxia* and *Strychnos,* dorsiventral in others. Simple, unicellular or uniseriate hairs reported. Extrafloral nectaries occur at the base of the petiole in *Fagraea* spp. Stomata variable; Cruciferous type in most genera; Rubiaceous type in species of *Gelsemium* and *Strychnos*.

Embryology.   Pollen grains 3- to 5-aperturate with a smooth exine and shed at 3-celled stage. Ovules anatropous, unitegmic, tenuinucellate. Polygonum type of embryo sac, 8-nucleate at maturity. Endosperm formation of the Nuclear type.

Chromosome Number. The basic chromosome number for *Gelsemium* is x = 4, for *Usteria* and *Strychnos* x = 11, and for others x = 8,10 or 12. The diploid chromosome number for *Nuxia floribunda* is 2n = 38 (Gadella 1966).

Chemical Features. Rich in alkaloids and seco-iridoids (Jensen et al. 1975); devoid of glycosides. Cyanogenic glucosides and morronisides are reported in various species of *Sambucus*[5]. (Jensen and Nielsen 1973, 1974). Strychnine and brucine are important alkaloids from *Strychnos nux-vomica* fruits. Iridoids isolated from fruit pericarp are loganin, loganic acid, deoxyloganin, ketologanin and secologanin (Bisset and Choudhury 1974).

Important Genera and Economic Importance. Some of the important genera of this family are: *Logania, Gelsemium, Fagraea* and *Strychnos.Gelsemium elegans* is a climbing shrub with golden-yellow flowers. *Strychnos nux-vomica* is a tree which occurs throughout tropical India. Seeds are a source of the drug nux-vomica, which is used as tonic, stimulant and in treatment of paralysis and nervous disorders (Singh et al. 1983). The seeds of *S. potatorum* are used to purify water for drinking. These are reported to be a rich source of polysaccharide gum suitable for use in paper and textile industries. Bark of *S. toxifera* produces curare to poison arrows. *Fagraea crenulata, F. elliptica, F. fragrans* and *S. nux-vomica* yield good timber.

Taxonomic Considerations. Most phylogenists treat the Loganiaceae as a member of the order Gentianales (Melchior 1964, Cronquist 1968, 1981, Dahlgren 1980a, 1983a, Takhtajan 1980, 1987). Lawrence (1951) places this family in the suborder Gentianineae of the order Contortae. Hutchinson (1969, 1973) erects a separate order, Loganiales, including Potaliaceae (= Desfontainiaceae), Buddlejaceae, Antoniaceae, Spigeliaceae, Strychnaceae and Oleaceae.

The Loganiaceae are distinguished from other families of the order Gentianales by the opposite stipulate leaves, bilocular, superior ovary with axile placentation and presence of internal phloem in the majority of the genera.

This family may be the most primitive in this order. It is justified to retain the family Loganiaceae in the order Gentianales because of: (1) presence of internal phloem, (2) unitegmic, tenuinucellate ovules, (3) the Nuclear type of endosperm formation, and (4) presence of alkaloids and seco-iridoids.

Desfontainiaceae

A unigeneric family with one species, *Desfontainia spinosa* (Takhtajan 1987), distributed in the tropics.

Vegetative Features. Trees or shrubs with opposite, simple, entire, or spinose-dentate (Fig. 49.2 A), often thickly coriaceous leaves, connected at the base by a stipule-like sheath.

Floral Features. Solitary terminal or a few-flowered cyme. Flowers ebracteate, ebracteolate, hermaphrodite, actinomorphic, hypogynous, pentamerous. Calyx of 5 sepals, polysepalous, imbricate; corolla of 5 petals, gamopetalous, contorted, infundibuliform (Fig. 49.2A). Stamens 5, epipetalous, inserted, introrse; anthers basifixed, with short filaments (Fig. 49.2 B). Gynoecium syncarpous, pentacarpellary, ovary superior with parietal placentation, ovules numerous; style 1, simple, stigma capitate. Fruit a berry; seeds numerous, unwinged and with endosperm.

Anatomy. Xylem (in young stems) consists of radial rows of thin-walled vessels and tracheids of small diameter; rays uniseriate; perforation plates scalariform. Leaves dorsiventral; stomata of Ranunculaceous type.

---

[5]Treated as member of Sambucaceae at present (Takhtajan 1987).

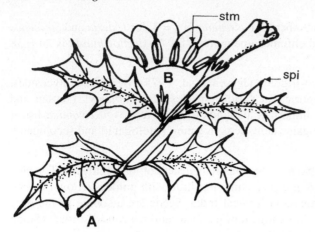

Fig. 49.2 Desfontainiaceae: **A, B** *Desfontainia spinosa.* **A** Twig bearing flower. **B** Corolla, opened. *spi* spine, *stm* stamen. (Adapted from Hutchinson 1969)

Chromosome Number. The basic chromosome number is x = 7.

Important Genera and Economic Importance. *Desfontainia spinosa*, the only member, has no economic importance.

Taxonomic Considerations. *Desfontainia* is included in Loganiaceae by Bentham and Hooker (1965b), Metcalfe and Chalk (1972), Lawrence (1951), Cronquist (1968, 1981) and Dahlgren (1975a, 1980a, 1983a). Melchior (1964) treats it as a unigeneric family because of its spinescent leaves and typically pentamerous, solitary, terminal flowers. Willis (1973) and Hutchinson (1969, 1973) include Desfontainiaceae in Potaliaceae. Takhtajan (1969, 1980, 1987) recognises Desfontainiaceae as a family distinct from Loganiaceae.

The genus *Desfontainia* is anatomically different from other members of Loganiaceae, and also in basic chromosome number, and some exomorphological features, such as solitary terminal and pentamerous flower, and pentacarpellary ovary. Under the circumstances, it is best treated as a distinct family.

Gentianaceae

A family of 70 genera and about 800 species of worldwide distribution but abundant in the arctic and alpine regions.

Vegetative Features. Mostly perennial herbs with rhizomes, rarely subshrubs or shrubs. The genus *Crawfurdia* is a twiner; some members are halophytes and saprophytes: *Cotylanthera, Leiphaimos, Obolaria* and *Voyria.* Stem soft, herbaceous; branching dichotomous, often winged or ribbed. Leaves opposite, decussate, often basally connate, simple, estipulate, usually entire.

Floral Features. Inflorescence dichasial or monochasial cyme. Flowers bracteate, sometimes bracteolate, hermaphrodite, actinomorphic, hypogynous, usually showy, 4- or 5-merous, rarely more, e.g. *Blackstonia.* Calyx of 4 or 5 sepals, gamosepalous, usually tubular, generally imbricate. Corolla of 4 or 5 gamopetalous petals, contorted, tubular, rotate, salverform or campanulate; scales or nectar pits occur in corolla tube. Stamens of the same number as the corolla lobes and alternate with them, epipetalous, distinct (syngenesious in *Voyria* and *Leiphaimos*); usually versatile; anthers bicelled, introrse, longitudinally dehiscent (poricidal in *Exacum*); pollen grains variable and distinctive; annular disc present or absent. Gynoecium syncarpous, bicarpellary, ovary usually unilocular, with two parietal placentae (bilocular with axile placentation in *Exacum*); ovules numerous. Style 1, simple, stigma simple or bilobed. Fruit a septicidal capsule; seeds with fleshy endosperm and a small embryo.

Anatomy. Xylem in the form of a closed cylinder without distinct medullary rays in *Blackstonia, Centaurium, Exacum* and *Gentiana.* Vessels not very distinct from the adjoining fibre cells; with simple perforations; parenchyma scanty, rays uniseriate. Intraxylary phloem present. Bundles collateral to bicollateral in saprophytic genera like *Voyria, Voyriella* and *Leiphaimos.* Acicular crystals occur in the cortex and pith of *Enicostema* and oil droplets in the epidermal cells of *Gentiana corymbifera.*

Embryology. Pollen grains tricolporate or tetracolpate with thick exine, 2- or 3-celled at the shedding stage and sometimes even in the same anther (*Canscora diffusa*). Pollen grains remain united in tetrads in species of *Adenolisianthus, Helia, Irlbachia* and others (Johri et al. 1992). Ovules anatropous, unitegmic, tenuinucellate. Embryo sac of the Polygonum type, 8-nucleate at maturity. Endosperm of more or less the Nuclear type. In the saprophytic genera—*Cotylanthera, Leiphaimos, Voyria,* and *Voyriella*—the embryos are 5- to 24-celled and undifferentiated at seed-shedding (Natesh and Rau 1984).

Chromosome Number. The basic chromosome numbers are x = 5–11, 13, 15, 17 and 19.

Chemical Features. Gentianaceae are very rich in seco-iridoids. Flavones and flavonols also occur (Gornall et al. 1979).

Important Genera and Economic Importance. The important genera of this family include *Gentiana, Swertia* and *Centaurium.* Many of the Gentianaceae are grown as ornamentals. An infusion of the dried roots and rhizome of *Gentiana lutea* is used for stimulating gastric secretion, *Swertia chirayita* of the Western Himalayas in India is the source of another digestive stimulant called 'chirata'.

Taxonomic Considerations. The family Gentianaceae is included in the order Gentianales along with 6 other families. It is a natural taxon, if the Menyanthaceae is treated as a separate family.

Menyanthaceae

A small family of 5 genera and 33 species distributed in north and south temperate zones and tropical Southeast Asia.

Vegetative Features. Perennial, aquatic or marshy herbs, often with creeping rhizomes as in *Menyanthes.* Stems erect or runner-like; leaves alternate or subopposite, simple, sometimes peltate or trifoliate, estipulate and with sheathing petiole.

Floral Features. Flowers on peduncles clustered at the nodes or on the apparent petioles close below the leaf blades (Fig. 49.4 A). Flowers hermaphrodite, actinomorphic, hypogynous. Calyx of 4 or 5 sepals, poly- or gamosepalous, oblong or lanceolate. Corolla of 5 petals, gamopetalous, rotate (Fig. 49.4 B), valvate or conduplicate-valvate; margins or interior of petals often fimbriate (Fig. 49.4 C). Stamens 5, alternipetalous, epipetalous; anthers bicelled, versatile and sagittate; filaments linear, short; nectaries present. Gynoecium syncarpous, bicarpellary, ovary superior, unilocular, placentation parietal, ovules numerous (Fig. 49.4 D); style simple, short; stigma bilobed. Heterostyly observed in *Menyanthes.* Fruit a globose or ellipsoid 2- to 4-valved capsule (Fig. 49.4 E, F), or fleshy and indehiscent. Seeds many, discoid (Fig. 49.4 G), sometimes winged, endospermous.

Anatomy. Stem cortex contains numerous vertically-elongated air cavities (Fig. 49.4 H, I) giving a laciniated appearance. Branched sclerenchymatous idioblasts occur (Fig. 49.4 I) in *Limnanthemum.* Vascular bundles dispersed, colateral; vessels spiral with simple (*Limnanthemum*) or sometimes scalariform perforation plates (*Menyanthes*). Intraxylary phloem absent. Leaves dorsiventral or isobilateral; stomata Ranunculaceous type confined to the upper surface only in the peltate leaves of *Limnanthemum.* Crystals of calcium oxalate absent.

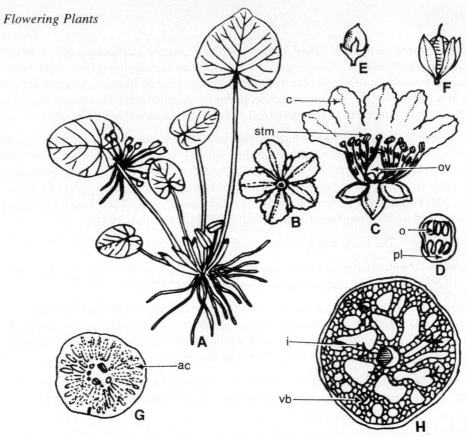

Fig. 49.4 Menyanthaceae: **A-H** *Nymphoides cristatum.* **A** Plant with inflorescence. **B** Flower, top view. **C** Corolla (opened). **D** Ovary, transection. **E, F** Fruit. **G, H** Transection of stem, note air cavities. *ac* air cavities, *c* corolla, *i* idioblasts, *o* ovule, *ov* ovary, *pl* placenta, *stm* stamen, *vb* vascular bundle. (**A-F** adapted from Subramanyam 1962, **G, H** after Metcalfe and Chalk 1972)

Embryology. Pollen grains triporate, spiraperturate and 3-celled at shedding stage. Ovules anatropous, unitegmic, tenuinucellate. Polygonum type of embryo sac, 8-nucleate at matuity. Endosperm of the Cellular type.

Chromosome Number. The basic chromosome numbers are x = 9, 27.

Chemical Features. Members of this family contain a bitter principle, meliatin; lack the flavones or C-glycoflavones of the Gentianaceae (Gornall et al. 1979).

Impotant Genera and Economic Importance. The family Menyanthaceae include the following genera: *Limnanthemum, Liparophyllum, Menyanthes, Nephrophyllidium* and *Villarsia. Limnanthemum indicum* plants are used as a substitute for chirata (*Swertia chirayita*) against fever and jaundice (Chopra et al. 1956).

Taxonomic Considerations. The placement of the family Menyanthaceae is highly controversial. It was included in the family Gentianaceae as a subfamily Menyanthoideae (Lawrence 1951, Rendle 1925). Morphological and anatomical differences justify raising it to a family level (Wettstein 1935, Lindsey 1938). Hutchinson (1959, 1969, 1973), Cronquist (1968, 1981), Dahlgren (1980a, 1983a) and Takhtajan (1980, 1987) also treat it as a distinct family. Cronquist (1968, 1981) does not even retain it in the

Gentianales. According to him: "the Menyanthaceae are phenetically perfectly at home in the Polemoniales". Enbryologically, also, this family is different from the Gentianaceae in having integumentary tapetum and Cellular type of endosperm (Vijayaraghavan and Padmanabhan 1969).

The morphological, anatomical and embryological characters fully justify a family rank for Menyanthaceae.

The generic name *Limnanthemum* Gmel. (1770) should be replaced by *Nymphoides* Hill (1756) on account of priority (Subramanyam 1962). According to Takhtajan (1987), the generic name *Nephrophyllidium* Gilg has been changed to *Fauria* Franch.

Apocynaceae

A large family with 300 genera and about 1300 species; cosmopolitan in distribution, but more common in the tropics.

Vegetative Features.   Members are herbs (*Catharanthus roseus*), shrubs (*Nerium indicum*), or trees (*Alstonia scholaris*); *Beaumontia grandiflora* and *Trachelospermum jasminoides* are large, stout, woody climbers. Stem woody in shrubs and trees; soft in herbs; usually contain milky latex (sap colour-less in *Catharanthus*). Leaves opposite, decussate, sometimes alternate, as in *Thevetia* or whorled, as in *Alstonia, Rauwolfia;* simple, entire, petiolate, estipulate.

Floral Features.   Flowers either solitary or in axillary pairs (*Catharanthus*) or in cymose inflorescence; axillary cymes in *Allamanda,* terminal cymes in *Beaumontia,* axillary or terminal cymes in *Mandevilla* and *Melodinus.* In *Carissa* the flowers are in corymbose cymes; in *Alstonia* and *Rauwolfia* in umbellate cymes.

Flowers bracteate, bracteolate, pedicellate, hermaphrodite, actinomorphic, complete, hypogynous, typically pentamerous. Calyx of 5 sepals, gamosepalous but deeply lobed, aestivation quincuncial, often with a hairy outer surface. Corolla of 5 petals, gamopetalous, contorted in bud, salverform or infundibuliform; throat of the corolla tube hairy or appendaged foming a corona-like structure. Stamens 5, free, alternate with the petals, epipetalous; anthers introrse, often sagittate, bicelled, free or adherent (by viscid exudates) to the stigma or 'clavuncle'; filaments short. Pollen grains separate or rarely in tetrads as in *Condylocarpon.* Gynoecium composed of 2 carpels, which may be either fully syncarpous as in *Carissa* or may remain free below with 2 distinct ovaries as in *Catharanthus,* but with a single style and stigma. Each ovary supeior to half-inferior, unilocular with parietal placentation; when syncapous, bilocular with axile placentation. Style simple, a thick or dumble-shaped stigma; ovules few to many. Fruit a follicle (*Plumeria, Nerium*), capsule (*Allemanda*), drupe (*Thevetia*) or a berry (*Carissa*). Seeds winged in *Plumeria,* with a tuft of hairs in *Alstonia.* Seeds with fleshy endosperm and a straight embryo.

Anatomy.   Vessels usually small; medium-sized in some species of *Alstonia, Aspidosperma, Kibatalia, Rauwolfia, Vallaris* and some others; and sometimes large, as in species of *Anodendron* and *Landolphia*; perforations simple. Parenchyma most commonly apotracheal, as scattered cells or narrow bands; sometimes paratracheal. Rays typically 2 or 3 cells wide or exclusively uniseriate, heterogeneous. Leaves dorsiventral, occasionally isobilateral as in *Nerium.* Laticiferous canals and intraxylary phloem are reported. Stomata mostly Ranunculaceous type but Rubiaceous type in some taxa. Twenty three different types of stomata have been reported in *Catharanthus roseus* (Baruah and Nath 1996).

Embryology.   Pollen grains triporate and 3-celled at shedding stage; often 2-celled as in *Catharanthus, Holarrhena* and *Plumeria.* Ovule hemianatropous or anatropous, rarely orthotropous (*Rauwolfia sumatrana*) (Johri et al. 1992). Embryo sac of the Polygonum type, 8-nucleate at maturity. Occasionally twin and abnormal embryo sacs with fewer or more than 8 nuclei occur. Degeneration of the embryo sacs common (Johri et al. 1992). Endosperm formation of the Nuclear type.

Chromosome Number. Basic chromosome number is variable: x = 8–12, 18 or 23.

Chemical Features. Apocynaceae members are rich in indole alkaloids. Cyclitols are reported in *Nerium* and *Catharanthus*.

Important Genera and Economic Importance. *Apocynum* is the type genus with all the usual features and is an erect herb or undershrub. The type species *A. androsaemifolium* is interesting. Its flowers emit a sweet honey-like fragrance which attracts insects. These insects, *Musca pipiens* in particular, enter the flower, are caught in the honey-like substance, and die. Some of the genera exhibit xerophytic features. *Adenium* and *Pachypodium* have fleshy stems and *Carissa* has spiny stems. Other interesting genera are: *Lepinia*, *Pleiocarpa* and *Notonerium* with 3- to 5-carpellary ovary and *Epigynum* and *Ichnocarpus* with half-inferior ovary. The fruits of *Alyxia concatenata* is a pair of moniliform follicles. In *Allemanda* an unusual type of gynoecial deveopment has been observed. The two carpels are free at initiation but fuse completely during development resulting in a unilocular ovary with parietal placentation (Fallen 1985).

Economically, the family has limited importance. *Rauwolfia serpentina* roots are used medicinally as a cure for epilepsy, high blood pressure, insanity and cardiac diseases. A number of alkaloids such as reserpine, reserpinine, serpentine, serpentinine, ajmalinine, isoajmaline and a few more alkaloids are reported from this plant and reserpine is the most important alkaloid amongst these. *R. tetraphylla* and *R. vomitoria* from America and Africa are also useful sources of reserpine. *Catharanthus roseus* is the source of anticancer drugs—Vincristin and Vinblastin.

Many plants such as *Plumeria* spp., *Nerium odorum*, *Thevetia peruviana*, *Tabernaemontana divaricata* and *Allemanda* sp. are grown as ornamentals. The wood of *Wrightia tinctoria* and *Alstonia scholaris* is soft and used for wood carvings. The fruits of *Carissa carandas* are sour and edible. *Landolphia comoriensis*, *L. kirkii* and *Funtumia elastica,* three vines from Africa, yield rubber from their latex.

Taxonomic Considerations. The family Apocynaceae is closely related to the more advanced family Asclepiadaceae. It differs from Asclepiadacae in: (a) single style, (b) absence of a corona, (c) mostly free pollen grains (never in pollinia), (d) stamens free from the stigma, and (e) in the absence of translators connecting the pollinia from the adjacent anther sacs.

Bentham and Hooker (1965b) treated this family as a member of the order Gentianales. Hallier (1912) combined the two families—Apocynaceae and Asclepiadaceae—and derived them from the Linaceae. Bessey (1915) and Rendle (1925) included these two families in the Gentianales. Engler and Diels (1936) placed it in the order Contortae. Hutchinson (1948) included the Apocynaceae alone in his Apocynales. In this order, Hutchinson (1969, 1973) also included three other families—Plocospermataceae, Periplocaceae and Asclepiadaceae—and derived Apocynales from the Loganiales. Cronquist (1968, 1981), Dahlgren (1980a, 1983a) and Takhtajan (1980, 1987) treat Apocynaceae as a member of the order Gentianales.

In the advanced family Apocynaceae, about 70% genera are apocarpous; the gynoecium consists of two carpels at the time of initiation. These are free above and usually united just at the base. During floral development, the uppermost part of the carpels undergoes a temporary postgenial fusion along their ventral flanks. According to Fallen (1986), the presence of the two fusion zones provide evidence for interpreting the apocarpous condition as phylogenetically secondary.

The family Apocynaceae is a natural taxon and is correctly placed in the order Gentianales.

Asclepiadaceae

A large family of 250 genera and about 3000 species distributed mainly in the Old World tropics and some in tropical America.

Vegetative Features.   The members are mostly herbs or climbers and a few shrubs (*Calotropis gigantea*); perennial (*Asclepias, Gymnema, Stapelia*) or annual (*Pergularia daemia*); the vegetative parts yield latex Stem herbaceous, climbing or twinning, sometimes covered with dense tomentum (*Calotropis*) or with wax (*Hoya parasitica*). The stem of *Stapelia* is succulent, green and the leaves are scale-like. Leaves opposite, simple, petiolate, estipulate, entire. In *Dischidia* the leaves are modified into pitchers which collect water and adventitious roots from the next upper node grow in this pitcher.

Floral Features.   Inflorescence di- or monochasial cyme or umbellate or racemose. Flowers bracteate, ebracteolate, hermaphrodite, actinomorphic, hypogynous, typically pentamerous except the carpels. Calyx of 5 free or basally connate sepals, quincuncial or open; corolla of 5 petals, gamopetalous, or only basally fused, contorted or valvate, salver-shaped or funnel-form, corona of 5 or more scales or appendages attached to the corolla tube or staminal tube. Stamens 5, usually adnate to the gynoecium to form a gynostegium; rarely free as in *Cryptostegia;* filaments flat and united to from a fleshy staminal tube around the ovary. The apical part is adnate to the pentangular stigmatic disc. Anthers bicelled, the pollen of each cell agglutinated in granular masses of tetrads as in *Cryptostegia* or waxy pollen masses or pollinia as in *Asclepias, Pergularia, Calotropis* and others. Pollinia of the adjoining cells of the contiguous anthers are united in pairs, either directly or by appendages (caudicles) to black dot-like glands (corpusculum) which lie at the angles of the pentangular stigmatic disc. The 10 pollen masses are united in 5 pairs forming five translators. Gynoecium apocarpous, bicapellary, ovary unilocular, placentation marginal, superior, ovules numerous; styles 2, distinct, stigmas fused to form a pentangular stgmatic disc. Fruit an etaerio of follicles, seeds many, usually flattened, crowned with a tuft of long silky hairs; endosperm thin and scanty, embryo large.

Anatomy.   Vessels small to medium-sized, with numerous radial multiples, sometimes ring-porous; perforations simple, intervascular pitting rather large and alternate. Parenchyma as scattered cells and irregular, uniseriate lines or scanty, paratracheal. Rays narrow and sometimes storied. Leaves usually dorsiventral; isobilateral in fleshy leaves of *Ceropegia* and *Hoya*. Hairs unicellular or uniseriate; stomata mostly Rubiaceous type but Cruciferous type in *Hoya* and *Stapelia* and Ranunculaceous type in *Sarcostemma, Solenostemma* and *Vincetoxicum*. Anomalous structure reported in climbing species.

Embryology.   Pollen grains are 3-celled and remain aggregated in pollinia. Ovules anatopous, unitegmic or ategmic and tenuinucellate. Ategmic ovules are reported in *Araujia, Asclepias, Gomphocarpus* and *Marsdenia* (Johri et al. 1992). Polygonum type of embryo sac, 8-nucleate at matuirty. Endospem formation of the Nuclear type.

Chromosome Number.   Basic chromosome number is x = 11.

Chemical Features.   Iridoids are totally absent. Seeds of various species of *Asclepias* are rich in linoleic acid (Shorland 1963).

Important Genera and Economic Importance.   *Asclepias* is the largest genus and includes some ornamentals like *A. curassavica* or 'blood' flower and *A. tuberosa* or 'butterfly' weed. *Oxypetalum caeruleum* (blue milkweed), *Hoya carnosa* (wax plant), and *Stapelia* spp. (carrion flower) are some other ornamentals. *Dischidia rafflesiana* is an interesting plant from the tropical rain forests of Malayan region, with its leaves modified into pitchers that collect rain water and adventitious roots from adjacent nodes grow into it. Economically, the members of this family have limited importance. The seed hairs of *Asclepias* are used as substitute for 'kapok'— another seed fiber from *Salmalia malabarica* (Bombacaceae). The hairs are fine, silky and light. The plants of *Tylophora indica* are used medicinally against asthma. Latex from *Cryptostegia grandiflora* is a source of rubber. Latex of *Gymnema laticiferum* and *Oxystelma secamore* can be consumed as 'milk'. Latex of *Matelea,* on the other hand, is used as an arrow poison. *Calotropis procera* and *Pergularia daemia* are some of the weeds.

Taxonomic Considerations.   The family Asclepiadaceae is closely allied to the family Apocynaceae, the differences between the two families are: (1) Stamens modified to pollinia and accompanied by translators, and (2) presence of a gynoestegium.

All taxonomists retain this family next to the Apocynaceae, in the order Gentianales. This is fully justified.

Rubiaceae

A large family of about 500 genera and 6000 species, distributed mostly in the tropics and a few in temperate and arctic regions.

Vegetative Features.   Trees (*Mitragyna parviflora, Anthocephalus cadamba*), shrubs (*Gardenia, Ixora*) or herbs (*Oldenlandia*); sometimes lianas (*Galium, Rubia*); mostly perennial, rarely annual (*Oldenlandia*). Leaves opposite, decussate (*Ixora*) or whorled (*Gardenia*), simple, subsessile or petiolate, entire or rarely toothed, stipulate, Stipules inter- or intrapetiolar, frequently united to one another and to the petioles forming a sheath around the stem. The two stipules—one from each leaf standing next to each other— are usually united in the tribe Rubieae; often leaf-like and as large as normal leaf. The entire structure appears as a whorl of leaves in the tribe Galieae; or stipules reduced to glandular setae, as in *Pentas*.

Floral Features.   Inflorescence basically axillary or terminal dichasial cyme—reduced to single flowers in *Gardenia;* sometimes aggregated into globose heads (Fig. 49.7 A) with the flowers basally adnate, e.g. *Morinda, Sarcocephala* (=*Nauclea*), *Anthocephalus, Uncaria* and others. Flowers hermaphrodite, actinomorphic, epigynous, 4- or 5-merous (Fig. 49.7 B). Calyx 4- or 5-lobed, aestivation open, persistent, enlarged in fruit in *Nematostylis*. Sometimes one sepal larger than others (Fig. 49.7 D-I) and brightly coloured, as in *Mussaenda*. Corolla of 4 or 5 petals, rarely 8–10 petals, gamopetalous, valvate, convolute or imbricate, rotate, salverform or funnel-form. Stamens 4 or 5, alternipetalous, epipetalous; anthers bicelled, dehiscence longitudinal, introrse. Gynoecium syncarpous, usually bicarpellary (pentacarpellary in *Hamelia patens*; monocarpellary with parietal placentation in *Gardenia*). Ovary bilocular (pentalocular in *Hamelia patens*), axile placentation, inferior, rarely superior as in *Gaertnera* and *Pagamea;* ovary half-inferior in *Synaptantha*. Ovules usually numerous per locule; uniovulate in *Pavetta* with 1 ovule sunken in fleshy funiculus. Style 1 and slender, often 2-branched, stigma linear, capitate or lobed. Distyly is sometimes reported, as in *Hedyotis salzmannii* (Riveros et al. 1995). Fruit a septi- or loculicidal capsule or indehiscent and separating into 1 seeded segments (*Galium, Rubia, Oldenlandia*); a fleshy berry in some, e.g. *Uncaria* (Fig. 49.7 C), *Coffea* and *Mitchella*. Seeds sometimes winged; embryo small in rich endosperm.

Anatomy.   Vessels typically small or medium-sized, solitary or rarely in multiples of four, numerous, rarely ring-porous; perforation plates simple, parenchyma apotracheal in species with non-septate fibers and absent in species with septate fibres. Rays mostly narrow, up to 2–3 cells wide. Leaves dorsiventral (centric in *Asperula cynanchica*). Hairs unicellular, uniseriate, tufted or rarely peltate; stomata nearly always on lower surface, Rubiaceous type. Raphides and crystal sand are reported.

Embryology.   The pollen grains are 2- or 3-colporate or 3-pororate with smooth exine. Pollen tetrad reported in *Gardenia.* Pollen trimorphism occurs in some genera (Mathew and Philip 1993). They are

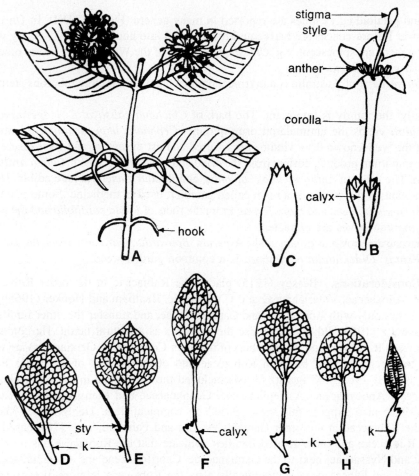

Fig. 49.7 Rubiaceae: **A-C** *Uncaria gambir*, **D** *Pinckneya pubescence*, **E** *Pogonopus tubulosus*, **F** *Capirona wurdeckii*, **G** *Calycophyllum spectabile*, **H** *C. candidissimum*, **I** *Cosmocalyx spectabilis*. **A** Twig with inflorescence. **B** Flower. **C** Fruit. **D-I** Genera with one enlarged and coloured sepal. *k* calyx, *sty* style. (Adapted from Hutchinson 1969)

2- or 3-celled at the dispersal stage. Ovules anatropous, unitegmic, tenuinucellate. Polygonum type of embryo sac, 8-nucleate at maturity. Twin embryo sacs reported in *Borreria hispida*, *B. stricta* and *Galium asperifolium* (Johri et al. 1992). Allium type of embryo sac in *Scyphiphora hydrophyllacea* and Peperomia type in *Crucianella* spp. (Davis 1966). Endosperm formation is of the Cellular type.

Chomosome Number.   Basic chromosome numbers are: x = 6–15, 17, x = 9 and 11 being more common.

Chemical Features.   Iridoids and complex alkaloids such as quinine and quinidine in *Cinchona* and caffeine in *Coffea* have been reported. Iridoid glucosides occur in *Gardenia jasminoides* fruits (Inouye et al. 1974); loganin in *Mitragyna* (Cordell 1974).

Important Genera and Economic Impotance.   An interesting family, both morphologically as well as economically. Bacterial nodules develop on the leaves of various species of *Pavetta*, *Chomelia asiatica* and *Psychotria bacteriophylla*. These bacteria have been shown to fix atmospheric nitrogen (Boodle 1923).

Enlarged and petaloid calyx lobes are reported in many genera (Fig. 49.7 D-I). In *Uncaria gambir* of Malaya the lower inflorescences are barren and transformed into hooks (Fig. 49.7 A) by which the plant climbs. Many genera are spinescent, e.g. *Catesbaea spinosa* of the West Indies; leaves modified to spines in *Phyllacantha grisbachiana*.

Myrmecodia beccarii of Australia is a myrmecophilous plant with a large tuberous stem which houses ants.

Economically, the family is important. The bark of *Cinchona calisaya, C. ledgeriana, C. officinalis* and *C. succirubra* yields the antimalarial drug quinine. *Cephaelis impecacuanha* is another medicinal plant yielding the well-known drug vinum ipecac, used against dysentery and liver diseases. The family yields another important product, coffee, from the dried and powdered seeds of *Coffea arabica, C. liberica* and *C. robusta.* The fruits of *Anthocephalas cadamba,* and *Randia dumetorum* are edible. *Uncaria gambir,* an Indonesian plant, is the source of a resin called gambier, used in medicine. Some plants yield timber: *Calycophyllum candidissimum* and *Ixora ferrea.* From the roots of *Rubia cordifolia* and the bark of *Morinda angustifolia,* important dyes are extracted.

Many Rubiaceae are grown as ornamentals: *Asperula, Bouvardia, Gardenia, Hamelia, Ixora, Mussaenda, Nertera* and *Pentas. Oldenlandia corymbosa* is a common garden weed.

Taxonomic Considerations.   Bessey (1915) placed the Rubiaceae in the order Rubiales along with Caprifoliaceae, Adoxaceae, Valerianaceae and Dipsacaceae. Bentham and Hooker (1965b) also placed it in the Rubiales, but only with Adoxaceae and Caprifoliaceae, and transfer the other families to the order Asterales. Wagenitz (1959) did not recognise the Rubiales as a natural taxon. He commented that the similarities between Rubiaceae and the members of the order Contortae (= Gentianales) are more important, as compared to the similarities between Rubiaceae and other families of the order Rubiales (sensu Bentham and Hooker 1965b). Wagenitz (1964) included the Rubiaceae in the order Gentianales along with Loganiaceae, Apocynaceae, Asclepiadaceae, Gentianaceae and Menyanthaceae. Takhtajan (1969) and Thorne (1968) also support this view. Based on chemical data, Dahlgren (1975a) supports the inclusion of the Rubiaceae in the order Gentianales. Lee and Faïrbrothers (1978), based on serological evidences on Rubiaceae and other related families, indicate that the Rubiaceae has maximal affinity to the Cornaceae and Nyssaceae; next to the Gentianaceae, Caprifoliaceae and Asclepiadaceae, and least to the Apocynaceae and Dipsacaceae. Cytologically also, the Rubiaceae is more akin to the Gentianales. The basic chromosome number for both the Rubiaceae and the Gentianales is x = 11. In both taxa, the pollen grains are eurypalynous, ovules mostly anatropous, unitegmic, tenuinucellate and endospem formation of the Nuclear type. Therefore, Wagenitz's placement of the Rubiaceae in the order Gentianales is justified. According to Bremekamp (1966), however, the absence of intraxylary phloem is very much against placing the Rubiaceae amongst the Gentianales. Also, Rubiaceae differs distinctly from the Gentianales in having inferior ovary. Bremekamp (1966) and Cronquist (1968, 1981), therefore, treat it as a monotypic order—Rubiales with more affinities to the Gentianales. Cronquist also regards Rubiales to be the connecting link between the Gentianales and the Dipsacales.

We also support this assignment because: (1) the basic chomosome number x = 11 is not so frequent in the Gentianales as in the Rubiaceae, (2) the extent of eurypalyny in the Rubiaceae is much more in contrast to that in the Gentianales which are moderately eurypalynous, (3) ovary inferior in Rubiaceae and supeior in all Gentianales members, and (4) intraxylary phloem is absent in the Rubiaceae.

The last two characters bring this family close to the order Dipsacales, and to the family Caprifoliaceae in particular. Taxonomists have expressed doubt that the two families, Rubiaceae and Caprifoliaceae, should be treated separately (see Cronquist 1968). However, the Dipsacales differ from the Rubiaceae in cellular type of endosperm, near-absence of the alkaloids and absence of the special glandular trichomes, colletéts on the inner surface of the stipules.

Rubiaceae is a large family (ca. 500 genera and 6000 species) and would be too dominating if included either in the order Gentianales or Dipsacales. It will be best to treat this family as a distinct order, the Rubiales.

Takhtajan (1987) includes six more families in this order—Saccifoliaceae, Carlemanniaceae, Spigeliaceae, Dialypetalanthaceae, Theligonaceae and Plocospermataceae.

Saccifoliaceae is a small unigeneric family with a single species of *Saccifolium* (*S. bandeirae*) from South America, Venezuella and Guyana Highlands, separated from the Gentianaceae by Maguire and Pires (1978). Thorne (1983, 1992a) includes Saccifoliaceae in the Gentianaceae as a distinct family. It differs from other Gentianaceae members by alternate leaves, scalariform perforation plates of the vessels and imbricate aestivation of corolla. *Saccifolium bandeirae* is a subshrub of cushion-like growth with entire leaves closely arranged in the upper part of stem. It contains iridoids but no alkaloids.

Flowers solitary, axillary, actinomorphic, dioecious, calyx and corolla 4- to 5-merous, more or less tubular; stamens isomerous, alternipetalous, adnate to corolla tube; gynoecium bicarpellary, syncarpous, bilocular, ovules numerous, anatropous (Nicholas and Baijnath 1994).

Carlemanniaceae, a family with only two genera, *Carlemannia* and *Silvianthus*, was earlier treated as a member of tribe Hedyotideae of the family Rubiaceae. Distributed from southeast Asia to Sumatra, these are perennial herbs or subshrubs with opposite, simple, sometimes oblique, dentate, estipulate leaves; petioles connected by a raised line. Flowers slightly zygomorphic, hermaphrodite, in dense terminal or axillary cymes. Calyx 4 or 5, more or less unequal, open, persistent; corolla 4 or 5, gamopetalous, imbricate or induplicate-valvate. Stamens 2, epipetalous, filaments short, anthers linear-oblong. Gynoecium syncarpous, bicarpellary, ovary inferior; style 1, long, stigma bifid, clavate or fusiform; ovules numerous, placentation axile. Fruit a membranous or fleshy, more or less globular, bilocular, 4- or 5-valved capsule. Seeds numerous with fleshy endosperm. Anatomically, these two genera—*Carlemannia* and *Silvianthus*—are very much alike.

Spigeliaceae includes 3 genera—*Spigelia, Mitreola* and *Polypremum*—and 90 species distributed in Madagascar, tropical Asia to north Australia, and warmer parts of North and South America. Perennial or annual herbs or rarely shrubby, occasionally with stellate hairs and opposite, simple, entire leaves; stipules small or obsolete. Inflorescence cymose or spicate. Flowers hemaphrodite, actinomorphic, hypogynous. Calyx of 2 to 5 sepals, poly- or gamosepalous; corolla of 4 or 5 petals, valvate. Stamens 4 or 5, alternipetalous, epipetalous, included. Gynoecium syncarpous, bicarpellary, ovary bilocular with axile placentation, inferior; bifid to bipartite style and capitate stigma; ovules numerous. Fruit a capsule, septicidal or transversely dehiscent; seeds unwinged, endospermous.

Plocospermataceae is a monotypic Central American family allied to the family Apocynaceae, and not Loganiaceae in which it is included by Bentham and Hooker (1965b). The only genus is *Plocosperma*.

Leaves dorsiventral, stomata of Ranunculaceous type.

The ovary of 2 united carpels is unilocular with parietal placentation; ovules 4 in two pairs, the lower pair erect and the upper pair pendulous. Style short, bilobed; fruit a capsule dehiscing along both sides; seed only one with a dense tuft of hairs at the tip.

Two other families (Takhtajan 1987)—Dialypetalanthaceae and Theligonaceae—have been included in the Myrtiflorae by Melchior (1964), and discussed under that order (in Polypetalae: pp. 385, 399, 400).

Taxonomic Considerations of the Order Gentianales

The order Gentianales (as defined here) comprises 7 families—Loganiaceae, Gentianaceae, Menyanthaceae, Desfontainiaceae, Apocynaceae, Asclepiadaceae and Rubiaceae.

The order Gentianales (sensu Bentham and Hooker 1965b) excludes the family Rubiaceae but includes the family Oleaceae. Engler and Diels (1936) and Lawrence (1951) designated this order Contortae comprising 2 suborders—Oleineae and Gentianineae. Rendle (1925) and Wettstein (1935) segregated the suborder Oleineae as a distinct order Oleales, and retained only the members of the Gentianineae in the order Gentianales.

Hutchinson (1969, 1973) assigns these families to four different orders—Loganiales (Loganiaceae, Oleaceae), Apocynales (Apocynaceae, Asclepiadaceae), Gentianales (Gentianaceae, Menyanthaceae), and Rubiales (Rubiaceae). He includes some more families in Loganiales and Apocynales.

Dahlgren (1980a) treats the Oleales as a separate order, and Gentianales comprises Loganiaceae (including Antoniaceae, Spigeliaceae, Strychnaceae and Potaliaceae), Menyanthaceae, Gentianaceae, Apocynaceae and Asclepiadaceae. According to Dahlgren, all Gentianales are rich in seco-iridoids, and endosperm formation is of the Nuclear type. The Oleales are rich in simple iridoids, and endosperm formation is of the Cellular type. Cronquist (1981) disagrees with most taxonomists and includes Oleaceae in the Scrophulariales, Rubiaceae and Theligonaceae in the Rubiales, and Loganiaceae, Retziaceae, Gentianaceae, Saccifoliaceae, Apocynaceae and Asclepiadaceae in the Gentianales. Takhtajan (1987) also recognizes the two orders, Oleales and Gentianales; however, he includes a number of small families in the order Gentianales (see p. 463) which may not deserve independent status as families.

The placement of Rubiaceae and Caprifoliaceae in the order Rubiales (Bentham and Hooker 1965b) is also questionable. Embryologically the two families do not have many common features. The reduction of nucellar tissue, occasional development of Peperomia type of embryo sac, Nuclear type of endosperm, and Solanad type of embryogeny with suspensor haustoria in Rubiaceae are quite distinct from those of the Caprifoliaceae (Johri et al. 1992). Therefore, on the basis of embryology, its assignment to a separate order Rubiales is justified (see Cronquist 1981).

The monotypic order Rubiales should be placed close to the Gentianales, due to certain similarities between the two: (a) well-developed stipules, (b) opposite leaves, (c) bicarpellate ovary, (d) anatropous, unitegmic, tenuinucellate ovule, (e) Nuclear type of endosperm formation, and (f) alkaloid content.

50

Order Tubiflorae

The Tubiflorae is a large order with 6 suborders and 26 families. The suborder Solanineae comprises 15 families, and the suborders Myoporineae and Phrymineae, 1 family each. The members are predominantly herbaceous. A few members of Bignoniaceae and Verbenaceae are trees and shrubs. All members of Lennoaceae and Orobanchaceae are parasites. Some members of Lentibulariaceae and the genus *Trapella* of Pedaliaceae are aquatic plants. Fouquieriaceae includes a xerophytic member, *Fouquieria*.

Inflorescences are of varied types. Flowers bisexual, mostly zygomorphic, bilipped, calyx persistent, corolla gamopetalous, floral parts in four whorls, stamens epipetalous. Ovary mostly superior (inferior in Columelliaceae), bicarpellary (tricarpellary in Polemoniaceae, pentacarpellary in Nolanaceae), syncarpous, bilocular (trilocular in Polemoniaceae, unilocular in Phrymaceae), placentation axile. Ovules numerous (only one in Phrymaceae), anatropous, unitegmic (bitegmic in Fouquieriaceae), tenuinucellate (rarely crassinucellate). Fruit mostly capsule, sometimes nutlets, drupe or berry.

Polemoniaceae

A medium-sized family comprising 13 genera and 265 species (17 genera, 300 species according to Takhtajan 1987), mostly confined to America.

Vegetative Features. Chiefly annual or perennial herbs, rarely shrubs, trees or twining vines (*Cantua* spp. of western South America are shrubs or small trees). Stem herbaceous, covered with hairs. Leaves alternate or opposite (*Phlox*), entire or palmately or pinnately compound (*Cobaea*), estipulate, often sessile, surface hairy.

Floral Features. Inflorescence usually cymose, corymbose (*Phlox*) or capitate (*Collomia*). Flowers rarely solitary axillary, bisexual, hypogynous, actinomorphic or weakly zygomorphic as in *Loselia* and *Bonplandia*, usually showy; calyx 5-lobed, gamosepalous, outer surface hairy, imbricate or valvate. Corolla lobes 5, contorted in bud, salverform or rotate or campanulate, as in *Cobaea*, hypogynous. Stamens 5, epipetalous, attached to the corolla tube at various heights. Gynoecium tricarpellary (rarely bi- or tetracarpellary), syncarpous, ovary trilocular, placentation axile with 1 to many ovules on each placenta, superior; style 1, filiform, terminal, stigma 3-partite (rarely 2). Fruit a loculicidal capsule (septicidal in *Cobaea*), rarely indehiscent. Seeds sometimes with mucilaginous coating, e.g. *Collomia* and *Gilia*; copious fleshy endosperm and a straight embryo.

Anatomy. Xylem vessels usually small (large in *Cobaea*), frequently arranged in radial rows. Intraxylary phloem absent. Vessel perforations mostly simple in species of *Cobaea*, *Collomia*, *Loselia* and *Phlox*, rarely scalariform. Leaves dorsiventral or isobilateral, stomata Ranunculaceous type, rarely Rubiaceous type.

Embryology. Pollen grains polyporate with a smooth exine and 2-celled at dispersal stage. Ovules hemianatropous, unitegmic, tenuinucellate. Polygonum type of embryo sac, 8-nucleate at maturity. Endosperm formation of the Nuclear type.

Chromosome Number. Basic chromosome number is x = 7, 8 or 9 (Takhtajan 1987). In *Cobaea* the haploid chromosome number is n = 26.

Chemical Features. Many members, like *Phlox* and *Gilia*, contain anthocyanin pigments of glucosidic type and pelargonidin (a glycone); *Gilia* also contains p-coumaric type of anthocyanins. Seed fat of *Cobaea scandens* is rich in various organic acids like palmitic, oleic and linoleic acids.

Important Genera and Economic Importance. *Phlox* is an important genus on which extensive hybridisation work has been conducted. Many of its members are ornamental and are cultivated as garden annuals: *Phlox, Polemonium, Gilia, Cobaea, Collomia* and *Linanthus*.

Taxonomic Considerations. Bentham and Hooker (1965b) included Polemoniaceae in Polemoniales. Engler and Diels (1936) considered it to be a member of the order Tubiflorae. Hallier (1905) showed its derivation from the Geraniaceae. Later, Hallier (1912) considered its possible derivation from the Linaceae. Bessey (1915) derived Polemoniaceae from Boraginaceous stock. Rendle (1925) considered it to have been derived from sympetalous members of the Rosaceae. Hutchinson (1959, 1969, 1973) includes Polemoniaceae in his Polemoniales and derives them from the Geraniales, with the Ranales ancestral to both. On the basis of floral morphology, Dawson (1936) stated that Polemoniaceae is closely related to both Caryophyllaceae and Geraniaceae. Embryologically also, Polemoniaceae and Geraniaceae are closely related. The tricarpellary condition in this family might have originated from the ancestral stock of Caryophyllaceae at a time before the reduction of trilocular ovary to unilocular ovary. Sunder Rao (1940), on the basis of embryological studies, derived Polemoniaeac from Geraniaceae. Cronquist (1968, 1981) regards it as a member of the order Polemoniales which has originated from Gentianales—both members of subclass Asteridae. Hutchinson, in his earlier work (1926), regarded it to be closely related to Geraniales, later Hutchinson (1969, 1973) derived it from Caryophyllaceous stock. Benson (1970) and Stebbins (1974) place it in Polemoniales. Dahlgren (1975a, b, 1980a, 1983a) includes Polemoniaceae in Solanales. Takhtajan (1980, 1987) attributes a much advanced position to this family including it in the order Polemoniales. He derives this order from Gentianales and particularly the family Loganiaceae. He, however, raises the genus *Cobaea* to family rank, Cobaeaceae.

From the views discussed above, it is clear that Polemoniaceae is an advanced family and has evolved from the Gentianales.

Fouquieriaceae

A monogeneric family with 11 species (Dahlgren et al. 1976) confined to Mexico.

Vegetative Features. Xerophytic shrubs or small trees covered with stout petiolar spines (Fig. 50.2.A, B). Stems thorny; at every node there is a spine formed from the petiole and midrib of a leaf and in the rainy season leaf clusters appear in the axils of the spines (Fig. 50.2B). Leaves simple, spirally arranged on long and short shoots, leaf bases decurrent, estipulate.

Floral Features. Inflorescence basically determinate, mostly a panicle varying from compound and drupiform or corymbiform to simple. Flowers showy, bisexual, hypogynous, actinomorphic, pentamerous (Fig. 50.2C). Sepals 5, unequal, polysepalous, imbricate. Petals 5, connate basally to form a tube and short spreading lobes imbricate in bud. Stamens 10–17 (Fig. 50.2C, D), in 1 or 2 whorls, hypogynous, with a stiff reddish basal part, puberulent on the outer surface; anthers cuspidate at apex, dorsifixed, tetrasporangiate, without appendages, introrse, dehiscing longitudinally. Gynoecium syncarpous, tricarpellary (Fig. 50.2E). Ovary unilocular, superior, base nectariferous; placenta parietal and intruded, ovules 4 to 6 on each placenta, styles 3, connate almost up to the middle, stigmas 3. Fruit a 3-valved, dry, loculicidal capsule (Fig. 50.2F), seeds broad, winged, wings made of unicellular trichomes (Fig. 50.2G) derived from epidermis of outer integument; embryo small, straight with well-developed cotyledons, endosperm present as a thin layer in the seed (Dahlgren et al. 1976).

Anatomy. Vessels small, often ring-porous; stomata Ranunculaceous type.

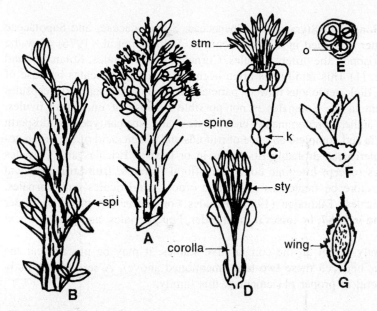

Fig. 50.2 Fouquieriaceae: A-G *Fouquieria splendens*. **A** Twig with inflorescence. **B** Part of sterile branch. **C** Flower. **D** Longisection. **E** Ovary, transection. **F** Fruit. **G** Seed. *k* calyx, *o* ovule, *spi* spine, *stm* stamen, *sty* style. (Adapted from Lawrence 1951)

Embryology.   Pollen grains tricolpate, elliptical, binucleate. Ovules anatropous, bitegmic, tenuinucellate. Polygonum type of embryo sac, 8-nucleate at maturity. Endosperm formation of the Cellular type (Govil 1970).

Chromosome Number. Basic chromosome number is x =12; diploids (2n = 24), tetraploids (2n = 48) and hexaploids (2n = 72) have been reported.

Chemical Features.   *Fouquieria* contains iridoids, ellagic acid, waxes, triterpenes and triterpene saponins (Hegnauer 1966), steroidal saponins (Gibbs 1974), tannins (Gibbs 1974) and coumarins; also caffeic acid, quercetin, kaempferol, leucocyanidin and other phenolic compounds. Seeds contain fatty acids.

Important Genera and Economic Importance.   *Fouquieria* sp. commonly known as boogum tree or curio of Baja California, and Sonora, Mexico, is often cultivated as a curiosity. This weird tree is like a giant carrot root turned upside down, and only small twigs appear at the apex of the trunk. *F. splendens*, the ocotillo or coach-whip plant grows in the desert areas from California to western Texas. The plant has a trunk only about 1 foot high and large wand-like branches as much as 20 feet long and 1 to 2.5 inches in diameter. The family is not of any economic importance.

Taxonomic Considerations.   Most taxonomists treat this family as a unigeneric one. However, Nash (1903) recognized 2 genera—*Fouquieria* with 6 species and the monotypic *Idria*. The phyletic position of this family is highly disputed. Wettstein (1935) included it in the Parietales, Engler and Diels (1936) in the Tubiflorae along with Convolvulaceae and Polemoniaceae, Bessey (1915) in Ebenales, whereas Hutchinson (1969, 1973) placed it in the Tamaricales. Humphrey (1935) reported possible affinities with Loasaceae and Polemoniaceae. Johnson (1936) mentioned the relationship between *Fouquieria* spp. and *Myricaria germanica* of the Tamaricaceae

According to Cronquist (1968), it is an isolated family, without evident close relatives. Although it is associated with the Polemoniales, there is hardly any relationship. Neither pollen morphology nor wood anatomy give any positive evidence. Ovules in this family are bitegmic, as in most of the members of Violales, and on this basis Cronquist (1968) included it in the Violales. Hutchinson (1969, 1973), Metcalfe and Chalk (1972) and Stebbins (1974) included Fouquieriaceae in the Tamaricales. Benson

(1970) includes it in the Ebenales along with Styracaceae, Ebenaceae, Symplocaceae, and Sapotaceae though he states that it would be better classified in the Polemoniales. Dahlgren et al. (1976) draw the following conclusions from a comparison of the orders Ericales, Cornales, Polemoniales, Solanales, and Tamaricales with the Fouquieriaceae: (1) This family has often been placed in Tamaricales because of its shrubby habit, sympetalous corolla, diplostemonous flowers, parietal placentation and septicidal capsules with winged seeds. (2) Its placement in the Solanales is not possible because of unitegmic ovules, haplostemonous flowers, nonapiculate anthers, predominantly bicarpellate ovary, Nuclear type of endosperm formation, lack of endosperm haustoria and complete absence of iridoids and ellagic acid of the Solanales including Polemoniaceae. (3) Fouquieriaceae probably fits better in or next to Ericales and Cornales because of its embryological features (except bitegmic condition), floral features, fruit structures and chemical features. (4) It should therefore be treated as a separate order near Ericales and Cornales. Dahlgren (1983a) places it near Ericales. Takhtajan (1980) includes Fouquieriaceae in the suborder Fouquierineae, order Tamaricales, and in 1987, he raises it to an order, Fouquieriales, and places it next to Tamaricales.

Fouquieriaceae is an isolated family. From all the comparative studies, it may be placed near the Ericales because of the resemblances between these two taxa (mentioned above), A serological study with the allied families might be useful in proper placement of this family.

Convolvulaceae

A medium-sized family including 50 genera and 1200 species, mostly distributed in the tropics and subtropics of the world. Some members extend up to north and south temperate zones also.

Vegetative Features.   Members vary in habit; may be herbs: trailers *Evolvulus, Convolvulus,* climbers *Ipomoea nil, I. pes-tigridis, Merremia aegyptiaca,* shrubs *Argyreia* or trees *Humbertia. Rivea hypocrateriformis* is a climbing shrub; *Porana paniculata* is a large woody climber. Aquatic plants like *I. aquatica* and xerophytes like *Hildebrandtia* are also known.

Root system well-developed, many creepers rooting at nodes: *Evolvulus nummularius*; storage roots: *I. batatas.* Stem prostrate, climbing or twining, rarely erect (*I. carnea*), herbaceous, rarely woody as in *Humbertia,* cylindrical, solid (hollow in *Argyria*), surface hairy in *I. pes-tigridis.* Vegetative parts often with milky latex. Leaves alternate, estipulate, petiolate, simple, as in *E. alsinoides, Convolvulus pluricaulis,* or pinnately (*Quamoclit* sp.) or palmately compound (*Merremia aegyptiaca*). Extrafloral nectaries present on petioles of *I. batatas.*

Floral Features.   Inflorescence usually axillary dichasial cymes or paniculate as in *Porana* or flowers solitary axillary as in *E. nummularius.* Flowers hermaphrodite, pentamerous; unisexual and tetramerous in *Hildebrandtia,* actinomorphic, pedicellate, hypogynous, bracteate, bracts in pairs. often form involucre; flowers cleistogamous in *Dichondra*[6]. Calyx of 5 sepals, polysepalous, imbricate, persistent. Corolla of 5 petals, gamopetalous, infundibuliform or salverform, rarely with appendages within as in *Cuscuta*[1]; usually induplicate-valvate or contorted, imbricate in *Cuscuta.* Stamens 5, distinct, epipetalous, alternate with corolla lobes, attached to the base of the corolla tube. Anthers bicelled, dorsifixed, dehiscence longitudinal, usually introrse. Gynoecium syncarpous, bicarpellary, ovary bilocular, placentation axile, 2 ovules in each locule or tetralocular by formation of false septum and 1 ovule per locule; superior, raised on a nectariferous disc, often with a dense covering of hairs; style 1, filiform, simple, sometimes 2 as in *Cressa* and *Cuscuta* or absent as in *Erycibe*; stigma usually terminal and capitate, 2 in *Convolvulus.* Fruit a loculicidal capsule, indehiscent and fleshy in *Argyria.* In *Erycibe* the fruit is ovoid, single-seeded berry.

---

[6]According to some authors, *Dichondra* belongs to a separate family, Dichondraceae, and *Cuscuta* to family Cuscutaceae.

Seeds smooth or hairy, endosperm hard, cartilaginous, embryo usually large with folded or bilobed cotyledons; in *Cuscuta* the coiled embryo is poorly developed, filiform and apparently without cotyledons and radicle.

Anatomy. Xylem vessels relatively small and arranged in radial rows. Intraxylary phloem present. Parenchyma mostly abundant, or very sparse and scanty, paratracheal. Leaves dorsiventral or isobilateral; stomata mostly Rubiaceous type, rarely Cruciferous type as in *Evolvulus* and *Hildebrandtia*, or Ranunculaceous type as in *Cressa*. Both glandular and nonglandular hairs present.

Embryology. Pollen grain structure varies in different genera and on this basis there are two groups:

  i. Convolvulus type—pollen grains ellipsoidal with longitudinal bands.
  ii. Ipomoea type—pollen grains spherical with warty exine.

Pollen grains shed at 2- or 3-celled stage. Ovules erect, anatropous, unitegmic, tenui- or crassinucellate and sessile. Polygonum type of embryo sac, 8-nucleate at maturity. Endosperm formation of the Nuclear type.

Chromosome Number. Basic chromosome numbers are very variable—may be x =7–15, 21, 25 and mainly 15. For *Cuscuta* x = 7, 14, 21.

Chemical Features. Tropane alkaloids are present in some members. Many Convolvulaceae members are capable of responding to fungal infection by synthesizing phytoalexins like sesquiterpene furanolactone-ipomeamarone (Harborne and Turner 1984).

Important Genera and Economic Importance. A number of species of *Convolvulus* and *Ipomoea* are weeds: *C. pluricaulis*, *C. arvensis*, *I. pes-tigridis*, *I. nil*, *I. scindica* and others. *Evolvulus nummularius* has small, almost round or oval-shaped leaves with apical notch, grows prostrate on the soil and bears adventitious roots at each node. *I. nil* (morning glory), *I. purpurea*, *I. palmata* (railway creeper), *I. tuberosa* (woodrose), *Jacquemontia pentantha*, *Quamoclit* sp. (Cypress vine), *Porana paniculata* (Christmas vine or bridal creeper), *Calonyction aculeatum* (moon-flower) are some of the ornamental plants of this family. *I. batatas* is cultivated for its edible roots rich in sugar and starch. The rootstock of *Calystegia sepium* is cooked and eaten in New Zealand, the green young shoots of *I. aquatica* are used as vegetables in India. Some are ecologically important as good soil-binders, e.g. *I. pes-caprae*, *Calystegia* sp., and *Convolvulus arvensis*. Seeds of *I. hederacea* and roots of *Covolvulus* spp. are used medicinally as purgative; *Meremmia tridentata* is used in rheumatism, piles and urinary disorders.

Taxonomic Considerations. Bentham and Hooker (1965b) placed the Convolvulaceae in Polemoniales of Bicarpellatae on the basis of sympetalous corolla and bicarpellary ovary. Engler and Diels (1936) included Convolvulaceae in the Tubiflorae (see also Hallier 1912). Bessey (1915) included it in Polemoniales. Wettstein (1935) separated it from Tubiflorae and raised it to the rank of an order, Convolvulales, on the basis of large embryo with folded cotyledons. Hutchinson (1959) retained this family in Solanales along with the families Solanaceae and Nolanaceae. These taxonomists included the genus *Cuscuta* in Convolvulaceae. Lawrence (1951) also retained this genus in Convolvulaceae and placed it in the suborder Convolvulineae of the order Tubiflorae. Some authors split up this family on the basis of parasitic habit of *Cuscuta* (see Lawrence 1951). Tiagi (1951) justified the separation of Cuscutaceae on embryological basis. Hutchinson (1969, 1973) prefered to place the small family Cuscutaceae in the order Polemoniales, along with Polemoniaceae and Hydrophyllaceae [see also (Stebbins 1974)]. Dahlgren (1975a, b, 1980a, 1983a) includes the two families Convolvulaceae and Cuscutaceae in Solanales. Takhtajan (1980, 1987) also recognises the two families but includes them in Polemoniales (1980) and later (1987) in a distinct order of its own, Convolvulales. The family Cuscutaceae was separated not only because of its parasitic nature but also because of its distinct embryological features: undifferentiated filiform embryo, and copious endosperm (see Johri et al. 1992). The separation of *Dichondra* is probably not necessary, as the distinction is based only on cleistogamous flowers.

Hydrophyllaceae

A medium-sized family of 20 genera and 265 species (18 genera and 250 species according to Takhtajan 1987), of wide distribution all over the world, except in Australia; occurs more commonly in North America.

Vegetative Features.   Mostly annual or perennial herbs, rarely shrubs, e.g. *Eriodictyon*. Stem herbaceous, often scabrid hairy, glandular hairy or bristly, densely prickly in *Codon* (Fig. 50.4A). Leaves alternate (*Romanzoffia, Codon*) or opposite (*Nemophila*) or in basal rosettes, as in some species of *Hydrophyllum*, simple as in *Wigandia* and *Phacelia viscida* or pinnately (*P. tanacetifolia*) or palmately lobed, often covered with hairs and dentate.

Floral Features.   Inflorescence mostly terminal or axillary, helicoid cyme, in *Hydrophyllum*, often circinnate, as in *P. tanacetifolia*, sometimes dichasial cyme as in *P. campanularis*, rarely solitary as in *Nemophila* and *Hesperochiron* or secund-raceme, i.e. one-sided raceme as in *Romanzoffia*. Flowers hermaphrodite, actinomorphic, usually pentamerous except in *Codon* with 8- to 12-merous flowers (Fig. 50.4B), mostly hypogynous, ebracteate, ebracteolate. Calyx 5-lobed (2 series of sepals in *Tricardia*), almost free in some genera, imbricate. Corolla 5-lobed, gamopetalous, rotate, or campanulate, or infundibuliform, imbricate or contorted. Stamens mostly 5, epipetalous, arise from the base of the corolla, alternipetalous, filaments equal or unequal, filiform, hairy; anthers bicelled, dehiscence longitudinal, versatile, introrse, often alternate with or subtended by scalelike or hairy appendages. Gynoecium syncarpous, bicarpellary, ovary superior or half-inferior as in *Nama* sp., unilocular with parietal placentation (free-central in *Codon*), ovules numerous (4 in *Hydrophyllum*); style 1 or shortly bifid at the apex (Fig. 50.4. C) or 2 as in *Wigandia* and *Hydrolea*; stigma capitate. Fruit mostly a loculicidal capsule, dehisces usually by 2 or rarely 4 valves, sometimes indehiscent. Seeds sometimes carunculate, variously pitted (Fig. 50.4E),

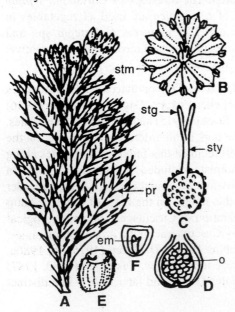

Fig. 50.4 Hydrophyllaceae: **A-F** *Codon royenii*. **A** Flowering twig with dense prickly stem and leaves. **B** Flower, top view. **C** Gynoecium. **D** Ovary, longisection. **E** Seed. **F** Longisection. *em* embryo, *o* ovule, *pr* prickle, *stg* stigma, *stm* stamen, *sty* style. (Adapted from Hutchinson 1969)

reticulate, sculptured or muricate; embryo small and straight, surrounded by copious endosperm (Fig. 50.4F). In *Hydrolea* 8 to 10 longitudinal ridges correspond to the lobes of the ruminate endosperm (Johri et al. 1992).

Anatomy. Vessels small to medium-sized, sometimes solitary, usually ring-porous; parenchyma apotracheal, rays up to 3–4 cells wide, hetero- or homogeneous. Leaves commonly dorsiventral, sometimes isobilateral; stomata Ranunculaceous type; unicellular glandular hairs present on leaves.

Embryology. Pollen grains are shed at 2-celled stage. Ovules anatropous, unitegmic and tenuinucellate. Polygonum type of embryo sac, 8-nucleate at maturity. Endosperm formation of the Nuclear or Cellular or sometimes intermediate between the two types (Johri et al. 1992).

Chromosome Number. Haploid chromosome number of *Phacelia* spp. is n = 11. Basic chromosome number is x = 5–9, 11–13, 19.

Chemical Features. Endosperm is often rich in lipids and proteins, as in *Nama jamaicense* (Johri et al. 1992).

Important Genera and Economic Importance. *Nemophila*, *Phacelia* and *Wigandia* are grown as ornamentals. A monotypic genus *Tricardia* reported from California, Arizona and Utah is very different from all other genera. It has 2 series of sepals, three are large and orbicular, cordate at the base and are membranous and reticulate in fruit; genus *Codon*, restricted to southwest Africa, has densely prickly branches and leaves, and pleiomerous flowers.

Taxonomic Considerations. The family Hydrophyllaceae is closely related to the families Convolvulaceae, Polemoniaceae and Boraginaceae. Bentham and Hooker (1965b) grouped this family in Polemoniales along with the other three families and Solanaceae. Engler and Diels (1936) placed it in the Tubiflorae along with 19 other families. Its affinity with Polemoniaceae and Boraginaceae has been pointed out. Wettstein (1935) and Rendle (1925) also grouped it in Tubiflorae but in the suborder Boragineae. Hallier (1912) placed it in Boraginaceae and treated it as a primitive member of the Campanulales; Bessey (1915) included the family Hydrophyllaceae in the Polemoniales. Hutchinson's (1926) Polemoniales comprised only Polemoniaceae and Hydrophyllaceae. Later, Hutchinson (1969) included Cuscutaceae also in the Polemoniales. According to him, Hydrophyllaceae is of somewhat mixed descent. Usually, it is regarded to have originated along with Polemoniaceae from the Geraniaceae and allied families. There is a strong possibility that a (small) part of it has originated from Saxifragaceae. *Nemophila insignis* and *Conanthus grandiflorus* resemble some members of Geraniaceae, but *Hydrophyllum*, *Phacelia* and *Romanzoffia* have a striking resemblance to *Saxifraga* spp. *Phacelia sericea* resembles *Echium* and *P. tanacetifolia* resembles *Heliotropium*—both members of Boraginaceae. Cronquist (1968, 1981) includes Hydrophyllaceae in the Polemoniales along with 6 other families. Benson (1970) and Stebbins (1974) also agree with this view. Dahlgren (1975a, b, 1980a, 1983a), however, treats it as a member of Solanales, which also includes Boraginaceae. Takhtajan (1980) regarded Hydrophyllaceae as a member of Polemoniales, between Polemoniaceae and Boraginaceae, but in 1987 he removed it to Boraginales.

In view of different placements, the correct assignment of Hydrophyllaceae would be between Polemoniaceae and Boraginaceae. Its members show resemblances to the members of both the families.

Boraginaceae

A large family of ca. 100 genera and 2000 species, widely distributed in both temperate and tropical countries.

Vegetative Features. Herbs annual (*Heliotropium indicum*) or perennial (*Arnebia*, *Cynoglossum*), shrubs (*Rhabdia lycioides*), lianas (*Tournefortia*), or sometimes trees (*Cordia dichotoma*).

Mostly normal tap root, root tuberous in *Symphytum*. Stem usually herbaceous except in woody plants. Stiff hairs occur all over the plant body (Fig. 50.5A, H, I). Leaves mostly alternate, lower ones sometimes opposite, as in *Trichodesma*, all leaves opposite in *Antiphytum*, simple, entire, estipulate and covered with stiff hairs. Cystoliths common in leaves.

Floral Features. Inflorescence determinate, usually composed of 1 or more scorpioid or helicoid cymes which uncoil as the flowers open (Fig. 50.5A). Sometimes the flowers are in terminal dense clustered racemes, as in *Cordia* and *Myosotidium*, or spikes as in *Lithospermum*, or loosely branched, a few-flowered cymes, as in *Mertensia*. Flowers hermaphrodite, actionomorphic (Fig. 50.5B), rarely zygomorphic as in *Lycopsis* and *Echium*; complete, hypogynous. Calyx of 5 sepals, all equal (Fig. 50.5B), rarely unequal as in *Myosotis azorica*; sepals 4 in *Myosotidium*, usually gamosepalous, sometimes only basally connate and free above, as in *Lithospermum*, imbricate, rarely valvate, persistent. Corolla 5-lobed, imbricate, or contorted in bud; rotate (*Borago*), salverform (*Myosotis*, *Lithospermum*), infundibuliform (*Moltkia*), campanulate (*Mertensia*) or tubular (*Onosma*); corolla tube sometimes partially closed by hairy appendages as in *Borago*, lobes unequal in *Echium*. Stamens 5, alternate with corolla lobes and inserted at the throat of the tube, i.e. epipetalous (Fig. 50.5L), equal or unequal as in *Echium*. Anthers bicelled, dehisce longitudinally, basifixed (Fig. 50.5C) or basally dorsifixed, introrse; annular nectariferous disc may or

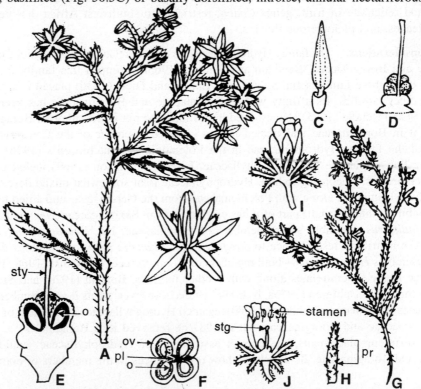

Fig. 50.5 Boraginaceae: **A-F** *Borago officinalis*, **G-J** *Harpagonella palmeri*. **A** Twig with flowers. **B** Flower. **C** Stamen. **D** Gynoecium. **E** Longisection. **F** Ovary, transection. **G** Twig with flowers in *H. palmeri*. **H** Part of stem shows prickles. **I** Flower. **J** Longisection shows epipetalous stamens. *o* ovule, *ov* ovary, *pl* placenta, *pr* prickle, *stg* stigma, *sty* style. (Adapted from Hutchinson 1969)

may not be present. Gynoecium syncarpous. bicarpellary, ovary bi- or tetralocular (Fig. 50.5D, F), superior, 1 ovule in each locule on axile placenta (Fig. 50.5E, F); style 1, simple gynobasic or terminal as in Heliotropioideae and Ehretioideae, stigma 1, simple, capitate (bilobed in *Anchusa, Pulmonaria*, 4-lobed in *Cordia*, 2 in *Echium*). Fruit of 4 nutlets (2 nutlets in *Cerinthe*) or a 1- to 4-seeded nut, or a durpe as in Cordioideae, Heliotropioideae and Ehretioideae; variously sculptured or glabrous and glossy. Seeds without or with scanty endosperm and a curved or straight embryo.

Anatomy. Vessels very small to large, often tend to be ring-porous, perforations simple; parenchyma varies from uniseriate to broad bands. Rays mostly 4–6 cells wide, heterogeneous. Leaves much variable, from isobilateral to dorsiventral within a genus, stomata usually Ranunculaceous type; thick-walled, stiff unicellular hairs cover leaf surfaces.

Embryology. The family is multipalynous. The pollen grains are triporate (e.g. *Coldenia procumbens*), tricolpate (*Heliotropium indicum*), multicolpate (*H. ovalifolium*) and dumbell-shaped (*Cynoglossum denticulatum*); 2- or 3-celled at the dispersal stage. Ovules ana-, hemiana- or ana-campylotropous, or orthotropous, as in *Cordia obliqua*; unitegmic, tenuinucellate or crassinucellate, as in *Coldenia, Ehretia* and *Rotula*; or pseudocrassinucellate, as in *Cordia alba* (Johri et al. 1992). Embryo sac of Polygonum type, 8-nucleate at maturity. Twin embryo sacs reported in some members (Johri et al. 1992).

Chromosome Number. Haploid chromosome number is n = 12, 14; in *Lappula microcarpa* n = 12. Basic chromosome number is x = 4–13.

Chemical Features. Anthocyanin pigments of glycoside type and aglycones such as petunidin are present (*Anchusa*). Most of the members of the subfamily Boraginoideae are rich in L-bornesitol, a monoethyl ether of myo-inositol (Plouvier 1963). Four pyrrolizidine alkaloids (intermedine, lycopsamine, O^7-acetylintermedine and O^7-acetyllycopsamine) have been detected in *Onosma alborosea, O. alborosea* x *sanguinolenta* and *O. arenaria* subsp. *pennina* (Röder et al. 1993).

Important Genera and Economic Importance. The genera *Patagonula* and *Caccinia* are characterized by the presence of much enlarged and stellately spreading calyx below the fruit. In *Harpagonella* the calyx lobes are deeply-divided and hooked (Fig. 50.5J, L). Corolla limb bilipped in *Echiochilon*. In *Caccinia* and *Heliocarya* 1 anther is much larger than the others. Anthers connivent into a cone-shaped structure in *Trichodesma* and *Borago*. *Heliotropium indicum*, a common wild plant with small flowers, is medicinally used against boils and swellings. Fruits of various species of *Cordia* are edible. *Arnebia hispidissima* is a prostrate herb of sandy places bearing yellow tubular flowers; its roots yield a purple dye. Roots of *Alkanna tinctoria* yield a red dye. *Myosotis palustris* bears bright-blue flowers commonly known as forget-me-not. *Ehretia acuminata* is a common tree from Indus to Sikkim and in the hills up to 700 m. It has sharply toothed leaves and white flowers, and drupes with 2 stones only. Its wood is very strong and used for making agricultural implements, sword-hilts, and scabbards; fruits edible. Leaves of *E. laevis* are used like betel leaf or 'paan', in some parts of India.

Taxonomic Considerations. The family Boraginaceae is divided into two groups with a number of subfamilies:

Group I—style terminal, fruit a drupe.
> Subfamily (i) Cordioideae—style 4-lobed, drupe 4-locular, seeds with plicate cotyledons and no endosperm as in *Cordia, Auxemma.*
> Subfamily (ii) Ehretioideae—style simple or bilobed. Fruit a small globose drupe divided into 2 bilocular or 4 unilocular pyrenes. Seeds with ovate cotyledons and scanty endosperm, as in *Ehretia, Coldenia.*

Subfamily (iii) Heliotropioideae—style simple or bilobed; a ring of hairs present below the stigma. Fruits sometimes dehiscent, splitting into nutlets as in *Heliotropium, Tournefortia.*

Group II—style gynobasic, fruit of nutlets.

Subfamily (iv) Boraginoideae—corolla with scales near the throat of the tube; fruit of 4 or fewer nutlets as in *Borago, Myosotis.*

Subfamily (v) Wellstedioideae—flowers 4-merous; ovary compressed, 2-celled; ovule 1 in each locule as in *Wellstedia.*

Bentham and Hooker (1965b) included Boraginaceae in Polemoniales along with 4 other families, on the basis of bicarpellary superior ovary, stamens as many as or fewer than corolla lobes and alternate with them. Engler and Diels (1936) placed Boraginaceae in the order Tubiflorae, because of gamopetalous corolla and epipetalous stamens. According to Hallier (1912), it is the most primitive member of his order Campanulinae and has given rise to Campanulaceae and Loasaceae. He derived it from the group Proberberideae. Bessey (1915) agreed with Bentham and Hooker. Hutchinson (1926, 1969) broke up Boraginaceae into 2 families—Ehretiaceae with woody members (except *Coldenia*), and Boraginaceae with herbaceous members. Cronquist (1968, 1981) considers Boraginaceae to be a single family and attaches importance to its resemblances with Verbenaceae and Labiatae. In both Boraginaceae and Verbenaceae-cum-Labiatae, the gynobasic style and the 4-nutlet fruit have evolved from the unlobed gynoecium with a terminal style that ripens into a 4-seeded drupe. If the Verbenaceae and the more primitive genera of Boraginaceae were to disappear, the characteristic gynoecium of the Labiatae and the rest of Boraginaceae could have originated from a common ancestor. Dahlgren (1975a, b, 1980a, 1983a) recognises two distinct families—Ehretiaceae and Boraginaceae—and includes them in the Solanales. Takhtajan (1980, 1987), however, treats this family as a consolidated one including 5 subfamilies— Cordioideae, Ehretioideae, Heliotropioideae, Boragoideae and Wellstedioideae. The Wellstedioideae is recognised as a distinct family by Dahlgren (1980a), and as a part of the family Boraginaceae by Hutchinson (1969). J.R.A. Lawrence (1937) concludes, from the study of floral anatomy of various members, that the subfamilies are best treated as components of a single family. Also, that Boragoideae members with their deeply cut nutlets and gynobasic style are the most highly evolved members. If the family is treated as a consolidated one, then the different trends in evolution can be studied more easily

Lennoaceae

A small family including only 3 genera and 4 species of root parasites, endemic to southwestern North America and Mexico.

Vegetative Features.   All the members are herbaceous, total parasites on the roots of *Clematis, Eriodictyon* and *Prosopis.* Stem simple or branched; leaves scale-like, erect-spreading. *Lennoa madreporoides* is parasitic on roots of *Stevia* sp. of the family Compositae (Fig. 50.6A, B).

Floral Features.   Inflorescence dense, subcorymbose cyme in *Lennoa*, spicate in *Pholisma* (Fig. 50.6F, G) and saucer-like and subcapitate in *Ammobroma.* Flowers hermaphrodite, actinomorphic, hypogynous. Sepals 5-10 (8 in *Lennoa*), polysepalous, or partly so, linear, filiform or subulate; puberulant or plumose. Corolla of 5 or 6 petals (Fig. 50.6H), 8-11 in *Lennoa* (Fig. 50.6C), gamopetalous, imbricate, tubular or infundibuliform, or salverform (*Lennoa*). Stamens 8 in 2 whorls in *Lennoa*, 5–10 in 1 whorl in *Pholisma* and *Ammobroma*, epipetalous, filaments very short, adnate to corolla tube for most of their length (Fig. 50.6D, I), emerging from it just at the mouth. Anthers bicelled, the two thecae parallel (Fig. 50.6.K), basally divergent in *Lennoa* (Fig. 50.6E), introrse. Gynoecium syncarpous, 6- to 14-carpellary, ovary superior, number of locules apparently twice the number of carpels by false septation, placentation axile

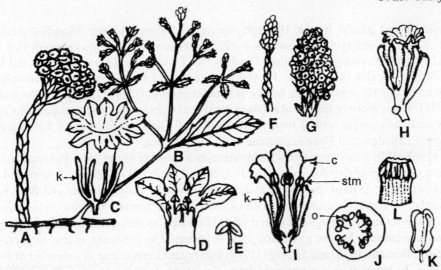

Fig. 50.6 Lennoaceae: **A, C-E** *Lennoa madreporoides*, **B** *Stevia* sp. of Compositae (host plant), **F-L** *Pholisma arenarium*. **A** Flowering twig of *L. madreporoides*. **B** Flowering twig of *Stevia* sp. **C** Flower. **D** Corolla opened up, note epipetalous stamens. **E** Anther. **F** Twig of *P. arenarium* with flowers. **G** Inflorescence. **H** Flower. **I** Longisection. **J** Ovary, transection. **K, L** Stamen (**K**) and stigma (**L**). *c* corolla, *k* calyx, *o* ovule, *stm* stamen. (Adapted from Hutchinson 1969)

(Fig. 50.6J), ovules 1 per locule; style 1, solid and simple, stigma subcapitate (Fig. 50.6L) or peltate-crenate. Fruit a fleshy capsule, irregularly circumscissile. Seeds enclosed in a bony endocarp formed from the locule wall, and pyrene-like, albuminous; embryo globose without any differentiation of cotyledons and radical at seed-shedding.

Anatomy. Vascular system in *Ammobroma* and *Lennoa* consists of a ring of large, isolated bundles accompanied by smaller cortical leaf traces. Xylem of *Ammobroma* is composed of pitted and reticulate vessels with simple perforations. Xylem of *Lennoa* includes only tracheids and no vessels. Stem epidermis bears stalked glandular hairs and stomata; stomata usually numerous and present on both surfaces of the scale leaves.

Embryology. Pollen grains single, 3-colporate in *Ammobroma* and *Lennoa*; 4-colporate predominant in *Pholisma* but 3- to 5-colporate grains also occur; 2-celled at shedding stage. Ovules anatropous, unitegmic, tenuinucellate. Embryo sac of Polygonum type, 8-nucleate at maturity. Endosperm formation of the Cellular type.

Chromosome Number. Basic chromosome number is x = 9.

Important Genera and Economic Importance. Both *Ammobroma* and *Pholisma* are monotypic genera with one species each—*A. sonorae* and *P. arenarium*. Both genera are small herbs, only a few inches above ground. Usually grow in sandy soil, have orange or brown, scale-covered stems and minute violet to purple flowers; *Lennoa* has two species, *L. madreporoides* and *L. arenaria*. The former is parasitic, mostly occurs on different members of Compositae.

Taxonomic Considerations. The family Lennoaceae has been placed in different groups. Bessey (1915) and Hutchinson (1926) placed it in the Ericales. Rydberg (1914) related Lennoaceae to the saprophytic group Monotropoideae of the Pyrolaceae. Hallier (1912, 1923), Copeland (1935) and Wettstein (1935), raised its position and placed it near Hydrophyllaceae and Boraginaceae. Morphological studies by Süssenguth (1927) and embryological studies by P. Maheshwari (1945) also support this assignment. Hutchinson (1969, 1973) recognised distinct families like Pyrolaceae, Monotropaceae and Lennoaceae, all members of his Ericales of Lignosae. However, neither the Lennoaceae nor the Monotropaceae are ligneous. Lawrence (1951) regarded this family as a member of Tubiflorae, following the work of Engler and Diels (1936). It has more or less tubular flowers with gamopetalous corolla, epipetalous stamens, free pollen grains and each carpel with 2 ovules. Metcalfe and Chalk (1972) placed Lennoaceae between Diapensiaceae of Ericales, and Plumbaginaceae of Plumbaginales.

Cronquist (1968, 1981), however, treats Lennoaceae as a member of the order Polemoniales. He points out that because of certain embryological peculiarities of the Ericales, it is a distinct group, and Lennoaceae cannot be retained there. Instead, they show resemblances with Hydrophyllaceae and Bignoniaceae. The increase in number of carpels is usually regarded as reflecting a secondary increase rather than an original high number. They resemble Boraginaceae and other Lamiales, as they produce twice as many 1-seeded nutlets as there are carpels, but the gynobasic style is absent. The similarity is due to parallelism and a common ancestry. Benson (1970) and Stebbins (1974) also regard Lennoaceae as a member of Polemoniales. Dahlgren (1975a, b) treats it as a member of Solanales along with the families like Boraginaceae and Hydrophyllaceae. Takhtajan regards Lennoaceae as a member of the suborder Boraginineae (1980) and order Boraginales (1987). In our opinion, family Lennoaceae is a member of the order Tubiflorae and closely related to the families Boraginaceae and Hydrophyllaceae.

Verbenaceae

A large family with ca. 98 genera and 2614 species, widely distributed all over the tropics and subtropics of the world. A few genera are known from the temperate regions.

Vegetative Features. Herbs (*Phyla*, *Verbena*), shrubs (*Duranta*, *Holmskioldia*), trees (*Tectona*) or sometimes climbing shrubs (*Clerodendrum thomsonae*, *Petrea volubilis*); mostly mesophytic but some halophytes such as *Avicennia*. Root system normal in most members, pneumatophores in the halophytes. Stem usually erect, woody or herbaceous, often bear tubercles on the surface, usually quadrangular. Leaves opposite or whorled, estipulate (stipulate in *Clerodendrum*), petiolate, usually simple (compound in *Vitex*).

Floral Features. Inflorescence variable: a raceme in *Duranta*, an umbel in *Lantana* and *Verbena*, a dichasial cyme as in *Clerodendrum*, or in trichotomous panicles as in *Tectona*. Flowers bisexual, zygomorphic, rarely actinomorphic as in *Duranta*, hypogynous, bracteate, bracteolate. Calyx of 5 sepals, gamosepalous, (4 sepals in *Lantana*), persistent, valvate, often enlarged in fruit; tubular in *Duranta*, bilipped in *Vitex*. Outer surface of sepals of *C. japonicum* bears extrafloral nectaries; calyx sometimes petaliferous. Corolla usually of 5 petals (6 to 16 petals in *Symphorema*), gamopetalous, lobes mostly unequal, usually salverform as in *Verbena* and *Lantana*, or tubular as in *Duranta*, rarely campanulate; sometimes bilipped as in *Vitex* and *Holmskioldia*; aestivation imbricate in bud. Stamens 5 in more primitive genera like *Tectona* and *Geunsia*, didynamous in others; the 5th stamen may or may not be represented by a staminode, often only 2 stamens and 3 staminodes, as in *Oxera*; epipetalous, either included as in *Lantana* or exerted, as in *Holmskioldia* and *Clerodendrum*. Anthers bicelled, longitudinal dehiscence, introrse. Gynoecium syncarpous, bi-, tetra- (*Duranta*) or pentacarpellary (*Geunsia*), ovary superior, locules as many as the number of carpels or twice the number by false septation, placentation axile (free-central in *Avicennia*), ovule 1 per locule, erect; style 1, terminal, stigma lobes usually as many as the carpels. Fruit usually a drupe with as many pyrenes as the number of ovules in the ovary or nutlets, as in *Verbena*, or a 2- to 4-valved capsule, as in *Avicennia*. Seeds without endosperm (present in *Avicennia*, *Stilbe* and *Chloanthes*), and a straight embryo.

Anatomy.   Vessels mostly medium-sized, commonly ring-porous, perforation plates simple but with a few scalariform or reticulate plates in some genera; parenchyma paratracheal, usually narrowly vasicentric, rays mostly 3–4 cells wide, hetero- to homogeneous. Included phloem develops in *Avicennia.* Leaves dorsiventral or isobilateral, surface waxy in halophytic species, hairs both glandular and nonglandular. Extrafloral nectaries occur on the lower surface of leaves in many genera, e.g. *Clerodendrum.* Stomata on both surfaces or only on the lower surface, Caryophyllaceous, Rubiaceous or Ranunculaceous type. Stomata on the upper surface of leaves of a few species of *Stachytarpheta* are arranged in groups.

Embryology.   The family is multipalynous. Pollen grains tricolpate and shed at 2- or 3-celled stage. Ovules anatropous, unitegmic, tenuinucellate, micropyle faces downwards. Polygonum type of embryo sac, 8-nucleate at maturity. Endosperm formation of the Cellular type. Both chalazal and micropylar haustorium develop.

Chromosome Number.   Basic chromosome number is x =5–7, 9, 11–18.

Chemical Features.   Anthocyanin pigments, cyanidin, delphinidin, pelargonidin and verbenalin, occur in many species of *Clerodendrum, Lantana* and *Verbena. Avicennia* wood is rich in lapachol, a naphthaquinone. Seed fat rich in stearic acid in *Tectona grandis* and *Lantana indica.* In *L. camara* the freshly opened flowers are yellow (rich in β-carotin) and undergo post-anthesis chromatic alteration to orange, scarlet and magenta due to synthesis of delphinidin monoglucoside (Mathur and Mohan Ram 1986).

Important Genera and Economic Importance.   The most important genus of this family is *Tectona,* the source of the valuable timber teak wood. The aromatic leaves of *Premna* are used as dinner plates (pattal) and the wood is used to manufacture combs. The wood of *Gmelina arborea* is extensively used in making musical instruments like drums. Many of the genera are grown as ornamentals: *Clerodendrum, Verbena, Lantana, Duranta, Petrea, Vitex, Holmskioldia* and *Callicarpa.* The bark of *Avicennia* is used in tanning.

Taxonomic Considerations.   Bentham and Hooker (1965b) and Bessey (1915) retaind Verbenaceae and Labiatae in the same order—Lamiales—because of certain resemblances in their corolla and gynoecium. Engler and Diels (1936) and Hallier (1912) included it in Tubiflorae along with many other families. Hutchinson (1926) retained Verbenaceae, Labiatae and also Phrymaceae in his Lamiales, but in 1948 he erected a separate order, Verbenales, deriving it from Rubiaceous stock. According to him, Verbenaceae is the most highly evolved family of the Lignosae (1969, 1973). Although in many classificatory systems Lamiaceae (= Labiatae) and Verbenaceae are placed together, neither has evolved from the other. The stem, leaves, inflorescence and gynoecium, in the two families, are rather similar; in their respective groups, both the families represent a climax of evolution.

Cronquist (1968, 1981) also includes Verbenaceae in Lamiales along with Labiatae and Phrymaceae. According to him, it is rather "difficult to know where to draw the line of distinction" between Verbenaceae and Labiatae. It may even be assumed that the Verbenaceae and Labiatae are the primitive and advanced segments of the same family. Phrymaceae, with a single genus and species, is also very closely related to the Verbenaceae. Phrymaceae are clearly a reduced pseudomonomerous type derived from this family. Benson (1970) and Stebbins (1974) also regard Verbenaceae as a member of Lamiales, and the latter includes Phrymaceae and Callitrichaceae as well.

Dahlgren (1975a, b, 1980a, 1983a) treats Verbenaceae as a member of Lamiales, but does not recognise Phrymaceae as a distinct family. Takhtajan (1980, 1987) also includes Verbenaceae in Lamiales embracing the families Verbenaceae, Lamiaceae and Callitrichaceae. Verbenaceae includes many subfamilies which are treated as distinct families: Avicenniaceae, Chloanthaceae, Dichrastylidaceae, Phrymaceae, Stilbaceae and Symphoremataceae. Other subfamilies included are: Viticoideae, Verbenoideae, Lithophytoideae, Phrymatoideae, Nyctanthoideae, and Caryopteridoideae (see Takhtajan 1987).

Verbenaceae and Labiatae (= Lamiaceae) should be placed together in the same order, as they can be derived from a common ancestral stock (see Cronquist 1968).

Callitrichaceae

A unigeneric family of cosmopolitan distribution. The genus *Callitriche* has about 26 species.

Vegetative Features.   Monoecious, aquatic or terrestrial, annual herbs. Stem filiform, herbaceous; leaves opposite, decussate, upper ones often form rosettes in the aquatic species.

Floral Features.   Flowers solitary, axillary, unisexual, actionmorphic, apetalous, bracteolate, hypogynous. Staminate flowers with single stamen, filament long, filiform; anther bicelled, dehisce laterally and longitudinally. Pistillate flowers with a single pistil; gynoecium syncarpous, bicarpellary, ovary superior, bilocular or tetralocular by formation of false septum, placentation axile, ovule 1 per locule; styles 2, filiform, papillose. Fruit a drupe splitting into 2 or 4 pyrenes or drupelets. Seeds with fleshy endosperm and a straight embryo.

Anatomy.   The vascular system of stem is highly reduced. Glandular hairs occur in terrestrial species. Stomata absent in aquatic species but their sporadic occurrence on the leaves of submerged forms are reported; numerous on both surfaces of the leaf and on stem in terrestrial species.

Embryology.   Pollen grains inaperturate or provided with thin irregular rifts; 3-celled at the time of shedding. The pollination and growth of pollen tube is very unusual (see Johri et al. 1992). Ovules pendulous, anatropous, unitegmic and tenuinucellate. Polygonum type of embryo sac, 8-nucleate at maturity. Endosperm formation of the Cellular type.

Chromosome Number.   Basic chromosome number is x = 3, 5, 6, 8.

Chemical Features.   The fleshy endosperm contains oily food reserve.

Important Genus and Economic Importance.   The only genus, *Callitriche*, has no economic importance.

Taxonomic Considerations.   Highly reduced flower has made it difficult to correctly determine the taxonomic position of Callitrichaceae. Even the study of vascular anatomy has not been of much help. Hegelmaier (1864) and Bentham and Hooker (1965b) included this family in the Haloragaceae. Hallier (1912) derived it from Linaceae and included it in the order Guttales. Bessey (1915) derived Callitrichaceae from Euphorbiaceae and placed it in his order Geraniales. Rendle (1925), Wettstein (1935) and Pax and Hoffman (1931b) considered it to be an advanced member of Euphorbiales, mainly because of its highly reduced, unisexal flowers. Hutchinson (1926) regarded it as a member of Lythrales including the families Lythraceae, Onagraceae, Trapaceae and Haloragidaceae. Hutchinson (1969, 1973) treated Callitrichaceae as a member of Onagrales (the new name of the order after removing Lythraceae). Benson (1970) included Callitrichaceae in a separate order, Callitrichales, and placed this order between Empetrales and Podostemales. Cronquist (1968, 1981), Stebbins (1974), Dahlgren (1975a, b, 1980a, 1983a) and Takhtajan (1980, 1987) place it in one of the most highly evolved orders, Lamiales, along with the families Verbenaceae and Labiatae (=Lamiaceae). According to Cronquist, althouth the floral structure is highly reduced, it has the characteristic unitegmic, tenuinuecellate ovule of the Asteridae and higher Dilleniidae. Callitrichaceae also has the fruit characteristic of the Verbenaceae-Labiatae, i.e. a drupe splitting into 4 drupelets; but, on the other hand, this family differs from most of the Lamiales in having well-developed endosperm and separate styles. Both these features are primitive as compared to other families in this order. Takhtajan (1980, 1987) is of the opinion that it is probably related to Lamiaceae and Verbenaceae.

Except for the embryological features, no other character supports its placement in such a highly evolved order as the Lamiales. The phyletic position of Callitrichaceae is sill uncertain.

Labiatae (= Lamiaceae)

A large family of about 200 genera and 3200 species, cosmopolitan in distribution; especially abundant in the Mediterranean region, the Old World and the mountains of subtropics.

Vegetative Features. Predominantly annual or perennial herbs, sometimes shrubs (*Lavandula dentata*), rarely trees (*Hyptis*, and *Leucosceptrum*), or erect, perennial marsh plant (*Scutellaria galericulata*). Stem herbaceous or woody as in tree members, glandular-pubescent, erect, sometimes prostrate and with suckers as in *Mentha viridis*, quadrangular. Leaves simple, opposite-decussate, estipulate, petiolate, with varied type of margins, surface glandular hairy. These glands contain volatile oil which makes the leaves aromatic.

Floral Features. Inflorescence is typical of the family. It is a spike or raceme of pairs of dichsial cymes at each node (Fig. 50.9A), rarely solitary, axillary as in *Scutellaria*. Sometimes the primary axis has short internodes and the pairs of biparous cymes are condensed together to form a globular head, as in *Lamium*, *Hyptis*, *Prunella* and *Monarda*. Flowers scattered in a raceme in a few species of *Scutellaria* and *Teucrium*. Each cyme at the axil is subtended by a foliaceous bract which may be longer than the cyme and coloured (*Salvia*). Flowers hermaphrodite, bracteolate or ebracteolate, zygomorphic (Fig. 50.9B) (actinomorphic in *Mentha* and *Elsholtzia*), pedicellate, complete, hypogynous, often showy. Calyx of 5 sepals, gamosepalous,

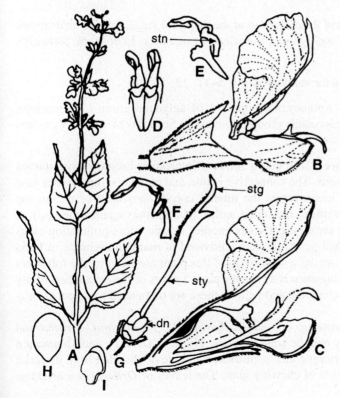

Fig. 50.9 Labiatae: A-E, G-I *Salvia urticifolia*, F *S. lyrata*. A Flowering plant. B Flower, side view. C Flower, longi-section. D Stamens front view E Side view. F Fertile stamen of *S. lyrata* with both the anther halves polliniferous. G Gynoecium, whole mount. H Nutlet. I Embryo (oriented as in nutlet). *dn* nectariferous disc, *stg* stigma, *stn* staminode, *sty* style. (Adapted from Radford 1987)

sometimes bilobed (Fig. 50.9C), as in *Alvesia, Ocimum*; sometimes the lobes are absent or appear to be only 2 with 5, 10, or 15 conspicuous ridges or ribs. Corolla of 5 petals, gamopetalous, imbricate, mostly bilabiate, the lower lip of 3 petals and upper lip of 2 petals. In *Ocimum* a single petal forms the lower lip and 4 petals together form the upper lip; the upper lip is rudimentary in *Teucrium*. Stamens 2 as in *Salvia* (Fig. 50.9D, E, F) or 4 as in *Ocimum*, didynamous, monadelphous in *Coleus*; staminodes rarely present, as in *Orthodon* and *Hypogomphia*, epipetalous. Anthers bicelled, longitudinal dehiscence, connective elongated with the two lobes at two ends as in *S. lyrata* (Fig. 50.9F), one lobe rudimentary in *S. urticifolia* (Fig. 50.9D, E). Hypogynous, nectariferous disc often present between the stamens and the ovary (Fig. 50.9C, G). Gynoecium syncarpous, bicarpellary; ovary superior, bilocular but tetralocular at maturity due to the intrusion of the ovary wall, ovule one in each of the 4 locules formed due to false septation of the ovary, placentation basal derived from axile; style gynobasic (Fig. 50.9G), rarely terminal as in *Ajuga*, mostly bifid, rarely tetrafid as in *Cleonia*, and stigmas minute at the end of these branches. Fruit typically a group of 4 nutlets, enclosed by the persistent calyx; drupaceous in *Stenogyne*. Sometimes the pericarp is fleshy, as in *Gomphostemma*, or develops into a wing-like membranous structure. In *Alvesia* of tropical Africa, the calyx enlarges and becomes bladder-like and reticulate. Seeds (Fig. 50.9H) non-endospermous or with scanty endosperm which is absorbed by the developing embryo; embryo straight (Fig. 50.9I).

Anatomy. Vessels small to minute, semi-ring-porous, with simple perforations, parenchyma paratracheal, rather sparse; rays 4 to 12 cells wide, sometimes heterogeneous. Hairs of various kinds—glandular, secrete essential oils, and uniseriate, tufted or branched nonglandular—occur on the surface of stem and leaves. Leaves dorsiventral, stomata on the lower surface, Caryophyllaceous type; crystals not very frequent.

Embryology. Pollen grains 3- to 6-colpate and 2- to 3-celled at the dispersal stage. Ovules anatropous, unitegmic, tenuinucellate. Polygonum type of embryo sac, 8-nucleate at maturity. Endosperm formation of the Cellular type.

Chromosome Number. The basic chromosome number is x = 5–11, 13, 17–20.

Chemical Features. Rich in volatile oils. Anthocyanin pigments of aglycone group like cyanidin, delphinidin and pelargonidin, and of acyl and glycosidic groups are present. Seed fats of various members are linolenic-rich.

Important Genera and Economic Importance. The genus *Salvia* is interesting because of its stamen structure. There are only two epipetalous stamens. The connective is thin and elongated, one anther lobe is fertile and the other rudimentary. As the insect enters the tubular corolla in search of nectar, the connective acts as a lever. Because of its movement, the fertile anther lobe brushes against the body of the insect, covering it with pollen. When the same insect visits another flower, cross-pollination takes place. *Ocimum sanctum* is a profusely branched perennial herb cultivated in many households in India (for religious purposes). Necklaces of beads from the woody twigs of this plant are worn by the followers of the Krishna cult and the Vaishnavas. *Mentha aquatica* is an aquatic plant. Leaves of *Coleus* are beautifully variegated and coloured with shades of red, violet, pink and yellow; these are often used for experimental work in laboratories.

Many genera are the source of volatile aromatic oil: *Salvia, Lavandula, Rosmarinus, Mentha* and *Pogostemon*. These oils are used in perfumery and the soap industry. Plants of *Ocimum* and *Mentha* are important medicinally. *M. piperata*, the peppermint plant, is cultivated and the oil distilled from its aerial parts is used medicinally and also in preparation of chewing gum. The leaves of *Orthosiphon aristatus* are used like tea leaves in Java. This drink is of medicinal value in kidney and bladder troubles. Some other members, like *Origanum* (marjoram), *Thymus* (thyme) and *Satureja* (savory), are important culinary herbs. The leaves of *Origanum vulgare* are used to make hair decorations in some parts of India. The ornamentals include *Salvia*, *Leonotis* (lion's head), *Dracocephalum* (dragonhead), *Nepeta* (catmint), *Scutellaria* (skull cap), *Coleus*, *Teucrium*, *Lavandula*, *Thymus* and *Pycnanthemum*.

Eremostachys superba grows in few and widely separated areas which are not connected by the normal dispersal ability of the plant. It is reported from Mohand, Dehra Dun in the Siwalik hills of Uttar Pradesh in India and from Peshawar in Pakistan. The plant is on the verge of extinction and may disappear within 4 to 5 years (Rao and Garg 1994).

Taxonomic Considerations.   Hallier (1912), Rendle (1925) and Wettstein (1935) included Labiatae in Tubiflorae. Bentham and Hooker (1965b) and Bessey (1915) retained it in Lamiales along with the family Verbenaceae. Hutchinson (1926) also placed the two families—Verbenaceae and Labiatae—in the order Lamiales, but later (1948) Hutchinson raised Verbenaceae to Verbenales. In 1969, he pointed out that the Labiatae is the most highly evolved family amongst the Herbaceae and Verbenaceae, amongst the Lignosae, and considered the two families unrelated. Benson (1970) placed only Verbenaceae and Labiatae in his order Lamiales, and the predominantly herbaceous family is treated as more advanced.

Cronquist (1968, 1981) and Stebbins (1974) group together Labiatae, Phrymaceae and Verbenaceae in the Order Lamiales. Dahlgren (1975a, b, 1980a, 1983a) considers Labiatae to belong to the order Lamiales, along with Verbenaceae. Takhtajan (1980, 1987) supports this assignment.

The family Labiatae has such distinctive features that it can easily be separated from others. However, the members of the two subfamilies—Ajugoideae and Prostantheroideae—resemble Verbenaceae in having terminal styles (not gynobasic). Similarly, there are some members of Verbenaceae which have a gynobasic style. Some Labiatae also resemble some members of the Boraginaceae from which they can be separated due to the character of ovules. In the ovule of Boraginaceae the micropyle points upward and the raphe is directed outward, whereas in Labiatae the micropyle points downward and the raphe is inwards.

The comparative data of these three families is given in Table 50.9.1.

Table 50.9.1   Comparative Data for the Families Verbenaceae, Labiatae and Boraginaceae

Verbenaceae	Labiatae	Boraginaceae
Herbs, shrubs and trees, stem quadrangular	Herbs and shrubs, stem quadrangular	Herbs, shrubs or trees, stem cylindrical
Inflorescence racemose or cymose	Verticillate or cymose	Cincinnus
Flowers zygomorphic, rarely bilipped	Zygomorphic, always bilipped	Actinomorphic, never bilipped
Calyx persistent	Calyx persistent	Calyx persistent
Stamens 2 + 2 or 2	Stamens 2 + 2 or 2	Stamens 5
Carpels (2), superior	Carpels (2), superior	Carpels (2), superior
Ovary 2- or 4-locular, placentation axile	Ovary 4-locular, placentation basal	Ovary 4-locular, placentation axile
Style simple, terminal, rarely gynobasic	Style gynobasic, rarely terminal	Style terminal, gynobasic in subfamily Boraginoideae
Fruit a drupe or berry	Fruit of 4-nutlets	Fruit a drupe

From the above data, it is clear that the two families Labiatae and Verbenaceae are closely related, and should be retained in the order Lamiales.

Nolanaceae

A small family of only 2 genera and ca. 63 species confined to Chile and Peru near the seashore, i.e. on the Pacific Coasts of South America and the Galapagos Islands.

Vegetative Features. More or less succulent herbs or shrubs. Stem woody or herbaceous; spotted and streaked with purple in *Nolana paradoxa*. Leaves simple, alternate in vegetative parts and opposite in the inflorescence region, estipulate, with or without a distinct petiole; often covered with glandular hairs.

Floral Features. Flowers solitary axillary, bisexual, actinomorphic, pentamerous. Sepals 5, gamosepalous; corolla of 5 petals, gamopetalous, plicate in bud, campanulate or infundibuliform. Stamens 5, usually equal, alternate with petals, epipetalous, arising from the base; anthers bithecous, dehiscence longitudinal. Gynoecium syncarpous, pentacarpellary, ovary superior, with as many locules as carpels or 2- or 3- times the number of carpels (due to transverse septation), placentation axile; style 1, terminal or gynobasic, stigma bilobed, more or less capitate. Fruit a schizocarp or 5 nutlets which may be 1 to many-seeded. Seeds albuminous, with a curved embryo.

Anatomy. Leaves dorsiventral, stomata mostly Cruciferous type, multicellular, nonglandular hairs occur; intraxylary phloem present. Crystal-sands present.

Embryology. Pollen grains trizonocolpate, oblate-spheroidal; 2-celled at dispersal stage. Ovules ana- to campylotropous, unitegmic, tenuinucellate. Polygonum type of embryo sac, 8-nucleate at maturity. Endosperm formation of the Cellular type.

Chromosome Number. Basic chromosome number is x = 12.

Chemical Features. *Nolana rostrata* is characterized by resinous specimens growing in deserts and semi-desert areas of Chile, and contains labdane diterpenoids, a secondary metabolite (Garbarino et al. 1986).

Important Genera and Economic Importance. The two genera of this family are *Nolana* with ca. 57 species and *Alona* with only 6 species. *N. paradoxa* and *N. acuminata* are cultivated as garden ornamentals.

Taxonomic Considerations. The family Nolanaceae has been placed in different taxonomical groups by different workers, because of varied morphological features. Bentham and Hooker (1965b) and Hutchinson (1969) included it in Convolvulaceae as a tribe, Nolaneae, because of its infundibuliform or funnel-shaped corolla, twisted aestivation and pentacarpellary ovary. The fruit bears resemblance to those of Boraginaceae. It has many characters in common with Solanaceae also. Hallier (1912) and Hutchinson (1948) placed Nolanaceae in the Solanaceae. Engler and Diels (1936) recognised it as a distinct family and its position in Tubiflorae just before Solanaceae. Hutchinson (1969, 1973) also raised it to the rank of a family (on the basis of the monographic study of the taxon by Johnston 1936). The pentacarpellary ovary is considered to be the persistence of generally primitive condition. As the family Nolanaceae shows close affinities with both Convolvulaceae and Solanaceae, Cronquist (1968, 1981) places it in his order Polemoniales, along with these two families and a few others. Benson (1970) and Stebbins (1974) also agreed with this assignment. According to Dahlgren (1975a, b), the Nolanaceae belongs to Solanales as a distinct family.

The Nolanaceae has been related to Convolvulaceae, Boraginaceae, and Solanaceae (Mesa 1976, 1981, Alfaro and Mesa 1979). Mesa (1986) comments that this family is closely related to Solanaceae, perhaps better placed between Convolvulaceae and Solanaceae, as both the families have intraxylary phloem (but not Boraginaceae). Common features of Nolanaceae and Solanaceae are: (i) crystal-sand of calcium oxalate, (ii) intraxylary phloem in the hypocotyl, (iii) hair types, (iv) pollen morphology (dimorphic pollen grain), and (v) capacity for increase in the number of carpels.

Dahlgren (1980a, 1983a) and Cronquist (1981) support the above assignment. Takhtajan (1980) considers the Nolanaceae as a subfamily of Solanaceae, on the basis that in Nolanaceae the fruits are derived from certain atypical baccate fruits present in some Solanaceae, such as *Lycium, Grabowskia* and *Nicandra. N. physaloides* has 5-merous gynoecium which is typical of Nolanaceae. However, Takhtajan (1987) treats Nolanaceae as a separate family near Solanaceae. Mesa (1986) retains this family as a distinct one, on the basis of: (i) area of distribution at the family level; Nolanaceae members are geographically isolated from Solanaceae being restricted to Peruvian-Chilean Atacama desert, (ii) mericarpic fruit with separate nutlets is typical of Nolanaceae, (iii) flowers are apparently heterostylous. This is also supported by pollen dimorphism—each species with subcircular and ovate pollen types.

D'Arcy (1986) presumes that *Nolana* should be included in Solanaceae as they also have the Solanaceous characters like alternate leaves, similar flower structure, and intraxylary phloem; but, at the same time, its geographical isolation, mericarpic fruits, and heterostylous condition support its treatment as a distinct family.

Solanaceae

A large family of ca. 85 genera and more than 2200 species which occur chiefly in Central and South America. The largest genus *Solanum* with about 1500 species occurs over most parts of the world.

Vegetative Features.   Predominantly herbaceous, shrubs, small trees, lianas and creepers are also known. Stem soft, herbaceous or woody, with hairy surface. Leaves alternate or opposite (near the inflorescence), simple or pinnatisect as in *Lycopersicon esculentum*, estipulate, often with oblique base (*Datura, Solanum nigrum*).

Floral Features.   Inflorescence axillary or extra-axillary cymes, often helicoid. Flowers solitary, axillary in *Datura, Atropa*, hermaphrodite, actinomorphic, or zygomorphic as in *Schizanthus*, hypogynous, ebracteate and ebracteolate. Calyx 5-lobed, gamosepalous but often connate only at base and free above, persistent and sometimes enlarged in fruit, as in *Physalis, Withania*. Corolla of 5 petals, gamopetalous, rotate (*Solanum*), infundibuliform (*Petunia, Datura*) or tubular (*Nicotiana tabacum*) , rarely bilabiate (*Schizanthus*). Aestivation usually plicate or convolute, rarely valvate. Stamens 5, epipetalous, alternate with corolla lobes, usually unequal; all perfect and inserted at the base of the corolla. Sometimes only 4 or even 2 stamens occur and then staminodes are present. Anthers bicelled (monothecous in *Browallia*), longitudinal dehiscence or dehisce by apical pores as in *Solanum*. Hypogynous disc usually present. Gynoecium syncarpous, bicarpellary (3- to 5-carpellary in *Nicandra*), ovary superior, bilocular or 3- to 5-locular by formation of false septa, rarely unilocular with only one ovule, as in *Henoonia*[7] or apically unilocular, as in *Capsicum*, placentation axile, placenta swollen, style 1, stigma bilobed, capitate. Fruit a berry, sometimes enclosed in an enlarged persistent calyx, as in *Physalis* and *Withania*; or septicidal capsule, as in *Datura*. Seeds smooth or pitted, albuminous and with a straight embryo.

Anatomy.   Vessels very small to medium-sized, few to numerous, perforations simple, intervascular pitting alternate; parenchyma either scanty, paratracheal or predominantly apotracheal. Rays uniseriate or up to 8 cells wide, almost homogeneous. Intraxylary phloem and crystal-sand are reported. Leaves usually dorsiventral, stomata Ranunculaceous type or sometimes Cruciferous or Caryophyllaceous type.

Embryology.   Pollen grains 3- to 5- or 6-colpate, colporate or non-aperturate; shed at 2-celled stage.

---

[7]Now included in Goetziaceae (Takhtajan 1987).

Ovules hemiana-, ana- or campylotropous, unitegmic, tenuinucellate. Polygonum type of embryo sac is common; Allium type occurs in *Capsicum frutescens* var. *tabasco*, *C. nigrum* and *C. pendulum* (Johri et al. 1992); 8-nucleate at maturity. Endosperm formation of the Nuclear, Cellular and intermediate type.

Chromosome Number.   Basic chromosome number is x = 7 to 12.

Chemical Features.   The Solanaceae are known for their tropane alkaloids and for the steroidal lactones (withanolides). The principal tropane alkaloids, hyoscine and hyoscyamine, are reported in at least 15 genera such as *Datura*, *Duboisia*, *Cyphanthera* and others (Evans 1986). Steroidal lactones are highly oxygenated C_{28} compounds and have been isolated from *Acnistus*, *Datura*, *Lycium*, *Jaborosa*, *Nicandra*, *Physalis* and *Withania*. In *W. somnifera* a number of chemotypes are known which are morphologically similar but differ in their withanolide content. Hybrids from such chemotypes often produce new withanolides not known in either parent (Eastwood et al. 1980, Nittala and Lavie 1981).

Important Genera and Economic Importance.   A number of genera of this family are important. The habit of different species of *Solanum* is much varied. *S. surattense* is a spiny xerophytic prostrate herb, *S. nigrum* a mesophytic erect herb, *S. tuberosum* a herb with underground stem modified for storage of food, *S. melongena* bears edible fruits that are used as vegetable; *S. verbascifolium*, a tall unarmed shrub or small tree, has 4- to 8-inches long, elliptic-lanceolate leaves which are woolly tomentose on the lower surface, and flowers in dichotomous corymbose subterminal cymes. *S. indicum* is an erect prickly undershrub about 1 to 6 feet tall. *Lycopersicon esculentum* is the tomato plant, bearing edible fruits rich in vitamins particularly vitamins A and B. *Capsicum frutescens* and *C. annuum* are the chillies. *Physalis peruviana* is the commonly known gooseberry in which the orange-yellow berries are completely enclosed in the enlarged, papery, persistent calyx. The fruits are eaten when ripe and also made into jams. *Nicotiana* is another important genus. The two species cultivated are *N. tabacum* and *N. rustica*, the leaves yield tobacco. *Datura alba*, *D. innoxia*, *D. metel* grow wild; *D. arborea* is a small tree from the Peruvian Andes bearing solitary axillary flowers with 6- to 9-inches long white corolla. *D. suaveolens* from Brazil is also a small tree. The seeds of these plants are poisonous if eaten in large quantities, but of medicinal value when used in minute doses. *Atropa belladona* is also of medicinal importance—source of the alkaloid atropin used for dilating the pupil of the eye. This, too, is poisonous when administered in higher doses. Other plants of medicinal importance are *Hyoscyamus niger* and *Mandragora officinarum*.

Ornamental plants include *Petunia violacea*, *Cestrum nocturnum*, *Nicandra physaloides* and many species of *Solandra*, *Schizanthus*, *Brunfelsia* and *Browallia*. *Lycium europaeum* is yet another interesting plant, a spiny xerophytic shrub.

Taxonomic Considerations.   The family Solanaceae has been placed variously by different authors. Bentham and Hooker (1965b) and Bessey (1915) treated it as a member of the order Polemoniales. Engler and Diels (1936), and Hallier (1912) placed it in Tubiflorae, whereas Hutchinson (1948, 1959, 1969) recognised a distinct order, Solanales, and derived it from Saxifragales. Cronquist (1968, 1981) considers it to be a member of his Polemoniales along with the families Nolanaceae, Convolvulaceae, Cuscutaceae, Polemoniaceae and others. Benson (1970) included Solanaceae and Nolanaceae in the same order, Solanales, and concluded that Solanaceae is the more advanced of the two. Varghese (1970) observed that various floral features are shared by the families Scrophulariaceae and Solanaceae. Zygomorphic flowers of Scrophulariaceae are present in *Salpiglossis*—a genus of Solanaceae. The oblique placenta as seen in Solanaceae is also present in *Scrophularia*—a member of Scrophulariaceae. However, embryologically, the two families are very distinct. In Solanaceae the endosperm formation is of the Cellular, Nuclear or Helobial type but in Scrophulariaceae it is only of the Cellular type. Embryogeny in Scrophulariaceae is of the Crucifer type and in Solanaceae it is of the Solanad type. Varghese (1970) concluded that the two families might have had a common origin but Scrophulariaceae has specialised more than Solanaceae. Stebbins (1974) included it in Polemoniales along with Nolanaceae. Dahlgren (1975a, b, 1980a, 1983a) places it in a separate order, Solanales, which also includes Nolanaceae. Takhtajan (1980) states that this family belongs to Scrophulariales but also has affinities with Convolvulaceae of the Polemniales. According to him, these two families probably had a common origin from Loganiaceous stock. In both the families intraxylary phloem is present—in Convolvulaceae it arises in the stem above the level of hypocotyl, and in Solanaceae it arises in the hypocotyl. At the same time, the seed structure of *Solanum* and *Lycopersicon* resembles the seed coat structure of *Strychnos* of the Loganiaceae (Corner 1976). Takhtajan (1987), however, removes Solanaceae to Solanales.

After reviewing the phylogeny of Solanaceae, it is best placed in the order Tubiflorae and close to the family Scrophulariaceae. There are certainly more resemblances of Solanaceae with Scrophulariaceae than with Convolvulaceae.

Duckeodendraceae

A unigeneric family from Brazil, with the genus *Duckeodendron* with only one species.

Vegetative Features.   Large trees with alternate, entire, simple, estipulate leaves.

Floral Features.   Inflorescence terminal or subterminal, few-flowered cymes. Calyx of 5 sepals, gamosepalous, persistent. Corolla of 5 petals, gamopetalous, with a long corolla tube and short imbricate lobes, greenish-white. Stamens 5, alternipetalous, epipetalous, exerted, with oblong, introrse, basally bilobed or sagittate, medianly fixed anthers; interstaminal disc present. Gynoecium syncarpous, bicarpellary, ovary may or may not be immersed in the disc, bilocular, with 1 anatropous ovule per locule; style 1, elongated, stigma dilated and shortly bilobed. Fruit a large, shiny, red drupe. Seed U-shaped with U-shaped embryo in scanty oily endosperm.

Embryology.   Embryology has not been studied.

Taxonomic Considerations.   The family Duckeodendraceae is probably closely related to Apocynaceae as the two families resemble each other in anatomical features, sagittate anthers, presence of disc and simple style (Willis 1973). Hutchinson (1969, 1973) includes it in the family Ehretiaceae of order Verbenales. Benson (1970), Cronquist (1968, 1981), Stebbins (1974) and Dahlgren (1975a, b, 1980a, 1983a) do not recognize it as a distinct family at all. According to Takhtajan (1980), this family is probably related to the family Solanaceae. Kuhlamann (1934) placed Duckeodendraceae in Solanaceae but on the basis of wood anatomy its resemblance is much more with Apocynaceae (Record 1933). Takhtajan (1987) places it in the order Solanales near the families Solanaceae and Nolanaceae.

Buddlejaceae

A small family of 8 genera and 160 species (7 genera and 120 species, according to Takhtajan 1987), distributed mainly in the tropical zones of both New and Old World.

Vegetative Features.   Trees, shrubs or woody vines, rarely herbs. Stem woody, leaves opposite or alternate, simple, covered with woolly or stellate indumentum, petiolate, sipulate, stipules deciduous.

Floral Features.   Inflorescence apical, compact panicle-like clusters or thyrses. Flowers hermaphrodite, actinomorphic, hypogynous, ebracteate, ebracteolate. Calyx 4- or 5-lobed, gamosepalous; corolla of 4 or 5 petals, gamopetalous, imbricate. Stamens 4, rarely 2 or 5, epipetalous, fused in the middle of the corolla tube, anthers bicelled, longitudinally dehiscent. Gynoecium syncarpous, bicarpellary, ovary superior, bilocular, placentation axile, ovules numerous. Fruit a septicidal, bivalved capsule or a drupe, as in

Adenoplea and *Adenoplusia*, or a berry as in *Nicodemia*. Seeds compressed, fusiform or discoid, often winged, endosperm fleshy, embryo straight.

Anatomy. Vessels very small to medium-sized, perforations simple; intraxylary phloem absent. Leaves dorsiventral, stomata confined to the lower surface, usually surrounded by many subsidiary cells.

Embryology. Pollen grains 3- or 4-colporate and shed at 2-celled stage. Ovules anatropous, unitegmic, tenuinucellate. Polygonum type of embryo sac, 8-nucleate at maturity. Endosperm formation of the Cellular type.

Chromosome Number. Basic chromosome number is x =7.

Chemical Features. *Buddleja* contains iridoids typical of Scrophulariales (Jensen et al. 1975).

Important Genera and Economic Importance. *Buddleja* is the type genus of this family, with about 100 species of trees, shrubs, or climbing shrubs distributed in the tropics and subtropics of America, Asia and South Africa. The plants are usually covered with stellate, glandular, or scaly pubescence. *B. globosa*, from Chile and Peru, bears long-peduncled globular, head-like inflorescences. Various species are cultivated for their ornamental value.

Taxonomic Considerations. Buddlejaceae owes its origin to Hutchinson (1959). Earlier, the genus *Buddleja* was included in the family Loganiaceae. The separation was mainly based on anatomical characters—the absence of intraxylary phloem and the presence of glandular, stellate or lepidote indumentum. In addition, the corolla lobes are imbricate. Hutchinson (1959, 1969, 1973) treats it as a member of the Loganiales. Cronquist (1968, 1981) removes it from Loganiales because of the absence of intraxylary phloem and presence of integumentary tapetum and Cellular type of endosperm. He, however, places this family in a much higher taxon—Scrophulariales. The relationship is suggested by the two genera of Buddlejaceae—*Sanango* and *Peltanthera*—both with 5 corolla lobes but only 4 functional stamens. In *Sanango* the 5th stamen is represented by a staminode, as in many Scrophulariaceae. According to Cronquist, the tetramerous flowers of other Buddlejaceae may have a pentamerous ancestry. Benson (1970) and Metcalfe and Chalk (1972) included it in the family Loganiaceae. Dahlgren (1975a, b, 1980a, 1983a) recognises it as a distinct family and, although he places it in Gentianales, he also suggests its alternate position in Scrophulariales. Takhtajan (1980) includes Buddlejaceae in the suborder Scrophulariineae of order Scrophulariales. Although it shows many similarities with Loganiaceae in external morphology and embryology, it differs in: (i) absence of true stipules, (ii) absence of intraxylary phloem, (iii) having glandular, stellate or lepidote indumentum, and (iv) having different chemical consitituents. Palynology and chemical features reveal a close relationship with the members of the Scrophulariales (Punt and Leenhouts 1967, Jensen et al. 1975). Takhtajan (1987) treats it as a distinct family, Buddlejaceae, in the order Scrophulariales.

Although Buddlejaceae was earlier included in the Loganiaceae, it should now be treated as a distinct family and should be placed near Scrophulariaceae (because of their similarities).

Scrophulariaceae

A large family of ca. 210 genera and nearly 3000 species, of cosmopolitan distribution and occurs in all the continents including Antarctica.

The members are mostly herbs or undershrubs (*Scoparia dulcis*), rarely trees (*Paulownia*), sometimes climbing shrubs (*Maurandya, Rhodochiton*), parasites (*Hyobanche, Harveya*), semi-parasites on roots (*Striga, Pedicularis*), or saprophytes (*Melampyrum, Castilleja*); *Hebe* spp. of New Zealand are xerophytes (resembling certain Coniferae).

Vegetative Features. Stem herbaceous or woody. Leaves simple, estipulate, alternate, opposite (Fig. 50.14A) or whorled, often lower opposite and upper alternate; entire or pinnately lobed or incised. In parasitic species the leaves are scale-like and devoid of chlorophyll.

Floral Features. Inflorescence variable—simple raceme or spike (indeterminate) or dichasial cyme (determinate). Solitary axillary flowers in *Linaria*. Flowers hermaphrodite (Fig. 50.14B, C), zygomorphic, hypogynous, usually bracteate and bracteolate. Bracts and bracteoles brightly coloured in *Castileja*. Calyx of usually 5 sepals (or 4), gamosepalous but often deeply cleft, persistent; posterior sepal is suppressed in *Veronica* and the two anterior sepals are fused in *Calceolaria*; aestivation imbricate or valvate. Corolla of 4 or 5 (rarely 6–8) petals, gamopetalous, aestivation imbricate or valvate. Corolla tube inconspicuous, as in *Veronica*, or prominent, as in *Digitalis*, usually bilabiate (Fig. 50.14B) and personate (e.g. *Antirrhinum*, *Linaria*), campanulate in *Digitalis*. Corolla spurred in *Linaria*, saccate in *Antirrhinum* or the limb develops into two unequal, inflated lips (*Calceolaria*); absent from 1 species of *Synthyris*. Stamens usually 4, sometimes

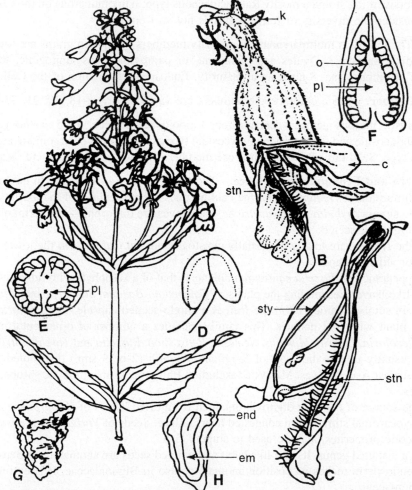

Fig. 50.14 Scrophulariaceae: **A-H** *Penstemon canescens*. **A** Twig with inflorescence. **B** Flower. **C** Longisection. **D** Anther. **E.** Ovary, transection. **F** Longisection. **G, H** Seed (**G**) and longisection (**H**). *c* corolla, *em* embryo, *end* endosperm, *k* calyx, *o* ovule, *pl* placenta, *stn* staminode, *sty* style. (Adapted from Radford 1987)

5 (*Verbascun*) or 2 (*Veronica, Linaria, Gratiola*), epipetalous and fused at the base of the corolla tube (Fig. 50.14C), alternating with the petals, didynamous, anthers bicelled (Fig. 50.14D), sometimes one cell is larger than the other, dehiscence longitudinal, rarely poricidal (e.g. *Seymeria*), introrse; nectariferous disc usually present at the base. Gynoecium bicarpellary, syncarpous; superior, bilocular ovary; placentation axile, ovules numerous on enlarged placentae (Fig. 50.14E, F); style 1, terminal, stigma bilobed. Fruit is a dry dehiscent capsule, normally septicidal but loculicidal in *Buchnera* and poricidal in *Antirrhinum*; rarely a berry, e.g. *Halleria, Teedia, Leucocarpus* and *Dermatocalyx*, or dry indehiscent capsule as in *Hebenstretia* or 1-seeded tardily dehiscent capsule as in *Tozzia*. Seeds smooth or with rugose surface (Fig. 50.14G), sometimes winged, as in *Mimulus*; endosperm fleshy (absent in *Wightia* and *Monttea*), embryo straight or slightly curved (Fig. 50.14H).

Anatomy.   Vessels very small and numerous, occasionally ring-porous, parenchyma usually sparse or absent, rarely abundant. Rays either absent or when present, 1 to 9 cells wide, homo- or heterogeneous. Leaves usually dorsiventral, stomata mostly Ranunculaceous type; hairs numerous on the vegetative parts and exhibit a considerable diversity of forms. Crystals not so frequent.

Embryology.   The family is multipalynous. In majority members the pollen grains are tricolporate and 2-celled at the dispersal stage. Ovules ana-, hemiana- or campylotropous, unitegmic, tenuinucellate. Polygonum type of embryo sac, 8-nucleate at maturity. Endosperm formation of the Cellular type.

Chromosome Number.   Basic chromosome numbers are variable: $x = 6$-18, 20, 21, 23-26 and 30.

Chemical Features.   Various members contain Group I iridoids (Jensen et al. 1975), anthocyanin pigments like rutinoside, diglucoside and glucosylside of glycosidic type and cyanidin, pelargonidin, and delphinidin of the aglycones type. Seed fats of Scrophulariaceae members are rich in linolenic acid (Shorland 1963).

Important Genera and Economic Importance.   The genus *Ixianthus* is unique in having verticillate leaves. Leaves dimorphic in *Hemiphragma*, the cauline ones are orbicular and the axillary ones linear. In the two aquatic genera *Ambulia* and *Hydrotriche*, too, leaves are dimorphic—the aerial ones entire and the submerged, highly dissected.

In *Vandellia* the flowers are totally or partially cleistogamous. In *Calceolaria* the lower corolla lip is entire, concave or slipper-shaped.

The general appearance of *Russelia equisetifolia* plants is that of a xerophyte with ridged chlorophyllous stems and scale-like leaves, resembling the plants of *Equisetum*. *Bacopa*, *Scoparia* and *Verbascum* have 5 stamens each. In semi-parasitic *Tozzia* the fruit is a single-seeded capsule. *Scrophularia nodosa* is a non-leguminous plant with root nodules. This family includes a number of ornamental plants such as *Antirrhinum, Calceolaria, Linaria, Mimulus, Nemesia, Penstemon, Russelia* and *Torenia*. *Striga densiflora*, a parasitic herb usually infects the roots of *Sorghum*. *Wightia* (2 or 3 spp.) distributed from Eastern Himalayas to southeast Asia and west Malaysia (excluding Philippines) are epiphytic shrubs; later become independent trees.

Digitalis is the source of the drug digitoxin obtained from dried leaves of *D. purpurea* and *D. lanata* and is used as myocardinal stimulant in congested heart failure. Seeds of *Verbascum thapsus* are to some extent, with narcotic properties, and are used to stupefy fish.

Paulwonia is a disputed genus. Its tree-like habit and winged seeds are anomalous in Scrophulariaceae, but its copious endosperm makes its position anomalous also in Bignoniaceae. It is extremely close to *Catalpa* of Bignoniaceae.

Taxonomic Considerations.   The family Scrophulariaceae belongs to the order Personales according to Bentham and Hooker (1965b). Engler and Diels (1936) and Hallier (1912) retained it in Tubiflorae.

Bessey (1915) and some other taxonomists placed this family in Scrophulariales, while Hutchinson (1948, 1969, 1973) included it in Personales and pointed out its derivation from Solanales through Salpiglossidaceae. According to Cronquist (1968, 1981), it belongs to Scrophulariales and is derived from Polemoniales. Benson (1970), Stebbins (1974), and Dahlgren (1975a, b, 1980a, 1983a) also placed it in Scrophulariales. Takhtajan (1980, 1987) also includes this family in Scrophulariales. According to him, it resembles the members of Solanaceae, especially the tribes Cestreae and Salpiglossideae, and also has close affinity with the family Buddlejaceae, particularly in embryological and chemical features. Thus, there is no controversy over the placement of Scrophulariaceae in Scrophulariales.

Globulariaceae

A small family of 3 genera and 23 species distributed mainly in the Mediterranean region, *Globularia* in southern Europe and Mediterranean to the Baltic, *Poskea* in Socotra and Somaliland and *Lytanthus* in Atlantic Islands.

Vegetative Features. Perennial herbs or shrubs. Leaves simple, alternate, estipulate.

Floral Features. Inflorescence a capitulum as in *Globularia* (Fig. 50.15A), in *Lytanthus* subtended by an involucre of bracts, or a spike, as in *Poskea*. Flowers hermaphrodite, zygomorphic (Fig. 50.15B), minute, on a scaly receptacle. Calyx tubular, 5-lobed, actinomorphic or bilabiate, persistent. Corolla 4- or 5-lobed, the upper lip of two petals is shorter than the 3-petalled lower lip (Fig. 50.15C). Stamens 4 (Fig. 50.15B, C), didynamous or only 2, epipetalous, arising from the upper portion of the corolla tube and alternate with the petals; anthers bicelled but fuse to form a single locule at the time of anthesis and dehisce by a single longitudinal slit (Fig. 50.15D), versatile. Gynoecium syncarpous, bicarpellary, ovary superior, unilocular with a single ovule, pendulous from locule apex (Fig. 50.15E); style 1, filiform (Fig. 50.15F), stigma capitate or bilobed. Fruit, a single-seeded nutlet enclosed within the persistent calyx. Seed with a straight embryo surrounded by a fleshy endosperm.

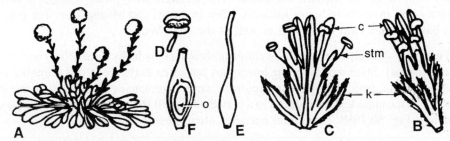

Fig. 50.15 Globulariaceae: **A-F** *Globularia aphyllanthus*. **A** Plant with flowering branches. **B, C** Flower, and longisection. **D** Monothecous anther. **E** Pistil. **F** Ovary, vertical section. *c* corolla, *k* calyx, *stm* stamen, *o* ovule. (After Lawrence 1951)

Anatomy. Vessels small, solitary or in a few short radial multiples, perforations simple, parenchyma scanty, rays only 1 or 2 cells wide. Leaves dorsiventral to centric, sometimes within the same genus; stomata usually Ranunculaceous type, sometimes intermixed with Caryophyllaceous or Rubiaceous type. The presence of calcareous glands is one of the important anatomical features of this family. These are small glandular hairs each with a short stalk cell and a head of 2 or more (rarely 4) convexly-arched cells separated from one another by vertical walls. These hairs secrete calcareous material in the form of scales.

Embryology. The embryology has not been studied in sufficient details. Ovules anatropous, unitegmic, tenuinucellate. Polygonum type of embryo sac, 8-nucleate at maturity. Endosperm formation of the Cellular type.

Chromosome Number. The haploid chromosome numbers are n = 8, 10.

Important Genera and Economic Importance. The genus *Globularia*, commonly known as globe daisy, probably because of its inflorescence bearing a resemblance to that of a Compositae, is cultivated for its ornamental value. About 22 herbs or small subshrubs belonging to this genus are cultivated. These are adapted to rock gardens. *G. aphyllanthes* and *G. nudicaulis* are small herbs with basal rosettes of leaves. *G. trichosantha, G. repens* and *G. meridionalis* are small subshrubs with stoloniferous stems.

Taxonomic Considerations. Bentham and Hooker (1965b) treated Globulariaceae as a member of the family Selaginaceae [not included in Melchior's (1964) work] of the Lamiales. Hallier (1912) merged it with Scrophulariaceae. However many taxonomists—Bessey (1915), Rendle (1925) and Wettstein (1935)—treated it as a distinct family and included it in Lamiales. Cronquist (1968, 1981) places Globulariaceae (including Selaginaceae) in Scrophulariales deriving it from the family Scrophulariaceae. Benson (1970), Stebbins (1974), and Dahlgren (1975a, b, 1980a, 1983a) accept this assignment. According to Metcalfe and Chalk (1972), Globulariaceae cannot have a family rank on the basis of anatomical features. Takhtajan (1980) includes it in his Scrophulariaceae as a subfamily Globularioideae but, later, in 1987, raises it to family rank, retaining it in the same order.

The family Globulariaceae shows the following resemblances to Scrophulariaceae: (i) predominantly herbaceous, (ii) bilabiate corolla. The differences between the two are: (i) typical calcareous glands, and (ii) unilocular ovary with a single pendulous ovule in Globulariaceae. Therefore a family rank is fully justified and it should be included in the order Scrophulariales.

Bignoniaceae

A large family with ca. 120 genera and 650 species. The members are typically tropical in nature. A large number of taxa occur in northern South America, a few in Africa and Madagascar. The genera *Campsis* and *Catalpa* have species both in the Old as well as the New World.

Vegetative Features. Mostly trees, shrubs or climbing shrubs, rarely herbs (*Incarvillea*) or with suffruticose habit (*Eccremocarpus*). Stem woody in the lower part but often herbaceous and climbing in the upper parts. Leaves estipulate, petiolate, opposite-decussate, rarely alternate as in *Kigelia*, simple or pinnately compound with the terminal leaflets reduced to a branched tendril (Fig. 50.16A), or disc-like suckers that help in climbing (Fig. 50.16H). Extrafloral nectaries often present on leaves.

Floral Features. Inflorescence usually a dichasial cyme, rarely monochasial as in *Eccremocarpus* (Fig. 50.16A). Flowers solitary in *Incarvillea compacta* var. *grandiflora*. Flowers bracteate, bracteolate, hermaphrodite, zygomorphic, hypogynous and showy. Calyx of 5 teeth or lobes, sometimes bilabiate or spathe-like, as in *Spathodea*. In *Tourrettia* the upper sterile flowers have coloured calyx and the lower fertile flowers have green calyx. Calyx double in *Amphilophium*. Corolla 5-lobed, campanulate or infundibuliform (Fig. 50.16H), sometimes bilabiate, imbricate, induplicate-valvate or sub valvate as in *Millingtonia*. Stamens usually 4 with 1 staminode, didynamous, epipetalous, rarely only 2, as in *Catalpa*; anthers bicelled, cells widely divergent (Fig. 50.16B, I) and sometimes appear to be one above the other; only one functional cell or locule in *Jacaranda*, the other is smaller and empty. Gynoecium syncarpous, bicarpellary, ovary superior, bilocular with axile placentation, unilocular with parietal placentation in *Eccremocarpus* (Fig. 50.16C, F) and *Kigelia*; ovules numerous; style 1, terminal (Fig. 50.16D, J), simple, stigma bilobed (Fig. 50.16D, J), often flat and broad. Fruit a loculicidal or septicidal capsule

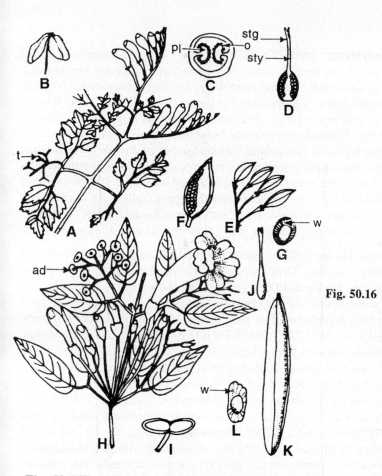

Fig. 50.16 Bignoniaceae: **A-G** *Eccremocarpus scaber*, **H-L** *Bignonia capreolata*. **A** Twig with inflorescence, leaflets modified into tendrils. **B** Stamen. **C, D** Ovary. **C** Transection shows parietal placentation. **D** Longisection. **E** Fruits. **F** Fruit, longisection. **G** Winged seed. **H** Twig of *B. capreolata* with flowers, leaflets modified into adhesive discs. **I** Stamen. **J** Pistil. **K** Fruit. **L** Winged seed. *ad* adhesive disc, *o* ovule, *pl* placenta, *stg* stigma, *sty* style, *t* tendril, *w* wing. (Adapted from Hutchinson 1969)

(Fig. 50.16E, K); seeds winged (Fig. 50.16G, L) or without wing as in *Kigelia*, *Parmentiera*, *Crescentia* and *Phyllarthron*.

Anatomy. Vessels small to medium-sized, sometimes ring-porous, perforations simple. Parenchyma paratracheal, often storied; rays 1–4 cells wide in arboreals and 5–13 cells wide in climbers, homogeneous. Several types of anomalous secondary growth is reported, particularly in climbing species. Hairs uni- and multiseriate, glandular and nonglandular reported. Leaves usually dorsiventral; stomata confined to the lower surface, surrounded by a large number of ordinary epidermal cells. A few genera have Rubiaceous and Caryophyllaceous type of stomata.

Embryology. Pollen grains tricolporate, and 2- or 3-celled at shedding stage. Ovules anatropous, unitegmic, tenuinucellate. Polygonum type of embryo sac, 8-nucleate at maturity. Endosperm formation of the Cellular type.

Chromosome Number. Basic chromosome number is x = 7 (Goldblatt and Gentry 1979). Haploid number is variable: n = 11, 13–15, 18–20 and 21.

Chemical Features. Anthocyanin pigments without a 3-hydroxyl group are rare. One such pigment, 'carajurin', is present in *Arrabidaea chica*. Seed oil of *Doxantha unguis-cati* is rich in palmitoleic acid, an unsaturated acid.

Important Genera and Economic Importance. Many of the members are ornamental trees and others yield valuable timber. *Kigelia pinnata*, an African species, commonly known as the sausage tree because of its large cylindrical fruits, is grown as an avenue tree and bears long, hanging inflorescence of dark-red flowers. *Jacaranda acutifolia* (= *J. mimosifolia*) is a Brazilian tree grown for its beautiful lilac flowers in loose pyramidal panicles. *Tabebuia* is also grown as an ornamental tree for its large, showy flowers of various colours. It is prized for its timber. *Spathodea campanlata* is the African tulip, with large scarlet flowers. Among the ornamental vines are *Bignonia capreolata*, various species of *Pyrostegia*, *Campsis*, *Doxantha* and *Tecomaria*. Some of these plants have remarkable structures to help them in climbing. In *B. capreolata* the upper leaflets of a pinnately compound leaf are transformed into small disc-like suckers which adhere to the support (Fig. 50.16H). In *Eccremocarpus scaber* the upper leaflets are transformed into tendrils (Fig. 50.16A). *Parmentiera cereifera* is the well-known candle tree of Panama. Large, elongated, cylindrical fruits of a yellow wax colour, resembling candles, hang from the stem and branches.

Taxonomic Considerations. Bentham and Hooker (1965b) placed the family Bignoniaceae in Personales along with Scrophulariaceae and other families: Orobanchaceae, Lentibulariaceae, Pedaliaceae, Columelliaceae, Gesneriaceae, Acanthaceae. Engler and Diels (1936), and Hallier (1912) included it in Tubiflorae along with other families with tubular corolla. Bessey (1915) treated Bignoniaceae as a member of Scrophulariales and derived it from Scrophulariaceae. Hutchinson (1948) retained it in Personales, between Gesneriaceae and Pedaliaceae. Later (1969, 1973), he raises it to a separate order, Bignoniales, along with the families Cobaeaceae, Pedaliaceae and Martyniaceae. Cronquist (1968, 1981) places it in Scrophulariales and points out that the two families—Scrophulariaceae and Bignoniaceae—are closely related, although the former is dominated by herbaceous members. Benson (1970), Stebbins (1974) and Dahlgren (1975a, b, 1980a, 1983a) also treat it as a member of Scrophulariales. Takhtajan (1980) reports that Bignoniaceae is very close to Scrophulariaceae, especially to the tribe Scrophularieae, and probably had a common origin with them. The two controversial genera, *Paulownia* and *Wightia*, are included in Bignoniaceae (Takhtajan 1980). According to Armstrong (1985), the relationship with Scrophulariaceae is uncertain but Takhtajan (1987) treats these two members in Scrophulariaceae as a subfamily, Paulownieae.

There is no controversy about the position of Bignoniaceae in the Tubiflorae and near the family Scroplulariaceae.

Henriqueziaceae

This family includes 2 genera, *Henriquezia* and *Platycarpum*, both South American taxa, and 13 species.

Vegetative Features. Small or moderate-sized trees. Leaves simple, entire, opposite or verticillate, stipulate.

Floral Features. Inflorescence terminal thyrses. Flowers large, zygomorphic, epigynous. Calyx of 4 or 5 sepals, polysepalous, deciduous; corolla lobes 5, gamopetalous, imbricate, campanulate. Stamens 5, unequally epipetalous, with curved filament and dorsifixed sagittate-based anthers, annular disc present. Gynoecium syncarpous, bicarpellary, ovary inferior, bilocular, with 2 to 4 ovules per locule, style 1, stigmas 2, apical. Fruit, a semi-superior or almost superior, transversely ovoid, subreniform or discoid, 2- to 4-seeded, loculicidal capsule. Seeds flattened but not winged, without endosperm.

Taxonomic Considerations. Cronquist (1968, 1981) points out that the peculiarities noted in these genera are not adequate to require familial segregation. He is also not convinced that the families of Tubiflorae are its possible allies. He presumes that these genera should tentatively be retained in the family Rubiaceae. The genus *Gleasonia* of Rubiaceae shows common feature of absence of endosperm with *Henriquezia* and *Platycarpum*. Metcalfe and Chalk (1972), on the basis of wood anatomy, include these genera in Rubiaceae. Hutchinson (1969, 1973) confirms this assignment. Dahlgren (1975a, b, 1980a, 1983a), however, places them in Scrophulariales near Bignoniaceae. Takhtajan (1980, 1987) includes them in the Rubiaceae.

The correct assessment of the phylogenetic position of the family Henriqueziaceae must await further investigations on anatomy, embryology, chromosome number and chemical features.

Acanthaceae

A large homogenous family of ca. 240 genera and 2200 species, distributed mostly in the tropical and subtropical areas of the world. The four tropical zones presumed to be the centres of distribution of the members of this family are: Indo-Malaya, Africa, Brazil and Central America.

Vegetative Features. Perennial herbs, shrubs or rarely medium-sized trees, sometimes lianas (*Mendoncia* and *Thunbergia*), some members are xerophytic herbs (*Acanthus* and *Blepharis molluginifolia*), undershrubs, e.g. *Barleria prionitis* and *Peristrophe bicalyculata*, rarely aquatics, e.g. *Acanthus ilicifolius*, a mangrove plant. Stem herbaceous, green, often ridged. Leaves opposite-decussate, simple, estipulate, petiolate, subsessile or sessile.

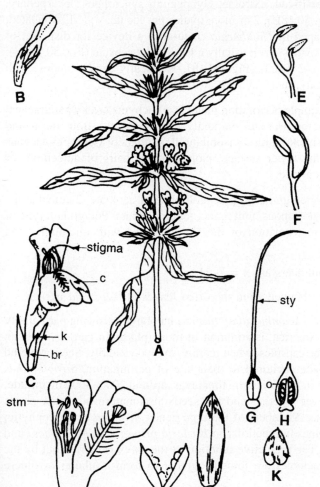

Fig. 50.18 Acanthaceae: **A-K** *Astercantha longifolia*. **A** Portion of plant. **B** Bud. **C** Flower. **D** Corolla (cut open). **E, F** Stamens. **G** Pistil. **H** Ovary, longisection. **I** Capsule. **J** Dehisced capsule. **K** Seed. *br* bract, *c* corolla, *k* calyx, *o* ovule, *stm* stamen, *sty* style. (Adapted from Hutchinson 1969)

Floral Features. Inflorescence terminal (Fig. 50.18A) or axillary dichasial cyme; rarely a raceme, e.g. *Beloperone*, or solitary axillary flowers, as in *Thunbergia*. Flowers bracteate and bracteolate (Fig. 50.18B, C), bracts and bracteoles often conspicuous and involucrate as in *Peristrophe*, sometimes spinescent as in *Acanthus mollis* and *Barleria prionitis*; zygomorphic, bisexual, pedicellate, often large and showy, e.g. *Thunbergia*. Calyx of 4 or 5 sepals, deeply lobed, often with spiny or bristly margin, e.g. *Barleria*, or highly reduced as in *Thunbergia*; imbricate, rarely contorted. Corolla of 4 or 5 petals, gamopetalous, usually bilabiate (Fig. 50.18C, D), exceptions *Thunbergia* and *Ruellia*, upper lip bifid and erect, lower lip almost horizontal and three-lobed (Fig. 50.18D). Inner surface of the corolla lobes hairy and the hairs often extend up to the mouth of the corolla; in *Acanthus* the upper lip is totally missing and the stamens are protected by the calyx; aestivation imbricate or contorted. Stamens 2 to 5, epipetalous, when 4, didynamous with the fifth one suppressed or represented by a staminode; when only 2, they are attached to the anterior petals. Filaments usually long and free above (Fig. 50.18E, F), and therefore the anthers are outside the floral tube (exception *Thunbergia* with short stamens within the floral tube). Form, position and number of anther lobes also vary; lobes 1 or 2, when 1-celled, there may be a rudimentary second lobe; when bicelled, the two lobes may be separated by an elongated connective as in *Peristrophe, Beloperone*; sometimes the 2 lobes are unequal; anthers dorsifixed, extrorse. Gynoecium syncarpous, bicarpellary, ovary superior, bilocular, placentation axile (Fig. 50.18H), 2 or more ovules per locule, style 1, elongated, filiform (Fig. 50.18G), stigma bifid, funnel-shaped in *Thunbergia*. Jaculator, a device for dispersal of seeds, may or may not be present just below the ovules. Fruit usually a loculicidal capsule (Fig. 50.18I, J), often elastically dehiscent; a drupe in *Mendoncia*. Seeds (Fig. 50.18J, K) mostly exalbuminous, with various types of testa such as mucilaginous, scaly, hairy or with an indurated funicle.

Anatomy. Vessels small with a radial pattern, simple perforation plates; parenchyma scanty paratracheal or vasicentric, rays 1 to 6 cells wide, uni- or multiseriate, markedly heterogeneous. Both intra- and interxylary phloem present. Leaves usually dorsiventral, rarely isobilateral, stomata of Caryophyllaceous type, occur on both the surfaces or only on the lower surface. Nonglandular hairs of unicellular or uniseriate type as well as glandular hairs are reported.

Embryology. The family is multipalynous; pollen grains colpate, colporate or acolpate, 2-celled at the shedding stage. Ovules ana-, hemiana- or campylotropous, unitegmic, tenuinucellate. Polygonum type of embryo sac, 8-nucleate at maturity. Endosperm formation of the Cellular type with micropylar and chalazal haustoria.

Chromosome Number. Basic chromosome numbers are x = 7–21.

Chemical Features. The alkaloid vasicine is obtained from the dried leaves of *Adhatoda vasica*.

Important Genera and Economic Importance. *Acanthus ilicifolius* is a halophyte, growing particularly in tidal swamps. *Ruellia tuberosa*, a native of America, is common in moist places in gardens. It is an erect annual herb with bluish-pink flowers and the capsules, when mature, explode audibly. Seeds covered with hygroscopic hairs that help them to become anchored to their site of germination. *Strobilanthes dalhausianus* is a common perennial undershrub in the Western Himalayas around 2000–3000 m altitude. These plants grow in abundance, come to flower together and the seeds also mature in all the plants together. The jungle birds feed upon these seeds. Members of the type genus *Acanthus* are xerophytic; the leaves and bracts are more or less spiny, spines interpetiolar in *Barleria prionitis*. In *Mendoncia* and *Thunbergia* the calyx is reduced to an annulus. The protective function of the calyx is carried out by the large leafy bracteoles. In *Boutonia* the two bracteoles are joined together to form a tubular involucre around each axillary flower.

Barleria, Ruellia, Justicia, Thunbergia and *Strobilanthes* are ornamentals. The spinescent plants are sometimes grown as a hedge. *Adhatoda vasica* is medicinally important. Its active principles vasicine and adhatodic acid, obtained from the dried leaves, are components of the cough mixture, glycodin, much used as an expectorant in India. Leaves of *Andrographis paniculata* are also used medicinally against liver ailments.

Taxonomic Considerations. Acanthaceae is divided into two subfamilies depending upon the presence or absence of jaculators, i.e. the curved retinacula which support the seeds:

Subfamily Thunbergioideae—seeds without jaculators: *Nelsonia, Mendoncia* and *Thunbergia,*
Subfamily Acanthioideae—seeds with jaculators: *Acanthus, Ruellia, Justicia* and *Adhatoda.*

Most taxonomists presume Acanthaceae to have been derived from Scrophulariaceae or stocks ancestral to them (Lawrence 1951). Hutchinson (1969, 1973) considers it to be the most advanced taxon of the Personales. Bessey (1915) agreed with this view and treated it as the most advanced amongst the members of Scrophulariales. Cronquist (1968, 1981) also includes this family in the same order as Bessey and comments on its relationship with the family Scrophulariaceae. There has been some controversy regarding the position of the two genera *Nelsonia* and *Elytraria.* The proposal for the transfer of these genera to tribe Rhinantheae under Scrophulariaceae have been rejected by Johri and Singh (1959), Mohan Ram and Masand (1963) and P. Maheshwari (1964). Presence of jaculator (though nonfunctional) in *Elytraria* and *Nelsonia* and asymmetric development of Cellular endosperm are Acanthaceous features and not reported in Rhinantheae (see Johri et al. 1992). In addition, presence of alternate leaves, parietal placentation, endothelium, funicular obturator and albuminous seed, support their inclusion in Acanthaceae. *Elytraria* forms a link between Acanthaceae and Scrophulariaceae. This genus differs from most Acanthaceae in well-developed endosperm and its funicle, though enlarged, does not develop into a typical jaculator. Benson (1970), Stebbins (1974) and Dahlgren (1975a, b, 1980a, 1983a) also retain it in Scrophulariales. Takhtajan (1980, 1987) reports that the Acanthaceae is closely related to the tribe Scrophularieae of the family Scrophulariaceae.

In our opinion too, the family Acanthaceae is closely related to the Scrophulariaceae and is an advanced taxon.

Pedaliaceae

A small family of 20 genera and ca. 60 species, distributed mainly in the Old World tropics and subtropics— South Africa, Madagascar, Indo-Malaya and Australia.

Vegetative Features. Annual or perennial herbs or rarely shrubs; sometimes aquatic, e.g. *Trapella.* Stem herbaceous, covered with mucilage-containing glandular hairs. Leaves opposite, entire or lobed, simple, estipulate, upper leaves may be alternate.

Floral Features. Flowers usually solitary axillary or in simple axillary dichasia, i.e. 3-flowered cymes with glands at the base of the stalks. Flowers ebracteate, ebracteolate, hermaphrodite, zygomorphic, hypogynous, rarely epigynous, e.g. *Trapella*, sessile or shortly pedicellate. Calyx of 5 or rarely 4 sepals, gamosepalous or basally connate, valvate. Corolla of 5 or 4 petals, tubular, obscurely bilipped (2 + 3), aestivation imbricate or valvate. Stamens 4, epipetalous, didynamous (only 2 in *Trapella*), the 5th represented by a staminode; anthers bicelled, dehisce longitudinally, introrse. Gynoecium syncarpous, bicarpellary, ovary superior or inferior as in *Trapella*, bilocular but tetralocular later due to formation of false septum, placentation axile, ovules 1 to many on each placenta; style 1, filiform, stigma bilobed. Fruit, a loculicidal capsule or nut, often spiny or with wings, hooks or thorns, may or may not be dehiscent.

Anatomy. Vessels irregularly distributed, with a wide range of size, with simple perforations. Leaves usually dorsiventral, stomata Ranunculaceous type on both the surfaces. Characteristic mucilage hairs occur all over the plant body.

Embryology. Pollen grains 6- (*Pedalium murex*) or 10-colpate (*Sesamum indicum*); shed at 3-celled stage. ovules anatropous, unitegmic, tenuinucellate. Polygonum type of embryo sac, 8-nucleate at maturity. Endosperm formation is of the Cellular type.

Chromosome Number. Haploid chromosome numbers are: n = 8, 13, 14.

Chemical Features. Seed fat of *Sesamum indicum* is rich in linolenic acid.

Important Genera and Economic Importance. *Sesamum indicum*, commonly called sesame or gingelly, is a rainy-season crop in India, grown for its oil-yielding seeds. The oil obtained is edible and is an accepted substitute for olive oil in European countries. *Ceratotheca triloba* is grown as an ornamental. Two species of the genus *Trapella* grow in China and Japan. These are aquatic herbs with broad floating leaves and epigynous flowers. Another outstanding genus is *Pterodiscus* with a 4-winged, indehiscent, unarmed fruit. In *Harpagophytum* the fruits are tardily dehiscent and flattened contrary to the septum and armed along the margins with 2 rows of long, horny, recurved spines.

Taxonomic Considerations. Bentham and Hooker (1965b) placed Pedaliaceae in Personales, and Engler and Diels (1936) in Tubiflorae. Hallier (1912) merged Pedaliaceae and Martyniaceae. Bessey (1915) recognised Pedaliaceae as a distinct family and included it in Scrophulariales. Lawrence (1951) treats it as a distinct family in the order Tubiflorae. Pollen morphological, and embryological studies are suggestive of their affinities with Scrophulariaceae as well as Bignoniaceae (Erdtman 1952, Crète 1951). Cronquist (1968, 1981) includes Pedaliaceae in his Scrophulariales. Benson (1970), Stebbins (1974), Dahlgren (1975a, b) and Takhtajan (1980) also treat Pedaliaceae as a member of the Scrophulariales. Hutchinson (1969, 1973), however, places Pedaliaceae in the Bignoniales.

Trapella is a controversial genus in the Pedaliaceae. It is different from other members of this family in morphology, palynology and embryology. G. Erdtman (1952), on the basis of pollen morphology, suggested the erection of a separate family, Trapellaceae, for this genus. This view has been supported by studies on floral morphology, endosperm, fruit and seed structure (Singh 1960) but Hutchinson (1969, 1973) is not in favour of declassifying *Trapella* as a separate family. Takhtajan (1987), however, treats it as a separate family, Trapellaceae. This view is supported by Willis (1973) and Dahlgren (1980a, 1983a).

Pedaliaceae undoubtedly is a distinct family closely related to both Scrophulariaceae and Bignoniaceae. The disputed genus *Trapella* is distinct from other members of the family Pedaliaceae and should be treated as a member of a separate family, Trapellaceae.

Martyniaceae

A small family of only 4 genera and ca. 16 species, mainly distributed in the New World tropics.

Vegetative Features. Mostly stout, annual or perennial herbs. Stem and leaves covered with glandular hairs. Leaves opposite, alternate towards the end of the branches, simple, margin undulate, estipulate.

Floral Features. Inflorescence a terminal raceme. Flowers bracteate, hermaphrodite, zygomorphic, hypogynous, pedicellate. Calyx spathaceous or of 5 distinct sepals subtended by 1 or 2 bracts which often become fleshy at maturity. Corolla always 5-lobed, gamopetalous, tube basally cylindrical, campanulate or infundibuliform above, often oblique. Stamens usually 4, didynamous with a posterior staminode or only 2 fertile stamens and 2 staminodes as in *Martynia*, epipetalous; anthers bicelled, divergent, dehiscence longitudinal, the 2 lobes of each pair of stamens are coherent, i.e. lie close together before anthesis. Gynoecium syncarpous, bicarpellary, ovary superior, unilocular, placentation parietal. Sometimes the 2 placentae are winged and coherent together to form a false septum; style 1, slender, stigmatic lobes 2. Fruit a horned capsule with fleshy exocarp, covered with viscid hairs and is deciduous. The single probosis-like persistent style splits into 2 horn-like processes when the fruit is mature. Seeds sculptured, embryo straight, endosperm present.

Anatomy.   Anatomically it is very similar to the members of Pedaliaceae.

Embryology.   Pollen grains spheroidal, 2-celled at dispersal stage. Ovules anatropous, unitegmic, tenuinucellate. Polygonum type of embryo sac, 8-nucleate at maturity. Endosperm formation of the Cellular type.

Chromosome Number.   Diploid chromosome number of *Martynia annua* is 2n = 36.

Important Genera and Economic Importance.   The family has no economic importance. These are known as unicorn plants. *Martynia louisiana* is sometimes grown for its tiger's claw-like fruits as a curiosity. *Craniolaria* (3 spp.) and *Proboscidea* (9 spp.) are the other two genera.

Taxonomic Considerations.   Bentham and Hooker (1965b), Hallier (1912) and Hutchinson (1948) placed this family in the Pedaliaceae. Martyniaceae was treated as a distinct family for the first time by Stapf (1895). Bessey (1915), Rendle (1925) and Wettstein (1935) treated it likewise. Cronquist (1968, 1981) does not recognise it as a separate family, nor does Stebbins (1974). Hutchinson (1969, 1973) recognizes it as a distinct family. Dahlgren (1975a, b, 1980a, 1983a) places Martyniaceae in the Scrophulariales next to Pedaliaceae. Takhtajan (1980, 1987) also agrees with this assignment and is of the opinion that Martyniaceae is closely related to Bignoniaceae and Pedaliaceae.

   Although Martyniaceae resembles the families Bignoniaceae and Pedaliaceae, it is separated from them by the presence of parietal placentation. Embryological studies support its greater affinities with Bignoniaceae than with Pedaliaceae.

Gesneriaceae

A large family of 85 genera and ca. 1200 species confined to the tropics and subtropics of both the hemispheres.

Vegetative Features.   Herbs, shrubs or rarely trees, and a few lianas and epiphytes. Stem herbaceous and hairy. Leaves opposite or in basal rosettes, when opposite, the two leaves may be equal or one is much reduced and stipule-like; usually decussate, and rarely alternate (due to complete suppression of the opposite leaf); surface hairy, coriaceous in epiphytes.

Floral Features.   Inflorescence cymose or flowers solitary axillary. Flowers ebracteate, ebracteolate, pedicellate, hermaphrodite, zygomorphic (actinomorphic in *Ramonda*), showy. Calyx 5-lobed (Fig. 50.21A), often basally connate, valvate or rarely imbricate. Corolla 5-lobed, gamopetalous (Fig. 50.21A, B), imbricate, rotate, campanulate, infundibuliform (Fig. 50.21E, F) or more or less bilabiate. Stamens 4 as in *Acanthonema* (Fig. 50.21G), didynamous or only 2 (Fig. 50.21A, B), 5 in *Ramonda* and *Sinningia*, one staminode, epipetalous; anthers coherent in pairs or in some genera connate, i.e. syngenesious stamens, bicelled, longitudinally dehiscent; annular nectariferous disc is sometimes 5-lobed and often 5 separate glands (Fig. 50.21C). Gynoecium syncarpous, bicarpellary (Fig. 50.21J), ovary superior, half-inferior or inferior, unilocular, placentation parietal; placentae often intruded and bifid, sometimes meeting in the center and then the ovary is bi- or tetralocular. Ovules numerous, anatropous; style 1, slender, stigma

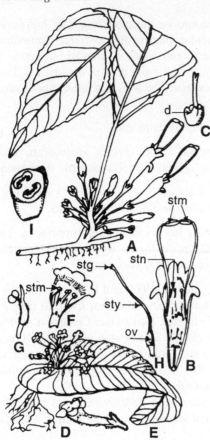

Fig. 50.21 Gesneriaceae: **A-C** *Agalmyla parasitica*, **D-I** *Acanthonema strigosum*. **A** Flowering twig. **B** Corolla (cut open) to show two stamens and three staminodes. **C** Gynoecium with disc (*d*) at the base. **D** Mature plant of *Acanthonema strigosum* consists of a single leaf—the foliaceous cotyledon and a flowering shoot. **E** Flower. **F** Corolla (cut open) to show four stamens. **G** Stamen. **H** Gynoecium. **I** Ovary, transection. *d* disc, *ov* ovary, *stg* stigma, *stm* stamen, *stn* staminode, *sty* style. (Adapted from Hutchinson 1969)

often bilobed (Fig. 50.21I). Fruit a loculicidal capsule (septicidal in *Ramonda*), rarely fleshy and berry-like as in *Cyrtandra*. Seeds numerous, minute, with massive endosperm and a small embryo.

Anatomy. Vessels typically small with a tendency to radial multiples that produce a radial pattern, perforations simple, parenchyma absent or scanty, paratracheal; rays 2–17 cells wide, with very few uniseriates. Leaves mostly dorsiventral, simple uniseriate hairs as well as stalked glandular hairs present; stomata Cruciferous type, distributed on both the surfaces.

Embryology. The triporate pollen grains are shed at 2-celled stage. Ovules anatropous, unitegmic, tenuinucellate. Polygonum type of embryo sac, 8-nucleate at maturity. Embryo sac is extra-micropylar in *Roettlera*[8] (Johri et al. 1992).

Chromosome Number. Haploid chromosome number is n = 4–17 or more.

Chemical Features. Aglycone anthocyanin pigments, apigenidin, cyanidin, delphinidin, luteolinidin, malvidin, pelargonidin, and petunidin occur in various members: *Achimenes*, *Columnea*, *Gesneria*, *Kohleria*, *Streptocarpus* and *Saintpaulia*. Gesnerin, a rare anthocyanin that lacks a 3-hydroxyl group, was first reported in *Gesneria fulgens* and later also in *Kohleria eriantha* (Harborne and Turner 1984).

---

[8]*Roettlera* Vahl. has been changed to *Didymocarpus* Wall.

Important Genera and Economic Importance.   An interesting plant of this family is *Aeschynanthus bracteata*, an epiphyte, with fleshy leaves, that grows at an elevation of 3048 m in the Darjeeling area in India. *Acanthonema strigosum* grows in the Cameroons mountains at 2000 ft in West Africa. The mature plant consists of a single leaf—an enlarged, foliaceous cotyledon and a flowering shoot (Fig. 50.21E). This peculiar feature is also seen in two other genera, *Platystemma violoides* and *Streptocarpus* sp. *Corallodiscus lanuginosus* is a scapigerous herb. A number of Gesneriaceae are cultivated as ornamentals, e.g. *Achimenes*, *Columnea*, *Episcia*, *Haberlea*, *Jancaea*, *Ramonda*, *Saintpaulia* and others.

Taxonomic Considerations.   Bentham and Hooker (1965b) placed the family Gesneriaceae in the order Personales along with other families: Scrophulariaceae, Lentibulariaceae, Orobanchaceae, Bignoniaceae, Acanthaceae and others. Engler and Diels (1936), and Hallier (1912) included it in the order Tubiflorae. Bessey (1915), Benson (1970) and Stebbins (1974) retained it in the order Scrophulariales. Hutchinson (1969, 1973) includes it in the order Personales along with five other families. Dahlgren (1975a, b, 1980a, 1983a) treats Gesneriaceae as a member of Scrophulariales. Takhtajan (1980) includes this family in the suborder Scrophulariineae of the order Scrophulariales but later, in 1987, treats it as a distinct family, Gesneriaceae. It is a highly advanced and presumably recent family and is very closely related to the families Scrophulariaceae, Bignoniaceae and Orobanchaceae (see Table 50.21.1).

Table 50.21.1   Comparative Data for Gesneriaceae, Bignoniaceae, Orobanchaceae and Scrophulariaceae

Gesneriaceae	Bignoniaceae	Orobanchaceae	Scrophulariaceae
Non-parasitic habit	Non-parasitic habit	Parasitic habit	Non-parasitic habit (with a few exceptions)
Parietal placentation	Axile or rarely parietal	Parietal placentation	Axile placentation
Unilocular ovary	Bi-, or rarely unilocular	Unilocular ovary	Bilocular ovary
Fruit a capsule	Fruit a fleshy siliqua	Fruit a capsule	Fruit a capsule
Seeds minute without wings	Seeds winged	Seeds minute without wings	Seeds minute without wings

Columelliaceae

A unigeneric (*Columellia*) family with 4 species distributed in the Andes in South America.

Vegetative Features.   Shrubs with evergreen, opposite, asymmetrical, estipulate, simple or compound leaves.

Floral Features.   Inflorescence cymose. Flowers hermaphrodite, epigynous, slightly zygomorphic. Calyx of 5 sepals, polysepalous, scarcely imbricate, persistent and adnate to the ovary. Corolla of 5 petals, gamopetalous, corolla tube very short, more or less campanulate, imbricate. Stamens 2, short and thick with irregular, broad connective and 1 twisted and plaited pollen chamber, inserted near the base of corolla, disc absent. Gynoecium syncarpous, bicarpellary, ovary inferior, incompletely bilocular, ovules numerous on 2 parietal placentae almost meeting in the middle of the ovary; style short and thick with broad, 2- to 4-lobed stigma. Fruit a capsule enclosed in persistent calyx. Seeds numerous with fleshy endosperm and minute straight embryo.

Anatomy.   This family shows primitive wood structure, because of scalariform perforation plates of vessels, and fibers with bordered pits. Embryology has not been studied.

The only genus included is *Columellia*, which occurs in Columbia and Peru and has no economic importance.

Taxonomic Considerations.   Columelliaceae is a highly controversial family. Hallier (1912) included the genus *Columellia* in tribe Philadelpheae of family Saxifragaceae. Cronquist (1968, 1981) places it in the order Rosales near the family Pittosporaceae. Its primitive wood anatomy makes Columelliaceae out of place in the Scrophulariales but the gamopetalous corolla, and only 2 stamens are unusual in the Rosales (except the Pittosporaceae). Hutchinson (1969, 1973) includes it in the family Euphorbiaceae; and Benson (1970) near Gesneriaceae in the order Scrophulariales, in spite of its primitive wood anatomy. Stebbins (1974), although includes it in the Scrophulariales, is not very certain about its relationship. Stern et al. (1969) conclude that the nearest relative of this family is Escalloniaceae of the order Cunoniales. The differences between the two families are: Columelliaceae has gamopetalous and slightly zygomorphic flowers and 2 stamens with a broad connective and 1 twisted pollen sac, whereas Escalloniaceae has polypetalous, actinomorphic flowers and 5 normal stamens. Willis (1973), too, is of the opinion that, despite sympetaly, slight zygomorphy and curious anthers, it is probably related to Escalloniaceae and Hydrangeaceae, and perhaps also to Loganiaceae.

Hutchinson (1973) considers the anatomical resemblance with Escalloniaceae is due to parallel evolution. Takhtajan (1980) includes Columelliaceae in the suborder Pittosporineae of the order Saxifragales but, later, in 1987, he placed it in the suborder Escalloniineae of the order Hydrangeales, as suggested by Willis (1973). Columelliaceae is included in the order Gentianales by Golbderg (1986). According to Nicholson and Baijnath (1994), Columelliaceae along with Desfontainiaceae is a bridging taxon between Cornales/Hydrangeales and Gentianales. The position of Columelliaceae can be ascertained only after further investigations.

Orobanchaceae

A family comprising 13 genera and 140 species. The members grow in north temperate regions as well as warm temperate parts of the Old World.

Vegetative Features.   Annual or perennial, fleshy herbs and total root parasites (lack chlorophyll). Leaves alternate, scale-like; stem fleshy.

Floral Features.   Flowers solitary in the axil of leaf or bract; hermaphrodite, zygomorphic, pedicellate, bracteolate. In *Epifagus* the upper flowers of a branch are sterile and the lower perfect and cleistogamous. Calyx of 2 to 5 sepals, lobed, divided or spathaceous, accordingly, the sepals are open or valvate. Corolla 5-lobed, straight or arcuate (= curved), usually bilabiate or obliquely so, imbricate with 2 adaxial lobes included. Stamens 4, didynamous, epipetalous, distinct, alternate with the lobes, the posterior 5th one is reduced to a staminode; anthers bicelled, dehiscence longitudinal, often coherent in pairs, sometimes one half-anther sterile. Gynoecium syncarpous, bicarpellary or rarely tricarpellary, ovary superior, unilocular or rarely bilocular as in *Christisonia*; placentation parietal, number of placental groups as many as the number of carpels or twice as many by the branches of each intruded placenta with reflexed and folded back against ovary wall; ovules numerous. Style 1, slender, stigma terminal, 2- to 4-lobed. Fruit a loculicidal capsule enclosed within the persistent calyx, bivalved and leathery. Seeds minute with a pitted or rough surface, albuminous, embryo undifferentiated.

Anatomy.   Leaves reduced to scales; stomata occur on leaf and stem but sometimes become disorganised. Members of this family show various degrees of reduction in relation to their parasitic mode of life.

Embryology.   Pollen grains tricolpate and shed at 2-celled stage. Ovules anatropous, unitegmic, tenuinucellate. Polygonum type of embryo sac, 8-nucleate at maturity. Endosperm formation of the Cellular type; endosperm soft, fleshy and oily, embryo undifferentiated, globular or ovoid at shedding stage of seed.

Chromosome Number. Haploid chromosome number of *Orobanche parishii* is n = 24.

Important Genera and Economic Importance. The members are total root parasites and have no significant economic importance, except that they infest many agricultural products.

Taxonomic Considerations. Bentham and Hooker (1965b) and Hutchinson (1969, 1973) include the family Orobanchaceae in the order Personales. Engler and Diels (1936) placed it in the order Tubiflorae. Bessey (1915) treated it as a member of Scrophulariales along with 9 other families. Lawrence (1951) stated that some phylogenists derive this family from Gesneriaceae and others from Scrophulariaceae. Hallier (1912), Benson (1970), Stebbins (1974), Thorne (1976), Dahlgren (1975a, b, 1980b, 1983a) as well as Takhtajan (1980) place the family Orobanchaceae in the order Scrophulariales. According to Takhatjan (1980), it is an advanced group that represents the final stage of the parasitic tendency exhibited in the subfamily Rhinanthoideae of Scrophulariaceae. This family is probably derived from Scrophulariaceae-Rhinanthoideae through forms like *Striga orobanchoides*. Embryological studies on this plant also confirm this view (Tiagi 1970). Takhtajan (1987) treats it as a subfamily Orobanchoideae in family Scrophulariaceae.

In our opinion, Orobanchaceae should be treated as a distinct family near Scrophulariaceae in Tubiflorae.

Lentibulariaceae

A small family with 5 genera and 260 species, widely distributed almost throughout the world.

Vegetative Features. Predominantly annual or perennial herbs, aquatic or terrestrial on moist land, insectivorous. When aquatic, the plants lack root system. Stem thin, delicate, herbaceous. Leaves variable, alternate or in basal rosettes, often dimorphic in aquatic plants, with submerged leaves finely dissected bearing insectivorous bladders of complex structure, and aerial leaves a floating rosette, or reduced to scales, or absent. In *Genlisea*, a terrestrial genus, rosettes of foliage leaves and tubular or pitcher-like insectivorous leaves are closely adpressed to the ground.

Floral Features. Inflorescence a scapose raceme or flowers are borne solitary on a scape (*Pinguicula*). Flowers bracteate, bracteolate, hermaphrodite, zygomorphic, hypogynous. Calyx of 2 to 5 sepals, gamosepalous, odd sepal posterior, lobes often distinct, imbricate. Corolla 5-lobed, gamopetalous, bilabiate, the lower lip saccate or spurred, aestivation imbricate. Stamens 2, inserted at the base of the corolla tube, staminodes 2; epipetalous, anthers 1-celled, the theca sometimes partially constricted in the middle, dehiscence longitudinal, disc absent. Gynoecium syncarpous, bicarpellary, ovary superior, unilocular with free-central placentation; ovules numerous (only 2 in *Biovularia*); style 1 or commonly obsolete with a bilobed sessile stigma. Fruit a capsule dehiscing by 2 to 4 valves or by an irregular splitting. Seeds minute with rough or bristly surface, non-endospermic, embryo poorly developed.

Anatomy. Stem structure is highly reduced with scanty mechanical tissue. Leaf structure different in various genera.

Embryology. Pollen grains 3- to polyaperturate, and of various shapes; exine reticulate or smooth, 3-celled at dispersal stage. In some species, they are shed in tetrads, e.g. *Utricularia arcuata* and *U. punctata*. Ovules many, often sunken into a globose placental mass; unitegmic, anatropous, occasionally ortho- or anacampylotropous. Polygonum type of embryo sac, 8-nucleate at maturity. Abnormal embryo sacs occur in different species of *Utricularia*. Endosperm formation of the Cellular type (Khan 1970).

Chromosome Number. Haploid chromosome numbers, as reported by Takhtajan (1987), are n = 6, 8, 9, 11, 16, 22, 32, 42 and the basic chromosome numbers are x = 6, 8, 9.

Important Genera and Economic Importance

1. *Utricularia.* Largest genus with ca. 200 species. Many species are aquatic, submerged, rootless plants with finely dissected leaves bearing insectivorous bladders. Opening of the bladder is very interesting. The bladder is provided with a valve, which opens only towards its inside in response to (some) external stimuli from any animalcule and closes immediately after the organism has been trapped. It is a sort of trapdoor for these minute aquatic insects. After capture, the insect is digested by digestive juices secreted from the glandular hairs lining the inner wall of the bladder.

2. *Pinguicula.* Commonly called butterwort, it has about 32 species distributed in the north temperate countries. Perennial herbs, grow in damp areas, leaves simple, entire, in rosettes, bear large number of glandular hairs which secrete a sticky substance. If any insect ventures to sit on this leaf, it inrolls and the victim is captured and digested.

3. *Genlisea.* Twelve species of this genus are land plants distributed in tropical America and Africa. It has two types of leaves—densely crowded simple foliage leaves in rosettes, and pitcher-like leaves adpressed to the ground. The insects can enter the pitcher but cannot come out.

4. *Polypompholyx.* This genus belongs to tropical Australia and South America, and resembles the land forms of *Utricularia*.

5. *Biovularia.* This genus, with 2 species, is native to western India and is a small floating aquatic plant. Economically, none of the members of this family is important.

Taxonomic Considerations. The family Lentibulariaceae resembles superficially some members of Scrophulariaceae, but it is distinguishable from them by the presence of only 2 stamens, 1-celled anther lobe, free-central placentation and insectivorous nature. Most taxonomists consider it to be closely related to the family Scrophulariaceae (Hutchinson 1973, Benson 1970, and Stebbins 1974). Takhtajan (1980, 1987) is of the opinion that it is near to and derived from Scrophulariaceae. Dahlgren (1980a, 1983a) also includes this family in Scrophulariales. Khan (1970) supported this view on the basis of morphological and embryological features. The position of Lentibulariaceae in the Tubiflorae near the family Scrophulariaceae is justified.

Myoporaceae

A family of 5 genera and ca. 110 species (3 genera and 150 species, according to Takhtajan 1987), mostly confined to the Old World. It is predominantly distributed in Australia and also shows disjunct distribution. *Myoporum* occurs in Australia, New Guinea and New Zealand and in western Asia with the Pacific Ocean inbetween. *Bontia* is spreading from the West Indies to parts of South America. Some species occur in the Pacific Islands and the islands of Mauritius and Rodriguez in the Indian Ocean. In the northern hemisphere, a few species occur in China, Japan and the Hawaiian Islands (Richmond 1994).

Vegetative Features. Mostly shrubs, rarely small trees; stem woody, herbaceous above, surface covered with stellate, glandular or plumose hairs. Leaves alternate, rarely opposite, simple, entire, estipulate, shortly petiolate, and surface covered with glandular hairs.

Floral Features. Inflorescence, a fascicular cyme or the flowers are solitary axillary (Fig. 50.25A, C). Flowers hermaphrodite, zygomorphic, rarely actinomorphic, hypogynous. Calyx 5-lobed, persistent, gamosepalous, valvate or imbricate. Corolla of 5 petals, gamopetalous (Fig. 50.25D), imbricate, sometimes bilabiate, e.g. *Bontia* sp. (Fig. 50.25B). Stamens 4, didynamous (Fig. 50.25B, E), the 5th sometimes

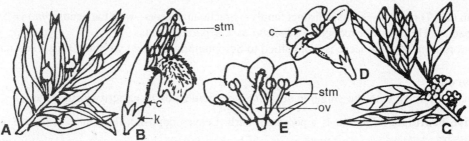

Fig. 50.25 Myoporaceae: **A, B** *Bontia daphnoides*, **C-E** *Myoporum lactum*. **A** Twig bearing fruits, **B** Flower, note hairs inside corolla, and didynamous stamens. **C** Twig of *M. lactum* bearing flowers. **D** Flower. **E** Corolla expanded to show four stamens and pistil. *c* corolla, *k* calyx, *ov* ovary, *stm* stamen. (After Lawrence 1951)

represented by a staminode, epipetalous; anthers bicelled, the 2 cells are often divergent and apically merge together; dehiscence longitudinal. Gynoecium syncarpous, bicarpellary, ovary superior, bilocular or more due to false septation, ovules superposed in pairs and 2 to 8 in each locule, pendulous; style 1, simple, stigma 1. Fruit a berry or drupe. Seeds with scanty or no endosperm.

Anatomy. Vessels small, mostly in radial multiples and clusters, perforations simple; parenchyma paratracheal, scanty, vasicentric, often storied. Rays 2–3 cells wide, heterogeneous. Leaves isobilateral, dorsiventral in *Oftia* and highly reduced in *Pholidia scoparia*. Stomata mostly Cruciferous type but Ranunculaceous type in *Oftia*, usually present on both the surfaces. Characteristic secretory cavities lined with epithelium occur in leaf and stem; hairs both glandular and nonglandular.

Embryology. Pollen grains 3-colporate (P.K.K. Nair 1970). Ovules unitegmic, tenuinucellate. Embryo sac of Polygonum type, 8-nucleate at maturity. Endosperm formation of the Cellular type (Johri et al. 1992). Embryology has not been studied in sufficient details.

Chromosome Number. The haploid chromosome number of *Myoporum* is n = 27.

Chemical Features. Inulin occurs in large quantities in the phloem parenchyma of stem and cortex of root of *Myoporum*.

Important Genera and Economic Importance. Two large genera are *Myoporum* with ca. 30 species and *Pholidia* with 60 species. *Myoporum* and *Bontia* are sometimes grown as ornamental shrubs. Genus *Eremophila* derives its name from 'eremophilous' meaning desert-loving, and occur in the semi-arid and arid regions in Western Australia. It is also known as emu-bush, as the fleshy fruits of *E. maculata* and *E. longifolia* are eaten by emu. The seeds are albuminous and contain stored lipids (Richmond 1994).

Taxonomic Considerations. According to Bentham and Hooker (1965b), Myoporaceae (= Myoporineae) is the first family, i.e. the most primitive amongst the members of Lamiales. Engler and Diels (1936) included it in their Tubiflorae. Hallier (1912) pointed out its derivation from the Solanaceae. Bessey (1915) treated this family as one of the members of the Lamiales. Lawrence (1951) placed it in the suborder Myoporineae of the order Tubiflorae. Hutchinson (1969, 1973) agrees with Bessey and includes it in Lamiales. However, Stebbins (1974), Cronquist (1981), Dahlgren (1983a) as well as Takhtajan (1987) include this family in the Scrophulariales. Takhtajan comments that probably Myoporaceae had a common origin with the members of Scrophulariaceae. The genus *Oftia* is different from the rest of the family in having intraxylary phloem in the sufficiently mature axis, and in the absence of secretory cavities and also in having a haploid chromosome number of n = 19. On the basis of these features,

Takhtajan (1987) recognises it as a distinct family—Spielmanniaceae—with the genus *Oftia* (= *Spielmannia*) and 2 species. He does not recognise *Pholidia* as a separate genus.

In our opinion, Myoporaceae is more allied to Scrophulariaceae and without *Oftia* it is a natural taxon.

Phrymaceae

A monotypic family with the single genus *Phryma*, indigenous to eastern North America.

Vegetative Features.   *Phryma* is a perennial herb. Leaves opposite, decussate, petiolate, estipulate.

Floral Features.   Inflorescence a slender raceme of a pair of flowers at each node on the spike. Flowers hermaphrodite, zygomorphic, hypogynous. Calyx of 5 sepals, gamosepalous, bilabiate, unequal lobes forming the two lips; strongly reflexed in fruit and appressed to the inflorescence. Corolla of 5 lobes, bilabiate with the lower lip 3-lobed and larger than the upper lip. Stamens 4, didynamous, epipetalous, fifth one absent. Gynoecium syncarpous, bicarpellary, ovary superior, unilocular, uniovulate, ovule erect. Fruit a 1-seeded capsule enclosed in persistent calyx. Seeds without endosperm, cotyledons convolute.

Embryology.   Embryology has not been investigated in sufficient details. Pollen grains tricolpate; ovule hemianatropous, unitegmic, tenuinucellate. Endosperm formation of the Cellular type.

Taxonomic Considerations.   *Phryma leptostachya* is the monotypic genus (of this family). It has doubtful taxonomic status. It resembles the family Verbenaceae, but is distinct from it in having unilocular and uniovulate ovary. Bentham and Hooker (1965b), Hallier (1912) and Hutchinson (1969, 1973) treat *Phryma* as belonging to the family Verbenaceae. Bessey (1915), Engler (1931) and Wettstein (1935) accepted Phrymaceae as a distinct family in the order Lamiales. Cronquist (1968, 1981) reports that the Phrymaceae are reduced pseudomonomerous type derived from the Verbenaceae. The single genus and species of Phrymaceae is habitually very much like *Verbena*, and has often been included in the Verbenaceae. As the gynoecium character is very much different, Cronquist prefers to place it in Lamiales as a distinct family and next to the family Verbenaceae. Benson (1970) included this family in Scrophulariales, although he emphasised its doubtful position. Hutchinson (1969, 1973) observes that this family occupies the ultimate place in his Lignosae, although it is a genus of herbs, and that it is clearly related to the Verbenaceae.

Therefore, it is doubtful that Phrymaceae is worth separating as a family from Verbenaceae. Its erect, orthotropous ovule may be considered an exception in the family Verbenaceae (having anatropous ovules). There is no transitional group between these two families.

Taxonomic Considerations of the Order Tubiflorae

The order as recognized by Engler and Diels (1936), comprised 8 suborders and 23 families. Rendle (1925) created a separate order, Convolvulales, for the family Convolvulaceae and did not include Fouquieriaceae and Lennoaceae in Tubiflorae. Wettstein (1935) removed the family Fouquieriaceae to the order Parietales and included Plantaginaceae in Tubiflorae. The arrangement of suborders and families in this book is in accordance with that of Melchior (1964).

The order Tubiflorae is a highly heterogeneous group and there are a few families of disputed position included in this order. Polemoniaceae is one such family for which different positions have been attributed by various workers (see also p. 466). The unigeneric family Fouquieriaceae also has a highly disputed position. The members of this family have bitegmic ovules, which is not true for the rest of the members of this order. The family Columelliaceae is yet another controversial taxon. Although it was placed in Tubiflorae by Engler and Diels (1936) and also by Melchior (1964), many later workers have placed it in the Rosales. Hallier (1912) included the genus *Columellia* in Saxifragaceae of the order Rosales. Cronquist

(1968, 1981) placed this family close to Pittosporaceae, also of the Rosales. Takhtajan (1980, 1987) treats Columelliaceae as a member of suborder Pittosporineae of the order Saxifragales. Its primitive wood anatomy does not support its placement in Tubiflorae and its gamopetalous, slightly zygomorphic flowers and two stamens are unusual in the Rosales (except in Pittosporaceae). Duckeodendraceae, another controversial family, is probably more closely related to the family Apocynaceae of the Gentianales.

Opinion is divided about the position of the family Henriqueziaceae. Cronquist (1968, 1981), Metcalfe and Chalk (1972) and Hutchinson (1969, 1973) are of the opinion that its members should be included in the family Rubiaceae, and Takhtajan (1987) did include it in the Rubiaceae. However, Dahlgren (1975a) placed it in the Scrophulariales near the family Bignoniaceae.

Another controversial family is Phrymaceae of the suborder Phrymineae. Taxonomic evaluation of this family shows clearly that it could have been considered as a part of the family Verbenaceae, as has been done by Takhtajan (1987).

Further research should be taken up to justify the position of the controversial families of this order.

In addition to these families, Takhtajan (1987) includes Sclerophylaceae and Goetzeaceae in Solanales, Cuscutaceae in Convolvulales, Cobaeaceae in Polemoniales, Ehretiaceae, Hoplestigmataceae, Cordiaceae and Wellstediaceae in Boraginales, Retziaceae, Stilbaceae, Trapellaceae, Spielmanniaceae, Thunbergiaceae and Mendonciaceae in Scrophulariales, and the order Hippuridales with Hippuridaceae in this group. Many of these are unigeneric and separated from a larger family, e.g. Spielmanniaceae (1 genus, *Oftia*, with 2 species) separated from Myoporaceae.

The separation into so many smaller families is not at all convincing. The features which justify the segregation may be regarded as exceptions in the larger families.

51

Order Plantaginales

The Order Plantaginales is a monotypic order comprising the only family, Plantaginaceae, with 3 genera.

Plantaginaceae

Plantaginaceae is a very small family including only 3 genera and 260 species (Takhtajan 1987). *Plantago*, a cosmopolitan genus, is the largest with ca. 200 species. The genus *Littorella* occurs in parts of Europe and the Antarctica; the monotypic genus *Bougueria* is reported only from Andes, South America.

Vegetative Features.   Herbs (Fig. 51.1A, C) or undershrubs; leaves mostly basal, alternate or rarely opposite; sometimes highly reduced; venation apparently parallel; base often sheathing; estipulate.

Floral Features.   Inflorescence scape-like, capitate or spicate. Flowers sessile (Fig. 51.1B) or shortly pedicellate, bracteate, ebracteolate, usually hermaphrodite (unisexual in *Littorella*; Fig. 51.1D, F), actinomorphic, hypogynous and tetramerous. Calyx tubular, sepals 4, membranous, hairy on the outer surface. Petals 4, gamopetalous (Fig. 51.1B, D), scarious, imbricate, persistent. Stamens 4, epipetalous, alternipetalous, all equal, exerted; anthers bicelled, large, versatile (Fig. 51.1E), dehiscence longitudinal.

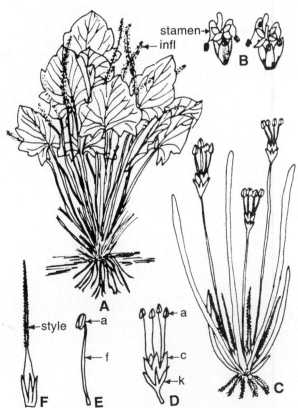

Fig. 51.1 Plantaginaceae: **A–B** *Plantago palmata*, **C–F** *Littorella uniflora*. **A** Plant with inflorescences. **B** Flowers. **C** Plant of *L. uniflora* with male and female flowers. **D** Male flower. **E** Stamen. **F** Female flower. *a* anther, *c* corolla, *f* filament, *infl* inflorescence, *k* calyx. (After Hutchinson 1969)

Gynoecium syncarpous, bicarpellary, ovary usually bilocular, rarely 1- to 4-locular, superior, placentation axile (free-central or basal in unilocular ovary); ovules 1 or more in each locule. Style 1, filiform, bears stigmatic hairs. Fruit a circumscissile capsule; seeds minute with small, straight embryo covered by copious endosperm.

Anatomy. Vessels very small, in radial lines; perforations simple, intervascular pitting alternate. Parenchyma absent or extremely sparse, rays absent. Leaves of *Plantago* isobilateral, dorsiventral or with the mesophyll wholly composed of palisade cells. Stomata variable, Caryophyllaceous type in *P. indica, P. lanceolata* and *P. psyllium;* stomata in *P. media* mostly surrounded by 3 and in *P. major* by 3–4 epidermal cells. In *Littorella uniflora* the stomata (in terrestrial forms) are arranged in longitudinal rows on both surfaces of the leaf; rare in aquatic forms. Mycorrhizal fungi have been reported in the roots of *P. coronopus* and *P. maritima.*

Embryology. Pollen grains spheroidal and panporate with a thin, smooth exine, 2-celled at dispersal stage. Ovules anacampylotropous, unitegmic and tenuinucellate. Polygonum type of embryo sac, 8-nucleate at maturity. Seed coat 2-layered, outer layer mucilaginous. Endosperm formation of the Cellular type; both micropylar and chalazal haustoria have been observed.

Chromosome Number. The basic chromosome number is x = 6,8,9.

Chemical Features. Iridoids of aucubin and plantagonine groups occur in *Plantago.* Trisaccharide planteose is characteristic of *Plantago* seeds (Harborne and Turner 1984). Thioglucoside is also reported, although sporadically, in members of this family (Kjaer 1963).

Important Genera and Economic Importance. Of the three genera of Plantaginaceae, the largest is *Plantago* with world-wide distribution. *P. indica* of southern Asia and *P. ovata,* originally of Europe, commonly known as Psyllium, have mucilaginous seed coat that is used medicinally as cathartic. The active constituent is xylin (Kochhar 1981). Another genus, *Bougueria,* has spicate-capitate inflorescence, a few upper flowers are bisexual and the rest female. In the third genus *Littorella* (Fig. 51.1C) the male and female flowers are distinct, the males on a long pedicel and the females sessile among the leaves at the base of the male flowers. *L. uniflora* has variable habit, depending on whether it grows on marshy ground or in water. The land forms have a basal rosette of leaves and the aquatic forms have long erect, cylindrical leaves. The last two genera do not have any economic importance.

Taxonomic Considerations. The position of the family Plantaginaceae has been considered in many different ways. Bentham and Hooker (1965b) and Metcalfe and Chalk (1972) consider Plantaginaceae to be an anomalous family. Engler and Diels (1936), Rendle (1925), Melchior (1964), Cronquist (1981), Dahlgren (1980a, 1983a) and Takhtajan (1987), consider it to be a highly advanced family.

Melchior (1964), in his revision of Engler's work, placed Plantaginaceae next to the order Tubiflorae. Several taxonomists presume that it is closely related to the order Scrophulariales (Cronquist 1968). This view is further supported by the biochemical studies on *Littorella* and *Plantago* by Bourdu et al. (1963). Hutchinson (1969, 1973), however, holds a different view. According to him, the Plantaginaceae is close to the Primulaceae and Plumbaginaceae. It is close to Primulaceae as in both the fruits show circumscissile dehiscence and both have Caryophyllaceous type of stomata. Dahlgren (1980a) and Takhtajan (1987) include Plantaginaceae in the order Scrophulariales.

It does appear more closely allied to the order Scrophulariales in view of the similarities in floral, embryological, and chemical features. However, as the flowers of this family are anemophilous, this may be treated as a member of a separate order Plantaginales placed close to the Scrophulariales.

Taxonomic Considerations of the Order Plantaginales

The order Plantaginales has been treated variously. It belongs to Asteridae and has small wind-pollinated flowers. It is a highly advanced family as its members are herbs and undershrubs (Cronquist 1968). Bentham and Hooker (1965b) and Metcalfe and Chalk (1972) consider this order to be an anomalous group. Bessey (1915) included it in his Primulales as derived from or allied to Plumbaginaceae. Most of the taxonomists agree that the family Plantaginaceae be included in the order Scrophulariales. The tetramerous anemophilous flowers, membranous corolla, and exerted stamens of this group support its being raised to a distinct order.

52

Order Dipsacales

The order Dipsacales as defined here consists of 4 families: Caprifoliaceae, Adoxaceae, Valerianaceae, and Dipsacaceae. The members of this order are mostly perennial herbs or shrubs with opposite leaves. Except for some Caprifoliaceae, they lack stipules. Corolla regular or irregular, stamens often fewer than corolla lobes. The filaments are attached to the corolla tube well above the base; pollen grains are trinucleate when shed. Ovules unitegmic, tenuinucellate. Endosperm formation is of the Cellular type. Dipsacales are rich in caffeic acid.

Caprifoliaceae

A small family of 18 genera and 290 species distributed mainly in the northern hemisphere, and some in tropical mountains. *Triosteum hirsutum* occurs at high altitudes in the Himalayas.

Vegetative Features. Mostly shrubs (*Sambucus*), sometimes lianas (*Lonicera* spp.), rarely herbs (*Triosteum* and *Sambucus ebulus*), or suffrutescent (*Linnaea*). Leaves opposite, connate-perfoliate in *Lonicera* (Fig. 52.1A), simple, rarely pinnately compound as in *Sambucus*, estipulate, stipulate in *Sambucus,* or stipules modified to nectariferous glands as in *Viburnum* and *Leycesteria;* sessile (*Lonicera tatarica*) or petiolate (*Viburnum tomentosum*); with reticulate venation.

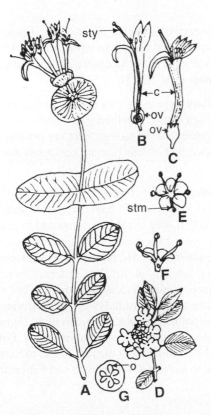

Fig. 52.1 Caprifoliaceae: **A–C** *Lonicera caprifolium,* **D–G** *Viburnum tomentosum.* **A** Twig with flowers. **B** Flower, longisection. **C** Bilabiate flower. **D** Twig of *V. tomentosum* with flowers. **E, F** Flower (**E**) and vertical section (**F**). **G** Ovary, transection. *c* corolla, *k,* calyx, *o* ovule, *ov* ovary, *stm* stamen, *sty* style. (**A–C** adapted from Hutchinson 1969, **D–G** after Lawrence 1951)

Floral Features. Inflorescence cymose (Fig. 52.1A) or various modifications of cymose. Flowers only a few in a cyme, usually hermaphrodite, sterile or neutral flowers known in some species of *Viburnum*, actinomorphic (*Sambucus* sp.) or zygomorphic (*Lonicera* sp.), epigynous. Calyx of 5 sepals, gamosepalous; corolla typically of 5 petals, variable in shape, often bilabiate as in *Lonicera* (Fig. 52.1B, C), rotate or salverform (Fig. 52.1D, E), imbricate. Stamens 5, sometimes 4, epipetalous, alternate with petals (Fig. 52.1E, F); anthers bicelled, dehiscence longitudinal, introrse. Gynoecium syncarpous, usually 3- to 5-carpellary, ovary 1- to 5-locular, placentation axile (Fig. 52.1G), sometimes with an approach towards parietal, inferior; ovules usually one in each locule (numerous in *Leycesteria*), pendulous; style 1, slender or sometimes minute or obscure, e.g. *Viburnum tomentosum* (Fig. 52.1F); stigmas as many as carpels. Fruit a berry or a drupe. Seeds with copious endosperm and usually a small straight embryo.

Anatomy. Vessels small, solitary, ring-porous or semi-ring-porous, with spiral thickenings; perforation plates simple or simple and scalariform. Parenchyma apotracheal, diffuse, rarely scanty paratracheal. Rays usually 3–4 cells wide, heterogeneous. Leaves dorsiventral with both glandular and nonglandular hairs. Glandular leaf teeth secreting resin and mucilage and extrafloral nectaries occur in different species. Stomata more common on the lower surface, sometimes on the upper surface also; Ranunculaceous or Rubiaceous type.

Embryology. Pollen grains 3-colporate and shed at 3-celled stage. Ovules anatropous, unitegmic, tenuinucellate. Polygonum type of embryo sac in *Viburnum*, Allium type in *Lonicera* and Adoxa type in *Sambucus nigra* (Johri et al. 1992). Endosperm formation is of the Cellular type.

Chromosome Number. The basic chromosome number is very variable. It may be x = 8–12, 16, or 18. The haploid chromosome number for *Sambucus* is n = 18, 19 or 21 and the basic chromosome number for *Viburnum* is x = 8 or 9

Chemical Features. Secologanins have been reported in *Diervilla*, *Dipelta*, *Kolkwitzia*, *Lonicera*, *Symphoricarpos*, *Weigela*. Loganin also occurs in 2 members—*Lonicera* and *Symphoricarpos*, and adoxoside in *Viburnum* (Jensen et al. 1975). Cyanogenic glucosides have been reported in *Sambucus nigra*, and seco-iridoid morroniside in species of *Lonicera* and *Sambucus* (Jensen et al. 1975).

Important Genera and Economic Importance. The important genera of this family are *Lonicera* (honeysuckle), *Sambucus* (commonly called elderberry), *Symphoricarpos* (coralberry or snowberry) and *Viburnum* (snowball). The family is economically important because most of its members are ornamental shrubs. However, *Triosteum* sp. and *Sambucus ebulus* are the two herbaceous plants. *Sambucus* berries are used in marking wine.

Taxonomic Considerations. The family Caprifoliaceae has often been included in the order Rubiales (Engler and Diels 1936, Lawrence 1951). Ferguson (1966a) and Porter (1967) have suggested that the two families Caprifoliaceae and Rubiaceae are so closely related that they should be combined into one. Hutchinson (1969, 1973) states that the comparison between these two families is scarcely possible owing to the disparity in their size. Their similarity is due to convergence. He places Caprifoliaceae in the Araliales. Melchior (1964) and Thorne (1968) retain the two families in different orders, Cronquist (1968, 1981) in two closely related orders. Takhtajan (1959) places Caprifoliaceae in Rubiales, but later (1980, 1987) in Dipsacales, in superorder Cornanae. According to Wilkinson (1949), Caprifoliaceae might be polyphyletic. On the basis of morphology and anatomy, he pointed out that *Sambucus* and *Viburnum* are derived from the Cornaceae, while the tribes Lonicereae and Linnaeae have a different origin. Pollen morphology and cytological data also distinguish *Sambucus* and *Viburnum* from the rest of the members. Serological studies report that Caprifoliaceae is uniformly dissimilar to Rubiaceae (Hillebrand and Fairbrothers 1970).

From the above discussion, therefore, it is clear that although there is superficial resemblance between the two families, they are not truly related. Caprifoliaceae enjoys a natural position in the order Dipsacales.

Adoxaceae

A monotypic family with the only genus *Adoxa* distributed in north temperate zone (subarctic and temperate regions) and with only one species, *A. moschatellina.*

Vegetative Features.   Small perennial herbs with creeping rhizomes, monopodial, at first filiform bearing 2 rows of alternate scale leaves and expanding towards its apex into a more or less flattened bulb-like body; bears 3–5 scale leaves, 1–3 long-petioled foliage leaves and a single flowering shoot. Flowering shoot or peduncle erect, tetragonal, arises from the bulb-like expansion of the rhizome in the axil of a scale leaf or a foliage leaf, about 8–10 cm long; with 1–3 radical leaves, and a pair of opposite 3-foliolate cauline leaves, variable in size, estipulate.

Floral Features.   Inflorescence a small head or a condensed dichotomous cyme of about 5 flowers. The terminal flower usually 4-merous, the laterals 5- or 6-merous, ebracteate, hermaphrodite, actinomorphic, greenish, inconspicuous, semi-epigynous. Perianth in 2 whorls, the outer usually 3-merous, open or valvate, persistent, often regarded as an involucre formed of bracts and bracteoles but probably a calyx, adnate to the ovary. The inner whorl (probably a corolla) of 5 or 4 lobes, gamopetalous, imbricate with odd lobe posterior, caducous. Stamens 4–6, alternate with petals, inserted on a raised ring at the mouth of the inner perianth; filaments short and divided and each segment bearing a half-anther; anthers extrorse, monothecous. Gynoecium syncarpous, 3- to 5-carpellary, rarely bicarpellary; ovary 3- to 5-loculed, half-inferior, with 1 pendulous ovule in each locule. At the base of each inner perianth lobe, on its upper surface is a nectary composed of about 30 minute glandular capitate hairs. Fruit a drupe with several stones or composed of 1–5 pyrenes, about 3 mm long, each surrounded by a slimy layer; endocarp coriaceous. Seeds albuminous, about 2 mm long, endosperm copious, cartilaginous (tough and hard, but not bony), embryo 0.3–0.5 mm long.

Anatomy.   The flowering stalk contains 4 vascular bundles of which 2 larger ones supply the cauline leaves and the 2 smaller ones lead to the flowers. Xylem when young, include spirally thickened vessels but scalariform thickenings arise later on. Parenchyma tissue contains abundant starch. Leaves dorsiventral, deciduous, tanniniferous secretory hairs on young leaves. Stomata Ranunculaceous type, confined to the lower surface.

Embryology.   Pollen grains 3-colporate and are shed at 3-celled stage, Pollen of *Adoxa moschatellina* is similar to that of *Sambucus* of Caprifoliaceae. Ovules anatropous, unitegmic, tenuinucellate. Embryo sac of Adoxa type, 8-nucleate at maturity. It is a tetrasporic embryo sac. Endosperm formation of the Cellular type.

Chromosome Number.   Haploid chromosome number of the genus *Adoxa* is n = 18.

Chemical Features.   Secologanin, including secologanic acid, foliamenthin, cantleyoside and other derivatives, and morroniside, including oliveridine, have been reported in *Adoxa* (Jensen et al. 1975).

Important Genera and Economic Importance.   *Adoxa* is the only genus with the species *A. moschatellina* known so far. One new species—*A. omeiensis* (Wu 1981, Hara 1981, 1983) and a new genus *Sinadoxa corydalifolia*—have also been recorded. Takhtajan (1987) records a third genus *Tetradoxa* and another species of *Adoxa*—*A. orientalis.*

Taxonomic Considerations.   The taxonomic position of *Adoxa* is much disputed. Engler and Diels

(1936) recognise Adoxaceae as a family of the Rubiales along with other related families. Hutchinson (1969, 1973) treated Adoxaceae as the climax family of his Saxifragales. Dahlgren (1980a) places it in the order Cornales. Cronquist (1981) and Takhtajan (1980, 1987) treat it as a member of the Dipsacales.

The herbaceous habit, bulb-like expansion of rhizome, floral morphology and anatomy of Adoxaceae suggest its alliance to the Saxifragaceae. On the other hand, cytological and serotaxonomical data support its relationship with Caprifoliaceae, Dipsacaceae and Rubiaceae (Sprague 1972). There is a similarity of floral structure between Adoxaceae and Caprifoliaceae. Also, the pollen grain and tetrasporic embryo sac of *Adoxa* are similar to those of *Sambucus* of Caprifoliaceae. Chemically, also, the Adoxaceae are similar to the Dipsacaceae in containing seco-iridoids.

The Adoxaceae is correctly grouped with Caprifoliaceae and Dipsacaceae in the same order, Dipsacales.

Valerianaceae

A small family of only 10 genera and about 370 species, mostly distributed in the north temperate region. Some species are reported from high altitude areas of the tropical zones also.

Vegetative Features.   Mostly annual or perennial herbs, rarely shrubs, woody at base; rootstock often produces suckers. Leaves opposite or in basal rosettes; cauline leaves entire or pinnately divided (Fig 52.3A), estipulate, often with sheathing leaf base.

Floral Features.   Inflorescence a monochasium, thyrse, or a many-flowered dichasial cyme, sometimes highly condensed into a capitulum (Fig. 52.3A). Flowers bracteate, bracteolate, hermaphrodite or unisexual and then the plants are usually dioecious as in *Valeriana* spp., irregular (almost regular in *Patrinia*), epigynous. Calyx usually develops tardily and is represented by an epigynous ring, rarely 2- to 4-toothed as in *Fedia* which develops into fruits also, adnate to ovary. Corolla usually tubular and 5-lobed (Fig. 52.3B), rarely 3- to 4-lobed, imbricate, often saccate or spurred basally; corolla bilabiate in *Centranthus*. Stamens vary in number from 1 to 4; 1 in *Centranthus,* 2 in *Fedia,* 3 in *Valeriana* and *Valarianella* and 4 in *Patrinia* and *Nardostachys;* epipetalous and alternipetalous; anthers bicelled, 4-locular, introrse, versatile, dehiscence longitudinal. Gynoecium syncarpous, tricarpellary, ovary basically trilocular but usually 2 locules suppressed and sterile and only one locule fertile (Fig. 52.3C), inferior; placentation basically axile but often appears to be parietal, ovule 1 per locule, pendulous (Fig. 52.3C); style 1, slender, stigma simple or 2- to 3-lobed. Fruit dry, indehiscent achene, calyx often develops into a winged, awned or plumose pappus. Seeds without endosperm, and a large, straight embryo.

Anatomy.   Stems frequently hollow. Vascular bundles individually distinct in very young stems but form a continuous cylinder in older stems. Vessels extremely small with simple, rather oblique perforations. Parenchyma absent; rays multiseriate, 4–6 cells wide. Leaves mostly dorsiventral, occasionally centric;

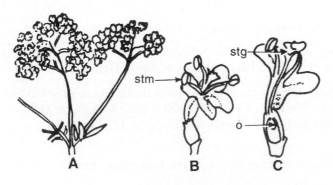

Fig. 52.3 Valerianaceae: **A–C** *Valeriana officinalis.* **A** Leaf, inflorescence. **B** Flower. **C** Vertical section. *o* ovule, *stg* stigma, *stm* stamen. (Adapted from Lawrence 1951)

both simple and glandular hairs common; stomata present on both sides or confined only to the lower surface, Ranunculaceous type.

Embryology.   Pollen grains monads, 3-colporate, 3-celled at the dispersal stage; exine echinate. Ovules anatropous, unitegmic, tenuinucellate. Polygonum type of embryo sac, 8-nucleate at maturity. Endosperm formation of the Cellular type.

Chromosome Number.   Basic chromosome number is quite variable for different genera: x = 7–13. The diploid number for *Centranthus* is 2n = 14, 32, for *Fedia* 2n = 32, for *Patrinia* 2n = 22 and for *Astrophia* 2n = 32.

Chemical Features.   A yellow flavone, 6-hydroxyluteolin has been reported from the leaves of 6 Mediterranean species of *Valerianella* (Gregor and Ernet 1973). Its exclusive occurrence in these species is a marker for them. Seed fats of the Valerianaceae contain α-olaeostearic acid (Shorland 1963). The family Valerianaceae occupies a singular position in producing certain iridoids called valeriana compounds, not reported in any other plant (Jensen et al. 1975).

The glandular hairs of aerial parts secrete a volatile oil which is not only lipophilic (Szentpetery et al. 1969) but also hydrophilic (Corsi and Pagni 1990). They could perhaps be a rubber-resin similar to that suggested by Bruni et al. (1987) for *Tamus communis* (family Dioscoreaceae). The excretions are found in rod-shaped deposits on the external surface of the glandular hairs.

Important Genera and Economic Importance.   *Valeriana,* the largest genus with about 210 species occur in both the Old and the New World. *Nardostachys jatamansi,* commonly called Jatamansi because of the bearded appearance of its rhizomes, is a perennial herb about 60 cm tall with woody, elongated rhizomes covered with fibres from petioles of withered leaves, and occurs in the alpine Eastern Himalayas between 3000 and 4000 m altitude. The dried rhizomes and roots are used medicinally in certain cardiac ailments. It is a substitute for the drug valerian from *Valeriana officinalis* which occurs in some regions of Kashmir at about 2500 m altitude. The fresh juice of the rhizomes and roots containing a volatile oil is used against nervous disorders and certain cardiac diseases. The efficacy of the drug is lost on drying.

Taxonomic Considerations.   Most taxonomists place the family Valerianaceae along with the Caprifoliaceae in the order Dipsacales (Melchior 1964, Cronquist 1981, Takhtajan 1980, 1987), or Rubiales (Lawrence 1951). Hutchinson (1969, 1973) treats it as a member of Valerianales. From the studies undertaken and data collected, its correct position is in the Dipsacales along with Caprifoliaceae and other families.

Dipsacaceae

A small family of 9 genera and 160 species (10 genera and 300 species according to Takhtajan 1987), distributed in the eastern Mediterranean and Balkan areas, across the steppes of Russia and in India.

Vegetative Features.   Annual, biennial or perennial herbs, rarely shrubs such as *Scabiosa* spp. Leaves opposite or rarely whorled, estipulate, usually simple or pinnatifid.

Floral Features.   Inflorescence a dense involucrate head (Fig. 52.4.1A) or spike or dense whorl-like subapical clusters surrounded by several bracts. Flowers hermaphrodite, zygomorphic, each with an involucre and an involucel, epigynous. Each flower surrounded by an epicalyx formed by the connation of 2 subtending bracteoles. Calyx small, cuplike or divided into 5–10 pappus-like segments (Fig. 52.4B, C). Petals 4 or 5, gamopetalous, imbricate. Nectaries usually situated at the base of corolla tube and consist of 1-celled epidermal hairs and an underlying nectar-secreting tissue. In *Cephalaria* and *Knautia,* one

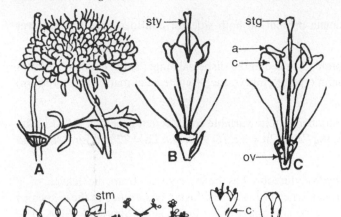

Fig. 52.4.1 Dipsacaceae: **A–C** *Scabiosa atropurpurea*. **A** Inflorescence. **B** Flower. **C** Vertical section. *a* anther, *c* corolla, *stg* stigma, *sty* style. (Adapted from Lawrence 1951)

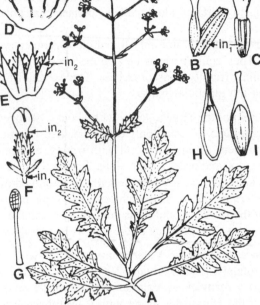

Fig. 52.4.2 Dipsacaceae: **A–I** *Triplostegia* sp. **A** Twig with inflorescence. **B, C** Flower open (**B**) and bud (**C**), note involucre (*in₁*) and corolla. **D** Corolla (cut open). **E** Involucel (*in₂*) opened out. **F** Flower bud with involucre (*in₁*) and involucel (*in₂*). **G** Glandular hair from involucel. **H, I** Vertical section (**H**) and entire fruit (**I**). *c* corolla, *in₁* involucre, *in₂* involucel, *stm* stamen. (After Hutchinson 1969)

continuous ring of nectariferous tissue has been observed (Wagenitz and Laing 1984). Stamens usually 4 (sometimes 2 or 3), epipetalous (Fig. 52.4.1C); anthers bicelled, dehiscence longitudinal, introrse. Gynoecium syncarpous, bicarpellary, one carpel suppressed; ovary unilocular (derived from one carpel), inferior; ovule solitary, pendulous from locule apex. Style 1, filiform, stigma simple or bilobed, lateral in *Dipsacus*. Fruit an achene, enclosed within the epicalyx, often crowned by the persistent calyx. Seeds with fleshy endosperm and a straight embryo.

Anatomy.   Xylem and phloem form a closed ring in most genera. Vessels in the primary xylem in radial rows, many with spiral thickenings and with simple perforation plates. Leaves dorsiventral or centric as in some species of *Dipsacus* and *Scabiosa*. Stomata generally present on both the surfaces of the leaves; mostly Ranunculaceous, rarely Cruciferous type.

Embryology.   Pollen grains tricolpate, 3-celled at the dispersal stage. Ovules anatropous, unitegmic, tenuinucellate. Embryo sac of Polygonum type, 8-nucleate at maturity. Endosperm formation of the Cellular type. The nuclear division in the zygote is followed by an oblique vertical wall.

Chromosome Number. Basic chromosome number of the family is x = 9. Diploid numbers are 2n = 10, 14, 16, 18, 20, 34, 36, 38, 42, 44, 46, 50 and 54. Euploidy, aneuploidy on the diploid level, polyloidy followed by aneuploidy, gene-chromosome and genome mutations have been important in the evolution of Dipsacaceae.

Chemical Features. C-glycoflavones occur in the members of this family. Iridoids of the natural methyl glucoside and secologanin and its derivatives have been reported in some genera (Jensen et al. 1975).

Important Genera and Economic Importance. Amongst the important genera are *Dipsacus, Scabiosa, Cephalaria,* and *Pterocephalus,* some of these are grown as ornamentals.

The genus *Triplostegia* with 2 species, *T. glandulifera* (Fig. 52.4.2A–I) from northern India and *T. repens* from New Guinea, is a disputed taxon. It was first ascribed to the Valerianaceae but transferred to the Dipsacaceae by Bentham and Hooker (1965b). It was subsequently placed back in the Valerianaceae (Melchior 1964, Cronquist 1981). The apparent morphology of the genus is that of Valerianaceae but there is a double involucel (Fig. 52.4.2F), unilocular and uniovulate ovary (Fig. 52.4.2H) and the seeds have copious endosperm. Therefore, Hutchinson (1959, 1969), Burtt (1977) and Thorne (1983) include it in Dipsacaceae. Airy Shaw (1965) raises the genus *Triplostegia* to a separate family Triplostegiaceae; Dahlgren (1983a) and Takhtajan (1987) support the erection of the new family, although Burtt (1977) strongly opposes it.

Taxonomic Considerations. The family Dipsacaceae has been placed in the order Rubiales by Engler and Diels (1936) and is a natural taxon with the exception of the genera *Morina* and *Triplostegia,* which are often treated in separate families—Morinaceae and Triplostegiaceae (Takhtajan 1987). The family Dipsacaceae is very close to the Valerianaceae but for the fact that it has: (a) unilocular ovary, (b) capitulum inflorescence, and (c) inflorescence surrounded by an involucre of bracts. Cronquist (1968, 1981), Takhtajan (1980, 1987), and Dahlgren (1983a) treat it as a member of Dipsacales. Hutchinson (1969, 1973) prefers to place it in Valerianales because of many similarities between Dipsacaceae and Valerianaceae.

The above discussion, however, supports its position in the Dipsacales, which is a distinct order.

Taxonomic Considerations of the Order Dipsacales

Bentham and Hooker (1965b) did not recognise Dipsacales as a distinct order. Its members were included partly in the Rubiales (Caprifoliaceae) and partly in Asterales (Valerianaceae and Dipsaceae). The family Adoxaceae was also not recognised by Bentham and Hooker (1965b). Engler and Diels (1936) placed all these families (including Adoxaceae) in Rubiales, although they grow best in temperate region, in contrast to the tropical Rubiaceae. Dipsacales members differ from the Rubiales also in Cellular type of endosperm, almost without any alkaloids and also lack coletters. Hutchinson (1969, 1973) places Caprifoliaceae in the order Araliales of Lignosae and the other two families in the Valerianales of Herbaceae, and Adoxaceae in another order, Saxifragales.

Melchior (1964) erects a separate order, Dipsacales, and includes all the four families (Caprifoliaceae, Adoxaceae, Valerianaceae, and Dipsacaceae) in this order. Cronquist (1968, 1981) and Takhtajan (1980, 1987) support this assignment. Dahlgren (1983a), however, places Caprifoliaceae and Adoxaceae in the order Cornales, and the rest of the families in Dipsacales, although both the orders are included in the Corniflorae. The common features are: (a) unitegmic, tenuinucellate ovule, (b) Cellular type of endosperm, (c) frequent sympetaly, and (d) endosperm with haustoria. Gornall et al. (1979) treat the Dipsacales as a member of the Gentianiflorae on the basis of chemical features. In most Gentianiflorae, the endosperm formation is of the Nuclear type (Gentianales) and they have stipulate leaves, and intraxylary phloem. These features are not common to the Dipsacales.

Within the order Dipsacales there is a progression from woody to herbaceous habit, from regular to irregular corolla, from isomerous to anisomerous stamens, from many-seeded condition to only one seed, a tendency for the endosperm to be consumed, and for the embryo to increase in size. The Caprifoliaceae is the most primitive family of the order, and has given rise to rest of the families.

The four families, Caprifoliaceae, Adoxaceae, Valerianaceae and Dipsacaceae of the order Dipsacales, form a natural group; and they are close to the Cornales.

53

Order Campanulales

An order comprising 8 families—Campanulaceae, unigeneric Sphenocleaceae and Pentaphragmataceae, Goodeniaceae, monotypic Brunoniaceae, Stylidiaceae, Calyceraceae and Compositae, the largest family of the angiosperms. These are characterised by pentamerous perianth, epipetalous stamens in a single whorl, bithecal, coherent to connate anthers, and usually unilocular ovary with a single ovule. In all the families, the ovules are anatropous, unitegmic and tenuinucellate. Iridoids are absent as a rule, except in Goodeniaceae. Some Compositae members also produce alkaloids.

The members of Campanulales are mostly perennial herbs or shrubs; sometimes may be annual; trees rare.

Campanulaceae

A family of 50–55 genera and 950 species (Takhtajan 1987), distributed in temperate, subtropical and tropical zones of both the Old and the New World.

Vegetative Features. Mostly perennial herbs and undershrubs, rarely trees, e.g. *Clermontia*; usually with latex. Leaves alternate or opposite; whorled in *Ostrowskia* and *Siphocampylus,* simple, stipulate.

Floral Features. Inflorescence terminal on the main shoot, or secondary shoots; generally racemose or thyrsiform; in some taxa, instead of single flowers in the axils of bracts of the raceme, small dichotomies may occur; cymose inflorescence in *Canarina*. Flowers bracteate and bracteolate, usually hermaphrodite, actino- or zygomorphic, pentamerous, epigynous. Calyx lobes 3–10, usually 5, polysepalous, imbricate or valvate. Corolla of 5 petals, basally connate, valvate, campanulate, tubuler or strongly bilabiate; number of petals variable, e.g. 6 in *Canarina* and *Michauxia,* 3–4 in *Wahlenbergia,* gamopetalous, often seemingly polypetalous, as in *Phyteuma*. Corolla sometimes absent as in the cleistogamous flowers of *Legenere, Heterocodon* and *Legousia*. Stamens as many as petals, alternipetalous and epipetalous at the extreme base of petals; anthers bicelled, introrse, longitudinally dehiscent, filament base often expanded, forms a dome-shaped chamber over the nectariferous disc. Gynoecium syncarpous, 2-, 3-, or 5-carpellary, ovary multilocular with axile placentation (unilocular with 2 parietal placentae in *Downingia, Legenere, Howellia* and *Apetahia;* deeply intruded placentae in *Siphocampylus*); usually inferior (half-inferior in *Wahlenbergia, Diastatea* and *Lobelia,* and superior in *Cyananthus*); ovules numerous, rarely few and apparently basal, as in *Merciera* and *Siphocodon*; style 1, sometimes 2- to 5-branched, stigmas as many as carpels. Fruit usually a capsule, dehisces apically by slits, as in *Lobelia* and *Wahlenbergia,* or circumscissilely, as in *Lysipomia*, or by apical or basal pores, as in *Campanula rapunculoides;* sometimes a berry as in *Canarina* and *Centropogon;* seeds with fleshy endosperm and a small, straight embryo.

Anatomy. Vessels small, solitary and in small radial multiples; perforations simple, intervascular pitting large and alternate; parenchyma absent. Rays broad, composed mostly of large upright cells. Leaves usually dorsiventral, unicellular and uniseriate hairs have been recorded. Stomata mostly present on both the surfaces; Ranunculaceous type. Hydathodes occur widely; crystals of various types reported.

Embryology. Pollen grains subprolate, oblate-sphaeroidal, sphaeroidal, prolate or prolate-sphaeroidal; 3- or 4-porate, 3-colporate or 7- to 10-colpate, shed at 3-celled or rarely 2-celled stage. Ovules anatropous, unitegmic, tenuinucellate. Polygonum type of embryo sac, 8-nucleate at maturity. Endosperm formation of the Cellular type; prominent micropylar and chalazal haustoria common.

Chromosome Number. Basic chromosome numbers are: x = 6-10, or 12–15,17 (Takhtajan 1987). The diploid numbers are: 2n = 14,16,18,24,28,30,36,54,64 or 72 (Kumar and Subramaniam 1987).

Chemical Features. Iridoids absent (Jensen et al. 1975); rich in delphinidin, an anthocyanin, inulins and polyacetylenes (Harborne 1963).

Important Genera and Economic Importance. *Campanula,* commonly called bell flower, is the largest genus with about 230 species; a number of these are cultivated. Other important genera include *Lobelia, Siphocampylus, Centropogon, Wahlenbergia, Phyteuma* and *Cyanea,* several of these are ornamentals.

Taxonomic Considerations. Campanulaceae comprises three subfmilies, according to Melchior (1964):

(a) Campanuloideae. Flowers actinomorphic, rarely zygomorphic, anthers free: *Campanula, Phyteuma, Wahlenbergia, Platycodon, Jasione.*

(b) Cyphioideae. Flowers zygomorphic, stamens sometimes united, anthers free: *Cyphia, Nemacladus.*

(c) Lobelioideae. Flowers zygomorphic, rarely actinomorphic, anthers united: *Centropogon, Siphocampylus, Lobelia.*

Hutchinson (1969, 1973) separated Lobeliaceae from the Campanulaceae and placed the two families in the same order, Campanales. Warming (1942) treated the subfamily Cyphioideae as a distinct family, Cyphiaceae. Cronquist (1968, 1981) includes Lobeliaceae in his Campanulaceae, Dahlgren (1975 a, b, 1980b, 1983a) treates Lobeliaceae as a family separate from Campanulaceae but includes the Sphenocleaceae in it. Takhtajan (1980) accounts for a single family, Campanulaceae; but in 1987 he breaks up Campanulaceae s.l. into Campanulaceae s.s. (50–55 genera), Cyphiaceae (1 genus), Nemocladaceae (3 genera), Lobeliaceae (30 genera) and Cyphocarpaceae (1 genus). According to Subramanyam (1970a), embryological characters strongly reflect the affinity of the Campanulaceae with the Lobeliaceae, and also their recognition as separate families.

It is, however, more appropriate that Lobeliaceae should remain as a subfamily of the family Campanulaceae. Even the breaking up of this family into so many smaller units (as was done by Takhtajan 1987) is not desirable.

Sphenocleaceae

A monotypic family of the genus *Sphenoclea* with 1 or 2 species. *S. zeylanica* occurs in India in moist areas and swampy places, especially near sea coast.

Vegetative Features. Herbaceous plants; stem soft, erect, more or less succulent, without latex, glabrous, hollow, branched. Leaves simple, alternate, entire, shortly petiolate or almost sessile, lanceolate or more or less elliptic, acute, estipulate.

Floral Features. Inflorescence a dense spike (Fig. 53.2 A, B). Flowers bracteate, hermaphrodite, actinomorphic, small, perigynous. Calyx of 5 sepals, polysepalous, lower portion fused with ovary, upper portion free, imbricate. Corolla lobes 5, sympetalous (Fig. 53.2 C, D), imbricate. Stamens as many as corolla lobes, alternipetalous, free or connate only by the anthers; anthers introrse, dorsifixed (Fig. 53.2 E), dehiscence longitudinal; filaments short, free from style. Gynoecium syncarpous, bicarpellary, ovary bilocular, half-inferior with multi-ovulate pendulous placentae (Fig. 53.2 F, G); style short, glabrous with a solitary bulbous stigma. Fruit a circumscissile capsule; seeds small with scanty endosperm and a large straight embryo.

Anatomy. The primary cortex of stem includes large intercellular air canals; pericyclic sclerenchyma

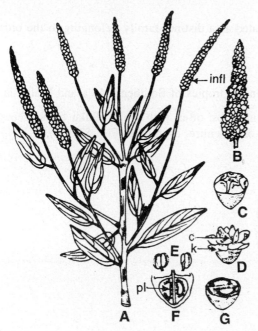

Fig. 53.2 Sphenocleaceae: **A-G** *Sphenoclea zeylanica.* **A** Flowering twig. **B** Inflorescence. **C, D** Flower bud closed (**C**) and open (**D**). **E** Stamens. **F, G** Vertical (**F**) and transverse (**G**) section of ovary. *c* corolla, *k* calyx, *infl* inflorescence, *pl* placenta. (After Hutchinson 1969)

cells present, xylem and phloem constitute continuous cylinders traversed by 1 or 2 wide medullary rays. Vessels thin-walled, angular with large, horizontal, bordered pits on lateral walls and simple perforations. Latex tubes absent in phloem, instead there are elongated cells with granular content. Leaves dorsiventral, stomata numerous on the lower and occasional on the upper surface, each surrounded by 4 epidermal cells—2 parallel and the other 2 at right angles to the guard cells. Raphides absent.

Embryology.   The mature pollen grains are tricolporate, subprolate, psilate with the stratification of the exine more or less obscure; 3-celled at shedding stage. Ovules anatropous, unitegmic, tenuinucellate. Polygonum type of embryo sac, 8-nucleate at maturity. Endosperm formation of the Cellular type; both micropylar and chalazal haustoria aggressive.

Chromosome Number.   The haploid chromosome number of *Sphenoclea* is n = 12 (Takhtajan 1987). The diploid numbers are 2n = 24, 32 and 42 (Kumar and Subramanian 1987).

Chemical Features.   Iridoids absent (Jensen et al. 1975).

Important Genera and Economic Importance.   *Sphenoclea,* the only genus, has no economic importance.

Taxonomic Considerations.   The genus *Sphenoclea* was placed in the family Campanulaceae by Bentham and Hooker (1965b), Wimmer (1943) and Hutchinson (1969, 1973). Airy Shaw (1948) suggested its removal from the Campanulaceae with the assumption that Sphenocleaceae is a 'peripheral Centrosperm' group and has possible connections to Phytolaccaceae. The typical circumscissile capsules occur in Portulaccaceae and Primulaceae also. Its resemblance to *Phytolacca*, according to Hutchinson (1959, 1969), "is only superficial and due to parallel evolution of the inflorescence". Fjodorow (1954) included *Sphenoclea* in the subfamily Sphenocleoideae of the Campanulaceae. Subramanyam (1970b), on the basis of embryological data, places this family in the order Campanulales. Cronquist (1968, 1981) comments that although *Sphenoclea* "is reminiscent of *Phytolacca*" in its habit, embryologically and palynologically it is more like the Campanulaceae, He includes Sphenocleaceae in the order Campanulales. Dahlgren (1975a,b, 1980a), however, retains *Sphenoclea* in the family Campanulaceae. Takhtajan (1987) includes the family Sphenocleaceae in the order Campanulales.

In our opinion also, the Sphenocleaceae is better treated as a distinct family belonging to the order Campanulales.

Pentaphragmataceae
A unigeneric family with about 30 species distributed in the tropics of Southeast Asia and Malaysia.

Vegetative Features.   Perennial, more or less succulent herbs, often with multicellular hairs. Leaves alternate, simple, usually asymmetrical, sinuate-dentate or subentire, estipulate (Fig. 53.3 A).

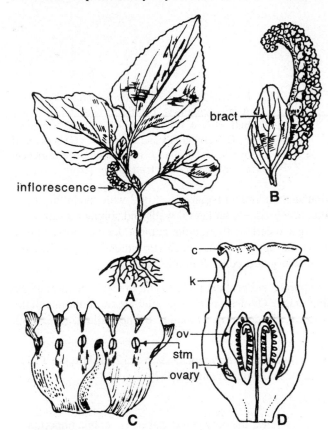

Fig. 53.3   Pentaphragmataceae: **A-D** *Pentaphragma horsefieldii*. **A** Plant with inflorescence. **B** Inflorescence. **C** Opened corolla, shows stamens and pistil. **D** Flower, vertical section. *c* corolla, *k* calyx, *n* nectary, *ov* ovary, *stm* stamen. (After Kapil and Vijayaraghavan 1965)

Floral Features.   Inflorescence a scorpioid cyme with numerous flowers arranged in acropetal succession (Fig. 53.3 A, B) and often subtended by a large, membranous bract. Flowers small, sessile, actinomorphic, hermaphrodite or rarely unisexual. Sepals 5, gamosepalous, forms a campanulate calyx tube, upper portion free, persistent. Petals 5, gamo- or polypetalous, valvate, fleshy or cartilaginous, persistent. Stamens 5, epipetalous, alternipetalous (Fig. 53.3 C); filaments persistent, short; anthers linear, introrse, basifixed. Gynoecium syncarpous, bicarpellary, ovary bilocular, with axile placentation, inferior; style short and thick, stigma massive, oblong; 5 nectaries at the base of the ovary (Fig. 53.3D); ovules numerous. Fruit baccate with persistent calyx and corolla; seeds numerous, small, ovoid, brown, with copious endosperm.

Anatomy.   Phloem and xylem (in transverse section of stem) appear as continuous rings, traversed by 1 or 2 cells wide medullary rays. Vessels numerous, angular, up to 60 μm in diameter, with scalariform lateral pitting and scalariform perforation plates. Leaves dorsiventral with long hairs on the lower surface. Stomata chiefly on the lower surface but present on the upper surface also, especially near marginal leaf teeth; each stoma surrounded by a ring of 3 or occassionally 4 cells.

Embryology. Mature pollen grains trilobate, tricolporate and at shedding stage, 2-celled with exine thickened at the poles (Subramanyam 1970c); surface without any sculpturing. Ovules anatropous, unitegmic, tenuinucellate. Embryo sac of Polygonum type, 8-nucleate at maturity. A peculiar feature is the micropylar protrusion of the embryo sac. Endosperm of the Cellular type; micropylar haustoria present.

Important Genera and Economic Importance. The only genus, *Pentaphragma* has no economic importance.

Taxonomic Considerations. The genus *Pentaphragma* is considered to be a member of the family Campanulaceae by Bentham and Hooker (1965b), Rendle (1925) and Hutchinson (1969, 1973). Airy Shaw (1941, 1954) advocated the creation of a new family, Pentaphragmataceae, as it shows resemblances to Begoniaceae, Rubiaceae, Gesneriaceae and Cucurbitaceae in vegetative features, and to Boraginaceae and Hydrophyllaceae in its inflorescence structure. However, he could not decide about its placement. Melchior (1964) placed this family in the order Campanulales, the embryological features support this position. According to Dunbar (1978), palynologically *Pentaphragma* should be placed in a separate family and not in Campanulaceae. Cronquist (1968, 1981) and Takhtajan (1980, 1987) also agree to treat Pentaphragmataceae as a distinct family in the order Campanulales. Dahlgren (1980a, 1983a), however, treats the genus *Pentaphragma* as a member of the family Campanulaceae.

In view of its resemblances to various families, and also some unique embryological features (such as extra-micropylar embryo sac, and uninucleate micropylar endosperm haustorium), and the palynological features, it is appropriate to retain this family as a distinct one, in the order Campanulales.

Goodeniaceae

A small family of 17 genera and 300 species distributed in the South Pacific region of Australia.

Vegetative Features. Perennial herbs or sometimes shrubs without any latex system. Leaves alternate, rarely opposite or basal, simple, estipulate.

Floral Features. Inflorescence variable: racemose, cymose or sometimes capitate, paniculate or even solitary (Fig. 53.4 A) as in *Selliera radicans*. Flowers hermaphrodite, usually zygomorphic, actinomorphic in *Selliera*, hypo- or perigynous. Calyx of 5 sepals, fused below, free above, imbricate; corolla of 5 petals, sometimes bilabiate, valvate or induplicate-valvate. Stamens 5, alternate with corolla lobes (Fig. 53.4 B), free or adnate to corolla base; anthers coherent and form a cylinder around the style, sometimes syngenesious; anthers bicelled, introrse, longitudinally dehiscent. Gynoecium syncarpous, bicarpellary, ovary uni- or bilocular, inferior or half-inferior, rarely superior as in *Velleia*; placentation basal or axile, ovules 1, 2 or many. Style 1, filiform, stigma simple or 2- to 3–times branched, ending in a cup-like structure for collecting pollen. Fruit a capsule, berry, drupe or a nut. Seeds with fleshy endosperm and straight embryo.

Anatomy. Vessels solitary, perforations simple; parenchyma vasicentric, diffuse. Rays up to 7 cells wide, high, heterogenebus. Fibres with bordered pits. Leaves dorsiventral, isobilateral or centric. Various types of glandular and nonglandular hairs recorded. Stomata mostly of Ranunculaceous type, sometimes Rubiaceous type; distributed on both the surfaces or only on the lower surface.

Embryology. Pollen grains subprolate or prolate-sphaeroidal, usually 3-colporate with thick exine and

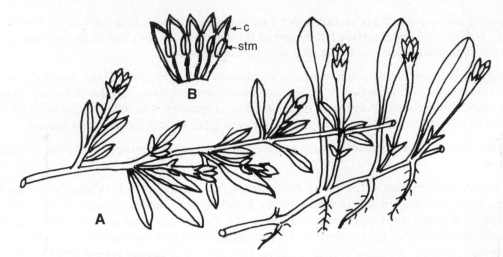

Fig. 53.4 Goodeniaceae: **A, B** *Selliera radicans*. **A** Flowering twigs. **B** Corolla (opened). *c* corolla, *stm* stamen. (Adapted from Hutchinson 1969)

shed at 2-celled stage. Ovules anatropous, unitegmic, tenuinucellate; hypostase present. Polygonum type of embryo sac, 8-nucleate at maturity. Endosperm formation of the Cellular type, haustorium absent.

Chromosome Number.   The basic chromosome number is x = 7–9.The haploid number is n = 8 in *Scaevola lobelia* and *S. sericea.*

Chemical Features.   The members of Goodeniaceae are rich in anthocyanidins including delphinidin and O-methyl derivatives. 3–O-methylflavonols are present (Gornall et al. 1979). The two genera *Scaevola suaveolens* and *Selliera radicans* contain seco-loganins (Jensen et al. 1975). Most members have inulin in the vegetative parts.

Important Genera and Economic Importance.   The largest genus of this family, *Scaevola*, is widely distributed in Australia, Asia, and Africa. *Selliera radicans* (Fig. 53.4 A), a herb, has a remarkable discontinuous distribution on both sides of the Pacific Ocean. It grows on the moist sandy soils of the sea-shore. *Calogyne* is another interesting genus with 2- to 3-lobed style. In *Leschenaultia* the indusium is deeply bilabiate. There is no economic importance.

Taxonomic Considerations.   Bentham and Hooker (1965b), Hallier (1912) and Bessey (1915) treated Goodeniaceae as a member of the order Campanulales. Hutchinson (1969, 1973) treated Goodeniaceae along with Brunoniaceae and Stylidiaceae as members of the order Goodeniales and placed it next to the Campanulales. Wagenitz (1964) and Thorne (1968) considered it as a member of the order Campanulales, and Cronquist (1968) retained it in the Campanulaceae itself. Dahlgren (1980a), Gornall et al. (1979), and Takhtajan (1980,1987) placed Goodeniaceae in a separate order, Goodeniales.

  The family Goodeniaceae does not deserve the rank of a separate order, but should be retained in the order Campanulales.

Brunoniaceae

A unigeneric family with only one species, endemic to Australia—*Brunonia australis*.

Vegetative Features. Plants perennial, herbaceous, about a foot high, with green, herbaceous, shortened stem. Leaves radical, entire, oblanceolate or spathulate, estipulate, in basal rosettes; latex absent.

Floral Features. Inflorescence a scapose capitate cyme or involucrate head. Flowers bracteate, ebracteolate, hermaphrodite, actinomorphic, hypogynous. Calyx lobes 5, gamosepalous, sepals subulate, margin densely hairy; corolla lobes 5, gamopetalous, free above, spathulate, valvate. Stamens 5, epipetalous, fused near the base; anthers coherent, introrse, filaments free. Gynoecium monocarpellary, ovary unilocular, with basal placentation, superior, ovule 1; style 1, stigma cup-like to store pollen grains. Fruit an one-seeded achene enclosed in persistent calyx. Seeds without endosperm.

Anatomy. Similar to that of the Goodeniaceae.

Embryology. Embryology has not been fully investigated. Pollen grains 2-celled at dispersal stage. Ovules anatropous, unitegmic, tenuinucellate. Polygonum type of embryo sac, 8-nucleate at maturity. Endosperm formation of the Cellular type. A multinucleate suspensor haustorium develops in *Brunonia australis* (Johri et al. 1992).

Chromosome Number. The basic chomosome number is x = 9.

Important Genera and Economic Importance. The only genus is *Brunonia* with a single species— *B. australis*. It is commonly known as 'blue pincushion' because of its capitate inflorescence of blue flowers, and is often cultivated as a hardy perennial.

Taxonomic Considerations. The earlier workers treated Brunoniaceae as a subfamily of Goodeniaceae. Even Dahlgren (1980a,1983a) includes this family in Goodeniaceae of the order Goodeniales, but most of the recent authors give it the rank of a separate family. Lawrence (1951), Melchior (1964) and Cronquist (1968, 1981) treat Brunoniaceae as a member of the order Campanulales and near Goodeniaceae, Hutchinson (1969, 1973) and Takhtajan (1980) include Brunoniaceae in Goodeniaceae and place it in the order Campanulales. Takhtajan (1987), however, raises it to the rank of a family and along with Goodeniaceae includes it in the order Goodeniales.

The family Brunoniaceae has many features common with the Goodeniaceae: (a) the indusiate pollen-collecting cup, (b) absence of milky sap, (c) similar anatomical features, (d) syngenesious anthers, and (e) basal, anatropous ovule. The differences are: (a) in their habit, (b) inflorescence, and (c) ovary inferior in Goodeniaceae, and superior in Brunoniaceae.

In view of these features, Brunoniaceae should be treated as a separate family and placed with Goodeniaceae in the same order, Campanulales.

Stylidiaceae

A small family of 5 genera and 164 species. Stylidiaceae is a distinctively Australian family: 134 of the 137 valid species of *Stylidium* occur in Australia, 129 are endemic. *Levenhookia* with 10 species is also endemic to Australia. Of the three smaller genera, *Phyllachne* and *Forstera* are represented by one species each in Tasmania. Only the monotypic *Oreostylidium* is absent from Australia. Species of *Forstera* (except *F. bellidifolia*), *Oreostylidium*, and two species of *Phyllacne* are endemic to New Zealand. *P. colensoi* occurs in Tasmania; *P. uliginosa* is endemic to Tierra del Fuego, in South America. The genus *Stylidium* extends beyond Australia only to the north and west: New Guinea, Indo-Malaysia and southeastern Asia (Carlquist 1969). Several species of *Phyllachne* are distributed in temperate South America across the South Pacific ocean (Subramanyam 1951).

The occurrence of *Stylidium* species in Indo-Malaysian region and India is due to long-distance dispersal from Australia. The long disjunction between southwestern and southeastern Australia for many species of *Stylidium* (such as *S. brachyphyllum, S. inundatum, S. beaugleholei*) and *Levenhookia dubia* and *L. pusilla* is also due to the same reason (Carlquist 1979), although Schuster (1976) speculates that Stylidiaceae might have been in India in the Upper Cretaceous. The members mostly grow on acidic and mineral-poor soils and often in sandy heaths.

Vegetative Features. Small herbs or undershrubs, more or less xerophytic and without latex; often perennial with short creeping rhizomes as in *S. squamosotuberosum.* Leaves usually in basal rosettes, simple, entire, estipulate, almost grass-like. Successive rosettes may be separated by a slight leafy piece of stem.

Floral Features. Inflorescence mostly scapose (Fig. 53.6A), racemose or corymbose. Flowers bracteolate, hermaphrodite or unisexual and then plants usually monoecious; zygomorphic, epigynous. Numerous stalked multicellular glands occur on inflorescence axis, bracts, sepals, petals and outer wall of ovary. Calyx of usually 5 sepals, sometimes 4–9, polysepalous or connate into a bilabiate cup-like structure; odd sepals posterior, often covered with stalked viscid glandular hairs. Corolla lobes 5, polypetalous, the anterior lobe or the labellum is either larger or smaller than other petals. Stamens 2, posterio-lateral, rarely three, adnate to the style to form a gynostegium; the anthers, therefore, are apparently sessile, one lying on each side just below the stigma. Stamens distinct and free in *Donatia*; free from corolla; anthers bicelled, extrorse, dehiscence longitudinal. Gynoecium syncarpous, bicarpellary (Fig. 53.6B), but sometimes the posterior carpel primitive; ovary inferior, bilocular in basal part, unilocular above. Placentation axile or free-central type, style elongated, stigma lobed (Fig. 53.6 C).

Two conspicuous nectaries, the anterior one larger, are present at the base of the column and immediately above the inferior ovary as in Lobeliaceae (Subramanyam 1949) and Goodeniaceae (Brough 1925).

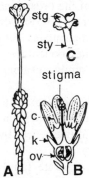

Fig. 53.6 Stylidiaceae: **A-C** *Forstera sedifolia*. **A** Twig bearing flower. **B** Flower, vertical section. **C** Stigmatic lobes. *c* corolla, *k* calyx, *ov* ovary, *stg stigma*, *sty* style. (Adapted from Hutchinson 1969)

Anatomy. Continuous circles of xylem and phloem occur in *Donatia, Forstera* and *Phyllachne;* vessels arranged in short radial rows. In *D. novae-zelandie* the phloem is in the form of strands; xylem comprises spiral, reticulate and annular tracheids and vessels with simple peforations. Leaves small, overlapping and adpressed. Stomata confined to the lower surface, more abundant towards the apex and may be Ranunculaceous or Rubiaceous type. Glandular hairs abundant and secrete mucilage.

Embryology. Mature pollen grains oval, (4)–5–(6)–colpate or –colporate and with 6 prominent bands around it; exine thick, closely and minutely echinulate; 3-celled at shedding stage and often germinate in situ, e.g. *Levenhookia* and *Stylidium.* Ovules anatropous, unitegmic, tenuinucellate. Polygonum type of embryo sac, 8-nucleate at maturity. Endosperm formation of the Cellular type; both micropylar and chalazal endosperm haustoria develop (Subramanyam 1970d).

Chemical Features.   Iridoids, monotropein and vaccinoside occur in *Stylidium* (Jensen et al. 1975). Tannins and leucoanthocyanins also occur (Hegnauer 1973). Inulin is reported.

Important Genera and Economic Importance.   Of the 6 genera comprising this family, *Stylidium* is the largest with about 137 species (Carlquist 1969) and have basal rosettes of grass-like leaves. *Donatia* and *Phyllachne* are more like herbs, with overlapping adpressed leaves and small axillary flowers borne near the apex of the short axis. Creeping or erect stems of *Forstera* also bear small overlapping leaves with flowers borne on a small leafless peduncle. *Levenhookia* includes small annuals with branching in the inflorescence region. *Oreostylidium* is a small, stemless rosette plant. None of these has any economic importance.

Taxonomic Considerations.   The family Stylidiaceae is interesting because of the discontinuous distribution and peculiarities of floral features. They resemble the Lobeliaceae in various structures, both morphological and embryological and also to some extent chemical (have inulin as reserve food). However, they differ from Campanulaceae and Lobeliaceae in: (a) presence of glandular hairs, (b) absence of laticiferous ducts, and (c) presence of iridoid—monotropein and vaccinoside. Lawrence (1951), Melchior (1964) and Cronquist (1968,1981) include Stylidiaceae in the Campanulales mainly on the basis of exomorphological resemblances. Hutchinson (1969,1973) places this family in the Goodeniales. Jensen et al. (1975) and Dahlgren (1975a, b) are of the opinion that the Stylidiaceae should be placed in the order Cornales, with which it shows resemblances (except the presence of inulin and fewer stamens). Thorne (1968) includes this family in the suborder Saxifragineae of Rosales, where many families of the Cornales (sensu Dahlgren 1975 a, b) are included.

   Takhtajan (1980) agrees with earlier workers and retains Stylidiaceae in the Campanulales; but later in 1987 he raises it to a separate order altogether, Stylidiales. It is preferable to retain this family in the order Campanulales if the micromolecular structures are not given much importance.

Calyceraceae

A small family of 4 genera and ca. 40 species endemic to South American tropics (6 genera and 60 spp., according to Takhtajan 1987).

Vegetative Features.   Dwarf, annual or perennial herbs, sometimes suffrutescent. Leaves alternate, cauline or in basal rosettes, entire or pinnately lobed, estipulate.

Floral Features.   Inflorescence heads with an involucre of bracts. Flowers usually hermaphrodite, rarely unisexual, actino- or zygomorphic, epigynous. Calyx 4- to 6-lobed, valvate. Stamens 4–6, in one whorl, alternipetalous, arising from near the mouth of the corolla tube; filaments basally or wholly connate; anthers bicelled, introrse, free or basally coherent or connate around the style; dehiscence longitudinal. Gynoecium monocarpellary, ovary unilocular, with a solitary, pendulous ovule; style 1, slender, stigma capitate. Fruit an achene crowned by the persistent calyx; adjoining achenes sometimes connate as in the fruits of the outer flowers of *Acicarpha*. Seeds with copious to scanty endosperm and a straight embryo.

Anatomy.   Anatomical studies of this family are incomplete. Vascular bundles are in a ring separated from each other by broad rays in *Acicarpha:* sometimes embedded in a ring of mechanical tissue as in *Boöpis.* Vessels moderately wide, circular with simple perforations and with bordered pits wherever in contact with the ray parenchyma; xylem fibre elements with broad lumen, thick walls and simple pits. Clustered crystals occur in cortex and pith of *Acicarpha.*

Embryology.   Pollen grains 2-celled at dispersal stage. Ovules anatropous, unitegmic, tenuinucellate. Polygonum type of embryo sac, 8-nucleate at maturity. Endosperm formation of the Cellular type.

Chromosome Number. The basic chromosome number is x = 8, 15, 18, 21.

Chemical Features. Calyceraceae contains simple seco-iridoids like seco-loganin and seco-loganic acid, foliaxenthin, cantleyoside and other derivatives (Gornall et al. 1979).

Important Genera and Economic Importance. None of the members (*Acicarpha, Boöpis, Calycera, Moschopsis*) has any economic importance.

Taxonomic Considerations. The family Calyceraceae resembles the family Compositae, particularly because of their inflorescence type, inferior, unilocular ovary and seemingly syngenesious stamens. The Calyceraceae have variously been referred to the Campanulales (Lawrence 1951, Melchior 1964), or Dipsacales (Cronquist 1968, 1981, Thorne 1968, Jensen et al. 1975), or Valerianales along with Dipsacaceae and Valerianaceae (Hutchinson 1969, 1973), or even to a separate order, Calycerales (Takhtajan 1987).

The pollen of the Calyceraceae is much like that of the Brunoniaceae and Goodeniaceae. The anthers surround the style and their introrse dehiscence is typical of the Campanulales members. The filaments of Calyceraceae are attached well up in the corolla tube, as in the Dipsacales and not at the base of the corolla, as in the Campanulales. Embryologically, the three families—Calyceraceae, Dipsacaceae and Compositae are closely related (Poddubnaja-Arnoldi 1964), but binucleate pollen grains, alternate leaves and centripetal (racemose) inflorescence of the Calyceraceae are all out of harmony with the rest of the Dipsacales.

Takhtajan (1969) excludes this family from Campanulales and Dipsacales because of its 2-celled pollen gains (3-celled in Dipsacales) and pendulous ovules and places it in a separate order Calycerales.

According to Jensen et al. (1975), the presence of simple seco-iridoids is not in accord with a position in or close to Campanulales or Asterales, but agrees well with the conditions in the orders of Gentianiflorae (Goodeniales, Oleales and Gentianales). It agrees with Dipsacales in having unilocular ovaries with 1 pendulous ovule, and disagrees in having alternate leaves, absence of glandular hairs, Asteraceae-like inflorescences and 2-celled pollen grain. Therefore, it would be appropriate to place this family in a separate order, Calycerales, close to the family Asteraceae.

Compositae (= Asteraceae)

It is the largest angiosperm family including 1250 to 1300 genera and 20 000 to 25 000 species (Takhtajan 1987) distributed all over the world and in almost all habitats.

Vegetative Features. Members are much diversified, may be annual or perennial; xerophytes, succulents or normal mesophytes; herbs, shrubs or less commonly trees or climbers. *Espeletia hartwegiana,* and *Senecio johnstonii.* from Kilimanjaro in Africa and *Vernonia arborea* are trees and *Mikania scandens* is a climber. *S. praecox* and *S. longiflorus* from southwest Africa are stem succulents. *Megalodonta beckii* is an aquatic; *S. hydrophilus* grows in wet ground or even in brackish water (Small 1919).

Normal tap root, branched and fibrous; root tubers produced in *Dahlia, Helianthus tuberosus* and *H. maximiliani.* Stem soft, erect or prostrate, rarely climbing, sometimes woody, usually hairy, often with milky or coloured sap (*Launaea, Sonchus*). Sometimes adventitious roots borne on stem surface, e.g. *Tagetes.* Leaves radical as in *Cichorium* or cauline, alternate, rarely opposite, as in *Dahlia*, estipulate, simple or pinnatisect, e.g. *Tagetes*; smooth or hairy.

Floral Features. Inflorescence, a capitulum consists of a few or large number of sessile flowers arranged on the variously shaped receptacle and surrounded by one or more than one whorl of involucral bracts, that are protective in function; the receptacle may be flat disc-like, convex, concave, conical or cylindrical. Florets are either tubular or strap-shaped, i.e. ligulate. A receptacle may comprise any one type of floret or of both types:

(i) with only tubular, bisexual flowers, e.g. *Ageratum*

(ii) with only ligulate, bisexual flowers, e.g. *Launaea.*

(iii) with both types, florets—outer ligulate, neutral, and inner tubular, bisexual, e.g. *Helianthus.*

Florets in the middle of the receptacle are disc florets and are mostly tubular. Each floret sessile, subtended by a scaly bract, usually hermaphrodite, actinomorphic, epigynous, pentamerous except the ovary. Calyx modified into a hairy, scaly or bristly pappus, persistent and forming a parachute-like structure in mature fruit. Corolla of 5 petals, gamopetalous, valvate. Stamens 5, epipetalous, syngenesious; anthers basifixed, connate or coherent forming a staminal column around the style, introrse, longitudinally dehiscent. Gynoecium syncarpous, bicarpellary, ovary unilocular, with one ovule on basal placenta, inferior; style 1, filiform, passing through the staminal column, stigma bifid, coiled. Fruit an achene or cypsella usually with the persistent pappus attached to it. Seeds without endosperm and with a straight embryo.

Ray florets on the margin of the receptacle are ligulate. Calyx similar to that of tubular flowers. Corolla of 5 petals, gamopetalous, zygomorphic and ligulate. Stamens and pistil similar to those of the tubular flowers. Mostly, the ray florets are neutral, i.e. without any stamen or pistil.

Anatomy. The anatomical structure of the family shows considerable diversity in correlation with their habit differences. Herbaceous stems in a transverse section usually exhibit a ring of collateral vascular bundles that are accompanied by pericyclic or bast fibers. Vessels smaller, often with radial multiples of 4 or more, sometimes with spiral thickenings; perforation plates simple and horizontal; parenchyma sparse and paratracheal. Rays 4–10 cells wide, uni- or multiseriate. Hairs both glandular and nonglandular. Presence of resin canals, secretory cavities and laticifierous vessels reported in various genera.

Leaf generally dorsiventral but variations occur such as scale leaves in *Helichrysum* and rolled leaves in *Olearia solandri*. Stomata variously distributed, mostly Ranunculaceous type, sometimes Cruciferous; absent from submerged leaves and present on both surfaces of aerial leaves in *Megalodonta beckii*, an aquatic member. Various types of anomalous secondary growth are also reported.

Embryology. Pollen grains 3- or 4-colporate, exine spinous or echinolophate; 3-celled at anthesis. Ovules anatropous, unitegmic, tenuinucellate. Polygonum type of embryo sac, 8-nucleate at maturity. Both Polygonum and Allium type of embryo sacs are seen in *Tridax trilobata, Sanvitalia procumbens* and *Vernonia cinerascens* (Johri et al. 1992). Fritillaria and Drusa type of embryo sac formation are also reported in many members. Pyrethrum parthenifolium type is seen in *Chrysanthemum maximum* (Johri et al. 1992).

Chromosome Number. The basic chromosome number is variable: $x = 2$–19 or more in various members, $x = 9$ being the most common number.

Chemical Features. Compositae members are rich in sesquiterpenes and polyacetylenes. Seco-iridoids are absent. An oleoresin produced by secretory ducts of *Artemisia campestris* subsp. *maritima* contains terpenoids, alkaloids, fatty acids and polyacetylenes (Ascensão and Pais 1988).

Important Genera and Economic Importance. A number of Compositae are obnoxious weeds, such as *Ageratum conyzoides, Blumea mollis, Eclipta prostrata, Erigeron bonariensis, Gnaphalium indicum, Launaea asplenifolia, L. nudicaulis, Parthenium hysterophorous, Sonchus asper, Sphaeranthus indicus, Tridax procumbens, Xanthium strumarium* and others.

Because of the large, showy capitula many genera are cultivated as ornamentals: *Aster, Calendula, Chrysanthemum, Dahlia, Helianthus, Helichrysum, Gerbera, Tagetes* and others. Compositae are also important for their food value. Many genera, e.g. *Scolymus hispanicus* in Spain, *Scorzonera hispanica* from Europe to Central Asia, and *Tragopogon porrifolius* of southern Europe are grown for their edible roots (Datta 1988). The roots of *Cichorium intybus* (commonly called chicory) are used as an adulterant of coffee in powdered form. Leaves of *Lactuca sativa* (lettuce) and *Cynara cardunculus* (cardoon) as also tubers of *Helianthus tuberosus* (Jerusalem artichoke) are edible.

The family is the source of a large number of drug plants also. Santonine used against intestinal worms is obtained from *Artemisia cina, A. maritima* and *A. nilagarica. Matricaria chamomilla* is the source of the medicine chamomile. Dried flower heads of *Chrysanthemum cinerariefolium* are the source of the insecticide Pyrethrum.

Oil is obtained from the seeds of *Carthamus oxycantha, Guizotia abyssinica* and *Helianthus annuus.* Latex of *Parthenium argentatum* of South America, *Solidago laevenworthii* and *Taraxacum kok-saghys* (in Russia) yield rubber.

Some of the tree members like *Brachylaena huillensis* and *Montanoa quadrangularis* (Colombia and Venezuela) and *Vernonia arborea* (Assam, India) furnish timber. The wood of *Tarchonanthus camphoratus* (South Aftrica) is used for musical instruments.

A red dye is extracted from the flowers of *Carthamus tinctorius, Tagetes erectus* and *Adenostemma tinctorium.*

A number of genera growing as weeds in pastures have poisonous effect on livestock, e.g. *Senecio, Xanthium* and *Solidago*. The pollen grains of *Ambrosia artemisifolia, A. trifida* and *Solidago* spp. are resonsible for hay fever.

Taxonomic Considerations.   Bentham and Hooker (1965b) placed Compositae in Asterales of Series Inferae in the Gamopetalae. Hutchinson (1969,1973), Cronquist (1968,1981), Dahlgren (1980a) and Takhtajan (1987) also agree to this assignment. However, Engler and Diels (1936) and Melchior (1964) include Compositae in Campanulales.

There has been diversified opinion about the phylogeny and evolution of this family. All taxonomists agree that it is the most advanced taxon of the dicots. It indicates similarity with the families Dipsacaceae and Valerianaceae, because of the similar type of inflorescences, inferior ovary, and solitary ovule, and also some embryological features like the embryo development (Asterad type in *Valeriana alitoris, Centranthus ruber, C. angustifolia* of Valerianaceae and *Scabiosa succisa* of Dipsacaceae), and presence of endothelium (Deshpande 1970). However, the pollen grains of Compositae are more akin to those of Brunoniaceae, Calyceraceae and Goodeniaceae and differ from those of Campanulaceae, Dipsacaceae, Stylidiaceae and Valerianaceae (G. Erdtman 1952).

Many taxonomists are of the opinion that the Compositae have evolved from the Campanulaceae through the subfamily Lobelioideae, but Campanulaceae differ from the Compositae in dichasial or monochasial cyme inflorescence, pentacarpellary (Campanuloideae) or bicarpellary (Lobelioideae) ovary, axile placentation and numerous ovules. On the other hand, their pollen presentation mechanism is similar. Also, both families contain polyacetylenes (Jenson et al. 1975, Dahlgren 1980a). According to Cronquist (1968,1981): "Only the Rubiales-Dipsacales complex has the characters necessary for a near ancestor of the Compositae". Comparative morphological studies show that the ancestral prototypes of Compositae must have been woody plants. Similar pollen dispersal mechanism is seen in some Rubiales (Rubiaceae) also. In many Rubiaceae, capitulum inflorescence occurs. Phenolic compounds widespread in Compositae are also present in many Rubiales and Dipsacales but not in Campanulales.

Stebbins (1977) concluded: "The Compositae cannot be regarded as descended from or closely related to any other modern family".

From the above discussion it is clear that the Compositae is no doubt the most highly evolved family and in all probability is polyphyletic with more than one ancestral form.

The important features of this family are:

1. Mainly herbaceaous; shrubs and trees make up for only 1.5% taxa.
2. 25000 species are approximately 10% of the dicots and occur in every conceivable spot on the earth's surface, though somewhat rare in tropical rain forests.
3. Flowers in a capitate inflorescence are a highly advantageous situation. A single insect can pollinate several flowers at a time. If, however, cross-pollination fails, the curling back of stigmas helps in self-pollination.
4. Pollen mechanism, pollen protection and nectar placement are such that any specialised insect is not required.
5. The ripening fruits are protected by the involucral bracts which bend inwards. A perfect mechanism for seed dispersal by wind is the presenc of hairy pappus.

Taxonomic Considerations of the Order Campanulales

The circumscription of this order is very variable. Companulales, according to Melchior (1964), comprises 8 families—Campanulaceae, Sphenocleaceae, Pentaphragmataceae, Goodeniaceae, Brunoniaceae, Stylidiaceae, Calyceraceae and Compositae. Bentham and Hooker (1965b) considered only Campanulaceae and Goodeniaceae as members of the order Campanulales. The families Sphenocleaceae and Pentaphragmataceae were included in the Campanulaceae, Brunoniaceae in Goodneniaceae and Stylidiaceae in Capparales. Cronquist (1968,1981) does not include Calyceraceae and Asteraceae in his Campanulales. He removes Calyceraceae to Dipsacales (1968) and later to Calycerales (1981), and Asteraceae to Asterales. Hutchinson (1969,1973) places Calyceraceae in Valerianales along with Dipsacaceae and Valerianaceae; Goodeniaceae, Brunoniaceae and Stylidiaceae together in Goodeniales, and Asteraceae in Asterales. His Campanulales, therefore, contained only Campanulaceae and Lobeliaceae; and Sphenocleaceae and Pentaphragmataceae were included in Campanulaceae. Dahlgren (1980a) treats Pentaphragmataceae, Campanulaceae (incl. Sphenocleaceae) and Lobeliaceae as members of his Campanulales, and Asteraceae as a member of the Asterales. He removed Goodeniaceae (incl. Brunoniaceae) to the order Goodeniales in Gentianiflorae because of similar chemical features (they contain iridoids not known to occur in Campanulales).

In this classification, the Stylidiaceae has a unique position, i.e. in the order Cornales, According to Dahlgren, these 3 families are placed incorrectly in the order Campanulales on the basis of superficial resemblances. About the position of the Calyceraceae, Jensen et al. (1975) proposed placing it in Dipsacales.

Takhtajan (1987) places Calyceraceae in a separate order, Calycerales. The anther filaments in Calyceraceae are attached well up in the corolla and not at the base. It also differs from other Campanulales in some embryological and chemical features. The suggestion of Dahlgren (1980a) to place it in Dipsacales was rejected by Takhtajan because of alternate leaves, lack of glandular hairs, the Asteraceae-like inflorescence and binucleate pollen grains (pollen grains 3-nucleate in Dipsacales). Takhtajan's Campanulales, therefore, includes only 3 of the families originally included because he separated Stylidiaceae to Stylidiales, Goodeniaceae and Brunoniaceae to Brunoniales, and Asteraceae to Asterales.

So much division, however, is not really necessary. There are also a number of features that are common to all these families, and on this basis, they may be retained in one order, Campanulales, as has been done by Melchior (1964).

Monocotyledons

(+ Concluding Remarks)

54

Order Helobiae

The order Helobiae comprises 4 suborders and 9 families:

1. Suborder Alismatineae—Alismataceae and Butomaceae.
2. Suborder Hydrocharitineae—Hydrocharitaceae.
3. Suborder Scheuchzeriineae—Scheuchzeriaceae.
4. Suborder Potamogetonineae—Aponogetonaceae, Juncaginaceae, Potamogetonaceae, Zannichelliaceae and Najadaceae.

Most of the members are aquatic herbs, submerged or of marshy habitats. Flowers solitary or in simple or compound inflorescences, more or less enclosed in a spathe; uni- or bisexual, regular, naked or with a single or double perianth. Stamens and carpels one to numerous, carpels superior or inferior, usually free. Seeds non-endospermous, embryo large, with a strongly developed hypocotyl. Oxalate raphides and silica bodies absent, vessels absent from stems. Pollen grains 3-celled at dispersal stage.

Alismataceae

A family of 13 genera and 100 species, cosmopolitan, but more common in temperate and tropical regions of the northern hemisphere.

Vegetative Features. Perennial or annual herbs, mostly grow in freshwater swamps and streams. Plants cauline and erect, stem generally a short, thick rhizome, but in *Sagittaria*, runners arise from the rhizome ending in tubers by which the plant propagates. Stem slender and floating in *Elisma*. Roots fibrous from stout rhizome. Leaves radical, erect, floating or submerged, long-petioled, sheathing basally; leaf blades variable from linear to strap-shaped to ovate or oblong; or when above water, with sagittate or hestate bases (e.g. *Sagittaria sagittifolia,* Fig. 54.1A); venation reticulate with few to many primary parallel veins converging apically and numerous close and parallel transverse veins; small, delicate, linear scales in leaf axils.

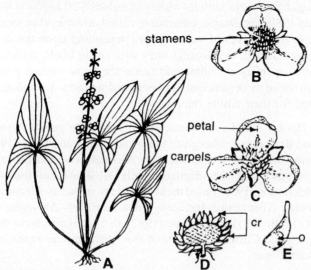

stamens

petal

carpels

cr

Fig. 54.1 Alismataceae: **A–E** *Sagittaria sagittifolia.* **A** Flowering plant. **B** Staminate flower. **C** Pistillate flower. **D** Vertical section (perianth removed). **E** Carpel. *cr* carpel, *o* ovule. (Adapted from Lawrence 1951)

Floral Features.   Inflorescence usually much branched, primary branching racemose, secondary often cymose; in *Alisma plantago* whorls of branches are borne on the inflorescence axis, on the ultimate branchlets the stalked flowers are borne. *Elisma* has a few flowers only, each on a long stalk, at the nodes of the slender floating stem. Flowers bracteate, ebracteolate, bi- or unisexual as in *Sagittaria* (Fig. 54.1B, C), pedicellate, actinomorphic. Calyx of 3 sepals, green, herbaceous, persistent, imbricate; corolla of 3 petals, larger, petaloid, deciduous, imbricate. Stamens 6 or more, rarely 3, free; anthers bicelled, extrorse, dehiscence by longitudinal slits. Gynoecium of 6 or more carpels, apocarpous; ovaries spirally arranged (Fig. 54.1D) or only in a single whorl, superior, unilocular; ovules solitary (Fig. 54.1E), rarely 2 or more as in *Damasonium,* in each carpel. Styles highly reduced or absent, acicular, persistent, with simple, terminal stigmas, scarcely distinguishable from style. Fruit a group of achenes, rarely follicular and dehiscing basally. In *Caldesia* and *Limnophyton* the 1-seeded fruits have a woody endocarp. Seeds non-endospermous, embryo horseshoe-shaped.

Anatomy.   Intercellular laticifer passages contain an oil emulsion in stem and leaf tissues. In the rhizome of *Alisma plantago* and tubers of *Sagittaria,* the laticifers form a network that is closely connected with the vascular bundles. Vessels absent in stem but vessels with simple perforation plates present in roots. Sieve tube plastids consistently have triangular protein bodies. Stomata either missing or paracytic (Rubiaceous).

Embryology.   Pollen grains pantoporate; exine reticulate, shed at 3-celled stage (Arguie 1975). Ovules anatropous, bitegmic, tenuinucellate; inner integument forms the micropyle. Allium type of embryo sac, 8-nucleate at maturity. Endosperm formation of the Helobial type with a small 1- or 2-nucleate chalazal chamber, or of the Nuclear type as in *Alisma, Damasonium, Elisma* and *Machaerocarpus*. Embryo curved, horseshoe-shaped.

Chromosome Number.   Basic chromosome number is x = 5–13.

Chemical Features.   Anthocyanin pseudobases absent. Oxalate raphides totally absent. C–glycoflavones, flavonols and flavonoid sulfates are common in leaf extracts.

Important Genera and Economic Importance.   Some of the important genera are *Alisma* (10 species) in north temperate regions and Australia, *Limnophyton* (3 species) distributed in tropical Africa, Madagascar, India to Indo-China, Java and Timor, *Caldesia,* with 3 of its 4 species, in the plains and marshes of India, *Echinodorus* (25 species) in tropical America and *Sagittaria* with its widely distributed 20 species (Jones and Luchsinger 1987, Kumar and Subramanian 1989). *Sagittaria*, commonly called arrowhead or swamp potato, is a water plant with a stout rhizome, bearing leaves of various types depending upon the depth of water. The fully submerged leaves are ribbon-shaped, the floating ones with ovate blade, and those projecting above the surface of water are sagitate; rhizomes edible and hence the name swamp potato. Some species of *Alisma* and *Sagittaria* are cultivated as ornamentals in pools and aquaria. *S. sagittifolia* and other Eurasian species are often cultivated for their edible rhizome.

Taxonomic Considerations.   According to Hooker (1894), Alismataceae comprise the genera *Alisma, Limnophyton, Sagittaria, Wisneria, Butomus* and *Butomopsis*. Marcgraf (1936) and Engler and Diels (1936) erected two suborders—Alismatineae and Butomineae, the former includes Alismataceae and the latter Butomaceae and Hydrocharitaceae. Rendle (1904) split the original family Alismataceae into two subfamilies—Alismoideae and Butomoideae. Hutchinson (1959) raised them to the rank of orders Alismatales and Butomales. Takhtajan (1966) recognised only Alismatales, including three families—Alismataceae, Butomaceae and Hydrocharitaceae. Cronquist (1968, 1981) and Thorne (1983) also recognise these three families. Dahlgren (1980a, 1983a) and Dahlgren and Clifford (1982) treat the Alismataceae as the only family of the Alismatales in superorder Alismatiflorae.

Floral anatomy (Singh and Sattler 1972) reveals that the flowers of Ranunculaceae are different from those of Alismataceae and there is no close link between the two, as presumed by Takhtajan (1966) and Cronquist (1968).

With biseriate perianth, apocarpous ovaries, and mature non-endospermous seeds, these aquatic or semiaquatic plants are the most primitive amongst the monocots. The family is very closely related to Limnocharitaceae, especially to the genus *Limnocharis*.

Butomaceae

A monotypic family of the only genus *Butomus* with only one species *B. umbellatus* distributed in temperate Eurasia.

Vegetative Features.   Aquatic herbs with creeping rhizome, from which spring erect, linear, twisted leaves with sheathing base.

Floral Features.   Inflorescence a scape, cymose umbel (Fig. 54.2A) with a involucre of bracts. Flowers bisexual (Fig. 54.2B), regular, trimerous, hypogynous or very slightly epigynous. Perianth lobes 6, free, imbricate, subequal, arranged in 2 series of 3 each, both whorls coloured, outer whorl sometimes with greenish tinge at the tips. Stamens 9, outer whorl of 6, inner whorl of 3; filaments flattened with basifixed, bithecous anthers opening by lateral slits. Gynoecium apocarpous, 6-carpellary; the 6 carpels appear to form a single whorl in a mature flower and have abaxially a common base for a short distance. The ventral margins of each pistil never fuse but are held together in the distal portion by interlocking hairs. Ovaries unilocular with numerous ovules scattered over the inner surface of the ovary wall except the margins and dorsal suture (Fig. 54.2C). No distinct style, margins of the pistils stigmatic in the upper portion (Fig. 54.2B, C) (Singh and Sattler 1974). Fruit a group of almost free, 6, turgid, leathery follicles which dehisce adaxially by pulling apart of the ventral margins. Seeds numerous, non-endospermous, embryo straight.

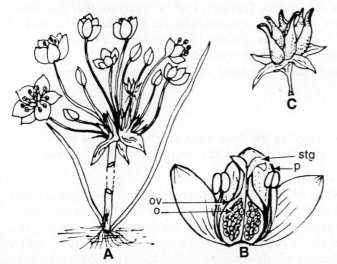

Fig. 54.2 Butomaceae: **A–C** *Butomus umbellatus.* **A** Flowering plant. **B** Flower, longisection. **C** Follicles. *o* ovule, *ov* ovary, *p* perianth, *stg* stigma. (Adapted from Lawrence 1951)

Anatomy.   Similar to that of the members of Alismataceae but without secretory canals.

Embryology.   Pollen monosulcate, exine sculpturing reticulate (Zavada 1983); oblate to suboblate (Arguie 1971); 3-celled at dispersal stage. Ovules anatropous, bitegmic with the inner integument forming the micropyle. Allium type of embryo sac, Polygonum type in *Butomus umbellatus* (Johri et al. 1992), 8-nucleate at maturity. Endosperm formation of the Helobial type.

Chromosome Number.   Basic chromosome number is x = 13.

Chemical Features.   Flavonols present, and also flavone-sulphates.

Important Genera and Economic Importance.   The only genus *Butomus* has a single species, *B. umbellatus.* It is an aquatic herb with attractive flowers, having petaloid perianth. No economic importance is reported.

Taxonomic Considerations.   The genus *Butomus* was included in the Alismataceae by Hooker (1894), suborder Butomineae of order Helobiae by Marcgraf (1936) and Engler and Diels (1936), and in the subfamily Butomoideae of Alismataceae by Rendle (1904). Hutchinson (1959, 1969, 1973) recognised a separate order Butomales for the family Butomaceae. Cronquist (1981), Thorne (1983) and Takhtajan (1987) include Butomaceae in the order Alismatales. Dahlgren (1980a, 1983a) and Dahlgren and Clifford (1982), however, treat this family as a member of the order Hydrocharitales.

According to Takhtajan (1966), the most primitive type of flower is seen in the Butomaceae and Limnocharitaceae. The biseriate flowers, apocarpous ovaries, laminar placentation are some of the primitive features of this family. Many earlier workers (Hallier 1905, Schaffner 1929, 1934, Eber 1934, Takhtajan 1954, 1966, 1969, Kimura 1956, Kaul 1967, 1968a, 1969, and Moseley 1971) suggested that this family, together with Alismataceae and Limnocharitaceae, have some definite similarities with the Nymphaeales of the dicotyledons. On the basis of xylary anatomy, however, Kosakai et al. (1970) conclude that the Nymphaeaceae are very unlikely ancestors of the monocots, and in all probability the Butomaceae and Alismataceae have arisen from terrestrial forms. Singh and Sattler (1974), on the basis of floral anatomy, reported considerable differences in the early phases of floral development between *Butomus* and Nymphaeaceae. Huber (1969), Kubitzki (1972) and Tomlinson (1970) also point out that the assumed relationship between the two taxa is rather far-fetched. Dahlgren (1983a) suggests that the family Cabombaceae of the dicotyledons have many features in common with the Butomaceae: (a) lack of vessels in the stem, (b) short-lived radicle, (c) trimerous flowers, (d) monosulcate pollen grains, (e) apocarpous ovary, (f) laminar placentation, and (g) Helobial type endosperm.

It is still controversial whether the two groups have a common ancestry lost in the geological history or one group is ancestral to the other.

Hydrocharitaceae

A family of 15 genera and 106 species, distributed in the fresh-water ponds, pools and lakes, and 3 genera, *Halophila, Enhalus* and *Thalassia* in sea water all over the tropics and subtropics, and a few spreading also to temperate zones.

Vegetative Features.   Partially (*Vallisneria*) (Fig. 54.3A, B) or completely submerged (*Halophila*) plants, rarely floating (*Hydrocharis*). Stem reduced and modified into stolons; in *Hydrocharis morsus-ranae* (Fig. 54.3C), from the reduced aerial stem, spreading fibrous root and stolons arise, each ending in a rosette of leaves. Leaves variable in shape and size; a rosette of stiff, tapering leaves with a spiny margin in *Stratiotes aloides;* a slender branching stem bearing whorls of narrow toothed leaves in *Elodea canadensis* and *Hydrilla verticillata;* a reduced stem bearing a tuft of roots and a crowded cluster of long, linear, grass-like leaves in *Vallisneria* (Fig. 54.3A, B).

Floral Features.   Inflorescence lateral, subtended by a bifid spathaceous bract or a pair of opposite bracts which often persist till the fruit is ripe. Spathes sessile or pedicellate or peduncled and the

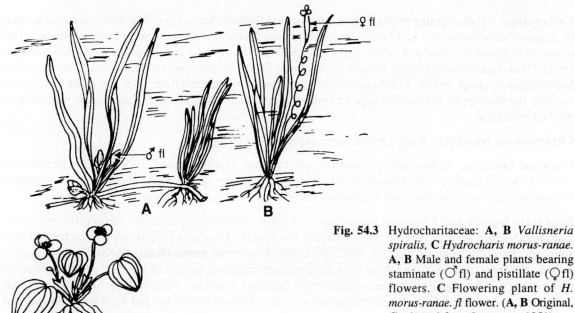

Fig. 54.3 Hydrocharitaceae: **A, B** *Vallisneria spiralis,* **C** *Hydrocharis morus-ranae.* **A, B** Male and female plants bearing staminate (♂fl) and pistillate (♀fl) flowers. **C** Flowering plant of *H. morus-ranae. fl* flower. (**A, B** Original, **C** adapted from Lawrence 1951)

peduncles coiled in some genera, e.g. *Vallisneria.* The female and bisexual flowers generally solitary, and male flowers solitary to many-flowered umbels. Only in *Halophila* are flowers of both sexes borne in the same inflorescence. Flowers usually unisexual (and then plants dioecious) or bisexual as in *Ottelia* or polygamous as in *Elodea;* actinomorphic, trimerous, epigynous. Perianth of 6 tepals, free, in two series of 3 each; the outer green and sepaloid, valvate, and inner petaloid, imbricate or convolute. Staminate flowers with 3 to many stamens in 1 to 5 whorls of which the innermost often reduced to staminodes; anthers bithecous, dehisce by parallel vertical slits; pistillodes sometimes present. Pistillate flowers with a similar perianth, have a syncarpous, 2- to 15-carpellary gynoecium. Ovary inferior, unilocular with 3 to 6 parietal placentae, which often intrude or are ill-defined and the ovules seemingly scattered over the inner surface of ovary wall, numerous. Style 1, usually divided into as many branches as the number of placentae. Staminodes occur in some (e.g. *Hydrocharis*). Fruit a berry, indehiscent, submerged or sometimes rupture irregularly as in *Hydrocharis.* Seeds numerous, endosperm absent, embryo straight.

Anatomy.   Stem (circular in transection) with a 1-layered epidermis of narrow elongated cells followed by a few subepidermal layers of collenchyma. Air cavities in the cortex are irregularly scattered (*Blyxa, Elodea*) or regularly disposed in 2–3 series with the smaller towards the centre (*Egeria, Hydrilla*). The central cylinder exhibits a central protoxylem lacuna, some metaxylem and phloem elements. Xylem and phloem show contradictory evolutionary trends (Ancibor 1979). Xylem is simple—the conducting elements are mostly spiral or reticulately thickened tracheids. Vessels have been seen only in roots—highly elongated elements with oblique, scalariform perforation plates. The leaves (*Stratiotes*), petioles (*Hydrocharis*) and stem (*Hydrilla*) have only protoxylem elements. The phloem consists of large, wide sieve-tube elements with horizontal or slightly oblique, simple sieve plates and a single companion cell per element. Stomata present only in the aerial floating leaves of *Hydrocharis, Limnobium* and *Ottelia,* and in some of the aerial leaves of *Stratiotes;* mostly of Rubiaceous type and sometimes of Ranunculaceous type.

Secretory tannin cells are present in the cortex and the central cylinder (Ancibor 1979).

Embryology.   Pollen grains of *Halophila*, often called filiform, are arranged in uniseriate chains; may be echinate or subechinate; in *Elodea, Hydrocharis* and *Stratiotes,* they are apparently acolpate; exine absent in *Vallisneria* . Shed at 3-celled stage, at 2-celled stage in *Ottelia* and *Blyxa octandra* (Lakshmanan 1961). Ovules anatropous (*Ottelia, Halophila, Stratiotes, Blyxa*) or orthotropous (*Vallisneria, Nechamandra, Lagerosiphon*) (Kaul 1968b; Lakshmanan 1970a), bitegmic, crassinucellate, with the inner integument forming the micropyle. Polygonum type of embryo sac, 8-nucleate at maturity. Endosperm formation of the Helobial type.

Chromosome Number.   Basic chromosome number is x = 7–12.

Chemical Features.   Anthocyanin pseudobases often present (Dahlgren and Clifford 1982). Gornall et al. (1979) reported flavones like luteonin/apigenin, cyanidin and/or pelargonidin, their common O-methyl derivatives, flavone bisulfates and C-glycosyl derivatives in various members of this family.

Important Genera and Economic Importance.   *Elodea, Hydrocharis, Hydrilla, Vallisneria, Halophila* and *Thalassia* are some of the important genera of this family. All genera (15) except *Ottelia* have at least some species with unisexual flowers which are borne on separate plants (Kaul 1968b). The two marine genera, *Halophila* and *Thalassia,* have only three perianth segments instead of biseriate perianth. *Hydrilla verticillata*, a submerged freshwater aquatic often becomes a serious weed in lakes and ponds. Many genera show interesting pollination mechanism, e.g. male flowers of *Elodea* and *Vallisneria spiralis* are free-floating and reach the female flowers with the water current, in *Halophila* the thread-like pollen grains are carried in a chain by ocean current, and there is explosive release of pollen by the anthers of *Hydrilla*. Datta and Biswas (1967) reported a number of form variations in *Ottelia alismoides*. The members of this family are of little or no economic importance. *Hydrilla verticillata* is often used as classwork material.

Taxonomic Considerations.   Hydrocharitaceae is distinct from the rest of the members of the Alismatales in having inferior, syncarpous ovary. Placentation is basically laminar, with the ovules scattered over the inner walls of the carpels. Laminar placentation is otherwise known in the Butomaceae, Limnocharitaceae and a few families of the Magnoliidae.

Bentham and Hooker (1965c) included Hydrocharitaceae in the order Microspermae, Engler and Diels (1936) in the Helobiae, Hutchinson (1959) in the Butomales and Thorne (1968, 1983) in the Alismatales. Cronquist (1968, 1981), Takhtajan (1969, 1980), Dahlgren (1980a, 1983a) and Dahlgren and Clifford (1982) treat this family as a member of the order Hydrocharitales, whereas Takhtajan (1987) considers it as a member of his Alismatales.

Cronquist (1968) and Takhtajan (1980) state that the Hydrocharitaceae have a common ancestor with the Butomaceae.

Embryological and chemical data support their placement with other members of the Helobiae (Johri 1970, Gornall et al. 1979).

Scheuchzeriaceae

A family of only one genus, *Scheuchzeria* with 2 species, distributed in north temperate regions.

Vegetative Features.   Slender, perennial herbs, occur in *Sphagnum* bogs; leaves alternate, linear, sheathing with ligules or squamulae intravaginalis in axils.

Floral Features.   Inflorescence terminal, bracteate racemes. Flowers actinomorphic, bisexual, greenish, anemophilous, protogynous. Perianth lobes 6 in 2 whorls of 3 each, homochlamydeous; stamens 3 + 3, with basifixed, extrorse anthers. Gynoecium 6- or 3-carpellary, shortly united below, with sessile stigmas and 2 to a few, basal, erect ovules per carpel. The carpel walls retain an open ventral suture throughout the early and mid-stages of development and even in late developmental stages, the suture does not fully fuse (Posluszny 1983). Fruit a schizocarp, with free, divaricate, 1- or 2-seeded, ventrally dehiscent follicles. Seeds non-endospermous, embryo chlorophyllous (Dahlgren 1980b).

Anatomy.   Vessels absent in leaf and stem; in root the vessels have scalariform perforation plates. Sieve tube plastids of the monocotyledonous type are common in roots.

Embryology.   Pollen grains inaperturate, globose, shed at 3-celled stage. Ovules anatropous, bitegmic crassinucellate; inner integument forms the micropyle. Polygonum type of embryo sac, 8-nucleate at maturity. Endosperm formation of the Helobial type.

Chromosome Number.   Basic chromosome number is x = 11.

Chemical Features.   Very rich in cyanogenic compounds.

Important Genera and Economic Importance.   *Scheuchzeria*, the only genus has no economic importance.

Taxonomic Considerations.   Rendle (1930) included *Scheuchzeria* in the Juncaginaceae, but most other workers recognise it as a distinct family Scheuchzeriaceae (Lawrence 1951, Melchior 1964, Cronquist 1968, 1981, Dahlgren 1975a,b, 1980a, 1983, Dahlgren and Clifford 1982, Takhtajan 1987). It has been included in different orders by different workers—in Helobiae by Engler and Diels (1936), in Najadales by Takhtajan (1980), in Zosterales by Dahlgren (1980a), and in Scheuchzeriales by Dahlgren and Clifford (1982) and Takhtajan (1987). Lakshmanan (1970b), on the basis of embryological features, presumes it to be a distinct family, Scheuchzeriaceae, close to *Triglochin* and *Lilaea,* of the Jancaginaceae. According to Takhtajan (1987), in spite of many specialised characters, this family is relatively archaic, has many features in common with the Alismatales, and has a common origin with it. One such primitive feature is that the carpel walls retain an open ventral suture throughout the early and mid-stages of development and even in late developmental stages the suture does not fully fuse (Posluszny 1983). Nevertheless, this family is closely linked with the members of the Alismataceae.

Aponogetonaceae

A monotypic family of the genus *Aponogeton* with 47 species distributed in the tropics and subtropics of Asia, Australia, Africa and in Madagascar.

Vegetative Features.   Submerged or floating (Fig. 54.5A), glabrous, laticiferous, aquatic herbs, rhizomes sympodial, tuberous or stoloniferous, usually with numerous fibrous roots. Leaves basal or radical, rarely sessile, petioles with sheathing base; usually floating, submerged in *Aponogeton fenestralis.* Leaf blade oblong to linear-lanceolate, membranous, with 3 to 7 or more longitudinal veins connected by numerous oblique, transverse veinlets. The entire leaf tissue between the veins breaks up as the leaf grows, leaving a network with holes inbetween.

Floral Features.   Inflorescence simple or 2- to 4-branched spike borne on long peduncles and project above the water surface (Fig. 54.5A), subtended by a caducous spathe. Flowers small, bisexual (Fig. 54.5B), actinomorphic, hypogynous, variously coloured. Perianth lobes usually 2 but may be 3 or 1 also, membranous, equal or unequal, normally persistent; stamens numerous, but usually 6 in 2 whorls of 3 each, free, hypogynous, persistent; filaments filiform or subulate; anthers small, bicelled, longitudinally dehiscent, basifixed (Fig. 54.5D), extrorse. Gynoecium apocarpous, 3- to 6-carpellary (Fig. 54.5B), ovary superior, unilocular and narrowed into a slender, simple stylodium; ovules 1 to 8 or more, basal (Fig. 54.5C) or 2-seriate on the ventral suture, ascending; septal nectaries present on lateral walls of free carpels. Fruits of 3 to 6 or more, free, inflated, coriaceous, beaked follicles, opening along ventral suture;

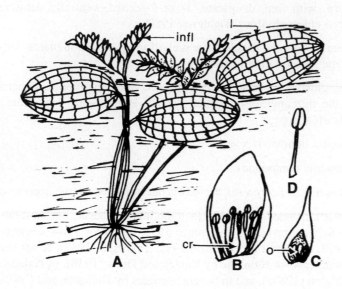

Fig. 54.5 Aponogetonaceae: **A–D** *Aponogeton distachyus.* **A** Habit (leaves float). **B** Flower. **C** Carpel, vertical section. **D** Stamen. *cr* carpel, *infl* inflorescence, *l* leaf, *o* ovule. (Adapted from Lawrence 1951)

1 to 8, erect, oblong or cylindrical, non-endospermous seeds; embryo straight, elongate, compressed or cylindrical.

Anatomy. Vessels absent in root, stem and leaves. Stomata of Rubiaceous type. Calcium oxalate crystals and silica bodies absent.

Embryology. Pollen grains subglobose or ellipsoid, monocolpate or trichotomocolpate; 2- or 3-celled at shedding stage. Ovules ana- or hemianatropous, bitegmic, crassinucellate; inner integument forms the micropyle. Polygonum type of embryo sac, 8-nucleate at maturity. Endosperm formation of the Helobial type. Embryo chlorophyllous (Dahlgren 1980b).

Chromosome Number. Basic chromosome number is x = 8.

Chemical Features. Common flavonoids present (Gornall et al. 1979).

Important Genera and Economic Importance. *Aponogeton* is the only genus. *A. natans* is an aquatic herb; its stoloniferous rootstock is edible.

Taxonomic Considerations. Bentham and Hooker (1965c) did not recognise Aponogetonaceae as a separate family and considered it under the family Najadaceae. Engler and Diels (1936) and Melchior (1964) treated it as a member of Helobiae. Cronquist (1968, 1981) and Takhtajan (1966, 1980) include this family in the Najadales, Dahlgren (1975a, 1980a) in Hydrocharitales, and Thorne (1983) in Zosterales. Takhtajan (1987) raised Aponogetonaceae to the rank of order Aponogetonales. This family is closely related to and probably originated from the Alismatales. As observed by Clifford and Williams (1980), Aponogetonaceae is similar to Butomaceae, Limnocharitaceae and Alismataceae in its reproductive biology.

Juncaginaceae

A small family of 3 genera, *Triglochin, Tetronchium* and *Cycnogeton* (Takhtajan 1987), or 5 genera including *Lilaea* and *Maundia* and 25 species, distributed in north and south temperate regions.

Vegetative Features. Annual or perennial, scapigerous herbs (Fig. 54.6A), grow in freshwater or saline marshy areas. Roots sometimes tuberous; stem highly reduced. Leaves radical, linear, with sheathing base.

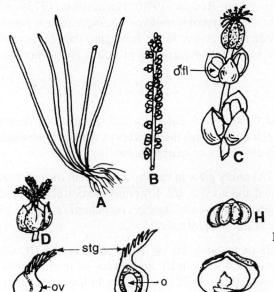

Fig. 54.6 Juncaginaceae: **A–H** *Triglochin maritimum.* A Habit. **B, C** Inflorescence. **D** Pistillate flower. **E** Carpel. **F** Vertical section of carpel. **G** Staminate flower. **H** Stamen. *fl* flower, *o* ovule, *ov* ovary, *stg* stigma. (Adapted from Lawrence 1951)

Floral Features. Inflorescence raceme or spike (Fig. 54.6B, C). Flowers small, ebracteate, actinomorphic, bi- or unisexual, the plants then dioecious or polygamodioecious, hypogynous. Perianth of 6 lobes in 2 whorls of 3 each, green, sepaloid, valvate; stamens 6 to 4, with subsessile extrorse anthers (Fig. 54.6H) and reduced filaments. Gynoecium 6- to 4-carpellary, syn- or apocarpous, ovary superior, trilocular, placentation basal or rarely apical, ovule 1 per locule. Style 1, stigma 1, plumose (Fig. 54.6D–F). Fruit cylindrical or obovate, dehiscent or indehiscent, sometimes conspicuously calcerate at base, sometimes 3 sterile carpels alternate with fertile ones. Seeds erect, non-endospermous.

Anatomy. Stems and leaves constantly devoid of vessels; in roots the vessels have scalariform perforation plates. Sieve tube plastids of monocotyledonous type reported in *Triglochin* (Behnke 1969).

Embryology. Pollen grains inaperturate, 2-celled at dispersal stage in *Triglochin striatum* (Gardner 1976) and *Lilaea subulata* (Lakshmanan 1970c). Ovules anatropous, bitegmic, crassinucellate; micropyle formed by inner integument only. Polygonum type of embryo sac, 8-nucleate at maturity. Endosperm formation of Nuclear type.

Chromosome Number. Basic chromosome numbers are x = 6, 8, 9.

Chemical Features. Common flavonoids and flavones present (Gornall et al. 1979).

Important Genera and Economic Importance. The chief genera are *Triglochin, Cycnogeton, Tetroncium, Lilaea* and *Maundia,* mostly marshy herbs without any economic importance. Only some species of *Triglochin—T. maritimum* leaves and *T. procerum* rhizomes are said to be edible.

Taxonomic Considerations. Juncaginaceae has been treated as a member of Najadaceae by Bentham and Hooker (1965c), in the order Helobiae by Engler and Diels (1936), in the order Juncaginales by

Hutchinson (1959) and Takhtajan (1987), in the order Najadales by Cronquist (1968, 1981) and Takhtajan (1966, 1980). Thorne (1968, 1983) includes this family in the Scheuchzeriaceae. The Juncaginaceae is closely allied to the family Najadaceae and is placed in it or close to it by most phylogenists.

The systematic position of the genus *Lilaea* is controversial. Benson (1970), Hutchinson (1973) and Takhtajan (1987) treat it as a distinct family. Takhtajan (1987) recognises even *Maundia* as a separate family Maundiaceae. Most phylogenists, however, include these genera in the Juncaginaceae. *Lilaea* has polygamic, monomerous flowers without perianth and is sometimes regarded as primitive and sometimes as advanced (Dahlgren and Clifford 1982).

Potamogetonaceae

A family of 8 genera (including *Zostera, Posidonia* and *Cymodocea*) and about 125 species, distributed widely all over the world and grow in oceanic coastal regions, brackish tidal waters as well as freshwater bodies such as ponds, pools, lakes, rivers, streams and also bogs or marshy areas.

Vegetative Features.   Perennial, aquatic herbs (Fig. 54.7A) rarely grow in marshy areas. Stem rhizomatous spreading on the ground surface under water, often jointed and nodose, the lower nodes root-bearing and upper ones foliaceous; leaves with sheathing base, sheath often apically ligulate, distichous, alternate or less often opposite; sessile or petiolate, often vaginate at the base, stipulate.

Floral Features.   Flowers solitary, spicate (54.7A, B) or cymose, bi- or unisexual (Fig. 54.7B, C), actinomorphic, often borne in a spathe. Perianth absent or of 4 to 6, small, herbaceous or membranous, free or fused segments; stamens 1 to 6, extrorse, sessile (Fig. 54.7E), mono- or bithecous anthers. Gynoecium of 1 to 6 carpels, free or rarely fused at base, each ovary superior, unilocular, uniovulate (Fig. 54.7D), ovule usually pendulous from the apex. Fruit coriaceous, subwoody or membranous, 1-seeded drupelets or nutlets; seeds without endosperm, embryo axile.

Anatomy.   Vessels absent in leaf and stem; in roots the vessels are with scalariform perforation plates. Sieve tube plastids of monocotyledonous type[9] occur in *Potamogeton* (Behnke 1969). Stomata of Rubiaceous type; occur only on upper surface of leaves.

Embryology.   Pollen grains inaperturate, tricolpate, globose with reticulate sculpturing, thread-like in *Cymodocea;* 3-celled at shedding stage. Ovules ana-, ortho- or campylotropous; bitegmic, crassinucellate; micropyle formed by inner integument only. Polygonum type of embryo sac, 8-nucleate at maturity. Endosperm formation of the Helobial type in *Potamogeton* and *Ruppia.* Endosperm formation precedes embryogenesis (Takaso and Bouman 1984).

Chromosome Number.   Basic chromosome number is $x = 13-15$ in *Potamogeton* and *Groenlandia;* $x = 8$ in *Ruppia.*

Important Genera and Economic Importance.   All the known genera are aquatic weeds. *Ruppia tuberosa*—a dark green, firmly rooted plant with its rhizome completely buried—occupies hypersaline habitats in western Australia (Davis and Tomlinson 1974). *Posidonia oceanica* is a bioindicator of mercury contamination in marine environments (Maserti et al. 1988).

Taxonomic Considerations.   Configuration of the family Potamogetonaceae has been rather varied. It is a heterogeneous group and has been treated variously by different authors. Bentham and Hooker

---

[9]All monocotyledons investigated have the same type of sieve tube plastids: a P-type with cuneate (triangular) crystalloid bodies, generally in a considerable number per plastid (Dahlgren and Clifford 1982).

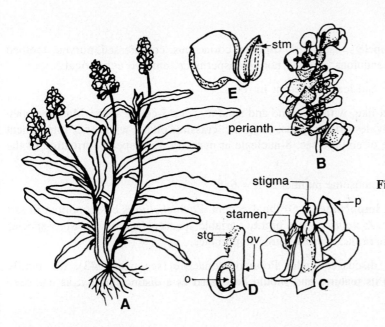

Fig. 54.7 Potamogetonaceae: **A–E** *Potamogeton crassipes* **A** Habit. **B** Inflorescence. **C** Flower. **D** Carpel, vertical section. **E** Perianth lobe with stamen attached. *o* ovule, *ov* ovary, *p* perianth, *stg* stigma, *stm* stamen. (Adapted from Lawrence 1951)

(1965c) included all the genera in the Najadaceae; Engler and Diels (1936) recognised Potamogetonaceae; Benson (1970) assigned all the genera to the Zosteraceae. Lawrence (1951) recognised 8 genera of the Potamogetonaceae; Hutchinson (1969, 1973) assigns them to 4 different families—*Cymodocea, Diplanthera* and *Zannichellia* to Zannichelliaceae, *Potamogeton* to Potamogetonaceae, *Zostera* to Zosteraceae, and *Ruppia* to Ruppiaceae. Huber (1969) also recognises 4 families: *Ruppia, Groenlandia* and *Potamogeton* in Potamogetonaceae, *Zostera, Heterozostera* and *Phyllospadix* in Zosteraceae, *Zannichellia, Althenia, Lepilaena* and *Vleisia* in Zannichelliaceae, and *Cymodocea, Diplanthera, Syringodium, Amphibolis, Halodule* and *Thalassodendron* in Cymodoceaceae. Dahlgren and Clifford (1982) also recognise these 4 families. Takhtajan (1987) raises them to the rank of order and includes 5 families: Potamogetonaceae, and Ruppiaceae in Potamogetonales, Zosteraceae in Zosterales, and Cymodoceaceae and Zannichelliaceae in Cymodoceales. The four families as recognised by Huber (1969) and Dahlgren and Clifford (1982) and also the fifth one—Ruppiaceae (Takhtajan 1987)—are so distinct from each other in their morphological (Singh 1965) as well as embryological features (Lakshmanan 1970d), and basic chromosome number, that it will be better to recognise them as distinct families under a separate order, i.e. Zosterales, as suggested by Dahlgren and Clifford (1982).

Zannichelliaceae

A family of 4 genera and 7 or 8 species, cosmopolitan in distribution.

Vegetative Features.   Perennial submerged aquatic herbs of fresh and brackish water. Stem slender, simple or cymosely branched and leafy; rootstock slender. Leaves alternate or opposite, linear to filiform, entire with sheathing base or free stipules.

Floral Features.   Flowers solitary axillary or cymose inflorescent. Minute, monoecious, naked flowers, both sexes enclosed together in a membranous deciduous spathe; one staminate and 2–5 pistillate flowers, often in the same axillary, cup-shaped involucre (Subramanyam 1962). Staminate flower without a perianth; stamens 1 to 3, sometimes connate; anthers 1 or 2 loculate, linear, filaments filiform. Pistillate flowers with a cupular, hyaline perianth. Gynoecium apocarpous, 1- to 9-carpellary; ovaries sessile or stipitate; each with a short or long, simple or 2- to 3-lobed style and peltate stigma; one pendulous ovule in each carpel. Fruit a group of drupelets, usually 4, stipitate, coriaceous, compressed, curved, toothed or entire along the margin; seeds pendulous, oblong, non-endospermous, embryo cylindrical.

Anatomy. Vessels absent in stem and leaves, present in roots.

Embryology. Pollen grains thread-like, without exine and shed at 2- or 3-celled stage. Mature pollen contains numerous starch grains. Ovules orthotropous, bitegmic, crassinucellate, with inner integument forming the micropyle. Allium type of embryo sac, 8-nucleate at maturity. Endosperm formation of the Helobial type (Lakshmanan 1970e).

Chromosome Number. Basic chromosome number is x = 6–8.

Important Genera and Economic Importance. Of the 4 genera, *Zannichellia* is cosmopolitan in salt marshes and at times in freshwater. *Z. palustris* is a highly variable species. *Lepilaena* with 3 species occur in Australia. None of the genera has any economic importance.

Taxonomic Considerations. As discussed under Potamogetonaceae (see page 543), the family Zannichelliaceae is very distinct in its features and should be treated as a distinct family, as has been done by almost all phylogenists.

Najadaceae

A monotypic family of the genus *Najas,* with 50 species; subcosmopolitan in distribution.

Vegetative Features. Fresh or brackish water aquatic herbs with opposite or pseudo-verticillate leaves that are dentate to entire; slender stem (Fig. 54.9A).

Floral Features. Flowers unisexual, male flowers unistaminate (Fig. 54.9B, C), stamen enclosed in an apically bilipped sheath and supported by two scales; anthers unilocular; quadrilocular in *N. graminea* and *N. lacerata* (Lakshmanan 1970f). The pistillate flowers naked or supported by a spathe (Fig. 54.9D), monocarpellate; ovary superior, unilocular, ovule solitary, erect (Fig. 54.9E); style cylindrical, stigma bifurcate, subulate.

Anatomy. Vessels absent in stems. Squamulae intravaginalis or axillary non-vascularized, scale-, gland- or finger-like trichomes, in pairs or in large number in the axils of vegetative leaves are present. Stomata absent.

Embryology. Pollen grains inaperturate, ellipsoidal or spherical, shed at 2-celled stage. Ovules anatropous, bitegmic, crassinucellate, micropyle formed by inner integument. Polygonum type of embryo sac, 8-nucleate at maturity. Endosperm formation of the Nuclear type. Polyembryony reported (Johri et al. 1992).

Chromosome Number. Basic chromosome number is x = 6,7.

Chemical Numbers. Glycoflavones occur in some members.

Important Genera and Economic Importance. *Najas* is the only genus of this family. *N. marina* growing in brackish water is a salt-tolerant plant. *N. graminea, N. indica, N. lacerata* and *N. minor* are some of the common species. In warmer regions, *Najas* species are reported to be a nuisance in irrigation ditches and rice fields. They are important food plants for fish; often used as packing material and also green fertilizer (Dutta 1988).

Taxonomic Considerations. This family is sometimes treated as a distinct order, Najadales (Takhtajan

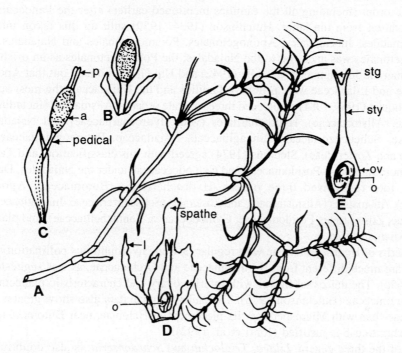

Fig. 54.9 Najadaceae: **A–E** *Najas* spp. **A** *N. minor*, a portion of the plant. **B** Staminate flower of *N. gracillima*. **C** Staminate flower of *N. flexilis*, the pedicel elongates before the dehiscence of anther. **D** Pistillate flower, vertical section. **E** Carpel, longisection. *a* anther, *l* leaf, *o* ovule, *ov* ovary, *p* perianth, *stg* stigma, *sty* style. (**A–C** after Rendle 1904, **D, E** after Lawrence 1951)

1987), separate from, and at other times included in, Zosterales. The two orders are closely allied to each other on the basis of exomorphological and embryological features (Davis 1966, Dahlgren and Clifford 1982). According to Lakshmanan (1970f), this family is allied to Lilaeaceae and Zanichelliaceae, on the basis of embryological characters. Najadaceae is probably best included in the Zosterales in sequence with the Scheuchzeriaceae and Juncaginaceae (Stebbins 1974).

Taxonomic Considerations of the Order Helobiae

Aquatic habit, hemi-cyclic to cyclic flowers, free carpels, seeds without endosperm, seedlings with a large hypocotyl and absence of vessels in stem and leaves are some of the characteristic features of this order.

It is regarded as the most primitive order among the monocotyledons by most phylogenists, on the basis of: (i) floral parts free and arranged in several whorls, (ii) numerous stamens and carpels showing spirocyclic to cyclic arrangement, (iii) prevalence of wide lamina-like filaments with basifixed anthers in some members such as Juncaginaceae and Potamogetonaceae, (iv) occurrence of laminar placentation in many members of the Alismataceae and Hydrocharitaceae, (v) more or less conduplicate carpels in some members of the Hydrocharitaceae, and (vi) absence of vessel elements from stem and leaves, often from roots as well.

Circumscription of the order is much variable. Engler and Diels (1936), Rendle (1904) and Lawrence (1951) placed the order (including all the families mentioned earlier) after the Pandanales, but did not consider its derivation from the latter. Hutchinson (1934, 1959) split up this taxon into six orders—Butomales, Alismatales, Juncaginales, Aponogetonales, Potamogetonales and Najadales. According to him, the Aponogetonales was ancestral to the Najadales; the Potamogetonales as an offshoot, somewhat more primitive than the Juncaginales. Cheadle (1942) and Uhl (1947) pointed out that Aponogetonaceae, Scheuchzeriaceae and Lilaeaceae were primitive families and the Najadaceae, the most advanced family of this order. Cronquist (1968, 1981) recognised three separate orders—Alismatales (including Butomaceae, Limnocharitaceae, Alismataceae), Hydrocharitales (with Hydrocharitaceae) and Najadales (including Aponogetonaceae, Scheuchzeriaceae, Juncaginaceae, Najadaceae, Potamogetonaceae, Ruppiaceae, Zannichelliaceae and Zosteraceae). Stebbins (1974) agreed with this classification, and Takhtajan (1969) recognised two more families—Posidoniaceae and Cymodoceaceae under the Najadales. Dahlgren (1975a, 1980a, 1983a), too, recognised three orders: Hydrocharitales (Butomaceae, Aponogetonaceae, Hydrocharitaceae), Alismatales (Alismataceae), and Zosterales (Scheuchzeriaceae, Juncaginaceae, Najadaceae, Potamogetonaceae, Zosteraccae, Posidoniaceae, Cymodoceacea, Zannichelliaceae), and placed them under the superorder Alismatiflorae.

The aquatic herbs of Hydrocharitaceae with regular flowers, hydrophillous pollination and indehiscent berry-like fruits are much different from Burmanniaceae and Orchidaceae, as was suggested by Bentham and Hooker (1965c). The unisexual flower with inferior ovary and characteristic plancentation indicates its alliance to Butomaceae (Hutchinson 1973). The embryological data also show greater similarity with the order Helobiae than with Microspermae. Its placement in Helobiae, near Butomaceae, in a separate suborder Hydrocharitineae is justified (Johri et al. 1992).

Relationship of the three genera, *Lilaea, Triglochin* and *Scheuchzeria,* is also doubtful. Many earlier taxonomists (Bentham and Hooker 1965c, Engler and Diels 1936, and Wettstein 1935) have placed them together in one family, Juncaginaceae. Hutchinson (1973) raises these three genera to three independent families—*Scheuchzeria* to Scheuchzeriaceae in Alismatales, *Lilaea* to Lilaeaceae and *Triglochin* to Juncaginaceae in Juncaginales. On the basis of embryological data, *Scheuchzeria* is distinct from the other two genera (Johri et al. 1992). Evidence from floral morphology and anatomy indicates that although the three genera are closely allied, *Scheuchzeria* is closer to *Aponogeton.*

Melchior's (1964) arrangement to retain *Scheuchzeria* in a separate suborder near the Aponogetonaceae and the other two genera in Juncaginaceae is, therefore, justified.

Suborder Potamogetonineae includes Aponogetonaceae, Juncaginaceae, Potamogetonaceae, Zannichelliaceae and Najadaceae. All these families include aquatic plants with spadix-type inflorescence. Although true bracts are absent, the petals are bract-like (Hutchinson 1973).

Aponogetonaceae resembles Scheuchzeriaceae more than any other family of this order.

Morphologically and embryologically, Potamogetonaceae are a heterogenous group and many taxonomists raise some of the genera to family level and recognise 4 families instead (Hutchinson 1973, Huber 1969, Dahlgren and Clifford 1982, Takhtajan 1987). Dahlgren and Clifford (1982) classified the group likewise.

The order Helobiae is often considered to be the most primitive group of the Monocotyledons. However, they are considered by Cronquist (1968) as a near basal side-branch, a relict group that has retained a number of primitive characters. Any primitive monocot should have binucleate pollen and endospermous seed, which is not observed here.

The apocarpous gynoecium of most members of this order and also numerous stamens and mostly uniaperturate pollen, indicate that if there exists any connection between the monocots and the dicots, it must be through the primitive subclass, Magnoliidae. The order Nymphaeales of this subclass is the closest to the Alismatales. Common features between the two orders are: (i) both are of aquatic habitat,

(ii) presence of apocarpous ovaries and numerous stamens, (iii) presence of laminar placentation, (iv) absence of vessels, and (v) herbaceous nature. It is presumed by many phylogenists that the Helobiales is the direct descendent of early dicots and direct ancestor of other monocots (Dutta 1988) but it is much more probable that both the Nymphaeales and the Alismatales originated from a common dicotyledonous ancestral form of terrestrial, herbaceous nature, extinct at present or lost in the geological history. The Alismatales or Helobiae is a heterobathomic group retaining primitive features like apocarpous ovaries and laminar placentation, as well as advanced features like seeds without endosperm. Instead of treating this group as ancestral to all the monocotyledons, this order should be regarded as an ancient side-branch of monocot development (cf. classificatory systems suggested by Takhtajan 1959,1969,1980).

55

Order Triuridales

A monotypic order of the family Triuridaceae. The plants are saprophytic herbs growing on decayed stem of trees or humus-rich soil. Flowers small, actinomorphic, usually unisexual, with apocarpous ovaries and fruits as small achenes. Seeds endospermous, rich in fats.

Triuridaceae

A small family of 7 genera and 80 species of tiny little saprophytic plants growing in the tropical forests of America, Asia and Africa.

Vegetative Features.   Small mycorrhizal saprophytes, lack chlorophyll, reddish, purplish or colourless; rhizomatous; grow on decayed, rotten wood or on humus-rich soil. Stems simple, very thin; leaves reduced in size, bract-like or absent (Fig. 55.1A).

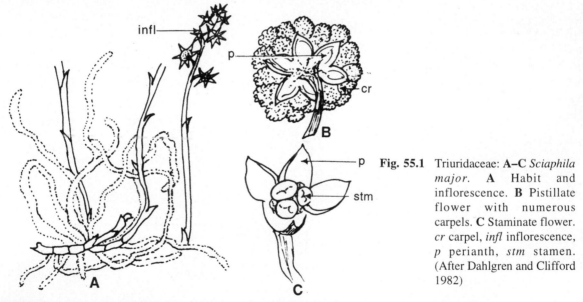

Fig. 55.1 Triuridaceae: **A–C** *Sciaphila major*. **A** Habit and inflorescence. **B** Pistillate flower with numerous carpels. **C** Staminate flower. *cr* carpel, *infl* inflorescence, *p* perianth, *stm* stamen. (After Dahlgren and Clifford 1982)

Floral Features.   Inflorescence terminal racemes (Fig. 55.1A) or rarely cymose. Flowers small, subtended by minute scale-shaped bracts, usually unisexual (Fig. 55.1B, C), sometimes bisexual, with 3 to 6 (rarely 4 or 10), equal or unequal, valvate tepals, sometimes extended as tails or with an apical knob or a tuft of hairs. Stamens 3–6 (Fig. 55.1C), rarely 2, with short or no filament; anthers bi- or tetrasporangiate, sometimes dehisce transversely. Carpels numerous (Fig. 55.1B), small, free, each with a single stylodium which may be terminal, lateral or gynobasic, and 1 erect, basal ovule in each carpel. Fruits small achenes, single-seeds; endospermous, with a minute undifferentiated globular embryo.

Anatomy.   Vascular system is highly reduced with vascular bundles arranged in one circle; vessels absent , stomata absent.

Embryology. Pollen grains smooth, inaperturate, monocolpate in *Sciaphila;* shed at 3-celled stage. Ovules solitary, basal, anatropous, bitegmic, tenuinucellate, the inner integument forms the micropyle. Polygonum type of embryo sac (probably Fritillaria type in *Triuris*), 8-nucleate at maturity. Endosperm formation of Nuclear type.

Chromosome Number. Basic chromosome numbers are x = 11, 12, 14.

Chemical Features. Seeds rich in albumen and fat.

Important Genera and Economic Importance. *Triuris, Hexuris, Andruris, Sciaphila, Soridium* are some of the genera of this family. They are not known to have any economic importance.

Taxonomic Considerations of Triuridaceae and the Order Triuridales

Bentham and Hooker (1965c) placed Triuridaceae in Apocarpae along with Alismataceae and Najadaceae. Engler and Diels (1936) and Hutchinson (1973) treated them as the only member of the order Triuridales. They are considered as an advanced but reduced form of Helobiae, and derived from it. Thorne (1977) traced the possibility of a common ancestry of Liliiflorae and Triuridales, on morphological basis. Dahlgren (1980b) assigns Triuridales to the superorder Triuridiflorae next to Alismatiflorae. Cronquist (1981) includes Petrosaviaceae and Triuridaceae in Triuridales in the subclass Alismatidae.

Triuridaceae is a very isolated family with possible affinity to the Alismataceae, to *Alisma* in particular because of apocarpy and gynobasic stylodia of the carpels (Dahlgren and Clifford 1982). It has many features in common with the Liliiflorae also, but in certain cases it is more archaic (Takhtajan 1987).

Because of its nonchlorophyllous plant body, terrestrial habit, endospermous seeds and 3-celled pollen grains, this family is quite distinct and it is justified to treat it as a member of a separate order of its own, i.e. Triuridales. Embryologically there are several resemblances between Triuridaceae and Burmanniaceae but they (Triuridaceae) resemble the Alismatiflorae (Alismatales, Helobiae) and Liliiflorae (Liliales) much more (Table 50.1).

Table 50.1 Comparative Data for Triuridaceae, Burmanniaceae, Liliales, and Alismatales

Feature	Triuridaceae	Burmanniaceae	Liliales	Alismatales
Anther	2-, 3- or 4-locular	4-locular	4-locular	4-locular
Tapetum	Secretory/plasmodial	Secretory	Secretory	Plasmodial
Pollen grains	Monads, 3-celled	Monads, dyads, or tetrads, 2-celled	Monads, 2-celled	Monads, 3-celled
Ovule	Anatropous, bitegmic, tenuinucellate	Anatropous, bitegmic tenuinucellate	Ana-, Ortho-, Hemianatropous, bitegmic tenui-nucellate	Ana-, Ortho-, Campylotropous, bitegmic, crassi-nucellate
Embryo sac	Polygonum type	Polygonum or Allium type	Polygonum type	Allium or Polygonum type
Endosperm	Nuclear	Helobial	Helobial/Nuclear	Helobial/Nuclear

On the basis of embryological features, Johri et al. (1992) support the placement of this order between the superorders/orders, Alismatiflorae and Liliiflorae (Liliales).

56

Order Liliiflorae

The order Liliiflorae comprises 5 suborders and 17 families:

1. Suborder Liliineae includes Liliaceae, Xanthorrhoeaceae, Stemonaceae, Agavaceae, Haemodoraceae, Cyanastraceae, Amaryllidaceae, Hypoxidaceae, Velloziaceae, Taccaceae and Dioscoreaceae.
2. Suborder Pontederiineae comprises the family Pontederiaceae.
3. Suborder Iridineae includes Iridaceae and Geosiridaceae.
4. Suborder Burmanniineae includes Burmanniaceae and Corsiaceae, and
5. Suborder Phylidrineae includes the family Philydraceae.

Herbs, rarely shrubs or woody, sparingly branched 'trees' without, or in several families with, secondary growth. Stems underground, mostly modified as rhizomes, corms or bulbs. Leaves usually alternate or rarely opposite or verticillate, linear to lanceolate, sessile, generally sheathing at base, lamina entire or rarely compound or digitately lobed as in Dioscoreaceae and Taccaceae; venation mostly parallel, reticulate in some.

Flowers hypo- or epigynous, actino- or zygomorphic, tepals normally petaloid. Stamens usually 6, sometimes 3, 2 or 1 as in Philydraceae, or rarely up to 9 or more as in *Vellozia* and *Pleea*. Gynoecium syncarpous, tricarpellary, placentation variable. Fruits generally capsules or berries, seeds numerous, endospermous.

Steroid saponins present.

Liliaceae

A large family of 240 genera and 4000 species widely distributed especially in the warm temperate and tropical regions of the world. Except for some xerophytic representatives, Liliaceae members do not form a dominant vegetation anywhere. *Astelia solanderi* is an epiphytic member from New Zealand (Ambasht 1990).

Vegetative Features. Mostly perennial, herbs (Fig. 56.1A,E,F) with rootstock modified to bulb, corm, tuber or rhizome; sometimes climbers (*Asparagus, Smilax*) and infrequently woody (*Dracaena, Yucca*); stem often prickly, as in *Smilax* and *Asparagus racemosus;* sometimes modified to subterranean storage organs or cladophylls (Fig. 56.1E). Leaves radical or cauline, alternate or in rosettes, opposite only in *Scolyopus,* mostly lamellate, sometimes reduced to spines (*Asparagus*) or scale leaves (*Ruscus*), sometimes fleshy (*Aloe*) or with prickly tip and margin (*Yucca*); commonly parallel veined (parallel reticulate in *Smilax*).

Floral Features. Inflorescence a scapigerous raceme, spike or umbel (*Allium cepa*), sometimes cymose. Flowers actinomorphic, weakly zygomorphic in *Gilliesia, Haworthia* and *Hemerocallis;* bisexual, rarely unisexual as in *Smilax, Lomandra, Ruscus,* hypogynous, bracteate, bracts small, scarious or spathaceous; ebracteate, often highly ornamental. Perianth mostly large and showy, of 6 tepals in 2 whorls of 3 each, rarely 4 or more than 6 as in *Paris* (Fig. 56.1 A-D), generally not distinguishable into calyx and corolla, free or basally connate into a tube, imbricate or the outer whorl valvate. Stamens 6, rarely 3 as in male flower of *Ruscus,* one whorl is suppressed; filaments free or adnate to tepals (Fig. 56.1G) often coloured as in *Gloriosa:* anthers bicelled, basifixed (Fig 56.1B,D) or versatile as in *Lilium canadense,* extrorse or

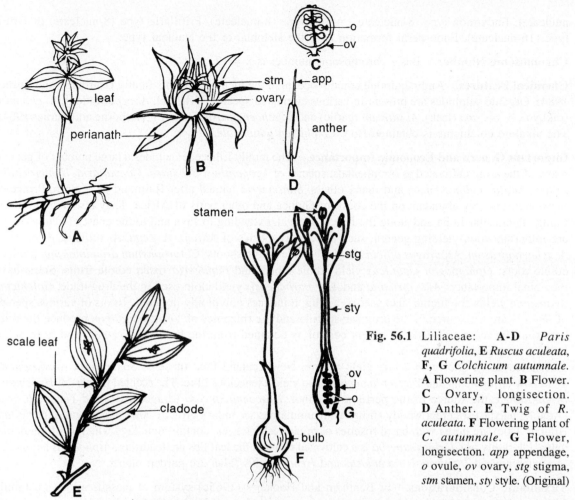

Fig. 56.1 Liliaceae: **A-D** *Paris quadrifolia*, **E** *Ruscus aculeata*, **F, G** *Colchicum autumnale*. A Flowering plant. B Flower. C Ovary, longisection. D Anther. E Twig of *R. aculeata*. F Flowering plant of *C. autumnale*. G Flower, longisection. *app* appendage, *o* ovule, *ov* ovary, *stg* stigma, *stm* stamen, *sty* style. (Original)

introrse, usually dehisce longitudinally. Gynoecium syncarpous, tricarpellary; ovary superior, half-inferior in *Mondo* (Dutta 1988), trilocular with axile placentation or rarely unilocular with parietal placentation; ovules numerous, biseriate; style usually 1, divided or trifid, stigmas 3 or 1 with 3 lobes. Fruit a loculi- or septicidal capsule or a berry. Seeds with small embryo and abundant endosperm.

Anatomy.   Based on the observations of Cheadle and Kosakai (1971), vessels in Liliaceae occur in roots of all species and in the stems of at least some species of the tribes Asphodeleae, Herrerieae, Johnsonieae, Polygonateae, Tricyrtideae and Asparageae. Vessels in stems are almost invariably primitive except in *Tricoryne elatior*, where some vessels are with simple perforation plates. In roots they vary from highly specialized forms to very primitive forms. Raphides present in some members like *Lilium* and *Lloydia* (Dutta 1988). Leaves isobilateral, mostly with parallel venation.

Embryology.   The family is multipalynous. Pollen grains trichotomocolpate, 1- or 2- or more-colpate or -porate and spiraperturate (P.K.K. Nair 1970); 3-celled at dispersal stage (2-celled in *Urginia indica* and some others). Ovules anatropous, bitegmic, tenui- or crassinucellate; usually inner integument forms the micropyle. Embryo sac (see Johri et al. 1992) of Polygonum type (8-nucleate), Allium type (8-

nucleate), Endymion type (8-nucleate), Adoxa type (8-nucleate), Fritillaria type (8-nucleate) or Drusa type (16-nucleate). Endosperm formation is of the Helobial or the Nuclear type.

Chromosome Number.   Basic chromosome number is x = 12.

Chemical Features.   Anthraquinone emodin occurs in the members of this family (Harborne and Turner 1984). Organic sulphides are present in various cultivated species of *Allium*: *A. cepa* (onion), *A. chinense* (rakkyo), *A. porrum* (leek), *A. sativum* (garlic) and *A. schoenoprasum* (chive) (Harborne and Turner 1984). The alkaloid colchicine is obtained from *Colchicum autumnale*.

Important Genera and Economic Importance.   The family Liliaceae includes a large number of genera, many of these are cultivated as ornamentals; species of *Agapanthus, Brodiaea, Convallaria, Hemerocallis, Lilium, Scilla, Tulipa, Yucca* and many others. *Sansevieria* named after Raimond de Sangro, Prince of Sanseviero, is very abundant on the coast of Guinea and other parts of Africa. They also abound in Sri Lanka, Peninsular India and along the Bay of Bengal, extending to Java and to the coast of China. Some are important food-yielding genera, such as various species of *Allium*: *A. cepa, A. sativum, A. porrum, A. schoenoprasum. Asparagus officinalis* yield edible young shoots; *Chlorophytum arundinaceum* produce edible roots; *Ophiopogon japonicus* yield edible tubers and *Lapageria rosea* edible fruits. Some have medicinal importance; *Aloe africana* and *Aloe barbadensis* yield aloin used in the drug trade; *Colchicum autumnale* yields colchicine, also used as a drug; it induces polyploidy in plants. Roots of various species of *Smilax* are the source of the drug sarsaparilla and the rhizomes of *Veratrum album* produce the drug veratrin. The red squill, used in rodent control, is obtained from the bulbs of *Urginea,* and from *Scilla* bulbs a rat poison is obtained.

The leaves of *Phormium tenax* yield a fibre, New Zealand flax, those of *Sansevieria roxburghiana* yield bow-string hemp and *Yucca filamentosa* also yield a tenacious fibre. The scented flowers of *Hyacinthus orientalis* are the source of the perfume hyacinth. *Dracaena spicata* (dragon tree) and *Yucca gloriosa* (Adam's needle) are erect woody shrubs with anomalous secondary growth. *Aloe indica* and *A. perfoliata* (Indian Aloe) are herbs with basal rosettes of thick fleshy leaves, contain mucilage which is of medicinal value. Glory lily or *Gloriosa superba* is a cultivated climber, the leaf tips are tendriller. *Asphodelus tenuifolius* is a winter weed. *Asparagus racemosus* and *Hemerocallis fulva* are garden plants.

Taxonomic Considerations.   In Bentham and Hooker's (1965c) system of classification, the family Liliaceae is included in the third series Coronarieae. Engler and Diels (1936) treated it as a member of suborder Liliineae of order Liliiflorae. Cronquist (1968, 1981), Dahlgren (1975a, 1980a, 1983a), Hutchinson (1959), Takhtajan (1980, 1987) and Thorne (1968,1983) include it in the Liliales.

Hutchinson (1959) considers Liliaceae to be the stock from which many other families of monocots evolved directly or indirectly. Although formerly regarded as a primitive family, it is the most typical family of the Monocotyledons.

Cronquist (1968,1981) includes the family Amaryllidaceae in the Liliaceae. The Liliaceae differ from the Amaryllidaceae in having superior ovary and 6 stamens. The two families are connected by the genus *Allium* (according to some Alliaceae), which has superior ovary, as in Liliaceae and umbellate inflorescence, as in Amaryllidaceae.

There is much controversy over the circumscription of this family. Hutchinson (1959, 1973) includes the tribes Agapanthieae, Allieae and Gillesieae in his Amaryllidaceae because of umbellate inflorescence. This view is substantiated by the palynological study of *Allium* by Maia (1941) and anatomical study of Monocotyledons by Cheadle (1942).

The Liliaceae have probably origniated from the Helobieae or its ancestors as some members show intermediate features, e.g. *Petrosavia* (Liliaceae) has partially-fused carpels, as in Helobieae.

Liliaceae, after the removal of Xanthorrhoeaceae, Agavaceae and Hypoxidaceae is a more natural taxon. However, further breakup into Smilacaceae, Alliaceae, Agapanthaceae and many others (Dahlgren et al. 1985, Takhtajan 1987) may not be necessary.

Xanthorrhoeaceae

A small family of 8 genera and 66 species (Willis 1973), distributed in Australia, New Caledonia and New Zealand.

Vegetative Features. Perennial, stout, woody subshrubs, rhizomatous or not. Stem sometimes tall and branched; leaves linear or not, simple, sometimes spiny, often sheathing.

Floral Features. Inflorescence spicate, capitate or paniculate; flowers actinomorphic, bi- or unisexual, usually dry and glumaceous, often persistent. Perianth of 6 tepals in 2 whorls of 3 each, sometimes shortly connate. Stamens 6 in 2 whorls of 3 each, outer usually free, inner adnate to inner whorl of tepals; anthers basifixed or versatile, introrse or latrorse. Gynoecium syncarpous, tricarpellary; ovary superior, trilocular with axile placentation or unilocular with basal placentation; ovule 1 to 3 or a few per locule, erect when basal, styles free or more or less completely connate. Fruit a loculicidal capsule or indehiscent nut; seed only 1 or a few, endospermous.

Anatomy. Roots with vessels have simple perforations (an advanced feature). Vessels sometimes present in leaves but missing in stems in species of *Xanthorrhoea*. Stomata paracytic (Rubiaceous type).

Embryology. Embryology has not been fully investigated. Ovules anatropous, bitegmic, crassinucellate with the inner integument forming the micropyle. Polygonum type of embryo sac, 8-nucleate at maturity.

Chromosome Number. Basic chromosome number is x = 11.

Chemical Features. Polysaccharide levan is reported as a storage carbohydrate in Xanthorrhoeaceae.

Important Genera and Economic Importance. The genus *Xanthorrhoea* with 15 species is endemic to Australia. *X. hastilis* commonly called the grass tree or black boy is a characteristic plant of Australian vegetation. It has the habit of *Aloe* or *Dasylirion*, with a long bulrush-like spike of flowers. From the bases of old leaves, a resin trickles which is used in making varnish, sealing wax, etc.

Taxonomic Considerations. Xanthorrhoeaceae were previously treated as a subfamily of the Liliaceae. Hutchinson (1973) removes this family to his Agavales, Embryologically, it is not fully investigated, to justify this separation (Johri et al. 1992). Morphologically, the two families are not very distinct but they differ in basic chromosome number and chemical features. Most modern taxonomists, however, recognise Xanthorrhoeaceae as a distinct family. We support this placement.

Stemonaceae

A monotypic family of the genus *Stemona* with 25 species distributed in Eastern Asia, Indo-Malaya and Australia.

Vegetative Features. Perennial, erect or climber, rhizomatous herbs, often with fasciculated tubers. Leaves alternate, opposite or verticillate, petiolate, often *Smilax*-like with parallel main veins and numerous closely parallel cross veins; sometimes reduced to scale.

Floral Features. Inflorescence axillary, few-flowered cymes. Flowers regular, bi- or unisexual, sometimes unpleasantly scented. Perianth in two whorls of two each, sepaloid or petaloid; stamens 2 + 2, filaments short , connate at base, anthers linear with long-produced, linear-lanceolate, sometimes lamellate connective.

Gynoecium syncarpous, bicarpellary, ovary unilocular, superior, with few to several basal ovules, compressed; style reduced, stigma small, subsessile. Fruit an ovoid, 2-valved capsule with a few to many seeds. Seeds endospermous, sometimes arillate.

Anatomy.   Similar to that of the Dioscoreaceae; stomata Ranunculaceous type.

Embryology.   The family Stemonaceae is unipalynous; pollen grains monocolpate (Zavada 1983). Ovules anatropous, bitegmic, crassinucellate; inner integument forms the micropyle. Polygonum type of embryo sac, 8-nucleate at maturity. Endosperm formation of the Nuclear type.

Chromosome Number.   Basic chromosome number is x = 7.

Chemical Features.   Steroid saponins abundant.

Important Genera and Economic Importance.   According to Takhtajan (1987), there are 4 genera— *Stemona, Stichoneuron*[10], *Croomia*[10] and *Pentastemona;* none has any economic importance.

Taxonomic Considerations.   Bentham and Hooker (1965c) included Stemonaceae in the Dioscoreaceae. Engler and Diels (1936) recognised this family as a member of the order Liliiflorae. Hutchinson (1934,1959) erected a separate family, Roxburghiaceae (= Stemonaceae), to include *Croomia, Stichoneuron* and *Stemona* (= *Roxburghia*) on the basis of morphological and anatomical features. However, Burkill (1960) disapproved Hutchinson's proposition. Melchior (1964), Cronquist (1968, 1981), Dahlgren (1975a, 1980a) and Takhtajan (1987), place it in Dioscoreales and treat Stemonaceae as a distinct family.

Agavaceae

A small family of 10 genera and 400 species (Takhtajan 1987), distributed in tropical and subtropical regions of the New World and in Central America.

Vegetative Features.   Xerophytic plants of arboreal habit. Leaves fibrous, radical or in basal rosettes with an erect, aerial scapigerous stem bearing inflorescence. Leaves borne on the scape (Fig. 56.4A) (*Polianthes*), often with spiny margin and/or tip (*Agave, Yucca*).

Floral Features.   Inflorescence racemose (*Polianthes*, Fig. 56.4A) or paniculate; large and consists of cymes arranged in a racemose manner in *Agave*; many flower buds modified into bulbils, which give rise to new plants. Flowers actinomorphic, bisexual, epigynous (Fig. 56.4 B), bracteate, bracts green, ebracteolate. Perianth of 6 segments, in 2 whorls of 3 each, gamophyllous to form a tube, upper portion free. Stamens 6, epiphyllous, filaments short and thick, anthers basifixed (Fig. 56.4D), introrse, longitudinally dehiscent. Gynoecium syncarpous, tricarpellary; ovary inferior, trilocular with many ovules arranged in 2 series on axile placentae (Fig. 56.4 C).

Anatomy.   Similar to Liliaceae.

Embryology.   Pollen grains 2-celled at dispersal stage. Ovules anatropous, bitegmic, crassinucellate; inner integument forms the micropyle. Polygonum type of embryo sac, 8-nucleate at maturity. Endosperm formation of the Helobial type. It is ruminate in *Yucca* spp. (Johri et al. 1992).

Chromosome Number.   Basic chromosome number is x = 30; karyotype shows 5 large and 25 small chromosomes (Takhtajan 1987).

---

[10]Included in Dioscoreaceae, in this work.

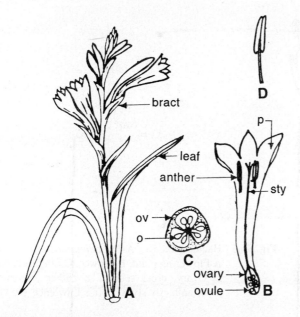

Fig. 56.4 Agavaceae: **A-D** *Polianthes tuberosa.*
A Flowering scape. **B** Flower, longisection.
C Ovary, cross section. **D** Anther. *o* ovule,
ov ovary, *p* perianth, *sty* style. (Original)

Chemical Features. The polysaccharide levan is reported in 10 species of this family (Pollard 1982). Steroidal saponin also occurs in *Agave*.

Important Genera and Economic Importance. Many members of this family are ornamentals. *Polianthes* is grown for its fragrant flowers; *Sanseviera* and *Dracaena* are grown for their foliage. Various species of *Agave* and *Furcraea* are the source of leaf fibres, used for cordage. In Latin America, a sugary exudate from *Agave* sp. is used for distilling liquors as 'mezcal', 'plulque' and 'tequila' (Dutta 1988).

Taxonomic Considerations. The family Agavaceae is treated in the order Liliales by Cronquist (1968, 1981) and Takhtajan (1980, 1987). Engler retained it in the Liliiflorae, Dahlgren (1980a, 1983a) in the order Asparagales and Hutchinson (1973) in the Agavales. Bentham and Hooker (1965c) and Thorne (1968, 1983) do not recognise Agavaceae as a distinct family.

This family is a heterogeneous assemblage of taxa, which could also have been retained in Liliaceae or Amaryllidaceae.

Haemodoraceae

A family of 14 genera and 75 species (Willis 1973), or only 9 genera and 40 species (Takhtajan 1987), largely distributed in the southern hemisphere—Africa, Australia, Tasmania, Central and South America; and also in some parts of North America.

Vegetative Features. Erect, terrestrial, often hairy herbs with subterranean rhizomes or tubers. Leaves radical, linear, distichous, often ensiform, parallel veined.

Floral Features. Inflorescence a panicle of a number of cymes arranged racemosely (Fig. 56.5A). Flowers hypo- or epigynous, actino- or zygomorphic (*Anigozanthos*) with 3 + 3 more or less petaloid tepals—sometimes fused into a tube (Fig. 56.5 B, C, D), imbricate or subvalvate, tube straight or curved. Stamens 6 or 3, inserted on inner perianth lobes; tetrasporangiate, basi- or dorsifixed, introrse anthers (Fig. 56.5 B, C). Gynoecium syncarpous, tricarpellary, ovary superior or inferior, trilocular, placentation axile (Fig. 56.5 D), style 1, stigma with dry or wet surface. Fruit a loculicidal capsule, seeds variously shaped, sometimes hairy with a small, globose-ovoid, non-chlorophyllous embryo and a starchy endosperm.

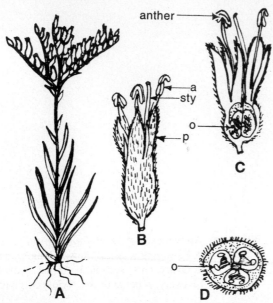

Fig. 56.5 Haemodoraceae: **A-D** *Lachnanthes tinctoria.*
A Flowering plant. **B** Flower. **C** Longisection.
D Ovary, cross section. *a* anther, *o* ovule,
p perianth, *sty* style. (After Lawrence 1951)

Anatomy.   Roots and sometimes stems with vessels, The leaves have unicellular or uniseriate, sometimes branched and multicellular hairs. Stomata Rubiaceous type; oxalate raphides common.

Embryology.   Pollen grains single, bicelled and usually sulcate or 2- (8)-foraminate; cylindrical, biporate and isopolar in *Phlebocarya* and apolar, globose with 7-8-porate apertures in *Tribonanthes australis* (Simpson 1983). Ovules hemianatropous or orthotropous, bitegmic, crassinucellate; micropyle formed either by inner or both integuments (Johri et al. 1992). Polygonum type of embryo sac, 8-nucleate at maturity. Endosperm formation of Helobial or Nuclear type.

Chemical Features.   Chelidonic acid present (Dahlgren and Clifford 1982); polysaccharide levan has also been reported (Pollard 1982).

Important Genera and Economic Importance.   *Haemodorum* with ca. 20 species is confined to Australia; 1 species of *Barberetta,* 2 species of *Dilatris* and 5 species of *Wachendorffia* are confined to South Africa, and one species of *Hagenbachia* and *Schiekia* and monotypic *Lachnanthes* occur in South America and along the Atlantic coastal plain from Florida to Cape Cod. *Conostylis* with 23 species occur in southwestern Australia. The position of the genus *Lophiola* is controversial. Once included in Haemodoaceae (Simpson and Dickison 1981), anatomical studies on its ensiform leaves show that it has a clear affinity with Liliaceae (Ambrose 1980, 1985). The family has no known economic importance.

Taxonomic Considerations.   The family Haemodoraceae is affiliated to the Cyanastraceae, Tecophilaeaceae and Hypoxidaceae (Dahlgren and Clifford 1982). Intrafamilial classification has varied considerably. Bentham and Hooker (1965c) originally proposed 4 tribes—Euhaemodoreae, Conostyleae, Conanthereae and Ophiopogoneae. These have been shifted in various combinations to other families. Melchior (1964) recognised only three tribes, i.e. Euhaemodoreae, Conostyleae and Conanthereae. Goernick (1969) and Hutchinson (1973) recognise only two tribes: Haemodoreae and Conostylideae. Palynologically, the two tribes are remarkably distinct (Simpson 1983). The genus *Cyanella* (Conanthereae) is embryologically so distinct that Dutt (1970a) and Johri et al. (1992) support its exclusion from Haemodoraceae and Amaryllidaceae, as suggested by Hutchinson (1973). Dutt and Johri et al. also support Hutchinson's view to raise Conanthereae to family Tecophilaeaceae. The position of the family Haemodoraceae, assigned by Melchior (1964), justifies its affiliation with the families Cyanastraceae (= Tecophilaeaceae), and Hypoxidaceae.

Cyanastraceae (= Tecophilaeaceae)

A small family of 6 genera and 22 species distributed mainly in southen hemisphere—Andes of South America, tropical and South Africa, and also in California.

Vegetative Features.   Perennial herbs with fibrous tunicated corms (Fig. 56.6 A) or thick, orbicular, flattened tubers. Leaves radical or towards the base of the flowering stems, alternate, linear to ovate-orbicular and cordate, glabrous.

Floral Features.   Inflorescence simple racemes borne separately from the tuber or corm, or panicle; bracteate, bracts large and membranous or small. Flowers bisexual (Fig. 56.6 B, C), actinomorphic or slightly zygomorphic, perigynous. Perianth in 2 whorls of 3 each, sometimes shortly connate to form a tubular structure; lobes spreading or reflexed, subequal, imbricate. Stamens 6, all perfect or 2 or 3 staminodes, inserted at the throat of the perianth, epipetalous; anthers bilocular, often connivent, the connective often produced at both ends, the base then swollen or spur-like; loculi open by a terminal pore (Fig. 56.6 C), rarely by a slit up to the base, and introrse. Gynoecium syncarpous, tricarpellary, ovary more or less semi-inferior, trilocular, placentation axile, ovules numerous, 2-seriate in each locule; style subulate or filiform, stigma 3-lobed or unlobed. Fruit a loculicidal capsule (Fig. 56.6 D); seeds one to numerous, with a fairly large embryo and a fleshy endosperm, chalazosperm is present in seeds of *Cyanastrum hostifolium* (Dahlgren and Clifford 1982); (Fig. 56.6 D).

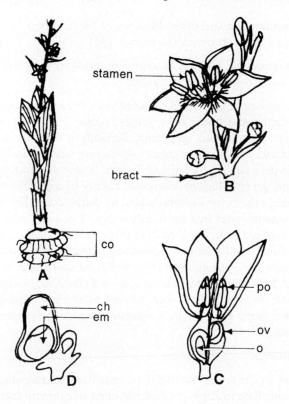

Fig. 56.6 Cyanastraceae: **A-D** *Cyanastrum hostifolium.* A Flowering plant. **B** Part of inflorescence. **C** Flower, longisection. **D** Fruit, longisection. *ch* chalazosperm, *co* corm, *em* embryo, *o* ovule, *ov* ovary, *po* pore. (After Dahlgren and Clifford 1982)

Anatomy. Vessels occur only in metaxylem of root systems (Cheadle 1969); with only scalariform or some variation of scalariform perforations. Number of bars may be few (*Calostemma hyacinthoides*) or numerous (*Conanthera bifolia*).

Embryology. Pollen grains are mono- and trichotomocolpate; 2-celled at the dispersal stage. Ovules anatropous, bitegmic, crassinucellate (Johri et al. 1992); the inner integument forms the micropyle. Polygonum type of embryo sac, 8-nucleate at maturity. In *Cyanastrum johnstoni* the endosperm degenerates following the formation of a few nuclei.

Chemical Features. The polysaccharide levan is reported in members of Cyanastraceae (Pollard 1982).

Important Genera and Economic Importance. *Cyanastrum* is a tropical African genus with 6 species. These are herbs with tubers or tuberous rhizome and racemes or panicles of actinomorphic, bisexual flowers. Fruit 1-seeded, perisperm present. No economic importance is known.

Taxonomic Considerations. The genus *Cyanastrum* has been placed in Haemodoraceae, as a subfamily Cyanastroideae (see Johri et al. 1992) or in a separate family, Cyanastraceae (Cronquist 1981, Takhtajan 1987). The members of Cyanastraceae are characterised by: (a) secretory anther tapetum, (b) Polygonum type of embryo sac, (c) Nuclear type of endosperm, (d) chalazosperm (in the absence of endosperm) present in mature seeds. Placement of this family in the order Liliiflorae, in the neighbourhood of Taccaceae, is justified. Both Taccaceae and Cyanastraceae members have broad leaves with a definite petiole and a more or less net-veined blade.

Amaryllidaceae

A family of 60–65 genera and 900 species, cosmopolitan in distribution. Majority of the members occur in the plains, plateaus and steppe areas of the tropics and subtropics (Lawrence 1951).

Vegetative Features. Perennial herbs with a rhizomatous, bulbous or cormous rootstock and aerial stem, a scape. Leaves mostly linear or lorate and basal rosettes, rarely cauline.

Floral Features. Inflorescence mostly umbellate, racemose or paniculate, sometimes reduced to single flower (*Zephyranthes*), subtended by a scarious spathe and borne on a leafless scape (Fig. 56.7 A). Flowers bisexual (Fig. 56.7 B, G), actino- or rarely zygomorphic, epigynous. Perianth of 6 tepals in 2 whorls of 3 each, sometimes polyphyllous (*Leuconium*) and in some genea (*Narcissus*) bear a cup-like corona (Fig. 56.7 F, G). Stamens 6, inserted on the perianth, in 2 whorls of 3 each, filaments short and thick or long and filamentous, usually free but in some genera filaments connected basally by a corolliform velamen or staminal corona. In *Vagaria* and *Urceolinia* the corona is represented by distinct or indistinct tooth on either side of filament base, in *Eurycles* by membranes that are shortly united (*Crinum*); anthers bicelled, basifixed or versatile, as in *Crinum,* or dorsifixed as in *Zephyranthes* (Fig. 56.7 D), dehiscence by vertical slits rarely by terminal pores, as in *Galanthus,* usually introrse, sometimes extrorse. Gynoecium syncarpous, tricarpellary, ovary inferior, trilocular with axile placentation (Fig. 56.7 E) or unilocular with basal placentation, as in *Calostemma* or unilocular with parietal placentation, as in *Leontochir;* ovules usually many but only 1 in *Choananthus*. Style 1, stigmas 3 or 1 and 3-lobed as in *Zephyranthes* (Fig. 56.7 B, C). Fruit usually a 3-celled capsule or a berry as in *Clivia, Cryptostephanus, Haemanthus;* seeds many with small embryo in fleshy endosperm.

Anatomy. Similar to that of the members of Liliaceae.

Embryology. Pollen grains monosulcate (disulcate in *Crinum*, Dutt 1970b); 2-celled at shedding stage. Ovules bitegmic, anatropous, crassinucellate, tenuinucellate in *Zephyranthes;* the inner integument forms

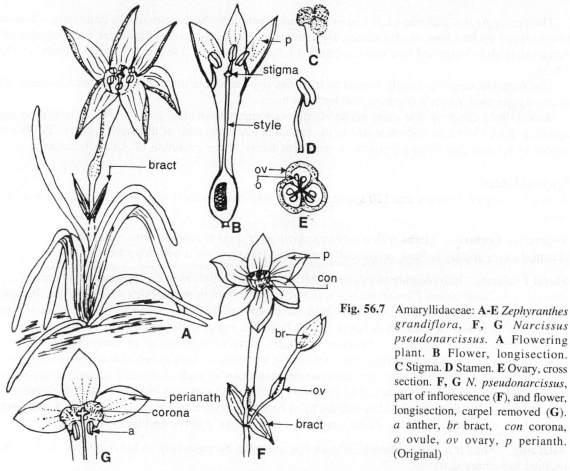

Fig. 56.7 Amaryllidaceae: **A-E** *Zephyranthes grandiflora*, **F, G** *Narcissus pseudonarcissus*. **A** Flowering plant. **B** Flower, longisection. **C** Stigma. **D** Stamen. **E** Ovary, cross section. **F, G** *N. pseudonarcissus*, part of inflorescence (**F**), and flower, longisection, carpel removed (**G**). *a* anther, *br* bract, *con* corona, *o* ovule, *ov* ovary, *p* perianth. (Original)

the micropyle. Embryo sac of Polygonum type, 8-celled at maturity. Endosperm formation of the Helobial or Nuclear type.

Chromosome Number. Basic chromosome numbers are x = 6–15, 23, 27, 29.

Chemical Features. *Narcissus pseudonarcissus* contains the galanthamine tazettine. Amryllidaceae members are particularly rich in steroidal saponins (Dahlgren and Clifford 1982, Harborne and Turner 1984). The polysaccharide levan is the storage carbohydrate in about 15 species of this family (Pollard 1982).

Important Genera and Economic Importance. The family consists of many well-known ornamental species such as *Crinum, Pancratium, Amaryllis, Clivia, Eucharis, Eurycles, Haemanthus, Hymenocallis, Lycoris, Narcissus, Nerine* and *Zephyranthes*.

Taxonomic Considerations. Bentham and Hooker (1965c) included Amaryllidaceae in the series Epigynae of the Monocotyledoneae, Engler and Diels (1936) in the order Liliiflorae, Hutchinson (1959) in the order Amaryllidales, Dahlgren (1980a, 1983a) in the order Asparagales and Takhtajan (1987) in the Liliales, Cronquist (1968) and Thorne (1968) do not recognise Amaryllidaceae as a distinct family and include its members in the Liliaceae.

The family Amaryllidaceae s.l. is a heterogeneous assemblage. Segregates of this family are Alliaceae (as suggested by Hutchinson), Agavaceae, Hypoxidaceae, Alstroemeriaceae, Taccaceae and Velloziaceae. Alliaceae is also recognised as a separate family by Takhtajan (1966, 1980, 1987) and Dahlgren (1975a, 1983a).

The Amaryllidaceae is closely related to Iridaceae (by the nature of ovary) and to the Liliaceae; the genus *Agapanthus* forms a common link between them.

Bose (1962) observes that gene mutation, chromosome repatterning, polyploidy, hybridization and apomixis have played an important role in the evolution and speciation of this family. Flory (1977) also agrees to the fact that hybridization is an important factor in the evolution of Amaryllidaceae.

Hypoxidaceae

A small family of 7 genera and 120 species, more or less cosmopolitan (except in Europe and northern Asia).

Vegetative Features.   Herbs with a tuberous rhizome or a corm covered with membranous or fibrous sheath. Leaves mostly radical, prominently veined and often clothed with long hairs.

Floral Features.   Inflorescence racemose or spicate or subumbellate as in *Campynemanthe*; often solitary flowers, e.g. *Campynema*. Flowers mostly white or yellow, actinomorphic, bisexual, epigynous. Perianth tube very short or consolidated into a large beak on the apex of the ovary, tepals 6 spreading, equal-sized and similar in colour. Stamens 6 or rarely 3 as in *Pauridia*, opposite the perianth segments and inserted at their base; anthers introrse or extrorse, bilocular, entire or sagittate as in *Campynema*, basifixed (*Campynemanthe*) or dorsifixed (*Hypoxis*), dehiscence longitudinal. Gynoecium syncarpous, tricarpellary, ovary inferior, trilocular, placentation axile, ovules numerous in each locule in 2 series, or rarely only a few. Fruit either a capsule, crowned by persistent perianth and variously dehiscent, such as opening by a circular slit or by short vertical slits near the tip or baccate and indehiscent, as in *Curculigo*. Seeds small, often black, with a distinct lateral hilum, sometimes carunculate, embryo enclosed in copious endosperm.

Anatomy.   Similar to that of Amaryllidaceae and Liliaceae. Stomata only on lower surface, without any defined subsidiary cells.

Embryology.   Pollen grains monocolpate and 2-celled at the dispersal stage. Ovules anatropous (*Campynema, Curculigo crassifolia*), hemi-anatropous (*Pauridia minida*) and campylotropous (*Empodium plicatum*); bitegmic and tenuinucellate (*Curculigo, Empodium, Hypoxis*) or crassinucellate (*Campynema*); micropyle is formed by inner integument only or by both. Polygonum type of embryo sac, 8-nucleate at maturity. Endosperm formation of the Helobial and/or Nuclear type.

Chromosome Number.   Haploid chromosome number is n = 6, 7, 9, 11, 14, 18, 22, 38, and 57.

Chemical Features.   The polysaccharide levan occurs in the members of this family (Pollard 1982).

Important Genera and Economic Importance.   Some of the important genera are; *Hypoxis, Curculigo* and *Molineria. C. latifolia* is an ornamental herb, occurs commonly in the Andamans. Fruits are edible and leaves are used for making fishing nets (Singh et al. 1990). *C. orchioides* is a perennial herb, distributed in the subtropical Himalayas, Khasi Hills and in Peninsular India. Its black root is ground and eaten like flour. *Hypoxis aurea,* a herb with tuberous rootstocks, occurs in hilly regions of India. The plant is used as tonic and aphrodisiac.

Taxonomic Considerations.   Bentham and Hooker (1965c) and Engler and Diels (1936) placed the members of Hypoxidaceae in Amaryllidaceae. Hutchinson (1934, 1959), Dahlgren (1975a, 1980a) and

Takhtajan (1987) include it in a separate order Haemodorales. Hutchinson (1959) presumes that Hypoxidaceae is closely related to the Orchidaceae through the genus *Curculigo*. Cronquist (1968, 1981) and Thorne (1983) include Hypoxidaceae in the Liliaceae.

According to Lawrence (1951), the two genera, *Campynema* and *Campynemanthe*, of Tasmania and New Caledonia, respectively, comprise the subfamily Campynematoideae of the Amaryllidaceae. The tribe Hypoxideae together with the subfamily Campynematoideae (of Amaryllidaceae) was later segregated and elevated as the family Hypoxidaceae by Pax and Hoffmann (1930), Hutchinson (1959, 1973), Melchior (1964) and Dutt (1970c). The position of these two genera is controversial. In contrast to features such as amoeboid anther tapetum, tenuinucellate ovule with micopyle formed by both integuments in most Hypoxidaceae, *Campynema* has secretory tapetum, a parietal cell in the ovule and the micropyle is formed by the inner integument. The secretory anther tapetum, and monosulcate, bicelled pollen grains show its alliance with Amaryllidaceae (Dutt 1970c, Johri et al. 1992). Occurrence of extrorse anthers and three free styles in these two genera—*Campynema* and *Campynemanthe*—also makes them distinct from other members of Hypoxidaceae (sensu Melchior 1964). Dutt (1970c), therefore, supports the retention of these two genera in the subfamily Campynematoideae within Amaryllidaceae.

Dahlgren and Lu (1985) recognise a distinct family Campynemataceae for *Campynema* and *Campynemanthe,* which together with Melanthiaceae comprise the order Melanthiales. They also observe that the two genera might be included in the family Melanthiaceae. Takhtajan (1987) does include the two genera in the family Melanthiaceae of order Liliales.

In our opinion also the genera *Campynema* and *Campynemanthe* should be segregated from the Hypoxidaceae. Its position in the order Liliales near the Amaryllidaceae is justified.

Velloziaceae

A small family of 5 or 6 genera and 260 species distributed in tropical South America, mainly Brazil, Africa, Madagascar and Arabia; mostly in rocky places and dry areas.

Vegetative Features.   Perennial, more or less xerophytic, occasionally arborescent with up to 6 m tall stem covered with persistent leaf sheaths and adventitious roots. Leaves usually spirally arranged, linear, parallel veined and sometimes with slightly dentate margin and often with spinous tips.

Floral Features.   Flowers solitary on each peduncle, white, yellow or blue, often very handsome; bisexual, epigynous, actinomorphic with 3 + 3 petaloid and basally connate tepals, segments equal, spreading, often covered with glandular hairs (Fig. 56.9 A, B). Stamens 3 + 3 or more by "dedoublement" in groups of 3 or more; anthers basi- or medifixed, linear, introrse, opening by longitudinal slits (Fig. 56.9 C, D). Gynoecium syncarpous, tricarpellary, ovary globose, inferior, trilocular (Fig. 56.9 E), placentation

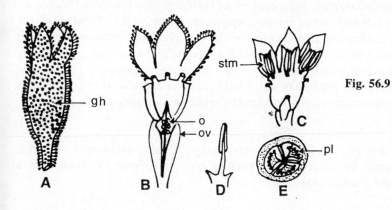

Fig. 56.9   Velloziaceae: **A-E** *Barbacenia bicolor.* **A** Flower, note glandular hairs on outer surface. **B** Longisection of flower without stamens. **C** Same, with stamens. **D** Stamen. **E** Ovary, cross section. *gh* glandular hair, *o* ovule, *ov* ovary, *pl* placenta, *stm* stamen. (Adapted from Lawrence 1951)

axile, style 1, simple, slender, stigma trilobate. Fruit a dry or hard capsule, often flat or concave on top; many-seeded, seeds with copious starchy endosperm and a small ovoid embryo.

Anatomy.   Vessels occur in roots and leaves, rarely in stem; only xylem tracheids present. Leaves dorsiventral, hairs unicellular; stomata of Rubiaceous type, predominantly on abaxial surface (Ayensu 1968). Vascular traces alternate with bands of sclerenchyma. Vessels short with simple perforation plates.

Embryology.   Pollen grains solitary or rarely in tetrads as in *Vellozia* (Ayensu and Skvarla 1974), sulcate, 2-celled at the dispersal stage. Ovules anatropous, bitegmic, tenuinucellate, with the micropyle formed by the inner integument. Polygonum type of embryo sac, 8-nucleate at maturity. Endosperm formation of the Helobial type.

Important Genera and Economic Importance.   *Barbecenia* occurs in tropical South America, tropical and south Africa and Madagascar. *Vellozia arabica* of Central Africa grows mainly on granite outcrops and is often host to small epiphytic orchids. The stem is thin but its coating of roots may be inches deep. Water poured over the roots disappears as if into a sponge and these plants are thus able to supply themselves from atmospheric precipitation like dew, during the dry season.

Taxonomic Considerations.   Most phylogenists—Bentham and Hooker (1965c), Engler and Diels (1936), Hallier (1912), Hutchinson (1973), Cronquist (1981)—include the family Velloziaceae in the Liliales; Thorne (1983) in the Commelinales; Huber (1969, 1977), Dahlgren et al. (1983) and Takhtajan (1987) in a distinct order Velloziales. Although this family is often associated with the Hypoxidaceae (Melchior 1964), this is not supported by recent studies.

    Vellozia exhibits a dichotomously branched woody stem bearing persistent leaf-bases and tufts of narrow, pointed leaves at the apex of the branches. The solitary terminal flowers with 'dedoubled' or branched stamens, and inferior trilocular ovaries and thick placentae are quite distinct. The family is sometimes considered closely related to the Bromeliaceae (Johri et al. 1992). The embryology of the two families is also comparable. Further investigations about embryogeny, chromosome number and chemical features may be helpful in deciding the phyletic position of the family Velloziaceae.

Taccaceae

A family of 2 genera, *Tacca* and *Schizocapsa,* and 31 species (Willis 1973) which occur in southeastern Asia and China.

Vegetative Features.   Perennial herbs with tubercular or creeping stach-rich rhizomes. Leaves radical, long-petiolate, with simple and entire (*Schizocapsa*) or palmately lobate to compound lamina.

Floral Features.   Inflorescence a pseudo-umbel, supported by an involucre of a few broad leaves and several long filiform bracts. Flowers bisexual, actinomorphic, epigynous, campanulate. Tepals 6 (3 + 3) subequal, dull-coloured, tube shortly and broadly cupular, segments broad, spreading, imbricate, persistent. Stamens 6, epipetalous; anthers broad, introrse, bicelled, dehisce longitudinally, with rounded connective; filaments short, petaloid. Gynoecium syncarpous, tricarpellary, ovary unilocular with parietal placentation, ovules many; style short, 3 reflexed stigmas, petaloid or not. Fruit baccate or capsular as in *Schizocapsa,* with prismatic, ex-arillate seeds, copious non-starchy endosperm, and minute embryo.

Anatomy.   Stem vesselless.

Embryology.   Pollen grains free, sulcate, 2-celled at dispersal stage. Ovules anatropous, bitegmic, crassinucellate; inner integument forms the micropyle. Polygonum type of embryo sac, 8-nucleate at maturity. Endosperm formation of the Nuclear type.

Chromosome Number. Basic chomosome number is x = 15.

Important Genera and Economic Importance. *Tacca* with 30 species is the larger genus of the two, *Tacca* and *Schizocapsa.* East Indian arrowroot is made from the rhizome of *T. leontopodioides* and other species. Monotypic *Schizocapsa* occurs in southeast China.

Taxonomic Considerations. Affinity of the family Taccaceae is probably closer to the Dioscoreales than to the Amaryllidaceae near which they are sometimes placed (Dahlgren and Clifford 1982). However, embrylogically, the two families are close (Johri et al. 1992). Melchior (1964) also places it near Dioscoreaceae. Takhtajan (1987), although includes it in a distinct order, Taccales, retains its position near Dioscoreales.

Dioscoreaceae

A large family of 8 genera[11] and 650 species, distributed in both the New and the Old World tropics.

Vegetative Features. Rhizome common in all genera but tubers only in a few. Evergreen climbers (*Avetra*), climbers (*Dioscorea, Rajania, Tamus, Stenomeris*), erect herb (*Croomia*), low shrub (*Stichoneuron*) or non-climber with short stem (*Trichopus*). Bulbils occur only in *Dioscorea*. Aerial stems annual, rounded or winged, with or without spines or hairs; 120–225 mm high in *D. barlettii* and over 396 m in *D. mangenotiana*. Leaves simple or compound, even within the same genus (*Dioscorea*), mostly distichous, sometimes tristichous or even irregular.

Floral Features. Inflorescence cymes, racemes or spikes in various species of *Dioscorea* and *Rajania*; solitary or racemes in *Tamus* and *Avetra;* only cymes in *Stenomeris;* cluster of solitary flowers in *Trichopus* and only racemes in *Stichoneuron.* Flowers unisexual (*Dioscorea, Rajania* and *Tamus*) or bisexual, epigynous, rarely hypogynous as in *Croomia* and *Stichoneuron.* Perianth mostly hexamerous, 6-lobed in *Stenomeris,* 6-partite in *Trichopus,* bell-shaped in *Avetra* and tetramerous in *Croomia* and *Stichoneuron.* In male flowers stamens mostly 6, but 4 in *Croomia* and *Stichoneuron,* in 2 whorls, the inner whorl sometimes reduced to staminodes; filaments free or briefly connate, anthers basifixed, introrse or extrorse, bicelled, the 2 cells confluent or contiguous, dehiscence longitudinal; the connective sometimes proliferated or appendaged. In female flowers the gynoecium is syncarpous, tricarpellary, ovary inferior, trilocular, placentation axile, ovules 2 to numerous in each locule; style 1 or more, stigmas 3- or sometimes 2-partite; staminodes 2. Bisexual flowers have a combination of staminate and pistillate characters. Fruit a capsule (*Dioscorea, Stenomeris, Croomia, Stichoneuron*) or samara (*Rajania, Avetra*) or berry (*Tamus, Trichopus*). Seeds compressed or ovoid, winged (*Dioscorea, Avetra*) or not winged.

Anatomy. Vessels small to medium-sized, pits on lateral walls, alternate and opposite; perforated scalariform plates oblique (Tan and Rao 1974). There are some very complex and peculiar assemblages of specialized vascular cells situated in the stems at the level of nodes. These have been described as 'glomeruli' (Ayensu 1972). Both xylem-glomerulus and phloem-glomerulus occur. The former is composed of short tracheids of variable shape closely fitted together, confined to nodes and have large bordered pits. Phloem-glomeruli are composed of somewhat funnel-shaped, thin-walled cells with numerous pits on the end walls. Stomata Ranunculaceous type, mostly confined to the lower surface but occasionally on upper surface also, as in *Dioscorea campestris* and *D. bulbifera.* Cuticular striations, uni or bicellular hairs, tannins and crystals occur in most members.

---

[11]Some of these genera have been included in other families or raised to the rank of a family (see p. 554).

Embryology.   Pollen grains heterogeneous (G. Erdtman 1952), monosulcate, 2- or 3-sulcate or 4- or 5-foraminoidate (Zavada 1983); 2-celled at dispersal stage (Johri et al. 1992). Ovules anatropous, bitegmic, crassinucellate, both integuments form the micropyle. Embryo sac of Polygonum type, 8-nucleate at maturity, or of Drusa type and 16-nucleate at maturity. Endosperm formation of the Nuclear type.

Chromosome Number.   Basic chromosome number of *Dioscorea* is x = 10 (Sharma and De 1956, Raghavan 1958); of *Tamus communis* x = 12 (Darlington and Wylie 1955); of *Trichopus* x = 14 (Ramachandran 1968)

Chemical Features.   *Dioscorea* spp. contain both sapogenins and diosgenins. In *D. deltoidea* and *D. prazeri* sapogenin concentration increases with the age of the tubers (Bindroo et al. 1985). Maximal diosgenin content is obtained late in the growing season. Full development of leaves is yet another factor for maximal concentration. Chelidonic acid is also reported in many members (Johri et al. 1992).

Important Genera and Economic Importance.   *Dioscorea* with ca. 600 species is the most important genus. Many of the species are well-known food plants. Yams are the starch-rich rhizomes used as food, particularly in the Far East. It is also the source of the drug diosgenin. The tuber is the potential source of precursors for cortisone synthesis. Young shoots of *Tamus communis* are eaten as asparagus.

Taxonomic Considerations.   Bentham and Hooker (1965c) treated Dioscoreaceae along with Scitamineae, Bromeliaceae, Haemodoraceae, Iridaceae, Amaryllidaceae and Taccaceae. Knuth (1930) included Dioscoreaceae in the suborder Liliineae of order Liliiflorae. He also recognized two tribes—Dioscoreae with dioecious flowers, and Stenomerideae with bisexual flowers. The unisexual flowers with inferior ovary suggest their advanced features. The members of this family show probable phylogeny with dicotyledons in reticulate venation and in the absence of leaf-sheath (Johri et al. 1992). Embryological and chemical data support its alliance with the family Liliaceae.

The genus *Petermannia*—formerly included in the tribe Stenomerideae—is now recognied as a distinct family, Petermanniaceae (Conran 1988). *P. cirrosa*, from eastern Australia, is the only member of this family. It has earlier been included in Dioscoreaceae (Lawrence 1951), Smilacaceae (Cronquist 1981, Schulze 1982, Conran and Clifford 1986) or in Philesiaceae (Schlittler 1949, Takhtajan 1980).

Hutchinson (1959) proposed raising a few other genera to the rank of family, such as Stemonaceae (including *Croomia, Stichoneuron* and *Stemona*) and Trichopodaceae (including *Avetra, Trichopus*). Burkill (1960), however, disapproved these assignments. At present, many phylogenists treat these families as separate taxa (Dahlgren and Clifford 1982, Takhtajan 1987), and it would be correct to do so to make Dioscoreaceae a more natural taxon.

Pontederiaceae

A family of 9 genera and 34 species, pantropically distributed.

Vegetative Features.   Rhizomatous and sometimes stoloniferous, free-floating or rooted herbs (Fig. 56.12A) in swamps and waters, roots fibrous; perennial or rarely annual, e.g. *Hydrothrix;* stem very short or erect, unbranched, mostly enveloped by sheaths of leaf base. Leaves alternate, distichous, generally differentiated into a leaf-sheath, petiole and the entire lamina. Ligules and stipule-like lobes often present.

Floral Features.   Inflorescence a raceme (Fig. 56.12A) or panicle, usually subtended by a spathe-like leaf-sheath. Flowers hypogynous, actinomorphic to zygomorphic, bisexual (Fig. 56.12B). Tepals mostly 6, imbricate, white or blue, often basally fused. Stamens usually 6 or 3 or 1, inserted on perianth, anthers tetrasporangiate, basi- or dorsifixed (Fig. 56.12 B, C), introrse, opening longitudinally or by pores. Gynoecium syncarpous, tricarpellary, ovary surperior (Fig. 56.12 D), trilocular, with axile placentation,

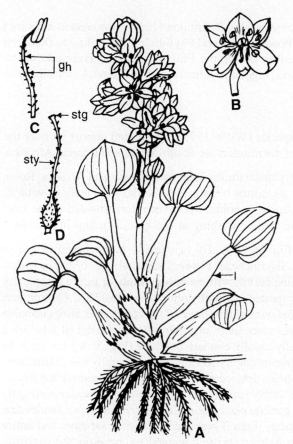

Fig. 56.12 Pontederiaceae: **A-D** *Eichhornia crassipes.*
A Flowering plant. **B** Flower. **C** Stamen.
D Carpel. *gh* glandular hair, *l* leaf, *stg* stigma,
sty style. (Original)

2 locules sometimes empty; style 1, stigma 1, dry; ovules 1 to numerous per locule. Fruit a loculicidal capsule or a nut, as in *Pontederia;* seeds with starchy endosperm and a linear straight embryo.

Anatomy. Vessels with scalariform perforation plates often present, laticifers absent; oxalate raphides sometimes present but styloides more dominant.

Embryology. Pollen grains free, sulcate or bi- or trisulcate, 2-celled at the dispersal stage. Ovules anatropous, bitegmic, crassinucellate; micropyle formed by both the integuments. Embryo sac of Polygonum type, 8-nucleate at maturity. Endosperm fomation of the Helobial type.

Chromosome Number. Basic chromosome number is x = 8, 14, 15.

Important Genera and Economic Importance. Of the 9 genera, *Hydrothrix* is monotypic and is reported from Brazil, *Reussia* with 3 species also occurs in South America, mainly in Brazil; *Eichhornia* with 5 species is tropical American but grows extensively in other tropical and subtropical parts of the world, often as an obnoxious weed. *E. crassipes* or water hyacinth is sometimes grown as an aquatic ornamental for its foliage and blue flowers. It is also known to absorb heavy metals present in the water as pollutants. *Pontederia* or pickerel weed is also grown as an ornamental.

Taxonomic Considerations. The Pontederiaceae has been included by most phylogenists in the order Liliales (or Liliiflorae). Dahlgren (1980a, 1983a), Dahlgren and Clifford (1982), Dahlgren et al. (1985)

and Takhtajan (1987) place it in a separate order, Pontederiales. Takhtajan (1987) also erected a distinct superorder, Pontederianae, including the two orders, Pontederiales and Phylidrales. According to Dahlgren and Clifford (1982), it probably has its closest relatives among the Philydraceae and Haemodoraceae. Johri et al. (1992), on the basis of embryologicial features, suggest its alliance with Liliaceae, Commelinaceae and Philydraceae.

Iridaceae

A moderately large family of 60 genera and 1800 species (Willis 1973), distributed almost all over the world except extremely cold regions; chief centres of distribution are South Africa and tropical America.

Vegetative Features. Perennial herbs or very rarely undershrubs such as *Witsenia* and *Nivenia*. Roots fibrous from the stem modified to bulbs (Fig. 56.13 A), corms or rhizomes. Aerial stem leafy or without leaves, herbaceous, in bunches from the bulb or rhizome or solitary. Leaves often crowded at the base of the stem, mostly narrowly linear, flattened at the sides, sheathing at the base, equitant in 2 ranks.

Floral Features. Terminal racemes (*Gladiolus*) (Fig. 56.13 A, E), or cymes—monochasial cymes in *Freesia;* only 1-flowered as in members of the tribe Sisyrincheae. Flowers bisexual, epi- or hypogynous, actinomorphic, with a straight perianth tube or the tube curved with an oblong limb, or rarely completely zygomorphic, usually ornamental with mottled or spotted perianth lobes (Fig. 56.13B,C,F). Perianth petaloid, of 6 tepals in 2 series, subequal and similar (*Belamcanda*) or those of the outer series distinguishable from those of the inner series by size, colour or shape; when limb (or tube) oblique, the dorsal lobe often the largest and hood-like, as in *Gladiolus,* all generally basally connate in a tube (Fig. 5.13 F) or the tube obsolete. Stamens 3, opposite outer perianth lobes, free or epiphyllous; filaments usually free, sometimes partially connate; anthers bicelled, extrorse or latrorse, dehiscence by vertical slits, mostly basifixed. Gynoecium syncarpous, tricarpellary; ovary inferior, rarely superior as in *Hewardia,* trilocular with axile placentation (Fig. 56.13 D) or rarely unilocular with parietal placentation, e.g. *Hermodactylus;* ovules few to many in each locule, rarely solitary; style 1, slender, often 3-branched, branches subulate and entire or deeply lobed, sometimes winged and petaloid; the style crests often winged and petaloid; the stigmatic

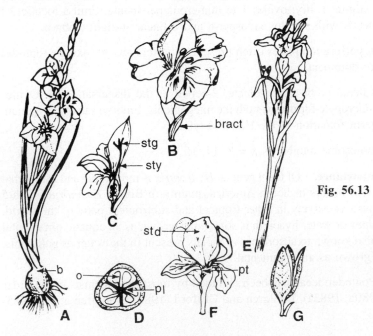

Fig. 56.13 Iridaceae: **A-D** *Gladiolus tristis*, **E-G** *Iris* spp. **A** Plant with inflorescence. **B** Flower. **C** Longisection. **D** Ovary, transection. **E** Inflorescence of *Iris xiphium.* **F** Flower of *I. germanica.* **G** Capsule of *I. sibirica. b* bulb, *o* ovule, *pl* placenta, *pt* perianth tube, *std* standard, *stg* stigma, *sty* style. (After Lawrence 1951)

surface terminal or adaxial. Fruit a loculicidal capsule dehiscing by 3 valves (Fig. 56.13 G). Seeds endospermous, sometimes arillate, embryo small.

Anatomy.   Similar to that of Amaryllidaceae and Liliaceae.

Embryology.   Pollen grains monosulcate or inaperturate, with reticulate exine (Zavada 1983); 2-celled at the dispersal stage. Ovules anatropous, bitegmic, tenuinucellate; the inner integument forms the micropyle. Embryo sac of Polygonum type, 8-nucleate at maturity. Endosperm formation of the Nuclear type.

Chromosome Number.   Basic chromosome number is $x = 3-19, 22, 25$.

Chemical Features.   Xanthones (a group of plant pigments) occur in *Iris* species (Harborne and Turner 1984); crocine, a yellow pigment, picrocine, and safranol (produces fragrance) are reported in *Crocus sativus* (Plessner et al. 1989).

Important Genera and Economic Importance.   Most genera of this family are ornamental with large beautiful showy flowers, e.g. *Belamcanda, Iris, Gladiolus, Cipura, Freesia, Ixia, Tigridia* and others. *Crocus sativus* is a sterile genotype propagated by annual replacement of corms. It has been cultivated for at least 3500 years, its red stigmatic lobes constitute saffron, valued since ancient times for its odoriferous, colouring and medicinal properties. Its active components—safranol for odour, picrocine for taste, and crocine (a carotenoid pigment)—are all located in the stigmatic lobes (Plessner et al. 1989). Hysteranthy (flowering before the appearanracre of foliage leaves) in *Crocus sativus* can be induced by manipulating temperature conditions of corm storage and planting (Plessner et al. 1989). The dried roots or rhizomes of *Iris florentina, I. pallida* and *I. germanica* var. *germanica* constitute orris root, a violet scented perfume. The rhizome of *I. foetidissima* serves as a specific for hysteria (Dutta 1988).

Taxonomic Considerations.   Iridaceae is a fairly large homogeneous and natural family related to Liliaceae. Bentham and Hooker (1965c) included it in the series Epignae, Engler and Diels (1936) in Liliiflorae, Cronquist (1968, 1981), Dahlgren (1975a, 1980a) and Thorne (1968, 1983) in the order Liliales, and Hutchinson (1959) and Takhtajan (1980, 1987) in the order Iridales, Bentham and Hooker (1965c) divided this family into 3 tribes and many subtribes which were raised to the rank of tribes by Hutchinson. He also included in this family. *Hewardia* (= *Isophysis*), a Tasmanian genus with superior ovary but otherwise Iridaceous. This genus is a link between Liliaceae and Iridaceae. The tribe Sisyrinchieae— included the genera *Libertia, Bobartia, Sisyrinchium* and *Belamcanda*— is a simple and primitive tribe. The more advanced tribes are Gladioleae and Antholyzeae with more or less zygomorphic flowers, mostly curved perianth tube with oblique limb and hood-like dorsal lobe. Tribe Antholyzieae is mostly confined to South Africa.

The general floral morphology of Iridaceae members resembles that of Liliaceae. Anatomical and embryological features are similar to those of Liliaceae and Amaryllidaceae.

This family has presumably originated from a Liliaceous stock, the genus *Hewardia* is the link between the two families.

Geosiridaceae

A monotypic family, the only genus *Geosiris,* with a single species which occurs in Madagascar.

Vegetative Features.   Small, colourless, saprophytic herbs arising from underground scaly rhizomes. Stems simple or branched; leaves alternate, squamiform (scale-like).

Floral Features.   Inflorescence loosely cymose, bracteate. Flowers actinomorphic, bisexual, epigynous. Perianth lobes 6 in 2 series, petaloid, shortly connate at the base, subequal; the outer 3 imbricate and the inner contorted, bluish coloured. Stamens 3, opposite the three outer tepals, filaments short; anthers large, oblique, basifixed, extrorse, dehiscence longitudinal. Gynoecium syncarpous, tricarpellary, ovary inferior, trilocular with numerous ovules, on dendriform axile placenta. Fruits trigono-obconic, crowned with an annulus at truncate apex. Seeds numerous, minute.

Anatomy.   Stems vesselless. Anatomical investigations are incomplete.

Embryology.   Pollen grains monosulcate. Ovules anatropous, bitegmic (Dahlgren and Clifford 1982). Embryology has not been investigated fully.

Important Genera and Economic Importance.   The only genus, *Geosiris,* is a parasitic herb, with no economic importance.

Taxonomic Considerations.   Most phylogenists place this family in the order Liliales (or Liliiflorae); Cronquist (1968, 1981) included it in the Orchidales. Hutchinson (1959) did not recognise Geosiridaceae as a distinct family and treated the genus *Geosiris* as a member of Burmanniaceae, which is allied to the Orchidaceae. Being saprophytic in nature, this family is allied to the members of Orchidales, but otherwise it is Liliaceous in other respects. Iridaceae and Geosiridaceae form the suborder Iridineae.

Burmanniaceae

A family of 17 genera and 125 species (Willis 1973); pantropic and grow mainly in rain forests of Japan, Tasmania and New Zealand.

Vegetative Features.   Annual or perennial, small, saprophytic and colourless, or rarely green and autotrophic herbs, the perennial taxa with rhizomes or tubers. Leaves linear and parallel veined, ligules and stipules absent, alternate or radical, simple, entire, more often reduced to scales.

Floral Features.   Inflorescence di- or monochasial cyme or racemose or flower solitary and terminal. Flowers actino- or zygomorphic, semi-epigynous to epigynous, generally with 2 whorls of tepals, similar or dissimilar, sometimes fused to a tube. Tepals coloured or white to hyaline, sometimes bizzare in shape; tube often winged, segments sometimes appendiculate; septal or placental nectaries sometimes present. Stamens 6 or 3, epipetalous, basifixed, tetrasporangiate, extrorse or introrse anthers. Gynoecium tricarpellary, carpels fused in the ovary region, with a single, filiform or conical style with 3 stigmas; placentation often parietal, rarely axile. Fruit capsular, sometimes fleshy with numerous, minute, exarillate seeds without phytomelan and a starchy endosperm, and a minute organ-less (undifferentiated) embryo (Johri et al. 1992).

Anatomy. Vessels present in roots, stem as well as leaves at least in one species of *Burmannia;* have scalariform peforation plates. Hairs absent or unicellular on leaf surface; stomata absent or without subsidiary cells. Oxalate raphides sometimes present.

Embryology.   Pollen grains free, monocolpate and 2- or 3-celled at the dispersal stage. Ovules anatropous, bitegmic, tenuinucellate. Embryo sac of Polygonum type, 8-nucleate at maturity. Endosperm formation of the Helobial type.

Important Genera and Economic Importance.   Amongst the members of this family, *Burmannia* and *Thismia* are well-known. *Burmannia* with its 57 species is a slender, saprophytic weed, most prevalent in damp tropical woods or savannahs. *B. coelestis* is a slender weed of waste places; *B. pusilla* bears bluish violet flowers (Dutta 1988). *Thismia americana* is a rare endemic known only from an open prairie near Chicago, Illinois. Neither of the genera has any economic importance, but they are of academic interest.

Taxonomic Considerations. Complexity of floral forms and reduction of floral members support the alliance of the family Burmanniaceae with the Orchidaceae.

Engler and Diels (1936), Rendle (1904), Wettstein (1935), Lawrence (1951), Cronquist (1968, 1981) and Hutchinson (1934, 1959, 1973) include Burmanniaceae in the Orchidales. Melchior (1964) and Thorne (1968, 1983), however, include it in the Liliales (or Liliiflorae) because of its probable resemblance with members of the Iridaceae (also of the Liliales). Dahlgren and Clifford (1982), Dahlgren et al. (1985), and Takhtajan (1987) recognise a separate order, Burmanniales, including Burmanniaceae and Corsiaceae. The family Burmanniaceae also shows some resemblance with members of Philydraceae in having a starchy endosperm, Helobial endosperm formation, zygomorphic flower, vessels with scalariform perforation plates, and with Amaryllidaceae in regular trimerous flowers. It is rather difficult to assess the true affinity of this family. Under the circumstances, it would probably be better to place it in an order of its own, i.e. Burmmaniales, and place it between the Liliales and Orchidales.

Corsiaceae

A small family of 2 genera, *Corsia* and *Arachnitis,* with 9 species, distributed in West Australia and Chile.

Vegetative Features. Small, erect, perennial, rhizomatous or tuberous saprophytes; leaves reduced to large scales.

Floral Features. Flowers solitary, terminal, zygomorphic, bisexual as in *Corsia,* or unisexual as in *Arachnitis,* epigynous. Perianth in 2 whorls of 3 each, posterior member of outer whorl large, coloured, cordate-ovate, margin laterally involute in bud, enclosing the remaining 5, sometimes with a large gland inside towards the base; other 5 linear-filiform and more or less reflexed. Stamens 6, exerted, with distinct but short and subequal filaments; anthers shortly ovoid, bicelled, extrorse, dehiscence longitudinal. Rudimentary ovary in staminate and staminodes in pistillate flowers. Gynoecium syncarpous, tricarpellary, ovary inferior, trilocular, globose or elongated with three parietal, much intruded and bifurcate placentae; style short with 3 short thick stigmas. Ovules numerous on each placenta. Fruit a short and broad (*Arachnitis*) or elongated capsule (*Corsia*); cylindrical, opening vertically by 3 valves. Seeds numerous, very small, testa reticulate and extends slightly (at each end) beyond the embryo.

Anatomy and Embryology have not been investigated.

Chromosome Number and Chemical Features. Have not been studied.

Important Genera and Economic Importance. The two genera, *Corsia* and *Arachnitis,* show disjunct distribution in West Australia and Chile. They have no economic importance.

Taxonomic Considerations. The family Corsiaceae has been treated as a tribe (Corsieae) of the Burmanniaceae by Bentham and Hooker (1965c). Jonker (1938), Hutchinson (1934, 1959), Lawrence (1951), Wettstein (1935), Cronquist (1968, 1981), Dahlgren (1975a, 1980a, 1983a), Dahlgren and Clifford (1982), Dahlgren et al. (1985), Takhtajan (1987) and Dutta (1988) treat Corsiaceae as a distinct family. It differs from the Burmanniaceae in zygomorphic flowers, and 6 stamens. It resembles Burmanniaceae in extremely minute seeds and absence of endosperm. Along with Burmanniaceae it is sometimes included in the Orchidales (Lawrence 1951, Cronquist 1981, Thorne 1983); more often in a separate order, Burmanniales (Dahlgren 1980a, 1983a, Takhtajan 1987), along with Burmanniaceae.

The position of Corsiaceae in the Orchidales is justified. All the members of Orchidales are similar to Liliiflorae, except that sometimes they lack chlorophyll and have numerous, tiny seeds, undifferentiated embryo, and endosperm scanty or none. Amongst the four members of Orchidales (sensu Cronquist 1981), Corsiaceae, Geosiridaceae and Burmanniaceae have remained comparatively less evolved.

Philydraceae

A small family of only 4 genera and 5 species, restricted to southeastern Asia, Indo-Malaya, New Guinea and parts of Australia.

Vegetative Features.   Erect, often large, perennial, marsh or aquatic herbs with rhizomes or tubers. Leaves alternate, distichous, linear, usually ensiform, flat, parallel veined, eligulate, estipulate.

Floral Features.   Inflorescence racemose. Flowers bracteate, hypogynous, zygomorphic, with petaloid tepals, the lateral members of the outer whorl and median of the inner whorl fuse to form a large upper lip, and the median member of the outer whorl forms a large lower lip; the lateral inner tepals are small. Septal nectaries absent. Androecium with a large functional stamen with a dorsifixed, basically introrse (in *Philydrum* helically twisted), anther with 4 microsporangia. Gynoecium syncarpous, tricarpellary, ovary tri- or unilocular, with simple style and stigma; placentation axile or intrusive-parietal. Fruit a capsule or a dry berry as in *Helmholtzia*; seeds carunculate; endosperm starchy but also contains oil; embryo small, straight.

Anatomy.   Vessels absent in stem and leaves. Glandular, uniseriate hairs with a long end-cell common. Stomata of Rubiaceous type or tetracytic; oxalate raphides in anther tapetal cells and styloids in vegetative parts common; silica bodies absent.

Embryology.   Pollen grains free or in tetrads. monosulcate, with reticulate exine; 2-celled at dispersal stage. The pollen grains usually germinate in situ, and many pollen tubes enter the stigma of the same flower (Johri et al. 1992). Ovules anatropous, bitegmic, crassinucellate; the inner integument forms the micropyle. Embryo sac of Polygonum type, 8-nucleate at maturity. Endosperm formation of the Helobial type (Hamann 1966).

Chromosome Number.   Basic chromosome number is x = 8, 17.

Chemical Features.   Flavonoid syringetin, a rare substance, reported in *Philydrum lanuginosum*.

Important Genera and Economic Importance.   *Philydrum lanuginosum* is one species of wide occurrence in Australia, Malaysia, East Asia and the Andamans; economic importance not known.

Taxonomic Considerations.   The family Philydraceae has been treated variously. It has been placed in Liliales (or Liliiflorae) by Melchior (1964) and Cronquist (1968, 1981), in Commelinales by Lawrence (1951), Thorne (1968, 1981) and Hamann (1966) and in the Philydrales by Dahlgren (1975a, 1980a, 1983a), Dahlgren and Clifford (1982) and Takhtajan (1987). According to Cronquist (1968), this family is a relatively unsuccessful offshoot from the primitive Liliid stock, when its differentiation from the Commelinid stock was not yet complete. Members of Philydraceae are much like the Commelinaceae members in general appearance, have starchy endosperm, and (apparently) lack nectaries. On the other hand, they also resemble many members of the Lliiflorae (Liliales); in the absence of silica bodies, presence of oxalate raphides, scalariform vessel perforations, petaline tepals, sulcate, binucleate pollen grains, axile or parietal placentation, capsular fruit and Helobial endosperm. According to Hamann (1966), the Philydraceae are closely related to the Pontederiaceae, and both of them are best treated as a peripheral subgroup within the Liliales. Dahlgren (1980a), while placing this family in a separate order, Philydrales, comments that they are transitional between the Commelinales and the Liliales. Under the circumstances, it would probably be better to treat this family in a separate order of its own, i.e. Philydrales, or in a suborder Philydrineae under the Liliiflorae (Liliales).

Taxonomic Considerations of the Order Liliiflorae

The Liliiflorae is a very large order comprising 17 families. A heterogeneous order, many of its suborders have been raised to the rank of distinct orders by Dahlgren (1980a, 1983a), Dahlgren and Clifford (1982), Dahlgren et al. (1985) and Takhtajan (1987); the suborders, Burmanniineae and Philydrineae, have been raised to Burmanniales and Philydrales, respectively. There are so many variations in exomorphological and cytological features that they might be treated as distinct orders; but, at the same time, all these families are bound together by the perennial habit of the plants and similar embryological and chemical features.

Therefore, it would be better to treat this order as comprising a number of suborders, as has been suggested by Melchior (1964).

Order Juncales

The order Juncales comprises two families—Juncaceae and the monotypic Thurniaceae. Perennial or annual herbs, mostly occur in wet and damp habitats of the temperate and arctic regions and tropical mountains.

The members have graminoid habit, rarely with distinct aerial stem; leaves mostly basal, small anemophilous flowers that are bracteate and bracteolate, in many-flowered, often head-like clusters; glumaceous perianth and superior ovary. Vessels present in stem and leaves; silica bodies mostly lacking in Juncaceae but present in Thurniaceae. Pollen grains in tetrads, ovules anatropous; seeds endospermous with a small straight embryo.

Juncaceae

A small family of 9 genera and 400 species, distributed worldwide but some genera restricted to southern hemisphere. The plants are most frequent in temperate and cold or montane regions, usually hygrophilous.

Vegetative Features.   Perennial or annual, often rhizomatous herbs with "graminoid" habit, rarely with a distinct aerial stem, as in *Prionium.* Rhizome erect or horizontal, or rarely absent. Stem mostly leafy at the base. Leaves mostly basal, alternate, usually tristichous, sometimes distichous, linear or filiform, flat, canaliculate, terete or laterally compressed, sheathing, often with stipule-like ears at the base of the lamina, parallel veined, glabrous or with ciliate margins. A median ligule is recorded in *Oxychloë* (Guédes 1967). Culm erect, leafless or leafy.

Floral Features.   Inflorescence mostly cymose—a cincinnus, as in *Juncus bufonius,* a monochasial cyme, as in *J. inflexus* or a small head-like cluster, as in *J. articulatus* (Fig. 57.1A), or even solitary flowers, as in *J. trifidus* and *Rostkovia.* Flowers uni- or bisexual (Fig. 57.1B), plants dioecious or not, actinomorphic, hypogynous; tepals in 2 whorls of 3 each or only 3, green, brown or hyaline, glumaceous or coriaceous, woody in *Marsippospermum,* petaloid in some species of *Juncas* and *Luzula*; free. Stamens as many and opposite the tepals (Fig. 57.1C) free; anthers bicelled, introrse, dehiscing longitudinally, basifixed. Gynoecium syncarpous, tricarpellary, ovary tri- or unilocular (Fig. 57.1D, E) (*Luzula*), superior, with axile or basal placentation. Style basally simple, 1, very short to linear, or 3, stigmas 3, linear to lanceolate, brush-like. tribracheate with dry stigma. Ovules 3 to numerous per locule, biseriate or only 1 and basal; in *Luzula* ovary unilocular with 3 basal ovules. Fruit a loculicidal capsule, rarely indehiscent as in *Oxychloë;* seeds sometimes with elaiosome, endosperm copious; embryo small, straight, basal, ovoid with a large cotyledon.

Anatomy.   Vessels in stem and leaves with scalariform, or simple-scalariform or scalariform-reticulate perforation plates. Lateral wall pitting reticulate, scalariform or alternate; perforation plates frequently oblique. Stomata Rubiaceous type. Oxalate raphides absent as also silica bodies.

Embryology.   Pollen grains in permanent tetrads, all functional, 3-celled at shedding stage. Ovules anatropous, bitegmic, crassinucellate; inner integument forms the micropyle. Polygonum type of embryo sac, 8-nucleate at maturity. Endosperm formation of the Helobial type.

Chromosome Number.   Basic chromosome numbers are x = 3–30.

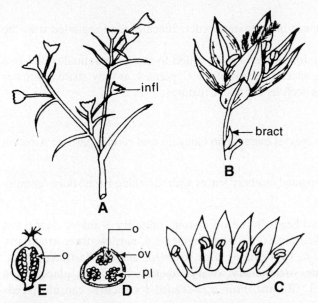

Fig. 57.1 Juncaceae: **A** *Juncus articulatus*, **B-E** *J. tenuis*. **A** Portion of plant with inflorescence. **B** Flower of *J. tenuis*. **C** Perianth and androecium. **D** Ovary, cross section. **E** Vertical section. *infl* inflorescence, *o* ovule, *ov* ovary, *pl* placentation. (**A** after S.C. Dutta 1988, **B-E** after Lawrence 1951)

Chemical Features. Luteolin and daphnetin derivatives are reported in Juncaceae members (Williams and Harborne 1975).

Important Genera and Economic Importance. *Juncus* with 300 species and *Luzula* with 80 species are the larger genera. The stems of *Juncus effusus* and *J. squarrosus* are sometimes used for weaving mats, hats, chair seats, etc. Many species of *Luzula* are used as a cure for kidney ailments (Dutta 1988). The tussocks or rushes (as species of *Juncus* are commonly known) often provide stepping-stones for crossing a wet meadow, as well as hiding places for many kinds of birds and mammals (Jones and Luchsinger 1987).

Taxonomic Considerations. The family Juncaceae has occasionally been placed in the Liliiflorae, because of the complete Lilüflorean flower construction and capsular fruits. Rendle (1904) commented that the relation between the two families, Juncaceae and Liliaceae s.l., is so close that several genera with membranous perianth like *Xanthorrhoea*, *Calectasia* and others were placed indifferently in either family.

The Juncaceae are a very old family, extending through the Tertiary and back into the Cretaceous. Different authors have different views about the centres of origin of this family. According to Buchenau (1906; see Lawrence 1951), it originated in the mountains of Eurasia; according to Vierhapper (1930; see Lawrence 1951), in the Old World tropics, and according to Weimarck (1946), it has probably spread northwards from Australia.

Though some taxonomists regarded Juncaceae as a group reduced from Liliaceous stock, others considered it as ancestral to the Liliaceae. Juncaceae is also allied to Cyperaceae and Gramineae in glumaceous perianth, feathery stigmas and wind pollination, but differs from them in a 6-parted perianth. From Liliaceae also they differ, as the Juncaceae characteristically have vessels throughout the plant body and stomates with two subsidiary cells, whereas the Liliaceae have no subsidiary cells, and have vessels confined chiefly to the roots (Cronquist 1968). In both these respects, they resemble the Restionaceae, Cyperaceae, and Gramineae. Hutchinson (1934, 1959) placed the Juncaceae in Juncales, a component of his Glumiflorae. Cronquist (1968, 1981) also included this family in Juncales and considered it to be a reduced Commelinideae, placing it between the Restionales and Cyperales. Thorne (1968, 1983) included it in the Commelinales as a suborder Juncineae. Dahlgren (1975a, 1980a, 1983a), Dahlgren and Clifford

(1982) and Takhtajan (1987) also treated Juncaceae in a separate order, Juncales, and regarded it as the progenitor of the Cyperales.

From the above discussion, it is evident that Juncaceae is more allied to the Commelinales and less so to the Liliales; it may also be regarded as the ancestral form of the Cyperales, as they share a number of morphological, embryological, anatomical as well as chemical features.

Thurniaceae

A monotypic family of the genus *Thurnia* with 3 species endemic to Guayana and certain parts of Amazon river valley.

Vegetative Features.   Sedge-like herbs, bear elongated, leathery leaves with sheathing bases; dorsiventrally flattened, margin smooth or spinulose-serrate.

Floral Features.   Inflorescence of one or more heads borne on a stout, obtusely 3-angled scape and subtended by more or less leafy, sessile or stalked bracts. Perianth segments 6, narrow, free, irregularly arranged below the ovary, persistent. Stamens 6, hypogynous, free, exerted; anthers basifixed, erect, dehiscence longitudinal. Gynoecium syncarpous, tricarpellary, ovary superior, trilocular, placentation axile; ovules 1 to a few per locule; stigmas 3, filiform. Fruit a 3-seeded loculicidal capsule; seeds endospermous.

Anatomy.   In stem metaxylem vessels small, with scalariform wall-pitting, and oblique scalariform perforation plates. Silica in the form of sand-like particles occur in certain parenchyma cells of *Thurnia jasminoides*. Tannin cells scattered. Stomata more frequent on lower surface, Rubiaceous type or tetracytic.

Embryology.   The embryology has not been investigated (Johri et al. 1992).

Chemical Features.   Steroid saponins and chelidonic acid absent.

Important Genera and Economic Importance.   *Thurnia*, the only genus, has no economic importance.

Taxonomic Considerations.   Formerly, this genus was included in the Rapateaceae or in the Juncaceae as a *genus anomalum* (Bentham 1965). Later, it was raised to the rank of a family, Thurniaceae. Hutchinson (1959, 1969,1973) placed it in the Juncales together with Juncaceae, Restionaceae and Centrolepidaceae. According to Cronquist (1968, 1981), although the Thurniaceae are not very distinct from the Juncaceae in morphological and palynological features, they would be a very anomalous group in the Juncaceae if their habit and anatomical features are taken into consideration. However, there is no second opinion, and Thurniaceae is a distinct family of the Juncales.

Taxonomic Considerations of the Order Juncales

Hutchinson (1934, 1959) considered the Juncales to be closely related to the Liliaceae and a reduced form derived directly from the Liliician stock. However, studies by Cronquist (1968, 1981), Dahlgren (1975a, 1980a, 1983a), Dahlgren and Clifford (1982) and Takhtajan (1987) have pointed out that it is a reduced form derived from the Commelinian stock. Like Commelinales members, they have vessels throughout the plant body, stomata with subsidiary cells, presence of silica bodies, and starchy endosperm. Juncales is also closely related to the Cyperales and shares a number of features pertaining to morphology, embryology, anatomy and chemistry. However, there are differences also (Table 57.3.1).

Table. 57.3.   Differences Between Juncales and Cyperales

Features	Juncales	Cyperales
1 Number of stamens	Mostly 6, rarely 3	Not more than 3
2 Pollen tetrads	All 4 pollen grains functional	Only 1 functional
3 Ovary	Trilocular, axile placentation, many ovules	Unilocular, basal placentation, 1 ovule
4 Endosperm formation	Helobial type	Nuclear type
5 Fruit	Capsule	Nutlet

From these differences, it is clear that the Cyperales are more advanced and Juncales are less so, but the two have originated from a common ancestor, the Commelinales.

58

Order Bromeliales

A monotypic order with the only family Bromeliaceae. These are terrestrial xerophytes or epiphytes, mostly grow in the American tropics; with actino- or rarely zygomorphic flowers with septal nectaries and inferior ovary. Seeds endospermous, endosperm starchy. Rich in flavonoids and myricetin is also reported in some genera.

Bromeliaceae

A family of ca. 45 genera and 2000 species, distributed chiefly in Southern and Central America, with one probably recent genus in Western Africa, *Pitcairnia feliciana* (Dahlgren and Clifford 1982). According to Takhtajan (1987), there are 51 genera and 2100 species.

Vegetative Features. Perennial herbs, rarely arborescent (*Puya*), terrestrial (*Ananas comosus*) or epiphytic (*Tillandsia*); often spinous xerophytes retaining water in internal storage tissue of leaf. The epiphytes are either facultative or frequently obligate with an unusual water economy involving either collection of free water in overlapping leaf bases or with the ability to absorb dew or free moisture over the whole surface. Extreme specialization is shown by the rootless, moss- or lichen-like epiphytes such as Spanish moss (*Tillandsia usneoides*). *Deuterocohnia* is a woody perennial. Leaves are spirally set in basal rosettes, generally stiff, succulent, sheathing at base, linear to ovate, often with dentate or spinulose margins. Axis often decumbent and branched to form cushion-like colonies. Leaves and stem glabrous or more often clothed with peltate or stellate multicellular water-absorbing hairs.

Floral Features. Inflorescence terminal spicate or paniculate; largest inflorescence of about 10.7 m in height and 2.4 m in diameter is reported in a Bolivian plant *Puya raimondii* (Nayar 1984). Flowers bracteate, bracts large, brightly coloured, actinomorphic, rarely zygomorphic as in *Pitcairnia*, hypo- or epigynous with 3 + 3 petaloid perianth, the outer generally much smaller than the inner, free or fused, often with fringed appendages between the stamens. Stamens 3 + 3, free or adnate to perianth segments, anthers versatile or dorsifixed, introrse, bicelled, longitudinally dehiscent. Septal nectaries present. Gynoecium syncarpous, tricarpellary, ovary inferior (*Ananas, Bromelia*) or superior (*Navia, Tillandsia*), rarely half-inferior, as in *Pitcairnia* and *Dyckia*; trilocular with axile placentation, ovules many per locule, style 1 with 3 often contorted stigmatic branches. Fruits mostly septicidal capsule or berry; seeds tailed or winged, with copious starchy endosperm and a small straight embryo.

Anatomy. Vessel elements mostly confined to roots, with scalariform perforations; sometimes occur in stem and leaves. Stomata often in furrows, with 2 narrow lateral and 2 short terminal subsidiary cells. Schizogenous ducts of mucilage occasionally present. Calcium oxalate raphides and rounded silica bodies widespread (Stebbins and Khush 1961).

Embryology. Pollen grains free (tetrads in *Cryptanthus*), monosulcate or as in the subfamily Bromelioideae, multiaperturate; exine reticulate; mostly 2-celled at shedding stage (Dahlgren and Clifford 1982, Zavada 1983). Ovules anatropous, bitegmic, crassinucellate; inner integument forms the micropyle. Embryo sac of Polygonum type, 8-nucleate at maturity. Endosperm formation of the Helobial type (Johri et al. 1992).

Chromosome Number. Basic chromosome number is x = 2–28.

Chemical Features. Common flavonols, flavones and their O-methyl derivatives and flavonoids are reported; myricetin is reported in two genera of this family; C-glycoflavones also occur (Gornall et al. 1979). Steroidal saponins present (Takhtajan 1980 but chelidonic acid absent.

Important Genera and Economic Importance. Some of the important genera are *Navia* with 60 species, *Bromelia* with 40 species, *Puya* with 140 species, *Pitcairnia*, a West African genus with 250 species, *Tillandsia* with 400 species and *Ananas* with 5 species. Many of the bromeliads grow as epiphytes in the tropical forests of South and Central America. They have adaptive strategies of occupying different habitats from mesic to xeric conditions. This, in addition to vegetative multiplication, has played an important role in the spread and dispersal of many members of this family. Interesting genera are *Catopis, Hechtia* and *Prionophyllum* with unisexual flowers, *Billbergia nutans* with tetramerous flowers and *Pitcairnia* with zygomorphic flowers due to the inner perianth segments forming a hooded structure over the stamens. *Tillandsia usneoides* or Spanish moss, a familiar epiphyte hanging in grey masses from trees and telephone/telegraph wires in various parts of tropical America. Economically, the family is important for the fruits of *Ananas comosus*, long cultivated by the American Indians and introduced to the Old World by the Spanish and Portuguese travellers. A fibre obtained from the stiff leaves is used for weaving Pina cloth. Leaves of *Bromelia* and *Neoglaziovia* also yield fibre. Leaf spines of *Puya chilensis* are used as hooks for fishing. Dried stems and leaves of *T. usneoides* are used in upholstery. Species of *Billbergia, Guzmania* and *Nidularium* are grown as ornamentals.

Taxonomic Considerations of Bromeliaceae and the Order Bromeliales

Bentham and Hooker (1965c) included the Bromeliaceae in the order Epignae. Engler and Diels (1936), Rendle (1904) and Lawrence (1951) placed it in the Farinosae, also Hamann (1961, 1962). Hutchinson (1934) considered the Bromeliaceae as representing the climax of a line of descent in which the calyx and corolla have remained distinct—probably a feature retained from the dicotyledonous stock. He treated it as a distinct order of its own Bromeliales, related to but more advanced than the Commelinales.

According to Cronquist (1968), Bromeliaceae are more closely allied to the Commelinales in well-differentiated calyx and corolla, several subsidiary cells associated with each stoma (Stebbins and Khush 1961) and starchy endosperm. Cronquist also suggests the Zingiberales to be close to this family, as both the taxa have retained floral nectaries and have achieved epigyny. However, he also observed that the mesophytic Zingiberales could not have evolved from the xeromorphic Bromeliaceae with hypogynous condition and that the epigyny in the two groups has evolved independently. In 1981, Cronquist placed the Bromeliales in Zingiberidae along with Zingiberales. The common features are: predominantly tropical distribution, regular to irregular flowers, septal nectaries, and inferior ovary. Some members of these two orders of this subclass have myricetin (Gornall et al. 1979).

Takhtajan (1980, 1987) considers the Bromeliaceae (of the Bromeliales) to be related to the Liliales on the one hand and the Commelinales on the other. The occurrence of raphides, steroidal saponins and flavonoids, mostly scalariform perforations in root vessels, septal nectaries, and Helobial endosperm make the Bromeliaceae more akin to the Liliaceous ancestry. They differ from the Liliales only in starchy endosperm, stomata usually with two narrow lateral and two short terminal subsidiary cells (Ziegenspeck 1944, Tomlinson 1969), and silica bodies.

Dahlgren (1983a) includes the Bromeliaceae in Bromeliales under superorder Bromeliiflorae, and indicates that this family represents one of the evolutionary lines from a Precommeliniflorean-Zingiberiflorean-Bromeliiflorean branch, which itself is derived from the Liliiflorean ancestors. Bromeliaceae with Liliiflorean appearance and starchy endosperm also suggests its derivation from Liliiflorean ancestors. From the above discussion it is evident that the Bromeliaceae (and Bromeliales) is an isolated group, a very natural and homogeneous order. The family has links with Liliiflorae (Agavaceae in particular),

Commeliniflorae (Rapateaceae in particular) and Zingiberiflorae (in floral nectaries, inferior ovary and chemicals like flavonoids and myricetin).

Within the family, there are four subfamilies:

Subfamily 1. Bromelioideae—Leaves toothed or spinose. Ovary inferior; fruit a berry, seeds neither tailed nor winged, e.g. *Ananas, Bromelia* and others.

Subfamily 2. Navioideae—Leaves dentate. Ovary superior, fruit a capsule, seeds without appendage or wing, e.g. *Navia*.

Subfamily 3. Pitcairnioideae—Leaves entire or spinose. Ovary superior to half-inferior, fruit a capsule, seeds tailed or winged, e.g. *Dyckia, Pitcairnia*, etc.

Subfamily 4. Tillandsioideae—Leaves entire. Ovary superior, fruit a capsule, seeds with long plumose, pappus-like appendages, e.g. *Tillandsia, Vriesia*.

59

Order Commelinales

The order Commelinales comprises four suborders and eight families:

1. Suborder Commelinineae—includes Commelinaceae, Mayacaceae, Xyridaceae and Rapateaceae.
2. Suborder Eriocaulineae—Eriocaulaceae.
3. Suborder Restionineae—Restionaceae and Centrolepidaceae.
4. Suborder Flagellariineae—Flagellariaceae.

Perennial or annual herbs, often of freshwater habitats (Mayacaceae, Eriocaulaceae), often succulent (Commelinaceae). Leaves simple, alternate or spirally arranged, with sheathing base. Inflorescence variable. Flowers actino- or zygomorphic, trimerous, hypogynous; with distinct sepals and petals; anthers basifixed, introrse, tetrasporangiate, longitudinally dehiscent (poricidal in Mayacaceae). Gynoecium syncarpous, tricarpellary, ovary 3- (rarely 2)- locular (unilocular in Mayacaceae), placentation axile or parietal. Fruit a capsule, rarely a berry; seeds sometimes arillate; embryo often conical, under a disc-like structure, "embryostega", beneath the seed coat; endosperm mealy. Stomata with 4–6 subsidiary cells. Silica bodies and calcium oxalate present. Endosperm formation of the Nuclear type.

Steroid saponins occur in some members (Commelinaceae).

Commelinaceae

A moderately large family of 47 genera and 700 species, tropical and subtropical in distribution. Tribe Tradescantieae is abundant in the New World and tribe Commelinieae is represented in tropical Africa.

Vegetative Features.   Perennial or annual, succulent herbs with jointed stems; roots fibrous or sometimes much thickened and tuber-like. Leaves alternate, flat, entire, sessile, linear or lanceolate to ovate, parallel-veined and with sheathing base (Fig. 59.1A).

Floral Features.   Inflorescence terminal, terminal and axillary or only axillary (*Rhoeo spathacea*), simple or compound helicoid cyme or thyrse, usually subtended by a boat-shaped or cymbiform spathe as in *Commelina* (Fig. 59.1A), and *Tradescantia* or foliaceous, often with coloured bracts. Flowers bisexual, actinomorphic (Fig. 59.1B, D) (subfamily Tradescantieae) or rarely zygomorphic, hypogynous. Perianth distinguished into calyx and corolla. Calyx usually green (petaloid in *Commelina* and *Rhoeo*), polysepalous, imbricate, rarely gamosepalous. Corolla of 3 petals, polypetalous, equal or unequal, rarely united basally, 1 petal often reduced in size or suppressed. Stamens usually 3 + 3 or only 3, the upper ones often staminodial or 1 whorl missing; anthers basifixed, introrse, tetrasporangiate, dehiscing longitudinally, rarely poricidal as in *Dichorisandra;* filaments free, often covered with coloured, moniliform hairs as in *Tradescantia virginiana* (Fig. 59.1D, E). Gynoecium syncarpous, tricarpellary, ovary superior, usually trilocular (Fig. 59.1C) (rarely bilocular), ovule 1 to a few per locule, placentation axile; style 1, stigma 1, capitate or trifid. Fruit a loculicidal capsule. Seeds with a punctiform to linear funicular scar, usually netted, muricate or ridged, sometimes arillate as in *Dichorisandra*, endospermous, endosperm mealy, embryo situated under a disc-like structure (embryostega) beneath the seed coat.

Anatomy.   Vessels in stems and leaves with simple perforation plates. Unicellular or uniseriate trichomes present; stomata on both surfaces, usually surrounded by 4–6 subsidiary cells. Raphides and silica bodies present.

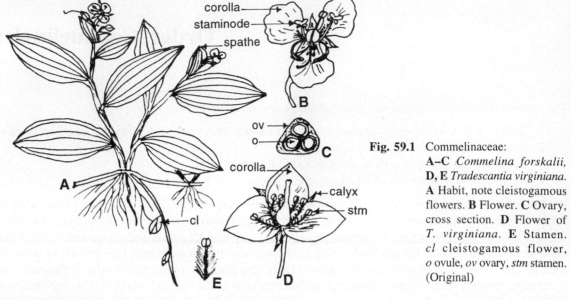

Fig. 59.1 Commelinaceae:
A–C *Commelina forskalii,*
D, E *Tradescantia virginiana.*
A Habit, note cleistogamous
flowers. **B** Flower. **C** Ovary,
cross section. **D** Flower of
T. virginiana. **E** Stamen.
cl cleistogamous flower,
o ovule, *ov* ovary, *stm* stamen.
(Original)

Embryology. Pollen grains simple, usually sulcate, 2- or rarely 3-celled at shedding stage. Ovules ortho-, ortho-campylo- or hemianatropous, bitegmic, tenui- or crassinucellate. Micropyle formed by inner integument or both (Johri et al. 1992). Embryo sac of Polygonum or Allium type, 8-nucleate at maturity. Endosperm formation of the Nuclear type.

Chromosome Number. Basic chromosome number is x = 5–19.

Chemical Features. Steroid saponins occur in some members of this family, such as *Cyanotis* (Dahlgren and Clifford 1982).

Important Genera and Economic Importance. Some of the important genera are *Aneilema, Commelina,* (commonly called dayflower), *Tradescantia, Cyanotis* and *Rhoeo. Tradescantia* is commonly known as spiderwort because of the soft mucilaginous substance that can be pulled from the broken ends of the stem and it hardens into a cobweb-like thread after exposure to the air (Jones and Luchsinger 1987). *Commelina benghalensis,* a common rainy season weed of the tropics, bears cleistogamous flowers on underground stems (Fig. 59.1A). The family is not of much economic importance except that various species of *Tradescantia, Commelina, Cyanotis, Zebrina* and *Rhoeo spathacea* are grown as ornamentals.

Taxonomic Considerations. There is no controversy about the position of family Commelinaceae in the Commelinales along with the families like Mayacaceae, Xyridaceae, Flagellariaceae and others. Hutchinson (1934, 1959) considered Commelinales to have been derived from the Butomales and Alismatales, and advanced over these orders in having syncarpous ovaries. Most taxonomists agree that the family is divided into 2 tribes: Tradescantieae and Commelineae.

Mayacaceae

A monotypic family of the genus *Mayaca* with 15 species distributed in tropical and subtropical America with a centre of development in the Amazonian basin, together with 1 species in Portuguese West Africa.

Vegetative Features. Small, perennial herbaceous plants with a *Lycopodium*-like habit (Fig. 59.2A), growing in swampy areas or even totally aquatic. Stem slender, creeping. Leaves alternate, simple, entire, linear, without sheathing base and shortly bifid at apex.

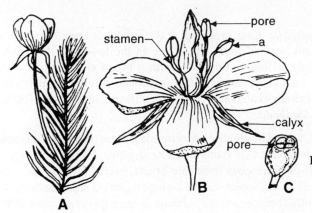

Fig. 59.2 Mayacaceae: **A–C** *Mayaca sellowiana.* **A** Plant with flower. **B** Flower. **C** Poricidal anther. *a* anther. (After Dahlgren and Clifford 1982)

Floral Features.   Solitary axillary or several subapical flowers, subtended by membranous bracts. Flowers bisexual (Fig. 59.2B), actinomorphic, hypogynous, partly cleistogamous, calyx and corolla distinguishable; calyx of 3 sepals, green, subvalvate. Corolla of 3 petals, corolline, imbricate, sometimes shortly clawed. Stamens 3, opposite the sepals (Fig. 59.2B), anthers basifixed, bithecous, dehiscing by apical pore (Fig. 59.2C). Gynoecium syncarpous, tricarpellary; ovary superior, unilocular, with numerous biseriate ovules on parietal placentae and simple or shortly trifid, filiform style. Fruit a 3-valved capsule, each valve bearing a median placenta. Seeds with strongly reticulate testa with "embryostega" and mealy endosperm.

Anatomy.   Vessels in stems and leaves with scalariform perforation plates. Trichomes unicellular or uniseriate; stomata usually with 4–6 subsidiary cells. Silica bodies and oxalate raphides present (Tomlinson 1969).

Embryology.   Pollen grains simple, monosulcate, with finely reticulate exine; shed at 2-celled stage. Ovules orthotropous, bitegmic, tenuinucellate; micropyle formed of both integuments. Embryo sac of Polygonum type, 8-nucleate at maturity. Endosperm formation of the Nuclear type (Dahlgren and Clifford 1982, Johri et al. 1992).

Important Genera and Economic Importance.   *Mayaca* (all but 1 species) occurs in tropical and subtropical America. Two species *M. fluviatilis* and *M. aubletii,* sometimes grow along margins of pools and streams of the Atlantic and Gulf coastal plains from as far north as Virginia, south to Florida and west to Texas (Lawrence 1951). The family has no known economic importance.

Taxonomic Considerations.   The family Mayacaceae is closely related to the Commelinaceae (Dahlgren and Clifford 1982). Takhtajan (1987) includes Mayacaceae in the suborder Commelinineae of the Commelinales along with the Commelinaceae. There is no controversy about the position of the family Mayacaceae.

Xyridaceae
A small family of 4 genera: *Abolboda, Achlyphila, Orectanthe* and *Xyris*, restricted to tropics and subtropics of North and South America; *Xyris* is widely distributed in the tropics and subtropics but abundant in America.

Vegetative Features.   Small, perennial herbs with corm-like or shortly creeping rhizome; roots fibrous. Leaves spirally or distichously arranged in a basal rosette below a long, usually naked, flowering scape (Fig. 59.3A). Axis (in *Xyris*) typically sympodial, rarely monopodial; branch system often complex but quite regular (Tomlinson 1969). Rhizome in *Achlyphila,* long, creeping, scaly; produces erect stems with distichously arranged leaves. Leaves equitant only in *Achlyphila* and *Xyris;* in the latter the leaves are keeled with sheathing base passing gradually into the blade. Eligulate, with a broad sheathing base in *Abolboda;* sheath not keeled or flattened in *Xyris.* Leaf blade narrow, either lanceolate or terete or subterete, sometimes spirally twisted in *Xyris;* blade dorsiventrally flattened in *Xyris reitzii* (Smith and Downs 1960). Leaf apex with a blunt terminal appendage in *Abolboda.*

Floral Features.   Inflorescence a pedunculate, terminal, globose or cylindrical head borne on a scape (Fig. 59.3B); axis naked in all genera except *Abolboda;* axis flattened in *Achlyphila.* Flowers bisexual, zygomorphic, subtended by an involucre of stiff or coriaceous imbricate bracts, each flower in the axil of an interfloral bract. Calyx and corolla distinct; sepals 3, 2 lower ones boat-shaped, keeled and membranous and dry; the inner 1 forms a hood over the corolla (before anthesis). Petals 3, gamopetalous, tube long or short, terminated by 3 uniformly spreading petals, usually yellow (Fig. 59.3C). Stamens 3, opposite corolla lobes, alternate with staminodes; staminodes often bifid, plumose or covered with moniliform hairs (Fig. 59.3C); anthers bicelled, extrorse, longitudinally dehiscent, filaments free, epipetalous. Gynoecium tricarpellary, syncarpous, ovary superior, unilocular, placentae 3 and parietal (Fig. 59.3D), or 1 and basal or 3 and free-central, ovules usually numerous. Style mostly 1 and elongated, stigmas 1 to 3, linear, spreading (Fig. 59.3C). Fruit an oblong, 3-valved loculicidal capsule, enclosed by the persistent corolla tube; seeds minute, apiculate, mostly striate, with a small apical embryo and copious mealy endosperm.

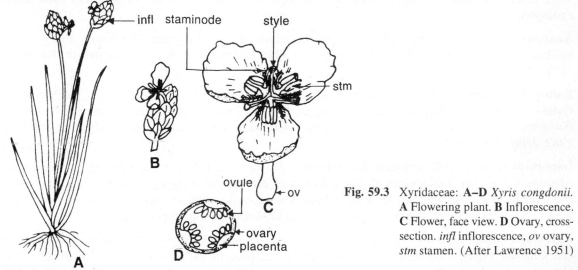

Fig. 59.3 Xyridaceae: **A–D** *Xyris congdonii.* **A** Flowering plant. **B** Inflorescence. **C** Flower, face view. **D** Ovary, cross-section. *infl* inflorescence, *ov* ovary, *stm* stamen. (After Lawrence 1951)

Anatomy.   Vessels with simple perforation plates recorded in all parts of the plants. Perforations often on oblique end walls. Vessel elements with more or less circular bordered pits in rhizome of *X. lanata;* otherwise vessel-pitting scalariform. Leaves isobilateral in *Xyris,* or dorsiventral as in *Abolboda* and *Orectanthe.* Uniseriate glandular hairs occur in *Xyris* but absent in others. Stomata usually occur on dorsal surface in *Abolboda, Orectanthe* but abundant on both surfaces in *X. fallax;* mostly of Rubiaceous type.

Embryology.   Pollen grains monocolpate; 2-celled at dispersal stage in *X. pauciflora* and 3-celled in *X. indica.* In *Abolboda grandis* the pollen grains are inaperturate and 2-celled with a large generative cell (Johri et al. 1992). Ovules orthotropous, bitegmic, tenuinucellate; both the integuments form the micropyle. Embryo sac of Polygonum type, 8-nucleate at maturity. Endosperm formation of the Nuclear type.

Chromosome Number. Basic chromosome numbers are x = 9, 13, 17.

Taxonomic Considerations. Xyridaceae is a heterogeneous family (Tomlinson 1969, Carlquist 1960). The four genera are divided into two distinct groups—*Achlyphila* and *Xyris* form one group and *Abolboda* and *Orectanthe* the other group. Members of the latter group are more closely related than the members of the former group; in certain respects *Achlyphila* resembles *Abolboda* more than to *Xyris*. This indicates that *Achlyphila* is a link between *Abolboda* and *Xyris,* and erection of two separate families—Abolbodaceae and Xyridaceae (Takhtajan 1959)—is not acceptable. Takhtajan (1987), however, includes Xyridaceae in suborder Xyridineae of the Commelinales and includes Abolbodaceae in it. Hutchinson (1934) placed it in a separate order, Xyridales, along with Rapateaceae. Lawrence (1951) observed that Xyridaceae is probably more closely related to Mayacaceae, Eriocaulaceae and Commelinaceae than to other members of this order. Dahlgren (1980a) and Dahlgren and Clifford (1982) place Xyridaceae in the order Eriocaulales along with Rapateaceae and Eriocaulaceae. Although this family is close to the Rapateaceae also, they differ in the absence of silica bodies and tannin-bearing cells and some anatomical features. Takhtajan (1987) states that the two families, Rapateaceae and Xyridaceae are closely related, and places them in the suborder Xyridineae.

The position of Xyridaceae in the Commelinales is, therefore, undoubted and it is closely related to Mayacaceae, Eriocaulaceae, Rapateaceae and Commelinaceae.

Rapateaceae

A small family of 16 genera and 90–100 species, very nearly endemic to the Guayana Highlands of South America (Maguire et al. 1958, 1965, Carlquist 1966). The genus *Epidryos* reaches as far west as Panama, and the monotypic *Maschalocephalus* is the only non-American taxon and occurs in Liberia (Carlquist 1966, Tomlinson 1969).

Vegetative Features. Herbs with short, usually simple and erect, subfleshy stems and linear or lanceolate leaves with blades rotated more or less 90 degrees; the lower portion of the leaf sheathing.

Floral Features. Inflorescence terminal spikes. Flowers bisexual, actinomorphic, hypogynous, bracteate (Fig. 59.4A). Perianth of 3 tepals; stamens 6, pedicel short and thick, anthers elongated, with a wrinkled lower part, tetrasporangiate, poricidal (Fig. 59.4B). Gynoecium syncarpous, tricarpellary, ovary trilocular, style 1, short, stigma trilobed.

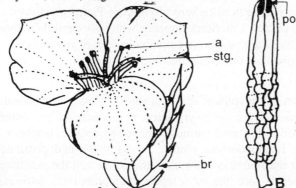

Fig. 59.4 Rapateaceae: **A, B** *Stegolepis neblinensis.* **A** Inflorescence. **B** Poricidal anther. *a* anther, *br* bract, *po* pore, *stg* stigma. (After Maguire et al. 1969)

Anatomy. Roots of Rapateaceae have a well-developed sclerenchymatous exodermis within which is a starch sheath, and internal to this is the central cortex which consists of two or three different types of cells (according to the tribe). The stems show a broad outer cortical tissue of thin walled cells containing starch. The endodermis consists of 2–5 layers of sclereids, that surrounds the vascular core

(Carlquist 1966). Vessels of root and stem with scalariform perforation plates; various degrees of sclereification occur in ground tissue of the vascular core, slime canals and silica bodies common in some genera. Leaves dorsiventral, without vessels, hairs variable, nonglandular, unicellular or uniseriate; stomata of Rubiaceous type, only on abaxial surface. Oxalate raphides absent but rounded druses of silica bodies are reported.

Embryology.   Pollen grains simple, monosulcate, zona- or bisulcate in *Spathanthus;* with reticulate, scabrate or psilate sculpturing; 2-celled at shedding stage. Ovules anatropous, bitegmic, crassinucellate, micropyle formed by both the integuments. Embryo sac of Polygonum type, 8–nucleate at maturity.

Chromosome Number.   Basic chromosome number is x = 11 in *Maschalocephalus* and 13 in *Spathanthus.*

Important Genera and Economic Importance.   The family is not of much economic importance; *Schoenocephalium maritianum* is sold as a cut flower in Bogota, Columbia.

Taxonomic Considerations.   According to Hamann (1961), Rapateaceae is close to Commelinaceae, Xyridaceae, Bromeliaceae and Liliaceae, in this order. Anatomical studies by Carlquist (1966) suggest that the closest affinities of the Rapateaceae are among the Bromeliaceae and Xyridaceae, and to a lesser extent to Commelinaceae, Eriocaulaceae and Mayacaceae. Rapateaceae, although similar to Xyridaceae in a number of features, differ in the presence of silica bodies and tannin cells, and the nature of vascular core of the roots (Tomlinson 1969). According to Dahlgren (1980a) and Dahlgren and Clifford (1982), Rapateaceae, along with Eriocaulaceae and Xyridaceae, belong to a separate order Eriocaulales and form a more or less allied, parallel line near the Commelinales. Takhtajan (1987), however, places only Xyridaceae and Rapateaceae in the suborder Xyridineae, and treats Eriocaulaceae as belonging to a separate suborder, Eriocaulineae.

From all available evidence it is apparent that Rapateaceae forms a natural family which shows numerous resemblances to the families Bromeliaceae and Xyridaceae.

Eriocaulaceae

A small family of 14 genera and 1180–1200 species (Takhtajan 1987), mainly tropical and subtropical, abundant in South America.

Vegetative Features.   Perennial, rhizomatous, scapose herbs (Fig. 59.5A), sometimes with a short vertical, subterranean stem and generally with a distinct basal rosette in distichous phyllotaxy. Plants mostly very small, e.g. *Philodice;* some species of *Paepalanthus* exceed 1 m. Some aquatics with only capitula above the surface of water are *Eriocaulon capillaceus* and *E. natans;* some are amphibious, e.g. *E. septangulare.* Leaves linear, sheathing, sometimes with a short pseudopetiole. Lamina linear to lanceolate, parallel-veined, ligule absent.

Floral Features.   Inflorescence capitate (Fig. 59.5A, B) or spicate; spikes always leafless, often ribbed. flowers unisexual (Fig. 59.5C–E), trimerous, actinomorphic or bisymmetric or because of the outer tepals, zygomorphic; numerous, sessile or shortly pedicelled, on a variously shaped receptacle; bracteate, bracts scarious, scale-like, coloured or colourless. In monoecious plants, the staminate and pistillate flowers are either mixed within the inflorescence or the staminate flowers in the centre and the pistillate flowers on the periphery. Perianth in 2 series of 3 tepals each, the outer tepals often bract-like, scarious, or membranous, polytepalous or coherent; the inner series often fused basally, stipitate and infundibuliform or cupular (Fig. 59.5E), sometimes thin and brightly coloured. Stamens 4–6 in 2 whorls or only 1 whorl, alternate with outer tepals; anthers basi- (Fig. 59.5F) or dorsifixed, introrse or extrorse, sometimes poricidal (1, 2 or 4 pores). Gynoecium bi- or tricarpellary, syncarpous, ovary superior, as many loculed

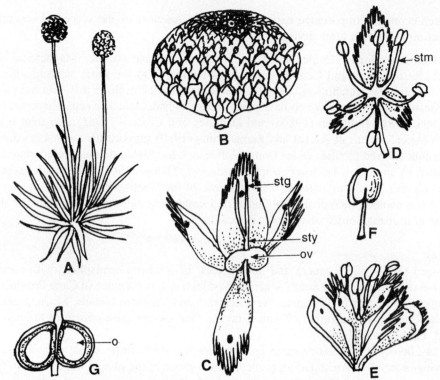

Fig. 59.5 ·Eriocaulaceae: **A–G** *Eriocaulon compressum.* **A** Flowering plant. **B** Inflorescence. **C** Pistillate flower. **D,**
E Staminate flowers. **F** Stamen. **G** Ovary, vertical section. *o* ovule, *ov* ovary, *stg* stigma, *stm* stamen, *sty*
style. **A, B, E–G** after Lawrence 1951, **C** after Rendle 1930, **D** after Dahlgren and Clifford 1982)

as the number of carpels, each locule with a single, pendulous ovule (Fig. 59.5G), placentation axile;
style 1 with as many branches as the number of locules, each branch simple or bifid. Pistillodes common
in staminate flowers but staminodes in pistillate flowers rare. Fruit a membranous, 2- to 3-seeded,
loculicidal capsule. Seeds with copious endosperm and a small, lenticular or conical embryo located at
the micropylar end.

Anatomy.   Vessels in all parts; in roots with simple, more or less transverse perforation plates; in other
parts either simple or less specialized as in leaf (Tomlinson 1969). Calcium oxalate common as needle-
like or prismatic crystals, sometimes druses. Hairs of various types.

Embryology.   Pollen grains spiraperturate (Thanikaimani 1965), or 3-celled at shedding stage (Arekal
and Ramaswamy 1980), sculpturing echinate. Ovules bitegmic, orthotropous, tenuinucellate; both
integuments form the micropyle. Polygonum type of embryo sac, 8-nucleate at maturity. Endosperm
formation of the Nuclear type.

Chromosome Number.   Basic chromosome number is x = 8, sometimes 10.

Important Genera and Economic Importance.   Important genera are *Eriocaulon* with 400 species,
Syngonanthus with 200 species, *Paepalanthus* with 500 species and *Lachnocaulon* with 10 species. South
America is the epicentre of distribution, with only one genus (*Syngonanthus*) in West Africa and Madagascar.
Mostly grow in bogs and along wet shores of water bodies in the tropics and subtropics.

Not of much economic importance except that the infloresences or the scapes of several genera are used in flower arrangements after drying.

Taxonomic Considerations. Engler and Diels (1936) treated the families Mayacaceae, Xyridaceae, Eriocaulaceae, Restionaceae and Centrolepidaceae as belonging to the same suborder, Enantioblastae, under the Commelinales. Hutchinson (1934) placed these families in three different but closely related orders, and considered Eriocaulaceae to be distinct enough to be treated as a separate order, Eriocaulales. Cronquist (1968, 1981), Dahlgren (1980a) and Dahlgren and Clifford (1982) also treat it as a distinct order. Embryological studies of Arekal and Ramaswamy (1980) support this view. Eriocaulaceae is one of the most highly evolved families under Commelinideae; it has highly derived pseudanthial, involucrate heads pollinated by insects. It is closer to the Xyridaceae. Thorne (1968, 1983) and Takhtajan (1987), however, include this family in suborder Eriocaulineae of the Commelinales.

Evidence from exomorphological, anatomical and embryological studies support the placement of Eriocaulaceae in a distinct order of its own, Eriocaulales.

Restionaceae

A medium-sized family of 37 genera and 400 species; in southern hemisphere with concentration in Australia, some representatives in South America, besides a rich occurrence in Cape Province in southern Africa. Members also occur in Tasmania. New Zealand, and Chatham Islands, Madagascar, the Malaya Peninsula, and Chile and Patagonia in South America. One species grows north of the equator in south Vietnam and another in Malawi.

The habitats favoured by Restionaceae are fairly moist, with winter rains; they are subjected to a period of drying, normally corresponding to the dormant phase of the plants' life cycle. No species occur in South Africa and only one genus (*Alexgeorgia*) in Australia.

Vegetative Features. Perennial or annual herbs, graminoid habit, with tufted growth, sometimes as scramblers; the aerial parts consist of several-noded, wiry, sterile or fertile shoots; simple or branched, erect or flexuose, terete, quadrangular or flattened, solid or fistular. The rhizome is tufted or creeping, covered with brown scarious scales. Deciduous or persistent leaf bases occur at each node of aerial stem, each sheath splits up to the base (Cutler 1969).

Leaves alternate, linear or lanceolate, sessile. The majority of the members have leaves reduced to scarious bracts.

Floral Features. Inflorescence spike-like. Flowers ebracteate, rarely bracteate and bracteolate as in *Lepyrodia* (Cutter and Shaw 1965), bi- or unisexual, actinomorphic or reduced, in most cases with a perianth of 3 or 3 + 3 or 0, free, scaly or scarious, similar or dissimilar tepals. In staminate flowers stamens 3, anthers usually bisporangiate or monothecous, rarely bithecous (Cutler and Shaw 1965); dorsifixed, introrse. Gynoecium in pistillate flowers syncarpous, tricarpellary; ovary superior, trilocular, with one pendulous ovule in each locule, placentation axile; style and stigma one, terminal. Fruit an indehiscent nut; seeds with copious endosperm and a small embryo.

Anatomy. Vessels in stems with single and scalariform perforation plates, oblique or transverse; usually with scalariform, sometimes alternate wall-pitting. Vessels absent in leaves. Uniseriate hairs occasionally present on the leaves, stomata generally of Rubiaceous type, oxalate raphides absent, silica bodies common only in *Lepyrodia*. Crystals (rhombic) present only in *Lyginia barbata* (Cutler and Shaw 1965).

Embryology. Pollen grains single, smooth, monoporate with Centrolepidoid type of pore (Zavada 1983), 2- or 3-celled at shedding stage. Ovules anatropous, bitegmic, crassinucellate, micropyle formed by both integuments. Embryo sac of Polygonum type, 8-nucleate at maturity. Endosperm formation of the Nuclear type.

Chromoseme Number. Basic chromosome number is x = 6–13.

Chemical Features. Flavonols dominant, together with 6- or 8-hydroxyflovonoids (Dahlgren and Clifford 1982). Glycoflavones are restricted only to South African members. Australasian taxa are rich in gossypetin, together with the related 8-hydroxyluteolin.

Important Genera and Economic Importance. Some of the important genera are *Restio, Chondropetalum, Lepyrodia, Leptocarpus,* and *Willdenowia.* Carlquist (1976) reported a new genus *Alexgeorgia* from Western Australia. These are perennial herbs with a wiry creeping underground rhizome covered with imbricate, whitish or brownish scales. Male flowers borne in bracteate spikes, provided with dry brown bracts. Perianth segments in 2 series, usually 6 (sometimes 5 or 4), scarious; stamens 3, filaments filiform, anthers oblong, monothecous. Female inflorescence sessile, in axils of rhizome bracts/leaves, apparently only one per fertile rhizomatous stem formed each year. Pistillate flower solitary, terminal on the inflorescence, sessile; perianth lobes 6 or less; styles 3, connate basally, filiform, highly elongated, purple. Fruit a large indehiscent nut; seeds with abundant endosperm and a small embryo. The genera *Anarthria* and *Ecdeiocolea* are so distinct in embryological, anatomical as well as some morphological and palynological features, that they deserve family rank (Cutler and Shaw 1965).

Taxonomic Considerations. In many earlier taxonomic works, Restionaceae has been included in Commelinales (Engler and Diels 1936, Melchior 1964, Thorne 1968, 1983). Most of the modern authors, however, treat it as belonging to a distinct order, Restionales (Cronquist 1968, 1981, Dahlgren 1975a, 1980a, 1983a, Takhtajan 1980, 1987). These authors recognise Anarthriaceae and Ecdeiocoleaceae as distinct families. According to Dahlgren and Clifford (1982), the presence of ligules in Restionaceae, as in Graminaceous members, the similarity of stomata type, particularly the occurrence of dumb-bell-shaped guard cells, similar pollen morphology and the occurrence of orthotropous ovules, indicate a close affinity between Gramineae and Restionaceae.

Takhtajan (1987) is of the opinion that the Restionaceae are nearer to the Commelinales and probably have originated from some primitive member of this taxon. Graminoid features (in this family), according to him, have originated separately.

The Restionaceae are wind-pollinated members of the Commelinales, with reduced flowers and a single, pendulous, orthotropous ovule in each locule. As in most of the families of the Commelinales, the embryo lies alongside the endosperm, rather than being surrounded by it. Under the circumstances, it appears that this family has originated from the Commelinaceae/Commelinales members and the resemblance with the Gramineae is due to parallelism. Also, it would be better to treat this taxon as a separate order, Restionales, comprising the families Restionaceae, Anarthriaceae, Ecdeiocoleaceae and Centrolepidaceae.

Centrolepidaceae

A predominantly Australasian family of 5 genera and 40 species; distributed also in New Guinea, Borneo, Philippines, Cambodia and South America.

Vegetative Features. Small, tufted perennial or annual herbs; often moss-like in appearance, mostly 1.5-5(-7.5) cm tall; some are submerged aquatics in shallow fresh water (*Hydatella*). The culm is short, basal part of the stem appear as a small upright rhizome. The leaves—linear or thread-like—are usually crowded or closely imbricate. Leaf bases inflated and partially sheathing, and often with acute hyaline apex, ligules present in several species.

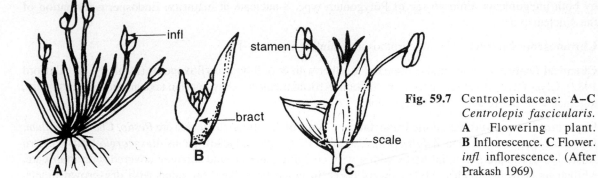

Fig. 59.7 Centrolepidaceae: **A–C** *Centrolepis fascicularis.* **A** Flowering plant. **B** Inflorescence. **C** Flower. *infl* inflorescence. (After Prakash 1969)

Floral Features. Inflorescences, interpreted by Hamann (1962) as condensed racemes, are borne on long stalks (Fig. 59.7A) and are subtended by a pair of opposite bracts bearing long, rigid hairs (Prakash 1969). The bracts enclose 10–15 bisexual pseudanthia of various ages (Fig. 59.7B). These consist of 1 stamen and 3 (2–4) carpels which show varying degrees of adnation and are enclosed (during younger stages) in 2 or 3 membranous scales (59.7C). Perianth absent. Stamen 1, filament elongated, anther bicelled, dorsifixed (basifixed in *Trithuria* and *Hydatella*). Gynoecium syncarpous (2-), 3- or (4)- carpellary; ovary superior, trilocular, placentation axile, ovules apical, 1 per locule, pendulous; monocarpellary, unilocular and uniovulate with some sessile, uniseriate stigmatic hairs in *Hydatella* and *Trithuria*. Mature carpels of each pseudanthium are obliquely superposed and often show different stages of growth (Prakash 1969). Usually, the first-matured carpel bears a seed; the others degenarate. Fruit indehiscent.

Anatomy. Vessels present in all vegetative parts; with scalariform or reticulate, lateral wall-pitting and oblique, scalariform or scalariform-reticulate perforation plates. Silica-like material and tannin present in some species, crystals absent. Stomata Rubiaceous in all species except in *Trithuria filamentosa,* where they are Ranunculaceous or absent altogether, as in *Hydatella*.

Embryolgoy. Pollen grains sulcate, irregularly globular, with thick sculptured exine, 3-celled as in *Centrolepis* or 2-celled as in *Hydatella* and *Trithuria* at shedding stage. Ovules orthotropous, bitegmic, tenuinucellate (anatropous and crassinucellate in *Hydatella*). Polygonum type of embryo sac, 8-nucleate at maturity. Endosperm formation of the Nuclear type.

Chromosome Number. Basic chromosome numbers are x = 10–13.

Important Genera and Economic Importance. *Hydatella,* a submerged aquatic and *Trithuria* are taxonomically important. They differ embryologically and palynologically (Bortenschlager et al. 1966, Prakash 1969) so much that they can be placed in a separate family, as suggested hy Hamann (1976). Economically, this family is not of any importance.

Taxonomic Considerations. The position of the family Centrolepidaceae is debatable. Originally, its members were included in the Restionaceae. Bentham and Hooker (1965c) considered them to belong to the Glumaceae along with Gramineae and Cyperaceae. Hutchinson (1934) included Centrolepidaceae in the Juncales along with Juncaceae and Thurniaceae. The unusual inflorescence and reduced nature of the plants have both contributed to taxonomic confusion (Cutler 1969). Hamann (1961, 1962) observed that the Centrolepidaceae is closely related to the Gramineae and the Restionaceae. Davis (1966) and Prakash (1969), on the basis of embryological studies, confirmed its close relationship with the Restionaceae and not with the Gramineae. Anatomically also, this family is closer to the Restionaceae (Cutler 1969). Dahlgren and Clifford (1982) include Restionaceae, Centrolepidaceae and Flagellariaceae in their Restionales, and show a close relationship between them. Melchior (1964) also considers these families to be related to each other but includes them in Commelinales as the advanced taxa.

It appears, however, that the erection of a separate order, Restionales, will be more appropriate than including these families in the Commelinales. Separation of *Hydatella* and *Trithuria* in a distinct family, Hydatellaceae (Dahlgren and Clifford 1982), is also supported in view of the distinct features of these two genera:

1. Annual herbs, submerged in shallow freshwater.
2. Leaves tufted, without a distinct sheath.
3. Inflorescence mono- or dioecious; flowers unisexual, minute, naked with only 1 stamen or a stipitate carpel.
4. Stamen with a massive filament, anthers basifixed.
5. Gynoecium monocarpellary, pseudomonomerous, ovary unilocular, uniovulate with some sessile, uniseriate stigmatic hairs.
6. Stomata anomocytic, Ranunculaceous type (*Trithuria*) or altogether absent (*Hydatella*).
7. Pollen grains shed at 2-celled stage.
8. Ovules anatropous, crassinucellate; endosperm formation of the Cellular type.

Flagellariaceae

A unigeneric family of the genus *Flagellaria* with 3 species, distributed from West Africa to Malaysia, Polynesia and north Australia. Earlier, *Hanguana* and *Joinvillea* were also included in this family.

Vegetative Features.   Erect, rhizomatous, perennial herbs with few to numerous shoots; scandent habit. Stems solid, slender, supported by leaf tendrils, frequently branched by equal dichotomy, internodes covered by leaf-sheath encircling the stem. Leaves elongated; with a tubular sheathing base, articulated to the indistinct petiole; lamina lanceolate, rolled in bud, the thickened midrib commonly extends into a circinately coiled tendril.

Floral Features.   Inflorescence terminal panicles. Flowers bi- or unisexual, actionomorphic, hypogynous, subsessile. Perianth lobes 6 in 2 series of 3 tepals each, imbricate, dry or somewhat petaloid, obtuse; Stamens 6, hypogynous or slightly adnate to the base of the perianth segments; anthers bicelled, introrse, longitudinally dehiscent. Gynoecium syncarpous, tricarpellary, ovary superior, trilocular, with 1 ovule per locule, placentation axile; styles 3, free or almost so, stigmas plumose or not. Fruits indehiscent, fleshy, drupaceous, more or less globose; seeds 1 or 2, with copious endosperm and a minute embryo.

Anatomy.   Vessels occur in all parts, mostly with simple perforation plates; in leaf with scalariform-reticulate perforation plates on more or less oblique end walls; simple perforation plates in leaf sheath (Tomlinson 1969). Leaf isobilateral, stomata uniformly scattered on both the surfaces, with dumb-bell-shaped guard cells. Calcium oxalate as druses and silica as small irregular bodies present.

Embryology.   Pollen grains spheroidal, monoporate, sculpturing verrucate-subgranulate and minutely punctate (Chanda 1966, Ladd 1977). Ovules pendulous, orthotropous, bitegmic, crassinucellate. The inner integument grows faster and forms the micropyle. Embryo sac of Allium type, 8-nucleate at maturity. Endosperm formation has not been investigated (Johri et al. 1992).

Chromosome Number.   Basic chromosome number is x = 19. In *Joinvillea* x = 18 (Newell 1969, Takhtajan 1987).

Important Genera and Economic Importance.   *Flagellaria* is the only genus recognised at present.

Earlier, *Joinvillea* and *Hanguana* were also included in this family but these have been raised to family rank. The family Flagellariaceae is hardly of any economic importance.

Taxonomic Considerations.   Earlier, the family Flagellariaceae included three genera: *Flagellaria, Joinvillea* and *Hanguana* Bl. (= *Susum* Bl.). These three genera were retained in the same family on the basis of: (a) similar paniculate inflorescence, (b) flowers conforming to the common Liliiflorous pattern, (c) similar fruits, and (d) an overlapping geographical distribution (Tomlinson and Smith 1970). They differ in vegetative morphology and anatomical features (Smithson 1957, Tomlinson 1969). Airy Shaw (1965) segregates *Hanguana* into a monotypic family, the Hanguanaceae, which (according to him) is allied to the Xanthorrhoeaceae. Cronquist (1981), Dahlgren (1980a, 1983a) and Takhtajan (1987) also support this view.

Morphological and anatomical studies show that even *Flagellaria* and *Joinvillea* do not form a natural unit (see Table 59.8.1) and should be treated separately as two distinct monotypic families (Tomlinson and Smith 1970).

Table 59.8.1   Differences Between *Flagellaria* and *Joinvillea*

	Flagellaria	*Joinvillea*
1	High-climbing lianas, the climbing stems arise from a diffuse sympodial rhizome; each rhizome segment bears only 2 renewal buds	Erect herbs, the aerial stems arise from a congested sympodial rhizome; each rhizome segment bears more than 2 renewal buds
2	Leaf rolled in bud, the adult leaf with a tendrillous tip, leaf sheath closed	Leaf plicate in bud, without tendrils, leaf sheath open
3	Lamina isobilateral or dorsiventral; epidermis without short cells and sinuous walls	Lamina dorsiventral; epidermis differentiated into long cells with sinuous walls and short cells
4	Hairs absent	Hairs common on aerial parts
5	Aerial stem solid, branching frequently by equal dichotomy	Aerial stem unbranched, with hollow internodes
6	Silica absent	Silica abundant in all parts except in roots
7	Secretory cells present in stem and leaf	Secretory cells absent

The position of the family Flagellariaceae is uncertain. Bentham and Hooker (1965c) placed it between the Rapateaceae and Juncaceae. Engler (1930e) observed that this family belongs to the first suborder of the order Farinosae. Hutchinson (1959) included Flagellariaceae in his order Commelinales together with Commelinaceae, Cartonemataceae and Mayacaceae. Thorne (1968, 1983) also agrees with this view. Cronquist (1968, 1981) and Takhtajan (1980, 1987) include this family in the Restionales and Dahlgren (1983a) in the order Poales. Hamann (1964) separates the Flagellariaceae after removing *Hanguana* to the probable affinity of Xanthorrhoeaceae, into a separate suborder of the Commelinales. Thorne (1968, 1983) includes it in the suborder Flagellariineae of the Commelinales, along with Restionaceae and Centrolepidaceae.

The Flagellariaceae are said to resemble grasses in their stomatal structure and pollen morphology (Tomlinson 1969) and have affinities with the Poales.

It is apparent from the above discussion that the Flagellariaceae s.l. should be segregated into 3 distinct, monotypic families—Flagellariaceae s.str., Joinvilleaceae and Hanguanaceae. Of these, the two former have close affinities with the Poales as well as Commelinales and the latter is close to Xanthorrhoeaceae of the Liliiflorae.

Taxonomic Considerations of the Order Commelinales

The order Commelinales (= Farinosae) of Engler and Diels (1936) is a large order of 13 small families, with mealy endosperm and syncarpous, superior ovary. Phylogenetically, this order is not homogeneous and Engler placed the included families in 6 suborders. Melchior (1964), while revising this work, retained only 8 families in 4 suborders, after removing Philydraceae to Philydrales, Bromeliaceae to Bromeliales, Thurniaceae to Juncales, and Cyanastraceae to Liliiflorae.

All later authors, however, did not accept Engler's or Melchior's views and these families have been redistributed in different orders. Hutchinson (1934, 1959) recognised only 3 families, Mayacaceae, Commelinaceae and Flagellariaceae in his Commelinales, and removed the rest of the families to different orders. Cronquist (1968, 1981), Dahlgren (1980a, 1983a) and Takhtajan (1980, 1987) include these families in different orders: Commelinales (with 4 families), monotypic Eriocaulales, Restionales (with 5 familes), monotypic Bromeliales, and Thurniaceae and Cyanastraceae in Juncales and Liliales, respectively.

The families included in the Commelinales of Engler and Diels (1936), and Melchior (1964) are ecologically highly diverse. Some of the families are wind-pollinated, associated with reduced flower structure, and some others are insect- or even bird-pollinated, associated with attractive flower and fruit structure. The members represent mesophytes to hydrophytes, xerophytes and epiphytes. Although basically terrestrial, some aquatics are also known, e.g. Mayacaceae. Apart from these, there are other differences also which advocate segregation of the Commelinales (sensu Melchior 1964) into 3 or 4 smaller orders— Commelinales s. str. including Rapateaceae, Xyridaceae, Mayacaceae and Commelinaceae; Eriocaulales including Eriocaulaceae, and Restionales including Flagellariaceae, Restionaceae and Centrolepidaceae. As discussed earlier, Flagellariaceae should now be treated as a monotypic family. Two other families are Joinvilleaceae and Hanguanaceae (formerly included in Flagellariaceae, see p. 590). Of these, Hanguanaceae has affinities with Xanthorrhoeaceae of the Liliales.

Similarly, the two small genera *Trithuria* and *Hydatella*, so far included in the Centrolepidaceae, should comprise a distinct family, Hydatellaceae. These two genera are submerged aquatic herbs from Australia and New Zealand, have bicelled or bilocular (not unilocular) anthers and a unique type of pollen which, according to G. Erdtman (1966), is difficult to describe. Zavada (1983) described the pollen grains of this family as monoporate with an ill-defined aperture, and sculpturing microverrucate to scabrate with minute punctae. According to Linder and Ferguson (1985), *Trithuria* and *Hydatella* are palynologically distinct from the Restionales, Poales and Cyperales group. Dahlgren et al. (1985) comment that: "It is so different from other monocotyledonous orders that its inclusion even in any superorder will be most strained". This family is therefore placed in the order Hydatellales of superorder Hydatellanae and precedes superorder Commelinanae (Takhtajan 1987).

60

Order Poales

A monotypic order including the only family. Poaceae (= Gramineae).

Poaceae (= Gramineae)

A very large family of 900 genera and 10500–11000 species (Takhtajan 1987), of cosmopolitan distribution.

Vegetative Features. Perennial or annual herbs or shrubs or trees, as in Bambusoideae. Rhizomes and stolons common; stem hollow (at internodes) or solid (at nodes) with prominent nodes; adventitious roots arising from nodes as in *Saccharum, Bambusa, Zea*. Leaves with sheathing base, usually linear or linear-lanceolate as in *Dendrocalamus, Setaria*, rarely petiolate as in Bambusoideae; ligulate, with parallel venation.

Floral Features. The inflorescence consists of distichous spikelets of determinate or indeterminate type, mostly in spikes (Fig. 60.1 C, D), sometimes in panicles, as in *Oryza sativa* and male inflorescence of *Zea mays* (Fig. 60.1 A); female inflorescence of *Zea mays* is cob-like (Fig. 60.1 B). Mostly bi- and rarely unisexual, enclosed within a pair of sterile glumes—lemma and palea (Fig. 60.1 E). Perianth modified to two small scale-like structures, the lodicules. Stamen usually 3, rarely 3 + 3, as in *Oryza*, two species of *Bambusa* and *Thyrsostachys oliveri* (Bhanwara 1988); more rarely 1 or 2 or numerous; up to 120 in *Ochlandra*. Filaments filiform, anthers bithecous, basifixed or versatile (Fig. 60.1 F), latrorse, dehiscence longitudinal, poricidal in *Bambusa*. Gynoecium bicarpellary (Fig. 60.1 G), tricarpellary in Bambuseae and the two carpels fused to form unicarpellary condition in *Zea*; ovary unilocular with only 1 ovule, placentation basal (Fig. 60.1 H, I); style and stigmas two, stigma papillate or plumose (Fig. 60.1 G), dry. Fruit generally a caryopsis; in some Bambusoideae, a nutlet or berry (*Melocanna*). Seeds with strach-rich endosperm, embryo lateral, the cotyledon forms a haustorial tissue and a tubular structure (scutellum and coleoptile).

Anatomy. Vessels in stems and leaves with simple or both simple and scalariform perforation plates. Leaves with Rubiaceous type of stomata having dumb-bell-shaped guard cells and often with unicellular or small bicellular hairs. Oxalate raphides absent, but silica bodies of variable shape present in epidermal cells.

Embryology. Pollen grains simple, spherical, smooth, ulcerate, shed at 3-celled stage. Ovules ana-, hemiana- or campylotropous, tenui-, crassi- or pseudo-crassinucellate, bitegmic. Inner integument grows well beyond the nucellus in most tribes (except in Panicoideae); outer integument degenerates after fertilization (Bhanwra 1988). Embryo sac of Polygonum type, 8-nucleate at maturity. Endosperm formation of the Nuclear type.

Chromosome Number. Basic chromosome number variable: x = 2–23.

Chemical Features. Cyanogenic compounds and alkaloids common; luteolin, glycoflavones, flavone 5-glucosides and sulphated flavonoids are also reported. Amongst anthocyanins, cyanidin and to some extent delphinidin occur. Dehydroquinate hydrolyase isoenzyme is present in members of Gramineae (Boudet et al. 1977). Ohmoto et al. (1970) reported the occurrence of triterpenoids in various members.

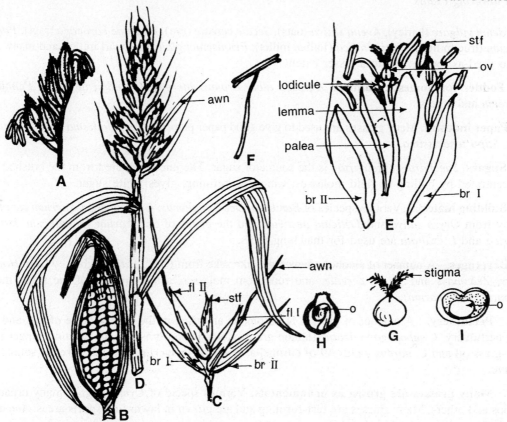

Fig. 60.1 Gramineae: **A, B** *Zea mays*, **C** *Avena sativa*, **D-I** *Triticum aestivum*. **A** Part of male inflorescence of *Z. mays*. **B** Female inflorescence of the same. **C** Dissected spikelet of *A. sativa*. **D** Inflorescence of *T. aestivum*. **E** Dissected spikelet of the same. **F** Stamen. **G** Carpel. **H, I** Vertical and cross section. *br* bract, *fl* flower (floret), *o* ovule, *ov* ovary, *stf* sterile flower. (Original)

Important Genera and Economic Importance. Although from the point of view of number of species, the grass family stands next to the Orchidaceae, it is the most widely distributed family of vascular plants. Its membes are known to occur in the arid and semi-arid zones (*Cenchrus, Duthiea*), as well as in the arctic and antarctic (*Deschampsia antarctica*); in areas where rainfall is neither very high nor very low, with moderately cold climate, forming the grassland vegetation like pampas, prairie, steppe and veldt.

Bambusa pallida and *Dendrocalamus strictus* are the arborescent members, *Cynodon dactylon* is the common lawn grass.

Economically, the members of this family of vascular plants are probably the most important. The three important civilisations of the world would not have grown without the members of the grass family. The Mediterranean and the Indus Valley Civilisations grew with domestication of wheat and other related plants like rye, barley and oats. Mayacan, Aztec and Incas Civilisations grew around maize. The Chinese Civilisation domesticated rice and millet. Grasses are grown for both human and animal consumption (Robbins et al. 1957).

The economically important plants of this family can be placed under the following categories:

(a) **Cereals.** Three most important cereals are: *Triticum aestivum* (Fig. 60.1 D) (wheat), *Oryza sativa* (rice) and *Zea mays* (Fig. 60.1 B) (maize), each used as staple food the world over.

Hordeum vulgare (barley), *Avena sativa* (oats), *Secale cereale* (rye), *Eleusine coracana* (ragi), *Panicum miliaceum* (true millet), *Setaria italica* (Italian millet), *Pennisetum glaucum* (pearl millet) and many others are also used as food grains to a lesser extent.

(b) **Fodder for domestic animals.** *Agrostis alba, Bromus inermis, Cynodon dactylon, Pennisetum purpureum* and others.

(c) **Paper industry.** Many grasses are used to give good paper pulp, e.g. *Ampelodesma tenax, Eulaliopsis binata, Stipa tenacissima*.

(d) **Sugars.** *Saccharum officinarum* is the source of sugar. The juice extracted from the crushed canes is concentrated by boiling to yield molasses, which, on refining, gives white sugar.

(e) **Building material.** Various species of *Bambusa, Dendrocalamus, Guadua* and *Gynerium sagittatum*. The hay from *Oryza sativa* and *Triticum aestivum* and the plants of *Saccharum spontaneum, Imperata cylindrica* and *I. exaltata* are used for thatching roofs.

(f) **Beverages.** A number of alcoholic beverages like sake from *Oryza sativa*, whiskey from *Hordeum vulgare, Zea mays* and *Secale cereale,* and rum from molasses, a bye-product of the sugar industry (*Saccharum officinarum*).

(g) **Perfumery.** A number of grasses are aromatic and used in the manufacture of oils and soaps and in perfumery. *Cymbopogon citratus* (lemon grass) yields lemon-grass oil, *C. martini* (ginger grass), ginger-grass oil and *C. nardus* yields oil of citronella. Roots of *Vetiveria zizanoides* are the source of oil of vetiver.

(h) **Many grasses are grown as ornamentals:** Various specie of *Cymbopogon,* many ornamental bamboos and others. Many grasses are turf-forming and are grown in lawns and sports areas. *Ammophila arenaria* is used as a sand-binder.

Taxonomic Considerations of Poaceae and the Order Poales

Taxonomically the Poaceae is very interesting and there are various opinions regarding its systematics. Arrangement of flowers in spikelets is an unique feature of this family, comparable to some extent only with the Cyperaceae members. The inflorescences are aggregates of spikelets arranged in spikes or racemes or panicles. The florets in each spikelet are highly reduced, enclosed within two glumes, lemma and palea, and the vestigial perianth represented by two or three lodicules (Rowlee 1898). There is controversy about the number of carpels in the florets. According to Bews (1929), Rendle (1904) and Engler and Diels (1936), the gynoecium is monocarpellary with 2 or 3 branched stigmas. Lotsy (1911), Weatherwax (1929), Arber (1934), and Randolf (1936) considered the gynoecium to be tricarpellary. According to Cronquist (1968, 1981), Jones and Luchsinger (1987), Takhtajan (1987) and Dahlgren et al. (1985), the gynoecia are mostly bicarpellary and rarely 3-carpellary. Floral anatomical studies support the latter view. Belk (1939) showed that there were 3 carpels joined edge to edge, and a single ovule of the ovary was always attached to the posterior wall of the only locule. The uniovulate and unilocular ovary of the Gramineae has therefore evolved from a tricarpellary condition. The original three carpels are often suggested by the presence of 3 stigmas in some genera (Bambuseae).

Bentham and Hooker (1965c) included the Poaceae in the series Glumaceae of the Monocotyledons, which also included the Eriocaulaceae, Centrolepideae, Restionaceae and Cyperaceae. Rendle (1930), on the basis of wind-pollinated and monochlamydeous flowers, placed it after the Triuridales. Hutchinson (1934) treated the Gramineae as the only member of his Graminales, and derived it from Liliaceous stock. Cronquist (1968, 1981) considered the Gramineae to be allied to the Restionales or "to a broadly defined

Commelinales, in which the Restionales is included". Like the members of Restionales, the members of Poales have orthotropous ovules, and position of embryo peripheral to the endosperm (also in most Commelinales). Takhtajan (1959) and Butzin (1965) suggested a relationship with Flagellariaceae, a member of the Restionales. G. Dahlgren (1989) includes this family in Poales along with Flagellariaceae, Joinvilleaceae, Restionaceae and Centrolepidaceae.

On the basis of ovular position in various families of the orders Commenilales, Restionales, Juncales and Poales, Cronquist (1968) suggests that "the Gramineae, Cyperaceae, Juncaceae and the families of the Restionales are derived from a common ancestry in the Commelinales near the Commelinaceae". Although the Restionales and the Poales are close to each other in many respects, they are separable from each other on the basis of pollen morphological data (Linder and Ferguson 1985, Kircher 1986).

Whether the two families, Cyperaceae and Poaceae, should be treated in separate orders or as members of one order is a matter of opinion. Cronquist (1968, 1981) observes that these two families are too close to be treated separately, but most others treat these in separate orders.

61

Order Principes

Perincipes comprises a single family, Palmae, which mostly occurs in moist, tropical regions of the world. Usually large, woody tree-like in habit; sometimes shrubby; rarely branching. Leaves few and large, palmate or pinnately compound; petiole stout, with a strong broad sheathing base. Flowers small, unisexual, trimerous, mostly on compound spadices of simple or branched rachillae; the inflorescence mostly enveloped by one or more, large spathaceous, often woody bracts. Gynoecium tricarpellary, syncarpous; ovary trilocular or unilocular, superior, ovules usually 3, often only 1 develops. Fruit a drupe or berry or nut, seeds endospermous, often ruminate.

Palmae

A large family of 212 genera and 3000 species (Takhtajan 1987), distributed throughout the tropics and also extending into the warmer parts of the temperate zones of both the New and the Old World. Mostly grows in hot and humid areas, sometimes semi-aquatic (*Nypa*); rarely xerophytic, e.g. *Hyphaene thebacia* of North Africa. *Rhopalostylis* is the only palm reported from New Zealand (Ambasht 1990).

Vegetative Features.   Perennial woody shrubs or trees, sometimes scrambling vines, e.g. *Calamus rotang,* mostly unbranched and attains great height. Stem stout or slender, often clothed with persistent bases of the leaves or with only the ring-like scars; rarely branched as in *Hyphaene*. In *Nypa*, the stem is underground, with prostrate, branched root stock. Leaves or fronds in a terminal cluster in arborescent species or scattered alternately in climbing and shrubby species; usually pinnately compound (feather plams), simple and flabellate in fan palms; simple and pinnately veined in *Calamus*; petiolate, leaf base persistent or not, sheathing, petiole smooth or margined with teeth or prickles or the lower leaflets regressively smaller and modified into long spines. Leaves show xerophytic adaptation, and have thick cuticle, sometimes even waxy, as in *Copernicia.* Young leaves appear in vertical fashion so that the thin delicate lamina is not exposed to the sun, till they are fully mature. The transpiring area is controlled by the appearance of new leaves only after the older ones have fallen off.

Floral Features.   Inflorescence simple or compound spadix, enclosed in one or more sheathing bracts or spathes, which may be hard and woody, opening by two valves; situated below (infrafoliar), or amongst the leaves (intrafoliar), or rarely above the crown of foliage (suprafoliar). When woody, the spathes are termed 'cymba' and in some genera, the cymba constricted basally into a handle-like claw called 'manubrium' (Lawrence 1951). Flowers small, actinomorphic, sessile or nearly so; mostly unisexual, and then plants mono- (*Cocos*) or dioecious (*Phoenix*), rarely bisexual as in *Livistona, Trachycarpus, Areca;* two types of flowers intermixed in *Geonoma;* male and female flowers arranged in upper and lower half of the branches in *Raphia farinifera*; in *Cocos* and *Caryota* 1 female flower occurs between 2 male flowers. Perianth of 6 tepals, in 2 whorls of 3 each, free or connate, coriaceous, white or petaloid, valvate, or imbricate, persistent. Stamens 6, in male flowers, in 2 whorls, free or united at base (*Roscheria*), or with petals; sometimes only 3 as in *Nypa* or many as in *Caryota* and *Pinango;* anthers bicelled, versatile, longitudinally dehiscent. Three staminodes present in female flowers. Gynoecium absent or rudimentary in male flower, tricarpellary, syncarpous; ovary superior, trilocular, unilocular in *Areca catechu;* each locule with 1 pendulous or erect ovule; stigma terminal and sessile. In *Cocos nucifera,* only 1 carpel develops, due to abortion of the other 2 carpels; ovules 1 or 3 in each loculus. Fruit a berry or a drupe or a nut. Seeds endospermic; embryo minute, endosperm ruminate.

Anatomy. Increase in girth of stem is affected by expansion of a (huge) apical meristem, which produces a number of vascular bundles and the ground parenchyma which separates them. In early stages there is very little elongation of the stem but the increase in diameter continues. Stomata tetracytic. Oxalate raphides and silica bodies present (Dahlgren 1979).

Embryology. The family is eurypalynous, the pollen grains are 1–2–4-colpate, mono- or trichotomo-sulcate (Ferguson et al. 1983) and 3-zonoporate (*Sclerosperma manii*; G. Erdtman and Singh 1957); exine smooth-walled or reticulate with various excrescences (P.K.K. Nair 1970). Exine of *Ammandra decasperma* is of reticulate type (Amwani and Kumar 1994); pollen grains 2-celled at dispersal stage (Johri et al. 1992) Ovules ana-, hemiana- or orthotropous, bitegmic, crassinucellate; both the integuments form the micropyle except in *Sabal* and *Washingtonia*. Embryo sac of Polygonum type, 8-nucleate at maturity. Endosperm formation of the Nuclear type.

Chromosome Number. Basic chromosome number is x = 8, 12–18.

Chemical Features. Negatively charged flavones and tricin are chemosystematic markers in the members of Palmae (Williams et al. 1973). Harborne (1975) suggests that the co-occurrence of flavone bisulphates and tricin may play a role in the maintenance of ionic balance in cells of plants growing in habitats with high salt concentration such as sea shores or coastal areas and very arid regions. According to Harris and Hartley (1980), ferulic acid and p-coumaric acid occur in the cell walls of the coconut palm.

Important Genera and Economic Importance. Important genera include *Areca* with 54 species, *Cocos* and *Lodoicea* with one species each, *Phoenix* with 17 species, *Calamus* with more than 300 species, *Copernicia* or wax palm of Brazil with 30 species, *Washingtonia* or fan palm with 2 species and *Roystonea* or royal plam with 17 species.

To residents of tropics, this family is second only to the grasses in economic importance. A large number of palms are used as important food materials. Various species of *Phoenix*—*P. dactylifera*, *P. sylvestris* and *P. acaulis*— are cultivated for the dates. Each pistillate tree of *P. dactylifera* can yield as much as 50 kg of dates each year. Other species are also tapped for a sweet sap which can be converted to plam sugar or jaggery or fermented into plam wine. *Cocos nucifera* or coconut is "man's most useful tree": food is obtained from the "meat" or endosperm of the fruit, a cooling drink from the "milk" or liquid endosperm, the juice tapped from the inflorescence spathe forms toddy, the dried kernel is the commercial copra that yields oil and the remaining oil cake is a cattle feed. *Borassus flabellifer* (Fan palm or Palmyra plam) yields toddy on tapping young inflorescences, which can be converted into sugar, alcohol or vinegar. *Caryota urens* (Toddy palm) also yields a sugary liquid on tapping its inflorescence. The pith of the stem is used in making sago. *Metroxylon laevis* and *M. rumphii* are commonly known as sago palm and yield sago starch. The terminal buds of a number of species of palms are used as food and are called 'hearts of palm'. Cabbage palm or *Sabal palmetto* is one example.

The oil palm, *Elaeis guineensis,* native of West Africa, yields a valuable oil that is both edible and used as lubricant. Oil is also obtained from dried kernels of *Cocos nucifera* (coconut oil), *Astrocaryum murumuru, A. tucuma. A. vulgare* and *Syagrus coronata* (palm carnel oil). *Attalea cohune* of Central and South America yields 'cohune' nut, which is a source of oil.

Copernicia cerifera supplies the natural wax with the higest melting point and is used in making candles and gramophone records. *Ceroxylon andicola* of northern Andes forms wax on the trunk which is scraped off and used commercially. Fibres are obtained from many species of palms. A red resin is extracted from the unripe fruits of *Daemonorops draco* of Indo-Malaysia. The pinnate and palmate leaves of many species are used for making bags, baskets, fans, mats, umbrellas and for thatching roofs. The stems of *Calamus* and *Daemonorops* are used in furniture making (canes).

Areca nut from *Areca catechu* is cultivated for its seeds, called betel nuts, which is used in betel leaf (*Piper betle*) chewing. The endosperm of ivory-nut palm (*Phytelephas*), is used as a substitute for ivory in making billiard balls, chessman and inlaid toilet articles.

Many palms are cultivated as ornamentals in parks and for landscaping, e.g. *Livistona chinensis, Chamaedorea elegans, Roystonea regia, Jubaea spectabilis, Washingtonia filifera,* and many others. *Nypa fruticans* is a stemless palm of the Sunderbans (West Bengal).

Taxonomic Considerations of Palmae and the Order Principes

The family Palmae has been treated variously by different workers. Bentham and Hooker (1965c) included it in the series Calycinae along with Juncaceae, because of the presence of sepaloid calycine peianth. Engler and Diels (1936) placed this family in the order Pincipes, due to the presence of homochlamydeous, trimerous, unisexual flowers. Rendle (1904) considered it as a member of his Spadiciflorae along with Araceae and Lemnaceae, because of the presence of spadix and unisexual flowers. Hutchinson (1934, 1973), placed this family in the Palmales near Agavales and traced its origin from Liliaceous ancestors. Family Palmae is arranged in the order Arecales by Cronquist (1968), Takhtajan (1966, 1980), Thorne (1968, 1983) and Dahlgren (1975a,b, 1980a, 1983a).

The 2-whorled trimerous perianth, mostly 6 stamens, tricarpellary, superior ovary and the habit of many Liliaceous genera like *Cordyline, Dracaena* and *Yucca* (similar to the habit of palms) bring Palmae closer to the Liliaceae. According to Takhtajan (1987), the order Arecales (including Palmae or Arecaceae) has a common origin with the Alismatidae and Liliidae. All these three groups are derived from a hypothetical extinct common ancestor. In all three groups, there are still forms with apocarpous gynoecium and primitive carpels. Amongst the Arecales (Palmae), some palms, like *Trithrinax, Trachycarpus, Chamaerops, Phoenix and Nypa,* have apocarpous ovaries and in some of them, carpels are conduplicate and stipitate, with open sutures and laminar or sublaminar placentation (Uhl and Moore 1971, Moore 1973). In *Trachycarpus fortunei,* trichomes are present along and to some extent within the unsealed suture (Uhl and Moore 1971). The septal nectaries have develoed independetly, as in the Liliaceae (Eames 1961).

The subfamilies Coryphoideae, Phoenicoideae, Borassoideae, Caryotoideae, Lepidocaryoideae, Arecoideae, Cocosoideae, Phytelephantoideae and Nypoideae are recognised under this family.

It is generally recognised that the palms are morphologically a unique and isolated group amongst all the monocots.Wettstein (1935) suggested that the Palmae are related to both the Pandanaceae and the Cyclanthaceae. Tomlinson (1960, 1961), on the basis of antomical studies, agrees that the Cyclanthaceae is one of the monocot families that is most closely related to the Palmae. Cronquist (1968) also reported that the seedling leaves of some palms, such as *Heterospathe*, are remarkably similar to the mature leaves of some Cyclanthaceae. On the other hand, floral structure and pollen morphology support the similarity with the Liliaceae. Palmae also resemble the Pandanaceae in having numerous, small, unisexual flowers and in being arborescent.

The placement of this family and the order Palmales or Principes or Arecales should be near to the Cyclanthales and the Pandanales. Dahlgren and Clifford (1982) place the three orders in the same superorder, Areciflorae. Takhtajan (1987) also included this order, Arecales, in subclass Arecidae, along with the orders Cyclanthales, Pandanales and also the Arales and Typhales.

62

Order Synanthe

A monotypic order with one family Cyclanthaceae. These are typically herbs of the forest floor, growing in the shady habitats. Herbs of palm-like habit, leaves often deeply lobed; flowers small, densely crowded into a spadix, subtended by large, more or less caducous, spathe-like bracts enveloping the spadix when young.

Cyclanthaceae

A small family of 11 genera and 180 species, restricted in its distribution to South and Central America, concentrated in the Amazon basin.

Vegetative Features.   Large, perennial stemless herbs or climbers, sometimes subepiphytic, rhizomatous, with mucilage cells in all parts. Leaves alternate, distichous or spirally set, petiolate and with a bifurcate or divided lamina, parallel venation. petiole sheathing at the base.

Floral Features.   Inflorescence a spadix or spike. Male and female flowers arranged as the squares of a chessboard on the thick spadices or in alternating, unisexual, superposed whorls, as in *Cyclanthus*. Male flowers naked or with cup-shaped perianth with short to obsolete lobes. Stamens numerous, with basally fused filaments. Anthers basifixed, latrorse. Female flowers tetramerous, without or with a 4-lobate carnose perianth with 4 staminodes and a 4-carpellate, unilocular, mostly inferior ovary with 4 sessile stigmatic crests. Ovules numerous, on 1 or 4 apical or 4 parietal placentae. Fruits baccate, often laterally coherent; seeds with a succulent, carnose seed coat and with copious, generally non-starchy endosperm and a minute, linear or curved embryo.

Anatomy.   Vessels in roots and leaves, with scalariform perforation plates.

Embryology.   Pollen grains ellipsoidal, smooth, mostly monosulcate; ulcerate in *Cyclanthus,* monoporate in *Carludovica* and biporate in *Thoracocarpus* (Zavada 1983); exine sculpturing varies from psilate to finely reticulate; dispersal at 2-celled stage. Ovules bitegmic, anatropous, tenuinucellate or pseudo-crassinucellate. Embryo sac of Polygonum type, 8-nucleate at maturity. Endosperm formation of the Helobial type.

Chromosome Number.   Basic chromosome numbers are x = 9–16.

Important Genera and Economic Importance.   The leaves of *Carludovica angustifolia* of Peru are used by local people for thatching huts. Panama hats are made from the leaves of *C. insignis* in Ecuador. In Guyana, brooms are made from the leaves of *C. sarmentosa* (Dutta 1988).

Taxonomic Considerations of the family Cyclanthaceae and the Order Synanthe

The family is divided into two subfamilies:

Subfamily Carludovicoideae—leaves bifid or entire, male and female flowers in groups that are spirally arranged. Fruiting spadix not screw-like, e.g. *Carludovica.*
Subfamily Cyclanthoideae—leaves deeply 2-partite, with forked costa. Male and female flowers in separate alternating whorls, sometimes partly spiral. Fruiting spadix screw-like, e.g. *Cyclanthus.*

Most authors agree that the family Cyclanthaceae is related to Araceae, Palmae, and Pandanceae. These families are, however, so different from each other that it is not possible to include any two of them in one order (Cronquist 1968). There is no controversy about the position of Cyclanthaceae and it is included in the order Cyclanthales of Class Arecidae by Cronquist (1968, 1981), Takhtajan (1980, 1987), and Thorne (1983). Dahlgren and Clifford (1982) and Dahlgren (1980a) include this order in superorder Areciflorae. Engler and Diels (1936) and later Melchior (1964) used the ordinal name Synanthae.

63

Oder Spathiflorae

This order comprises only two families, Araceae and Lemnaceae. Mostly rhizomatous herbs or climbers, sometimes aquatic; perennial. Leaves often differentiated into petiole and blade, sometimes digitate or palmately compound (*Anthurium*, *Arisaema*). Flowers minute, highly reduced, on a thickened spadix that is subtended by a single, large spathe; without or with much reduced perianth. Ovary usually superior, 1- to 3-carpellary; fruit a berry. Plants rich in calcium oxalate raphides. Seeds mostly endospermous.

Araceae

A fairly large family with 115 genera and 2000 species, distributed widely in the tropics and subtropics and a few in temperate regions also, e.g. *Arisaema* and *Lysichiton* extending from Alaska southwards to the Santa Cruz mountains of California.

Vegetative Features. Rhizomatous or tuberous, perennial, herbaceous terrestrials (Fig. 63.1 A), rarely aquatic, e.g. *Pistia* (Fig. 63.1 C); sometimes epiphytic with aerial roots, often large climbers such as *Pothos* and *Philodendron*. When aquatic, the plants are floating and with root pockets (Fig. 63.1 C, D). Vegetative parts often with milky latex or with watery or pungent sap. Leaves simple or compound,

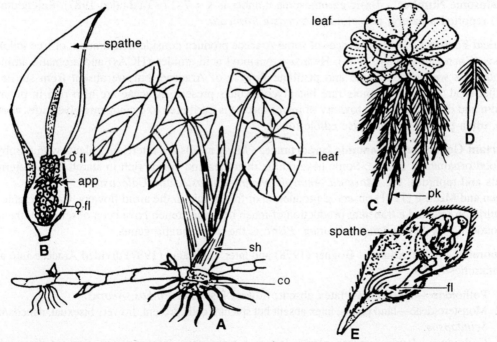

Fig. 63.1 Araceae: **A** *Colocasia antiquorum*, **B** *Arum maculatum*, **C-E** *Pistia stratiotes*. **A** Habit of *C. antiquorum*. **B** Inflorescence of *A. maculatum*. **C** Habit of *P. stratiotes*. **D** Part of root. **E** Inflorescence. *app* appendage, *fl* flower, *co* corm, *pk* root pocket, *sh* leaf sheath. (**A** original, **B** after Gangulee et al. 1982, **C** after G.L. Chopra 1973, **D, E** after Lawrence 1951)

sometimes solitary; basal or cauline and alternate, petiolate, with a membranous sheathing base. Leaf blades ensiform and parallel veined or digitate or palmately lobed with pinnately or palmately reticulate venation. Ligule present in *Calla*, intravaginal squamulae occur in some species of *Philodendron*.

Floral Features.   Inflorescence a simple spadix (Fig. 63.1 B, E) subtended by a large, caducous, herbaceous, often brightly coloured (*Calla*) or white (*Anthurium*) spathe. Flowers bi- or unisexual (the plants then mono- or rarely dioecious), small, often with foetid smell; perianth absent in unisexual flowers. present in bisexual flowers, of 4 to 6 tepals, free or basally connate. Stamens 2, 4 or 8 or rarely 1, with basifixed, bithecous, mostly extrorse, poricidal or longitudinally dehiscent anthers; filaments free or connate. Gynoecium 1- to 3-carpellary, ovary with as many locules, superior or inferior and then embedded in spadix; ovules 1 to many, placentation basal, parietal, axile, or apical; style and stigma highly reduced or absent. Fruit a berry, seeds mostly endospermous, often starchy, embryo straight or curved, frequently macropodus.

Anatomy.   Stem and leaves vesselless; secondary growth absent. Leaves dorsiventral, stomata mostly with 4 or more, rarely 2 subsidiary cells. Laticifers often present; silica bodies absent but calcium oxalate raphides widely distributed. *Pistia* has S-type sieve-element plastids; all others have monoctyledon-specific plastids (Behnke 1995).

Embryology.   Pollen grains inaperturate (*Monstera*), monosulcate (*Pistia, Pothos*), or 2–4-sulcate (Thanikaimoni 1969) or 3–4-forate; 2- or 3-celled at shedding stage. The genus *Syngonium* has both 2- and 3-celled pollen grains (Grayum 1986). Ovules ana-, hemiana- or orthotropous, bitegmic, pseudo-crassi- or tenuinucellate (*Acorus, Arisaema, Typhonodendron*). Embryo sac of Polygonum type, 8-nucleate at maturity. Endosperm formation of the Cellular type (Johri et al. 1992).

Chromosome Number.   Basic chromosome number is x = 7–17 (Takhtajan 1987). Rajendran and Jos (1972) reported a natural polyploid in *Alocasia fornicata*.

Chemical Features.   Inflorescences of some Araceae produce considerable amount of free indole during anthesis (Chen and Meeuse 1971). Hydroxycinnamoyl acid amides (HCAs) and aromatic amines occur in spathes as well as staminate and pistillate flowers of Araceae, but are absent from sterile flowers (Ponchet et al. 1982). Flavonols rare but glycoflavones present. Araceae are also rich in pelargonidin, cyanidin and delphinidin. The toxicity of many species is attributed to cyanogenic glycosides (Dring et al. 1995), often present in otherwise edible tubers.

Important Genera and Economic Importance.   Many aroids are interesting plants with climbing habit and good ornamental foliage. Some have edible rhizomes and corms rich in starch and calcium oxalate crystals and raphides, e.g. *Maranta, Amorphophalus, Alocasia,* and *Colocasia*.

Chen and Meeuse (1971) observed production of free indole in the aroid flowers. Considerable changes in respiration rate and a transient production of much heat and stench have been observed in the appendix of voodoo lily or *Sauromatum guttatum. Pistia* is the only aquatic genus.

Taxonomic Considerations.   Bogner (1978) and later Takhtajan (1987) divided Araceae into a number of subfamilies/tribes:

1 Pothoideae—land plants, latex absent, flowers bisexual: *Pothos, Acorus.*
2 Monsteroideae—land plants, latex absent but spicular cells present, flowers bisexual, naked: *Monstera, Scindapsus.*
3 Calloideae—land or marsh plants, leaves not sagittate, latex present, flowers bisexual, naked: *Calla.*
4 Lasioideae—land or marsh plants. leaves sagittate, often lobed and reticulately veined, latex present, flowers bi- or unisexual: *Amorphophallus, Lasia.*

5 Philodendroideae—land plants often on wet habitat, leaves always parallel-veined, flowers bi- or unisexual: *Richardia, Philodendron.*

6 Colocasioideae—land plants, sometimes on wet habitat, leaves always reticulate, never sagittate; latex sac present, branched. Flowers unisexual, naked, stamens in synandria; *Alocasia, Colocasia.*

7 Aroideae—land or marsh plants, leaves reticulately veined, latex sacs straight. Flowers with or without perianth; stamens free or in synandria: *Arum, Typhonium.*

8 Pistioideae—aquatic plants, leaves parallel veined. Flowers highly reduced: *Pistia.*

Bentham and Hooker (1965c) placed the Araceae in the Nudiflorae as the perianth is absent in many genera of this family. Engler and Diels (1936), Lawrence (1951) and Melchior (1964) treated this family as a member of the Spathiflorae along with Lemnaceae, Rendle (1904) in Spadiciflorae on the basis of spadix inflorescence; Hutchinson (1959, 1973) in his Arales, and Dahlgren (1975 a,b, 1980a,1983a) and Takhtajan (1987) in the superorder Arecanae.

Araceae, according to many taxonomists, are associated with Arecales (= Palmae), Pandanales, Cyclanthales and Typhales, and even the members of Liliiflorae like Dioscoreaceae (Tables 63.1.1 and 63.1.2)

Table 63.1.1 Similarities between Dioscoreaceae and Araceae

1 Petiolate leaves with broad, often cordate and sometimes compound lamina; reticulate venation
2 Occurrence of calcium oxalate raphides
3 Spike inflorescence subtended by spathe; flowers small or highly reduced, greenish
4 Basifixed stamens
5 Stigmas dry
6 Mostly anatropous ovules

Table 63.1.2 Comparison between Araceae, Arecaceae, Pandanaceae and Cyclanthaceae

Features	Araceae	Arecaceae	Pandanaceae	Cyclanthaceae
Habit	Small to giant herbs	Large woody plants or lianas	Large woody plants	Giant herbs
Leaf base	Alisma type	Bambusa type	Bambusa type	Bambusa type
Laticifers	Present	Absent	Absent	Absent
Silica bodies	Absent	Occasional	Occasional	Occasional
Vessels	Absent in the shoot	Present in leaf and stem	Present in leaf and stem	Absent in stem
Anthers	Extrorse	Latrorse or introrse	Latrorse or introrse	Latrorse or introrse
Endosperm formation	Cellular type	Nuclear type	Nuclear type	Helobial type
Chemicals:				
Luteolin/apigenin	Absent	Present	Absent	Absent
Tricin	Absent	Present	Absent	Absent
Flavonol sulphates	Present in a few taxa	Abundant	Absent	Absent
Flavones	Absent	Present	Absent	Absent
Glycoflavones and flavonols	Rare	Present	Absent	Absent

The Araceae also resembles members of Piperales of the Dicotyledoneae (Lotsy 1911, Emberger 1960, Dahlgren 1975a, b, 1980a, Huber 1976, 1977, Burger 1977). Dahlgren and Clifford (1982) list a number of similarities and dissimilarities between the two taxa. According to these taxonomists, "the differences are more significant and the two groups exhibit extraordinarily good examples of convergent evolution".

The differences are quite significant to consider Araceae and Piperales closely allied at all. In all probability, Araceae is derived from an established monocotyledonous stock.

The family Araceae is characterised by the spadix inflorescence of non-bracteate flowers, abundant calcium oxalate crystals and raphides in different parts of the plant body, extrorse stamens, Cellular type of endosperm formation and production of free indole during anthesis. Table 63.1.2 indicates that the Araceae is quite distinct and cannot be grouped together with these families.

Lemnaceae

This family of 6 genera and 30 species is distributed almost throughout the world in freshwater habitats.

Vegetative Features. Floating or submerged perennial herbs; plants without or with roots or with unbranched rhizoids. The shoot is highly reduced to a minute, oval or oblong, flat or globose thallus, leafless, often purplish underneath. Asexual reproduction by buds common. Daughter shoots arise from pockets near or at the base of the parent shoot, and may become separated or remain attached to it. Daughter shoots develop only on one side or on both sides, and accordingly the branches are helicoid or dichasial. *Wolffia* is rootless and shows no such differentiation of shoot.

Floral Features. Flowers rarely develop, particularly in temperate countries. In *Lemna* and *Spirodela* the simple inflorescences arise in the pocket on the less vigorously growing side of the shoot. Flowers unisexual, naked as in *Wolffia* or initially enclosed in a leaf-like spathe; staminate flowers solitary or in pairs of 1 or rarely 2 stamens, anthers mono- or bithecous; filaments absent or present and then stout. Pistillate flowers solitary, usually subtended by a membranous sheathing spathe; gynoecium monocarpellary, ovary unilocular, sessile, ovules 1–7, placentation basal; style and stigma 1. Fruit a 1-4-seeded utricle, seeds with or without endosperm, when present, fleshy, embryo straight.

Anatomy. The internal structure is of a spongy nature, consists of parenchyma cells and air spaces of various size. Vascular tissue absent in *Wolffia* and in others a single median strand of very simple structure.

Embryology. Pollen grains spherical, covered with small warts; monosulcate, striate, monoporate with echinate exine in *Lemna minor* (de Sloover 1961); 3-celled at dispersal stage. Ovules ana- (*Spirodela*), hemiana- (*L. paucicostata*) or orthotropous (*Wolffia*); bitegmic, crassinucellate, both integuments form the micropyle. Embryo sac of Polygonum type; Allium type in *Lemna* and *Wolffia,* 8-nucleate at maturity. Endosperm formation of the Cellular type (Johri et al. 1992).

Chromosome Number. Basic chromosome number is x = 8, 10, 11, 21.

Chemical Features. McClure and Alston (1964, 1966) report the occurrence of flavonoids in all the 6 genera of this family.

Important Genera and Economic Importance. These plants are said to be the source of food for water fowls and fish, sometimes cultivated as aquarium plant. *Wolffia arrhiza* is the smallest known flowering plant; the individuals are barely visible to the naked eye, and masses of plants are often mistaken for green algae on the surface of water.

Taxonomic Considerations. The family Lemnaceae is included in the order Nudiflorae by Bentham and Hooker (1965c), in the Spathiflorae by Engler and Diels (1936) and in the Arales by Hutchinson (1959, 1973), Cronquist (1968, 1981), Dahlgren (1975a, b, 1980a, 1983a), Dahlgren and Clifford (1982) and Takhtajan (1987).

This family is characterised by free-floating, aquatic plants with undifferentiated thalloid vegetative shoots, unisexual flowers, one or two male flowers with a single stamen each and a unicarpellate female flower with 1 to 6 basal, erect ovules.

Most phylogenists presume that Lemnaceae is an offshoot of the Araceae with origin probably from *Pistia* or an ancestral stock of close affinity. Brooks (1940), on the basis of morphological, anatomical and cytological studies, concluded that the Lemnacae is derived from the Araceae; it has highly reduced structure and show a reduction series from *Spirodela* through *Lemna* to *Wolffia* and *Wolffiella*. The *Spirodela*-like species may have given rise to *Lemna*-like species and the latter to the most highly reduced *Wolffia* and *Wolffiella*. S.C. Maheshwari (1958) considered *Spirodela polyrrhiza* to be the link between the Araceae and the Lemnaceae. This has also been coroborated by chemotaxonomic studies (McClure and Alston 1966). The chemical data is interesting, pointing to a phyletic reduction series—*Spirodela* contains anthocyanins, flavones, glycoflavones and flavonols; *Lemna* contains anthocyanins, flavones and glycoflavones; *Wolffia*, flavones and glycoflavones; and *Wolffiella*, only flavonols. The flavonoid data also suggests that *Wolffia* is biphyletic; some of the species are derived through *Lemna*-like line and another from *Wolffiella*-like element.

Taxonomic Considerations of the Order Spathiflorae

This herbaceous order is closely related to Cyclanthaceae and the Zingiberales in being tropical and subtropical herbs of forest floors; and also to Palmae and Pandanaceae. It is primitive to all these families as the vessels are confined only to roots and not in other parts of the plant body.

Bentham and Hooker (1965c) termed this order Nudiflorae, including the families Araceae and Lemnaceae. Engler and Diels (1936), Lawrence (1951) and Melchior (1964) retained the two families in the order Spathiflorae. Hutchinson (1959, 1973), Cronquist (1968, 1981), Thorne (1983) and Takhtajan (1980, 1987) have included these two families in the Arales, a highly evolved order of the class Arecidae. There is no controversy about its position. Dahlgren (1980a), however, treats this order in superorder Ariflorae and suggested its alliance to Alismatiflorae.

Aquatic habitat, highly reduced plant body as well as floral structure and the occurrence of a large variety of chemicals, point to advanced nature of this order although the absence of vessels from stem and leaves is a primitive feature. Hence, Spathiflorae is an advanced order of the surperorder Arecidae.

64

Order Pandanales

Pandanales is a small order of only 3 families, Pandanaceae, Sparganiaceae, and Typhaceae, mostly distributed in the tropics.

Perennial aquatics or marsh herbs, usually rhizomatous and sparingly branched (in Pandanaceae only). Leaves elongated, sessile, sheathing at the base, generally tough and grass-like, often with spiny margin (e.g. *Pandanus*). Inflorescence racemes or dense spikes; unisexual, anemophilous, minute, perianth often reduced; number of stamens vary, anthers basifixed. Gynoecium syncarpous, ovary unilocular, with one ovule on basal or parietal placentation. Fruits drupe-like; seeds endospermous.

Pandanaceae

A small family of 3 genera and 800 species, distributed in the Old World tropics, *Freycinetia*, *Pandanus* and *Sararanga*.

Vegetative Features. Dioecious trees, often with considerable trunk, and generally sparingly branched stems; also shrubs, climbers and large herbs; stem supported by aerial roots (Fig. 64.1 A). Leaves distinctly tristichous or spirally arranged, long, linear, sessile with sheathing base, stiff, generally tough and grass-like, occasionally with lateral spines (*Pandanus*).

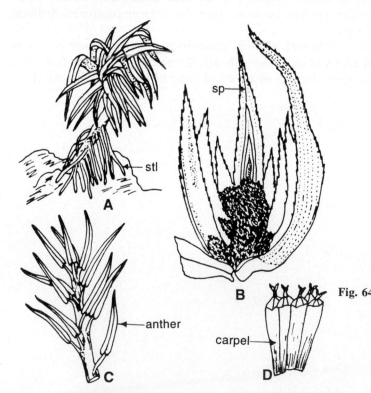

Fig. 64.1 Pandanaceae: **A-C** *Pandanus fascicularis*, **D** *P. canaranus*. **A** Habit. **B** Male inflorescence. **C** Stamens. **D** Drupes of *P. canaranus*. *sp* spathe, *stl* stilt root. (**A** Original, **B-D** adapted from Saldanha and Nicolson 1976)

Floral Features. Inflorescence racemose, spadices or panicles as in *Sararanga,* subtended by white or coloured spathes (Fig. 64.1 B). Flowers unisexual, naked or with reduced perianth. Male flowers with numerous, variously arranged, free or basally fused stamens (Fig. 64.1 C); anthers erect, basifixed, bicelled, dehiscence by longitudinal slits. Female flowers naked; gynoecium 1- to many-carpellary, syncarpous; carpels situated in rows or rings to form 'phalanges'. Ovary unilocular or many-loculed, with 1 or numerous ovules on basal or parietal placentae; stylodia in rows free, fused or obsolete. Fruit multilocular drupes (Fig. 64.1 D) with 1-seeded locules or of berries containing small seeds with fleshy, presumably non-starchy endosperm and a minute linear embryo.

Anatomy. Vessels present in roots, stems and leaves, with scalariform perforation plates. Oxalate raphides widespread but silica bodies apparently absent.

Embryology. Pollen grains inaperturate, exine-sculpturing of small spines or verrucose (Zavada 1983); 2–celled at dispersal stage. Ovules anatropous, bitegmic, crassinucellate; the micropyle formed by the inner integument. Embryo sac of Polygonum type, 8-nucleate at maturity. Endosperm formation of the Nuclear type.

Chromosome Number. Basic chromosome number is x = 30.

Chemical Features. Oxalate raphides absent (Dahlgren 1980a).

Important Genera and Economic Importance. *Pandanus* with 600 species is an important genus. Most of the species are evergren trees or shrubs, often grow along the coastal regions (*P. odoratissimus*) of India. 'Kewra' essence is obtained from the spadices, a very popular perfume used in India since ancient times. Fruits of *P. andamanensium* and *P. lerum* are edible. Leaves of many species are used for making mats, baskets and as a source of fibre. *P. utilis* a native of Malagasy, is grown in Indian gardens as a decorative plant (Singh et al. 1983).

Taxonomic Considerations. Pandanaceae have been included by all phylogenists along with Typhaceae and Sparganiaceae, in the Pandanales. They differ from the other two families in having tree-like habit and coastal habitat, and also in chemical features (Dutta 1988). There is no controversy about its position in the Pandanales.

Typhaceae

A monotypic family of the genus *Typha* with 15 species, cosmopolitan in distribution.

Vegetative Features. Amphibian plants, grow in marsh or water; perennial, rhizomatous, monoecious, with erect or floating stem. Leaves radical or cauline, distichous, elongated, linear, flat (Fig. 64.2A) or triangular in transection, sheathing at base, thick and spongy.

Floral Features. Inflorescence compound, globose or cylindrical (Fig. 64.2 B) with secondary and sometimes tertiary minute axes. Flowers unisexual, male flowers usually yellow, in upper part and female flowers brown in lower part of the inflorescences, actinomorphic. Perianth of 1, 3, 4, or 6 tepals, reduced to hairs situated at some distance below the stamens. Stamens in male flowers 2 to 5, sometimes with fused filaments (Fig. 64.2 C); anthers linear-oblong, basifixed, broadening apically, tetrasporangiate, connective often projecting above. Gynoecium normally monomerous, ovary with one locule and 1 styloidal branch and elongated stigmatic surface (Fig. 64.2 D); ovule one, pendulous. Fruit dry, nut-like or achenes but ultimately dehiscent, covered with long threads or scales. Seeds with striate testa, copious endosperm, partly starchy; embryo linear or cuneate-fusiform, nearly as long as the seed.

Anatomy. Vessels with scalariform perforation plates in stems and leaves. Hairs generally absent; stomata of Rubiaceous type. Oxalate raphides present in the vegetative parts but silica bodies absent.

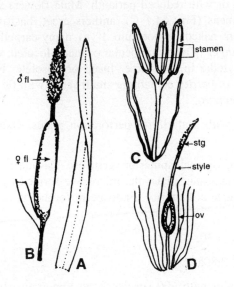

Fig. 64.2 Typhaceae: **A-D** *Typha latifolia*. **A** Leaf. **B** Inflorescence. **C** Staminate flower. **D** Pistillate flower. *fl* flower, *ov* ovary, *stg* stigma. (Adapted from Saldanha and Nicolson 1976)

Embryology. Pollen grains usually in tetrads, exine finely reticulate (Nilsson et al. 1977, Zavada 1983), 2-celled at dispersal stage (Johri et al. 1992). Ovules anatropous, bitegmic, crassinucellate, micropyle formed by the inner integument. Embryo sac of Polygonum type, 8-nucleate at maturity. Endosperm formation of the Helobial type.

Chromosome Number. Diploid chromosome number is 2n = 30 (Kumar and Subramanian 1987).

Chemical Features. Phenolic compounds like kaempferol, cyanidin, sinapic and ferulic acids occur in *Typha* (Gibbs 1974).

Important Genera and Economic Importance. The only genus, *Typha,* is an aquatic or marsh herb. Leaves of *T. latifolia* (red mace) are used for making chair bottoms, hassocks, mats, baskets, etc.

Taxonomic Considerations. The family Typhaceae is usually included in the Pandanales (Lawrence 1951, Melchior 1964). Hutchinson (1934, 1959, 1973) interpreted the Typhaceae as a derivative of the Liliaceae and as a reduced series with aquatic habit. Cronquist (1968, 1981) and Stebbins (1974) include this family along with Sparganiaceae in the Commelinidae. According to Cronquist, these families represent a line in Commelinidae with reduced, unisexual, wind-pollinated flowers with uniovulate, pseudomonomerous ovary, and apical, pendulous, anatropous ovule. Anatomically also, they have some features in common with the Commelinidae members, e.g. stomata with two subsidiary cells and occurrence of vessels in all vegetative organs. These authors consider the Pandanaceae to be too specialised to be considered as probable ancestors of this family.

But cytological studies reveal that there is homogeneity in the number and morphology of chromosomes, characterized by very small chromosomes, with nearly identical types of constrictions. Sharma (1964) observed that *Pandanus* (of Pandanaceae), *Typha* and *Sparganium* are not easily distinguishable on the

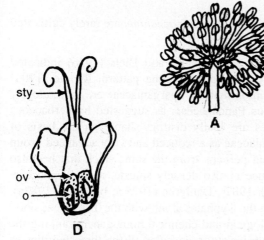

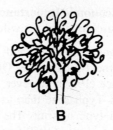

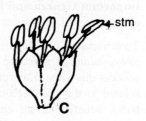

Fig. 64.3 Sparganiaceae: **A-D** *Sparganium eurycarpum.* **A, B** Male and female inflorescence. **C, D** Staminate and pistillate flower. *o* ovule, *ov* ovary, *stm* stamen, *sty* style. (Adapted from Lawrence 1951)

basis of their karyotypes as they look very similar to each other. Chemically, they are closely associated (Gibbs 1974).

The above discussion reveals that the Typhaceae are closely linked to both Sparganiaceae and Pandanaceae. Berger and Uwe (1989) report these two families to be very closely related serologically.

Sparganiaceae

A monotypic family of the genus *Sparganium* with 15 species, distributed mainly in the temperate and frigid regions of the northern hemisphere and in Australia and New Zealand.

Vegetative Features.   Aquatic perennial herbs with a creeping rhizome; stems simple or branched; leafy; leaves elongated, alternate, distichous, stiff or flaccid, erect or floating, sessile and sheathing at base.

Floral Features.   Inflorescence dense, axillary, globose, sessile or pedunculate clusters (Fig. 64.3 A, B), the upper ones are staminate and more crowded, and the lower pistillate and less crowded. Perianth of a few (3–6) membranous, elongated sepaloid scales. Stamens 3–6, in male flowers (Fig. 64.3 C), alternate with perianth lobes (when equal in number); anthers oblong or cuneate, basifixed, longitudinally dehiscent. In pistillate flowers, the gynoecium mono- or bicarpellary (Fig. 64.3 D), syncarpous, ovary superior, mostly unilocular, sessile, narrow at base, with one pendulous ovule; styles simple or forked, with a simple unilateral stigma. Fruits indehiscent, drupaceous with a narrow obconic base; exocarp spongy, endocarp hard and bony; in muricate globular heads. Seeds with a thin testa and embryo in the middle of a mealy endosperm.

Anatomy.   Similar to that of Typhaceae members.

Embryology.   Pollen grains monoporate, with finely reticulate exine (Nilsson et al. 1977), globose, 2- or 3-celled at dispersal stage. Ovules anatropous, bitegmic, crassinucellate, micropyle formed by both integuments. Embryo sac of Polygonum type, 8-nucleate at maturity. Endosperm formation of the Helobial type.

Chromosome Number.   Basic chromosome number is x = 15.

Chemical Features.   Close to the Typhaceae in chemical constituents, particularly in the phenolic compounds.

Important Genera and Economic Importance. Species of the genus *Sparganium* are rarely cultivated and are useful only as source of food for wild animals.

Taxonomic Considerations. Bentham and Hooker (1965c) and Engler and Diels (1936) indicated *Sparganium* as allied to the Pandanaceae. On the basis of similar branching pattern, Rendle (1904) accepts this view. Lawrence (1951), however, is of the opinion that the Sparganiaceae are more closely related to the monoecious Typhaceae than to the dioecious Pandanaceae, as suggested by herbaceous habit, aquatic habitat and type of fruit. The two families are easily distinguishable on the basis of inflorescence structure. Hutchinson (1959) regards Sparganiaceae as a reduced and very advanced group derived from the Liliaceous stock, not through Araceae but perhaps from the same stock that has also given rise to the Xanthorrhoeaceae, where the inflorescence is also densely spicate, as in Typhaceae. Cronquist (1968, 1981), Stebbins (1974), Takhtajan (1980, 1987), Dahlgren (1975 a, b, 1980a, 1983a), and Dahlgren and Clifford (1982) include Sparganiaceae in the Typhales along with the Typhaceae only.

However, as discussed earlier, the cytological, embryological and chemical homogeneity among the three families cannot be ignored and it would be more appropriate to consider all the three families as belonging to the order Pandanales.

Taxonomic Considerations of the Order Pandanales

Bentham and Hooker (1965c), Engler and Diels (1936) and Melchior (1964) treated the three families—Pandanaceae, Typhaceae and Sparganiaceae—in the order Pandanales. In later works, however, two orders have been recognised—Pandanales with only Pandanaceae, and Typhales include Typhaceae and Sparganiaceae (Cronquist 1968, 1981, Takhtajan 1987, Dahlgren 1980a, 1983a). Cronquist (1968) included Typhales in the Commelinidae and the Pandanales in Arecidae. Takhtajan (1987) retains both the orders in the Arecidae but in two separate superorders—Pandananae and Typhanae. Hutchinson (1973) considered Typhales to be related to the Liliaceae, and Cronquist (1968, 1981) to the Commelinidae, on the basis of reduced flower, wind pollination, stomatal organisation and vessel distribution. Reduction to this extent has not been observed among the Liliaceae members, according to him. Pandanales, a separate order comprising only Pandanaceae, is considered as woody Arecidae with numerous, firm, narrow, parallel veined leaves, usually with spiny margin.

Cronquist (1981), Dahlgren (1980a, 1983a) and Takhtajan (1980, 1987) relate the Pandanales (including only Pandanaceae) to the Cyclanthales, Arecales and Arales, all members of the Arecidae. Dahlgren and Clifford (1982) and Dahlgren et al. (1985) also recognise Pandanales and Typhales as two distinct orders. Sparganiaceae is sometimes included in Typhaceae within the Typhales (D. Müller-Doblies and U. Müller-Doblies 1977). G. Dahlgren (1989) supports this suggestion and includes Typhales in superorder Bromenianae; Pandanaceae of the Pandanales is placed in the most highly evolved superorder, Pandananae.

However, chromosomal and karyotypic studies (Sharma 1964, Mallik and Sharma 1966) make it evident that the three families—Pandanaceae, Typhaceae and Sparganiaceae are homogeneous. Chromosome numbers are $x = 30$ for Pandanaceae and $x = 15$ for Typhaceae and Sparganiaceae. In all the three families, the chromosomes are characteristically small, with nearly identical types of constrictions. Chemotaxonomic studies also coroborate this view.

Although, in many earlier classificatory systems the order Pandanales has been regarded as a primitive one, the highly reduced flowers, wind pollination, syncarpous ovary, point to its advanced nature. The Pandanaceae is certainly sufficiently distinct from the other two to be treated in a separate order of its own.

65

Order Cyperales

A monotypic order of the only family, Cyperaceae; distribution cosmopolitan. Mostly perennial, grass-like herbs, often grow in wet and marshy habitats. Stems or culms leafy or leafless, triangular. Leaves linear, in basal tufts, estipulate.

Inflorescence spikelets, variously arranged. Flowers bi- or unisexual, highly reduced. Stamens 3, ovary superior, fruit nut-like, seeds endospermous.

Cyperaceae

A very large family of 120 genera and 5600 species, distributed throughout the world, more common in the subarctic and temperate regions of both the hemispheres. A number of genera, *Cyperus, Scirpus, Carex,* abundant in the tropics.

Vegetative Features.   Perennial, or rarely annual, grass- or rush-like herbs (Fig. 65.1 A), often grow in wet, marshy or riparian habitats. Frequently rhizomatous, tufted herbs; fibrous roots arise from the base of the short or elongated and creeping rhizome. Stems and culms are leafy or leafless, mostly solid, terete, biconvex or triangular (Fig. 65.1 B), generally unbranched below the inflorescence. Leaves linear, in basal tufts or 3-ranked, with sheathing bases, estipulate; venation parallel, ligule usually absent.

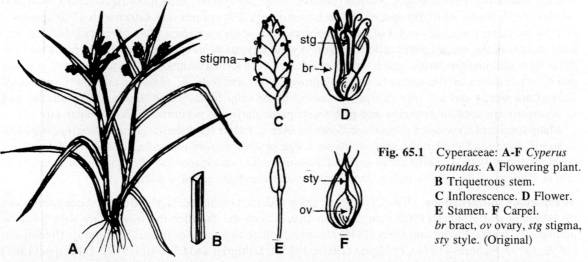

Fig. 65.1   Cyperaceae: **A-F** *Cyperus rotundas*. **A** Flowering plant. **B** Triquetrous stem. **C** Inflorescence. **D** Flower. **E** Stamen. **F** Carpel. *br* bract, *ov* ovary, *stg* stigma, *sty* style. (Original)

Floral Features.   Inflorescence mostly spikelets arranged in spikes, racemes, panicles or umbels (Fig. 65.1 A, C). Flowers bi- or unisexual (and then plants monoecious or dioecious), highly reduced in structure. Florets in the axils of glume-like, closely imbricated bracts (also called scales or glumes) (Fig. 65.1 D, F). Perianth 0 or reduced to bristles or hairs, rarely present as 3+3 bractlike scales, e.g. *Oreobalus*. Stamens 1 to 6, mostly 3, filaments thin, elongated, anthers bicelled, basifixed (Fig. 65.1 E), introrse, longitudinally dehiscent. Gynoecium syncarpous, bi- or tricarpellry, ovary superior, sometimes subtended and enveloped by a single, posterior prophyll, as in *Carex* (Fig. 65.1 F), unilocular, with a single, basal, erect ovule; style basally 1, with 2 or 3 long stylodjal branches. In the *Mapania* group, there are several anthers in a flower, each subtended by a scale-like bract, Their interpretation is undecided (Dahlgren and Clifford 1982). Fruit nut-like (achene or nutlet), sometimes enclosed in a flask-shaped utricle; seeds with copious endosperm and a broad embryo.

Anatomy. Stem and leaves with vessels; perforation plates simple and/or scalariform. Oxalate raphides absent. Silica bodies present in epidermal cells, generally conical, simple or compound. Leaves isobilateral, stomata usually in parallel rows, equally distributed on both surfaces.

Embryology. Pollen grains in tetrads, with only one functional, i.e. pseudomonads; smooth, 3-celled at dispersal stage. Ovules anatropous, bitegmic, crassinucellate, inner integument forms the micropyle. Embryo sac of the Polygonum type, 8-nucleate at maturity. Endosperm formation of the Nuclear type.

Chromosome Number. Basic chromosome number is $x = 5-13$. Some of the Cyperous genera, like *Carex*, have diffuse centromeres (Sharma and Bal 1956).

Chemical Features. Flavones like tricin, luteolin, hydroxyflavonoids like 6-hydroxyluteolin, glycoflavones and flavone 5-glucoside have been reported. Chalcones and flavanones are also present. Luteolin 5-methyl ether is an interesting marker of this family. Isoenzymes of dehydroquinate (DHQ-ase) are present in Cyperaceae members (Boudet et al. 1977).

Fruit and inflorescence are coloured due to aurone aurensidin, together with the characteristic 3-deoxyanthocyanidin carexidin. Leaf flavonoids are luteolin, tricin and flavone C-glycosides (Harborne et al. 1982). Flavonoid pigmentation (in sedges) has been reported by Clifford and Harborne (1969), and presence of quinones in the genus *Cyperus* is helpful in classification (Allen et al. 1978).

Important Genera and Economic Importance. Some of the chief genera include *Cyperus* with 600 species, commonly called sedges, *Scirpus* (bulrush) with 250 species, *Eleocharis* (spike rush), with 200 species, *Fimbristylis* with 200 species, *Rhynchospora* with 250 species and *Carex* with 1100 species.

Economically, the family is not of much importance. *Cyperus papyrus,* a plant of riverine habitat, was used in making the special paper called 'papyrus' by the Egyptians, as early as 2400 B.C. The triquetrous stems were split into thin strips, which were pressed together to form a continuous structure. Both *C. papyrus* and *C. alternifolius* or the umbrella plant are grown as ornamentals. *C. rotundus* and many species of *Scirpus* are weeds and are very difficult to control in the crop fields. Dried tubers of *C. rotundus* and *C. scariosus* are used in medicine and perfumery, particularly the perfumed sticks or 'agarbatti'.

Many species of *Cyperus, Kyllinga* and *Carex* are used as fodder for domestic animals. *Remirea maritima* is an effective sand binder. Perianth of cotton sedge or *Eriophorum* is used as a stuffing material for cushions and pillows. *Cyperus esculentus* and *Scirpus grossus* var. *kysoor* are often grown for their edible tubers. Tubers of *Eleocharis dulcis*, the Chinese water chestnut, are also edible.

Taxonomic Considerations. The Cyperaceae, along with the Gramineae, is placed in the series Glumaceae by Bentham and Hooker (1965c) on the basis that in both the families the flowers are subtended by glumaceous bracts. Engler and Diels (1936) also placed these two families in the Glumiflorae. Hutchinson (1959, 1973), Stebbins (1974), Dahlgren (1980a, 1983a), Dahlgren and Clifford (1982), Takhtajan (1987) and G. Dahlgren (1989) recognise two separate orders for the two families. As discussed earlier (see p. 573, 574) this family has a number of features common with the Juncaceae, but are advanced over Juncaceae. Possibly, the two families have a common ancester.

Taxonomic considerations of the Order Cyperales

Cyperales, with its reduced floral structure and mostly wind pollination, are closer to the Commelinales and Juncales (sometimes included in the Commelinidae). Like these orders, Cyperales members have vessels in all vegetative organs and stomates with subsidiary cells and also 3-celled pollen grains.

The order, according to earlier authors (Bentham and Hooker 1965c, Engler and Prantl 1931, Cronquist 1968, 1981), comprised of two families—Cyperaceae and Gramineae or Poaceae—and has also been termed Glumiflorae, Graminales and Poales. In recent years, some taxonomists treat these two families in two separate monotypic orders—Cyperales and Poales. The Cyperales are more closely affiliated to the Juncales, and the Poales to the Restionales (Cronquist 1968). The chemical data, however, indicate a close relationship between Juncaceae, Cyperaceae, Gramineae and Restionaceae. Keeping this in view, Cronquist (1968, 1981) points out that the chemical similarities between Gramineae and Cyperaceae are more suggestive of phyletic unity than convergence, but the occurrence of anatropous ovule and pollen tetrads in Cyperales make them distinct from the Gramineae with orthotropous ovules and single pollen grains. In addition, in Cyperales the embryo is embedded in the endosperm, and in Graminales the embryo is peripheral. There are other morphological distinctions and different methods of germination, which sharply delimit the two orders (Table 65.1.1).

Table 65.1.1   Comparison between Cyperales and Poales

	Characters	Cyperales	Poales
1	Florets	Axillary	Terminal
2	Perianth	Reduced to bristles or hairs	Reduced to lodicules
3	Leaves	Tristichous, without ligule	Distichous, with ligule
4	Stem	Solid, triangular	Hollow, cylindrical
5	Pollen grains	In tetrads	As monads
6	Ovule	Anatropous	Orthotropous
7	Fruit	Nutlet or achene	Caryopsis
8	Germination of seed	Cotyledon comes out of the seed	Cotyledon does not come out of seed

The two orders Cyperales and Poales, each including one family, are distinct from each other. They also do not indicate any phyletic relationship. The order Cyperales is more closely related to the Juncales.

66

Order Scitamineae

An order comprising 5 families: Musaceae, Zingiberaceae, monotypic Cannaceae, Marantaceae, and Lowiaceae distributed mainly in the tropics of both the New and the Old World.

Small to very large, generally perennial herbs, rarely shrubs (*Maranta*) or "trees" (*Ravenala*) with starch-rich rhizomes), rarely aquatic (*Thalia*). Vertical aerial stem often short; inflorescence-bearing stems covered with bracteate leaves. Leaves alternate, frequently distichous, with a sheathing base, usually petiolate with a large and simple or secondarily split, broad and pinnately veined lamina. Ligules present in many Zingiberaceae. Intravaginal squamules absent.

Flowers epi- or rarely perigynous, mostly with 3+3 petaloid, basally connate tepals; in Musaceae, 5 tepals fused to from a sheath, and one tepal free. Stamens 3+3 or fewer, only one in Zingiberaceae. Gynoecium syncarpous, tricarpellary, ovary mostly tri- and rarely unilocular with axile or parietal placentation. Fruit usually a loculicidal capsule, rarely a berry, nut or schizocarp; seeds arillate, endospermous with linear, capitate or curved embryo.

Vessels present in roots mostly and rarely in stems. Pollen grains in monads, 2- or 3-celled. Ovules mostly anatropous, crassinucellate; endosperm formation of the Helobial or Nuclear type.

Saponins and steroid saponins are reported in some families; some others are rich in flavonoids.

Musaceae

A small family of only 2 genera and 47 species, occur in tropical Asia, Africa and Australia with a large number of cultivars.

Vegetative Features.   Large herbs, often apparently tree-like in appearance (Fig. 66.1 A) and the "stem" sometimes semi-ligneous; "stem" unbranched and covered with leaf sheaths that are rolled round each other, concealing the short conical axis. Stem modified into an underground rhizome, the aerial stem from between the leaves elongates to form the inflorescence. Leaves large, alternate or distichous (*Ravenala*) or spirally arranged; entire or lacerated due to the effect of wind; with a strong midrib from which numerous parallel veins run to the margins.

Floral Features.   Inflorescence a spike or panicle or sometimes capitate, subtended by spathaceous bracts that are brilliantly coloured (Fig. 66.1 B), coriaceous or semisucculent and cymbiform. Flowers bi- or unisexual (Fig. 66.1 D-F); when unisexual, the plants monoecious with staminate flowers in the upper bracts and the pistillate ones within the lower bracts, zygomorphic, epigynous. Perianth of 6 segments, mostly in 2 series; tepals unequal in size and shape free or variously connate. Stamens 6, all fertile in *Ravenala* and 5 fertile ones with a staminode in *Musa*; anthers bithecous. Linear, basifixed, dehisce by longitudinal slits; filaments distinct and filiform. Gynoecium syncarpous, tricarpellary, ovary inferior, trilocular, placentation axile (Fig. 66.1 G), ovules numerous per locule; style 1, filiform, stigma capitate or trilobed. Fruit a 3-celled capsule or a berry, seeds endospermous, often arillate; embryo straight.

Anatomy.   Vessels present in roots but absent in stem and leaves.

Embryology.   Pollen grains simple, inaperturate, exine sculpturing psilate; shed at 3-celled stage. Ovules anatropous, bitegmic, crassinucellate, both the integuments form the micropyle. Embryo sac of Polygonum type, 8-nucleate at maturity. Endosperm formation of the Nuclear type (Johri et al. 1992).

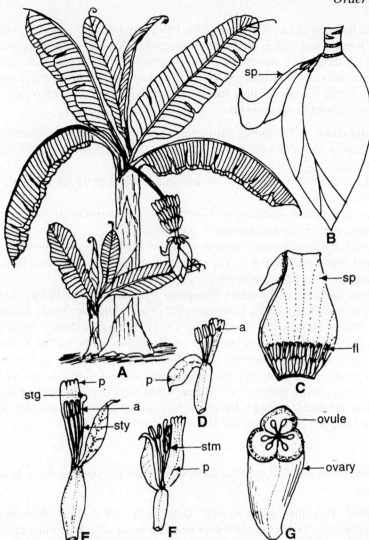

Fig. 66.1 Musaceae: **A-G** *Musa paradisiaca*. **A** Plant bearing inflorescence. **B** Inflorescence. **C** A spathe with flowers attached at the base on the inner surface. **D** Staminate flower. **E, F** Bisexual flowers. **G** Ovary, cross section. *a* anther, *fl* flower, *p* perianth, *sp* spathe, *stg* stigma, *stm* stamen, *sty* style. (Adapted from G.L. Chopra 1973)

Chromosome Number.   Basic chromosome number is x = 9–11, 16, 17.

Chemical Features.   Musaceae contains 3-deoxyanthocyanidins, a character rare in the monocots (Gornall et al. 1979).

Important Genera and Ecomomic Importance.   The genus *Musa* with 35 species is the most important member of this family. It has numerous cultivars and most of them yield edible fruits. In addition, banana fruit is a good source of starch, banana powder can be used as baby food and also in chocholate and biscuit industries. Raw banana is used in dysentry and diarrhoea. Banana leaves are used as plates for serving food in many parts of India. *M. superba* is a wild seeded banana, cultivated as an ornamental.

M. textilis or Manila hemp or abaca is a native of the Philippines but cultivated in southern parts of India also. Fibre obtained from its leaf stalk is used for marine ropes, twine, and wrapping paper. *Ensete ventricosa* is yet another fibre-yielding plant. *Ravenala madagascariensis* is the Madagascar traveller's tree. It is a large ornamental tree, native of Madagascar, so called as the water accumulated at the leaf bases is used as drinking water by thirsty travellers. Other species, like *E. superbum, Heliconia metallica, M. nepalensis Strelitzia reginae,* are grown as ornamentals.

Taxonomic Considerations. The family Musaceae s.l. is divided into 3 subfamilies (Dutta 1988):

1 Lowioideae—leaves half-alternate, flowers bisexual, largest median petal forms a labellum, e.g. *Orchidantha* (= *Lowia*).

2 Musoideae—leaves spiral, flowers uni- or bisexual, fruit a berry or leathery, indehiscent; seeds without aril, e.g. *Musa, Ensete.*

3 Strelitzioideae—leaves half-alternate or distichous; flowers bisexual, fruit a woody schizocarpic capsule, seeds arillate, e.g. *Heliconia, Ravenala* and *Strelitzia.*

At present, however, most taxonomists recognise only Musoideae as true Musaceae. i.e. Musaceae s. s. with only *Ensete* and *Musa* included in it. The other two subfamilies have also been raised to the rank of a family, i.e. Lowiaceae and Strelitziaceae.

The Musaceae was included in the series Epigyneae by Bentham and Hooker (1965c), in the order Scitamineae by Engler and Diels (1936), Lawrence (1951) and Melchior (1964). Cronquist (1968, 1981), Dahlgren (1980a, 1983a), Hutchinson (1934, 1959, 1973). Takhtajan (1980, 1987) and Thorne (1983) place Musaceae in the Zingiberales with the same configuration as the earlier Scitamineae.

Musaceae is the least specialised member of the Zingiberales, according to Dahlgren (1983a), and is characterised by the presence of oxalate raphides, 5 functional stamens (6 only in *Ravenala*), septal nectaries, and seeds with copious starchy endosperm. Although amongst the members of the Zingiberales, the Musaceae is the least advanced, amongst the entire Monocotyledons, it is a highly advanced family, as is evidenced by the presence of zygomophic flowers and inferior ovary.

Zingiberaceae

A large family of 46 genera and 850 species, distributed in the tropics, chiefly in Indo-Malaysia and the Indian sub-continent.

Vegetative Features. Perennial, small to large herbs with fleshy, short, distichous, scaly rhizomes, branching sympodially from axils of scale leaves near the bases of erect stems; each branch terminates in an unbranched leafy shoot or an inflorescence. Leaves radical or cauline, sometimes petiolate, alternate or distichous, sheaths overlap to form a pseudostem surrounding a thin true stem (Rogers 1984); ligule present at the insertion of petiole.

Floral Features. Inflorescence terminal, borne on a leafy shoot, open raceme or compact spike or uniflorous, subtended by prominent spirally arranged bracts. Flowers bisexual, zygomorphic, epigynous, highly diverse in appearance and usually last only for a day. Perianth biseriate, outer whorl of 3 sepals connate into a tube, usually split along one side; inner whorl of 3 petals fused basally with the androecium into a narrow tube, the dorsal lobe larger than the other 2 lobes, sometimes appendaged. Stamens highly modified—the 2 lateral ones modified into staminodes; 1 median stamen of inner whorl is fertile and 2 lateral ones of the same whorl united to form a labellum; anthers bithecous. Gynoecium syncarpous, tricarpellary, ovary inferior, trilocular, with axile placentation, or unilocular with parietal, basal or free-central placentation, rarely bilocular; ovules numerous. Style thin, passing through a channel in the filament and between the two locules of the anther, stigmas ciliate, often sunken (Rogers 1984). Fruit a capsule, seeds arillate.

Anatomy. Vessels confined to roots; stomata with asymmetrical guard cells; plants without raphide sacs.

Embryology. Pollen grains (Punt 1968) inaperturate or monosulcate (*Zingiber, Dimerocostus*), spiraperturate (*Tapeinocheilos*) or polyporate (*Costus, Monocostus*); exine sinulose to psilate (Zavada 1983, Takhtajan 1987). Pollen grains 2-celled at dispersal stage and rich in starch. Ovules anatropous, bitegmic, crassinucellate, micropyle formed by inner integument only (Panchaksharappa 1966). Embryo sac of Polygonum type, 8-nucleate at maturity. Endosperm formation of the Helobial type; seeds with both perisperm and endosperm.

Chromosome Number. Basic chromosome numbers vary x = 9, 11, 12, 9–26.

Chemical Features. The aromatic oils of Zingiberaceae members are rich in monoterpenoids. Sesquiterpenoids are common and sometimes predominant in the oil. Phenylpropane compounds are also reported. Flavonoids are abundant, quercetin and kaempferol common, some unusual ones like myricetin, isorhamnetin and syringetin flavonols occur in glycosidic combination with glucuronic acid, rhamnose or glucose (Gibbs 1974, William and Harborne 1977). The genus *Alpinia* is similar to Juncaceae and Cyperaceae in making 5-0-methyl flavones (Gornall et al. 1979).

Important Genera and Economic Importance. Some of the important genera include *Alpinia* with 250 species, *Amomum* with 150 species, *Curcuma* with 50 species distributed in Indo-Malaya, China and North Australia, *Elettaria* with 7 species, occurs in Western Ghats in India, *Hedychium* with 50 species, *Costus* with 1 wild and 1 cultivated species, and *Zingiber* with 80–90 species distributed in East Asia, Indo-Malaya and North Australia. *Curcuma amada* is an interesting plant with its rhizome smelling of mango and hence the common name 'mango ginger'. *Globba bulbifera* is a plant of marshy places and river banks, has a terminal panicle, the lower flowers are modified into bulbils. *Costus speciosus* is a rhizomatous plant growing in shady moist places. Many Zingiberaceae members have long been used as spices and condiments, in perfumes and in medicines. The rhizomes of *Zingiber officinale* yield ginger; turmeric from rhizomes of *Curcuma longa* and *C. domestica*. The seeds of *Aframomum melegueta* give melegueta pepper. *Amomum aromaticum, A. dealbatum* and *A. subulatum* yield large cardamoms, the seeds are used as flavouring agent for sweetmeats and flavouring beverages; mostly grown in swampy areas of Bengal, Assam, Sikim and Tamil Nadu, in India. The capsules and seeds of *Elettaria cardamomum* are used as a spice and also in medicine. *Hedychium coronarium* or ginger lily is an ornamental, rhizomatous herb, occurs throughout the moist regions of India. An essential oil is obtained from the rhizome; powdered rhizomes are used in medicines. Rhizomes of *Alpinia galanga* and *A. officinarum* are also used medicinally as stomachic, stimulant and carminative. An essential oil obtained from the former has insecticidal properties (Singh et al. 1983). Arrowroot is extracted from rhizomes of *Curcuma angustifolia* (Burtt 1977a, b, 1980, Ilyas 1978). Several taxa are grown as ornamentals—*Alpinia, Hedychium, Kaempferia* and *Roscoea* (Rogers 1984).

Taxonomic Considerations. The famly Zingiberaceae is divided into two subfamilies (Loesner 1930, Dutta 1988):

Subfamily 1 Costoideae—plants not aromatic, leaves spirally arranged, lateral staminodes absent or reduced; nectaries replaced by septal glands, *Costus*.

Subfamily 2 Zingiberoideae—plants aromatic; leaves distichous, outer and inner perianth markedly distinct; the three tribes are Hedychieae, Globbeae and Zingibereae.

Cronquist (1968, 1981), Takhtajan (1987) and Dahlgren (1975 a, b 1980a, 1983a) recognise two distinct families—Costaceae and Zingiberaceae s.s. The Zingiberaceae is regarded as one of the highly advanced monocot families due to the presence of inferior ovary and zygomorphic flowers with highly complicated structure.

Cannaceae

A monotypic family with 55 species, native to Central America and the West Indies; a few are cultivars.

Vegetative Features. Large, perennial herbs with a tuberous rhizome and unbranched aerial stem. Leaves large, with open, eligulate sheaths, cauline, parallel veined.

Floral Features. Inflorescence terminal on leafy shoots, raceme or panicle (Fig. 66.3 A). Flowers large, showy, bisexual, irregular, epigynous (Fig. 66.3 A, B), each flower or a pair of flowers subtended by a bract and rarely with scale-like bracteoles. Perianth biseriate, distinguishable into calyx and corolla (Fig. 66.3 B). Calyx of 3 sepals, polysepalous, persistent, lanceolate or elliptic, more or less green to purple. Corolla of 3 petals, usually unequal, erect, basally connate and adnate to androecium and style to form a tube, deciduous. The staminal column consists of 4 to 6 petaloid stamens in 2 series. The outer 3 petaloid stamens are sterile, the largest forms the labellum; inner series consists of 1 or 2 petaloid staminodes and a free petaloid fertile stamen with a monothecous anther adnate to the petaloid margin (Fig. 66.3 B). Gynoecium tricarpellary, syncarpous, ovary inferior, trilocular, with axile placentation (Fig. 66.3 D); ovules numerous per locule; style 1, petaloid, stigma 1, represented by a stigmatic crest on apical margin (Fig. 66.3 C). Fruit a warty capsule, seeds small, numerous, subglobose, with very hard endosperm and a straight embryo.

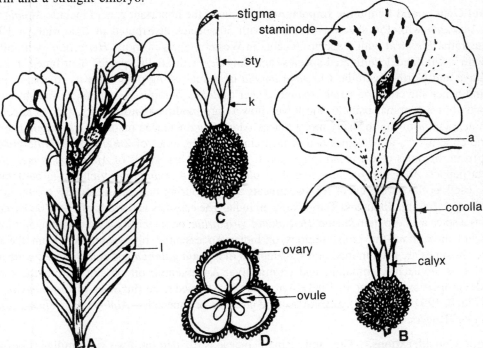

Fig. 66.3 Cannaceae: **A-D** *Canna indica.* **A** Inflorescence. **B** Flower. **C** Carpel with calyx (*k*) attached. **D** Ovary, cross section. *a* anther, *k* calyx, *l* leaf, *sty* style. (Original)

Anatomy. Aerial stems with mucilage canals or cavities. Stomata predominantly of Rubiaceous type.

Embryology. Pollen grains mostly spherical, usually with small spinules, inaperturate (P.K.K. Nair 1960), 3-celled at shedding stage (Davis 1966). Ovules anatropous, bitegmic, crassinucellate, micropyle formed by inner integument. Embryo sac of Polygonum type, 8-nucleate at maturity. Endosperm formation of the Nuclear type, but reduced to a thin layer in mature seed.

Chromosome Number.   Basic chromosome number is x = 9.

Chemical Features.   Flavonols present.

Important Genera and Economic Importance.   The only genus *Canna* with its 55 species and many cultivars is grown as an ornamental. *Canna edulis* is used as a source of food for livestock; rhizomes rich in starch. Burning the plants is said to produce an insecticidal smoke (Rogers 1984). Extracts from *C. indica* and other species have molluscicidal activity (Mahran et al. 1977a, b).

Taxonomic Considerations.   The family Cannaceae has been included in Scitamineae (or Zingiberales) by all phylogenists. It is a highly advanced family of this order.

Marantaceae

A moderately large family of 30 genera and 400 species distributed chiefly in the American tropics and Indian subcontinent.

Vegetative Features.   Perennial, erect or scrambler or climber, small to large herbs with sympodially branched rhizome with short internodes. Branches sometimes clustered on or above ground, often widely divergent, each bearing a series of petiolate leaves and terminal inflorescence. Leaves distichous, basal or cauline, with open, often auriculate, rarely ligulate sheaths and often assymmetric blades; homotropous or antitropous. Petiole with a swollen, pubescent pulvinous at the junciton of blade and sheath.

Floral Features.   Inflorescence a panicle or spike, subtended and surrounded by spathaceous bracts. Flowers bisexual, asymmetric, usually in mirror-image pairs, each pair or group of pairs subtended by a bract. Perianth of 6 segments in 2 series of 3 each. Calyx of three sepals, free, green; corolla of 3 petals, unequal in size, connate basally, adnate to androecium and gynoecium to form a tube. Outer staminal whorl totally absent or represented by 1 or 2 petaloid staminodes; inner whorl represented by 1 fertile stamen which is often petaloid; only a half-anther, the other two are petaloid staminodes. Gynoecium tricarpellary, syncarpous, ovary inferior, tri- or unilocular, ovules solitary in each locule, style twisted, lobed or apically dilated. Fruit a capsule, seeds arillate, embryo curved or invaginated, endosperm copious.

Anatomy.   Vessels confined to roots. Plants glabrous or with unicellular hairs, each surrounded at the base by a cluster of inflated epidermal cells. Stomata mostly paracytic (Rubiaceous type) (Rogers 1984).

Embryology.   Pollen grains inaperturate or uniaperturate (Rogers 1984). Ovules campylotropous, bitegmic, crassinucellate, micropyle formed by inner integument. Embryo sac of Polygonum type, 8-nucleate at maturity. Endosperm formation of the Nuclear type.

Chromosome Number.   Basic chromosome numbers are x = 4, 6, 9, 11–13.

Chemical Features.   Rich in flavonols; C-glycoflavones and flavone bisulfates are also reported (Gornall et al. 1979).

Important Genera and Ecnonomic Importance.   Amongst the members of Marantaceae, *Maranta* with 23 species, *Calathea* with 150 species and *Phrynium* with 30 species are important genera. Of these, *Calathea* spp. and species of *Maranta* are cultivated sporadically (Kumar and Subramanian 1989). *M. arundinacea* (commonly called West Indian arrowroot) is a herb or shrub native to tropical America but cultivated widely for its edible rhizome; arrowroot starch is used in preparation of biscuits, cakes, puddings and jellies (Singh et al. 1983). Tubers of *Calathea allouia* are also edible. *Schumannianthus dichotomus* is a woody shrub occurring in Assam and Bengal in India and also in Bangla Desh; its split stems are used for weaving mats. *Thalia dealbata* is a favourite garden aquatic.

Taxonomic Considerations. Phylogenists agree that Marantaceae is the most highly advanced family in this order, as evidenced by the highly complex androecium reduced to a single stamen, which is again monothecous; only 1 ovule per locule and in many genera, only 1 of the 3 locules bears ovules.

There is much resemblance between the Cannaceae and the Marantaceae, particularly in solitary 1-celled anther and numerous staminodes, but Marantaceae differs from the Cannaceae in only a 1- to 3-loculed ovary with only 1 ovule per locule and seeds with curved or invaginated or folded embryo.

It may be concluded that although the two families Cannaceae and Marantaceae are quite closely related, the differences between the two do not support their placement in the same family.

Lowiaceae

A monotypic family of the genus *Orchidantha* with two species, distributed in South China and West Malaysia.

Vegetative Features. Stemless, rhizomatous, perennial herbs; leaves radical, distichous, lanceolate or ovate-lanceolate, with long sheathing petiole and conspicuous tansverse venation.

Floral Features. Short, radical, few-flowered cymes, subtended by bracts. Flowers bisexual, bracteate, bracts oblong, highly zygomorphic. Sepals 3, united below into a long and slender stalk-like tube, limbs free, linear or lanceolate. Corolla of 3 petals, unequal, the two lateral ones small, and the middle one forms a large coloured elliptic or spathulate lip or labellum. Stamens 5, inserted with the petals, anthers bithecous, loculi parallel, opening by longitudinal slits. Gynoecium syncarpous, tricarpellary, ovary inferior, at the base of the calyx tube, trilocular, placentation axile, ovules numerous in each locule; style as long as the stamens, trilobed, lobes laciniate. Fruit 3-locular, seeds endospermous, enclosed in a 3-lobed aril.

Anatomy. Vessels confined to roots. Oxalate raphides present. Mucilage canals absent.

Chromosome Number. Basic chromosome number is x = 9.

Chemical Features. Rich in flavonols (Gornall et al. 1979).

Important Genera and Economic Importance. The only genus, *Lowia* (or *Ochidantha*) has no economic importance.

Taxonomic Considerations. The family Lowiaceae is related to the Musaceae and Strelitziaceae of the Zingiberales, on the basis of: (a) occurrence of 5 stamens, (b) presence of oxalate raphides, and (c) presence of abundant starchy endosperm. It has been treated as a subfamily, Lowioideae, of the Musaceae (Lawrence 1951). Hutchinson (1959, 1973), Melchior (1964), Cronquist (1968, 1981), Dahlgren (1975a, b, 1983a) and Takhtajan (1980, 1987) treat Lowiaceae as a distinct family, more akin to the Musaceae.

In our opinion also, Lowiacaae should be treated as a distinct family.

Taxonomic Considerations of the Order Scitamineae

The order Scitamineae (Zingiberales) comprises 5 families (according to Melchior 1964). Some of the subfamilies have been raised to the rank of families and 8 families are recognised.

This order is quite advanced on the basis of floral structure—zygomorphic or even asymmetric (Cannaceae and Marantaceae), and the androecium shows progressive specialisation and reduction from 5 or 6 to 1 or even 1/2 anther, and also adaptation for cross-pollination.

Bentham and Hooker (1965c) included this order in the series Epignyae. Rendle (1904) and Lawrence (1951) placed it just before the Orchidales (probably considering it as ancestral to this order). According to Cronquist (1968, 1981), the Zingiberales are equal to Commelinidae with irregular flowers, an inferior ovary, and pinnately veined leaves. They resemble Commelinidae members in: distinct sepals and petals, starchy endosperm, and two or more subsidiary cells for each stomata. Hutchinson (1969, 1973) associated the Zingiberales with the Bromeliales and the Commelinales. The co-occurrence of the chemical myricetin in Zingiberales, Typhales and Bromeliales also support this view (Gornall et al. 1979). A detailed chemical study by these authors has revealed an intermediate position of the Zingiberales with Liliales-Typhales on the one hand, and Commelinales-Arecales on the other.

Takhtajan (1980, 1987) observes that the order Scitamineae is related to the Liliales and Bromeliales, and probably has a common origin with the Bromeliales from ancient Lilealian stock. Chemically, Zingiberales and Liliales are alike and both have raphides and chelidonic acid (Gibbs 1974), in both the vessels are confined to the roots (Wagner 1977), and both have septal nectaries and cylindrical embryo. According to Tomlinson (1962), the family Strelitziaceae may be considered to represent the nearest approach to the ancestral stock from which the entire order has evloved. Dahlgren (1980a, 1983a) agrees that it is a very well-defined and natural order with the combination of features like arillate and frequently perispermous seeds, petiolate leaves and a tendency of some of the stamen homologues to be petaloid. The less-evolved families of this order (Strelitziaceae, Musaceae, Heliconiaceae, Lowiaceae) have 5–6 stamens, no staminode or only one, oxalate raphides and seeds without perisperm but copious endosperm. He, too, is of the opinion that "these families approach some Liliiflorean orders (Pontederiales, Bromeliales, Philydrales)".

The most highly-evolved family of this order is the Marantaceae with asymmetrical flowers, petaloid staminodes, a half-petaloid fertile stamen (with only half-anther) and an inferior ovary.

From studies on vessel distribution, it is apparent that the Zingiberales/Scitaminae might have diverged from the main line Commelinidae before that line had achieved a general distribution of vessels throughout the plant body.

The above discussion brings to light that the Scitaminae/Zingiberales is a natural taxon with some very distinct features of its own, such as; perennial herbs, shrubs or trees with starchy rhizomes, inflorescence-bearing stems covered with sheathing leaf-bases, leaves often large and simple, petiolate and with broad and pinnately-veined lamina, assymetrical and epigynous flowers, petaloid staminodes (Cannaceae, Marantaceae) and arillate and perispermous or endospermous seeds. It can also be separated from other allied orders very easily. Amongst monocots, order Scitaminae/Zingiberales is a highly evolved order and is derived from Commelinales stock rather than from a Liliales stock.

67

Order Microspermae

A monotypic order with the only family Orchidaceae. The members are mostly epiphytic or terrestrial herbs of moist tropical forests. Irregular, ornamental flowers are much prized; unilocular, inferior ovary with numerous minute seeds are characteristic features. Seeds non-endospermous, often with chlorophyllous embryo. Alkaloids common.

Orchidaceae

Largest angiosperm family with 750 genera and 20 000–25 000 species, widely distributed but dominant in moist tropical, evergreen forests of tropical America, southeastern Asia and the Eastern Himalayas. Often species with similar flowers grow in different geographical areas, such as damp mountain sides of Norway and New Zealand to the dry savannah grasslands of tropical Africa and South America; or from the coastal regions of the South Pacific to the dense forests of the Eastern Himalayan range, to the plains of north Bengal in India.

Vegetative Features.   Mostly terrestrial (*Geodorum, Zeuxine*) or epiphytic (*Dendrobium, Vanda*) herbs, rarely a climber, e.g. *Vanilla,* sometimes saprophytic; perennial. Terrestrial species with fibrous roots or with thickened tuberous or cord-like roots; epiphytes with aerial roots (Fig. 67.1G) with a covering of velamen tissue; normal roots may store food reserves. Stems often swollen and fleshy (pseudo-bulbs, Fig. 67.1 A), leafy or scapose with sympodial or monopodial growth; saprophytes without chlorophyll. Leaves alternate to subopposite, distichous or not, rarely whorled or opposite; occasionally reduced to scales, membranous, coriaceous or succulent; linear to broadly ovate or circular, rarely ensiform; sometimes falsely petiolate, never compound, sheathing at the base, sheath generally closed and envelop the stem; stipules and ligules absent.

Floral Features.   Inflorescence a spike, raceme or a panicle or flowers solitary (Fig. 67.1 A). Flowers zygomorphic, epigynous (Fig. 67.1 B), bracteate, sessile or pedicellate, usually resupinated by twisting of pedicel or ovary. Perianth of 6 tepals in 2 series of 3 each, the outer series of 3 sepals green or petaloid, all similar, or the median sepal more prominent due to size or coloration, mostly imbricate. The inner series is petaloid, variously ornamental, upper median one usually enlarged to form a labellum and frequently spurred. Nectariferous glands often present at the base of tepals. Stamens 1 or 2, joined to the style in a unique structure, the column (Fig. 67.1 C, D) or gynandrium; when 1, terminal on the column, when 2, lateral subtending the terminal stigma. Anthers bicelled, introrse, longitudinally dehiscent, pollen granular or agglutinated into polliinia (Fig. 67.1 D, E) which may be mealy, waxy or bony; pollinia 2–8 per anther. The distal end of pollinium sometimes attenuated into a sterile filamentous strand or caudicle; staminodes often present as glandular or dentiform or ovate processes. Gynoecium tricarpellary, syncarpous, ovary unilocular, inferior, with parietal placentation and numerous small ovules (Fig. 67.1 F). Fruit a many-seeded capsule; seeds without endosperm and with a small, often chlorophyllous embryo (Dahlgren 1980b, Dahlgren and Clifford 1982).

Anatomy.   Vessels present in roots and rarely in stems (e.g. *Vanilla*) but not known in leaves. Stomata variable, with or without subsidiary cells; spherical silica bodies sometimes present and oxalate raphides widely distributed. Laticifers absent. Multicellular, or unicellular hairs frequent, but many taxa glabrous.

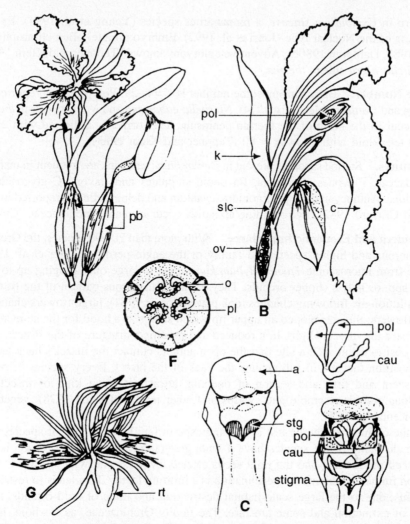

Fig. 67.1 Orchidaceae: **A-F** *Cattleya leuddemanniana*, **G** *Vanda roxburghii*. **A** Habit, showing pseudobulbs and flower. **B** Flower, vertical section. **C** Column. **D** Open anther to show pollinia. **E** A pair of pollinia. **F** Fruit, cross section. **G** Plant of *V. roxburghii* with clinging roots. *a* anther, *c* corolla, *cau* caudicle, *k* calyx, *o* ovule, *ov* ovary, *pb* pseudobulb, *pl* placenta, *pol* pollinium, *rt* roots, *stc* stylar canal, *stg* stigma. (**A**, **B** and **F** after Lawrence 1951, **C-E** after Radford 1986, **G** after G.L. Chopra 1973)

Embryology. Pollen grains may be monosulcate, in monads and with reticulate exine, as in subfamily Apostasieae; or sulcate, ulcerate or porate, in monads or tetrads, with reticulate or scabrate exine as in Cypripedieae (Zavada 1983); ulcerate, porate, inaperturate, monads, tetrads or pollinia (Balogh 1970), with reticulate exine in subfamily Neottieae (Ackerman and Williams 1980, 1981); or porate or ulcerate-inaperturate, in pollinia, with variously sculptured exine or exine-less, as in subfamilies—Orchideae and Epidendreae (N. Caspers and L. Caspers 1976, Dulieu 1973, Williams and Broome 1976). Ovules anatropous, bitegmic, tenuinucellate, with the inner integument forming the micropyle; undifferentiated at the time of pollination. There is variation in embryo sac types—Polygonum type in Monandrae, Allium type in Diandrae; 4-, 5- or 6-nucleate embryo sacs common in these groups. Embryo sac development is of bisporic pattern in *Cymbidium sinense,* a monandrous species (Yeung et al. 1994). Endosperm absent, when present, is of the Nuclear type (Johri et al. 1992). Embryo minute, often chlorophyllous (Dahlgren and Clifford 1982, Dahlgren 1980b). Adventive embryony known (Teppner and Klein 1993) in tetraploid species, *Nigritella nigra* subsp. *iberica.*

Chromosome Number. Basic chromosome number is x = 6–29. Polyploidy is reported in *Cymbidium, Paphiopedilum* and *Cattleya* (Mehlquist 1974). *Nigritella gabasiana* and *N. nigra* subsp. *iberica* with brown-red flowers occur in the north of the Iberian peninsula. The former is a diploid with 2n = 4x = 40 and the latter is a tetraploid with 2n = 4x = 80 (Teppner and Klein 1993).

Chemical Features. Steriod saponins present in *Dendrobium*; saponins are frequent in many other members of the Orchidaceae. Flavones, flavonols, flavonoid sulphates and flavone C-glycosides also occur in varying quantities. Anthocyanidins, pelargonidin, cyanidin and delphinidin are reported in different species (Dahlgren and Clifford 1982). Pyrrolizidine alkaloids occur in many Orchidaceae (Culvenor 1978).

Important Genera and Economic Importance. With more than 20 000 species, the Orchidaceae represnt the most numerous and highly specialised family of the angiosperms (Wolter et al. 1988). The orchid flowers range from microscopic *Taeniophyllum khasianum* to large ones ranging up to 18 cm across in *Cypripedium* spp. or lady's slipper orchids. They exhibit a trimerous pattern of the flowers of Liliaceae but during evolution the following changes took place (Corner 1964): large flowers changed to bilaterally symmetrical flowers; they developed an upper lip which serves as a hood for the stamens and stigma and a lower lip where the insects alight. In a reduced and modified structure of the flower, stamens reduced to 1 or 2 are positioned at such a site that the open anthers contact the insect's back and the stigma too is in such a position that it will rub against the back of the insect. Every species of orchid, by colour, size, shape, scent and time and season of opening, attracts its own kind of insect for pollination. Myrmecophilous orchids resemble ants, wasps and other insects (Horich 1978); vegetative mimicry is also known (Kennedy 1975).

Although the orchids are the largest of the angiosperm families (number of species), they are not so abundant, nor do many of the species have a wide geographical range. The reason is that, in spite of voluminous seed production, and the light seeds carried away far and wide by wind, the physiological dependence on fungal symbionts for germination is a limiting factor. This leads to restricted distribution. Along with this, due to the large-scale habitat destruction and trade of wild orchids, many species are on the verge of extinction and some are rare. The family Orchidaceae, as a whole, is included in the Appendix II of CITES or Convention on International Trade of Endangered Specied (R.W. Read and Desautels 1983).

A rare saprophytic orchid *Didymoplexis pallens* is reported from Indian Botanic Garden, Howrah (Rao and Rathore 1981). *Eulophia mannii* is a scarcely-known ground orchid from Assam (Rao and Hajra 1977). A list of 63 orchid species has been prepared that are rare and endangered (Nayar 1984) in India.

Some of the well-known orchids are: *Orchis* (35 spp.), *Cypripedium* (50 spp.) or lady's slipper, *Goodyera* (40 spp.) or rattlesnake plantain, *Ophyris* (30 spp.) or bee orchid, *Habenaria* (600 spp.) or fringe orchid, *Epidendrum* (400 spp.) or greenfly orchid, *Cattleya* (60 spp.) or florist's orchid, *Cymbidium* (40 spp.), *Vanilla* (90 spp.), *Odontoglossum* (200 spp.), *Rhynchostylis* (15 spp.), *Vanda* (60 spp.) and many others. Economically, the family is most important in florist's trade. Russell (1958) comments: "For use in corsages and as an ornamental the orchids are more sought after than any other type of plant, and not only is an industry built upon orchid culture, but growing orchids is also a widespread hobby". The leaves of *Calanthe veratrifolia* contain a glycoside, indican which, (on hydrolysis) yields indigo blue. The capsules of *Vanilla planifolia* yield commercial vanilla, a flavouring agent.

Taxonomic Considerations of Orchidaceae and the Order Orchidales

The family Orchidaceae is included in the order Microspermae by Bentham and Hooker (1965c) and Engler and Diels (1936). Cronquist (1968, 1981), Dahlgren (1975a, b, 1980a, 1983a), Hutchinson (1934, 1959) and Takhtajan (1980, 1987) treat Orchidaceae as a member of the Orchidales. Thorne (1983), however, included it under the Liliales. Most phylogenists agree that this family is closely related and derived from the Liliales. It shows closest relationship with the Hypoxidaceae (Rolfe 1909, 1910, Hallier 1912, Gobi 1916, Hutchinson 1934, 1959, Garay 1960, Rao 1969), especially with *Hypoxis* and *Curculigo* (Liliaceae). *C. orchioides* resembles orchids (Takhtajan 1980). Chromosomes of the Hypoxidaceae resemble those of the Orchidaceae (Sharma 1969). The connecting link between Hypoxidaceae and Orchidaceae is the most primitive subfamily of Orchidaceae—the Apostasioideae. Garay (1960) also considers Apostasioideae as a part of the Ochidaceae, whereas Hutchinson (1934, 1959) and Dahlgren (1980a, 1983a) regard it as a distinct family, Apostasiaceae.

Orchidaceae is assumed to be an old family derived from ancestors related to the Liliaceae, Amaryllidaceae and Burmanniaceae—a polyphyletic origin, according to Garay (1960). Cronquist (1968, 1981) includes Geosiridaceae, Burmanniaceae and Corsiaceae also in this order. According to him, the evolutionary advances are towards specialised adaptations to potential pollinators so that the pollen can be transferred en masse. Only the Orchidaceae have efficiently exploited the evolutionary opportunities of the order. The other three families could not adapt to these situations. Takhtajan (1980, 1987) observes that there is a direct relationship between the Orchidales and the Burmanniales, and the similarities between the two are the result of parallel evolution. The two taxa differ in the aestivation of perianth segments and the structure of their androecia—epipetalous in the Burmanniaceae, that is, adnate to perianth lobes. In the Orchidaceae, they are not epipetalous and there is adnation between stamen and style to form a gynoestegium.

There is no controversy about the derivation of this family from the Liliales and amongst them, Amaryllidaceae has the characters from which those of the Orchidales might have risen.

68

Concluding Remarks: Taxonomy—Yesterday, Today and Tomorrow

Taxonomy was one of the earliest disciplines of plant sciences, and deals with the identification, nomenclature and classification of living organisms. Various definitions have been advanced, of which the best could be: "a study aimed at producing a system of classification which best reflects the totality of similarities and differences" (see Cronquist 1968). This, however, is an overwhelming task, and taxonomic studies have grown and developed over the last several decades. As we do not, yet possess all the information about all the genera and species, this discipline will continue to develop as long as the researchers (all over the world) keep accumulating useful data from different expanding fields of biology. More mathematical data, biosystematic, chemical and serological data will make taxonomy a creative and dynamic subject. The new taxonomy will provide a sound and strong foundation for the progress of biological sciences.

The need for better and more objective methods of ascertaining relationships amongst various plants and plant groups was realised by the beginning of this century. The comparative studies of various plant groups from which taxonomic conclusions can be drawn, should ideally include all the characters of all the species, and from as many individual plants as possible, from the entire geographical range of the species. In this type of study, evidence from all related fields—vegetative and floral morphology, micromorphological observations on anatomy, embryology, palynology, cytology and genetics, and also geographical distribution, habitat requirement, serology and chemical analysis—are used for effective diagnosis of various taxa. The base of taxonomy is historical, but the application of modern applied science has made it an ever-changing system.

The ideal situation of using "all the characters of all the individuals of a species" is yet to be attained. The availability of information is time-consuming and demands more than just an individual effort. Presently, there are numerous "gaps" in the information available. Much research in systematic botany still remains to be taken up. A (large) number of families have not been studied at all, or have been incompletely investigated. A few examples may be cited.

For the family Dipentodontaceae, the data on embryology, chromosome number and chemical analysis are not available.

For the taxon *Medusagyne,* detailed data are not available (except for exomorphology and some anatomy). Takhtajan (1987) observes that: "Medusagynaceae is an almost extinct family and occupies such an isolated position that it is necessary to place it in an independent order". Medusagynaceae has certain features common with Paracryphiaceae, Theaceae and Ochnaceae and is, therefore, tentatively included in the suborder Dilleniidae of the order Guttiferales (until additional data, particularly in serology, become available).

The monotypic Strasburgeriaceae is yet another family with incomplete data. On the basis of the scanty data available, it has been grouped with the Ochnaceae in the order Guttiferales.

The Ancistrocladaceae has been included either in the Theales or the Violales. Takhtajan (1987) erected a separate order, Ancistrocladales, to accommodate this family. In the absence of any close relative in either Theales or Violales, it seems appropriate to place it in a distinct order of its own.

For the families Pentaphylacaceae, Dialypetalanthaceae, Lecythidaceae and Cynomoriaceae, the micromorphological data are scanty.

Lactoridaceae, with its only member *Lactoris fernandeziana*, is endemic to the Juan Fernandez Islands off the western coast of South America. Although its macro- and micromorphology have been studied, its chemical features are unknown, nor has any serological study been done. With only six taxa surviving (Zãvada and Taylon 1986), it is an isolated group in a distinct order, Lactoridales, between Laurales and Piperales.

A few families are of unknown relationship. The family Grubbiaceae includes the South African genus *Grubbia*, which is an autotrophic shrub of somewhat ericoid habit. It has been placed in such diverse groups as Santalales, Ericales, Rosales and Thymelaeales. Further comparative studies are necessary to assign *Grubbia* to its correct position.

The unigeneric family Balanophoraceae includes the total parasite *Balanophora*. Its nearest relative appears to be *Cynomorium*, which, according to Melchior (1964), belongs to the highly evolved order, Myrtales. *Cynomorium* is more primitive to Balanophoraceae due to its well-developed ovular integument, but more advanced in its inferior ovary.

The Hydrostachyaceae is a monotypic family of aquatic plants from tropical East Africa and Madagascar. The reduced or highly modified reproductive and vegetative morphology of these plants has made their placement difficult. Dahlgren (1980a) and Takhtajan (1987) treat this family as member of the Lamiales and the Scrophulariales, respectively. Thorne (1992b) places them in the Bruniales and Cronquist (1981) in the Callitrichales (because of their adaptation to aquatic habit). Absence of any diagnostic chemical compound makes it difficult to show any close link of Hydrostachyaceae to another group. Presently, however, the rbcL (Rubisco or Ribulose, 1,5 bisphosphate carboxylase enzyme) (see Glossary, p. 689) sequence data used to assess the position of this family brings it close to the Cornales s.l. (Hempel et al. 1995).

The disputed Picrodendraceae is a monotypic family, with the only genus *Picrodendron* from the West Indies. Bentham and Hooker (1965c), Engler and Diels (1936) and Takhtajan (1987) do not recognise it as a distinct family. Trifoliate leaves and drupaceous fruits bring it closer to the Rutales. Cronquist (1968) includes Picrodendraceae in the Juglandales along with Rhoipteleaceae and Juglandaceae. Hayden et al. (1984) and Takhtajan (1987) include the genus *Picrodendron* in the Euphorbiaceae. The absence of data on embryology, chromosome number, and chemical features makes it difficult to ascertain the correct taxonomic position of Picodendraceae.

Loasaceae is yet another problematic family. Although traditionally placed in the Violales because of their parietal placentation, numerous stamens and centrifugal stamen initiation, some taxonomists have suggested different alliances on the basis of various anatomical, embryological and chemical features. Dahlgren (1975a) was the first to suggest that the Loasaceae are related to the Cornales on the basis of a chemical feature (presence of iridoids) and embryology (unitegmic, tenuinucellate ovules). Hufford (1992) also suggests the same assignment on a morphological and chemical basis which is supported by rbcL sequence data provided by Hempel et al. (1995).

The families with incomplete data are listed (see p. 628). Extensive comparative data for such families are essential for their appropriate assignment is the current classificatory systems. As new relationships are discovered with the incorporation of new ideas and data, a new system may also emerge. Dahlgren (1983b) also listed a number of families in which serological investigations should be taken up. This may reveal new connections between them.

Families with Incomplete Data (see also Johri et al. 1992):

Dicotyledons: Archichlamydeae

Balanopaceae
Rhoipteleaceae
Dipentodontaceae
Grubbiaceae
Balanophoraceae
Medusandraceae
Sargentodoxaceae
Lactoridaceae
Medusagynaceae
Dioncophyllaceae
Strasburgeriaceae
Ancistrocladaceae
Cephalotaceae
Cunoniaceae
Neuradaceae
Hydrostachyaceae
Akaniaceae

Tremandraceae
Bretschneideraceae
Sabiaceae
Pentaphyllaceae
Staphyleaceae
Sarcolaenaceae
Scytopetalaceae
Penaeaceae
Peridiscaceae
Stachyuraceae
Scyphostegiaceae
Achariaceae
Sphaerosepalaceae
Datiscaceae
Crypteroniaceae
Dialypetalanthaceae
Theligonaceae
Cynomoriaceae

Dicotyledons: Sympetalae

Plumbaginaceae
Sarcospermataceae
Lissocarpaceae
Hoplestigmataceae
Callitrichaceae
Duckeodendraceae
Henriqueziaceae
Columelliaceae
Phrymaceae
Brunoniaceae

Monocotydelons

Velloziaceae
Geosiridaceae
Burmanniaceae
Corsiaceae
Mayacaceae

The integration and disintegration of taxa (at all levels—species, genera, families and orders) will continue ad infinitum as long as new taxa are discovered in explored and unexplored areas, and additional data become available; consequently inter- and intrafamilial and inter- and intra-ordinal assignments will also change. The evolutionary tendencies will also become clearer.

Appendix 1
Procedures for Field and Laboratory Study

Diagnostic Features

In taxonomic studies a *character* is defined as any attribute or feature of a taxon which can be related to form, function or behaviour of that taxon. Characters are abstractions and their expression is important. For example, stout stem is a character, but stem 20"–40" in diameter is an expression. Similarly, flower colour is a character which can be expressed as blue, red, etc. Unless a character is expressed, it has no value. A particular feature or a group of features can distinguish one taxon from the related taxa. These are *diagnostic* features.

Each organism has a large number of potential characters, and any one may be used as diagnostic feature. Some of these characters are present singly, and others in groups of two or three, are known as *correlated characters*, e.g. if in a taxon the characters such as syngenesious stamens, inferior ovary, single basal ovule and capitulum inflorescence occur together, it may be concluded that the taxon belongs to the family Asteraceae (= Compositae). Similarly, a taxon belonging to Solanaceae has gamopetalous corolla, epipetalous stamens, obliquely placed placenta and persistent calyx. Such *correlated characters* are, therefore, very useful for indentification. The two species *Cassia tora* and *C. obtusifolia* are very much alike, and have often been treated as only one species, i.e. *C. tora*. However, they can be distinguished on the basis of a few correlated characters (Table 1.1).

Correlated characters can be used very effectively in solving taxonomic problems, not only when the taxa are very closely related, but also for identification of the unknown specimens. The classificatory system of George Bentham and Joseph Dalton Hooker (1862–1883) is also based on correlated characters.

Table 1.1   Correlated Characters to distinguish *Cassia tora* from *C. obtusifolia*

	Cassia tora	*Cassia obtusifolia*
1	Leaf apex obtuse	Leaf apex mucronate
2	Extrafloral nectary between both pairs of leaflets	Extrafloral nectary between lower pair of leaflets only
3	Nectaries greenish-yellow	Nectaries deep-orange
4	Nectaries narrow at both end	Nectaries club-shaped

When a large collection of flowering plants has been made, one would like to know their names. The first step is to study their characters and to express them in a co-ordinated manner. The description of the plant and its flower is necessary to identify the undetermined taxon correctly. A critical study of the external morphology of different organs of the plants is necessary for a detailed description of a specimen.

Given below is a glossary of various morphological terms used in describing a plant specimen.

General Morphology

Aquatic plants: Grow in aquatic habitat, have slender stem and petiole, dissected leaves and poorly developed root system.

Autotrophic plants: Synthesize their own "food" from the inorganic raw materials in the presence of sunlight.

Carnivorous plants: Have variously modified leaves to trap the insects and digest them.

Epiphyte: These plants need a support and grow on the stems, branches and leaves of other plants (but are not parasitic on them). They have two sets of roots: the "clinging roots" are narrow and cling to the surface on which they grow. They also aquire nutrients from the debris collected on the bark. The other set of roots is much softer and thicker and remains suspended in the air. These are green and covered with a special tissue called velamen which absorbs moisture from the atmosphere. Most orchids are epiphytes.

Heterotrophic plants: The heterotrophs depend for their nutrition upon autotrophic plants.

Parasites: These plants grow on other plants and obtain their nutrition from the host plant through haustoria. Partial parasites are green and can prepare the carbohydrate needed for their nutrition. Total parasites are devoid of the green pigment. Root parasites are attached to the roots of the host plant; stem parasites are attached to the stems and branches.

Saprophytes: These plants grow on dead plant or animal debris or decayed products such as rotten leaves, humus etc.

Symbionts: These plants grow together without any dependence on each other.

Terrestrial: Plants grow on the surface of land, do not show any modification or adaptation due to habitat: *Euphorbia hirta.*

Xerophytic plant: Plants grow under xeric conditions and have modified green stem, flattened phylloclades or cladoes; leaves reduced to scales or spines, and stomata in stomatal pits to reduce transpiration: *Opuntia, Asparagus, Nerium.*

The Root
The root is the descending part of the plant body; positively geotropic and negatively phototropic. Root surface normally non-green, without nodes and internodes; branches or secondary roots borne at random.

Annulated tuberous roots: Appear to be formed of numerous swollen discs placed one above the other: *Cephaëlis ipecacuanha,* (Fig. 1G).

Assimilatory roots: These roots are green and can carry on carbon assimilation; epiphytic roots are assimilatory.

Climbing roots: Adventitious roots which develop from nodes and help in climbing (Fig. 2D).

Clinging roots: The epiphytes have clinging roots which enter the crevices of the support to keep the plant in place as in orchids.

Conical: Broad at the upper end gradually tapering towards the lower end, e.g. *Daucus carota* (Fig. 1B).

Fascicled roots: Tuberous roots develop in clusters in plants like *Asparagus* (Fig. 1F) and *Dahlia* (Fig. 1E). These are both for perennation and propagation. Palmately branched tuberous roots occur in *Orchis* (an orchid).

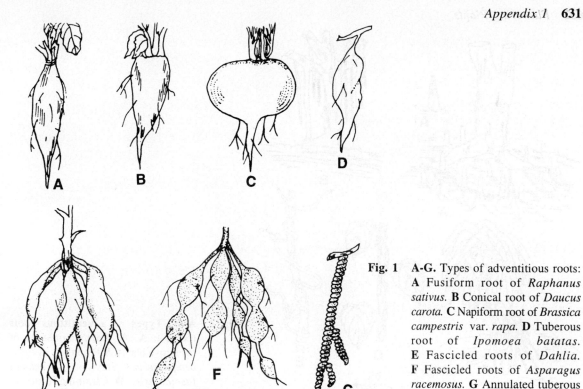

Fig. 1 **A-G.** Types of adventitious roots: **A** Fusiform root of *Raphanus sativus*. **B** Conical root of *Daucus carota*. **C** Napiform root of *Brassica campestris* var. *rapa*. **D** Tuberous root of *Ipomoea batatas*. **E** Fascicled roots of *Dahlia*. **F** Fascicled roots of *Asparagus racemosus*. **G** Annulated tuberous root of *Cephaëlis ipecacuanha*.

Fibrous roots: Common in monocots. The radicle becomes arrested in growth and additional roots develop close to the base of the radicle. These "seminal" roots and some adventitious roots grow from the base of the plumule and form the fibrous root system of the monocots.

Floating roots: Grow at the nodes of certain aquatic plants, are spongy and help to keep the plant buoyant. They function as air floats, as in *Jussiaea repens* (Fig. 2F).

Fusiform roots: A type of storage root, swollen in the middle and gradually tapering towards both ends, e.g. *Raphanus sativus.* (Fig. 1A)

Haustoria: These structures develop in parasites and obtain nourishment from the host plant (Fig. 2E, G–I).

Holdfast roots or haptera: Help in attaching the plant to the surface on which it is growing. The adhesive discs at the tip of climbing roots of *Hedera helix* are an example. The thalloid branched roots of Podostemaceae attach the plants to the rock surface with such haptera.

Moniliform roots: These are tuberous roots with alternate swollen and constricted parts which gives a beaded appearance, e.g. *Asparagus racemosus* (Fig. 1F).

Mycorrhiza: Some roots are infested with fungal mycelium and may be ecto- or endotrophic. The fungal hyphae act as root hairs do and help in absorption. Mycorrhiza is essential for seeds of orchids to germinate and growth of seedlings.

Napiform root: Storage roots swollen above and abruptly tapering towards the lower end, e.g. *Brassica campestris* var. *rapa* (Fig. 1C).

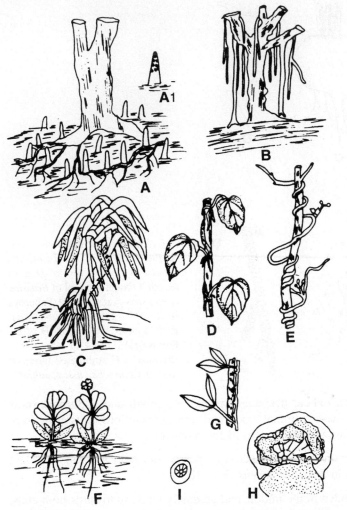

Fig. 2   **A-H.** Types of adventitious roots (contd.): **A, A₁** Pneumatophores of mangrove plants. **B** Prop roots of *Ficus benghalensis.* **C** Stilt roots of *Pandanus fascicularis.* **D** Climbing roots of *Scindapsus aureus.* **E** Haustoria of *Cuscuta reflexa.* **F** Floating roots of *Jussiaea repens.* **G** Haustorial roots of *Phrygilanthus celastroides.* **H** Section through swollen junction of the host *Pittosporum tenuifolium* and parasite (stippled). **I** Section through normal stem of the host. (**G-I** after Johri and Bhatnagar 1972).

Nodulose tuberous roots: Only the apices of the adventitious roots are swollen, as in *Costus speciosus.*

Pneumatophores, respiratory roots: Roots of plants which grow in saline coastal marshy regions (mangroves) are negatively geotropic and grow vertically upward. These roots have small pores (lenticels) through which gaseous exchange takes place.

Prop roots: Grow adventitiously from the horizontal branches of some tropical trees such as *Ficus,* and hang vertically downward. They reach the soil, get anchored, and become thick and almost as strong as the main stem. They give support to the heavy branches and sometimes also replace the main stem (Fig. 2B).

Root buttresses: These are broad and wing-like, radiating from the base of the stem; partly root and partly stem, and may develop in very old plants; *Adansonia digitata.*

Root tubers: These are formed at the nodes of running stems (runners); swollen in the middle and tapering at both ends, e.g. *Ipomoea batatas* (Fig. 1D).

Root pocket: In some water plants, the root tip is protected by a different type of root cap (called root pocket) which does not regenerate.

Stilt roots: These roots provide additional support to certain shrubs and small trees which grow along the edges of water bodies (*Pandanus*; Fig. 2C) or marshes, where anchorage of the main stem is not very strong. Adventitious roots which grow from the lower nodes of *Zea mays* or *Saccharum officinarum* also have a similar function.

Storage roots: Are swollen and fleshy due to the accumulation of stored food; may also act as perennating organs (see also Fusiform, Conical, and Napiform roots).

The Stem

The stem is the main portion of the ascending axis and develops from the plumule, bears leaves, branches, flowers and fruits. It is positively phototropic and negatively geotropic.

Acaulescent: Plants without a stem (only apparently), such as the monocots.

Adhesive climber: Plants provided with adhesive discs at the end of their climbing organs; with their help they can adhere to any surface, e.g. *Bignonia capreolata* (Fig. 50.16H).

Annual: Short-lived plants, complete their life cycle in one season.

Biennial: Complete their life cycle in two seasons/years.

Bulb: An underground modified bud with highly reduced convex or conical disc-like stem and fleshy scale leaves arising from it, e.g. *Allium cepa* (Fig. 3D). The apical bud grows into a flowering scape and the axillary buds grow into daughter bulbs. The fleshy leaves store food materials. When the scale leaves are arranged in concentric rings, as in *Allium cepa, Hyacinthus* and *Polianthes,* it is known as a *tunicated bulb.* When the scale leaves arise from the disc loosely and appear as petals of a flower, possibly overlapping one another only at the margin, as in *Allium sativum.* These individual buds are called the *cloves,* which are enclosed by a whitish, skinny scale leaf.

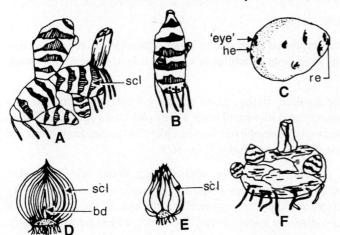

Fig. 3 **A-F.** Underground modifications of stem: **A** Rhizome of *Zingiber officinale.* **B** Rootstock of *Alocasia indica.* **C** Tuber of *Solanum tuberosum.* **D** Bulb of *Allium cepa.* **E** Bulb of *A. sativum.* **F** Corm of *Amorphophallus campanulatus. bd* bud, *he* heel end, *re* rose end, *scl* scale leaf.

Bulbil: These are modifications of axillary vegetative or floral buds, often borne on large scapiferous inflorescence, e.g. *Agave.*

Caudex or columnar stem: Usually do not branch at all. The lateral buds on the long columnar trunk remain dormant or collapse. There is a crown of leaves on the top.

Cladode: Cladode is a phylloclade of one internode only, e.g. *Asparagus racemosus* (Fig. 5D).

Climber: Weak-stemmed, mostly herbaceous, annual or perennial plants.

Corm: It is a highly condensed vertical rootstock with a large apical bud and many axillary buds. In *Amorphophallus campanulatus* (Fig. 3F) it is a huge, condensed single internode with numerous adventitious buds and roots arising from its surface.

Creeper: Perennial, herbaceous plants; to start with the plant is erect and soon gives rise to branches which, after travelling a short distance horizontally, become rooted and give rise to another plant which behaves in a similar way and covers a large area within a short time. *Runners, stolons* and *suckers* are the examples.

Decumbent: When the branches try to straighten up after growing parallel with soil surface for some distance.

Deliquescent stem: Stem with well-developed lateral buds but a weak apical bud; such trees have a spreading habit.

Diffuse: Some trailing plants produce a number of branches in all directions; these are termed as *diffuse.*

Excurrent stem: Grows indefinitely, branches are borne at a higher level and are in acropetal order.

Herbs: Rather small, weak-stemmed and short-lived plants. May be annual, biennial or perennial.

Lianas: Perennial, woody climbers which wind round tall trees to reach the top and form the uppermost strata of the forest; mostly occur in tropical rain forests.

Offsets: These are runner-like branches of aquatic plants but are shorter and thicker, e.g. *Eichhornia crassipes* (Fig. 4F).

Perennials: Grow for very long periods.

Procumbent: Branches lying parallel with soil surface.

Pseudobulb: Fleshy, tuberous structures metamorphosed from one or more internode of stem as seen in many aerial orchids (Fig. 5G). These organs store large quantities of water to tide over the unfavourable season.

Phylloclade: Stem modified to a flattened or swollen, fleshy, green structure which can function as a photosynthetic organ. The leaves are metamorphosed into small scale leaves (*Ruscus,* Fig. 5A) or spines (*Opuntia,* Fig. 5B). Phylloclades occur commonly in xerophytic families like Cactaceae, Euphorbiaceae and one member of Polygonaceae, *Muehlenbeckia platiclados* (Fig. 5C).

Rhizome: Dorsiventral, underground stems growing horizontally, with distinct nodes and internodes, brown scale leaves at the nodes and apical as well as axillary buds, e.g. *Zingiber officinale* (Fig. 3A), *Curcuma longa.* Adventitious roots develop from the lower surface and the apical and axillary buds produce new shoots in favourable season. *Rootstock* is a special type of rhizome which grows vertically, e.g. *Alocasia indica* (Fig. 3B).

Runner: These are axillary branches arising from the lower leaves which grow along the surface of soil and give rise to new daughter plant at a distance from the mother plant, e.g. *Oxalis corniculata* (Fig. 4A), *Colocasia antiquorum* (Fig. 4B), *Hydrocotyle asiatica, Cynodon dactylon,* etc. A special type of underground runner or sobole is seen in some grasses like *Agropyron* (Fig. 4C), *Saccharum spontaneum,* etc. These grasses are very difficult to eradicate and are good soil-binders.

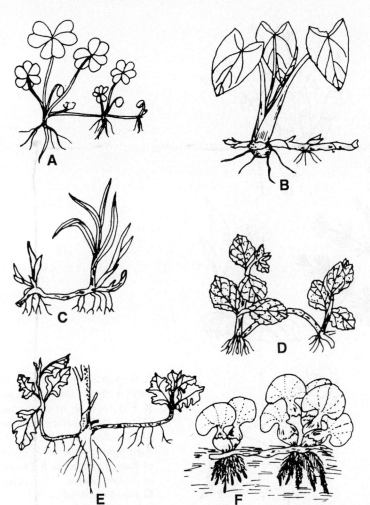

Fig. 4 **A-F** Subaerial modifications of stem: **A** Runner of *Oxalis corniculata.* **B** Runner of *Colocasia antiquorum.* **C** Sobole of *Agropyron.* **D** Stolon of *Mentha viridis.* **E** Sucker of *Chrysanthemum.* **F** Offsets of *Eichhornia crassipes.*

Scape: The stalk-bearing inflorescence in plants with basal rosette of leaves, e.g. *Narcissus.*

Scrambler: These plants grow on other plants and retain their position with the help of small prickle-like outgrowths on their stem surface. These are common in tropical rain forests, where they form thick, impenetrable bushes because of their profuse growth, e.g. *Calamus rotung.*

Shrubs: Shorter than trees in height, and branching occurs close to the level of soil. The woody branches are not as strong as those of a tree.

Stem tendril: Tendrils are climbing organs and may develop from any part of the plant. Axillary branches in *Passiflora,* inflorescence axes of *Antigonon leptopus* (Fig. 5F) and *Cardiospermum helicacabum* are some examples of stem tendril.

Stolon: Here the lower axillary branches grow upwards in the beginning and then arch down to meet the ground and root, where daughter plants are formed, e.g. *Mentha* (Fig. 4D).

Suckers: These are underground runners just below the surface of the soil and form daughter plants some distance away from the mother plant after forming adventitious roots. Such daughter plants are also formed from axillary buds on the underground runner, e.g. *Chrysanthemum* (Fig. 4E).

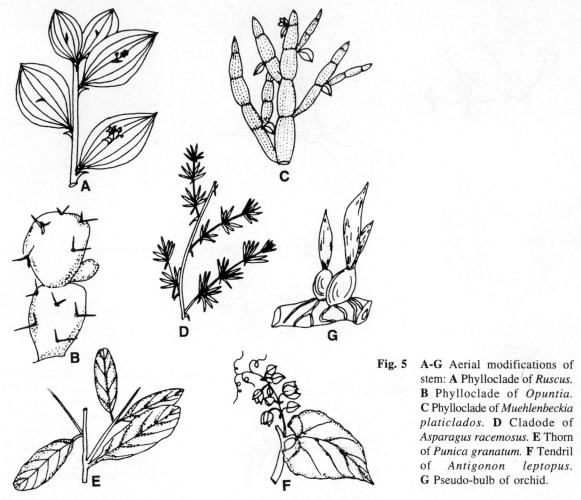

Fig. 5 **A-G** Aerial modifications of stem: **A** Phylloclade of *Ruscus*. **B** Phylloclade of *Opuntia*. **C** Phylloclade of *Muehlenbeckia platiclados*. **D** Cladode of *Asparagus racemosus*. **E** Thorn of *Punica granatum*. **F** Tendril of *Antigonon leptopus*. **G** Pseudo-bulb of orchid.

Tendril: Thin, wiry climbing organs which twine around any support they come in contact with. Any part of the plant body can be modified to a tendril (see also Stem tendril).

Thorn: These are axillary branches transformed into stiff pointed structures, e.g. *Punica granatum* (Fig. 5E), *Duranta*. Thorns are different from spines and prickles, are very deep-seated and have vascular connection. Prickles are only superficial outgrowths on the epidermis and can be easily broken off, as in *Rosa*. Spines are hard, pointed structures usually modified stipules or any other part of a plant, e.g. *Acacia nilotica*. Thorns are often branched, as in *Carissa carandus*.

Tuber: Axillary or adventitious branches arising from underground part of stem swell up at the apices and form the tubers, e.g. *Solanum tuberosum*. The nodes and internodes on the tubers are not so prominent, the nodal points being marked by the presence of scale leaves at younger stages and rudimentary buds or "eyes" in the axils (Fig. 3C). The eyes are more crowded towards the rose end (or the distal end) of the tubers. The point of attachment with the branches is the heel end.

Twiners: These are weak-stemmed plants which coil round any support. Coiling may be anti-clockwise from the top (sinistrorse climbers) or clockwise (dextrorse climbers). *Clitoria ternatea* is a sinistrorse and *Dolichos lablab* is a dextrorse climber.

The Leaf
Foliage leaf is the green flattened structure borne at nodes on stem and branches subtending a bud in its axil. Leaves develop from leaf primordia, seen as exogenic protuberances on the shoot apex.

Leaf apex: Apex of the lamina.

Acuminate: A type of acute apex whose sides are more or less concave (Fig. 9A).

Acute: Ending in a sharp point; the sides of the tapered apex essentially straight or slightly convex (Fig. 9B).

Aristate: Tapering to a narrow, finely elongated apex (Fig. 9D).

Caudate: Bearing a tail-like appendage (Fig. 9E).

Cirrhose: Slender coiled apex, e.g. *Gloriosa* (Fig. 9F).

Cuspidate: Acute but coriaceous and stiff (Fig. 9H).

Emarginate: An obtuse apex that is deeply notched as in *Bauhinia* (Fig. 9I).

Mucronate: Obtuse apex with a sharp, pointed tip as in *Catharanthus roseus* (Fig. 9J).

Mucronulate: When the mucron in broader than long, straight (Fig. 9K).

Obtuse: Margins straight to convex, forming a terminal angle of more than 90°, as in *Ficus benghalensis* (Fig. 9H).

Retuse: When the obtuse apex is slightly notched, as in *Pistia* (Fig. 9N).

Leaf Base
Basal portion of the lamina.

Attenuate: Showing a long gradual taper.

Auriculate: An ear-shaped appendage at the base, particularly in sessile leaves.

Connate: Basal portions of opposite leaves fused together, giving the appearance of a single leaf through the centre of which the stem passes as in *Swertia chirayita, Canscora diffusa* (Fig. 7G).

Cordate: Rounded lobes at the base.

Cuneate: Wedge-shaped or triangular, with the narrow end at the point of attachment e.g. *Pistia stratiotes* (Fig. 63.1C).

Oblique: Slanting; or unequal-sided, as in Solanaceae members.

Obtuse: Blunt or rounded base.

Perfoliate: Basal lobes fused together after completely clasping the stem as in *Bupleurum.*

Truncate: Appearing as if cut off abruptly at the end.

Leaf Lamina
The flat, green part of the leaf, carrying out the most important metabolic activities like photosynthesis, respiration, and transpiration.

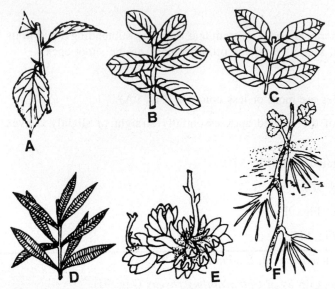

Fig. 6 A-F Phyllotaxy of leaf: **A-D** Cauline leaves, **E** Radical leaves, **F** Heterophylly. **A** Alternate, *Hibiscus rosa-sinensis.* **B** Opposite-decussate, *Calotropis procera.* **C** Opposite-superposed, *Quisqualis indica.* **D** Verticillate, *Nerium odorum.* **E** Radical, *Saxifraga macnabiana.* **F** Heterophylly in *Ranunculus aquatilis.*

Abaxial: Surface of the leaf away from the stem, also called dorsal or lower surface.

Adaxial: Surface of the leaf facing the stem, also called ventral or upper surface.

Cordate: Heart-shaped; with a sinus and rounded lobes at the base and ovate in general outline.

Cuneate: Wedge-shaped; triangular, with the narrow end at point of attachment.

Dorsiventral: If the two surfaces, dorsal and ventral, differ in structure.

Elliptical: Oval in outline with narrowed to rounded ends and broadest almost in the middle.

Hastate: Triangular, with the basal lobes directed outwards, i.e. away from the petiole: *Alocasia fornicata.*

Isobilateral: The two surfaces, dorsal and ventral, are similar.

Lanceolate: Lance–shaped, as in *Nerium.*

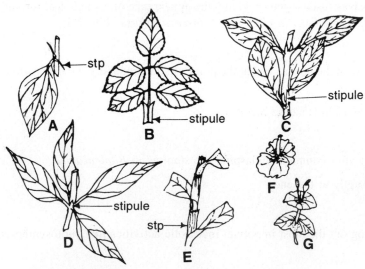

Fig. 7 A-G Stipules **A-E,** Leaf base **F, G: A** Free-lateral stipules, *Hibiscus rosa-sinensis.* **B** Adnate, *Rosa indica.* **C** Intrapetiolar, *Gardenia.* **D** Interpetiolar, *Ixora.* **E** Ochreate, *Polygonum hydropiper.* **F** Perfoliate leaf base, *Bupleurum.* **G** Connate leaf base, *Canscora diffusa.* *stp* stipule

Linear: Long and slightly broad.

Lyrate: Pinnatifid, but with enlarged terminal lobe and smaller lower lobes.

Obcordate: Deeply lobed at the apex; opposite of cordate.

Oblong: More or less rectangular, as in *Musa.*

Obovate: The terminal half of the lamina is broader than the basal.

Ovate: Egg-shaped, i.e. broader towards the base and the narrower end above the middle.

Peltate: When the petiole is attached to the centre of the leaf lamina.

Pinnatifid: Cleft or parted in a pinnate manner.

Pnnatisect: Split up to the midrib in a pinnate manner.

Reniform: Kidney-shaped.

Rotand: Orbicular, leaf is circular or disc-shaped as in *Nelumbo.*

Sagittate: Triangular, with the basal lobes pointing downward or incurved, as in *Sagittaria sagittifolia.*

Spathulate: Spoon-shaped as in *Drosera burmannii.*

Subulate: Long and narrow, tapering gradually from base to apex.

Leaf Margin

Aculate: Prickly (Fig. 10A).

Ciliate: With cilia or trichomes protruding from margin (Fig. 10B).

Cleft: Indentations up to 1/4 to 1/2 distance to the midrib (Fig. 10C).

Crenate: Shallowly ascending, round-toothed (Fig. 10D).

Crenulate: The rounded teeth smaller than in crenate (Fig. 10E).

Dentate: Teeth pointed and sharp, pointing outward at right angles to the midvein (Fig. 10F).

Denticulate: Diminutive of dentate (Fig. 10G).

Entire: With a continuous margin, may or may not be ciliate (Fig. 10H).

Incised: Margin sharply and deeply cut at random (Fig. 10L).

Lacerate: Margin irregularly, not sharply cut (Fig. 10M).

Laciniate: Margin cut into ribbon-like segments (Fig. 10N).

Parted: Cut or cleft 1/2 to 3/4 distance to the midrib (Fig. 10R).

Palmatifid: Cut palmately almost up to the tip of the petiole (Fig. 10P).

Pinnatifid: Cut pinnately almost up to the midrib (Fig. 10O).

Revolute: Margin rolled towards lower side (Fig. 10Q).

Serrate: Sharp and ascending teeth (Fig. 10I).

Serrulate: Diminutive of serrate (Fig. 10J).

Sinuate: Wavy in a horizontal plane; shallowly and smoothly lobed, without distinctive teeth or lobes (Fig. 10T).

Undulate: Wavy in a vertical plane (Fig 10S).

Leaf Types

Leaves may be simple (Fig. 8A, B), i.e. not divided into leaflets, or compound, i.e. divided into two or more leaflets.

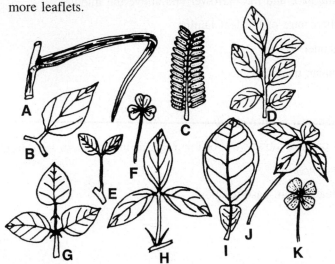

Fig. 8 **A-K** Leaf types: **A** Simple leaf of Gramineae (monocot). **B** Of *Digera arvensis* (dicot).' **C** Paripinnate, *Sesbania*. **D** Imparipinnate, *Murraya exotica*. **E** Bipinnate, *Hardwickia binata*. **F** Trifoliate, *Oxalis corniculata*. **G** Pinnate, trifoliate, *Dolichos lablab*. **H** Palmate, trifoliate, *Aegle marmelos*. **I** Palmately unifoliate, *Citrus*. **J, K** Palmate, quadrifoliate, *Paris* (**J**) and *Marsilea* (**K**).

Bifoliate palmate or pinnate: Two leaflets articulated on the top of the petiole, e.g. *Balanites roxburghii, Hardwickia binnata* (Fig. 8E).

Bipinnate: The rachis (or midrib) is branched once and the pinnules are borne on the rachila as in *Prosopis juliflora, Mimosa pudica*.

Decompound: In this type, the branching of the rachis is of still higher order than tripinnate, as in *Daucas carota, Foeniculum vulgare* and other members of the Umbelliferae.

Imparipinnate: Rachis terminates in an unpaired or odd leaflet, e.g. *Murraya exotica* (Fig. 8D).

Imparipinnate trifoliate: Leaflets three; an elongated rachis present between the lower pair of pinnae and the odd pinna as in *Dolichos lablab* (Fig. 8G).

Digitate: More than four leaflets borne at the same point on the top of the petiole, as in *Cleome viscosa, C. gynandra, Salmalia malabaricum*.

Palmate compound leaf: The rachis does not develop at all and the leaflets are articulated at one common point on the apex of the petiole.

Pinnate compound leaf: Leaflets (pinnae) borne on the rachis (midrib) or on its branches (rachilla) or on its branchlets.

Quadrifoliate palmate: A rare type, common in *Paris quadrifolia* (Fig. 8J) and is with four leaflets attached at the same point on the tip of the petiole; also seen in *Marsilia quadrifolia* (Fig. 8K).

Trifoliate palmate or ternate: Three leaflets articulated on the tip of the petiole as in *Trifolium, Oxalis* (Fig. 8F), *Aegle marmelos* (Fig. 8H).

Tripinnate: Third order branching of the rachis with pinnules borne on these secondary branches, e.g. *Moringa oliefera.*

Unifoliate: A single leaflet borne on the tip of the petiole. Presence of the joint or the articulation proves that it is not a simple leaf, as in *Citrus* (Fig. 8I).

Unipinnate: Leaflets or pinnae borne directly on the rachis. When rachis terminates in an unpaired or odd leaflet as in *Murraya exotica* (Fig. 8D), it is *imparipinnate.* When the pinnae are borne in pairs as in *Sesbania sesban,* it is *paripinnate* (Fig. 8C).

Petiole: Stalk of the leaf; leaf with petiole is *petiolate,* without it is *sessile.* The petiole is usually solid and cylindrical but may be *hollow* as in *Carica papaya* or *grooved* as in many Fabaceae or *spongy and grooved* as in *Musa, Canna; spongy and swollen* as in *Eichhornia; winged* as in *Citrus; tendrillar* as in *Clematis; flattened* and *leaflike* or *phyllodial* as in *Acacia auriculiformis.* In sessile leaves of Gramineae, a membranous outgrowth, *ligule* occurs at the junction of the leaf-sheath and the lamina, on the inner surface. Sometimes the sheathing base surrounds the stem completely and then it is called *amplexicaule* as in *Polygonum orientale.* When petiole, leaf-base as well as the stem are winged, and these wings extend up to the next lower node, the condition is called *decurrent.*

Pulvinus: The swollen base of the petiole, as in many leguminous plants.

Stipules: The two lateral appendages at the base of the petiole; give protection to axillary buds. When stipules are present, the leaves are *stipulate,* when absent, they are *estipulate.* Stipules may be caducous i.e. shed very early (*Michelia champaka*), or deciduous, i.e. shed after one season (*Dillenia indica*) or persistent (*Rosa indica*). Following types of stipules are known:

Adnate: Stipules attached to petiole, e.g. *Rosa indica* (Fig. 7B).

Free-lateral: Two, free, filiform stipules on both sides of the petiole as in *Hibiscus rosa-sinensis* (Fig. 7A).

Foliaceous: Leafy and green, carry out photosynthesis, as in *Lathyrus aphaca.*

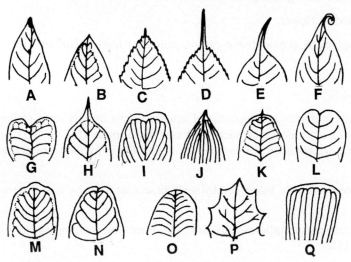

Fig. 9 A-Q Leaf apex: **A** Acuminate. **B** Acute. **C** Apiculate. **D** Aristate. **E** Caudate. **F** Cirrhose. **G** Cleft. **H** Cuspidate. **I** Emarginate. **J** Mucronate. **K** Mucronulate. **L** Ob-cordate. **M** Obtuse. **N** Retuse. **O** Rounded. **P** Spinose. **Q** Truncate. (Adapted from Radford 1987)

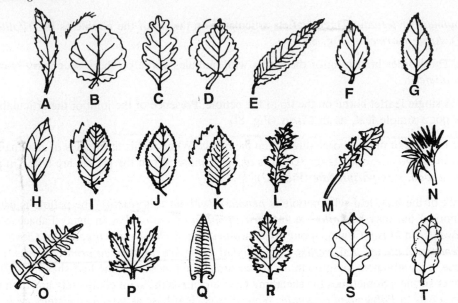

Fig. 10   A-T Leaf margin: **A** Aculeate. **B** Ciliate. **C** Cleft. **D** Crenate. **E** Crenulate. **F** Dentate. **G** Denticulate. **H** Entire. **I** Serrate. **J** Serrulate. **K** Double-serrate. **L** Incised. **M** Lacerate. **N** Laciniate. **O** Pinnatifid. **P** Palmatifid. **Q** Revolute. **R** Parted. **S** Undulate. **T** Sinuate. (Adapted from Radford 1987)

Interpetiolar: Two stipules of two opposite leaves coherent by their outer margins and appear on two sides of the stem between the petioles of two opposite leaves, e.g. *Ixora* (Fig. 7D).

Intrapetiolar: Two stipules coherent by their inner margin, resulting in a single fused stipule in the axil of a leaf, e.g. *Gardenia* (Fig. 7C).

Ochreate: Two stipules fuse to form a tube-like structure, covering the internode up to a certain height (Fig. 7E), e.g. *Polygonum hydropiper.*

Tendrillar: Stipules modified to tendrils as in *Smilax;* helps in climbing.

Spiny: Modified to spine serving as defensive organ, e.g. *Carissa carandus.*

Sheathing or protective: Enclosing a bud or flower.

Stipel: Stipule-like structures (scales, spines or glands) borne on the petiole of leaflets (or petiolules).

Inflorescence
Arrangement of flowers on the floral axis.

Capitulum or head inflorescence: A dense, compact inflorescence of usually sessile or only short-stalked flowers arranged on a flattened, convex or conical receptacle, in a centripetal order (Fig. 11G). Each capitulum is surrounded by whorls of bracts called *involucral bracts,* and the individual flowers are subtended by a scaly bract each.

Catkin: A modified spike with a thin and weak axis, subtended by scaly bracts, as in many amentiferous families (Fig. 11H).

Corymb: A more or less flat-topped indeterminate inflorescence with the outer flowers opening first, as in many Cruciferae members (Fig. 11D).

Cyathium: A cup-shaped involucre formed by the union of bracts, enclosing a group of highly reduced staminate flowers surrounding the single, centrally placed stigmatic flower. Each flower has its individual pedicel and is subtended by a bract (Fig. 12E–H). This is characteristic of some members of the family Euphorbiaceae (*Euphorbia, Poinsettia*).

Cymose: A determinate inflorescence in which the opening of the flowers is basipetal.

Dichasial cyme: The most common type of cymose inflorescence. Two lateral branches develop on the two sides of the terminal apical flower which is the oldest. These branches also terminate in a flower each and give out two branches which also behave in the same way (Fig. 11O).

Helicoid cyme: A type of monochasial cyme in which the lateral branches develop on the same side forming a helix as in members of the family Boraginaceae (Fig. 11M).

Hypanthodium: A cup-shaped receptacle with a small opening on the top and minute unisexual flowers arranged on the inner surface (Fig. 12B–D) as in *Ficus* (Moraceae). In *Dorstenia* (Moraceae) the receptacle is saucer-shaped with slightly curved-up margin and is termed *Coenanthium* (see Ganguli et al. 1972).

Monochasial cyme: The main axis terminates in a flower and one lateral branch develops from near the base which also ends in a flower after producing another branch.

Panicle: An indeterminate branching raceme in which the branches of the primary axis are racemose and the flowers pedicellate as in *Oryza sativa* (Fig. 11C, L).

Polychasial cyme. In this, more than two branches are formed from the base of the apical flower. These branches often end in monochasial cymes, as in *Hamelia patens* (Rubiaceae) (Fig. 12A).

Raceme: A simple, elongated, indeterminate inflorescence with pedicelled flowers, as in *Antirrhinum, Brassica* (Fig. 11A)

Racemose: An indeterminate inflorescence with older flowers near the base.

Scape: It is the flower-bearing shoot coming out from rosette of radical leaves of acaulescent, annual or perennial, sometimes bulbous plants, e.g. *Allium cepa, Amaryllis;* flowers may be sessile or pedicellate (Fig. 11F).

Scorpioid cyme: A type of monochasial cyme in which the lateral branches arise alternately on opposite side of the axis; the whole structure sometimes appear as a racemose inflorescence. Within this type there may be: (i) *rhipidium,* with the lateral branches lying in the same plane as the main axis (Fig. 11N), or (ii) *cincinnus,* with the lateral branches in angular plane, as in *Commelina.*

Spadix: A modified spike in which the rachis is thick and fleshy and the whole inflorescence is covered by one or more spathe-like bract; flowers unisexual (Fig. 11I), e.g. *Arum.*

Spike: Similar to a raceme but the flowers are sessile, as in *Achyranthes aspera* (Fig. 11B).

Spikelet: A secondary spike; a part of a compound inflorescence which itself is spicate as in *Triticum aestivum.* The floral unit is composed of highly reduced flowers and their subtending bracts (Fig. 11Ja, b).

Strobile: A type of spike in which the individual flowers are subtended by a persistent membranous bract, as in *Humulus lupulus* (Fig. 11K).

Thyrse: Compact, more or less compound panicle.

Fig. 11 **A-O** Types of inflorescence: **A** Raceme. **B** Spike. **C** Panicle. **D** Corymb. **E** Umbel. **F** Scape. **G** Capitulum. **H** Catkin. **I** Spadix. **Ja** Spike of spikelets. **Jb** Spikelet. **K** Strobile. **L** Panicle of spikelets. **M** Helicoid cyme. **N** Scorpioid cyme. **O** Dichasial cyme.

Umbel: A more or less flat-topped, indeterminate inflorescence in which the pedicels and peduncles arise from a common point, as in members of Umbelliferae (Fig. 11E). Umbels may be primary or secondary.

Verticillaster: A raceme of pairs of dichasial cymes at each node, characteristic of family Labiatae. As the flowers are sessile and clustered together, they appear to form false whorls or *verticels* round the inflorescence axis. Each dichasium is subtended by a large bract (Fig. 12I).

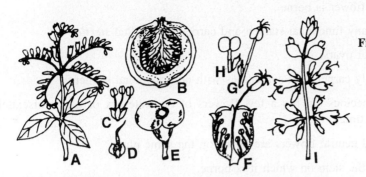

Fig. 12 **A-I** Some special types of inflorescences: **A** Polychasial cyme. **B** Hypanthodium. **C, D** Staminate (**C**) and Pistillate (**D**) flower from a hypanthodium. **E** Cyathium. **F** Vertical section. **G, H** Staminate (**G**) and Pistillate (**H**) flower from a cyathium. **I** Verticillaster inflorescence.

The Flower

The flower is the most conspicuous part of an angiospermous plant. The floral parts or the sporophylls of different types are borne on the *thalamus* or the floral axis, spirally or in a cyclic or in a spirocyclic manner. Normally, the primitive families show spiral or spirocyclic arrangement and advanced families show cyclic arrangement of the floral parts. A typical flower has four succesive floral whorls arranged on the thalamus: (i) outermost *calyx* composed of sepals, (ii) *corolla* of petals, (iii) *androecium* of stamens, and (iv) *gynoecium* of carpels. When all the four whorls are present, the flower is *complete,* if not, it is *incomplete.*

Achlamydeous: Flowers without the whorls of calyx and corolla; may be staminate, or pistillate or bisexual, subtended by bract(s).

Actinomorphic: When a flower can be cut into two equal halves through any vertical plane; also termed *regular.*

Anterior: Side of the flower facing the subtending bract, or the side opposite the stem on which it is borne.

Asymmetrical or irregular: Flowers which cannot be divided into equal halves in any plane, e.g. flower of *Canna.*

Bisexual: Flowers with both stamens and carpels (functional in each flower); also called *perfect* or *hermaphrodite.*

Bracteate: With a subtending leaf-like structure or bract at the base of the flower; *ebracteate,* when no such structure is present.

Bracteolate: With leaf-like structure(s) on the pedicel of the flower; *ebracteolate,* when no such structure is present on the pedicel of the flower.

Dichlamydeous: Flowers with both accessory whorls, i.e. calyx and corolla.

Dioecious: If male and female flowers are borne on separate plants, the plant is dioecious.

Epigynous: Flowers with inferior ovary.

Hypogynous: Flowers with superior ovary.

Monochlamydeous or *Haplochlamydeous:* With only one accessory whorl, i.e. calyx, corolla or perianth.

Monoecious: If male and female flowers are borne on the same plant, the plant is monoecious.

Mother axis: The axis on which the flower is borne.

Neuter or Neutral: Flowers without any functional stamen and carpel; also called *sterile.*

Perigynous: Flowers with half-inferior ovary.

Pistillate: Unisexual flowers with only carpels and sometimes with nonfunctional stamens.

Polygamodioecious: Functionally dioecious but has a few flowers of opposite sex or a few bisexual flowers on all plants at flowering time.

Polygamous: When male, female and neutral flowers are borne on the same plant.

Posterior: Side of the flower facing the stem on which it is borne.

Staminate: Unisexual flowers with only stamens, and sometimes with nonfunctional carpels.

Unisexual: Flowers with only stamens or carpels in one flower; also called *imperfect.*

Zygomorphic: The flower which can be cut into two equal halves only through one particular vertical plane.

The Calyx

The outermost (or lowermost) floral whorl; usually green, rarely petaloid, e.g. *Delphinium;* protects the flower in bud stage.

Bilabiate: Two-lipped or bilipped.

Campanulate: Bell-shaped.

Cleft: Sepals fused up to the middle.

Entire: Sepals completely united.

Gamosepalous: Sepals united to form a cup-like structure.

Partite: Sepals fused only at the base and free above.

Polysepalous: Sepals free.

Sepal: Individual part of a calyx.

Toothed: Sepals almost completely fused, only the tips are free.

Tubular: Tube-like.

Urceolate: Urn-shaped.

Modifications of Calyx

Calyptra: Calyx forms a cap-like structure, *calyptra,* which falls off at the time of anthesis, as in *Eucalyptus.*

Pappus: Calyx modified into hairy or bristly structures, as in members of Compositae.

Spines: Modified to spines, as in *Trapa bispinosa.*

Spur: One of the calyx lobes is modified into an elongated, tubular structure, i.e. spur, as in *Delphinium ajacis, Impatiens* spp. and *Tropaeolum majus.*

The Corolla

The second floral whorl, inner to the calyx; usually brightly coloured due to the presence of various pigments like anthocyanin, anthoxanthin and carotenoids. It protects the inner whorls and makes the flowers attractive to pollinating agents. Occasionally, petals are sepaloid. When the individual parts of corolla, the *petals* are free, the corolla is *polypetalous,* and when fused, it is *gamopetalous* or *sympetalous*

Aestivation: Arrangement of the sepals, petals or perianth lobes (tepals) in relation to one anther, in bud condition, may be of following types:

Contorted or twisted or convolute: One edge of every member overlapped by the edge of the next member in one direction, as in *Nerium, Thevetia.*

Imbricate: Floral lobes are not in one whorl; one member is completely in or overlapped from both margins and one member is completely out, i.e. overlapping the next members' margin from one side and the other three with one side in and one side out, as in *Clerodendrum.*

Quincuncial: This is a variation of Imbricate type, with two members completely in and two others completely out and one member with one side overlapped and one side overlapping, as in *Psidium guajava.*

Valvate: Meeting only at the edges without overlapping, as in *Mimosa pudica.*

Vexillary: This is typical of papilionaceous corolla. The posterior largest petal is the vexillum, which overlaps the two laterals, wings which overlap the paired anterior petals, carina or keel.

Shape

Bilabiate: Corolla two-lipped; gamopetalous with a posterior upper lip, of usually two petals and an anterior lower lip, of usually three petals; characteristic of the families Labiatae, Acanthaceae and Scrophulariaceae (Fig. 13I).

Campanulate: Bell-shaped, e.g. *Cucurbita moschata* (Fig. 13B).

Caryophyllaceous: Five free petals, with the limbs at right angles to the claws, as in the family Caryophyllaceae.

Corona: Appendages of various types such as scales or hairs developing between the corolla and stamens or on the corolla, as in some *Amaryllis,* or as outgrowths of staminal part as in *Calotropis;* function is to make the flower more attractive.

Cruciform: Four free petals arranged in the form of a Christian cross, as in family Cruciferae.

Hypocrateriform or *Salverform:* Slender corolla tube with an abruptly expanded flat limb, as in *Phlox drumondii* (Fig. 13D).

Infundibuliform: Funnel-shaped, as in *Petunia* (Fig. 13C).

Ligulate: Strap-shaped; a gamopetalous corolla in which the petals are united to form a very short tube,

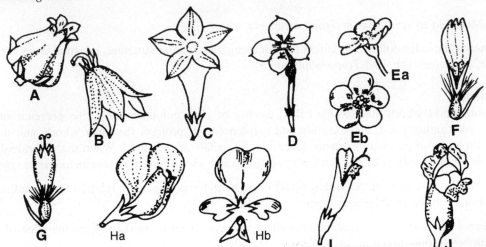

Fig. 13 **A-J** Types of corolla: **A** Urceolate. **B** Campanulate. **C** Infundibuliform. **D** Hypocrateriform. **Ea, Eb** Rotate. **F** Ligulate. **G** Tubular. **Ha, Hb** Papilionaceous. **I** Bilabiate. **J** Personate.

split on one side, forming a flat, ribbon-like structure, as the ray florets of many Compositae (Fig. 13F).

Papilionaceous: See aestivation vexillary; characteristic of family Papilionaceae.

Personate: A variation of bilabiate corolla, the throat is closed by a projection from the lower lip, called the palate; seen in some Scrophulariaceae members (Fig. 13J).

Rotate: Wheel-shaped; a gamopetalous corolla with a flat and circular limb at right angles to a short or obsolete tube as in *Brunnera* (Fig. 13Ea, Eb).

Saccate or *gibbous:* Corolla tube with a pouch at the base as in *Antirrhinum* (Fig. 13J).

Spurred: Gamopetalous (*Linaria*) or polypetalous (*Aquilegia*) corolla, with one or all the petals appendaged into spurs.

Tubular: Corolla tube more or less cylindrical, limbs not spreading, e.g. the disc florets of Compositae (Fig. 13G).

Urceolate: Urn-shaped; a gamopetalous corolla which is swollen in the middle and tapers towards both the base and the apex, as in *Pieris, Bryophyllum* (Fig. 13A).

The Androecium

The third whorl of floral organs comprises the *stamens*. East stamen consists of a stalk, termed *filament*, and a pollen-bearing *anther*.

Adnate: Filament and the connective appear to be attached to the entire length of the back of the two anther lobes, as in some Magnolialian families (Fig. 15K).

Alternipetalous: Stamens alternate with the petals.

Antipetalous: Stamens opposite the petals.

Antisepalous: Stamens opposite the sepals.

Apostemonous: Flower with separate stamens.

Appendiculate: Stamens with connective bearing appendage, e.g. a prolonged feathery structure in *Nerium odorum* (Fig. 15D).

Basifixed: Filament attached to the base of the anther (Fig. 15L).

Bithecous: Bilobed anthers.

Connective: The tissue connecting the two anther lobes;

Diadelphous: Two groups of stamens fused by their filaments (Fig. 14B).

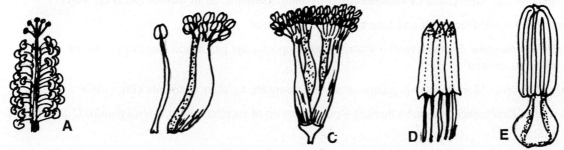

Fig. 14   A-E Types of androecium: **A** Monadelphous. **B** Diadelphous. **C** Polyadelphous. **D** Syngenesious. **E** Synandrous.

Didynamous: Two long and two short stamens.

Diplostemonous: Stamens in two whorls, outer opposite the sepals and inner opposite the petals.

Discrete: The connective tissue is almost nil and the anther lobes remain very close together, as in *Euphorbia* (Fig. 15A).

Distractile: The connective is highly elongated and bears a fertile lobe on one end and the other end is sterile, e.g. *Salvia officinalis* (Fig. 15C).

Divaricate: The connective appears to have bifurcated and the two anther lobes are separated from each other, as in *Tilia* (Fig. 15B).

Dorsifixed: Filament attached firmly to the back of the anther (Fig. 15M).

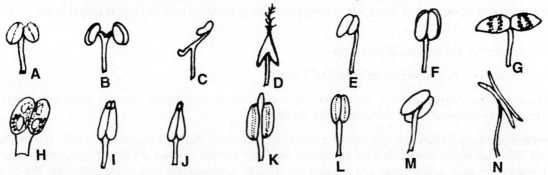

Fig. 15   A-N Anthers. **A-D** Types of anthers, **E-J** Types of dehiscence, **K-N** Types of attachment of anthers: **A** Discrete. **B** Divaricate. **C** Distractile. **D** Appendiculate. **E** Oblique. **F** Longitudinal. **G** Transverse. **H** Valvular. **I, J** Poricidal. **K** Adnate. **L** Basifixed. **M** Dorsifixed. **N** Versatile.

Epipetalous: Stamens attached to the base of the petals of a gamopetalous corolla.

Extrorse: Dehiscence of anthers longitudinal and face the outside of the flower or the petals.

Gynandril or *Gynostaminal:* Stamens and carpels fused, as in Asclepiadaceae and Orchidaceae.

Introrse: Dehiscence longitudinal and face the inside of the flower or the carpels.

Latrorse: Dehiscence longitudinal and lateral.

Longitudinal: When the thecae (anther lobes) dehisce along longitudinal sutures (Fig. 15F).

Monadelphous: One group of stamens fused by their filaments, as in Malvaceae (Fig. 14A).

Monothecous: Anthers with one lobe only, e.g. Malvaceae.

Obdiplostemonous: Stamens in two whorls, outer opposite the petals and inner opposite the sepals, e.g. *Murraya exotica.*

Polyadelphous: More than two groups of stamens connate by their filaments (Fig. 14C).

Poricidal: Dehiscence of anthers through a pore at the tip of each lobe as in *Cassia fistula, C. occidentalis* (Fig. 15I, J).

Synandrous: Stamens in three bundles of 2 + 2 + 1; each pair of stamens has completely fused filaments and the anthers are sinuous or S-shaped, e.g. *Momordica charantia, Luffa cylindrica* (Cucurbitaceae) (Fig. 14E).

Syngenesious: Stamens with free filaments but fused anthers as in Compositae (Fig. 14D).

Tetradynamous: Four long and two short stamens.

Transverse: Dehiscence at right angles to the long axis of theca (Fig. 15G).

Valvular: Dehiscence through pore(s) on the surface of the thecae covered by a flap of tissue as in *Berberis, Laurus* (Fig. 15H).

Versatile: Filament attached merely at a point, about the middle of the connective (Fig. 15N), to make the anthers swing, e.g. Gramineae.

The Gynoecium

The innermost or the central floral whorl comprises one or more carpels. A typical carpel has three parts: ovary, style, and stigma.

Bicarpellary: A gynoecium of two carpels.

Monocarpellary: A gynoecium of only one carpel.

Polycarpellary: A gynoecium of more than one or two carpels: may be *apocarpous* when the carpels are free, or *syncarpous* when the carpels are fused.

Syncarpous: Carpels fused; a syncarpous gynoecium may show different degrees of union. For example, in *Hibiscus rosa-sinensis*, ovaries and styles are fused but the stigmas are free. In *Dianthus,* only the ovaries are united, the styles and stigmas are free. In Apocynaceae and Asclepiadaceae, the ovaries and styles are free and the stigmas are fused to form a stigmatic disc.

The Ovule

Apical: Ovule pendulous or suspended, with the placenta at the apex of the ovary, as in *Coriandrum sativum.*

Basal: Erect, with the micropyle facing downward as in Compositae, or ascending, i.e. rising obliquely from the side of the ovary wall, as in some Ranunculaceae.

Horizontal: Ovule arising almost from the middle of the side wall, as in *Podophyllum.*

Placenta: Ovules in the locules are attached to a tissue called placenta which runs along the margin of the carpels. Placenta may also develop on a direct prolongation of the thalamus, at the base of the ovary. Depending upon the distribution of placental tissue, there may be various types of placentation (Fig. 16).

Axile: Placentae along the central axis of a syncarpous, multilocular ovary, as in members of Malvaceae, Solanaceae, Liliaceae, Commelinaceae.

Basal: Placenta at the base of an unilocular ovary, apparently placed on the tip of the thalamus at the floor of the ovary, as in Compositae.

Free Central: Placenta along the central axis of the ovary and the ovary is polycarpellary and unilocular. This condition develops in two different ways:

(a) In an ovary with axile placentation, the walls forming the locules may break down at a later stage, so that the ovary is apparently unilocular, e.g. *Dianthus.*
(b) The thalamus may extend into the ovary for a short distance and the placenta develop on this extended portion, as in members of Primulaceae.

Laminate: Placenta all over the inner surface of the ovary wall, as in *Nymphaea.* Here, the ovary is multicarpellary and multilocular.

Marginal: Placenta along the margin of a unilocular ovary, e.g. Leguminales.

Parietal: Placentae on the wall or the intruding partitions of a syncarpous but unilocular ovary, rarely bi- or trilocular, as in Cruciferae and Cucurbitaceae. Bilocular condition in Cruciferae is due to the formation of a false septum called replum.

Superficial: Same as *laminate.*

The stigma

The stigma is the uppermost part of the carpel, may be of the following types:

Capitate: Head-like or knob-like, e.g. *Petunia.*

Clavate: Club-shaped, e.g. *Dendrophthoë falcata, Musa paradisiaca.*

Discoid: Disc-like, e.g. *Hibiscus rosa-sinensis.*

Diffuse: Spread over a wide surface, e.g. *Elmerillia* sp.

Fimbriate: Fringed, e.g. *Najas* sp.

Lobed: Divided into parts, such as bifid (Bignoniaceae), trifid (Cucurbitaceae).

Plumose: Feather-like, as in Gramineae.

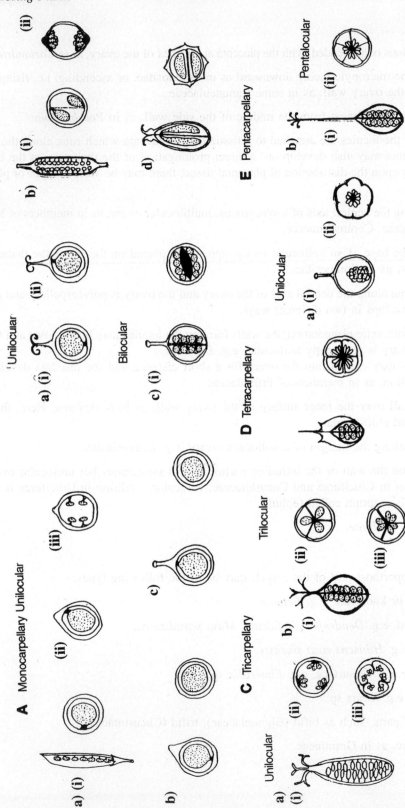

Fig. 16   A-E Types of placentation: A Monocarpellary. **a** (i), (ii), (iii) Marginal. **b** Basal. **c** Apical. B Bicarpellary—Unilocular **a** (i) Basal. **a** (ii) Apical. **b** (i), (ii), (iii). Parietal. Bilocular **c** Axile. **d** Apical. C Tricarpellary—Unilocular **a** (i), (ii), (iii). Parietal Trilocular. **b** (i), (ii), (iii). Axile. D Tetracarpellary—Unilocular **a** (i), (ii). Free-central. Pentalocular **b** (i), (ii). Axile. **E** Pentacarpellary—Unilocular **a** (i), (ii), Free-central. Pentalocular **b** (i), (ii). Axile.

Striate: Star-like, or radiate, as in *Papaver somniferum.*

Terete: Cylindrical and elongate, e.g. *Casuarina equisetifolia.*

Truncate: Cut abruptly at right angles to the long axis, as in *Acacia nilotica.*

In *Begonia* the stigma is highly branched, in Euphorbiaceae all the three stigmas are bifid, and in *Crocus sativus,* the stigmas are funnel-shaped.

Stylopodium: The swollen base of the style as in Umbelliferae

The Style

Gynobasic: Arising from the centre of the depressed apex of the ovary, as in *Salvia, Ocimum.*

Lateral: Arising from the side of the ovary, as in *Mangifera indica.* In *Gloriosa superba* the style is positioned at right angles to the ovary axis.

Terminal: Arising from the tip of the ovary, common in most plants.

Appendix 2
Herbarium Techniques

What is a Herbarium?

The original term herbarium described a book about medicinal plants (Radford et al. 1974), and not a mere collection of plants. However, at present, this word means a collection of dried specimens arranged according to a well-known system of classification.

Hence, a herbarium is a collection of dry, pressed and preserved specimens of plants arranged systematically for the purpose of reference and identification. For preservation of every detail in the specimen, and also for the length of time they are preserved, the procedure of pressing and drying has been very useful; such specimens have been and always will be helpful for various taxonomic research.

A herbarium may also be regarded as a museum which is well represented as regards the geographical distribution of various species. In addition, it can also be used as a rich data bank for various specimens because each specimen is accompanied by as much detailed data as possible, collected from different fields, with the result that a modern herbarium is more or less a great filing system and can furnish data in any field. A herbarium has many functions, some of the important ones are:

1 It may serve as a reference collection, with its help newly collected plants can be verified or identified.
2 It may serve as an institution offerring courses on herbarium techniques to graduate and undergraduate students.
3 It may also serve as a reference collection for plant taxonomy and other relevant courses.
4 Help can be sought from a herbarium for writing the flora of a particular region.
5 Classification or nomenclatural problems can be sorted out with the help of herbarium material.
6 A herbarium is helpful in providing data on various fields such as vegetative and reproductive morphology, pollen samples, leaf samples for chemical analysis, anatomical study after clearing, data for distribution maps, economic value, etc. (from the herbarium labels).
7 Serves as a repository for Type specimens, chromosomal, chemosystematic and experimental voucher specimens.

Based on the purposes they serve, Davis and Heywood (1963) divided all herbaria into three Categories:

1. Major or National herbaria,
2. Regional or local herbaria,
3. Working herbaria.

According to Heywood's definition, National herbaria are major herbaria where there is a large staff working on a wide range of taxonomic and floristic projects covering various areas, and indeed countries, as opposed to small university or local herbaria, whose programmes are related primarily to local floras or a small range of activities. Functions of National herbaria can be grouped into four categories: (a) various types of research programmes of their own, (b) serving as a repository for Type specimens and other historically important specimens, (c) giving specimens on loan to other institutions, and (d) training graduate students.

The types of research work done in a National herbarium may not be possible anywhere else, mainly because of limited resources. Such studies often require the examination of large number of specimens from a wide geographical distribution. Not that it is impossible to obtain specimens on loan from various herbaria or to go to different herbaria for research, but this requires ample funding resources and also time. It is always better to have all the materials for study under one roof. Work on major Floras or Manuals or an effective broad-scale study of a family is possible only in a National herbarium.

Regional floras are often written at a National herbarium. It is not expected of a writer of such a flora to provide extensive data for every genus, but working in such a herbarium, provides more authentic information. Floristic collections from many countries have been written with the help of National herbaria, and have also broad-based studies of families, where attention is paid to generic limits and the relationships among genera, are chiefly carried out. Biosystematic studies are also carried out at National herbaria, especially those which are also attached to a university.

The importance of such herbaria as repositories of types and other historical material is also well known.

Another function of National herbaria is to give specimens on loan to other institutions for research. Most National herbaria are also associated with universities providing graduate training in taxonomy.

According to Davis and Heywood (1963), there are regional, local and working herbaria. In regional herbaria, the collections concentrate on one region; in local herbaria, the concentration is on a more local area within the region, and working herbaria are essentially for the identification of collections for a particular area or group.

Regional herbaria usually have a good collection of specimens, and offer various types of research work, such as population studies (Hedberg 1961), can be conducted in addition to the two main functions: (a) correct identification, and (b) alpha (α) taxonomic research.

The work of Hedberg (1961) is proof of the use of herbarium material for biosystematic and cytotaxonomic work. However, it is not always possible to use herbarium material for cytological studies, unless indirectly by raising seedlings using seeds from herbarium sheets. Of course, other micro-characters, such as pollen grain or stomatal size (Aalders and Hall 1962), can be studied; seed morphological (Ball and Heywood 1962) and even anatomical studies (Schwabe 1961) can be carried out with herbarium material.

In addition, it is also now known that many chemical constituents of plants remain detectable in herbarium material even from very old collections (Bate-Smith 1965, Harborne 1967). Such functions of the regional herbaria could be taken over by major herbaria as well, but if such foraging expeditions are taken up by the biosystematists or phytochemists, the major herbaria may not be able to retain their status for long.

Local herbaria also have an active role in identification and distributional studies, but they are not involved in taxonomic research on a large scale. However, both regional and local herbaria can serve in different ways. Both can test the application of new technological developments in herbarium practices. This is not possible in major herbaria, as (a) they are more conservative, and (b) their Type specimens are irreplaceable. An important new technology which deserves special mention is the application of electronic data-processing equipment in herbaria (Sokal and Sneath 1966). If the experiment is conducted in a major herbarium, it may take 8 to 10 years to obtain the result. Any time-consuming new method of operation should be well-proven before it is adopted for a major herbarium.

Methods of Preservation

The essential requirement for the preparation of herbarium specimens is a wooden press or plant press. With the help of this equipment, it is possible to press the fresh specimens flat and to dry them quickly. The press is made of two wooden frames, in which a number of thin but strong pieces of wood are screwed at the two ends. Between the frames is kept the moisture-absorbing material, usually blotting paper or corrugated sheets, and between these, folded newspaper sheets, used for pressing the specimens.

Once the specimens have been placed in the folded newspapers, the press is tightly bound with the help of two leather straps.

A good specimen should be:

1. Of appropriate size, i.e. ca. 10" to 12" long. If the specimen itself is small, then more than one may be mounted on one sheet. If it is a large specimen, it should be folded in the shape of a V or W.
2. It should be at the flowering and/or fruiting stage.
3. Its leaves should be healthy and not infected by fungus or eaten by insects.
4. Pressing should be done in such a way that parts do not overlap and some of the leaves should show the dorsal and others the ventral surface.
5. The press containing the plants should be kept in the sun if possible, or suspended over moderate heat. The blotters should be dried at least once daily till the plants are completely dry.

If, however, it is not possible to press the material on the spot, then it can be brought back to the camping site and then pressed. To keep the material fresh for pressing after the lapse of a few hours, one can make use of another piece of equipment called the *vasculum*. This is a container made of sheet metal, oval at both ends and with a hinged lid on one surface. The lid is provided with a device to keep the interior of the container air-tight. Of course, no specimen remains as fresh if pressing is delayed for a long period. Plastic bags are also excellent containers for this purpose, if they are closed properly.

Other instruments required for plant collection include a digger and a pair of scissors or secateurs. When a herbaceous plant is collected, its underground parts also should be pressed. For this purpose, a digging instrument is essential. On the other had, for cutting a woody branch, it is equally essential to have a pair of scissors or secateurs. A pair of forceps and a good hand lens are also indispensable when collecting for a herbarium.

Once the specimens have been dried completely, they are mounted on sheets and then preserved. The mounting sheets are white and of standard size i.e. 11.5" × 16.5". The quality of paper should be such that it can be stored for a long time.

Specimens are attached to the sheets in various ways. One common method is to use strips of white gummed cloth or Holland cloth (Porter 1969) and a suitable non-staining gum solution. The gum solution is spread on the surface of the plant body and then pieces of white gummed cloth are used. However, care should be taken to avoid placing these stripes over important features of the specimen. Use of cellotape should be avoided, as it comes off after some time. Archer (1950) devised a new technique, whereby ethyl cellulose and a resin are dissolved in a mixture of toluene and methyl alcohol. The result, a thick, syrupy, adhesive liquid, can be applied with the help of a bottle with a nozzle. The mounts appear very neat and can be treated within a very short time.

Herbarium labels. A specimen without a label is useless. The label provides all the data about the plant which cannot be determined by examining the specimen. It should be ca. 2.75" × 4.25" in size, of high quality paper and either typewritten or printed. It should contain;

1. *Heading:* Country, state or province name, name of the institution.
2. *Scientific name* of the specimen followed by the author's name and the name of the family.
3. *Locality:* Specific locality should be mentioned, so that if another person wants to collect the same specimen, he should be able to reach the exact site without much difficulty.
4. *Habitat:* Vegetation type, moisture content of the soil and atmosphere, soil type, elevation, direction of slope, etc. should also be mentioned.
5. *Date of collection* should include the exact month and year to indicate when to collect the specimen.

6. *Name of collector*
7. *Collection number:* The literature on plant systematics identifies and refers to the specimen by the collector's name and collection number. Hence, the collection number is a must for any collector.

The label should be pasted on the lower right corner of the sheet. A short note may be added on the label itself, to provide additional information about the specimen, such as the flower colour, scented or not, height of the plant, abundance of the plant in the area, etc.

Filing: Filing is the process of placing the specimens in a proper sequence. The mounted specimens have to be handled with care. They are usually kept in closed steel almirahs, in which each shelf is further divided into smaller areas. In each of these cavities, a family cover containing a number of genera, or a genus cover containing a number of species is placed, depending upon the size of the family or genus or species. On the front edge of each of these folders, the scientific name of the specimen is written in bold letters.

In all major and regional herbaria, specimens are arranged according to any well-known system of classification. In smaller herbaria, like local or working herbaria, specimens are often arranged alphabetically.

Maintenance

Specimens thus kept and handled with care, can remain as they are for an indefinite period of time. Additional care is taken against insect and fungal attacks. Specimens may be fumigated by paradichlorobenzene or may be poisoned by smearing with a solution of mercuric chloride. Fumigants can be kept even inside the cabinets, the doors remaining closed when not in use.

Specimens should be handled with extreme care so that they are not damaged and become worthless. Each and every broken part (leaf or petal) should be placed in a small packet in the same folder. If there is name change, an additional label with all the new data, i.e. the changed name, author's name and date, should be attached to the sheets above the regular labels. These are called 'annotation' labels.

Preventive measures against fire should also be taken. To achieve this the best method would be to house the specimens in fire-proof cabinets in a fire-proof building.

Insect damage to herbarium specimens is a serious problem in the tropics. There are many chemicals which have been used as fumigants. However, most fumigants and insect repellants are also health hazards for the staff working in a herbarium. Low heating or fumigation of the new entries to a herbarium and a climate in the herbarium itself suitable to ward off insects, is a good practice. Amongst the most serious insect pests are the three types of beetles—cigarette beetle, drugstore beetle and black carpet beetle. Various methods have been used for the control of these insects:

1. Incoming specimens are given a preliminary heat treatment at 60 °C for 6 h in a specially constructed cabinet.
2. Insect repellants like paradichlorobenzene or naphthalene are used, as these chemicals can ward off insects by their offensive odour and taste. Usually a 2- or 3-oz cloth bag refilled once a year is sufficient for a standard case. However, workers in herbaria should not be exposed to the fumes of these chemicals.
3. Fumigants are another group of chemicals used in the form of gas. Fumigation of a herbarium is necessary and also a common practice in many of them. Recent information indicates that many of these chemicals are extremely hazardous towards human health. The Federal Insecticide, Fungicide and Rodenticide Act (of USA) prohibits the sale of fumigants except for precisely those uses for which they are labelled. Herbaria are required to comply with the Occupational Safety and Health Act (Jones and Luchsinger 1987). Dowfume-75 is one fumigant which has been permitted to be used against insect pests such as beetles. However, only a trained fumigator, wearing a gas-mask,

should do the fumigation. All new entries to a herbarium are fumigated with ethylene dioxide, methyl bromide or ethylene dibromide at a temperature of 38 °C for 3 to 4 h.

In many herbaria, ethylene dichloride or 1,2 dichloroethane mixed with carbon tetrachloride, CCl_4, is used as one of the fumigants. Six ounces of the mixture is taken in a beaker and kept on the uppermost shelf of the herbarium case. The case is kept tightly closed for 4 or 5 days. Ethylene dichloride alone is a dangerous explosive, and excessive exposure to it causes injury to human liver and kidney. Carbon tetrachloride is also toxic, causing damage to the liver (Monro 1969).

Type specimens are the invaluable specimens of any herbarium, being the specimens on which the description of the name of the taxon is based, and also designated by the author himself. These are, therefore, irreplaceable. In most herbaria, such specimens are kept separately from general and more recent collections. Extra precautionary measures should be observed, so that they are not destroyed because of unnecessary handling. In some herbaria, these are filed separately in folders of different colours.

Loan and Exchange of Herbarium Materials

Giving specimens on loan for research purposes is one of the functions of major and regional herbaria. It is not possible for an individual to move from one herbarium to another with a research project, as this is both costly and impractical. On the other hand, one can borrow specimens from different herbaria and work in one place.

A loan is always made to other institutions and arrangements for giving and receiving the loans are conducted by the curators of the two institutions.

The specimens taken on loan must be handled with care. Any new information due to revision of the specimen should be annotated on the sheets. These sheets may be marked lightly with a pencil so that it is easier to send back the specimens to different herbaria without any difficulty.

Herbarium Ethics

Every herbarium should provide information on working days and hours, filing arrangements, loan procedures, availability of microscopes, use of library collections, etc. A visitor should be aware of these regulations beforehand.

Even otherwise, one should always practice the following rules while working in a herbarium:

1. Specimens are fragile; the sheets on which they are mounted should be kept flat.
2. Folders and sheets should be lifted one at a time.
3. While carrying the specimens from the cabinet to the working table, a supporting sheet should be kept below.
4. Heavy books or elbows should not be placed on the sheets.
5. Any broken part should be kept in a small packet in the same folder.
6. Materials for dissection should be taken only sparingly and with permission.
7. Damaged specimens should be kept aside for repair.

Identification

Identification is an essential aspect of taxonomy. A plant specimen cannot remain as A, B or C for an unspecified length of time. It has to be given its correct name for future reference. The process of determination of an unknown specimen, as being identical with another, already known specimen is called *identification*. Identification and determination of the correct name go together, and this is known as *specimen determination* (Jones and Luchsinger 1987).

Some of the Major Herbaria of the World and the Number of Specimens in them

Royal Botanic Gardens, Kew	6.5 million
Museum of Natural History, Paris	6.5 million
Komarov Botanical Institute, Leningrad	over 5 million
Conservatory and Botanical Garden, Geneva	5 million
Combined Herbaria, Harvard University, Cambridge	4.5 million
New York Botanical Garden, Bronx	4.3 million
British Museum of Natural History, London	4 million
U.S. National Herbarium (Smithsonian), Washington, D.C.	4.1 million
Natural History Museum, Vienna	3.5 million
Missouri Botanical Garden, St. Louis	2.9 million
Central National Herbarium, Calcutta	2.5 million
Royal Botanic Garden, Edinburgh	1.7 million
National Botanical Garden of Belgium, Brussels	over 2 million
National Botanic Garden Herbarium, Lucknow	over 1 million
National Herbarium, Melbourne	1.5 million
Arnold Arboratum, Boston	700 thousand
Forest Research Institute, Dehradun	300 thousand
Blatter Herbarium, St. Xavier's College, Bombay	100 thousand

Source: (a) Jones and Luchsinger (1987), and (b) Jain and Rao (1977).

For identifying plants, one must:

1. Observe the specimens in the field prior to identification and determine the important or distinguishing features.
2. Have a knowledge of the resources for identification, such as floras, manuals or relevant herbarium.

While making field trips for plant collection, one must carry notebooks for making field notes for each plant. These books should have enough writing space, and the collection number printed on them. At the foot there should be three or four detachable collection numbers which can be tagged with the specimens.

The following items about the plant are always desirable for identification purposes:

(a) The plant is a herb, shrub or tree, or any other.
(b) Root adventitious or taproot.
(c) Stem woody or herbaceous, and other detailed morphological features that can be seen with the naked eyes.
(d) Altitude of the area; type of soil, amount of annual rainfall; type of vegetation, i.e. if it is a thorny scrub vegetation of mainly thorny xerophytic plants, if it is a marshy land vegetation with improper drainage, if it is a tropical rain forest area or any other.
(e) Flower colour and scent, which cannot be retained in pressed specimens.
(f) Any typical agent of pollination noticed.
(g) Any other special feature which is not discernable from herbarium specimens.

All these are macroscopic features and give detailed information about the specimen.

Identification with Keys

It is far more efficient to identify a specimen by the use of key than to shuffle through a stack of herbarium sheets for comparison. Analytical keys have long been used for identification of plant taxa.

Keys are mechanical devises by which an unknown taxon may be identified. It is *artificial* in the sense that pairs of contrasting characters are selected, of these one is to be accepted and the others rejected. Such a pair of contrasting or contradictary characters is known as a *couplet*, and each statement of a couplet is known as *lead*. The two leads of a couplet are arranged in *yokes*, and these two leads of a couplet are also identified by the same number or letters.

There are various types of keys. A key may be useful in assessing the natural relationship or affinities amongst various taxa, or it may be purely artificial, i.e. without keeping the natural relationship in view.

Key can be as simple as only one pair of leads or very complex. They may be based only on floral characters, or only on vegetative characters, or on both.

---

1. Corolla actinomorphic; stamens numerous......Mimosaceae
1. Corolla zygomorphic; stamens few
   2. Stamens diadelphous, all fertile ......Papilionaceae
   2. Stamens free, staminodes often present ... Caesalpiniaceae

---

Depending upon the arrangement of the couplets, there can be a *bracket* key or an *indented* key. In *bracket* key (also known as parallel keys), the couplets are arranged either with alternate indentation or all starting from the common margin. Numbering or lettering is consecutive or continuous. In this type of key, the two leads of a couplet are always together.

---

1. Corolla zygomorphic, stamens ten .... (2)
1. Corolla actinomorphic, stamens numerous .... Mimosaceae
   2. Stamens diadelphous .... Papilionaceae
   2. Stamens free .... Caesalpiniaceae

---

In this type of key, all leads are more or less of same length.

In *indented* (or yoked) keys, each successive couplet is indented under the one preceding it. In this, the two leads of a couplet may not remain together all the time. This key is better understood because similar elements are grouped in such a way that they are visible at a glance.

For preparing an identification key, certain rules are to be followed:

1. Only contradictory statements should be taken into account, so that one of these fits and the other is rejected.
2. The smaller group of the two should be dealt with first, followed by the larger group.
3. The same character should not be used repeatedly.
4. As far as possible, a negative statement should not be chosen, e.g. instead of saying "stamens not diadelphous" we may say, "stamens free".
5. Each lead of a couplet must start with the same word.
6. Use of discontinuous characters is more beneficial than continuous or overlapping characters. Generalised characters must also be avoided, e.g. leaves 5–10 cm broad and leaves 3–5 cm broad, or pedicel short, leaves with broad, white margin and pedicel longer, leaves with narrower margin.
7. Only those morphological features should be considered that can be seen with the naked eye, or at best with a hand lens. If microscopic features like chromosome number, or anatomical or embryological characters are used to form a key, it will not be possible to identify unknown herbarium specimens.

8. While making a key for dioecious plants, characters of both the male and female plants must be incorporated, as the male and female plant may not both be available at one time.

Keys have their own limitations. Even at their best, they are only selected assortments of a few characters of the groups of plants or the plant families involved. There is a possibility of coming across exceptions to the statements included in the key. Variation is one of the significant laws of biology. A key usually includes the most outstanding or the most usual feature, and it may not be possible or even practical to include all the exceptions encountered.

Appendix 3

Comparative Placement of Orders and Families in Various Classificatory Systems

This is a comparison, in tabular form, of the flowering plant classificatory systems of Melchior (1964), Bentham and Hooker (1965a, b, c), Hutchinson (1973), Dahlgren (1980a), Cronquist (1981) and Takhtajan (1987). All families accepted in any one of these six systems have been accounted for. The linear sequence of the orders and families is that presented in Melchior (1964).

Notes in Appendix:

1. Note a): The family Corylaceae (no. 63) of Hutchinson's system (1973) is a member of Fagales, but according to the systems of Melchior (1964), Cronquist (1981) and Takhtajan (1987), Corylaceae is included in family Betulaceae and not treated as a distinct family. Similarly, the family Barbeyaceae of Cronquist (1981) is included in Ulmaceae of Melchior (1964) and as a member of the order Berbeyales in Takhtajan's system (1987).

2. Notes b), c) .... If there is more than one such family in any order, it has been marked b, c, d, up to z, and then aa, bb, cc, up to pp.

3. A dash (-) indicates that the taxon has not been treated in some of the systems.

4. A question mark (?) indicates that the position of the taxon is undecided.

5. The numbers before the family/order is the number given by the Author.

6. *in* before the family/order indicates the placement of the taxon.

Melchior (1964)	Benth. & Hook. (1965a,b,c)	Hutchinson (1973)	Dahlgren (1980a)	Cronquist (1981)	Takhtajan (1987)
Order Casuarinales					
1. Casuarinaceae	158 Unisexuales	67 Casuarinales	Casuarinales	63 Casuarinales	92 Casuarinales
Order Juglandales					
2. Myricaceae	157 Unisexuales	59 Myricales	Myricales	59 Myricales	95 Myricales
3. Juglandaceae	156 Unisexuales	65 Juglandales	Juglandales	58 Juglandales	97 Juglandales
(a) —	—	66 Picrodendraceae (Juglandales)	—	—	*in* Euphorbiaceae
Order Balanopales					
4. Balanopaceae	152 Unisexuales	60 Balanopales	Balanopales	60 Fagales	86 Balanopales
Order Leitneriales					
5. Leitnericeae	155 Unisexuales	58 Leitneriales	Sapindales	56 Leitneriales	272 Leitneriales
6. Didymelaceae	near Leitneriaceae	81 Bixales	Euphorbiales (?)	48 Didymelales	87 Didymelales
Order Salicales					
7. Salicaceae	160 Ordines anomali	57 Salicales	Salicales	131 Salicales	158 Salicales
Order Fagales					
8. Betulaceae	159 Unisexuales	61 Fagales	Fagales	62 Fagales	94 Betulales
9. Fagaceae	159 Unisexuales	62 Fagales	Fagales	61 Fagales	93 Fagales
a) *in* Betulaceae	*in* Unisexuales	63 Corylaceae (Fagales)		*in* Betulaceae	*in* Betulaceae
Order Urticales					
10. Rhoipteleaceae	—	64 Juglandales	Juglandales	57 Juglandales	96 Rhoipteleales
11. Ulmaceae	*in* Urticaceae	68 Urticales	Urticales	51 Urticales	184 Urticales
12. Eucommiaceae	—	73 Urticales	Eucommiales	49 Eucommiales	80 Eucommiales
a) *in* Ulmaceae	—	72 Urticales	Urticales	50 Barbeyaceae (Urticales)	189 Barbeyales
13. Moraceae	*in* Unisexuales	70 Urticales	Urticales	51 Urticales	185 Urticales
b) *in* Moraceae	*in* Unisexuales	69 Cannabaceae (Urticales)	Urticales	52 Urticales	186 Urticales
c) —	—	—	Cecropiaceae (Urticales)	54 Urticales	187 Urticales
14. Urticaceae	153 Unisexuales	71 Urticales	Urticales	55 Urticales	188 Urticales
Order Proteales					
15. Proteaceae	144 Daphnales	89 Proteales	Proteales	184 Proteales	322 Proteales
Order Santalales					
16. Olacaceae	Olacaceae	196 Olacales	Santalales	209 Santalales	311 Santalales
17. Dipentodontaceae	—	200 Olacales	Violales	208 Santalales	148 Violales
18. Opiliaceae	*in* Olacaceae	197 Olacales	Santalales	210 Santalales	312 Santalales

Melchior (1964)	Benth. & Hook. (1965a,b,c)	Hutchinson (1973)	Dahlgren (1980a)	Cronquist (1981)	Takhtajan (1987)
19. Grubbiaceae	in Santalaceae	203 Santalales	—	141 Ericales	131 Ericales
20. Santalaceae	Santalales	204 Santalales	Santalales	211 Santalales	315 Santalales
21. Mysodendraceae (or Misodendraceae)	in Santalaceae	205 Santalales	Santalales	212 Santalales	316 Santalales
22. Loranthaceae	Achlamydosporae	202 Santalales	Santalales	213 Santalales	317 Santalales
a) in Olacaceae	—	198 Olacales	in Olacaceae	in Olacaceae (?)	313 Octoknemaceae
b) in Olacaceae	in Olacaceae	199 Aptandraceae (Olacales)	in Olacaceae	in Olacaceae	in Olacaceae
c) in Loranthaceae	in Loranthaceae	in Loranthaceae	in Loranthaceae	214 Santalales	318 Viscaceae (Santalales)
—	—	—	—	215 Santalales	319 Eremolepidaceae (Santalales)
Order Balanophorales					
23. Balanophoraceae	Achlamydosporae	206 Santalales	Balanophorales	216 Santalales	37 Balanophorales
a) in Balanophoraceae	—	in Balanophoraceae	in Balanophoraceae	in Balanophoraceae	33 Dactylanthaceae
b) "	—	"	"	"	34 Sarcophytaceae
c) "	—	"	"	"	35 Latraeophilaceae
d) "	—	"	"	"	36 Lophophytaceae
e) in Balanophoraceae	in Balanophoraceae	in Balanophoraceae	Cynomoriaceae	in Balanophoraceae	364 Cynomoriales
Order Medusandrales					
24. Medusandraceae	—	201 Olacales	—	207 Santalales	314 Santalales
Order Polygonales					
25. Polygonaceae	134 Curvembryeae	286 Polygonales	Polygonales	76 Polygonales	74 Polygonales
Order Centrospermae					
26. Phytolaccaceae	132 Curvembryeae	289 Chenopodiales	Caryophyllales	64 Caryophyllales	57 Caryophyllales
27. Gyrostemonaceae	in Phytolaccaceae	290 Chenopodiales	Capparales (?)	137 Batales	168 Batales
28. Achatocarpaceae	in Amaranthaceae	79 Bixales	in Phytolaccaceae	64 Caryophyllales	58 Caryophyllales
29. Nyctaginaceae	128 Curvembryeae	88 Thymelaeales	Caryophyllales	66 Caryophyllales	60 Caryophyllales
30. Molluginaceae	in Aizoaceae	282 Caryophyllales	Caryophyllales	74 Caryophyllales	63 Caryophyllales
31. Aizoaceae	79 Ficoidales (as Ficoideae)	284 Caryophyllales	Caryophyllales	67 Caryophyllales	61 Caryophyllales
32. Portulacaceae	24 Caryophyllineae	285 Caryophyllales	Caryophyllales	72 Caryophyllales	65 Caryophyllales
33. Basellaceae	in Chenopodiaceae	298 Chenopodiales	Caryophyllales	73 Caryophyllales	67 Caryophyllales
34. Caryophyllaceae	in Curvembryeae	283 Caryophyllales	Caryophyllales	75 Caryophyllales	71 Caryophyllales
35. Dysphaniaceae	in Illecebraceae	294 Chenopodiales	in Chenopodiaceae	in Chenopodiaceae	in Chenopodiaceae
36. Chenopodiaceae	131 Curvembryeae	293 Chenopodiales	Caryophyllales	70 Caryophyllales	73 Caryophyllales
37. Amaranthaceae	130 Curvembryeae	295 Chenopodiales	Caryophyllales	71 Caryophyllales	72 Caryophyllales

Melchior (1964)	Benth. & Hook. (1965a,b,c)	Hutchinson (1973)	Dahlgren (1980a)	Cronquist (1981)	Takhtajan (1987)
38. Didiereaceae	—	232 Sapindales	Caryophyllales	68 Caryophyllales	70 Caryophyllales
a) *in* Phytolaccaceae	*In* Phytolaccaceae (?)	288 Chenopodiales	*in* Phytolaccaceae	*in* Phytolaccaceae	59 Barbeuiaceae (Caryophyllales)
b) *in* Aizoaceae	*in* Aizoaceae	*in* Aizoaceae	*in* Aizoaceae	*in* Aizoaceae	62 Tetragoniaceae (Caryophyllales)
c) *in* Phytolaccaceae	*in* Phytolaccaceae	Pittosporales	Caryophyllales	*in* Phytolaccaceae	64 Stegnospermataceae (Caryophyllales)
d) *in* Caryophyllaceae	*in* Portulacaceae & Caryophyllaceae	*in* Brassicaceae	Caryophyllales	—	66 Hectorellaceae
e) *in* Chenopodiaceae	—	*in* Chenopodiaceae	Caryophyllales	*in* Chenopodiaceae	68 Halophytaceae
f) *in* Phytolaccaceae	*in* Phytolaccaceae	291 Agdestidaceae (Chenopodiales)	*in* Phytolaccaceae	*in* Phytolaccaceae	*in* Phytolaccaceae
g) *in* Phytolaccaceae	*in* Phytolaccaceae	292 Petiveriaceae (Chenopodiales)	—	*in* Phytolaccaceae	*in* Phytolaccaceae
h) *in* Caryophyllales	129 Curvembryeae	287 Illecebraceae (Polygonales)	*in* Caryophyllaceae	*in* Caryophyllaceae	*in* Caryophyllaceae
i) *in* Aizoaceae	*in* Aizoaceae	*in* Aizoaceae	Mesembryanthemaceae *in* Aizoaceae	*in* Aizoaceae	*in* Aizoaceae
Order Cactales					
39. Cactaceae	78 Ficoidales	115 Cactales	Caryophyllales	69 Caryophyllales	69 Caryophyllales
Order Magnoliales					
40. Magnoliaceae	4 Ranales	1 Magnoliales	Magnoliales	6 Magnoliales	3 Magnoliales
41. Degeneriaceae	—	*in* Winteraceae	Magnoliales	2 Magnoliales	1 Magnoliales
42. Himantandraceae	*in* Magnoliales	6 Magnoliales	Magnoliales	3 Magnoliales	2 Magnoliales
43. Winteraceae	*in* Magnoliales	3 Magnoliales	Magnoliales	1 Magnoliales	8 Winterales
44. Annonaceae	5 Ranales	10 Annonales	Annonales	8 Magnoliales	5 Annonales
45. Eupomatiaceae	*in* Annonaceae	11 Annonales	Annonales	4 Magnoliales	4 Eupomatiales
46. Myristicaceae	141 Microembryeae	18 Laurales	Annonales	9 Magnoliales	7 Annonales
47. Canellaceae	16 Parietales	4 Magnoliales	Annonales	10 Magnoliales	6 Annonales
48. Schisandraceae	*in* Magnoliaceae	5 Magnoliales	Illiciales	24 Illiciales	10 Illiciales
49. Illiciaceae	*in* Magnoliaceae	2 Magnoliales	Illiciales	23 Illiciales	9 Illiciales
50. Austro-baileyaceae	—	13 Laurales	Laurales	5 Magnoliales	11 Austrobaileyales
51. Trimeniaceae	*in* Monimiaceae	14 Laurales	Laurales	12 Magnoliales	13 Laurales
52. Amborellaceae	*in* Monimiaceae	*in* Monimiaceae	Laurales	11 Magnoliales	12 Laurales
53. Monimiaceae	142 Microembryeae	12 Laurales	Laurales	13 Magnoliales	14 Laurales

Melchior (1964)	Benth. & Hook. (1965a,b,c)	Hutchinson (1973)	Dahlgren (1980a)	Cronquist (1981)	Takhtajan (1987)
a) *in* Monimiaceae	*in* Monimiaceae	*in* Monimiaceae	*in* Monimiaceae	*in* Monimiaceae	15 Atherospermataceae (Laurales)
b) *in* Monimiaceae	*in* Monimiaceae	*in* Monimiaceae	*in* Monimiaceae	*in* Monimiaceae	16 Siparunaceae (Laurales)
54. Calycanthaceae	3 Ranales	26 Rosales	Laurales	15 Magnoliales	19 Laurales
55. Gomortegaceae	*in* Euphorbiaceae (?)	16 Laurales	Laurales	14 Magnoliales	17 Laurales
c) —	—	—	*in* Calycanthaceae	16 Magnoliales	20 Idiospermaceae (Laurales)
56. Lauraceae	143 Daphnales	15 Laurales	Laurales	17 Magnoliales	21 Laurales
57. Hernandiaceae	*in* Lauraceae	17 Laurales	Laurales	18 Magnoliales	18 Laurales
d) *in* Hernandiaceae	*in* Combretaceae	*in* Hernamdiaceae	in Hernandiaceae	in Hernandiaceae	22 Gyrocarpaceae (Laurales)
58. Tetracentraceae	—	49 Hamamelidales	Trochodendrales	40 Trochodendrales	77 Trochodendrales
59. Trochodendraceae	*in* Magnoliaceae	8 Magnoliales	Trochodendrales	41 Trochodendrales	76 Trochodendrales
60. Eupteleaceae	*in* Magnoliaceae	*in* Trochodendraceae	Trochodendrales	43 Hamamelidales	79 Eupteleales
61. Cercidiphyllaceae	—	9 Magnoliales	Trochodendrales	42 Hamamelidales	78 Cercidiphyllales
Order Ranunculales					
62. Ranunculaceae	1 Ranales	257 Ranales	Ranunculales	30 Ranunculales	47 Ranunculales
a) *in* Ranunculaceae	*in* Ranunculaceae	266 Berberidales	Ranunculales	31 Ranunculales	48 Circaeasteraceae (Ranunculales)
b) *in* Ranunculaceae	*in* Ranunculaceae	*in* Helleboraceae	*in* Ranunculaceae	*in* Ranunculaceae	49 Hydrastidaceae (Ranunculales)
c) *in* Ranunculaceae	*in* Ranunculaceae	*in* Circaeasteraceae	Kingdoniaceae (Ranunculales)	*in* Circaeasteraceae	*in* Ranunculaceae
d) *in* Ranunculaceae	*in* Ranunculaceae	256 Helleboraceae (Ranales)	*in* Ranunculaceae	*in* Ranunculaceae	*in* Ranunculales
e) *in* Berberidaceae	*In* Berberidaceae	259 Podophyllaceae (Ranales)	*in* Berberidaceae	*in* Ranunculaceae	*in* Ranunculales
f) 63. Berberidaceae	7 Ranales	267 Berberidales	Ranunculales	32 Ranunculales	50 Ranunculales
g) *in* Berberidaceae	*in* Berberidaceae	265 Nandinaceae (Berberidales)	*in* Berberidaceae	*in* Berberidaceae	51 Ranunculales
64. Sargentodoxaceae	—	262 Berberidales	Ranunculales	33 Ranunculales	45 Ranunculales
65. Lardizabalaceae	*in* Berberidaceae	263 Berberidales	Ranunculales	34 Ranunculales	44 Ranunculales
66. Menispermaceae	6 Ranales	264 Berberidales	Ranunculales	35 Ranunculales	46 Ranunculales
67. Nymphaeaceae	8 Ranales	258 Ranales	Nymphaeales	26 Nymphaeales	40 Nymphaeales
h) *in* Nymphaeaceae	*in* Nymphaeaceae	*in* Nymphaeaceae	*in* Nymphaeaceae	25 Nelumbonaceae (Nymphaeales)	43 Nelumbonales

Melchior (1964)	Benth. & Hook. (1965a,b,c)	Hutchinson (1973)	Dahlgren (1980a)	Cronquist (1981)	Takhtajan (1987)
i) *in* Nymphaeaceae	*in* Nymphaeaceae	*in* Nymphaeaceae	*in* Nymphaeaceae	27 Barclayaceae (Nymphaeales)	41 Nymphaeales
j) *in* Nymphaeaceae	*in* Nymphaeaceae	261 Cabombaceae	*in* Nymphaeaceae	28 Cabombaceae (Nymphaeales)	39 Nymphaeales.
68. Ceratophyllaceae	163 Ordines anomali	260 Ranales	Nymphaeales	29 Nymphaeales	42 Ceratophyllales
Order Piperales					
69. Piperaceae	139 Microembryeae	272 Piperales	Piperales	21 Piperales	26 Piperales
70. Saururaceae	*in* Piperaceae	273 Piperales	Piperales	20 Piperales	25 Piperales
71. Chloranthaceae	140 Microembryeae	274 Piperales	Magnoliales	19 Piperales	24 Chloranthales
72. Lactoridaceae	*in* Piperaceae	7 Magnoliales	Magnoliales	7 Magnoliales	23 Lactoridales
Order Aristolochiales					
73. Aristolochiaceae	138 Multiovulatae-terrestris	268 Aristolochiales	Aristolochiales	22 Aristolochiales	27 Aristolochiales
74. Rafflesiaceae	Multiovulatae-terrestris	270 Aristolochiales (= Cytinaceae)	Rafflesiales	219 Rafflesiales	29 Rafflesiales
—	—	*in* Rafflesiaceae	*in* Rafflesiaceae	*in* Rafflesiaceae	30 Aponodontaceae (Rafflesiales)
in Rafflesiaceae	*in* Rafflesiaceae	*in* Rafflesiaceae (1969)	*in* Rafflesiaceae	218 Rafflesiales	31 Mitrastemonaceae (Rafflesiales)
—	—	*in* Rafflesiaceae	*in* Rafflesiaceae	—	32 Cytinaceae (Rafflesiales)
75. Hydnoraceae	*in* Rafflesiaceae	269 Aristolochiales	Rafflesiales	*in* Rafflesiaceae ?	28 Hydnorales
Order Guttiferales					
76. Dilleniaceae	2 Ranales	19 Dilleniales	Dilleniales	78 Dilleniales	98 Dilleniales
77. Paeoniaeeae.	*in* Ranunculaceae	255 Ranales	Paeoniales	79 Dilleniales	53 Paeoniales
a) *in* Ranunculaceae	*in* Ranunculaceae	*in* Helleboraceae	—	*in* Ranunculaceae	52 Glaucidiaceae (Glaucidiales)
78. Crossosomataceae	*in* Dilleniaceae (?)	21 Dilleniales	Rosales	176 Rosales	220 Crossosomatales
79. Eucryphiaceae	—	164 Guttiferales	Cunoniales	158 Rosales	198 Cunoniales
80. Medusagynaceae	—	145 Theales	Theales	96 Theales	114 Medusagynales
a) *in* Theaceae	*in* Theaceae	136 Bonnetiaceae (Theales)	—	*in* Theaceae	112 Theales
b) *in* Actinidiaceae	*in* Theaceae	138 Saurauriaceae (Theales)	*in* Actinidiaceae	*in* Actinidiaceae	?
81. Actinidiaceae	*in* Theaceae	139 Theales	Ericales	86 Theales	99 Actinidiales
a) *in* Theaceae	*in* Theaceae	140 Pellicierales (Theales)	—	*in* Theaceae	111 Theales

Melchior (1964)	Benth. & Hook. (1965a,b,c)	Hutchinson (1973)	Dahlgren (1980a)	Cronquist (1981)	Takhtajan (1987)
82. Ochnaceae	Geraniales	148 Ochnales	Theales	80 Theales	115 Theales
83. Dioncophyllaceae	—	in Flacourtiaceae	Theales	118 Violales	159 Dioncophyllales
84. Strasburgeriaceae	—	147 Ochnales	Theales	—	119 Ochnales
—	—	146 Diegodendraceae (Ochnales)	—	—	118 Ochnales
85. Dipterocarpaceae	Guttiferales	151 Ochnales	Malvales	83 Theales	177 Malvales
86. Theaceae	Guttiferales	137 Theales	Theales	85 Theales	102 Theales
87. Caryocaraceae	in Theaceae	144 Theales	Theales	84 Theales	109 Theales
88. Marcgraviaceae	in Theaceae	143 Theales	Theales	92 Theales	105 Theales
89. Quiinaceae	in Guttiferae	165 Guttiferales	Theales	93 Theales	120 Ochnales
90. Guttiferae (= Clusiaceae)	27 Guttiferales	163 Guttiferales	Theales	97 Theales	113 Theales
—	—	162 Hypericaceae	(= Clusiaceae) Theales		
91. Ancistrocladaceae	in Dipterocarpaceae	152 Ochnales	Theales	119 Violales	122 Ancistrocladales
Order Sarraceniales					
92. Sarraceniaceae	9 Parietales	318 Sarraceniales	Sarraceniales	104 Nepenthales	125 Sarraceniales
93. Nepenthaceae	136 Multiovulatae terrestris	271 Aristolochiales	Theales	105 Nepenthales	38 Nepenthiales
94. Droseraceae	61 Rosales	317 Sarraceniales	Droserales	106 Nepenthales.	215 Droserales
Order Papaverales					
95. Papaveraceae	10 Parietales	275 Rhoeadales	Papaverales	38 Papaverales	54 Papaverales
a) in Papaveraceae	in Papaveraceae	in Fumariaceae	in Fumariaceae	in Fumariaceae	55 Hypecoaceae (Papaverales)
b) in Papaveraceae	in Papaveraceae	276 Rhoeadales	Papaverales	39 Papaverales	56 Fumariaceae (Papaverales)
96. Capparaceae	12 Parietales	95 Capparales	Capparales	133 Capparales	163 Capparales
—	—	277 Cleomaceae (Rhoeadales)	in Capparaceae	—	in Capparaceae
c) in Capparaceae	in Simaroubaceae	180 Celastrales	Koeberliniaceae (Capparales)	in Capparaceae	Capparales (1969)
d) in Capparaceae	—	187 Celastrales	Pentadiplandraceae (Capparaceae)	in Capparaceae	Capparales (1969)
e) in Capparaceae	in Capparaceae	279 Oxystylidaceae (Brassicales)	in Capparaceae	in Capparaceae	in Capparaceae
97. Cruciferae	11 Parietales	278 Brassicales	Capparales	134 Capparales	164 Capparales
98. Tovariaceae	—	97 Capparales	Capparales	132 Capparales	165 Capparales
99. Resedaceae	13 Parietales	280 Resedales	Capparales	136 Capparales	166 Capparales

Melchior (1964)	Benth. & Hook. (1965a,b,c)	Hutchinson (1973)	Dahlgren (1980a)	Cronquist (1981)	Takhtajan (1987)
100. Moringaceae	55 Sapindales ? (Anomalous genera)	96 Capparales	Capparales	135 Capparales	167 Moringales
Order Batales.					
101. Bataceae	133 Curvembryeae	297 Chenopodiales	Capparales	138 Batales	169 Batales
Order Rosales					
102. Platanaceae	—	52 Hamamelidales	Hamamelidales	44 Hamamelidales	84 Hamamelidales
103. Hamamelidaceae	62 Rosales	50 Hamamelidales.	Hamamelidales	45 Hamamelidales	81 Hamamelidales
a) in Hamamelidaceae	in Hamamelidaceae	in Hamamelidaceae	in Hamamelidaceae	in Hamamelidaceae	82 Rhodoleiaceae (Hamamelidales)
b) in Hamamelidaceae	in Hamamelidaceae	in Hamamelidaceae	in Hamamelidaceae	in Hamamelidaceae	83 Altingiaceae (Hamamelidales)
104. Myrotha-mnaceae	—	51 Hamamelidales	Hamamelidales	46 Hamamelidaceae	88 Myrothamnales
105. Crassulaceae	60 Rosales	308 Saxifragales	Saxifragales	171 Rosales	203 Saxifragales
106. Cephalotaceae	in Saxifragaceae	309 Saxifragales	Saxifragales	172 Rosales	204 Saxifragales
107. Saxifragaceae	59 Rosales	310 Saxifragales	Saxifragales	173 Rosales	205 Saxifragales
a) in Saxifragaceae	in Saxifragaceae	34 Cunoniales	—	166 Rosales	206 Grossulariaceae
b) in Saxifragaceae	in Saxifragaceae	in Escalloniaceae	Saxifragales	in Grossulariaceae	207 Iteaceae
c) in Saxifragaceae	in Saxifragaceae	312 Saxifragales	Saxifragales	in Saxifragaceae	208 Vahliaceae
d) in Saxifragaceae	in Saxifragaceae	311 Saxifragales	—	in Saxifragaceae	209 Eremosynaceae
					210 Rousseaceae
e) in Melianthaceae (Sapindales)	in Sapindaceae (Sapindales)	36 Cunoniales	Saxifragales	167 Rosales	211 Greyiaceae
f) in Saxifragaceae	in Saxifragaceae	313 Saxifragales	Cunoniales	in Saxifragaceae	212 Frankoaceae
g) in Saxifragaceae	in Saxifragaceae	315 Saxifragales	Droserales	in Saxifragaceae	213 Parnassiaceae
h) in Saxifragaceae	in Saxifragaceae	in Saxifragaceae	Droserales	in Saxifragaceae	214 Lepuropeta-laceae
108. Brunelliaceae	in Simaroubaceae	22 Dilleniales	Cunoniales	156 Rosales	199 Cunoniales
109. Cunoniaceae	in Saxifragaceae	31 Cunoniales	Cunoniales	159 Rosales	195 Cunoniales
a) in Saxifragaceae	in Rosaceae	30 Pterostemonaceae (Cunoniales)	Cornales	in Grossulariaceae	340 Hydrangeales
b) in Saxifragaceae	in Saxifragaceae	32 Philadelphaceae (Cunoniales)	in Saxifragaceae	in Hydrangeaceae	—
c) in Saxifragaceae	in Saxifragaceae	33 Hydrangeaceae (Cunoniales)	Cornales	164 Rosales	337 Hydrangeales
d) in Saxifragaceae	in Saxifragaceae	38 Baueraceae (Cunoniales)	Cunoniales	in Cunoniaceae	196 Cunoniales

Melchior (1964)	Benth. & Hook. (1965a,b,c)	Hutchinson (1973)	Dahlgren (1980a)	Cronquist (1981)	Takhtajan (1987)
110. Davidsoniaceae	—	*in* Cunoniaceae	Cunoniales	160 Rosales	197 Cunoniales
111. Pittosporaceae	Polygalinae	90 Pittosporales	Pittosporales	162 Rosales	354 Pittosporales
112. Byblidaceae	*in* Droseraceae	91 Pittosporales	Pittosporales	163 Rosales	355 Byblidales
113. Roridulaceae	*in* Droseraceae	*in* Byblidaceae	Ericales	in Byblidaceae	338 Roridulaceae (Hydrangeales)
114. Bruniaceae	63 Rosales	56 Hamamelidales	Cunoniales	168 Rosales	200 Bruniales
115. Rosaceae	58 Rosales	24 Rosales	Rosales	174 Rosales	217 Rosales
116. Neuradaceae	*in* Rosaceae	*in* Rosaceae	Rosales	175 Rosales	219 Rosales
117. Chrysobalanaceae	*in* Rosaceae	*in* Rosaceae	Rosales	177 Rosales	218 Rosales
a) *in* Rosaceae	*in* Rosaceae	*in* Rosaceae	Malaceae (Rosales)	*in* Rosaceae	*in* Rosaceae
b) *in* Rosaceae	*in* Rosaceae	*in* Rosaceae	Amygdalaceae (Rosales)	*in* Rosaceae	*in* Rosaceae
118. Connaraceae	56 Rosales	20 Dilleniales	Sapindales	157 Rosales	242 Connarales
119. Leguminosae (or Fabaceae)	57 Rosales	29 Papilionaceae (Leguminales)	Fabales	180 Fabales	241 Fabales
a) *in* Leguminosae	*in* Leguminosae	27 Caesalpiniaceae (Leguminales)	Fabales	181 Fabales	*in* Fabaceae (as Subfamily Caesalpinioideae)
b) *in* Leguminosae	*in* Leguminosae	28 Mimosaceae (Leguminales)	Fabales	182 Fabales	*in* Fabaceae (as Subfamily Mimosoideae)
120. Krameriaceae	*in* Polygalaceae	103 Polygalales	Polygalales	249 Polygalales	297 Polygalales
Order Hydrostachyales 121 Hydrostachyaceae	—	319 Podostemales	Hydrostachyales	291 Callitrichales	417 Hydrostachyales
Order Podostemales 122. Podostemaceae	125 Multiovulatae-aquaticae	320 Podostemales	Podostemales	185 Podostemales	221 Podostemales
Order Geraniales					
123. Limnanthaceae	*in* Geraniaceae	343 Geraniales	Tropaeolales	267 Geraniales	292 Limnanthales
124. Oxalidaceae	*in* Geraniaceae	344 Geraniales	Geraniales	265 Geraniales	280 Geraniales
a) *in* Oxalidaceae	*in* Geraniaceae	220 Averrhoaceae (Rutales)	*in* Oxalidaceae	*in* Oxalidaceae	*in* Oxalidaceae
125. Geraniaceae	38 Geraniales	342 Geraniales	Geraniales	266 Geraniales	284 Geraniales
a) *in* Oxralidaceae	*in* Geraniaceae	*in* Oxalidaceae		*in* Oxalidaceae	282 Hypseocharitaceae (Geraniales)
b) *in* Geraniaceae	—	116 Tiliales	Geraniales	*in* Geraniaceae	285 Dirachmaceae (Geraniales)

Melchior (1964)	Benth. & Hook. (1965a,b,c)	Hutchinson (1973)	Dahlgren (1980a)	Cronquist (1981)	Takhtajan (1987)
c) *in* Geraniaceae	*in* Geraniaceae	132 Malpighiales	Geraniales	*in* Geraniaceae	281 Lepidobotrya-ceae (Geraniales)
d) *in* Geraniaceae	*in* Geraniaceae	*in* Geraniales	Geraniales	*in* Geraniaceae	283 Biebersteiniaceae
e) *in* Geraniaceae	*in* Geraniaceae	129 Malpighiales	Geraniales	—	286 Ledocarpaceae
—	—				287 Rhyncothecaceae
f) *in* Geraniaceae	*in* Geraniaceae	93 Pittosporales	Geraniales	—	289 Vivianiaceae
126. Tropaeolaceae	*in* Geraniaceae	345 Geraniales	Tropaeolales	268 Geraniales	291 Tropaeolales
127. Zygophyllaceae	Geraniales	134 Malpighiales	Geraniales	264 Sapindales	262 Rutales
128. Linaceae	Geraniales	126 Malpighiales	Geraniales	242 Linales	275 Linales
a) *in* Linaceae	*in* Linaceae	123 Ixonanthaceae (Malpighiales)	Geraniales	240 Linales	277 Linales
b) *in* Linaceae	Geraniales	125 Humiriaceae (Malpighiales)	Geraniales	239 Linales	278 Linales
c) *in* Simaroubaceae	*in* Simroubaceae (?)	127 Irvingiaceae (Malpighiales)	—	*in* Simaroubaceae	260 Rutales
d) *in* Linaceae	*in* Linaceae	*in* Linaceae	—	241 Linales	274 Hugoniaceae (Linales)
e) *in* Linaceae	—	131 Ctenolophonaceae (Malpighiales)	Geraniales	*in* Linaceae	276 Linales
f) *in* Zygophyllaceae	*in* Simaroubaceae	133 Balanitaceae (Malpighiales)	Geraniales	*in* Simaroubaceae	264 Rutales
g) *in* Zygophyllaceae	*in* Zygophyllaceae	*in* Zygophyllaceae	Geraniales	*in* Zygophyllaceae	263 Nitrariaceae (Rutales)
h) *in* Zygophyllaceae	*in* Rutaceae	*in* Zygophyllaceae	Geraniales	*in* Zygophyllaceae	265 Peganaceae (Rutales)
i) *in* Simaroubaceae	*in* Simaroubaceae	*in* Simaroubaceae	Rutales	178 Rosales	261 Surianaceae (Rutales)
j) —	—	—	—	—	257 Tetradichidaceae (Rutales)
k) —	—	—	*in* Rutaceae	179 Rosales near Centrospermae	256 Rhabdodendra-ceae (Rutales)
129. Erythroxylaceae	*in* Linaceae	130 Malpighiales	Geraniales	238 Linales	279 Linales
130. Euphorbiaceae	Unisexualis	135 Euphorbiales	Euphorbiales	234 Euphorbiales	190 Euphorbiales
131. Daphniphyl- laceae	*in* Euphorbiaceae	55 Hamamelidales	Buxales	47 Daphniphyllales	85 Daphniphyllales
Order Rutales					
132. Rutaceae	39 Geraniales	217 Rutales	Rutales	263 Rutales	255 Rutales

Melchior (1964)	Benth. & Hook. (1965a,b,c)	Hutchinson (1973)	Dahlgren (1980a)	Cronquist (1981)	Takhtajan (1987)
133. Cheoraceae	*in Simaroubaceae*	181 Celastrales	Rutales	261 Rutales	258 Rutales
134. Simaroubaceae	40 Geraniales	218 Rutales	Rutales	260 Rutales	259 Rutales
135. Picrodendraceae	*in Simaroubaceae* (?)	66 Juglandales	*in Euphorbiaceae*	—	—
136. Burseraceae	42 Geraniales	219 Rutales	Rutales	257 Rutales	270 Rutales
137. Meliaceae	43 Geraniales	221 Meliales	Rutales	262 Rutales	266 Rutales
a) *in Simaroubaceae*	—	*in Simaroubaceae*		*in Simaroubaceae*	267 Kirkiaceae
b) —	—	*in Simaroubaceae*		[*in/near* Meliaceae, Sapindaceae]	268 Ptaeroxylaceae (Rutales)
c) —	—	—			269 Tupeianthaceae (Rutales)
d) *in Meliaceae*	—	*in Sapindaceae*	*in Meliaceae*		Aitoniaceae (Rutales)?
			in Meliaceae	[*in/near* Meliaceae, Rutaceae]	Flindersiaceae (Rutales)?
138. Akaniaceae	*in Sapindaceae*	228 Sapindales	Sapindales	253 Rutales	250 Sapindales
139. Malpighiaceae	Geraniales	124 Malpighiales	Polygalales	243 Polygalales	293 Polygalales
140. Trigoniaceae	*in Vochysiaceae*	104 Polygalales	Polygalales	245 Polygalales	294 Polygalales
141. Vochysiaceae	Polygalinae	105 Polygalales	Polygalales	244 Polygalales	295 Polygalales
142. Tremandraceae	19 Polygalinae	94 Pittosporales	Pittosporales (?)	246 Polygalales	298 Polygalales
143. Polygalaceae	20 Polygalinae	102 Polygalales	Pittosporales	247 Polygalales	296 Polygalales
a) *in Polygalaceae*	*in Polygalaceae*	*in Polygalaceae*	*in Polygalaceae*	248 Xanthophyllaceae	*in Polygalaceae*
Order Sapindales					
144. Coriariaceae	54 Anomalous family	23 Coriariales	Sapindales	36 Ranunculales	273 Coriariales
145. Anacardiaceae	53 Sapindales	226 Sapindales	Sapindales incl. Pistasciaceae	258 Sapindales	271 Rutales incl. Spondiaceae, Terebinthaceae
146. Aceraceae	*in Sapindaceae*	227 Sapindales	Sapindales	256 Sapindales	246 Sapindales
147. Bretschneideraceae	—	*in Sapindaceae*	Sapindales (?)	252 Sapindales	248 Sapindales
148. Sapindaceae	Sapindales	223 Sapindales	Sapindales	254 Sapindales	245 Sapindales
b) —	—	—	—	—	244 Tapisciaceae (1987)
c) *in Chrysobalanaceae*	*in Rosaceae*	*in Rosaceae* (1969)	*in Sapindaceae*	?	251 Stylobasiaceae (= Stylobasidiaceae)
d) *in Capparaceae*	*in Capparaceae*	*in Flacourtiaceae* (1969)	Sapindales	*in Capparaceae*	252 Emblingiaceae
e) —	—	—	—	—	254 Physenaceae (1985)

Melchior (1964)	Benth. & Hook. (1965a,b,c)	Hutchinson (1973)	Dahlgren (1980a)	Cronquist (1981)	Takhtajan (1987)
149. Hippocastanaceae	*in* Sapindaceae	228 Sapindales	Sapindales	255 Sapindales	247 Sapindales
in Anacardiaceae?	*in* Sapindaceae	224 Podoaceae (Sapindales)	Sapindales	?	?
150. Sabiaceae	Sapindales	225 Sapindales	Sapindales	37 Ranunculales	253 Sapindales
151. Melianthaceae	*in* Sapindaceae	222 Sapindales	Sapindales	251 Sapindales	249 Sapindales
152. Aextoxicaceae	*in* Euphorbiaceae	185 Colastrales	*in* Euphorbiaceae	227 Celastrales	193 Euphorbiales
153. Balsaminaceae	*in* Geraniaceae	346 Geraniales	Balsaminales	269 Geraniales	290 Balsaminales
Order Julianiales					
154. Julianiaceae	*in* Anacardiaceae (?)	231 Sapindales	*in* Anacardiaceae	259 Sapindales	271 Rutales
Order Celastrales					
155. Cyrillaceae	Olacales	182 Celastrales	Ericales	139 Ericales	129 Ericales
156. Pentaphyllaceae	*in* Theaceae	141 Theales	—	88 Theales	—
157. Aquifoliaceae	Olacales	178 Celastrales	Cornales	225 Celastrales	310 Celastrales
158. Corynocarpaceae	*in* Anacardiaceae	189 Celastrales	Celastrales	229 Celastrales	191 Euphorbiales
159. Pandaceae	—	177 Celastrales	Euphorbiales	233 Euphorbiales	305 Celastrales
160. Celastraceae	Celastrales	188 Celastrales	Celastrales incl. Tripterygiaceae	221 Celastrales	
161. Staphyleaceae	*in* Sapindaceae	228 Sapindales	Sapindales	250 Sapindales	243 Sapindales
162. Hippocrateaceae	*in* Celastraceae	192 Celastrales	*in* Celastraceae	222 Celastrales	308 Celastrales
163. Stackhousiaceae	Celastrales	190 Celastrales	Celastrales	223 Celastrales	309 Celastrales
164. Salvadoraceae	Gentianales	179 Celastrales	Salvadorales	224 Celastrales	
165. Buxaceae	*in* Euphorbiaceae	54 Hamamelidales	Buxales	231 Euphorbiales	89 Buxales
166. Icacinaceae	*in* Olacaceae	184 Celastrales	Cornales	226 Celastrales	301 Celastrales
167. Cardiopteridaceae	*in* Olacaceae	182 Celastrales	Celastrales (?)	228 Celastrales	303 Celastrales
a) *in* Celastraceae	*in* Celastraceae	191 Celastrales	*in* Celastrales	*in* Celastraceae	306 Goupiaceae (Celastrales)
b) —	—	—	Celastrales	—	307 Lophopyxidaceae (Celastrales)
c) *in* Saxifragaceae	*in* Saxifragaceae	*in* Escalloniaceae	—	*in* Grossulariaceae (Rosales)	304 Brexiaceae (Celastrales)
d) *in* Aquifoliaceae	*in* Rutaceae (?)	*in* Aquifoliaceae	Cornales	*in* Aquifoliaceae	300 Phellinaceae (Celastrales)
e) *in* Aquifoliaceae	—	—	—	—	302 Sphenostemonaceae (Celastrales)
f) *in* Celastraceae	*in* Celastraceae	Celastrales	Siphonodontaceae (*in* Celastraceae)	?	—
g) —	—	Erythropalaceae	—	—	—

Melchior (1964)	Benth. & Hook. (1965a,b,c)	Hutchinson (1973)	Dahlgren (1980a)	Cronquist (1981)	Takhtajan (1987)
Order Rhamnales					
168. Rhamnaceae	Celastrales	209 Rhamnales	Rhamnales	235 Rhamnales	310 Rhamnales
169. Vitaceae	Celastrales	210 Rhamnales	Vitidales	237 Rhamnales	323 Vitaceae
170. Leeaceae	Celastrales	in Vitaceae	in Vitidaceae	236 Rhamnales	324 Leeaceae
Order Malvales					
171. Elaeocarpaceae	in Tiliaceae	in Tiliaceae	Malvales	98 Malvales	173 Malvales
172. Sarcolaenaceae	Guttiferales	149 Ochnales	Malvales	82 Theales	178 Malvales
173. Tiliaceae	Malvales	118 Tiliales	Malvales	99 Malvales	175 Malvales
174. Malvaceae	Malvales	122 Malvales	Malvales	102 Malvales	182 Malvales
175. Bombaceae	in Malvaceae	120 Tiliales	Malvales	101 Malvales	181 Malvales
176. Sterculiaceae	Malvales	119 Tiliales	Malvales	100 Malvales	180 Malvales
177. Scytopetalaceae	in Olacaceae	117 Tiliales	Theales	87 Theales	121 Ochnales
a) —	—	—	—	—	174 Plagiopteraceae
b) —	—	—	—	—	176 Monotaceae
c) in Styracaceae	—	128 Malpighiales	Malvales	111 Violales	183 Huaceae (Malvales)
Order Thymeleales					
178. Geissolomataceae	in Penaeaceae	85 Thymeleaeales	Hamamelidales	220 Celastrales	201 Geissolomatales
179. Penaeaceae	Daphnales	86 Thymelaeales	Myrtales	190 Myrtales	229 Myrtales
180. Dichapetalaceae	Geraniales	25 Rosales	—	230 Celastrales	192 Euphorbiales
181. Thymelaeaceae	Daphnales	87 Rosales	Thymelaeales	192 Myrtales	194 Thymelaeales
182. Elaeagnaceae	Daphnales	208 Rhamnales	Elaeagnales	180 Proteales	321 Elaeagnales
Order Violales					
183. Flacourtiaceae	Parietales	76 Bixales	Violales	107 Violales	143 Violales
184. Peridiscaceae	in Bixineae	80 Bixales	Violales (?)	108 Violales	146 Violales
185. Violaceae	Parietales	101 Violales	Violales	115 Violales	147 Violales
186. Stachyuraceae	in Theaceae	53 Hamamelidales	Theales	114 Violales	101 Theales
187. Scyphostegiaceae	—	195 Celastrales	Violales	113 Violales	149 Violales
188. Turneraceae	Passiflorales	106 Loasales	Violales	120 Violales	151 Violales
189. Malesherbiaceae	in Passifloraceae	108 Passiflorales	Violales	121 Violales	152 Violales
190. Passifloraceae	Passiflorales	109 Passiflorales	Violales	122 Violales	150 Violales
191. Achariaceae	in Passifloraceae	110 Passiflorales	Violales	123 Violales	153 Violales
192. Cistaceae	Parietales	75 Bixales	Malvales	110 Violales	172 Bixales
193. Bixaceae	Parietales	74 Bixales	Malvales	109 Violales	170 Bixales
194. Sphaerosepalaceae (=Rhopalocarpaceae)	—	150 Ochnales	Malvales	—	179 Malvales

Melchior (1964)	Benth. & Hook. (1965a,b,c)	Hutchinson (1973)	Dahlgren (1980a)	Cronquist (1981)	Takhtajan (1987)
195. Cochlospermaceae	*in* Bixineae	77 Bixales	Malvales	81 Theales	171 Bixales
196. Tamaricaceae	Caryophyllinae	99 Tamaricales	Tamaricales	116 Violales	155 Tamaricales
197. Frankeniaceae	Caryophyllinae	98 Tamaricales	Tamaricales	117 Violales	156 Tamaricales
198. Elatinaceae	—	281 Caryophyllales	Theales	94 Theales	123 Elatinales
199. Caricaceae	*in* Passifloraceae	114 Cucurbitales	Violales	124 Violales	154 Violales
200. Loasaceae	Passiflorales	107 Loasales	Loasales	130 Violales	
201. Datiscaceae	Passiflorales	113 Cucurbitales	Violales	128 Violales	
202. Begoniaceae	Passiflorales	112 Cucurbitales	Violales	129 Violales	
a) —	—	—	—	—	141 Berberidopsidaceae (Violales) 1985
b) —	—	—	—		142 Aphloiaceae (Violales) 1985
c) —	—	—	—		144 Kiggelariaceae (Violales)
d) *in* Flacourtiaceae	Anomalous family	82 Bixales	*in* Flacourtiaceae	112. Violales	145 Lacistemataceae (Violales)
Order Cucurbitales					
203. Cucurbitaceae	Passiflorales	111 Cucurbitales	Violales	127 Violales	160 Cucurbitales
Order Myrtiflorae					
204. Lythraceae	Myrtales	170 Myrtales	Myrtales	189 Myrtales	234 Myrtales
205. Trapaceae	*in* Onagraceae	300 Onagrales	Myrtales	193 Myrtales	239 Myrtales
206. Crypteroniaceae	*in* Lythraceae	39 Cunoniales	Myrtales	191 Myrtales	232 Myrtales
207. Myrtaceae	Myrtales	166 Myrtales	Myrtales incl. Heteropyxidaceae	194 Myrtales	226 Myrtales
208. Dialypetalanthaceae	—	244 Rubiales	Cornales	161 Rosales	371 Gentianales
209. Sonneratiacae	*in* Lythraceae	169 Myrtales	Myrtales	188 Myrtales	237 Myrtales
210. Punicaceae	*in* Lythraceae	174 Myrtales	Myrtales	195 Myrtales	235 Myrtales
211. Lecythidaceae	—	172 Myrtales	Theales	103 Lecythidales	124 Lecythidales
212. Melastomataceae	Myrtales	176 Myrtales	Myrtales incl. Memecylaceae	200 Lecythidales	233 Myrtales
213. Rhizophoraceae	Myrtales	171 Myrtales	Rhizophorales	202 Rhizophorales	223 Rhizophorales
214. Combretaceae	Myrtales	173 Myrtales	Myrtales	201 Myrtales	231 Myrtales
215. Onagraceae	Myrtales	299 Onagrales	Myrtales	196 Myrtales	238 Myrtales
216. Oliniaceae	*in* Lythraceae	35 Cunoniales	Myrtales	199 Myrtales	230 Myrtales
217. Haloragaceae	Rosales	301 Onagrales	Haloragales	186 Haloragales	240 Haloragales

Melchior (1964)	Benth. & Hook. (1965a,b,c)	Hutchinson (1973)	Dahlgren (1980a)	Cronquist (1981)	Takhtajan (1987)
218 Theligonaceae	*in* Haloragaceae	296 Chenopodiales	*in* Rubiaceae	314 Rubiales	369 Gentianales
219. Hippuridaceae	*in* Urticaceae	*in* Haloragaceae	Hippuridales	291 Callitrichales	413 Hippuridales
220. Cynomoriaceae	*in* Balanophoraceae	*in* Balanophoraceae	Balanophorales	—	364 Cynomoriales
a) —	—	—	—	—	227 Alzateaceae (Myrtales) 1986
b) —	—	—	—	—	228 Rhynchocalycaceae (Myrtales) 1986
c) —	—	—	—	—	236 Duabangaceae (Myrtales) 1986
d) —	—	Myrtales	Myrtales	*in* or near Myrtaceae	224 Psiloxylaceae (Myrtales)
e) *in* Myrtaceae	*in* Lythraceae	207 Rhamnales	*in* Myrtaceae	*in* Myrtaceae	225 Heteropyxidaceae (Myrtales)
f) —	—	Barringtoniaceae	*in* Lecythidaceae	—	—
g) —	—	Napoleonaceae	*in* Lecythidaceae	—	—
h) *in* Rhizophoraceae	*in* Rhizophoraceae	Anisophylleaceae	Cornales (?)	*in* Rhizophoraceae	222 Rhizophorales
Order Umbelliflorae					
221. Alangiaceae	*in* Cornaceae	44 Araliales	Cornales	203 Cornales	348 Cornales
222. Nyssaceae	*in* Cornaceae	46 Araliales	Cornales	204 Cornales	342 Cornales
223. Davidiaceae	Umbellales	*in* Nyssaceae	Cornales	—	341 Cornales
224. Cornaceae	*in* Cornaceae	43 Araliales	Cornales	205 Cornales	343 Cornales
225. Garryaceae	Umbellales	45 Araliales	Cornales	206 Cornales	347 Cornales
226. Araliaceae	Umbellales	47 Araliales	Araliales	270 Apiales	352 Apiales
227. Umbelliferae	Umbellales	321 Umbellales	Araliales (= Apiaceae)	271 Apiales	353 Apiales
a) *in* Cornaceae	*in* Araliaceae	*in* Araliaceae	Araliales	*in* Cornaceae	351 Helwingiaceae (Apiales) (1986)
b) —	—	—	—	—	344 Curtisiaceae (Cornales) (1987)
c) *in* Cornaceae	*in* Cornaceae	*in* Cornaceae	—	*in* Cornaceae	345 Mastixiaceae (Cornales)
d) —	—	—	Cornales	—	346 Aucubaceae (Cornales)
e) —	—	—	—	—	349 Aralidiaceae (Aralidiales)
f) *in* Cornaceae	*in* Cornaceae	*in* Cornaceae	—	*in* Cornaceae	350 Toricelliaceae (Toricelliales)

Melchior (1964)	Benth. & Hook. (1965a,b,c)	Hutchinson (1973)	Dahlgren (1980a)	Cronquist (1981)	Takhtajan (1987)
Order Diapensiales					
228. Diapensiaceae	98 Ericales	158 Ericales	Ericales	147 Diapensiales	132 Diapensiales
Order Ericales					
229. Clethraceae	*in* Ericaceae	153 Ericales	Ericales	140 Ericales	126 Ericales
230. Pyrolaceae	*in* Ericaceae	154 Ericales	Ericales	145 Ericales	*in* Ericaceae
231. Ericaceae	94 Ericales	155 Ericales	Ericales	144 Ericales	127 Ericales
232. Empetraceae	Anomalous fam.	184 Celastrales	Ericales	142 Ericales	130 Ericales
233. Epacridaceae	97 Ericales	157 Ericales	Ericales	143 Ericales	128 Ericales
a) *in* Ericaceae and Epacridaceae	*in* Ericaceae & Epacridaceae	156 Prionotaceae (Ericales)	?	*in* Ericaceae & Epacridaceae	*in* Ericaceae & Epacridaceae
b) *in* Pyrolaceae	96 Ericales	159 Monotropaceae	Ericales	146 Ericales	*in* Ericaceae
c) *in* Ericaceae	Ericales	161 Vacciniaceae	*in* Ericaceae (?)	*in* Ericaceae	*in* Ericaceae
d) *in* Guttiferales	—	—	Actinidiaceae (Ericales) *incl* Saururiaceae	—	*in* Actinidiales
Order Primulales					
234. Theophrastaceae	*in* Myrsiniaceae	212 Myrsinales	Primulales	153 Primulales	137 Primulales
235. Myrsinaceae	Primulales	211 Myrsinales	Primulales	154 Primulales	138 Primulales
236. Primulaceae	Primulales	305 Primulales	Primulales	155 Primulales	140 Primulales
a) *in* Myrsinaceae	*in* Myrsinaceae	213 Aegicerataceae (Myrsinales)	Primulales	*in* Myrsinaceae	139 Primulales
b) —	—	—	Coridaceae (Primulales)	—	*in* Primulaceae
Order Plumbaginales					
237. Plumbaginaceae	Primulales	306 Primulales	Plumbaginales	77 Plumbaginales	75 Plumbaginales
Order Ebenales					
238. Sapotaceae	Ebenales	215 Ebenales	Ebenales	148 Ebenales	136 Sapotales
239. Sarcospermataceae	*in* Sapotaceae	216 Ebenales	Ebenales	*in* Sapotaceae	
240. Ebenaceae	Ebenales	214 Ebenales	Ebenales	149 Ebenales	135 Ebenales
241. Styracaceae	Ebenales	41 Styracales	Ebenales	150 Ebenales	133 Ebenales
242. Lissocarpaceae	*in* Styracaceae	40 Styracales	Ebenales	151 Ebenales	134 Ebenales
243. Symplocaceae	*in* Styracaceae	42 Styracales	Cornales	152 Ebenales	108 Theales
244. Hoplestigmataceae	—	78 Bixales	Boraginales (?)	126 Violales	391 Boraginales
Order Oleales					
245. Oleaceae	Gentianales	239 Loganiales	Oleales	295 Scrophulariales	378 Oleales

Melchior (1964)	Benth. & Hook. (1965a,b,c)	Hutchinson (1973)	Dahlgren (1980a)	Cronquist (1981)	Takhtajan (1987)
Order Gentianales					
246. Loganiaceae	Gentianales	234 Loganiales	Gentianales	272 Gentianales	366 Gentianales
a) *in* Loganiaceae	*in* Loganiaceae	233 Potaliaceae (Loganiales)	*in* Loganiaceae	*in* Loganiaceae	
b) *in* Loganiaceae	*in* Loganiaceae	236 Antoniaceae (Loganiales)	*in* Loganiaceae	*in* Loganiaceae	
c) *in* Loganiaceae	*in* Loganiaceae	237 Spigeliaceae (Loganiales)	*in* Loganiaceae	*in* Loganiaceae	367 Gentianales
d) *in* Loganiaceae	*in* Loganiaceae	238 Strychnaceae (Loganiales)	*in* Loganiaceae	*in* Loganiaceae	
247. Desfontainiaceae	*in* Loganiaceae	*in* Potaliaceae	—	*in* Loganiaceae	365 Gentianales
248. Gentianaceae	Gentianales	303 Gentianales	Gentianales	274 Gentianales	372 Gentianales
249. Menyanthaceae	*in* Gentianaceae	304 Gentianales	Gentianales	283 Solanales	374 Gentianales
250. Apocynaceae	Gentianales	241 Apocynales	Gentianales	276 Gentianales	376 Gentianales
e) *in* Loganiaceae	*in* Loganiaceae	240 Plocospermataceae (Apocynales)	*in* Apocynaceae	*in* Apocynaceae	375 Gentianales
f) *in* Asclepiadaceae	*in* Asclepiadaceae	242 Periplocaceae (Apocynales)	*in* Asclepiadaceae	*in* Asclepiadaceae	*in* Asclepiadaceae
251. Asclepiadaceae	Gentianales	243 Apocynales	Gentianales	277 Gentianales	377 Gentianales
252. Rubiaceae	Rubiales	245 Rubiales	Gentianales	313 Rubiales	368 Gentianales
g) —	—	—	—	—	370 Carlemanniaceae (Gentianales)
h) *in* Loganiaceae	*in* Solanaceae	332 Solanales	Scrophulariales	273 Retziaceae (Gentianales)	399 Scrophulariales
i) —	—	—	—	275 Gentianales	373 Saccifoliaceae (Gentianales)
Order Tubiflorae					
253. Polemoniaceae	Polemoniales	347 Polemoniales	Solanales	284 Solanales	388 Polemoniales
254. Fouquieriaceae	*in* Tamaricaceae	100 Tamaricales	Fouquieriales	125 Violales	157 Fouquieriales
a) *in* Polemoniaceae	*in* Polemoniaceae	346 Bignoniales	Solanales	*in* Polemoniaceae	387 Cobaeaceae (Polemoniales)
255. Convolvulaceae	Polemoniales	347 Polemoniales	Solanales incl. Humbertiaceae	281 Solanales	385 Convolvulales
256. Hydrophyllaceae	Polemoniales	333 Solanales	Solanales	285 Solanales	389 Boraginales
257. Boraginaceae	Polemoniales	350 Boraginales	Boraginales	287 Lamiales	394 Boraginales
b) *in* Convolvulaceae	*in* Convolvulaceae	349 Cuscutaceae (Polemoniales)	Solanales	282 Solanales	386 Convolvulales

Melchior (1964)	Benth. & Hook. (1965a,b,c)	Hutchinson (1973)	Dahlgren (1980a)	Cronquist (1981)	Takhtajan (1987)
c) —	—	—	—	*in* Boraginaceae	393 Cordiaceae (Boraginales)
d) *in* Boraginaceae	*in* Boraginaceae	250 Ehretiaceae (Verbenales)	Boraginales	*in* Boraginaceae	392 Boraginales
e) —	—	*in* Boraginaceae (?)	Boraginales	*in* Boraginaceae (?)	395 Wellstediaceae (Boraginales)
258. Lennoaceae	Ericales	160 Ericales	Boraginales	286 Lamiales	390 Boraginales
259. Verbenaceae	Lamiales	251 Verbenales	Lamiales	288 Lamiales	414 Lamiales
f) *in* Verbenaceae	*in* Verbenaceae	252 Stilbaceae (Verbenales)	Scrophulariales	*in* Verbenaceae	400 Scrophulariales
g) ?	?	253 Dirachstylidaceae	?	?	?
260. Callitrichaceae	*in* Haloragaceae	302 Onagrales	Lamiales	291 Callitrichales	416 Lamiales
261. Labiatae (= Lamianae)	Lamiales	354 Lamiales	Lamiales	289 Lamiales	415 Lamiales
262. Nolanaceae	*in* Convolvulaceae	334 Solanales	*in* Solanaceae	279 Solanales	381 Solanales
263. Solanaceae	Polemoniales	331 Solanales	Solanales	280 Solanales	380 Solanales
264. Duckeo-dendraceae	—	*in* Ehretiaceae	*in* Solanaceae	278 Solanales	382 Solanales
h) —	—	—	—		383 Sclerophylaca-ceae (Solanales)
i) —	—	—	—		384 Goetzeaceae (Solanales)
265. Buddlejaceae	*in* Loganiaceae	235 Loganiales	Scrophulariales	294 Scrophulariales	396 Scrophulariales
266. Scrophulariaceae	Personales	335 Personales	Scrophulariales	296 Scrophulariales	397 Scrophulariales
j) *in* Solanaceae	*in* Solanales	336 Salpiglossidaceae (Personales)	—	*in* Solanaceae	*in* Solanaceae
k) —	—	—	Nelsoniaceae (Scrophulariales)	—	—
267. Globulariaceae	*in* Selaginaceae	353 Lamiales	Scrophulariales	297 Scrophulariales	398 Scrophulariales
268. Bignoniaceae	Personales	247 Bignoniales	Scrophulariales	303 Scrophulariales	401 Scrophulariales
269. Henriqueziaceae	*in* Rubiaceae	*in* Rubiaceae	Scrophulariales (?)	*in* Rubiaceae	*in* Rubiaceae
270. Acanthaceae	Personales	337 Personales	Scrophulariales	301 Scrophulariales	410 Scrophulariales
271. Pedaliaceae	Personales	248 Bignoniales	Scrophulariales	302 Scrophulariales	402 Scrophulariales
l) —	—	—	—	*in* Acanthaceae	411 Thunbergiaceae (Scrophulariales)
m) —	—	—	Scrophulariales	304 Scrophulariales	412 Mendonciaceae (Scroplulariales)

Melchior (1964)	Benth. & Hook. (1965a,b,c)	Hutchinson (1973)	Dahlgren (1980a)	Cronquist (1981)	Takhtajan (1987)
n) *in Pedaliaceae*	—	*in Pedaliaceae*	Scrophulariales	*in Pedaliaceae*	403 Trapellaceae (Scrophulariales)
272. Martyniaceae	*in Pedaliaceae*	249 Bignoniales	Scrophulariales	*in Pedaliaceae*	404 Scrophulariales
273. Gesneriaceae	Personales	338 Personales	Scrophulariales	300 Scrophulariales	405 Scrophulariales
274. Columelliaceae	Personales	341 Personales	Cornales	—	333 Hydrangeales
275. Orobanchaceae	Personales	339 Personales	*in Scrophulariaceae*	299 Scrophulariales	407 Scrophulariales
276. Lentibulariaceae	Personales	340 Personales	Scrophulariales	305 Scrophulariales	408 Scrophulariales
277. Myoporaceae	Personales	351 Lamiales	Scrophulariales	298 Scrophulariales	*in Scrophulariaceae*
o) *in Scrophulariaceae*	Lamiales	352 Selaginaceae (Lamiales)	Scrophulariales	*in Globulariaceae*	409 Spielmanniaceae (Scrophulariaceae)
p) —	—	—	—	—	
278 Phrymaceae	*in Verbenaceae*	254 Verbenales	*in Verbenaceae*	*in Verbenaceae* (?)	*in Verbenaceae* (?)
Order Plantaginales					
279. Plantaginaceae	Anomalous family	307 Plantaginales	Scrophulariales	293 Plantaginales	406 Scrophulariales
Order Dipsacales					
298. Caprifoliaceae	Rubiales	48 Araliales	Cornales	315 Dipsacales	356 Dipsacales
299. Adoxaceae	*in Caprifoliaceae*	316 Saxifragales	—	316 Dipsacales	359 Dipsacales
300. Valerianaceae	Asterales	322 Valerianales	—	317 Dipsacales	360 Dipsacales
301. Dipsacaceae	Asterales	323 Valerianales	Cornales	318 Dipsacales	362 Dipsacales
a) —	—	—	—	—	[357 Viburnaceae (Dipsacales)]
b) ?	?	?	—	?	358 Sambucaceae (Dipsacales)
c) —	—	—	—	*in Dipsacaceae*	361 Triplostegiaceae (Dipsacales)
d) —	—	—	—	*in Dipsacaceae*	363 Morinaceae (Dipsacales)
Order Campanulatae					
302. Campanulaceae	Campanales	325 Campanales	Campanulales	308 Campanulales	420 Campanulales
303. Sphenocleaceae	*in Campanulaceae*	*in Campanulaceae*	*in Campanulaceae*	307 Campanulales	419 Campanulales
304. Pentaphrag-mataceae	*in Campanulaceae*	*in Campanulaceae*	Campanulales	306 Campanulales	418 Campanulales
a) —	—	—	—	—	421 Cyphiaceae
b) —	—	—	—	—	422 Nemacladaceae
in Campanulaceae	*in Campanulaceae*	326 Campanales	—	*in Campanulaceae*	423 Lobeliaceae
c) —	—	—	—	—	424 Cyphocarpaceae

Melchior (1964)	Benth. & Hook. (1965a,b,c)	Hutchinson (1973)	Dahlgren (1980a)	Cronquist (1981)	Takhtajan (1987)
305. Goodeniaceae	Campanales	327 Goodeniales	Goodeniales	312 Campanulales	427 Goodeniales
306. Brunoniaceae	in Goodeniaceae	328 Goodeniales	in Goodeniaceae	311 Campanulales	428 Goodeniales
307. Stylidiaceae	Capparales	329 Goodeniales	Cornales	309 Campanulales	426 Stylidiales
d) in Stylidiaceae	in Saxifragaceae	315 Saxifragales	—	310 Campanulales	425 Donatiaceae (Stylidiales)
308. Calycyceraceae	Asterales	324 Valerianales	Asterales	319 Calycerales	429 Calycerales
309. Compositae (= Asteraceae)	Asterales	330 Asterales	Asterales	320 Asterales	430 Asterales
Order Helobiae					
310. Alismataceae	Apocarpae	357 Alismatales	Alismatales *in Alismataceae*	323. Alismatales / 322 Limnocharitaceae	436 Alismatales / 435 Alismatales
311. Butomaceae	in Alismaceae	355 Butomales	Hydrocharitales	321 Alismatales	431 Butomales
312. Hydrocharitaceae	Microspermae	356 Butomales	Hydrocharitales	324 Hydrocharitales	432 Hydrocharitales
a) —	—	—	in Hydrocharitaceae	—	433 Thalassiaceae
b) —	—	—	in Hydrocharitaceae	—	434 Halophilaceae
313. Scheuchzeriaceae	in Najadaceae	358 Alismatales	Zosterales	326 Najadales	438 Scheuchzeriales
314. Aponogetonaceae	in Najadaceae	364 Aponogetonales	Hydrocharitales	325 Najadales	437 Aponogetonales
315. Juncaginaceae	in Najadaceae	361 Juncaginales	Zosterales	327 Najadales	439 Juncaginales
316. Potamogetonaceae	in Najadaceae	366 Potamogetonales	Zosterales	328 Najadales	442 Potamogetonales
317. Zannichelliaceae	in Najadaceae	368 Najadales	Zosterales	331 Najadales	446 Cymodoceales
318. Najadaceae	Apocarpae	369 Najadales	Zosterales	330 Najadales	448 Najadales
c) in Liliaceae	in Liliaceae (?)	359 Petrosaviaceae (Najadales)	Liliaceae	335 Triuridales	—
d) in Juncaginaceae	in Najadaceae	362 Lilaeaceae (Najadales)	in Juncaginaceae	—	440 Juncaginales
e) in Potamogetonaceae	in Najadaceae	365 Najadales	Zosteraceae (Zosterales)	334 Najadales	445 Zosterales
f) in Potamogetonaceae	in Najadaceae	363 Posidoniaceae (Najadales)	Zosterales	332 Najadales	444 Posidoniales
g) in Potamogetonaceae	in Najadaceae	367 Ruppiaceae (Najadales)	in Potamogetonaceae	329 Najadales	443 Potamogetonales
h) in Zannichelliaceae	in Najadaceae	in Zannichelliaceae	—	333 Cymodoceaceae (Najadales)	447 Cymodoceales (Najadales)
i) —	—	—	—	—	441 Maundiaceae (Juncaginales)

Melchior (1964)	Benth. & Hook. (1965a,b,c)	Hutchinson (1973)	Dahlgren (1980a)	Cronquist (1981)	Takhtajan (1987)
Order Triuridales					
319. Triuridaceae	Apocarpae	360 Triuridales	Triuridales	336 Triuridales	449 Triuridales
Order Liliiflorae					
320. Liliaceae	Coronarieae	384 Liliales	Liliales	371 Liliales	458 Liliales
321. Xanthorrho-eaceae	*in* Juncaceae	403 Agavales	Asparagales	376 Liliales	462 Amaryllidales
322. Stemonaceae	Coronarieae	401 Dioscoreales (*as* Roxburghiaceae)	Dioscoreales	379 Liliales	492 Dioscoreales
324 Agavaceae	*in* Amaryllidaceae and Liliaceae	404 Agavales	Asparagales	375 Liliales	469 Amaryllidales
325. Haemodoraceae	Epigynae	408 Haemodorales	Haemidorales *in* Haemodoraceae	369 Liliales	496 Haemodorales
a) —					497 Conostylidaceae (Haemodorales)
326. Cyanastraceae	—	*in* Tecophilaeaceae	Asparagales	370 Liliales	455 Liliales
327. Amaryllidaceae	Epigynae	397 Amaryllidales	Asparagales	*in* Liliaceae	474 Amaryllidales
328. Hypoxidaceae	*in* Amaryllidaceae	409 Haemodorales	Asparagales	*in* Liliaceae	498 Dioscoreales
b) —			Liliales		450 Melanthiaceae (Liliales)
c) —			Liliales		451 Calochortaceae (Liliales)
d) *in* Haemodoraceae	*in* Haemodoraceae	Liliales	Asparagales	*in* Haemodoraceae	454 Tecophilaeaceae (Liliales)
e) —			Asparagales		456 Eriospermaceae (Liliales)
f) ?	?	?	?	?	[457 Medeolaceae (Liliales) 1987]
g) —			Asparagales incl. Aloeaceae		461 Asphodelaceae (Amaryllidales)
h) —			Asparagales		463 Dasypogonaceae (Amaryllidales)
i) *in* Liliaceae	*in* Liliaceae	*in* Liliaceae	Asparagales	—	464 Aphyllanthaceae (Amaryllidales)
j) *in* Liliaceae	*in* Liliaceae	*in* Liliaceae	Asparagales	*in* Liliaceae	465 Hyacinthaceae (Amaryllidales)
k) *in* Liliaceae	*in* Liliaceae	*in* Amaryllidaceae	Asparagales incl. Agapanthaceae & Gilliesiaceae	*in* Liliaceae	466 Alliaceae

	Melchior (1964)	Benth. & Hook. (1965a,b,c)	Hutchinson (1973)	Dahlgren (1980a)	Cronquist (1981)	Takhtajan (1987)
l)						467 Hesperocallidaceae
m)				Asparagales		468 Funkiaceae
n)				Asparagales		470 Hemerocallidaceae
o)				Asparagales		471 Phormiaceae
p)				Asparagales		472 Blandfordiaceae / 473 Doryanthaceae
q)				Asparagales		475 Ixioliriaceae (Amaryllidales)
r)				Asparagales		476 Convallariaceae (Asparagales)
s)	*in Liliaceae*	*in Liliaceae*	389 Liliales	Asparagales	*in Liliaceae*	477 Ruscaceae
t)	*in Liliaceae*	*in Liliaceae*	*in Liliaceae*	Asparagales	*in Liliaceae*	478 Asparagaceae (Asparagales)
u)				Asparagales		479 Dracaenaceae (Asparagales)
v)				Asparagales		480 Nolinaceae (Asparagales)
w)				Asparagales		481 Herreriaceae (Asparagales)
x)				Asparagales		482 Asteliaceae (Asparagales)
y)				Asparagales (?)		483 Hanguanaceae (Asparagales)
z)	*in Liliaceae*	*in Liliaceae*	392 Alstroemeriales	Philesiaceae (Asparagales)	*in Liliaceae*	485 Smilacales
aa)	—	—	—	Luzuriagaceae (Asparagales)	—	484 Smilacales
bb)	—	—	—	Getonoplesiaceae (Asparagales)	—	—
cc)	*in Liliaceae*	*in Liliaceae*	Liliales	Smilacaceae (Asparayales)	380 Liliales	488 Smilacales
dd)	*in Liliaceae*	*in Dioscoreaceae*	391 Alstroemeriales	Petermanniaceae (Asparayales)	*in Liliaceae*	486 Smilacales
ee)	—	—	—	Dianellaceae (Asparagales)	—	—

Melchior (1964)	Benth. & Hook. (1965a,b,c)	Hutchinson (1973)	Dahlgren (1980a)	Cronquist (1981)	Takhtajan (1987)
ff) —	—	—	Anthericaceae (Asparagales)	—	—
gg) *in* Liliaceae	*in* Amaryllidaceae	390 Alstroemericeae (Alstroemeriales)	Liliales	*in* Liliaceae	495 Alstroemeriales
hh) *in* Liliaceae	—	—	Colchicaceae (Liliales)	—	—
ii) —	—	—	Tricyrtidaceae (Liliales)	—	—
jj) ?	?	?	?	?	487 Ripogonaceae (1987) (Smilacales)
kk) *in* Dioscoreaceae	*in* Dioscoreaceae	400 Dioscoreales	Dioscoreales	*in* Dioscoreaceae	489 Trichopodaceae (Dioscoreales)
ll) *in* Dioscoreaceae	*in* Dioscoreaceae	399 Dioscoreales	*in* Dioscoreaceae	*in* Dioscoreaceae	490 Stenomeridaceae (Dioscoreales)
mm) *in* Liliaceae	*in* Liliaceae	386 Liliales	Dioscoreales	*in* Liliaceae	493 Trilliaceae (Dioscoreales)
329. Velloziaceae	*in* Amaryllidaceae	410 Haemodorales	Velloziales	373 Liliales	503 Velloziales
330. Taccaceae	Epigynae	412 Haemodorales	Dioscoreales	378 Liliales	494 Taccales
nn) *in* Orchidaceae	Microspermae	411 Apostasiaceae (Haemodorales)	Orchidales	*in* Orchidaceae	*in* Orchidaceae
331. Dioscoreaceae	Epigynae	402 Dioscoreales	Dioscoreales	381 Liliales	491 Dioscoreales
oo) —	—	—	*in* Haemodoraceae	—	497 Conostylidaceae (1987) (Haemodorales)
332. Pontederiaceae	Coronarieae	387 Liliales	Pontederiales	368 Liliales	500 Pontederiales
333. Iridaceae	Epigynae	398 Iridales	Liliales	372 Liliales	452 Liliales
334. Geosiridaceae	—		Liliales	382 Orchidales	453 Liliales
335. Burmanniaceae	Microspermae	414 Burmanniales	Burmanniales	383 Orchidales	459 Burmanniales
336. Corsiaceae	*in* Burmanniaceae	416 Burmanniales	Burmanniales	384 Orchidales	460 Burmanniales
pp) *in* Burmanniaceae	*in* Burmanniaceae	415 Thismiaceae	Burmanniales	*in* Burmanniaceae	*in* Burmanniaceae
337. Philydraceae	Coronarieae	413 Haemodorales	Philydrales	367 Liliales	501 Philydrales
Order Juncales					
338. Juncaceae	Calycinae	418 Juncales	Juncales	351 Juncales	512 Juncales
339. Thurniaceae	*in* Juncaceae	419 Juncales	Juncales	352 Juncales	513 Juncales
Order Bromeliales					
440. Bromeliaceae	Epigynae	376 Bromeliales	Bromeliales	358 Bromeliales	502 Bromeliales
Order Commelinales					
441. Commelinaceae	Coronarieae	370 Commelinales	Commelinales	345 Commelinales	516 Commelinales

Melchior (1964)	Benth. & Hook. (1965a,b,c)	Hutchinson (1973)	Dahlgren (1980a)	Cronquist (1981)	Takhtajan (1987)
a) —	—	371 Cartonemataceae	*in* Commelinaceae	—	—
442. Mayacaceae	Coronarieae	373 Commelinales	Commelinales	344 Commelinales	517 Commelinales
443. Xyridaceae	Coronarieae	374 Xyridales	Eriocaulales	343 Commelinales	518 Commelinales
444. Rapateaceae	Coronarieae	375 Xyridales	Eriocaulales	342 Commelinales	519 Commelinales
445. Eriocaulaceae	Glumaceae	376 Eriocaulales	Eriocaulales	346 Eriocaulales	520 Commelinales
446. Restionaceae	Glumaceae	421 Juncales	Poales	349 Restionales	523 Restionales
447. Centrolepidaceae	Glumaceae	420 Juncales	Poales	350 Restionales	526 Restionales
448. Flagellariaceae	Calycinae	372 Commelinales	Poales	347 Restionales	521 Restionales
			Poales	348 Joinvilleaceae (Restionales)	522 Restionales
b) *in* Restionaceae	*in* Restionaceae	*in* Restionaceae	*in* Restionaceae	*in* Restionaceae (??)	524 Anarthriaceae
c) *in* Restionaceae	*in* Restionaceae	*in* Restionaceae	*in* Resionaceae	*in* Restionaceae (??)	525 Ecdeiocoleaceae
Order Poales					
449. Gramineae (= Poaceae)	Glumaceae	423 Graminales	Poales	354 Cyperales	527 Poales
Order Principes					
450. Palmae (= Arecaceae)	Calycinae	405 Palmales	Arecales	337 Arecales	528 Arecales
Order Synanthae					
451. Cyclanthaceae	Nudiflorae	407 Cyclanthales	Cyclanthales	338 Cyclanthales	529 Cyclanthales
Order Spathiflorae					
452. Araceae	Nudiflorae	393 Arales	Arales	340 Arales	531 Arales
453. Lemnaceae	Nudiflorae	394 Arales	Arales	341 Arales	532 Arales
Order Pandanales					
454. Pandanaceae	Nudiflorae	406 Pandanales	Pandanales	339 Pandanales	530 Pandanales
455. Sparganiaceae	*in* Typhaceae	395 Typhales	Typhales	356 Typhales	533 Typhales
456. Typhaceae	Nudiflorae	396 Typhales	Typhales	357 Typhales	534 Typhales
Order Cyperales					
457. Cyperaceae	Glumaceae	422 Cyperales	Cyperales	353 Cyperales	514 Cyperales
Order Scitamineae					
458. Musaceae	*in* Epigyneae as Scitamineae	378 Zingiberales	Zingiberales	361 Zingiberales	505 Zingiberales
459. Zingiberaceae	*as* Scitamineae	381 Zingiberales	Zingiberales	363 Zingiberales	508 Zingiberales
460. Cannaceae		382 Zingiberales	Zingiberales	365 Zingiberales	510 Zingiberales
461. Marantaceae		383 Zingiberales	Zingiberales	366 Zingiberales	511 Zingiberales
462. Lowiaceae		380 Zingiberales	Zingiberales	362 Zingiberales	507 Zingiberales
a) *in* Musaceae	*in* Epigyneae as Scitamineae	379 Zingiberales	Zingiberales	359 Zingiberales	504 Strelitziaceae

Melchior (1964)	Benth. & Hook. (1965a,b,c)	Hutchinson (1973)	Dahlgren (1980a)	Cronquist (1981)	Takhtajan (1987)
b) *in Musaceae*	in Epigyneae as Scitamineae	*in* Strelitziaceae	Zingiberales	360 Zingiberales	506 Heliconiaceae
c) *in Zingiberaceae*	in Epigyneae as Scitamineae	*in* Zingiberaceae	Zingiberales	364 Zingiberales	509 Costaceae
d) —	—	—	Hydatellales (?)	355 Hydatellales	515 Hydatellaceae (Hydatellales)
Order Microspermae					
449. Orchidaceae	Microspermae	417 Orchidales	Orchidales	385 Orchidales	499 Orchidales
a) —	—	—	Cypripediaceae (Orchidales)	—	—

Glossary of Terms

Acropetal: Proceeding from basal or proximal end to the distal end.

Adnate: Different organs attached with each other.

Aestivation: Arrangement of sepals and/or petals with respect to each other.

Amentiferae: A group of dicotyledonous plants with inflorescence an ament or catkin.

Androgynophore: An axis which arises from the receptacle and bears both androecium and gynoecium.

Anthesis: Flowering period when pollination takes place.

Anthocyanin: Flavonoid pigments, colour ranges from blue or violet to purple or red, occur in the central vacuole of a cell, especially in petals.

Anthotaxy: A term used as against phyllotaxy for the floral parts.

Anthoxanthin: Closely allied to anthocyanin, colour ranges from yellow or orange to orange-red.

Antitropous: Broader halves of leaves on alternate side of midrib.

Betacyanin: A chemical class of nitrogenous, water-soluble pigments, colour ranges from blue or violet to purple or red.

Betalain: Nitrogenous, water-soluble pigments—betacyanins and betaxanthins.

Betaxanthin: Pigments allied to betacyanin, but colour ranges from yellow or orange to orange-red.

Bolster-like: Swollen like a pillow, or cushion-like.

Cauliflory: Flowers borne on the main stem or older branches, as in *Cocoa*.

Caruncle: A growth near the hilum of some seeds such as castor.

Confluent: Blending or merging together.

Connate: Attached to each other, but the organs are similar (like-organs).

Cupule: Cup-like structure at the base of some fruits formed by the dry and enlarged floral envelopes.

Cymbiform: Boat-shaped

Cymule: Diminutive of cyme; of usually a few flowers.

Domatia: Depressions, pockets, sacs or tufts of hairs on the principal vein, vein-axils where they occur exclusively on the abaxial surface of leaves. Predominantly seen in woody plants of humid tropical or subtropical regions (Metcalfe and Chalk 1979).

Embryo sac: The female gametophyte. Depending upon the number of megaspore nuclei taking part in the development, these may be monosporic, bisporic or tetrasporic.

Monosporic embryo sacs; Polygonum type—8-nucleate at maturity; Oenothera type—4-nucleate at maturity.

Bisporic embryo sacs: Allium type—8-nucleate at maturity; Podostemon type—4-nucleate at maturity.

Tetrasporic embryo sacs: Peperomia type—16-nucleate at maturity; Penaea type—16-nucleate at maturity; Drusa type—16-nucleate at maturity; Fritillaria type—8-nucleate at maturity; Plumbagella type—4-nucleate at maturity; Plumbago type—8-nucleate at maturity, tetrapolar; Adoxa type—8-nucleate at maturity; bipolar; Chrysanthemum cinerariaefolium type—10- to 12-nucleate at maturity.

Endosperm: A starch- or oil-filled tissue in mature seeds. There are three types: *Cellular*—Cell walls develop immediately after mitosis of primary endosperm nucleus so that there is no initial free-nuclear stage. *Nuclear*—Endosperm which has a free-nuclear phase throughout or only during early ontogeny. Wall formation takes place only after several mitotic divisions have occurred. *Helobial*—Endosperm in which the first division of the triple-fusion-nucleus is followed by the formation of a partition wall; afterwards the micropylar chamber becomes cellular and the other remains free-nuclear. Many variations in each type.

Equitant: Overlapping of leaves in two ranks, as in members of the family Iridaceae.

Exine: The outer wall layer of a pollen grain.

Farinose: Surface covered with a mealy (starch or starch-like materials) coating, as leaves of some *Primula* spp.

Fetid: With a disagreeable odour.

Follicetum: An aggregate of follicles; product of a multipistillate, apocarpous gynoecium.

Foveolate: Pitted.

Geniculate: Bent knee-like.

Gibbous: Swollen on one side, usually at the base as in *Antirrhinum majus*.

Glabrous: Surface without hairs.

Glaucous: Surface covered with very fine hairs.

Glutinous: Sticky

Haustoria: The nutrition absorbing extensions of pollen, megaspore, embryo sac, endosperm and embryo suspensor. Also in roots and shoots of parasitic plants.

Heterobathmy: Unequal rate of evolution of different features within one lineage.

Hippocrepiform: Horseshoe-shaped.

Homotropous: Broader halves of leaves on the same side of midrib, right or left.

Hypanthium: The cup-like receptacle derived usually from the fusion of floral envelopes and androecium, and on which the calyx, corolla and stamens are apparently borne.

Hypocotyl: The axis of the embryo below the cotyledons which—on seed germination—develops into the radicle.

Indumentum: A thick pubescence.

Inflated: Bladder-like or balloon-like.

Introrse: Turned inward as an anther with the line of dehiscence towards the center of the flower.

Involucel: A secondary involucre.

Involucre: One or more whorls of small leaves or bracts borne underneath an inflorescence or a cluster or flowers.

Keel: The two anterior united petals of a papilionaceous corolla.

Labellum: Lip; part of a perianth as in the flower of orchids.

Lactiferous: Producing latex.

Lax: Loose.

Lemma: The fertile bract or glume as in grass florets.

Lepidote: Covered with small scurfy scales (exfoliating scaly incrustations).

Leucoanthocyanins: Colourless anthocyanins.

Ligneous: Woody.

Lorate: Strap-shaped (apex not acute or acuminate).

Macropodous or Macropodial: With food reserve stored in hypocotyl.

Mycorrhiza: A symbiotic association of a fungus and the root of a vascular plant.

Myrosin: An enzyme involved in the formation of mustard oil.

Nut: An indehiscent, 1-seeded hard and bony fruit.

Nutlet: A diminutive or small nut.

Obdiplostemonous: Stamens in 2 whorls, outer opposite petals, inner opposite sepals.

Ochreate: Formation of a nodal sheath by fusion of two stipules.

Ovule: Embryonic seed consisting of integument(s) and nucellus.

Ovule types: *Amphitropous*—An ovule with the body half-inverted so that the funiculus is attached near the middle. The micropyle points at right angles to the funiculus.

Anatropous—The ovule with the body fully inverted so that the micropyle is basal, adjoining the funiculus.

Campylotropous—An ovule curved by uneven growth so that its axis is approximately at right angles to its funiculus (stalk).

Hemianatropous—Same as amphitropous.

Circinotropous—The ovule after attaining fully inverted position (anatropous ovule), continues the curvature until it has turned over completely and the micropylar end again points upwards as in *Opuntia*.

Orthotropous— Straight or unbent ovule with the micropyle at the opposite end from the stalk or funiculus.

Epitropous—An erect ovule with dorsal raphe, or a pendulous ovule with ventral raphe.

Crassinucellate—Ovule with the nucellus several cells thick at least at the micropylar end.

Tenuinucellate—Ovule with the nucellus of a single layer of cells.

Perisperm: Food storage tissue in seed, derived from nucellus.

Phenetic: Pertaining to the expressed characteristics of an individual, irrespective of its genetic nature.

Phylad: A natural group of any rank, considered from the standpoint of its evolutionary history.

Pollen grains: Young male gametophytes contained in the anther. Pollen grain types:

Multipalynous (syn. *eurypalynous*)—Taxa with more than one pollen morphoforms.

Unipalynous (syn. stenopalynous)—Taxa with one pollen morphoform.

Inaperturate—Without any aperture.

Pantoporate—With many apertures.

Triporate—With three apertures.

Uniporate—With one aperture.

Pollinia: Pollen grains in uniform coherent masses.

Polygamodioecious: Functionally dioecious but with a few flowers of opposite sex or a few bisexual flowers at the time of flowering.

Prophyll: Bracteole.

Pseudanthia: Cluster of small or reduced flowers, collectively appearing as a single flower.

Pseudomonomerous: An ovary apparently composed of a single carpel but phyletically derived from a compound or polycarpellary ovary.

Psychomimetics: Act as hallucinogens.

Rbc L or Rubisco: Rubisco or Ribulose, 1,5 bisphosphate carboxylase enzyme is located on the stromal surface of the thylakoid membrane. It is the most abundant protein in the biosphere, particularly in the chloroplast and is involved in CO_2 fixation. This enzyme consists of 2 subunits, each encoded by a separate gene, the smaller subunit rbc S (encoded by a nuclear gene) and the larger subunit rbc L (encoded by a chloroplast gene).

Raphe: The portion of the funiculus of an ovule that is adnate to the integument, usually represented by a ridge; present in anatropous ovules.

Raphides: Acicular or needle-shaped crystals of calcium oxalate.

Retrorse: Bent or turned over backward or downward.

Resupinated: Twisted by 180°.

Rugose: Covered with wrinkles.

Ruminate: Coarsely wrinkled or mottled. A ruminate endosperm is formed due to invaginations of outer tissues, which penetrate deeper and deeper and appear as dark wavy bands in mature seeds as in *Areca* nut, Annonaceae and Myristicaceae members.

Scarious: Thin, dry and membranaous.

Squamiform: Scale-like.

Stipel: Stipule of a leaflet.

Stomatal types:

Anomocytic or *Ranunculaceous* type—Stoma remains surrounded by a limited number of cells which cannot be distinguished from other epidermal cells.

Anisocytic or *Cruciferous* type—Stoma remains surrounded by three subsidiary cells of which one is distinctly smaller than the other two.

Diacytic or *Caryophyllaceous* type—Stoma remains enclosed by two subsidiary cells whose common wall is at right angles to the guard cells.

Paracytic or *Rubiaceous* type—Stoma is surrounded by one or more subsidiary cells on either side which lie parallel to the long axis of the stomatal pore.

Taxon: A general term used for any taxonomic group irrespective of its rank.

Tumid—Swollen.

Vasicentric—Concentrated around the vessels.

Verticillate—Arranged in a whorl.

Verrucose—Warty.

References

Aalders LE, Hall IV (1962) New evidence on the cytotaxonomy of *Vaccinium* species as revealed by stomatal measurements from herbarium specimens. Nature 196:694

Abbe EC (1935) Studies in the phylogeny of the Betulaceae. 1. Floral and inflorescence anatomy and morphology. Bot Gaz 97: 1–67

Abbe EC, Earle TT (1940) Inflorescence, floral anatomy and morphology of *Leitneria floridana*. Bull Torrey Bot Club 67: 173–193

Ackerman JD, Williams NH (1980) Pollen morphology of the tribe Neottieae and its impact on the classification of the Orchidaceae. Grana Palynol 19: 7–18

Ackerman JD, Williams NH (1981) Pollen morphology of the Chloraeinae (Orchidaceae: Diurideae) and related subtribes. Am J Bot 68: 1392–1402

Adams JE (1949) Studies in the comparative anatomy of the Cornaceae. J Elisha Mitchell Scient Soc 56: 218–244

Agarwal Saroj (1961) The embryology of *Strombosia* Blume. Phytomorphology 11: 269–272

Agarwal Saroj (1963) Morphological and embryological studies in the family Olacaceae 1. *Olax* L. Phytomorphology 13: 185–196

Airy Shaw HK (1941) Additions to the flora of Borneo and other Malay Islands. The Pentaphragmataceae of the Oxford University Expedition of Sarawak, 1932. Bull Misc Inf R Bot Gdns Kew 1941: 233–236

Airy Shaw HK (1948) Sphenocleaceae. *In:* Van Steenis CGGJ (ed) Flora Malesiana 4: 27–28

Airy Shaw HK (1952) On the Dioncophyllaceae. Kew Bull 3: 327–347

Airy Shaw HK (1954) Pentaphragmataceae. *In:* Van Steenis CGGJ (ed) Flora Malesiana 4: 517–528

Airy Shaw HK (1965) Diagnoses of new families, new names, etc. for the seventh edition of Willis's Dictionary. Kew Bull 18: 249–273

Al-Eisawi DM (1988) Resedaceae in Jordan. Bot Jb 110: 17–39

Aleykutty KM, Inamdar JA (1978) Structure, ontogeny and taxonomic significance of trichomes and stomata in some Capparidaceae. Feddes Report 89: 19–30

Alfaro ME, Mesa A (1979). El origen morfológico del floema intraaxilar en Nolanáceas y la posición sistemática de esta familia. Bol Soc Argent Bot 18: 123–126

Allan RD, Wells RJ, Correll RL, Macleod JK (1978) The presence of quinones in the genus *Cyperus* as an aid to classification. Phytochem 17: 263–266

Alston RE, Turner BL (1963) Natural hybridization among four species of *Baptisia*. Am J Bot 50: 159–173

Ambasht RS (1990) A Textbook of Plant Ecology, pp 1–358. Students' Friends Publ, Varanasi (India)

Ambrose JD (1985) *Lophiola,* familial affinity with the Liliaceae. Taxon 34: 149–150

Ambwani K, Kumar Madhav (1994) Pollen morphology of *Ammandra decasperma* Cook., Phytelephantoideae. Phytomorphology 44: 89–93

Ancibor E (1979) Systematic anatomy of vegetative organs of the Hydrocharitaceae. Bot J Linn Soc London 78: 237–266

Arachi JX (1968) Pictorial Presentation of Indian Flora, pp 1–190. Higginbothams (P) Ltd. Madras Bangalore Ooty Trivandram, India

Arber A (1925) Monocotyledons: A Morphological Study, pp 1–258. Univ Press, Cambridge (England)

Arber A (1942) Studies in flower structure 7. On the gynoecium of *Reseda,* with a consideration of paracarpy. Ann Bot (London) 6: 43–48

Archer WA (1950) New plastic aid in mounting herbarium sheets. Rhodora 52: 298–299

Arekal GD, Ramaswamy SN (1980) Embryology of *Eriocaulon hookerianum* Stapf. and the systematic position of Eriocaulaceae. Bot Notiser 133: 295–309

Arguie CL (1971) Pollen of Butomaceae and Alismataceae. Grana 11: 131–144

Armstrong JE (1985) The delimitation of Bignoniaceae and Scrophulariaceae based on floral anatomy, and the placement of problem genera. Am J Bot 72: 755–766

Arora N (1953) The embryology of *Zizyphus rotundifolia* Lamk. Phytomorphology 3: 88–98

Ascensão L, Pais MS (1988) Ultrastructure and histochemistry of secretory ducts in *Artemisia campestris* ssp. *maritima* (Compositae). Nordic J Bot 8: 283–292

Ayensu ES (1968) The anatomy of *Barbaceniopsis*: A new genus recently described in the Velloziaceae. Am J Bot 55: 399–405

Ayensu ES (1972) Dioscoreales. *In*: Metcalfe CR (ed) Anatomy of the Monocotyledons, pp 1–182. Vivian Ridler, Clarendon Press, Oxford (UK)

Ayensu ES, Skvarla JJ (1974) Fine structure of Velloziaceae pollen. Bull Torrey Bot Club 101: 250–256

Baas P (1972) Anatomical contributions to plant taxonomy. 2. The affinities of *Hua* Pierre and *Afrostyrax* Perkins et Gilg. Blumea 20: 161–192

Baas P (1975) Vegetative anatomy and the affinities of Aquifoliaceae: *Sphenostemon, Phelline* and *Oncotheca*. Blumea 22: 311–407

Baas P (1981) A note on stomatal types and crystals in the leaves of Melastomataceae. Blumea 27: 475–479

Baas P (1984) Vegetative anatomy and the taxonomic status of *Ilex collina* and *Nemopanthus* (Aquifoliaceae). J Arnold Arbor 65: 243–250

Baas P, Zweypfenning RCVJ (1979) Wood anatomy of the Lythraceae. Acta Bot Neerl 28: 117–155

Balle S, Dandy JE, Gilmour JSL, Holttum RE, Stern WT, Thoday D (1960) *Loranthus*. Taxon 9: 208–210

Balogh P (1979) Pollen morphology of the tribe Cranichideae Endlicher, subtribe Spiranthinae Bentham (Orchidaceae). Orquidea 7: 241–260

Bailey IW (1966) The significance of the reduction of vessels in the Cactaceae. J Arnold Arbor 47: 288–292

Bailey IW, Nast CG (1945a) The comparative morphology of the Winteraceae. 7. Summary and Conclusions. J Arnold Arbor 26: 37–47

Bailey IW, Nast CG (1945b) Morphology and relationships of *Trochodendron* and *Tetracentron*. 1. Stem, root and leaf. J Arnold Arbor 26: 143–154

Bailey IW, Nast CG (1945c) Morphology and relationships of *Trochodendron* and *Tetracentron*. 2. Inflorescence, flower and fruit. J Arnold Arbor 26: 267–276

Bailey IW, Smith AC (1942) Degeneriaceae: A new family of flowering plants from Fiji. J Arnold Arbor 23: 356–365

Bailey IW, Swamy BGL (1949) The morphology and relationships of *Austrobaileya*. J Arnold Arbor 30: 211–226

Bailey IW, Swamy BGL (1951) The conduplicate carpel of dicotyledons and its initial trends of specialization. Am J Bot 38: 373–379

Bailey IW, Nast CG, Smith AC (1943) The family Himantandraceae. J Arnold Arbor 24: 190–206

Bailey IW, Nast CG, Smith AC (1948) Morphology and relationships of *Illicium, Schisandra* and *Kadsura*. J Arnold Arbor 29: 77–89

Baker HG (1976) Mistake pollination as a reproductive system in the special reference to the Caricaceae, pp 161–169. *In:* Burley J, Styles BT (eds) Tropical Trees: Variation, Breeding and Conservation, pp 1–243. Academic Press, London

Baker W, Ollis WD (1961) Biflavonyls. *In:* Ollis WD (ed) Recent Developments in the Chemistry of Natural Phenolic Compounds. Oxford London

Ball PW, Heywood VH (1962) A revision of the genus *Petrorhagia*. Bull Br Mus Natu Hist Bot 3: 121–172

Bamber CJ (1916) Plants of the Punjab: A Descriptive Key to the Flora of the Punjab, North-West Frontier Province and Kashmir, pp 1–285. Govt Printing Press, Lahore (Punjab; Br India)

Bandaranayake WM, Karunanayake S, Subramanian S, Sultanbawa MUS, Balasubramanian S (1977) Triterpenoid taxonomic markers for *Stemonoporous* and other genera of the Dipterocarpaceae. Phytochem 16: 699–701

Barfod A (1988) Inflorescence morphology of some South American Anacardiaceae and the possible phylogenetic trends. Nordic J Bot 8: 3–11

Barlow BA (1959) Chromosome numbers in the Casuarinaceae. Aust J Bot 7: 230–237

Barlow BA (1964) Classification of the Loranthaceae and Viscaceae. Proc Linn Soc NSW 89: 268–272

Barlow BA (1983) Biogeography of Loranthaceae and Viscaceae, pp 19–46. *In:* Calder M, Bernhardt P (eds) The Biology of Mistletoes, pp 1–348. Harcourt Brace Jovanovich, Academic Press, Sydney Australia

Barlow BA, Wiens D (1971) The cytogeography of the Loranthaceous mistletoes. Taxon 20: 291–312

Baruah A, Nath SC (1996) Stomatal diversities in *Catharanthus roseus* (L.) G. Don—Some additional information. Phytomorphology 46: 365–369

Basinger JF, Dilcher DL (1983) Fruits of *Cercidiphyllum* from the early Tertiary of Ellesmere Islands, arctic Canada. Am J Bot 70: 67

Bate-Smith EC (1965) Investigation of the chemistry and taxonomy of subtribe Quillajeae of the Rosaceae, using comparisons of fresh and herbarium material. Phytochem 4: 127–133

Bate-Smith EC (1974) Systematic distribution of ellagitannins in relation to the phylogeny and classification of the angiosperms, pp 93–102. *In:* Bendz G, Santesson J (eds) Chemistry in Botanical Classification, pp 1–320. Nobel Symposium 25. Academic Press, London

Bate-Smith EC, Swain T (1966) The asperulosides and the aucubins, pp 159–174. *In:* Swain T (ed) Comparative Phytochemistry, pp 1–360. Academic Press, London New York

Bate-Smith EC, Whitmore TC (1959) Chemistry and taxonomy in the Dipterocarpaceae. Nature 184: 795–796

Bate-Smith EC, Davenport SM, Harborne JB (1967) Comparative biochemistry of the flavonoids. 3. A correlation between chemistry and plant geography in the genus *Eucryphia.* Phytochem 6: 1407–1413

Bate-Smith EC, Ferguson IK, Hutson K, Jensen SR, Nielsen BJ, Swain T (1975) Phytochemical interrelationship in the Cornaceae. Biochem Syst Ecol 3: 79–89

Battaglia E (1955) A consideration of a new type of meiosis (mis-meiosis) in Juncaceae (*Luzula*) and *Hemiptera.* Bull Torrey Bot Club 82: 383–396

Batygina TB, Kravtsova TI, Shamrov II (1980) The comparative embryology of some representatives of the orders Nymphaeales and Nelumbonales (in Russian). Bot Zh (Moscow Leningrad) 65: 1071–1087

Becker HF (1973) The York Ranch flora of the Upper Ruby River Basin, south-western Montana. Palaeontographica B Palaeophytol 143: 18–93

Behnke H–D (1969) Die Siebröhren-Plastiden der Monocotyledonen. Vergleichende Untersuchungen über Feinbau und Verbreitung eines characteristischen Plastidentyps. Planta 84: 174–184

Behnke H–D (1974) P- und S-Typ Siebelement-Plastiden bei Rhamnales. Beitr Biol Pfl 50: 457–464

Behnke H–D (1975) The bases of angiosperm phylogeny: Ultrastructure. Ann Mo Bot Gard. 62: 647–663

Behnke H–D (1976a) Ultrastructure of sieve-element plastids in Caryophyllales (Centrospermae): Evidence for the delimitation and classification of the order. Pl Syst Evol 126: 31–54

Behnke H–D (1976b) A tabulated survey of some characters of systematic importance in Centrospermous families. Plant Syst Evol 126: 95–98

Behnke H–D (1981) Sieve-element characters. Nordic J Bot 1: 381–400

Behnke H–D (1982) Sieve-element plastids, exine sculpturing and the systematic affinities of the Buxaceae. Pl Syst Evol 139: 257–266

Behnke H–D (1985) Contributions to the knowledge of P-type sieve-element plastids in dicotyledons. 2. Eucryphiaceae. Taxon 34: 607–610

Behnke H–D (1986) Contributions to the knowledge of sieve-element plastids in Gunneraceae and allied families. Pl Syst Evol 151: 215–222

Behnke H.–D (1995) P-type sieve-element plastids and the systematics of the Arales. Pl Syst Evol 195: 87–119

Behnke H–D, Turner BL (1971) On specific sieve-tube plastids in Caryophyllales: Further investigations with special reference to Bataceae. Taxon 20: 731–737

Benson L (1962) Plant Taxonomy: Methods and Principles. Ronald Press, New York

Benson L (1970) Plant Classification, pp 1–688. Oxford & IBH Publ, New Delhi

Bentham G (1870) Flora Australiensis. 5: 164–165

Bentham G (1965) Juncaceae. *In:* Bentham G, Hooker JD (eds) Genera Plantarum 3: 861–869

Bentham G, Hooker JD (1965a) Genera Plantarum, Vol 1. L. Reeve, London (Reprint edition).

Bentham G, Hooker JD (1965b) Genera Plantarum, Vol 2. L. Reeve, London (Reprint edition)

Bentham G, Hooker JD (1965c) Genera Plantarum, Vol 3. L. Reeve, London (Reprint edition)

Bergner I, Jensen U (1989) Phytoserological contribution to the systematic placement of the Typhales. Nordic J Bot 8: 447–456

Berridge EM (1914) The structure of the flower of the Fagaceae and its bearing on the affinities of the group. Ann Bot (London) 28: 509–526

Bessey CE (1915) The phylogenetic taxonomy of flowering plants. Ann Mo Bot Gard 2: 109–164

Bews JW (1929) The world's grasses. Their differentiation, distribution, economics and ecology, pp 1–408. Longmans, Green and Co., London New York Toronto

Bhandari NN (1971) Embryology of the Magnoliales and comments on their relationships. J Arnold Arbor 52: 1–39, 285–304

Bhandari NN, Vohra SCA (1983) Embryology and affinities of Viscaceae, pp 69–86. *In:* Calder M, Bernhardt P (eds) The Biology of Mistletoes, pp 1–348. Harcourt Brace Jovanovich, Academic Press, Sydney Australia

Bhanwra RK (1988) Embryology in relation to systematics of Gramineae. Ann Bot (London) 62: 215–233

Bhatnagar AK, Garg M (1977) Affinities of *Daphniphyllum:* Palynological approach. Phytomorphology 27: 92–97

Bhatnagar AK, Kapil RN (1982) Seed development in *Daphniphyllum himalayense* with a discussion on taxonomic position of Daphniphyllaceae. Phytomorphology 32: 66–81

Bhatnagar SP (1965) Studies in Angiospermic Parasites. 2. *Santalum album*—The Sandalwood Tree. Bull Natl Bot Gard (Lucknow, India) No. 112: 1–90

Bhatnagar SP (1970) Santalaceae. *In:* Proc Symp Comparative Embryology of Angiosperms. Bull Indian Natl Sci Acad No. 41: 15–18

Bhatnagar SP, Johri BM (1983) Embryology of Loranthaceae, pp 47–67. *In:* Calder M, Bernhardt P (eds) The Biology of Mistletoes, pp 1–348. Harcourt Brace Jovanovich, Academic Press, Sydney Australia

Bhaumik PK, Mukherjee B, Juneau JP, Bhacca NS, Mukherjee R (1979) Alkaloids from leaves of *Annona squamosa.* Phytochem 18: 1584–1586

Bindroo BB, Bhat BK, Kachroo P (1985) Seasonal periodicity of diosgenin production in *Dioscorea deltoidea* Wall. Beitr Biol Pfl 60: 333–339

Bisset NG, Choudhury AK (1974) Alkaloids and iridoids from *Strychnos nux-vomica* fruits. Phytochem 13: 265–269

Bisset NG, Diaz-Parra MA, Ehret C, Ourisson G (1967) Études chimiotaxonomiques dans la famille des Diptérocarpacées. 3. Constituants des genres *Anisoptera* Korth., *Cotylelobium* Pierre, *Dryobalanops* Gaertn. f. et *Upuna* Sym. Phytochem 6: 1395–1405

Boesewinkel FD (1984a) Development of ovule and seed coat in *Cneorum tricoccum* (Cneoraceae). Acta Bot Neerl 33: 61–70

Boesewinkel FD (1984b) Ovule and seed structure in Datiscaceae. Acta Bot Neerl 33: 419–429

Boesewinkel FD (1988) The seed structure and taxonomic relationships of *Hypseocharis.* Acta Bot Neerl 37: 111–120

Boesewinkel FD, Bouman F (1984) The seed: Structure, pp 567–610. *In:* Johri BM (ed) Embryology of Angiosperms, pp 1–830. Springer, Berlin Heidelberg New York Tokyo

Bogner J (1978) A critical list of the aroid genera. Aroideana 13: 63–73

Bohlmann F, Burkhardt T, Zdero C (1973) Naturally-occurring Acetylenes. Academic Press, London

Bohm BA, Ornduff R (1981) Leaf flavonoids and ordinal affinities of Coriariaceae. Syst Bot, New York 6: 15–26

Boke N (1964) The Cactus gynoecium: A new interpretation. Am J Bot 51: 598–610

Boothroyd LE (1930) The morphology and anatomy of the inflorescence and flower of the Platanaceae. Am J Bot 17: 678–693

Bortenschlager S (1967) Vorlanfige Mitteilungen zur Pollen morphologie in der Familie der Geraniaceen und ihre systematische Beudeutung. Grana 7: 400–468

Bortenschlager S, Erdtman G, Praglowski J (1966) Pollen morphologische Notizen über einige Blütenflanzen incertae sedis. Bot Notiser 119: 160–168

Bose S (1962) Cytotaxonomy of Amaryllidaceae. Bull Bot Surv India 4: 27–38

Bottomley WB (1911) The structure and physiological significance of the root nodules of *Myrica gale.* Proc R Soc London 84: 215–216

Bottomley WB (1912) The root nodules of *Myrica gale.* Ann Bot (London) 26: 111–117

Boudet AM, Boudet A, Bouyssou H (1977) Taxonomic distribution of isoenzymes of dehydroquinate hydrolyase in the angiosperms. Phytochem 16: 919–922

Bouman F (1984) The Ovule, pp 123–157. *In:* Johri BM (ed) Embryology of Angiosperms, pp 1–830. Springer, Berlin Heidelberg New York Tokyo

Bourdu R, Cartier D, Gorenfolt R (1963) Biochemical affinities of the genera *Littorella* and *Plantago.* Bull Soc Bot Fr 110: 107–109

Bremekamp CEB (1966) Remarks on the position, the delimitation and the subdivision of the Rubiaceae. Acta Bot Neerl 15: 1–53

Brenan JMP (1952) Plants of the Cambridge Expedition 1947–1948. 2. A new order of flowering plants from the British Cameroons. Kew Bull 2: 227–236

Brenan JMP (1953) *Soyauxia,* a second genus of Medusandraceae. Kew Bull 4: 507–511

Brewbaker JL (1967) The distribution and phylogenetic significance of binucleate and trinucleate pollen grains in the angiosperms. Am J Bot 54: 1069–1083

Bridle P, Stott KG, Timberlake CF (1973) Anthocyanins in *Salix* species: A new anthocyanin in *S. purpurea.* Phytochem 12: 1103–1106

Briggs BG, Johnson LAS (1979) Evolution in the Myrtaceae: Evidence from inflorescence structure. Proc Linn Soc NSW 102: 157–272

Britton NL, Rose JN (1919) Descriptions and illustrations of the Cactus family. Carnegie Inst Washington 248, 1: 1–236

Britton NL, Rose JN (1920) Descriptions and illustrations of the Cactus family. Carnegie Inst Washington 248, 2: 1–239

Britton NL, Rose JN (1922) Descriptions and illustrations of the Cactus family. Carnegie Inst Washington 248, 3: 1–255

Britton NL, Rose JN (1923) Descriptions and illustrations of the Cactus family. Carnegie Inst Washington 248, 4: 1–318

Brizicky GK (1961a) The genera of Turneraceae and Passifloraceae in the south-eastern United States. J Arnold Arbor 42: 204–218

Brizicky GK (1961b) The genera of Violaceae in the south-eastern United States. J Arnold Arbor 42: 321–333

Brizicky GK (1962) The Genera of Simaroubaceae and Burseraceae in the south-eastern United States. J Arnold Arbor 43: 173–186

Brizicky GK (1963) The genera of the Sapindales in the south-eastern United States. J Arnold Arbor 44: 462–501

Bronckers F, Stainer F (1972) Contribution a l'étude morphologique du pollen de la familie des Stylidiaceae. Grana 12: 1–22

Brough P (1927) Studies in the Goodeniaceae. 1. The life history of *Dampiera stricta* R.Br. Proc Linn Soc NSW 52: 471–498

Brown R (1814) General remarks, geographical, systematical on the botany of Terra Australis. *In:* Finders M (ed) A Voyage to Terra Australis. Vol 2: 1–550

Brown RW (1962) Palaeocene flora of the Rocky Mountains and Great plains, pp 1–375. Profess Pap US Geol Surv

Brown WH (1938) The bearing of nectaries on the phylogeny of flowering plants. Proc Am Phil Soc 79: 549–595

Bruni A, Tosi B, Modenesi P (1987) Morphology and secretion of glandular trichomes in *Tamus communis.* Nordic J Bot 7: 79–84

Buchenau F (1906) Juncaceae. *In:* Engler A (ed) Das Pflanzenreich 25 (IV 36): 1–284

Bukowiecki H, Furmanowa M, Oledzka H (1972) The numerical taxonomy of Nymphaeaceae Bentham et Hooker. Acta Polon Pharm 29: 319–327

Burger WC (1978) The Piperales and the monocots. Alternative hypothesis for the origin of monocotyledonous flowers. Bot Rev 43: 345–393

Burger WC (1981) Heresay revived: The monocot theory of angiosperm origin. Evol Theory (Chicago) 5: 189–225

Burkill IH (1960) The organography and the evolution of Dioscoreaceae, the family of the yams. Bot J Linn Soc 56: 319–412

Burtt BL (1977a) Classification above the genus, as exemplified by Gesneriaceae, with parallels from other groups. Pl Syst Evol Suppl 1: 97–109

Burtt BL (1977b) *Curcuma zedoaria.* Gard Bull (Singapore) 30: 59–62

Burtt BL (1977c) The nomenclature of turmeric and other Ceylon Zingiberaceae. Notes R Bot Gard Edinb 35: 209–215

Burtt BL (1980) Cardamoms and other Zingiberaceae in Hortus Malabaricus, pp 139–148. *In:* Manilal KS (ed) Botany and History of Hortus Malabaricus. Rotterdam

Buxbaum F (1944) Untersuchungen zur Morphologie der Kakteenblüte. Bot Arch 45: 190–247

Buxbaum F (1961) Vorläufige Untersuchungen über Umfang systematische Stellung und Gliederung der Caryophyllales (Centrospermae). Beitr Biol Pfl 36: 1–56

Calder DM (1983) Mistletoes in Focus: An Introduction, pp 1–18. *In:* Calder DM, Bernhardt P (eds) The Biology of Mistletoes, pp 1–348. Harcourt Brace Jovanovich, Academic Press, Sydney Australia

Camp WH, Hubbard MM (1963) Vascular supply and structure of the ovule and aril in peony and of the aril in nutmeg. Am J Bot 50: 174–178

Canright JE (1962) Comparative morphology of pollen of Annonaceae. Pollen Spores (Paris) 4: 338–339

Carlquist S (1960) Anatomy of Guayana Xyridaceae: *Abolboda, Orectanthe* and *Achlyphila.* Mem NY Bot Gard 10: 65–117

Carlquist S (1964) Pollen morphology of Sarcolaenaceae. Brittonia 16: 231–254

Carlquist S (1966) Anatomy of Rapateaceae—Roots and stems. Phytomorphology 16: 17–38

Carlquist S (1969) Studies in Stylidiaceae: New taxa, field observations, evolutionary tendencies. Aliso 7: 13–64

Carlquist S (1976a) Wood anatomy of Byblidaceae. Bot Gaz 137: 35–38

Carlquist S (1976b) Wood anatomy of Roridulaceae, with ecological and phylogenetic implications. Am J Bot 63: 1003–1008

Carlquist S (1976c) Wood anatomy and relationships of the Geissolomataceae. Bull Torrey Bot Club 102: 128–134

Carlquist S (1976d) *Alexgeorgia*, a bizarre new genus of Restionaceae from Western Australia. Aust J Bot 24: 281–295

Carlquist S (1977a) A revision of Grubbiaceae. J S Afr Bot 43: 115–128

Carlquist S (1977b) Wood anatomy of Grubbiaceae. J S Afr Bot 43: 129–144

Carlquist S (1979) *Stylidium* in Arnhem Land: New species, modes of speciation on the sandstone plateau, and comments on floral mimicry. Aliso 9: 411–461

Carlquist S (1985) Wood anatomy of Coriariaceae: Phylogenetic and ecological implications. Syst Bot, New York 10: 174–183

Carlquist S, De Buhr L (1977) Wood anatomy of Penaeaceae (Myrtales): Comparative phylogenetic and ecological implications. Bot J Linn Soc 75: 211–227

Carolin RC (1982) The trichomes of the Chenopodiaceae and Amaranthaceae. Bot Jb 103: 451–466

Carolin RC (1987) A review of the family Portulacaceae. Aust J Bot 35: 383-412

Caspers N, Caspers L (1976) Zür Oberflaechenskulpturierung der Pollinien Mediterraner Orchis- und Ophyris-arten. Pollen Spores (Paris) 18: 203–215

Chaudefaud M, Emberger L (eds) (1960) Traité de Botanique Systematique. Vol 2: 1–1539. Masson et Cie, Paris

Chakravarty HL (1966) Monograph on the Cucurbitaceae of Iraq. Tech Bull No. 133. Mins Agric, Baghdad

Challice JS (1974) Rosaceae chemotaxonomy and the origin of the Pomoideae. Bot J Linn Soc 69: 239–259

Chanda S (1966) On the pollen morphology of the Centrolepidaceae, Restionaceae and Flagellariaceae with special reference to taxonomy. Grana Palynol 6: 355–415

Chanda S (1969) A contribution to the palynotaxonomy of Casuarinaceae, pp 191–208. *In:* Santapau H, Ghosh AK, Roy SK, Chanda S, Chaudhuri SK (eds) J Sen Memorial Volume. J Sen Memorial Committee and Botanical Society of Bengal, Calcutta

Chandler MEJ (1964) The Lower Tertiary Floras of Southern England. 4. A Summary and Survey of Findings in the light of Recent Botanical Observations, pp 1–150. Br Museum (Natu Hist), London

Chandrasekharan A (1974) Megafossil flora from the Genesse locality, Alberta (Canada). Palaeontographica B Palaeophytol 147: 1–41

Cheadle VI (1942) The occurrence and types of vessels in the various organs of the plant in the Monocotyledoneae. Am J Bot 29: 441–450

Cheadle VI (1968) Vessels in Haemodorales. Phytomorphology 18: 412–420

Cheadle VI (1969) Vessels in Amaryllidaceae and Tecophylaeaceae. Phytomorphology 19: 8–16

Cheadle VI, Kosakai H (1971) Vessels in Liliaceae. Phytomorphology 21: 320–333

Chen J, Meeuse BJD (1971) Production of free indole by some aroids. Acta Bot Neerl 20: 627–635

Chopra RN, Nayar SL, Chopra IC (1956) Glossary of Indian Medicinal Plants, pp 1–330. Council Sci Indus Res (CSIR), New Delhi

Chupov VS (1978) The comparative immunoelectrophoretic investigations of pollen proteins of some Amentiferous taxa (in Russian). Bot Zh (Moscow, Leningrad) 63: 1579–1584

Clifford HT, Harborne JB (1969) Flavonoid pigmentation in the sedges, Cyperaceae. Phytochem 8: 123–126

Clifford HT, Williams WT (1980) Interrelationships amongst the Liliatae: A Graph Theory approved. Aust J Bot 28: 261–268

Cochrane TS (1978) *Podandrogyne formosa* (Capparidaceae), a new species from Central America. Brittonia 30: 405–410

Conran JG (1988) Embryology and possible relationships of *Petermannia cirrosa* (Petermanniaceae). Nordic J Bot 8: 13–17

Conran JC, Clifford HT (1986) Smilacaceae. *In:* George AS (ed) Flora of Australia. Govt Printers, Canberra 46: 180–196

Copeland HF (1935) The structure of the flower of *Pholisma arenarium.* Am J Bot 22: 366–383

Copeland HF (1953) Observations on the Cyrillaceae particularly on the reproductive structures of the North American species. Phytomorphology 3: 405–411

Copeland HF (1955) The reproductive structures of *Pistacia chinensis* (Anacardiaceae). Phytomorphology 5: 440–448

Cordell GA (1974) The biosynthesis of indole-alkaloids. Lloydia 37: 219–298

Corner EJH (1946) Centrifugal stamens. J Arnold Arbor 27: 423–437

Corner EJH (1964) The Life of Plants, pp 1–315. Oxford Univ Press, London

Corner EJH (1976) The Seeds of Dicotyledons. Vol 1; pp 1–311, Vol 2: pp 1–552. Univ Press, Cambridge London New York Melbourne

Corsi G, Pagni AM (1990) The glandular hairs of *Valeriana officinalis* subsp. *collina.* 1. Some unusual features in their development and differentiation. Bot J Linn Soc 104: 381–388

Crane PR (1978) Angiosperm leaves from the lower Tertiary of Southern England Cour Forsch.—Inst Senckenberg 30: 126–136

Crane PR, Stockey RA (1985) Growth and reproductive biology of *Joffrea speirsii* gen. et sp. nov., A *Cercidiphyllum*-like plant from the Late Paleocene of Alberta, Canada. Can J Bot 63: 340–364

Crane PR, Friis Else Marie, Kaj RP (1995) The origin and early diversification of angiosperms. Nature 374: 27–33

Cribb AB, Cribb JW (1975) Wild Food in Australia. W. Collins, Sydney, Australia

Croizat L (1940) Notes on Dilleniaceae and their allies: Austrobaileyeae subfam. nov. J Arnold Arbor 21: 397–404

Cronquist A (1968) The Evolution and Classification of Flowering Plants, pp 1–396. Thomas Nelson, London Edinburgh

Cronquist A (1981) An Integrated System of Classification of Flowering Plants, pp 1–1262. Columbia Univ Press, New York

Cronquist A (1982) Basic Botany, pp 1–536. Harper and Row, Cambridge (USA)

Cronquist A (1983) Some realignments in the dicotyledons. Nordic J Bot 3: 75–83

Culvenor CCJ (1978) Pyrrolizidine alkaloids—occurrence and systematic importance in angiosperms. Bot Notiser 131: 473–486

Cutler DF (1969) Juncales. *In:* Melcalfe CR (ed) Anatomy of the Monocotyledons. Oxford Univ Press, London

Cutler DF, Shaw KHA (1965) Anarthriaceae and Ecdeiocoleaceae: Two new Monocotyledonous families separated from the Restionaceae. Kew Bull 19: 489–499

Dahlgren G (1989) An updated angiosperm classification. Bot J Linn Soc 100: 197–203

Dahlgren R (1967a) Studies on Penaeaceae. 3. The genus *Glischrocolla.* Bot Notiser 120: 57–68

Dahlgren R (1967b) Studies on Penaeaceae. 4. The genus *Endonema.* Bot Notiser 120: 68–83

Dahlgren R (1971) Studies on Penaeaceae. 6. The genus *Penaea* L. Op bot čech 29: 1–58

Dahlgren R (1975a) A system of classification of the angiosperms to be used to demonstrate the distribution of characters. Bot Notiser 128: 119–147

Dahlgren R (1975b) The distribution of characters within an angiosperm system. 1. Some embryological characters. Bot Notiser 128: 181–197

Dahlgren R (1977a) A commentary on a diagrammatic presentation of the angiosperms in relation to the distribution of character states. Pl Syst Evol Suppl 1: 253–283

Dahlgren R (1977b) A note on the taxonomy of the 'Sympetalae' and related groups. Bull Cairo Univ Herb 7–8: 83–102

Dahlgren R (1980a) A revised system of classification of the angiosperms. Bot J Linn Soc 80: 91–124

Dahlgren R (1980b) The taxonomic significance of chlorophyllous embryos in angiosperm seeds. Bot Notiser 133: 337–341

Dahlgren R (1983a) General aspects of angiosperm evolution and macrosystematics. Nordic J Bot 3: 119–149

Dahlgren R (1983b) The importance of modern serological research for angiosperm classification, pp 371–394. *In:* Jensen U, Fairbrothers DE (eds) Protein and Nucleic Acids in Plant Systematics, pp 1–408. Springer, Berlin Heidelberg New York Tokyo

Dahlgren R, An-ming Lu (1985) *Campynemanthe* (Campynemataceae): Morphology, microsporogenesis, early ovule ontogeny and relationships. Nordic J Bot 5: 321–330

Dahlgren R, Clifford HT (1982) The Monocotyledons—A Comparative Study, pp 1–378. Academic Press, London

Dahlgren R, Rao VS (1969) A study of the family Geissolomataceae. Bot Notiser 122:207–227

Dahlgren R, Thorne RF (1984) The order Myrtales: Circumscription, variation and relationships. Ann Mo Bot Gard 71: 633–699

Dahlgren R, Clifford HT, Yeo PF (1985) The Families of the Monocotyledons, pp 1–520. Springer, Berlin

Dahlgren R, Jensen SR, Nielsen BJ (1976) Iridoid compounds in Fouquieriaceae and notes on its possible affinities. Bot Notiser 129: 207–212

Dahlgren R, Jensen SR, Nielsen BJ (1981) A revised classification of the angiosperms with comments on the correlation between chemical and other characters, pp 149–204. *In:* Young DA, Seigler DS (eds) Phytochemistry and Angiosperm Phylogeny. Praeger Scientific, New York

Danser BH (1929) On the taxonomy and nomenclature of the Loranthaceae of Asia and Australia. Bull Jard Bot Buitenz 19: 291–373

D'Arcy WG (1986) Taxonomy and biogeography, pp 1–4. *In:* D'Arcy WG (ed) Solanaceae: Biology and Systematics. Columbia Univ Press, New York

Darwin C (1859) The Origin of Species. J Murray, London

Dathan ASR, Singh D (1972) Development of embryo sac and seed of *Bixa* L. and *Cochlospermum* Kunth. J Indian Bot Soc 51: 254–266

Datta RM, Mitra JN (1947) The systematic position of the family Moringaceae based on a study of *Moringa pterygosperma* Gaertn (*M. oleifera* Lam.). J Bombay Natu Hist Soc 47: 355–357

Davidson C (1973) An anatomical and morphological study of Datiscaceae. Aliso 8: 49–110

Davis GL (1966) Systematic Embryology of the Angiosperms, pp 1–526. John Wiley New York London Sydney

Davis TS, Tomlinson PB (1974) A new species of *Ruppia* in high salinity in Western Australia. J Arnold Arbor 55: 59–66

Dawson ML (1936) The floral morphology of Polemoniaceae. Am J Bot 23: 501–511

De Beer GR (1958) Embryos and Ancestors. Oxford (UK)

De Buhr LE (1975) Phylogenetic relationships of the Sarraceniaceae. Taxon 24: 297–306

De Buhr LE (1977) Wood anatomy of the Sarraceniaceae: Ecological and evolutionary implications. Pl Syst Evol 128: 159–169

Decker JM (1966) Wood anatomy and phylogeny of Luxemburgiaceae (Ochnaceae). Phytomorphology 16: 39–55

Dement WA, Mabry TJ (1972) Flavonoids of North American species of *Thermopsis*. Phytochem 11: 1089–1093

Dement WA, Raven PH (1973) Distribution of the chalcone isosalipurposide in the Onagraceae. Phytochem 12: 807–808

Den Hartog C (1970) *Odinea,* a new genus of Nymphaeaceae. Blumea 18: 413–416

Den Hartog RM (neé Van Tholen T), Baas P (1978) Epidermal characters of the Celastraceae sensu lato. Acta Bot Neerl 27: 355–388

Den Outer RW, Van Veenendaal WLH (1981) Wood and bark anatomy of *Azima tetracantha* Lam. (Salvadoraceae) with description of its included phloem. Acta Bot Neerl 30: 199–207

Dermen H (1931) A study of chromosome number in two genera of Berberidaceae, *Mahonia* and *Berberis.* J Arnold Arbor 12: 281–287

Dermen H (1932) Cytological studies of *Cornus*. J Arnold Arbor 13: 410–415

De Sloover J–L (1961) Note sur le pollen d'*Lemna minor* L. Pollen Spores (Paris) 3: 5–10

Deshpande PK (1970) Compositae. *In:* Proc Symp Comparative Embryology of Angiosperms. Bull Indian Natl Sci Acad No. 41: 325–333

De Wit HCD (1963) Plants of the World: The Higher Plants. Vol 2: 1–340. Thames and Hudson, London

Dickison WC (1967a) Comparative morphological studies in Dilleniaceae. 1. Wood anatomy. J Arnold Arbor 48: 1–29

Dickison WC (1967b) Comparative morphological studies in Dilleniaceae. 2. Pollen. J Arnold Arbor 48: 231–240

Dickison WC (1975) Studies on the floral anatomy of the Cunoniaceae. Am J Bot 62: 433–447

Dickison WC (1978) Comparative anatomy of Eucryphiaceae. Am J Bot 65: 722–735

Dickison WC (1981) Contributions to the morphology and anatomy of *Strasburgeria* and a discussion of the taxonomic position of the Strasburgeriaceae. Brittonia 33: 564–580

Dickison WC, Nowicke JW, Skvarla JJ (1982) Pollen morphology of the Dilleniaceae and Actinidiaceae. Am J Bot 69: 1055–1073

Dickson J (1936) Studies in floral anatomy. 3. An interpretation of the gynoecium in the Primulaceae. Am J Bot 23: 385–393

Diels L (1914) Diapensiaceen—Studien. Bot Jb 50 (suppl): 304–330

Diels L (1930a) Byblidaceae. *In:* Engler A, Prantl K (eds) Die natürlichen Pflanzenfamilien. Edn 2. 18a:286–288

Diels L (1930b) Roridulaceae. *In:* Engler A, Prantl K (eds) Die natürlichen Pflanzenfamilien. Edn 2. 18a: 346–348

Dnyansagar VR (1970) Leguminosae. *In:* Proc Symp Comparative Embryology of Angiosperms. Bull Indian Natl Sci Acad No. 41: 93–103

Dollo L (1893) Les lois de l'évolution. Bull Soc Belg Géol 7: 164–166

Don G (1838) General history of Dichlamydeous plants, vol 4. London

Douglas GE (1936) Studies in the vascular anatomy of the Primulaceae. Am J Bot 23: 199–212

Dorasami LS, Gopinath DM (1945) An embryological study of *Linum mysorens* Hyn. Proc Indian Acad Sci (Pl Sci) 22: 6–9

Dring JV, Kite GC, Nash RJ, Reynolds T (1995) Chemicals in aroids: A survey, including new results for polyhydroxy alkaloids and alkylresorcinols. Bot J Linn Soc 117: 1–12

Drude O (1889) Pirolaceae. *In:* Engler A, Prantl K (eds) Die natürlichen Pflanzenfamilien. Edn 1. 4(1): 3–11

Drude O (1897–1898) Umbelliferae. *In:* Engler A, Prantl K (eds) Die natürlichen Pflanzenfamilien. Edn 1. 3(8): 63–250

Dulien D (1973) Étude morphologique de la surface pollinique de *Ponthieve maculata* Lindl. Orchidaceae en microscopie électronique a balagage. Adansonia 13:229–234

Dunbar A (1975) On pollen of Campanulaceae and related families with special reference to the surface ultrastructure. Bot Notiser 128: 73–101

Dunbar A (1978) Pollen morphology and taxonomic position of the genus *Pentaphragma* Wall. (Pentaphragmataceae). Grana 17: 141–147

Dutt BSM (1970a) Haemodoraceae. *In:* Proc Symp Comparative Embryology of Angiosperms. Bull Indian Natl Sci Acad No. 41: 358–361

Dutt BSM (1970b) Cyanastraceae. *In:* Proc Symp Comparative Embryology of Angiosperms. Bull Indian Natl Sci Acad No. 41: 362–364

Dutt BSM (1970c) Amaryllidaceae. *In:* Proc Symp Comparative Embryology of Angiosperms. Bull Indian Natl Sci Acad No. 41: 365–367

Dutt BSM (1970d) Hypoxidaceae. *In:* Proc Symp Comparative Embryology of Angiosperms. Bull Indian Natl Sci Acad No. 41: 368–372

Dutt BSM (1970e) Velloziaceae. *In:* Proc Symp Comparative Embryology of Angiosperms. Bull Indian Natl Sci Acad No. 41: 373–374

Dutta SC (1988) Systematic Botany, pp 1-645. Wiley Eastern, New Delhi Bangalore Bombay Calcutta Hyderabad Madras

Dutta SC, Biswas KK (1967) Form-variation in *Ottelia alismoides* (L.) Pers. Brotéria 36: 63–69

Eames AJ (1961) Morphology of the Angiosperms, pp 1–518. McGraw-Hill, New York Toronto London

Earle FR, Glass GA, Gessinger GC, Wolff IA, Bagby MO, Jones A (1960) Search for new industrial oils. J Am Oil Chem Soc 37:440–447

Eastwood FW, Kirson I, Lavie D, Abraham A (1980) New withanolides from a cross of a South African Chemotype by Chemotype II (Israel) in *Withania somniferum.* Phytochem 19: 1503–1507

Eber E (1934) Karpelbau und Plazentationsverhältnisse in der Reiche der Helobiae. Flora (Jena, Germany) 127: 273–330

Eckardt T (1964) Centrospermae. *In:* Melchior H (ed) A Engler's Syllabus der Pflanzenfamilien. 2: 79–101 Borntraeger, Berlin

Eckardt T (1967) Vergleich von *Dysphania* mit *Chenopodium* und mit Illecebraceae. Bauhinia 3: 327–344

Eckardt T (1968) Zur Blütenmorphologie von *Dysphania plantaginella* F. vM. Phytomorphology 17: 165–172

Eckardt T (1976) Classical morphological features of Centrospermous families. Pl Syst Evol 126: 5–25

Edlin HLA (1935) A critical revision of certain taxonomic groups of the Malvales. New Phytol 34: 1–20, 122–143

Eglinton G, Hamilton RJ (1963) The distribution of alkanes, pp 187–217. *In:* Swain T (ed) Chemical Plant Taxonomy, pp 1–543. Academic Press, London New York

Ehrendorfer F (1976a) Chromosome numbers and differentiaion of Centrospermous families. Pl Syst Evol 126: 27–30

Ehrendorfer F (1976b) Closing remarks: Systematics and Evolution of Centrospermous families. Pl Syst Evol 126: 99–105

Ehrendorfer F (1977) New ideas about the early differentiation of angiosperms. Pl Syst Evol Suppl 1: 227–234

Ehrendorfer F (1983a) Summary statement—New evidence of relationships and modern systems of classifications of the angiosperms. Nordic J Bot 3:151–155

Ehrendorfer F (1983b) Angiospermae, pp 796–915. *In:* Denffer DV et al. (eds) Lehrbuch der Botanik. 32nd Edn. Fischer, Stuttgart New York

Ehrendorfer F, Krendle F, Habeler E, Sauer W (1968) Chromosome number and evolution in primitive angiosperms. Taxon 17: 337–353

Ehrendorfer F, Morawetz W, Dawe J (1984) The neotropical angiosperm families Brunelliaceae and Caryocaraceae: First karyosystematical data and affinities. Pl Syst Evol 145: 183–191

Ekambaram T, Panje RR (1935) Contributions to our knowledge of *Balanophora* 2: Life history of *B. dioica.* Proc Indian Acad Sci (Pl Sci) 12B: 522–542

Endress PK (1977a) Evolutionary trends in the Hamamelidales-Fagales group. Pl Syst Evol Suppl 1: 321–347

Endress PK (1977b) Über Blütenbau und Verwandtschaft der Eupomatiaceae und Himantandraceae. Ber Deutsch Bot Ges 90: 83–103

Endress PK (1979) Noncarpellary pollination and "hyperstigma" in an angiosperm (*Tambourissa religiosa*, Monimiaceae). Experientia 35 : 45

Endress PK (1980a) The reproductive structures and systematic position of the Austrobaileyaceae. Bot Jb 101: 393–433

Endress PK (1980b) Floral structure and relationships of *Hortonia* (Monimiaceae). Pl Syst Evol 133: 199–221

Endress PK (1980c) Ontogeny, function and evolution of extreme floral construction in Monimiaceae. Pl Syst Evol 134: 79–120

Endress PK (1983) Dispersal and distribution in some small archaic relic angiosperm families (Austrobaileyaceae, Eupomatiaceae, Himantandraceae, Idiospermoideae—Calycanthaceae). Sonderbd Naturwiss (Hamburg) 7: 201–217

Endress PK (1986) Floral structure, systematics and phylogeny in Trochodendrales. Ann Mo Bot Gard 73: 297–324

Endress PK (1987) The Chloranthaceae: Reproductive structures and phylogenetic position. Bot Jb 109: 153–226

Endress PK, Honegger R (1980) The pollen of the Austrobaileyaceae and its phylogenetic significance. Grana 19: 177–182

Endress PK, Sampson FB (1983) Flower structure and relationships of Trimeniaceae (Laurales). J Arnold Arbor 64: 447–473

Engel T, Barthlott W (1988) Epicuticular waxes of Centrosperms. Pl Syst Evol 161: 71–85

Engler A (1909) Syllabus der Pflanzenfamilien. Edn 6. Borntraeger, Berlin

Engler A (1925a) Quiinaceae. *In:* Engler A, Prantl K (eds) Die natürlichen Pflanzenfamilien. Edn 2. 21: 106–108

Engler A (1925b) Guttiferae. *In:* Engler A, Prantl K (eds) Die natürlichen Pflanzenfamilien. Edn 2. 21: 154–237

Engler A (1930a) Podostemonaceae. *In:* Engler A, Prantl K (eds) Die natürlichen Pflanzenfamilien. Edn 2. 18a: 3–68

Engler A (1930b) Saxifragaceae. *In:* Engler A, Prantl K (eds) Die natürlichen Pflanzenfamilien. Edn 2. 18a: 74–225

Engler A (1930c) Flagellariaceae. *In:* Engler A, Prantl K (eds) Die natürlichen Pflanzenfamilien. Edn 2. 15a: 6–8

Engler A, Diels L (1936) Syllabus der Pflanzenfamilien, pp 1–419. Gebruder Borntraeger, Berlin

Engler A, Gilg E (1924) Syllabus der Pflanzenfamilien, pp 1–420. Gebruder Borntraeger, Berlin

Engler A, Melchior H (1925) Medusagynaceae. *In:* Engler A, Prantl K (eds) Die natürlichen Pflanzenfamilien. Edn 2. 21: 50–52

Eramian EN (1971) Palynological data on systematics and phylogeny of the Cornaceae Dumort. and related families, pp 235–272. *In:* Kuprianova LA, Yakovlev MS (eds) Morphology of the pollen of Cucurbitaceae, Thymeleaceae and Cornaceae. "Nauka", Leningrad

Erber C, Leins P (1982) Zur spirale in Magnolien-Blüten. Beitr Biol Pfl 56: 225–241

Erber C, Leins P (1983) Zur Sequenz von Blütenorganen bei einigen Magnoliiden. Bot Jb 103: 433–449

Erdtman G (1952) Pollen Morphology and Plant Taxonomy, pp 1–539. Chronica Botanica, Waltham, Mass., USA

Erdtman G (1954) Pollen morphology and plant taxonomy. Bot Notiser 1954: 65–81

Erdtman G (1966) Pollen morphologische Notizen über einige Blütenpflanzen incertae sedis. Bot Notiser 119: 160–168

Erdtman G, Singh G (1957) On the pollen morphology of *Sclerosperma manii*. Bull Jard Bot Buitenz 27: 217–220

Erdtman G, Leins P, Melville R, Metcalfe CR (1969) On the relationships of *Emblingia*. Bot J Linn Soc 62: 169–186

Erdtman H (1963) Some aspects of chemotaxonomy, pp 89–125. *In:* Swain T (ed) Chemical Plant Taxonomy, pp 1–543. Academic Press, London New York

Ernst WR, Thompson HJ (1963) The Loasaceae in the south-eastern United States. J Arnold Arbor 44: 138–142

Ettlinger MG, Kjaer A (1968) Sulfur compound in plants: Recent Advances in Phytochemistry 1: 59–144

Evans CS, Bell EA (1978) Uncommon amino acids in the seeds of 64 species of Caesalpinieae. Phytochem 17: 1127–1129

Evans WC (1986) Hybridization and secondary metabolism in the Solanaceae, pp 179–186. *In:* D'Arcy WG (ed) Solanaceae: Biology and Systematics, pp 1–603. Columbia Univ Press, New York

Eyde RH (1963) Morphological and palaeobotanical studies of the Nyssaceae. 1. A survey of the modern species and their fruits. J Arnold Arbor 44: 1–60

Eyde RH (1966) Systematic anatomy of the flower and fruit of *Corokia*. Am J Bot 53: 833–847

Eyde RH (1967) The peculiar gynoecial vasculature of Cornaceae and its systematic significance. Phytomorphology 17: 172–182

Fagerlind F (1947) Die systematische Stellung der Familie Grubbiaceae. Svensk Bot Tidskr 41: 315–320

Fagerlind F (1948) Beiträge zur kenntnis der Gynäceummorphologie und phylogenie der Santalales-Familien. Svensk Bot Tidskr 42: 195–229

Fahselt D (1971) Flavonoid components of *Dicentra canadensis* (Fumariaceae). Can J Bot 49: 1559–1563

Fahselt D, Ownbey M (1968) Chromatographic comparison of *Dicentra* species and hybrids. Am J Bot 55: 334–345

Fairbairn JW, El-Muhtadi FJ (1972) Chemotaxonomy of anthraquinones in *Rumex*. Phytochem 11: 263–268

Fairbairn JW, Williamson EM (1978) Meconic acid as a chemotaxonomic marker in the Papaveraceae. Phytochem 17: 2087–2089

Fairbrothers DE (1968) Chemosystematics, with emphasis on systematic serology, pp 141–174. *In:* Heywood VH (ed) Modern Methods in Plant Taxonomy, pp 1–312. Academic Press, New York

Fairbrothers DE (1983) Evidence from nucleic acid and protein chemistry, in particular serology, in angiosperm classification. Nordic J Bot 3: 35–41

Fairbrothers DE, Johnson MA (1964) Comparative serological studies within the families Cornaceae (dogwood) and Nyssaceae (sour gum), pp 305–318. *In:* Leone CA (ed) Taxonomic Biochemistry and Serology. Ronald Press, New York

Fairbrothers DE, Peterson FD (1983) Serological investigation of the Annoniflorae (Magnoliiflorae, Magnoliidae), pp 301–310. *In:* Jensen U, Fairbrothers DE (eds) Protein and Nucleic Acids in Plant Systematics, pp 1-408. Springer, Berlin Heidelberg New York Tokyo

Fallen ME (1985) The gynoecial development and systematic position of *Allamanda* (Apocynaceae). Am J Bot 72: 572–579

Fallen ME (1986) Floral structure in Apocynaceae: Morphological, functional and evolutionary aspects. Bot Jb 106: 245–286

Favre-Duchartre M (1984) Homologies and phylogeny, pp 697–734. *In:* Johri BM (ed) Embryology of Angiosperms, pp 1–830. Springer, Berlin Heidelberg New York Tokyo

Fedde F (1936) Papaveraceae Unterfamilie Fumarioideae. *In:* Engler A, Prantl K (eds) Die natürlichen Pflanzenfamilien. Edn 2. 17b: 121–145

Fenzl E (1841) Die Gattung *Tetradiclis* Steven und ihre Stellung im natürlichen Systeme. Linnaea 15: 289–299

Ferguson IK (1966a) The genera of Caprifoliaceae in South eastern United States. J Arnold Arbor 47: 33–59

Ferguson IK (1966b) Notes on the nomenclature of *Cornus*. J Arnold Arbor 47: 100–105

Ferguson IK (1966C) The Cornaceae in the south-eastern United States. J Arnold Arbor 47: 106–116

Ferguson IK (1977) Cornaceae. *In:* Nilsson S (ed) World pollen and spore flora 6: Angiospermae, pp 1–34. Almquist and Wiksell, Stockholm

Ferguson IK, Dransfield J, Page FC, Thanikaimoni G (1983) Notes on the pollen morphology of *Pinanga* with special reference to *P. aristata* and *P. pilosa* (Palmae: Arecoideae). Grana 22: 65–72

Feuer S (1981) Pollen morphology and relationships of the Myzodendraceae. Nordic J Bot 1: 731–734

Feuer S, Kuijt J (1978) Fine structure of mistletoe pollen, 1. Eremolepidaceae, *Lepidoceras* and *Tupeia*. Can J Bot 56: 2853–2864

Feuer S, Kuijt J (1979) Pollen morphology and evolution in *Psittacanthus* (Lorantháceae). Bot Notiser 132: 295–309

Fey BS, Endress PK (1983) Development and morphological interpretation of the cupule in Fagaceae. Flora 173: 451–468

Finet A, Gagnepain F (1905) Contribution à la flore de l'Asie orientale (Fasc 2). Bull Soc Bot Fr 52: Mém 4: Magnoliacées, pp 23–54

Fischer MJ (1928) The morphology and anatomy of flowers of Salicaceae. 1. Am J Bot 15:307–326

Fjodorow AA (1954) *Sphenoclea*. *In*: Flora der USSR 24:449–450

Flory WS (1977) Overview of chromosome evolution in the Amaryllidaceae. Nucleus 20:70–88

Forman LL (1965) A new genus of Ixonanthaceae with notes on the family. Kew Bull 19:517–526

Forman LL (1966) The reinstatement of *Galearia* Zoll. et Mor. and *Microdesmis* Hook. f. in Pandaceae. Kew Bull 20:309–321

Foster AS (1950) Morphology and venation of the leaf in *Quiina acutangula*. Am J Bot 37:159–171

Foster AS (1951) Heterophily and foliar venation in *Lacunaria*. Bull Torrey Bot Club 78:382–400

Franz E (1908) Beiträge zur Kenntnis der Portulacaceen und Basellaceen. Bot Jb 97:1–46

Fritsch FE (1908) The anatomy of the Julianiaceae considered from the systematic point of view. Trans Linn Soc London Bot Ser 2, 7:129–151

Gadella WJ (1966) Some cytological observations in the Loganiaceae. 3. Acta Bot Neerl 15:490–491

Gagnepain F, Boureau E (1946) Une nouvelle famille de Gymnosperms: Sarcopodaceae. Bull Soc Bot Fr 93: 313–320

Ganders FR (1979) Heterostyly in *Erythroxylon coca* (Erythroxylaceae). Bot J Linn Soc 78:11–20

Gangulee HC, Das KS, Datta CT (1972) College Botany. Vol I, pp 1–920. New Central Book Agency, Calcutta

Garay LA (1960) On the origin of the Orchidaceae. Bull Mus Leafl Harv Univ 19:57–96

Garbarino JA, Chamy MC, Gambaro V (1986) Labdane diterpenoids from *Nolana rostrata*. Phytochem 25: 2833–2836

Garcia V (1962) Embryological studies in the Loasaceae with special reference to endosperm haustoria, pp 157–161. *In*: Symp Plant Embryology, Council Sci Indus Res (CSIR), pp1–261. New Delhi

Gardner RO (1975) A survey of the distribution of binucleate and trinucleate pollen in New Zealand Flora. N Z J Bot 13:361–366

Gardner RO (1976) Binucleate pollen in *Triglochin* L. N Z J Bot 14:115–116

Garg M (1981) Pollen morphology and systematic position of *Coriaria*. Phytomorphology 30:5–10

Garrat GA (1933) Bearing of wood anatomy on the relationships of the Myristicaceae. Trop Woods 36:20–44

Geernick D (1969) Genere des Haemodoraceae et des Hypoxidaceae. Bull Jard Bot Natt Belg 39:47–82

George AS (1982) Gyrostemonaceae. *In*: George AS (ed) Flora of Australia, Lecythidales to Batales. 8:362–379

Gershenzon J, Mabry TJ (1983) Secondary metabolites and the higher classification of angiosperms. Nordic J Bot 3:5-34

Giannasi DE (1978) Generic relationships in the Ulmaceae based on flavonoid chemistry. Taxon 27:331–344

Giannasi DE, Niklas KJ (1977) Flavonoid and other chemical constituents of fossil Miocene *Celtis* and *Ulmus*. Science 197:765–767

Gibbs RD (1958) Chemical evolution in plants. Bot J Linn Soc 56:49–57

Gibbs RD (1963) History of chemical taxonomy, pp 41–88. *In*: Swain T (ed) Chemical Plant Taxonomy, pp 1–543. Academic Press, London New York

Gibbs RD (1974) Chemotaxonomy of Flowering Plants, Vol I:1–680; Vol II:681–1274: Vol III:1275–1982; Vol IV: 1983–2372. McGill-Queen's Univ Press, Montreal

Gilg E (1914) Zür Frage der Verwandschaft der Salicaceae mit den Flacourtiaceae. Bot Jb 50:424–434

Gill LS, Hawksworth FG (1961) The mistletoes: A literature review. Bull US Dep Agric No. 1242:1–87

Glad J (1976) Taxonomy of *Mentzelia mollis* and allied species (Loasaceae). Madroño (San Fransisco) 23:283–292

Glazner JT, Devlin B, Ellstrand NC (1988) Biochemical and morphological evidence for host race evolution in desert mistletoe, *Phoradendron californicum* (Viscaceae). Pl Syst Evol 161:13–21

Goldberg A (1986) Classification, evolution and phylogeny of the families of dicotyledons. Smithsonian Contr Bot 58:1–314

Goldblatt P (1971) Cytological and morphological studies in the Southern African Iridaceae. J S Afr Bot 37: 317–460

Goldblatt P (1976) Chromosome numbers and its significance in *Batis maritima* (Bataceae). J Arnold Arbor 57: 526–530

Goldblatt P, Endress PK (1977) Cytology and evolution in Hamamelidaceae. J Arnold Arbor 58:67–71

Goldblatt P, Gentry AH (1979) Cytology of Bignoniaceae. Bot Notiser 132:475–482

Gornall RJ, Bohn BA, Dahlgren R (1979) The distribution of flavonoids in the angiosperms. Bot Notiser 132:1–30

Gottlieb OR, Kalpan MAC, Kubitzki K, Barros Toledo JR (1989) Chemical dichotomies in the Magnolialean complex. Nordic J Bot 8:437–444

Gottwald H, Parameswaran N (1966) Das sekundäre Xylem der Familie Dipterocarpaceae, anatomische Untersuchungen zur Taxonomie und Phylogenie. Bot Jb 85:410–508

Gottwald H, Parameswaran N (1967) Beiträge zur Anatomie und Systematik der Quiinaceae. Bot Jb 87:361–381

Govil CM (1970) Fouquieriaceae. *In*: Proc Symp Comparative Embryology of Angiosperms. Bull Indian Natl Sci Acad No. 41:244–245

Graham SA, Graham A (1971) Palynology and systematics of *Cuphea* (Lythraceae). 2. Pollen morphology and infrageneric classification. Am J Bot 58:844–857

Graham S, Wood Jr CE (1965) The genera of Polygonaceae in the south-eastern United States. J Arnold Arbor 46:91–121

Gray A (1848) Remarks on the structure and affinities of the order Ceratophyllaceae. Ann Lyceum Natu Hist 4: 41–50

Grayum MH (1986) Phylogenetic implications of pollen nuclear number in the Araceae. Pl Syst Evol 151:145–161

Grund C, Jensen U (1981) Systematic relationships of the Saxifragales revealed by serological characteristics of seed proteins. Pl Syst Evol 137:1–22

Greger H, Ernet D (1973) Flavonoid–Muster, Systematik und Evolution bei *Valerianella*. Phytoechem 12:1693–1699

Greuter W (1981) "XIII International Botanical Congress". Taxon 30:904–912

Greuter W (1988) International Code of Botanical Nomenclature. Regnum Veg 118:1–328

Grootjen CJ, Bouman F (1988) Seed structure in Cannaceae: Taxonomic and ecological implications. Ann Bot (London) 61:363–371

Guédès M (1967) Stipules médianes et stipules ligulaires chez quelques Liliacées, Joncacées et Cypéracées. Beitr Biol Pfl 43:59–103

Guérin P (1923) Les Urticées: Cellules à mucilage, laticifères et canaux sécréteurs. Bull Soc Bot Fr 70: 125–136, 207–215, 255–263

Gundersen A (1927) The Frankeniaceae as a link in the classification of dicotyledons. Torreya 27:65–71

Gundersen A (1950) Families of Dicotyledons, pp 1–237. Chronica Bot, Waltham (Mass., USA)

Gurni AA, Kubitzki K (1981) Flavonoid chemistry and systematics of the Dilleniaceae. Biochem Syst Ecol 9: 109–114

Gurni AA, Konig WA, Kubitzki K (1981) Flavonoid glycosides and sulfates from the Dilleniaceae. Phytochem 20:1057–1059

Gutzwiller M-A (1961) Die phylogenetische Stellung von *Suriana maritima* L. Bot Jb 81:1–49

Haber TM (1959) The comparative anatomy and morphology of flowers and inflorescence of Proteaceae 1. Some Australian taxa. Phytomorphology 9:325–358

Haber TM (1961) The comparative anatomy and morphology of flowers and inflorescence of Proteaceae 2. Some American taxa. Phytomorphology 11:1–61

Hagerup O (1953) The morphology and systematics of the leaves in Ericales. Phytomorphology 3:459–464

Haines RW, Lye KA (1975) Seedlings of Nymphaeaceae. Bot J Linn Soc 70:255–265

Hallier H (1905) Provisional scheme of the natural (phylogenetic) system of flowering plants. New Phytol 4: 151–162

Hallier H (1912) L'origine et le système phylétique des Angiosperms exposés à l'aide de leur arbre gènéralogique. Arch Neerl Sci Exact Natu Ser 1:146–234

Hallier H (1923) Beiträge zur Kenntnis der Linaceae. Beih Bot Zbl 39:1–178

Hallock FA (1930) The relationship of *Garrya*: The development of the flowers and seeds of *Garrya* and its bearing on the phylogenetic position of the genus. Ann Bot (London) 44:771–813

Halse RR, Mishaga R (1988) Seed germination in *Sidalcea nelsoniana*. Phytologia 64:179–184

Hamann U (1961) Merkmalsbestand und Verwandtschaftsbeziehungen der 'Farinosae'. Willdenowia 2:639–768

Hamann U (1962a) Weiteres über Merkmalsbestand und Verwandtschaftsbeziehungen der 'Farinosae'. Willdenowia 3:169–207

Hamann U (1962b) Beiträge zur Embryologie der Centrolepidaceae mit Bemerkungen über den Bau der Blüten und Blütenstände und die systematische Stellg der Familie. Ber Deutsch Bot Ges 75:153–171

Hamann U (1964) Commelinales. *In*: Melchior HA (ed) Engler's Syllabus der Pflanzenfamilien 2:549–561

Handel-Mazzetti H (1932) Rhoiptelaeaceae, eine nene Familie der Monochlamydeae. Repert Spec Nov Reg Vég 30: 75–80

Hansen B (1972) Balanophoraceae. *In*: Smitinand T, Kai Larsen, Hansen B (eds) Flora Thailand. Pt 2:177–182

Hansen L, Boll PM (1986) Polyacetelenes in Araliaceae: Their chemistry, biosythesis and biological significance. Phytochem 25:285–293

Hara H (1981) A new species of the genus *Adoxa* from Mt Oci of China. J Jap Bot 56:271–274

Hara H (1983) A revision of Caprifoliaceae of Japan with reference to allied plants in other districts and to Adoxaceae, pp 1–360. Academic Scientific Book, Tokyo

Harborne JB (1963) Distribution of anthocyanins in higher plants, pp 359–388. *In*: Swain T (ed) Chemical Plant Taxonomy, pp 1–543. Academic Press, London New York

Harborne JB (1967a) Comparative Biochemistry of the Flavonoids, pp1–383. Academic Press, London New York

Harborne JB (1967b) Correlations between chemistry, pollen morphology and systematics in the family Plumbaginaeae. Phytochem 6:1415–1428

Harborne JB (1968) Correlations between flavonoid pigmentation and systematics in the family Primulaceae. Phytochem 7:1215–1230

Harborne JB (1975a) Flavonoid sulphates: A new class of sulphur compounds in higher plants. Phytochem 14: 1147–1155

Harborne JB (1975b) Flavonoid bisulphates and their co-occurrence with ellagic acid in the Bixaceae, Frankeniaceae and related families. Phytochem 14:1331–1337

Harborne JB, Green PS (1980) A chemotaxonomic survey of flavonoids in leaves of the Oleaceae. Bot J Linn Soc 81:155–167

Harborne JB, Turner BL (1984) Plant Chemosystematics, pp 1–562. Academic Press, London

Harborne JB, Williams CA, Wilson KL (1982) Flavonoids in leaves and inflorescences of Australian *Cyperus* species. Phytochem 21:2491–2507

Hardman R, Benjamin TV (1976) The co-occurrence of ecdyosomes with bufadienolides and steroidal saponins in the genus *Helleborus*. Phytochem 15:1515–1516

Harms (1935) Balanophoraceae. *In*: Engler A, Prantl K. (eds) Die natürlichen Pflanzenfamilien. Edn 2. 16b:296–339

Harris PJ, Hartley RD (1980) Phenolic constituents of the cell walls of monocotyledons. Biochem Syst Evol 8: 153–160

Hart H't, Koek-Noorman J (1989) The origin of woody Sedoideae (Crassulaceae). Taxon 38:535–544

Hartley RD, Harris PJ (1981) Phenolic constituents of the cell walls of dicotyledons. Biochem Syst Ecol 9:189–203

Hartley TG (1973) A survey of New Guinea plants for alkaloids. Lloydia 36:217–319

Hauman L (1951) Contribution a l'étude des *Chrysobalanoides africaines*. Bull Jard Bot L'Etat Bruxelles 21: 167–198

Hayden WJ (1977) Comparative anatomy and systematics of *Picrodendron*, genus incartae sedis. J Arnold Arbor 58:257–279

Hayden WJ, Gillis WT, Stone DE, Broome CR, Webster GL (1984) Systematics and palynology of *Picrodendron*: Further evidence for relationship with the Oldfieldioideae (Euphorbiaceae). J Arnold Arbor 65:105–127

Hedberg I (1961) Cytotaxonomic studies in *Anthoxanthum odoratum* s. lat. l. Morphologic analysis of herbarium specimens. Svensk Bot Tidskr 55:118–128

Hegelmaier F (1864) Monographie der Gatung *Callitriche*, pp 1–64 Verlag von Ebner and Seubert, Stuttgart

Hegnauer R (1963) The taxonomic significance of alkaloids, pp 389–427. *In*: Swain T (ed) Chemical Plant Taxonomy, pp 1–543. Academic Press, London

Hegnauer R (1964) Chemotaxonomie der Pflanzen. 3:1–743. Birkhauser, Basel Stuttgart

Hegnauer R (1969) Chemical evidence for the classification of some plant taxa, pp 121–138. *In*: Harborne JB, Swain T (eds) Perspective in Phytochemistry, pp 1–235. Academic Press. London

Hegnauer R (1971) Chemical patterns and relationships of the Umbelliferae, pp 267–277. *In*: Heywood VH (ed) The Biology and Chemistry of the Umbelliferae, pp 1–438. Academic Press, London

Hegnauer R (1973) Chemotaxonomie der Pflanzen. 6:1–882. Birkhauser, Basel Stuttgart

Hegnauer R (1977) Cyanogenic compounds as systematic markers in Tracheophytes. Pl Syst Evol Suppl 1:191–209

Heimerl A (1934) Achatocarpaceae. *In*: Engler A, Prantl K (eds) Die natürlichen Pflanzenfamilien. Edn 2. 16c: 174–178

Heimsch Jr C (1940) Wood anatomy and pollen morphology of *Rhus* and allied genera. J Arnold Arbor 21:279–291

Heimsch Jr C (1942) Comparative anatomy of the secondary system in the "Gruniales" and "Terebinthales" of Wettstein, with reference to taxonomic grouping. Lilloa 8:83–198

Heinig KH (1951) Studies in the floral morphology of the Thymelaeaceae. Am J Bot 38:113–132

Hempel Alice L, Patrick A, Reeves R, Olmstead G, Jansen RK (1995) Implications of rbcL sequence data for higher order relationships of the Loasaceae and the anomalous aquatic plant *Hydrostachys* (Hydrostachyaceae). Pl Syst Evol 194:25–37

Herr Jr JM (1984) Embryology and taxonomy, pp 647–696. *In*: Johri BM (ed) Embryology of Angiosperms, pp 1–830. Springer, Berlin Heidelberg New York Tokyo

Hershkovitz MA (1989) Phylogenetic studies in Centrospermae: A brief appraisal. Taxon 38:602–610

Hickey LJ (1973) Classification of the architecture of dicotyledonous leaves. Am J Bot 60:17–33

Hickey LJ, West RM, Dawson MR, Choi DK (1983) Arctic terrestrial biota: Palaeomagnetic evidence of age disparity with mid-Northern latitudes during the late Cretaceous and early Tertiary. Science 221:1153–1156

Hideux MJ, Ferguson IK (1976) The stereostructure of the exine and its evolutionary significance in Saxifragaceae sensu lato, pp 327–377. *In*: Ferguson IK, Müller J (eds) The Evolutionary Significance of the Exine. pp 1–591. Linn Symp Ser I, Academic Press London New York

Hiepko P (1965) Vergleichende–morphologische und entwicklungsgeschichtliche Untersuchungen über das Perianth bei den Polycarpicae. Bot Jb 84:359–508

Hill RJ (1976 Taxonomic and phylogenetic significance of seed-coat microsculpturing in *Mentzelia* (Loasaceae) in Wyoming and adjacent Western States. Brittonia 28:86–112

Hill RJ (1977) Variability of soluble seed proteins in populations of *Mentzelia* L. (Loasaceae) from Wyoming and adjacent States. Bull Torrey Bot Club 104:93–101

Hillebrand GR, Fairbrothers DE (1970) Serological investigation of the Caprifoliaceae. 1. Correspondence with selected Rubiaceae and Cornaceae. Am J Bot 57:810–815

Hilsenbeck RA, Levin DA, Mabry TJ, Raven PH (1984) Flavonoids of *Gaura triangulata*. Phytochem 23:1077–1079

Hocking PJ, Fineran BA (1983) Aspects of the nutrition of root parasitic Loranthaceae, pp 230–258. *In*: Calder M, Bernhardt P (eds) The Biology of Mistletoes, pp 1–348. Academic Press, Sydney, Australia

Hogg RW, Gillan FT (1984) Fatty acids, Sterols and hydrocarbons in the leaves from 11 species of mangrove. Phytochem 23:93–97

Hohn MF, Meinschein WG (1976) Seed oil fatty acids: Evolutionary significance in the Nyssaceae and Cornaceae. Biochem Syst Ecol 4:193–199

Holm L (1979) Some problems in angiosperm taxonomy in the light of the rust data, pp 177–181. In: Hedberg I (ed) Parasites as Plant Taxonomists—Proceedings of a Symposium held in Uppsala, 1978. Symb Bot Upsal, Uppsala, pp 1–221

Hooker JD (1965) Chenopodiaceae. *In*: Bentham G, Hooker JD (eds) Genera Plantarum (Reprinted) 3:76–78

Hooker JD (1894) The Flora of British India. Vol 6:1–792. L. Reeve and Co., London

Horich CK (1978) Ants, wasps and orchids—Myrmecophilous orchids. Florida Orchidist 21:154–162

Hou D (1967) *Sarawakodendron*: A new genus of Celastraceae. Blumea 15:139–143

Hou D (1969) Pollen of *Sarawakodendron* (Celastraceae) and some related genera. Blumea 17:97–120

Hsu C (1967) Preliminary chromosome studies on the vascular plants of Taiwan 1. Taiwania 13:117–129

Hu Shiu-Ying (1960) A revision of the genus *Clethra* in China. J Arnold Arbor 41:164–190

Huber H (1963) Die Verwandtsschaftsverhältnisse der Rosifloren. Mitt Bot Staatssamml München 5:1–48

Huber H (1969) Die Samenmerkmale und Verwandschaftsverhältnisse der Liliifloren. Mitt Bot Staatssamml, Münich 8:219–538

Huber H (1977) The treatment of the monocotyledons in an evolutionary system of classification. Pl Syst Evol Suppl 1:285–298

Hufford LD (1992) Rosidae and their relationship to the nonmagnoliid dicotyledons: A phylogenetic analysis using morphological and chemical data. Ann Mo Bot Gard 79:219–248

Hummel A (1971) The genus *Cercidiphyllum* in the Tertiary flora of Poland and of the neighbouring regions. Roczn Sekc Dendrol Pol Tow Bot 25:63–75

Humphrey RR (1935) A study of *Idria columnaris* and *Fouquieria splendens*. Am J Bot 22:184–206

Hutchinson J (1926) The Families of Flowering Plants. Vol 1, Dicotyledons, pp 1–328. Macmillan, London

Hutchinson J (1934) The Families of Flowering Plants. Vol 2, Monocotyledons, pp 1–243. Macmillan, London

Hutchinson J (1948) British Flowering Plants, pp 1–374. P.R. Gawthorn Ltd., London (UK)

Hutchinson J (1959) The Families of Flowering Plants. Edn 2. Dicots, Vol 1: pp 1–510. Clarendon Press, Oxford (UK)

Hutchinson J (1959) The Families of Flowering Plants, Edn 2. Monocots, Vol 2: pp 511–792. Clarendon Press, Oxford (UK)

Hutchinson J (1969) Evolution and Phylogeny of Flowering Plants, pp 1–717. Academic Press, London New York

Hutchinson J (1973) The Families of Flowering Plants arranged according to a New System based on their Probable Phylogeny, pp 1–968. Edn 3. Oxford Univ Press, London (UK)

Iljinskaja IA (1972) Correction of the volume of the genus *Trochodendroides* and new fossil species of *Cocculus* (in Russian). Bot Zh (Moscow, Leningrad) 57:17–30

Ilyas M 1978 (1979) The spices of India 2. Econ Bot 32:238–263

Inamdar JA, Aleykutty KM (1979) Studies on *Cabomba aquatica* (Cabombaceae). Pl Syst Evol 132:161–166

Inouye H, Takeda Y, Nishimura H (1974) Two new iridoid glucosides from *Gardenia jasminoides* fruits. Phytochem 13:2219–2224

Ito M (1987) Phylogenetic systematics of the Nymphaeales. Bot Mag Tokyo 10:17–35

Iwashina T, Ootani S, Hayashi K (1986) Determination of minor flavonol-glycosides and sugar-free flavonols in the tepals of several species of Cereoideae (Cactaceae). Bot Mag Tokyo 99:53–62

Jacques F (1965) Morphologie du pollen et des ovules de *Couroupita guianensis* Aubl. (Lecythidaceae). Pollen Spores (Paris) 7:175–180

Jain SK (1983) Medicinal Plants, pp 1–180. National Book Trust, New Delhi.

Jäger-Zürn I (1966) Infloreszenz- und blütenmorphologische sowie embryologische Untersuchungen an *Myrothamnus* Welw. Beitr Biol Pfl 42:241–271

Jähnichen H, Mai DH, Walther H (1980) Blätter und Früchte von *Cercidiphyllum* Siebold & Zuccarini im mitteleuropäischen Tertiar. Schriftenreihe Geol Wiss 16:357–399

Jeffrey C (1980) A review of the Cucurbitaceae. Bot J Linn Soc 81:233–247

Jensen SR, Nielsen BJ (1973) Cyanogenic glucosides in *Sambucus nigra* L. Acta Chem Scand 27:2661–2662

Jensen SR, Nielsen BJ, Dahlgren R (1975) Iridoid compounds, their occurrence and systematic importance in the Angiosperms. Bot Notiser 128:148–180

Johansen DA (1936) Morphology and embryology of *Fouquieria*. Am J Bot 23:95–99

Johnson DS (1935) The development of the shoot, male flower and seedling of *Batis maritima* Linn. Bull Torrey Bot Club 62:19–31

Johnson LAS, Briggs BG (1975) On the Proteaceae: The evolution and classification of a southern family. Bot J Linn Soc 70:83–182

Johnson LAS, Briggs BG (1985) Myrtales and Myrtaceae: A phylogenetic analysis. Ann Mo Bot Gard 71:700–756

Johnson MA (1958) The epiphyllous flowers of *Turnera* and *Helwingia*. Bull Torrey Bot Club 85:313–323

Johnston IM (1936) A study of Nolanaceae. Contr Gray Herb Harv 112:1–83

Johri BM (1970a) Limnanthaceae. *In*: Proc Symp Comparative Embryology of Angiosperms. Bull Natl Sci Acad No. 41:110–113

Johri BM (1970b) Alismataceae and Butomaceae. *In*: Proc Symp Comparative Embryology of Angiosperms. Bull Natl Sci Acad No 41:334–335

Johri BM, Ambegaokar KB (1984) Embryology: Then and Now, pp 1–52. *In*: Johri BM (ed) Embryology of Angiosperms, pp 1–830. Springer, Berlin Heidelberg New York Tokyo

Johri BM, Bhatnagar SP (1955) A contribution to the morphology and life history of *Aristolochia*. Phytomorphology 5:123–137

Johri BM, Bhatnagar SP (1960) Embryology and taxonomy of the Santalales 1. Proc Natl Inst Sci India B 26: 199–220

Johri BM, Bhatnagar SP (1972) Loranthaceae. Botanical Monograph No. 8:1–155. Council Sci Indus Res (CSIR), New Delhi

Johri BM, Ambegaokar KB, Srivastava PS (1992) Comparative Embryology of Angiosperms. Vol 1:1–614, 2:615–1221. Springer, Berlin Heidelberg New York Tokyo

Jones GN (1955) Leguminales: A new ordinal name. Taxon 4:188–189

Jones SB, Luchsinger AE (1987) Plant Systematics, pp 1–512. McGraw Hill, New York

Jonker FP (1938) A monograph of the Burmanniaceae. Meded Bot Mus Herb Rijks (Utrecht) 51:1–279

Jørgensen LB (1981) Myrosin cells and dilated cistarnae of the endoplasmic reticulum in the order Capparales. Nordic J Bot 1:433–445

Joshi AC (1946) A note on the development of pollen of *Myristica fragrans* van Houtten and the affinities of the family Myristicaceae. J Indian Bot Soc 25:139–143

Kak AM, Durani S (1986) A contribution to the seed anatomy of *Nelumbium nuciferum* Gaertn. J Pl Anat Morphol (Jodhpur, Rajasthan, India) 3:59–64

Kalkman C (1988) The phylogeny of the Rosaceae. Bot J Linn Soc 98:37–59

Kanis A (1968) A revision of the Ochnaceae of the Indo-Pacific area. Blumea 16:1–83

Kapil RN, Ahluwalia K (1963) Embryology of *Peganum harmala* L. Phytomorphology 13:127–140

Kapil RN, Jalan SS (1964) *Schisandra* Michaux: Its embryology and systematic position. Bot Notiser 117:285–306.

Kapil RN, Mohana Rao PR (1966a) Embryology and systematic position of *Theligonum* Linn. Proc Natl Inst Sci B 32:218–232

Kapil RN, Mohana Rao PR (1966b) Studies of the Garryaceae 2. Embryology and systematic position of *Garrya* Douglas ex Lindley. Phytomorphology 16:564–578

Kapil RN, Sethi S Bala (1963) Development of male and female gametophytes in *Camellia sinensis* (L.) O. Kuntze. Proc Natl Inst Sci India B 29:567–574

Kasapligil B (1951) Morphological and ontogenetic studies of *Umbellularia californica* Nutt. and *Laurus nobilis* L. Univ Calif Publ Bot 25 3:115–240

Kaul RB (1967) Ontogeny and anatomy of the flower of *Limnocharis flava* (Butomaceae). Am J Bot 54:1223–1230

Kaul RB (1968a) Floral development and vasculature in *Hydrocleys nymphoides* (Butomaceae). Am J Bot 55: 236–242

Kaul RB (1968b) Floral maophology and phylogeny in the Hydrocharitaceae. Phytomorphology 18:13–35

Kaul RB (1969) Morphology and development of the flowers of *Bootia cordata*, *Ottelia alismoides* and their synthetic hybrid (Hydrocharitaceae). Am J Bot 56:951–959

Kawasaki M, Kanomata T, Yoshitama K (1986) Flavonoids in the leaves of 28 Polygonaceous plants. Bot Mag Tokyo 99:63–74

Keating RC (1969) Comparative morphology of Cochlospermaceae. 1. Synopsis of the family and wood anatomy. Phytomorphology 19:379–392

Keating RC (1970) Comparative morphology of the Cochlospermaceae. 2. Anatomy of the young vegetative shoots. Am J Bot 57:889–898

Keating RC (1972a) The comparative morphology of the Cochlospermaceae. 3. The flower and pollen. Ann Mo Bot Gard 59:282–296

Keating RC (1972b) The pollen morphology and systematics of the Flacourtiaceae. Brittonia 24:121–122

Keating RC (1974) Trends of specialization of pollen of Flacourtiaceae with comparative observations of Cochlospermaceae and Bixaceae. Grana 15:29–49

Keefe JM, Moseley MF (1978) Wood anatomy and phylogeny of *Paeonia* section Moutan. J Arnold Arbor 59: 274–297

Keighery GJ (1975) Chromosome numbers in the Gyrostemonaceae Endl. and the Phytolaccaceae Lindl.: A comparison. Aust J Bot 23:335–338

Keighery GJ (1979) Chromosome counts in *Cephalotus* (Cephalotaceae). Pl Syst Evol 133:103–104

Keighery GJ (1985) *Walternanthus*: A new genus of Gyrostemonaceae from Western Australia. Bot Jb 106:107–113

Kelsey HP, Dayton WA (1942) (eds) Standardized Plant Names, pp 1–675. Edn 2. Am Joint Com Hort Nom Harrisburg. J Horace McFarland Co., Harrisburg, Pa.

Kennedy GC (1975) Vegetative mimicry in orchid plants. Orchid Dig 39:95–97

Khan R (1970) Lentibulariaceae. *In*: Proc Symp Comparative Embryology of Angiosperms. Bull Indian Natl Sci Acad No. 41:290–297

Khanna P (1965) Morphological and embryological studies in Nymphaeaceae 2. *Brasenia schreberei* Gmel. and *Nelumbo nucifera* Gaertn. Aust J Bot 13:379–387

Kimler L, Mears J, Mabry TJ, Rosler H (1970) On the question of the mutual exclusiveness of betalains and anthocyanins. Taxon 19:875–878

Kimura Y (1956) Système et phylogénie des monocotylèdones. Notulae Systematicae (Paris) 15:137–159

Kircher P (1986) Untersuchungen zur Bluten- und Infloreszenmorphologie, Embryologie und Systematik der Restionaceen im Vergleich mit Graineen und verwandten Familien. Diss Bot 94:1–218

Kjaer A (1963) The distribution of sulphur compounds, 453–473. *In*: Swain T (ed) Chemical Plant Taxonomy, pp 1–543. Academic Press, London

Kjaer A, Malver O (1979) Glucosinolates in *Tersonia brevipes*, Gyrostemonaceae. Phytochem 18:1565

Kochhar SL (1981) Economic Botany in the Tropics, pp 1–476. MacMillan India, New Delhi

Koek-Noorman J, Hogeweg P, Van Maanen WHM, Ter Welle BJH (1979) Wood anatomy of the *Blakeeae* (Melastomataceae). Acta Bot Neerl 28:21–43

Kolbe K-P (1978) Serologischer Beotrag zur Systematik der Capparales. Bot Jb 99:468–489

Kolbe K-P, John T (1979) Serologische Untersuchungen zur Systematik der Violales. Bot Jb 101:3–15

Kooiman K (1971) Ein phytochemischer Beitrag zur Lösung des Verwandtschaftsprobleme der Theligonaceae. Öst Bot Z 119:395–398

Kooiman K (1974) Iridoid glycosides in the Loasaceae and the taxonomic position of the family. Acta Bot Neerl 23:677–679

Kool RA (1980) A taxonomic revision of the genus *Ixonanthus* (Linaceae). Blumea 26:191–204

Kosakai H, Moseley MF, Cheadle VI (1970) Morphological studies of the Nymphaeaceae 5: Does *Nelumbo* have vessels? Am J Bot 57:487–494

Kostermans AJGH (1960) Miscellaneous botanical notes. Reinwardtia (Bogor, Indonesia) 5:233–254

Kostermans AJGH (1978) *Pakaraimaea dipterocarpacea* belongs to Tiliaceae. Taxon 27:357–359

Kostermans AJGH (1985) Family status for the Monotoideae Gilg and the Pakaraimoideae Ashton, Maquire and de Zeeuw (Dipterocarpaceae). Taxon 34:426–435

Krach JE (1977) Seed characters in and affinities among Saxifragineae. Pl Syst Evol Suppl 1:141–153

Kramer PR (1939) The woods of *Billia*, *Cashalia*, *Henoonia* and *Juliania*. Trop Woods 58:1–5

Krause J (1942a) Corynocarpaceae. *In*: Engler A, Prantl K (eds) Die natürlichen Pflanzenfamilien. Edn 2.20b: 22–35

Krause J (1942b) Staphyleaceae. *In*: Engler A, Prantl K (eds) Die natürlichen Pflanzenfamilien. Edn 2. 20b:255–361

Kshetrapal S (1970) A contribution to the vascular anatomy of the flower of certain species of the Salvadoraceae. J Indian Bot Soc 49:92–98

Kubitzki K (1968) Flavonoide und Systematik der Dilleniaceen. Ber Deutsch Bot Ges 81:238–251

Kubitzki K (1972) Probleme der Grossgliederung der Blütenpflazen. Ber Deutsch Bot Ges 85:259–277

Kubitzki K (1987) Origin and significance of trimerous flowers. Taxon 16:21–28

Kuhlmann JG (1934) Notas sobra o genera *Duckeodendron*. Archos Inst Biol Veg (Rio de Janero) 1:35–37

Kuijt J (1968) Mutual affinities of Santalalean families. Brittonia 20:136–147

Kuijt J (1969) The Biology of Parasitic Flowering Plants, pp 1–245. Univ Calif Press. Berkeley Los Angelos

Kuijt J (1981) Inflorescence morphology of Loranthaceae: An evolutionary synthesis. Blumea 27:1–73

Kuijt J (1986) Morphology, biology and systematic relationships of *Desmaria* (Loranthaceae). Pl Syst Evol 151: 121–130

Kuijt J, Bruns D (1987) Roots in *Corynaea* (Balanophoraceae). Nordic J Bot 7:539–542

Kumar V, Subramanian B (1987) Chromosome Atlas of Flowering Plants of the Indian Subcontinent. Vol 1:1–464, Dicotyledons; Vol 2:465–1095, Monocotyledons. Bot Surv India (Calcutta)

Kuprainova LA (1965) The palynology of the Amentiferae. Komarov Bot Instt (Acad Sci URSS) 1:1–214 Nauka, Moscow Leningrad

Ladd PG (1977) Pollen morphology of some members of the Restionaceae and related families, with notes on fossil record. Grana palynol 16:1–14

Lakshmanan KK (1961) Embryological studies in the Hydrocharitaceae. 1. *Blyxa octandra* Planch. J Madras Univ B 31:133–142

Lakshmanan KK (1970a) Hydrocharitaceae. *In*: Proc Symp Comparative Embryology of Angiosperms. Bull Indian Natl Sci Acad No. 41:336–341

Lakshmanan KK (1970b) Scheuchzeriaceae. *In*: Proc Symp Comparative Embryology of Angiosperms. Bull Indian Natl Sci Acad No. 41:342–343

Lakshmanan KK (1970c) Juncaginaceae. *In*: Proc Symp Comparative Embryology of Angiosperms. Bull Indian Natl Sci Acad No. 41:344–347

Lakshmanan KK (1970d) Potamogetonaceae. *In*: Proc Symp Comparative Embryology of Angiosperms. Bull Indian Natl Sci Acad No. 41:348–351

Lakshmanan KK (1970e) Zannichelliaceae. *In*: Proc Symp Comparative Embryology of Angiosperms. Bull Indian Natl Sci Acad No. 41:352–353

Lakshmanan KK (1970f) Najadaceae. *In*: Proc Symp Comparative Embryology of Angiosperms. Bull Indian Natl Sci Acad No. 41:354–357

Lam HJ (1939) On the system of the Sapotaceae with some remarks on taxonomical methods. Recl Trav Bot Neerl 36:509–525

Langlet O (1928) Einige Beobachtungen über Zytologie der Berberidaceae. Svensk Bot Tidskr 22:169–184

Langlet O (1932) Über chromosomenverhältnisse und Systematik der Ranunculaceae. Svensk Bot Tidskr 26:381

Lawrence GHM (1951) Taxonomy of Vascular Plants, pp 1–823. MacMillan, New York

Lawrence JRA (1937) Correlation of the taxonomy and the floral anatomy of certain Boraginaceae. Am J Bot 24:433–444

Leandri J (1937) Sur l'aire et la position systématique du genre malgache *Didymeles* Thouars. Ann Sci Natn Bot Ser 10. 19:309–317

Lee HM (1987) The biology of *Hakea epiglottis* Labill. (Proteaceae). Aust J Bot 35:689–699

Leboeuf M, Cavé A, Bhaumik PK, Mukherjee B, Mukherjee R (1982) The phytochemistry of the Annonaceae. Phytochem 21:2783–2813

Lee YS, Fairbrothers DE (1978) Serological approaches to the Rubiaceae and related families. Taxon 27:159–185

Leenhouts PW (1959) Revision of the Burseraceae of the Malaysian area in a wider sense. 10a. *Canarium* Stickm. Blumea 9:275–475

Leinfellner W (1971) Das Gynözeum von *Krameria* und sein Vergleich mit jenem der Leguminosæ und der Polygalaceae. Öst Bot Z 119:102–117

Lemesle R (1936) Les vaisseaux à perforations scalariformes de l'*Eupomatia* et leur importance dans la phylogénie des Polycarpes. C R Acad Sci Paris 203:1538–1540

Lemesle R (1937) Etude microchemique des divers tannoids de l'*Eupomatia*. Bull Soc Bot Fr 84:535–538

Lemesle R (1943) Les trachéides a ponctuations aréolées des *Sargentodoxa cuneata* Rehd et Wils. et leur importance dans la phylogénie des Sargentodoxacées. Bull Soc Bot Fr 90:104–107

Lemesle R (1945) Les ponctuations aréolées des fibres des genres *Schizandra* L., *Kadsura* J., *Illicium* L. et leurs rapports avec la phylogénie. C R Acad Sci Paris 221:113–115

Les DH (1988) The origin and affinities of the Ceratophyllaceae. Taxon 37:326–345

Le Thomas A (1981) Ultrastructural characters of the pollen grains of African Annonaceae and their significance for the phylogeny of primitive angiosperms (2nd part). Pollen Spores (Paris) 23:5–36

Li HL (1954) *Davidia* as the type of a new family Davidiaceae. Lloydia 17:329–331

Li HL (1955) Classification and phylogeny of Nymphaeaceae and allied families. Am Midl Naturalist 54:33–41

Linder HP, Ferguson IK (1985) On the pollen morphology and phylogeny of the Restionales and Poales. Grana 24:65–76

Lindsey AA (1938) Anatomical evidence for the Menyanthaceae. Am J Bot 25:480–485

Lindsey AA (1940) Floral anatomy of the Gentianaceae. Am J Bot 27:640–651

Lobreau D (1969) Les limites de l' "Order" de Célastrales d'après le pollen. Pollen Spores (Paris) 11:499–555

Lobreau-Callen D (1982) Structures et affinités polliniques des Cardiopterygaceae, Dipentodontaceae, Erythropalaceae et Octoknemaceae. Bot Jb 103:371–412

Lobreau-Callen D, Nilsson S, Albers F, Straka H (1978) Les Cneoraceae (Rutales): Étude taxonomique palynologique et systématique. Grana 17:125–139

Loesener T (1942) Celastraceae. *In*: Engler A, Prantl K (eds) Die natürlichen Pflanzenfamilien. Edn 2. 20b:87–197

Lotsy JP (1911) Vorträge über botanische Stammesgeschichte. Vol 3. Cormophyta Siphonogamia. Gustav Fischer, Jena

Lu S.-Y, Hsu K.-S, Fan F.-H (1986) Bretschneideraceae, a new family recorded for the flora of Taiwan (in Chinese). Quarterly Chinese Forestry 19:115–119

Lundin R (1983) Taxonomy of *Sonerila* (Melastomataceae) of Ceylon. Nordic J Bot 3:633–656

Luza JG, Polito VS (1991) Porogamy and chalazogamy in walnut (*Juglans regia* L.). Bot Gaz 152:100–106

Mabry TJ (1976) Pigment dichotomy and DNA-RNA hybridization data for Centrospermous families. Pl Syst Evol 126:79–94

Mabry TJ, Behnke H-D (1976) Betalains and P-type sieve element plastids: The systematic position of *Dysphania* R.Br. (Centrospermae). Taxon 25:109–111

Mabry TJ, Turner BL (1964) Chemical investigations of the Batidaceae: Betaxanthin and their systematic implications. Taxon 13:197–200

Mabry TJ, Taylor A, Turner BL (1963) The betacyanins and their distribution. Phytochem 2:61–64

Mabry TJ, Eifert IJ, Chang C, Mabry H, Kidd C, Behnke H-D (1975) Theligonaceae: Pigment and ultrastructural evidence which excludes it from the order Centrospermae. Biochem Syst Ecol 3:53–55

Madhav R, Seshadri TR, Subramanian GBV (1967) Identity of the polyphenol of *Shorea* species with Hopeaphenol. Phytochem 6:1155–1156

Magda O, Weber-El Ghobary (1984) The sytematic relationships of *Aegialitis* (Plumbaginaceae) as revealed by pollen morphology. Pl Syst Evol 144:53–58

Maguire B, Pires JM (1978) Saccifoliaceae: A new monotypic family of the Gentianales. *In*: Maguire B, Wurdack JJ (eds) The Botany of the Guayana Highlands. Pt 10. Mem N Y Bot Gdn No. 29

Maguire B, Ashton PS, Zeew de C, Giannasi DE, Nicklas KJ (1977) Pakaraimoideae, Dipterocarpaceae of the Western Hemisphere. Taxon 26:341–385

Maguire B, Wurdack JJ (eds) (1958) The Botany of the Guayana Highland. Pt 3. Mem N Y Bot Gdn 10:1–156

Maguire B, Wurdack JJ (eds) (1965) The Botany of the Guayana Highland. Pt 6. Mem N Y Bot Gdn 12:1–285

Maheshwari JK (1965) Illustrations to the Flora of Delhi, pp 1–282. Council Sci Indus Res (CSIR) New Delhi

Maheshwari P (1945) The place of angiosperm embryology in research and teaching. J Indian Bot Soc 24:25–41

Maheshwari P (1950) An Introduction to the Embryology of Angiosperms, pp 1–453. McGraw-Hill, New York London

Maheshwari P (1954) Embryology and systematic botany. Proc VIIIth Intl Bot Congr, Paris Sect 7–8:254–255

Maheshwari P (1958) Embryology and taxonomy. Mem Indian Bot Soc 1:1–9

Maheshwari P, Johri BM (1956) The morphology and embryology of *Floerkea proserpinacoides* Willd. with a discussion on the systematic position of the family Limnanthaceae. Bot Mag Tokyo (Ogura Comm Vol) 69:410–423

Maheshwari P, Kapil RN (1966) Some Indian contributions to the embryology of angiosperms. Phytomorphology 16:339–391

Maheshwari P, Singh B (1952) Embryology of *Macrosolen cochinchinensis*. Bot Gaz 114:20–32

Maheshwari P, Johri BM, Dixit SN (1957) The floral morphology and embryology of the Loranthoideae (Loranthaceae). J Madras Univ B 27:121–136

Maheshwari SC (1958) *Spirodela polyrrhiza*: The link between the aroids and the duckweeds. Nature 181:1745–1746

Maheswari Devi, H (1972) Salvadoraceae: A study of its embryology and systematics. J Indian Bot Soc 51:56–62

Mahran GH, El-Hossary GA, Saleh M, Motawe HM (1977a) Isolation and identification of certain molluscicidal substances in *Canna indica* L. J Afr Med Pl 1:107–119

Mahran GH, El-Hossary GA, Saleh M, Mohamed AM, Motawe HM (1977b) Contribution to the molluscicidal activity of *Canna* species growing in Egypt. J Afr Med Pl 1:147–155

Maia LD'O (1941) Le grain de pollen dans l'identification et la classification des plantes. 1. Sur la position systématique du genre *Allium*. Bull Bot Soc Portugaise Sci Natu 13:135–147

Maksoud SA, El Hadidi MN (1988) The flavonoids of *Balanites aegyptiaca* (Balanitaceae) from Egypt. Pl Syst Evol 160: 153–158

Mallick R, Sharma AK (1966) Chromosome studies in Indian Pandanales. Cytologia 31:402–410

Markgraf Fr (1936) Blutenbau under Verwandtschaft bei den einfachsten Helobial. Ber Deutsch Bot Ges 54: 191–229

Markgraf Fr (1963) Die Phylogenetische Stellung der Gattung *Davidia*. Ber Deutsch Bot Ges 76 (1 Generalversammlungsheft): 63–69

Markham KR, Mabry TJ, Swift WJ (1970) Distribution of flavonoids in the genus *Baptisia*. Phytochem 9:2359–2364

Martin AC (1946) The comparative internal morphology of seeds. Am Midl Naturalist 36:513–660

Martin HA (1977) The history of *Ilex* (Aquifoliaceae) with special reference to Australia: Evidence from pollen. Aust J Bot 25:655–673

Mary TN, Malik CP (1973) Cytogenetic studies in *Papaver*, 3. Induced polyploids in some *Papaver* species. Chromosome Inform Ser (Calcutta) No. 15:27–29

Masand P (1970) Zygophyllaceae. *In*: Proc Symp Comparative Embryology of Angiosperms. Bull Indian Natl Sci Acad No. 41:123–126

Maserti BE, Ferrara R, Paterno P (1988) *Posidonia oceanica* (L.) Delile as indicators of mercury contamination in seafood. Proc Intl Symp Environ Life Elements and Health, p 77. Chinese Academy of Sciences in association with Third world Academy of Sciences and WHO, Beijing, China

Mathew PM, Philip O (1983) Studies in the pollen morphology of South Indian Rubiaceae. Advances in Pollen Spore Res, Lucknow 10:1–80

Mathur G, Mohan Ram HY (1986) Floral biology and pollination in *Lantana camara*. Phytomorphology 36:76–100

Mauritzon J (1933) Über die systematische Stellung der Familien Hydrostachyaceae and Podostemaceae. Bot Notiser 1933:172–180

Mauritzon J (1936) Zur Embryologie und systematischen Abgrenzung der Reihen Terebinthales und Celastrales. Bot Notiser 1936:161–212

McClure JW, Alston RE (1964) Patterns of selected chemical components of *Spirodela oligorhiza* formed under various conditions of axenic culture. Nature 201:311–313

McClure JW, Alston RE (1966) Chemotaxonomy of the Lemnaceae. Am J Bot 53:849–859

McLaughlin RP (1933) Systematic anatomy of the woods of the Magnoliales. Trop Woods 34:3–37

McLaughlin J (1959) The woods and flora of the Florida Keys: Wood anatomy and phylogeny of Batidaceae. Trop Woods 110:1–15

Meacham CA (1980) Phylogeny of the Berberidaceae with an evaluation of classification. Syst Bot (New York) 5:149–172

Meeuse ADJ (1975a) Floral evolution of Hamamelidae. 3. Hamamelidales and associated groups including Urticales and final conclusions. Acta Bot Neerl 24:181–191

Meeuse ADJ (1975b) Taxonomic relationships of Salicaceae and Flacourtiaceae: Their bearing on interpretative floral morphology and dilleniid phylogeny. Acta Bot Neerl 24:437–457

Meher-Homiji VM (1979) Distribution of the Dipterocarpaceae: Some phytogeographic considerations on India. Phytocoenologia 6:85–93

Mehlquist GAL (1974) Some aspects of polyploidy in orchids with particular reference to *Cymbidium*, *Paphiopedilum* and the *Cattleya* alliance, pp 393–409. *In*: Withner CL (ed) The Orchids: Scientific studies, pp 1–604. John Wiley, New York

Meijer W (1972) The genus *Axinandra*—Melastomataceae: A missing link in Myrtales? Ceylon J Sci Biol 10: 72–74

Meijer W (1976) A note on *Podostemum ceratophyllum* Michx. as an indicator of clean streams in and around the Appalachian Mountains. Castanea 41:319–324

Melchior H (1964) A. Engler's Syllabus der Pflanzenfamilien (Revised). Vol 2: 368–666. Borntraeger, Berlin

Melville R (1983) The affinity of *Paeonia* and a second genus of Paeoniaceae. Kew Bull 38:87–105

Mesa A (1981) Nolanaceae. *In*: Flora Neotrópica Bronx New York. N Y Bot Gdn Monograph No. 26

Mesa A (1986) The classification of the Nolanaceae, pp 86–90. *In*: D'Arcy WG (ed) Solanaceae: Biology and systematics, pp 1–603. Columbia Univ Press, New York

Metcalfe CR (1935) The structure of some sandalwoods and of their substitutes and of some other little known woods. Kew Bull 4:165–194

Metcalfe CR (1952a) *Medusandra richardiana* Brenan: Anatomy of leaf, stem and wood. Kew Bull 2:237–244

Metcalfe CR (1952b) The anatomical structure of the Dioncophyllaceae, in relation to the taxonomic affinities of the family. Kew Bull 2:351–368

Metcalfe CR (1956) *Scyphostegia borneensis* Stapf.: Anatomy of stem and leaf in relation to its taxonomic position. Reinwardtia (Bogor, Indonesia) 4:99–104

Metcalfe CR (1962) Notes on the systematic anatomy of *Whittonia* and *Peridiscus*. Kew Bull 15:472–475

Metcalfe CR, Chalk L (1972) Anatomy of the Dicotyledons. Vol 1: 1–724, Vol 2: 725–1500. Clarendon Press, Oxford (UK)

Metcalfe CR, Chalk L (1979) Anatomy of the Dicotyledons. Vol 1:1–276. Clarendon Press, Oxford (UK)

Meyer FG (1976) A revision of the genus *Koelreuteria* (Sapindaceae). J Arnold Arbor 57:127–166

Meylan BA, Butterfield BG (1975) Occurrence of simple, multiple and combination perforation plates in the vessels of New Zealand woods. N Z J Bot 13:1–18

Mez C (1926) Die Bedeutung der Serodiagnostik für stammesgeschichtliche Forschung. Bot Arch 16:1–23

Milby TH (1971) Floral anatomy of *Krameria lanceolata*. Am J Bot 58:569–576

Mirande M (1922) Sur l'origine morphologique de liber interne des Nolanaceés et la position systématique de cette familie. C R Acad Sci Paris 175:375–376

Mitra JN (1956) On the systematic position of the family Cactaceae as based from the studies of morphological characters of the flowers of *Selenicereus grandiflorus* Brit. et Rose and *Opuntia dillenii* How. Sci Cult 21: 460–461

Mitra K, Mondal M, Saha S (1977) The pollen morphology of Burseraceae. Grana 16:75–79

Miyaji Y (1930) Beitrage zur Chromosomen phylogenie der Berberidaceen. Planta 11:650–659

Mohana Rao PR (1963) Suspensor polyembryony in *Garrya veatchii* Kell. Curr Sci 32:468–469

Mohana Rao PR (1983) Seed and fruit anatomy of *Trochodendron aralioides*. Phytomorphology 31:18–23

Money LL, Bailey IW, Swamy BGL (1950) The morphology and relationships of the Monimiaceae. J Arnold Arbor 31:372–404

Monro HAU (1969) Manual of Fumigation for Insect Control. Edn 2. FAO Agric Studies No. 79:1–289

Moore Jr HE (1973) The major groups of palms and their distribution. Genetes Herb 1:27–140

Moore PG, Inamdar JA (1976) *Dendrophthoe falcata* (L.f.) Ettings: A parasite on the leaf of *Mangifera indica* Linn. Curr Sci 45:305

Morton CM, Dickison WC (1992) Comparative pollen morphology of the Styracaceae. Grana 31:1–15

Moseley MF (1948) Comparative anatomy and phylogeny of the Casuarinaceae. Bot Gaz 110:232–280

Moseley MF (1958) Morphological studies in the Nymphaeaceae, 1. The nature of stamens. Phytomorphology 8: 1–29

Moseley MF (1971) Morphological studies of Nymphaeaceae 6. Development of flower of *Nuphar*. Phytomorphology 21:253–283

Moseley MF, Beeks RM (1955) Studies of the Garryaceae. 1. The comparative morphology and phylogeny. Phytomorphology 5: 314–346

Mudaliar CR, Rao JS (1951) A contribution to the taxonomy of the genus *Turnera*. Madras Agric J 38:369–371

Mueller J (1975) Note on the pollen morphology of Crypteroniaceae. Blumea 22:275–294

Murgai P (1962) Embryology of *Paeonia* together with a discussion on its systematic position, pp 215–223. *In*: Symp Plant Embryology, pp 1–261. Council Sci Indus Res (CSIR), New Delhi

Murray KE, Shipton J, Whitfield FB (1972) Volatile constituents of passion fruit, *Passiflora edulis*. Aust J Chem 25:1921–1933

Musselman LJ (1977) Seed germination and seedlings of *Krameria lanceolata.* Sida 7:224–225

Musselman LJ, Mann Jr WF (1978) Root parasites of southern forests. US For Serv Gen Tech Rep 20:1–76

Müller-Doblies D, Müller-Doblies U (1977) Ordnung Typhales. *In*: Hegi G (ed) Illustrierte Flora von Mitteleuropa. 2 (1):275–317

Nagendran CR, Arekal GD, Swamy BGL (1980) Facultative stomata in *Griffithella* (Podostemaceae). Curr Sci 49:561

Nair NC (1970a) Meliaceae. *In*: Proc Symp Comparative Embryology of Angiosperms. Bull Indian Natl Sci Acad No. 41:151–155

Nair NC (1970b) Rhamnaceae. *In*: Proc Symp Comparative Embryology of Angiosperms. Bull Indian Natl Sci Acad No. 41:168–173

Nair NC (1970c) Vitaceae *In*: Proc Symp Comparative Embryology of Angiosperms. Bull Indian Natl Sci Acad No. 41:174–179

Nair NC (1970d) Leeaceae. *In*: Proc Symp Comparative Embryology of Angiosperms. Bull Indian Natl Sci Acad No. 41:180–184

Nair NC, Nathawat KS (1958) Vascular anatomy of the flower of some species of Zygophyllaceae. J Indian Bot Soc 37:172–180

Nair PKK (1960) Pollen grains of cultivated plants. 1. *Canna* L. J Indian Bot Soc 39:373–381

Nair PKK (1970) Pollen Morphology of Angiosperms: A Historical and Phylogenetic Study, pp 1–160. Barnes and Noble, New York

Nakai T (1943) Ordines, familiae, tribe, genera, sectiones, species, varietas, formae et combinationes novae a Prof. Nakai-Takenoschin adhuc ut novis edita. Appendix: Questiones characterium naturalism plantarum, etc. Tokyo

Narang N (1953) The life history of *Stackhousia linariaefolia* A. Cunn. with a discussion on its systematic position. Phytomorphology 3:485–493

Narasimha Rao VL, Ravindranath V (1964) A further contribution to the host range of *Dendrophthoë falcata* (L.f.) Ettings. Bull Bot Surv India 6:103

Narayana HS (1962) Postfertilization study on *Moringa oleifera* Lamk.: A reinvestigation. Phytomorphology 12: 65–69

Narayana HS (1970) Moringaceae. *In*: Proc Symp Comparative Embryology of Angiosperms. Bull Indian Natl Sci Acad No. 41:78–83

Narayana LL (1960a) Studies in Burseraceae 1. J Indian Bot Soc 39:204–209

Narayana LL (1960b) Studies in Burseraceae 2. J Indian Bot Soc 39:402–409

Narayana LL (1970a) Oxalidaceae. *In*: Proc Symp Comparative Embryology of Angiosperms. Bull Indian Natl Sci Acad No. 41:114–116

Narayana LL (1970b) Geraniaceae. *In*: Proc Symp Comparative Embryology of Angiosperms. Bull Indian Natl Sci Acad No. 41:117–120

Narayana LL (1970c) Tropaeolaceae. *In*: Proc Symp Comparative Embryology of Angiosperms. Bull Indian Natl Sci Acad No. 41:121–122

Narayana LL (1970d) Linaceae. *In*: Proc Symp Comparative Embryology of Angiosperms. Bull Indian Natl Sci Acad No. 41:127–132

Narayana LL (1970e) Balsaminaceae. *In*: Proc Symp Comparative Embryology of Angiosperms. Bull Indian Natl Sci Acad No. 41:158–162

Narayana LL, Rao D (1971) Contribution to the floral anatomy of Linaceae. 2. Phytomorphology 21:64–67

Nash GV (1903) Revision of the family Fouquieriaceae. Bull Torrey Bot Club 31:449–454

Nast GG, Bailey IW (1946) Morphology of *Euptelea* and comparison with *Trochodendron*. J Arnold Arbor 27: 186–192

Natesh S, Rau MA (1984) The Embryo, pp 377–443. *In*: Johri BM (ed) Embryology of Angiosperms, pp 1–830. Springer, Berlin Heidelberg New York Tokyo

Nayar MP (1984a) Bromeliaceae. *In*: Flora of India (Ser IV). Key works to the Taxonomy of Flowering Plants of India. Bot Surv India, Calcutta 1:187–191

Nayar MP (1984b) Cardiopteridaceae. *In*: Flora of India (Ser IV). Key works to the Taxonomy of Flowering Plants of India. Bot Surv India, Calcutta 1:255

Nayar MP (1984c) Ceratophyllaceae. *In*: Flora of India (Ser IV). Key works to the Taxonomy of Flowering Plants of India. Bot Surv India, Calcutta 1:286–288

Nayar MP (1984d) Connaraceae. *In*: Flora of India (Ser IV). Key works to the Taxonomy of Flowering Plants of India. Bot Surv India, Calcutta 1:385–387

Nayar MP (1984e) Myrtaceae. *In*: Flora of India (Ser IV). Key works to the Taxonomy of Flowering Plants of India. Bot Surv India, Calcutta 4:123–130

Nayar MP (1984f) Neuradaceae. *In*: Flora of India (Ser IV). Key works to the Taxonomy of Flowering Plants of India. Bot Surv India, Calcutta 4:141

Nayar MP (1984g) Nymphaeaceae. *In*: Flora of India (Ser IV). Key works to the Taxonomy of Flowering Plants of India. Bot Surv India, Calcutta 4:147–154

Nayar MP (1984h) Orchidaceae. In: Flora of India (Ser IV). Key works to the Taxonomy of Flowering Plants of India. Bot Surv India, Calcutta 4:184–268

Nayar MP (1985) Loranthaceae. *In*: Flora of India (Ser IV). Key works to the Taxonomy of Flowering Plants of India. Bot Surv India, Calcutta 3:83–86

Nayar MP (1986a) Pittosporaceae. *In*: Flora of India (Ser IV). Key works to the Taxonomy of Flowering Plants of India. Bot Surv India, Calcutta 5:72–75

Nayar MP (1986b) Podostemaceae. *In*: Flora of India (Ser IV). Key works to the Taxonomy of Flowering Plants of India. Bot Surv India, Calcutta 5:242–248.

Némejc F (1956) On the problems of the origin and phylogenetic development of the angiosperms. Sb nár Mus Praze Sect B 12:59–143

Newell TK (1969) A study of the genus *Joinvillea* (Flagellariaceae). J Arnold Arbor 50:527–555

Nicholas A, Baijnath H (1994) A consensus classification for the order Gentianales with additional details on the suborder Apocynineae. Bot Rev 60:440–482

Niklas KJ, Giannasi DE (1977a) Flavonoids and other chemical constituents of fossil Miocene *Zelkovia* (Ulmaceae). Science 196:877–878

Niklas KJ, Giannasi DE (1977b) Geochemistry and thermolysis of flavonoids. Science 197:767–769

Nilsson S, Praglowski J, Nilsson L (1977) Atlas of Airborne Pollen Grains and Spores in Northern Europe, pp 1–139. Naturoch Kultur, Stockholm

Nittala SS, Lavie D (1981) Chemistry and genetics of withanolides in *Withania somniferum* hybrids. Phytochem 20: 2741–2748

Nooteboom HP (1966) Flavonols, leuco-anthocyanins, cinnamic acids and allkaloids in dried leaves of some Asiatic and Malesian Simaroubaceae. Blumea 14:253–356

Nooteboom HP (1967) The taxonomic position of Irvingioideae, *Allantospermum* Forman and *Cyrillopsis* Kuhlm. Adansonia Ser 2. 7:161–168

Nowicke JW (1970) Pollen morphology in the Nyctaginaceae. Grana 10:79–88

Nowicke JW (1975) Preliminary survey of pollen morphology in the order Centrospermae. Grana 15:51–77

Nowicke JW, Skvarla JJ (1981) Pollen morphology and phylogenetic relationships of the Berberidaceae. Smithson Contr Bot 50:1–83

Nowicke JW, Skvarla JJ (1983) Pollen morphology and the relationships of the Corynocarpaceae. Taxon 32: 176–183

Nowicke JW, Miller JS (1989) Pollen morphology and the relationships of Hoplestigmataceae. Taxon 38:12–16

Ohmoto T, Ikuse M, Natori S (1970) Triterpenoides of the Gramineae. Phytochem 9:2137–2148

Okada H, Ueda K (1984) Cytotaxonomical studies on Asian Annonaceae. Pl Syst Evol 144:165–177

Oliver D (1895) *Eucommia ulmoides* Oliv. *In*: Hooker WJ. Icon Pl Ser 24:2361

Ozenda P (1946) Sur l'anatomie liberoligneuse des Schizandracées. C R Acad Sci Paris 217:31–33

Panchaksharappa MG (1966) Embryological studies in some members of Zingiberaceae. 2. *Elettaria cardamomum, Hitchenia caulina* and *Zingiber macrostachyum*. Phytomorphology 16:412–417

Paris R (1963) The distribution of plant glycosides, pp 337–358. *In*: Swain T (ed) Chemical Plant Taxonomy, pp 1–543. Academic Press, London New York

Parkin J (1914) The evolution of the inflorescence. Bot J Linn Soc 42:511–582

Parks CR, Kondo K (1974) Breeding systems in *Camellia* (Theaceae). 1. A chemosystematic analysis of synthetic hybrids and back crosses involving *Camellia japonica* and *C. salvenensis*. Brittonia 26:321–322

Parulekar NK (1970) Annonaceae. *In*: Proc Symp Comparative Embryology of Angiosperms. Bull Indian Natl Sci Acad No. 41:38–41

Patel RN (1975) Wood anatomy of the dicotyledons indigenous to New Zealand. 8. *Corynocarpus*. N Z J Bot 13: 89–109

Patel VS, Skvarla JJ, Raven PH (1985) Pollen characters in relation to the delimitation of the Myrtales. Ann Mo Bot Gard 71:558–969

Pauzé F, Sattler R (1979) La placentation axillaire chez *Ochna atropurpurea*. Canad J Bot 57:100–107

Pax F (1927) Zur Phylogenie der Caryophyllaceae. Bot Jb 61:223–241

Pax F, Hoffman K (1931) Callitrichaceae. *In*: Engler A, Prantl K (eds) Die natürlichen Pflanzenfamilien. Edn 2. 2:236–246

Pax F, Hoffman K (1934) Dysphaniaceae. *In*: Engler A, Prantl K (eds) Die natürlichen Pflanzenfamilien. Edn 2. 16c:272–274

Periasamy K (1962) Studies on seeds with ruminate endosperm. 2. Development of rumination in the Vitaceae. Proc Indian Acad Sci B 56:13–26

Peterson BR (1961) Studies of floral morphology in the Epacridaceae. Bot Gaz 122:259–279

Peterson FP, Fairbrothers DE (1979) Serological investigation of selected amentiferous taxa. Syst Bot (New York) 4:230–241

Peterson FP, Fairbrothers DE (1983) A serotaxonomic appraisal of *Amphipterygium* and *Leitneria*: Two amentiferous taxa of Rutiflorae (Rosidae). Syst Bot (New York) 8:134–148

Philipson WR (1974) Ovular morphology and the major classification of the dicotyledons. Bot J Linn Soc 68: 89–108

Philipson WR (1975) Evolutionary lines within the dicotyledons. N Z J Bot 13:73–91

Philipson WR (1987) *Corynocarpus* J.R. & G. Frost. : An isolated genus. Bot J Linn Soc 95:9–18

Pichon M (1948) Les Monimiacées, familie Hétérogene. Bull Mus Hist Natu Paris, Ser 2. 20:383–384

Pilger R (1925) Caryocaraceae. *In*: Engler A, Prantl K (eds) Die natürlichen Pflanzenfamilien. Edn 2. 21:90–93

Plessner O, Negbi M, Ziv M, Basker D (1989) Effects of temperature on the flowering of the saffron crocus (*Crocus sativus* L.): Induction of hysteranthy. Israel J Bot 38:1–7

Plouvier V (1963) Distribution of aliphatic polyols and cyclitols, pp 313–336. *In*: Swain T (ed) Chemical Plant Taxonomy, pp 1–543. Academic Press, London New York

Poddubnaya-Arnoldi VA (1964) The General Embryology of Angiosperms (in Russian). Moscow

Pollard CJ (1982) Fructose oligosaccharides in the Monocotyledons: A possible delimitation of the order Liliales. Biochem Syst Ecol 10:245–250

Poppendick HH (1980) A monograph of the Cochlospermaceae. Bot Jb 10:191–265

Porter CL (1967) Taxonomy of Flowering Plants, pp 1–452. W H Freeman, San Fransisco

Ponchet M, Martin-Tanguy J, Marais A, Martin C (1982) Hydroxycinnamoyl acid amides and aromatic amines in the inflorescences of some Araceae species. Phytochem 21:2865–2869

Posluszny U (1983) Re-evaluation of certain key relationships in the Alismatales. Floral organogenesis of *Scheuchzeria palustris*. Am J Bot 70:925–933

Posluszny U, Tomlinson PB (1977) Morphology and development of floral shoots and organs in certain Zannichelliaceae. Bot J Linn Soc 75:21–46

Praglowski J (1970) The pollen morphology of the Haloragaceae with reference to taxonomy. Grana 10:159–239

Prakash N (1969) The floral development and embryology of *Centrolepis fascicularis*. Phytomorphology 19:285–291

Prakash N, Lim AL, Manurung R (1977) Embryology of duku and langsat varieties of *Lansium domesticum*. Phytomorphology 27:50–59

Prance GT (1970) The genera of Chrysobalanaceae in the southeastern United States. J Arnold Arbor 51:521–528

Prance GT (1972) Chrysobalanaceae, pp 1–409, Flora Neotrópica. N Y Bot Gard Monogr No 9. Hafner Publ, New York

Prantl K (1888) Trochodendraceae. *In*: Engler A, Prantl K (eds) Die natürlichen Pflanzenfamilien. Edn 1. 3 (2): 21–23

Preisner RM, Shamma M (1980) The spirobenzylisoquinoline alkaloids. J Natu Prod 43:305–318

Presting D, Straka H, Friedrich B (1983) Palynologia Madagassica et Mascarenica Familien. Trop Subtrop Pflanzen 44:1–93

Price JR (1963) The distribution of alkaloids in the Rutaceae, pp 429–452. *In*: Swain T (ed) Chemical Plant Taxonomy, pp 1–543, Academic Press, London New York

Punt W (1968) Morphology of the American species of the subfamily Costoideae (Zingiberaceae). Rev Paleobot Palynol 7:31–43

Radford AE, Dickison WC, Massey JR, Bell CR (1974) Vascular Plant Systematics, pp 1–891. Harper & Row, New York

Raghavan RS (1958) A chromosome survey of Indian Dioscoreas. Proc Indian Acad Sci B 48:59

Raj B (1970) Morphological and embryological studies in the family Loranthaceae. 13. *Amylotheca dictyophleba* Van Tiegh. Öst Bot Z 118:417–430

Rajendran PG, Jos JS (1972) A natural pentaploid in *Alocasia fornicata* Schott. Curr Sci 41:612–613

Ram Manasi (1956) Floral morphology and embryology of *Trapa bispinosa* with a discussion on the systematic position of the genus. Phytomorphology 6:4–19

Ram Manasi (1959) Morphological and embryological studies in the family Santalaceae. 2. *Exocarpus* with a discussion on its systematic position. Phytomorphology 9:4–19

Ram Manasi (1970) Misodendraceae. *In*: Proc Symp Comparative Embryology of Angiosperms. Bull Indian Natl Sci Acad No. 41:19–21

Ramachandran K (1968) Cytological studies in Dioscoreaceae. Cytologia 33:401–410

Randolf LF (1936) Developmental morphology of the caryopsis in maize. J Agric Res 53:881–916

Rangaswamy NS, Chakrabarty B (1966) The Leguminosae of Delhi: Some studies on their morphology and taxonomy. Bull Bot Surv India (Calcutta) 8:25–41

Rao AS, Hajra PK (1977) *Eulophia mannii* Hook. f.: A scarcely known ground orchid from Assam. Bull Bot Surv India (Calcutta) 16:156–157

Rao AS, Rathore SR (1981) *Didymoplexis pallens* Griff.: Rediscovery of a rare saprophytic orchid in the Indian Botanic Garden, Howrah. Bull Bot Surv India (Calcutta) 21:151–155

Rao RR, Garg Arti (1994) Can *Eremostachys superba* be saved from extinction? Curr Sci 67:80–81

Rao VS (1969) The floral anatomy and relationship of the rare Apostasias. J Indian Bot Soc 68:374–385

Rau MA (1962) Review of recent work on the embryogeny of some families and genera of disputed systematic position, pp 75–80. *In*: Symp Plant Embryology, pp 1–261. Council Sci Indus Res (CSIR), New Delhi

Rau MA, Sharma VK (1970a) Coriariaceae. *In*: Proc Symp Comparative Embryology of Angiosperms. Bull Indian Natl Sci Acad No. 41:156–157

Rau MA, Sharma VK (1970b) Elaeagnaceae. *In*: Proc Symp Comparative Embryology of Angiosperms. Bull Indian Natl Sci Acad No. 41:185–187

Raven PH (1975) The bases of angiosperm phylogeny: Cytology. Ann Mo Bot Gard 62:724–764

Raven PH, Kyhos DW (1965) New evidence concerning the original basic chromosome number of Angiosperms. Evolution 19:244–248

Raynal A (1965) Les espèces africaines du genre *Laurembergia* Berg. (Haloragaceae) et leur répartition. Webbia 19:683–695

Read RW, Desantels PE (1983) CITES: Orchids as a family endangered by legislation. Am Orchid Soc Bull 52: 15–21

Record SJ (1921) *Lignum-vitae*: A study of the woods of the Zygophyllaceae with reference to the true Lignum-vitae of commerce. Bull Yale Sch For No 6: 1–48

Record SJ (1933) The woods of *Rhabdodendron* and *Duckeodendron*. Trop Woods 33:6–10

Record SJ (1938) The American woods of the family Euphorbiaceae. Trop Woods 54:7–40

Rendle AB (1904) The Classification of Flowering Plants. Vol 1: 1–412. Gymnosperms and Monocotyledons. Cambridge Univ Press, Cambridge London

Rendle AB (1925) The Classification of Flowering Plants. Vol 2: 1–636. Dicotyledons. Cambridge Univ Press, Cambridge London

Retallack G, Dilcher DL (1981) Early angiosperm reproduction: *Prisca reynoldsii* gen. et sp. nov. from mid-cretaceous coastal deposits in Kansas (USA). Palaeontographica B Paläophytol 179:103–137

Richardson PM (1978) Flavonols and C-glycosylflavonoids of the Caryophyllales. Biochem Syst Ecol 6:283–286

Richardson PM (1981) Flavonoids of some controversial members of the Caryophyllales (Centrospermae). Pl Syst Evol 138:227–233

Richmond GS (1994) Seed germination of the Australian shrub *Eremophila* (Myoporaceae). Bot Rev 60:483–503

Riveros GM, Barria O Rosa, Humaña Ana Maria (1995) Self-compatibility in distylous *Hedyotis salzmannii* (Rubiaceae). Pl Syst Evol 194:1–8

Rizk AFM (1987) The chemical constituents and economic plants of the Euphorbiaceae. Bot J Linn Soc 94:293–326

Robbins WW, Weier TE, Stocking CR (1957) An Introduction to Plant Science. John Wiley, New York

Roberts ML, Haynes RR (1983) Ballistic seed dispersal in *Illicium* (Illiciaceae). Pl Syst Evol 143:227–232

Rodriguez RL (1971) The relationships of the Umbellales. Bot J Linn Soc 64 (Suppl 1): 63–91

Rogers GK (1984) The Zingiberales (Cannaceae, Marantaceae and Zingiberaceae) in the south-eastern United States. J Arnold Arbor 65:5–55

Rogers GK (1985) The genera of Phytolaccaceae in the south-eastern United States. J Arnold Arbor 67:1–37

Rolfe RA (1909) The evolution of the Orchidaceae. Orchid Rev 17:129–132, 193–196, 289–292, 353–356

Rolfe RA (1910) The evolution of the Orchidaceae. Orchid Rev 18:33–36, 87–99, 129–132, 162–166, 289–294, 321–325

Rosatti T (1986) The genera of Sphenocleaceae and Campanulaceae in the south-eastern United States. J Arnold Arbor 67: 1–64

Rosén W (1935) Beiträge zur Embryologie der Stylidiaceen. Bot Notiser 1935: 273–278

Rosén W (1949) Endosperm development in Campanulaceae and closely related families. Bot Notiser 1949: 137–147

Roth I (1977) Fruits of Angiosperms. Encycl Pl Anat 10: 1–675

Rousseau D (1928) Contribution á l'anatomie comparée des Piperacées. Arch Inst Bot, Univ Liége 7:45

Ruijgrok HWL (1966) The distribution of ranunculin and cyanogenetic compounds in the Ranunculaceae, pp 175–186. *In:* Swain T (ed) Comparative Phytochemistry, pp 1–360. Academic Press, London

Röder E, Wiedenfeld H, Kröger R, Teppner H (1993) Pyrrolizidialkaloide dreier *Onosma*-Sippen (Boraginaceae-Lithospermae) Phyton (Horn, Austria) 33: 41–49

Rowlee WW (1898) The morphological significance of the lodicules of grasses. Bot Gaz 15: 199–203

Rüdenberg L (1967) The chromosomes of *Austrobaileya*. J Arnold Arbor 48: 241–244

Salasoo I (1985) Rimuene in Epacridaceae and Ericaceae. Existence of chemophytes. Aust J Bot 33: 239–243

Saldanha CJ, Nicolson DH (1976) Flora of Hasan district, pp 1-915. Amerind Publ. New Delhi

Sampson FB (1969) Studies on the Monimiaceae. 2. Floral morphology of *Laurelia novae-zelandiae* A. Cunn. N Z J Bot 7: 214–240

Sampson FB, Endress PK (1984) Pollen morphology in the Trimeniaceae. Grana 23: 129–137

Samuelsson G (1913) Studien über die Entwicklungsgeschichte der Blüten einiger Bicornes-typen. Svensk Bot Tidskr 7: 97–188

Saupe SG (1981) Cyanogenic compounds and angiosperm phylogeny, pp 80–116. *In:* Young DA, Seigler DS (eds) Phytochemistry and Angiosperm Phylogeny. Praeger Scientific, New York

Savile DBO (1979) Fungi as aids in higher plant classification. Bot Rev 45: 377–503

Sax K (1933) Chromosome behaviour in *Calycanthus*. J Arnold Arbor 14: 279–282

Schaffner JH (1929) Principles of plant taxonomy 8. Ohio J Sci 29: 243–252

Schaffner JH (1934) Phylogenetic taxonomy of plants. Quart Rev Biol 9: 129–160

Schellenberg G (1938) Connaraceae. *In:* Engler A (ed) Das Pflanzenreich (IV 127) 103: 317–326

Schill R, Rauh W, Wieland HP (1974) Weitere Untersuchungen an Didiereacean 4, pp 1–14. Die chromosomenzahlen den einzelnen Arten. *In:* Rauh H (ed) Tropische und Subtropische Pflanzenwelt 2. Akad Wissench u Lit, Mainz (Germany)

Schinz H (1934) Amaranthaceae. *In:* Engler A, Prantl K (eds) Die natürlichen Pflanzenfamilien. Edn. 16c: 7–85

Schlittler J (1949) Die systematische Stellung der Gattung *Petermannia* F.v. Muell. und ihre phylogenetischen Bezeihungen zu den Luzuriagoideae Engl. und den Dioscoriaceae Lindl. Vjschr naturf Ges Zürich 94: 1–28

Schmid R (1964) Die systematische Stellung der Dioncophyllaceae. Bot Jb 83: 1–56

Schmid R, Carlquist S, Hufford LD, Webster GL (1984) Systematic anatomy of *Oceanopapaver:* A monotypic genus of the Capparaceae from New Caledonia. Bot J Linn Soc 89: 119–152

Schodde R (1970) Two new suprageneric taxa in the Monimiaceae alliance (Laurales). Taxon 19: 324–328

Schulze W (1982) Beitäge zur Taxonomie der Liliifloren 7. Philesiaceae. Wiss. Z Friedrich Schiller Univ Jena 31: 285–289

Schulze-Menz GK (1964) Rosaceae. *In:* Melchior H (ed) A Engler's Syllabus der Pflanzenfamilien. Vol 2: 209–218. Borntraeger, Berlin

Schuster RM (1976) Plate tectonics and its bearing on the geographical origin and dispersal of angiosperms, pp 48–138. *In:* Beck CB (ed) Origin and early Evolution of Angiosperms, pp 1–341. Columbia Univ Press, New York London

Schwabe H (1961) Glicol de etilene: Un nuevo método en histologia para ablander material de herbario. Bol Soc Argent Bot 9: 383–394

Scora RW, Bergh BO, Hopfinger JA (1975) Leaf alkanes in *Persea* and related taxa. Biochem Syst Ecol 3: 215–218

Scott PJ, Day RT (1983) Diapensiaceae: A review of the taxonomy. Taxon 32: 417–423

Scott RA, Wheeler EA (1982) Fossil woods from the Eocene Clarno formation of Oregon. Intl Assoc Wood Anat Bull 3: 135–154

Sebsebe D (1985) The genus *Maytenus* (Celastraceae) in the NE tropical Africa and tropical Arabia. Acta Univ Upsal Symp Bot Upsala 15: 2–101

Secor TB, Conn EC, Dunn JE, Seigler DS (1976) Detection and identification of cyanogenic glucosides in six species of *Acacia.* Phytochem 15: 1703–1706

Sehgal A, Mohan Ram HY (1981) Comparative developmental morphology of two populations of *Ceratophyllum* L. (Ceratophyllaceae) and their taxonomy. Bot J Linn Soc 82: 343–356

Sehgal A, Mohan Ram HY, Bhatt JR (1993) In vitro germination, growth, morphogenesis and flowering of an aquatic angiosperm, *Polypleurum stylosum.* Aqua Bot (Amsterdam) 45: 269–283

Seibert M, Williams G, Folger G, Milne T (1986) Fuel and chemical co-production from tree crops. Biomass 9: 49–66

Seigler DS (1977) The naturally occurring cyanogenic glycosides. Progr Phytochem 4: 83–120

Seigler DS (1981a) Secondary metabolites and plant systematics, pp 139–176. *In:* Conn EE (ed) The Biochemistry of Plants, pp 1–798. Academic Press, New York

Seigler DS (1981b) Terpenes and plant phylogeny, pp 117–148. *In:* Young DA, Seigler DS (eds) Phytochemistry and Angiosperm Phylogeny. Praeger Scientific, New York

Seigler DS, Kawahara W (1976) New reports of cyanolipids from Sapindaceous plants. Biochem Syst Ecol 4: 263–265

Seigler DS, Dunn JE, Conn EE, Holstein GL (1978) Acacipetalin from six species of *Acacia* of Mexico and Texas. Phytochem 17: 445–446

Seigler DS, Simpson BB, Martin C, Neff JL (1978) Free 3-acetoxy fatty acids in floral glands of *Krameria* species. Phytochem 17: 995–996

Seneviratne AS, Fowden L (1968) The amino acids of the genus *Acacia.* Phytochem 7: 1039–1045

Seshavataram V (1970) Onagraceae. *In:* Proc Symp Comparative Embryology of Angiosperms. Bull Indian Natl Sci Acad No. 41: 220–225

Sethi S Bala (1965) Structure and development of seed in *Camellia sinensis* (L.) O. Kuntze. Proc Natl Inst Sci India B 31: 24–33

Shamanna S (1954) A contribution to the embryology of *Olax wightiana* Wall. Proc Indian Acad Sci B 39: 249–256

Shamanna S (1955) A contribution to the embryology of *Opilia amentacea* Roxb. Curr Sci 24: 154–157

Shamanna S (1961) A contribution to the embryology of *Strombosia ceylanica* Gardn. Proc Indian Acad Sci B 54: 12–16

Shamma M, Moniot JL (1978) Isoquinoline Alkaloid Research 1972–1977. Plenum, New York

Sharma AK (1964) Cytology as an aid in taxonomy. Bull Bot Soc Bengal (Calcutta) 18: 1–4

Sharma AK (1969) Evolution and taxonomy of monocotyledons. *In:* Darlington CD (ed) Chromosome today 2: 241–249

Sharma AK, Bal AK (1956) A cytological investigation of some members of the family Cyperaceae. Phyton (Buenos Aires) 6: 7–22

Sharma AK, De DN (1956) Polyploidy in *Dioscorea.* Genetica 28: 112–120

Sharma BD (1969) Pollen morphology of Tiliaceae in relation to plant taxonomy. J Palynol 5:7–27

Sharma BD (1970) Contribution to the pollen morphology and plant taxonomy of the family Bombacaceae. Proc Indian Natl Sci Acad B 36: 175–191

Sharma VK (1966) Embryology of *Elaeagnus conferta* Roxb. Curr Sci 35: 185–186

Sharma VK (1968) Floral morphology, anatomy and embryology of *Coriaria nepalensis* Wall. with a discussion on the interrelationships of the family Coriariaceae. Phytomorphology 18: 143–153

Shorland FB (1963) The distribution of fatty acids in plant lipids, pp 253–311. *In:* Swain T (ed) Chemical Plant Taxonomy, pp 1–543. Academic Press, London New York

Shulgin AT (1966) Possible implications of myristicin as a psychotropic substance. Nature 210: 380–383

Simon JP (1970) Comparative serology of the order Nymphaeales. 1. Preliminary survey on the relationships of *Nelumbo.* Aliso 7: 243–261

Simpson BB, Skvarla JJ (1981) Pollen morphology and ultrastructure of *Krameria* (Krameriaceae): Utility in questions of intrafamilial and interfamilial classification. Am J Bot 68: 277–294

Simpson BB, Neff JL, Seigler D (1977) *Krameria,* free fatty acids and oil-collecting bees. Nature 267: 150–151

Simpson MG (1983) Pollen ultrastructure of the Haemodoraceae and its taxonomic significance. Grana 22: 79–103

Simpson MG, Dickison WC (1981) Comparative anatomy of *Lachnanthes* and *Lophiola* (Haemodoraceae). Flora 171: 95–113

Singh B (1944) A contribution to the anatomy of *Salvadora persica* L. with special reference to the origin of the included phloem. J Indian Bot Soc 23: 71–78

Singh B (1952) The embryology of *Dendrophthoë falcata* (Linn. f.) Ettings. Bot J Linn Soc 53: 449–473

Singh B (1962) Studies in angiospermic parasites 1. *Dendrophthoë falcata* (L.f.) Ettings., its life history, list of hosts and control measures. Bull Natl Bot Gard (Lucknow) No. 69: pp 1–75

Singh U, Wadhwani AM, Johri BM (1983) Dictionary of Economic Plants in India, pp 1–288. Second enlarged and revised Edition. Reprinted 1990, 1996. Indian Council Agricultural Research, New Delhi

Singh V (1965) Morphological and anatomical studies in Helobiae. 2. Vascular anatomy of the flower of Potamogetonaceae. Bot Gaz 126: 137–144

Singh V, Sattler R (1972) Floral development of *Alisma triviale.* Can J Bot 50: 619–627

Singh V, Sattler R (1974) Floral development of *Butomus umbellatus.* Can J Bot 52: 223–230

Skottsberg C (1935) Myzodendraceae. *In:* Engler A, Prantl K (eds) Die natürlichen Pflanzenfamilien. Edn 2. 16b: 92–97

Skvarla JJ, Nowicke JW (1976) Ultrastructure of pollen exine in Centrospermous families. Pl Syst Evol 126: 55–78

Sleumer H (1942) Icacinaceae. *In:* Engler A, Prantl K (eds) Die natürlichen Pflanzenfamilien. Edn 2. 20b: 322–396

Sleumer H (1980) A taxonomic account of the Olacaceae of Asia, and the adjacent areas. Blumea 26: 145–168

Small E, Bassett IJ, Crompton CW, Lewis H (1971) Pollen phylogeny in *Clarkia.* Taxon 20: 739–746

Small J (1919) Origin and development of the Compositae. New Phytol 18: 201–234

Smith AC (1940) The American species of Hippocrateaceae. Brittonia 3: 341–555

Smith AC (1945) A taxonomic review of *Trochodendron* and *Tetracentron.* J Arnold Arbor 26: 123–142

Smith AC (1946) A taxonomic review of *Euptelea.* J Arnold Arbor 27: 175–185

Smith AC (1947) The families Illiciaceae and Schisandraceae. Sargentia (Jamaica Plain, Massachussets, USA) 7: 1–224

Smith AC (1949) Additional notes on *Degeneria vitiensis.* J Arnold Arbor 30: 1–9

Smith AC (1971) An appraisal of the orders and families of primitive extant angiosperms. J Indian Bot Soc (Golden Jubillee Vol.) 50A: 215–226

Smith FH, Smith Elizabeth C (1942) Floral anatomy of the Santalaceae and some related forms. Ore St Monogr Stud Bot No. 5: 1–93

Smith LB, Downs RJ (1960) Xyridaceae from Brazil 2. Proc Biol Soc Wash 73: 245–260

Smith PM (1976) The Chemotaxonomy of Plants, pp 1–313. Edward Arnold Publ, London

Smithson E (1957) The comparative anatomy of the Flagellariaceae. Kew Bull 18: 491–501

Sokal RR, Sneath PHA (1966) Efficiency in Taxonomy. Taxon 15: 1–21

Souégès R (1953) Embryogénie des Peganacées: Développement de l'embryon chez le *Peganum harmala* L. C R Acad Sci (Paris) 236:2186–2188

Sørensen NA (1963) Chemical taxonomy of acetylenic compounds, pp 219–252. *In:* Swain T (ed) Chemical Plant Taxonomy, pp 1–543. Academic Press, London New York

Spencer KC, Seigler KS (1984) Cyanogenic glycosides of *Carica papaya* and its phylogenetic position with respect to the Violales and Capparales. Am J Bot 71: 1444–1447

Sponberg S (1971) The Staphyleaceae in south-eastern United States. J Arnold Arbor 52: 196–203

Sporne KR (1972) Some observations on the evolution of pollen types in dicotyledons. New Phytol 71: 181–185

Sprague TA (1927) The morphology and taxonomic position of the Adoxaceae. Bot J Linn Soc 47: 471–487

Stähl B (1987) The genus *Theophrasta* (Theophrastaceae), foliar structures, floral biology and taxonomy. Nordic J Bot 7: 529–538

Stapf O (1894) On the flora of Mount Kinabalu, in North Borneo. Trans Linn Bot Soc London 2: 69–263

Stapf O (1895) Martyniaceae. *In:* Engler A, Prantl K (eds) Die natürlichen Pflanzenfamilien. Edn 1. 4(3b): 265–269

Stebbins GL (1972) Ecological distribution of centres of major adaptive radiation in angiosperms, pp 7–34. *In:* Valentine DH (ed) Taxonomy and Phytogeography and Evolution, pp 1–431. Academic Press, London

Stebbins GL (1974) Flowering Plants: Evolution above the species level, pp 1–399. Arnold Press, London

Stebbins GL (1977) Development and comparative anatomy of the Compositae, pp 91–109. *In:* Heywood VH, Harborne JB, Turner BL (eds) The Biology and Chemistry of the Compositae. Vol 1: 1–619. Academic Press, London New York

Stebbins GL, Khush GS (1961) Variation in the organization of the stomatal complex in the leaf epidermis of monocotyledons and its bearing on their phylogeny. Am J Bot 48: 51–59

Stern KR (1962) The use of pollen morphology in the taxonomy of *Dicentra*. Am J Bot 49: 362–368

Stern WL (1955) Xylem anatomy and relationships of Gomortegaceae. Am J Bot 42: 874–885

Stern WL, Brizicky GK, Eyde RH (1969) Comparative anatomy and relationship of Columelliaceae. J Arnold Arbor 50: 36–75

Stockey RA, Crane PR (1983) In situ *Cercidiphyllum*-like seedlings from the Paleocene of Alberta, Canada. Am J Bot 70: 1564–1568

Stone DE (1968) Cytological and morphological notes on the south-eastern endemic *Schisandra glabra* (Schisandraceae). J Elisha Mitchell Scient Soc 84: 351–356

Stone DE, Freeman JL (1968) Cytotaxonomy of *Illicium floridanum* and *I. parviflorum* (Illiciaceae). J Arnold Arbor 49: 41–51

Subrahmanyam GV, Khoshoo TN (1984) Evolution of garden Nymphaeas. Curr Sci 53: 360–363

Subramanyam K (1949) An embryological study of *Lobelia pyramidalis* Wall. with special reference to the mechanism of nutrition of the embryo in the family Lobeliaceae. New Phytol 48: 365–373

Subramanyam K (1951) A morphological study of *Stylidium graminifolium*. Lloydia 14: 65–81

Subramanyam K (1962) Aquatic Angiosperms, pp 1–190. Council Sci Indus Res (CSIR), New Delhi

Subramanyam K (1970a) Campanulaceae. *In:* Proc Symp Comparative Embryology of Angiosperms. Bull Indian Natl Sci Acad No. 41: 306–312

Subramanyam K (1970b) Sphenocleaceae. *In:* Proc Symp Comparative Embryology of Angiosperms. Bull Indian Natl Sci Acad No. 41: 313–316

Subramanyam K (1970c) Stylidiaceae. *In:* Proc Symp Comparative Embryology of Angiosperms. Bull Indian Natl Sci Acad No. 41: 317–320

Subramanyam K (1970d) Pentaphragmataceae. *In:* Proc Symp Comparative Embryology of Angiosperms. Bull Indian Natl Sci Acad No. 41: 321–324

Subramanyam K, Sreemadhavan CP (1971) A conspectus of the families Podostemaceae and Tristicaceae. Bull Bot Surv India (Calcutta) 11: 161–168

Sunder Rao Y (1940) Male and female gametophytes of *Polemonium caeruleum* with a discussion of the affinities of the family Polemoniaceae. Proc Natl Inst Sci India B 61: 695–704

Sundari KT, Radharkri M, Narayana LL (1942) Chemotaxonomy of *Ceratophyllum*. Acta Bot Indica 10: 304

Süssenguth K (1927) Über die Gattung *Lennoa*. Ein Beitrag zur Kenntnis exotischer Parasitien. Flora (Jena) 122: 264–305

Süssenguth K (1935) Leeaceae. *In:* Engler A, Prantl K (eds) Die natürlichen Pflanzenfamilien. Edn 2. 20d: 372–390

Swamy BGL (1949) The comparative embryology of the Santalaceae: Node, secondary xylem and pollen. Am J Bot 36: 661–673

Swamy BGL (1953a) On Chloranthaceae. J Arnold Arbor 34: 375–408

Swamy BGL (1953b) On the floral structure of *Scyphostegia*. Proc Natl Inst Sci India B 19: 127–142

Swamy BGL (1960) Contributions to the embryology of *Cansjera rheedii*. Phytomorphology 10: 397–409

Swamy BGL, Bailey IW (1949) The morphology and relationships of *Cercidiphyllum*. J Arnold Arbor 30: 187–210

Swingle DB (1962) A Textbook of Systematic Botany, pp 1–343. McGraw-Hill, New York London

Szentpetery RG, Kovacs A, Sarkany S (1969) Volatile oil excretion of the differentiating epidermis on the developing leaf of *Valeriana collina* Wallr. Acta Agro Acad Sci Hungaricae 18: 287–296

Takahashi H (1987) Pollen morphology and its taxonomic significance of the Monotropoideae (Ericaceae). Bot Mag Tokyo 100: 385–405

Takashi S (1987) Chromosome number of *Saruma* Oliver (Aristolochiaceae). Bot Mag Tokyo 100: 99–101

Takaso T, Bouman F (1984) Ovule ontogeny and seed development in *Potamogeton natans* (Potamogetonaceae) with a note on the campylotropous ovule. Acta Bot Neerl 33: 519–533

Takhtajan A (1959) Die evolution der Angiospermen, pp 1–344. Gustav Fischer, Jena

Takhtajan A (1966) Systema et phylogenia Magnoliophytorum, pp 1–610. Soviet Sciences Press, Moscow Leningrad

Takhtajan A (1969) Flowering Plants: Origin and Dispersal, pp 1–310. Oliver & Boyd, Edinburgh

Takhtajan A (1980) Outline of the Classification of Flowering plants (Magnoliophyta). Bot Rev 46: 225–359

Takhtajan A (1987) Systema Magnoliophytorum (in Russian), pp 1–439. Nauka Publ, Moscow

Tamura M (1963) Morphology, ecology and phylogeny of the Ranunculaceae 1. Sci Rep Osaka Univ 11: 115–126

Tamura M (1972) Morphology and phyletic relationship of the Glaucidiaceae. Bot Mag Tokyo 85: 29–41

Tan AS, Rao AN (1974) Studies on the developmental anatomy of *Dioscorea sansibarensis* Pax (Dioscoreaceae). Bot J Linn Soc 69: 211–227

Tang CS, Syed MM, Hamilton RA (1972) Benzyl isothiocyanate content as a possible chemotaxonomic criterion in the Caricaceae. Phytochem 11: 2531–2533

Tang Y (1932) Timber studies of Chinese trees: Timber anatomy of Rhoipteleaceae. Bull Fan Inst Biol Peking ? 127–131

Taylor H (1972) The secondary xylem of the Violaceae: A comparative study. Bot Gaz 133: 230–242

Teppner H, Klein E (1993) *Nigritella gabasiana* spec. nova, *N. nigra* subsp. *iberica* subsp. nova (Orchidaceae-Orchideae) und deren Embryologie. Phyton (Horn, Austria) 33: 179–209

Ter Welle BJH, Koek-Noorman J (1981) Wood anatomy of the Neo-tropical Melastomataceae. Blumea 27: 335–394

Thanikaimoni G (1965) Contribution to the pollen morpholgy of Eriocaulaceae. Pollen Spores Paris 7: 181–191

Thanikaimoni G (1969) Esquisse palynologique des Aracées. Trav Sect Scient Tech Inst Fr Pondichery 11: 1–286

Thien LB, Heimermann W, Holman RT (1975) Floral odours and quantitative taxonomy of *Magnolia* and *Liriodendron*. Taxon 24: 557–568

Thomas JL (1961) The genera of the Cyrillaceae and Clethraceae of the south-eastern United States. J Arnold Arbor 42: 96–106

Thomas V, Mercykutty VC, Saraswathy Amma CK (1996) Seed biology of Para Rubber tree (*Hevea brasiliensis* Muell. Arg., Euphorbiaceae): A Review. Phytomorphology 46: 335–342

Thornber CW (1970) Alkaloids of the Menispermaceae. Phytochem 9: 157–187

Thorne RF (1963) Some problems and guiding principles of angiosperm phylogeny. Am Natu 97: 287–305

Thorne RF (1968) Synopsis of a putatively phylogenetic classification of the flowering plants. Aliso 6: 57–66

Thorne RF (1973) Inclusion of the Apiaceae (Umbelliferae) in the Araliaceae. Notes R Bot Gard Edinb 32: 161–165

Thorne RF (1976) A phylogenetic classification of the Angiospermae. Evolut Biol 9: 35–106

Thorne RF (1977) Some realignments in the Angiospermae. Pl Syst Evol Suppl 1: 299–319

Thorne RF (1981) Phytochemistry and angiosperm phylogeny: A summary statement, pp 233–295. *In:* Young DA, Seigler DS (eds) Phytochemistry and Angiosperm Phylogeny, Praeger Scientific, New York

Thorne RF (1983) Proposed new realignments in the angiosperms. Nordic J Bot 3: 85–117

Thorne RF (1992a) An updated phylogenetic classification of the families of dicotyledons. Aliso 13: 365–389

Thorne RF (1992b) Classification and geography of the flowering plants. Bot Rev 58: 225–348

Thorne RF, Scogin R (1978) *Forsellesia* Greene (*Glossopetalon* Gray): A third genus in the Crossosomataceae, Rosineae, Rosales. Aliso 9: 171–178

Thulin M (1978) *Cyphia* (Lobeliaceae) in tropical Africa. Bot Notiser 131: 455–471

Tiagi B (1951) A contribution to the morphology and embryology of *Cuscuta hyalina* Roth and *C. planiflora* Tenore. Phytomorphology 1: 9–21

Tiagi B (1970) Orobanchaceae. *In:* Proc Symp Comparative Embryology of Angiosperms. Bull Indian Natl Sci Acad No. 41: 282–289

Tippo O (1938) Comparative anatomy of the Moraceae and their presumed allies. Bot Gaz 100: 1–99

Tippo O (1940) The comparative anatomy of the secondary xylem and the phylogeny of the Eucommiaceae. Am J Bot 27: 832–838

Titman PW (1949) Studies in the wood anatomy of the family Nyssaceae. J Elisha Mitchell Soc 65: 245–261

Tobe H (1981) Embryological studies in *Glaucidium palmatum* Sieb. et Zucc. with discussion on the taxonomy of the genus. Bot Mag Tokyo 94: 207–224

Tobe H, Ching-I Peng (1990) The embryology and taxonomic relationships of *Bretschneidera* (Bretschneideraceae). Bot J Linn Soc 103: 139–152

Tobe H, Raven PH (1983a) An embryological analysis of the Myrtales: Its definition and characteristics. Ann Mo Bot Gard 70: 71–94

Tobe H, Raven PH (1983b) The embryology of *Axinandra zeylanica* (Crypteroniaceae) and the relationships of the genus. Bot Gaz 144: 426–432

Tobe H, Raven PH (1984) An embryological contribution to systematics of the Chrysobalanaceae 1. Tribe Chrysobalaneae. Bot Mag Tokyo 97: 397–411

Tomlinson PB (1960) Seedling leaves in palms and their morphological significance. J Arnold Arbor 41: 414–428

Tomlinson PB (1961) Anatomy of the Monocotyledons 2. Palmae. Clarendon Press, Oxford (UK)

Tomlinson PB (1962) Phylogeny of the Scitamineae: Morphological and anatomical considerations. Evolution 16: 192–213

Tomlinson PB (1969) Commelinales—Zingiberales. *In:* Metcalfe CR (ed) Anatomy of the Monocotyledons, Vol 3: 1–446. Clarendon Press, Oxford (UK)

Tomlinson PB, Posluszny U (1976) Generic limits in the Zannichelliaceae (sensu Dumortier). Taxon 25:273–279

Tomlinson PB, Smith AC (1970) Joinvilleaceae; A new family of Monocotyledons. Taxon 10: 887–889

Tomlinson PB, Zimmerman MH (1969) Vascular anatomy of monocotyledons with secondary growth: An introduction. J Arnold Arbor 50: 159–179

Troll W (1964) Die Infloreszenzen; Typologie und Stellung in Aufleau des Vegetation skörpers. 1; Fischer, Stuttgart

Treub M (1891) Sur les Casuarinées et leur place dans le systeme naturel. Ann Jard Bot Buitenz 10: 145–231

Tucker SC (1960) Ontogeny of the floral apex of *Michelia fuscata*. Am J Bot 47: 266–277

Tucker SC, Sampson FB (1979) The gynoecium in Winteraceous plants. Science 203: 920–921

Turner BL (1958) Chromosome numbers in the genus *Krameria:* Evidence for familial status. Rhodora, Boston Mass. (USA) 60: 101–106

Ühl NW, Moore Jr HE (1971) The palm gynoecium. Am J Bot 58: 945–992

Van Bakhuizen B (1943) A contribution to the knowledge of the Melastomataceae occurring in the Malaya Archipelago, especially in the Netherlands Indies. Recl Trav Bot Neerl 40: 1–391

Van Balgooy MMJ (1982) A revision of *Sericolea* Schlechter (Elaeocarpaceae). Blumea 28: 103–141

Van Beusekom–Osinga RJ (1977) Crypteroniaceae. *In:* Van Steenis CGGJ (ed) Flora Malesiana Ser 1. 8:187–204

Van Beusekom-Osinga RJ, Van Beusekom CF (1975) Delimitation and subdivision of the Crypteroniaceae (Myrtales). Blumea 22: 255–266

Van Borssum Waalkes J (1966) Malesian Malvaceae revised. Blumea 14; 1–213

Van Heel WA (1966) Morphology of the androecium in Malvales. Blumea 13: 177–394

Van Heel WA (1967) Anatomical and ontogenetic investigations on the morphology of the flowers and the fruit of *Scyphostegia borneensis* Stapf. (Scyphostegiaceae). Blumea 15: 107–125

Van Steenis CGGJ (1959) Reduction of the Papuan genus *Kaernbachia* Schltr. (Cunoniaceae) to *Turpinia* (Staphyleaceae). Nova Guinea 10: 211, 212

Van Steenis CGGJ (1960) Staphyleaceae. *In:* Van Steenis CGGJ (ed) Flora Malesiana Ser 1. 6: 49–59

Van Tieghem P (1869) Anatomie d la fleur femelle et du fruit du noyer. Bull Soc Bot Fr 16: 412

Van Tieghem P (1896) Classification des Lorathinées. Bull Soc Bot Fr 30: 246–286

Van Tieghem P (1899) Sur les genres *Actinidie* et *Sauravie* considérés comme types d'une famille nouvelle, les Actinidiacées. J Bot (Paris) 13: 170–173

Van Tieghem P (1900) Sur les dicotylédons du groupe des Homoxylées. J Bot (Paris) 14: 259–297, 330–361

Van Vilet GJCM (1975) Wood anatomy of Crypteroniaceae sensu lato. J Microscop 104: 65–82

Van Vilet GJCM, Baas P (1975) Comparative anatomy of Crypteroniaceae sensu lato. Blumea 22: 173–195

Van Vilet GJCM, Koek-Noorman J, Ter Welle BJH (1981) Wood anatomy, classification and phylogeny of the Melastomataceae. Blumea 27: 463–473

Vani Hardev (1972) Systematic embryology of *Roridula gorgonius* Planch. Beitr Biol Pfl 48: 339–351

Varghese TM (1970) Solanaceae. *In:* Proc Symp Comparative Embryology of Angiosperms. Bull Indian Natl Sci Acad No. 41: 255–258

Venkata Rao C (1949) Contributions to the embryology of Sterculiaceae. 1. J Indian Bot Soc 28: 180–197

Venkata Rao C (1953a) Floral anatomy and embryology of two species of *Elaeocarpus*. J Indian Bot Soc 32: 21–33

Venkata Rao C (1953b) Contributions to the embryology of Sterculiaceae. 5. J Indian Bot Soc 32: 208–238

Venkata Rao C (1954a) Embryological studies in Malvaceae. 1. Development of gametophytes. Proc Natl Inst Sci India B 20: 127–150

Venkata Rao C (1954b) A contribution to the embryology of Bombacaceae. Proc Indian Acad Sci B 39: 51–57

Venkateswarlu J, Lakshminarayana L (1957) A contribution to the embryology of *Hydrocera triflora* W. & A. Phytomorphology 7:194–203

Verkerke W (1985) Ovule ontogeny and seed-coat development in *Krameria* Roefling (Krameriaceae). Beitr Biol Pfl 60: 341–352

Vestal PA (1937) The significance of comparative anatomy in establishing the relationships of the Hypericaceae to the Guttiferae and their allies. Philipp J Sci 64: 199–256

Vidyashankari B, Mohan Ram HY (1987) In vitro germination and origin of thallus in *Griffithella hookeriana* (Podostemaceae). Aqua Bot (Amsterdam) 28: 161–169

Vierhapper F (1930) Juncaceae. *In:* Engler A, Prantl K (eds) Die natürlichen Pflanzenfamilien. Edn 2. 15a: 192–224

Vijayaraghavan MR (1965) Morphology and embryology of *Actinidia polygama* Franch et Sav. and systematic position of the family Actinidiaceae. Phytomorphology 15: 224–235

Vijayaraghavan MR (1969) Studies in the family Cyrillaceae. 1. Development of male and female gametophytes in *Cliftonia monophylla* (Lam.) Britton ex Sarg. Bull Torrey Bot Club 96: 484–489

Vijayaraghavan MR (1970a) Actinidiaceae. *In:* Proc Symp Comparative Embryology of Angiosperms. Bull Indian Natl Sci Acad No. 41: 69–74

Vijayaraghavan MR (1970b) Cyrillaceae. *In:* Proc Symp Comparative Embryology of Angiosperms. Bull Indian Natl Sci Acad No. 41: 163–167

Vijayaraghavan MR, Dhar Usha (1976) *Scytopetalum tieghmii:* Embryologically unexplored taxon and affinities of the family Scytopetalaceae. Phytomorphology 26: 10–22

Vijayaraghavan MR, Dhar Usha (1978) Embryology of *Cyrilla* and *Cliftonia* (Cyrillaceae). Bot Notiser 131: 127–138

Vijayaraghavan MR, Kaur Davinder (1966) Morphology and embryology of *Turnera ulmifolia* L. and affinities of the family Turneraceae. Phytomorphology 16: 539–553

Vijayaraghavan MR, Malik U (1972) Morphology and embryology of *Scaevola frutescens* K. and affinities of the family Goodeniaceae. Bot Notiser 125: 241–254

Vijayaraghavan MR, Padmanaban Usha (1969) Morphology and embryology of *Centaurium ramosissimum* Druce and affinities of the family Gentianaceae. Beitr Biol Pfl 46: 15–37

Vink W (1978) The Winteraceae of the Old World. 3. Notes on the ovary of *Takhtajania*. Blumea 24: 521–525

Voroshilov VN, Nekrasov AA (1954) The Far-Eastern *Euryale* (in Russian). Priroda 43: 108–109

Wagenitz G (1959) Die systematische Stellung der Rubiaceae. Bot Jb 79: 17–35

Wagenitz G (1964a) Ebenales. *In:* Melchior H (ed) A Engler's Syllabus der Pflanzenfamilien. Vol 2: 396–403. Borntraeger, Berlin

Wagenitz G (1964b) Rubiaceae. *In:* Melchior H (ed) A Engler's Syllabus der Pflanzenfamilien. Vol 2: 417–424. Borntraeger, Berlin

Wagenitz G (1975) Blütenreduktion als ein zentrales problem der Angiospermen-systematik. Bot Jb 96: 448–470

Wagenitz G (1976) Systematics and phylogeny of the Compositae (Asteraceae). Pl Syst Evol 125: 29–46

Wagenitz G, Laing B (1984) Die Nektarien den Dipsacales und ihre systematische Bedeutung. Bot Jb 104: 448–507

Wanger P (1977) Vessel types of the monocotyledons: A survey. Bot Notiser 130: 383–402

Walia Karvita, Kapil RN (1965) Embryology of *Frankenia* Linn. with some comments on the systematic position of the Frankeniaceae. Bot Notiser 118: 412–429

Walker JW (1976) Comparative pollen morphology and phylogeny of the Ranalean complex, pp 241–299. *In:* Beck CB (ed) Origin and Early Evolution of Angiosperms, pp 1–341. Columbia Univ Press, New York

Walker RL (1950) Megasporogenesis and development of megagametophyte in *Ulmus*. Am J Bot 37: 47–52

Wallaart RAM (1980) Distribution of sorbitol in Rosaceae. Phytochem 19: 2603–2610

Walters JL (1956) Spontaneous meiotic chromosome breakage in natural populations of *Paeonia californica*. Am J Bot 43: 343–354

Wangerin W (1910a) Nyssaceae. *In:* Engler A (ed) Das Pflanzenreich 41(4) 220a: 1–20

Wangerin W (1910b) Alangiaceae. *In:* Englar A (ed) Das Pflanzenreich 41(4) 220b: 1–25

Wangerin W (1910c) Cornaceae. *In:* Engler A (ed) Das Pflanzenreich 41(4) 229: 1–110

Warburg O (1897) Monographie der Myristicaceen. Nova Acta Acad Leop-Carol 68: 1–680

Warming W (1942) Handbook of Systematic Botany, pp 1–506. 3rd Edn. Berlin

Weatherwax P (1929) The morphology of the spikelet of six genera of Oryzeae. Am J Bot 16: 547–555

Webb IJ, Tracey JG (1981) Australian rain forests: Patterns and change, pp 605–694. *In:* Keast A (ed) Ecological Biogeography of Australia. W Junk, Hague (Netherlands)

Webber IE (1941) Systematic anatomy of the woods of the Burseraceae. Lilloa 6: 441–465

Webster GL (1967) The genera of Euphorbiaceae in the southeastern United States. J Arnold Arbor 48: 303–430

Webster GL (1975) Conspectus of a new classification of the Euphorbiaceae. Taxon 24: 593–601

Weimarch H (1946) Studies in Juncaceae, with special reference to the species in Ethiopia and the Cape. Svensk Bot Tidskr 40: 141–178

Wernham HF (1911) Floral evolution: With particular reference to the sympetalous dicotyledons. 3. The Pentacyclidae. New Phytol 10: 145–159

Wettstein R (1935) Handbuch der Systematischen Botanik. Edn 4. pp 1–1152. Leipzing Wien

Whitaker TW (1933) Chromosome number and relationship in the Magnoliales. J Arnold Arbor 14: 376–385

White CT (1933) Ligneous plants collected for the Arnold Arboretum in North Queensland by SF Kajewski in 1929. Contr Arnold Arbor 4: 1–113

White CT (1948) A new species of *Austrobaileya* (Austrobaileyaceae) from Australia. J Arnold Arbor 29: 255–256

Wilber RL (1963) The Leguminous Plants of North Carolina. Tech Bull North Carolina Agric Exp Stn No, 151: 1–294

Wilkinson AM (1949) Floral anatomy and morphology of *Triosteum* and of Caprifoliaceae in general. Am J Bot 36: 481–489

Williams NH, Broome CR (1976) Scanning electron microscope studies of orchid pollen. Am Orchid Soc Bull 45: 699–707

Williams CA, Harborne JB, (1975) Luteolin and daphnetin derivatives in the Juncaceae and their systematic significance. Biochem Syst Ecol 3: 181–190

Williams CA, Harborne JB, Clifford HT (1973) Negatively charged flavones and tricin as chemosystematic markers in the Palmae. Phytochem 12: 2417–2430

Willis JC (1973) A Dictionary of the Flowering Plants and Ferns. Edn 8 (revised by Airy Shaw HK), pp 1-1245. Univ Press, Cambridge (England)

Wilson TK (1964) Comparative morphology of the Canellaceae. 3. Pollen. Bot Gaz 125: 192–197

Wilson TK (1966) Comparative morphology of the Canellaceae. 4. Floral morphology and conclusions. Am J Bot 53: 336–343

Wimmer FE (1943) Campanulaceae—Lobelioideae. *In:* Engler A (ed) Das Pflanzenreich. 106 (IV, 276b): 1–260

Wink M (1985) Chemische Verteidigung der Lupinen: Zur biologischen Bedeutung der chinolizidinalkaloide. Pl Syst Evol 150: 65–81

Withner CL (1941) Stem anatomy and phylogeny of the Rhoipteleaceae. Am J Bot 28: 872–878

Wodehouse RP (1931) Pollen grains in the identification and classification of plants. 6. Polygonaceae. Am J Bot 18: 149–764

Wollenweber E (1982) Flavones and flavonols, pp 189–259. *In:* Harborne JB, Mabry TJ (eds) The Flavonoids: Advances in Research from 1975–1981. Chapman & Hall, London

Wolter M, Seuffert C, Schill R (1988) The ontogeny of pollinia and elastoviscin in the anther of *Doritis pulcherrima* (Orchidaceae). Nordic J Bot 8: 77–88.

Worsdell WC (1908a) A study of vascular system in certain orders of the Ranales. Ann Bot (London) 22: 561–682

Worsdell WC (1908b) The affinities of *Paeonia*. J Bot (London) 46: 114–116

Wu CY (1981) Another new genus of Adoxaceae, with special references on the infrafamiliar evolution and the systematic position of the family. Acta Bot Yunn (China) 3: 383–388

Wu CY, Wu ZL, Huang RF (1981) *Sinadoxa,* genus novum familiae Adoxacearum. Acta Phytotax Sinica 19: 203–210

Wunderlich R (1967) Some remarks on the taxonomic significance of the seed coat. Phytomorphology 17: 301–311

Wunderlich R (1971) Die systematische Stellung von *Theligonum.* Öst Bot Z 119: 239–394

Yakovlev MS, Yoffe MD (1957) On some peculiar features in the embryology of *Paeonia* L. Phytomorphology 7: 74–87

Yakovlev MS, Yoffe MD (1961) Further studies of the new type of embryogenesis in angiosperms. Bot Zh 46: 1402–1421

Yang D.-Q, Hu C.-M (1985) The chromosomes of *Bretschneidera* Hemsl. Notes R Bot Gard Edinb 42: 347–349

Yeo PF (1985) Fruit-discharge type in *Geranium* (Geraniaceae): Its use in classification and its evolutionary implications. Bot J Linn Soc 89: 1–36

Yeung EC, Zee SY, Ye XL (1994) Embryology of *Cymbidium sinense:* Ovule development. Phytomorphology 44: 55–63

Young DA (1981) The usefulness of flavonoids in angiospermous phylogeny: Some selected examples, pp 205–232. *In:* Young DA, Siegler DS (eds) Phytochemistry and Angiosperm Phylogeny. Praeger Scientific, New York

Youngken HW (1919) The comparative morphology, taxonomy and distribution of the Myricaceae of the eastern United States. Contr Bot Lab Univ Paris 4: 339–400

Yu Jin, Xiao Pei-Gen (1987) A preliminary study of the chemistry and systematics of Paeoniaceae (in Chinese). Acta Phytotaxa Sinica 25: 172–179

Zavada MS (1983) Comparative morphology of monocot pollen and evolutionary trends of apertures and wall structures. Bot Rev 49: 331–379

Zavada MS, Taylon TN (1986) Pollen morphology of Lactoridaceae. Pl Syst Evol 154: 31–39

Ziegenspeck H (1944) Das Vorkommen von Öl in den Stomata der Monokotyledonen und die Bedeutung des Konstitutionalen Vorkommens für die Systematic derselben. Repert Spec Nov 53: 151–173

Zweifel R (1939) Cytologische-embryologische Untersuchungen an *Balanophora abbreviata* Blume und *B. indica* Wall. Diss Zürich

*Johri BM, Singh H (1959) The morphology, embryology and systematic position of *Elytraria acaulis* (Linn.f.). Bot Notiser 112: 227–251

*Maheshwari P (1964) Embryology in relation to taxonomy. In: Turrill WB (ed) Vistas in Botany, vol 4 pp 55–97 Pergamon Oxford

*Mohan Ram HY, Masand P (1963) The embryology of *Nelsonia campestris* R. Br. Phytomorphology 13: 82–91

---

*Added afterwards.

Plant Index